3,000 Solved Problems in Chemistry

David E. Goldberg, Ph.D.
Brooklyn College

Schaum's Outline Series

New York Chicago San Francisco Lisbon London Madrid
Mexico City Milan New Delhi San Juan Seoul
Singapore Sydney Toronto

The McGraw-Hill Companies

David E. Goldberg, Ph.D., who is the former department chairman at Brooklyn College, has taught chemistry for many years and has more recently entered the field of computer languages. He is the author of 8 textbooks, as well as *Schaum's Outline of Beginning Chemistry*.

4 5 6 7 8 9 10 CUS/CUS 1 9 8 7 6 5 4 3

ISBN 978-0-07-175500-9
MHID 0-07-175500-4

CONTENTS

To the Student

The best way to ensure that you understand the concepts of General Chemistry is to solve many problems on each topic. You should attempt a large number of different problems, rather than merely reworking the same problems again and again, since you might have a tendency to memorize the solution in the latter case. Be sure to read each problem carefully, since a small difference in the wording of a problem can make a large difference in its solution.

Since there is no set order of topics in general chemistry texts, you will have to consult the Table of Contents in this book to find the problems you wish to do. The problems in each section start with the more basic ones and progress to those that are more difficult. In some chapters, you may find problems based on material which you have not yet covered in your course. Do not attempt to do these before you cover the material on which they are based. For example, some texts cover equilibrium before thermodynamics, while others cover it afterward. Do not attempt the section on the thermodynamics of equilibrium until you have been introduced to both topics in your course.

There are numerous methods to solve most problems. The solution methods are usually related to each other but may seem very different. In this book, some related problems are solved using one method, and some with another. Many of the problems, especially in the early chapters, use several solution methods. You should attempt to do the problems yourself before looking at the solutions given. If you get the correct answer using a reasonable method, you need not worry that you did not use the method selected here. If the method presented is clearer that the one you used, however, you might consider adopting it for future similar problems.

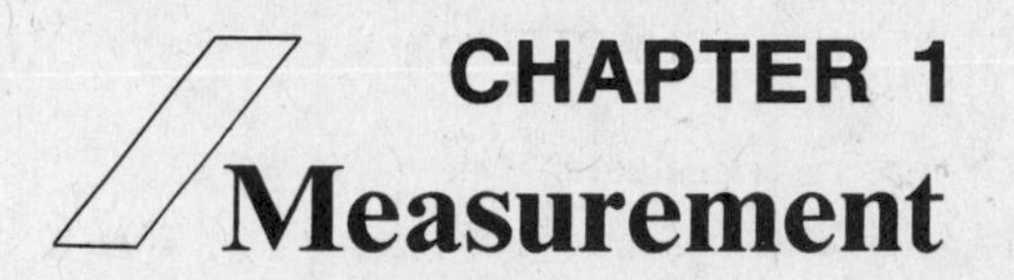

CHAPTER 1 Measurement

1.1 EXPONENTIAL NUMBERS

1.1 List the powers of ten, from 10^{-4} to 10^6, together with their explicit meanings.

▌

$$10^0 = 1 \qquad 10^{-1} = \frac{1}{10} = 0.1$$

$$10^1 = 10$$

$$10^2 = 10 \times 10 = 100 \qquad 10^{-2} = \frac{1}{10^2} = \frac{1}{100} = 0.01$$

$$10^3 = 10 \times 10 \times 10 = 1000$$

$$10^4 = 10 \times 10 \times 10 \times 10 = 10\,000 \qquad 10^{-3} = \frac{1}{10^3} = \frac{1}{1000} = 0.001$$

$$10^5 = 10 \times 10 \times 10 \times 10 \times 10 = 100\,000$$

$$10^6 = 10 \times 10 \times 10 \times 10 \times 10 \times 10 = 1\,000\,000 \qquad 10^{-4} = \frac{1}{10^4} = \frac{1}{10\,000} = 0.0001$$

In the expression 10^5, the *base* is 10 and the *exponent* is 5.

1.2 Express the following numbers in standard exponential form:

(*a*) 22 400 (*b*) 7 200 000 (*c*) 454 (*d*) 0.454
(*e*) 0.0454 (*f*) 0.000 06 (*g*) 0.003 06 (*h*) 0.000 000 5

▌ Any number may be expressed as an integral power of 10, or as the product of two numbers one of which is an integral power of 10 (e.g., $300 = 3 \times 10^2$).

(*a*) $22\,400 = 2.24 \times 10^4$ (*b*) $7\,200\,000 = 7.2 \times 10^6$ (*c*) $454 = 4.54 \times 10^2$
(*d*) $0.454 = 4.54 \times 10^{-1}$ (*e*) $0.0454 = 4.54 \times 10^{-2}$ (*f*) $0.000\,06 = 6 \times 10^{-5}$
(*g*) $0.003\,06 = 3.06 \times 10^{-3}$ (*h*) $0.000\,000\,5 = 5 \times 10^{-7}$

Moving the decimal point one place to the right is equivalent to multiplying a number by 10; moving the decimal point two places to the right is equivalent to multiplying by 100, and so on. Whenever the decimal point is moved to the right by n places, compensation can be achieved by *dividing* at the same time by 10^n; the value of the number remains unchanged. Thus

$$0.0325 = \frac{3.25}{10^2} = 3.25 \times 10^{-2}$$

Moving the decimal point one place to the left is equivalent to dividing by 10. Whenever the decimal point is moved to the left n places, compensation can be achieved by *multiplying* at the same time by 10^n; the value of the number remains unchanged. For example,

$$7296 = 72.96 \times 10^2 = 7.296 \times 10^3$$

1.3 Evaluate:

(*a*) $a^3 \times a^5$ (*b*) $10^2 \times 10^3$ (*c*) 10×10
(*d*) $10^7 \times 10^{-3}$ (*e*) $(4 \times 10^4)(2 \times 10^{-6})$ (*f*) $(2 \times 10^5)(3 \times 10^{-2})$

▌ In multiplication, exponents of like bases are added.

(*a*) $a^3 \times a^5 = a^{3+5} = a^8$ (*b*) $10^2 \times 10^3 = 10^{2+3} = 10^5$
(*c*) $10 \times 10 = 10^{1+1} = 10^2$ (*d*) $10^7 \times 10^{-3} = 10^{7-3} = 10^4$
(*e*) $(4 \times 10^4)(2 \times 10^{-6}) = 8 \times 10^{4-6} = 8 \times 10^{-2}$ (*f*) $(2 \times 10^5)(3 \times 10^{-2}) = 6 \times 10^{5-2} = 6 \times 10^3$

1.4 Evaluate:

(*a*) $\dfrac{a^5}{a^3}$ (*b*) $\dfrac{10^2}{10^5}$ (*c*) $\dfrac{8 \times 10^2}{2 \times 10^{-6}}$ (*d*) $\dfrac{5.6 \times 10^{-2}}{1.6 \times 10^4}$

▌ In division, exponents of like bases are subtracted.

(*a*) $\dfrac{a^5}{a^3} = a^{5-3} = a^2$ (*b*) $\dfrac{10^2}{10^5} = 10^{2-5} = 10^{-3}$

(*c*) $\dfrac{8 \times 10^2}{2 \times 10^{-6}} = \dfrac{8}{2} \times 10^{2+6} = 4 \times 10^8$ (*d*) $\dfrac{5.6 \times 10^{-2}}{1.6 \times 10^4} = \dfrac{5.6}{1.6} \times 10^{-2-4} = 3.5 \times 10^{-6}$

1.5 Evaluate the following expressions: (*a*) a^0 (*b*) 10^0 (*c*) $(3 \times 10)^0$ (*d*) 7×10^0 (*e*) 8.2×10^0

▌ (*a*) $a^0 = 1$ (*b*) $10^0 = 1$ (*c*) $(3 \times 10)^0 = 1$ (*d*) $7 \times 10^0 = 7$ (*e*) $8.2 \times 10^0 = 8.2$

1.6 Express as radicals: (*a*) $10^{2/3}$ (*b*) $10^{3/2}$ (*c*) $10^{1/2}$ (*d*) $4^{3/2}$

▌ (*a*) $10^{2/3} = \sqrt[3]{10^2}$ (*b*) $10^{3/2} = \sqrt{10^3}$ (*c*) $10^{1/2} = \sqrt{10}$ (*d*) $4^{3/2} = \sqrt{4^3} = \sqrt{64} = 8$

1.7 Simplify: (*a*) $(10^3)^2$ (*b*) $(10^{-2})^3$ (*c*) $(a^3)^{-2}$

▌ (*a*) $(10^3)^2 = 10^{3 \times 2} = 10^6$ (*b*) $(10^{-2})^3 = 10^{-2 \times 3} = 10^{-6}$ (*c*) $(a^3)^{-2} = a^{-6}$

1.8 Take the square root of each of the following numbers, using exponential notation as an aid: (*a*) 90 000 (*b*) 3.6×10^3 (*c*) 4.9×10^{-5} Take the cube roots of the following numbers: (*d*) 8×10^9 (*e*) 1.25×10^5

▌ To extract the square root of a power of 10, divide the exponent by 2. If the exponent is an odd number it should be increased or decreased by 1, and the coefficient adjusted accordingly. To extract the cube root of a power of 10, adjust so that the exponent is divisible by 3; then divide the exponent by 3. The coefficients are treated independently.

(*a*) $\sqrt{90\,000} = \sqrt{9 \times 10^4} = \sqrt{9} \times \sqrt{10^4} = 3 \times 10^2$ or 300
(*b*) $\sqrt{3.6 \times 10^3} = \sqrt{36 \times 10^2} = \sqrt{36} \times \sqrt{10^2} = 6 \times 10^1$ or 60
(*c*) $\sqrt{4.9 \times 10^{-5}} = \sqrt{49 \times 10^{-6}} = \sqrt{49} \times \sqrt{10^{-6}} = 7 \times 10^{-3}$ or 0.007
(*d*) $\sqrt[3]{8 \times 10^9} = \sqrt[3]{8} \times \sqrt[3]{10^9} = 2 \times 10^3$ or 2000
(*e*) $\sqrt[3]{1.25 \times 10^5} = \sqrt[3]{125 \times 10^3} = \sqrt[3]{125} \times \sqrt[3]{10^3} = 5 \times 10$ or 50

1.9 Calculate the value of $\dfrac{(4.0 \times 10^{-100}) + (2.0 \times 10^{-101})}{2.0 \times 10^{-200}}$

▌ You must know the rules for handling exponents, because not all electronic calculators do this type of problem.

$$\frac{(4.0 \times 10^{-100}) + (2.0 \times 10^{-101})}{2.0 \times 10^{-200}} = \frac{(4.0 \times 10^{-100}) + (0.20 \times 10^{-100})}{2.0 \times 10^{-200}} = \frac{4.2 \times 10^{-100}}{2.0 \times 10^{-200}} = 2.1 \times 10^{100}$$

1.10 Units can also be expressed with exponents, both positive and negative. Show the dimensions of the result two ways when a mass in grams (g) is divided by a volume in cubic meters (m^3).

▌ g/m^3 or $g \cdot m^{-3}$

1.11 Compute and state the answers in standard exponential form:

(*a*) $(2.0 \times 10^{13}) + (1.5 \times 10^{14})$ (*b*) $(8.0 \times 10^{-14})/(4.0 \times 10^{-13})$
(*c*) $(5.0 \times 10^{17})(2.0 \times 10^{-4})$ (*d*) $(6.6 \times 10^{15}) - (3.0 \times 10^{16})$

▌ (*a*) 1.7×10^{14} (*b*) 2.0×10^{-1} (*c*) $10 \times 10^{13} = 1.0 \times 10^{14}$ (*d*) -2.3×10^{16}

1.12 Evaluate: (*a*) $10^4 \times 10^{-2}$ (*b*) $10^4/10^{-3}$ (*c*) $2.5 \times 10^7 \times 4.0 \times 10^3$

▌ (*a*) In multiplication, add exponents algebraically: 10^2 (*b*) In division, subtract exponents algebraically: 10^7 (*c*) Multiply coefficients and exponential parts separately and simplify: $10 \times 10^{10} = 10^{11}$.

1.13 Simplify the following expressions:

(*a*) $\dfrac{48\,000\,000}{1200}$ (*b*) $\dfrac{0.0078}{120}$

(*c*) $(4 \times 10^{-3})(5 \times 10^{4})^2$ (*d*) $\dfrac{(6\,000\,000)(0.00004)^4}{(800)^2(0.0002)^3}$

(*e*) $(\sqrt{4.0 \times 10^{-6}})(\sqrt{8.1 \times 10^{3}})(\sqrt{0.0016})$ (*f*) $(\sqrt[3]{6.4 \times 10^{-2}})(\sqrt[3]{27\,000})(\sqrt[3]{2.16 \times 10^{-4}})$

▌ (*a*) $\dfrac{48\,000\,000}{1200} = \dfrac{48 \times 10^6}{12 \times 10^2} = 4.0 \times 10^{6-2} = 4.0 \times 10^4$ or 40 000

(*b*) $\dfrac{0.0078}{120} = \dfrac{7.8 \times 10^{-3}}{1.2 \times 10^2} = 6.5 \times 10^{-5}$ or 0.000 065

(*c*) $(4 \times 10^{-3})(5 \times 10^4)^2 = (4 \times 10^{-3})(5^2 \times 10^8) = 4 \times 5^2 \times 10^{-3+8} = 100 \times 10^5 = 1 \times 10^7$

(*d*) $$\frac{(6\,000\,000)(0.000\,04)^4}{(800)^2(0.000\,2)^3} = \frac{(6 \times 10^6)(4 \times 10^{-5})^4}{(8 \times 10^2)^2(2 \times 10^{-4})^3} = \frac{6 \times 4^4}{8^2 \times 2^3} \times \frac{10^6 \times 10^{-20}}{10^4 \times 10^{-12}}$$
$$= \frac{6 \times 256}{64 \times 8} \times \frac{10^{6-20}}{10^{4-12}} = 3 \times \frac{10^{-14}}{10^{-8}} = 3 \times 10^{-6}$$

(*e*) $$(\sqrt{4.0 \times 10^{-6}})(\sqrt{8.1 \times 10^3})(\sqrt{0.0016}) = (\sqrt{4.0 \times 10^{-6}})(\sqrt{81 \times 10^2})(\sqrt{16 \times 10^{-4}})$$
$$= (2.0 \times 10^{-3})(9.0 \times 10^1)(4.0 \times 10^{-2})$$
$$= 72 \times 10^{-4} = 7.2 \times 10^{-3} \text{ or } 0.0072$$

(*f*) $$(\sqrt[3]{6.4 \times 10^{-2}})(\sqrt[3]{27\,000})(\sqrt[3]{2.16 \times 10^{-4}}) = (\sqrt[3]{64 \times 10^{-3}})(\sqrt[3]{27 \times 10^3})(\sqrt[3]{216 \times 10^{-6}})$$
$$= (4.0 \times 10^{-1})(3.0 \times 10^1)(6.0 \times 10^{-2})$$
$$= 72 \times 10^{-2} \text{ or } 0.72$$

1.14 Find the logarithm of each of the following numbers: (*a*) 4.56 (*b*) 1.70 (*c*) 9.75 (*d*) 1.07 (*e*) 3.16 (*f*) 1.00

▌ (*a*) 0.6590 (*b*) 0.2304 (*c*) 0.9890 (*d*) 0.0294 (*e*) 0.4997 (*f*) 0.0000

1.15 Find the antilogarithm of each of the following: (*a*) 0.4502 (*b*) 0.8579 (*c*) 0.7042 (*d*) 0.6080 (*e*) 0.9695

▌ (*a*) 2.82 (*b*) 7.21 (*c*) 5.06 (*d*) 4.055 (*e*) 9.322

1.16 Determine the antilog of each of the following: (*a*) 2.6170 (*b*) 7.42 (*c*) −2.0057 (*d*) −0.4776

▌ (*a*) $4.14 \times 10^2 = 414$ (*b*) 2.6×10^7 (*c*) 9.87×10^{-3} (*d*) $3.33 \times 10^{-1} = 0.333$

1.2 METRIC SYSTEM

1.17 Show the relationships among the metric units of length, volume, and mass and the common English units of these quantities.

▌

1 meter = 100 cm = 1000 mm = 0.001 km = 39.37 in.

1 kilogram = 1000 g = 2.2 lb

1 inch = 2.54 cm = 0.0254 m = 25.4 mm = 2.54×10^7 nm

1 foot = 12 in. = 12 × 2.54 cm = 30.48 cm = 0.3048 m = 304.8 mm

1 liter = 1 dm^3 = 10^{-3} m^3 = 1.06 quarts

1 yard = 3 ft = 91.44 cm = 914.4 mm = 0.9144 m

1 mile = 5280 ft = 6.336×10^4 in. = 1.609×10^5 cm = 1.609×10^3 m = 1.609×10^6 mm

1 pound = 0.4536 kg = 453.6 g = 4.536×10^5 mg

1 ounce = $\dfrac{1}{16}$ lb = $\dfrac{1}{16} \times 453.6$ g = 28.35 g = 0.02835 kg

1 metric ton = 1000 kg = 10^6 g

1.18 (*a*) Express 3.69 m in kilometers, in centimeters, and in millimeters. (*b*) Express 36.24 mm in centimeters and in meters.

▌ (*a*) $(3.69\ \text{m})\left(\frac{1\ \text{km}}{10^3\ \text{m}}\right) = 3.69 \times 10^{-3}\ \text{km}$ (*b*) $(36.24\ \text{mm})\left(\frac{1\ \text{cm}}{10\ \text{mm}}\right) = 3.624\ \text{cm}$

$(3.69\ \text{m})\left(\frac{100\ \text{cm}}{\text{m}}\right) = 369\ \text{cm}$ $(36.24\ \text{mm})\left(\frac{1\ \text{m}}{10^3\ \text{mm}}\right) = 3.624 \times 10^{-2}\ \text{m}$

$(3.69\ \text{m})\left(\frac{10^3\ \text{mm}}{\text{m}}\right) = 3.69 \times 10^3\ \text{mm}$

1.19 (*a*) How many cubic centimeters are there in 1 m^3? (*b*) How many liters are there in 1 m^3? (*c*) How many cubic centimeters are there in 1 L?

▌ (*a*) $1\ \text{m}^3 = (1\ \text{m})^3 = (100\ \text{cm})^3 = (10^2\ \text{cm})^3 = 10^6\ \text{cm}^3$
(*b*) $1\ \text{m}^3 = (10\ \text{dm})^3 = 10^3\ \text{dm}^3 \times 1\ \text{L/dm}^3 = 10^3\ \text{L}$ (*c*) $1\ \text{L} = 1\ \text{dm}^3 = (10\ \text{cm})^3 = 10^3\ \text{cm}^3$

1.20 Find the capacity in liters of a tank 0.6 m long, 10 cm wide, and 50 mm deep.

▌ Convert to decimeters, since $1\ \text{L} = 1\ \text{dm}^3$.

$$\text{Volume} = 0.6\ \text{m} \times 10\ \text{cm} \times 50\ \text{mm} = 6\ \text{dm} \times 1\ \text{dm} \times 0.5\ \text{dm} = 3\ \text{dm}^3 = 3\ \text{L}$$

1.21 Convert: (*a*) 2.0×10^3 g to milligrams (*b*) $1.6 \times 10^{-3}\ \text{cm}^3$ to liters.

▌ (*a*) $(2.0 \times 10^3\ \text{g})\left(\frac{10^3\ \text{mg}}{\text{g}}\right) = 2.0 \times 10^6\ \text{mg}$ (*b*) $(1.6 \times 10^{-3}\ \text{cm}^3)\left(\frac{1\ \text{L}}{10^3\ \text{cm}^3}\right) = 1.6 \times 10^{-6}\ \text{L}$

1.22 How many millimeters are there in 15.0 cm?

▌
$$(15.0\ \text{cm})\left(\frac{1\ \text{m}}{100\ \text{cm}}\right)\left(\frac{1000\ \text{mm}}{1\ \text{m}}\right) = 150\ \text{mm}$$

1.23 Convert: (*a*) 1.47 km to millimeters (*b*) 1.42 mL to cubic centimeters (*c*) 1.7×10^7 mg to kilograms (*d*) 1.54×10^{-3} L to milliliters (*e*) 70.5 g/L to grams per milliliter (*f*) 4.66 kg/L to grams per milliliter

▌ (*a*) 1.47×10^6 mm (*b*) 1.42 cm^3 (*c*) 17 kg
(*d*) 1.54 mL (*e*) 0.0705 g/mL (*f*) 4.66 g/mL

1.24 Calculate the volume, in liters, of a rectangular bar which measures 0.10 m long, 2.0 cm thick, and 4.0 cm wide.

▌
$$(0.10\ \text{m})(2.0\ \text{cm})(4.0\ \text{cm}) = (10\ \text{cm})(2.0\ \text{cm})(4.0\ \text{cm}) = (80\ \text{cm}^3)\left(\frac{1\ \text{L}}{1000\ \text{cm}^3}\right) = 0.080\ \text{L}$$

1.25 Convert: (*a*) 1.6×10^{-2} km to centimeters (*b*) 20.0 mL to liters (*c*) 16.2 g/cm^3 to kilograms per liter

▌ (*a*) $(1.6 \times 10^{-2}\ \text{km})\left(\frac{10^3\ \text{m}}{\text{km}}\right)\left(\frac{10^2\ \text{cm}}{1\ \text{m}}\right) = 1.6 \times 10^3\ \text{cm}$ (*b*) $(20.0\ \text{mL})\left(\frac{1\ \text{L}}{1000\ \text{mL}}\right) = 0.0200\ \text{L}$

(*c*) $\left(\frac{16.2\ \text{g}}{\text{cm}^3}\right)\left(\frac{1\ \text{kg}}{1000\ \text{g}}\right)\left(\frac{1000\ \text{cm}^3}{\text{L}}\right) = 16.2\ \text{kg/L}$

1.26 Perform the following calculations: (*a*) $(2.0 \times 10^2\ \text{cm}) + (1.12 \times 10^{-1}\ \text{m})$
(*b*) $(0.10\ \text{m})(1.0 \times 10^{-4}\ \text{km})(1.0 \times 10^2\ \text{mm})$

▌ (*a*) $(2.0 \times 10^2\ \text{cm}) + (0.112 \times 10^2\ \text{cm}) = 2.1 \times 10^2\ \text{cm}$ (*b*) $(10\ \text{cm})(10\ \text{cm})(10\ \text{cm}) = 10^3\ \text{cm}^3 = 1.0\ \text{L}$

1.27 The color of light depends on its wavelength. The longest visible rays, at the red end of the visible spectrum, are 7.8×10^{-7} m in length. Express this length in micrometers, in nanometers, and in angstroms.

▌ $(7.8 \times 10^{-7}\ \text{m})\left(\frac{10^6\ \mu\text{m}}{\text{m}}\right) = 0.78\ \mu\text{m}$, $(7.8 \times 10^{-7}\ \text{m})\left(\frac{10^9\ \text{nm}}{\text{m}}\right) = 780\ \text{nm}$, $(7.8 \times 10^{-7}\ \text{m})\left(\frac{10^{10}\ \text{Å}}{\text{m}}\right) = 7800\ \text{Å}$

1.28 When a sample of healthy human blood is diluted to 200 times its initial volume and microscopically examined in a layer 0.10 mm thick, an average of 30 red corpuscles are found in each 100×100 micrometer square. (*a*) How many red cells are in a cubic millimeter of blood? (*b*) The red blood cells have an average life of 1 month, and the adult blood volume is about 5 L. How many red cells are generated every second in the bone marrow of the adult?

■ **(a)** $(100\ \mu m) \times (100\ \mu m) \times (0.10\ mm) = (0.10\ mm)^3 = 1.0 \times 10^{-3}\ mm^3$

$$\frac{30\ \text{corpuscles}}{1.0 \times 10^{-3}\ mm^3} \times 200 = 6 \times 10^6\ \text{corpuscles undiluted}$$

(b) $$(5\ L)\left(\frac{10^3\ cm^3}{L}\right)\left(\frac{10^3\ mm^3}{cm^3}\right)\left(\frac{6 \times 10^6\ \text{cells}}{mm^3}\right) = 3 \times 10^{13}\ \text{cells}$$

Since 3×10^{13} cells are required per month,

$$\left(\frac{3 \times 10^{13}\ \text{cells}}{\text{month}}\right)\left(\frac{1\ \text{month}}{30\ \text{days}}\right)\left(\frac{1\ \text{day}}{24\ h}\right)\left(\frac{1\ h}{3600\ s}\right) = 1 \times 10^7\ \text{cells/s}$$

1.3 SIGNIFICANT FIGURES

1.29 How many significant figures are there in each of the following numbers: **(a)** 17 **(b)** 103 **(c)** 1.035 **(d)** 0.0010 **(e)** 1.00×10^6 **(f)** π

■ **(a)** two **(b)** three **(c)** four **(d)** two **(e)** three **(f)** an infinite number

1.30 Perform the following operations:

$$\begin{array}{r} 12.01\ cm \\ 17.3\ cm \\ +\ 0.11\ cm \\ \hline \end{array} \qquad \begin{array}{r} 133\ \ g \\ -\ \ 2.2\ g \\ \hline \end{array}$$

■ The answers are 29.4 cm and 131 g rather than 29.42 cm and 130.8 g. The 2 in the hundredths column of the sum is farther to the right than the 3 of 17.3, and so it cannot be significant. It is dropped because it is less than 5. The 8 of the second example is not significant for the same reason, but it is over 5, so the answer is rounded up to the next higher integer.

1.31 **(a)** $12.7 \times 11.2 = ?$ **(b)** $108/7.2 = ?$

■ In the first calculation, three significant figures may be retained since each factor has three. In the second, only two significant figures are retained in the answer. **(a)** 142 **(b)** 15

1.32 Underline each significant digit in the numbers below. If the digit is uncertain, place a question mark below it.

(a) 1.066 **(b)** 750 **(c)** 0.050 **(d)** 0.2070 **(e)** 50.0

■ **(a)** $\underline{1.066}$ **(b)** $\underline{75}0$ (? below the 0) **(c)** $0.0\underline{50}$ **(d)** $0.\underline{2070}$ **(e)** $\underline{50.0}$

1.33 Calculate to the proper number of significant digits:

(a) $(4.50 \times 10^2\ m) + (3.00 \times 10^6\ mm)$ **(b)** $(4.50 \times 10^2\ cm)(2.00 \times 10^6\ cm)$

(c) $(4.50 \times 10^2\ mL) - (0.0225\ L)$

■ **(a)** $(4.50 \times 10^2\ m) + (3.00 \times 10^3\ m) = 3.45 \times 10^3\ m$ **(b)** $9.00 \times 10^8\ cm^2$

(c) $(4.50 \times 10^2\ mL) - (0.225 \times 10^2\ mL) = 4.28 \times 10^2\ mL$

1.34 Explain why in calculations involving more than one arithmetic operation, rounding off to the proper number of significant figures may be done once at the end if all the operations are multiplications and/or divisions or if they are all additions and/or subtractions, but not if they are combinations of additions or subtractions with multiplications or divisions.

■ There are different rules for the number of significant digits in the answer to an addition and to a multiplication, and so they must be applied separately when a mixed calculation is made.

1.35 Add the following quantities expressed in grams, to the proper number of significant digits.

$$\textbf{(a)}\ \begin{array}{r} 25.340 \\ 5.465 \\ 0.322 \\ \hline \end{array} \qquad \textbf{(b)}\ \begin{array}{r} 58.0 \\ 0.0038 \\ 0.00001 \\ \hline \end{array} \qquad \textbf{(c)}\ \begin{array}{r} 4.20 \\ 1.6523 \\ 0.015 \\ \hline \end{array} \qquad \textbf{(d)}\ \begin{array}{r} 415.5 \\ 3.64 \\ 0.238 \\ \hline \end{array}$$

▮ (*a*)
$$\begin{array}{r} 25.340 \\ 5.465 \\ 0.322 \\ \hline 31.127\text{ g} \end{array}$$
(*b*)
$$\begin{array}{l} 58.0 \\ 0.0038 \\ 0.00001 \\ \hline 58.00381 = 58.0\text{ g} \end{array}$$
(*c*)
$$\begin{array}{l} 4.20 \\ 1.6523 \\ 0.015 \\ \hline 5.8673 = 5.87\text{ g} \end{array}$$
(*d*)
$$\begin{array}{r} 415.5 \\ 3.64 \\ 0.238 \\ \hline 419.378 = 419.4\text{ g} \end{array}$$

1.36 (*a*) What is the usual method for correcting a calculated answer to the proper number of significant digits? (*b*) Do the following calculations: 2.48/1.24, 17 790/2.0. (*c*) Explain why the usual method given in part (*a*) is not used with either calculation in part (*b*).

▮ (*a*) Drop the extra digits and round the last digit retained according to the value of the first digit dropped. (*b*) 2.00 and 6400. (*c*) In 2.00, digits were added to obtain the proper number. In 6400, the last two digits were not dropped (or the value 64 would have resulted) but instead were changed to nonsignificant zeros.

1.37 Convert each of the following measurements to the basic unit (with no prefix), and express the result in standard exponential notation to the proper number of significant digits. (*a*) 9.50×10^{-1} kg (*b*) 4.40×10^{3} mm (*c*) 0.00102 cm (*d*) 400.0 mL

▮ (*a*) 9.50×10^{2} g (*b*) 4.40 m (*c*) 1.02×10^{-5} m (*d*) 4.000×10^{-1} L

1.38 $(1.20 \times 10^{-6}) + (6.00 + 10^{-5}) = ?$

▮
$$(0.120 \times 10^{-5}) + (6.00 \times 10^{-5}) = 6.12 \times 10^{-5}$$

1.39 Find the sum to the proper number of significant digits: 14.90 + 0.0070 + 1.0 + 0.091

▮
$$\begin{array}{l} 14.90 \\ 0.0070 \\ 1.0 \\ 0.091 \\ \hline 15.9_{980} \rightarrow 16.0 \end{array}$$

1.40 Calculate to the correct number of significant digits: (14.90)(0.0070)/(0.091).

▮
$$(14.90)(0.0070)/(0.091) = 1.1$$

1.41 Calculate to the proper number of significant digits:

(*a*) (1.0042 − 0.0034)(1.23) (*b*) (1.0042)(0.0034)(1.23) (*c*) (1.0042)(−0.0034)/1.23

▮ (*a*) 1.23 (*b*) 4.2×10^{-3} (*c*) -2.8×10^{-3} When 0.0034 is subtracted from 1.0042, a number with five significant digits results. When this number is multiplied by 1.23, with only three significant digits, the result must be expressed to only three significant digits.

1.42 A solid has a volume of 1.23 cm³. Its mass plus that of a piece of weighing paper is 10.024 g; the paper weighs 0.03 g. Calculate the density of the solid to the proper number of significant digits.

▮
$$\begin{array}{r} 10.024\text{ g} \\ -\ 0.03\ \text{ g} \\ \hline 9.99\not{4}\text{ g} \end{array} \qquad \frac{9.99\text{ g}}{1.23\text{ cm}^3} = 8.12\text{ g/cm}^3$$

The 4 must be dropped from the difference since the mass of the weighing paper is known only to the second decimal place.

1.43 Perform the following calculations to the proper number of significant digits. (*a*) $(2.00 \times 10^{-2}\text{ km}) + (4.2 \times 10^{2}\text{ cm})$ (*b*) $(1.5 \times 10^{1}\text{ cm})(8.0 \times 10^{2}\text{ cm})(0.0100\text{ m})$

▮ (*a*) (20.0 m) + (4.2 m) = 24.2 m (*b*) $(1.5 \times 10^{1}\text{ cm})(8.0 \times 10^{2}\text{ cm})(1.00\text{ cm}) = 1.2 \times 10^{4}\text{ cm}^3$

1.4 CALCULATIONS WITH METRIC QUANTITIES

1.44 What volume will 300 g of mercury occupy? Density of mercury is 13.6 g/cm³.

▮
$$\text{Volume} = \frac{\text{mass}}{\text{density}} = \frac{300\text{ g}}{13.6\text{ g/cm}^3} = 22.1\text{ cm}^3$$

1.45 Find the density of ethyl alcohol if 80.0 mL weighs 63.3 g.

∎

$$\frac{63.3\text{ g}}{80.0\text{ mL}} = 0.791\text{ g/mL}$$

1.46 Find the volume of 40 kg of carbon tetrachloride, whose density is 1.60 g/cm³.

∎

$$(40\text{ kg})\left(\frac{10^3\text{ g}}{\text{kg}}\right)\left(\frac{1\text{ cm}^3}{1.60\text{ g}}\right)\left(\frac{1\text{ L}}{10^3\text{ cm}^3}\right) = 25\text{ L} \quad \text{or else} \quad (40\text{ kg})\left(\frac{1\text{ L}}{1.60\text{ kg}}\right) = 25\text{ L}$$

1.47 An important physical quantity has the value 1.987 cal or 0.08206 L·atm. What is the conversion factor from liter·atmospheres to calories?

∎

$$\frac{1.987\text{ cal}}{0.08206\text{ L}\cdot\text{atm}} = 24.21\text{ cal/L}\cdot\text{atm}$$

1.48 Calculate the density, in grams per cubic centimeter, of a body that weighs 420 g (has a mass of 420 g) and has a volume of 52 cm³.

∎

$$\text{Density} = \frac{\text{mass}}{\text{volume}} = \frac{420\text{ g}}{52\text{ cm}^3} = 8.1\text{ g/cm}^3$$

1.49 Calculate the volume of 400 g of gold (density = 19.3 g/cm³).

∎

$$(400\text{ g})\left(\frac{1\text{ cm}^3}{19.3\text{ g}}\right) = 20.7\text{ cm}^3$$

1.50 The density of a metal is 9.50 g/cm³. Calculate the number of ***(a)*** kilograms per cubic meter ***(b)*** cubic centimeters per gram.

∎ ***(a)***

$$\left(\frac{9.50\text{ g}}{\text{cm}^3}\right)\left(\frac{1\text{ kg}}{10^3\text{ g}}\right)\left(\frac{10^2\text{ cm}}{\text{m}}\right)^3 = \frac{9.50\times 10^3\text{ kg}}{\text{m}^3}$$

(b)

$$\frac{1\text{ cm}^3}{9.50\text{ g}} = 0.105\text{ cm}^3/\text{g}$$

1.51 Fool's gold is so called because it bears a visual similarity to real gold. A block of fool's gold which measures 1.50 cm by 2.50 cm by 3.00 cm has a mass of 56.25 g. How can this material be distinguished from real gold by means of its physical properties?

∎ It can be distinguished, among other ways, by its density, 5.00 g/cm³. (It would be extremely coincidental if the density of fool's gold were the same as that of real gold, 19.3 g/cm³.)

1.52 The density of platinum is 21.45 g/cm³. Calculate its density in kilograms per cubic meter.

∎

$$\left(\frac{21.45\text{ g}}{\text{cm}^3}\right)\left(\frac{1\text{ kg}}{1000\text{ g}}\right)\left(\frac{100\text{ cm}}{\text{m}}\right)^3 = \frac{21.45\times 10^3\text{ kg}}{\text{m}^3}$$

1.53 A block of platinum 6.00 cm long, 3.50 cm wide, and 4.00 cm thick has a mass of 1802 g. What is the density of platinum?

∎ The volume, V, of the block is determined by multiplying its length, l, times its width, w, times its thickness, t: $V = lwt = (6.00\text{ cm})(3.50\text{ cm})(4.00\text{ cm}) = 84.0\text{ cm}^3$. The density is the mass per unit volume:

$$d = \frac{m}{V} = \frac{1802\text{ g}}{84.0\text{ cm}^3} = 21.5\text{ g/cm}^3$$

1.54 What is the density of a steel ball which has a diameter of 7.50 mm and a mass of 1.765 g? [Volume of a sphere of radius r is $\frac{4}{3}\pi r^3$.]

∎

$$V = \left(\frac{4\pi}{3}\right)\left(\frac{7.50\text{ mm}}{2}\right)^3 = 221\text{ mm}^3$$

$$d = \frac{m}{V} = \frac{1.765\text{ g}}{221\text{ mm}^3} = \frac{1.765\times 10^{-3}\text{ kg}}{221\times 10^{-9}\text{ m}^3} = 7.99\times 10^3\text{ kg/m}^3$$

1.55 An alloy was machined into a flat disk, 31.5 mm in diameter and 4.5 mm thick, with a hole 7.5 mm in diameter drilled through the center. The disk weighed 20.2 g. What is the density of the alloy?

▌ The volume of the disk is

$$V = \pi r_1^2 h - \pi r_2^2 h = \pi h(r_1^2 - r_2^2)$$
$$= \pi(4.5 \text{ mm})\left[\left(\frac{31.5}{2}\right)^2 \text{mm}^2 - \left(\frac{7.5}{2}\right)^2 \text{mm}^2\right] = 3308 \text{ mm}^3 = 3.308 \text{ cm}^3$$
$$d = \frac{m}{V} = \frac{20.2 \text{ g}}{3.308 \text{ cm}^3} = 6.11 \text{ g/cm}^3 = 6110 \text{ kg/m}^3$$

1.56 A glass vessel weighed 20.2376 g when empty and 20.3102 g when filled to an etched mark with water at 4 °C. The same vessel was then dried and filled to the same mark with a solution at 4 °C. The vessel was now found to weigh 20.3300 g. What is the density of the solution?

▌
$$\text{Mass of water} = (20.3102 \text{ g}) - (20.2376 \text{ g}) = 0.0726 \text{ g}$$
$$\text{Mass of solution} = (20.3300 \text{ g}) - (20.2376 \text{ g}) = 0.0924 \text{ g}$$

The density of water is 1.000 g/cm³ at 4 °C. Hence,

$$V = (0.0726 \text{ g})\left(\frac{1 \text{ cm}^3}{1.000 \text{ g}}\right) = 0.0726 \text{ cm}^3 \quad \text{and} \quad \text{density} = \frac{0.0924 \text{ g}}{0.0726 \text{ cm}^3} = 1.27 \text{ g/cm}^3$$

1.57 A sample of concentrated sulfuric acid is 95.7% H_2SO_4 by weight and its density is 1.84 g/mL. ***(a)*** How many grams of pure H_2SO_4 are contained in 1.00 L of the acid? ***(b)*** How many milliliters of acid contains 100 g of pure H_2SO_4?

▌
(a) $(1.00 \text{ L acid})\left(\frac{1.84 \text{ kg acid}}{\text{L acid}}\right)\left(\frac{95.7 \text{ kg } H_2SO_4}{100 \text{ kg acid}}\right) = 1.76 \text{ kg } H_2SO_4$

(b) $(100 \text{ g } H_2SO_4)\left(\frac{100 \text{ g acid}}{95.7 \text{ g } H_2SO_4}\right)\left(\frac{1 \text{ mL acid}}{1.84 \text{ g acid}}\right) = 56.8 \text{ mL acid}$

1.58 Analysis shows that 20.0 mL of concentrated hydrochloric acid of density 1.18 g/mL contains 8.36 g HCl. ***(a)*** Find the mass of HCl per milliliter of acid solution. ***(b)*** Find the percent by weight (mass) of HCl in the concentrated acid.

▌
(a) $\frac{8.36 \text{ g HCl}}{20.0 \text{ mL}} = 0.418 \text{ g/mL}$ *(b)* $\frac{8.36 \text{ g HCl}}{(20.0 \text{ mL})(1.18 \text{ g/mL})} \times 100\% = 35.4\% \text{ HCl}$

1.59 A piece of gold leaf (density 19.3 g/cm³) weighing 1.93 mg can be beaten further into a transparent film covering an area of 14.5 cm². ***(a)*** What is the volume of 1.93 mg of gold? ***(b)*** What is the average thickness of the transparent film, in angstroms?

▌
(a) $(1.93 \text{ mg})\left(\frac{1 \text{ g}}{10^3 \text{ mg}}\right)\left(\frac{1 \text{ cm}^3}{19.3 \text{ g}}\right) = 1.00 \times 10^{-4} \text{ cm}^3$

(b) $\frac{1.00 \times 10^{-4} \text{ cm}^3}{14.5 \text{ cm}^2} = (6.90 \times 10^{-6} \text{ cm})\left(\frac{1 \text{ Å}}{10^{-8} \text{ cm}}\right) = 690 \text{ Å}$

1.60 A piece of capillary tubing was calibrated in the following manner. A clean sample of the tubing weighed 3.247 g. A thread of mercury, drawn into the tube, occupied a length of 23.75 mm, as observed under a microscope. The weight of the tube with the mercury was 3.489 g. The density of mercury is 13.60 g/cm³. Assuming that the capillary bore is a uniform cylinder, find the diameter of the bore.

▌
$$m = 3.489 \text{ g} - 3.247 \text{ g} = 0.242 \text{ g}$$
$$V = (0.242 \text{ g})\left(\frac{1 \text{ cm}^3}{13.60 \text{ g}}\right) = 0.0178 \text{ cm}^3$$
$$A = \frac{V}{l} = \frac{0.0178 \text{ cm}^3}{2.375 \text{ cm}} = 0.00749 \text{ cm}^2 = \pi d^2/4$$
$$d = \sqrt{\frac{4(0.00749 \text{ cm}^2)}{3.14159}} = 0.0976 \text{ cm}$$

1.61 The General Sherman tree, located in Sequoia National Park, is believed to be the most massive of living things. If the overall density of the tree trunk is assumed to be 850 kg/m^3, calculate the mass of the trunk by assuming that it may be approximated by two right conical frustra having lower and upper diameters of 11.2 and 5.6 m, and 5.6 and 3.3 m, respectively, and respective heights of 2.4 and 80.6 m (see Fig. 1-1). A frustrum is a portion of a cone bounded by two planes, both perpendicular to the axis of the cone. The volume of a frustrum is given by $\pi h(r_1^2 + r_2^2 + r_1 r_2)/3$, where h is the height and r_1 and r_2 are the radii of the cone at the bounding planes.

▌

$$V_1 = \tfrac{1}{3}\pi(80.6\text{ m})[1.65^2 + 2.8^2 + (1.65)(2.8)] = 1281\text{ m}^3$$
$$V_2 = \tfrac{1}{3}\pi(2.4\text{ m})[5.6^2 + 2.8^2 + (5.6)(2.8)] = 138\text{ m}^3$$

So $V = 1419\text{ m}^3$ and $m = (1419\text{ m}^3)\left(\dfrac{850\text{ kg}}{\text{m}^3}\right) = 1.21 \times 10^6\text{ kg} = 1.21 \times 10^3$ metric tons

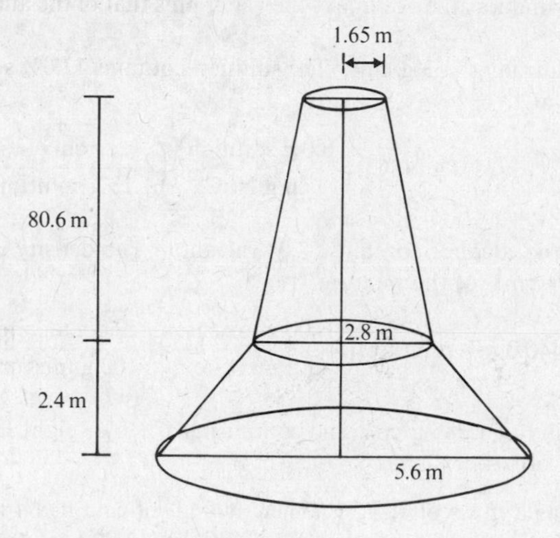

Fig. 1.1

1.62 Calculate the percent sodium in a breakfast cereal which is advertised to contain 110 mg of sodium per 100 g of cereal.

▌

$$\frac{110 \times 10^{-3}\text{ g Na}}{100\text{ g total}} \times 100\% = 0.110\%\text{ Na}$$

1.63 **(a)** Calculate the mass of pure HNO_3 per mL of the concentrated acid which assays 69.8% by weight HNO_3 and has a density of 1.42 g/mL. **(b)** Calculate the mass of pure HNO_3 in 60.0 mL of concentrated acid. **(c)** What volume of the concentrated acid contains 63.0 g of pure HNO_3?

▌ **(a)** 1.00 mL of acid has a mass of 1.42 g. Since 69.8% of the total mass of the acid is pure HNO_3, then the number of grams of HNO_3 in 1.00 mL of acid is $0.698 \times 1.42\text{ g} = 0.991\text{ g}$. **(b)** Mass of HNO_3 in 60.0 mL of acid = 60.0 mL × 0.991 g/mL = 59.5 g HNO_3. **(c)** 63.0 g HNO_3 is contained in

$$\frac{63.0\text{ g}}{0.991\text{ g/mL}} = 63.6\text{ mL acid}$$

1.64 The density of gold is 19.3 g/cm^3. Calculate the diameter of a solid gold sphere having a mass of 422 g.

▌ The result sought is a diameter, D, in centimeters. The data are a mass of gold (in grams), its density, d (in grams per cubic centimeter), and the shape of the object (spherical). The radius, r, of a sphere is related to its volume, V, by $r = \sqrt[3]{3V/4\pi}$, and the volume of this sphere is 422 g/(19.3 g/cm^3). Thus

$$D = 2r = 2\sqrt[3]{\frac{3(422\text{ g})}{4\pi(19.3\text{ g/cm}^3)}} = 3.47\text{ cm}$$

1.65 The density of aluminum is 2.70 g/cm^3. An irregularly shaped piece of aluminum weighing 40.0 g is added to a 100-mL graduated cylinder containing exactly 50.0 mL of water. To what height in the cylinder will the water level rise?

▌ The water level will change by an amount which reflects the volume of the piece of aluminum. The volume of the piece may be determined from the density of aluminum and its mass by using either the equation for density or the factor-label method.

the equation	the factor-label method
$d = \frac{m}{V}$ $V = \frac{m}{d} = \frac{40.0\ \text{g}}{2.70\ \text{g/cm}^3} = 14.8\ \text{cm}^3$	$\overbrace{40.0\ \text{g}}^{\text{quantity given}}\underbrace{\left(\frac{1\ \text{cm}^3}{2.70\ \text{g}}\right)}_{\text{ratio}} = 14.8\ \text{cm}^3$

The water level in the cylinder is the volume of the water plus that of the aluminum: 50.0 mL + 14.8 mL = 64.8 mL.

1.66 The density of a salt solution is 1.13 g/cm³. The solution contains 17.0% sodium chloride. What volume of solution will contain 35.0 g of NaCl?

▌
$$(35.0\ \text{g NaCl})\left(\frac{100\ \text{g solution}}{17.0\ \text{g NaCl}}\right)\left(\frac{1\ \text{cm}^3}{1.13\ \text{g solution}}\right) = 182\ \text{cm}^3$$

1.67 A mixture of gasoline and alcohol contains 22.0% alcohol. The density of the mixture is 0.800 g/mL. What mass of alcohol is there in 40.0 mL of the mixture?

▌
$$(40.0\ \text{mL mixture})\left(\frac{0.800\ \text{g mixture}}{\text{mL mixture}}\right)\left(\frac{22.0\ \text{g alcohol}}{100\ \text{g mixture}}\right) = 7.04\ \text{g alcohol}$$

1.68 Battery acid has a density of 1.285 g/cm³ and contains 38.0% by weight H_2SO_4. How many grams of pure H_2SO_4 are contained in a liter of battery acid?

▌ 1.000 cm³ of acid has a mass of 1.285 g. Then 1.000 L of acid has a mass of 1285 g. Since 38.0% by weight of the acid is pure H_2SO_4, the number of grams of H_2SO_4 in 1.000 L of battery acid is $0.380 \times 1285\ \text{g} = 488\ \text{g}$.
Formally, the above solution can be written as follows:

$$\text{Mass of } H_2SO_4 = 1285\ \text{g acid} \times \frac{38.0\ \text{g } H_2SO_4}{100\ \text{g acid}} = 488\ \text{g } H_2SO_4$$

Here the conversion factor $38.0\ \text{g } H_2SO_4/100\ \text{g acid}$ is taken to be equal to 1. Although the condition $38.0\ \text{g } H_2SO_4 = 100\ \text{g acid}$ is not a universal truth in the same sense that 1 in. always equals 2.54 cm, the condition is a rigid one of association of 38.0 g H_2SO_4 with every 100 g acid for this particular acid preparation. Mathematically, these two quantities may be considered to be equal for this problem.

1.69 A nugget of gold and quartz weighs 100 g. The densities of gold, quartz, and the nugget are 19.3, 2.65, and 6.4 g/cm³, respectively. Determine the weight of gold in the nugget.

▌ Let x = grams of gold in nugget; then $100\ \text{g} - x$ = grams of quartz in nugget.

Volume of nugget = (volume of gold in nugget) + (volume of quartz in nugget)

$$\frac{100\ \text{g}}{6.4\ \text{g/cm}^3} = \left(\frac{x}{19.3\ \text{g/cm}^3}\right) + \left(\frac{100\ \text{g} - x}{2.65\ \text{g/cm}^3}\right)$$

from which $x = 68$ g gold.

1.70 A clay was partially dried and then contained 50% silica and 7% water. The original clay contained 12% water. What is the percentage of silica in the original sample?

▌ The original and dried clays have the compositions

	% water	% silica	% other
original	12	x	$88 - x$
dried	7	50	43

The ratio of silica to the other dry constituents must be the same in both clays; hence

$$\frac{x}{88 - x} = \frac{50}{43}$$

Solving, $x = 47$; i.e. there is 47% silica in the original clay.

1.71 Determine the value of each of the following expressions:

(*a*) $x = 0.0592 \log \frac{10^2/L}{10^{-1}/L}$ (*b*) $z = 0.0296 \log \frac{(10^{-5})^2(10^{-1})^2}{(10^3)^2}$

(*c*) $y = 0.0296 \log (3.0 \times 10^7)$ (*d*) $w = \log \frac{(10^{-5})^3(10^{-1})^3}{3 \times 10^9}$

▮ (*a*) 0.178 (*b*) −0.533 (*c*) 0.221 (*d*) −27.5

1.72 A bullet, of density 5.40 g/cm³ and mass 80.0 g, is dropped into a 100-mL graduated cylinder containing exactly 50.0 mL of water. To what height will the water level rise in the cylinder?

▮

$$\text{Volume of bullet} = (80.0\text{ g})\left(\frac{1\text{ cm}^3}{5.40\text{ g}}\right) = 14.8\text{ cm}^3 = 14.8\text{ mL}$$

$$\text{Total volume} = (50.0\text{ mL}) + (14.8\text{ mL}) = 64.8\text{ mL}$$

1.73 On the "dry basis" a sample of coal analyzes as follows: combustibles, 21.06%; fixed carbon, 71.80%; ash, 7.14%. If the moisture present in the coal is 2.49%, what is the analysis on the "wet basis"?

▮ On the wet basis, 2.49% of the sample is water, and the remaining 100% − 2.49% = 97.5% of the sample has the "dry" composition.

$$\%\text{ water} = 2.5\% \qquad \%\text{ carbon} = (71.80\%)(0.975) = 70.0\%$$

$$\%\text{ combustibles} = (21.06\%)(0.975) = 20.5\% \qquad \%\text{ ash} = (7.14\%)(0.975) = 7.0\%$$

Check: 2.5 + 20.5 + 70.0 + 7.0 = 100.0%

1.74 A Pennsylvania bituminous coal is analyzed as follows: Exactly 2.500 g is weighed into a fused silica crucible. After drying for 1 h at 110 °C, the moisture-free residue weighs 2.415 g. The crucible next is covered with a vented lid and strongly heated until no volatile matter remains. The residual coke button weighs 1.528 g. The crucible is then heated without the cover until all specks of carbon have disappeared, and the final ash weighs 0.245 g. What are the percents of moisture, volatile combustible matter (VCM), fixed carbon (FC), and ash?

▮

$$\text{moisture} = 2.500\text{ g} - 2.415\text{ g} = 0.085\text{ g} \qquad \text{FC} = 1.528\text{ g} - 0.245\text{ g} = 1.283\text{ g}$$

$$\text{VCM} = 2.415\text{ g} - 1.528\text{ g} = 0.887\text{ g} \qquad \text{ash} = 0.245\text{ g}$$

Total: 0.085 + 0.887 + 1.283 + 0.245 = 2.500 g coal

$$\text{Fraction of moisture} = \frac{0.085\text{ g}}{2.500\text{ g}} = 0.034 = 3.4\%$$

Similarly, the other percentages are calculated to be: 35.5% VCM, 51.3% FC, 9.8% ash.

1.75 A granulated sample of aircraft alloy (Al, Mg, Cu) weighing 8.72 g was first treated with alkali to dissolve the aluminum, then with very dilute HCl to dissolve the magnesium, leaving a residue of copper. The residue after alkali-boiling weighed 2.10 g, and the acid-insoluble residue from this weighed 0.69 g. What is the composition of the alloy?

▮

$$\text{wt Al} = 8.72\text{ g} - 2.10\text{ g} = 6.62\text{ g} \qquad \text{wt Mg} = 2.10\text{ g} - 0.69\text{ g} = 1.41\text{ g} \qquad \text{wt Cu} = 0.69\text{ g}$$

$$\text{fraction Al} = \frac{6.62\text{ g}}{8.72\text{ g}} = 0.759 = 75.9\% \qquad \text{fraction Cu} = \frac{0.69\text{ g}}{8.72\text{ g}} = 0.079 = 7.9\%$$

$$\text{fraction Mg} = \frac{1.41\text{ g}}{8.72\text{ g}} = 0.162 = 16.2\%$$

Check: 75.9 + 16.2 + 7.9 = 100.0%

1.76 How much 58.0% sulfuric acid solution is needed to provide 150 g of H_2SO_4?

▮

$$\left(\frac{0.580\text{ g } H_2SO_4}{\text{g soln}}\right)x = 150\text{ g } H_2SO_4 \qquad \text{so} \qquad x = 259\text{ g soln}$$

1.77 A liter flask is filled with two liquids (A and B) of specific gravity (density relative to that of water) 1.4 together. The specific gravity of liquid A is 0.80 and of liquid B, 1.80. What volume of each exists in the mixture? Assume no change of volume on mixing.

Let x = volume of A in milliliters; then $1000 - x$ = volume of B in milliliters, and

$$0.80x + 1.80(1000 - x) = 1.4(1000)$$

Solving, $x = 400$ mL A and $1000 - x = 600$ mL B

1.78 A clay contains 45% silica and 10% water. What is the percentage of silica in the clay on a dry (water-free) basis?

Assume 100.0 g of wet clay. Then there will be 90.0 g of dry clay, containing 45.0 g silica.

$$\frac{45.0 \text{ g silica}}{90.0 \text{ g dry clay}} \times 100\% = 50.0\% \text{ silica in dry clay}$$

1.79 A coal contains 2.4% water. After drying, the moisture-free residue contains 71.0% carbon. Determine the percentage of carbon on the "wet basis."

Assume a 100.0 g sample of wet coal. Then there is 97.6 g of dry coal.

$$(97.6 \text{ g})(0.710) = 69.3 \text{ g C} \quad \text{or} \quad 69.3\% \text{ C in the sample}$$

1.80 A household cement gave the following analytical date: A 28.5 g sample, on dilution with acetone, yielded a residue of 4.60 g of aluminum powder. The filtrate, on evaporation of the acetone and solvent, yielded 3.2 g of plasticized nitrocellulose which contained 0.80 g of benzene-soluble plasticizer. Determine the composition of this cement.

$$(28.5 \text{ g}) - (3.2 \text{ g}) - (4.6 \text{ g}) = 20.7 \text{ g solvent}$$

$$\frac{20.7 \text{ g solvent}}{28.5 \text{ g sample}} \times 100\% = 72.6\% \text{ solvent} \qquad \frac{2.4 \text{ g nitrocellulose}}{28.5 \text{ g sample}} \times 100\% = 8.4\% \text{ nitrocellulose}$$

$$\frac{4.60 \text{ g Al}}{28.5 \text{ g sample}} \times 100\% = 16.1\% \text{ Al} \qquad \frac{0.80 \text{ g plasticizer}}{28.5 \text{ g sample}} \times 100\% = 2.8\% \text{ plasticizer}$$

Check: $72.6 + 16.1 + 8.4 + 2.8 = 99.9\%$

1.81 A cold cream sample weighing 8.41 g lost 5.83 g of moisture on heating to 110 °C. The residue on extracting with water and drying lost 1.27 g of water-soluble glycerol. The balance was oil. Calculate the composition of this cream.

$$\text{Mass of oil} = 8.41 \text{ g} - 5.83 \text{ g} - 1.27 \text{ g} = 1.31 \text{ g oil}$$

$$\frac{5.83 \text{ g water}}{8.41 \text{ g total}} \times 100\% = 69.3\% \text{ water}$$

$$\frac{1.27 \text{ g glycerine}}{8.41 \text{ g total}} \times 100\% = 15.1\% \text{ glycerine}$$

$$\frac{1.31 \text{ g oil}}{8.41 \text{ g total}} \times 100\% = 15.6\% \text{ oil}$$

Check: 100.0%

1.82 Fatty acids spread spontaneously on water to form a monomolecular film. A benzene solution containing 0.10 mm^3 of stearic acid is dropped into a tray full of water. The acid is insoluble in water but spreads on the surface to form a continuous film area of 400 cm^2 after all of the benzene has evaporated. What is the average film thickness in angstroms?

$$1 \text{ mm}^3 = (10^{-3} \text{ m})^3 = 10^{-9} \text{ m}^3 \qquad 1 \text{ cm}^2 = (10^{-2} \text{ m})^2 = 10^{-4} \text{ m}^2$$

$$\text{Film thickness} = \frac{\text{volume}}{\text{area}} = \frac{(0.10)(10^{-9} \text{ m}^3)}{(400)(10^{-4} \text{ m}^2)} = 2.5 \times 10^{-9} \text{ m} = (2.5 \times 10^{-9} \text{ m})(10^{10} \text{ Å/m}) = 25 \text{ Å}$$

1.83 A porous catalyst for chemical reactions has an internal surface area of 800 m^2 per cubic centimeter of bulk material. Fifty percent of the bulk volume consists of the pores (holes), while the other 50% of the volume is made up of the solid substance. Assume that the pores are all cylindrical tubules of uniform diameter d and length l, and that the measured internal surface area is the total area of the curved surfaces of the tubules. What is the diameter of each pore? [The volume of a cylinder is $V = \frac{1}{4}\pi d^2 l$.]

▮ Consider 1.0 cm³ of the catalyst.

$$V_{\text{holes}} = 0.50\ \text{cm}^3 = (0.50\ \text{cm}^3)\left(\frac{10^8\ \text{Å}}{\text{cm}}\right)^3 = 0.50 \times 10^{24}\ \text{Å}^3 = \frac{1}{4}\pi d^2 l$$

$$A_{\text{holes}} = (800\ \text{m}^2)\left(\frac{10^{10}\ \text{Å}}{\text{m}}\right)^2 = 800 \times 10^{20}\ \text{Å}^2 = \pi d l$$

$$\frac{\frac{1}{4}\pi d^2 l}{\pi d l} = \frac{d}{4} = \frac{0.50 \times 10^{24}\ \text{Å}^3}{8.0 \times 10^{22}\ \text{Å}^2} = \frac{50}{8.0}\ \text{Å} \quad \text{so} \quad d = 4\left(\frac{50}{8.0}\right)\ \text{Å} = 25\ \text{Å}$$

1.5 ENGLISH-METRIC CONVERSIONS

1.84 Convert 5.00 in. to (*a*) centimeters (*b*) millimeters (*c*) meters

▮ (*a*) $5.00\ \text{in.} = (5.00\ \text{in.})(2.54\ \text{cm/in.}) = 12.7\ \frac{\text{in.}\cdot\text{cm}}{\text{in.}} = 12.7\ \text{cm}$

The procedure can be understood easily in terms of the word definition of the conversion factor: 2.54 is the number of centimeters *per* inch; that is, the number of centimeters in 1 in. Thus the number of centimeters in 5 in. is 5×2.54.
More formally, the conversion factor may be considered to be a statement of equality between 2.54 cm and 1 in. Since $2.54\ \text{cm} = 1\ \text{in.}$,

$$2.54\ \text{cm per in.} = 2.54\ \text{cm/in.} = \frac{2.54\ \text{cm}}{1\ \text{in.}} = \frac{2.54\ \text{cm}}{2.54\ \text{cm}} = 1$$

Thus the conversion factor, 2.54 cm/in., is mathematically equal to 1, so that any quantity may be multiplied or divided by the conversion factor without changing the essential value of the quantity.

(*b*) $12.7\ \text{cm} = (12.7\ \text{cm})(10\ \text{mm/cm}) = 127\ \frac{\text{cm}\cdot\text{mm}}{\text{cm}} = 127\ \text{mm}$

(*c*) $12.7\ \text{cm} = \frac{12.7\ \text{cm}}{100\ \text{cm/m}} = 0.127\ \frac{\text{cm}\cdot\text{m}}{\text{cm}} = 0.127\ \text{m}$

Here it was appropriate to divide by the conversion factor. If by mistake we had multiplied 12.7 cm by 100 cm/m, the answer would have been expressed in square centimeters per meter, since

$$\text{cm} \times \frac{\text{cm}}{\text{m}} = \frac{\text{cm}^2}{\text{m}}$$

We would immediately realize our error, seeing that the answer is not expressed in meters.

1.85 Convert (*a*) 14.0 cm and (*b*) 7.00 m to inches.

▮ (*a*) $14.0\ \text{cm} = \frac{14.0\ \text{cm}}{2.54\ \text{cm/in.}} = 5.51\ \text{in.}$ or $14.0\ \text{cm} \times 0.3937\ \text{in./cm} = 5.51\ \text{in.}$

Note that the reciprocal of a conversion factor is also a conversion factor (since the reciprocal of 1 is 1).

$$2.54\ \text{cm/in.} = 1 \qquad \frac{1}{2.54\ \text{cm/in.}} = 0.3937\ \text{in./cm} = 1$$

(*b*) $7.00\ \text{m} = \frac{700\ \text{cm}}{2.54\ \text{cm/in.}} = 276\ \text{in.}$ or $7.00\ \text{m} \times 39.37\ \text{in./m} = 276\ \text{in.}$

1.86 Determine the mass of 66 lb of sulfur in (*a*) kilograms and (*b*) grams. (*c*) Find the mass in pounds of 3.4 kg of copper.

▮ (*a*) $66\ \text{lb} = 66\ \text{lb} \times 0.454\ \text{kg/lb} = 30\ \text{kg}$ or $66\ \text{lb} = \frac{66\ \text{lb}}{2.2\ \text{lb/kg}} = 30\ \text{kg}$

(*b*) $30\ \text{kg} = 30\,000\ \text{g}$ or $66\ \text{lb} = 66\ \text{lb} \times 454\ \text{g/lb} = 30\,000\ \text{g}$

(*c*) $3.4\ \text{kg} = 3.4\ \text{kg} \times 2.2\ \text{lb/kg} = 7.5\ \text{lb}$

1.87 A tennis ball was observed to travel at a speed of 95.0 miles per hour. Express this in meters per second.

$$95.0\ \text{mi/h} = \left(95.0\,\frac{\text{mi}}{\text{h}}\right)\left(1.609\times10^3\,\frac{\text{m}}{\text{mi}}\right)\left(\frac{1\ \text{h}}{3.60\times10^3\ \text{s}}\right) = 42.5\ \text{m/s}$$

The conversion factor relating the meter to the mile was taken from Prob. 1.17.

1.88 Determine the number of (*a*) millimeters in 10.0 in., (*b*) feet in 5.00 m, (*c*) centimeters in 4 ft 3 in.

(*a*) $$(10.0\ \text{in.})\left(\frac{2.54\ \text{cm}}{\text{in.}}\right)\left(\frac{10\ \text{mm}}{\text{cm}}\right) = 254\ \text{mm}$$

(*b*) $$(5.00\ \text{m})\left(\frac{100\ \text{cm}}{\text{m}}\right)\left(\frac{1\ \text{in.}}{2.54\ \text{cm}}\right)\left(\frac{1\ \text{ft}}{12\ \text{in.}}\right) = 16.4\ \text{ft}$$

(*c*) $$4\ \text{ft}\ 3\ \text{in.} = (4\ \text{ft})\left(\frac{12\ \text{in.}}{\text{ft}}\right) + (3\ \text{in.}) = 51\ \text{in.} \qquad \text{and} \qquad (51\ \text{in.})\left(\frac{2.54\ \text{cm}}{\text{in.}}\right) = 130\ \text{cm}$$

1.89 Express the weight (mass) of 32.0 g of oxygen in milligrams, in kilograms, and in pounds.

$$(32.0\ \text{g})\left(\frac{10^3\ \text{mg}}{\text{g}}\right) = 32\,000\ \text{mg} \qquad (32.0\ \text{g})\left(\frac{1\ \text{kg}}{10^3\ \text{g}}\right) = 0.0320\ \text{kg} \qquad (32.0\ \text{g})\left(\frac{1\ \text{lb}}{454\ \text{g}}\right) = 0.0705\ \text{lb}$$

1.90 How many grams are there in 5.00 lb of copper sulfate? How many pounds are there in 4.00 kg of mercury? How many milligrams are there in 1 lb 2 oz of sugar?

$$(5.00\ \text{lb})\left(\frac{454\ \text{g}}{\text{lb}}\right) = 2270\ \text{g} \qquad (4.00\ \text{kg})\left(\frac{2.202\ \text{lb}}{\text{kg}}\right) = 8.81\ \text{lb} \qquad (1.125\ \text{lb})\left(\frac{454\ \text{g}}{\text{lb}}\right)\left(\frac{10^3\ \text{mg}}{\text{g}}\right) = 5.1\times10^5\ \text{mg}$$

1.91 Convert the weight (mass) of 500 lb of coal to (*a*) kilograms (*b*) metric tons (*c*) U.S. tons (1 ton = 2000 lb)

(*a*) $$(500\ \text{lb})\left(\frac{454\ \text{g}}{\text{lb}}\right)\left(\frac{1\ \text{kg}}{10^3\ \text{g}}\right) = 227\ \text{kg}$$ (*b*) $$(227\ \text{kg})\left(\frac{1\ \text{metric ton}}{10^3\ \text{kg}}\right) = 0.227\ \text{metric ton}$$

(*c*) $$(500\ \text{lb})\left(\frac{1\ \text{ton}}{2000\ \text{lb}}\right) = 0.250\ \text{ton}$$

1.92 How many square inches are there in one square meter?

$$1\ \text{m} = \frac{100\ \text{cm}}{2.54\ \text{cm/in.}} = 39.37\ \text{in.} \qquad 1\ \text{m}^2 = (1\ \text{m})^2 = (39.37\ \text{in.})^2 = 1550\ \text{in.}^2$$

1.93 A wood block, 10 in. × 6.0 in. × 2.0 in., weighs 3 lb 10 oz. What is the density of the wood in kilograms per cubic meter?

$$V = (10\ \text{in.})(6.0\ \text{in.})(2.0\ \text{in.}) = 120\ \text{in.}^3 = (120\ \text{in.}^3)\left(\frac{2.54\ \text{cm}}{\text{in.}}\right)^3 = 1966\ \text{cm}^3$$

$$m = (58\ \text{oz})\left(\frac{28.35\ \text{g}}{\text{oz}}\right) = 1644\ \text{g} \qquad \text{and} \qquad d = \frac{1644\ \text{g}}{1966\ \text{cm}^3} = 0.84\ \text{g/cm}^3 = 840\ \text{kg/m}^3$$

1.94 A uniform steel bar is 16.0 inches long; its mass is 6 lb 4 oz. Determine the mass per unit length of the bar in grams per centimeter.

$$16.0\ \text{in.} = (16.0\ \text{in.})(2.54\ \text{cm/in.}) = 40.6\ \text{cm} \qquad 6.25\ \text{lb} = (6.25\ \text{lb})(454\ \text{g/lb}) = 2840\ \text{g} \qquad \frac{2840\ \text{g}}{40.6\ \text{cm}} = 69.9\ \text{g/cm}$$

1.95 A pressure of one atmosphere is equal to 101.3 kPa. Express this pressure in pounds force per square inch. The pound force is 4.448 N and 1 Pa = 1 N/m^2.

$$1\ \text{atm} = 101.3\ \text{kPa} = (101.3\times10^3\ \text{N/m}^2)\left(\frac{1\ \text{lbf}}{4.448\ \text{N}}\right)\left(\frac{2.540\times10^{-2}\ \text{m}}{1\ \text{in.}}\right)^2 = 14.69\ \text{lbf/in.}^2$$

Notice that the conversion factor between m and in. is squared to give the conversion factor between m^2 and in.^{2}.

1.96 New York City's 7.9 million people have a daily per capita consumption of 173 gal of water. How many tons of sodium fluoride (45% fluorine by weight) would be required per year to give this water a tooth-strengthening dose of 1 part (by weight) fluorine per million parts water? One U.S. gallon of water at normal room temperature weighs 8.34 lbf (i.e., has a mass of 8.34 lb).

▮ The mass of water, in tons, required per year is

$$[(7.9 \times 10^6) \times 173 \times 365]\left(\frac{\text{gal water}}{\text{yr}}\right)\left(\frac{8.34\ \text{lb water}}{1\ \text{gal water}}\right)\left(\frac{1\ \text{ton}}{2000\ \text{lb}}\right) = 2.08 \times 10^9\ \frac{\text{tons water}}{\text{yr}}$$

so the mass of sodium fluoride, in tons, required per year is

$$\left(2.08 \times 10^9\ \frac{\text{tons water}}{\text{yr}}\right)\left(\frac{1\ \text{ton fluorine}}{10^6\ \text{tons water}}\right)\left(\frac{1\ \text{ton sodium fluoride}}{0.45\ \text{ton fluorine}}\right) = 4.6 \times 10^3\ \frac{\text{tons sodium fluoride}}{\text{yr}}$$

1.97 The density of cast iron is 7200 kg/m³. Calculate its density in pounds per cubic foot.

▮

$$\text{Density} = \left(7200\ \frac{\text{kg}}{\text{m}^3}\right)\left(\frac{1\ \text{lb}}{0.454\ \text{kg}}\right)\left(\frac{0.3048\ \text{m}}{1\ \text{ft}}\right)^3 = 449\ \text{lb/ft}^3$$

The two conversion factors were taken from Prob. 1.17.

1.98 A casting of an alloy in the form of a disk weighed 50.0 g. The disk was 0.250 in. thick and had a circular cross section of diameter 1.380 in. What is the density of the alloy, in grams per cubic centimeter?

▮

$$\text{Volume of the cylinder} = \frac{\pi d^2 h}{4} = \left[\frac{\pi(1.380)^2(0.250)}{4}\ \text{in.}^3\right]\left(\frac{2.54\ \text{cm}}{1\ \text{in.}}\right)^3 = 6.13\ \text{cm}^3$$

$$\text{Density of the alloy} = \frac{\text{mass}}{\text{volume}} = \frac{50.0\ \text{g}}{6.13\ \text{cm}^3} = 8.15\ \text{g/cm}^3$$

1.99 The density of zinc is 455 lb/ft³. Find the mass of 9.00 cm³ of zinc.

▮ First express the density in grams per cubic centimeter:

$$\left(455\ \frac{\text{lb}}{\text{ft}^3}\right)\left(\frac{1\ \text{ft}}{30.48\ \text{cm}}\right)^3\left(\frac{454\ \text{g}}{\text{lb}}\right) = 7.29\ \frac{\text{g}}{\text{cm}^3}$$

Thus 1.00 cm³ of zinc has a mass of 7.29 g, so 9.00 cm³ has a mass of $9.00 \times 7.29\ \text{g} = 65.6\ \text{g}$.

1.100 Convert 22.4 L to cubic centimeters, to cubic meters, and to cubic feet.

▮

$$(22.4\ \text{L})\left(\frac{10^3\ \text{cm}^3}{\text{L}}\right) = 2.24 \times 10^4\ \text{cm}^3 \qquad (22.4\ \text{L})\left(\frac{1\ \text{m}^3}{10^3\ \text{L}}\right) = 2.24 \times 10^{-2}\ \text{m}^3$$

$$(2.24 \times 10^4\ \text{cm}^3)\left(\frac{1\ \text{in.}}{2.54\ \text{cm}}\right)^3\left(\frac{1\ \text{ft}}{12\ \text{in.}}\right)^3 = 0.791\ \text{ft}^3$$

Note that when a conversion factor is cubed, *each number and unit* within it is cubed.

1.101 Determine the number of **(a)** cubic centimeters in a cubic inch, **(b)** cubic inches in a liter, **(c)** cubic feet in a cubic meter.

▮ **(a)** $(1.00\ \text{in.}^3)\left(\frac{2.54\ \text{cm}}{\text{in.}}\right)^3 = 16.4\ \text{cm}^3$ **(b)** $(1.00\ \text{L})\left(\frac{1000\ \text{cm}^3}{\text{L}}\right)\left(\frac{1\ \text{in.}}{2.54\ \text{cm}}\right)^3 = 61.0\ \text{in.}^3$

(c) $(1.00\ \text{m}^3)\left(\frac{100\ \text{cm}}{\text{m}}\right)^3\left(\frac{1\ \text{in.}}{2.54\ \text{cm}}\right)^3\left(\frac{1\ \text{ft}}{12\ \text{in.}}\right)^3 = 35.3\ \text{ft}^3$

1.102 In a crystal of platinum, centers of individual atoms are 2.8 Å apart along the direction of closest packing. How many atoms would lie on a 1.0-in. length of a line in this direction?

▮

$$(1.0\ \text{in.})\left(\frac{2.54\ \text{cm}}{\text{in.}}\right)\left(\frac{10^8\ \text{Å}}{\text{cm}}\right)\left(\frac{1\ \text{atom}}{2.8\ \text{Å}}\right) = 9.1 \times 10^7\ \text{atoms/in.}$$

1.103 The density of water is 1.000 kg/L at 4 °C. Calculate the density of water in pounds per cubic foot at the same temperature.

▮
$$\left(\frac{1000\text{ g}}{1000\text{ cm}^3}\right)\left(\frac{1\text{ lb}}{454\text{ g}}\right)\left(\frac{2.54\text{ cm}}{1\text{ in.}}\right)^3\left(\frac{12\text{ in.}}{\text{ft}}\right)^3 = 62.4\text{ lb/ft}^3$$

1.104 What is the average speed, in miles per hour, of a sprinter in doing the 100-m dash in 10.1 s?

▮
$$\left(\frac{100.0\text{ m}}{10.1\text{ s}}\right)\left(\frac{10^2\text{ cm}}{\text{m}}\right)\left(\frac{1\text{ in.}}{2.54\text{ cm}}\right)\left(\frac{1\text{ ft}}{12\text{ in.}}\right)\left(\frac{1\text{ mi}}{5280\text{ ft}}\right)\left(\frac{3600\text{ s}}{\text{h}}\right) = 22.1\text{ mi/h}$$

1.105 The silica gel which is used to protect sealed overseas shipments from moisture seepage has a surface area of 6.0×10^5 m^2/kg. What is this surface area in square feet per gram?

▮
$$\left(\frac{6.0\times 10^5\text{ m}^2}{\text{kg}}\right)\left(\frac{1\text{ ft}}{0.3048\text{ m}}\right)^2\left(\frac{1\text{ kg}}{10^3\text{ g}}\right) = 6.5\times 10^3\text{ ft}^2\text{/g}$$

1.106 There is reason to think that the length of the day, determined from the earth's period of rotation, is increasing uniformly by about 0.0010 s every century. What is this variation in parts per billion?

▮
$$\left(\frac{0.0010\text{ s}}{100\text{ y}}\right)\left(\frac{1\text{ y}}{365\text{ days}}\right)\left(\frac{1\text{ day}}{24\text{ h}}\right)\left(\frac{1\text{ h}}{3600\text{ s}}\right) = 3.2\times 10^{-13} = \frac{3.2\times 10^{-4}\text{ s}}{10^9\text{ s}} = 3.2\times 10^{-4}\text{ ppb}$$

1.107 Determine the mass of 20.0 ft^3 of aluminum (density = 2.70 g/cm^3).

▮
$$(20.0\text{ ft}^3)\left(\frac{12\times 2.54\text{ cm}}{\text{ft}}\right)^3\left(\frac{2.70\text{ g}}{\text{cm}^3}\right) = 1.53\times 10^6\text{ g} = 1.53\times 10^3\text{ kg}$$

1.108 Air weighs about 8 lbf per 100 ft^3. Find its density in ***(a)*** grams per cubic foot, ***(b)*** grams per liter, ***(c)*** kilograms per cubic meter.

▮ ***(a)***
$$\left(\frac{8\text{ lb}}{100\text{ ft}^3}\right)\left(\frac{454\text{ g}}{\text{lb}}\right) = 36\text{ g/ft}^3$$

(b)
$$\left(\frac{36\text{ g}}{\text{ft}^3}\right)\left(\frac{1\text{ ft}}{12\times 2.54\text{ cm}}\right)^3\left(\frac{10^3\text{ cm}^3}{\text{L}}\right) = 1.3\text{ g/L}$$

(c)
$$\left(\frac{1.3\text{ g}}{\text{L}}\right)\left(\frac{10^3\text{ L}}{\text{m}^3}\right)\left(\frac{1\text{ kg}}{10^3\text{ g}}\right) = 1.3\text{ kg/m}^3$$

1.109 An electrolytic tin-plating process gives a coating 30 millionths of an inch thick. How many square meters can be coated with 1 kg of tin, density 7300 kg/m^3?

▮
$$V = (1.00\text{ kg})\left(\frac{1\text{ m}^3}{7300\text{ kg}}\right) = 1.37\times 10^{-4}\text{ m}^3 \quad\text{and}\quad t = (30\times 10^{-6}\text{ in.})\left(\frac{0.0254\text{ m}}{\text{in.}}\right) = 7.62\times 10^{-7}\text{ m}$$

$$A = \frac{V}{t} = \frac{1.37\times 10^{-4}\text{ m}^3}{7.62\times 10^{-7}\text{ m}} = 180\text{ m}^2$$

1.110 There is available 10 tons of a coal containing 2.5% sulfur, and also supplies of two coals containing 0.80% and 1.10% sulfur. How many tons of each of the latter should be mixed with the original 10 tons to give 20 tons containing 1.7% sulfur?

▮ Let x = mass of 0.80% S; then $10.0 - x$ = mass of 1.10% S.

$$\text{Mass of sulfur} = (0.25\text{ ton}) + 0.0080x + 0.0110(10.0 - x) = 0.34\text{ ton}$$

Solving, x = 6.7 tons 0.80% S and $10.0 - x$ = 3.3 tons 1.10% S.

1.111 A typical formulation for a cationic asphalt emulsion calls for 0.5% tallow amine emulsifier and 70% asphalt; the rest consists of water and water-soluble ingredients. How much asphalt can be emulsified per pound of the emulsifier?

▮ On the basis of 100 lb emulsion,

$$\frac{70.0\text{ lb asphalt}}{0.50\text{ lb emulsifier}} = \frac{140\text{ lb asphalt}}{\text{lb emulsifier}}$$

1.112 Two unblended manganese ores contain 40 and 25% manganese, respectively. How many pounds of each ore must be mixed to give 100 lb of blended ore containing 35% manganese?

▌ Let x = pounds of 40% ore required; then $100 \text{ lb} - x$ = pounds of 25% ore required.

$$\text{Mn from 40\% ore} + \text{Mn from 25\% ore} = \text{total Mn in 100 lb of mixture}$$

$$(0.40)x + (0.25)(100 \text{ lb} - x) = (0.35)(100 \text{ lb})$$

Solving, $x = 67$ lb of 40% ore. Then $100 \text{ lb} - x = 33$ lb of 25% ore.

1.113 A fusible alloy is made by melting together 10.6 lb bismuth, 6.40 lb lead, and 3.00 lb tin. ***(a)*** What is the percentage composition of the alloy? ***(b)*** How much of each metal is required to make 70.0 g of alloy? ***(c)*** What weight of alloy can be made from 4.2 lb of tin?

▌ *(a)* $(10.6 \text{ lb}) + (6.40 \text{ lb}) + (3.00 \text{ lb}) = 20.0 \text{ lb total}$

$$\% \text{ Bi} = \frac{10.6 \text{ lb Bi}}{20.0 \text{ lb total}} \times 100\% = 53.0\% \text{ Bi} \qquad \% \text{ Pb} = \frac{6.40 \text{ lb Pb}}{20.0 \text{ lb total}} \times 100\% = 32.0\% \text{ Pb}$$

$$\% \text{ Sn} = 100.0\% - 53.0\% - 32.0\% = 15.0\% \text{ Sn}$$

(b) $$(70.0 \text{ g total})\left(\frac{53.0 \text{ g Bi}}{100 \text{ g total}}\right) = 37.1 \text{ g Bi} \qquad (70.0 \text{ g total})\left(\frac{32.0 \text{ g Pb}}{100 \text{ g total}}\right) = 22.4 \text{ g Pb}$$

$$(70.0 \text{ g}) - (37.1 \text{ g}) - (22.4 \text{ g}) = 10.5 \text{ g Sn}$$

(c) $$(4.2 \text{ lb Sn})\left(\frac{100 \text{ g total}}{15.0 \text{ g Sn}}\right) = 28 \text{ lb total}$$

1.114 The blue iridescence of butterfly wings is due to striations which are 0.15 μm apart, as measured by the electron microscope. ***(a)*** What is this distance in millionths of an inch? ***(b)*** How does this spacing compare with the wavelength of blue light, about 4500 Å?

▌

(a) $$(0.15 \ \mu\text{m})\left(\frac{100 \text{ cm}}{10^6 \ \mu\text{m}}\right)\left(\frac{1 \text{ in.}}{2.54 \text{ cm}}\right) = 5.9 \times 10^{-6} \text{ in. or 5.9 millionths of an inch}$$

(b) $$(4500 \text{ Å})\left(\frac{1 \text{ m}}{10^{10} \text{ Å}}\right) = 4.5 \times 10^{-7} \text{ m} = 0.45 \times 10^{-6} \text{ m} = 0.45 \ \mu\text{m}$$

$$\frac{0.15 \ \mu\text{m}}{0.45 \ \mu\text{m}} = \frac{1}{3} \qquad \text{(one-third of the wavelength)}$$

1.115 The thickness of a soap bubble film at its thinnest (bimolecular) stage is about 60 Å. ***(a)*** What is this thickness in inches? ***(b)*** How does this thickness compare with the wavelength of yellow sodium light, which is 0.5890 μm?

▌

(a) $$(60 \text{ Å})\left(\frac{1 \text{ cm}}{10^8 \text{ Å}}\right)\left(\frac{1 \text{ in.}}{2.54 \text{ cm}}\right) = 2.4 \times 10^{-7} \text{ in.}$$

(b) $$(0.5890 \ \mu\text{m})\left(\frac{10^{10} \text{ Å}}{10^6 \ \mu\text{m}}\right) = 5890 \text{ Å} \qquad \frac{60 \text{ Å}}{5890 \text{ Å}} \approx \frac{1}{100} \qquad \text{(about one-hundredth the wavelength)}$$

1.116 An average man requires about 2.00 mg of riboflavin (vitamin B_2) per day. How many pounds of cheese would a man have to eat per day if this were his only source of riboflavin and if the cheese contained 5.5 μg riboflavin per gram?

▌

$$(2.00 \text{ mg B}_2)\left(\frac{10^3 \ \mu\text{g}}{\text{mg}}\right)\left(\frac{1 \text{ g cheese}}{5.5 \ \mu\text{g}}\right)\left(\frac{1 \text{ lb}}{454 \text{ g}}\right) = 0.80 \text{ lb cheese}$$

1.117 The bromine content of average ocean water is 65 parts by weight per million. Assuming 100% recovery, how many cubic meters of ocean water must be processed to produce 1.0 lb of bromine? Assume that the density of seawater is 1.0×10^3 kg/m^3.

▌

$$(1.0 \text{ lb Br})\left(\frac{454 \text{ g}}{\text{lb}}\right)\left(\frac{10^6 \text{ g water}}{65 \text{ g Br}}\right)\left(\frac{1 \text{ m}^3 \text{ water}}{10^6 \text{ g water}}\right) = 7.0 \text{ m}^3$$

1.6 TEMPERATURE SCALES

1.118 Change ***(a)*** 40 °C and ***(b)*** −5 °C to the Kelvin scale.

▌ *(a)* $T = (40 + 273) \text{ K} = 313 \text{ K}$ *(b)* $T = (-5 + 273) \text{ K} = 268 \text{ K}$

1.119 Convert (*a*) 220 K and (*b*) 498 K to the Celsius scale.

▮ (*a*) $t = (220 - 273)\ °C = -53\ °C$ (*b*) $t = (498 - 273)\ °C = 225\ °C$

1.120 (*a*) Express 10 °C and 20 °C in kelvins. (*b*) Calculate the difference between these two temperatures in both scales. What is the relationship between a temperature *change* in Celsius and in Kelvin?

▮ (*a*) 283 K and 293 K. (*b*) The difference (or change) is $(20\ °C) - (10\ °C) = 10\ °C$, or $(293\ K) - (283\ K) = 10\ K$. The difference (or change) in temperature is the same on the two scales since their degree-sizes are the same.

1.121 The temperature of Dry Ice (sublimation temperature at normal pressure) is −109 °F. Is this higher or lower than the temperature of boiling ethane (a component of bottled gas), which is −88 °C?

▮ $$\tfrac{5}{9}(-109\ °F - 32°) = -78.3\ °C \qquad \text{(higher)}$$

1.122 Gabriel Fahrenheit in 1714 suggested for the zero point on his scale the lowest temperature then obtainable from a mixture of salt and ice, and for his 100° point he suggested the highest known normal animal temperature. Express these "extremes" in degrees Celsius.

▮ $$\tfrac{5}{9}(0\ °F - 32°) = -17.8\ °C \qquad \tfrac{5}{9}(100\ °F - 32°) = 37.8\ °C$$

1.123 Convert 300 K, 760 K, and 180 K to degrees Celsius.

▮ $$300\ K - 273° = 27\ °C \qquad 760\ K - 273° = 487\ °C \qquad 180\ K - 273° = -93\ °C$$

1.124 Express 8 K and 273 K in degrees Fahrenheit.

▮ $$8\ K - 273° = -265\ °C \qquad \tfrac{9}{5}(-265\ °C) + 32 = -445\ °F$$
$$273\ K - 273° = 0\ °C \qquad \tfrac{9}{5}(0\ °C) + 32 = 32\ °F$$

1.125 Convert 14 °F to degrees Celsius and kelvins.

▮ $$\tfrac{5}{9}(14\ °F - 32°) = -10\ °C = 263\ K$$

1.126 At what temperature have the Celsius and Fahrenheit readings the same numerical value?

▮ $$x = \tfrac{5}{9}(x - 32) \qquad \text{or} \qquad x = -40\ °C = -40\ °F$$

1.127 A water-stabilized electric arc was reported to have reached a temperature of 25 600 °F. On the absolute scale, what is the ratio of this temperature to that of an oxyacetylene flame, 3500 °C?

▮ $$\tfrac{5}{9}(25\,600\ °F - 32°) = (14\,204\ °C) + 273 = 14\,477\ K \qquad \text{and} \qquad 3500\ °C + 273 = 3773\ K$$
$$\text{Ratio} = \frac{14\,477\ K}{3773\ K} = 3.84$$

1.128 Construct a temperature scale on which the freezing and boiling points of water are 100° and 400°, respectively, and the degree interval is a constant multiple of the Celsius degree interval. (*a*) What is the absolute zero on this scale, and (*b*) what is the boiling point of sulfur, which is 444.6 °C?

▮ (*a*) $3(-273\ °C) + 100° = -719°$ (*b*) $3(444.6\ °C) + 100° = 1434°$

1.129 (*a*) Convert 68 °F to °C; 5 °F to °C; 176 °F to °C. (*b*) Convert 30 °C to °F; 5 °C to °F; −20 °C to °F.

▮ (*a*) $\tfrac{5}{9}(68\ °F - 32°) = 20\ °C$ $\tfrac{5}{9}(5\ °F - 32°) = -15\ °C$ $\tfrac{5}{9}(176\ °F - 32°) = 80\ °C$
(*b*) $\tfrac{9}{5}(30\ °C) + 32° = 86\ °F$ $\tfrac{9}{5}(5\ °C) + 32° = 41\ °F$ $\tfrac{9}{5}(-20\ °C) + 32° = -4\ °F$

1.130 Convert the following temperatures: −195.5 °C to °F; −430 °F to °C; 1705 °C to °F.

▮ $\tfrac{9}{5}(-195.5\ °C) + 32° = -319.9\ °F$ $\tfrac{5}{9}(-430\ °F - 32°) = -256.7\ °C$ $\tfrac{9}{5}(1705\ °C) + 32° = 3101\ °F$

1.131 During the course of an experiment, laboratory temperature rose 0.800 °C. Express this rise in degrees Fahrenheit.

▮ Temperature *intervals* are converted differently than temperature *readings*. For intervals, 100 °C = 180 °F, or 5 °C = 9 °F; hence

$$0.800\ °C = (0.800\ °C)\left(\frac{9\ °F}{5\ °C}\right) = 1.44\ °F$$

1.132 Mercury (*a*) boils at 675 °F and (*b*) solidifies at −38.0 °F, at 1 atm pressure. Express these temperatures in degrees Celsius.

(*a*) $t = \frac{5}{9}(675 - 32)\ ^\circ\text{C} = \frac{5}{9}(643)\ ^\circ\text{C} = 357\ ^\circ\text{C}$ (*b*) $t = \frac{5}{9}(-38.0 - 32.0)\ ^\circ\text{C} = \frac{5}{9}(-70.0)\ ^\circ\text{C} = -38.9\ ^\circ\text{C}$

1.133 Ethyl alcohol (*a*) boils at 78.5 °C and (*b*) freezes at −117 °C, at 1 atm pressure. Convert these temperatures to the Fahrenheit scale.

(*a*) $t = [\frac{9}{5}(78.5) + 32]\ ^\circ\text{F} = (141 + 32)\ ^\circ\text{F} = 173\ ^\circ\text{F}$

(*b*) $t = [\frac{9}{5}(-117) + 32]\ ^\circ\text{F} = (-211 + 32)\ ^\circ\text{F} = -179\ ^\circ\text{F}$

CHAPTER 2

Structure of Matter

2.1 ELEMENTS, COMPOUNDS, MIXTURES

2.1 Define each of the following terms: (*a*) element (*b*) molecule (*c*) polyatomic ion (*d*) electrolyte (*e*) mass number (*f*) monatomic molecule.

▌ (*a*) An element is one of the hundred or so building blocks of the universe. A pure sample of an element cannot be broken down into simpler substances by chemical or physical means. (*b*) A molecule is an uncharged collection of one or more atoms joined by covalent bonds if more than one. CH_4, CO_2, and He are examples. (*c*) A polyatomic ion is a collection of two or more atoms joined by covalent bonds which has either an excess or a deficiency of electrons compared to protons, and therefore has an electric charge. CO_3^{2-} and NH_4^+ are examples. (*d*) An electrolyte is a substance which will conduct an electric current when melted or dissolved. NaCl and HCl are electrolytes because their water solutions conduct electricity well. (*e*) The mass number of an isotope is the sum of its number of protons and number of neutrons. The mass number of the most common isotope of carbon is 12, the sum of its 6 protons and 6 neutrons. (*f*) A monatomic molecule is an uncharged atom which is not bonded to any other atom. The noble gases He, Ne, etc., generally are found as monatomic molecules.

2.2 Give the symbol for (*a*) sodium (*b*) phosphorus (*c*) silver (*d*) iodine (*e*) manganese (*f*) lead

▌ (*a*) Na (*b*) P (*c*) Ag (*d*) I (*e*) Mn (*f*) Pb

2.3 Write the symbol for (*a*) lead (*b*) gold (*c*) tin (*d*) iodine

▌ (*a*) Pb (*b*) Au (*c*) Sn (*d*) I

2.4 Name: (*a*) K (*b*) Co (*c*) Ag (*d*) Cs

▌ (*a*) Potassium (*b*) cobalt (*c*) silver (*d*) cesium

2.5 Name: (*a*) H (*b*) Mg (*c*) Ne (*d*) Se (*e*) N (*f*) K

▌ (*a*) Hydrogen (*b*) magnesium (*c*) neon (*d*) selenium (*e*) nitrogen (*f*) potassium

2.6 Write the symbols for (*a*) iron (*b*) calcium (*c*) cobalt (*d*) bromine (*e*) phosphorus (*f*) chlorine

▌ (*a*) Fe (*b*) Ca (*c*) Co (*d*) Br (*e*) P (*f*) Cl

2.7 (*a*) List the elements whose symbols start with a letter that is different from the first letter of the name of the element. (*b*) Write the symbol for the element copper. (*c*) Give the group of the periodic table for each of the elements listed in (*a*) and (*b*).

▌ (*a*) Ag silver, Au gold, Fe iron, Hg mercury, K potassium, Na sodium, Pb lead, Sb antimony, Sn tin, W tungsten (*b*) Cu (*c*) Ag IB, Au IB, Fe VIIIB, Hg IIB, K IA, Na IA, Pb IVA, Sb VA, Sn IVA, W VIB, Cu IB

2.8 Using the periodic table, fill in the following:

	symbol	AW
Fluorine	_____	_____
Magnesium	_____	_____
Calcium	_____	_____

▌

	symbol	AW
Fluorine	F	19.0
Magnesium	Mg	24.3
Calcium	Ca	40.1

2.9 Write the symbol of each element and its atomic weight (AW) to three significant digits.

	symbol	AW		symbol	AW
Hydrogen	____	____	Cesium	____	____
Lithium	____	____	Fluorine	____	____
Sodium	____	____	Chlorine	____	____
Potassium	____	____	Bromine	____	____
Rubidium	____	____	Iodine	____	____

▮

	symbol	AW		symbol	AW
Hydrogen	H	1.01	Cesium	Cs	133
Lithium	Li	6.94	Fluorine	F	19.0
Sodium	Na	23.0	Chlorine	Cl	35.5
Potassium	K	39.1	Bromine	Br	79.9
Rubidium	Rb	85.5	Iodine	I	127

2.10 A combination of iron and sulfur is partially soluble in excess CS_2. The combination may be described as a ______.

▮ Different parts of the combination have different properties; it must be a *mixture*.

2.11 Zinc metal "dissolves" in aqueous HCl. Solid NaCl dissolves in water. Solid sucrose, ordinary table sugar ($C_{12}H_{22}O_{11}$), dissolves in water. How can it be demonstrated that only the last two of these are actually cases of dissolving, while the first is a chemical reaction?

▮ (1) Evaporation of the last two solutions to dryness will regenerate the NaCl and sucrose, respectively. Evaporation of the first solution will not regenerate the zinc metal. (2) $ZnCl_2$ is recovered from the first solution; NaCl from the second; sucrose from the third.

2.12 For each pair of materials listed, state several properties which can be used to distinguish between the two: (*a*) steel and aluminum (*b*) water and a solution of sodium chloride in water (*c*) water and motor oil (*d*) FeS and a mixture of iron and sulfur

▮ (*a*) Density, magnetic attraction, luster, tendency to rust (*b*) density, taste, boiling point (*c*) viscosity, density, volatility, lubricating properties (*d*) density, ability for the elements to be separated by a magnet, partial solubility in carbon disulfide, appearance.

2.13 A certain homogeneous material has a melting point of 94.6 °C. When 10 g of the material is placed in 20 mL of water, only 2 g of the material dissolves. Suggest two further experiments which could be used to determine whether the material is a mixture or a pure substance.

▮ Among other methods, one could determine if more sample dissolves to a similar degree in a second portion of water; that is, whether the first solution was saturated. Alternatively, one could determine if both the dissolved material and the undissolved material melt at 94.6 °C.

2.14 When 10.0 g of A was heated, 4.4 g of B was given off, leaving 5.6 g of E. The same quantity of B can also be prepared by combination of 1.2 g of C and 3.2 g of D. The E can be electrolyzed, after melting, to yield 4.0 g of F and 1.6 g of G, neither of which can be further decomposed by ordinary chemical means. E combines with water to give 1.3 g of J per gram of E. A combination of E with 1.5 times its mass of water yields a homogeneous material, L. Identify as far as possible each lettered material as element, compound, or mixture.

▮ A and E can be decomposed; B can be formed by a combination of C and D. Hence A, B, and E are compounds or mixtures, probably the former. F and G are elements, being undecomposable. J is probably a compound of E with hydrogen or oxygen or both. L is probably a solution (or else a second compound of the same elements of which J is composed).

2.15 A 10-g sample of material A was placed in water, whereupon 4 g of it, B, dissolved. The remaining material, C, was placed in a second sample of water, but no change took place. The sample of B had a sharp melting point and, after melting, was electrolyzed to yield 1.5 g of D and 2.5 g of E, neither of which could be further broken down. When the sample of C was heated in air, it reacted completely to give 22 g of a gas, F. Upon cooling to −100°C, material F solidified, and upon warming it sublimed at precisely −78°C. Identify each of the lettered materials as either definitely or probably an element, a compound, or a mixture.

/ Some but not all of A is soluble in water, so it must be a mixture. B decomposes by electrolysis; it cannot be an element. Its sharp melting point suggests that it is a compound. D and E are elements, because they cannot be further broken down. F has a mass greater than that of C; hence, it must be a combination of elements. Its sharp sublimation point makes it likely that F is a compound, most likely a compound of C and an element in the air. C, since it produces a compound (F) rather than a mixture, is probably a pure substance—element or compound—but not enough information is provided to differentiate.

2.16 In its elementary form, oxygen exists as diatomic molecules. Look up the formulas of all the other nonmetals in their elementary forms in a reference book such as the *Handbook of Chemistry and Physics* (CRC Press, Cleveland, Ohio). Which, if any, occur(s) as monatomic molecules?

/ Only the noble gases exist as monatomic molecules.

2.2 ELEMENTARY ATOMIC STRUCTURE

TABLE 2.1 Nuclidic Masses

^{1}H	1.00783 u	^{12}C	12.00000 u	^{17}O	16.9991 u	^{35}Cl	34.9689 u
^{2}H	2.01410	^{13}C	13.00335	^{18}O	17.9992	^{37}Cl	36.9659
^{3}H	3.01605	^{14}C	14.00324	^{18}F	18.00094	^{36}Ar	35.9676
^{4}He	4.00260	^{16}C	16.01470	^{18}Ne	18.00571	^{38}Ar	37.9627
^{6}He	6.01889	^{14}N	14.00307	^{28}Si	27.9769	^{40}Ar	39.9624
^{6}Li	6.01512	^{15}N	15.0001	^{29}Si	28.9765	^{87}Rb	86.9092
^{7}Li	7.01600	^{16}N	16.00610	^{30}Si	29.9738		
^{7}Be	7.01693	^{16}O	15.9949	^{32}S	31.97207		

2.17 (*a*) What is the charge on a sodium ion? (*b*) What is the charge on a sodium nucleus? (*c*) What is the charge on a sodium atom?

/ (*a*) 1+ (*b*) 11+ (*c*) 0 Note that it is extremely important to read the question carefully, since a small difference in wording can make an entirely different question.

2.18 (*a*) What is the atomic number of sodium? (*b*) How many protons are there in the sodium nucleus? (*c*) How many protons are there in the sodium atom?

/ The answer to each question is 11. These seemingly different questions are all really the same question in different forms.

2.19 (*a*) For the ion $^{39}_{19}K^+$, state how many electrons, how many protons, and how many neutrons are present. (*b*) Which of these particles—electron, proton, neutron—has the smallest mass?

/ (*a*) 18 electrons, 19 protons, 20 neutrons (*b*) electron

2.20 An atom has a net charge of −1. It has 18 electrons and 20 neutrons. Give: (*a*) its isotopic symbol (*b*) its atomic number (*c*) its mass number (*d*) the charge on its nucleus (*e*) the number of protons

/ (*a*) $^{37}_{17}Cl^-$ (*b*) 17 (*c*) 37 (*d*) 17+ (*e*) 17

2.21 Which of the following are isotopes:

I Atoms of an element having different numbers of electrons

II Atoms of an element having different numbers of neutrons

III $^{40}_{19}K$ and $^{40}_{20}Ca$

IV $^{90}_{38}Sr$ and $^{88}_{38}Sr$

/ II and IV are each isotopes.

2.22 What is the number of electrons in $^{40}_{19}K^+$?

/ 18; the positive charge denotes the loss of one electron.

2.23 Determine the number of protons, electrons, and neutrons in (*a*) $^{80}_{35}Br^-$ (*b*) $^{79}_{34}Se$

▮

	protons	neutrons	electrons
(*a*)	35	45	36
(*b*)	34	45	34

2.24 Complete the following table.

symbol	atomic number	mass number	number of protons	number of electrons	number of neutrons	net charge
$^{90}Sr^{2+}$						
	11			10	12	
		82	35			1−

▮

symbol	atomic number	mass number	number of protons	number of electrons	number of neutrons	net charge
$^{90}Sr^{2+}$	38	90	38	36	52	2+
$^{23}Na^+$	11	23	11	10	12	1+
$^{82}Br^-$	35	82	35	36	47	1−

2.25 Complete the following table.

isotopic symbol	atomic number	mass number	number of protons	number of neutrons	number of electrons	charge
^{15}N						
	19			20		1+
			1	2	0	

▮

isotopic symbol	atomic number	mass number	number of protons	number of neutrons	number of electrons	charge
^{15}N	7	15	7	8	7	0
$^{39}K^+$	19	39	19	20	18	1+
$^{3}H^+$	1	3	1	2	0	1+

2.26 Tabulate the symbols and the values of the atomic number (Z), mass number (A), number of protons, number of neutrons, and number of electrons for (*a*) $^{23}Na^+$ (*b*) a species with two electrons, three protons, and four neutrons (*c*) a species with $Z = 92$ and with 88 electrons and 146 neutrons

▮

		Z	A	protons	neutrons	electrons
(*a*)	$^{23}Na^+$	11	23	11	12	10
(*b*)	$^{7}Li^+$	3	7	3	4	2
(*c*)	$^{238}U^{4+}$	92	238	92	146	88

2.27 Complete the following table.

isotope of element	Z	number of protons	number of neutrons	A
^{13}C				
	17		18	
		26		56
			2	3
	52			128
		50	70	

▌

isotope of element	Z	number of protons	number of neutrons	A
$^{13}_{6}C$	6	6	7	13
$^{35}_{17}Cl$	17	17	18	35
$^{56}_{26}Fe$	26	26	30	56
$^{3}_{1}H$	1	1	2	3
$^{128}_{52}Te$	52	52	76	128
$^{120}_{50}Sn$	50	50	70	120

2.28 State the number of protons, neutrons, and electrons in an atom of *(a)* ^{235}U *(b)* ^{90}Sr *(c)* D *(d)* ^{34}S

▌

	protons	neutrons	electrons
(a)	92	143	92
(b)	38	52	38
(c)	1	1	1
(d)	16	18	16

2.29 Carbon occurs in nature as a mixture of atoms of which 98.89% have a mass of 12.0000 u and 1.11% have a mass of 13.00335 u. Calculate the atomic weight of carbon.

▌

$$\text{AW} = (12.0000\ \text{u} \times 0.9889) + (13.00335\ \text{u} \times 0.0111) = 12.011\ \text{u}$$

2.30 Complete each line in the table. There are only two naturally occurring isotopes of each element.

	isotope A			isotope B			
	isotope	percent	mass (u)	isotope	percent	mass (u)	atomic weight (u)
(a)	^{191}Ir	37.30	190.9609	^{193}Ir	62.70	192.9633	______
(b)	^{121}Sb	57.25	______	^{123}Sb	42.75	122.9041	121.75
(c)	^{107}Ag	51.82	106.9041	^{109}Ag	48.18	______	107.868
(d)	^{79}Br	______	78.9183	^{81}Br	______	80.9163	79.904
(e)	^{12}C	98.89	______	^{13}C	1.11	______	12.011

▌ (*a*) $(0.3730)(190.9609) + (0.6270)(192.9633) = 192.2\ \text{u}$

(*b*) Let x = mass of ^{121}Sb

$$0.5725x + 0.4275(122.9041) = 121.75 \quad \text{hence} \quad x = \frac{121.75 - 52.54}{0.5725} = 120.9\ \text{u}$$

(*c*) Let x = mass of ^{109}Ag

$$0.5182(106.9041) + 0.4818x = 107.868 \quad \text{hence} \quad x = 108.9\ \text{u}$$

(*d*) Let x = fraction of ^{79}Br

$$x(78.9183) + (1 - x)(80.9163) = 79.904 \quad \text{hence} \quad x = 0.5067 \text{ or } 50.67\%$$

(*e*) The mass of ^{12}C, by definition of the atomic weight scale, is 12.0000 u. Hence

$$0.9889(12.0000) + 0.0111x = 12.011 \quad \text{hence} \quad x = 13.0\ \text{u}$$

2.31 Naturally occurring argon consists of three isotopes, the atoms of which occur in the following abundances: 0.34% ^{36}Ar, 0.07% ^{38}Ar, and 99.59% ^{40}Ar. Calculate the atomic weight of argon from these data.

▌ The isotopic weight data are from Table 2.1.

$$\frac{0.34}{100}(35.9676) + \frac{0.07}{100}(37.9627) + \frac{99.59}{100}(39.9624) = 39.95\ \text{u}$$

2.32 Naturally occurring boron consists of 80% ^{11}B (nuclidic mass = 11.01) and 20% another isotope. To account for the atomic weight, 10.81, what must be the nuclidic mass of the other isotope?

▌

$$\frac{80}{100}(11.01) + \frac{20}{100}(x) = 10.81 \quad \text{so} \quad x = 10.0 \quad (^{10}\text{B is the isotope})$$

2.33 ^{35}Cl and ^{37}Cl are the only naturally occurring chlorine isotopes. What percentage distribution accounts for the atomic weight, 35.453?

▌ Let x = fraction ^{35}Cl; then $(1.00 - x)$ = fraction ^{37}Cl

$$x(34.9689) + (1.00 - x)(36.9659) = 35.453$$

$x = 0.7576$ ^{35}Cl is 75.76% and ^{37}Cl is 24.24% of natural chlorine.

2.34 To account for nitrogen's atomic weight of 14.0067, what must be the ratio of ^{15}N to ^{14}N atoms in natural nitrogen?

▌ Let x = fraction ^{14}N; then $1.000 - x$ = fraction ^{15}N

$$x(14.00307) + (1.000 - x)(15.0001) = 14.0067$$

$$x = 0.9964 \qquad 1 - x = 0.0036$$

$$\text{Ratio} = \frac{0.0036}{0.9964} = 0.0036$$

2.35 At one time there was a chemical atomic weight scale based on the assignment of the value 16.0000 to naturally occurring oxygen. What would have been the atomic weight, on such a table, of silver, if current information had been available? The atomic weights of oxygen and silver on the present table are 15.9994 and 107.868.

▌ Since the old atomic weight of oxygen was 16.0000/15.9994 times as large as the new value, the old atomic weight of silver must be

$$(107.868)\left(\frac{16.0000}{15.9994}\right) = 107.872$$

2.36 It has been found by mass spectrometric analysis that in nature the relative abundances of the various isotopic atoms of silicon are: 92.23% ^{28}Si, 4.67% ^{29}Si, and 3.10% ^{30}Si. Calculate the atomic weight of silicon from this information and from the nuclidic masses.

▌ The atomic weight is the average mass of the three nuclides, each weighted according to its own relative abundance. The nuclidic masses are given in Table 2.1.

$$\text{AW} = (0.9223)(27.977) + (0.0467)(28.976) + (0.0310)(29.974) = 25.803 + 1.353 + 0.929 = 28.085$$

2.37 Naturally occurring carbon consists of two isotopes, ^{12}C and ^{13}C. What are the percent abundances of the two isotopes in a sample of carbon whose atomic weight is 12.01112?

Let $x = \%$ abundance of ^{13}C; then $100 - x$ is % of ^{12}C.

$$AW = 12.01112 = \frac{(12.00000)(100 - x) + (13.0034)x}{100}$$

$$= 12.0000 + \frac{(13.0034 - 12.0000)x}{100} = 12.0000 + (0.010034)x$$

Thus $$x = \frac{12.01112 - 12.00000}{0.010034} = \frac{0.01112}{0.010034} = 1.109\%\ ^{13}C \quad \text{and} \quad 100 - x = 98.891\%\ ^{12}C$$

2.38 Before 1961, a physical atomic weight scale was used whose basis was an assignment of the value 16.00000 to ^{16}O (compare Prob. 2.35). What would have been the physical atomic weight of ^{12}C on the old scale?

The ratio of the masses of any two nuclides must be independent of the reference point chosen.

$$\frac{\text{old AW}(^{12}C)}{\text{old AW}(^{16}O)} = \frac{\text{new AW}(^{12}C)}{\text{new AW}(^{16}O)} = \frac{12.0000}{15.9949}$$

or $$\text{old AW}(^{12}C) = (16.0000)\left(\frac{12.0000}{15.9949}\right) = 12.0038$$

2.39 The nuclidic mass of ^{90}Sr had been determined on the old physical scale ($^{16}O = 16.0000$) as 89.936. Recompute this to the present atomic weight scale, on which ^{16}O is 15.9949.

$$(89.936)\left(\frac{15.9949}{16.0000}\right) = 89.907$$

2.40 In a chemical atomic weight determination, the tin content of 3.7692 g of $SnCl_4$ was found to be 1.7170 g. If the atomic weight of chlorine is taken as 35.453, what is the value for the atomic weight of tin determined from this experiment?

$$\text{Mass of chlorine} = (3.7692\ g) - (1.7170\ g) = 2.0522\ g\ Cl$$

$$\text{Moles Cl} = \frac{2.0522\ g\ Cl}{35.453\ g/mol} = 0.057885\ mol\ Cl$$

$$\text{Moles Sn} = \frac{1.7170\ g\ Sn}{x} = \tfrac{1}{4}(0.057885)\ mol\ Sn$$

$$x = \frac{4(1.7170)}{0.057885} = 118.65\ g/mol$$

2.3 IONIC AND COVALENT BONDING

2.41 Construct a table comparing metals with nonmetals in terms of (*a*) the sign of the charges possible on monatomic ions, (*b*) the possibility of reaction with other elements of the same class, (*c*) the range of possible numbers of valence electrons, and (*d*) the ability of the elements to conduct electricity in the elementary state.

	(*a*)	(*b*)	(*c*)	(*d*)
Metals	+	unusual	1–5	all
Nonmetals	–	normal	3–8	few

2.42 In which of the following compounds is the bonding essentially ionic, in which is the bonding essentially covalent, and in which are both types of bonding represented? (*a*) PCl_3 (*b*) $(NH_4)_2S$ (*c*) $Ba(CN)_2$ (*d*) NaBr (*e*) CH_3CH_2OH

(*a*) Covalent (*b*) both (*c*) both (*d*) ionic (*e*) covalent. One should recognize the ammonium ion and the cyanide ion as having covalently bonded atoms within the ions.

2.43 (*a*) State the number of atoms of each element indicated in each of the following formulas. (*b*) State the formula of each of the ions and the number of each.

	(*a*) atoms		(*b*) ions	
	element	no. of atoms	formula	no. of ions
$Co(ClO_3)_2$	___	___	___	___
	___	___	___	___
	___	___		
$(NH_4)_2CO_3$	___	___	___	___
	___	___	___	___
	___	___		
	___	___		

▌

	(*a*) atoms		(*b*) ions	
	element	no. of atoms	formula	no. of ions
$Co(ClO_3)_2$	Co	1	Co^{2+}	1
	Cl	2	ClO_3^-	2
	O	6		
$(NH_4)_2CO_3$	N	2	NH_4^+	2
	H	8	CO_3^{2-}	1
	C	1		
	O	3		

2.44 Write the formula for a compound of Cl which contains (*a*) ionic bonds only (*b*) ionic and covalent bonds (*c*) covalent bonds only

▌ One possible formula is given for each. (*a*) NaCl (*b*) NaClO (*c*) NCl_3

2.45 Write the formula of a (the) compound expected when each of the following pairs of elements is combined: (*a*) carbon and chlorine (*b*) sodium and sulfur (*c*) nitrogen and lithium

▌ (*a*) CCl_4 (*b*) Na_2S (*c*) Li_3N

2.46 What is the formula of the compound corresponding to the combination of each of the following pairs: (*a*) Al and S (*b*) PO_4^{3-} and Mg^{2+} (*c*) ClO_3^- and Co^{3+} (*d*) Na and Cl_2

▌ (*a*) Al_2S_3 (*b*) $Mg_3(PO_4)_2$ (*c*) $Co(ClO_3)_3$ (*d*) NaCl

2.47 Write the formulas of the ions which constitute Na_2Se.

▌ Na^+ and Se^{2-}

2.48 Write the formula of the compound formed by each of the following pairs of elements: (*a*) aluminum and selenium (*b*) magnesium and nitrogen (*c*) magnesium and chlorine

▌ (*a*) Al_2Se_3 (*b*) Mg_3N_2 (*c*) $MgCl_2$

2.49 Write the formula of the compound formed by:

(*a*) Ca^{2+} and PO_4^{3-} (*b*) Li^+ and SO_3^{2-} (*c*) Mn^{2+} and O^{2-}
(*d*) Sr and S (*e*) N^{3-} and Mg^{2+} (*f*) Cu^+ and O^{2-}

▌ (*a*) $Ca_3(PO_4)_2$ (*b*) Li_2SO_3 (*c*) MnO (*d*) SrS (*e*) Mg_3N_2 (*f*) Cu_2O

2.50 Pure liquid H_2SO_4 solidifies below 10.4 °C. Neither the pure liquid nor the solid conducts electricity; however, aqueous solutions of H_2SO_4 conduct electricity well. Solid Na_2SO_4, which melts at 884 °C, does not conduct electricity, but molten Na_2SO_4 as well as aqueous solutions of Na_2SO_4 conduct electricity well. Explain the difference in properties between pure Na_2SO_4 and H_2SO_4.

Pure H_2SO_4 is essentially covalent; Na_2SO_4 is ionic, with Na^+ and SO_4^{2-} ions. H_2SO_4 does not conduct because there are no ions to carry the current. $Na_2SO_4(s)$ does not conduct because the ions are not free to move. When H_2SO_4 dissolves, it reacts with water to form ions, and the solution conducts well. When solid Na_2SO_4 is melted or dissolved, the ions are freed and the material conducts well.

2.51 Determine the charges of the ions in parentheses [or brackets] in the following formulas:

(*a*) $Na_2(MnO_4)$ (*b*) $K_4[Fe(CN)_6]$ (*c*) $NaCd_2(P_3O_{10})$ (*d*) $Na_2(B_4O_7)$
(*e*) $Ca_3(CoF_6)_2$ (*f*) $Mg_3(BO_3)_2$ (*g*) $(UO_2)Cl_2$ (*h*) $(SbO)_2SO_4$

(*a*) The charge of (MnO_4) must balance that of two Na^+, i.e., $(MnO_4)^{2-}$. (This ion is called *manganate* and is different from permanganate.)
(*b*) The ion in brackets must balance the charge of four K^+, i.e., $[Fe(CN)_6]^{4-}$.
(*c*) The charge of (P_3O_{10}) must balance the charge of Na^+ and two Cd^{2+}, i.e., $(P_3O_{10})^{5-}$.
(*d*) The charge of (B_4O_7) must balance the charge of two Na^+, i.e., $(B_4O_7)^{2-}$.
(*e*) The charge of two (CoF_6) ions must balance the charge of three Ca^{2+}, i.e., $(CoF_6)^{3-}$.
(*f*) The charge of two (BO_3) ions must balance the charge of three Mg^{2+}, i.e., $(BO_3)^{3-}$.
(*g*) The charge of the (UO_2) ion must balance the charge of two Cl^-, i.e., $(UO_2)^{2+}$.
(*h*) The charge of two (SbO) ions must balance the charge of SO_4^{2-}, i.e., $(SbO)^+$.

2.52 Determine the ionic charges of the groups in parentheses: (*a*) $Ca(C_2O_4)$ (*b*) $Ca(C_2H_3O_2)_2$ (*c*) $Mg_3(AsO_3)_2$ (*d*) $(MoO)Cl_3$ (*e*) $(CrO_2)F_2$ (*f*) $(PuO_2)Br$ (*g*) $(PaO)_2S_3$

(*a*) 2−, to balance the 2+ on the Ca^{2+} ion.
(*b*) 1−, the two 1− ions balance the charge on the Ca^{2+} ion.
(*c*) 3−, the two 3− balance the three 2+ on the Mg^{2+} ions.
(*d*) 3+, to balance the three 1− on the Cl^- ions.
(*e*) 2+, to balance the two 1− on the F^- ions.
(*f*) 1+, to balance the 1− on the Br^- ion.
(*g*) 3+, the two 3+ balance the three 2− on the S^{2-} ions.

2.53 The formula of potassium arsenate is K_3AsO_4. The formula of potassium ferrocyanide is $K_4Fe(CN)_6$. Write the formulas of (*a*) calcium arsenate (*b*) iron(III) arsenate (*c*) barium ferrocyanide (*d*) aluminum ferrocyanide

(*a*) $Ca_3(AsO_4)_2$ (*b*) $FeAsO_4$ (*c*) $Ba_2Fe(CN)_6$ (*d*) $Al_4[Fe(CN)_6]_3$

2.54 The formula of calcium pyrophosphate is $Ca_2P_2O_7$. Determine the formulas of sodium pyrophosphate and iron(III) pyrophosphate.

The charge of the pyrophosphate ion must be 4− to balance the charge of two Ca^{2+}. We can then write $Na_4P_2O_7$ and $Fe_4(P_2O_7)_3$.

2.4 ELECTRON DOT STRUCTURES AND THE OCTET RULE

2.55 Write all possible octet structural formulas for (*a*) CH_4O (*b*) C_2H_3F

(*a*) Since the valence of hydrogen is saturated with two electrons, each hydrogen can form only one covalent bond. Thus hydrogen cannot serve as a bridge between C and O. The only possibility of providing four bonds to the C is to have three H's and the O bonded directly to it. Only one structure is possible, as shown in Fig. 2.1. Note that the total number (14) of valence electrons in the structure is the sum of the numbers of valence electrons in the free component atoms: 4 (in C) + 6 (in O) + 4 (in 4 H).

```
      H
      |    ..
  H—C—O—H
      |    ..
      H
```

Fig. 2.1

(*b*) The only way of providing four pairs for each C within the limitation of 18 valence electrons is to have a C=C bond, as in Fig. 2.2. The reader should determine by trial and error that no other structure is possible.

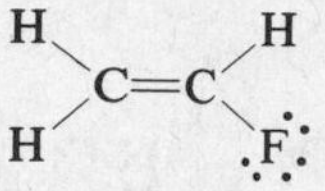

```
H       H
 \     /
  C = C
 /     \
H       F:
```

Fig. 2.2

2.56 Draw electron dot diagrams for (*a*) SO_3 (*b*) SO_3^{2-} (*c*) Na_2SO_3 (*d*) H_2SO_3 (*e*) Name each of these species.

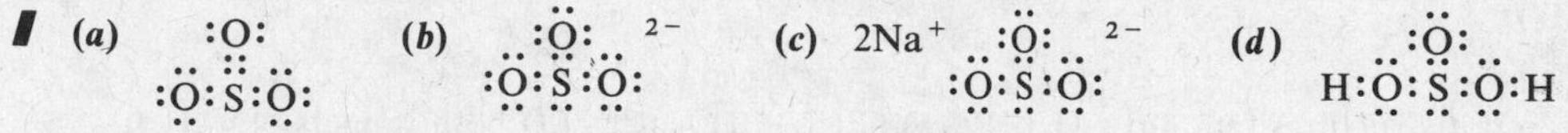

▌ (*a*)
```
   :O:
:O:S:O:
```
(*b*)
```
   :O:   2-
:O:S:O:
```
(*c*)
```
2Na+  :O:   2-
   :O:S:O:
```
(*d*)
```
    :O:
H:O:S:O:H
```

(*e*) Sulfur trioxide, sulfite ion, sodium sulfite, sulfurous acid.

2.57 Draw an electron dot diagram for (*a*) fluoride ion (*b*) calcium ion (*c*) calcium fluoride (*d*) fluorine gas

▌ (*a*) :F:⁻ (*b*) Ca^{2+} (*c*) Ca^{2+} 2[:F:⁻] (*d*) :F:F:

2.58 Write electron dot and line structures for (*a*) phosphorus trichloride, PCl_3 (*b*) carbon monoxide, CO (*c*) hydroxide ion, OH^-

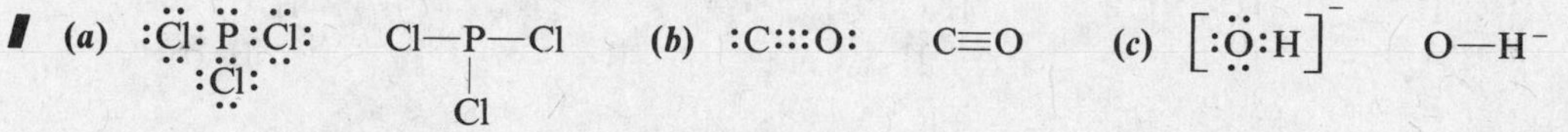

▌ (*a*)
```
:Cl:P:Cl:      Cl—P—Cl
   :Cl:           |
                  Cl
```
(*b*) :C:::O: C≡O (*c*) [:O:H]⁻ O—H⁻

In (*a*) each atom has an octet, formed by sharing one pair of electrons with each neighboring atom in single bonds. In (*b*) the sharing of three pairs of electrons constitutes a triple bond. In (*c*) the hydroxide ion has a negative charge, indicating one electron in excess of those provided by the hydrogen and oxygen atoms. The hydrogen atom does not have an octet, but by sharing a pair of electrons it attains a configuration similar to that of helium.

2.59 Write electron dot diagrams for chlorine, Cl_2, and calcium chloride, $CaCl_2$. Explain the difference between the method of satisfying the octet of chlorine in the two cases.

▌ :Cl:Cl: Ca^{2+} 2[:Cl:⁻] In the element, the chlorine atoms are covalently bonded to each other; in the compound, they exist as chloride ions.

2.60 Write an electron dot diagram for SeO_3^{2-}

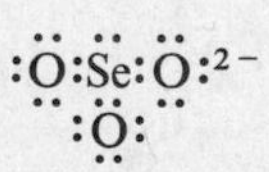

▌
```
:O:Se:O:2-
   :O:
```

2.61 Write electron dot diagrams for (*a*) $AsCl_3$ (*b*) NO_3^-

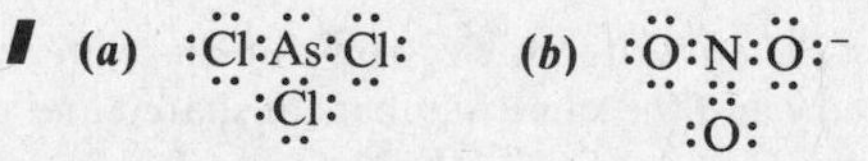

▌ (*a*)
```
:Cl:As:Cl:
   :Cl:
```
(*b*)
```
:O:N:O:⁻
   :O:
```

2.62 Write electron dot diagrams for (*a*) CCl_4 (*b*) elementary iodine (*c*) Li_3PO_4 (*d*) NCl_3 (*e*) ClO_2^- (*f*) H_2O (*g*) CS_2

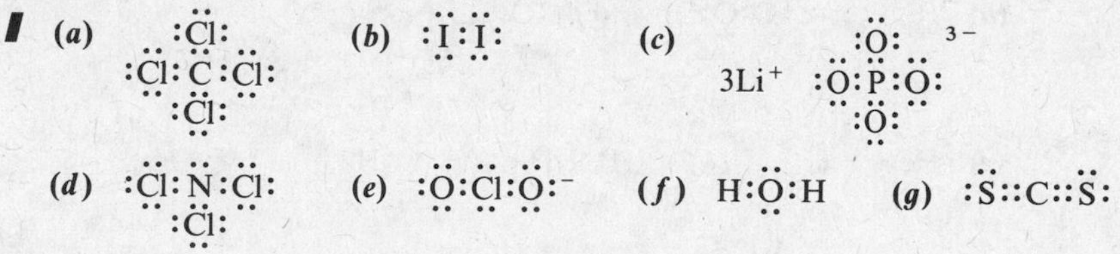

▌ (*a*)
```
   :Cl:
:Cl:C:Cl:
   :Cl:
```
(*b*) :I:I:

(*c*)
```
        :O:   3-
3Li+ :O:P:O:
        :O:
```
(*d*)
```
:Cl:N:Cl:
   :Cl:
```
(*e*) :O:Cl:O:⁻ (*f*) H:O:H (*g*) :S::C::S:

2.63 Write electron dot diagrams for (*a*) $COCl_2$ (the central atom is C) (*b*) NH_4^+

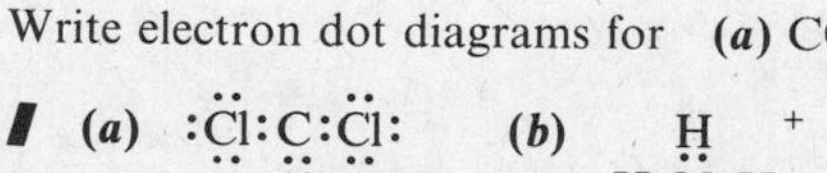

▌ (*a*)
```
:Cl:C:Cl:
   :O:
```
(*b*)
```
  H   +
H:N:H
  H
```

2.64 Write electron dot diagrams for (*a*) CN^- (*b*) H_2CO

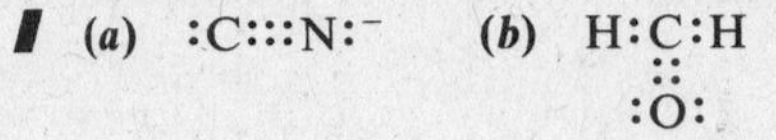

(*a*) :C:::N:$^-$ (*b*)
```
H:C:H
  ::
 :O:
```

2.65 Draw electron dot diagrams for (*a*) PO_4^{3-} (*b*) SF_2 (*c*) BrO_2^-

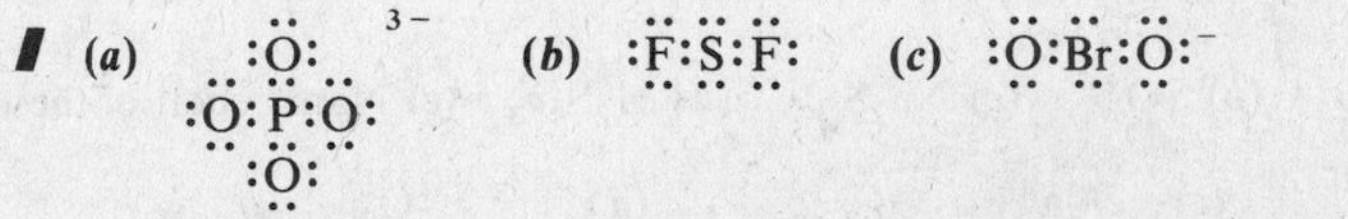

(*a*)
```
   :O:   3−
 :O:P:O:
   :O:
```
(*b*) :F:S:F: (*c*) :O:Br:O:$^-$

2.66 Write electron dot structures for (*a*) CH_2Cl_2 (*b*) C_2Cl_4 (*c*) NaCl (*d*) both isomers of $C_2H_4Cl_2$

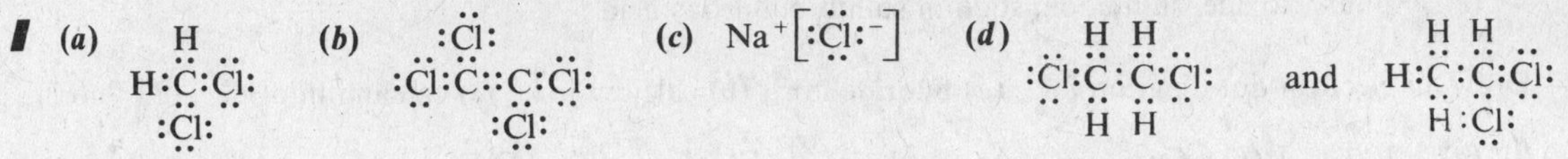

(*a*)
```
  H
H:C:Cl:
 :Cl:
```
(*b*)
```
     :Cl:
:Cl:C::C:Cl:
     :Cl:
```
(*c*) Na$^+$[:Cl:$^-$]

(*d*)
```
     H H                H H
:Cl:C:C:Cl:   and     H:C:C:Cl:
     H H                H:Cl:
```

2.67 Write an electron dot structure for (*a*) CH_4 (*b*) CH_2O (*c*) CH_4O (*d*) C_2Cl_2 (*e*) CH_5N

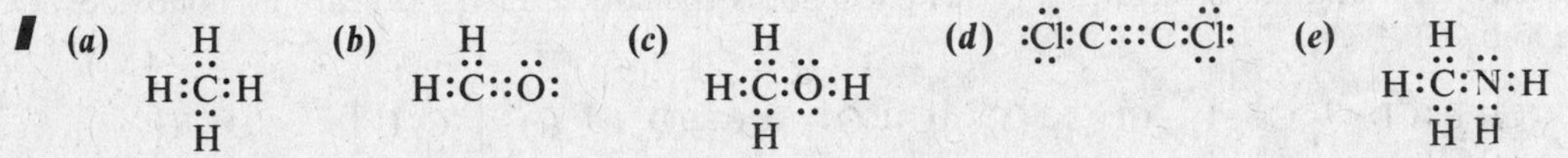

(*a*)
```
  H
H:C:H
  H
```
(*b*)
```
  H
H:C::O:
```
(*c*)
```
  H
H:C:O:H
  H
```
(*d*) :Cl:C:::C:Cl:

(*e*)
```
  H
H:C:N:H
  H H
```

2.68 Draw electron dot structures for (*a*) CH_3OCH_3 (*b*) CH_3CH_2OH (*c*) CH_3COCH_3 (*d*) C_2H_4 (*e*) C_2H_2 (*f*) HCO_2H

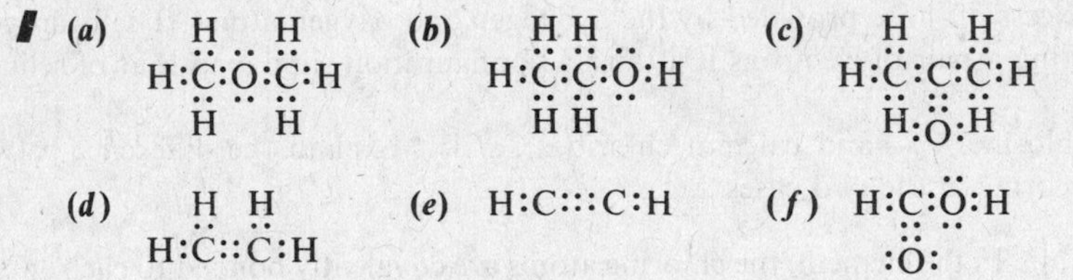

(*a*)
```
  H   H
H:C:O:C:H
  H   H
```
(*b*)
```
  H H
H:C:C:O:H
  H H
```
(*c*)
```
  H   H
H:C:C:C:H
  H:O:H
```
(*d*)
```
  H  H
H:C::C:H
```
(*e*) H:C:::C:H

(*f*)
```
H:C:O:H
  ::
 :O:
```

2.69 Write electron dot formulas for each of the following substances, none of which follows the octet rule: (*a*) BF_3 (*b*) PF_5 (*c*) ICl_3 (*d*) SF_6.

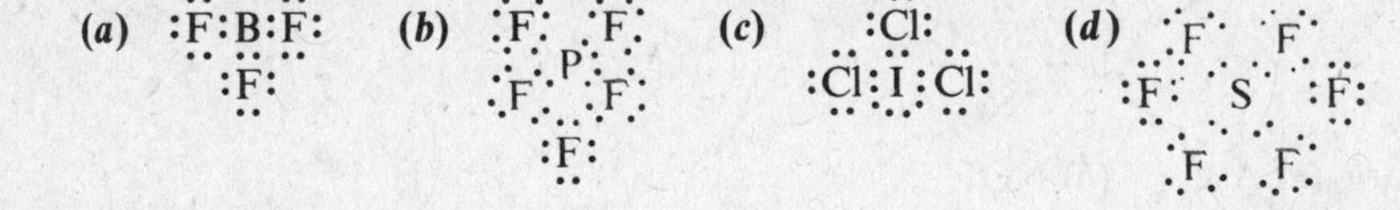

(*a*)
```
:F:B:F:
  :F:
```
(*b*)
```
 :F:  :F:
    P
 :F:  :F:
   :F:
```
(*c*)
```
   :Cl:
:Cl:I:Cl:
```
(*d*)
```
   :F:  :F:
:F:   S   :F:
   :F:  :F:
```

2.70 Write electron dot diagrams for each of the following molecules. Then write an electron dot diagram for an ion which is isostructural with each, that is, which has the same geometry and the same numbers of shared and unshared electrons: (*a*) CH_4 (*b*) F_2 (*c*) HNO_3 (*d*) CF_4 (*e*) He (*f*) NH_3 (*g*) SO_2 (*h*) N_2O_4 (*i*) N_2 (*j*) CS_2

One isostructural ion of the many possibilities is presented for each case.

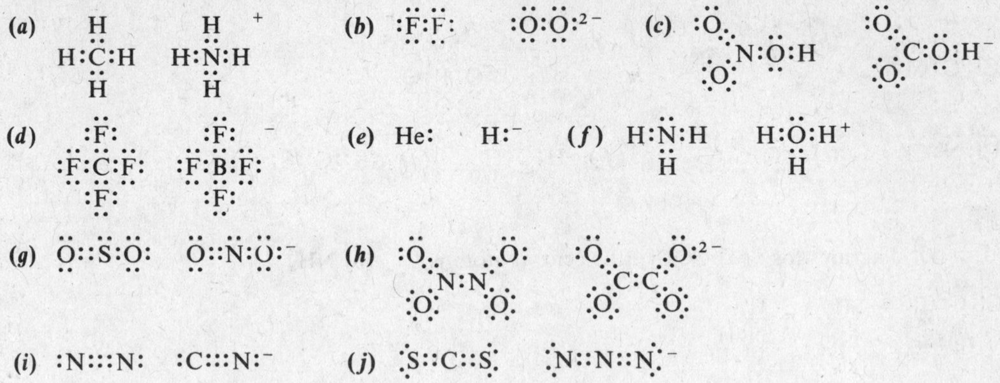

(*a*)
```
  H       H   +
H:C:H   H:N:H
  H       H
```
(*b*) :F:F: :O:O:$^{2-}$

(*c*)
```
:O.              :O.
   N:O:H            C:O:H−
:O:              :O:
```
(*d*)
```
  :F:        :F:  −
:F:C:F:    :F:B:F:
  :F:        :F:
```
(*e*) He: H:$^-$

(*f*)
```
H:N:H     H:O:H+
  H         H
```
(*g*) O::S:O: O::N:O:$^-$

(*h*)
```
:O.   .O:     :O.   .O:2−
   N:N           C:C
:O:   :O:     :O:   :O:
```
(*i*) :N:::N: :C:::N:$^-$ (*j*) :S::C::S: :N::N::N:$^-$

2.71 Draw an electron dot diagram (Lewis structure) for BF_4^-.

▮

$$\left[\begin{array}{c} :\ddot{\text{F}}: \\ :\ddot{\underset{\cdot\cdot}{\text{F}}}:\ddot{\underset{\cdot\cdot}{\text{B}}}:\ddot{\underset{\cdot\cdot}{\text{F}}}: \\ :\underset{\cdot\cdot}{\text{F}}: \end{array}\right]^-$$

2.72 Write electron dot diagrams for (*a*) CN^- (*b*) BrO_3^-

▮ (*a*) $:C:::N:^-$ (*b*)

$$\left[\begin{array}{c} :\ddot{\text{O}}: \\ :\ddot{\underset{\cdot\cdot}{\text{O}}}:\ddot{\underset{\cdot\cdot}{\text{Br}}}:\ddot{\underset{\cdot\cdot}{\text{O}}}: \end{array}\right]^-$$

CHAPTER 3
Periodic Table

3.1 PERIODIC TRENDS

3.1 Give examples of each of the following: (*a*) alkali metal (*b*) noble gas (*c*) halogen (*d*) alkaline earth metal

▌ (*a*) Li, Na, K, Rb, Cs, or Fr (*b*) He, Ne, Ar, Kr, Xe, or Rn
(*c*) F, Cl, Br, I, or At (*d*) Be, Mg, Ca, Sr, Ba, or Ra

3.2 Which is the most nonmetallic of the following: (*a*) Be (*b*) B (*c*) Al (*d*) Ga (*e*) Mg

▌ B (boron). It is farthest right and highest of the elements listed.

3.3 (*a*) Which main group elements form monatomic ions in their compounds? (*b*) Give the charge on these ions, as a function of the periodic group.

▌ (*a*) Groups I, II, III, V, VI, and VII of the main groups form monatomic ions. (*b*) The charges are, respectively, 1+, 2+, 3+, 3−, 2−, and 1−. Examples are Na^+, Mg^{2+}, Al^{3+}, P^{3-}, S^{2-}, and Cl^-. The heaviest elements of group IV form 4+ or 2+ ions, such as Pb^{4+} and Pb^{2+}.

3.4 Use the words at the right to complete the sentences at the left:

(*a*) Bonds between nonmetal atoms are ______.
(*b*) Bonds between a metal and a nonmetal atom are ______.
(*c*) In some compounds, ______ are formed.

ionic
covalent
both types
neither

▌ (*a*) covalent (*b*) ionic (*c*) both types

3.5 In the periodic table, note that for the main group elements, if the periodic group of the element is even, the monatomic ion and the oxyanions of the element have charges that are even numbers, while if the periodic group number is odd, the respective charges are odd. Is this generalization true for elements which are not in a main group of the periodic table? List several exceptions.

▌ No. (Fe^{2+} and Fe^{3+} both exist, for example, despite the fact that iron is a member of an even-numbered group.) Exceptions are Fe^{3+}, Mn^{2+}, Cr^{3+}, and Cu^{2+}, among many others.

3.6 Write the formula of a binary compound of fluorine with each main group element in the fourth period of the periodic table.

▌ KF, CaF_2, GaF_3, GeF_4, AsF_3 or AsF_5, SeF_2 or SeF_6, BrF or BrF_5, KrF_2

3.7 Which one of the following periodic groups consists entirely of metals: (*a*) IIA (*b*) IIIA (*c*) IVA (*d*) VIA (*e*) VIIA

▌ (*a*) IIA. The alkaline earth elements.

3.8 Which one of the following elements has chemical properties most like those of sulfur: (*a*) Cl (*b*) F (*c*) P (*d*) N (*e*) Se

▌ (*e*) Se, which is in the same periodic group.

3.9 Indicate the element which (*a*) is the alkali metal with the fewest protons (*b*) has atoms with 7 outermost electrons each and is in the third period (*c*) is the most variable in its properties, sometimes acting as a metal and other times as a nonmetal.

▌ (*a*) Li (*b*) Cl (*c*) H

3.10 Which main group element(s) has (have) a different number of outermost electrons than its (their) group number?

▌ The noble gases, group 0, have 8 valence electrons.

3.11 Look up and tabulate the following properties of silicon, tin, gallium, and arsenic: color, density, melting point (mp), ability to conduct electricity. Write the formula for the chloride of each element. Look up and tabulate the density, melting point, and boiling point (bp) of each of these compounds. Using these data, predict analogous properties of the element germanium and those of the compound of germanium with chlorine.

▮

	color	density	mp	conductance
Si	steel gray	2.33	1410	nonconductor
Sn	gray[a]	5.75	232	conductor
Ga	gray black	5.904	30	nonconductor
As	gray	5.727	613 sublimes	nonconductor
Ge	gray white	5.35	937	nonconductor

	density	mp	bp
$SiCl_4$	1.48	−70	58
$SnCl_4$	2.226	−33	114.1
$GaCl_3$	2.47	77.9	201.3
$AsCl_3$	2.163	−8.5	63
$GeCl_4$	1.8443	−49.5	84

[a] One allotropic form.

3.2 INORGANIC NOMENCLATURE

3.12 How are binary nonmetal-nonmetal compounds named? How are metal-nonmetal compounds named?

▮ Nonmetal-nonmetal compounds are usually named by giving the name of the nonmetal farthest to the left and/or farthest down in the periodic table, then giving the name of the other element with its ending changed to *ide*. The number of atoms of the second element is denoted by a prefix—mono, di, tri, tetra, penta, hexa, ... Metal-nonmetal compounds are named by giving the name of the metal first. If the metal is in group IA or IIA or is Zn, Al, or Cd, its name is not further described. If it is another metal, its oxidation state (equal to the charge on the monatomic ion) is given in parentheses attached to the name. Then the name of the nonmetal is given, with its ending changed to *ide*.

3.13 (*a*) Name: $AlCl_3$, PCl_3, $CoCl_3$. (*b*) Explain why their names are so different.

▮ (*a*) Aluminum chloride, phosphorus trichloride, cobalt(III) chloride. (*b*) The compounds are named using different systems. Aluminum is a metal which never varies from +3 oxidation state in its compounds, so aluminum chloride implies aluminum with a charge of 3+. Phosphorus is a nonmetal, and nonmetal-nonmetal compounds are named using the prefixes (tri in this case). Cobalt is a metal which exhibits several oxidation states in its compounds; we must be specific about which cobalt chloride we are referring to in this compound by affixing a roman numeral to denote the oxidation state. An older terminology uses cobaltic chloride (as opposed to cobaltous chloride) to denote the cobalt ion with the higher oxidation state.

3.14 Name: (*a*) NaCl (*b*) KBr (*c*) $MgCl_2$ (*d*) NaOH (*e*) $Ca(OH)_2$ (*f*) LiCN

▮ (*a*) sodium chloride (*b*) potassium bromide (*c*) magnesium chloride
(*d*) sodium hydroxide (*e*) calcium hydroxide (*f*) lithium cyanide

3.15 Write formulas for (*a*) barium nitrate (*b*) aluminum sulfate (*c*) iron(II) hydroxide.

▮ (*a*) Since two negative charges are required to balance the charge on one barium ion, Ba^{2+}, and each NO_3^- ion has only one negative charge, the formula must be written $Ba(NO_3)_2$. (*b*) To achieve equal numbers of positive and negative charges, the formula must be $Al_2(SO_4)_3$. (*c*) Since iron(II) is specified, it takes two OH^- ions to supply the appropriate number of negative charges—hence the formula is $Fe(OH)_2$.

3.16 Explain why it is necessary to use the prefixes mono, di, tri, etc., more often for covalent compounds than for ionic compounds.

▮ The nonmetal-nonmetal compounds, for which the prefixes are mainly used, are covalent.

3.17 Name: (*a*) Mg_3P_2 (*b*) $Hg_2(NO_3)_2$ (*c*) NH_4TcO_4

▌ (*a*) The ion of phosphorus must be an anion, since magnesium forms only a cation. Application of the usual formula for naming binary compounds gives the name *magnesium phosphide*. (*b*) Since the total charge on two NO_3^- (nitrate ions) is 2−, the total cationic charge must be 2+. Since the average charge per Hg is 1+, the name is *mercury(I) nitrate* (or *mercurous nitrate*). (*c*) The charge on the anion must be 1− in order to balance the 1+ charge on the ammonium ion. Since technetium belongs to the same group of the periodic table as Mn, TcO_4^- is analogous to MnO_4^-, permanganate. The name is thus *ammonium pertechnetate*.

3.18 How can you tell from the formula of a compound if it is an acid?

▌ The formulas of acids have hydrogen written first. Thus HCl is an acid and NH_3 is not. H_2O is an exception. (As one gets more familiar with chemical compounds, this practice is deviated from in part; thus organic chemists write acids in accordance with their structures, such as CH_3COOH.)

3.19 Name: (*a*) SO_3^{2-} (*b*) SO_3 (*c*) ClO^- (*d*) H_2SO_3 (*e*) $HClO_4$ (*f*) PBr_3 (*g*) Na_2SO_3 (*h*) $NaClO_2$ (*i*) $Ba(ClO)_2$ (*j*) HCl (*k*) $KClO_4$ (*l*) $Al(ClO_2)_3$

▌ (*a*) sulfite ion (*b*) sulfur trioxide (*c*) hypochlorite ion
(*d*) sulfurous acid (*e*) perchloric acid (*f*) phosphorus tribromide
(*g*) sodium sulfite (*h*) sodium chlorite (*i*) barium hypochlorite
(*j*) hydrochloric acid (*k*) potassium perchlorate (*l*) aluminum chlorite

3.20 Write formulas for:

(*a*) hydroiodic acid (*b*) hypoiodous acid (*c*) iodous acid (*d*) iodic acid
(*e*) periodic acid (*f*) iodide ion (*g*) hypoiodite ion (*h*) iodite ion
(*i*) iodate ion (*j*) periodate ion

▌ (*a*) HI (*b*) HIO (*c*) HIO_2 (*d*) HIO_3 (*e*) HIO_4
(*f*) I^- (*g*) IO^- (*h*) IO_2^- (*i*) IO_3^- (*j*) IO_4^-

3.21 Write formulas for: (*a*) barium oxide (*b*) aluminum chloride (*c*) magnesium phosphate

▌ (*a*) The formula is BaO since the 2+ charge on one barium ion just balances the 2− charge on the oxide ion. (*b*) Three chloride ions, 1− charge on each, are needed to balance the 3+ charge of one aluminum ion. The formula is $AlCl_3$. (*c*) Since neither 2 (the positive charge on magnesium ion) nor 3 (the negative charge on phosphate) is an integral multiple of the other, the smallest number must be found which is a multiple of both. This number is 6. The formula, $Mg_3(PO_4)_2$, shows 6 units of positive charge (3 × 2) and 6 units of negative charge (2 × 3) per formula unit.

3.22 Give the formula for (*a*) sulfite ion (*b*) iron(III) bicarbonate (*c*) hydrochloric acid (*d*) sodium nitrite (*e*) barium hydroxide

▌ (*a*) SO_3^{2-} (*b*) $Fe(HCO_3)_3$ (*c*) HCl (*d*) $NaNO_2$ (*e*) $Ba(OH)_2$

3.23 Name: (*a*) $Mg(IO)_2$ (*b*) $Fe_2(SO_4)_3$ (*c*) $CaMnO_4$ (*d*) $KReO_4$ (*e*) $CaWO_4$ (*f*) $CoCO_3$

▌ (*a*) magnesium hypoiodite (*b*) iron(III) sulfate or ferric sulfate (*c*) calcium manganate
(*d*) potassium perrhenate (*e*) calcium tungstate (*f*) cobalt(II) carbonate

Answers (*d*) and (*e*) illustrate the usefulness of the periodic table in naming compounds. In (*d*), ReO_4^- is related to MnO_4^-, a common ion, because the two central elements are in the same periodic group. In (*e*), WO_4^{2-} is related to SO_4^{2-} because both central elements are in periodic group VI (albeit one is a main group element and the other a transition element).

Roman numerals attached to the metal ion name are not necessary for alkali metals, alkaline earth metals, zinc, aluminum, and cadmium, because these elements in their compounds always have charges equal to their group numbers.

3.24 Name: (*a*) $(NH_4)_2CrO_4$ (*b*) $NaHSO_4$ (*c*) Hg_2I_2 (*d*) ClO_4^- (*e*) $Cu_3(PO_4)_2$

▌ (*a*) ammonium chromate (*b*) sodium hydrogen sulfate or sodium bisulfate

(*c*) mercury(I) iodide or mercurous iodide (*d*) perchlorate ion
(*e*) copper(II) phosphate or cupric phosphate

3.25 Write formulas for: (*a*) lithium hydride (*b*) calcium bromate (*c*) chromium(II) oxide (*d*) thorium(IV) perchlorate (*e*) nickel(II) phosphate (*f*) zinc sulfate

▌ (*a*) LiH. Equal magnitudes of positive and negative charges require equal numbers of positive and negative ions. (*b*) $Ca(BrO_3)_2$. Two mononegative anions are needed to balance the charge on each dipositive ion. (*c*) CrO. See (*a*). (*d*) $Th(ClO_4)_4$. Each ClO_4^- ion has one negative charge. (*e*) $Ni_3(PO_4)_2$. Three dipositive cations balance two trinegative anions. (*f*) $ZnSO_4$. See (*a*).

3.26 For each of the following pairs of elements, write the formula for and name the binary compound which they form, and state whether the bonding in each compound will be ionic or covalent: (*a*) potassium and phosphorus (*b*) carbon and fluorine (*c*) hydrogen and sulfur (*d*) potassium and hydrogen (*e*) fluorine and nitrogen

▌ (*a*) K_3P, ionic, potassium phosphide (*b*) CF_4, covalent, carbon tetrafluoride (*c*) H_2S, covalent, hydrogen sulfide (*d*) KH, ionic, potassium hydride (*e*) NF_3, covalent, nitrogen trifluoride. Note that the element farther to the right in the periodic table is named last. The nonmetal-nonmetal compounds (except for the hydrogen compound) use the prefixes from Prob. 3.12.

3.27 Write the formula for sodium nitrate.

▌ $NaNO_3$

3.28 Write the formula for sodium nitride.

▌ Na_3N

3.29 Name $CoCl_3$.

▌ cobalt(III) chloride

3.30 Write formulas for (*a*) chlorate (*b*) permanganate (*c*) dichromate (*d*) sulfite

▌ (*a*) ClO_3^- (*b*) MnO_4^- (*c*) $Cr_2O_7^{2-}$ (*d*) SO_3^{2-}

3.31 Write formulas for (*a*) sodium carbonate (*b*) chromium(III) nitrite (*c*) barium chloride

▌ (*a*) Na_2CO_3 (*b*) $Cr(NO_2)_3$ (*c*) $BaCl_2$

3.32 Name: (*a*) $HgCl_2$ (*b*) Ag_2SO_4 (*c*) NH_4CN

▌ (*a*) mercury(II) chloride (*b*) silver sulfate (*c*) ammonium cyanide

3.33 Write formulas for (*a*) manganese(II) chlorite (*b*) iron(II) hydroxide (*c*) sulfur hexafluoride (*d*) sulfurous acid

▌ (*a*) $Mn(ClO_2)_2$ (*b*) $Fe(OH)_2$ (*c*) SF_6 (*d*) H_2SO_3

3.34 Name: (*a*) KCN (*b*) PCl_5 (*c*) $Co(ClO_4)_3$ (*d*) $CoCl_2$ (*e*) PbO_2

▌ (*a*) potassium cyanide (*b*) phosphorus pentachloride (*c*) cobalt(III) perchlorate
(*d*) cobalt(II) chloride (*e*) lead(IV) oxide

3.35 Name: (*a*) $Ca(OH)_2$ (*b*) VCl_3 (*c*) Cu_2O (*d*) $Ba(ClO)_2$ (*e*) H_2SO_3

▌ (*a*) calcium hydroxide (*b*) vanadium(III) chloride (*c*) copper(I) oxide
(*d*) barium hypochlorite (*e*) sulfurous acid

3.36 Name: (*a*) K_2S (*b*) BCl_3

▌ (*a*) potassium sulfide (*b*) boron trichloride

3.37 Write formulas for (*a*) ammonium carbonate (*b*) cobalt(II) cyanide

▌ (*a*) $(NH_4)_2CO_3$ (*b*) $Co(CN)_2$

3.38 Name: ***(a)*** NH_4Br ***(b)*** Cu_2S ***(c)*** CuS ***(d)*** $Ba(OH)_2$

▌ *(a)* ammonium bromide *(b)* copper(I) sulfide *(c)* copper(II) sulfide *(d)* barium hydroxide

3.39 State the formula for ***(a)*** chromium(III) phosphate ***(b)*** nickel(II) nitrate ***(c)*** sodium bromide ***(d)*** magnesium oxide

▌ *(a)* $CrPO_4$ *(b)* $Ni(NO_3)_2$ *(c)* $NaBr$ *(d)* MgO

3.40 Name: ***(a)*** CO ***(b)*** CO_2 ***(c)*** PCl_5 ***(d)*** SF_6 ***(e)*** $AlCl_3$

▌ *(a)* carbon monoxide *(b)* carbon dioxide *(c)* phosphorus pentachloride
(d) sulfur hexafluoride *(e)* aluminum chloride

3.41 Name: SF_2.

▌ sulfur difluoride

3.42 Name: BaI_2.

▌ barium iodide

3.43 Write formulas for: ***(a)*** nickel(II) hydroxide ***(b)*** ammonium sulfate.

▌ *(a)* $Ni(OH)_2$ *(b)* $(NH_4)_2SO_4$

3.44 Write the formula for: aluminum bromide.

▌ $AlBr_3$

3.45 Name: ***(a)*** $CuSO_4$ ***(b)*** $Fe(NO_3)_3$ ***(c)*** FeI_2 ***(d)*** $Sr(ClO_3)_2$ ***(e)*** $K_2Cr_2O_7$ ***(f)*** Hg_2Cl_2 ***(g)*** $(NH_4)_2SO_3$

▌ *(a)* copper(II) sulfate *(b)* iron(III) nitrate *(c)* iron(II) iodide
(d) strontium chlorate *(e)* potassium dichromate *(f)* mercury(I) chloride
(g) ammonium sulfite

3.46 Name: CS_2.

▌ carbon disulfide

3.47 Name: ***(a)*** $Cu(ClO_2)_2$ ***(b)*** $(NH_4)_2C_2O_4$ ***(c)*** $Al_2(CO_3)_3$ ***(d)*** $Ba(CN)_2$

▌ *(a)* copper(II) chlorite *(b)* ammonium oxalate *(c)* aluminum carbonate *(d)* barium cyanide

3.48 For SCl_2, $HClO_3$, Li_3N, and NH_4ClO_2: ***(a)*** state if the compound is an acid. ***(b)*** If it is not an acid, state whether there is any ionic bonding in the compound. ***(c)*** Name the compound.

▌

	(a)	*(b)*	*(c)*
SCl_2	no	no	sulfur dichloride
$HClO_3$	yes		chloric acid
Li_3N	no	yes	lithium nitride
NH_4ClO_2	no	yes	ammonium chlorite

3.49 Name the ion which is produced when each of the following acids is neutralized by a base: ***(a)*** hydrochloric acid ***(b)*** nitric acid ***(c)*** sulfurous acid

▌ *(a)* chloride ion *(b)* nitrate ion *(c)* sulfite ion

3.50 Write the chemical formula for ***(a)*** potassium periodate ***(b)*** iron(II) phosphate ***(c)*** selenate ion

▌ *(a)* K_5IO_6 or KIO_4 *(b)* $Fe_3(PO_4)_2$ *(c)* SeO_4^{2-}

3.51 Name: ***(a)*** $Ba(HSO_4)_2$ ***(b)*** $Ba(H_2PO_2)_2$

▌ *(a)* barium hydrogen sulfate or barium bisulfate *(b)* barium dihydrogen hypophosphite

3.52 Using the periodic table as a guide, if necessary, write formulas for the following compounds: (*a*) ammonium perrhenate (*b*) lithium selenate (*c*) copper(I) arsenide (*d*) strontium iodate

▌ (*a*) NH_4ReO_4 (analogous to permanganate) (*b*) Li_2SeO_4 (analogous to sulfate)
(*c*) Cu_3As (*d*) $Sr(IO_3)_2$

3.53 Write formulas for (*a*) ammonium dichromate (*b*) lead(II) chlorate (*c*) sodium acetate (*d*) copper(I) oxide (*e*) barium arsenate (*f*) silver sulfate

▌ (*a*) $(NH_4)_2Cr_2O_7$ (*b*) $Pb(ClO_3)_2$ (*c*) $NaC_2H_3O_2$
(*d*) Cu_2O (*e*) $Ba_3(AsO_4)_2$ (*f*) Ag_2SO_4

3.54 Use the periodic table as an aid to write formulas for (*a*) radium sulfide (*b*) magnesium astatide (*c*) potassium phosphide

▌ (*a*) RaS (*b*) $MgAt_2$ (*c*) K_3P

3.55 Name: (*a*) Li_2SO_4 (*b*) $KReO_4$ (*c*) Na_2MoO_4

▌ (*a*) lithium sulfate (*b*) potassium perrhenate (*c*) sodium molybdate. These names are analogous to sodium sulfate, potassium permanganate, and sodium chromate, respectively.

CHAPTER 4

Chemical Formulas

4.1 PERCENT COMPOSITION

4.1 Carbon monoxide is 43% carbon by weight. Express that fact using pounds, kilograms, grams, atomic mass units.

▌ $$\frac{43\ \text{lb C}}{100\ \text{lb CO}} \qquad \frac{43\ \text{kg C}}{100\ \text{kg CO}} \qquad \frac{43\ \text{g C}}{100\ \text{g CO}} \qquad \frac{43\ \text{u C}}{100\ \text{u CO}}$$

4.2 A certain whiskey is 45% alcohol by volume. Express that fact using at least four different volume units.

▌ $$\frac{45\ \text{m}^3\ \text{alcohol}}{100\ \text{m}^3\ \text{whiskey}} \qquad \frac{45\ \text{L alcohol}}{100\ \text{L whiskey}} \qquad \frac{45\ \text{pt alcohol}}{100\ \text{pt whiskey}} \qquad \frac{45\ \text{mL alcohol}}{100\ \text{mL whiskey}}$$

4.3 Sulfur trioxide is 25 mol % sulfur. Which of the following ratios correctly expresses that percentage?

$$\frac{25\ \text{mol S}}{100\ \text{mol SO}_3} \qquad \frac{25\ \text{kg S}}{100\ \text{kg SO}_3} \qquad \frac{25\ \text{L S}}{100\ \text{L SO}_3}$$

▌ Only the first ratio, since a mole of sulfur has a different mass than a mole of sulfur trioxide and has a different volume than a mole of sulfur trioxide.

4.4 Determine the percent composition of $Mg_3(PO_4)_2$.

▌ In 1 mol of $Mg_3(PO_4)_2$ there are

$$3 \times 24.3 = 72.9\ \text{g Mg} \qquad \frac{72.9\ \text{g Mg}}{262.9\ \text{g total}} \times 100\% = 27.7\%\ \text{Mg}$$

$$2 \times 31.0 = 62.0\ \text{g P} \qquad \frac{62.0\ \text{g P}}{262.9\ \text{g total}} \times 100\% = 23.6\%\ \text{P}$$

$$8 \times 16.0 = 128.0\ \text{g O} \qquad \frac{128.0\ \text{g O}}{262.9\ \text{g total}} \times 100\% = 48.7\%\ \text{O}$$

$$\text{Total} = 262.9\ \text{g} \qquad\qquad \text{Total} = 100.0\%$$

You may check your answers by adding all the percentages. If the total is 100% within 0.1%, your answer *may* be correct. If the total is not 100%, you have made an error somewhere. (If the answer is 100%, you still *may* have made a mistake. Try using the wrong atomic weight for one of the elements, for example.)

4.5 Calculate the percent composition of ethyl alcohol, CH_3CH_2OH.

▌ Each molecule contains two carbon atoms, six hydrogen atoms, and one oxygen atom. Hence

$$\text{MW} = (2 \times 12.01\ \text{u}) + (6 \times 1.008\ \text{u}) + (16.00\ \text{u}) = 46.07\ \text{u}$$

$$\%\ \text{carbon} = \frac{24.02\ \text{u}}{46.07\ \text{u}} \times 100\% = 52.14\%\ \text{C}$$

$$\%\ \text{hydrogen} = \frac{6.048\ \text{u}}{46.07\ \text{u}} \times 100\% = 13.13\%\ \text{H}$$

$$\%\ \text{oxygen} = \frac{16.00\ \text{u}}{46.07\ \text{u}} \times 100\% = 34.73\%\ \text{O}$$

4.6 Given the formula K_2CO_3, determine the percent composition of potassium carbonate.

▌ 1 FW(K_2CO_3) contains

$$2\ \text{AW(K)} = 2 \times 39.098 = 78.196\ \text{parts K by weight}$$
$$1\ \text{AW(C)} = 1 \times 12.011 = 12.011\ \text{parts C by weight}$$
$$3\ \text{AW(O)} = 3 \times 15.999 = 47.998\ \text{parts O by weight}$$
$$\text{FW}(K_2CO_3) = 138.205\ \text{parts by weight}$$

$$\text{Fraction of K in } K_2CO_3 = \frac{78.196}{138.205} = 0.5658 = 56.58\%\ \text{K}$$

$$\text{Fraction of C in } K_2CO_3 = \frac{12.011}{138.205} = 0.0869 = 8.69\%\ \text{C}$$

$$\text{Fraction of O in } K_2CO_3 = \frac{47.998}{138.205} = 0.3473 = 34.73\%\ \text{O}$$

Check 100.00%

4.7 Using the following atomic weights, calculate the percent by mass of oxygen in $Ca(ClO_3)_2$: Ca 40.0 Cl 35.5 O 16.0.

▮

$1 \times 40.0 = 40.0$

$2 \times 35.5 = 71.0$

$6 \times 16.0 = 96.0$

Total = 207.0

$$\frac{96.0}{207.0} \times 100\% = 46.4\%\ \text{O}$$

4.8 Calculate the percentage of carbon in $Ca(HCO_3)_2$.

▮

$1 \times 40.0 = 40.0$ g Ca

$2 \times 1.0 = 2.0$ g H

$2 \times 12.0 = 24.0$ g C

$6 \times 16.0 = 96.0$ g O

Total = 162.0 g

$$\frac{24.0\ \text{g C}}{162\ \text{g total}} \times 100\% = 14.8\%\ \text{C}$$

4.9 Calculate the percent composition of $CsClO_2$.

▮

Cs	132.9 g	Cs:	(132.9/200.4) × 100% =	66.3% Cs
Cl	35.5 g	Cl:	(35.5/200.4) × 100% =	17.7% Cl
2O	32.0 g	O:	(32.0/200.4) × 100% =	16.0% O
Total	200.4 g		Total =	100.0%

4.10 Calculate the percent composition of each of the following compounds: ***(a)*** benzene, C_6H_6 ***(b)*** acetylene, C_2H_2. ***(c)*** Compare and explain the results obtained in *(a)* and *(b)*. ***(d)*** What additional data could be used to distinguish between benzene and acetylene?

▮

(a) $$\%\text{H} = \frac{6(1.008\ \text{u})}{6(12.011\ \text{u}) + 6(1.008\ \text{u})} \times 100\% = 7.74\%\ \text{H} \qquad \%\text{C} = 100.0\% - 7.74\% = 92.26\%\ \text{C}$$

(b) $$\%\text{H} = \frac{2(1.008\ \text{u})}{2(12.011\ \text{u}) + 2(1.008\ \text{u})} \times 100\% = 7.74\%\ \text{H}$$ Again, 92.26% C.

(c) The percent of each element is the same in both C_2H_2 and C_6H_6, since they both have the same empirical formula—CH. ***(d)*** The molecular weight would allow distinction between the two; C_6H_6 has a molecular weight of 78 u; C_2H_2 has a molecular weight of 26 u.

4.11 A strip of electrolytically pure copper weighing 3.178 g is strongly heated in a stream of oxygen until it is all converted to 3.978 g of the black oxide. What is the percent composition of this oxide?

▮

Total weight of black oxide = 3.978 g

Weight of copper in oxide = 3.178 g

Weight of oxygen in oxide = 0.800 g

$$\text{Fraction of copper} = \frac{\text{weight of copper in oxide}}{\text{total weight of oxide}} = \frac{3.178\ \text{g}}{3.978\ \text{g}} = 0.799 = 79.9\%$$

$$\text{Fraction of oxygen} = \frac{\text{weight of oxygen in oxide}}{\text{total weight of oxide}} = \frac{0.800\ \text{g}}{3.978\ \text{g}} = 0.201 = 20.1\%$$

Check 100.0%

4.12 Calculate the percent composition of $Co_2(CO_3)_3$.

In each mole of compound there is

$$\begin{aligned}(2\text{ mol Co})(58.9\text{ g/mol}) &= 117.8\text{ g Co}\\(3\text{ mol C})(12.0\text{ g/mol}) &= 36.0\text{ g C}\\(9\text{ mol O})(16.0\text{ g/mol}) &= 144.0\text{ g O}\\\text{Total} &= 297.8\text{ g}\end{aligned}$$

The percent of each element is thus

$$\frac{117.8\text{ g Co}}{297.8\text{ g total}} \times 100\% = 39.6\%\text{ Co} \qquad \frac{144.0\text{ g O}}{297.8\text{ g total}} \times 100\% = 48.4\%\text{ O}$$

$$\frac{36.0\text{ g C}}{297.8\text{ g total}} \times 100\% = 12.1\%\text{ C} \qquad \text{Total} = 100.1\%$$

4.13 Uranium hexafluoride, UF_6, is used in the gaseous diffusion process for separating uranium isotopes. How many kg of elementary uranium can be converted to UF_6 per kg of combined fluorine?

In 1.00 mol of UF_6 there is

$$\frac{238\text{ g U}}{6(19.0)\text{ g F}} = 2.09\text{ g U/g F} = 2.09\text{ kg U/kg F}$$

4.14 Determine the weight of sulfur required to make 1.00 metric ton of H_2SO_4.

$$(1.00 \times 10^6\text{ g }H_2SO_4)\left(\frac{1\text{ mol }H_2SO_4}{98.0\text{ g }H_2SO_4}\right)\left(\frac{1\text{ mol S}}{\text{mol }H_2SO_4}\right)\left(\frac{32.06\text{ g S}}{\text{mol S}}\right) = 327 \times 10^3\text{ g S} = 327\text{ kg S}$$

4.15 A 10.00-g sample of a crude ore contains 2.80 g of HgS. What is the percentage of mercury in the ore?

HgS:	
Hg	200.6 u
S	32.1
Total	232.7 u

$$\%\text{ Hg} = \frac{200.6\text{ u}}{232.7\text{ u}} \times 100\% = 86.21\%\text{ Hg in the compound}$$

$$(2.80\text{ g HgS})\left(\frac{86.21\text{ g Hg}}{100.0\text{ g HgS}}\right) = 2.41\text{ g Hg} \qquad \frac{2.41\text{ g Hg}}{10.0\text{ g ore}} \times 100\% = 24.1\%\text{ Hg in the ore}$$

4.16 How much phosphorus is contained in 5.00 g of the compound $CaCO_3 \cdot 3Ca_3(PO_4)_2$?

$CaCO_3 \cdot 3Ca_3(PO_4)_2$:	
10Ca	400.8 u
C	12.0
27O	432.0
6P	186.0
Total	1030.8 u

$$\%\text{ P} = \frac{186.0\text{ u}}{1030.8\text{ u}} \times 100\% = 18.0\%\text{ P}$$

$$(5.00\text{ g compound})\left(\frac{18.0\text{ g P}}{100\text{ g compound}}\right) = 0.900\text{ g P}$$

4.17 How many kg of metallic sodium and of liquid chlorine can be obtained from 1.00 metric ton of salt?

NaCl:	
Na	23.0 u
Cl	35.5
Total	58.5 u

$$\%\text{ Na} = \frac{23.0\text{ u}}{58.5\text{ u}} \times 100\% = 39.3\%\text{ Na}$$

$$\%\text{ Cl} = 100.0 - 39.3 = 60.7\%\text{ Cl}$$

$$(1.00\text{ metric ton NaCl})\left(\frac{10^3\text{ kg}}{\text{metric ton}}\right)\left(\frac{39.3\text{ kg Na}}{100\text{ kg NaCl}}\right) = 3.93 \times 10^2\text{ kg} = 393\text{ kg Na}$$

$$(1000\text{ kg}) - (393\text{ kg}) = 607\text{ kg Cl}$$

4.18 Calculate the percent composition of each of the following compounds: ***(a)*** H_2O ***(b)*** H_2O_2 ***(c)*** vinyl chloride, C_2H_3Cl ***(d)*** $Al_2(SO_4)_3$ ***(e)*** $NH_4C_2H_3O_2$

(a) $\%\text{ H} = \dfrac{2(1.008\text{ u})}{2(1.008\text{ u}) + (15.999\text{ u})} \times 100\% = 11.19\%\text{ H} \qquad \%\text{ O} = 100.00\% - 11.19\% = 88.81\%\text{ O}$

(*b*) $\%\,H = \dfrac{2(1.008\text{ u})}{2(1.008\text{ u}) + 2(15.999\text{ u})} \times 100\% = 5.927\%\text{ H} \qquad \%O = 100.000\% - 5.927\% = 94.073\%O$

(*c*) Formula weight = 2(12.011 u) + 3(1.008 u) + 35.453 u = 62.499 u

$$\%\,C = \frac{2(12.011\text{ u})}{62.499\text{ u}} \times 100\% = 38.44\%\text{ C} \qquad \%\,H = \frac{3(1.008\text{ u})}{62.499\text{ u}} \times 100\% = 4.84\%\text{ H}$$

$$\%\,Cl = \frac{35.453\text{ u}}{62.499\text{ u}} \times 100\% = 56.73\%\text{ Cl}$$

(*d*) Formula weight = 2(26.98 u) + 3(32.06 u) + 12(16.00 u) = 342.14 u

$$\%\,Al = \frac{53.96\text{ u}}{342.14\text{ u}} \times 100\% = 15.77\%\text{ Al} \qquad \%\,S = \frac{96.18\text{ u}}{342.14\text{ u}} \times 100\% = 28.11\%\text{ S}$$

$$\%\,O = \frac{192.00\text{ u}}{342.14\text{ u}} \times 100\% = 56.12\%\text{ O}$$

(*e*) The compound has an empirical formula $NH_7C_2O_2$, with a formula weight = 77.09 u

$$\frac{14.01\text{ u}}{77.09\text{ u}} \times 100\% = 18.17\%\text{ N} \qquad \frac{7.056\text{ u}}{77.09\text{ u}} \times 100\% = 9.15\%\text{ H}$$

$$\frac{24.02\text{ u}}{77.09\text{ u}} \times 100\% = 31.16\%\text{ C} \qquad \frac{32.00\text{ u}}{77.09\text{ u}} \times 100\% = 41.51\%\text{ O}$$

4.19 What weight of CuO will be required to furnish 200 kg copper?

▮

CuO:	Cu	63.5 u
	O	16.0
	Total	79.5 u

$$\%\,Cu = \frac{63.5\text{ u}}{79.5\text{ u}} \times 100\% = 79.9\%\text{ Cu} \qquad (200\text{ kg Cu})\left(\frac{100.0\text{ kg CuO}}{79.9\text{ kg Cu}}\right) = 250\text{ kg CuO}$$

4.20 What weight of silver is present in 3.45 g Ag_2S?

▮

Ag_2S:	2Ag	2(107.9 u) = 215.8 u
	S	32.0
	Total	247.8 u

$$\%\,Ag = \frac{215.8\text{ u}}{247.8\text{ u}} \times 100\% = 87.1\%\text{ Ag} \qquad (3.45\text{ g }Ag_2S)\left(\frac{87.1\text{ g Ag}}{100.0\text{ g }Ag_2S}\right) = 3.00\text{ g Ag}$$

4.21 The purest form of carbon is prepared by decomposing pure sugar, $C_{12}H_{22}O_{11}$ (driving off H_2O). What is the maximum number of grams of carbon that could be obtained from 1.00 lb of sugar?

▮ The maximum mass of carbon available is the number of grams present:

$C_{12}H_{22}O_{11}$:	12C	144.0 u
	22H	22.0 u
	11O	176.0
	Total	342.0 u

$$\frac{144\text{ u C}}{342\text{ u total}} \times 100\% = 42.1\%\text{ C}$$

$$(1.00\text{ lb})\left(\frac{454\text{ g}}{\text{lb}}\right)\left(\frac{42.1\text{ g C}}{100.0\text{ g}}\right) = 191\text{ g C}$$

4.22 Determine the percent composition of (*a*) UO_2F_2 (*b*) $C_3Cl_2F_6$

▮

(*a*)

UO_2F_2:	U	238.0 u
	2O	32.0
	2F	38.0
	Total	308.0 u

$$\%\,U = \frac{238.0\text{ u}}{308.0\text{ u}} \times 100\% = 77.27\%\text{ U}$$

$$\%\,O = \frac{32.0\text{ u}}{308.0\text{ u}} \times 100\% = 10.4\%\text{ O}$$

$$\%\,F = \frac{38.0\text{ u}}{308.0\text{ u}} \times 100\% = 12.3\%\text{ F}$$

(*b*) $C_3Cl_2F_6$:

3C	36.0 u	$\%\ C = \dfrac{36.0\ u}{220.9\ u} \times 100\% = 16.3\%\ C$
2Cl	70.9	$\%\ Cl = \dfrac{70.9\ u}{220.9\ u} \times 100\% = 32.1\%\ Cl$
6F	114.0	$\%\ F = \dfrac{114.0\ u}{220.9\ u} \times 100\% = 51.6\%\ F$
Total	220.9 u	

4.23 Determine the percent composition of (*a*) silver chromate, Ag_2CrO_4 (*b*) calcium pyrophosphate, $Ca_2P_2O_7$

▮ (*a*)

2Ag	2(107.9 u) = 215.8 u	$\%\ Ag = \dfrac{215.8\ u}{331.8\ u} \times 100\% = 65.0\%\ Ag$
Cr	52.0	$\%\ Cr = \dfrac{52.0\ u}{331.8\ u} \times 100\% = 15.7\%\ Cr$
4O	4(16.0) = 64.0	$\%\ O = \dfrac{64.0\ u}{331.8\ u} \times 100\% = 19.3\%\ O$
	Total = 331.8 u	

(*b*)

2Ca	2(40.1 u) = 80.2 u	$\%\ Ca = \dfrac{80.2\ u}{254.2\ u} \times 100\% = 31.5\%\ Ca$
2P	2(31.0) = 62.0	$\%\ P = \dfrac{62.0\ u}{254.2\ u} \times 100\% = 24.4\%\ P$
7O	7(16.0) = 112.0	$\%\ O = \dfrac{112.0\ u}{254.2\ u} \times 100\% = 44.1\%\ O$
	Total = 254.2 u	

4.24 What is the nitrogen content (fertilizer rating) of (*a*) NH_4NO_3 (*b*) $(NH_4)_2SO_4$?

▮ (*a*)

2N	2(14.0) = 28.0 u
4H	4(1.008) = 4.0
3O	3(16.0) = 48.0
	Total = 80.0 u

$$\%\ N = \frac{28.0\ u}{80.0\ u} \times 100\% = 35.0\%\ N$$

(*b*)

2N	28.0 u
8H	8.0
S	32.0
4O	64.0
Total	132.0 u

$$\%\ N = \frac{28.0\ u}{132.0\ u} \times 100\% = 21.2\%\ N$$

4.25 Calculate the percentage of copper in (*a*) cuprite, Cu_2O (*b*) copper pyrites, $CuFeS_2$ (*c*) malachite, $CuCO_3 \cdot Cu(OH)_2$. (*d*) How many tons of cuprite will give 500 tons of copper?

▮ (*a*)

2Cu	2(63.5) = 127 u
O	16
	Total = 143 u

$$\%\ Cu = \frac{127\ u}{143\ u} \times 100\% = 88.8\%\ Cu$$

(*b*)

Cu	63.5 u
Fe	55.8
2S	2(32.06) = 64.1
	Total = 183.4 u

$$\%\ Cu = \frac{63.5\ u}{183.4\ u} \times 100\% = 34.6\%\ Cu$$

(*c*)

2Cu	2(63.5) = 127 u
C	12
5O	5(16) = 80
2H	2(1) = 2
	Total = 221 u

$$\%\ Cu = \frac{127\ u\ Cu}{221\ u\ total} \times 100\% = 57.5\%\ Cu$$

(*d*) $$(500\ tons\ Cu)\left(\frac{100\ tons\ total}{88.8\ tons\ Cu}\right) = 563\ tons$$

4.26 Find the percentage arsenic in a polymer having the empirical formula C_2H_8AsB.

▮

2C	24.0 u	As	74.9
8H	8.0	B	10.8

Total 117.7 u

$$\% \text{ As} = \frac{74.9 \text{ u}}{117.7 \text{ u}} \times 100\% = 63.6\% \text{ As}$$

4.27 The specifications for a transistor material called for one boron atom in 10^{10} silicon atoms. What would be the boron content of 1 kg of such material?

▮

$$\text{Total mass} = (10^{10} \text{ Si})(28.1 \text{ u/Si}) = 28.1 \times 10^{10} \text{ u} \qquad \text{(The one B atom is negligible.)}$$

$$\% \text{ B} = \frac{10.8 \text{ u B}}{28.1 \times 10^{10} \text{ u total}} \times 100\% = 3.84 \times 10^{-9}\% \text{ B}$$

$$(1 \text{ kg total})\left(\frac{3.84 \times 10^{-9} \text{ kg B}}{100 \text{ kg total}}\right) = 4 \times 10^{-11} \text{ kg B}$$

4.28 Calculate the percent composition of polyvinyl chloride, $(C_2H_3Cl)_n$. Compare the answer to that for vinyl chloride, Prob. 4.18(*c*).

▮ Polyvinyl chloride has the same percent composition as vinyl chloride.

$$\frac{2n \text{ moles C} \times 100\%}{n \text{ moles } C_2H_3Cl} = \frac{2n(12.01 \text{ u}) \times 100\%}{n(62.50 \text{ u})} = \frac{2(12.01 \text{ u}) \times 100\%}{62.50 \text{ u}} = 38.44\%$$

Since the unknown n drops out of the equation because it appears in both numerator and denominator, the degree of polymerization (n) does not affect the percent composition.

4.29 (***a***) Determine the percentages of iron in $FeCO_3$, Fe_2O_3, and Fe_3O_4. (***b***) How many kilograms of iron could be obtained from 2.000 kg of Fe_2O_3?

▮ (***a***) Formula weight of $FeCO_3$ is 115.86; of Fe_2O_3, 159.70; of Fe_3O_4, 231.55. Consider 1 mol of each compound.

$$\text{Fraction of Fe in } FeCO_3 = \frac{(1 \text{ mol Fe})}{(1 \text{ mol } FeCO_3)} = \frac{55.85 \text{ g}}{115.86 \text{ g}} = 0.4820 = 48.20\%$$

$$\text{Fraction of Fe in } Fe_2O_3 = \frac{(2 \text{ mol Fe})}{(1 \text{ mol } Fe_2O_3)} = \frac{2(55.85 \text{ g})}{159.70 \text{ g}} = 0.6994 = 69.94\%$$

$$\text{Fraction of Fe in } Fe_3O_4 = \frac{(3 \text{ mol Fe})}{(1 \text{ mol } Fe_3O_4)} = \frac{3(55.85 \text{ g})}{231.55 \text{ g}} = 0.7237 = 72.37\%$$

(***b***) From (*a*), the weight of Fe in 2.000 kg Fe_2O_3 is $0.6994 \times 2.000 \text{ kg} = 1.399 \text{ kg Fe}$.

4.30 (***a***) Calculate the percentage of CaO in $CaCO_3$. (***b***) How many pounds of CaO can be obtained from 1.000 ton of limestone that is 97.0% $CaCO_3$?

▮ (***a***) The quantitative factor can be written by considering the conservation of Ca atoms. 1 mol $CaCO_3$ contains 1 mol Ca, which is the same amount of Ca as is contained in 1 mol CaO. Therefore,

$$\text{Fraction of CaO in } CaCO_3 = \frac{\text{FW(CaO)}}{\text{FW}(CaCO_3)} = \frac{56.1}{100.1} = 0.560 = 56.0\%$$

(***b***) Weight of $CaCO_3$ in 1.000 ton limestone $= 0.970 \times 2000 \text{ lb} = 1940 \text{ lb } CaCO_3$

$$\text{Weight of CaO} = (\text{fraction of CaO in } CaCO_3)(\text{weight of } CaCO_3)$$
$$= 0.560 \times 1940 \text{ lb} = 1090 \text{ lb CaO per ton limestone}$$

4.31 Compute the amount of zinc in a ton of ore containing 60.0% zincite, ZnO.

▮

ZnO:		
	Zn	65.4 u
	O	16.0
		81.4 u

$$\% \text{ Zn} = \frac{65.4 \text{ u}}{81.4 \text{ u}} \times 100\% = 80.3\% \text{ Zn in ZnO}$$

$$(1.00 \text{ ton ore})\left(\frac{2000 \text{ lb}}{\text{ton}}\right)\left(\frac{60.0 \text{ lb ZnO}}{100.0 \text{ lb ore}}\right)\left(\frac{80.3 \text{ lb Zn}}{100 \text{ lb ZnO}}\right) = 964 \text{ lb Zn}$$

4.32 The arsenic content of an agricultural insecticide was reported as 28% As_2O_5. What is the percentage arsenic in this preparation?

▮

As_2O_5:	2As	148.8 u
	5O	80.0
		228.8 u

$$\% \text{ As} = \frac{148.8 \text{ u}}{228.8 \text{ u}} \times 100\% = 65.0\% \text{ As in } As_2O_5$$

$$(28\% \text{ } As_2O_5)\left(\frac{65.0 \text{ g As}}{100 \text{ g } As_2O_5}\right) = 18\% \text{ As in insecticide}$$

4.33 Express the potassium content of a fertilizer in % K_2O if its elementary potassium content is 4.5%.

▮

$$\left(\frac{4.5 \text{ g K}}{100 \text{ g sample}}\right)\left(\frac{1 \text{ mol K}}{39.1 \text{ g K}}\right)\left(\frac{1 \text{ mol } K_2O}{2 \text{ mol K}}\right)\left(\frac{94.2 \text{ g } K_2O}{\text{mol } K_2O}\right) = \frac{5.4 \text{ g } K_2O}{100 \text{ g sample}} = 5.4\% \text{ } K_2O$$

4.34 A typical analysis of a Pyrex glass showed 12.9% B_2O_3, 2.2% Al_2O_3, 3.8% Na_2O, 0.4% K_2O, and the balance SiO_2. Assume that the oxide percentages add up to 100%. What is the ratio of silicon to boron atoms in the glass?

▮

$$\% \text{ Si} = 100.0\% - 12.9\% - 2.2\% - 3.8\% - 0.4\% = 80.7\% \text{ } SiO_2$$

SiO_2:	Si	28.1 u
	2O	32.0
		60.1 u

$$\% \text{ Si} = \frac{28.1 \text{ u}}{60.1 \text{ u}} \times 100\% = 46.8\% \text{ Si in } SiO_2$$

B_2O_3:	2B	21.6 u
	3O	48.0
		69.6 u

$$\% \text{ B} = \frac{21.6 \text{ u}}{69.6 \text{ u}} \times 100\% = 31.0\% \text{ B in } B_2O_3$$

Per 100 u of glass,

$$(80.7 \text{ u } SiO_2)\left(\frac{46.8 \text{ u Si}}{100 \text{ u } SiO_2}\right) = (37.8 \text{ u Si})\left(\frac{1 \text{ atom Si}}{28.1 \text{ u Si}}\right) = 1.35 \text{ atoms Si}$$

$$(12.9 \text{ u } B_2O_3)\left(\frac{31.0 \text{ u B}}{100 \text{ u } B_2O_3}\right) = (4.00 \text{ u B})\left(\frac{1 \text{ atom B}}{10.8 \text{ u B}}\right) = 0.370 \text{ atom B}$$

$$\frac{\text{Si}}{\text{B}} = \frac{1.35}{0.370} = 3.6 \text{ atoms Si/atom B}$$

4.35 A piece of plumber's solder weighing 3.00 g was dissolved in dilute nitric acid, then treated with dilute H_2SO_4. This precipitated the lead as $PbSO_4$, which after washing and drying weighed 2.93 g. The solution was then neutralized to precipitate stannic acid, which was decomposed by heating, yielding 1.27 g SnO_2. What is the analysis of the solder as % Pb and % Sn?

▮

SnO_2:	Sn	118.7 u
	2O	32.0
	Total	150.7 u

$$\% \text{ Sn} = \frac{118.7 \text{ u}}{150.7 \text{ u}} \times 100 = 78.8\% \text{ Sn in } SnO_2$$

$$(1.27 \text{ g } SnO_2)\left(\frac{78.8 \text{ g Sn}}{100 \text{ g } SnO_2}\right) = 1.00 \text{ g Sn in sample}$$

$PbSO_4$:	Pb	207.2 u
	S	32.1
	4O	64.0
	Total	303.3 u

$$\% \text{ Pb} = \frac{207.2 \text{ u}}{303.3 \text{ u}} \times 100 = 68.3\% \text{ Pb in } PbSO_4$$

$$(2.93 \text{ g } PbSO_4)\left(\frac{68.3 \text{ g Pb}}{100 \text{ g } PbSO_4}\right) = 2.00 \text{ g Pb}$$

$$\frac{2.00 \text{ g Pb}}{3.00 \text{ g total}} \times 100\% = 66.7\% \text{ Pb} \qquad \frac{1.00 \text{ g Sn}}{3.00 \text{ g total}} \times 100\% = 33.3\% \text{ Sn}$$

4.36 A sample of impure cuprite, Cu_2O, contains 66.6% copper. What is the percentage of pure Cu_2O in the sample?

▮

$$\left(\frac{66.6 \text{ g Cu}}{100 \text{ g sample}}\right)\left(\frac{1 \text{ mol Cu}}{63.5 \text{ g Cu}}\right)\left(\frac{1 \text{ mol } Cu_2O}{2 \text{ mol Cu}}\right)\left(\frac{143 \text{ g } Cu_2O}{\text{mol } Cu_2O}\right) = \frac{75.0 \text{ g } Cu_2O}{100 \text{ g sample}} = 75.0\% \text{ } Cu_2O$$

4.37 A taconite ore consisted of 35.0% Fe_3O_4 and the balance siliceous impurities. How many tons of the ore must be processed in order to recover a ton of metallic iron (*a*) if there is 100% recovery, (*b*) if there is only 75% recovery?

▌ (*a*)
$$(1.00 \text{ ton Fe})\left(\frac{2000 \text{ lb}}{\text{ton}}\right)\left(\frac{454 \text{ g}}{\text{lb}}\right)\left(\frac{1 \text{ mol Fe}}{55.85 \text{ g Fe}}\right)\left(\frac{1 \text{ mol } Fe_3O_4}{3 \text{ mol Fe}}\right)\left(\frac{231.6 \text{ g } Fe_3O_4}{\text{mol } Fe_3O_4}\right)$$
$$\times\left(\frac{1 \text{ lb}}{454 \text{ g}}\right)\left(\frac{1 \text{ ton}}{2000 \text{ lb}}\right)\left(\frac{100 \text{ tons ore}}{35.0 \text{ tons } Fe_3O_4}\right) = 3.95 \text{ tons ore}$$

(As usual in these problems, we could have solved in grams and used the results in tons. As can be seen above, the conversion factors from tons to grams and back merely cancel out.)

(*b*)
$$\frac{3.95 \text{ tons}}{0.75} = 5.27 \text{ tons}$$

4.38 A 5.82 g silver coin is dissolved in nitric acid. When sodium chloride is added to the solution, all the silver is precipitated as AgCl. The AgCl precipitate weighs 7.20 g. Determine the percentage of silver in the coin.

▌
$$\text{Fraction of Ag in AgCl} = \frac{\text{AW(Ag)}}{\text{FW(AgCl)}} = \frac{107.9}{143.3} = 0.753$$
$$\text{Weight of Ag in 7.20 g AgCl} = (0.753)(7.20 \text{ g}) = 5.42 \text{ g Ag}$$

Hence the 5.82 g coin contains 5.42 g Ag.

$$\text{Fraction of Ag in coin} = \frac{5.42 \text{ g}}{5.82 \text{ g}} = 0.931 = 93.1\% \text{ Ag}$$

4.39 A sample of impure sulfide ore contains 42.34% Zn. Find the percentage of pure ZnS in the sample.

▌ The formula ZnS shows that 1 FW(ZnS) contains 1 AW(Zn).

$$\text{Fraction of Zn in ZnS} = \frac{\text{AW(Zn)}}{\text{FW(ZnS)}} = \frac{65.38}{97.44} = 0.6710 = 67.10\%$$

If the sample were 100% ZnS, it would contain 67.10% Zn. But since the sample contains only 42.34% Zn, it is

$$\frac{42.34}{67.10} \times 100\% = 63.10\% \text{ pure ZnS}$$

4.40 When the *Bayer process* is used for recovering aluminum from siliceous ores, some aluminum is always lost because of the formation of an unworkable "mud" having the following average formula: $3Na_2O \cdot 3Al_2O_3 \cdot 5SiO_2 \cdot 5H_2O$. Since aluminum and sodium ions are always in excess in the solution from which this precipitate is formed, the precipitation of the silicon in the "mud" is complete. A certain ore contained 13% (by weight) kaolin ($Al_2O_3 \cdot 2SiO_2 \cdot 2H_2O$) and 87% gibbsite ($Al_2O_3 \cdot 3H_2O$). What percent of the total aluminum in this ore is recoverable in the Bayer process?

▌ Consider 100 g ore, which contains 13 g kaolin and 87 g gibbsite.

$$\text{Weight of Al in 13 g kaolin} = (13 \text{ g kaolin})\left(\frac{2 \text{ mol Al}}{1 \text{ mol kaolin}}\right) = 13 \times \frac{54.0}{258} = 2.7 \text{ g Al}$$
$$\text{Weight of Al in 87 g gibbsite} = (87 \text{ g gibbsite})\left(\frac{2 \text{ mol Al}}{1 \text{ mol gibbsite}}\right) = 87 \times \frac{54.0}{156} = 30.1 \text{ g Al}$$
$$\text{Total weight of Al in 100 g ore} = 2.7 \text{ g} + 30.1 \text{ g} = 32.8 \text{ g}$$

Kaolin has equal numbers of Al and Si atoms, and 13 g kaolin contains 2.7 g Al. The mud takes 6 Al atoms for 5 Si atoms. Hence the precipitation of all the Si from 13 g kaolin involves the loss of $(6/5)(2.7 \text{ g}) = 3.2$ g Al.

$$\text{Fraction of Al recoverable} = \frac{\text{recoverable Al}}{\text{total Al}} = \frac{(32.8 - 3.2) \text{ g}}{32.8 \text{ g}} = 0.90 = 90\%$$

4.2 THE MOLE; FORMULA CALCULATIONS

4.41 ^{12}C is the standard for the atomic weights of atoms. What is the standard for the molecular weights of molecules? Explain.

▌ ^{12}C. The same standard is used for all atomic, molecular, and formula weights.

4.42 If the atomic weight of carbon were set at 100 u, what would be the value of Avogadro's number? Is Avogadro's number a fundamental physical constant?

▌ Avogadro's number would be the number of atoms in 100 g of carbon. The number of atoms in 100 g of carbon may be calculated as follows:

$$(100 \text{ g C})\left(\frac{6.02 \times 10^{23} \text{ atoms}}{12.01 \text{ g C}}\right) = 5.01 \times 10^{24} \text{ atoms}$$

Obviously, Avogadro's number depends on the basis of the atomic weight scale. (It changed very little in 1961, when the basis of the atomic weight scale was changed from the naturally occurring mixture of oxygen isotopes at 16.0000 u to ^{12}C, which put oxygen at 15.9994 u.)

4.43 Determine **(*a*)** the formula weight of potassium hexachloroiridate(IV), K_2IrCl_6, and **(*b*)** the molecular weight of trifluorosilane, $SiHF_3$.

▌ Potassium hexachloroiridate(IV) does not exist as discrete molecules represented by the empirical formula, but trifluorosilane does. The term *formula weight* may be used in every case to describe the relative weight of the indicated formula unit; further chemical experience is needed to define the applicability of the term *molecular weight*. Some books use the two terms interchangeably.

(*a*)	2K = 2(39.10) = 78.20	**(*b*)**	Si = (28.086)	= 28.086
	Ir = (192.22) = 192.22		H = (1.008)	= 1.008
	6Cl = 6(35.453) = 212.72		3F = 3(18.9984)	= 56.995
	Formula weight = 483.14		Molecular weight	= 86.086

Note that the atomic weights are not all known to the same number of significant figures or to the same number of decimal places in u. In (*a*) there is no point in writing the atomic weight of K as 39.098, since the value for Ir is known to only 0.01 u. Note also that in order to express 6 times the atomic weight of Cl to 0.01 u, it was necessary to use the atomic weight to 0.001 u. Similarly, an extra figure was used in the atomic weight for fluorine to give the maximum significance to the last digit in the sum column.

4.44 What is the formula weight of Na_2S?

▌ 2 × AW(Na) = 45.98 u   AW(S) = 32.06 u   Formula weight = total = 78.04 u

4.45 What is the molecular weight of glucose, $C_6H_{12}O_6$?

▌

6 × AW(C) = 72.06 u	6 × AW(O) = 96.00 u
12 × AW(H) = 12.10 u	Molecular weight = total = 180.16 u

4.46 Determine the molecular weights or formula weights to 0.01 u for **(*a*)** NaOH **(*b*)** HNO_3 **(*c*)** F_2 **(*d*)** S_8 **(*e*)** $Ca_3(PO_4)_2$ **(*f*)** $Fe_4[Fe(CN)_6]_3$ **(*g*)** $TiO_{1.12}$

▌ All noninteger quantities are in u.

(*a*)	Na	22.99	**(*b*)**	H	1.008	**(*c*)**	2F	2(19.00)	**(*d*)**	8S	8(32.06)
	O	16.00		N	14.006		F_2	38.00		S_8	256.48
	H	1.008		3O	48.00						
	NaOH	40.00		HNO_3	63.01						

(*e*)	3Ca	3(40.08) = 120.24	**(*f*)**	7Fe	7(55.85) = 390.95	**(*g*)**	Ti	47.90
	2P	2(30.97) = 61.94		18C	18(12.01) = 216.18		1.12O	1.12(16.00) = 17.92
	8O	8(16.00) = 128.00		18N	18(14.01) = 252.18		$TiO_{1.12}$	65.82
	$Ca_3(PO_4)_2$	= 310.18		$Fe_4[Fe(CN)_6]_3$	= 859.31			

Note that $TiO_{1.12}$ is an unusual type of compound which is called nonstoichiometric. Its numbers of moles are not in a small integral ratio.

4.47 **(*a*)** What is the mass of 4.00×10^{-3} mol of glucose, $C_6H_{12}O_6$? **(*b*)** How many carbon atoms are there in 4.00×10^{-3} mol of glucose?

▌ **(*a*)** $(4.00 \times 10^{-3} \text{ mol})\left(\frac{180 \text{ g}}{\text{mol}}\right) = 0.720 \text{ g}$

(*b*) $$(4.00 \times 10^{-3} \text{ mol } C_6H_{12}O_6)\left(\frac{6 \text{ mol C}}{\text{mol } C_6H_{12}O_6}\right)\left(\frac{6.02 \times 10^{23} \text{ C atoms}}{\text{mol C}}\right) = 1.44 \times 10^{22} \text{ C atoms}$$

4.48 How many moles of $C_2H_4O_2$ contains 6.02×10^{23} atoms of hydrogen?

$$(1.00 \text{ mol H})\left(\frac{1 \text{ mol } C_2H_4O_2}{4 \text{ mol H}}\right) = 0.250 \text{ mol } C_2H_4O_2$$

4.49 (*a*) How many atoms of oxygen are there in 1.00 mol of Ce_2O_3? (Ce is element number 58.) (*b*) How many grams of Ce_2O_3 are in 0.400 mol of Ce_2O_3?

(*a*) $$(1.00 \text{ mol } Ce_2O_3)\left(\frac{3 \text{ mol O}}{\text{mol } Ce_2O_3}\right)\left(\frac{6.02 \times 10^{23} \text{ O atoms}}{\text{mol O}}\right) = 1.81 \times 10^{24} \text{ O atoms}$$

(*b*) $$(0.400 \text{ mol } Ce_2O_3)\left(\frac{328 \text{ g } Ce_2O_3}{\text{mol } Ce_2O_3}\right) = 131 \text{ g } Ce_2O_3$$

4.50 Calculate the number of oxygen atoms in 25.0 g of $CaCO_3$.

$$(25.0 \text{ g } CaCO_3)\left(\frac{1 \text{ mol } CaCO_3}{100 \text{ g } CaCO_3}\right)\left(\frac{3 \text{ mol O atoms}}{1 \text{ mol } CaCO_3}\right) = 0.750 \text{ mol O}$$

$$(0.750 \text{ mol O})\left(\frac{6.02 \times 10^{23} \text{ O atoms}}{\text{mol O}}\right) = 4.52 \times 10^{23} \text{ O atoms}$$

4.51 How many molecules of water are there in 36.0 g of H_2O?

$$(36.0 \text{ g } H_2O)\left(\frac{1 \text{ mol } H_2O}{18.0 \text{ g } H_2O}\right)\left(\frac{6.02 \times 10^{23} \text{ molecules } H_2O}{\text{mol } H_2O}\right) = 1.20 \times 10^{24} \text{ molecules}$$

4.52 How many atoms of hydrogen are there in 25.0 g of NH_4Cl?

$$(25.0 \text{ g } NH_4Cl)\left(\frac{1 \text{ mol } NH_4Cl}{53.5 \text{ g } NH_4Cl}\right)\left(\frac{4 \text{ mol H}}{\text{mol } NH_4Cl}\right) = 1.87 \text{ mol H}$$

$$(1.87 \text{ mol H})\left(\frac{6.02 \times 10^{23} \text{ H atoms}}{\text{mol H}}\right) = 1.13 \times 10^{24} \text{ H atoms}$$

4.53 How many molecules of H_2 are present in 7.5 g of H_2? How many H atoms?

$$(7.5 \text{ g } H_2)\left(\frac{1 \text{ mol } H_2}{2.0 \text{ g } H_2}\right)\left(\frac{6.02 \times 10^{23} \text{ molecules } H_2}{\text{mol } H_2}\right) = 2.2 \times 10^{24} \text{ molecules } H_2$$

$$(2.2 \times 10^{24} \text{ molecules } H_2)\left(\frac{2 \text{ atoms H}}{\text{molecule } H_2}\right) = 4.4 \times 10^{24} \text{ atoms H}$$

4.54 Calculate the number of oxygen atoms in 300 g of $CaCO_3$.

$$(300 \text{ g } CaCO_3)\left(\frac{1 \text{ mol } CaCO_3}{100 \text{ g } CaCO_3}\right)\left(\frac{3 \text{ mol O}}{\text{mol } CaCO_3}\right)\left(\frac{6.02 \times 10^{23} \text{ O atoms}}{\text{mol O}}\right) = 5.42 \times 10^{24} \text{ O atoms}$$

4.55 How many grams of H_2O are there in 2.50 mol of H_2O?

$$(2.50 \text{ mol } H_2O)\left(\frac{18.0 \text{ grams}}{\text{mol}}\right) = 45.0 \text{ g}$$

4.56 Which one of the following, if any, contains the greatest number of oxygen atoms? the greatest number of molecules? 1.0 g of O atoms, 1.0 g of O_2, or 1.0 g of ozone, O_3.

All have the same number of atoms. The 1.0-g sample of O has the largest number of molecules (albeit monatomic).

4.57 How many moles of atoms of each element are there in 1.0 mol of each of the following compounds? (*a*) Fe_3O_4 (*b*) $AsCl_5$ (*c*) $Mg(C_2H_3O_2)_2$ (*d*) $CuSO_4 \cdot 5H_2O$

(*a*) 3 mol Fe, 4 mol O (*b*) 1 mol As, 5 mol Cl

(*c*) 1 mol Mg, 4 mol C, 6 mol H, 4 mol O (*d*) 1 mol Cu, 1 mol S, 9 mol O, 10 mol H

4.58 How many moles of oxygen atoms are there in each of the following? (*a*) 0.17 mol of O_2 (*b*) 6.02×10^{24} molecules of CO (*c*) 1.0 mol of $BaS_2O_8 \cdot 4H_2O$ (*d*) 20.0 g of O_2 (*e*) 1.6 g of CO_2 (*f*) 1.0 mol of $Ba(NO_3)_2 \cdot H_2O$

(*a*) $(0.17 \text{ mol } O_2)\left(\frac{2 \text{ mol O}}{\text{mol } O_2}\right) = 0.34 \text{ mol O}$

(*b*) $(6.02 \times 10^{24} \text{ CO molecules})\left(\frac{1 \text{ mol CO}}{6.02 \times 10^{23} \text{ molecules}}\right)\left(\frac{1 \text{ mol O}}{\text{mol CO}}\right) = 10.0 \text{ mol O}$

(*c*) 12 mol (*d*) $(20.0 \text{ g } O_2)\left(\frac{1 \text{ mol } O_2}{32.0 \text{ g } O_2}\right)\left(\frac{2 \text{ mol O}}{\text{mol } O_2}\right) = 1.25 \text{ mol O}$

(*e*) $(1.6 \text{ g } CO_2)\left(\frac{1 \text{ mol } CO_2}{44 \text{ g } CO_2}\right)\left(\frac{2 \text{ mol O}}{\text{mol } CO_2}\right) = 0.073 \text{ mol O}$ (*f*) 7.0 mol O

4.59 How many grams of each of the constituent elements are contained in 1.000 mol of (*a*) CH_4 (*b*) Fe_2O_3 (*c*) Ca_3P_2? How many atoms of each element are contained in the same amount of compound?

(*a*) 1 mol C = 12.01 g C — 6.02×10^{23} atoms
4 mol H = 4.032 g H — $4(6.02 \times 10^{23}) = 2.41 \times 10^{24}$ atoms
(*b*) 2 mol Fe = 111.70 g Fe — $2(6.02 \times 10^{23}) = 1.20 \times 10^{24}$ atoms
3 mol O = 48.00 g O — $3(6.02 \times 10^{23}) = 1.81 \times 10^{24}$ atoms
(*c*) 3 mol Ca = 120.24 g Ca — 1.81×10^{24} atoms
2 mol P = 61.95 g P — 1.20×10^{24} atoms

4.60 Calculate the number of g in a mol of each of the following common substances: (*a*) calcite, $CaCO_3$ (*b*) quartz, SiO_2 (*c*) cane sugar, $C_{12}H_{22}O_{11}$ (*d*) gypsum, $CaSO_4 \cdot 2H_2O$ (*e*) white lead, $Pb(OH)_2 \cdot 2PbCO_3$

The method is the same as the previous problem. The numbers of grams of each element are merely totaled. (*a*) 100.09 g (*b*) 60.09 g (*c*) 342.3 g Parts (*d*) and (*e*) have one additional facet. (*d*) The 2 coefficient within the formula multiplies everything after it until the next centered dot or until the end of the formula.

1 mol Ca	40.08 g	
1 mol S	32.06 g	
6 mol O	96.00 g	(2 mol from the H_2O of hydration)
4 mol H	4.03 g	(from the H_2O of hydration)
Total	172.17 g	

(*e*) 775.7 g

4.61 What is the average weight in kg of (*a*) a hydrogen atom (*b*) an oxygen atom (*c*) a uranium atom?

(*a*) $\frac{1.008 \times 10^{-3} \text{ kg/mol}}{6.02 \times 10^{23} \text{ atoms/mol}} = 1.67 \times 10^{-27} \text{ kg/atom}$ (*b*) $\frac{16.00 \times 10^{-3} \text{ kg/mol}}{6.02 \times 10^{23} \text{ atoms/mol}} = 2.66 \times 10^{-26} \text{ kg/atom}$

(*c*) $\frac{238.03 \times 10^{-3} \text{ kg/mol}}{6.02 \times 10^{23} \text{ atoms/mol}} = 3.95 \times 10^{-25} \text{ kg/atom}$

4.62 How many mmol of iron are there in 500 mg of iron?

$$\text{Number of mmol} = \frac{500 \text{ mg Fe}}{55.85 \text{ mg Fe/mmol Fe}} = 8.95 \text{ mmol Fe}$$

4.63 How many moles of nitrogen gas, N_2, are there in 35.7 g of nitrogen?

$$\text{Number of moles} = \frac{35.7 \text{ g}}{28.0 \text{ g/mol}} = 1.28 \text{ mol}$$

4.64 How many hydrogen atoms are present in 0.235 g of NH_3?

$$\text{Number of moles of } NH_3 = \frac{0.235 \text{ g } NH_3}{17.0 \text{ g/mol}} = 0.0138 \text{ mol } NH_3$$

$$(0.0138 \text{ mol } NH_3)\left(\frac{3 \text{ mol H}}{1 \text{ mol } NH_3}\right)\left(\frac{6.02 \times 10^{23} \text{ H atoms}}{\text{mol H}}\right) = 2.49 \times 10^{22} \text{ H atoms}$$

4.65 The atomic weight of sulfur was determined by decomposing 6.2984 g of Na_2CO_3 with sulfuric acid and weighing the resultant Na_2SO_4 formed. The weight was found to be 8.4380 g. Taking the atomic weights of C, O, and Na as 12.011, 15.999, and 22.990, respectively, what value is computed for the atomic weight of sulfur?

▌ Each mole of Na_2CO_3 is converted to 1 mol of Na_2SO_4. The formula weight of Na_2CO_3 is

$$2(22.990\text{ u}) + (12.011\text{ u}) + 3(15.999\text{ u}) = 105.988\text{ u}$$

The number of moles of Na_2CO_3 (and thus Na_2SO_4) is

$$\frac{6.2984\text{ g } Na_2CO_3}{105.988\text{ g/mol}} = 0.0594256\text{ mol}$$

FW(Na_2SO_4) is therefore

$$\frac{8.4380\text{ g } Na_2SO_4}{0.0594256\text{ mol } Na_2SO_4} = 141.993\text{ g/mol}$$

Of that, 2 mol Na and 4 mol O constitute

$$2(22.990\text{ u}) + 4(15.999\text{ u}) = 109.976\text{ u}$$

leaving (141.993 u) − (109.976 u) = 32.017 u for S

4.66 A 12.5843-g sample of $ZrBr_4$ was dissolved and, after several chemical steps, all of the combined bromine was precipitated as AgBr. The silver content of the AgBr was found to be 13.2160 g. Assume the atomic weights of silver and bromine to be 107.868 and 79.904. What value was obtained for the atomic weight of Zr from this experiment?

▌

$$(13.2160\text{ g Ag})\left(\frac{1\text{ mol Ag}}{107.868\text{ g Ag}}\right)\left(\frac{1\text{ mol Br}}{\text{mol Ag}}\right)\left(\frac{1\text{ mol } ZrBr_4}{4\text{ mol Br}}\right) = 0.030630\text{ mol } ZrBr_4$$

$$\frac{12.5843\text{ g}}{0.030630\text{ mol}} = 410.8\text{ g/mol } ZrBr_4 \qquad 410.8 - 4(79.904) = 91.2\text{ g/mol}$$

4.67 How many mol of atoms are contained in *(a)* 32.7 g Zn *(b)* 7.09 g Cl *(c)* 95.4 g Cu *(d)* 4.31 g Fe *(e)* 0.378 g S?

▌ *(a)* $(32.7\text{ g Zn})\left(\frac{1\text{ mol Zn}}{65.4\text{ g Zn}}\right) = 0.500\text{ mol Zn}$ *(b)* $(7.09\text{ g Cl})\left(\frac{1\text{ mol Cl}}{35.45\text{ g Cl}}\right) = 0.200\text{ mol Cl}$

(c) $(95.4\text{ g Cu})\left(\frac{1\text{ mol Cu}}{63.55\text{ g Cu}}\right) = 1.50\text{ mol Cu}$ *(d)* $(4.31\text{ g Fe})\left(\frac{1\text{ mol Fe}}{55.85\text{ g Fe}}\right) = 0.0772\text{ mol Fe}$

(e) $(0.378\text{ g S})\left(\frac{1\text{ mol S}}{32.06\text{ g S}}\right) = 0.0118\text{ mol S}$

4.68 What is the weight of one molecule of *(a)* CH_3OH *(b)* $C_{60}H_{122}$ *(c)* $C_{1200}H_{2000}O_{1000}$?

▌ *(a)* $12.0 + 4.0 + 16.0 = 32.0\text{ g/mol}$ $\quad \frac{32.0\text{ g/mol}}{6.02 \times 10^{23}\text{ molecules/mol}} = 5.32 \times 10^{-23}\text{ g/molecule}$

(b) $60(12.0) + 122(1.0) = 842\text{ g/mol}$ $\quad \frac{842\text{ g/mol}}{6.02 \times 10^{23}\text{ molecules/mol}} = 1.40 \times 10^{-21}\text{ g/molecule}$

(c) $\frac{32\,400\text{ g/mol}}{6.02 \times 10^{23}\text{ molecules/mol}} = 5.38 \times 10^{-20}\text{ g/molecule}$

4.69 How many moles are in *(a)* 24.5 g H_2SO_4 *(b)* 4.00 g O_2?

▌ *(a)* $(24.5\text{ g } H_2SO_4)\left(\frac{1\text{ mol } H_2SO_4}{98.0\text{ g } H_2SO_4}\right) = 0.250\text{ mol } H_2SO_4$ *(b)* $(4.00\text{ g } O_2)\left(\frac{1\text{ mol } O_2}{32.0\text{ g } O_2}\right) = 0.125\text{ mol } O_2$

Note that O_2 has a *molecular* weight of 32.0 g/mol.

4.70 *(a)* How many mol of Ba and of Cl are contained in 107.0 g of $Ba(ClO_3)_2 \cdot H_2O$? *(b)* How many molecules of water of hydration are in this same amount?

▌ *(a)* $[107.0\text{ g } Ba(ClO_3)_2 \cdot H_2O]\left(\frac{1\text{ mol } Ba(ClO_3)_2 \cdot H_2O}{322\text{ g } Ba(ClO_3)_2 \cdot H_2O}\right) = 0.332\text{ mol}$

This fraction of a mole of compound contains 0.332 mol Ba and 0.664 mol Cl, as well as 0.332 mol H_2O.

(*b*) $(0.332 \text{ mol } H_2O)\left(\frac{6.02 \times 10^{23} \text{ molecules}}{\text{mol}}\right) = 2.00 \times 10^{23} \text{ molecules } H_2O$

4.71 How many mol of Fe and of S are contained in (*a*) 1 mol of FeS_2 (pyrite), (*b*) 1.00 kg of FeS_2? (*c*) How many kg of S are contained in 1.00 kg of FeS_2?

▌ (*a*) 1 mol Fe and 2 mol S

(*b*) $(1.00 \text{ kg } FeS_2)\left(\frac{10^3 \text{ g}}{\text{kg}}\right)\left(\frac{1 \text{ mol } FeS_2}{120 \text{ g } FeS_2}\right) = 8.33 \text{ mol } FeS_2$ which contains 8.33 mol Fe and 16.7 mol S

(*c*) $(16.7 \text{ mol S})\left(\frac{32.06 \text{ g S}}{\text{mol S}}\right)\left(\frac{1 \text{ kg}}{10^3 \text{ g}}\right) = 0.535 \text{ kg S}$

4.72 How many (*a*) g of H_2S (*b*) mol of H and of S (*c*) g of H and of S (*d*) molecules of H_2S (*e*) atoms of H and of S are contained in 0.400 mol H_2S?

▌ Atomic weight of H is 1.01; of S, 32.06. Molecular weight of H_2S is $2(1.01) + 32.06 = 34.08$.
Note that it was not necessary to express the molecular weight to 0.001 u, even though the atomic weights are known to this significance. Since the limiting factor in this problem is $n(H_2S)$, known to one part in 400, the value 34.08 (expressed to one part in over 3000) for the molecular weight is more than adequate.

(*a*) Number of grams of compound = (number of mol) × (weight of 1 mol)

Number of grams of H_2S = 0.400 mol × 34.08 g/mol = 13.63 g H_2S

(*b*) One mol of H_2S contains 2 mol H and 1 mol S. Then 0.400 mol H_2S contains

$0.400 \times 2 = 0.800$ mol H and 0.400 mol S.

(*c*) Number of grams of element = (number of mol) × (weight of 1 mol)

Number of grams of H = 0.800 mol × 1.008 g/mol = 0.806 g H

Number of grams of S = 0.400 mol × 32.06 g/mol = 12.82 g S

(*d*) Number of molecules = (number of mol) × (number of molecules in 1 mol)

$= 0.400 \text{ mol} \times 6.02 \times 10^{23} \text{ molecules/mol} = 2.41 \times 10^{23} \text{ molecules}$

(*e*) Number of atoms of element = (number of mol) × (number of atoms per mol)

Number of atoms of H = $0.800 \text{ mol} \times 6.02 \times 10^{23} \text{ atoms/mol} = 4.82 \times 10^{23}$ atoms H

Number of atoms of S = $0.400 \text{ mol} \times 6.02 \times 10^{23} \text{ atoms/mol} = 2.41 \times 10^{23}$ atoms S

4.73 How many mol of atoms are contained in (*a*) 10.02 g calcium, (*b*) 92.91 g phosphorus? (*c*) How many moles of molecular phosphorus are contained in 92.91 g phosphorus if the formula of the molecule is P_4? (*d*) How many atoms are contained in 92.91 g phosphorus? (*e*) How many molecules are contained in 92.91 g phosphorus?

▌ Atomic weights of Ca and P are 40.08 and 30.974. Hence 1 mol Ca = 40.08 g Ca 1 mol P = 30.974 g P

(*a*) $n(\text{Ca}) = \frac{\text{weight of Ca}}{\text{AW(Ca)}} = \frac{10.02 \text{ g}}{40.08 \text{ g/mol}} = 0.250 \text{ mol Ca atoms}$

(*b*) $n(\text{P}) = \frac{\text{weight of P}}{\text{AW(P)}} = \frac{92.91 \text{ g}}{30.974 \text{ g/mol}} = 3.000 \text{ mol P atoms}$

(*c*) $\text{MW}(P_4) = 4 \times 30.974 = 123.90$. Then $n(P_4) = \frac{\text{weight of P}}{\text{MW}(P_4)} = \frac{92.91 \text{ g}}{123.90 \text{ g/mol}} = 0.7500 \text{ mol } P_4 \text{ molecules}$

(*d*) Number of atoms of P = $3.000 \text{ mol} \times 6.023 \times 10^{23} \text{ atoms/mol} = 1.807 \times 10^{24}$ atoms P

(*e*) Number of molecules of P_4 = $0.7500 \text{ mol} \times 6.023 \times 10^{23} \text{ molecules/mol} = 4.517 \times 10^{23}$ molecules P_4

4.74 How many mol are represented by (*a*) 9.540 g SO_2 (*b*) 85.16 g NH_3 (*c*) 25.02 g $TiS_{1.85}$?

▌ Atomic weight of S, 32.06; of O, 16.00; of N, 14.007; of H, 1.008; of Ti, 47.90. From these,

$\text{MW}(SO_2) = 32.06 + 2(16.00) = 64.06 \qquad \text{MW}(NH_3) = 14.007 + 3(1.008) = 17.031 \qquad \text{FW}(TiS_{1.85}) = 107.21$

Hence, 1 mol SO_2 = 64.06 g SO_2, 1 mol NH_3 = 17.031 g NH_3, 1 mol $TiS_{1.85}$ = 107.21 g $TiS_{1.85}$.

(*a*) $\text{Amount of } SO_2 = \dfrac{\text{mass of } SO_2}{\text{mass per mol of } SO_2} = \dfrac{9.540\text{ g}}{64.06\text{ g/mol}} = 0.1489\text{ mol } SO_2$

(*b*) $\text{Amount of } NH_3 = \dfrac{\text{mass of } NH_3}{\text{mass per mol of } NH_3} = \dfrac{85.16\text{ g}}{17.031\text{ g/mol}} = 5.000\text{ mol } NH_3$

(*c*) Titanium sulfides belong to a relatively small class of nonstoichiometric solid compounds, whose structures allow a limited variability in composition. The actual atomic ratios of titanium and sulfur may vary by as much as 10 percent, depending on the details of preparation. The formula given here describes a particular preparation. The mole concept may be applied just as well to entities with nonintegral formulas as to those with integral formulas.

$$\text{Amount of } TiS_{1.85} = \frac{\text{mass of } TiS_{1.85}}{\text{mass per mol of } TiS_{1.85}} = \frac{25.02\text{ g}}{107.21\text{ g/mol}} = 0.2334\text{ mol } TiS_{1.85}$$

4.75 Calculate the number of mol of $Cu(C_2H_3O_2)_2$ present in 200 g.

▮

$$(200\text{ g})\left(\frac{1\text{ mol}}{182\text{ g}}\right) = 1.10\text{ mol}$$

4.76 Verify the law of multiple proportions for an element, X, which forms oxides having percentages of X equal to 77.4%, 63.2%, 69.6%, and 72.0%.

▮

	division by 1.72	multiplication by 6
$\dfrac{77.4\text{ g X}}{22.6\text{ g O}} = 3.42\text{ g X/g O}$	1.99	12
$\dfrac{63.2\text{ g X}}{36.8\text{ g O}} = 1.72\text{ g X/g O}$	1.00	6
$\dfrac{69.6\text{ g X}}{30.4\text{ g O}} = 2.29\text{ g X/g O}$	1.33	8
$\dfrac{72.0\text{ g X}}{28.0\text{ g O}} = 2.57\text{ g X/g O}$	1.49	9

Per (6/1.72) g O, there are 12 g X, 6 g X, 8 g X, and 9 g X in the four compounds. Since the masses of X in the compounds, for a given mass of O, are in integral ratio, they obey the law of multiple proportions.

4.77 A certain element, X, forms three different binary compounds with chlorine, containing 59.68%, 68.95%, and 74.75% chlorine, respectively. Show how these data illustrate the law of multiple proportions.

▮ According to the law of multiple proportions, the relative amounts of an element combining with some fixed amount of a second element in a series of compounds are the ratios of small whole numbers. We could proceed by fixing the mass of either X or Cl and computing for each compound the corresponding mass of the other element. Once we make this decision, we can choose any arbitrary amount as the fixed amount.

The following tabulations based on 100 g of each compound may be helpful. The percentage of X in each compound is equal to 100 minus the percentage of Cl.

compound A	compound B	compound C
59.68 g Cl	68.95 g Cl	74.75 g Cl
40.32 g X	31.05 g X	25.25 g X
100.00 g A	100.00 g B	100.00 g C

Compute for each compound the amount of X that combines with 1.0000 g of Cl. For compound A, the amount is 1/59.68 times the amount that is combined in 100 g A, i.e., combined with 59.68 g Cl. Formally, this can be expressed by an equation which recognizes the fixed relationship between the combining amounts of two elements in a particular compound.

$$m(\text{X}) = (1.0000\text{ g Cl})\left(\frac{40.32\text{ g X}}{59.68\text{ g Cl}}\right) = 0.6756\text{ g X}$$

Similarly, for compounds B and C,

$$m(\text{X}) = (1.0000\text{ g Cl})\left(\frac{31.05\text{ gX}}{68.95\text{ Cl}}\right) = 0.4503\text{ g X}$$

$$m(\text{X}) = (1.0000\text{ g Cl})\left(\frac{25.25\text{ g X}}{74.75\text{ g Cl}}\right) = 0.3378\text{ g X}$$

The relative amounts of X in the three cases are not affected if all three amounts are divided by the smallest of them.

$$0.6756:0.4503:0.3378 = \frac{0.6756}{0.3378}:\frac{0.4503}{0.3378}:\frac{0.3378}{0.3378} = 2.000:1.333:1.000$$

These relative amounts are indeed the ratios of small whole numbers, 6:8:3, within the precision of the analyses.

Had a fixed amount of X been chosen, inverse ratios would have been found. Let us compute, for example, the amounts of Cl combined with 1.0000 g X in each of the compounds.

$$m(\text{Cl}) = (1.0000\text{ g X})\left(\frac{59.68\text{ g Cl}}{40.32\text{ g X}}\right) = 1.480\text{ g Cl}$$

$$m(\text{Cl}) = (1.0000\text{ g X})\left(\frac{68.95\text{ g Cl}}{31.05\text{ g X}}\right) = 2.221\text{ g Cl}$$

$$m(\text{Cl}) = (1.0000\text{ g X})\left(\frac{74.75\text{ g Cl}}{25.25\text{ g X}}\right) = 2.960\text{ g Cl}$$

and

$$1.480:2.221:2.960 = \frac{1.480}{1.480}:\frac{2.221}{1.480}:\frac{2.960}{1.480} = 1.000:1.501:2.000$$

These relative amounts are again the ratios of small whole numbers, 2, 3, and 4, and the law of multiple proportions is verified.

4.78 A 1.5276-g sample of $CdCl_2$ was converted to metallic cadmium and cadmium-free products by an electrolytic process. The weight of the metallic cadmium was 0.9367 g. If the atomic weight of chlorine is taken as 35.453, what must be the atomic weight of Cd from this experiment?

▮

$$\begin{aligned} \text{Weight of } CdCl_2 &= 1.5276\text{ g} \\ \text{Weight of Cd in } CdCl_2 &= \underline{0.9367\text{ g}} \\ \text{Weight of Cl in } CdCl_2 &= 0.5909\text{ g} \end{aligned} \qquad n(\text{Cl}) = \frac{0.5909\text{ g}}{35.453\text{ g/mol}} = 0.016667\text{ mol}$$

From the formula $CdCl_2$ we see that the number of moles of Cd is exactly half the number of moles of Cl.

$$n(\text{Cd}) = \tfrac{1}{2}n(\text{Cl}) = \tfrac{1}{2}(0.016667) = 0.008333\text{ mol}$$

The atomic weight is the weight per mole.

$$\text{AW(Cd)} = \frac{0.9367\text{ g}}{0.008333\text{ mol}} = 112.41\text{ g/mol}$$

4.79 In a chemical determination of the atomic weight of vanadium, 2.8934 g of pure $VOCl_3$ was allowed to undergo a set of reactions as a result of which all the chlorine contained in this compound was converted to AgCl. The weight of the AgCl was 7.1801 g. Assuming the atomic weights of Ag and Cl are 107.868 and 35.453, what is the experimental value for the atomic weight of vanadium?

▮ This problem is similar to Prob. 4.78 except that n(Cl) must be obtained by way of n(AgCl). The three Cl atoms of $VOCl_3$ are converted to 3 formula units of AgCl, the formula weight of which is 143.321 (the sum of 107.868 and 35.453).

$$n(\text{AgCl}) = \frac{7.1801\text{ g}}{143.321\text{ g/mol}} = 0.050098\text{ mol}$$

From the formula AgCl,

$$n(\text{Cl}) = n(\text{AgCl}) = 0.050098\text{ mol Cl}$$

Also, from the formula $VOCl_3$,

$$n(\text{V}) = \tfrac{1}{3}n(\text{Cl}) = \tfrac{1}{3}(0.050098) = 0.016699\text{ mol V}$$

To find the weight of vanadium in the weighed sample of $VOCl_3$, we must subtract the weights of chlorine and oxygen contained. If we designate the mass of any substance or chemical constituent X by *m*(X), then

$$m(\mathrm{X}) = n(\mathrm{X}) \times \mathrm{FW(X)}$$

where FW(X) is the formula weight of X.

$$m(\mathrm{Cl}) = n(\mathrm{Cl}) \times \mathrm{AW(Cl)} = (0.050098\ \mathrm{mol})(35.453\ \mathrm{g/mol}) = 1.7761\ \mathrm{g\ Cl}$$

Noting from the formula $VOCl_3$ that $n(\mathrm{O}) = n(\mathrm{V})$,

$$m(\mathrm{O}) = n(\mathrm{O}) \times \mathrm{AW(O)} = (0.016699\ \mathrm{mol})(15.999\ \mathrm{g/mol}) = 0.2672\ \mathrm{g\ O}$$

By difference,

$$m(\mathrm{V}) = m(\mathrm{VOCl_3}) - m(\mathrm{O}) - m(\mathrm{Cl}) = (2.8934 - 0.2672 - 1.7761)\ \mathrm{g} = 0.8501\ \mathrm{g}$$

Then

$$\mathrm{AW(V)} = \frac{m(\mathrm{V})}{n(\mathrm{V})} = \frac{0.8501\ \mathrm{g}}{0.016699\ \mathrm{mol}} = 50.91\ \mathrm{g/mol}$$

4.80 A certain public water supply contained 0.10 ppb (part per billion) of chloroform, $CHCl_3$. How many molecules of $CHCl_3$ would be contained in a 0.050 mL drop of this water?

▮

$$(0.050\ \mathrm{mL\ water})\left(\frac{1.0\ \mathrm{g\ water}}{\mathrm{mL\ water}}\right)\left(\frac{0.10\ \mathrm{g\ CHCl_3}}{10^9\ \mathrm{g\ water}}\right)\left(\frac{1\ \mathrm{mol\ CHCl_3}}{119.5\ \mathrm{g\ CHCl_3}}\right)\left(\frac{6.02\times10^{23}\ \mathrm{molecules}}{\mathrm{mol}}\right) = 2.5\times10^{10}\ \mathrm{molecules}$$

4.81 If 2.00 mol of calcium carbonate (formula weight = 100) occupies a volume of 67.0 mL, what is its density?

▮

$$(2.00\ \mathrm{mol})\left(\frac{100\ \mathrm{g}}{\mathrm{mol}}\right) = 200\ \mathrm{g} \qquad d = \frac{200\ \mathrm{g}}{67.0\ \mathrm{mL}} = 2.99\ \mathrm{g/mL}$$

4.82 How many tons of $Ca_3(PO_4)_2$ must be treated with carbon and sand in an electric furnace to make 1.00 ton of phosphorus?

▮ The formula $Ca_3(PO_4)_2$ indicates that 2 mol P (2×30.97 g = 61.94 g P) is contained in 1 mol $Ca_3(PO_4)_2$ [310.2 g $Ca_3(PO_4)_2$]. Then, changing grams to tons in the weight ratio,

$$\text{Weight of } \mathrm{Ca_3(PO_4)_2} = (1.00\ \mathrm{ton\ P})\left(\frac{310.2\ \mathrm{tons\ Ca_3(PO_4)_2}}{61.94\ \mathrm{tons\ P}}\right) = 5.01\ \mathrm{tons\ Ca_3(PO_4)_2}$$

4.83 (*a*) How much H_2SO_4 could be produced from 500 kg of sulfur? (*b*) How many kg of Glauber's salt, $Na_2SO_4 \cdot 10H_2O$, could be obtained from 1.000 kg H_2SO_4?

▮ (*a*) The formula H_2SO_4 indicates that 1 mol S (32.06 g S) will give 1 mol H_2SO_4 (98.08 g H_2SO_4). Then, since the *ratio* of any two constituents is independent of the mass units,

$$\text{Weight of } \mathrm{H_2SO_4} = (500\ \mathrm{kg\ S})\left(\frac{98.08\ \mathrm{kg\ H_2SO_4}}{32.06\ \mathrm{kg\ S}}\right) = 1530\ \mathrm{kg\ H_2SO_4}$$

(*b*) 1 mol H_2SO_4 (98.08 g H_2SO_4) will give 1 mol $Na_2SO_4 \cdot 10H_2O$ (322.2 g $Na_2SO_4 \cdot 10H_2O$), since each substance contains one sulfate (SO_4) group. Then

$$\text{Weight of } \mathrm{Na_2SO_4 \cdot 10H_2O} = (1.000\ \mathrm{kg\ H_2SO_4})\left(\frac{322.2\ \mathrm{kg\ Na_2SO_4 \cdot 10H_2O}}{98.08\ \mathrm{kg\ H_2SO_4}}\right) = 3.285\ \mathrm{kg\ Na_2SO_4 \cdot 10H_2O}$$

4.84 How much calcium is in the amount of $Ca(NO_3)_2$ that contains 20.0 g of nitrogen?

▮ It is not necessary to find the weight of $Ca(NO_3)_2$ containing 20.0 g of nitrogen. The relationship between two component elements of a compound may be found directly from the formula.

$$\text{Weight of Ca} = (20.0\ \mathrm{g\ N})\left(\frac{1\ \mathrm{mol\ Ca}}{2\ \mathrm{mol\ N}}\right) = (20.0\ \mathrm{g\ N})\left(\frac{40.0\ \mathrm{g\ Ca}}{2 \times 14.0\ \mathrm{g\ N}}\right) = 28.6\ \mathrm{g\ Ca}$$

4.85 A procedure for analyzing the oxalic acid content of a solution involves the formation of the insoluble complex $Mo_4O_3(C_2O_4)_3 \cdot 12H_2O$. (*a*) How many g of this complex would form per g of oxalic acid, $H_2C_2O_4$, if 1 mol of the complex results from the reaction with 3 mol of oxalic acid? **(b)** How many g of molybdenum are contained in the complex formed by reaction with 1 g of oxalic acid?

▌ The molecular weight of oxalic acid is 90.0 g/mol. 1 g is therefore (1/90.0) mol, which would form (1/270) mol of complex. The molecular weight of the complex is 912 g/mol:

4Mo	4(95.94) =	383.8 u
27O	27(16.0) =	432.0
6C	6(12.0) =	72.0
24H	24(1.0) =	24.0
	Total =	911.8 u

(a) $(912\ \text{g/mol})(1/270)\ \text{mol} = 3.38\ \text{g complex}$

(b) $(3.38\ \text{g complex})\left(\dfrac{384\ \text{u}}{912\ \text{u}}\right) = 1.42\ \text{g Mo}$

4.3 EMPIRICAL FORMULAS

***Warning*: Use at least three significant digits in empirical-formula calculations.**

4.86 The formula $2PbCO_3 \cdot Pb(OH)_2$ could be confusing. Write this formula with parentheses to remove the ambiguity.

▌ $(PbCO_3)_2Pb(OH)_2$

4.87 How many atoms of each element are there in each formula unit listed?

(a) $NaC_2H_3O_2 \cdot Mg(C_2H_3O_2)_2 \cdot 3UO_2(C_2H_3O_2)_2 \cdot 6H_2O$ *(b)* $(NH_4)_3PO_4 \cdot 12MoO_3$
(c) $FeSO_4 \cdot (NH_4)_2SO_4 \cdot 6H_2O$

▌ *(a)* 1 Na, 1 Mg, 3 U, 18 C, 39 H, 30 O *(b)* 3 N, 12 H, 1 P, 40 O, 12 Mo *(c)* 1 Fe, 2 S, 14 O, 2 N, 20 H

4.88 What is the empirical formula of a compound which contains 60.0% oxygen and 40.0% sulfur by mass?

▌ Since no specific mass of sample has been given, one may choose any total mass. A 100-g sample is a convenient choice, because then the mass of each of the elements is numerically equal to the percentage of that element in the compound. Then

$$\frac{60.0\ \text{g O}}{16.0\ \text{g O/mol O}} = 3.75\ \text{mol O} \qquad \frac{40.0\ \text{g S}}{32.0\ \text{g S/mol S}} = 1.25\ \text{mol S}$$

Since formulas are "always" expressed in terms of whole numbers of atoms, the mole ratio must be expressed as a ratio of integers. Therefore to obtain an integral ratio, divide the numbers of moles of the elements present by the number present in smallest amount.

$$\frac{3.75\ \text{mol O}}{1.25\ \text{mol S}} = \frac{3.00\ \text{mol O}}{1.00\ \text{mol S}}$$

The ratio of 3 mol of oxygen atoms to 1 mol of sulfur atoms corresponds to the formula SO_3.

4.89 Derive the empirical formula of a hydrocarbon that on analysis gave the following percent composition: $C = 85.63\%$, $H = 14.37\%$.

▌ A solution based on 100 g of compound is as follows:

E	*m*(E)	AW(E)	$n(E) = \dfrac{m(E)}{AW(E)}$	$\dfrac{n(E)}{7.129\ \text{mol}}$
C	85.63 g	12.011 g/mol	7.129 mol	1.000
H	14.37 g	1.008 g/mol	14.26 mol	2.000

where E = element; *m*(E) = mass of element per 100 g of compound; AW(E) = atomic weight of element; *n*(E) = amount of element per 100 g of compound, expressed in moles of atoms.

The procedure of dividing each *n*(E) by *n*(C) is equivalent to finding the number of atoms of each element for every atom of carbon. The ratio of H to C atoms is 2:1. Hence, the empirical formula is CH_2. (The formulas C_2H_4, C_3H_6,

C_4H_8, etc., imply the same percentage composition as does CH_2, but for the empirical formula the smallest possible integers are chosen.)

4.90 A compound gave on analysis the following percent composition: K = 26.57%, Cr = 35.36%, O = 38.07%. Derive the empirical formula of the compound.

▌ A tabular solution, as applied to 100 g of compound, follows.

(1) E	(2) $m(E)$	(3) AW(E)	(4) $n(E) = \frac{m(E)}{AW(E)}$	(5) $\frac{n(E)}{0.6800 \text{ mol}}$	(6) $\frac{n(E)}{0.6800 \text{ mol}} \times 2$
K	26.57 g	39.10 g/mol	0.6800 mol	1.000	2
Cr	35.36 g	52.00 g/mol	0.6800 mol	1.000	2
O	38.07 g	16.00 g/mol	2.379 mol	3.499	7

In contrast to Prob. 4.89, the numbers in column (5) are not all integers. The ratio of the numbers of atoms of two elements in a compound must be the ratio of small whole numbers, in order to satisfy one of the postulates of Dalton's atomic theory. Allowing for experimental and calculational uncertainty, we see that the entry for oxygen in column (5), 3.499, is, to within the allowed error, 3.500 or $\frac{7}{2}$, indeed the ratio of small whole numbers. By rounding off in this way and multiplying each of the entries in column (5) by 2, we arrive at the set of smallest integers that correctly represent the relative numbers of atoms in the compound, as tabulated in column (6). The formula is thus $K_2Cr_2O_7$.

4.91 Calculate the empirical formula for an oxide of iron with composition Fe 69.94%, O 30.06% using only one significant digit. Repeat the calculation using three significant digits. Comment on the results.

▌

	one significant digit	three significant digits
$(69.94 \text{ g Fe})\left(\frac{1 \text{ mol Fe}}{55.85 \text{ g Fe}}\right)$	= 1 mol Fe	1.25 mol Fe
$(30.06 \text{ g O})\left(\frac{1 \text{ mol O}}{16.0 \text{ g O}}\right)$	= 2 mol O	1.88 mol O
Mole ratio $\frac{O}{Fe}$	$\frac{2}{1}$	$\frac{1.88}{1.25} = \frac{1.50}{1.00} = \frac{3}{2}$
Empirical formula	FeO_2 (incorrect)	Fe_2O_3 (correct)

The use of one significant digit causes roundoff errors and results in an incorrect empirical formula.

4.92 The insecticide DDT has the following composition by mass: 47.5% C, 2.54% H, and 50.0% Cl. Determine the empirical formula of DDT.

▌

$$\frac{47.5 \text{ g C}}{12.0 \text{ g/mol}} = 3.95 \text{ mol C} \qquad \frac{2.54 \text{ g H}}{1.008 \text{ g/mol}} = 2.52 \text{ mol H} \qquad \frac{50.0 \text{ g Cl}}{35.5 \text{ g/mol}} = 1.41 \text{ mol Cl}$$

Chlorine is present in the smallest number of moles. Per mole of chlorine, there are

$$\frac{3.95 \text{ mol C}}{1.41 \text{ mol Cl}} = 2.80 \text{ mol C/mol Cl} \qquad \frac{2.52 \text{ mol H}}{1.41 \text{ mol Cl}} = 1.79 \text{ mol H/mol Cl}$$

Analytical data are not likely to be 10% in error; therefore the ratio of 1.79:1 should not be merely rounded off to 2:1. Instead, the number of moles of each element per mole of Cl is multiplied by the same small integer to obtain nearly integral ratios:

$$\frac{\text{mol C}}{\text{mol Cl}} = \frac{2.80}{1.00} = \frac{2.80 \times 5}{1.00 \times 5} = \frac{14.0 \text{ mol C}}{5.00 \text{ mol Cl}} \qquad \frac{\text{mol H}}{\text{mol Cl}} = \frac{1.79}{1.00} = \frac{1.79 \times 5}{1.00 \times 5} = \frac{8.95 \text{ mol H}}{5.00 \text{ mol Cl}}$$

The last ratio is close enough to round off to 9:5, and the resulting 14:9:5 ratio corresponds to the formula $C_{14}H_9Cl_5$.

4.93 A 2.500-g sample of uranium was heated in the air. The resulting oxide weighed 2.949 g. Determine the empirical formula of the oxide.

▌ 2.949 g of the oxide contains 2.500 g U and 0.449 g O. A calculation on the basis of 2.949 g of the oxide shows 2.672 mol oxygen atoms per mol uranium atoms. The smallest multiplying integer that will give whole numbers is 3.

$$\frac{n(\text{O})}{n(\text{U})} = \frac{2.672 \text{ mol O}}{1.000 \text{ mol U}} = \frac{3(2.672 \text{ mol O})}{3(1.000 \text{ mol U})} = \frac{8.02 \text{ mol O}}{3.00 \text{ mol U}}$$

The empirical formula is U_3O_8.

Emphasis must be placed on the importance of carrying out the computations to as many significant figures as the analytical precision requires. If numbers in the ratio 2.67:1 had been multiplied by 2 to give 5.34:2 and these numbers had been rounded off to 5:2, the wrong formula would have been obtained. This would have been unjustified because it would have assumed an error of 34 parts in 500 in the analysis of oxygen. The weight of oxygen, 0.449 g, indicates a possible error of only a few parts in 500. When the multiplying factor 3 was used, the rounding off was from 8.02 to 8.00, the assumption being made that the analysis of oxygen may have been in error by 2 parts in 800; this degree of error is more reasonable.

4.94 An oxide of nitrogen contains 30.4% nitrogen. What is its empirical formula?

▌
$$(30.4 \text{ g N})\left(\frac{1 \text{ mol N}}{14.0 \text{ g N}}\right) = 2.17 \text{ mol N} \qquad (69.6 \text{ g O})\left(\frac{1 \text{ mol O}}{16.0 \text{ g O}}\right) = 4.35 \text{ mol O}$$

The mole ratio is 1:2, and the empirical formula is NO_2.

4.95 A *borane* (a compound containing only boron and hydrogen) analyzed 88.45% boron. What is its empirical formula?

▌ The percent hydrogen is 100.00% − 88.45% = 11.55%.

$$(88.45 \text{ g B})\left(\frac{1 \text{ mol B}}{10.81 \text{ g B}}\right) = 8.182 \text{ mol B} \qquad (11.55 \text{ g H})\left(\frac{1 \text{ mol H}}{1.008 \text{ g H}}\right) = 11.46 \text{ mol H}$$

$$\frac{8.182 \text{ mol B}}{8.182} = 1.00 \text{ mol B} \qquad \frac{11.46 \text{ mol H}}{8.182} = 1.40 \text{ mol H}$$

Multiplying these numbers by 5 yields an integer ratio of 5:7; the empirical formula is B_5H_7.

4.96 Determine the simplest formula of a compound that has the following composition: Cr = 26.52%, S = 24.52%, O = 48.96%.

▌
$$(26.52 \text{ g Cr})\left(\frac{1 \text{ mol Cr}}{52.0 \text{ g Cr}}\right) = 0.510 \text{ mol Cr} \qquad (24.52 \text{ g S})\left(\frac{1 \text{ mol S}}{32.06 \text{ g S}}\right) = 0.765 \text{ mol S}$$

$$(48.96 \text{ g O})\left(\frac{1 \text{ mol O}}{16.0 \text{ g O}}\right) = 3.06 \text{ mol O}$$

Dividing each of these by 0.510 and then multiplying them by 2 yields the integral ratio 2:3:12. The empirical formula is $Cr_2S_3O_{12}$, or $Cr_2(SO_4)_3$.

4.97 A compound contains 63.1% carbon, 11.92% hydrogen, and 24.97% fluorine. Derive its empirical formula.

▌
$$(63.1 \text{ g C})\left(\frac{1 \text{ mol C}}{12.0 \text{ g C}}\right) = 5.26 \text{ mol C} \qquad (11.92 \text{ g H})\left(\frac{1 \text{ mol H}}{1.008 \text{ g H}}\right) = 11.8 \text{ mol H}$$

$$(24.97 \text{ g F})\left(\frac{1 \text{ mol F}}{19.0 \text{ g F}}\right) = 1.31 \text{ mol F}$$

Division by 1.31 yields the integer ratio 4:9:1; the empirical formula is C_4H_9F.

4.98 A compound contains 21.6% sodium, 33.3% chlorine, 45.1% oxygen. Derive its empirical formula.

▌ *Warning*: We are calculating moles of Cl and O atoms in the compound. The fact that these elements form diatomic molecules in their elementary states is not relevant in this problem.

$$(21.6\text{ g Na})\left(\frac{1\text{ mol Na}}{23.0\text{ g Na}}\right) = 0.939\text{ mol Na} \qquad (33.3\text{ g Cl})\left(\frac{1\text{ mol Cl}}{35.45\text{ g Cl}}\right) = 0.939\text{ mol Cl}$$

$$(45.1\text{ g O})\left(\frac{1\text{ mol O}}{16.0\text{ g O}}\right) = 2.82\text{ mol O}$$

$$\frac{0.939\text{ mol Na}}{0.939} = 1.00\text{ mol Na} \qquad \frac{0.939\text{ mol Cl}}{0.939} = 1.00\text{ mol Cl} \qquad \frac{2.82\text{ mol O}}{0.939} = 3.00\text{ mol O}$$

The mole ratio is 1:1:3; the empirical formula is $NaClO_3$.

4.99 When 1.010 g of zinc vapor is burned in air, 1.257 g of the oxide is produced. What is the empirical formula of the oxide?

▌ The mass of oxygen in the compound is

$$1.257\text{ g} - 1.010\text{ g} = 0.247\text{ g O}$$

$$(1.010\text{ g Zn})\left(\frac{1\text{ mol Zn}}{65.38\text{ g Zn}}\right) = 0.0154\text{ mol Zn} \qquad (0.247\text{ g O})\left(\frac{1\text{ mol O}}{16.0\text{ g O}}\right) = 0.0154\text{ mol O}$$

The mole ratio is 1:1; the empirical formula is ZnO.

4.100 A sample of a pure compound contains 2.04 g of sodium, 2.65×10^{22} atoms of carbon, and 0.132 mol of oxygen atoms. Find the empirical formula. (*Hint*: Think in terms of moles.)

▌

$$(2.04\text{ g Na})\left(\frac{1\text{ mol Na}}{23.0\text{ g Na}}\right) = 0.0887\text{ mol Na} \qquad (2.65 \times 10^{22}\text{ C atoms})\left(\frac{1\text{ mol C}}{6.02 \times 10^{23}\text{ C atoms}}\right) = 0.0440\text{ mol C}$$

and we are given 0.132 mol O. Dividing each number of moles by 0.0440 yields the integral ratio 2:1:3; the empirical formula is Na_2CO_3.

4.101 Derive the empirical formulas of the substances having the following percentage compositions: ***(a)*** Fe = 63.53%, S = 36.47% ***(b)*** Fe = 46.55%, S = 53.45% ***(c)*** Fe = 53.73%, S = 46.27%

▌

(a) $$(63.53\text{ g Fe})\left(\frac{1\text{ mol Fe}}{55.85\text{ g Fe}}\right) = 1.138\text{ mol Fe} \qquad (36.47\text{ g S})\left(\frac{1\text{ mol S}}{32.06\text{ g S}}\right) = 1.138\text{ mol S}$$

Divide each number of moles by 1.138 to get a 1:1 mole ratio. The empirical formula is FeS.

(b) $$(46.55\text{ g Fe})\left(\frac{1\text{ mol Fe}}{55.85\text{ g Fe}}\right) = 0.8335\text{ mol Fe} \qquad (53.45\text{ g S})\left(\frac{1\text{ mol S}}{32.06\text{ g S}}\right) = 1.667\text{ mol S}$$

$$\frac{0.8335\text{ mol Fe}}{0.8335} = 1\text{ mol Fe} \qquad \frac{1.667\text{ mol S}}{0.8335} = 2\text{ mol S}$$

The mole ratio is 1:2; the empirical formula is FeS_2.

(c) $$(53.73\text{ g Fe})\left(\frac{1\text{ mol Fe}}{55.85\text{ g Fe}}\right) = 0.9620\text{ mol Fe} \qquad (46.27\text{ g S})\left(\frac{1\text{ mol S}}{32.06\text{ g S}}\right) = 1.443\text{ mol S}$$

$$\frac{0.9620\text{ mol Fe}}{0.9620} = 1\text{ mol Fe} \qquad \frac{1.443\text{ mol S}}{0.9620} = 1.5\text{ mol S}$$

The mole ratio is 1:1.5 = 2:3; the empirical formula is Fe_2S_3.

4.102 An experimental catalyst used in the polymerization of butadiene is 23.3% Co, 25.3% Mo, and 51.4% Cl. What is its empirical formula?

▌

$$(23.3\text{ g Co})\left(\frac{1\text{ mol Co}}{58.9\text{ g Co}}\right) = 0.396\text{ mol Co} \qquad (25.3\text{ g Mo})\left(\frac{1\text{ mol Mo}}{95.94\text{ g Mo}}\right) = 0.264\text{ mol Mo}$$

$$(51.4\text{ g Cl})\left(\frac{1\text{ mol Cl}}{35.45\text{ g Cl}}\right) = 1.45\text{ mol Cl}$$

Dividing each of these numbers of moles by 0.264 yields a ratio of 1.5:1.0:5.5, or 3:2:11. The empirical formula is $Co_3Mo_2Cl_{11}$.

4.103 (*a*) Calculate the empirical formula of a compound containing 23.3% Mg, 30.7% S, and 46.0% O. (*b*) Name the compound.

▌ (*a*) $(23.3\text{ g Mg})\left(\frac{1\text{ mol Mg}}{24.3\text{ g Mg}}\right) = 0.959\text{ mol Mg}$ $\quad (30.7\text{ g S})\left(\frac{1\text{ mol S}}{32.1\text{ g S}}\right) = 0.956\text{ mol S}$

$$(46.0\text{ g O})\left(\frac{1\text{ mol O}}{16.0\text{ g O}}\right) = 2.88\text{ mol O}$$

Dividing by the smallest of these yields the ratio 1 mol Mg:1 mol S:3 mol O. The empirical formula is $MgSO_3$.

(*b*) The compound is magnesium sulfite. Since ionic compounds do not form molecules, we recognize them by their empirical formulas. It is easy to recognize that the compound is ionic since it has a group IIA metal in it, and in this compound the SO_3^{2-} ion is necessary to balance the charge on Mg^{2+}.

4.104 Calculate the empirical formula of the compound formed when 7.30 g of iron powder reacts completely with 6.30 g of powdered sulfur.

▌

$$(7.30\text{ g Fe})\left(\frac{1\text{ mol Fe}}{55.85\text{ g Fe}}\right) = 0.131\text{ mol Fe} \qquad (6.30\text{ g S})\left(\frac{1\text{ mol S}}{32.06\text{ g S}}\right) = 0.197\text{ mol S}$$

Dividing by the smaller:

$$\frac{0.131\text{ mol Fe}}{0.131} = 1.0\text{ mol Fe} \qquad \frac{0.197\text{ mol S}}{0.131} = 1.5\text{ mol S}$$

Doubling these last numbers gives a whole number ratio of 2 mol Fe:3 mol S. The empirical formula is therefore Fe_2S_3.

4.105 Determine the empirical formula of a compound which consists of 1.8% H, 56.1% S, and 42.1% O.

▌

$$(1.8\text{ g H})\left(\frac{1\text{ mol H}}{1.0\text{ g H}}\right) = 1.8\text{ mol H} \qquad (56.1\text{ g S})\left(\frac{1\text{ mol S}}{32.1\text{ g S}}\right) = 1.75\text{ mol S} \qquad (42.1\text{ g O})\left(\frac{1\text{ mol O}}{16.0\text{ g O}}\right) = 2.63\text{ mol O}$$

Dividing these by the smallest yields:

1.0 mol H 1.0 mol S 1.5 mol O

Doubling each of these last numbers of moles yields a whole number ratio of

2 mol H:2 mol S:3 mol O Empirical formula: $H_2S_2O_3$

4.106 Calculate the empirical formula of a compound containing 52.9% carbon and the rest oxygen.

▌

$$(52.9\text{ g C})\left(\frac{1\text{ mol C}}{12.0\text{ g C}}\right) = 4.41\text{ mol C} \qquad (47.1\text{ g O})\left(\frac{1\text{ mol O}}{16.0\text{ g O}}\right) = 2.94\text{ mol O}$$

$$\frac{4.41\text{ mol C}}{2.94} = 1.50\text{ mol C} \quad (\times 2 = 3\text{ mol C}) \qquad \frac{2.94\text{ mol O}}{2.94} = 1.00\text{ mol O} \quad (\times 2 = 2\text{ mol O})$$

Empirical formula C_3O_2

4.107 Determine the empirical formula of a compound containing 40.6% carbon, 5.1% hydrogen, and 54.2% oxygen. What are possible molecular formulas for this compound, based on these data alone?

▌ In 100 g of compound there is

$$(40.6\text{ g C})\left(\frac{1\text{ mol C}}{12.0\text{ g C}}\right) = 3.38\text{ mol C} \qquad (5.1\text{ g H})\left(\frac{1\text{ mol H}}{1.0\text{ g H}}\right) = 5.1\text{ mol H} \qquad (54.2\text{ g O})\left(\frac{1\text{ mol O}}{16.0\text{ g O}}\right) = 3.39\text{ mol O}$$

The mole ratio:

$$\frac{3.38}{3.38} = 1.0\text{ mol C} \qquad \frac{5.1}{3.38} = 1.5\text{ mol H} \qquad \frac{3.39}{3.38} = 1.0\text{ mol O}$$

The mole ratio is simplified by multiplication by 2:

$1.0 \times 2 = 2$ mol carbon $1.5 \times 2 = 3$ mol hydrogen $1.0 \times 2 = 2$ mol oxygen

Empirical formula $C_2H_3O_2$

The possible molecular formulas include any multiple of $C_2H_3O_2$, such as $C_2H_3O_2$, $C_4H_6O_4$, $C_6H_9O_6$, ... With only percent composition data, it is impossible to distinguish among them.

4.108 An oxide of arsenic was analyzed and found to contain 75.74% As. What is the empirical formula of the compound?

$$(75.74 \text{ g As})\left(\frac{1 \text{ mol As}}{74.92 \text{ g As}}\right) = 1.011 \text{ mol As} \qquad (24.26 \text{ g O})\left(\frac{1 \text{ mol O}}{16.00 \text{ g O}}\right) = 1.516 \text{ mol O}$$

$$\frac{1.516 \text{ mol O}}{1.011 \text{ mol As}} = \frac{1.50 \text{ mol O}}{1 \text{ mol As}} = \frac{3 \text{ mol O}}{2 \text{ mol As}}$$

Hence the formula is As_2O_3.

4.109 A 3.245-g sample of a titanium chloride was reduced with sodium to metallic titanium. After the resultant sodium chloride was washed out, the residual titanium metal was dried, and weighed 0.819 g. What is the empirical formula of this titanium chloride?

$$(3.245 \text{ g}) - (0.819 \text{ g}) = 2.426 \text{ g Cl}$$

$$(0.819 \text{ g Ti})\left(\frac{1 \text{ mol Ti}}{47.9 \text{ g Ti}}\right) = 0.0171 \text{ mol Ti} \qquad (2.426 \text{ g Cl})\left(\frac{1 \text{ mol Cl}}{35.45 \text{ g Cl}}\right) = 0.0684 \text{ mol Cl}$$

$$\frac{0.0171 \text{ mol Ti}}{0.0171} = 1.00 \text{ mol Ti} \qquad \frac{0.0684 \text{ mol Cl}}{0.0171} = 4.00 \text{ mol Cl}$$

The empirical formula is $TiCl_4$.

4.110 An organic compound was found on analysis to contain 47.37% carbon and 10.59% hydrogen. The balance was presumed to be oxygen. What is the empirical formula of the compound?

$$\%O = 100.00\% - 47.37\% - 10.59\% = 42.04\%$$

$$(47.37 \text{ g C})\left(\frac{1 \text{ mol C}}{12.01 \text{ g C}}\right) = 3.944 \text{ mol C} \qquad (10.59 \text{ g H})\left(\frac{1 \text{ mol H}}{1.008 \text{ g H}}\right) = 10.51 \text{ mol H}$$

$$(42.04 \text{ g O})\left(\frac{1 \text{ mol O}}{16.00 \text{ g O}}\right) = 2.628 \text{ mol O}$$

$$\frac{3.944 \text{ mol C}}{2.628} = 1.501 \text{ mol C} \qquad \frac{10.51 \text{ mol H}}{2.628} = 4.000 \text{ mol H} \qquad \frac{2.628 \text{ mol O}}{2.628} = 1.000 \text{ mol O}$$

The mole ratio is 3:8:2; the empirical formula is $C_3H_8O_2$.

4.111 A hydrate of iron(III) thiocyanate, $Fe(SCN)_3$, was found to contain 19.0% H_2O. What is the empirical formula for the hydrate?

In 100 g of compound, there are 19.0 g of H_2O and 81.0 g of $Fe(SCN)_3$.

$$(19.0 \text{ H}_2\text{O})\left(\frac{1 \text{ mol H}_2\text{O}}{18.0 \text{ g H}_2\text{O}}\right) = 1.05 \text{ mol H}_2\text{O} \qquad (81.0 \text{ g})\left(\frac{1 \text{ mol}}{230 \text{ g}}\right) = 0.352 \text{ mol Fe(SCN)}_3$$

There is 3 mol of water per mol of $Fe(SCN)_3$. The empirical formula is $Fe(SCN)_3 \cdot 3H_2O$.

4.112 A 15.00-g sample of an unstable hydrated salt, $Na_2SO_4 \cdot xH_2O$, was found to contain 7.05 g of water. Determine the empirical formula of the salt.

Hydrates are compounds containing water molecules loosely bound to the other components. H_2O may usually be removed intact by heating such compounds and may then be replaced by wetting. The Na_2SO_4 and H_2O groups may thus be considered as the units of which the compound is made, and their formula weights are used in place of atomic weights. It is more convenient to base the tabular solution in this case on 15.00 g of compound (which contains 15.00 − 7.05 = 7.95 g Na_2SO_4).

X	$m(X)$	FW(X)	$n(X) = \frac{m(X)}{FW(X)}$	$\frac{n(X)}{0.0559 \text{ mol}}$
Na_2SO_4	7.95 g	142.1 g/mol	0.0559 mol	1.00
H_2O	7.05 g	18.02 g/mol	0.391 mol	6.99

The mole ratio of H_2O to Na_2SO_4 is, to within the allowed error, 7:1, and the empirical formula is $Na_2SO_4 \cdot 7H_2O$.

4.113 A 1.367-g sample of an organic compound was combusted in a stream of air to yield 3.002 g CO_2 and 1.640 g H_2O. If the original compound contained only C, H, and O, what is its empirical formula?

It is necessary to use quantitative factors for CO_2 and H_2O to find how much C and H are present in the combustion products and, thus, in the original sample.

$$(3.002 \text{ g } CO_2)\left(\frac{1 \text{ mol C}}{1 \text{ mol } CO_2}\right) = (3.002 \text{ g } CO_2)\left(\frac{12.01 \text{ g C}}{44.01 \text{ g } CO_2}\right) = 0.819 \text{ g C}$$

$$(1.640 \text{ g } H_2O)\left(\frac{2 \text{ mol H}}{1 \text{ mol } H_2O}\right) = (1.640 \text{ g } H_2O)\left(\frac{2 \times 1.008 \text{ g H}}{18.02 \text{ g } H_2O}\right) = 0.1835 \text{ g H}$$

The amount of oxygen in the original sample cannot be obtained from the weight of combustion products, since the CO_2 and H_2O contain oxygen that came partly from the combined oxygen in the compound and partly from the air stream used in the combustion process. The oxygen content of the sample can be obtained, however, by difference.

$$m(\text{O}) = m(\text{compound}) - m(\text{C}) - m(\text{H}) = (1.367 \text{ g}) - (0.819 \text{ g}) - (0.184 \text{ g}) = 0.364 \text{ g}$$

The problem can now be solved by the usual procedures. The numbers of moles of the elementary atoms in 1.367 g of compound are found to be: C, 0.0682; H, 0.1820; O, 0.0228. These numbers are in the ratio 3:8:1, and the empirical formula is C_3H_8O.

4.114 Derive the empirical formulas of the minerals that have the following compositions: (*a*) $ZnSO_4 = 56.14\%$, $H_2O = 43.86\%$ (*b*) $MgO = 27.16\%$, $SiO_2 = 60.70\%$, $H_2O = 12.14\%$ (*c*) $Na = 12.10\%$, $Al = 14.19\%$, $Si = 22.14\%$, $O = 42.09\%$, $H_2O = 9.48\%$.

Here, the simple compounds and their formula weights are treated just as the elements and their atomic weights are usually treated.

(*a*) $$(56.14 \text{ g } ZnSO_4)\left(\frac{1 \text{ mol } ZnSO_4}{161.4 \text{ g } ZnSO_4}\right) = 0.3478 \text{ mol } ZnSO_4 \qquad (43.86 \text{ g } H_2O)\left(\frac{1 \text{ mol } H_2O}{18.0 \text{ g } H_2O}\right) = 2.44 \text{ mol } H_2O$$

$$\frac{0.3478 \text{ mol } ZnSO_4}{0.3478} = 1.00 \text{ mol } ZnSO_4 \qquad \frac{2.44 \text{ mol } H_2O}{0.3478} = 7.00 \text{ mol } H_2O$$

The empirical formula is $ZnSO_4(H_2O)_7$ or $ZnSO_4 \cdot 7H_2O$.

(*b*) $$(27.16 \text{ g MgO})\left(\frac{1 \text{ mol MgO}}{40.30 \text{ g MgO}}\right) = 0.6739 \text{ mol MgO} \qquad (60.70 \text{ g } SiO_2)\left(\frac{1 \text{ mol } SiO_2}{60.09 \text{ g } SiO_2}\right) = 1.010 \text{ mol } SiO_2$$

$$(12.14 \text{ g } H_2O)\left(\frac{1 \text{ mol } H_2O}{18.02 \text{ g } H_2O}\right) = 0.6737 \text{ mol } H_2O$$

The mole ratio is 1:1.5:1 = 2:3:2. The empirical formula is $(MgO)_2(SiO_2)_3(H_2O)_2$, or $2MgO \cdot 3SiO_2 \cdot 2H_2O$.

(*c*) $$(12.10 \text{ g Na})\left(\frac{1 \text{ mol Na}}{23.0 \text{ g Na}}\right) = 0.526 \text{ mol Na} \qquad (22.14 \text{ g Si})\left(\frac{1 \text{ mol Si}}{28.1 \text{ g Si}}\right) = 0.788 \text{ mol Si}$$

$$(42.09 \text{ g O})\left(\frac{1 \text{ mol O}}{16.0 \text{ g O}}\right) = 2.63 \text{ mol O} \qquad (9.48 \text{ g } H_2O)\left(\frac{1 \text{ mol } H_2O}{18.0 \text{ g } H_2O}\right) = 0.527 \text{ mol } H_2O$$

Dividing each number of moles by 0.526, the smallest of them, then multiplying the results by 2 yields a mole ratio of 2:3:10:2. The empirical formula is $Na_2Si_3O_{10} \cdot 2H_2O$.

4.115 A 1.500-g sample of a compound containing only C, H, and O was burned completely. The only combustion products were 1.738 g CO_2 and 0.711 g H_2O. What is the empirical formula of the compound?

The mass of the carbon and hydrogen in the compound can be determined from the masses of the products:

$$(1.738 \text{ g } CO_2)\left(\frac{1 \text{ mol } CO_2}{44.0 \text{ g } CO_2}\right)\left(\frac{1 \text{ mol C}}{\text{mol } CO_2}\right)\left(\frac{12.0 \text{ g C}}{\text{mol C}}\right) = 0.474 \text{ g C}$$

$$(0.711 \text{ g } H_2O)\left(\frac{1 \text{ mol } H_2O}{18.0 \text{ g } H_2O}\right)\left(\frac{2 \text{ mol H}}{\text{mol } H_2O}\right)\left(\frac{1.008 \text{ g H}}{\text{mol H}}\right) = 0.0796 \text{ g H}$$

The mass of the oxygen in the compound cannot be determined that way, since some of the oxygen in the products came from the O_2. The oxygen in the original compound is determined by difference.

$$(1.500 \text{ g total}) - (0.474 \text{ g C}) - (0.079 \text{ g H}) = 0.947 \text{ g O}$$

$$(0.474 \text{ g C})\left(\frac{1 \text{ mol C}}{12.0 \text{ g C}}\right) = 0.0395 \text{ mol C} \qquad (0.0796 \text{ g H})\left(\frac{1 \text{ mol H}}{1.008 \text{ g H}}\right) = 0.0790 \text{ mol H}$$

$$(0.947 \text{ g O})\left(\frac{1 \text{ mol O}}{16.0 \text{ g O}}\right) = 0.0592 \text{ mol O}$$

Division of these numbers of moles by 0.0395, followed by multiplication by 2 yields the integer ratio 2:4:3. The empirical formula is $C_2H_4O_3$.

4.116 Elementary analysis showed that an organic compound contained C, H, N, and O as its only elementary constituents. A 1.279-g sample was burned completely, as a result of which 1.60 g of CO_2 and 0.77 g of H_2O were obtained. A separately weighed 1.625-g sample contained 0.216 g nitrogen. What is the empirical formula of the compound?

▌ The masses of carbon and hydrogen in the organic compound:

$$(1.60 \text{ g } CO_2)\left(\frac{1 \text{ mol } CO_2}{44.0 \text{ g } CO_2}\right)\left(\frac{1 \text{ mol C}}{1 \text{ mol } CO_2}\right)\left(\frac{12.0 \text{ g C}}{\text{mol C}}\right) = 0.436 \text{ g C}$$

$$(0.77 \text{ g } H_2O)\left(\frac{1 \text{ mol } H_2O}{18.0 \text{ g } H_2O}\right)\left(\frac{2 \text{ mol H}}{\text{mol } H_2O}\right)\left(\frac{1.0 \text{ g H}}{\text{mol H}}\right) = 0.086 \text{ g H}$$

The mass of N in the 1.279-g sample:

$$(1.279 \text{ g sample})\left(\frac{0.216 \text{ g N}}{1.625 \text{ g sample}}\right) = 0.170 \text{ g N}$$

The mass of O, by difference:

$$(1.279 \text{ g}) - (0.436 \text{ g}) - (0.086 \text{ g}) - (0.170 \text{ g}) = 0.587 \text{ g O}$$

$$(0.436 \text{ g C})\left(\frac{1 \text{ mol C}}{12.0 \text{ g C}}\right) = 0.0364 \text{ mol C} \qquad (0.086 \text{ g H})\left(\frac{1 \text{ mol H}}{1.008 \text{ g H}}\right) = 0.086 \text{ mol H}$$

$$(0.587 \text{ g O})\left(\frac{1 \text{ mol O}}{16.0 \text{ g O}}\right) = 0.0367 \text{ mol O} \qquad (0.170 \text{ g N})\left(\frac{1 \text{ mol N}}{14.0 \text{ g N}}\right) = 0.0121 \text{ mol N}$$

Division by 0.0121 yields a mole ratio 3:7:3:1. The empirical formula is $C_3H_7O_3N$.

4.117 Manganese forms nonstoichiometric oxides having the general formula MnO_x. Find the value of x for a compound that analyzed 63.70% Mn.

▌

$$(63.70 \text{ g Mn})\left(\frac{1 \text{ mol Mn}}{54.94 \text{ g Mn}}\right) = 1.159 \text{ mol Mn} \qquad (36.30 \text{ g O})\left(\frac{1 \text{ mol O}}{16.00 \text{ g O}}\right) = 2.269 \text{ mol O}$$

Division of the numbers of moles by 1.159 yields a 1:1.958 mole ratio. The value of x is 1.958.

4.118 A 23.2-g sample of an organic compound containing carbon, hydrogen, and oxygen was burned in excess oxygen and yielded 52.8 g of CO_2 and 21.6 g of water. Determine the empirical formula of the compound.

▌

$$(52.8 \text{ g } CO_2)\left(\frac{1 \text{ mol}}{44.0 \text{ g}}\right)\left(\frac{1 \text{ mol C}}{1 \text{ mol } CO_2}\right) = 1.20 \text{ mol C} \qquad (1.20 \text{ mol C})\left(\frac{12.0 \text{ g}}{\text{mol}}\right) = 14.4 \text{ g C}$$

$$(21.6 \text{ g } H_2O)\left(\frac{1 \text{ mol } H_2O}{18.0 \text{ g } H_2O}\right)\left(\frac{2 \text{ mol H}}{\text{mol } H_2O}\right) = 2.40 \text{ mol H} \qquad (2.40 \text{ mol H})(1.00 \text{ g/mol}) = 2.40 \text{ g H}$$

$$\text{Mass O in compound} = (23.2 \text{ g}) - (14.4 \text{ g}) - (2.4 \text{ g}) = 6.4 \text{ g O} \qquad (6.4 \text{ g O})\left(\frac{1 \text{ mol}}{16 \text{ g}}\right) = 0.40 \text{ mol O}$$

The C/H/O mole ratio is thus 1.2:2.4:0.40 or 3:6:1, and the empirical formula is therefore C_3H_6O.

4.119 A certain compound was known to have a formula which could be represented as $[PdC_xH_yN_z](ClO_4)_2$. Analysis showed that the compound contained 30.15% carbon and 5.06% hydrogen. When converted to the corresponding thiocyanate, $[PdC_xH_yN_z](SCN)_2$, the analysis was 40.46% carbon and 5.94% hydrogen. Calculate the values of x, y, and z.

▮ Let the formula weight of the first compound be represented by F. That of the second compound is therefore equal to F minus twice the formula weight of ClO_4^- plus twice the formula weight of SCN^-:

$$F - 2(99.5) + 2(58.0) = F - 83.0$$

The percent carbon in the first compound is

$$\frac{12.0x}{F} \times 100\% = 30.15\%$$

In the second compound, which has 2 mol of carbon in the anions per mol of compound, the percent carbon is

$$\frac{12.0(x+2)}{F - 83.0} \times 100\% = 40.46\%$$

Solving by simultaneous equations

$$1200x = 30.15F$$
$$1200x + 2400 = 40.46F - (40.46)(83.0)$$
$$1200x + 2400 = 40.46\left(\frac{1200x}{30.15}\right) - 3358$$
$$x = 14$$
$$F = \frac{1200x}{30.15} = \frac{1200(14)}{30.15} = 557 \text{ u}$$

The percent hydrogen in the first compound is

$$\frac{1.008y}{557} \times 100\% = 5.06\% \qquad y = 28$$

The total formula weight of all the elements other than nitrogen is 501 u; therefore 56 u must represent the nitrogen in one formula unit. There are four nitrogen atoms per formula unit, and the complete formula is

$$[PdC_{14}H_{28}N_4](ClO_4)_2$$

4.4 MOLECULAR FORMULAS

4.120 A compound has the following percent composition: C = 40.0%, H = 6.67%, O = 53.3%. Its molecular weight is 60.0. Derive its molecular formula.

▮
$$(40.0 \text{ g C})\left(\frac{1 \text{ mol C}}{12.0 \text{ g C}}\right) = 3.33 \text{ mol C} \qquad (6.67 \text{ g H})\left(\frac{1 \text{ mol H}}{1.008 \text{ g H}}\right) = 6.62 \text{ mol H}$$
$$(53.3 \text{ g O})\left(\frac{1 \text{ mol O}}{16.0 \text{ g O}}\right) = 3.33 \text{ mol O}$$

The empirical formula is CH_2O. The weight of one empirical formula unit is 12.0 + 2(1.0) + 16.0 = 30.0 u.

$$\frac{60.0 \text{ g/mol}}{30.0 \text{ g/empirical formula unit}} = 2 \text{ empirical formula units/mol}$$

The molecular formula is $(CH_2O)_2$ or $C_2H_4O_2$.

4.121 Determine the empirical formula and the molecular formula of a hydrocarbon which has a molecular weight of 84 u and which contains 85.7% carbon.

▮
$$(85.7 \text{ g C})\left(\frac{1 \text{ mol C}}{12.0 \text{ g C}}\right) = 7.14 \text{ mol C} \qquad (14.3 \text{ g H})\left(\frac{1 \text{ mol H}}{1.008 \text{ g H}}\right) = 14.2 \text{ mol H}$$

The empirical formula is CH_2. Its formula weight is 14 u.

$$\frac{84 \text{ u}}{14 \text{ u}} = 6$$

Hence the molecular formula is C_6H_{12}.

4.122 A compound with molecular weight about 175 u consists of 40.0% carbon, 6.7% hydrogen, and 53.3% oxygen. What is its molecular formula?

▮ In 100 g of compound there are

$$(40.0\text{ g C})\left(\frac{1\text{ mol C}}{12.0\text{ g C}}\right) = 3.33\text{ mol C} \qquad (6.7\text{ g H})\left(\frac{1\text{ mol H}}{1.0\text{ g H}}\right) = 6.7\text{ mol H} \qquad (53.3\text{ g O})\left(\frac{1\text{ mol O}}{16.0\text{ g O}}\right) = 3.33\text{ mol O}$$

to give a mole ratio

$$1\text{ mol C}:2\text{ mol H}:1\text{ mol O}$$

The empirical formula is CH_2O. That formula has a formula weight of 30 u. There are thus

$$\frac{175\text{ u}}{30\text{ u}} = 6\text{ empirical formula units per molecule}$$

The molecular formula is therefore $C_6H_{12}O_6$.

4.123 A certain compound consists of 93.71% C and 6.29% H. Its molecular weight is approximately 130 u. What is its molecular formula?

▮ For a 100-g sample:

$$(93.71\text{ g C})\left(\frac{1\text{ mol}}{12.01\text{ g}}\right) = 7.803\text{ mol C} \qquad (6.29\text{ g H})\left(\frac{1\text{ mol}}{1.008\text{ g}}\right) = 6.24\text{ mol H}$$

$$\frac{7.803\text{ mol C}}{6.24\text{ mol H}} = \frac{1.25\text{ mol C}}{\text{mol H}} = \frac{5\text{ mol C}}{4\text{ mol H}}$$

C_5H_4 has an empirical formula weight of 64 u.

$$\frac{130\text{ u}}{64\text{ u}} = 2 \qquad (C_5H_4)_2 = C_{10}H_8$$

4.124 A compound consisting of 82.66% carbon and 17.34% hydrogen has a molecular weight 58.1 u. Determine its molecular formula.

▮ The empirical formula is determined first:

$$\frac{82.66\text{ g C}}{12.01\text{ g/mol}} = 6.883\text{ mol C} \qquad \frac{17.34\text{ g H}}{1.008\text{ g/mol}} = 17.20\text{ mol H}$$

$$\frac{17.20\text{ mol H}}{6.883\text{ mol C}} = \frac{2.50\text{ mol H}}{1.00\text{ mol C}} = \frac{5\text{ mol H}}{2\text{ mol C}}$$

The empirical formula is C_2H_5. This formula has a formula weight of 29.06 u. The number of empirical formula units per molecule is

$$\frac{58.1\text{ u}}{29.06\text{ u}} = 2$$

The molecular formula is $(C_2H_5)_2$ or C_4H_{10}.

4.125 A sample of polystyrene prepared by heating styrene with tribromobenzoyl peroxide in the absence of air has the formula $Br_3C_6H_3(C_8H_8)_n$. The number n varies with the conditions of preparation. One sample of polystyrene prepared in this manner was found to contain 10.46% bromine. What is the value of n?

▮ In 100 g of compound,

$$(10.46\text{ g Br})\left(\frac{1\text{ mol Br}}{79.90\text{ g Br}}\right) = 0.1309\text{ mol Br}$$

Since there are 3 mol Br per mol of compound, the number of mol of compound is $0.1309/3 = 0.04364$ mol. The molecular weight is

$$\frac{100\text{ g}}{0.04364\text{ mol}} = 2.29 \times 10^3\text{ g/mol}$$

$$3(79.9) + 6(12.0) + 3(1.0) + n(104) = 2290$$

$$n = 19$$

4.126 A saturated hydrocarbon contains 82.66% carbon. What is its empirical formula? its molecular formula?

▌ For a 100-g sample:

$$(82.66\text{ g C})\left(\frac{1\text{ mol C}}{12.01\text{ g C}}\right) = 6.883\text{ mol C} \qquad (17.34\text{ g H})\left(\frac{1\text{ mol H}}{1.008\text{ g H}}\right) = 17.20\text{ mol H}$$

$$\frac{17.20\text{ mol H}}{6.883\text{ mol C}} = \frac{2.5\text{ mol H}}{\text{mol C}} = \frac{5\text{ mol H}}{2\text{ mol C}}$$

C_2H_5 is the empirical formula. A saturated hydrocarbon has a formula of the form C_nH_{2n+2}. Therefore, the molecular formula is C_4H_{10}.

4.127 One of the earliest methods for determining the molecular weight of proteins was based on chemical analysis. A hemoglobin preparation was found to contain 0.335% iron. (*a*) If the hemoglobin molecule contains one atom of iron, what is its molecular weight? (*b*) If it contains 4 atoms of iron, what is its molecular weight?

▌ (*a*) $n(\text{Fe}) = n(\text{hemoglobin})$ $\qquad (0.335\text{ g Fe})\left(\frac{1\text{ mol Fe}}{55.85\text{ g Fe}}\right) = (100\text{ g hemoglobin})\left(\frac{1\text{ mol}}{\text{MW}}\right)$

$$\text{MW(hemoglobin)} = \frac{(100)(55.85)}{0.335} = 16\,700\text{ g/mol}$$

(*b*) $n(\text{Fe}) = \frac{n(\text{hemoglobin})}{4}$ $\qquad \text{MW(hemoglobin)} = 4(16\,700\text{ g/mol}) = 66\,800\text{ g/mol}$

4.128 A polymeric substance, tetrafluoroethylene, can be represented by the formula $(C_2F_4)_x$, where x is a large number. The material was prepared by polymerizing C_2F_4 in the presence of a sulfur-bearing catalyst that served as a nucleus upon which the polymer grew. The final product was found to contain 0.012% S. What is the value of x if each polymeric molecule contains (*a*) 1 sulfur atom, (*b*) 2 sulfur atoms? In either case, assume that the catalyst contributes a negligible amount to the total mass of the polymer.

▌ Per 100-g sample,

$$(0.012\text{ g S})\left(\frac{1\text{ mol S}}{32\text{ g S}}\right) = 3.7_5 \times 10^{-4}\text{ mol S}$$

(*a*) Hence, there is $3.7_5 \times 10^{-4}$ mol of polymer

$$\frac{100\text{ g}}{3.7_5 \times 10^{-4}\text{ mol}} = 2.7 \times 10^5\text{ g/mol} \qquad C_2F_4\text{ has a formula weight 100}$$

$$x = \frac{2.7 \times 10^5\text{ g/mol}}{100\text{ g/formula unit}} = 2.7 \times 10^3\text{ formula units/mol}$$

(*b*) 1.9×10^{-4} mol polymer, hence 5.3×10^5 g/mol and $x = 5.3 \times 10^3$ formula units per mol

4.129 A peroxidase enzyme isolated from human red blood cells was found to contain 0.29% selenium. What is the minimum molecular weight of the enzyme?

▌

$$(0.29\text{ g Se})\left(\frac{1\text{ mol Se}}{78.96\text{ g Se}}\right) = 3.67 \times 10^{-3}\text{ mol Se}$$

Assuming only 1 mol of Se per mol of enzyme,

$$\frac{100\text{ g enzyme}}{3.67 \times 10^{-3}\text{ mol enzyme}} = 2.7 \times 10^4\text{ g/mol}$$

4.130 A purified cytochrome protein isolated from a bacterial preparation was found to contain 0.376% iron. What can be deduced about the molecular weight of the protein?

▌ The protein must contain at least 1 atom of iron per molecule. If it contains only one, of weight 55.8 u, then the molecular weight is given by

$$0.00376\text{ MW} = 55.8\text{ u}$$

$$\text{MW} = 14\,800\text{ u}$$

If the protein molecule contained x atoms of Fe, the molecular weight would be $14\,800x$ u.

The method given here is useful for determining the *minimum* molecular weight of a macromolecular substance when an analysis can be done for one of the minor components.

4.131 A purified pepsin isolated from a bovine preparation was subjected to an amino acid analysis. The amino acid present in smallest amount was lysine, $C_6H_{14}N_2O_2$, and the amount of lysine was found to be 0.43 g per 100 g protein. What is the minimum molecular weight of the protein?

▮ Proteins do not contain free amino acids, but they do contain chemically linked forms of amino acids which on degradative analysis can be reconverted to the free amino acid form. As in Prob. 4.130, the protein molecule must be at least heavy enough to contain one lysine residue (146 u).

$$0.0043\ \text{MW}_{\text{min}} = 146\ \text{u}$$

$$\text{MW}_{\text{min}} = 34\,000\ \text{u}$$

4.132 A sample of potato starch was ground in a ball mill to give a starchlike molecule of lower molecular weight. The product analyzed 0.086% phosphorus. If each molecule is assumed to contain one atom of phosphorus, what is the average molecular weight of the material?

▮ Instead of using the method of Probs. 4.130 and 4.131, let us here work in terms of moles. By assumption, 1 mol of phosphorus atoms (31.0 g P) is contained in 1 mol of the material. Since 0.086 g P is contained in 100 g of material, then 31.0 g P is contained in

$$\frac{31.0}{0.086}(100\ \text{g}) = 36\,000\ \text{g of material}$$

Hence the average molecular weight of the material is 36 000 u.

4.133 An organic compound was prepared containing at least 1 and no more than 2 sulfur atoms per molecule. The compound had no nitrogen, but oxygen could have been present. The mass-spectrometrically determined weight of a fragment ion was 111.028 u. **(a)** What are the allowable formulas consistent with the mass number 111 and with the facts about the elementary composition? **(b)** What is the formula of the ion?

▮ **(a)** The nonhydrogen skeleton would be made up of the elements C, O, and S. The number of possible skeletons can be reduced by the following considerations. (i) The maximum number of carbon atoms is 6, since the mass number for 7 carbons plus 1 sulfur would be 116, in excess of the given value. (ii) The maximum number of hydrogen atoms is $2n(\text{C}) + 2 = 14$. (iii) The (C, O, S) skeleton must therefore contribute between 97 and 111, inclusive, to the mass number. A rather short list, Table 4.1, will now suffice.

(b) Of the four formulas consistent with the known mass number, only C_6SH_7 is consistent with the precise weight of the ion.

TABLE 4.1

(1) (C, O, S) skeleton	(2) mass number of skeleton	(3) 111−(2)	(4) n(H, max)	(5) formula	(6) nuclidic weight
CO_4S	108	3	4	CO_4SH_3	110.975
CO_2S_2	108	3	4	$CO_2S_2H_3$	110.957
C_2O_3S	104	7	6		
C_2OS_2	104	7	6		
C_3O_2S	100	11	8		
C_3OS_2	100	11	8		
C_4OS	96	15	10		
C_5OS	108	3	12	C_5OSH_3	110.993
C_6S	104	7	14	C_6SH_7	111.027

4.134 An intermediate in the synthesis of a naturally occurring alkaloid had a mass-spectrometrically determined molecular weight of 205.147. The compound is known to have no more than 1 nitrogen atom and no more than 2 oxygen atoms per molecule. **(a)** What is the most probable molecular formula of the compound? **(b)** What must the precision of the measurement be to exclude the next to most probable formula?

▌ (*a*) Possible formulas, along with their nuclidic molecular weights, are tabulated. The probable formula is $C_{13}H_{19}NO$. (Nuclidic molecular weight is 205.147.)

formula	MW	formula	MW
$C_{12}H_{15}NO_2$	205.110	$C_{14}H_{21}O$	205.159
$C_{13}H_3NO_2$	205.016	$C_{14}H_{23}N$	205.183
$C_{13}H_{17}O_2$	205.123	$C_{15}H_{11}N$	205.089
$C_{13}H_{19}NO$	205.147	$C_{15}H_9O$	205.065
$C_{14}H_5O_2$	205.029	$C_{16}H_{13}$	205.102

(*b*) The next closest molecular weight is 205.159 for $C_{14}H_{21}O$. The range of uncertainty in the experimental value should not exceed half the difference between 205.147 and 205.159; i.e., it should be less than 0.006, or about 1 part in 35 000.

4.135 An organic ester was decomposed inside a mass spectrometer. An ionic decomposition product had the formula weight 117.090. What is the formula of this product, if it is known in advance that the only possible constituent elements are C, O, and H, and that no more than 4 oxygen atoms are present in the molecule?

▌ Note that this information concerns a fragment of the ester. The possible formulas are tabulated, along with the precise nuclidic formula weights.

formula	FW	formula	FW
C_9H_9	117.070	$C_6H_{13}O_2$	117.092
C_8H_5O	117.034	$C_5H_9O_3$	117.055
C_7HO_2	116.997	$C_4H_5O_4$	117.019

The correct formula is $C_6H_{13}O_2$.

4.136 An alkaloid was extracted from the seed of a plant and purified. The molecule was known to contain 1 atom of nitrogen, no more than 4 atoms of oxygen, and no other elements besides C and H. The mass-spectrometrically determined nuclidic molecular weight was found to be 297.140. (*a*) How many molecular formulas are consistent with mass number 297 and with the other known facts except the precise molecular weight? (*b*) What is the probable molecular formula?

▌ Formulas with the correct total mass number are tabulated below, along with their precise nuclidic molecular weights. There are 17 such formulas, the most probable one being $C_{18}H_{19}O_3N$.

formula	MW	formula	MW	formula	MW
$C_{16}H_{27}O_4N$	297.194	$C_{19}H_7O_3N$	297.043	$C_{21}H_{15}ON$	297.115
$C_{17}H_{15}O_4N$	297.100	$C_{19}H_{23}O_2N$	297.173	$C_{21}H_{31}N$	297.246
$C_{17}H_{31}O_3N$	297.230	$C_{19}H_{39}ON$	297.303	$C_{22}H_3ON$	297.021
$C_{18}H_3O_4N$	297.006	$C_{20}H_{11}O_2N$	297.079	$C_{22}H_{19}N$	297.152
$C_{18}H_{19}O_3N$	297.137	$C_{20}H_{27}ON$	297.209	$C_{23}H_7N$	297.058
$C_{18}H_{35}O_2N$	297.267	$C_{20}H_{43}N$	297.340		

CHAPTER 5
Modern Structure of the Atom

5.1 PHYSICAL BACKGROUND

Unit conversions:

$$1\ \text{J} = 1\ \text{kg}\cdot\text{m}^2/\text{s}^2 = 10^7\ \text{ergs} \qquad 1\ \text{J} = 1\ \text{V} \times 1\ \text{C} \qquad 1\ \text{eV} = 1.60 \times 10^{-19}\ \text{J}$$

If a particle being accelerated by a potential has a charge equal in magnitude to the charge on an electron, the number of electron volts of energy is numerically equal to the potential in volts.

$$1\ \text{eV} = \text{energy of 1 electron being accelerated by 1 V}$$
$$x\ \text{eV} = \text{energy of 1 electron being accelerated by } x\ \text{V}$$

5.1 Convert 1 atomic mass unit (u) to kg.

▌ By definition, 1 mol of ^{12}C has a mass of 0.0120000 kg. Since there are 6.03×10^{23} atoms in 1 mol, each of mass 12.00 u,

$$12.00\ \text{u} = \frac{0.0120000\ \text{kg}}{6.03 \times 10^{23}} \qquad \text{or} \qquad 1\ \text{u} = \frac{0.0120000\ \text{kg}}{12.00(6.03 \times 10^{23})} = 1.66 \times 10^{-27}\ \text{kg}$$

5.2 What did Rutherford's alpha-particle-scattering experiment prove?

▌ Atoms contain massive, positively charged centers—the nuclei.

5.3 In an experiment to measure the charge on the electron, the following values of charge were found on oil droplets (in arbitrary units):

$$-1.6 \times 10^{-19} \qquad -2.4 \times 10^{-19} \qquad -4.0 \times 10^{-19}$$

What value of the electronic charge would be indicated by these results (in the same units)?

▌ -0.8×10^{-19}, the largest common factor of the values given.

5.4 An electron and a body with a $+1.0$ C charge on it are 2.0 m apart. Calculate the force of attraction between them. The charge on the electron is -1.6×10^{-19} C.

▌

$$f = k\frac{q_1q_2}{d^2} = (9.0 \times 10^9\ \text{N}\cdot\text{m}^2/\text{C}^2)\left(\frac{(-1.6 \times 10^{-19}\ \text{C})(1.0\ \text{C})}{(2.0\ \text{m})^2}\right) = -3.6 \times 10^{-10}\ \text{N}$$

The negative sign implies an attractive force.

5.5 How much energy will be released when a sodium ion and a chloride ion, originally at infinite distance, are brought together to a distance of 2.76 Å (the shortest distance of approach in a sodium chloride crystal)? $1\ \text{Å} = 1 \times 10^{-10}$ m. Assume that the ions act as point charges, each with a magnitude of 1.60×10^{-19} C (the electronic charge).

▌

$$E = k\frac{q_1q_2}{d} = \left(\frac{9.0 \times 10^9\ \text{J}\cdot\text{m}}{\text{C}^2}\right)\left(\frac{(1.60 \times 10^{-19}\ \text{C})^2}{2.76 \times 10^{-10}\ \text{m}}\right) = 8.3 \times 10^{-19}\ \text{J}$$

This energy corresponds to 119 kcal per mol of Na^+, Cl^- ion pairs. Of course in a crystal of NaCl, there are attractions between a given ion and several ions of opposite charge as well as repulsions between ions of like charge; hence for a mole of ion pairs in a crystal the energy of attraction is 185 kcal/mol, approximately 1.5 times as great.

5.6 What experiment negated Thomson's model of the atom as an intimate mixture of negative and positive particles?

▌ Rutherford's experiments with large-angle scattering of alpha particles by a metal foil.

5.7 The magnitude of the charge on the electron is 4.8×10^{-10} esu. What is the magnitude of the charge on the proton? on the nucleus of a helium atom?

▌ Since the charge on the proton is equal in magnitude to that on the electron, the charge is also 4.8×10^{-10} esu. Since a helium nucleus contains 2 protons, its charge is 9.6×10^{-10} esu.

5.8 Calculate the force between two bodies 2.00 cm apart, each having a 1.0×10^{-5} C charge.

▌
$$f = k\frac{q_1q_2}{d^2} = (9.0 \times 10^9\ \text{J}\cdot\text{m/C}^2)\frac{(1.0 \times 10^{-5}\ \text{C})^2}{(0.0200\ \text{m})^2} = 2.2 \times 10^3\ \text{J/m} = 2.2 \times 10^3\ \text{N}$$

5.9 Two 1.0-g carbon disks 1.00 cm apart have opposite charges of equal magnitude such that there is a 1.00×10^{-5} N force between them. Calculate the ratio of excess electrons to total atoms on the negatively charged disk.

▌
$$f = kq_1q_2/r^2 \qquad \text{hence} \qquad q_1q_2 = fr^2/k$$
$$q_1 = q_2 = \sqrt{fr^2/k} = \sqrt{(1.00 \times 10^{-5}\ \text{N})(0.0100\ \text{m})^2/(9.0 \times 10^9\ \text{J}\cdot\text{m/C}^2)}$$
$$= 3.3 \times 10^{-10}\ \text{C on each disk}$$
$$(3.3 \times 10^{-10}\ \text{C})\left(\frac{1\ \text{electron}}{1.60 \times 10^{-19}\ \text{C}}\right) = 2.1 \times 10^9\ \text{electrons}$$
$$(1.0\ \text{g})\left(\frac{6.02 \times 10^{23}\ \text{atoms}}{12.0\ \text{g}}\right) = 5.0 \times 10^{22}\ \text{atoms}$$
$$\frac{2.1 \times 10^9\ \text{excess electrons}}{5.0 \times 10^{22}\ \text{atoms}} = 4.2 \times 10^{-14}\ \text{electron/atom}$$

Notice the relatively low number of electrons which can make a measurable force. By no means can the carbon atoms be considered ions.

5.10 Calculate the energy required to move a 1.0×10^{-10} C negatively charged body from infinite distance to a point (*a*) 1.0 cm from a 1.0 C negatively charged body, (*b*) 0.10 cm from a 1.0 C negatively charged body, (*c*) from position (*a*) to position (*b*). (*d*) How would the answer to (*a*) be changed if the first body were positively charged?

▌
$$E = k\frac{q_1q_2}{d}$$

(*a*) $E = (9.0 \times 10^{11}\ \text{J}\cdot\text{cm/C}^2)\dfrac{(1.0 \times 10^{-10}\ \text{C})(1.0\ \text{C})}{1.0\ \text{cm}} = 90\ \text{J}$

(*b*) $E = (9.0 \times 10^{11}\ \text{J}\cdot\text{cm/C}^2)\dfrac{(1.0 \times 10^{-10}\ \text{C})(1.0\ \text{C})}{0.10\ \text{cm}} = 900\ \text{J}$

(*c*) $\Delta E = (900\ \text{J}) - (90\ \text{J}) = 810\ \text{J}$

(*d*) The 810 J of energy would be *liberated* instead of *required.*

5.11 (*a*) Would it take more energy to move a negatively charged body from 3.0 to 2.0 cm from a second negatively charged body, or from 2.0 to 1.0 cm from the second body? Explain. (*b*) Would it take more energy to move a charged body from midway between two charged parallel plates 1.0 cm toward the positive plate or from this second (off center) position to a position 1.0 cm closer to the positive plate? Explain.

▌ (*a*) The energy required to move the charged body from infinite distance to a given distance may be expressed in terms of the distance d by the equation

$$E = \text{constant}/d$$

The energy in moving the charged body from 3.0 cm to 2.0 cm from a second charged body is merely the difference:

$$\frac{\text{constant}}{2.0\ \text{cm}} - \frac{\text{constant}}{3.0\ \text{cm}} = \text{constant}\left(\frac{1}{2.0} - \frac{1}{3.0}\right) = \text{constant}/6.0$$

Similarly for the difference between 2.0 and 1.0 cm:

$$\frac{\text{constant}}{1.0\ \text{cm}} - \frac{\text{constant}}{2.0\ \text{cm}} = \text{constant}/2.0$$

Since $\frac{1}{2}$ is greater than $\frac{1}{6}$, the energy required to move the charged body from 2.0 cm to 1.0 cm is greater than that to move from 3.0 cm to 2.0 cm. **(*b*)** In contrast, a charge in a uniform electric field experiences a force which is not dependent on its position in the field; hence both of the moves described require equal energies.

5.12 What, if any, is the effect of a magnetic field on **(*a*)** a static electric charge in the field? **(*b*)** an electric charge moving through the field?

▌ **(*a*)** None. **(*b*)** An electric charge in motion generates its own magnetic field which the external magnetic field will interact with, creating a force perpendicular to its original direction of motion.

5.13 Why does the charge-to-mass ratio of positive rays depend on the residual gas in the discharge tube? Why is the charge-to-mass ratio of all cathode rays the same?

▌ Cathode rays, no matter what their source, are composed of electrons—all of which have the same charge-to-mass ratio. The ions remaining after the loss of the electrons might have the same magnitude of charge, but different masses. Hence they will have different charge-to-mass ratios.

5.14 State at least four ways in which positive (canal) rays differ from cathode rays.

▌

	canal rays	**cathode rays**
Sign of charge	Positive	Negative
e/m	Variable, depends on ions	Definite value
Mass	Variable, depends on ions	Definite value
Magnitude of charge	Mostly 1+, but sometimes 2+, 3+, . . .	Always 1−

5.15 Calculate the ratio of protons in the atomic nuclei to sodium atoms in a sample of NaCl. Is it necessary to know the size of the sample, the total number of atoms in the sample, or the total number of protons in the sample? To determine the charge-to-mass ratio of the electron, is it necessary to know either the charge or the mass of the electron? Under what conditions do the strengths of the magnetic and electric fields applied in a cathode ray tube permit the determination of the charge-to-mass ratio of the electron?

▌ Per mol of NaCl there are

$$\frac{(11 + 17)\text{ mol of protons}}{1\text{ mol of Na atoms}} = \frac{28}{1}\text{ mole ratio}$$

The size of the sample is immaterial to determining the mole ratio; for example, if one doubles the sample size, one doubles both the number of protons and the number of sodium atoms, and the ratio remains the same.

The e/m ratio for electrons was known before either e or m; obviously, it is not necessary to know the individual values to determine e/m. The e/m ratio is determined using equal magnetic and electric forces on the moving electrons.

5.16 An electron volt (eV) is the energy necessary to move an electronic charge (e) through a potential of exactly 1 V. Express this energy in **(*a*)** J **(*b*)** kcal per mole of electrons **(*c*)** kJ per mole of electrons

▌ **(*a*)** $1.00\text{ eV} = (1.60 \times 10^{-19}\text{ C})(1.00\text{ V}) = 1.60 \times 10^{-19}\text{ J}$

(*b*) $$(1.60 \times 10^{-19}\text{ J})\left(\frac{1\text{ kcal}}{4184\text{ J}}\right)\left(\frac{6.023 \times 10^{23}}{\text{mol}}\right) = 23.0\text{ kcal/mol}$$

(*c*) $$(1.60 \times 10^{-19}\text{ J})\left(\frac{1\text{ kJ}}{10^3\text{ J}}\right)\left(\frac{6.023 \times 10^{23}}{\text{mol}}\right) = 96.4\text{ kJ/mol}$$

5.17 In an oil drop experiment, the following charges (in arbitrary units) were found on a series of oil droplets: 2.30×10^{-15}, 6.90×10^{-15}, 1.38×10^{-14}, 5.75×10^{-15}, 3.45×10^{-15}, 1.96×10^{-14}. Calculate the magnitude of the charge on the electron (in the same units).

▌ 1.15×10^{-15}, which is the largest number which divides all the listed charges evenly. (The smallest of the charges listed is 2.30×10^{-15}, but this charge does not divide into all the others an even number of times; hence 2.30×10^{-15} must represent the charge (in arbitrary units) of 2 electrons.)

5.18 Assuming that the oil was sufficiently nonvolatile, would it be possible to perform Millikan's oil drop experiment in an evacuated apparatus? Explain.

▮ No, the oil droplets must reach a terminal velocity to be able to estimate their weights by Stokes' law. In the absence of air, there would be no good way to determine their weights, and the force of the electric field would then not be precisely known.

5.19 In an oil drop experiment, the terminal velocity of an oil droplet was observed to be 1.00 mm/s. The density of the oil is 0.850 g/cm³, and the viscosity of air (η) is 1.83×10^{-5} N·s/m². Calculate the mass and the radius of the oil droplet.

▮

$$d = \frac{m}{V} = \frac{m}{\frac{4}{3}\pi r^3} \qquad m = \tfrac{4}{3}\pi d r^3 \qquad v = \frac{mg}{6\pi\eta r} = \frac{(\frac{4}{3}\pi d r^3)g}{6\pi\eta r} = \frac{2}{9}\frac{dr^2 g}{\eta}$$

$$r = \sqrt{\frac{9v\eta}{2dg}} = \sqrt{\frac{9(0.00100\ \text{m/s})(1.83 \times 10^{-5}\ \text{N·s/m}^2)}{2(0.850 \times 10^3\ \text{kg/m}^3)(9.80\ \text{m/s}^2)}} = 3.14 \times 10^{-6}\ \text{m}$$

$$m = \tfrac{4}{3}\pi d r^3 = (1.33)(3.14)(0.850 \times 10^3\ \text{kg/m}^3)(3.14 \times 10^{-6}\ \text{m})^3 = 1.10 \times 10^{-13}\ \text{kg}$$

5.20 Although alpha particles have a larger charge, beta particles are deflected more than alpha particles in a given electric field. Explain this observation.

▮ Beta particles are much less massive, and therefore their charge-to-mass ratio is larger despite their lower charge.

5.21 The radius of a nucleus, in cm, can be estimated by

$$R \approx 1.4 \times 10^{-13} A^{1/3}$$

where A is the mass number of the atom. Calculate the approximate density of a polonium-210 nucleus.

▮

$$R \approx (1.4 \times 10^{-13})(210)^{1/3} = 8.3 \times 10^{-13}\ \text{cm}$$

5.22 The characteristic x-ray wavelengths for the lines of the K_α series in magnesium and chromium are 9.87 and 2.29 Å, respectively. Using these values, determine the constants a and b in Moseley's equation and predict the wavelengths of x-rays in this series for strontium and for chlorine.

▮

$$\sqrt{\nu} = a(Z - b)$$

$$\nu_{Mg} = \frac{c}{\lambda} = \left(\frac{3.00 \times 10^8\ \text{m/s}}{9.87\ \text{Å}}\right)\left(\frac{10^{10}\ \text{Å}}{\text{m}}\right) = 3.04 \times 10^{17}/\text{s} = 3.04 \times 10^{17}\ \text{Hz}$$

$$\nu_{Cr} = 1.31 \times 10^{18}\ \text{Hz}$$

$$\sqrt{\nu_{Mg}} = 5.51 \times 10^8$$

$$\sqrt{\nu_{Cr}} = 1.14 \times 10^9$$

$$5.51 \times 10^8 = a(12 - b)$$

$$1.14 \times 10^9 = a(24 - b)$$

Dividing the second of these equations by the first to eliminate a yields

$$\frac{1.14 \times 10^9}{5.51 \times 10^8} = \frac{24 - b}{12 - b} = 2.07$$

$$24 - b = 2.07(12 - b) = 24.8 - 2.07b$$

$$b = 0.75$$

Substitution of this value of b into the first equation above yields

$$5.51 \times 10^8 = a(12 - 0.75) = 11.25a$$

$$a = \frac{5.51 \times 10^8}{11.25} = 4.90 \times 10^7$$

The values may be checked using the second equation

$$1.14 \times 10^9 = 4.90 \times 10^7(24 - 0.75) = 1.14 \times 10^9$$

With the known values of a and b:

$$\sqrt{\nu_{Sr}} = 4.90 \times 10^7(38 - 0.75) = 1.825 \times 10^9$$
$$\sqrt{\nu_{Cl}} = 4.90 \times 10^7(17 - 0.75) = 7.96 \times 10^8$$
$$\nu_{Sr} = 3.33 \times 10^{18} \text{ Hz}$$
$$\nu_{Cl} = 6.34 \times 10^{17} \text{ Hz}$$
$$\lambda_{Sr} = c/\nu = \left(\frac{3.00 \times 10^8 \text{ m/s}}{3.33 \times 10^{18} \text{ s}^{-1}}\right)\left(\frac{10^{10} \text{ Å}}{\text{m}}\right) = 0.901 \text{ Å}$$
$$\lambda_{Cl} = \left(\frac{3.00 \times 10^8 \text{ m/s}}{6.34 \times 10^{17} \text{ s}^{-1}}\right)\left(\frac{10^{10} \text{ Å}}{\text{m}}\right) = 4.73 \text{ Å}$$

5.2 LIGHT

5.23 Determine the frequencies of light of the following wavelengths: ***(a)*** 1.0 Å ***(b)*** 5000 Å ***(c)*** 4.4 μm ***(d)*** 89 m ***(e)*** 562 nm

▌ The basic equation for all these problems is $\nu = c/\lambda = (2.998 \times 10^8 \text{ m/s})/\lambda$.

(a) $$\nu = \frac{3.0 \times 10^8 \text{ m/s}}{(1.0 \text{ Å})(10^{-10} \text{ m/Å})} = 3.0 \times 10^{18} \text{ s}^{-1} = 3.0 \times 10^{18} \text{ Hz}$$

(b) $$\nu = \frac{2.998 \times 10^8 \text{ m/s}}{(5000 \text{ Å})(10^{-10} \text{ m/Å})} = 5.996 \times 10^{14} \text{ Hz} = 599.6 \text{ THz}$$

(c) $$\nu = \frac{3.00 \times 10^8 \text{ m/s}}{4.4 \times 10^{-6} \text{ m}} = 6.8 \times 10^{13} \text{ Hz} = 68 \text{ THz}$$

(d) $$\nu = \frac{3.00 \times 10^8 \text{ m/s}}{89 \text{ m}} = 3.4 \times 10^6 \text{ Hz} = 3.4 \text{ MHz}$$

(e) $$\nu = \frac{2.998 \times 10^8 \text{ m/s}}{562 \times 10^{-9} \text{ m}} = 5.33 \times 10^{14} \text{ Hz} = 533 \text{ THz}$$

5.24 ***(a)*** What change in energy per mol of atoms would be associated with an atomic transition giving rise to radiation at 1 Hz? ***(b)*** What is the relationship between the electron volt and the wavelength in nm of the energetically equivalent photon?

▌ ***(a)*** If each of N_A atoms gives off one 1-Hz photon,

$$\Delta E = N_A(h\nu) = (6.022 \times 10^{23} \text{ mol}^{-1})(6.626 \times 10^{-34} \text{ J}\cdot\text{s})(1.00 \text{ s}^{-1}) = 3.990 \times 10^{-10} \text{ J}\cdot\text{mol}^{-1}$$

Since ΔE and ν are proportional, we may treat the ratio

$$\frac{3.990 \times 10^{-10} \text{ J}\cdot\text{mol}^{-1}}{1 \text{ Hz}}$$

as a "conversion factor" between Hz and $\text{J}\cdot\text{mol}^{-1}$.

(b) First let us find the frequency equivalent of 1 eV from the Planck equation, and then find the wavelength from the frequency.

$$\nu = \frac{\varepsilon}{h} = \frac{1.6022 \times 10^{-19} \text{ J}}{6.626 \times 10^{-34} \text{ J}\cdot\text{s}} = 2.4180 \times 10^{14} \text{ s}^{-1}$$
$$\lambda = \frac{c}{\nu} = \frac{2.998 \times 10^8 \text{ m}\cdot\text{s}^{-1}}{2.4180 \times 10^{14} \text{ s}^{-1}} = (1.2398 \times 10^{-6} \text{ m})(10^9 \text{ nm/m}) = 1239.8 \text{ nm}$$

Because of the inverse proportionality between wavelength and energy, the relationship may be written

$$\lambda\varepsilon = hc = 1239.8 \text{ nm}\cdot\text{eV}$$

5.25 What is the equivalent of the energy unit 1.00 cm^{-1} in ***(a)*** J per photon ***(b)*** kcal per mol of photons ***(c)*** kJ per mol of photons?

▌ ***(a)*** $E = h\nu = hc/\lambda = hc\bar{\nu} = (6.62 \times 10^{-34} \text{ J}\cdot\text{s})(3.00 \times 10^{10} \text{ cm/s})(1.00 \text{ cm}^{-1}) = 1.99 \times 10^{-23} \text{ J}$

(b) $\left(\frac{1.99 \times 10^{-23}\ \text{J}}{\text{photon}}\right)\left(\frac{6.02 \times 10^{23}\ \text{photons}}{\text{mol}}\right)\left(\frac{1\ \text{cal}}{4.184\ \text{J}}\right)\left(\frac{1\ \text{kcal}}{10^3\ \text{cal}}\right) = 2.86 \times 10^{-3}\ \text{kcal/mol}$

(c) $\left(\frac{1.99 \times 10^{-23}\ \text{J}}{\text{photon}}\right)\left(\frac{6.02 \times 10^{23}\ \text{photons}}{\text{mol}}\right)\left(\frac{1\ \text{kJ}}{10^3\ \text{J}}\right) = 1.20 \times 10^{-2}\ \text{kJ/mol}$

5.26 The wavelength of a beam of light is 24.0 μm. What is ***(a)*** its wavelength in cm ***(b)*** its frequency ***(c)*** its wave number ***(d)*** the energy of one of its photons?

▌ ***(a)*** $24.0\ \mu\text{m} = 24.0 \times 10^{-6}\ \text{m} = 2.40 \times 10^{-3}\ \text{cm}$

(b) $\nu = \frac{c}{\lambda} = \frac{3.00 \times 10^{10}\ \text{cm/s}}{2.40 \times 10^{-3}\ \text{cm}} = 1.25 \times 10^{13}\ \text{Hz}$

(c) $\bar{\nu} = \frac{1}{\lambda} = \frac{1}{2.40 \times 10^{-3}\ \text{cm}} = 4.17 \times 10^{2}\ \text{cm}^{-1}$

(d) $E = h\nu = (6.63 \times 10^{-34}\ \text{J}\cdot\text{s})(1.25 \times 10^{13}\ \text{s}^{-1}) = 8.29 \times 10^{-21}\ \text{J}$

5.27 How many photons of light having a wavelength of 4000 Å are necessary to provide 1.00 J of energy?

▌

$$E_{\text{photon}} = h\nu = h\frac{c}{\lambda} = \frac{(6.63 \times 10^{-34}\ \text{J}\cdot\text{s})(3.00 \times 10^{8}\ \text{m/s})}{4000 \times 10^{-10}\ \text{m}} = 4.97 \times 10^{-19}\ \text{J}$$

$$\frac{1.00\ \text{J}}{4.97 \times 10^{-19}\ \text{J/photon}} = 2.01 \times 10^{18}\ \text{photons}$$

5.28 Calculate for the longest and the shortest wavelengths of visible light: ***(a)*** wavelength ***(b)*** wave number ***(c)*** J/mol ***(d)*** ergs/photon ***(e)*** frequency ***(f)*** kcal/mol of photons.

▌ ***(a)*** The longest and shortest wavelengths of visible light are 7000 Å and 4000 Å, respectively.

(b) $\bar{\nu} = \frac{1}{\lambda} = \left(\frac{1}{7000\ \text{Å}}\right)\left(\frac{1 \times 10^{8}\ \text{Å}}{\text{cm}}\right) = 1.428 \times 10^{4}\ \text{cm}^{-1}$

$\bar{\nu} = \frac{1}{\lambda} = \left(\frac{1}{4000\ \text{Å}}\right)\left(\frac{1 \times 10^{8}\ \text{Å}}{\text{cm}}\right) = 2.500 \times 10^{4}\ \text{cm}^{-1}$

(c) $(1.428 \times 10^{4}\ \text{cm}^{-1})\left(\frac{1.20 \times 10^{-2}\ \text{kJ/mol}}{\text{cm}^{-1}}\right)\left(\frac{10^3\ \text{J}}{\text{kJ}}\right) = 1.71 \times 10^{5}\ \text{J/mol}$

↑ See Prob. 5.25(*c*).

$(2.500 \times 10^{4}\ \text{cm}^{-1})\left(\frac{1.20 \times 10^{-2}\ \text{kJ/mol}}{\text{cm}^{-1}}\right)\left(\frac{10^3\ \text{J}}{\text{kJ}}\right) = 3.00 \times 10^{5}\ \text{J/mol}$

(d) $(1.428 \times 10^{4}\ \text{cm}^{-1})\left(\frac{1.99 \times 10^{-16}\ \text{erg}}{\text{cm}^{-1}}\right) = 2.84 \times 10^{-12}\ \text{erg}$

For 4000 Å light, 4.97×10^{-12} erg

(e) $c = \nu\lambda$

$\nu = \frac{c}{\lambda} = \left(\frac{3.00 \times 10^{8}\ \text{m/s}}{7000\ \text{Å}}\right)\left(\frac{10^{10}\ \text{Å}}{\text{m}}\right) = 4.29 \times 10^{14}\ \text{Hz}$

For 4000 Å light, 7.50×10^{14} Hz

(f) $(1.428 \times 10^{4}\ \text{cm}^{-1})\left(\frac{2.86 \times 10^{-3}\ \text{kcal/mol}}{\text{cm}^{-1}}\right) = \frac{40.8\ \text{kcal}}{\text{mol}}$

↑ See Prob. 5.25(*b*).

For 4000 Å light, 71.5 kcal/mol.

5.29 Which of the following relate to light as wave motion, to light as a stream of particles, or to both? ***(a)*** diffraction ***(b)*** interference ***(c)*** photoelectric effect ***(d)*** $E = mc^2$ ***(e)*** $E = h\nu$

▌ ***(a)*** Wave motion ***(b)*** wave motion ***(c)*** particles ***(d)*** particles ***(e)*** both. (In $E = h\nu$, the E refers to the energy of each photon of light, while the ν refers to the frequency of the waves of light.)

5.30 Which has the greater energy—a photon of violet light or a photon of green light?

▮ Violet, which has a shorter wavelength and thus a higher energy.

$$E = h\nu = hc/\lambda$$

5.31 How many photons of light of 7000 Å wavelength are equivalent to 1.00 J of energy?

▮

$$\left(\frac{1.00\ \text{J}}{1.71 \times 10^5\ \text{J/mol}}\right)\left(\frac{6.02 \times 10^{23}\ \text{photons}}{\text{mol}}\right) = 3.52 \times 10^{18}\ \text{photons}$$

↑ See Prob. 5.28(*c*).

5.32 If the second hand on a watch rotates 60 cycles/h, what is the period (the time required for 1 cycle)? What is the relationship between frequency and period? between their units?

▮

$$\frac{60\ \text{cycles}}{\text{h}} = \frac{60}{\text{h}}\left(\frac{1\ \text{h}}{60\ \text{min}}\right) = 1\ \text{min}^{-1}$$

$$\text{Period} = T = \frac{1}{\nu} = \frac{1}{1\ \text{min}^{-1}} = 1\ \text{min}$$

Period and frequency are reciprocals of one another; for example, their units are s and s^{-1} = Hz.

5.33 When white light that has passed through sodium vapor is viewed through a spectroscope, the observed spectrum has a dark line at 5890 Å. Explain this observation.

▮ The light of 5890 Å wavelength has precisely the energy absorbed by the sodium atom to promote its outermost electron to a higher level from its ground state level.

5.34 Using a sound wave as an example, describe in familiar terms the disturbance of the medium, the velocity, the amplitude, and the frequency. Explain why the pitch seems higher when the source of sound is traveling toward you, but lower when the source is traveling away.

▮ The medium usually is air, but sound travels also through solids and liquids as well as other gases. It does not travel through a vacuum. The velocity in air, the speed at which the wave motion is propagated, is about 800 ft/s. Lightning can be seen before the corresponding thunder is heard because of the relatively low speed of sound compared to the almost instantaneous speed of light. The amplitude of sound is its loudness. The frequency of sound determines its pitch; the higher the frequency, the higher the tone. The pitch seems higher when a source of sound moves toward you because the sound emitted after the source has moved a certain distance has less far to travel and thus arrives somewhat before it otherwise would. In other words, the frequency is increased. Look up the Doppler effect for a more complete explanation.

5.35 It has been found that gaseous iodine molecules dissociate into separated atoms after absorption of light at wavelengths less than 4995 Å. If each quantum is absorbed by one molecule of I_2, what is the minimum input, in kcal/mol, needed to dissociate I_2 by this photochemical process?

▮

$$E(\text{per mol}) = N_A(h\nu) = \frac{N_A hc}{\lambda} = \frac{(6.022 \times 10^{23}\ \text{mol}^{-1})(6.626 \times 10^{-34}\ \text{J}\cdot\text{s})(2.998 \times 10^8\ \text{m}\cdot\text{s}^{-1})}{4995 \times 10^{-10}\ \text{m}}$$

$$= (239.5\ \text{kJ/mol})\left(\frac{1\ \text{kcal}}{4.184\ \text{kJ}}\right) = 57.2\ \text{kcal/mol}$$

5.36 Find the wavelength λ in the indicated units for light of the following frequencies: **(*a*)** 55 MHz (λ in m) **(*b*)** 1000 Hz (λ in cm) **(*c*)** 7.5×10^{15} Hz (λ in Å)

▮

(*a*) $55\ \text{MHz} = 55 \times 10^6/\text{s}$ $\qquad \lambda = \frac{c}{\nu} = \frac{3.0 \times 10^8\ \text{m/s}}{55 \times 10^6/\text{s}} = 5.5\ \text{m}$

(*b*) $\lambda = \frac{3.0 \times 10^{10}\ \text{cm/s}}{1000/\text{s}} = 3.0 \times 10^7\ \text{cm}$

(*c*) $\lambda = \frac{3.0 \times 10^{18}\ \text{Å/s}}{7.5 \times 10^{15}/\text{s}} = 400\ \text{Å}$

5.37 In a measurement of the quantum efficiency of photosynthesis in green plants, it was found that 8 quanta of red light at 6850 Å were needed to evolve 1 molecule of O_2. The average energy storage in the photosynthetic process is 112 kcal/mol O_2 evolved. What is the energy conversion efficiency in this experiment?

▮ $$E = h\nu = hc/\lambda = (6.63 \times 10^{-34}\ \text{J·s})(3.00 \times 10^{8}\ \text{m/s})/(6850 \times 10^{-10}\ \text{m}) = 2.90 \times 10^{-19}\ \text{J}$$

Eight quanta provide $8(2.90 \times 10^{-19}\ \text{J}) = 2.32 \times 10^{-18}\ \text{J}$

Required: $$\left(\frac{112\ \text{kcal}}{\text{mol}}\right)\left(\frac{4.184 \times 10^{3}\ \text{J}}{\text{kcal}}\right)\left(\frac{1\ \text{mol}}{6.02 \times 10^{23}\ \text{molecules}}\right) = \frac{7.78 \times 10^{-19}\ \text{J}}{\text{molecule}}$$

$$\left(\frac{7.78 \times 10^{-19}}{2.32 \times 10^{-18}}\right) \times 100\% = 33.5\%\ \text{efficiency}$$

5.38 O_2 undergoes photochemical dissociation into 1 normal oxygen atom and 1 oxygen atom 1.967 eV more energetic than normal. The dissociation of O_2 into 2 normal oxygen atoms is known to require 498 kJ/mol O_2. What is the maximum wavelength effective for the photochemical dissociation of O_2?

▮ For normal O atoms:

$$\left(\frac{498 \times 10^{3}\ \text{J}}{\text{mol O}_2}\right)\left(\frac{1\ \text{mol O}_2}{6.02 \times 10^{23}\ \text{molecules O}_2}\right) = 8.27 \times 10^{-19}\ \text{J}$$

Extra energy for excited O atom:

$$(1.967\ \text{eV})\left(\frac{1.60 \times 10^{-19}\ \text{J}}{1\ \text{eV}}\right) = 3.15 \times 10^{-19}\ \text{J}$$

Total energy for excited O atom $= 1.142 \times 10^{-18}$ J

$$\lambda = \frac{hc}{E} = \frac{(6.63 \times 10^{-34}\ \text{J·s})(3.00 \times 10^{8}\ \text{m/s})}{1.142 \times 10^{-18}\ \text{J}} = 1.74 \times 10^{-7}\ \text{m} = 174\ \text{nm}$$

5.39 The dye acriflavine, when dissolved in water, has its maximum light absorption at 4530 Å and its maximum fluorescence emission at 5080 Å. The number of fluorescence quanta is, on the average, 53% of the number of quanta absorbed. Using the wavelengths of maximum absorption and emission, what percentage of absorbed energy is emitted as fluorescence?

▮ Energy of each photon absorbed:

$$E = h\nu = hc/\lambda = (6.63 \times 10^{-34}\ \text{J·s})(3.00 \times 10^{8}\ \text{m/s})/(4530 \times 10^{-10}\ \text{m}) = 4.39 \times 10^{-19}\ \text{J}$$

Energy of each photon emitted:

$$E = h\nu = hc/\lambda = (6.63 \times 10^{-34}\ \text{J·s})(3.00 \times 10^{8}\ \text{m/s})/(5080 \times 10^{-10}\ \text{m}) = 3.92 \times 10^{-19}\ \text{J}$$

Energy fraction emitted:

$$\frac{0.53(3.92 \times 10^{-19})}{4.39 \times 10^{-19}} \times 100\% = 47\%$$

5.40 The prominent yellow line in the spectrum of a sodium vapor lamp has a wavelength of 590 nm. What minimum accelerating potential is needed to excite this line in an electron tube containing sodium vapor?

▮ $$E = h\nu = hc/\lambda = (6.63 \times 10^{-34}\ \text{J·s})(3.00 \times 10^{8}\ \text{m/s})/(590 \times 10^{-9}\ \text{m})$$
$$= (3.37 \times 10^{-19}\ \text{J})\left(\frac{1\ \text{eV}}{1.60 \times 10^{-19}\ \text{J}}\right) = 2.11\ \text{eV} = (2.11\ \text{V})(\text{charge on 1 electron})$$

Hence 2.11 V is needed to accelerate the electron away from the atom. When an electron fills the vacated location from outside the atom, a 590-nm photon will be emitted.

5.41 Distinguish between ***(a)*** a proton and a photon ***(b)*** a photon and a quantum.

▮ ***(a)*** A proton is a positively charged nuclear particle; a photon is a particle of light energy. ***(b)*** A quantum is a bundle of energy of definite magnitude, but not necessarily light energy; a photon is a quantum of light (electromagnetic energy).

5.3 PHOTOELECTRIC EFFECT

5.42 Light of wavelength λ shines on a metal surface with intensity X, and the metal emits Y electrons per second, of average energy Z. What will happen to Y and Z (*a*) if λ is halved (*b*) if X is doubled?

▮ (*a*) If the wavelength of incident light is halved, the energy of each photon is doubled and the energy of each emitted electron rises. (*b*) If the intensity of the light is doubled, the number of electrons emitted per second (Y) will be doubled. The average energy (Z) will remain the same.

5.43 The energy required to remove an electron from metal X is $E = 3.31 \times 10^{-20}$ J. Calculate the maximum wavelength of light that can photoeject an electron from metal X.

▮
$$\lambda = \frac{hc}{E} = \frac{(6.63 \times 10^{-34}\ \text{J}\cdot\text{s})(3.00 \times 10^{8}\ \text{m/s})}{3.31 \times 10^{-20}\ \text{J}} = 6.01 \times 10^{-6}\ \text{m}$$

5.44 For silver metal, the threshold frequency ν_0 is 1.13×10^{17} Hz. What is the maximum kinetic energy of the photoelectrons produced by shining ultraviolet light of 15.0 Å wavelength on the metal?

▮
$$\nu = c/\lambda = \left(\frac{3.00 \times 10^{8}\ \text{m/s}}{15.0\ \text{Å}}\right)\left(\frac{10^{10}\ \text{Å}}{\text{m}}\right) = 2.00 \times 10^{17}\ \text{Hz}$$
$$\text{KE} = h\nu - h\nu_0 = (6.63 \times 10^{-34}\ \text{J}\cdot\text{s})[(2.00 - 1.13) \times 10^{17}\ \text{s}^{-1})] = 5.8 \times 10^{-17}\ \text{J}$$

5.45 When a certain metal was irradiated with light of frequency 3.2×10^{16} Hz, the photoelectrons emitted had twice the kinetic energy as did photoelectrons emitted when the same metal was irradiated with light of frequency 2.0×10^{16} Hz. Calculate ν_0 for the metal.

▮
$$\text{KE} = h\nu - h\nu_0 = h(\nu - \nu_0) \qquad \nu - \nu_0 = \text{KE}/h$$
$$\text{KE}_2 = 2\text{KE}_1 \qquad \nu_2 - \nu_0 = \text{KE}_2/h \qquad \nu_1 - \nu_0 = \text{KE}_1/h$$

Dividing these equations yields:

$$\frac{\nu_2 - \nu_0}{\nu_1 - \nu_0} = \frac{\text{KE}_2/h}{\text{KE}_1/h} = 2$$
$$\nu_2 - \nu_0 = 2\nu_1 - 2\nu_0$$
$$\nu_0 = 2\nu_1 - \nu_2 = 2(2.0 \times 10^{16}\ \text{Hz}) - (3.2 \times 10^{16}\ \text{Hz}) = 8 \times 10^{15}\ \text{Hz}$$

5.46 One photon of ultraviolet light can excite an electron from the surface of a certain metal. When the same metal surface is irradiated with 2 photons of red light having a total energy equal to that of the ultraviolet photon, no photoelectrons are produced. Explain these facts in terms of Einstein's theory of the photoelectric effect.

▮ The red photons are below the threshold frequency, and 1 electron can interact with only 1 photon.

5.47 In a photoelectric effect experiment, irradiation of a metal with light of frequency 2.00×10^{16} Hz yields electrons with maximum kinetic energy 7.5×10^{-18} J. Calculate ν_0 for the metal.

▮
$$\text{KE} = h(\nu - \nu_0)$$
$$7.5 \times 10^{-18}\ \text{J} = (6.63 \times 10^{-34}\ \text{J}\cdot\text{s})(2.00 \times 10^{16}\ \text{s}^{-1} - \nu_0)$$
$$2.0 \times 10^{16}\ \text{s}^{-1} - \nu_0 = \frac{7.5 \times 10^{-18}\ \text{J}}{6.63 \times 10^{-34}\ \text{J}\cdot\text{s}} = 1.13 \times 10^{16}\ \text{s}^{-1}$$
$$\nu_0 = 8.7 \times 10^{15}\ \text{s}^{-1} = 8.7 \times 10^{15}\ \text{Hz}$$

5.48 Calculate the kinetic energy of the electron emitted when light of frequency 3.0×10^{15} Hz is shone on a metal surface which has a threshold frequency 1.0×10^{15} Hz.

▮
$$\text{KE} = h\nu - h\nu_0 = h(\nu - \nu_0) = (6.63 \times 10^{-34}\ \text{J}\cdot\text{s})(2.0 \times 10^{15}\ \text{s}^{-1}) = 1.3 \times 10^{-18}\ \text{J}$$

5.49 The minimum energy necessary to overcome the attractive force between the electron and the surface of silver metal is 7.52×10^{-19} J. What will be the maximum kinetic energy of the electrons ejected from silver which is being irradiated with ultraviolet light having a wavelength 360 Å?

▮
$$\text{KE} = h\nu - h\nu_0 = \frac{hc}{\lambda} - h\nu_0 = \frac{(6.63 \times 10^{-34}\ \text{J}\cdot\text{s})(3.00 \times 10^{8}\ \text{m/s})}{3.60 \times 10^{-8}\ \text{m}} - (7.52 \times 10^{-19}\ \text{J}) = 4.77 \times 10^{-18}\ \text{J}$$

5.50 The critical wavelength for producing the photoelectric effect in tungsten is 2600 Å. **(a)** What is the energy of a quantum at this wavelength in J and in eV? **(b)** What wavelength would be necessary to produce photoelectrons from tungsten having twice the kinetic energy of those produced at 2200 Å?

▌ (*a*) 2600 Å = 2600×10^{-10} m

$$E = h\nu = \frac{hc}{\lambda} = \frac{(6.63 \times 10^{-34}\ \text{J}\cdot\text{s})(3.00 \times 10^{8}\ \text{m/s})}{2600 \times 10^{-10}\ \text{m}} = 7.7 \times 10^{-19}\ \text{J}$$

(*b*) (1) $\text{KE}_{2200} = h\nu - h\nu_0 = hc\left(\frac{1}{\lambda_{2200}} - \frac{1}{\lambda_0}\right)$

(2) $\text{KE}_{\text{new}} = h\nu - h\nu_0 = hc\left(\frac{1}{\lambda_{\text{new}}} - \frac{1}{\lambda_0}\right)$

Dividing (2) by (1)

$$2 = \frac{\dfrac{1}{\lambda} - \dfrac{1}{2600}}{\dfrac{1}{2200} - \dfrac{1}{2600}} \qquad \frac{2}{2200} - \frac{1}{2600} = \frac{1}{\lambda} \qquad \lambda = 1900\ \text{Å}$$

5.51 In the photoelectric effect, an absorbed quantum of light results in the ejection of an electron from the absorber. The kinetic energy of the ejected electron is equal to the energy of the absorbed photon minus the energy of the longest-wavelength photon that causes the effect. Calculate the kinetic energy of a photoelectron produced in cesium by 400 nm light. The critical (maximum) wavelength for the photoelectric effect in cesium is 660 nm.

▌ Using the result of Prob. 5.24(*b*),

$$\text{Kinetic energy of electron} = h\nu - h\nu_{\text{crit}} = \frac{hc}{\lambda} - \frac{hc}{\lambda_{\text{crit}}} = \frac{1240\ \text{nm}\cdot\text{eV}}{400\ \text{nm}} - \frac{1240\ \text{nm}\cdot\text{eV}}{660\ \text{nm}} = 1.22\ \text{eV}$$

5.4 BOHR THEORY

5.52 Calculate the wavelengths of the first line and the series limit for the Lyman series for hydrogen.

▌

$$\text{First line:} \quad \frac{1}{\lambda} = 109\,678\left(\frac{1}{1^2} - \frac{1}{2^2}\right) = 82\,259\ \text{cm}^{-1}$$

$$\lambda = 1.2157 \times 10^{-5}\ \text{cm}$$

$$\text{Series limit:} \quad \frac{1}{\lambda} = 109\,678\left(\frac{1}{1^2} - \frac{1}{\infty}\right) = 109\,678\left(\frac{1}{1^2} - 0\right) = 109\,678\ \text{cm}^{-1}$$

$$\lambda = 9.1176 \times 10^{-6}\ \text{cm}$$

5.53 Identify each parameter in the Bohr energy expression

$$E_n = -k^2\frac{2\pi^2 mZ^2e^4}{n^2h^2}$$

▌ m = mass of electron, Z = atomic number of element, e = charge on the electron, n = integer (quantum number), h = Planck's constant, k = Coulomb's law constant.

5.54 Calculate the radius of the first allowed Bohr orbit for hydrogen.

▌ For the first orbit, $n = 1$. Substituting the values of the other constants in the equation

$$r = \frac{n^2h^2}{4\pi^2kme^2}$$

yields

$$r = \frac{1^2(6.63 \times 10^{-34}\ \text{J}\cdot\text{s})^2}{4(3.14)^2(9.00 \times 10^9\ \text{J}\cdot\text{m/C}^2)(9.109 \times 10^{-31}\ \text{kg})(1.60 \times 10^{-19}\ \text{C})^2} = 0.529 \times 10^{-10}\ \text{m} = 0.529\ \text{Å}$$

Experimental methods of determining the effective radius of the hydrogen atom yield the value 0.53 Å.

5.55 Calculate the energy of an electron in the first Bohr orbit of hydrogen.

▌

$$E_1 = -\frac{ke^2}{2r} = -\frac{(9.00 \times 10^9\ \text{J}\cdot\text{m/C}^2)(1.60 \times 10^{-19}\ \text{C})^2}{2(0.529 \times 10^{-10}\ \text{m})} = -2.178 \times 10^{-18}\ \text{J}$$

This value agrees closely with the experimentally determined energy required to remove an electron from a gaseous hydrogen atom.

5.56 Show that a_0, the radius of the first Bohr orbit of hydrogen, is 5.29×10^{-11} m, using the value of the permittivity of free space instead of the Coulomb's law constant.

▌

$$a_0 = \frac{h^2\varepsilon_0}{\pi me^2} = \frac{(6.626 \times 10^{-34}\ \text{J}\cdot\text{s})^2(8.854 \times 10^{-12}\ \text{C}^2/\text{N}\cdot\text{m}^2)}{(3.1416)(9.109 \times 10^{-31}\ \text{kg})(1.602 \times 10^{-19}\ \text{C})^2} = 5.29 \times 10^{-11}\ \text{m} = 0.529\ \text{Å}$$

5.57 If the energy difference between the ground state of an atom and its excited state is 4.4×10^{-19} J, what is the wavelength of the photon required to produce this transition?

▌

$$\lambda = \frac{hc}{\Delta E} = \frac{(6.63 \times 10^{-34}\ \text{J}\cdot\text{s})(3.00 \times 10^8\ \text{m/s})}{4.4 \times 10^{-19}\ \text{J}} = 4.5 \times 10^{-7}\ \text{m}$$

5.58 The third line in the Balmer series corresponds to an electronic transition between which Bohr orbits in hydrogen?

▌

$$5 \rightarrow 2$$

The lines of the Balmer series are first $3 \rightarrow 2$, second $4 \rightarrow 2$, third $5 \rightarrow 2$, fourth $6 \rightarrow 2$

5.59 Evaluate the quotient from the Bohr theory, $2\pi^2k^2me^4/ch^3$. Compare the result to the Rydberg constant, R.

▌

$$\frac{2\pi^2k^2me^4}{ch^3} = \frac{2(3.14)^2(9.00 \times 10^9\ \text{J}\cdot\text{m/C}^2)^2(9.10 \times 10^{-31}\ \text{kg})(1.60 \times 10^{-19}\ \text{C})^4}{(3.00 \times 10^8\ \text{m/s})(6.63 \times 10^{-34}\ \text{J}\cdot\text{s})^3} = 1.09 \times 10^7\ \text{m}^{-1} = 1.09 \times 10^5\ \text{cm}^{-1}$$

The result is, to 3 significant figures, equal to the Rydberg constant.

5.60 Express the Rydberg constant, $R = 109\,678\ \text{cm}^{-1}$, in **(*a*)** J/mol **(*b*)** J/atom.

▌

(*a*) $$(109\,678\ \text{cm}^{-1})\left(\frac{12.0\ \text{J/mol}}{\text{cm}^{-1}}\right) = 1.32 \times 10^6\ \text{J/mol}$$

↑ from Prob. 5.25(*c*)

(*b*) $$(109\,678\ \text{cm}^{-1})\left(\frac{1.99 \times 10^{-23}\ \text{J}}{\text{cm}^{-1}}\right) = 2.18 \times 10^{-18}\ \text{J/atom}$$

↑ from Prob. 5.25(*a*)

5.61 Calculate the frequency of light emitted for an electron transition from the sixth to the second orbit of the hydrogen atom. In what region of the spectrum does this light occur?

▌

$$\frac{1}{\lambda} = R\left(\frac{1}{2^2} - \frac{1}{6^2}\right) = (109\,678\ \text{cm}^{-1})(0.2222) = 24\,373\ \text{cm}^{-1}$$

$$\nu = \frac{c}{\lambda} = c\left(\frac{1}{\lambda}\right) = (3.00 \times 10^{10}\ \text{cm/s})(24\,373\ \text{cm}^{-1}) = 7.31 \times 10^{14}\ \text{Hz}$$

The line is in the visible spectrum of hydrogen. It is a part of the Balmer series because the electron has had a transition to the second orbit from a higher orbit.

5.62 Calculate the energy of an electron in the second Bohr orbit of a hydrogen atom.

▌

$$E_2 = -k^2\left(\frac{2\pi^2me^4}{n^2h^2}\right)$$

$$= -(9.00 \times 10^9\ \text{J}\cdot\text{m/C}^2)^2\left(\frac{2(3.14)^2(9.109 \times 10^{-31}\ \text{kg})(1.60 \times 10^{-19}\ \text{C})^4}{(2)^2(6.63 \times 10^{-34}\ \text{J}\cdot\text{s})^2}\right)$$

$$= -5.44 \times 10^{-19}\ \text{J}$$

The k is the constant in Coulomb's law.

5.63 For the hydrogen atom, $E_n = -(1/n^2)R_H$, where $R_H = 2.179 \times 10^{-18}$ J. Find the wavelength of the transition from the ground state to the $n = 2$ state.

▮

$$\Delta E = (2.179 \times 10^{-18}\ \text{J})\left(\frac{1}{1^2} - \frac{1}{2^2}\right) = 1.63 \times 10^{-18}\ \text{J}$$

$$\lambda = \frac{hc}{\Delta E} = \frac{(6.63 \times 10^{-34}\ \text{J·s})(3.00 \times 10^8\ \text{m/s})}{1.63 \times 10^{-18}\ \text{J}} = 1.22 \times 10^{-7}\ \text{m}$$

5.64 (*a*) Calculate the radii of the first two Bohr orbits of Li^{2+}. (*b*) Calculate the difference in potential energy between these two orbits. (*c*) Calculate the difference in total energy between these orbits.

▮

(*a*)
$$r_1 = \frac{n^2h^2}{4\pi^2 kmZe^2}$$

$$= \frac{1^2(6.63 \times 10^{-34}\ \text{J·s})^2}{4(3.14)^2(9.00 \times 10^9\ \text{J·m/C}^2)(9.109 \times 10^{-31}\ \text{kg})(3)(1.60 \times 10^{-19}\ \text{C})^2}$$

$$= 1.77 \times 10^{-11}\ \text{m} = 0.177\ \text{Å}$$

(Alternatively, the value of r_1 may be obtained by dividing r_1 for hydrogen, 0.529 Å, by 3, the atomic number of lithium. Explain why.)

$$r_2 = n_2^2 r_1 = 4r_1 = 7.08 \times 10^{-11}\ \text{m} = 0.708\ \text{Å}$$

(*b*) The potential energy of charged particles is given by

$$\text{PE} = k\frac{q_1q_2}{r}; \quad \text{so}$$

$$\text{PE}_1 = \frac{9.00 \times 10^9\ \text{J·m}}{\text{C}^2}\,\frac{(-1.60 \times 10^{-19}\ \text{C})(3 \times 1.60 \times 10^{-19}\ \text{C})}{1.77 \times 10^{-11}\ \text{m}} = -3.90 \times 10^{-17}\ \text{J}$$

$$\text{PE}_2 = \text{PE}_1/4 = -9.75 \times 10^{-18}\ \text{J}$$

$$\Delta\text{PE} = \text{PE}_2 - \text{PE}_1 = (-9.75 \times 10^{-18}\ \text{J}) - (-3.90 \times 10^{-17}\ \text{J}) = 2.92 \times 10^{-17}\ \text{J}$$

(*c*)
$$E_n = -k^2\left(\frac{2\pi^2 mZ^2e^4}{n^2h^2}\right)$$

$$= -(9.00 \times 10^9\ \text{J·m/C}^2)^2\left(\frac{2(3.14)^2(9.109 \times 10^{-31}\ \text{kg})(3)^2(1.60 \times 10^{-19}\ \text{C})^4}{1^2(6.63 \times 10^{-34}\ \text{J·s})^2}\right)$$

$$= -1.95 \times 10^{-17}\ \text{J}$$

(Alternatively, $E_1 = 9E_1$ for hydrogen.)

$$E_2 = E_1/4 = -4.88 \times 10^{-18}\ \text{J}$$

$$\Delta E = (-4.88 \times 10^{-18}\ \text{J}) - (-1.95 \times 10^{-17}\ \text{J}) = 1.46 \times 10^{-17}\ \text{J}$$

Note that the difference in total energy, part (*c*), is exactly half that of the potential energy.

5.65 If the lowest energy x-rays have $\lambda = 4.0 \times 10^{-8}$ m, estimate the minimum difference in energy between two Bohr orbits such that an electronic transition would correspond to the emission of an x-ray. Assuming that the electrons in other shells exert no influence, at what Z (minimum) would a transition from the second energy level to the first result in the emission of an x-ray?

▮

$$E = h\nu = hc/\lambda = \frac{(6.63 \times 10^{-34}\ \text{J·s})(3.00 \times 10^8\ \text{m/s})}{4.0 \times 10^{-8}\ \text{m}} = 5.0 \times 10^{-18}\ \text{J}$$

$$\Delta E_H = \tfrac{3}{4}(2.178 \times 10^{-18}\ \text{J}) = 1.63 \times 10^{-18}\ \text{J}$$

$$\Delta E = \Delta E_H(Z^2)$$

$$Z^2 = \Delta E/\Delta E_H = (5.0 \times 10^{-18})/(1.63 \times 10^{-18}) = 3.06$$

$$Z = 2 \quad \text{(helium)}$$

5.66 The energy of an electron in the first Bohr orbit for hydrogen is −13.6 eV. Which one(s) of the following is (are) possible excited state(s) for electrons in Bohr orbits of hydrogen? (*a*) −3.4 eV (*b*) −6.8 eV (*c*) −1.7 eV (*d*) +13.6 eV

▮

$$E_n = -k^2\frac{2\pi^2 mZ^2e^4}{n^2h^2}$$

Excited states therefore have energies

$$E_n = E_1/n^2$$

The only energy which is equal to -13.6 eV divided by the square of an integer is (*a*) $-3.4\text{ eV} = -13.6\text{ eV}/4$. The electron with positive energy value (*d*) must be out of the atom.

5.67 Analogous to the accomplishment of Balmer, fit the following three series of numbers into an equation involving integers:

series 1 wavelength, Å	series 2 wavelength, Å	series 3 wavelength, Å
68.26	22.76	11.38
91.02	34.13	18.20
102.40	40.96	22.76
109.22	45.51	26.00

▮ The equation, with n_1 and n_2 integral, is

$$\lambda = k\left(\frac{1}{n_1} - \frac{1}{n_2}\right) \qquad \text{where } k = 136.52\text{ Å}$$

One can guess a reciprocal relationship by the steady drop in the differences between successive numbers.

5.68 A photon was absorbed by a hydrogen atom in its ground state, and the electron was promoted to the fifth orbit. When the excited atom returned to its ground state, visible and other quanta were emitted. In this process, radiation of what wavelength *must* have been emitted? Explain.

▮ The emission of visible light implies some transition to the second orbit. After that, there *must* be a transition $2 \rightarrow 1$ for the atom to return to its ground state. λ is given by

$$\frac{1}{\lambda} = 109\,678\left(\frac{1}{1^2} - \frac{1}{2^2}\right)$$

5.69 The wave number of the first line in the Balmer series of hydrogen is 15 200 cm^{-1}. What is the wave number of the first line in the Balmer series of Be^{3+}?

▮

$$\Delta E_{\text{Be}} = Z^2(\Delta E_{\text{H}})$$

$$\bar{\nu}_{\text{Be}} = 16\bar{\nu}_{\text{H}} = 16(15\,200\text{ cm}^{-1}) = 2.43 \times 10^5\text{ cm}^{-1}$$

5.70 Derive an expression for the velocity of an electron in any Bohr orbit of a hydrogenlike atom. Calculate the velocity of an electron in the first orbit of a hydrogen atom. What is the ratio of this velocity to the velocity of light in a vacuum?

▮

$$\text{KE} = kZe^2/2r = mv^2/2$$

$$\frac{kZe^2}{r} = mv^2$$

$$v = \frac{\sqrt{kZe}}{\sqrt{mr}} = \frac{\sqrt{kZe}}{\sqrt{m}\sqrt{\dfrac{n^2h^2}{4\pi^2kme^2}}} = \frac{2\pi k\sqrt{Z}e^2}{nh}$$

For hydrogen:

$$v_1 = \frac{2(3.14)(9.00 \times 10^9\text{ J}\cdot\text{m/C}^2)(1)(1.60 \times 10^{-19}\text{ C})^2}{1(6.63 \times 10^{-34}\text{ J}\cdot\text{s})} = 2.18 \times 10^6\text{ m/s}$$

$$\frac{c}{v_1} = \frac{3.00 \times 10^8\text{ m/s}}{2.18 \times 10^6\text{ m/s}} = 137$$

Light travels 137 times as fast as the electron in this orbit.

5.71 Transitions between which orbits in He^+ ions would result in the emission of visible light?

▮

$$E_n = -k^2\left(\frac{2\pi^2 mZ^2e^4}{n^2h^2}\right)$$

Since Z for helium is 2, the corresponding energy transformations will be $2^2 = 4$ times as energetic as in hydrogen. The visible region ranges from 2.84×10^{-19} J to 4.97×10^{-19} J (Prob. 5.28(*d*)). For helium

$$E_1 = -8.712 \times 10^{-18}\ \text{J} \qquad (4 \text{ times } E_1 \text{ for H})$$

For any transition

$$\Delta E = E_1\left(\frac{1}{n_1^2} - \frac{1}{n_2^2}\right)$$

Thus

$$\frac{1}{n_1^2} - \frac{1}{n_2^2} = \frac{\Delta E}{E_1} = \frac{2.84 \times 10^{-19}}{-8.712 \times 10^{-18}} = 0.0325 \quad \text{to} \quad \frac{4.97 \times 10^{-19}}{-8.712 \times 10^{-18}} = 0.0570$$

Trial and error shows the values of n_1 and n_2 can be $4 \to 3$ and from $6 \to 4$ up to $13 \to 4$, and no others.

5.72 Assuming that it were possible for such an atom to exist, calculate the energy of a positron (a positive electron) in the first Bohr orbit of a hydrogen atom.

▮ +13.6 eV. A positron has characteristics the same as an electron except for the sign of its charge, which would cause its energy in the atom to be opposite that of the electron.

5.73 Calculate the energy of transition of an electron from the fourth to the second orbit of a helium ion. (The comparable value for hydrogen is 20 437 cm^{-1}.)

▮

$$(20\,437)(4) = 81\,748\ \text{cm}^{-1} \qquad (\text{For He}^+, \quad Z = 2 \quad \text{and} \quad Z^2 = 4.)$$

5.74 The Rydberg constant for deuterium (^{2}H) is 109 707 cm^{-1}. (This value reflects a refinement of the simple Bohr theory, wherein the Rydberg constant and the orbital radii depend on the so-called *reduced mass* rather than the electron mass. The reduced mass, in turn, varies slightly with the mass of the nucleus.) Calculate **(*a*)** the shortest wavelength in the absorption spectrum of deuterium, **(*b*)** the ionization potential of deuterium, and **(*c*)** the radii of the first three Bohr orbits.

▮ **(*a*)** The shortest-wavelength transition would correspond to the highest frequency and to the highest energy. The transition would thus be from the lowest energy state (the *ground state*), for which $n = 1$, to the highest, for which $n = \infty$.

$$\bar{\nu} = R\left(\frac{1}{1^2} - \frac{1}{\infty^2}\right) = R = 109\,707\ \text{cm}^{-1}$$

$$\lambda = \frac{1}{109\,707\ \text{cm}^{-1}} = (0.91152 \times 10^{-5}\ \text{cm})(10^7\ \text{nm/cm}) = 91.152\ \text{nm}$$

(*b*) The transition computed in (*a*) is indeed the ionization of the atom in its ground state. From the result of Prob. 5.24(*b*),

$$\text{IP} = \frac{1239.8\ \text{nm}\cdot\text{eV}}{91.152\ \text{nm}} = 13.601\ \text{eV}$$

This value is slightly greater than the value for ^{1}H.

(*c*) From the equation with $(Z = 1)$

$$r = n^2 a_0 = n^2(5.29 \times 10^{-11}\ \text{m})$$

the radii are 1, 4, and 9 times a_0, or 0.529, 2.116, and 4.76 Å. The reduced-mass correction, involving an adjustment of 3 parts in 10^4, is not significant, and the a_0 for the first Bohr orbit of ^{1}H is a perfectly satisfactory substitution.

5.75 **(*a*)** Neglecting reduced-mass effects, what optical transition in the He^+ spectrum would have the same wavelength as the first Lyman transition of hydrogen $(n = 2$ to $n = 1)$? **(*b*)** What is the second ionization potential of He? **(*c*)** What is the radius of the first Bohr orbit for He^+?

▮ **(*a*)** He^+ has only one electron. It is thus classified as a hydrogenlike species with $Z = 2$, and the Bohr equations may be applied. From the equation

$$\bar{\nu} = RZ^2\left(\frac{1}{n_1^2} - \frac{1}{n_2^2}\right)$$

the first Lyman transition for hydrogen would be given by

$$\bar{\nu} = R\left(\frac{1}{1^2} - \frac{1}{2^2}\right)$$

The assumption regarding mass effects is equivalent to considering R for He^+ the same as for ^1H. The Z^2 term can just be compensated by increasing n_1 and n_2 by a factor of 2 each.

$$\bar{\nu} = R(2)^2\left(\frac{1}{2^2} - \frac{1}{4^2}\right)$$

The transition in question is thus the transition from $n = 4$ to $n = 2$.

(*b*) The second ionization potential for He is the same as the ionization potential for He^+, and the Bohr equations may be applied to the ground state of He^+, for which $Z = 2$ and $n = 1$. From the Z^2 dependence of the energy of a Bohr level, and from the treatment of deuterium in Prob. 5.74, for which the mass correction is closer to that for He^+ than to that for ^1H,

$$\text{IP}(\text{He}^+) = Z^2 \times \text{IP}(^2\text{H}) = 4 \times 13.6 \text{ eV} = 54.4 \text{ eV}$$

(*c*) $r = \dfrac{n^2 a_0}{Z} = \dfrac{0.529 \text{ Å}}{2} = 0.264 \text{ Å}$

5.5 ELECTRON DIFFRACTION

5.76 Given Einstein's relationship, $E = mc^2$, and Planck's hypothesis, $E = h\nu$, derive the relationship between the wavelength of a photon and its mass and velocity. Compare this relationship to that derived by de Broglie for the wavelength of an electron.

▮

$$E = mc^2 = h\nu = hc/\lambda$$

$$\lambda = \frac{hc}{mc^2} = \frac{h}{mc} \quad \text{analogous to} \quad \lambda = \frac{h}{mv} \quad \text{for electrons}$$

5.77 Show that de Broglie's hypothesis applied to an electron moving in a circular orbit leads to Bohr's postulate of quantized angular momentum.

▮

$$\text{de Broglie's hypothesis:} \quad \lambda = h/mv$$

An electron in a circular orbit must have its path length equal to an integral number of wavelengths for reinforcement to occur. The path length, $2\pi r$, is thus equal to $n\lambda$:

$$2\pi r = n\lambda \qquad \lambda = \frac{2\pi r}{n} = \frac{h}{mv} \qquad mvr = \frac{nh}{2\pi}$$

5.78 An electron in a hydrogen atom in its ground state absorbs 1.50 times as much energy as the minimum required for it to escape from the atom. What is the wavelength of the emitted electron?

▮ Since 13.6 eV is needed for ionization, 20.4 eV must have been absorbed. Of this, 6.8 eV is converted to kinetic energy. The velocity of the electron may be calculated as follows:

$$6.8 \text{ eV} = 6.8(1.60 \times 10^{-19} \text{ C})(1 \text{ V}) = 1.09 \times 10^{-18} \text{ J}$$

$$\text{KE} = \tfrac{1}{2}mv^2$$

$$v = \sqrt{\frac{2\text{KE}}{m}} = \sqrt{\frac{2(1.09 \times 10^{-18} \text{ J})}{9.109 \times 10^{-31} \text{ kg}}} = 1.55 \times 10^6 \text{ m/s}$$

According to the de Broglie equation

$$\lambda = \frac{h}{mv} = \frac{6.63 \times 10^{-34} \text{ J·s}}{(9.109 \times 10^{-31} \text{ kg})(1.55 \times 10^6 \text{ m/s})} = 4.70 \times 10^{-10} \text{ m}$$

5.79 What accelerating potential must be imparted to a proton beam to give it an effective wavelength of 0.050 Å?

▮

$$v = \frac{h}{m\lambda} = \frac{6.63 \times 10^{-34} \text{ J·s}}{[1.008/(6.02 \times 10^{26}) \text{ kg}](0.050 \times 10^{-10} \text{ m})} = 7.92 \times 10^4 \text{ m/s}$$

$$\text{KE} = \tfrac{1}{2}mv^2 = \tfrac{1}{2}\left(\frac{1.008}{6.02 \times 10^{26}}\right)(7.92 \times 10^4)^2 = 5.25 \times 10^{-18} \text{ J}$$

$$= (5.25 \times 10^{-18} \text{ J})\left(\frac{1 \text{ eV}}{1.60 \times 10^{-19} \text{ J}}\right) = 33 \text{ eV}$$

Since the proton has a charge equal in magnitude to the electronic charge, the potential needed is equal in magnitude to the number of electron volts, 33 V.

5.80 What accelerating potential is needed to produce an electron beam with an effective wavelength of 0.090 Å?

▮

$$v = \frac{h}{m\lambda} = \frac{6.63 \times 10^{-34}\ \text{J}\cdot\text{s}}{(9.109 \times 10^{-31}\ \text{kg})(0.090 \times 10^{-10}\ \text{m})} = 8.09 \times 10^{7}\ \text{m/s}$$

$$\text{KE} = \tfrac{1}{2}mv^2 = \tfrac{1}{2}(9.109 \times 10^{-31}\ \text{kg})(8.09 \times 10^{7}\ \text{m/s})^2 = 2.98 \times 10^{-15}\ \text{J}$$

$$= (2.98 \times 10^{-15}\ \text{J})\left(\frac{1\ \text{eV}}{1.60 \times 10^{-19}\ \text{J}}\right) = 1.86 \times 10^{4}\ \text{eV} = 18.6\ \text{keV}$$

Potential = 18.6 kV

5.81 A 1.0-g projectile is shot from a gun with a velocity of 100 m/s. What is the de Broglie wavelength?

▮

$$\lambda = \frac{h}{mv} = \frac{6.63 \times 10^{-34}\ \text{J}\cdot\text{s}}{(1.0 \times 10^{-3}\ \text{kg})(100\ \text{m/s})} = 6.63 \times 10^{-33}\ \text{m}$$

5.82 An electron diffraction experiment was performed with a beam of electrons accelerated by a potential difference of 10.0 kV. What was the wavelength of the electron beam?

▮ We can use the de Broglie equation, taking the mass of an electron as 0.911×10^{-30} kg. The velocity of the electron is found by equating its kinetic energy, $\frac{1}{2}mv^2$, to its loss of electrical potential energy, 10.0 keV.

$$\tfrac{1}{2}mv^2 = (1.00 \times 10^{4}\ \text{eV})(1.602 \times 10^{-19}\ \text{J/eV}) = 1.602 \times 10^{-15}\ \text{J} = 1.602 \times 10^{-15}\ \text{kg}\cdot\text{m}^2\cdot\text{s}^{-2}$$

$$v = \left(\frac{2 \times 1.602 \times 10^{-15}\ \text{kg}\cdot\text{m}^2\cdot\text{s}^{-2}}{0.911 \times 10^{-30}\ \text{kg}}\right)^{1/2} = (35.17 \times 10^{14})^{1/2}\ \text{m/s} = 5.93 \times 10^{7}\ \text{m/s}$$

Now the de Broglie equation gives

$$\lambda = \frac{h}{mv} = \frac{6.63 \times 10^{-34}\ \text{J}\cdot\text{s}}{(0.911 \times 10^{-30}\ \text{kg})(5.93 \times 10^{7}\ \text{m/s})}$$

$$= \frac{1.23 \times 10^{-11}\ \text{kg}\cdot\text{m}^2\cdot\text{s}^{-1}}{\text{kg}\cdot\text{m}\cdot\text{s}^{-1}} = (1.23 \times 10^{-11}\ \text{m})(10^{10}\ \text{Å/m}) = 0.123\ \text{Å}$$

5.83 A beam of electrons accelerated through 4.64 V in a tube containing mercury vapor was partly absorbed by the vapor. As a result of absorption, electronic changes occurred within a mercury atom and light was emitted. If the full energy of a single incident electron was converted into light, what was the wave number of the emitted light?

▮ Using the result of Prob. 5.24(*b*),

$$\bar{\nu} = \frac{1}{\lambda} = \frac{\nu}{c} = \frac{h\nu}{hc} = \frac{4.64\ \text{eV}}{1240\ \text{nm}\cdot\text{eV}} = 0.00374\ \text{nm}^{-1} = 37\,400\ \text{cm}^{-1}$$

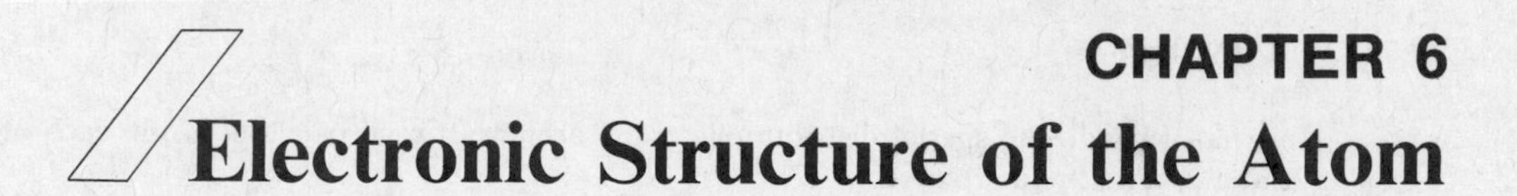

CHAPTER 6
Electronic Structure of the Atom

6.1 SHELLS, SUBSHELLS, ORBITALS

6.1 Describe briefly the contribution to the present theory of atomic structure made by each of the following: **(a)** Bohr **(b)** de Broglie **(c)** Heisenberg **(d)** Hund **(e)** Pauli **(f)** Schrödinger

(a) Bohr was first to postulate a theory which included the concept of discrete energy levels in atoms. **(b)** de Broglie predicted wave properties for a stream of electrons. **(c)** Heisenberg postulated the uncertainty principle, which states that there is an uncertainty of position and/or momentum for small particles such as electrons. **(d)** Hund showed that electrons in degenerate orbitals would remain unpaired as much as possible. **(e)** Pauli stated that no two electrons in a given atom can have the same set of four quantum numbers. **(f)** Schrödinger postulated the wave equation, which describes the properties of electrons in atoms better than the Bohr theory does.

6.2 What are the possible values of l for an electron with $n = 3$?

l can be 0, 1, or 2.

6.3 What are the possible values of m_l for an electron with $l = 2$?

The value of m_l can vary from $-l$ to $+l$; in this case m_l can have values -2, -1, 0, $+1$, and $+2$.

6.4 What are the possible values of m_s for an electron with $m_l = 0$?

m_s must be $-\frac{1}{2}$ or $+\frac{1}{2}$. These are the only permitted values, regardless of the values of the other quantum numbers.

6.5 Give the set of quantum numbers that describe an electron in a $3p$ orbital.

$$n = 3; \quad l = 1; \quad m_l = -1, 0, \text{ or } 1; \quad m_s = +\tfrac{1}{2} \text{ or } -\tfrac{1}{2}.$$

6.6 How many electrons can be placed **(a)** in the shell with $n = 2$ **(b)** in the shell with $n = 3$ **(c)** in the shell with $n = 3$ before the first electron enters the shell with $n = 4$?

(a) $2n^2 = 2(2)^2 = 8$ **(b)** $2(3)^2 = 18$ **(c)** 8. (The $4s$ electrons precede the $3d$ electrons.)

6.7 Match each of the following quantum number values with the proper letter designation (K, L, M, N, s, p, d, f): **(a)** $n = 1$ **(b)** $l = 2$ **(c)** $l = 0$ **(d)** $n = 3$

(a) K **(b)** d **(c)** s **(d)** M

6.8 What is the maximum number of electrons which can be accommodated **(a)** in the shell with $n = 4$ **(b)** in the $4f$ subshell **(c)** in an *atom* in which the highest principal quantum number value is 4?

(a) $2n^2 = 2(4)^2 = 32$ **(b)** 14 (the maximum in any f subshell) **(c)** 36, Kr. The atom can have no $5s$ electrons, precluding any atom with a regular electronic configuration beyond Kr. Note that the fourth shell is not filled in Kr. (Pd, element 46, has an exceptional electronic configuration—$4s^2\ 4p^6\ 4d^{10}$—and certainly is a correct answer to this question.)

6.9 **(a)** What values are permitted for the orbital angular momentum (azimuthal) quantum number l for an electron with principal quantum number $n = 4$? **(b)** How many different values for the magnetic quantum number are possible for an electron with orbital angular momentum quantum number $l = 3$? **(c)** How many electrons can be placed in each of the following subshells: s, p, d, f? **(d)** What is the maximum number of electrons which can be placed in an atomic orbital which has an orbital angular momentum quantum number $l = 3$? **(e)** What is the lowest shell which has an f subshell?

(a) 3, 2, 1, 0 **(b)** seven (-3, -2, -1, 0, $+1$, $+2$, $+3$) **(c)** s: 2, p: 6, d: 10, f: 14 **(d)** 2 (The maximum number of electrons in any orbital is 2.) **(e)** fourth

6.10 Which shell would be the first to have a g subshell?

Since the angular momentum quantum number, l, must be less than the principal quantum number, n, and since the g subshell designation represents an l value of 4, the minimum n value possible is 5. In literal notation ($KLMNO$), the fifth shell is designated the O shell.

6.11 In what ways do the spatial distributions of the orbitals in each pair differ from each other? (*a*) 1*s* and 2*s* (*b*) 2*s* and 2*p* (*c*) $2p_x$ and $2p_z$

▮ (*a*) The 2*s* orbital is larger, and it has a spherical node, which the 1*s* orbital does not have. (*b*) The 2*p* orbital is directed in space, aligned along one of the *x*, *y*, or *z* axes. It has a planar node which includes the other two axes. (*c*) Each orbital is aligned along the axis corresponding to its subscript.

6.12 Describe the term *penetration* as it applies to electronic configuration. The properties of which one of the following elements are most modified by penetration, and the properties of which one are least modified: Zn, Ca, Br, H?

▮ Penetration is the finite probability of finding an electron with a higher principal quantum number closer to the nucleus for a brief time than an electron with a smaller value of its principal quantum number. H is least affected, since it has only one electron. Zn is most affected since its 4*s* electrons can penetrate to a nuclear charge of 30, which causes a greater effect than the penetration of the 4*s* electrons of Ca (to a nucleus with a charge of only 20) or the penetration of the 4*p* electrons of Br.

6.13 The nucleus of an atom is located at $x = y = z = 0$. (*a*) If the probability of finding an *s*-orbital electron in a tiny volume around $x = a$, $y = z = 0$, is 1.0×10^{-5}, what is the probability of finding the electron in the same sized volume around $x = z = 0$, $y = a$? (*b*) What would be the probability at the second site if the electron were in a p_z orbital? Explain.

▮ (*a*) 1.0×10^{-5}. Since the distance from the nucleus is the same, and the *s* orbital is spherically symmetrical, the probability in each of the two volumes is the same. (*b*) Since the p_z orbital has a node at $z = 0$, the probability of finding the electron in this volume is 0.

6.14 Draw a representation of a 3*p* orbital, including in your sketch the information that it has the planar node characteristic of *p* orbitals but in addition has a spherical node.

▮ See Fig. 6.1.

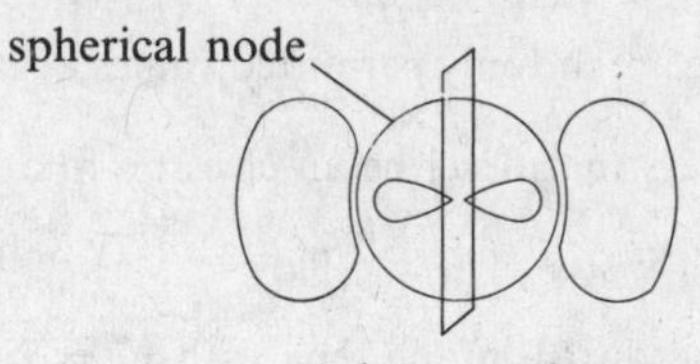

Fig. 6.1

6.15 (*a*) Suppose a particle has four quantum numbers such that the permitted values are those given below. How many particles could be fitted into the $n = 1$ shell? into the $n = 2$ shell? into the $n = 3$ shell?

n:	1, 2, 3, . . .
l:	$(n-1), (n-3), (n-5), \ldots$ but no negative number
j:	$(l + \frac{1}{2})$ or $(l - \frac{1}{2})$ if the latter is not negative
m:	$-j$ in integral steps to $+j$

(*b*) These quantum numbers apply to protons and neutrons in atomic nuclei. Explain the stability of $^{4}_{2}He$, $^{16}_{8}O$, and $^{40}_{20}Ca$.

▮ (*a*)

$n = 1$
$l = 0$
$j = \frac{1}{2}$
$m = -\frac{1}{2}, +\frac{1}{2}$

Two nucleons

$n = 2$	
$l = 1$	
$j = 1\frac{1}{2}$	$\frac{1}{2}$
$m = -1\frac{1}{2}, -\frac{1}{2}, +\frac{1}{2}, +1\frac{1}{2}$	$-\frac{1}{2}, +\frac{1}{2}$

Six nucleons

$n = 3$		
$l = 2$		0
$j = 2\frac{1}{2}$	$1\frac{1}{2}$	$\frac{1}{2}$
$m = -2\frac{1}{2}, -1\frac{1}{2}, -\frac{1}{2}, +\frac{1}{2}, +1\frac{1}{2}, +2\frac{1}{2}$	$-1\frac{1}{2}, -\frac{1}{2}, \frac{1}{2}, 1\frac{1}{2}$	$-\frac{1}{2}, \frac{1}{2}$

Twelve nucleons

(b) ^4_2He has 2 protons and 2 neutrons, with both $n = 1$ shells (protons and neutrons) filled. $^{16}_8O$, with 8 protons and 8 neutrons, has the first 2 shells of protons and the first 2 shells of neutrons filled $(2 + 6 = 8)$. $^{40}_{20}Ca$ has 3 proton shells and 3 neutron shells filled.

6.16 If there were three possible values $(-\frac{1}{2}, 0, +\frac{1}{2})$ for the spin magnetic quantum number m_s, how many elements would there be in the second period of the periodic table? (Quantum numbers n, l, and m_l are defined as usual.) Construct a periodic table showing the first 54 elements in such a hypothetical situation.

Twelve:

$n =$	2			
$l =$	1			0
$m_l =$	-1	0	$+1$	0
$m_s =$	$-\frac{1}{2}, 0, \frac{1}{2}$	$-\frac{1}{2}, 0, \frac{1}{2}$	$-\frac{1}{2}, 0, \frac{1}{2}$	$-\frac{1}{2}, 0, \frac{1}{2}$

$1s^3$		
$2s^3$		$2p^9$
$3s^3$		$3p^9$
$4s^3$	$3d^{15}$	$4p^9$

6.17 The uncertainty principle may be stated mathematically

$$\Delta(mv)\,\Delta x \simeq h/4\pi$$

where $\Delta(mv)$ represents the uncertainty in the momentum of a particle and Δx represents the uncertainty in its position. **(a)** If a 1.00-g body is traveling along the x axis at 100 cm/s within 1 cm/s, what is the theoretical uncertainty in its position? **(b)** If an electron is traveling at 100 m/s within 1 m/s, what is the theoretical uncertainty in its position? Explain why the uncertainty principle is not important for macroscopic bodies.

(a) The velocity has an uncertainty of 2 cm/s (from 99 to 101 cm/s). Therefore

$$\Delta x \simeq \frac{h}{4\pi\Delta(mv)} = \frac{6.63 \times 10^{-27}\ \text{erg}\cdot\text{s}}{4(3.14)(2\ \text{g}\cdot\text{cm/s})} = 3 \times 10^{-28}\ \text{cm} = 3 \times 10^{-30}\ \text{m}$$

$$\textbf{(b)}\ \Delta x \simeq \frac{6.63 \times 10^{-34}\ \text{J}\cdot\text{s}}{4(3.14)(2 \times 9.109 \times 10^{-31}\ \text{kg}\cdot\text{m/s})} = 3 \times 10^{-5}\ \text{m} = 30\ \mu\text{m}$$

It may be seen that the small mass of the electron causes a very significant uncertainty in its position, whereas the position of the macroscopic body is known *very* precisely.

6.18 The solution to the Schrödinger equation for an electron in the ground state of the hydrogen atom is

$$\psi_{1s} = \frac{1}{\sqrt{\pi a_0^3}}\, e^{-r/a_0}$$

where r is the distance from the nucleus and a_0 is 0.529×10^{-8} cm. The probability of finding an electron at any *point* in space is proportional to $|\psi|^2$. Using calculus, show that the maximum probability of finding the electron in the 1s orbital of hydrogen occurs at $r = a_0$.

The probability of finding the electron within a shell between r and $r + dr$ is simply the probability of finding it in a unit volume, ψ^2, times the volume of the shell, $2\pi r^2\, dr$. From calculus, the maximum is determined to be the radius at which the derivative of the probability per unit volume with respect to the radius is equal to 0. Thus,

$$\text{Probability} = 2\pi r^2 \psi^2 = \frac{2r^2 e^{-2r/a_0}}{a_0^3}$$

$$0 = \frac{2}{a_0^3}\frac{d(r^2 e^{-2r/a_0})}{dr} = \frac{2}{a_0^3}\left(2r(e^{-2r/a_0}) - \frac{2r^2}{a_0}(e^{-2r/a_0})\right)$$

Since ψ^2 can be 0 only at $r = \infty$

$$2r - \frac{2r^2}{a_0} = 0$$

$$r\left(1 - \frac{r}{a_0}\right) = 0$$

$$r = a_0$$

6.2 ELECTRONIC STRUCTURES OF ATOMS AND IONS

6.19 What is the detailed electronic configuration of phosphorus?

▮ $1s^2\ 2s^2\ 2p^6\ 3s^2\ 3p_x^1\ 3p_y^1\ 3p_z^1$

6.20 (*a*) List all the elements whose atoms have only one electron in a *p* subshell. (*b*) List all the elements whose atoms have only one electron in an *s* subshell. (*c*) Which of these two sets of elements contains elements in only one periodic group? (*d*) How many periodic groups are represented in the other set?

▮ (*a*) B, Al, Ga, In, Tl—the group IIIA elements (*b*) H, Li, Na, K, Rb, Cs, Fr, Cu, Ag, Au, Cr, Nb, Mo, Tc, Ru, Rh, Pt—the group IA and IB elements plus many elements with "exceptional" electronic configurations (*c*) The elements listed in set (*a*). (*d*) 6 (Note that Ru, Rh, and Pt are all in one group.)

6.21 Write the electronic configuration and identify the periodic group for the elements whose atomic numbers are 6, 16, 26, 36, 56.

▮

$1s^2\ 2s^2\ 2p^2$	group IVA
$1s^2\ 2s^2\ 2p^6\ 3s^2\ 3p^4$	group VIA
$1s^2\ 2s^2\ 2p^6\ 3s^2\ 3p^6\ 4s^2\ 3d^6$	group VIIIB
$1s^2\ 2s^2\ 2p^6\ 3s^2\ 3p^6\ 4s^2\ 3d^{10}\ 4p^6$	group 0
$1s^2\ 2s^2\ 2p^6\ 3s^2\ 3p^6\ 4s^2\ 3d^{10}\ 4p^6\ 5s^2\ 4d^{10}\ 5p^6\ 6s^2$	group IIA

6.22 Write the detailed electronic configuration of each of the following elements: Fe, F, Li, Na, K, Rb.

▮

Fe	$1s^2\ 2s^2\ 2p^6\ 3s^2\ 3p^6\ 4s^2\ 3d^6$
F	$1s^2\ 2s^2\ 2p^5$
Li	$1s^2\ 2s^1$
Na	$1s^2\ 2s^2\ 2p^6\ 3s^1$
K	$1s^2\ 2s^2\ 2p^6\ 3s^2\ 3p^6\ 4s^1$
Rb	$1s^2\ 2s^2\ 2p^6\ 3s^2\ 3p^6\ 4s^2\ 3d^{10}\ 4p^6\ 5s^1$

6.23 Several experimenters have attempted to synthesize superheavy elements by bombarding atoms of the actinoid series with heavy ions. Pending confirmation and acceptance of the results, some investigators in the early 1970s referred to elements 104 and 105 as *eka*-hafnium and *eka*-tantalum, respectively. Why were these names chosen?

▮ They are in the same periodic groups as hafnium and tantalum, respectively. Mendeleev had used the prefix *eka*- (Sanskrit word for first) to name elements whose existence he predicted, applying the prefix to a known element in the same periodic group as the predicted element.

6.24 Write the electronic configurations of S^{2-} and Ni^{2+}.

▮ The configurations of ions are derived from the configurations of their neutral atoms:

S^{2-}	$1s^2\ 2s^2\ 2p^6\ 3s^2\ 3p^6$	(from S $1s^2\ 2s^2\ 2p^6\ 3s^3\ 3p^4$)
Ni^{2+}	$1s^2\ 2s^2\ 2p^6\ 3s^2\ 3p^6\ 4s^0\ 3d^8$	(from Ni $1s^2\ 2s^2\ 2p^6\ 3s^2\ 3p^6\ 4s^2\ 3d^8$)

6.25 Write detailed electronic configurations for the following atoms and ions: Br^-, Ca, Fe^{2+}, P.

▮

Br^-	$1s^2\ 2s^2\ 2p^6\ 3s^2\ 3p^6\ 4s^2\ 3d^{10}\ 4p^6$
Ca	$1s^2\ 2s^2\ 2p^6\ 3s^2\ 3p^6\ 4s^2$
Fe^{2+}	$1s^2\ 2s^2\ 2p^6\ 3s^2\ 3p^6\ 4s^0\ 3d^6$
P	$1s^2\ 2s^2\ 2p^6\ 3s^2\ 3p^3$

6.26 Give the symbol for an atom (if any) and also for an ion (if any) whose ground state corresponds to each of the following electronic configurations:

(*a*) $1s^2\ 2s^2\ 2p^6\ 3s^2\ 3p^5$
(*b*) $1s^2\ 2s^2\ 2p^6\ 3s^2\ 3p^6\ 4s^2\ 3d^{10}\ 4p^6\ 4d^8$
(*c*) $1s^2\ 2s^2\ 2p^6\ 3s^2\ 3p^6\ 4s^2\ 3d^{10}\ 4p^6\ 5s^2\ 4d^{10}\ 5p^6\ 4f^7\ 6s^2$
(*d*) $1s^2\ 2s^2\ 2p^6\ 3s^2\ 3p^6\ 4s^2\ 3d^{10}\ 4p^6\ 5s^2\ 4d^{10}$
(*e*) $1s^2\ 2s^2\ 2p^6$

▮ (*a*) Cl (*b*) Pd^{2+} ion (Note that no 5*s* electrons are present, so no neutral atom is involved.) (*c*) Eu (The 6*s* electrons which are present preclude the possibility that any ion is involved.) (*d*) Cd, Sn^{2+}, or Sb^{3+} (*e*) Ne, F^-, O^{2-}, N^{3-}, Na^+, Mg^{2+}, Al^{3+}

6.27 Write the detailed electronic configuration for Cl^- and Ni^{2+}.

▮

Cl^- $1s^2\ 2s^2\ 2p^6\ 3s^2\ 3p^6$
Ni^{2+} $1s^2\ 2s^2\ 2p^6\ 3s^2\ 3p^6\ 4s^0\ 3d^8$

6.28 How many unpaired electrons are there in the Ni^{2+} ion?

▮ From Prob. 6.27: $3d^8$ $\underline{\uparrow\downarrow}\ \underline{\uparrow\downarrow}\ \underline{\uparrow\downarrow}\ \underline{\uparrow}\ \underline{\uparrow}$ (3*d*), so there are two unpaired electrons.

6.29 Write electronic structures for (*a*) Ar and S^{2-} (*b*) Fe and Ni^{2+} (*c*) Which pair is isoelectronic? Explain. (*d*) Are any ions of a transition element isoelectronic with free elements?

▮ (*a*) Both are $1s^2\ 2s^2\ 2p^6\ 3s^2\ 3p^6$ (*b*) Fe: $1s^2\ 2s^2\ 2p^6\ 3s^2\ 3p^6\ 4s^2\ 3d^6$; Ni^{2+}: $1s^2\ 2s^2\ 2p^6\ 3s^2\ 3p^6\ 3d^8$ (*c*) The first pair is isoelectronic. For the latter, the last electrons added are not the same as the ones removed on formation of the positive ion. (*d*) Only the exceptional Pd is isoelectronic with Ag^+ and Cd^{2+}.

6.30 Explain, using copper atoms and copper(II) ions as examples, why the electronic configurations of some ions are rather easily predictable despite the fact that the electronic configurations of the corresponding atoms do not obey the aufbau principle.

▮ The aufbau configuration for copper and the actual configuration for copper are as follows: $4s^2\ 3d^9$ and $4s^1\ 3d^{10}$. No matter which of these configurations is assumed to be the one on which the configuration of the ion is based, the removal of the 4*s* electron(s) before the 3*d* electrons results in the configuration for copper(II) ion: $4s^0\ 3d^9$.

6.31 Write the electronic configuration for each of the following ions: (*a*) Co^{3+} (*b*) Ni^{4+} (*c*) Zn^{2+}.

▮ (*a*) $[Ar]3d^6$ (*b*) $[Ar]3d^6$ (*c*) $[Ar]3d^{10}$

6.32 Write the electronic configurations of (*a*) Ti^{4+} (*b*) V^{3+}

▮ (*a*) $1s^2\ 2s^2\ 2p^6\ 3s^2\ 3p^6$ (*b*) $1s^2\ 2s^2\ 2p^6\ 3s^2\ 3p^6\ 3d^2$

6.33 Write the electronic configuration of each of the following atoms or ions: (*a*) Sc (*b*) Pd^{2+} (*c*) Os^{2+}

▮ (*a*) $1s^2\ 2s^2\ 2p^6\ 3s^2\ 3p^6\ 4s^2\ 3d^1$
(*b*) $1s^2\ 2s^2\ 2p^6\ 3s^2\ 3p^6\ 4s^2\ 3d^{10}\ 4p^6\ (5s^0)\ 4d^8$ ($5s^0$ circled: may be omitted)
(*c*) $1s^2\ 2s^2\ 2p^6\ 3s^2\ 3p^6\ 4s^2\ 3d^{10}\ 4p^6\ 5s^2\ 4d^{10}\ 5p^6\ (6s^0)\ 4f^{14}\ 5d^6$

6.34 Write the electronic configuration for each of the following atoms or ions. State the number of unpaired electrons contained in each. (*a*) C (*b*) Cu^{2+} (*c*) Zn (*d*) Eu (*e*) Gd^{3+} (*f*) Tl^+ (*g*) Fe^{3+} (*h*) U (*i*) Au

▮

(*a*)	$1s^2\ 2s^2\ 2p^2$	two
(*b*)	$1s^2\ 2s^2\ 2p^6\ 3s^2\ 3p^6\ 3d^9$	one
(*c*)	$1s^2\ 2s^2\ 2p^6\ 3s^2\ 3p^6\ 4s^2\ 3d^{10}$	zero
(*d*)	$1s^2\ 2s^2\ 2p^6\ 3s^2\ 3p^6\ 4s^2\ 3d^{10}\ 4p^6\ 5s^2\ 4d^{10}\ 5p^6\ 6s^2\ 4f^7$	seven
(*e*)	$1s^2\ 2s^2\ 2p^6\ 3s^2\ 3p^6\ 4s^2\ 3d^{10}\ 4p^6\ 5s^2\ 4d^{10}\ 5p^6\ 4f^7$	seven
(*f*)	$1s^2\ 2s^2\ 2p^6\ 3s^2\ 3p^6\ 4s^2\ 3d^{10}\ 4p^6\ 5s^2\ 4d^{10}\ 5p^6\ 4f^{14}\ 5d^{10}\ 6s^2$	zero
(*g*)	$1s^2\ 2s^2\ 2p^6\ 3s^2\ 3p^6\ 3d^5$	five
(*h*)	$[Rn]\ 7s^2\ 6d^1\ 5f^3$	four
(*i*)	$[Xe]\ 6s^1\ 5d^{10}\ 4f^{14}$	one

6.35 Write the electronic structure of each of the following ions: ***(a)*** Pb^{2+} ***(b)*** Tl^{+} ***(c)*** Sn^{2+} ***(d)*** Explain on the basis of these ions what is meant by the term *inert pair*. Is the pair truly inert, or merely low in reactivity?

▌ ***(a)*** [Xe] $6s^2\,5d^{10}\,4f^{14}$ ***(b)*** [Xe] $6s^2\,5d^{10}\,4f^{14}$ ***(c)*** [Kr] $5s^2\,4d^{10}$ ***(d)*** Each of these ions has retained the pair of electrons in its outermost *s* subshell—the inert pair. This pair of electrons is more difficult to remove than the *p* electrons, which have already been lost by these ions, but they are not totally unreactive. When lost, they form Pb^{4+}, Tl^{3+}, and Sn^{4+}, respectively.

6.36 Nickel has the electron configuration [Ar] $3d^8\,4s^2$. How do you account for the fact that the configuration of the next element, Cu, is [Ar] $3d^{10}\,4s$?

▌ In the hypothetical procedure of making up the electronic complement of Cu by adding 1 electron to the configuration of the preceding element, Ni, one might have expected only a ninth $3d$ electron in Cu. For atomic number 19, the $3d$ subshell is decidedly of higher energy than $4s$; thus potassium has a single $4s$ electron and no $3d$ electron. After the $3d$ subshell begins to fill, however, beginning with element 21 (after $4s$ has already been filled), the addition of each succeeding $3d$ electron is accompanied by a lowering of the average energy of the $3d$ level. This is because each succeeding element has an increased nuclear charge which is only partly screened from a test $3d$ electron by the additional electron in the same subshell. The energy of the $3d$ subshell decreases gradually as the subshell undergoes filling and drops below the level of the $4s$ toward the end of the transition series.

Another factor is that the configuration $3d^{10}\,4s^1$ represents one filled and one half-filled subshell, a stabilizing arrangement, while $3d^9\,4s^2$ would have 1 filled subshell, $4s$, but no specially favorable complement in the $3d$ to impart extra stabilization.

6.37 Write the full electronic configuration for ***(a)*** Mg^{2+} ***(b)*** V

▌ ***(a)*** Mg^{2+} $1s^2\,2s^2\,2p^6$ ***(b)*** V $1s^2\,2s^2\,2p^6\,3s^2\,3p^6\,4s^2\,3d^3$

6.38 What would you predict for the atomic number of the noble gas beyond Rn, if such an element had sufficient stability to be prepared or observed? Assume that *g* orbitals are still not occupied in the ground states of the preceding elements.

▌ The electronic configuration of the element would be [Rn] $7s^2\,5f^{14}\,6d^{10}\,7p^6$. Thus the element is 32 electrons beyond atomic number 86, or atomic number 118.

6.39 Explain why chromium has only one electron in its $4s$ subshell.

▌ The $3d^5\,4s^1$ configuration has two half-filled subshells, which lends a slightly greater stability than the (expected) $3d^4\,4s^2$ configuration, which has only one filled subshell.

6.40 ***(a)*** Write the electron configuration for the ground state of Pr^{3+}. ***(b)*** How many unpaired electrons would there be?

▌ ***(a)*** Pr^{3+} has 56 electrons, 3 fewer than the neutral Pr atom. Beyond Xe, the previous rare gas, the order of filling for the next period of elements is $6s^2$, then one $5d$ electron, then the whole $4f$ subshell, then the rest of the $5d$ subshell, then the $6p$ subshell. There are frequent replacements of the first $5d$ assignment with an additional $4f$, or of one of the $6s$ with an additional $5d$, but these irregularities are of no consequence in the assignment of electrons in Pr^{3+}. The 3 electrons removed from the neutral atom to form the ion follow the general rule of removal first from the outermost shell and then from the next-to-outermost; in this case the $6s$ electrons are removed first and no $5d$ electrons would remain even if there were one in the neutral atom.

$$1s^2\,2s^2\,2p^6\,3s^2\,3p^6\,3d^{10}\,4s^2\,4p^6\,4d^{10}\,4f^2\,5s^2\,5p^6 \quad \text{or} \quad [\text{Xe}]4f^2$$

(b) Pr^{3+} has 2 unpaired electrons, the two $4f$ electrons which singly occupy 2 of the 7 available $4f$ orbitals.

6.41 What are the electron configurations of Re^{3+} and Ho^{3+}? How many unpaired electron spins are in each of these ions?

▌ The electronic configurations of the transition metal ions and the inner transition metal ions are obtained from those of the neutral atoms by removal of the electrons with the highest principal quantum number(s) first.

Re^{3+}	$1s^2\,2s^2\,2p^6\,3s^2\,3p^6\,4s^2\,3d^{10}\,4p^6\,5s^2\,4d^{10}\,5p^6\,6s^0\,4f^{14}\,5d^4$	4 unpaired
Ho^{3+}	$1s^2\,2s^2\,2p^6\,3s^2\,3p^6\,4s^2\,3d^{10}\,4p^6\,5s^2\,4d^{10}\,5p^6\,6s^0\,4f^{10}\,5d^0$	4 unpaired

6.3 CONSEQUENCES OF ELECTRONIC STRUCTURE

6.42 Which properties of the elements depend on the electronic configuration of the atoms and which do not?

Chemical and physical properties depend on electronic configuration; nuclear properties do not.

6.43 Explain in terms of electronic configuration why the halogens have similar chemical properties. Why do they not have identical properties?

Their properties are similar because of similar outermost electronic configurations ($ns^2\ np^5$). The differences stem from their differences in size and the magnitude of electronic shielding available for the outermost electrons.

6.44 Account for the great chemical similarity of the lanthanoid elements (atomic numbers 57 to 71).

Their outermost two shells of electrons are alike—$6s^2\ 5d^1$—or that configuration is very close to the ground state in stability. In any event, all the lanthanoids act as if they had this configuration to a great extent.

6.45 (*a*) Select the largest species in each group: Ti^{2+}, Ti^{3+}, Ti; F^-, Ne, Na^+
(*b*) Select the species with the largest ionization potential in each group: Na, K, Rb; F, Ne, Na
(*c*) Explain why Pd and Pt have such similar sizes.

(*a*) Ti and F^- (*b*) Na and Ne (*c*) Because of the lanthanoid contraction. (The increase in size in going from one period to the next is negated by the 14 extra decreases in size corresponding to the 14 "extra" lanthanoid elements.)

6.46 Which ion has the smallest radius: Li^+ Na^+ K^+ Be^{2+} Mg^{2+}?

Be^{2+}. Be is a second period element, thus smaller than higher period elements. The Be^{2+} ion is isoelectronic with Li^+ but has a greater nuclear charge.

6.47 Which ion or atom has the largest radius: Mg Na Na^+ Mg^{2+} Al?

Na. The size decreases as one goes to the right in the periodic table, and also as one removes electrons.

6.48 From the following list, choose the element that has the lowest ionization potential and the element that has the highest ionization potential: K Ca Se Br Kr

K has the smallest ionization potential; Kr has the largest. (Ionization potential increases across a period, being very small for the alkali metals and very large for the noble gases.)

6.49 Select from each of the following groups the one which has the largest radius: (*a*) Co, Co^{2+}, Co^{3+} (*b*) S^{2-}, Ar, K^+ (*c*) Li, Na, Rb (*d*) C, N, O (*e*) Ne, Na, Mg (*f*) La, Lu (*g*) Cu, Ag, Au (*h*) Ba, Hf

(*a*) Co (The others have the same nuclear charge, but fewer electrons.) (*b*) S^{2-} (All have the same electronic configuration, but the sulfur nucleus has the smallest positive charge.) (*c*) Rb (It is in the largest period.) (*d*) C (It is the farthest left in the periodic table.) (*e*) Na (It is the first element of a new period.) (*f*) La (Lu is smaller because of the lanthanoid contraction.) (*g*) Au (Ag is almost the same size, because of the effect of the lanthanoid contraction on Au.) (*h*) Ba (Hf is much smaller because of the effect of the lanthanoid contraction.)

6.50 The ionic radii of S^{2-} and Te^{2-} are 1.84 and 2.21 Å, respectively. What would you predict for the ionic radius of Se^{2-} and for P^{3-}?

One can estimate the ionic radius of Se^{2-} as the average of those of S^{2-} and Te^{2-}; that is, 2.02 Å. The actual value is 1.98 Å.

One should expect the radius of P^{3-} to be greater than that of S^{2-}, since the two ions are isoelectronic but phosphorus has one fewer proton. Its actual radius is 2.12 Å.

6.51 In the ionic compound KF, the K^+ and F^- ions are found to have practically identical radii, about 1.34 Å each. What do you predict about the relative atomic radii of K and F?

The atomic radius of K should be much greater than 1.34 Å, and that of F much smaller, since atomic cations are smaller than their parent atoms, while atomic anions are bigger than their parents. The observed atomic radii of K and F are 2.31 and 0.64 Å, respectively.

6.52 The single covalent radius of P is 0.11 nm. What do you predict for the single covalent radius of Cl?

▮ P and Cl are members of the same period. Cl should have a smaller radius in keeping with the usual trend across a period. The experimental value is 0.10 nm.

6.53 From the data for the F atom and the trend shown in the radii for F^-, Ne, and Na^+, estimate the radius of Mg^{2+}. Estimate a reasonable radius for Na^{2+}.

▮ One can plot the sizes of the isoelectronic species F^-, Ne, and Na^+ and extrapolate to determine the estimated size of Mg^{2+}. To determine a reasonable size for the hypothetical Na^{2+}, one can take the same fraction of the size of Na^+ as the fraction of the sizes F/F^-. Mg^{2+} is estimated to be about 0.75 Å, while Na^{2+} would be about 0.45 Å on this basis.

6.54 Ionization potential is an old term for ionization energy. Explain why the two are synonymous.

▮ The ionization energy of an electron in electron volts is numerically equal to the ionization potential in volts. (See introduction to Sec. 5.1.)

6.55 The first ionization energy of C is 11.2 eV. Would you expect the first ionization energy of Si to be greater or less than this amount?

▮ Less. Ionization energy falls as one progresses down a group (largely due to increasing atomic size, but other factors also play a significant part). The observed value is 8.1 eV.

6.56 The first ionization energies of Al, Si, and S are 6.0, 8.1, and 10.3 eV, respectively. What do you predict for the first ionization energy of P?

▮ An electron from phosphorus should be slightly harder to ionize than one from sulfur because of the added stability of the half-filled *p* subshell. In general, atoms to the right have higher ionization energies than their left neighbors. You should predict between 10.3 and 11.0 eV. The observed value is 10.9 eV.

6.57 The ionization energies of Li and K are 5.4 and 4.3 eV, respectively. What do you predict for the ionization energy of Na?

▮ The IE of Na should be intermediate between that of Li and K. The IE of Na should be close to the arithmetic average of the two, or 4.9 eV. (The observed value is 5.1 eV.)

6.58 The ionization energies of Li, Be, and C are 5.4, 9.3, and 11.3 eV. What do you predict for the ionization energies of B and N?

▮ There is a general increasing trend of IE with increasing atomic number in a given period. Note, however, a larger increase in going from Li $(Z = 3)$ to Be $(Z = 4)$ than in going from Be to C $(Z = 6)$. The filling of the 2*s* subshell gives Be a greater stability than would be suggested by a smooth progression across the periodic table. The next element, B $(Z = 5)$, would have an IE that represents a balancing of the two oppositely directed factors, an increase with respect to Be because of the increased *Z* and a decrease because a new subshell is beginning to be filled in the case of B. One might guess that the IE for B might actually be less than for Be, and this turns out to be the case. The observed IE for B is 8.3 eV.

Nitrogen has a half-filled 2*p* subshell and should have extra stability for this reason. The increase in going from $Z = 5$ to $Z = 6$ is 3.0 eV, and the additional increase should be greater than this; thus the IE for N should exceed 14.3 eV. The experimental value is 14.5 eV.

6.59 Which of these elements is expected to have the lowest first ionization potential: Sr As Xe S F?

▮ Sr. It is a metal.

6.60 Explain why for sodium the second ionization potential is so much larger than the first, whereas for magnesium the magnitude of the difference between the first and second ionization potentials is much less.

▮ The large differences in successive ionization potentials occur at the points when electrons are first removed from completed *s* plus *p* subshells.

6.61 Without consulting any table, select from each of the following groups the element which has the largest ionization potential: **(*a*)** Na, P, Cl **(*b*)** He, Ne, Ar **(*c*)** O, F, Ne

▮ **(*a*)** Cl (Cl is farthest to the right of the three elements, all of the third period.) **(*b*)** He **(*c*)** Ne

6.62 Arrange the species in each group in order of increasing ionization potential, and in each case explain the reason for the sequence: **(*a*)** K^+, Ar, Cl^- **(*b*)** Fe, Fe^{2+}, Fe^{3+} **(*c*)** Na, Mg, Al **(*d*)** K, Ca, Sc **(*e*)** N, O, F **(*f*)** C, N, O **(*g*)** Cu, Ag, Au **(*h*)** Be, B, C **(*i*)** K, Rb, Cs

▌ **(*a*)** $Cl^- < Ar < K^+$ (All have the same electronic configuration, and the nuclear charge increases in the order listed.) **(*b*)** $Fe < Fe^{2+} < Fe^{3+}$ (All have the same nuclear charge, and the number of electrons decreases in the order listed.) **(*c*)** Na < Al < Mg (It is easiest to remove an electron from Na because of its large size. It is more difficult to remove an electron from Mg because the electron being removed is a 3*s* electron from a filled subshell. The 3*p* electron of Al is more easily removed.) **(*d*)** K < Ca < Sc (In each of these cases, a 4*s* electron is being removed, and the order is that shown.) **(*e*)** O < N < F (The half-filled *p* subshell of N imparts enough extra stability to make its ionization potential greater than that of the next element—O.) **(*f*)** C < O < N (Again the stability of the half-filled subshell is evident.) **(*g*)** Ag < Cu < Au (The lanthanoid contraction makes Ag and Au about equal in size, but Au has a much greater nuclear charge.) **(*h*)** B < Be < C (The 2*s* electron is more difficult to remove than the 2*p*.) **(*i*)** Cs < Rb < K (The trend is to smaller ionization potentials for elements in the same periodic group as one goes to heavier elements.)

6.63 Explain why the first ionization potential for copper is higher than that for potassium, whereas the second ionization potentials are in the reverse order.

▌ Copper has 10 more protons and 10 more electrons than does K, but the electrons do not screen the nucleus perfectly. Hence the first ionization potential of Cu is higher. The second ionization potential for K involves the loss of an electron from an octet—an inner, complete *s* plus *p* subshell arrangement, whereas that for Cu involves the more easily ionized d^{10} configuration.

6.64 Compare the first ionization potentials of sodium, magnesium, and aluminum with those of potassium, calcium, and scandium. Explain the difference in trend of ionization potential in the two neighboring series.

▌ See Prob. 6.62(*c*) and (*d*).

6.65 Account for the difference in ionization potential **(*a*)** between Cu and K **(*b*)** between K^+ and Ca^+ **(*c*)** between Cu^+ and Zn^+

▌ **(*a*)** Cu has 10 additional protons, and also 10 additional electrons which screen incompletely. **(*b*)** K^+ loses an electron from its $3p^6$ subshell; Ca^+ from its $4s^1$ subshell. The latter takes much less energy. **(*c*)** Zn^+ loses an electron from $4s^1$ more easily than Cu^+ loses an electron from $3d^{10}$.

6.66 The frequency of a line in the spectrum of an element enables one to determine which of the following? **(*a*)** One of the energy levels for electrons in the atoms. **(*b*)** The ionization potential of the atom. **(*c*)** The principal quantum number of the atom. **(*d*)** The difference between two electronic energy levels of the atom. **(*e*)** The difference between the ionization potential and the electron affinity of the atom.

▌ **(*d*)** (The change in energy level in the atom is associated with the emitted light.)

6.67 Determine the *total* ionization energy for the first three electrons of aluminum and for the first two electrons of sodium. Prove on the basis of these data that ionization energy is not the sole factor governing the chemical stability of aluminum(III).

▌ For Al: 53.03 eV; for Na: 52.20 eV. Since there is a stable Al(III) oxidation state but no stable Na(II) oxidation state, despite the greater energy required to remove three electrons from Al than two from Na, there must obviously be other factors involved.

6.68 **(*a*)** What differences are there between the ionization potential of an element and the oxidation potential of the element? **(*b*)** Determine the value (including the sign) of the electron affinity of Na^+.

▌ **(*a*)** Ionization potential refers to the loss of an electron from a gaseous atom to form a gaseous ion; it is not a common chemical process. Oxidation potential is the loss of one electron or more from an element in its normal state at 25 °C to an ion in its normal state. Ionization potential, despite its name, is an energy term, whereas oxidation potential is a potential. The former can occur without any reduction reaction occurring concurrently, while the latter cannot. **(*b*)** The electron affinity of a positive ion is merely the energy of addition of an electron to that ion; therefore it is the opposite of the ionization potential of the species with one lower positive charge. In this case, the reaction is

$$Na^+(g) + e^- \longrightarrow Na(g)$$

This reaction is the opposite of that corresponding to the ionization potential of sodium, and therefore has a value −5.1 eV.

6.69 PtF_6 is a powerful oxidizing agent, capable of effecting the following reaction:

$$O_2(g) + PtF_6(s) \longrightarrow O_2PtF_6(s)$$

Compare the first ionization energy of O_2 (12.5 eV) with those of the noble gases (Table 6.1). Which, if any, of the noble gases might undergo a comparable reaction with PtF_6?

▍ Xe has an ionization energy comparable to O_2. The "inert gas" configuration was thought to prevent reaction of the noble gases before the experiment was actually performed by Bartlett.

TABLE 6.1 Ionization Energies of Noble Gases

gas	helium	neon	argon	krypton	xenon	radon
IE, eV	24.46	21.47	15.68	13.93	12.08	10.70

6.70 (*a*) Explain in words the meaning of a negative value of electron affinity. Does any element have a negative ionization potential? (*b*) Write equations to demonstrate the difference between electron affinity and the *reverse* of ionization potential.

▍ (*a*) Negative electron affinity means that energy is *required* rather than liberated during the addition of an electron to a gaseous atom to form a gaseous ion. No element has a negative ionization potential—that is, all elements require energy to remove an electron from a gaseous atom to form a gaseous ion.

(*b*) Electron affinity: $A + e^- \longrightarrow A^-$

Reverse of ionization potential: $A^+ + e^- \longrightarrow A$

6.71 Explain each of the following observations: (*a*) The radius of Cd^{2+} is less than that of Sr^{2+}. (*b*) It is easy to separate V from Nb in a mixture, but difficult to separate Nb from Ta in a mixture. (*c*) Eu can be separated from the other lanthanoid elements much more easily than Gd can. (*d*) Sc and Y have higher ionization potentials than do Ga and Tl. (*e*) The ionization potentials increase in the series Ru, Rh, Pd but decrease in the series Fe, Co, Ni. (*f*) The members of the second period of the periodic table are not typical representatives of their respective groups. (*g*) The difference in size between Hf and Ba is much greater than that between Zr and Sr. (*h*) The electronic configurations of the positive ions of transition elements are more regular than those of the neutral atoms.

▍ (*a*) Cd^{2+} has a nuclear charge 10 units greater than that of Sr^{2+}, and its 10 extra electrons are not 100% effective in screening the nucleus. (*b*) Because of the lanthanoid contraction, Nb and Ta have almost exactly the same size (which is one of the main factors in such properties as solubility). (*c*) Eu has a +2 oxidation state (stabilized by its f^7 half-filled subshell); most of the other lanthanoid elements do not. (A difference in oxidation state makes separation rather easy.) (*d*) Ga and Tl are losing *p* electrons; Sc and Y are losing *s* electrons from a filled subshell. (*e*) The Pd atom, with its $4d^{10}\,5s^0$ configuration, involves ionization of an electron from a fully filled subshell. Ru and Rh are virtually equal in ionization potential. (*f*) The second group elements have no low-energy *d* orbitals to interact with orbitals of other atoms, and they are rather small. (*g*) Hf and Ba differ in atomic number by 16 units; Zr and Sr differ by only 2 units. Thus the lanthanoid contraction causes a greater reduction in size between Ba and Hf. (*h*) See Prob. 6.30.

6.72 Explain why the +2 oxidation states of tin and lead are more stable than those of carbon and silicon. Explain why that of lead is more stable than that of tin.

▍ The $5s^2$ electrons of Sn and the $6s^2$ electrons of Pb are relatively stable—they are called the "inert pair." The outer *s* electrons of the lighter group IV elements do not share this stability.

6.73 Explain why so few transition elements have +1 oxidation states, and list those which do.

▍ Only a few transition elements have configurations including ns^1 outermost subshells, and of these, only the IB elements form compounds with +1 oxidation states.

6.74 All the lanthanoid elements form stable compounds containing the 3+ cation. Of the few other ionic forms known, Ce forms the stablest 4+ series of ionic compounds and Eu the stablest 2+ series. Account for these unusual ionic forms in terms of their electronic configurations.

▍ Ce^{4+} has the stable electronic configuration of the rare gas Xe. Eu^{2+}, with 61 electrons, could have the configuration [Xe] $4f^7$, with the added stability of a half-filled $4f$ subshell.

6.75 Predict the magnetic moment for Co^{3+}.

▌ Co^{3+}, a $3d^6$ ion, has 4 unpaired electrons: ↑↓ ↑ ↑ ↑ ↑

$\mu = \sqrt{n(n+2)} = \sqrt{4(4+2)} = 4.9$ B.M. (The value of the Bohr magneton is 9.27×10^{-24} J/T.)

6.76 Predict the magnetic moment of the Cu^{2+} ion.

▌ The electronic configuration of Cu^{2+} is [Ar] $3d^9$. The $3d$ orbitals are occupied as represented in the following diagram, where each electron is represented by an arrow pointing either up or down to denote its spin.

Cu^{2+} ↑↓ ↑↓ ↑↓ ↑↓ ↑ (3d) __ (4s)

Hence Cu^{2+} has one unpaired electron, and its moment is given by

$$\mu = \sqrt{1(1+2)} = 1.73 \text{ B.M.}$$

6.77 Give the electronic configuration, the number of unpaired electrons, and the type of magnetic behavior for (*a*) Co^{3+} (*b*) Se^{2-} (*c*) Gd (*d*) Ni (*e*) Gd^{3+}

▌

	Configuration	Unpaired	Behavior
(*a*)	[Ar] $3d^6$	4	paramagnetic
(*b*)	[Ar] $3d^{10}\,4s^2\,4p^6$	0	diamagnetic
(*c*)	[Xe] $6s^2\,5d^1\,4f^7$	8	paramagnetic
(*d*)	[Ar] $4s^2\,3d^8$	2	paramagnetic
(*e*)	[Xe] $4f^7$	7	paramagnetic

6.78 Which of the following substances would be drawn most strongly into a magnetic field: $TiCl_4$ VCl_3 $FeCl_2$? Briefly state why.

▌ $FeCl_2$, since it has the most unpaired electrons.

6.79 Explain how the magnetic properties of $CsAuCl_3$ might be used to determine whether it contained gold(II) or an equal number of moles of gold(I) and gold(III).

▌ $CsAuCl_3$ is diamagnetic. It could not consist of simple gold(II) ions, because gold(II) has an odd number of electrons. (It is conceivable that there could be Au—Au bonds.) The actual structure consists of an equal number of $[AuCl_2]^-$ ions and $[AuCl_4]^-$ ions, each balanced by a Cs^+ ion. Thus there are equal numbers of gold(I) and gold(III) atoms in the compound.

6.80 Complete the following statements:

(*a*) Two electrons in the same ______ must have opposite spins.
(*b*) The presence of unpaired electrons in an atom gives rise to ______.
(*c*) When $l = 3$, m_l may have values from ______ to ______.
(*d*) The neutral fourth-period atom having a total of six *d* electrons is ______.
(*e*) The ionization potentials of elements in the same periodic group ______ as the atomic number increases.
(*f*) Orbitals with the same energy are said to be ______.
(*g*) Ca^{2+} has a smaller radius than K^+ because it has ______.
(*h*) The electronic configuration of Sn is [Kr] ______.
(*i*) The $2p$ orbitals of an atom have identical shapes but differ in their ______.
(*j*) A nodal surface is one at which the probability of finding an electrons is ______.

▌ (*a*) orbital (*b*) paramagnetism (*c*) -3 to $+3$ (*d*) Fe (*e*) decrease (*f*) degenerate (*g*) a higher nuclear charge (with the same number of electrons) (*h*) $5s^2\,4d^{10}\,5p^2$ (*i*) orientations in space (*j*) zero.

6.81 The first ionization energy for Li is 5.4 eV and the electron affinity of Cl is 3.61 eV. Compute ΔH for the reaction

$$Li(g) + Cl(g) \longrightarrow Li^+ + Cl^-$$

carried out at such low pressures that the resulting ions do not combine with each other.

▌ The overall reaction may be decomposed into two partial reactions

$$(1)\quad Li(g) \longrightarrow Li^+ + e^- \qquad \Delta E = N_A(\text{IE})$$

$$(2)\quad Cl(g) + e^- \longrightarrow Cl^- \qquad \Delta E = N_A(-\text{EA})$$

where e^- stands for an electron. Note that for the addition of an electron to an atom having a positive electron affinity, the reaction is exothermic. Although ΔH for each of the above partial reactions differs slightly from ΔE (by the term $P\,\Delta V$), the ΔH of the overall reaction is the sum of the two ΔE's (the overall volume change is zero).

$$\Delta H(\text{total reaction}) = \Delta E(1) + \Delta E(2) = N_A(\text{IE} - \text{EA})$$
$$= (6.02 \times 10^{23})(1.8\ \text{eV})(1.60 \times 10^{-19}\ \text{J/eV}) = 170\ \text{kJ}$$

6.82 For the gaseous reaction $K + F \rightarrow K^+ + F^-$, ΔH was calculated to be 19 kcal under conditions where the cations and anions were prevented by electrostatic separation from combining with each other. The ionization energy of K is 4.3 eV. What is the electron affinity of F?

▮

$$K + F \xrightarrow{19\ \text{kcal}} F^-(g) + K^+(g)$$

(cycle: $-\text{EA}$ from F to $F^-(g)$; IE = 4.3 eV from K to $K^+(g)$)

$$\Delta H = \left(\frac{19\ \text{kcal}}{\text{mol}}\right)\left(\frac{4.18 \times 10^3\ \text{J}}{\text{kcal}}\right)\left(\frac{1\ \text{eV}}{1.60 \times 10^{-19}\ \text{J}}\right)\left(\frac{1\ \text{mol}}{6.02 \times 10^{23}\ \text{atoms}}\right) = 0.82\ \text{eV}$$

$$4.3\ \text{eV} - \text{EA} = 0.82\ \text{eV}$$
$$\text{EA} = 3.5\ \text{eV}$$

Another method:

$$K(g) + F(g) \longrightarrow F^-(g) + K^+(g) \qquad \Delta H = 19\ \text{kcal} = 0.82\ \text{eV}$$
$$K(g) \longrightarrow K^+(g) + e^- \qquad \text{IE} = 4.3\ \text{eV}$$
$$F(g) + e^- \longrightarrow F^-(g) \qquad \text{EA} = \Delta H - \text{IE} = 4.3 - 0.82 = 3.5\ \text{eV}$$

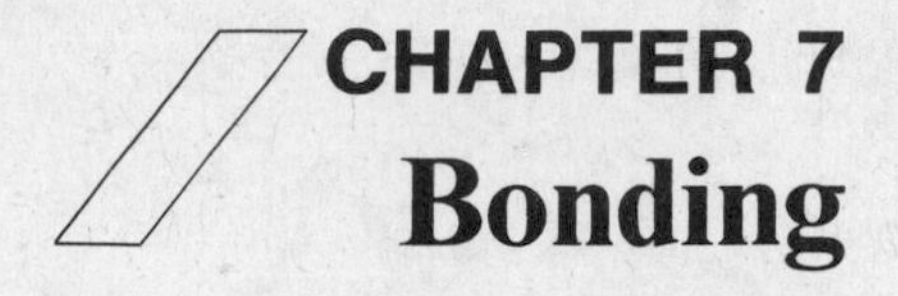

CHAPTER 7
Bonding

7.1 BOND LENGTHS AND BOND ENERGIES

Note: Before working the problems of this chapter, you should familiarize yourself with Table 7.1, below, and Table 7.2 on p. 101.

(The problems at the end of this section require prior study of enthalpy change.)

TABLE 7.1 Covalent Bond Radii, Å

single bond radii						multiple bond radii	
H	0.28	P	1.10	Te	1.37	C=	0.67
C	0.77	As	1.21	F	0.64	C≡	0.61
Si	1.17	Sb	1.41	Cl	0.99	N=	0.63
Ge	1.22	O	0.66	Br	1.14	N≡	0.55
Sn	1.40	S	1.04	I	1.33		
N	0.70	Se	1.17				

7.1 Using Table 7.1, calculate the bond lengths in the molecules H_2O and H_2. Compare the results with the observed values, 0.94 Å and 0.75 Å, respectively, and explain any discrepancy.

▌ The sum of the H— and O— radii is 0.94 Å. The actual bond distance in H_2O is 0.94 Å. The sum of the radii for two hydrogen atoms is 0.56 Å, but the observed bond distance is 0.75 Å. The discrepancy in the latter case is due to the experimental covalent radius of hydrogen being determined using molecules other than H_2. H_2 is unique; there is no inner electron cloud. In other covalent molecules the proton "buries" itself in the surrounding electron cloud, causing a somewhat smaller effective radius.

7.2 Calculate the bond lengths in ***(a)*** NH_3 ***(b)*** SCl_2 ***(c)*** CH_2Cl_2 ***(d)*** HOCl ***(e)*** H_3PO_4 ***(f)*** HCN ***(g)*** CH_3NH_2

▌ Covalent bond radii from Table 7.1 are merely added. Be sure that data for the correct bond orders are used.

(a) 0.70 Å + 0.28 Å = 0.98 Å
(b) 1.04 Å + 0.99 Å = 2.03 Å
(c) C—H 0.77 Å + 0.28 Å = 1.05 Å C—Cl 0.77 Å + 0.99 Å = 1.76 Å
(d) H—O 0.94 Å, O—Cl 1.65 Å
(e) H—O 0.94 Å, O—P 1.76 Å
(f) H—C 1.05 Å, C≡N 0.61 Å + 0.55 Å = 1.16 Å
(g) C—H 1.05 Å, C—N 1.47 Å, N—H 0.98 Å

7.3 Arrange C—C, C=C, and C≡C in order of ***(a)*** increasing bond energy ***(b)*** increasing bond length

▌ ***(a)*** C—C < C=C < C≡C ***(b)*** C≡C < C=C < C—C

7.4 The As—Cl bond distance in $AsCl_3$ is 2.20 Å. Estimate the single-bond covalent radius of As.

▌

$$\text{Internuclear distance} - \text{radius of chlorine atom} = \text{radius of As atom}$$
$$2.20\ \text{Å} - 0.99\ \text{Å} = 1.21\ \text{Å}$$

7.5 The platinum-chlorine distance has been found to be 2.32 Å in several crystalline compounds. If this value applies to both of the compounds shown in Fig. 7.1, what is the chlorine-chlorine distance in ***(a)*** structure *(a)*, ***(b)*** structure *(b)*?

▌ ***(a)*** 2(2.32 Å) = 4.64 Å ***(b)*** Using the Pythagorean theorem, $\sqrt{2}(2.32\ \text{Å}) = 3.28\ \text{Å}$

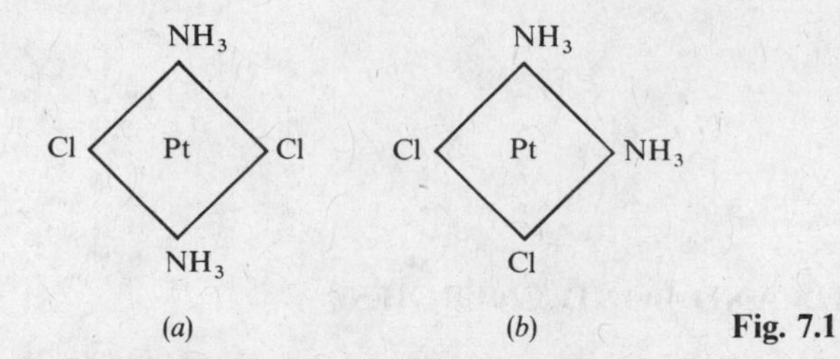

Fig. 7.1

7.6 Two substances having the same molecular formula, C_4H_8O, were examined in the gaseous state by electron diffraction. The carbon-oxygen distance was found to be 1.43 Å in compound A and 1.24 Å in compound B. What can you conclude about the structures of these two compounds?

▮ In compound A, the carbon-oxygen distance is the sum of the single-bond covalent radii of carbon and oxygen,

$$0.77\ \text{Å} + 0.66\ \text{Å} = 1.43\ \text{Å}$$

The oxygen must therefore be nonterminal. One such structure, the actual one selected for this experiment, is the heterocyclic compound tetrahydrofuran [Fig. 7.2(*a*)].

In compound B, the carbon-oxygen distance is close to that predicted for a double bond, 1.22 Å. The oxygen must therefore be terminal. One such structure, the actual one selected for this experiment, is butanone [Fig.7.2(*b*)].

$H_2C—CH_2$ (ring: H_2C, CH_2, O) $\qquad CH_3CCH_2CH_3$ (with $=O$ on the second carbon)

(*a*) (*b*) Fig. 7.2

7.7 **(*a*)** Using data from Table 7.1, calculate the molecular lengths of C_2H_2 and HCN. **(*b*)** From the observed oxygen-to-oxygen distance in the CO_2 molecule of 2.323 Å and data from Table 7.1, estimate the covalent radius of double-bonded oxygen atoms.

▮ **(*a*)** H—H≡C—H The length of the molecule is the sum of two hydrogen atom radii, two carbon single-bond radii, and two carbon triple-bond radii:

$$2(0.28\ \text{Å}) + 2(0.77\ \text{Å}) + 2(0.61\ \text{Å}) = 3.32\ \text{Å}$$

(*b*) The total distance is equal to the two oxygen double-bond radii plus two carbon double-bond radii.

$$2x + 2(0.67\ \text{Å}) = 2.323\ \text{Å}$$

$$x = 0.49\ \text{Å}$$

7.8 The C—C single-bond distance is 1.54 Å. What is the distance between the terminal carbons in propane, C_3H_8? Assume that the four bonds of any carbon atom are pointed toward the corners of a regular tetrahedron.

▮ With reference to Fig. 7.3, two terminal carbons can be thought of as lying at *A* and *B*, while the central carbon is at *O*. Then

$$\overline{AB} = 2\overline{AP} = 2\left(\overline{AO}\sin\frac{\theta}{2}\right) = 2(1.54\ \text{Å})(\sin 54°44') = 2.51\ \text{Å}$$

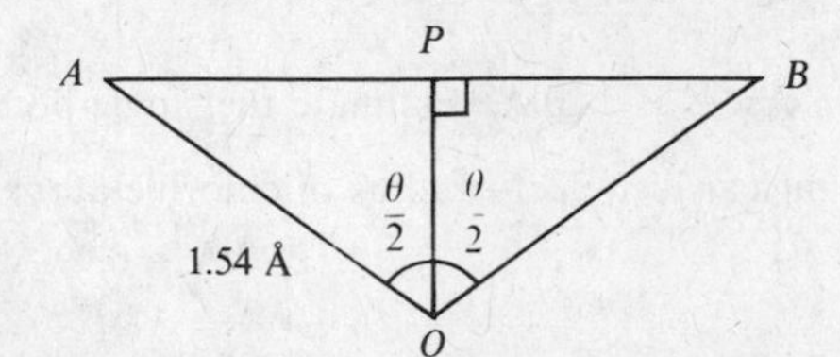

Fig. 7.3

7.9 What is the length of a polymer molecule containing 4001 carbon atoms singly bonded in a line, if the molecule could be stretched to its maximum length consistent with maintenance of the normal tetrahedral angle within any C—C—C group?

▮ With reference to Fig. 7.3,

$$d = \overline{AP} = 1.54 \sin(54.75^\circ) = 1.26 \text{ Å}$$

4001 atoms are separated by 4000 distances, $\overline{AP}$.

$$\text{Total distance} = 4000d = 5040 \text{ Å}$$

7.10 A plant virus was examined by the electron microscope and was found to consist of uniform cylindrical particles 150 Å in diameter and 5000 Å long. The virus has a specific volume of 0.75 cm³/g. If the virus particle is considered to be one molecule, what is its molecular weight?

▮

$$V = \pi r^2 h = (3.14)\left(\frac{150 \text{ Å}}{2}\right)^2 (5000 \text{ Å}) = 8.83 \times 10^7 \text{ Å}^3 = (8.83 \times 10^7 \text{ Å}^3)\left(\frac{10^{-8} \text{ cm}}{\text{Å}}\right)^3 = 8.83 \times 10^{-17} \text{ cm}^3$$

$$(8.83 \times 10^{-17} \text{ cm}^3)\left(\frac{1 \text{ g}}{0.75 \text{ cm}^3}\right) = 1.18 \times 10^{-16} \text{ g/molecule}$$

$$(1.18 \times 10^{-16} \text{ g/molecule})\left(\frac{6.02 \times 10^{23} \text{ molecules}}{\text{mol}}\right) = 7.10 \times 10^7 \text{ g/mol}$$

7.11 Compute the iodine-to-iodine distance in each of the isomeric compounds $C_2H_2I_2$. Use Table 7.1 for bond distances.

▮ The starting point of the solution is a set of diagrams, Fig. 7.4, based on the 120° angle in double-bond compounds.

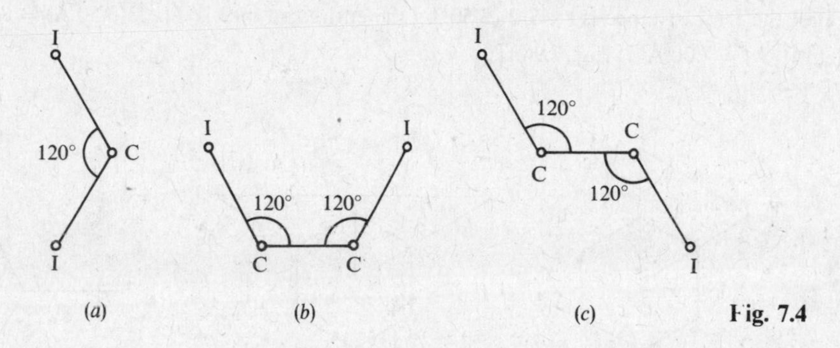

Fig. 7.4

(a) From Table 7.1, we find the C—I bond length to be 2.10 Å. This is the hypotenuse AB of a 60° right triangle, ABD in Fig. 7.5, whose side AD is half the nonbonded iodine-iodine distance AC. Then $\overline{AC} = 2\overline{AD} = 2(2.10 \sin 60^\circ) = 3.64$ Å.

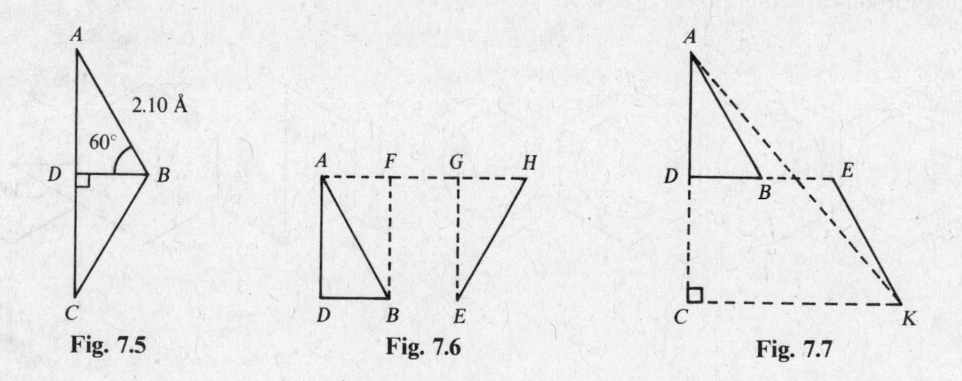

Fig. 7.5 **Fig. 7.6** **Fig. 7.7**

(b) This is related to the preceding case. The iodine-iodine distance (Fig. 7.6) is

$$\overline{AH} = \overline{AF} + \overline{FG} + \overline{GH}$$

Now, $\overline{AF} = \overline{GH} = \overline{BD} = 2.10 \cos 60^\circ = 1.05$ Å; $\overline{FG}$ is the C=C double bond distance, 1.33 Å. Thus $\overline{AH} = 1.05 + 1.33 + 1.05 = 3.43$ Å.

(c) This can be solved by reference to parts (*a*) and (*b*). The iodine-iodine distance is AK, the hypotenuse of right triangle ACK. The legs of this triangle ACK (Fig. 7.7) are AC and CK. AC was evaluated in (*a*) to be 3.64 Å. $\overline{CK}$ equals

$\overline{AH}$, evaluated in (*b*) to be 3.43 Å. Then

$$\overline{AK} = \sqrt{(3.64)^2 + (3.43)^2} = 5.00 \text{ Å}$$

7.12 Assuming the additivity of covalent radii in the C—I bond, what would be the iodine-iodine distance in each of the three diiodobenzenes (Fig. 7.8)? Assume that the ring is a regular hexagon and that each C—I bond lies on a line through the center of the hexagon. The distance between adjacent carbons is 1.40 Å.

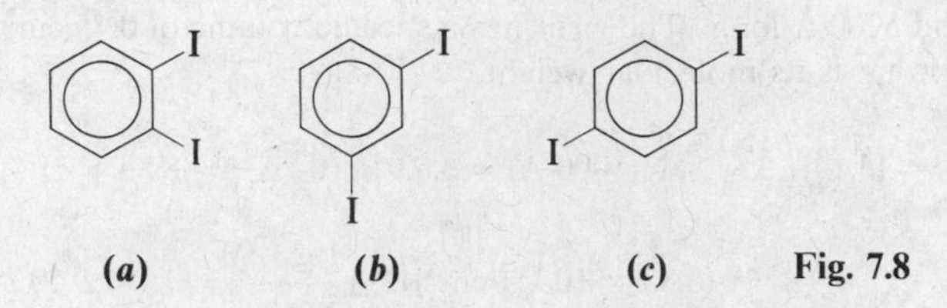

Fig. 7.8

▮ The distance from the center of the hexagon to any corner (apex) is 1.40 Å. Since the hexagon is regular, the lines from the center to two adjacent apices form an equilateral triangle.

(*a*) The distance from the center of the hexagon to an iodine atom is (1.40 Å) + (0.77 Å) + (1.33 Å) = 3.50 Å. Since this distance is also the side of an equilateral triangle, the distance from one adjacent iodine to another is also 3.50 Å [Fig. 7.9(*a*)].

(*b*) Half the I—I distance is ($\sqrt{3}/2$)(3.50 Å); the entire distance is ($\sqrt{3}$(3.50 Å) = 6.06 Å [Fig. 7.9(*b*)].

(*c*) 2(3.50 Å) = 7.00 Å [Fig. 7.9(c)].

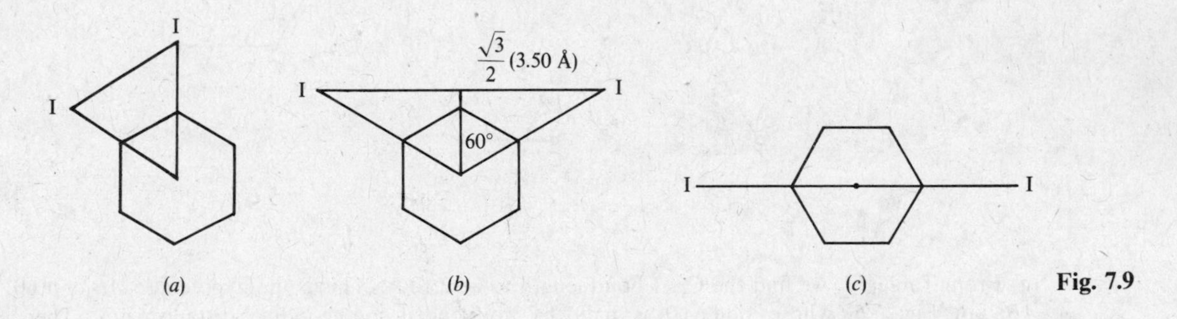

Fig. 7.9

7.13 Estimate the length and width of the carbon skeleton in naphthacene [Fig. 7.10(*a*)]. Assume hexagonal rings with equal carbon-carbon distances of 1.40 Å.

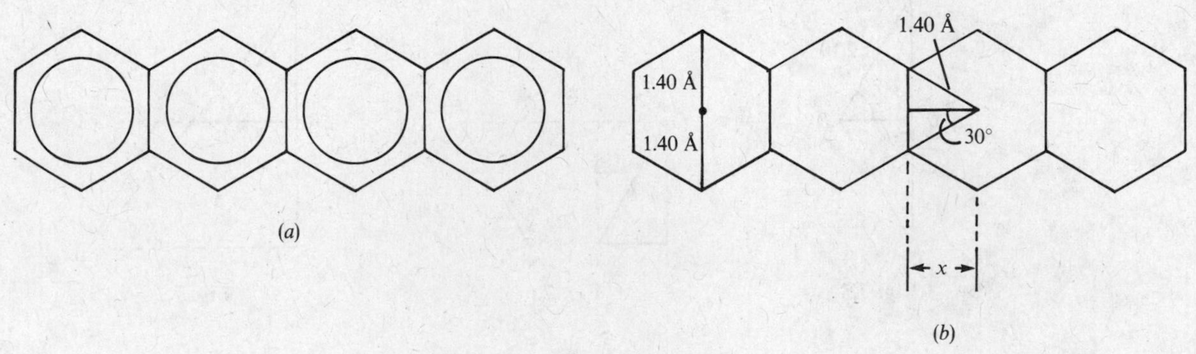

Fig. 7.10

▮ The width of the skeleton is merely the sum of two center-apex distances. The length is equal to 8 times the length shown as x in Fig. 7.10(*b*). That length is

$$x = 1.40(\sqrt{3}/2) = 1.40(0.866) = 1.21 \text{ Å}$$
$$8x = 9.68 \text{ Å}$$

7.14 There are two structural isomers of $C_2H_4I_2$. **(*a*)** In the isomer in which both iodine atoms are attached to the same carbon, find the iodine-iodine distance. **(*b*)** In the other isomer find the minimum and maximum distances between the two I atoms as one CH_2I group rotates about the C—C bond as an axis. Assume tetrahedral angles and additivity of covalent bond radii. (*Hint*: Refer to Probs. 7.8 and 7.11).

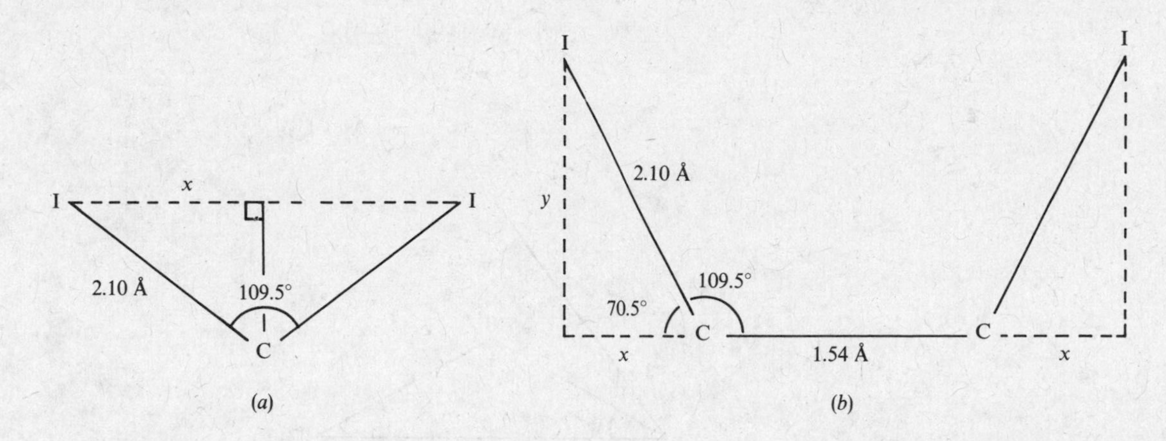

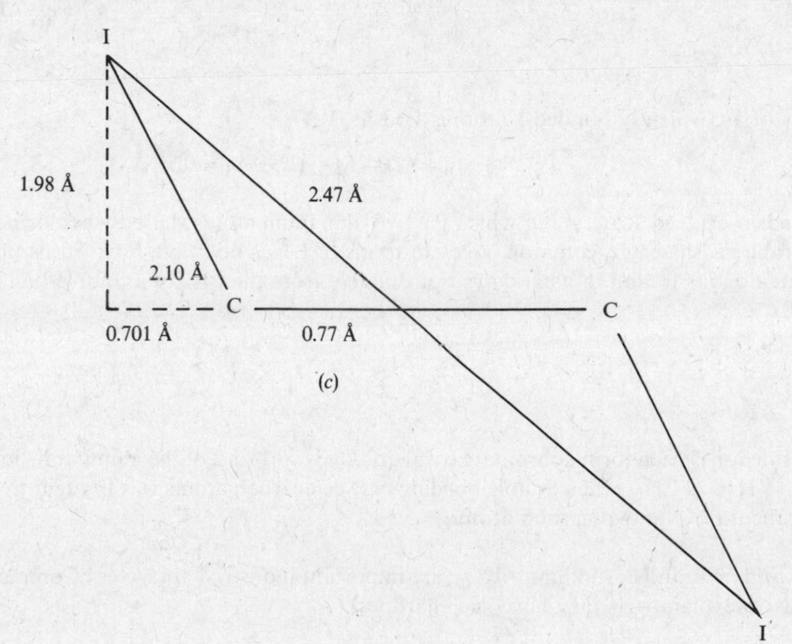

Fig. 7.11

(*a*) The iodine atoms are located 2.10 Å from the carbon atom at an angle of 109.5° [Fig. 7.11(*a*)].

$$x = 2.10 \sin(54.75°) = 1.71 \text{ Å}$$

$$2x = \text{I—I distance} = 3.42 \text{ Å}$$

(*b*) From Fig. 7.11(*b*),

$$x = 2.10 \cos(70.5°) = 0.701 \text{ Å}$$

$$\text{I—I distance} = 2x + 1.54 \text{ Å} = 2.94 \text{ Å}$$

(*c*) From Fig. 7.11(*c*),

$$y = 2.10 \sin(70.5°) = 1.98 \text{ Å}$$

The distance from the iodine atom to the center of the molecule may be calculated from the Pythagorean theorem:

$$d = \sqrt{(1.98)^2 + (0.77 + 0.701)^2} = 2.47 \text{ Å}$$

The iodine-iodine distance is twice that distance, 4.94 Å.

7.15 BI_3 is a symmetrical planar molecule, all B—I bonds lying at 120° to each other. The distance between I atoms is found to be 3.54 Å. From this fact and from information in Table 7.1, estimate the covalent radius of boron, assuming that the bonds are all single bonds.

▌ In Fig. 7.12,

$$x = d_{I-B} = 1.77/(\sin 60°) = 2.04\ \text{Å}$$

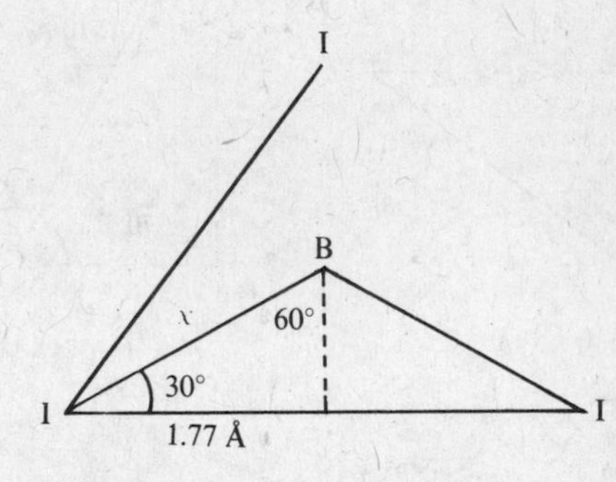

Fig. 7.12

The radius of a covalently bonded I atom is 1.33 Å.

$$r_B = (2.04\ \text{Å}) - (1.33\ \text{Å}) = 0.71\ \text{Å}$$

7.16 $Al_2Cl_6(g)$ has a bridged structure in which the two aluminum atoms share two chlorine atoms, and the effect is that of two tetrahedra sharing a common edge. In terms of Lewis notation, how is this possible? In what ways is this structure analogous to and in what ways is it different from the case of a double-bonded molecule such as C_2Cl_4?

▌

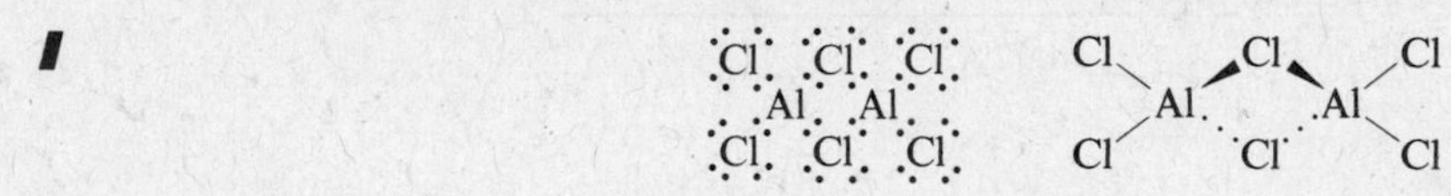

The "chloride ions" each form coordinate covalent bonds to 1 or 2 of the aluminum "ions." Historically, an analogy was made to $H_2C{=}CH_2$, when double bonding between carbon atoms was thought to be the sharing of two corners of the tetrahedra of the two carbon atoms.

7.17 Carborundum, SiC, and corundum, Al_2O_3, are important industrial abrasives. Comment on the structures for these compounds to explain why they have such hardness.

▌ The hardness results from the fact that the atoms are bonded together in three-dimensional arrays by strong bonds. SiC has the diamond structure. The bonds in Al_2O_3 have ionic character as well as covalent.

7.18 The average C—C bond energy is 343 kJ/mol. What do you predict for the Si—Si single-bond energy?

▌ Covalent bonds between third-period nonmetals are generally weaker than bonds between atoms of their second-period neighbors. Si—Si bonds are expected to be weaker than C—C bonds; their bond strength should be less than 300 kJ/mol.

7.19 Using bond energy data (Table 7.2), estimate whether CH_4 or C_8H_{18} would produce more heat *per gram* upon complete combustion. Assume all reactants and products to be in the gaseous state.

▌ Per mole of CH_4:

$$\begin{array}{cccc} CH_4 & +\ 2O_2 & \longrightarrow\ CO_2 & +\ 2H_2O \\ 4C{-}H & 2O_2 & 2C{=}O & 4H{-}O \end{array}$$

$$4(100) + 2(119.2) - 2(177) - 4(111) = -160\ \text{kcal/mol}$$

$$\Delta H = (-160\ \text{kcal/mol})\left(\frac{1\ \text{mol}}{16.0\ \text{g}}\right) = -10.0\ \text{kcal/g}$$

Per mole of C_8H_{18}:

$$C_8H_{18} + \tfrac{25}{2}O_2 \longrightarrow 8CO_2 + 9H_2O$$

$$18(\text{C—H}) \quad 7(\text{C—C}) \quad \tfrac{25}{2}(O_2) \quad 16(\text{C=O}) \quad 18(\text{H—O})$$

$$18(100) + 7(82) + \tfrac{25}{2}(119.2) - 16(177) - 18(111) = -966 \text{ kcal/mol}$$

$$\Delta H = (-966 \text{ kcal/mol})\left(\frac{1 \text{ mol}}{114 \text{ g}}\right) = -8.47 \text{ kcal/g}$$

Methane produces more heat per gram.

TABLE 7.2 Bond Energies at 298 K

single bond	D kcal/mol	D kJ/mol	single bond	D kcal/mol	D kJ/mol	multiple bond	D kcal/mol	D kJ/mol
H—H	104.2	436.0	N—N	32	134	C=C	146	611
H—C	100	418	N—F	56	234	C≡C	200	837
H—N	93	389	N—Cl	37	155	C=N	147	615
H—O	111	464	O—O	33	138	C≡N	213	891
H—F	135	565	O—F	45	188	C=O	177	741
H—S	81	339	O—Si	106	444	C≡O	256	1070
H—Cl	103.2	431.8	O—Cl	50	209	N=N	100	418
H—Br	88	368	O—S	124	519	N≡N	226	946
H—I	71	297	F—F	37.8	158	bond in O_2	119.2	498.7
C—C	82	343	P—Cl	78	326			
C—N	64	268	S—S	49	205			
C—O	83	347	Cl—Cl	58	243			
C—F	102	427	Cl—I	50	209			
C—S	61	255	Br—Br	46	192			
C—Cl	79	330	I—I	36	151			
C—Br	66	276						
C—I	52	218						

7.20 Calculate the electronegativity of chlorine from the bond energy of ClF (61 kcal/mol) and the data of Table 7.2.

▮

$$\Delta = D(\text{ClF}) - \sqrt{D(F_2)D(Cl_2)} = 61 - \sqrt{(38)(58)} = 14 \text{ kcal/mol}$$

$$\text{EN(F)} - \text{EN(Cl)} = 0.208\sqrt{\Delta}$$

$$\text{EN(Cl)} = \text{EN(F)} - 0.208\sqrt{\Delta} = 4.0 - 0.78 = 3.2$$

The electronegativity of 3.0 tabulated for chlorine is a rounded-off average of values from many compounds.

7.21 In what ways does the periodic variation of the electronegativities of the elements differ from the periodic variation of ionization potentials?

▮ Electronegativity is only semiquantitative, and therefore the variations from element to element are less detailed. Moreover, electronegativity for some noble gas elements is undefined.

7.22 Distinguish between electronegativity and electron affinity.

▮ Electronegativity denotes the relative attraction of covalently bonded atoms for the bonding electron pair(s). Electron affinity is the energy released when an electron is added to a *gaseous* atom or ion to form a gaseous ion.

7.23 Using bond energy data from Table 7.2 and taking the electronegativity of hydrogen as 2.1, estimate the electronegativities of sulfur and chlorine.

▮

$$\Delta = D(\text{H—S}) - \sqrt{D(\text{H—H})D(\text{S—S})} = 81 - \sqrt{(104.2)(49)} = 9.5 \text{ kcal/mol}$$

$$\text{EN(S)} = \text{EN(H)} + 0.208\sqrt{9.5} = 2.1 + 0.64 = 2.74$$

$$\Delta = D(\text{H—Cl}) - \sqrt{D(\text{H—H})D(\text{Cl—Cl})} = 103.2 - \sqrt{(104.2)(58)} = 25 \text{ kcal/mol}$$

$$\text{EN(Cl)} = \text{EN(H)} + 0.208\sqrt{25} = 2.1 + 1.04 = 3.1$$

7.24 Calculate the electronegativity of nitrogen from the bond energies given in Table 7.2 and the electronegativity of fluorine. Compare the result to the tabulated electronegativity of nitrogen, and suggest a reason for any difference.

▌ Note that single bond energies are used.

$$\Delta = D(\text{N—F}) - \sqrt{D(\text{N—N})D(\text{F—F})} = 56 - \sqrt{(32)(37.8)} = 21 \text{ kcal/mol}$$

$$\text{EN(N)} = \text{EN(F)} - 0.208\sqrt{\Delta} = 3.05$$

The tabulated value of 3.0 for nitrogen is an average value from many compounds.

7.25 In 1934, R. S. Mulliken proposed that a value of electronegativity for an element could be defined as the average of its ionization potential and its electron affinity. Using units of electron volts for the ionization potential and electron affinity, determine the electronegativity of each halogen using this system. Compare the relative electronegativities of these elements on this scale with that given by the Pauling scale.

	IP	EA
F	17.41	3.34
Cl	13.01	3.61
Br	11.84	3.36
I	10.45	3.06

▌

	average	Pauling EN	ratio
F	10.38	4.0	2.6
Cl	8.31	3.0	2.8
Br	7.60	2.8	2.7
I	6.76	2.5	2.7

Dividing the average of the ionization potential and the electron affinity by the Pauling electronegativity gives approximately the same value each time, showing that the two measures are directly proportional.

7.26 From the bond energies given in Table 7.2, calculate the extra ionic resonance energy in the HF molecule.

▌
$$\Delta(\text{HF}) = 135 - \sqrt{(104)(38)} = 72 \text{ kcal/mol}$$

7.27 Given that $\Delta H_f(\text{H}) = 218$ kJ/mol, express the H—H bond energy in kJ/mol and in kcal/mol.

▌ The bond energy is the energy needed to dissociate gaseous H_2 into separated atoms.

$$\text{H}_2 \longrightarrow 2\text{H}$$

ΔH for this reaction is twice ΔH_f for 1 mol of H.

$$\Delta H = 2(218 \text{ kJ/mol}) = 436 \text{ kJ/mol} = \frac{436 \text{ kJ/mol}}{4.184 \text{ kJ/kcal}} = 104 \text{ kcal/mol}$$

7.28 Estimate the enthalpy of formation of ammonia from bond energy data (Table 7.2).

▌
$$\text{N}_2 + 3\text{H}_2 \longrightarrow 2\text{NH}_3$$

Noting that there are three N—H bonds in each NH_3 molecule, one obtains

$$\begin{array}{rlrr} 3\text{H}(g) + \text{N}(g) & \longrightarrow \text{NH}_3(g) & -3D(\text{N—H}) = & -279 \text{ kcal} \\ \tfrac{1}{2}\text{N}_2(g) & \longrightarrow \text{N}(g) & \tfrac{1}{2}D(\text{N}_2) = & 113 \text{ kcal} \\ \tfrac{3}{2}\text{H}_2(g) & \longrightarrow 3\text{H}(g) & \tfrac{3}{2}D(\text{H}_2) = & 156 \text{ kcal} \\ \hline \tfrac{1}{2}\text{N}_2(g) + \tfrac{3}{2}\text{H}_2(g) & \longrightarrow \text{NH}_3(g) & \Delta H_f = & -10 \text{ kcal} \end{array}$$

The estimated enthalpy of formation is of the order of magnitude of the experimental value, −11 kcal/mol.

7.29 Estimate the C—H bond energy in methane, CH_4. The required enthalpy data are the enthalpy of formation of methane, −17.9 kcal/mol; the bond energy of hydrogen, 104.2 kcal/mol; and the enthalpy of sublimation of carbon, 171.7 kcal/mol. Illustrate the calculation with an enthalpy diagram.

▌ The C—H bond energy in methane is *defined* as one-fourth of the energy required to break the four C—H bonds.

$$CH_4(g) \longrightarrow C(g) + 4H(g)$$

To determine the enthalpy change of this reaction the data are combined in the following manner:

$$\begin{array}{ll} C(s) \longrightarrow C(g) & \Delta H_{sub} = \quad 171.7\text{ kcal} \\ 2H_2(g) \longrightarrow 4H(g) & 2D(H_2) = 2(104.2)\text{ kcal} \\ CH_4(g) \longrightarrow C(s) + 2H_2(g) & -\Delta H_f = \quad 17.9\text{ kcal} \\ \hline CH_4(g) \longrightarrow C(g) + 4H(g) & \Delta H = \quad 398.0\text{ kcal} \end{array}$$

The C—H bond energy is one-fourth of the energy required to break all four bonds, 99.5 kcal/mol. The enthalpy diagram is shown in Fig. 7.13.

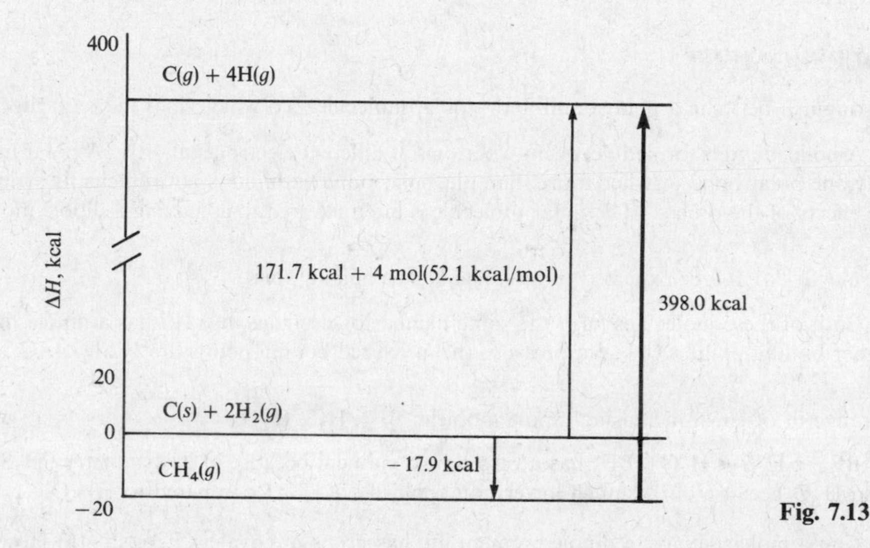

Fig. 7.13

7.30 The enthalpy of combustion of gaseous cyclohexane is −944.4 kcal/mol, while that of gaseous cyclopropane is −496.8 kcal/mol. In each case calculate the enthalpy change per mole of O_2 reacted and explain the difference.

▌ For cyclohexane, the enthalpy change is −104.9 kcal/mol of O_2; for cyclopropane, the enthalpy change is −110.4 kcal/mol of O_2. The higher energy state of the strained three-membered ring of cyclopropane is responsible for the relatively greater release of energy when this substance reacts with oxygen.

7.31 Using the bond energies of H_2 and Cl_2 and the enthalpy of formation of HCl(*g*), −22.1 kcal/mol, calculate the energy of the HCl bond, *D*(HCl).

▌

$$\begin{array}{ll} HCl(g) \longrightarrow \tfrac{1}{2}H_2(g) + \tfrac{1}{2}Cl_2(g) & -\Delta H_f^\circ(HCl) = 22.1\text{ kcal} \\ \tfrac{1}{2}H_2(g) \longrightarrow H(g) & \tfrac{1}{2}D(H_2) = 52.1\text{ kcal} \\ \tfrac{1}{2}Cl_2(g) \longrightarrow Cl(g) & \tfrac{1}{2}D(Cl_2) = 29.0\text{ kcal} \end{array}$$

The bond energy of HCl, *D*(HCl), is obtained by combining these equations and the respective enthalpy changes.

$$HCl(g) \longrightarrow H(g) + Cl(g) \qquad D(HCl) = -\Delta H_f^\circ(HCl) + \tfrac{1}{2}D(H_2) + \tfrac{1}{2}D(Cl_2)$$
$$= 22.1\text{ kcal} + 52.1\text{ kcal} + 29.0\text{ kcal} = 103.2\text{ kcal}$$

7.32 From the enthalpies of formation of PH_3 and P(*g*) (5.5 kcal/mol and 79.8 kcal/mol), calculate the average P—H bond energy. Then, using appropriate electronegativity values, estimate the bond energy of a P_2 molecule.

▌

$$\begin{array}{ll} \tfrac{1}{4}P_4 + \tfrac{3}{2}H_2 \longrightarrow PH_3 & \Delta H_f = \ 5.5\text{ kcal} \\ \tfrac{1}{4}P_4 \longrightarrow P(g) & \Delta H = 79.8\text{ kcal} \\ \tfrac{1}{2}H_2 \longrightarrow H(g) & \Delta H = 52.1\text{ kcal} \quad \text{(Table 7.2)} \end{array}$$

Subtraction of the second equation and three times the third equation from the first equation, and subtraction of the corresponding enthalpy changes, results in the following equation and enthalpy change:

$$P(g) + 3H(g) \longrightarrow PH_3 \qquad \Delta H = 5.5 - 79.8 - 3(52.1) = -230.6 \text{ kcal}$$

The average P—H bond energy has a magnitude of one-third that value, 76.9 kcal.

$$EN(P) - EN(H) = 2.1 - 2.1 = 0.0 = 0.208\sqrt{\Delta}$$
$$\Delta = 0.0 = D(P{-}H) - \sqrt{D(P_2)D(H_2)}$$
$$D(P{-}H) = \sqrt{D(P_2)D(H_2)}$$
$$D(P_2) = [D(P{-}H)]^2/D(H_2) = (76.9)^2/(104.2) = 57 \text{ kcal/mol}$$

7.33 Is ΔH_f(Br atoms) sufficient data for the evaluation of the Br—Br bond energy?

▮ No; the enthalpy of formation of Br is given with respect to the standard state for Br_2, which is the liquid and not the gas.

7.2 DIPOLE MOMENT

7.34 Distinguish between a polar bond and a polar molecule. To which does the term dipole refer?

▮ A polar bond is formed between two atoms of different electronegativity. A polar molecule results when there is only one polar bond or when more than one polar bond within it is not sufficiently symmetrically oriented to cancel the effects of the others. The polar molecule is often referred to as having a dipole moment.

7.35 Would Br_2 or ICl be expected to have the higher boiling point?

▮ Both of these molecules have the same number of electrons, but ICl has a dipole moment. ICl should have the higher boiling point. At 1 atm pressure the measured boiling points are ICl, 97.4 °C, and Br_2, 58.78 °C.

7.36 Arrange in order of increasing dipole moment: BF_3, H_2S, H_2O.

▮ $BF_3 < H_2S < H_2O$ BF_3 has a zero dipole moment because of its symmetry. H_2S has a lower dipole moment than H_2O because of the much lower bond polarity of H—S compared to H—O.

7.37 (*a*) Can a molecule have a dipole moment if it has no polar covalent bonds? (*b*) How is it possible for a molecule to have polar bonds but no dipole moment?

▮ (*a*) No. (*b*) If the various polar bonds are arranged symmetrically, as in CCl_4 and CO_2, the effects of one bond are cancelled by the effects of the other(s).

7.38 Arrange in order of decreasing polarity of the bonds: SbH_3, AsH_3, PH_3, NH_3.

▮ Decreasing electronegativity difference between H (2.1) and Sb (1.9), As (2.0), P (2.1), and N (3.0) cause the order

$$PH_3 < AsH_3 < SbH_3 < NH_3.$$

7.39 The dipole moments of SO_2 and CO_2 are 5.37×10^{-30} C·m and zero, respectively. What can be said about the shapes of the two molecules?

▮ Oxygen is considerably more electronegative than either sulfur or carbon. Thus each sulfur-oxygen and carbon-oxygen bond should be polar, with a net negative charge residing on the oxygen.

Since CO_2 has no net dipole moment, the two C—O bond moments must exactly cancel. This can occur only if the two bonds are in a straight line. (The net moment of the molecule is the vector sum of the bond moments.) The existence of a dipole moment in SO_2 must mean that the molecule is not linear, but bent.

7.40 The dipole moments of NH_3, AsH_3, and BF_3 are 4.97×10^{-30}, 0.60×10^{-30} and 0.00×10^{-30} C·m, respectively. What can be concluded about the shapes of these molecules?

▮ NH_3 and AsH_3 are both pyramidal and BF_3 is planar. From the dipole moments alone, nothing can be concluded about the relative flatness of the NH_3 and AsH_3 pyramids, because the electronegativities of N and As are different.

7.41 Dipole moments are sometimes expressed in *debyes* (D), where

$$1\ \text{D} = 10^{-18}\ (\text{esu of charge})\cdot\text{cm}$$

The electrostatic unit (esu) of charge is defined by $1\ \text{C} = 2.998 \times 10^9\ \text{esu}$. What is the value of 1 D in SI units?

▮

$$1\ \text{D} = (10^{-18}\ \text{esu}\cdot\text{cm})\left(\frac{1\ \text{C}}{2.998 \times 10^9\ \text{esu}}\right)\left(\frac{1\ \text{m}}{10^2\ \text{cm}}\right) = 3.336 \times 10^{-30}\ \text{C}\cdot\text{m}$$

7.42 Using data from Table 7.1 and the dipole moments of HCl (103 D) and HI (0.38 D), compare the magnitude of the partial charge on the hydrogen atom in HCl (*a*) with that on the hydrogen atom of HI and (*b*) with that of a singly charged positive ion, which is 4.80×10^{-10} esu.

▮ (*a*) If it is assumed that the partial positive and negative charges are centered on the hydrogen and halogen atoms, respectively, the distances between charge centers are calculated to be 1.27 and 1.61 Å for HCl and HI, respectively, from data in Table 7.1. The charge, δ, equals the dipole moment divided by the separation, d.

$$\text{For HCl:}\quad \delta = \frac{\text{dipole moment}}{d} = \frac{1.03 \times 10^{-18}\ \text{esu}\cdot\text{cm}}{1.27 \times 10^{-8}\ \text{cm}} = 8.11 \times 10^{-11}\ \text{esu}$$

$$\text{For HI:}\quad \delta = \frac{0.38 \times 10^{-18}\ \text{esu}\cdot\text{cm}}{1.61 \times 10^{-8}\ \text{cm}} = 2.4 \times 10^{-11}\ \text{esu}$$

The charge on the hydrogen atom of HCl is over 3 times that on the hydrogen atom of HI, in accordance with the greater electronegativity difference between hydrogen and chlorine compared to the difference between hydrogen and iodine.

(*b*) The charge on the hydrogen atom of HCl is about one-sixth the charge on a monopositive ion.

7.43 The dipole moment of HBr is $2.60 \times 10^{-30}\ \text{C}\cdot\text{m}$, and the interatomic spacing is 1.41 Å. What is the percent ionic character of HBr?

▮ The dipole moment of a 100% ionic "molecule" at the given internuclear distance would be

$$(1.60 \times 10^{-19}\ \text{C})(1.41 \times 10^{-10}\ \text{m}) = 2.26 \times 10^{-29}\ \text{C}\cdot\text{m}$$

The actual dipole is less; the percent ionic character is given by

$$\frac{2.60 \times 10^{-30}\ \text{C}\cdot\text{m}}{2.26 \times 10^{-29}\ \text{C}\cdot\text{m}} \times 100\% = 11.5\%$$

7.44 A diatomic molecule has a dipole moment of 1.2 D. If its bond distance is 1.0 Å, what fraction of an electronic charge, e, exists on each atom?

▮

$$\delta = \frac{\text{dipole moment}}{d} = \frac{1.2\ \text{D}}{1.0 \times 10^{-8}\ \text{cm}} = \frac{1.2 \times 10^{-18}\ \text{esu}\cdot\text{cm}}{1.0 \times 10^{-8}\ \text{cm}} = 1.2 \times 10^{-10}\ \text{esu}$$

The fraction of an electronic charge is

$$\frac{1.2 \times 10^{-10}\ \text{esu}}{4.8 \times 10^{-10}\ \text{esu}/e} = 0.25e = 25\%\ \text{of}\ e$$

7.45 The dipole moment of LiH is $1.964 \times 10^{-29}\ \text{C}\cdot\text{m}$, and the interatomic distance between Li and H in this molecule is 1.596 Å. What is the percent ionic character (Prob. 7.43) in LiH?

▮ Let us calculate the dipole moment of a hypothetical, fully ionized Li^+H^- ion pair with a separation of 1.596 Å.

$$\mu(\text{hypothetical}) = (1\ \text{electronic charge})(\text{separation}) = (1.602 \times 10^{-19}\ \text{C})(1.596 \times 10^{-10}\ \text{m}) = 2.557 \times 10^{-29}\ \text{C}\cdot\text{m}$$

The fractional ionic character is the actual dipole moment divided by the hypothetical.

$$\text{Fractional ionic character} = \frac{1.964 \times 10^{-29}\ \text{C}\cdot\text{m}}{2.557 \times 10^{-29}\ \text{C}\cdot\text{m}} = 0.768$$

The bond is computed to be 76.8% ionic.

7.46 In water, the H—O—H bond angle is 105°. Using the dipole moment of water and the covalent radii of the atoms, determine the magnitude of the charge on the oxygen atom in the water molecule.

$$\mu = 1.85 \text{ D} = 1.85 \times 10^{-18} \text{ esu·cm} = \delta d$$

H, 105°, δ+, Oδ−, H; 0.94 Å, 52.5°, d

$$\cos 52.5° = d/0.94 \text{ Å}$$

$$d = (0.609)(0.94 \text{ Å}) = 0.572 \text{ Å}$$

$$\delta = \frac{\mu}{d} = \frac{1.85 \text{ D}}{0.572 \text{ Å}} = \frac{1.85 \times 10^{-18} \text{ esu·cm}}{0.572 \times 10^{-8} \text{ cm}} = 3.2 \times 10^{-10} \text{ esu}$$

or 0.67 times the electronic charge $(4.8 \times 10^{-10}$ esu).

7.47 At a given potential, a certain pair of parallel plates can hold charges in a vacuum, and when HCl is between them, as tabulated. The concentration of HCl at the two temperatures is the same. Explain these data.

	charge at 0 °C	charge at 100 °C
With HCl	5×10^{-8} C	3×10^{-8} C
In vacuum	1×10^{-8} C	1×10^{-8} C

The higher the temperature, the more thermal agitation there will be and the less well aligned the HCl molecules will be. The greater the alignment, the greater the effect.

7.48 The percentage ionic character of a single bond may be estimated from the ratio of the observed dipole moment to the calculated moment, assuming oppositely charged ions located at a distance equal to the bond length. The observed dipole moments of HCl, HBr, and HI are 1.03, 0.79, and 0.38 D, respectively. Calculate the percentage ionic character of the bond in each of these compounds. Do the results obtained parallel the magnitudes of the extra ionic resonance energies, Δ, for these molecules?

The bond distances are calculated from the data of Table 7.1.

H—Cl	0.28 Å + 0.99 Å = 1.27 Å
H—Br	0.28 Å + 1.14 Å = 1.42 Å
H—I	0.28 Å + 1.33 Å = 1.61 Å

For HCl:
$$d = \frac{\mu}{d} = \frac{1.03 \times 10^{-18} \text{ esu·cm}}{1.27 \times 10^{-8} \text{ cm}} = 8.11 \times 10^{-11} \text{ esu}$$

$$[(8.11 \times 10^{-11} \text{ esu})/(4.80 \times 10^{-10} \text{ esu})] \times 100\% = 16.9\% \text{ ionic}$$

For HBr:
$$\delta = \frac{0.79 \times 10^{-18} \text{ esu·cm}}{1.42 \times 10^{-8} \text{ cm}} = 5.6 \times 10^{-11} \text{ esu or } 12.0\% \text{ ionic}$$

For HI:
$$\delta = \frac{0.38 \times 10^{-18} \text{ esu·cm}}{1.61 \times 10^{-8} \text{ cm}} = 2.4 \times 10^{-11} \text{ esu or } 5.0\% \text{ ionic}$$

$$\Delta = \left[\frac{\text{EN(X)} - \text{EN(H)}}{0.208}\right]^2 = \left(\frac{0.9}{0.208}\right)^2 = 18 \text{ for HCl}$$

$$= \left(\frac{0.7}{0.208}\right)^2 = 11 \text{ for HBr}$$

$$= \left(\frac{0.4}{0.208}\right)^2 = 4 \text{ for HI}$$

The results parallel the values of Δ rather well in these cases.

7.3 OTHER INTERMOLECULAR FORCES

7.49 Arrange the following types of interactions in order of increasing stability: covalent bond, van der Waals force, hydrogen bonding, dipole attraction.

van der Waals < dipole < hydrogen bonding < covalent

7.50 List properties of water that stem from hydrogen bonding.

High melting point, high boiling point, high heat of vaporization, low density of ice compared to water, high specific heat, high ionic conductance of hydronium and hydroxide ions and many others.

7.51 Predict the order of increasing boiling points of the noble gases.

The greater the atomic number, the greater the number of electrons in each atom, and the greater is the van der Waals force. The greater the forces between the molecules, the higher should be the boiling point. The actual boiling points of the nobles gases increase with increasing atomic number, as expected: He, 4 K; Ne, 27 K; Ar, 87 K; Kr, 120 K; Xe, 166 K; Rn, 211 K.

7.52 The two molecules indicated below are capable of intramolecular hydrogen bonding. Which is likely to form more stable hydrogen bonds? Suggest a reason for your choice.

H
H₃C—C(=O)—C(H)=C(—O—H)—CH₃   H₃C—C(=O)—CH₂—CH(—O—H)—CH₃

The first of the molecules has two resonance forms, which make the hydrogen bonding more stable. The second molecule has no such resonance stabilization.

H₃C—C(=O)—CH=C(—O—H)—CH₃   H₃C—C(—O—H)=CH—C(=O)—CH₃

7.53 Which of the following has the highest boiling point: H_2 He Ne Xe CH_4?

Xe. All are nonpolar molecules, but Xe has the greatest van der Waals forces because it has the most electrons.

7.54 At 300 K and 1.00 atm pressure, the density of gaseous HF is 3.17 g/L. Explain this observation, and support your explanation by calculations.

In each liter volume the number of molecules is given by

$$n = \frac{PV}{RT} = \frac{(1.00\text{ atm})(1.00\text{ L})}{(0.0821\text{ L}\cdot\text{atm/mol}\cdot\text{K})(300\text{ K})} = 0.0406\text{ mol}$$

$$\frac{3.17\text{ g}}{0.0406\text{ mol}} = 78.1\text{ g/mol}$$

The formula weight of HF is 19 g/mol. The large apparent formula weight from the gas density data means that the gas is appreciably associated even in the gas phase, presumably by hydrogen bonding. The average cluster of HF molecules is about 4.

7.55 From Fig. 7.14, estimate the boiling point which water would have if there were no hydrogen bonding.

Extrapolation yields a value about −90 °C.

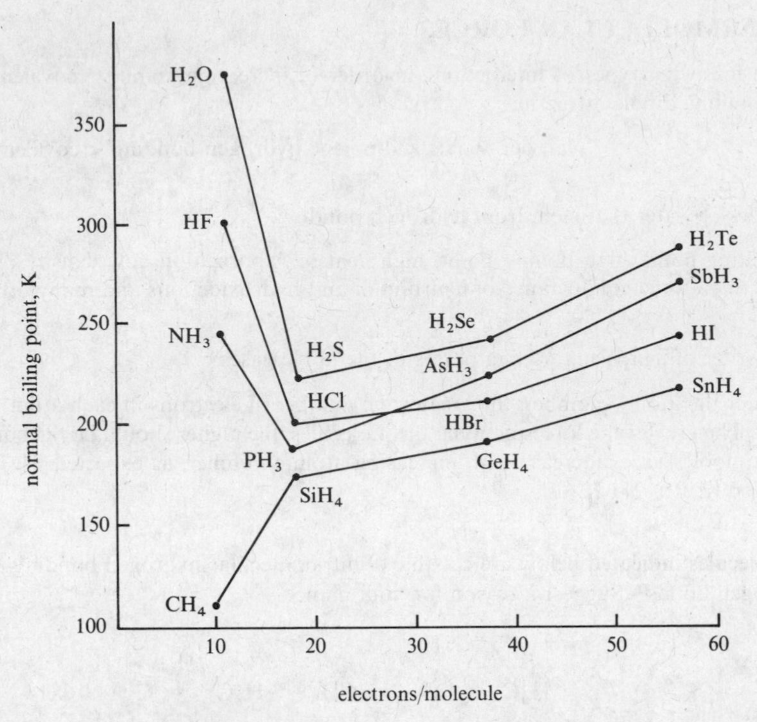

Fig. 7.14

7.56 Which one in each of the following pairs is expected to exhibit hydrogen bonding? ***(a)*** CH_3CH_2OH and CH_3OCH_3 ***(b)*** CH_3NH_2 and CH_3SH ***(c)*** CH_3OH and $(CH_3)_3N$

▌ ***(a)*** CH_3CH_2OH (H is connected to O) ***(b)*** CH_3NH_2 (S does form hydrogen bonds) ***(c)*** CH_3OH (No H on the nitrogen atom of $(CH_3)_3N$.)

7.57 Which of the following is expected to have the highest melting point: PH_3 NH_3 $(CH_3)_3N$? Explain why.

▌ NH_3 has the strongest intermolecular forces, thus it is expected to have the highest melting point. (Actual melting points are NH_3 −77.7 °C, PH_3 −133 °C, $(CH_3)_3N$ −117 °C.)

7.58 In which molecule is the van der Waals force likely to be the most important in determining the melting point and boiling point: ICl Br_2 HCl H_2S CO

▌ Br_2. Each of the other molecules has a dipole in addition to van der Waals forces. Only in Br_2 are the van der Waals forces the only intermolecular forces.

7.59 Boric acid, $B(OH)_3$, forms hexagonal crystals that cleave easily into thin layers, indicating weak interplanar forces. Diagram a possible structure for a layer of crystalline $B(OH)_3$.

▌

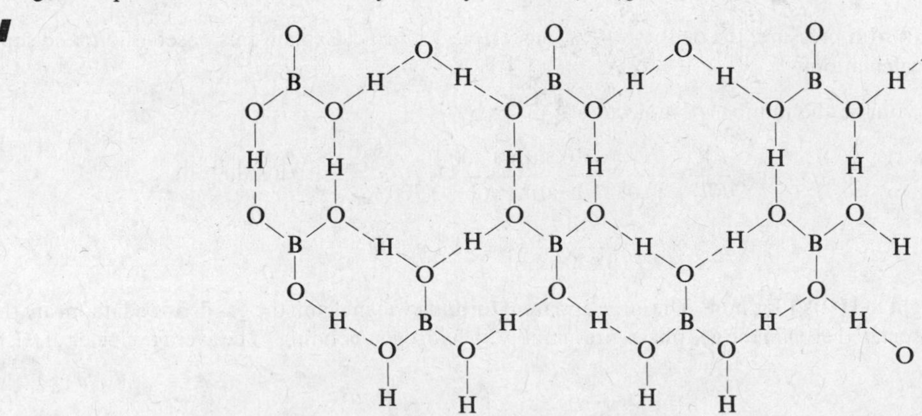

Hydrogen bonding maintains the planar H_3BO_3 units in layers.

7.60 The HF_2^- ion exists in the solid state and also in liquid HF solutions, but not in dilute aqueous solutions. Explain.

In the solid state and in liquid HF, the HF_2^- ion is held together by hydrogen bonding. In aqueous solutions, there is hydrogen bonding, but each HF molecule hydrogen bonds with the much more prevalent H_2O present instead of other HF molecules, and H_3O^+ and F^- are much more likely to be formed. H_2F^+ is also possible, but is not as prevalent since HF is a stronger acid than H_2O.

7.4 RESONANCE

7.61 Draw two or more electron dot diagrams showing resonance in the CO_3^{2-} ion.

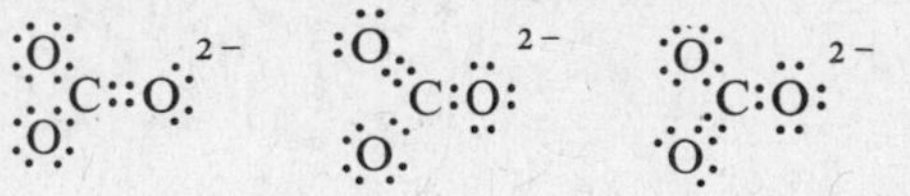

$$\left[\begin{matrix}:\ddot{O}\\ \quad C::\ddot{O}\\ :\ddot{O}\end{matrix}\right]^{2-} \quad \left[\begin{matrix}:\ddot{O}\\ \quad C:\ddot{O}:\\ ::\ddot{O}\end{matrix}\right]^{2-} \quad \left[\begin{matrix}:\ddot{O}\\ \quad C:\ddot{O}:\\ \ddot{O}\end{matrix}\right]^{2-}$$

7.62 Draw the complete Lewis electron dot diagram for each of the following, showing resonance structures where appropriate.

(*a*) MgO (*b*) CN^- (*c*) SO_3 (S in center) (*d*) BrF_3 (Br in center)

(*a*) $[Mg^{2+}][:\ddot{O}:^{2-}]$ (*b*) $:C:::N:^-$

(*c*) $:\ddot{O}:S:\ddot{O}: \longleftrightarrow :\ddot{O}:S::O: \longleftrightarrow :O::S:\ddot{O}:$ (each with $:\ddot{O}:$ below S) (*d*) $:\ddot{F}-\ddot{Br}-\ddot{F}:$ (with $:\ddot{F}:$ below Br)

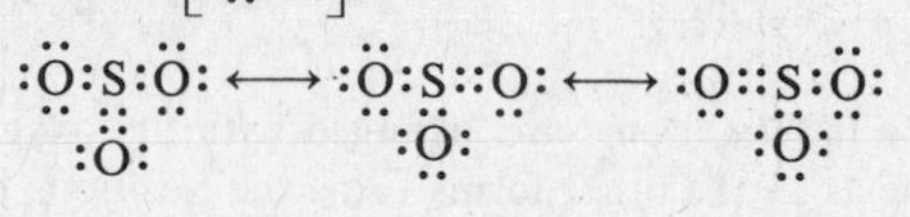

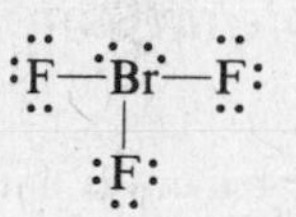

7.63 NO_2 gas is paramagnetic at room temperature. When a sample of the gas is cooled below 0 °C, its molecular weight increases and it loses its paramagnetism. When it is reheated, the behavior is reversed. (*a*) Using electron dot structures, write an equation which accounts for these observations. (*b*) How does this phenomenon differ from resonance?

(*a*) $:\ddot{O}\!:\!\dot{N}\!:\!\ddot{O}: + :\ddot{O}\!:\!\dot{N}\!:\!\ddot{O}: \rightleftharpoons$ $\begin{matrix}:\ddot{O} & & \ddot{O}:\\ & N:N & \\ :\ddot{O} & & \ddot{O}:\end{matrix}$

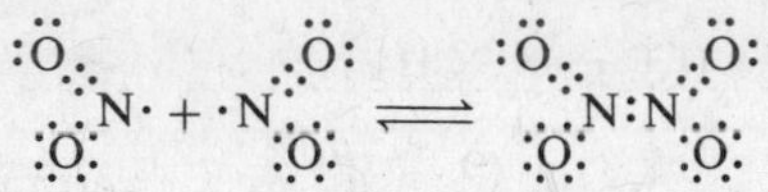

(*b*) This is an actual chemical reaction—an equilibrium. Resonance forms are only multiple representations of the same structure.

7.64 In each of the following pairs, select the species having the greater resonance stabilization: (*a*) HNO_3 and NO_3^- (*b*) $H_2C{=}O$ and $HC(=O)-O^-$.

(*a*) NO_3^- has three equal-energy resonance forms; HNO_3 has only two, since the oxygen atom bonded to the hydrogen atom is not equivalent to the others.
(*b*) $HC(=O)-O^-$ has two equal resonance forms, the other being $HC(-O^-)=O$. $H_2C{=}O$ does not have other resonance forms as low in energy. ($H_2\ddot{C}^-{-}\ddot{O}^+$ contributes little; $H_2\overset{+}{C}{-}\ddot{O}:^-$ contributes some.)

7.65 Draw all possible octet structural formulas for N_3^-. Which ones are possible resonance forms?

The possible linear structures are shown in Fig. 7.15.

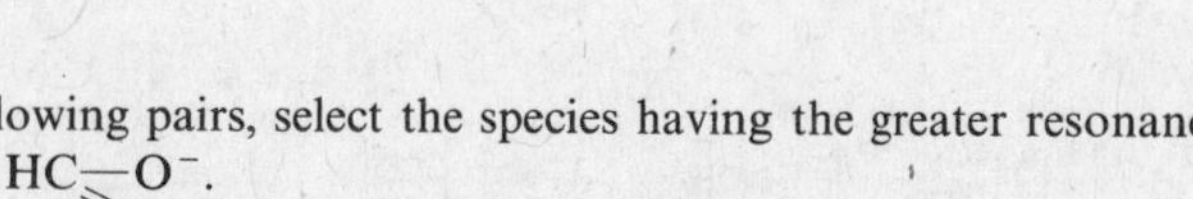

$[:\ddot{N}{=}N{=}\ddot{N}:]^-$ (*a*) $[:\ddot{N}{-}N{\equiv}N:]^-$ (*b*) $[:N{\equiv}N{-}\ddot{N}:]^-$ (*c*)

Fig. 7.15

$\left[\begin{matrix}:N{=}N:\\ \diagdown\ \diagup\\ :N:\end{matrix}\right]^-$

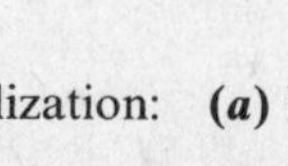

Fig. 7.16

Two resonance structures are written with the triple bond because there is no reason why one of the terminal nitrogens in an isolated azide ion should be different from the other. A different type of structure, involving a ring, also satisfies the octet rule. This structure (Fig. 7.16) is ruled out as a resonance form because the atoms are in different positions. Besides, the bond angles of 60° or less demanded by the structure require considerable strain from the normal angles of bonding.

7.66 Would resonance stabilization be greater in (*a*) CO_3^{2-} or (*b*) H_2CO_3?

▌ The possible resonance hybrids are

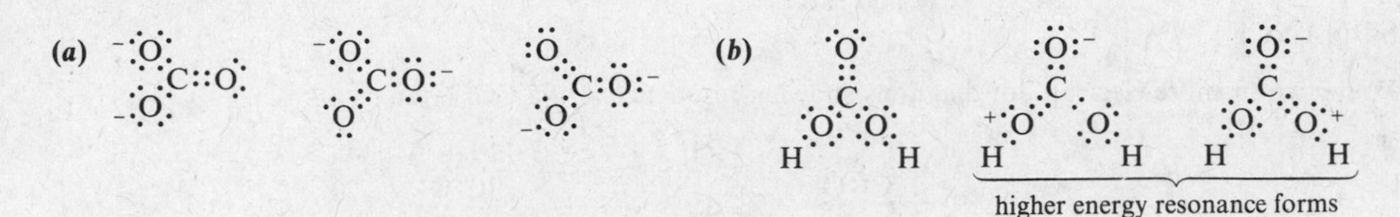

In CO_3^{2-}, three hybrids of equal energy contribute equally to the resonance structure. In H_2CO_3, the last two forms are of higher energy than the first and therefore are less important. Because it has three forms of equal energy, CO_3^{2-} has a much greater resonance stabilization than does H_2CO_3.

7.67 Which one(s) of the following structures *cannot* represent resonance forms for (diamagnetic) NNO? (*a*) :N::N::O: (*b*) :N:::N:O: (*c*) :N:N:::O: (*d*) :N::O::N: (*e*) :N::N::O:

▌ The forms shown as (*d*) and (*e*) cannot be used. In (*d*) the atoms are arranged with the oxygen atom between the nitrogen atoms. This arrangement is not the same as in the other forms or as the positions found experimentally. The form shown as (*e*) has four unpaired electrons, and this is not possible, since the molecule is diamagnetic.

7.68 Draw octet structural formulas for (*a*) C_2HCl (*b*) C_2H_6O (*c*) C_2H_4O (*d*) NH_3O (*e*) NO_2^- (both oxygens terminal) (*f*) N_2O_4 (all oxygens terminal) (*g*) OF_2

▌ (*a*) H—C≡C—Cl (*b*) H_3C—CH_2—Ö—H *or* H_3C—Ö—CH_3

(*c*) H_3C—CH=Ö *or* H_2C=CH—Ö—H

(*d*) H_2N—Ö—H (*e*) $[\ddot{O}$=N—Ö:$]^-$ ⟷ $[$:Ö—N=Ö$]^-$ *resonance*

(*f*) O_2N—NO_2 (four forms with alternate N=O double bond locations) *resonance*

(*g*) :F—O—F:

7.69 The sulfate ion is tetrahedral, with four equal S—O distances of 1.49 Å. Draw a reasonable structural formula consistent with these facts.

▌ It is possible to place the 32 valence electrons (6 for each of the five Group VI atoms plus 2 for the net negative charge) in an octet structure having only single bonds. There are two objections to this formula (Fig. 7.17). (1) The predicted bond distance, $r_S + r_O = 1.04 + 0.66 = 1.70$ Å, is much too high. (2) The calculated formal charge on the sulfur, 2+, is rather high. An alternative is to write resonance structures containing double bonds. A structure like that in Fig. 7.18 places zero formal charge on the sulfur and 1− on each of the singly bonded oxygens. The shrinkage in bond length due to double bond formation helps to account for the low observed bond distance. Other resonance structures with alternate locations of the double bonds are, of course, implied. Structures like this, with an expanded valence level beyond the octet, involve d orbitals of the central atom. This is the reason that second-period elements (C, N, O, F) do not form compounds requiring more than 8 valence electrons per atom; there is no $2d$ subshell.

Fig. 7.17　　Fig. 7.18

7.70 Write resonance structures for the phenolate ion, $C_6H_5O^-$.

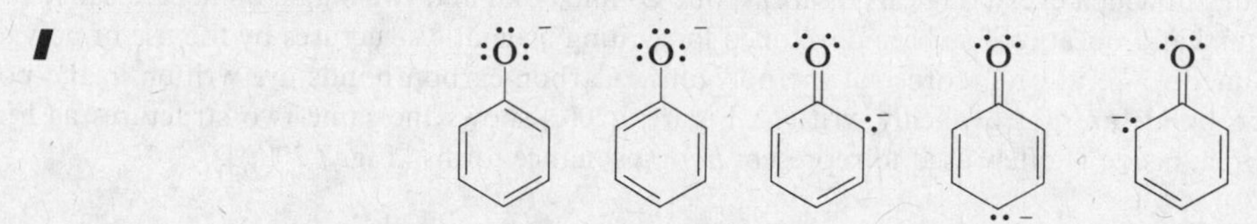

The following form may be substituted for the first two:

7.71 By completing the following structures, adding unshared electron pairs when necessary, evaluate the formal charges.

(*a*) N≡C—C≡N　(*b*) N=C=C=N　(*c*) Cl—C≡N
(*d*) Cl—C=N　(*e*) N=N=O　(*f*) N≡N—O
(*g*) Cl_2C=O　(*h*) Cl=C(Cl)—O　(*i*) Cl—N(=O)—O

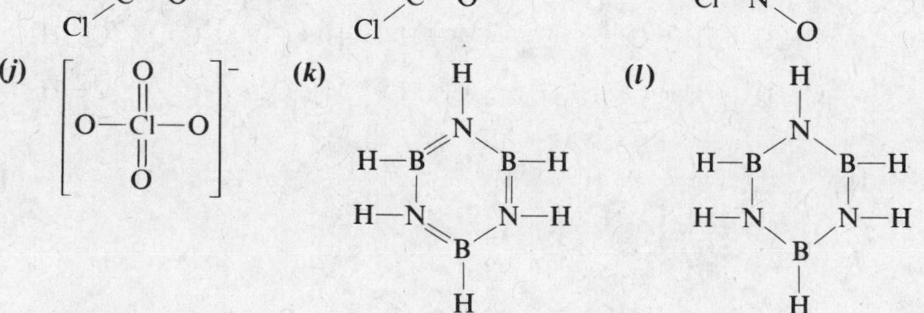

(*j*)　(*k*)　(*l*)

(*a*) All zero (*b*) 1+ on one N (which does not have an octet), 1− on the other (*c*) all zero (*d*) 1+ on Cl, 1− on N (*e*) 1− on terminal N, 1+ on central N (*f*) 1+ on central N, 1− on O (*g*) all zero (*h*) 1+ on doubly bonded Cl, 1− on O (*i*) 1+ on N, 1− on singly bonded O (*j*) 1+ on Cl, 1− on each singly bonded O (*k*) 1+ on each N, 1− on each B (*l*) all zero

7.72 The structure of 1,3-butadiene is often written as $H_2C{=}CH{-}CH{=}CH_2$. The distance between the central carbon atoms is 1.46 Å. Comment on the adequacy of the assigned structure.

There must be nonoctet resonance structures involving double bonding between the central carbons, such as $\overset{+}{C}H_2{-}CH{=}CH{-}\ddot{C}H_2^-$.

7.73 Draw all the octet resonance structures of (*a*) benzene, C_6H_6, and (*b*) naphthalene, $C_{10}H_8$. Benzene is known to have hexagonal symmetry, and the carbon framework of naphthalene consists of two coplanar fused hexagons. Invoke bonding only between adjacent atoms.

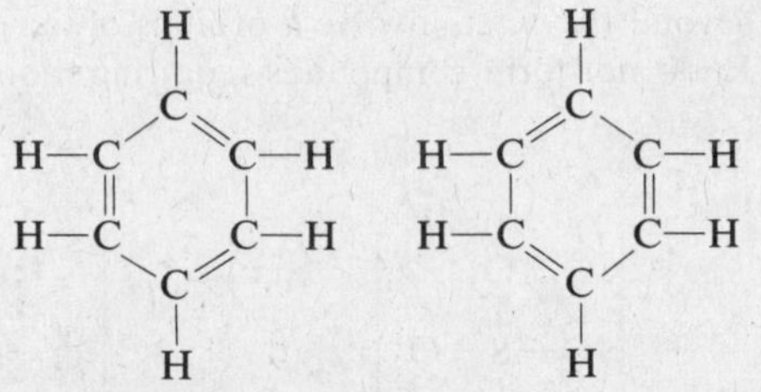

Fig. 7.19

(*a*) The hydrogens are distributed uniformly, one on each carbon, to conform with the hexagonal symmetry (Fig. 7.19). This leaves three electron pairs to bond at each carbon to bring its total covalence to four. Alternating single and double bonds form the only scheme for writing formulas within all of the above restrictions. Hydrocarbons containing planar rings in which every ring carbon forms one double bond and two single bonds are called *aromatic* hydrocarbons. A shorthand notation has been developed for writing aromatic structures by the use of polygons. A carbon atom is assumed to lie at each corner of the polygons. Carbon-carbon bonds are written in the polygons but carbon-hydrogen bonds are not explicitly written. Figure 7.20(*a*) shows the same two structures as Fig. 7.19. A hexagon with a circle inside is often used to represent *both* resonance forms [Fig. 7.20(*b*)].

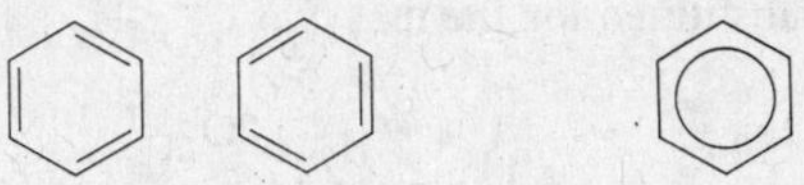

(*a*) (*b*) Fig. 7.20

(*b*) The planarity of naphthalene indicates its aromatic character. Note in Fig. 7.21 that the two carbons at the fusion of the two rings reach their covalence of four without bonding to hydrogen. The shorthand notation of these structures is indicated in Fig. 7.22.

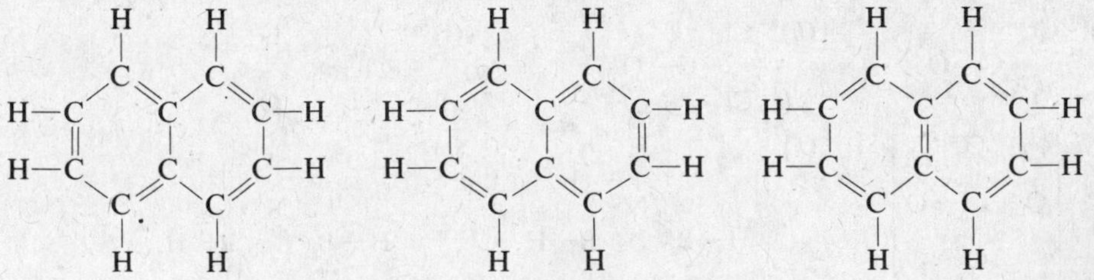

Fig. 7.21

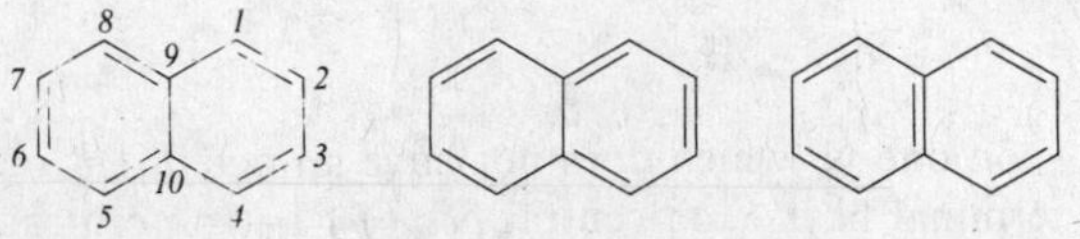

(*a*) (*b*) (*c*) Fig. 7.22

A C—H bond is assumed at each of those carbons sharing only three electron pairs within the carbon skeleton.

7.74 Enthalpies of hydrogenation of ethylene (C_2H_4) and benzene (C_6H_6) have been measured, where all reactants and products are gases.

$$C_2H_4 + H_2 \longrightarrow C_2H_6 \qquad \Delta H = -32.7 \text{ kcal}$$
$$C_6H_6 + 3H_2 \longrightarrow C_6H_{12} \qquad \Delta H = -49.2 \text{ kcal}$$

Estimate the resonance energy of benzene.

If C_6H_6 had three isolated carbon-carbon double bonds, ΔH of hydrogenation would be close to three times ΔH of hydrogenation of C_2H_4, with one double bond, or -98.1 kcal. The fact that benzene's hydrogenation is less exothermic by

$$98.1 \text{ kcal} - 49.2 \text{ kcal} = 48.9 \text{ kcal}$$

means that benzene has been stabilized by resonance to the extent of 48.9 kcal/mol, or 205 kJ/mol.

7.75 Show resonance structures for the following aromatic hydrocarbons

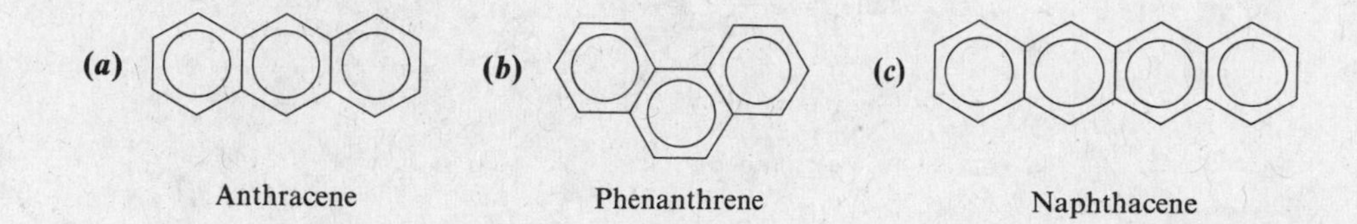

Consider double bonds only between adjacent carbons. (The circle inside the hexagon is a shorthand notation used to designate an aromatic ring without having to write out all the valence-bond structures.)

▌ See Fig. 7.23.

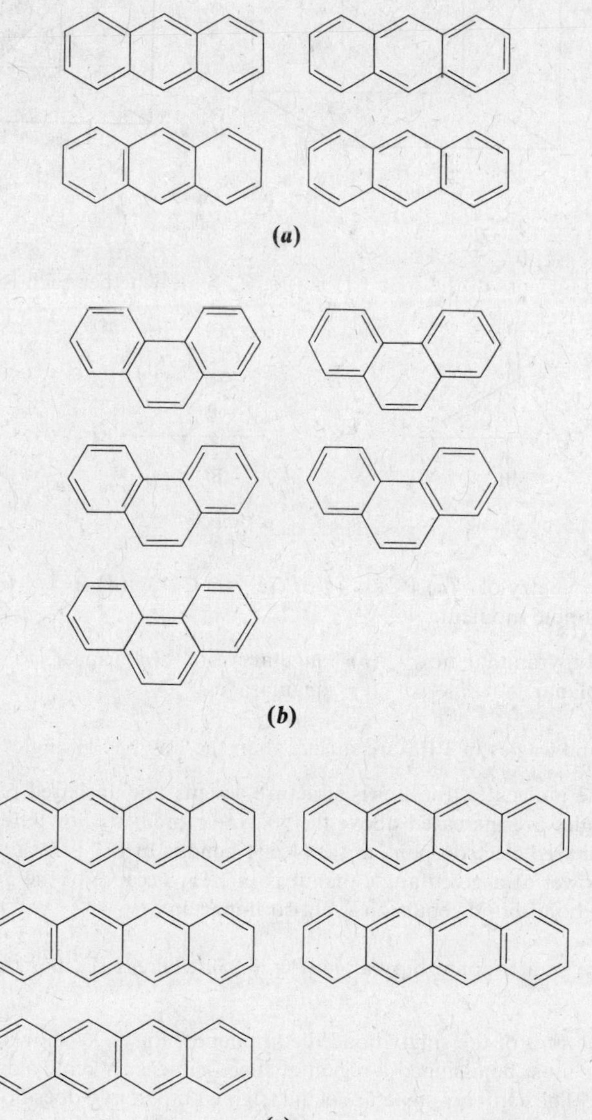

Fig. 7.23

7.5 GEOMETRY OF MOLECULES

7.76 Verify the value $\theta = 109°28'$ for the central angles in a regular tetrahedron.

▮ A simple way of constructing a regular tetrahedron is to select alternating corners of a cube and to connect each of the selected corners with each of the other three, as in Fig. 7.24(*a*). Figure 7.24(*b*) shows triangle OAB, determined by the center of the cube, which is also the center of the tetrahedron, and two corners of the tetrahedron. If P is the midpoint of AB, we see from right triangle OPA that

$$\tan\frac{\theta}{2} = \frac{a\sqrt{2}/2}{a/2} = \sqrt{2}$$

$$\frac{\theta}{2} = 54°44'$$

$$\theta = 109°28'$$

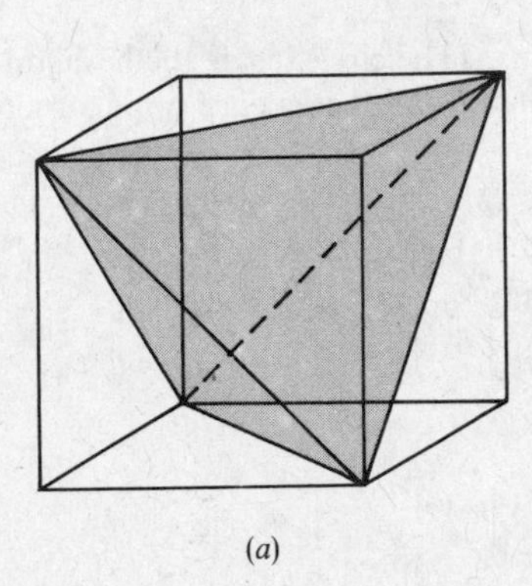

(*a*)

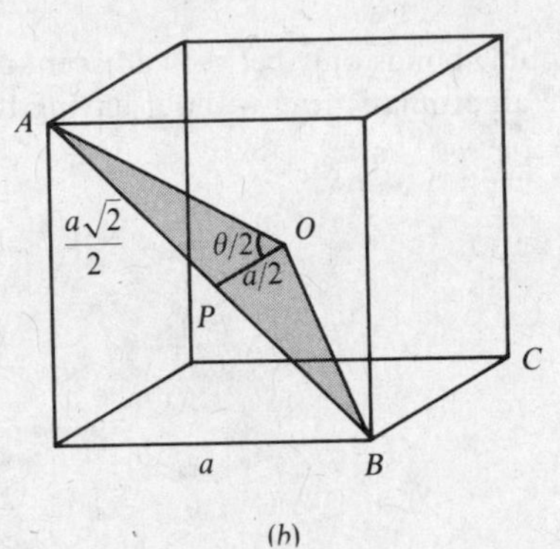

(*b*)

Fig. 7.24

7.77 Draw all geometric isomers of the PBr_2Cl_3 molecule. State whether each isomer has a dipole moment.

▮

Br
Cl
P
Cl Cl
Br

no dipole

Cl
Cl
P
Cl Br
Br

dipole

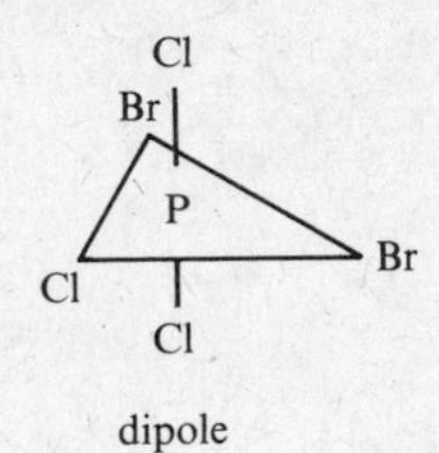

dipole

7.78 Characterize the geometry of **(*a*)** PCl_5 **(*b*)** H_2Se **(*c*)** CO_2 **(*d*)** BCl_3 **(*e*)** H_2O and state if each molecule has a finite (nonzero) dipole moment.

▮ **(*a*)** Trigonal bipyramidal, no **(*b*)** angular, yes **(*c*)** linear, no
(*d*) trigonal planar, no **(*e*)** angular, yes

7.79 The H—P—H bond angles in PH_3 are smaller than the H—N—H angles in NH_3. Explain.

▮ In each of these molecules, the Lewis structure assigns one unshared pair of electrons to the central atom. In NH_3 the bond angles are increased above the 90° value predicted for pure *p* bonding in order to provide greater separation of the shared electron pairs in the N—H bonds. In PH_3, the intrinsic electron repulsion between P—H bonds is already lower at a given angle than it is in NH_3 because of the greater size of the P atom, so that angle widening to avoid bond-bond repulsion is not so important.

7.80 PCl_5 has the shape of a trigonal bipyramid (Fig. 7.25), whereas IF_5 has the shape of a square pyramid. Account for this difference.

▮ The Lewis structures of the singly bonded compounds, Fig. 7.26, show an unshared electron pair on the iodine. This unshared pair must be assigned a region of space sufficiently removed from the I—F bonds so as to minimize electron repulsion. The compact structure of the trigonal bipyramid does not allow room for the unshared pair. The square pyramid structure of IF_5 may be thought of as an octahedron with the unshared pair pointing to one of the corners.

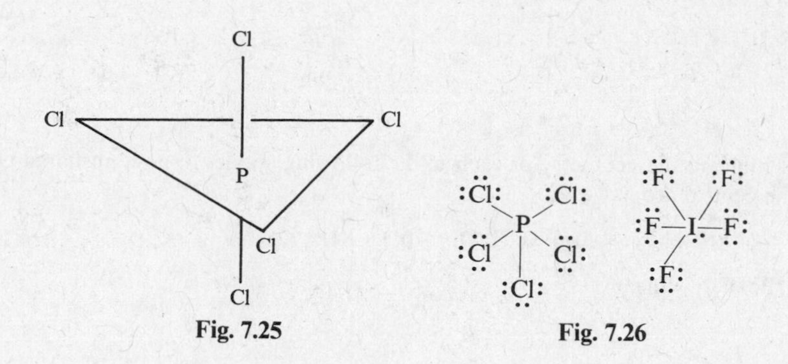

Fig. 7.25 Fig. 7.26

Iodine could be prepared for octahedral ($sp^3 d^2$) hybridization by promoting its ground-state configuration, $5s^2\ 5p^5$, to $5s^2\ 5p^3\ 5d^2$.

The trigonal bipyramid geometry is achieved by $sp^3 d$ hybridization. Phosphorus can be prepared for this hybridization by promoting its ground-state configuration from $3s^2\ 3p^3$ to $3s^1\ 3p^3\ 3d^1$.

7.81 Write electron dot structures for and predict the geometry of (*a*) H_3O^+ (*b*) $CH_2{=}NH$ (*c*) ClO_2^- (*d*) NH_4^+ (*e*) N_2H_4

▮ (*a*) [H:Ö:H with H above]$^+$ pyramidal

(*b*) H₂C::N̈—H (H, H on C; H on N) planar The carbon atom is trigonal (planar); the nitrogen atom is angular. The whole molecule lies in one plane.

(*c*) [:Ö:C̈l:Ö:]$^-$ angular

(*d*) [H:N̈:H with H above and below]$^+$ tetrahedral

(*e*) H:N̈:N̈:H (H H below) Both nitrogen atoms are pyramidal; the molecule is nonplanar.

7.82 Deduce the shape of (*a*) SO_3 (*b*) SO_3^{2-} (*c*) BF_3 (*d*) BF_4^- (*e*) NF_3

▮ (*a*) Trigonal (*b*) pyramidal (*c*) trigonal (*d*) tetrahedral (*e*) pyramidal

7.83 Which one of each of the following pairs is expected to have the larger bond angle? (*a*) H_2O and NH_3 (*b*) SF_2 and BeF_2 (*c*) BF_3 and BF_4^- (*d*) PH_3 and NH_3 (*e*) NH_3 and NF_3

▮ (*a*) NH_3 (Lone pairs expand more, crowding bonding pairs together. NH_3 has only one lone pair; H_2O has two.) (*b*) BeF_2 (*sp* hybrids are linear.) (*c*) BF_3 (*d*) NH_3 (*e*) NH_3

7.84 Two different bond lengths are observed in the PF_5 molecule, but only one bond length is observed in SF_6. Explain the difference.

▮ Trigonal bipyramidal structures have inherently two different types of bonding—axial, 90° away from a plane, and equatorial, in the plane. The equatorial bonds are 120° away from each other. Since the bonding is different, the bond lengths are not expected to be exactly the same. SF_6 is octahedral; all 6 bonds are exactly the same.

7.85 Draw an electron dot structure for Br_3^-. Deduce an approximate value for the bond angle, and explain your deduction.

▮ $:\ddot{Br}:\ddot{Br}:\ddot{Br}:^-$ The molecule should be linear, utilizing two axial dsp^3 orbitals, with the three other dsp^3 orbitals occupied by lone pairs.

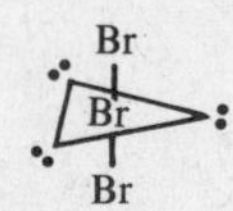

7.86 Sketch and indicate the geometry of each of the following species. Show unshared electron pairs only when they occur on the central atom.

(*a*) BeF_2 (*b*) NO_2^- (*c*) CH_2O (*d*) XeF_4

▮ (*a*) F:Be:F linear (*b*) $\ddot{N}$ with O and O, angular

(*c*) $H_2C::O$ trigonal (*d*) XeF_4 (Xe with four F and two lone pairs) square planar

7.87 Predict the geometry of each of the following molecules: BeH_2, BF_3, CH_4, PF_5, SF_6, NH_3, H_2O, XeF_4, and CO_2.

▮ For BeH_2, the two electron pairs will be located on opposite sides of the beryllium atom, yielding a linear molecule. For BF_3, the 3 electron pairs will be located at angles of 120° with respect to each other, giving the molecule a planar, triangular shape.

For CH_4, so that the 4 electron pair bonds will be located as far as possible from each other, they must assume a regular tetrahedral orientation, with angles of 109.5° with respect to each other. The structures of PF_5 and SF_6 are trigonal bipyramidal and octahedral, respectively.

The electron dot formula for ammonia suggests a tetrahedral arrangement of the four electron pairs. However, the unshared pair of electrons exerts a comparatively larger repulsion, and therefore the angles between the bonded pairs are somewhat smaller than 109.5° (actually 107°). The molecule has the shape of a triangular pyramid, with the nitrogen at the apex. The electron pair which is not involved in bonding is not included in the description of the geometry of the molecule.

In the H_2O molecule there are two sets of unshared electron pairs. The arrangement of bonding electron pairs is more distorted from the tetrahedral angle than in NH_3, yielding an angular molecule (with a bond angle of 105°). The electron dot formula for XeF_4 contains two pairs of unshared electrons and four pairs of shared electrons on the xenon atom. The former repel other electron pairs most and therefore lie on opposite sides of the xenon atom. The other four pairs are arranged at 90° angles about the xenon atom, all in one plane. A square planar molecule results.

The electron dot formula of carbon dioxide is written with two double bonds—two groups of electrons. Since there are no other electron pairs in the outermost shell of carbon, these double bonds will be located on opposite sides of the carbon atom, and the molecule is linear.

7.88 The observed H—C—H angle in the molecule $H_2C{=}O$ is 111°, rather than 120°. Explain this observation.

▮ The electron dot formula, with three groups of electrons, suggests a trigonal planar molecule:

$H_2C::\ddot{O}:$

The electron pairs in the double bond repel the others more than the electron pairs of the single bonds repel each other. The angle between the single bonds is expected to be somewhat less than 120°.

7.89 The bond angle in $SnCl_2$ is close to 120°, but the angle in I_3^- is 180°. Explain in terms of electron repulsions.

▮ Sn has one unshared pair in $SnCl_2$, allowing a trigonal structure with the unshared pair pointing to one of the corners of the triangle. The central iodine in I_3^- has three unshared pairs, and a trigonal bipyramidal shape is needed to accommodate each unshared pair at a corner. Among the possible trigonal bipyramidal arrangements shown in Fig. 7.27, (*a*) is most stable because no two of the orbitals occupied by an unshared pair are at 90° to each other. (The smaller the angle with respect to the central atom, the greater the repulsion between the pairs and the lower the stability.)

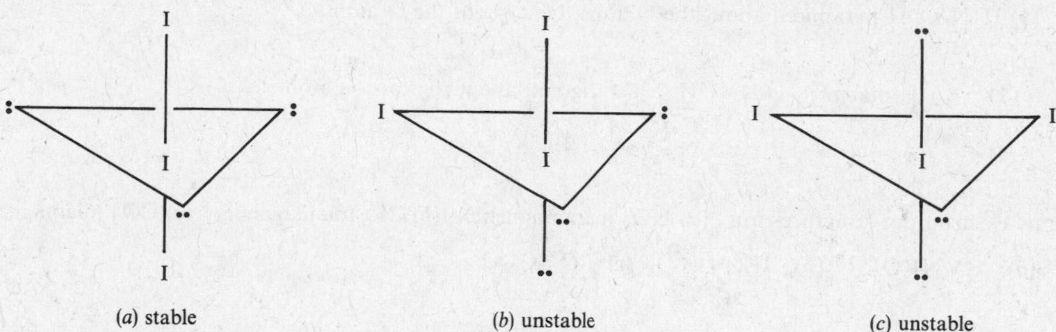

Fig. 7.27

7.90 Predict the shapes of **(a)** $TeCl_4$ **(b)** ClO_3 **(c)** ICl_2^+

▮ See Fig. 7.28: **(a)** Seesaw shaped **(b)** trigonal pyramid with angles less than 109.5° **(c)** angular, with bond angle less than 109.5°

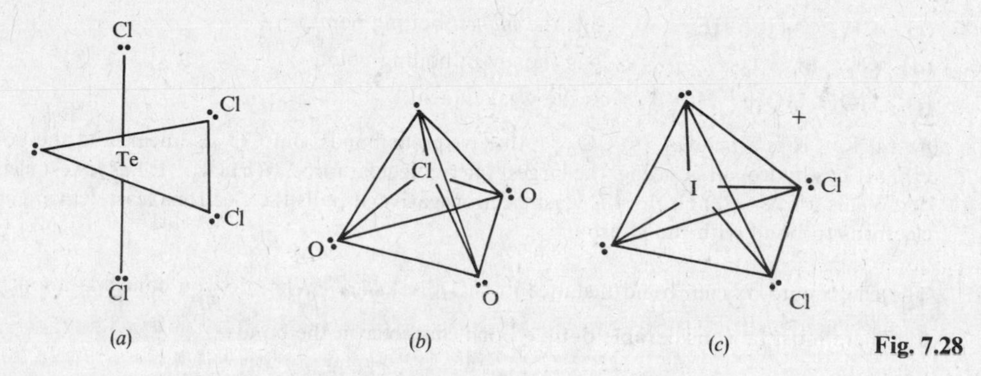

Fig. 7.28

7.91 Experimentally the azide ion, N_3^-, is found to be linear, with each adjacent nitrogen-nitrogen distance equal to 1.16 Å. **(a)** Evaluate the formal charge at each nitrogen in each of the 3 linear octet structures depicted in Fig. 7.15. **(b)** Predict the relative importance of these 3 resonance structures in N_3^-.

▮ **(a)** In structure (a) of Fig. 7.15, the central N is assigned half of the 4 shared pairs, or 4 electrons. This number is one less than the number of valence electrons in a free N atom, and thus this atom has a formal charge of 1+. Each terminal N is assigned 4 unshared electrons plus half of the two shared pairs, or 6 altogether. The formal charge is thus 1−. The net charge of the ion, 1−, is the sum $2(-1)+1$.

In structures (b) and (c) of Fig. 7.15, the central N again has 4 assigned electrons with a resulting formal charge of 1+. The terminal triple bonded N has 2 plus half of 3 pairs, or a total of 5, with a formal charge of 0. The singly bonded terminal N has 6 plus half of 1 pair, or a total of 7, with a formal charge of 2−. The net charge of the ion, 1−, is the sum of +1 and −2.

(b) For structure (a) the nitrogen-nitrogen bond distance is predicted to be 1.26 Å. The observed bond length, 1.16 Å, is shorter, possibly because of a contribution from structures (b) and (c), which would give shortening because of the triple bond.

7.92 Deduce the geometry of each of the following molecules: **(a)** XeF_6 **(b)** XeO_3 **(c)** XeF_4 **(d)** BrF_5

▮ **(a)** The 14 valence electrons constitute 6 single bonds and 1 lone pair. The geometry is related to an octahedron, with an extra pair of electrons in one face center, which distorts the regularity of the symmetry. The geometry is termed "distorted octahedral." **(b)** The three oxygen atoms can bond by providing an empty (sp) orbital to overlap with a filled sp^3 orbital of Xe. The expected geometry is pyramidal. **(c)** The molecule is square planar. The 12 electrons are octahedral, with bonding pairs at the corners of the square. **(d)** Square pyramidal. The 12 valence electrons are at the corners of an octahedron, but one of the 6 pairs has no F atom to bond with.

7.93 Write electron dot structures and describe the geometry of the following molecules: **(a)** hydrazine, NH_2NH_2 **(b)** hydroxylamine, NH_2OH **(c)** methylenimine, $CH_2{=}NH$ **(d)** acetyl chloride, $CH_3C(=O)—Cl$.

▌ (*a*) H:N:N:H pyramidal about each N atom
H H

(*b*) H:N:O:H pyramidal about the N atom, bent about the O atom
H

(*c*) H C::N H planar (*d*) H:C:C O: trigonal about the carbon atom
H H Cl

7.94 Draw electron dot structures for (*a*) NO_2 (paramagnetic) (*b*) PF_3 (diamagnetic) (*c*) CO_2 (diamagnetic)

▌ (*a*) :O:N::O: (*b*) :F:P:F: (*c*) :O::C::O:
:F:

7.95 Select the species which is best described by the statement to the right.

(*a*)	Cl Ar K	has the smallest ionization potential.
(*b*)	CO_2 NH_3 CO	has a zero dipole moment.
(*c*)	CH_4 NH_3 HF	has the *highest* boiling point.
(*d*)	Cl_2 Br_2 I_2	has the *lowest* boiling point.
(*e*)	HOI HOBr HOCl	is the weakest acid.

▌ (*a*) K. K is a metal. (*b*) CO_2. All have polar bonds, but CO_2 is linear, and the bond moments cancel. (*c*) HF. It has hydrogen bonding, the largest intermolecular force. (*d*) Cl_2. It has fewest electrons, therefore least van der Waals forces. (*e*) HOI. I is least electronegative; it pulls the electrons least, leaving a greater fraction of the O electrons to bond with the H atom.

7.96 The chlorine-to-oxygen bond distance in ClO_4^- is 1.44 Å. What do you conclude about the structure of this ion?

▌ There must be considerable double bond character in the bonds.

7.97 The $POCl_3$ molecule has the shape of an irregular tetrahedron with the P atom located centrally. The Cl—P—Cl angle is found to be 103.5°. Give a qualitative explanation for the deviation of this structure from a regular tetrahedron.

▌ The Lewis structure for $POCl_3$ would show some double bond character between P and O. (P is allowed to exceed the octet because of the availability of 3*d* orbitals.) The increased electron density in the P═O bond would make the intrinsic repulsion between the P═O bond and a P—Cl bond greater than between two P—Cl bonds. Thus the Cl—P—Cl angle is lowered and the Cl—P═O angle raised, as compared with a regular tetrahedron.

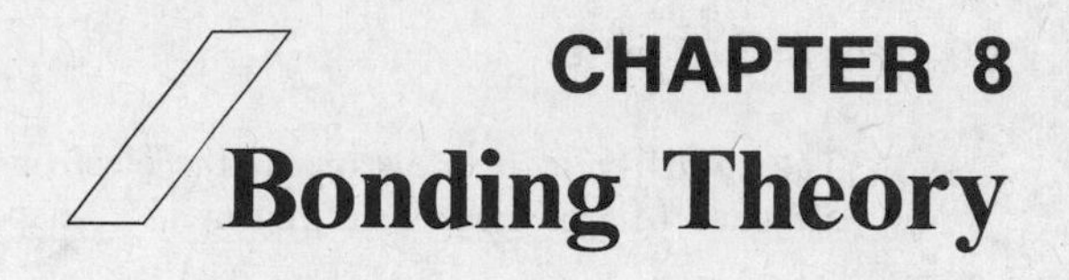

CHAPTER 8
Bonding Theory

8.1 VALENCE BOND THEORY

8.1 Compare the shapes of a *p* orbital and an *sp* hybrid orbital. Which one has a greater directional orientation? Explain.

▌ The *p* orbital has equal sized lobes; most of the *sp* probability density is on one side, making the latter more directional in character.

8.2 Complete the following table:

	hybrid type	geometry
(*a*)	______	linear
(*b*)	dsp^2	______
(*c*)	______	trigonal bipyramidal
(*d*)	______	octahedral
(*e*)	sp^2	______

▌ (*a*) *sp* (or *ds*) (*b*) square planar (*c*) dsp^3 (*d*) d^2sp^3 or sp^3d^2 (*e*) trigonal planar

8.3 Which of the sets of hybridized orbitals—*sp*, sp^2, sp^3, dsp^2, dsp^3, d^2sp^3, sp^3d^2—do (does) not maximize the angles between electron pairs?

▌ The sp^3d hybrids have 90° and 120° angles. Rearrangement of the bond angles could produce angles greater than 90° but at the expense of the highly symmetrical structure.

8.4 Describe the promotion and the hybrid orbitals in (*a*) each carbon atom in acetylene, HC≡CH (*b*) SF_6 (*c*) ICl_3

▌ (*a*) The ground-state carbon atom configuration is

↑↓	↑	↑	
2*s*		2*p*	

One electron is promoted to increase the number of unpaired electrons in *s* and *p* subshells to the maximum, in this case 4.

↑	↑	↑	↑
2*s*		2*p*	

Since there are two atoms bonded to each carbon atom, and no lone pairs, 2 hybrid orbitals are required. The hybrid orbitals formed are *sp*, which are used to form σ bonds to the bonded atoms. The other two unpaired electrons on each atom pair up in π bonds with the similar electrons on the other carbon atom. A linear molecule results.

(*b*) The sulfur atom has a ground state

↑↓	↑↓	↑	↑					
3*s*		3*p*				3*d*		

Since 6 uncharged fluorine atoms are to be added, 6 unpaired electrons are required in 6 hybrid orbitals. The sp^3d^2 hybrids which result cause formation of an octahedral molecule.

(*c*) The iodine atom has a ground state

↑↓	↑↓	↑↓	↑					
5*s*		5*p*				5*d*		

Since there are to be 3 atoms bonded, together with 2 lone pairs, there must be 3 unpaired electrons. After promotion, the configuration is

↑↓	↑↓	↑	↑	↑				
5*s*		5*p*				5*d*		

The sp^3d hybrid orbitals orient the electron pairs toward the corners of a trigonal bipyramid; the atoms will be attached to 3 of these, resulting in a "T-shaped" molecule.

8.5 Indicate the type of hybrid orbitals of the underlined atom and the molecular geometry. **(a)** $\underline{Be}Cl_2$ **(b)** $\underline{C}Cl_4$ **(c)** $\underline{C}_2F_4$ **(d)** $\underline{S}F_6$ **(e)** $\underline{B}Cl_3$ **(f)** $H\underline{C}N$

▌ **(a)** sp, linear **(b)** sp^3, tetrahedral **(c)** sp^2, planar
(d) sp^3d^2, octahedral **(e)** sp^2, trigonal planar **(f)** sp, linear

8.6 Deduce the hybridization of the central atom and the geometry of each of the following molecules: **(a)** NH_3 **(b)** C_2H_4 **(c)** ClO_3^-

▌ **(a)** sp^3, pyramidal **(b)** sp^2, planar (trigonal about each C) **(c)** sp^3, pyramidal: ↑↓ ↑↓ ↑↓ ↑↓ (O O O above the last three)

8.7 The carbon-carbon double bond energy in C_2H_4 is 615 kJ/mol and the carbon-carbon single bond energy in C_2H_6 is 347 kJ/mol. Why is the double bond energy appreciably less than twice the single bond energy?

▌ A σ orbital has greater electron overlap between the atoms because its component p atomic orbitals are directed toward each other, whereas the component p orbitals making up the π orbital are directed perpendicular to the internuclear axis.

8.8 The bond angle in H_2O is 105° and in H_2S it is 92°. Explain this difference.

▌ The larger size of the S atom as compared with O minimizes electron repulsions and allows the bonds in H_2S to be more purely p-type.

8.9 The two $—CH_2$ groups in C_2H_4 do not rotate freely around the bond connecting them, although the two $—CH_3$ groups in C_2H_6 have almost an unhindered rotation around the C—C bond. Why?

▌ C_2H_4 has a π bond. The overlap of the atomic p orbitals making up this bond would be destroyed if the $—CH_2$ groups did rotate.

8.10 Which of the four carbon-carbon bond types in naphthalene would you predict to be the shortest? Refer to Fig. 7.22.

▌ The four different kinds of carbon-carbon bonds are represented by *1–2*, *1–9*, *2–3*, and *9–10*. (Every other carbon-carbon bond is equivalent to one of these four. For example, *6–7* is equivalent to *2–3*, *7–8* is equivalent to *1–2*, and so on.) The bond with the greatest double bond character should be the shortest. Among the three resonance structures shown in Fig. 7.22, the frequency of double bonds for the various bond types is as follows: 2 in *1–2* [in (*a*) and (*c*)], 1 in *1–9* [in (*b*)], 1 in *2–3* [in (*b*)], and 1 in *9–10* [in (*c*)]. Bond *1–2* is expected to have the greatest double bond character and thus the shortest length. This prediction is found to be true experimentally. The above four bonds are found to have lengths of 1.365, 1.425, 1.404, and 1.393 Å, respectively.

Note that the method of counting the number of resonance structures containing a double bond between a given pair of carbon atoms is very crude and cannot distinguish among the last three of the listed bond types, each of which shows a double bond in only one resonance structure. Even within the framework of the limited resonance theory, it would be necessary to know the relative weighting of each of the two equivalent structures (*a*) and (*b*) with the nonequivalent structure (*c*).

8.11 In benzene, what is the valence bond hybridization on each carbon atom?

▌ sp^2

8.12 What are the hybridization states of each carbon atom in the following molecules?
(a) $CH_3CH_2CH_2CH_3$ **(b)** $CH_2{=}CH{-}CH{=}CH_2$ **(c)** $CH_3CH{=}CHCH_3$ **(d)** $H{-}C{\equiv}C{-}H$

▌ **(a)** All sp^3 **(b)** all sp^2 **(c)** sp^3 sp^2 sp^2 sp^3 **(d)** both sp

8.13 What hybridization is expected on the central atom of each of the following molecules? **(a)** BeH_2 **(b)** CH_2Br_2 **(c)** PF_6^- **(d)** BF_3

▌ **(a)** sp **(b)** sp^3 **(c)** sp^3d^2 **(d)** sp^2

8.14 **(a)** Which molecule, AX_3, AX_4, AX_5, AX_6, is most likely to have a trigonal bipyramidal structure? **(b)** If the central atom, A, has no lone pairs, what type of hybridization will it have?

▮ (*a*) AX_5 is the only molecule with 5 atoms bonded to a central atom—it is the only possible molecule listed which could be trigonal bipyramidal. (*b*) In the absence of lone pairs, it must be dsp^3 or sp^3d. (An inner or a valence *d* orbital may be involved.)

8.15 Determine the geometry of each of the following molecules and the hybridization about the central atom in each: (*a*) $BeF_2(g)$ (*b*) AlH_3 (*c*) NH_3 (*d*) $HC{\equiv}CH$

▮ (*a*) Linear, *sp* (*b*) trigonal (planar), sp^2 (*c*) pyramidal, sp^3 (*d*) linear, *sp* on each carbon atom

8.16 Predict the shapes of the following species and describe the type of hybrid orbitals on the central atom: (*a*) $PbCl_4$ (*b*) SbF_6^- (*c*) BH_4^- (*d*) PCl_3 (*e*) N_2Cl_4

▮ (*a*) Tetrahedral, sp^3 (*b*) octahedral, sp^3d^2 (*c*) tetrahedral, sp^3 (*d*) pyramidal, sp^3 (a lone pair occupies the fourth tetrahedral position) (*e*) two pyramidal nitrogen atoms, with sp^3 hybridization on each, yield a nonplanar molecule.

8.17 What hybrid orbitals are ascribed to carbon in the (short-lived) CH_2^{2+} ion? What is the geometry of this ion?

▮

	2s	2p		
Ground state of carbon	↑↓	↑	↑	__
Ground state of C^{2+}	↑↓	__	__	__
Promotion would yield	↑	↑	__	__

Only two bonds (and no lone pairs) are required for each carbon atom, so *sp* hybridization will occur. Thus a linear molecule is expected. There will be no multiple bonding, since the hydrogen atoms have no orbitals available and the carbon atom has lost the extra electrons.

8.18 Show schematically, according to valence bond theory, the orbital occupancy of electrons in the chlorine atom in (*a*) ClO_3^- (*b*) ClO_4^- (*c*) Describe the geometries of these two species.

▮

	3s	3p		
Cl	↑↓	↑↓	↑↓	↑
Cl^-	↑↓	↑↓	↑↓	↑↓
	sp^3 ↑↓	sp^3 ↑↓	sp^3 ↑↓	sp^3 ↑↓

The oxygen atoms can bond to the chloride ion utilizing fully occupied Cl orbitals, with no electrons furnished by the oxygen atoms.

(*a*)

	:Ö:	:Ö:	:Ö:	
ClO_3^-	↑↓	↑↓	↑↓	↑↓
	sp^3	sp^3	sp^3	sp^3

(*b*)

	:Ö:	:Ö:	:Ö:	:Ö:
ClO_4^-	↑↓	↑↓	↑↓	↑↓
	sp^3	sp^3	sp^3	sp^3

(*c*) ClO_3^- is pyramidal; ClO_4^- is tetrahedral.

8.19 Describe and compare the geometries of the molecules and the hybridization of the carbon and boron atoms in $F_2C{=}C{=}CF_2$ and $F_2B{-}C{\equiv}C{-}BF_2$. Compare the relative orientations of the sets of fluorine atoms in the two cases. In which case is it impossible for all four fluorine atoms to lie in the same plane?

▮ In $F_2C{=}C{=}CF_2$ the carbon atoms are linear owing to the *sp* hybridization of the center carbon atom. The other carbon atoms are sp^2, one utilizing the "horizontal" plane and the other the "vertical" plane of the center carbon atom. The molecule will have two fluorine atoms on each side in different planes (Fig. 8.1). In $F_2B{-}C{\equiv}C{-}BF_2$, the B and C atoms are again linear as a result of the *sp* hybridization of each carbon atom. The B atoms are sp^2 hybridized and are trigonal. No set orientation of the fluorine atoms with respect to each other is expected, since there is free rotation about the B—C single bonds.

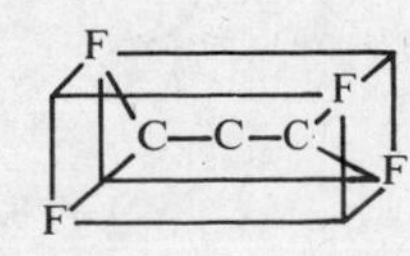

Fig. 8.1

8.20 Deduce the geometry and diagram the electronic structure of **(a)** I_3^- **(b)** ClO_3^- **(c)** ClO_3^+ **(d)** F_2SeO **(e)** $IO_2F_2^-$

▮ **(a)** Linear sp^3d The electrons are directed toward the corners of a trigonal bipyramid [Fig. 8.2(*a*)].

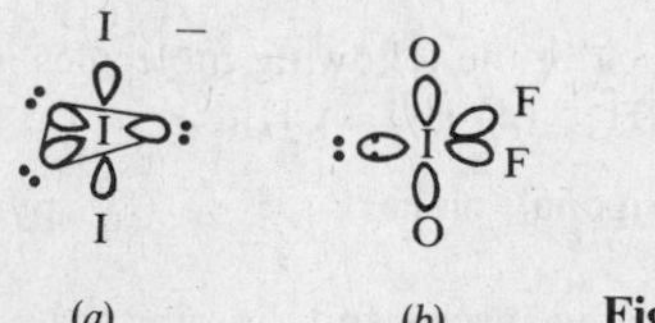

Fig. 8.2

(b) pyramidal sp^3 The electrons are arranged in a tetrahedron. See Prob. 8.18.

(c) trigonal sp^2 In this unstable ion, Cl—O double bonding is expected:

$$\mathrm{O{-}\underset{\underset{\displaystyle O}{|}}{Cl}{=}O^+}$$

(d) pyramidal sp^3 The structure is analogous to SO_3^{2-}.

$$:\ddot{\mathrm{F}}:\ddot{\mathrm{Se}}:\ddot{\mathrm{F}}: \quad :\ddot{\underset{\cdot\cdot}{\mathrm{O}}}:$$

(e) seesaw sp^3d The electrons are located at the corners of a trigonal bipyramid, but one of the equatorial pairs is unshared [Fig. 8.2(*b*)].

8.21 Either of the hybridizations **(a)** dsp^2 **(b)** sp^3d^2 of a central atom can lead to a square planar molecule. Give one example of each.

▮ **(a)** $[Ni(CN)_4]^{2-}$ (see Chap. 27) **(b)** XeF_4, which has 6 pairs of electrons at the corners of an octahedron, of which only 4, at the corners of a square, are bonding.

8.2 MOLECULAR ORBITAL THEORY

8.22 Sketch the orbital configuration (boundary surface diagram) for the orbital types **(a)** $3d_{xy}$ **(b)** σ_{2p_x} **(c)** $\pi^*_{2p_x}$ **(d)** sp. Clearly label all axes and nuclei.

▮ See Fig. 8.3.

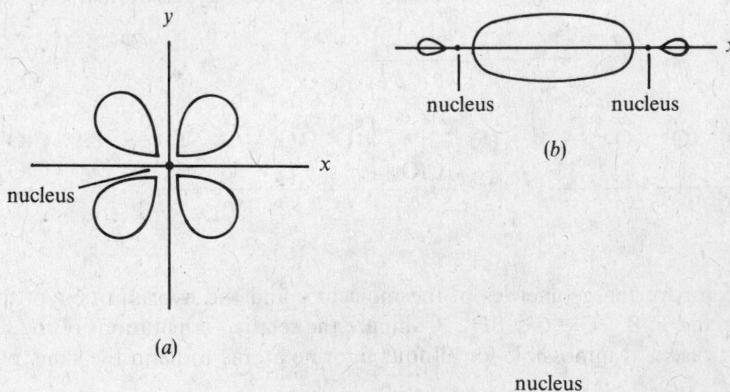

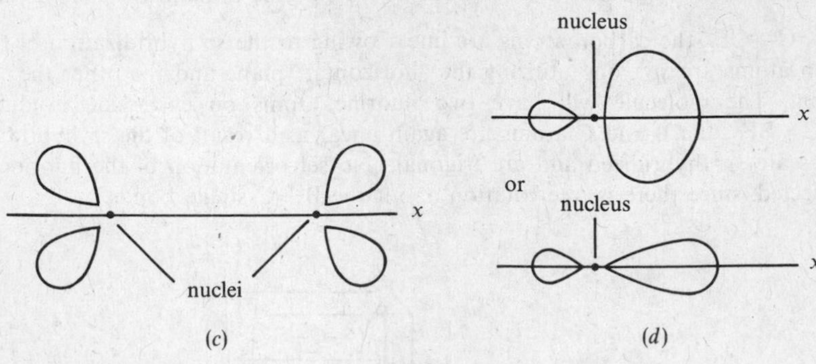

Fig. 8.3

8.23 Draw a diagram showing the formation of a σ_{2p} bonding orbital from the atomic orbitals.

▌ See Fig. 8.4.

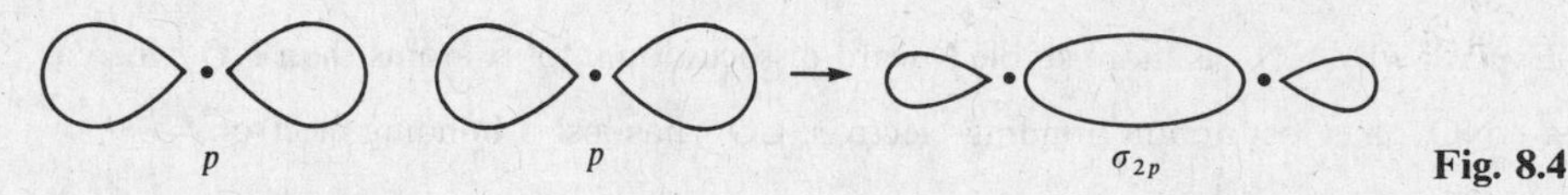

Fig. 8.4

8.24 (*a*) Draw and label an energy level diagram for molecular orbitals up to σ^*_{2p} for a molecule with 14 electrons. (*b*) Use this diagram to give a molecular orbital description of the CO molecule. (*c*) Briefly compare the CO molecule to the CN^- ion.

▌ (*a*)

	___	σ^*_{2p}
___	___	π^*_{2p}
	___	σ_{2p}
___	___	π_{2p}
	___	σ^*_{2s}
	___	σ_{2s}

(*b*) The 10 second-shell electrons of the two atoms fill the lowest 5 orbitals of the molecule, creating a triply bonded species: $\sigma^2_{2s}\sigma^{*2}_{2s}\pi^4_{2p}\sigma^2_{2p}$ (*c*) The CN^- ion is isoelectronic with CO.

8.25 What is the bond order in NO?

▌ The electronic configuration of NO is shown in Fig. 8.5. The 6 bonding and 1 antibonding electrons yield a net of 5 bonding electrons, for a bond order of $2\frac{1}{2}$.

N atom	NO	O atom
	___ σ^*	
	↑ π^* ___ π^*	
↑ ↑ ↑ (2*p*)		↑↓ ↑ ↑ (2*p*)
	↑↓ σ	
	↑↓ π ↑↓ π	

Fig. 8.5

8.26 (*a*) Use diagrams such as the one shown below to write the molecular orbital configurations of F_2 and OF.

	___		σ^*_{2p}
___		___	π^*_{2p}
___		___	π_{2p}
	___		σ_{2p}
	___		σ^*_{2s}
	___		σ_{2s}

(*b*) Which, if any, of these molecules is (are) paramagnetic? Why? (*c*) Which should be more stable toward dissociation into atoms, OF or F_2?

▌ (*a*)

F_2					OF	
	___		σ^*_{2p}		___	
↓↑		↓↑	π^*_{2p}	↓		↓↑
↓↑		↓↑	π_{2p}	↓↑		↓↑
	↓↑		σ_{2p}		↓↑	
	↓↑		σ^*_{2s}		↓↑	
	↓↑		σ_{2s}		↓↑	

(*b*) OF is paramagnetic, since it has one unpaired electron. (With an odd number of electrons, it must be paramagnetic.) (*c*) OF should be more stable, since it has one fewer antibonding electron.

8.27 Explain why NO^+ is more stable toward dissociation into its atoms than NO, whereas CO^+ is less stable than CO.

▌ NO^+ has lost an antibonding electron; CO^+ has lost a bonding electron.

8.28 Represent by electron dot structures and by molecular orbital boundary surface diagrams the electronic structures of (*a*) CO (*b*) CO_2 (*c*) NO_2 (*d*) NO_2^- (*e*) NO_2^+ (*f*) SO_3

▌ See Fig. 8.6.

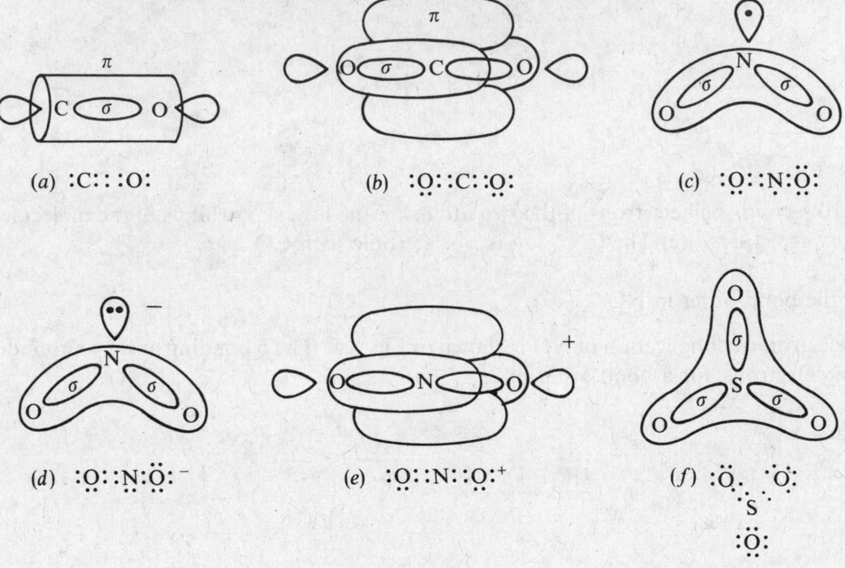

Fig. 8.6

8.29 What would be the expected electronic arrangement and magnetic moment in (*a*) the superoxide ion, O_2^- (*b*) the peroxide ion, O_2^{2-}?

▌ (*a*) O_2^-

	__	
↑↓		↑
↑↓		↑↓
	↑↓	

1 unpaired electron, hence $\mu = \sqrt{n(n+1)} = \sqrt{1(1+2)} = 1.73$ B.M.

(*b*) O_2^{2-}

	__	
↑↓		↑↓
↑↓		↑↓
	↑↓	

no unpaired electrons, hence $\mu = 0$

8.30 Has the peroxide ion, O_2^{2-}, a longer or shorter bond length than O_2? Explain.

▌ O_2^{2-} has a bond order of 1; O_2 has a bond order of 2. Hence O_2^{2-} will have a longer and weaker bond.

8.31 Predict whether the He_2^+ ion in its electronic ground state is stable toward dissociation into He and He^+.

▌ It should be stable because it has one more bonding electron than antibonding (Fig. 8.7).

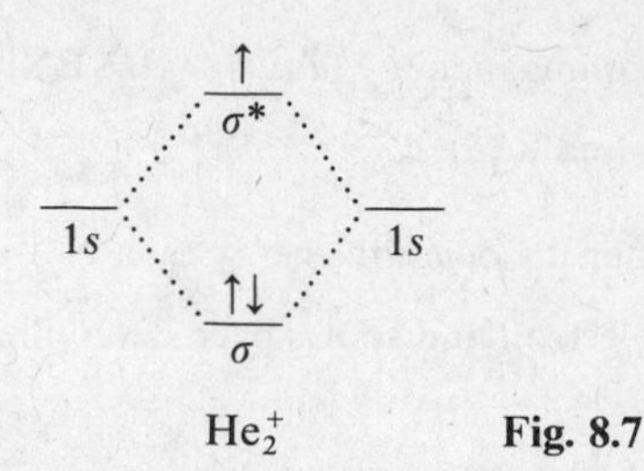

Fig. 8.7

8.32 Compare and contrast the concepts of hybrid orbitals and molecular orbitals with respect to **(*a*)** the number of atoms involved, **(*b*)** the number of orbitals produced from a given number of ground state orbitals, **(*c*)** the energies of the resulting orbitals with respect to one another.

▮

	hybrid	**molecular**
(*a*)	One central atom	more than one atom
(*b*)	Same number	same number
(*c*)	All the same	bonding orbitals lower in energy than antibonding orbitals

8.33 Distinguish between nonbonding orbitals and antibonding orbitals.

▮ Nonbonding orbitals have the same energy as the atomic orbitals from which they are formed; antibonding orbitals have higher energies than the highest energy atomic orbital from which they are formed.

8.34 Draw a molecular orbital energy level diagram for each of the species Li_2, Be_2, B_2, C_2, N_2, O_2, F_2, and Ne_2. Verify that each has the number of bonding and antibonding electrons listed.

▮

Li_2	Be_2	B_2	C_2	N_2	O_2	F_2	Ne_2
							↑↓
					↑ ↑	↑↓ ↑↓	↑↓ ↑↓
				↑↓	↑↓ ↑↓	↑↓ ↑↓	↑↓ ↑↓
		↑ ↑	↓↑ ↑↓	↑↓ ↑↓	↑↓	↑↓	↑↓
—	↑↓	↑↓	↑↓	↑↓	↑↓	↑↓	↑↓
↑↓	↑↓	↑↓	↑↓	↑↓	↑↓	↑↓	↑↓

8.35 State the bond order and indicate whether the species is paramagnetic: **(*a*)** B_2 **(*b*)** C_2 **(*c*)** N_2 **(*d*)** O_2 **(*e*)** CN^- **(*f*)** Br_2 **(*g*)** Cl_2^+ **(*h*)** NO **(*i*)** NO^+ **(*j*)** CO

▮

	bond order	**magnetic behavior**		**bond order**	**magnetic behavior**
(*a*)	1	paramagnetic	(*f*)	1	diamagnetic
(*b*)	2	diamagnetic	(*g*)	1.5	paramagnetic
(*c*)	3	diamagnetic	(*h*)	2.5	paramagnetic
(*d*)	2	paramagnetic	(*i*)	3	diamagnetic
(*e*)	3	diamagnetic	(*j*)	3	diamagnetic

8.36 Which of the following molecules has the highest bond order? **(*a*)** BN **(*b*)** CO **(*c*)** NO **(*d*)** Ne_2 **(*e*)** F_2

▮ **(*b*)** CO has a bond order of 3. [The others have bond orders of **(*a*)** 2 **(*c*)** 2.5 **(*d*)** 0 and **(*e*)** 1.]

8.37 Which of the following is paramagnetic? **(a)** O_2^{2-} **(b)** BN

▮ Neither (see Probs. 8.34 and 8.35)

8.38 Explain why N_2 has a greater dissociation energy than N_2^+, whereas O_2 has a lower dissociation energy than O_2^+.

▮ N_2^+ has lost a bonding electron and so has been destabilized; O_2^+ has lost an antibonding electron and has been stabilized.

8.39 Deduce the bond order in each of the following: CN^+, CN, CN^-, NO.

▮

	CN^+	CN	CN^-	NO
	—	—	—	—
	— —	— —	— —	↑ —
	—	↓	↓↑	↓↑
	↓↑ ↓↑	↓↑ ↓↑	↓↑ ↓↑	↓↑ ↓↑
Bond order	2	$2\frac{1}{2}$	3	$2\frac{1}{2}$

8.40 Make a table giving (i) number of orbitals with a given energy, (ii) maximum number of electrons per orbital, and (iii) maximum number of electrons at a given energy for the following types of orbitals: **(a)** s **(b)** p **(c)** sp^2 **(d)** sp^3 **(e)** σ **(f)** σ^* **(g)** π^*

▮

	(i)	(ii)	(iii)		(i)	(ii)	(iii)
(a)	1	2	2	**(e)**	1	2	2
(b)	3	2	6	**(f)**	1	2	2
(c)	3	2	6	**(g)**	2	2	4
(d)	4	2	8				

8.41 Which supposed homonuclear diatomic molecules of the second-period elements should have zero bond order?

▮ Be$_2$ and Ne$_2$ would have as many antibonding as bonding electrons and thus would not be stable.

Be_2	Ne_2
—	↑↓
— —	↑↓ ↑↓
—	↑↓ ↑↓
— —	↑↓
↑↓	↑↓
↑↓	↑↓

8.42 Which homonuclear diatomic molecule(s) of second-period elements, besides O_2, should be paramagnetic?

▮ B_2. Paramagnetism arises from unpaired electrons. Each diatomic molecule has an even number of electrons, and the only way any electrons can be unpaired is if they occupy π orbitals singly.

σ_{2p}	—
π_{2p}	↑ ↑
σ_{2s}^*	↑↓
σ_{2s}	↑↓
σ_{1s}^*	↑↓
σ_{1s}	↑↓

8.43 **(a)** What are the bond orders for CN^-, CN, and CN^+? **(b)** Which of these species should have the shortest bond length?

▮

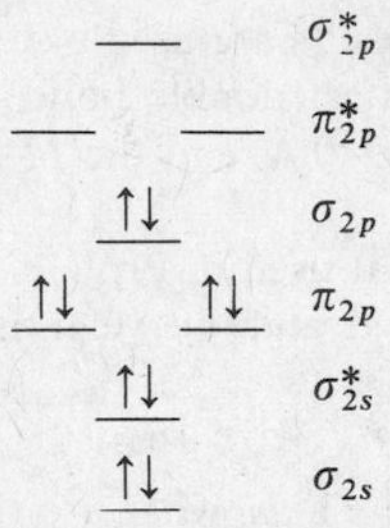

—		σ^*_{2p}
—	—	π^*_{2p}
↑↓		σ_{2p}
↑↓	↑↓	π_{2p}
↑↓		σ^*_{2s}
↑↓		σ_{2s}

CN^-

(a) CN has 4 bonding pairs and 1 antibonding pair, for a net bond order of 3. CN has one fewer bonding electron, for a bond order of $2\frac{1}{2}$, and CN^+ has still one fewer, for a bond order of 2.

(b) CN has the shortest bond because it has the highest bond order.

8.44 By the use of molecular orbital considerations, account for the fact that oxygen gas is paramagnetic. What is the bond order in O_2?

▮ Each oxygen atom has the $1s^2\ 2s^2\ 2p^4$ configuration in the ground state. Aside from the K electrons of the two atoms in O_2, which are so deeply imbedded in their respective atoms as not to overlap with other electrons, the remaining 12 electrons (6 from each of the atoms) will fill the lowest of the available molecular orbitals.

$$\sigma_s^2\sigma_s^{*2}\pi_{x,y}^4\sigma_{p_z}^2\pi_x^{*1}\pi_y^{*1}$$

The last two electrons go into the antibonding equienergetic π^* orbitals, one into π^*_x and one into π^*_y, so as to maximize electron spin in accordance with Hund's rule. These two unpaired electrons confer paramagnetism upon the molecule.

$$\text{Bond order} = \frac{(\text{number of electrons in bonding orbitals}) - (\text{number of electrons in antibonding orbitals})}{2}$$

$$= \frac{8-4}{2} = 2$$

8.45 Explain the observations that the bond length in N_2^+ is 0.02 Å greater than in N_2, while the bond length in NO^+ is 0.09 Å less than in NO.

▮ The electron configurations should be written for the 4 molecules according to the buildup principle:

N_2	(K electrons)$\sigma_s^2\sigma_s^{*2}\pi_{x,y}^4\sigma_{p_z}^2$
N_2^+	(K electrons)$\sigma_s^2\sigma_s^{*2}\pi_{x,y}^4\sigma_{p_z}^1$
NO	(K electrons)$\sigma_s^2\sigma_s^{*2}\pi_{x,y}^4\sigma_{p_z}^2\pi_{x,y}^{*1}$
NO^+	(K electrons)$\sigma_s^2\sigma_s^{*2}\pi_{x,y}^4\sigma_{p_z}^2$

The computed bond orders for N_2 and N_2^+ are 3 and $2\frac{1}{2}$, respectively. N_2 therefore has the stronger bond and should have the shorter bond length. The computed bond orders for NO and NO^+ are $2\frac{1}{2}$ and 3, respectively. NO^+ has the stronger bond and should have the shorter bond length. As opposed to ionization of N_2, which involves the loss of an electron in a *bonding* orbital, ionization of NO involves the loss of an electron in an antibonding orbital.

8.46 Consider the aromatic molecule anthracene, $C_{14}H_{10}$, shown in Fig. 8.8.

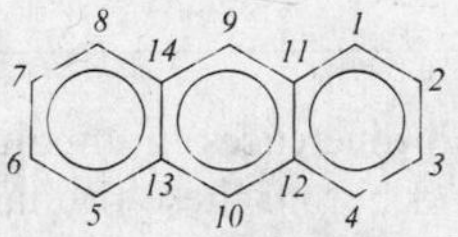

Fig. 8.8

Approximate calculations of the π-bond order for carbon-carbon bonds yield the following results:

bond	π-bond order
1–2	0.738
1–11	0.535
2–3	0.586
9–11	0.606
11–12	0.485

Which carbon-carbon bond would be the shortest and which the longest?

▌ This problem is similar to Prob. 8.10, except that molecular orbital results are used instead of simple resonance theory. The bond with the greatest π-bond order should have the greatest double bond character and thus the shortest distance, and vice versa. The *1–2* bond is indeed the shortest, 1.370 Å; and the *11–12* bond is the longest, 1.436 Å.

The method used in this problem is intrinsically more accurate than that used in Prob. 8.10 because, for one reason, the π-bond order is a calculated parameter for the molecule as a whole, and the weighting of various resonance structures is not required.

8.47 If the internuclear axis in the diatomic molecule AB is designated as the z axis, what are the various pairs of s, p, or d atomic orbitals that can be combined to form **(*a*)** π_x and **(*b*)** π_y orbitals?

▌ **(*a*)** (p_x, p_x), (p_x, d_{xz}), (d_{xz}, d_{xz}) **(*b*)** (p_y, p_y), (p_y, d_{yz}), (d_{yz}, d_{yz}) [Fig. 8.9 indicates the combinations in xz-projection]

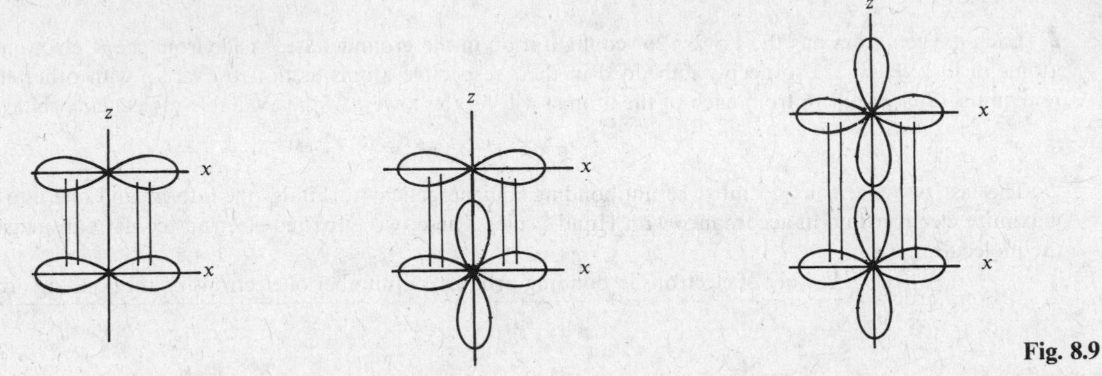

Fig. 8.9

8.48 Using energy-level diagrams in which **(*a*)** the π_{2p} orbitals lie below the σ_{2p} orbital, and **(*b*)** the opposite order is followed, write the electronic configuration of O_2 and F_2. Repeat the process for B_2 and C_2. In which case(s) does (do) the choice of energy level diagram make a difference in the number of unpaired electrons? Explain.

▌

	O_2	F_2	B_2	C_2
(*a*)	___	___	___	___
	↑ ↑	↑↓ ↑↓	___ ___	___ ___
	↑↓	↑↓	___	___
	↑↓ ↑↓	↑↓ ↑↓	↑ ↑	↑↓ ↑↓
(*b*)	___	___	___	___
	↑ ↑	↑↓ ↑↓	___ ___	___ ___
	↑↓ ↑↓	↑↓ ↑↓	___ ___	↑ ↑
	↑↓	↑↓	↑↓	↑↓

The oxygen and fluorine cases have no differences in unpaired electrons, since all the orbitals which have reversed order are filled anyway. In carbon and boron, these orbitals are only partially filled, and it makes a difference in the number of unpaired electrons as to which order actually occurs.

8.49 The bonding σ_{2s} orbital has a higher energy than the antibonding σ_{1s}^* orbital. Why is the former a bonding orbital while the latter is antibonding?

▌ The σ_{2s} orbital originates from atomic orbitals of higher energy than itself, and so is bonding. The σ_{1s}^* originates from atomic orbitals of lower energy than itself, and so is antibonding. The relative energies of the two σ orbitals is immaterial.

8.50 Is the HHe molecule apt to be stable toward dissociation into atoms? Given two hydrogen atoms and two helium atoms, which one of the following combinations (if any) has the lowest energy? **(*a*)** 2HHe **(*b*)** $H_2 + He_2$ **(*c*)** $He_2 + 2H$ **(*d*)** $H_2 + 2He$ **(*e*)** 2H + 2He

▮ The electronic structure of such a molecule would be as shown to the right and should be stable (corresponding to half a single bond). However, the species would be unstable with respect to dissociation into $H_2 + 2He$.

H	HHe	He
	↑	
↑		↑↓
	↑↓	

8.51 Show that if two atoms bond along their z axes, the various d orbitals can combine to form σ, π, or δ (delta) type molecular orbitals.

▮ See Fig. 8.10.

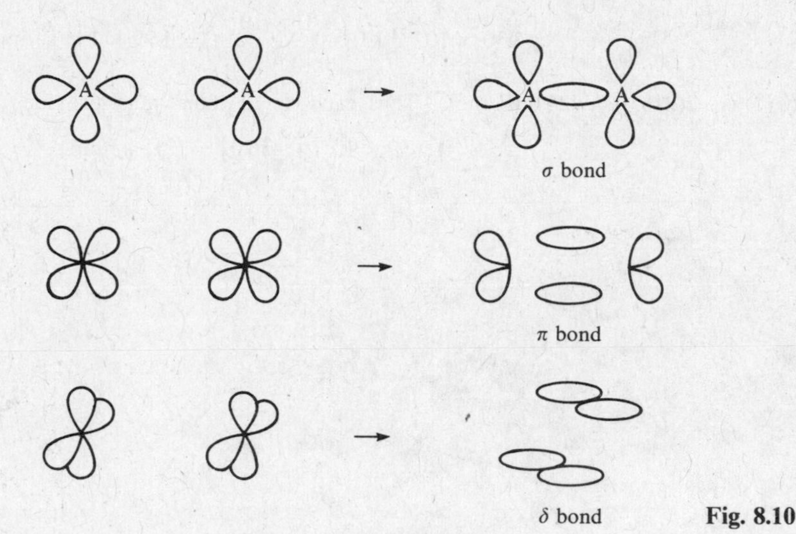

Fig. 8.10

8.52 Analogous to the formation of π_{2p} orbitals from separate p orbitals, diagram the approach of the d_{xy} orbitals of two atoms along their z axes, to form bonding δ and antibonding δ^* orbitals.

▮ See bottom of Fig. 8.10 for the δ orbital. The δ^* orbital would have 8 lobes—two on each side of each δ lobe shown.

8.53 An electronic excited state of He_2 has been observed, but no ground state has been found. Explain why He_2 molecules in an excited state could be stable toward dissociation into atoms.

▮

He	He_2	He
2s		2s
	↑ σ	
	↑ σ*	
1s		1s
	↑↓ σ	

The excited state must have more bonding than antibonding electrons. The preponderance of electrons between the nuclei makes the excited molecule stable toward dissociation into atoms (while it remains in its excited electronic state).

8.54 In terms of molecular orbital theory, using the F_2 molecule as an example, criticize the postulate that a single bond consists of one σ bond only.

▮ F_2 has a single bond consisting of one bonding pair of electrons in a σ_{2p} orbital, two bonding pairs in π_{2p} orbitals, and two antibonding pairs in π^*_{2p} orbitals.

8.55 Helium can be excited to the $1s^1\,2p^1$ configuration by light of 58.44 nm. The lowest excited singlet state, with the configuration $1s^1\,2s^1$, lies 4857 cm^{-1} below the $1s^1\,2p^1$ state. What would the average He—H bond energy have to be in order that HeH_2 could form nonendothermically from He and H_2? Assume that the compound would form from the lowest excited singlet state of helium, neglect any differences between ΔE and ΔH, and take $\Delta H_f(\text{H}) = 218.0$ kJ/mol.

▮ Formation of 2 mol of He—H bonds must provide (1) energy of excitation from $1s^2$ to $1s^1\,2s^1$ and (2) energy to produce 2 mol of H atoms.

Energy (1) is the difference between the promotion energy to $1s^1\ 2p^1$ and the difference between that level and the $1s^1\ 2s^1$ level:

$$E(1s^2 \rightarrow 1s^1\ 2p^1) = \frac{hc}{\lambda} = \frac{hc}{\lambda} = \frac{(6.63 \times 10^{-34}\ \text{J·s})(3.00 \times 10^8\ \text{m/s})}{58.44 \times 10^{-9}\ \text{m}} = 3.40 \times 10^{-18}\ \text{J}$$

$$E(1s^1\ 2p^1 \rightarrow 1s^1\ 2s^1) = hc\left(\frac{1}{\lambda}\right) = (6.63 \times 10^{-34}\ \text{J·s})(3.00 \times 10^8\ \text{m/s})(485\,700\ \text{m}^{-1}) = 9.66 \times 10^{-20}\ \text{J}$$

$$E(\text{singlet}) = (3.40 \times 10^{-18}\ \text{J}) - (9.66 \times 10^{-20}\ \text{J}) = 3.30 \times 10^{-18}\ \text{J} = 3.30 \times 10^{-21}\ \text{kJ}$$

Per mole:

$$(3.30 \times 10^{-21}\ \text{kJ})\left(\frac{6.02 \times 10^{23}}{\text{mol}}\right) = 1.99 \times 10^3\ \text{kJ/mol}$$

Energy (2) is $2(218.0) = 436.0$ kJ/mol. Thus, each He—H bond must provide

$$\tfrac{1}{2}[(1.99 \times 10^3) + 436.0] = 1215\ \text{kJ/mol}$$

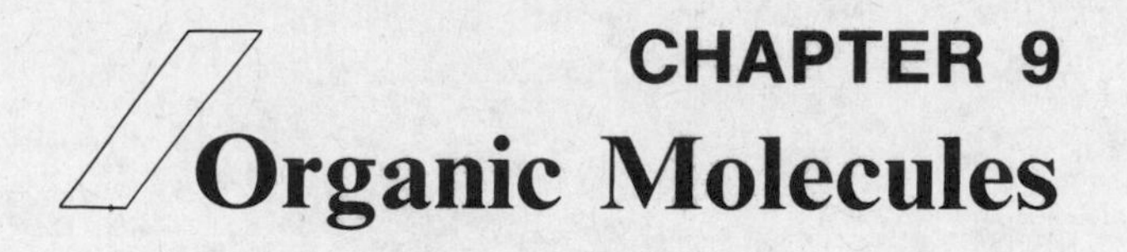

CHAPTER 9
Organic Molecules

9.1 ORGANIC NOMENCLATURE AND CLASSIFICATION

9.1 Name the following compounds:

(*a*) $CH_3{-}CH(CH_3){-}CH(CH_3){-}CH_3$ (*b*) $CH_3{-}C(CH_3)_2{-}CH_2{-}CH_3$ (*c*) $CH_3{-}CH_2{-}C(CH_3)_2{-}CH_3$

I (*a*) 2,3-dimethylbutane (*b*) and (*c*) 2,2-dimethylbutane

9.2 Write formulas for (*a*) 2-methyl-3-ethylhexane (*b*) 2,2-dimethylhexane (*c*) 2,3,6-trimethylheptane

I (*a*) $CH_3{-}CH(CH_3){-}CH(CH_2CH_3){-}CH_2{-}CH_2{-}CH_3$ (*b*) $CH_3{-}C(CH_3)_2{-}CH_2{-}CH_2{-}CH_2{-}CH_3$

(*c*) $CH_3{-}CH(CH_3){-}CH(CH_3){-}CH_2{-}CH_2{-}CH(CH_3){-}CH_3$

9.3 Explain why 2-ethylheptane is not a standard IUPAC name.

I The longest continuous chain would include the ethyl group, and the compound would have the standard name 3-methyloctane.

$$CH_3\overset{3}{C}H(\underset{2}{C}H_2\underset{1}{C}H_3)\overset{4}{C}H_2\overset{5}{C}H_2\overset{6}{C}H_2\overset{7}{C}H_2\overset{8}{C}H_3$$

9.4 Envision the "ball and stick" models of methane, ethane, methyl alcohol, dimethyl ether (CH_3OCH_3), ethyl alcohol, and water. (*a*) Which one(s) of these substances can be considered as "derived" from methane? (*b*) Which one(s) can be considered as being "derived" from water? (*c*) Which ones are isomers of each other? (*d*) Which pairs are in the same class of organic compounds?

I (*a*) Methyl alcohol, dimethyl ether, ethane (*b*) methyl alcohol, ethyl alcohol, and dimethyl ether (*c*) ethyl alcohol and dimethyl ether (*d*) methane and ethane (alkanes) and methyl alcohol and ethyl alcohol (alcohols)

9.5 What is the maximum number of other atoms to which an atom of each of the following can be bonded in organic compounds? (*a*) hydrogen (*b*) carbon (*c*) nitrogen (*d*) oxygen (*e*) chlorine

I (*a*) 1 (*b*) 4 (*c*) 3 (*d*) 2 (*e*) 1

9.6 Justify the rule

$$n(\mathrm{H,max}) = 2n(\mathrm{C}) + n(\mathrm{N}) + 2$$

which relates the maximum number of hydrogen atoms in an organic molecule to the numbers of atoms of carbon and nitrogen.

I Assume that the molecule contains only C, N, O, and H, the most commonly occurring elements in organic compounds. The rule may be explained in terms of the maximum covalences for C, N, O, and H, which are 4, 3, 2, and 1, respectively. (A fourth covalence for N, arising from coordinate covalence, occurs mainly in ion formation, such as NH_4^+, rather than in neutral molecules.)

Consider the skeleton structures in Fig. 9.1, containing many of the commonly occurring features in organic compounds. Hydrogen atoms are not shown; they are assumed to be attached to any terminal valence, but they cannot

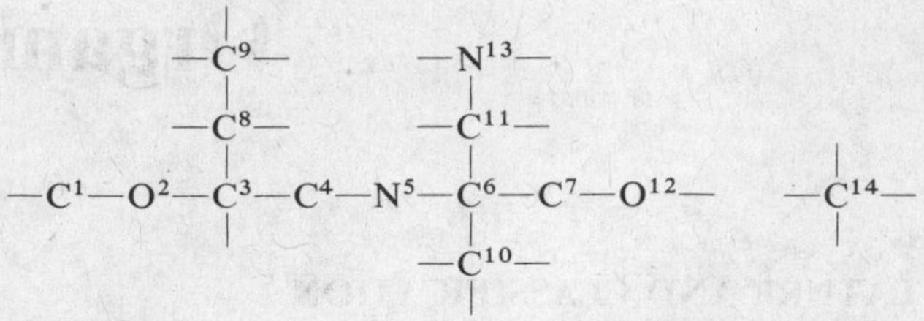

```
              |
            —C⁹—          —N¹³—
              |              |
            —C⁸—          —C¹¹—
              |              |
—C¹—O²—C³—C⁴—N⁵—C⁶—C⁷—O¹²—        —C¹⁴—
              |              |               |
                          —C¹⁰—
                             |
```

Fig. 9.1

bridge two different atoms. Double bonds are not used, since they reduce the number of bonded hydrogens and we are to calculate the maximum number of hydrogens. These structures have almost all the possibilities for these elements consistent with the assumptions: chain-terminating carbon (*1*, *9*, and *10*); bridging carbon with zero (*4*, *7*, *8*, *11*), one (*3*), or two (*6*) branches; carbon unbonded to C, N, or O (*14*); bridging (*5*) and terminal (*13*) nitrogen; bridging (*2*) and terminal (*12*) oxygen.

For type *4*, *7*, *8*, *11* carbon, 2 bonds to hydrogen are available per carbon, justifying the first term on the right side of the stated equation, $2n(\mathrm{C})$. For type *1*, *9*, *10* carbon, 3 bonds to hydrogen are available per carbon; 2 of these 3 carbons (e.g., *9* and *10*) may be considered as the beginning and end of the longest carbon chain in the molecule. The two ends, accounting for one more hydrogen each than the $2n(\mathrm{C})$ prescription, are responsible for the additive term, 2, in the equation. The additive term 2 is also appropriate for CH_4 (*14*). The other terminal carbon, *1*, also has a third bond to hydrogen, but 1 of the 3 bonds replaces 1 of the bonds that the main-chain carbon *3* had to give up in order to initiate the branch off the main chain toward carbon *1*. Branching oxygen *2* has no available bonds to hydrogen; thus $n(\mathrm{O})$ does not appear in the formula. Terminal oxygen *12* has one available bond to hydrogen, but this replaces one of the bonds to hydrogen that carbon *7* had to give up in order to initiate the branch toward oxygen *12*. Bridging nitrogen *5* has one bond available for hydrogen, thus justifying the second term on the right, $n(\mathrm{N})$. Terminal nitrogen *13* has 2 bonds available, but one of them replaces one bond to hydrogen which carbon *11* had to give up in order to initiate the branch toward nitrogen *13*. Thus, the $n(\mathrm{N})$ term is justified.

Double bonds between C and C, N, or O would merely have reduced the number of hydrogens below the calculated maximum. Halogen atoms, being monovalent, would also diminish the hydrogen atoms. Atoms of other Group IV or V elements would have to be treated separately, often analogously to C and N.

9.7 Give the systematic name of the following compound:

```
CH3CH2CHCH2CH2CH2CH3
        |
        CH—CH3
        |
        CH3
```

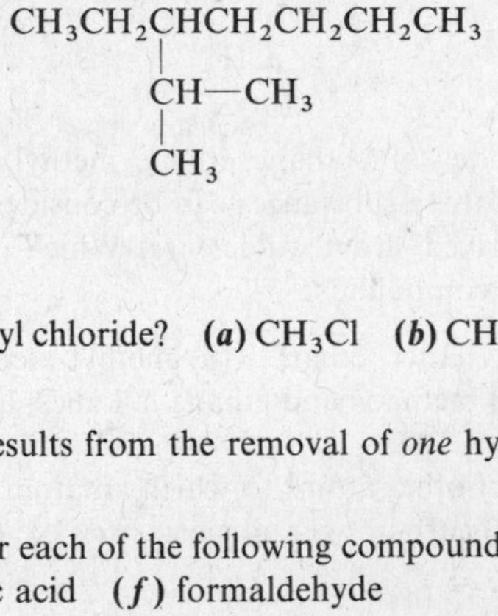

▌ 2-Methyl-3-ethylheptane

9.8 Which of the following is (are) called methyl chloride? (*a*) CH_3Cl (*b*) CH_2Cl_2 (*c*) $CHCl_3$ (*d*) CCl_4 (*e*) Explain your choice.

▌ (*a*) The methyl group is CH_3—. It results from the removal of *one* hydrogen atom from methane.

9.9 Give the systematic name and formula for each of the following compounds: (*a*) phenyl ethyl ketone (*b*) ethylene (*c*) acetone (*d*) acetaldehyde (*e*) acetic acid (*f*) formaldehyde

▌ (*a*) 1-Phenyl-1-propanone C_6H_5—C(=O)—CH_2CH_3 (*b*) ethene $CH_2{=}CH_2$

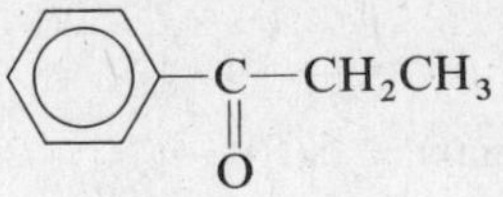

(*c*) propanone CH_3COCH_3 (*d*) ethanal CH_3CHO

(*e*) ethanoic acid CH_3CO_2H (*f*) methanal HCHO

9.10 What class of compounds is represented by the type formula ROR′? Using CH_3 and C_6H_5 as the radicals, write formulas for three compounds which correspond to this type formula.

▌ Ethers. CH_3OCH_3, $CH_3OC_6H_5$, $C_6H_5OC_6H_5$. (The fact that the radicals are denoted R and R′ means that they *may* be different, not that they *must* be different.)

9.11 Name the following compounds:

(*a*) $$\begin{array}{ccccccccc} CH_3 & - & CH_2 & - & CH & - & CH_2 & - & CH_3 \\ & & & & | & & & & \\ & & & & CH_2 & & & & \\ & & & & | & & & & \\ & & & & OH & & & & \end{array}$$

(*b*) $$\begin{array}{ccccccccc} HO & - & CH_2 & - & CH_2 & - & CH & - & CH_3 \\ & & & & & & | & & \\ & & & & & & CH_3 & & \end{array}$$

(*c*) $$\begin{array}{ccccccccc} HO & - & CH & - & CH_2 & - & CH_2 & - & CH_3 \\ & & | & & & & & & \\ & & CH_2 & & & & & & \\ & & | & & & & & & \\ & & CH_3 & & & & & & \end{array}$$

(*d*) $$\begin{array}{ccccccccc} CH_2 & = & CH & - & CH_2 & - & CH & - & CH_3 \\ & & & & & & | & & \\ & & & & & & CH_3 & & \end{array}$$

(*e*) $CH_3{-}CH{=}CH{-}CH_2{-}CH_3$

(*f*) $$\begin{array}{c} CH_3CH_2CHCH_2CH_3 \\ | \\ CH_2 \\ | \\ Cl \end{array}$$

■ (*a*) 2-Ethyl-1-butanol (*b*) 3-methyl-1-butanol (*c*) 3-hexanol
(*d*) 4-methyl-1-pentene (*e*) 2-pentene (*f*) 3-(chloromethyl)pentane

9.12 Name the following compounds:

(*a*) $$\begin{array}{c} CH_3CCH_2CH_2Cl \\ \| \\ CH_2 \end{array}$$ (*b*) $CH_3(CH_2)_6CO_2H$ (*c*) $$\begin{array}{c} CH_3CH_2CHCH_2CH_2CH_3 \\ | \\ CHO \end{array}$$

■ (*a*) 4-Chloro-2-methyl-1-butene (*b*) octanoic acid (*c*) 2-ethyl-1-pentanal

9.13 (*a*) Explain why the name *butanol* is not specific, whereas the name *butanone* represents one specific compound. (*b*) Is the name *pentanone* specific?

■ (*a*) Butanol may have the OH group on the first or second carbon atom. Butanone cannot have the carbonyl group on the end (where it would be an aldehyde functional group). (*b*) Pentanone could refer to 2-pentanone or to 3-pentanone, so the name is not specific.

9.14 Name the following compounds: (*a*) $CH_2{=}CH{-}CH{=}CH{-}CH{=}CH_2$ (*b*) $(C_6H_5)_2O$ (*c*) $(C_2H_5)_2NH$ (*d*) $CH_3CH_2CONHCH_3$ (*e*) $(C_2H_5)_2NH_2^+Cl^-$

■ (*a*) 1,3,5-hexatriene (*b*) diphenyl ether (*c*) diethyl amine
(*d*) methyl propanamide (*e*) diethylammonium chloride

9.15 Write formulas for (*a*) ethylamine (*b*) propionaldehyde (*c*) butanone (*d*) ethyl propionate (*e*) butyl formate (*f*) bromobenzene (*g*) acetylene (*h*) phenylacetylene

■ (*a*) $CH_3CH_2NH_2$ (*b*) CH_3CH_2CHO (*c*) $CH_3COCH_2CH_3$
(*d*) $CH_3CH_2CO_2CH_2CH_3$ (*e*) $HCO_2CH_2CH_2CH_2CH_3$ (*f*) C_6H_5Br
(*g*) $HC{\equiv}CH$ (*h*) $C_6H_5C{\equiv}CH$

9.16 Write formulas for (*a*) 2-butene (*b*) methyl ethyl ether (*c*) propanal (*d*) 2-propanol (*e*) 2,4-pentandione (acetylacetone)

■ (*a*) $CH_3CH{=}CHCH_3$ (*b*) $CH_3OCH_2CH_3$ (*c*) CH_3CH_2CHO
(*d*) $CH_3CHOHCH_3$ (*e*) $CH_3COCH_2COCH_3$

9.17 Write a structural formula for each of the following:

(*a*) 4-ethylheptane (*b*) 4-propylheptane (*c*) 4-(1-methylethyl)heptane

■ (*a*) $$\begin{array}{c} CH_3CH_2CH_2CHCH_2CH_2CH_3 \\ | \\ CH_2 \\ | \\ CH_3 \end{array}$$ (*b*) $$\begin{array}{c} CH_3CH_2CH_2CHCH_2CH_2CH_3 \\ | \\ CH_2 \\ | \\ CH_2 \\ | \\ CH_3 \end{array}$$

(*c*) $CH_3CH_2CH_2CHCH_2CH_2CH_3$ with $CH-CH_3$ attached to the fourth carbon, and CH_3 attached to that CH

$$\begin{array}{l} CH_3CH_2CH_2\underset{|}{C}HCH_2CH_2CH_3 \\ \qquad\qquad\quad \underset{|}{CH}-CH_3 \\ \qquad\qquad\quad CH_3 \end{array}$$

9.18 Molecules containing which of the functional groups act in aqueous solution (*a*) as Brønsted bases (*b*) as Brønsted acids?

▌ (*a*) Amines (*b*) carboxylic acids

9.19 Write an equation for the reaction with HCl(*aq*) of (*a*) $CH_3CH_2CH_2NH_2$ (*b*) $(CH_3)_3N$

▌ (*a*) $HCl(aq) + CH_3CH_2CH_2NH_2 \longrightarrow CH_3CH_2CH_2NH_3^+Cl^-$

(*b*) $HCl(aq) + (CH_3)_3N \longrightarrow (CH_3)_3NH^+Cl^-$

9.20 Name each of the following compounds: (*a*) CH_3CH_2OH (*b*) $CH_3C_6H_5$ (*c*) CH_3COCH_3 (*d*) CH_3OCH_3 (*e*) CH_3CHO (*f*) $CH_3CHOHCH_3$ (*g*) CH_3CO_2H (*h*) $HOCH_2CH_3$ (*i*) C_6H_5OH (*j*) $CH_3CO_2CH_3$ (*k*) $CH_3CHClCH_2CH_3$

▌ (*a*) Ethyl alcohol (*b*) toluene (*c*) acetone (*d*) dimethyl ether
(*e*) acetaldehyde (*f*) 2-propanol (*g*) acetic acid (*h*) ethyl alcohol
(*i*) phenol (*j*) methyl acetate (*k*) 2-chlorobutane

9.21 Write formulas for the following compounds: (*a*) butyl alcohol (*b*) propyl phenyl ether (*c*) ethyl acetate (*d*) diphenylamine

▌ (*a*) $CH_3CH_2CH_2CH_2OH$ (*b*) $CH_3CH_2CH_2OC_6H_5$
(*c*) $CH_3\underset{\underset{O}{\|}}{C}OC_2H_5$ (often written $CH_3CO_2C_2H_5$) (*d*) $(C_6H_5)_2NH$

9.22 Classify the following molecules according to functional group:

(*a*) $CH_3CH_2CH_2OCH_3$ (*b*) cyclic: H_2C-CH_2 / H_2C $CHOH$ / H_2C-CH_2 (*c*) $CH_3-\underset{\underset{O}{\|}}{C}-CH_2CH_3$

(*d*) $CH_2{=}CHCH_2C_6H_5$ (*e*) cyclic: H_2C-CH_2 / H_2C O / H_2C-CH_2 (*f*) cyclic: H_2C-CH_2 / H_2C $C{=}O$ / H_2C-CH_2

(*g*) $CH_2{=}CH\underset{\underset{CH_2}{\|}}{C}H$

▌ Molecules (*a*) and (*e*) have the ROR′ type formula and are ethers; (*b*), having the form ROH, is an alcohol; (*c*) and (*f*) are ketones because they contain the $C-\underset{\underset{O}{\|}}{C}-C$ group; (*d*) is an alkene which also contains an aromatic group; and (*g*) is a diene, with two double bonds connecting carbon atoms.

9.23 Write line formulas for

(*a*) $CH_3\underset{\underset{O}{\|}}{C}-O-CH_3$ (*b*) $CH_3\underset{\underset{O}{\|}}{C}-NH_2$ (*c*) $\begin{matrix}CH_3\\ \quad\rangle CHCH_3\\ CH_3\end{matrix}$ (*d*) $CH_3\underset{\underset{O}{\|}}{C}-CH_3$

▌ (*a*) $CH_3CO_2CH_3$ (*b*) CH_3CONH_2 (*c*) $(CH_3)_3CH$ or $(CH_3)_2CHCH_3$ (*d*) CH_3COCH_3

9.24 Explain why the alkene and the cycloalkane series of hydrocarbons both have the same general formula, C_nH_{2n}. Explain why the dienes and the alkynes have the same general formula, C_nH_{2n-2}. What is the number of carbon atoms in the smallest member of each group?

▮ Either insertion of a double bond or formation of a ring reduces the number of hydrogen atoms of the corresponding alkane by 2. A triple bond, or 2 double bonds, reduces the number of hydrogen atoms by 4. Alkenes and alkynes must have at least 2 carbon atoms to form a multiple bond. Dienes must have at least 3 carbon atoms to form 2 double bonds ($H_2C{=}C{=}CH_2$). Cycloalkanes must have at least 3 carbon atoms in order to form a ring.

9.25 The carbon atoms of the benzene ring may be numbered for identification of substituent groups, just as continuous chains of carbon atoms are numbered. Again, the smallest set of numbers designating the substituents is the preferred set. Draw structures for each of the following:

(a) 1,3-dichlorobenzene *(b)* 2,4,6-trinitrotoluene *(c)* 1,4-diethylbenzene

▮ *(a)* Cl, Cl *(b)* CH_3, O_2N, NO_2, NO_2 *(c)* CH_2CH_3, CH_2CH_3

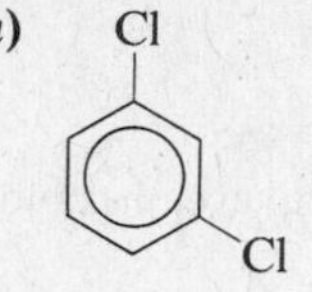

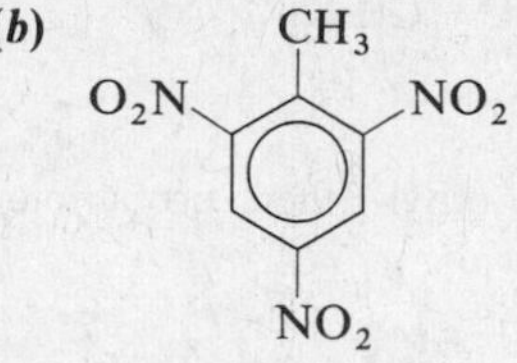

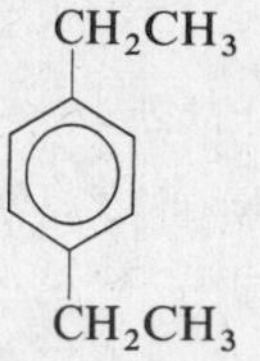

9.26 Write the formula for 1-phenyl-1-butanone. Explain why the designation "1-butanone" is somewhat unusual.

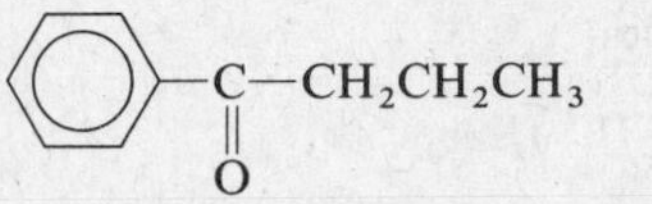

▮ $C_6H_5{-}C({=}O){-}CH_2CH_2CH_3$ The carbonyl group positioned on an end (number 1) carbon atom usually is a part of an aldehyde group.

9.27 In the following representation of a hydrocarbon molecule, designate the primary, secondary, tertiary, and quaternary carbon atoms:

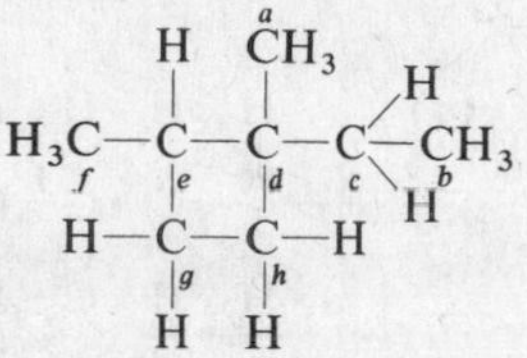

```
        H  CH3 (a)
        |  |      H
H3C — C — C — C ⟨ CH3
 f    |e |d  c   \  b
      |  |        H
  H — C — C — H
      |g |h
      H  H
```

▮ Atoms a, b, and f are primary; c, g, and h are secondary; e is tertiary; and d is quaternary.

9.28 For each class of simple organic compounds, *(a)* write a formula for the compound in the class which contains the fewest possible carbon atoms, *(b)* name each compound in *(a)*, *(c)* determine the oxidation number of the carbon atoms in the compound.

▮

	(a)	*(b)*	*(c)*
Halide	CH_3F	fluoromethane (methyl fluoride)	-2
Alcohol	CH_3OH	methanol (methyl alcohol)	-2
Ether	$(CH_3)_2O$	dimethyl ether	-2
Aldehyde	CH_2O	methanal (formaldehyde)	0
Ketone	CH_3COCH_3	propanone (acetone)	$-\frac{4}{3}$
Acid	HCO_2H	formic acid	$+2$
Ester	HCO_2CH_3	methyl formate	0
Amine	CH_3NH_2	aminomethane (methyl amine)	-2
Amide	$HCONH_2$	formamide	$+2$
Alkene	$H_2C{=}CH_2$	ethene (ethylene)	-2
Alkyne	$HC{\equiv}CH$	ethyne (acetylene)	-1
Aromatic	C_6H_6	benzene	-1

9.29 From the formulas listed below, select all examples of *(a)* primary amines *(b)* secondary amines *(c)* secondary alcohols

(i) $CH_3CHNH_2CH_3$, (ii) $CH_3CHOHCH_2NHCH_3$, (iii) $(CH_3)_2COHCH_2NH_2$;
(iv) $(CH_3)_2NCH(CH_3)_2$.

▮ (*a*) i and iii. (The NH_2 group, with only one hydrogen of NH_3 replaced by an R group, is the primary amine, even when it is attached to a secondary or tertiary carbon atom.) (*b*) ii. (*c*) ii. (The OH group is attached to a secondary carbon atom.)

9.30 (*a*) Classify each of the following compounds as an alkane, alkene, alcohol, ether, ester, acid, amine, amide, aldehyde, or ketone.

```
CH3CH2OCH3    CH3CH2CO2H     H2C—CH2       CH3CONH2    CH3CO2CH3
                            H2C     C=O
                             H2C—CH2
```

(*b*) Name the compounds in (*a*).

▮ (*a*) Ether, acid, ketone, amide, ester (*b*) methyl ethyl ether, propanoic acid, cyclohexanone, acetamide, methyl acetate

9.2 STRUCTURAL ISOMERISM

9.31 Which of the following formulas represent different molecules?

```
(a)                                  (c)
              H                           H  H  H  H  H
            H | H                         |  |  |  |  |
     H  H  H  C  H                     H—C—C—C—C—C—H
     |  |  |  |  |                        |  |  |  |
  H—C—C—C—C—C—H                         H  C  H  H  H
     |  |  |  |  |                         /|\
     H  H  H  H  H                        H H H

(b)    H                             (d)
     H | H                                H  H  H  H  H
       C    H    H                        |  |  |  |  |
       |    |    |                     H—C—C—C—C—C—H
  H—C———C———C—H                          |  |  |  |  |
       |    |    |                        H  H  C  H  H
       C    H    C                              /|\
     /|\        /|\                            H C H
    H H H      H H H
```

▮ (*a*), (*b*), (*c*) are formulas of the same compound. Each has a five-carbon chain with a branch at the second carbon atom from one end. (*d*) represents a different molecule, since the branch is on the third carbon atom.

9.32 In organic compounds the functional group —OH is characteristic of alcohols. Write structures and common names for four alcohols derived from the first *three* paraffin hydrocarbons. Which are isomers?

▮

```
     H            H  H            H  H  H             H  H  H
     |            |  |            |  |  |             |  |  |
  H—C—O—H      H—C—C—H        H—C—C—C—O—H         H—C—C—C—H
     |            |  |            |  |  |             |  |  |
     H            H  O            H  H  H             H  O  H
                     |                                   |
                     H                                   H
methyl alcohol  ethyl alcohol    propyl alcohol      isopropyl alcohol
```

The last two are isomers of each other.

9.33 Distinguish between the terms *isotope* and *isomer*.

▮ An isotope is an atom with the same atomic number but a different mass number as another atom. An isomer is a molecule with the same molecular formula as a different molecule.

9.34 What continuous-chain hydrocarbon is isomeric with 2-methyl-3-ethylhexane?

▮ Nonane

9.35 Two isomeric compounds, A and B, have the formula C_4H_8. Compound A reacts to decolorize a solution of bromine in CCl_4; compound B does not react with bromine. (Bromine reacts with double bonded carbon atoms.) Diagram possible structures for A and B.

I Compound A must have a double bond. It could be *cis* or *trans* $CH_3CH{=}CHCH_3$, $CH_2{=}CHCH_2CH_3$, or $(CH_3)_2C{=}CH_2$. Compound B must have no double bond. It could be

$$\begin{array}{l} H_2C \\ \ \ | \quad \rangle CH{-}CH_3 \\ H_2C \end{array} \quad \text{or} \quad \begin{array}{c} H_2C{-}CH_2 \\ |\qquad | \\ H_2C{-}CH_2 \end{array}$$

9.36 Write structural formulas for all the structural isomers of cyclobutane.

I $CH_2{=}CHCH_2CH_3 \qquad CH_3CH{=}CHCH_3 \qquad \begin{array}{c} H_2C{=}CCH_3 \\ | \\ CH_3 \end{array} \qquad \begin{array}{c} H_2C{-}CHCH_3 \\ \backslash\ / \\ CH_2 \end{array}$

9.37 Give a straight-chain hydrocarbon that is isomeric with 2,2,4-trimethylpentane.

I Octane

9.38 How many isomers are there corresponding to the formula $C_4H_{10}O$?

I There are 7—1-butanol, 2-butanol, 2-methyl-1-propanol, 2-methyl-2-propanol, diethyl ether, methyl propyl ether, and methyl isopropyl ether.

9.39 Draw the structures of all isomers of pentane, C_5H_{12}.

I

$$CH_3CH_2CH_2CH_2CH_3 \qquad \begin{array}{c} CH_3CH_2CHCH_3 \\ | \\ CH_3 \end{array} \qquad \begin{array}{c} CH_3 \\ | \\ CH_3CCH_3 \\ | \\ CH_3 \end{array}$$

9.40 Draw structures of all isomers of C_3H_5Cl.

I $CH_2{=}CHCH_2Cl \qquad \begin{array}{ccc} Cl & & H \\ & C{=}C & \\ H & & CH_3 \end{array} \qquad \begin{array}{ccc} Cl & & CH_3 \\ & C{=}C & \\ H & & H \end{array} \qquad \begin{array}{ccc} H & & Cl \\ & C{=}C & \\ H & & CH_3 \end{array} \qquad \begin{array}{c} H_2C{-}CH_2 \\ \backslash\ / \\ C \\ /\ \backslash \\ H \quad Cl \end{array}$

9.41 Draw formulas for all structural isomers of $C_4H_8I_2$.

I

$$\begin{array}{c} H\ \ H\ \ H\ \ H \\ I{-}C{-}C{-}C{-}C{-}H \\ I\ \ H\ \ H\ \ H \end{array} \qquad \begin{array}{c} H\ \ H\ \ H\ \ H \\ I{-}C{-}C{-}C{-}C{-}H \\ H\ \ I\ \ H\ \ H \end{array} \qquad \begin{array}{c} H\ \ H\ \ H\ \ H \\ I{-}C{-}C{-}C{-}C{-}H \\ H\ \ H\ \ I\ \ H \end{array}$$

$$\begin{array}{c} H\ \ H\ \ H\ \ H \\ I{-}C{-}C{-}C{-}C{-}I \\ H\ \ H\ \ H\ \ H \end{array} \qquad \begin{array}{c} H\ \ I\ \ H\ \ H \\ H{-}C{-}C{-}C{-}C{-}H \\ H\ \ I\ \ H\ \ H \end{array} \qquad \begin{array}{c} H\ \ H\ \ H\ \ H \\ H{-}C{-}C{-}C{-}C{-}H \\ H\ \ I\ \ I\ \ H \end{array}$$

$$\begin{array}{c} H\ \ H\ \ H \\ I{-}C{-}C{-}C{-}I \\ H\ \ |\ \ H \\ C \\ H\ \ H\ \ H \end{array} \qquad \begin{array}{c} H\ \ H\ \ H \\ I{-}C{-}C{-}C{-}H \\ I\ \ |\ \ H \\ C \\ H\ \ H\ \ H \end{array} \qquad \begin{array}{c} H\ \ I\ \ H \\ I{-}C{-}C{-}C{-}H \\ H\ \ |\ \ H \\ C \\ H\ \ H\ \ H \end{array}$$

9.42 How many structural isomers are possible for $C_2H_4Cl_2$?

I Two—1,1-dichloroethane and 1,2-dichloroethane

9.43 If the double bonds in dichlorobenzene, $C_6H_4Cl_2$, were localized between specific carbon atoms, how many isomers of this compound would exist? How many isomers actually exist?

▮ If the bonds were localized, there would be 4 isomers; actually there are only 3. Of the following, the first two are identical, because the bonds are not localized.

Cl Cl Cl Cl Cl Cl Cl Cl

9.44 Write all the structural isomeric formulas for C_4H_9Cl.

▮ The molecular composition resembles butane, C_4H_{10}, with the exception that one H is replaced by Cl. The formulas can be found by picking all the differently located hydrogens in the two isomeric formulas of C_4H_{10} (Fig. 9.2). The results are shown in Fig. 9.3.

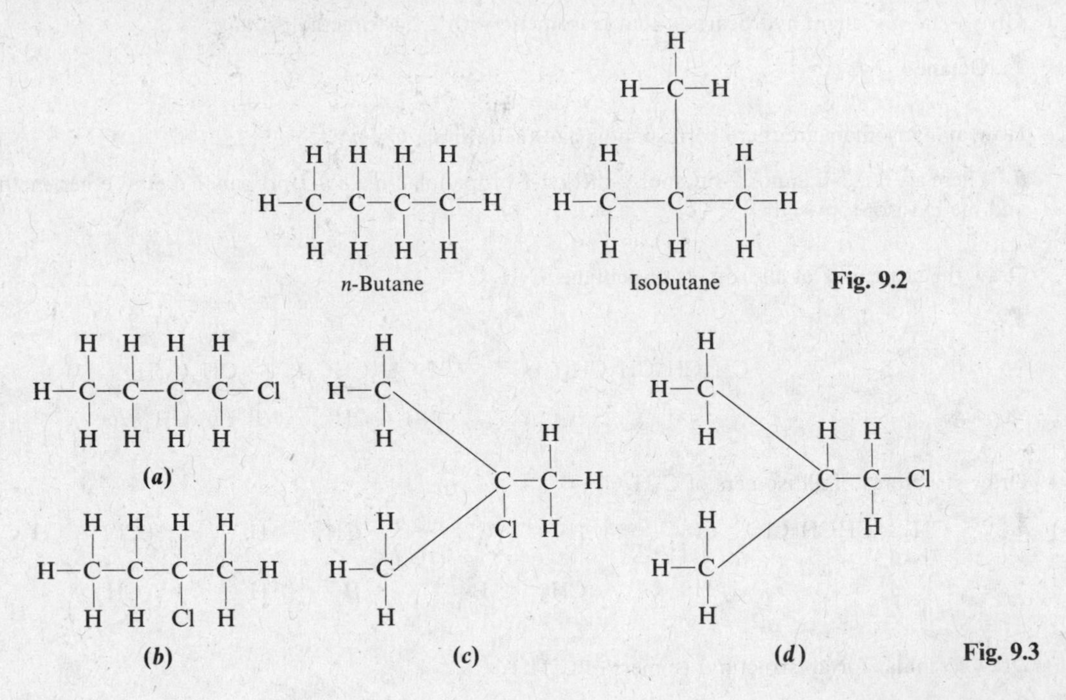

Fig. 9.2

Fig. 9.3

Note that the two end carbons of *n*-butane are identical and the two interior carbons are identical. Thus a chlorine substituted on the left-hand carbon would have given the same compound as (*a*), only viewed from a different end. Similarly, substitution on the carbon next to the left would have given a compound identical with (*b*). In isobutane, the three terminal carbons are identical. Because of the free rotation around the C—C single bond axis, it is immaterial which of the several hydrogens attached to a given C is substituted for; all positions become averaged in time anyway, because of the free rotation.

9.45 Write structural formulas for all the compounds which have the molecular formula C_5H_{10}.

▮ $CH_2{=}CHCH_2CH_2CH_3$ $CH_2{=}CHCH(CH_3)_2$ $CH_3CH{=}CHCH_2CH_3$

$CH_3CH{=}C(CH_3)_2$ $CH_3CH_2(CH_3)C{=}CH_2$

$H_2C{-}CH_2$, H_2C, CH_2, CH_2 (cyclopentane) $H_2C{-}CH{-}CH_3$, $H_2C{-}CH_2$ $H_2C{-}CH{-}CH_2CH_3$, CH_2 $H_3C{-}CH{-}CH{-}CH_3$, CH_2

9.46 Give formulas for all compounds with the molecular formula $C_5H_{11}Cl$.

▮ $ClCH_2CH_2CH_2CH_2CH_3$ $CH_3CHClCH_2CH_2CH_3$ $CH_3CH_2CHClCH_2CH_3$

$ClCH_2CH_2CH(CH_3)_2$ $CH_3CHClCH(CH_3)_2$ $CH_3CCl(CH_3)CH_2CH_3$

$ClCH_2C(CH_3)_3$ $ClCH_2CH(CH_3)CH_2CH_3$

9.47 Write formulas for all structural isomers corresponding to the formula $C_5H_{11}I$.

```
CH3CH2CH2CH2CH2I    CH3CH2CH2CHCH3    CH3CH2CHCH2CH3    ICH2CH2CHCH3
                              |                |                   |
                              I                I                  CH3

CH3CH—CHCH3            I              ICH2CHCH2CH3          CH2I
   |    |              |                   |                 |
   I   CH3         CH3CCH2CH3             CH3          CH3—C—CH3
                       |                                     |
                      CH3                                   CH3
```

9.48 Write all possible structural formulas for compounds with the molecular formula $C_5H_{10}I_2$.

```
I—CHCH2CH2CH2CH3     I—CH2CHCH2CH2CH3     I—CH2CH2CHCH2CH3
  |                         |                         |
  I                         I                         I

I—CH2CH2CH2CHCH3     I—CH2CH2CH2CH2CH2    CH3CHCHCH2CH3
               |                     |        | |
               I                     I        I I

     I               CH3CHCH2CHCH3                I
     |                  |     |                   |
CH3CCH2CH2CH3           I     I            CH3CH2CCH2CH3
     |                                            |
     I                                            I

I—CHCHCH2CH3         I—CH2CHCH2CH3        I—CH2CHCHCH3
  | |                       |                   | |
  I CH3                    CH2I                H3C I

I—CH2CHCH2CH2I             I              I I
       |                   |              | |
      CH3            I—CH2C—CH2CH3        CH3C—CHCH3
                           |                 |
                          CH3               CH3

     I                        I                  I I
     |                        |                  | |
CH3C—CH2CH2I          CH3CH—C—CH3         CH3CH—CHCH2
     |                   |  |                |
    CH3                 H3C I               CH3

          I                  CH3                    CH3
          |                  |                      |
CH3CHCH2CH            I—CH2—C—CH2I          I—CH—C—CH3
   |     |                   |                |   |
  CH3    I                  CH3               I  CH3
```

9.49 Draw the carbon skeleton and any chlorine atoms for all structural isomers of the following molecules. You need not bother drawing the hydrogen atoms, but see that the correct number are present for each isomer. (*a*) C_5H_{12} (*b*) C_3H_7Cl (*c*) $C_3H_6Cl_2$ (*d*) $C_5H_{11}Cl$ (*e*) C_6H_{14} (*f*) C_7H_{16}

```
(a)  C—C—C—C—C     C—C—C—C          C
                     |              |
                     C            C—C—C
                                    |
                                    C

(b)  C—C—C—Cl      C—C—C
                     |
                     Cl

(c)  C—C—C     Cl          C—C—C        Cl
     |  |      |           |   |        |
     Cl Cl     C—C—C       Cl  Cl     C—C—C
               |                        |
               Cl                       Cl
```

(*d*)

```
C—C—C—C—C     C—C—C—C—C     C—C—C—C—C     C—C—C—C—Cl
|               |               |             |
Cl              Cl              Cl            C

C—C—C—C           Cl          C—C—C—C             C
  |  |            |             |  |              |
  C  Cl         C—C—C—C         Cl C            C—C—C—Cl
                  |                               |
                  C                               C
```

(*e*)

```
C—C—C—C—C—C     C—C—C—C—C     C—C—C—C—C          C
                  |                 |              |
                  C                 C            C—C—C—C
                                                   |
                                                   C

C—C—C—C
  |  |
  C  C
```

(*f*)

```
C—C—C—C—C—C—C     C—C—C—C—C—C     C—C—C—C—C—C
                    |                   |
                    C                   C

    C             C—C—C—C—C       C—C—C—C—C
    |               |  |            |     |
  C—C—C—C—C         C  C            C     C
    |
    C

      C              C            C—C—C—C—C
      |              |                |
  C—C—C—C—C        C—C—C—C            C
      |              |  |             |
      C              C  C             C
```

9.50 Demonstrate that CH_2Cl_2 could exist in more than one isomeric form if the bonds about the carbon atom in saturated compounds were (*a*) square pyramidal or (*b*) trigonal pyramidal, both of which are incorrect.

▮ The (incorrect) structures would be

```
(a)      C              C           (b)     Cl              H
     H···/\···H     H···/\···Cl             |  .H           |  .Cl
     Cl/    \Cl     Cl/    \H           Cl—C·:          Cl—C·:
                                            \H              \H
```

9.51 Using Table 7.2, calculate ΔH° for the reaction at 25 °C

$$C_2H_4(g) + 3O_2(g) \longrightarrow 2CO_2(g) + 2H_2O(g)$$

▮

$$\begin{aligned}\Delta H^\circ &= 4D(\text{C—H}) + D(\text{C=C}) + 3D(O_2) - 4D(\text{C=O}) - 4D(\text{H—O}) \\ &= 4(100) + 146 + 3(119.2) - 4(177) - 4(111) = -248 \text{ kcal/mol}\end{aligned}$$

9.52 Using Table 7.2, estimate whether CH_4 or C_8H_{18} would yield more energy per gram upon complete combustion.

▮

$$CH_4 + 2O_2 \longrightarrow CO_2 + 2H_2O$$

Per mol of CH_4

$$\begin{aligned}\Delta H &= 4D(\text{C—H}) + 2D(O_2) - 2D(\text{C=O}) - 4D(\text{H—O}) \\ &= 4(100) + 2(119.2) - 2(177) - 4(111) = -160 \text{ kcal/mol}\end{aligned}$$

Per g of CH_4:

$$\left(\frac{-160 \text{ kcal}}{\text{mol}}\right)\left(\frac{1 \text{ mol}}{16.0 \text{ g}}\right) = -10.0 \text{ kcal/g}$$

Per mol of C_8H_{18}:

$$C_8H_{18} + \tfrac{25}{2}O_2 \longrightarrow 8CO_2 + 9H_2O$$

$$\begin{aligned}\Delta H &= 7D(\text{C—C}) + 18D(\text{C—H}) + \tfrac{25}{2}D(O_2) - 16D(\text{C=O}) - 18D(\text{H—O}) \\ &= 7(82) + 18(100) + 12.5(119.2) - 16(177) - 18(111) = -966 \text{ kcal/mol}\end{aligned}$$

Per g of C_8H_{18}:

$$\left(\frac{-966\ \text{kcal}}{\text{mol}}\right)\left(\frac{1\ \text{mol}}{114\ \text{g}}\right) = -8.47\ \text{kcal/g}$$

CH_4 yields more energy per gram.

9.53 Using Table 7.2, and assuming a Kekule (localized double bond) structure, calculate the enthalpy of combustion of benzene to CO_2 and water. Compare this value with that experimentally determined for benzene, -782.3 kcal/mol, and account for the difference.

▮

```
      H       H
       \     /
        C=C
       /    \
   H—C       C—H  +  15/2 O2 ⟶ 6CO2 + 3H2O
      \\    //
        C—C
       /    \
      H      H
```

$$\Delta H = 6D(\text{C—H}) + 3D(\text{C—C}) + 3D(\text{C=C}) + \tfrac{15}{2}D(\text{O}_2) - 12D(\text{C=O}) - 6D(\text{H—O})$$
$$= 6(100) + 3(82) + 3(146) + 7.5(119.2) - 12(177) - 6(111) = -612\ \text{kcal/mol}$$

The value of ΔH_{comb} given is -782.3 kcal/mol. The difference includes the delocalization energy of benzene. (Other differences are expansion against the atmosphere, included in ΔH_{comb} but not in the bond energy calculation and production of liquid versus gaseous water in the two cases.)

9.3 GEOMETRIC AND OPTICAL ISOMERISM

9.54 Name the isomeric forms of C_6H_{14} [Prob. 9.49(*e*)] according to the IUPAC system. Which one(s), if any, is (are) optically active?

▮

```
CH3CH2CH2CH2CH2CH3     CH3CHCH2CH2CH3     CH3CH2CHCH2CH3
                          |                     |
                          CH3                   CH3
      hexane            2-methylpentane      3-methylpentane

CH3CH—CHCH3              CH3
   |    |                 |
  H3C  CH3           H3C—C—CH2CH3
                          |
                          CH3
2,3-dimethylbutane   2,2-dimethylbutane
```

None of these isomers is optically active.

9.55 Write formulas for all the structural and geometrical isomers of C_4H_8.

▮ If the four carbons are not in a ring there must be one double bond to satisfy the tetracovalence of carbon. The double bond occurs either in the center of the molecule or toward an end. In the former case, two geometrical isomers occur with different positions of the terminal carbons relative to the double bond; in the latter case, two structural isomers occur, differing in the extent of branching within the skeleton. Additional possibilities are ring structures without double bonds.

```
H3C      CH3     H3C      H      H3C      H
   \    /           \    /          \    /
    C=C              C=C             C=C
   /    \           /    \          /    \
  H      H         H      CH3     H3C      H

CH3CH2      H     H2C—CH2          CH2
      \    /       |   |          /  \
       C=C        H2C—CH2      H2C—C—CH3
      /    \                        |
     H      H                       H
```

Note the shorthand notation of grouping 2 or 3 hydrogens with the carbon to which they are bonded. It is understood, of course, that each hydrogen is bonded to the carbon of its group and that the carbon of the group is bonded to the next carbon or to the carbon of the adjoining group or groups.

9.56 Which of the following compounds can exist as geometric isomers? CH_2Cl_2, $CH_2Cl{-}CH_2Cl$, $CHBr{=}CHCl$, $CH_2Cl{-}CH_2Br$.

▌ Only CHBr=CHCl can exist as geometric isomers:

```
Br        Cl              Br        H
  \      /                  \      /
   C == C        and         C == C
  /      \                  /      \
 H        H                H        Cl
```

In $CH_2Cl—CH_2Cl$ and $CH_2Cl—CH_2Br$, the carbon atoms are connected by a single bond about which the groups can rotate relatively freely. Thus any conformation of the halogen atoms may be converted into any other simply by rotation about the single bond. In CH_2Cl_2, the configuration of the molecule is tetrahedral, and all interchanges of atoms yield exactly equivalent configurations.

9.57 Which of the isomeric C_4H_9Cl compounds would you expect to be optically active? Refer to Fig. 9.3.

▌ Compound (*b*) is the only one which would exist in optically active isomeric forms, because it is the only one which has a carbon atom (the one bonded to Cl) which is bonded to four different groups. Every other carbon atom in this or the other structures is bonded to at least two hydrogen atoms or to two CH_3 groups.

9.58 Among the paraffin hydrocarbons (C_nH_{2n+2}, where *n* is an integer), what is the formula of the compound of lowest molecular weight which could demonstrate optical activity in at least one of its structural isomers?

▌ For an alkane to exhibit optical activity, at least one carbon atom must be bonded to 4 different groups. The smallest possible groups are H, CH_3, C_2H_5, and C_3H_7. The molecular formula would be C_7H_{16}.

9.59 Select from the following the pairs of **(*a*)** geometric isomers **(*b*)** optical isomers **(*c*)** structural isomers

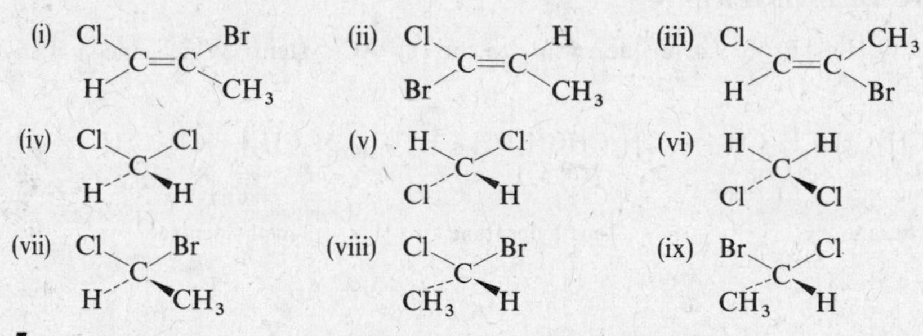

▌ **(*a*)** i and iii **(*b*)** viii with vii or ix, which are the same **(*c*)** ii with i and iii

9.60 Are all optical isomers necessarily optically active? Explain.

▌ Meso isomers are optically inactive, despite being in the class of optical isomers.

9.61 Explain the difference between a meso isomer and a racemic mixture. What characteristic(s) do they have in common?

▌ A meso isomer is a single compound; a racemic mixture contains an equimolar mixture of two compounds. The meso isomer is inherently inactive because of *internal* compensation; the racemic mixture does not rotate the plane of polarized light because the effect of one of the isomers negates the effect of the other. The racemic mixture can be separated into two optically active compounds by physical means (as Pasteur did). The meso isomer and the racemic mixture have in common their inability to rotate the plane of polarized light. Both have two centers of (opposite) chirality.

9.62 Write a structural formula for an alcohol which has the formula $C_4H_{10}O$ and which can exist as optically active enantiomers.

▌

```
   H  H  H  H
   |  |  |  |
H—C—C—C—C—H
   |  |  |  |
   H  H  O  H
         |
         H
```

Carbon 2 is asymmetric.

9.63 How many optical isomers can exist for 2,3-butanediol? Would all of these be optically active?

▌ There are three optical isomers, one of which is inactive (the meso form) because of the two carbon atoms having the same groups arranged in opposite configurations.

9.64 Tartaric acid is a dihydroxydicarboxylic acid, $HOCOCHOHCHOHCO_2H$. Identify any chiral centers, draw all optical isomers, and explain the total number of such isomers of tartaric acid in terms of the 2^n rule.

▌ The two middle carbon atoms in each structure are chiral centers. They are starred, to indicate their chirality.

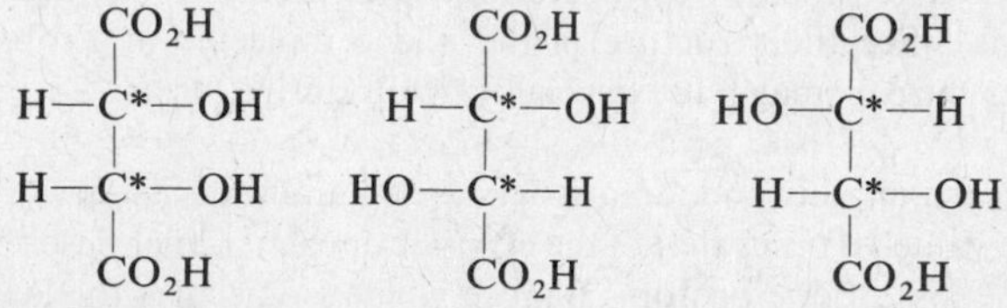

```
   CO2H            CO2H            CO2H
    |               |               |
 H—C*—OH        H—C*—OH        HO—C*—H
    |               |               |
 H—C*—OH       HO—C*—H          H—C*—OH
    |               |               |
   CO2H            CO2H            CO2H
   meso
```

There are 3 isomers instead of $2^2 = 4$, one a meso isomer.

9.65 Identify the asymmetric carbon atom(s) in each of the following substances, and identify each of the molecules which has a meso isomer.

(i) $CH_3CHNH_2CONH_2$ (ii) $HOCH_2CHOHCH_2CH_3$ (iii) $HOCH_2CHOHCH_2OH$

```
(iv)    H3C                    (v)     H2C—CH2
          \                           /       \
          HC—CH2                  H3CCH        CH2
         /      \                     \       /
     H3CCH       CH2                   HC=CH
          \     /
          H2C—CH2
```

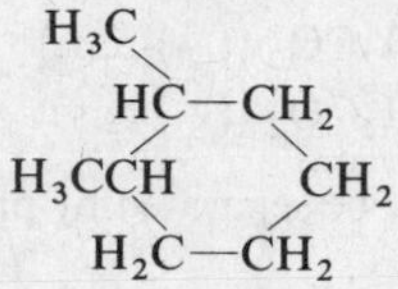

▌ The asymmetric carbon atoms are identified by asterisks. The compound which forms a meso isomer is labeled.

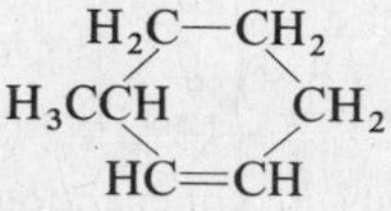

$CH_3C^*HNH_2CONH_2$ $HOCH_2C^*HOHCH_2CH_3$ $HOCH_2CHOHCH_2OH$ (none)

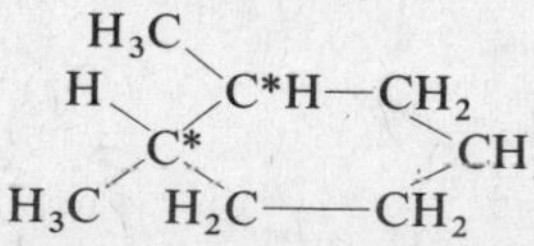

```
        H3C
     H     \C*H—CH2                                H2C—CH2
      \C*           CH2   (forms meso isomer)      /      \
     /   \         /                        H3C—C*H        CH2
  H3C    H2C——CH2                                 \      /
                                                   HC=CH
```

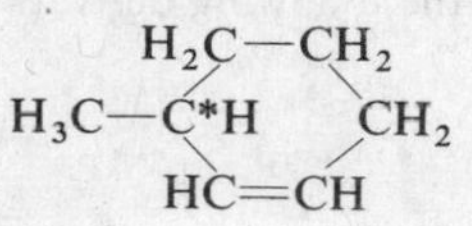

9.66 Draw structures for all isomers of C_3H_4ClBr. State which of them, if any, will be optically active.

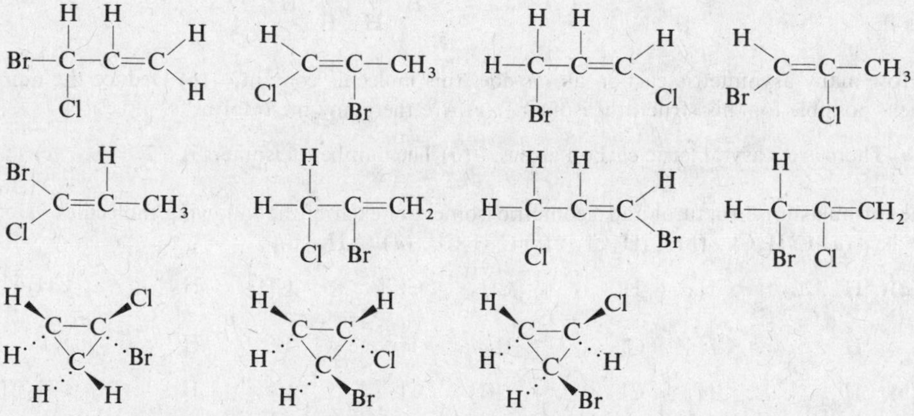

▌

```
    H  H                                              H  H
    |  |    H      H                                  |  |    H      H
Br—C—C=C             C=C—CH3                    H—C—C=C              C=C—CH3
    |       H      Cl    |                            |     Cl     Br    |
    Cl                   Br                           Br                 Cl

     H             H                          H  H               H
Br   |             |                          |  |    H          |
  C=C—CH3      H—C—C=CH2                  H—C—C=C           H—C—C=CH2
Cl                 |  |                       |     Br           |  |
                   Cl Br                      Cl                 Br Cl

H       Cl        H       H               H       Cl
  C—C               C—C                     C—C
H   C   Br        H   C   Cl              H   C   H
  H   H             H   Br                  H   Br
                  (+) and (−)             (+) and (−)
```

9.67 Construct molecular models of 1,2-dichlorocyclobutane. *(a)* How many asymmetric carbon atoms does this molecule contain? *(b)* Can this substance exist as a meso isomer? *(c)* What is the total number of isomers containing a four-membered ring corresponding to the formula $C_4H_6Cl_2$?

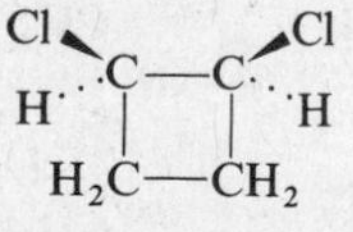

▌

```
  Cl       Cl
    C——C
 H  |  |  H
   H2C—CH2
```

(a) Two *(b)* yes, as shown above *(c)* 6. The isomers are the 1,1-dichlorocyclobutane, 1,2-dichloro isomers in (+), (−), and meso forms, 1,3-dichloro in cis and trans forms.

9.68 Construct a molecular model of 2,3-pentadiene. **(*a*)** Does this molecule contain any asymmetric carbon atoms? **(*b*)** Does it contain a chiral center? **(*c*)** Can it exist in optically active isomeric forms?

▮ **(*a*)** There is no asymmetric carbon atom. **(*b*)** The combination of the 3 middle carbon atoms gives the molecule as a whole an elongated tetrahedral configuration. The molecule has a chiral center consisting of the 3 double bonded carbon atoms. (See the structure of the allene molecule in Prob. 8.19). **(*c*)** Since the molecule is not superimposable on its mirror image, it exists in optically active forms.

9.69 The dicarboxylic acids maleic acid and fumaric acid are cis and trans isomers, respectively, of $HOCOCH{=}CHCO_2H$. One of them has a K_1 value 10 times that of the other. For each isomer describe an appropriate structure of the anion which results from the loss of one proton. Deduce which isomer has the larger K_1 value, and suggest a reason for the difference. On this basis, which K_2 value should be larger?

▮ The cis isomer can form intramolecular hydrogen bonds, making the mononegative ion more stable than that of the trans isomer. Thus K_1 for the cis isomer is greater, but its K_2 value is lower.

9.70 Describe a reaction sequence which might be used to resolve a racemic mixture of 3-chlorobutanoic acid.

▮ A reaction sequence such as the following might be employed:

$$\left.\begin{array}{l}(+)\text{-}CH_3CHClCH_2CO_2H\\(-)\text{-}CH_3CHClCH_2CO_2H\end{array}\right\} + 2((-)\text{-base}) \longrightarrow \left\{\begin{array}{l}(+)\text{-}CH_3CHClCH_2CO_2\text{-}((-)\text{-base})\\(-)\text{-}CH_3CHClCH_2CO_2\text{-}((-)\text{-base})\end{array}\right.$$

Since the pair of compounds formed in the reactions are not enantiomers, they may be separated by physical means. After separation, each is treated with excess HCl to restore the original acid.

$$(+)\text{-}CH_3CHClCH_2CO_2\text{-}((-)\text{-base}) + HCl \longrightarrow (+)\text{-}CH_3CHClCH_2CO_2H + H((-)\text{-base})Cl$$
$$(-)\text{-}CH_3CHClCH_2CO_2\text{-}((-)\text{-base}) + HCl \longrightarrow (-)\text{-}CH_3CHClCH_2CO_2H + H((-)\text{-base})Cl$$

9.71 Consider the following derivative of cyclohexane:

```
            Cl  H
            |   | /H
   CH3\     C — C       H
       C  /     H \  C/
   H/     \ C — C /    \Cl
          H/ |   | \H
            H   H
```

(*a*) How many asymmetric carbon atoms does this molecule contain? **(*b*)** Deduce the number of optically active isomers possible for this structural isomer. **(*c*)** Are there any meso forms?

▮ **(*a*)** There are 3 asymmetric carbon atoms. **(*b*)** The number of isomers is $2^3 = 8$. **(*c*)** There are no meso forms.

9.72 Write formulas for structural and geometric isomers of each of the following molecules. Do not include ring compounds. **(*a*)** C_3H_5Cl **(*b*)** $C_3H_4Cl_2$ **(*c*)** C_4H_7Cl **(*d*)** C_5H_{10}

▮ **(*a*)**

H, CH_3	H, CH_3	Cl, CH_3	H, CH_2Cl
C=C	C=C	C=C	C=C
H, Cl	Cl, H	H, H	H, H

(*b*)

H, CH_3	Cl, CH_3	Cl, CH_3	H, CH_2Cl
C=C	C=C	C=C	C=C
Cl, Cl	H, Cl	Cl, H	H, Cl

H, CH_2Cl	Cl, CH_2Cl	H, $CHCl_2$
C=C	C=C	C=C
Cl, H	H, H	H, H

(*c*)

Cl, CH_2CH_3	CH_3, CH_3	CH_3, CH_2Cl	H, CH_2CH_3
C=C	C=C	C=C	C=C
H, H	Cl, H	H, H	Cl, H

Cl, CH_3	H, CH_2Cl	H, CH_2CH_3	H, CH_3
C=C	C=C	C=C	C=C
CH_3, H	CH_3, H	H, Cl	Cl, CH_3

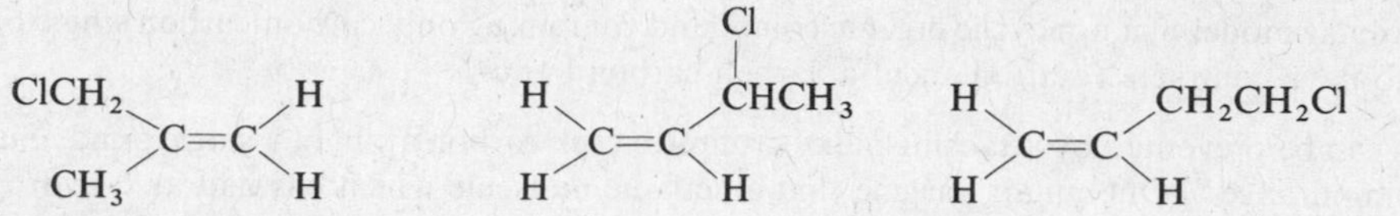

$$\underset{CH_3}{\overset{ClCH_2}{}}\!\!>C{=}C<\!\!\underset{H}{\overset{H}{}} \qquad \underset{H}{\overset{H}{}}\!\!>C{=}C<\!\!\underset{H}{\overset{CHCH_3(Cl)}{}} \qquad \underset{H}{\overset{H}{}}\!\!>C{=}C<\!\!\underset{H}{\overset{CH_2CH_2Cl}{}}$$

(*d*)

$$\underset{H}{\overset{H}{}}\!\!>C{=}C<\!\!\underset{H}{\overset{CH_2CH_2CH_3}{}} \qquad \underset{H}{\overset{CH_3}{}}\!\!>C{=}C<\!\!\underset{H}{\overset{CH_2CH_3}{}} \qquad \underset{CH_3}{\overset{H}{}}\!\!>C{=}C<\!\!\underset{H}{\overset{CH_2CH_3}{}}$$

$$\underset{H}{\overset{H}{}}\!\!>C{=}C<\!\!\underset{CH_3}{\overset{CH_2CH_3}{}} \qquad \underset{H}{\overset{H}{}}\!\!>C{=}C<\!\!\underset{H}{\overset{CHCH_3(CH_3)}{}} \qquad \underset{CH_3}{\overset{CH_3}{}}\!\!>C{=}C<\!\!\underset{H}{\overset{CH_3}{}}$$

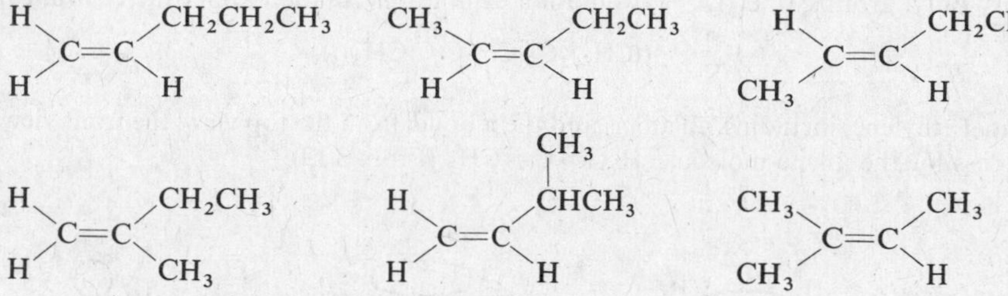

9.4 MORE ADVANCED TOPICS

9.73 Draw a two-dimensional formula for two different conformations of 2-methylbutane.

▮

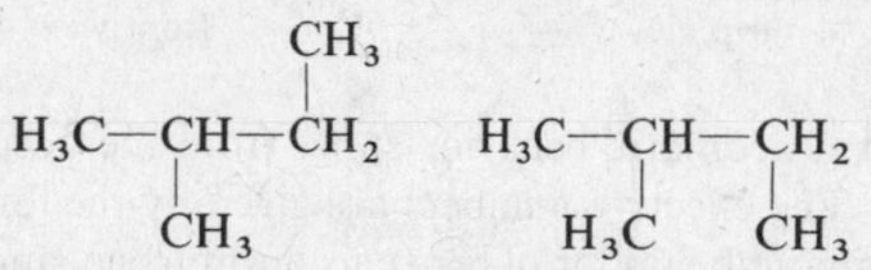

$$H_3C{-}\underset{CH_3}{CH}{-}\overset{CH_3}{CH_2} \qquad H_3C{-}\underset{H_3C}{CH}{-}\underset{CH_3}{CH_2}$$

9.74 In aqueous solution, tetramethylammonium hydroxide, $(CH_3)_4NOH$, is a strong base; aqueous solutions of trimethylamine are weakly basic. Explain the difference in terms of the structures of these substances.

▮ $(CH_3)_3N^+OH^-$ is an ionic compound. R_3H is basic because of the reaction with water forming some OH^- ions in an equilibrium reaction.

9.75 Choose from the following list those molecules which are (*a*) Brønsted acids (*b*) Brønsted bases (*c*) Lewis bases

(i) C_2H_5OH (ii) C_5H_5N (iii) $C_2H_5OCH_3$ (iv) C_6H_5OH (v) $(C_2H_5)_2NH_2^+$
(vi) $C_6H_5CO_2H$ (vii) $(C_2H_5)_2NH$

▮ (*a*) iv, v, and vi (v is the conjugate acid of $(C_2H_5)_2NH$) (*b*) ii and vii (*c*) ii, iii, and vii

9.76 What kinds of hybrid orbitals are formed on the carbon atoms in cyclohexane? In cyclopropane?

▮ sp^3 in each

9.77 In terms of electron delocalization, explain why amides are much weaker bases than amines.

▮ The third form represents the electron delocalization of the two resonance forms, shown first:

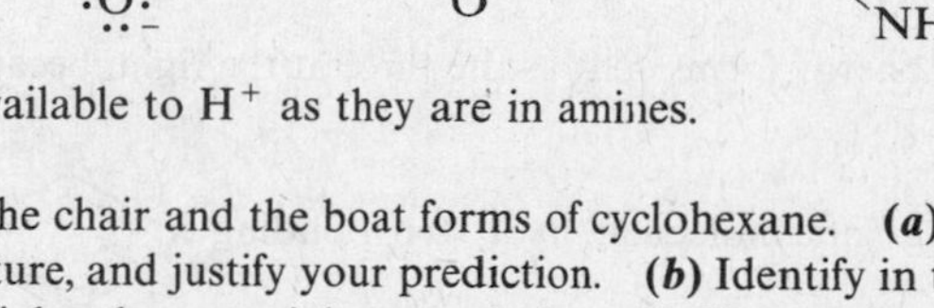

$$R{-}\underset{:\ddot{O}:^-}{C}{=}NH_2^+ \qquad R{-}\underset{O}{\overset{\|}{C}}{-}\ddot{N}H_2 \qquad R{-}C\!\overset{O}{\underset{NH_2}{\vdots}}$$

The electrons are thus not as available to H^+ as they are in amines.

9.78 Construct molecular models of the chair and the boat forms of cyclohexane. (*a*) Predict which form predominates in cyclohexane at room temperature, and justify your prediction. (*b*) Identify in the chair form the sets of hydrogen atoms which are referred to as axial and equatorial, respectively. (*c*) Change the conformation of the chair form into the boat form and then back into a chair form which is not the same as the original conformation. What positions do the hydrogen atoms which were originally axial now occupy? (*d*) How many isomers are there of chlorocyclohexane?

▮ (*a*) The chair form of cyclohexane predominates at room temperature. (*b*) and (*c*) The axial hydrogen atoms change into equatorial hydrogen atoms, and vice versa, when the original chair form is changed through the boat form into another conformation of the chair form. (*d*) Since the positions are interconvertible, there is only one form of chlorocyclohexane.

9.79 Construct a molecular model of a noncyclic organic compound containing only carbon-carbon single bonds in which it is impossible to have complete rotation about a carbon-carbon bond.

▎ Free rotation can be prevented by attaching bulky groups to each carbon atom of a single bond. Such a situation is termed "steric hindrance." One might imagine that an ethane molecule which has had its 6 hydrogen atoms replaced by tertiary butyl groups, $(CH_3)_3C$—, would not exhibit free rotation about the central carbon-carbon bond.

$$((CH_3)_3C)_3C—C(C(CH_3)_3)_3$$

9.80 Draw a diagram of ethylene, including all atoms and the π bond, from the top view, the front view, and the side view. Repeat this process for the allene molecule, $H_2C{=}C{=}CH_2$ (Prob. 8.19).

▎ See Fig. 9.4.

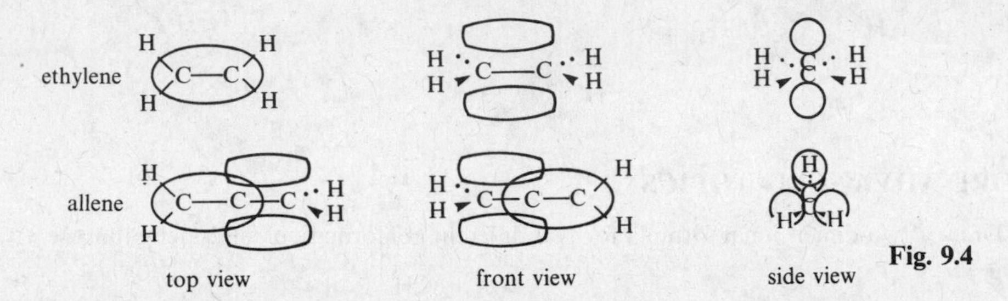

Fig. 9.4

9.81 (*a*) It has been suggested that aromatic behavior stems from the occurrence of certain numbers of electrons in delocalized π orbital systems. The effective numbers are given by the formula $4n + 2$, where n is an integer. Based on this fact, explain the aromatic behavior of benzene, anthracene, and azulene. (*b*) Draw resonance structures for azulene:

▎ (*a*) Benzene has 6 electrons in delocalized molecular orbitals, naphthalene has 10, and azulene has 10. Each of these values fits the expression $4n + 2$.

(*b*)

9.82 Show why the phenolate ion, $C_6H_5O^-$, has a greater resonance stabilization than phenol, C_6H_5OH.

▎

Phenol does not have low energy forms such as the three at the right, because the oxygen atom would have a partial positive charge.

9.83 Draw a structure showing intramolecular hydrogen bonding in [structure: benzene ring with —O—H and —C(H)=O substituents]. How many atoms are in the additional ring formed?

▎ The extra ring consists of 6 atoms.

9.84 Draw resonance structures for

■

Normal resonance forms may be drawn for the first 3 structures. See also Prob. 9.83.

9.85 Draw structures representing the resonance in each of the following: (*a*) benzene (*b*) naphthalene (*c*) anthracene

■ (*a*) (*b*)

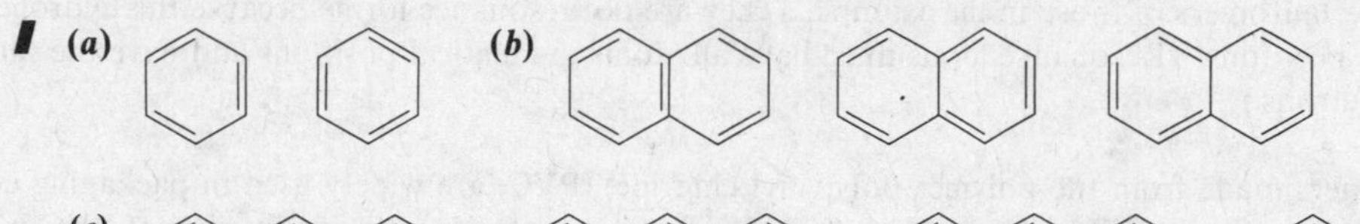

(*c*)

9.86 (*a*) Calculate the oxidation number of the carbon atom in formaldehyde and in methandiol, $HOCH_2OH$. (*b*) Using Table 7.2 and $\Delta H_{sub}(C) = 171.7$ kcal/mol, calculate the enthalpies of formation of these two substances. (*c*) Determine ΔH° for the following reaction, and predict if one of the compounds would be stable with respect to reaction to produce the other.

$$HCHO + H_2O \longrightarrow HOCH_2OH$$

■ (*a*) Zero in each. (*b*) For HCHO:

$$C(s) \longrightarrow C(g) \qquad \Delta H_{sub} = 171.7 \text{ kcal}$$
$$H_2 \longrightarrow 2H(g) \qquad D(H_2) = 104.2$$
$$\tfrac{1}{2}O_2 \longrightarrow O(g) \qquad \tfrac{1}{2}D(O_2) = \tfrac{1}{2}(119.2)$$
$$C(g) + 2H(g) + O(g) \longrightarrow HCHO \qquad -2D(C{-}H) - D(C{=}O) = -2(100) - 177$$
$$C(s) + H_2 + \tfrac{1}{2}O_2 \longrightarrow HCHO$$
$$\Delta H = \Delta H_{sub}(C) + D(H_2) + \tfrac{1}{2}D(O_2) - 2D(C{-}H) - D(C{=}O)$$
$$= 171.7 + 104.2 + \tfrac{1}{2}(119.2) - 2(100) - 177 = -41.5 \text{ kcal/mol}$$

For $HOCH_2OH$:

$$C(s) \longrightarrow C(g) \qquad \Delta H_{sub} = 171.7 \text{ kcal}$$
$$2H_2 \longrightarrow 4H(g) \qquad 2D(H_2) = 2(104.2)$$
$$O_2 \longrightarrow 2O(g) \qquad D(O_2) = 119.2$$
$$C(g) + 4H(g) + 2O(g) \longrightarrow HOCH_2OH \qquad -2D(H{-}O) - 2D(C{-}O) - 2D(C{-}H) = -2(111) - 2(83) - 2(100)$$
$$C(s) + 2H_2 + O_2 \longrightarrow HOCH_2OH$$
$$\Delta H = \Delta H_{sub}(C) + 2D(H_2) + D(O_2) - 2D(H{-}O) - 2D(C{-}O) - 2D(C{-}H)$$
$$= 171.7 + 2(104.2) + 119.2 - 2(111) - 2(83) - 2(100) = -88.7 \text{ kcal/mol}$$

(*c*) $\Delta H^\circ = \Delta H_f(HOCH_2OH) - \Delta H_f(HCHO) - \Delta H_f(H_2O) = -88.7 - (-41.5) - (-68.32) = +21.1$ kcal/mol

Since ΔH° is positive and ΔS for the combination is expected to be negative, $\Delta G = \Delta H^\circ - T\Delta S$ is positive for the reaction as written. Thus, HCHO is stable with respect to transformation by water into $HOCH_2OH$.

9.87 Textbooks represent the formula of a pyrimidine-type base as one of the following:

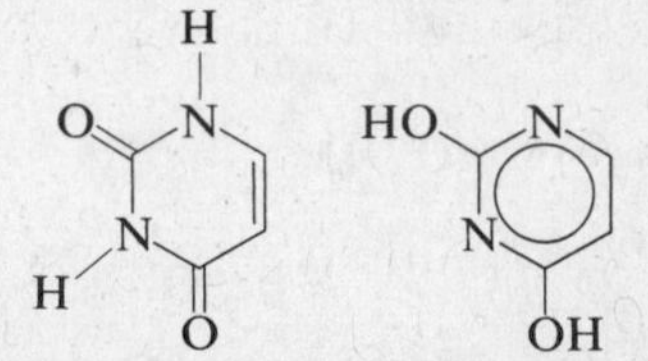

Noting that an OH group attached directly to an aromatic ring is acidic, write the formula for still another possible representation of the base. What relationships exist among these three representations of pyrimidine?

Any of the following might have been chosen:

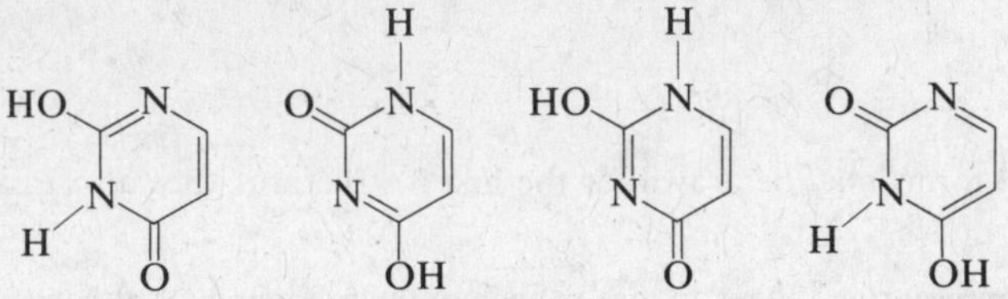

These molecules are tautomers of those in the example. They are not resonance forms because the hydrogen atoms are not in the same positions. (Resonance forms must have all atoms in identical positions and have the same number of unpaired electrons.)

9.88 Throwaway containers made from the polymer poly(vinyl chloride) (PVC) are widely used in packaging consumer goods. The formula for polyvinyl chloride is $(C_2H_3Cl)_n$. Write an equation for the combustion of PVC in oxygen. Suggest why incineration is not a satisfactory method for disposing of used packaging materials made of PVC.

$$(C_2H_3Cl)_n + \tfrac{5}{2}nO_2 \longrightarrow 2nCO_2 + nH_2O + nHCl$$

The HCl produced on combustion is a strong acid. When breathed from the atmosphere, it causes choking and is a corrosive acid which reacts with inanimate objects.

9.89 Explain why each of the following equations represents an oxidation of the organic reactant:

$$CH_3CHO + \tfrac{1}{2}O_2 \longrightarrow CH_3CO_2H$$

$$CH_3CH_3 \xrightarrow{\text{Pt catalyst}} CH_2{=}CH_2 + H_2$$

	average oxidation state of carbon
CH_3CHO	−1
CH_3CO_2H	0
CH_3CH_3	−3
$CH_2{=}CH_2$	−2

9.90 A solution of bromine in benzene is stable indefinitely, but when an iron nail is put into the solution, bromination of the benzene occurs fairly rapidly. Explain the function of the iron. Write equations for all the reactions.

The iron reacts with some bromine to form $FeBr_3$, a catalyst for the reaction of bromine with benzene:

$$2Fe + 3Br_2 \longrightarrow 2FeBr_3$$

$$Br_2 + C_6H_6 \xrightarrow{FeBr_3} C_6H_5Br + HBr$$

9.91 The presence of a mass/charge ratio for the parent peak at an odd value in the mass spectrum of an organic compound is indicative of an amine or amide. Explain why amines and amides exhibit this type of behavior, whereas alcohols, hydrocarbons, ethers, etc., do not.

Of the common elements, except for hydrogen, which constitute the majority of organic compounds, all have even atomic weights, but only nitrogen has an odd total bond order. Hence compounds with odd numbers of nitrogen atoms are most apt to have odd mass/charge ratios for the parent ion. (Try counting the hydrogen atoms in aminoethane and diaminoethane to see the effect of bond orders.)

9.92 Select from compounds (i) to (vi) below the one(s) which has (have) ***(a)*** a major peak of 30 in its mass spectrum, ***(b)*** a major mass spectral peak at 71, ***(c)*** a parent peak in the mass spectrum having an odd value of mass/charge.

(i) $CH_3CH_2—C_6H_5$ (ii) $C_6H_5—CHO$ (iii) $NH_2CH_2CH_2NH_2$

(iv) $CH_3CH_2NH_2$ (v) $NH_2CH_2CH{=}NCH_2CH_3$ (vi) $C_6H_4(CHO)(OH)$

▌ ***(a)*** iii, iv, and v, all of which have the NH_2CH_2 group ***(b)*** v; $NH_2CH_2CH{=}NCH_2$ ***(c)*** iv (See the answer to Prob. 9.91.) Compounds iii and v do not have odd parent peaks because each has *two* nitrogen atoms.

9.93 Write structural formulas for all isomers corresponding to the molecular formula C_2H_7N. Indicate how to distinguish each of these using mass spectrometry.

▌

$CH_3CH_2NH_2$ (H—C(H)(H)—C(H)(H)—N(H)—H): peaks at 15, 16, 29, 30

CH_3NHCH_3 (H—C(H)(H)—N(H)—C(H)(H)—H): peaks at 15, 30

9.94 What major differences would be expected between the mass spectra of the isomeric compounds CH_3CH_2OH and CH_3OCH_3?

▌ The mass spectrum of CH_3CH_2OH would contain a peak at the mass/charge ratio of 29, $CH_3CH_2^+$, while the spectrum of the ether should not contain this peak. Both spectra should contain a peak at $m/e = 31$.

9.95 Write formulas for all the possible structural isomers of $C_2H_8N_2$. Which isomer would *not* have a peak at a mass/charge ratio of 45?

▌ ***(a)*** $H_2NCH_2CH_2NH_2$ ***(b)*** $CH_3NHCH_2NH_2$ ***(c)*** $CH_3NHNHCH_3$
(d) $CH_3—N(NH_2)—CH_3$ ***(e)*** $CH_3CH_2NHNH_2$

Isomer *(a)* has no CH_3 group and could not yield a significant peak at a mass/charge ratio of 45 (15 less than the molecular weight), representing the loss of CH_3.

9.96 Explain how to distinguish the isomers of $C_2H_4Cl_2$ by mass spectroscopy.

▌ There are two possible isomers:

(a) Cl—C(H)(Cl)—C(H)(H)—H ***(b)*** Cl—C(H)(H)—C(H)(H)—Cl

The mass spectrum of *(a)* will contain a peak at 15 (plus others corresponding to the two major isotopes of Cl). That of *(b)* will contain no peak at 15.

9.97 What peaks are to be expected in the mass spectrum of CH_3Cl?

▌ There should be major peaks at 15, 35, 37, 50, and 52. The second and third correspond to the major isotopes of chlorine, ^{35}Cl and ^{37}Cl, which exist as 75.5% and 24.5% of natural chlorine, respectively. (There is no peak at 35.45.)

9.98 Show how to use mass spectroscopy to differentiate between the two isomers in each of the following sets. Explain how each isomer would be identified.

(*a*) $(CH_3)_2CHCONH_2$ and $CH_3CH_2CONHCH_3$ (*b*) CH_3CH_2OH and CH_3OCH_3
(*c*) $CH_3CH_2CH_2CH_3$ and $(CH_3)_2CHCH_3$ (*d*) $CH_2{=}CICH_2CH_3$ and $CH_3CI{=}CHCH_3$

▮ (*a*) The mass spectrum of the first compound would not contain a major peak at 29; that of the second compound would. (*b*), (*c*), and (*d*) The mass spectrum of the first compound contains a peak at 29, the second does not.

9.99 Identify the peak occurring at a mass/charge ratio of 46 in the mass spectrum of $HSCH_2CH_2OH$

▮ The peak must be due to SCH_2.

9.100 Two isomeric compounds, A and B, have the molecular formula C_3H_9N. The important peaks in their mass spectra are tabulated. Identify the compounds.

compound A		compound B	
mass peak	**relative abundance**	**mass peak**	**relative abundance**
59	100	59	100
58	10	58	5
44	23	44	40
43	30	30	30
16	38	29	30
15	20	15	30

▮ Compound A must contain the NH_2 group, which causes the peak at 16. Since it has no peak at 29, there can be no CH_3CH_2 group. Compound A is $(CH_3)_2CHNH_2$. Compound B contains a CH_3CH_2 group (29) and a CH_3NH group (30). The second group cannot be CH_2NH_2 or there would be a peak at 16 (NH_2). Compound B is $CH_3CH_2NHCH_3$.

9.101 Explain why carbon-carbon multiple bonds are reactive centers in chemical reactions, but are relatively stable in mass spectrometry.

▮ The π bonding electrons are susceptible to "attack" by electrophilic (electron-seeking) reactants, but these same π bonding electrons increase the strength of the bond between the atoms. The question "stability toward what?" should be asked whenever the word stable is used. (For example, carbon is stable alone at 500 °C, but in the presence of O_2 at 500 °C it is not stable.)

9.102 A compound contains 54.55% carbon, 9.09% hydrogen, and 36.36% oxygen, and has a parent ion with a mass/charge ratio of 88 u/*e*. What are its empirical formula and its molecular formula?

▮

$$(54.55\text{ g C})\left(\frac{1\text{ mol C}}{12.01\text{ g C}}\right) = 4.542\text{ mol C}$$

$$(9.09\text{ g H})\left(\frac{1\text{ mol H}}{1.008\text{ g H}}\right) = 9.02\text{ mol H}$$

$$(36.36\text{ g O})\left(\frac{1\text{ mol O}}{16.00\text{ g O}}\right) = 2.272\text{ mol O}$$

The empirical formula is C_2H_4O, which has a formula weight of 44 g/mol. From the mass spectroscopic data, the molecular formula must be $C_4H_8O_2$.

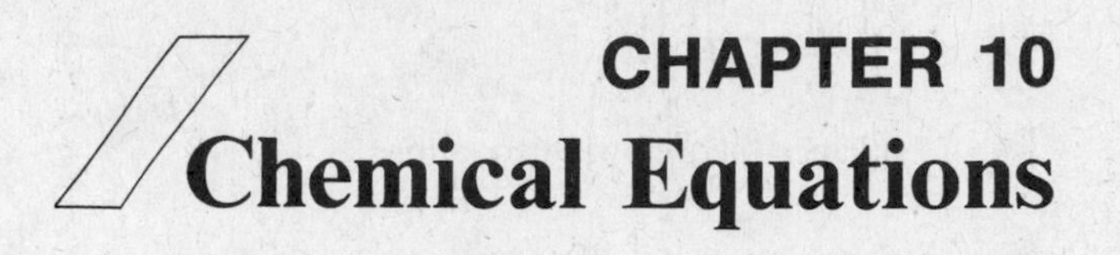

CHAPTER 10
Chemical Equations

10.1 BALANCING CHEMICAL EQUATIONS

10.1 In a *balanced* chemical equation, what does the absence of any coefficient imply?

▌ The absence implies a coefficient of 1. Before the species is balanced, however, the lack of a coefficient might merely signify that the balancing has not yet been done. Take care to distinguish between these meanings.

10.2 Balance the following skeleton equations with the smallest integral coefficients:

(*a*) $FeS_2 + O_2 \longrightarrow Fe_2O_3 + SO_2$ (*b*) $C_7H_6O_2 + O_2 \longrightarrow CO_2 + H_2O$

▌ (*a*) There are no fixed rules for balancing simple equations. Often a trial-and-error procedure is used. It is commonly helpful to start with the most complex formula. Fe_2O_3 has two different elements and a greater total number of atoms than any of the other substances, so we might start with it. We note that oxygen atoms occur in pairs in the molecules O_2 and SO_2 but not in the formula unit Fe_2O_3. If we write the equation with symbols representing the *integral* coefficients,

$$w\text{FeS}_2 + x\text{O}_2 \longrightarrow y\text{Fe}_2\text{O}_3 + z\text{SO}_2$$

then the total number of oxygen atoms on the left, $2x$, is even for any integral value of x. The total number on the right, $3y + 2z$, could be even or odd, depending on whether y is even or odd. We conclude from the required equality of $2x$ with $3y + 2z$ that y must be even. We can now try the smallest even number, 2, and proceed from there.

$$w\text{FeS}_2 + x\text{O}_2 \longrightarrow 2\text{Fe}_2\text{O}_3 + z\text{SO}_2$$

To balance iron atoms, w must equal 4.

$$4\text{FeS}_2 + x\text{O}_2 \longrightarrow 2\text{Fe}_2\text{O}_3 + z\text{SO}_2$$

To balance sulfur, z must equal 8.

$$4\text{FeS}_2 + x\text{O}_2 \longrightarrow 2\text{Fe}_2\text{O}_3 + 8\text{SO}_2$$

Finally, to balance oxygen, $2x = 6 + 16$, or $x = 11$.

$$4\text{FeS}_2 + 11\text{O}_2 \longrightarrow 2\text{Fe}_2\text{O}_3 + 8\text{SO}_2$$

Note that the coefficient of the simplest substance, elementary oxygen in this case, was evaluated last. This is the usual consequence of beginning the balancing procedure with the most complex substance.

(*b*) The most complex substance in this equation is $C_7H_6O_2$. We may assume 1 molecule of this substance and immediately write the coefficients for CO_2 and H_2O that will lead to balance of C and H, respectively.

$$\text{C}_7\text{H}_6\text{O}_2 + x\text{O}_2 \longrightarrow 7\text{CO}_2 + 3\text{H}_2\text{O}$$

Balancing oxygen atoms is saved for last because an adjustment of x would not interfere with the balance of any other element. An arithmetic balance now demands that x equal $\frac{15}{2}$, giving an equation

$$\text{C}_7\text{H}_6\text{O}_2 + \tfrac{15}{2}\text{O}_2 \longrightarrow 7\text{CO}_2 + 3\text{H}_2\text{O}$$

that, while balanced, violates the stated requirement for integral coefficients. The correct ratio is preserved and the fraction eliminated by multiplying each coefficient by 2.

$$2\text{C}_7\text{H}_6\text{O}_2 + 15\text{O}_2 \longrightarrow 14\text{CO}_2 + 6\text{H}_2\text{O}$$

10.3 Write balanced chemical equations for the following reactions:

Zinc sulfide + oxygen gas ⟶ zinc oxide + sulfur dioxide

Nitric acid + copper(II) carbonate ⟶ water + carbon dioxide + copper(II) nitrate

▌

$$2\text{ZnS} + 3\text{O}_2 \longrightarrow 2\text{ZnO} + 2\text{SO}_2$$

$$2\text{HNO}_3 + \text{CuCO}_3 \longrightarrow \text{H}_2\text{O} + \text{CO}_2 + \text{Cu(NO}_3)_2$$

Balance the following equations.

10.4 $BCl_3 + P_4 + H_2 \longrightarrow BP + HCl$

▌ $$4BCl_3 + P_4 + 6H_2 \longrightarrow 4BP + 12HCl$$

10.5 $C_2H_2Cl_4 + Ca(OH)_2 \longrightarrow C_2HCl_3 + CaCl_2 + H_2O$

▌ $$2C_2H_2Cl_4 + Ca(OH)_2 \longrightarrow 2C_2HCl_3 + CaCl_2 + 2H_2O$$

10.6 $(NH_4)_2Cr_2O_7 \longrightarrow N_2 + Cr_2O_3 + H_2O$

▌ $$(NH_4)_2Cr_2O_7 \longrightarrow N_2 + Cr_2O_3 + 4H_2O$$

10.7 $Zn_3Sb_2 + H_2O \longrightarrow Zn(OH)_2 + SbH_3$

▌ $$Zn_3Sb_2 + 6H_2O \longrightarrow 3Zn(OH)_2 + 2SbH_3$$

10.8 $HClO_4 + P_4O_{10} \longrightarrow H_3PO_4 + Cl_2O_7$

▌ $$12HClO_4 + P_4O_{10} \longrightarrow 4H_3PO_4 + 6Cl_2O_7$$

10.9 $C_6H_5Cl + SiCl_4 + Na \longrightarrow (C_6H_5)_4Si + NaCl$

▌ $$4C_6H_5Cl + SiCl_4 + 8Na \longrightarrow (C_6H_5)_4Si + 8NaCl$$

10.10 $Sb_2S_3 + HCl \longrightarrow H_3SbCl_6 + H_2S$

▌ $$Sb_2S_3 + 12HCl \longrightarrow 2H_3SbCl_6 + 3H_2S$$

10.11 $IBr + NH_3 \longrightarrow NI_3 + NH_4Br$

▌ $$3IBr + 4NH_3 \longrightarrow NI_3 + 3NH_4Br$$

10.12 $KrF_2 + H_2O \longrightarrow Kr + O_2 + HF$

▌ $$2KrF_2 + 2H_2O \longrightarrow 2Kr + O_2 + 4HF$$

10.13 $Na_2CO_3 + C + N_2 \longrightarrow NaCN + CO$

▌ $$Na_2CO_3 + 4C + N_2 \longrightarrow 2NaCN + 3CO$$

10.14 $K_4Fe(CN)_6 + H_2SO_4 + H_2O \longrightarrow K_2SO_4 + FeSO_4 + (NH_4)_2SO_4 + CO$

▌ $$K_4Fe(CN)_6 + 6H_2SO_4 + 6H_2O \longrightarrow 2K_2SO_4 + FeSO_4 + 3(NH_4)_2SO_4 + 6CO$$

10.15 $Fe(CO)_5 + NaOH \longrightarrow Na_2Fe(CO)_4 + Na_2CO_3 + H_2O$

▌ $$Fe(CO)_5 + 4NaOH \longrightarrow Na_2Fe(CO)_4 + Na_2CO_3 + 2H_2O$$

10.16 $H_3PO_4 + (NH_4)_2MoO_4 + HNO_3 \longrightarrow (NH_4)_3PO_4 \cdot 12MoO_3 + NH_4NO_3 + H_2O$

▌ $$H_3PO_4 + 12(NH_4)_2MoO_4 + 21HNO_3 \longrightarrow (NH_4)_3PO_4 \cdot 12MoO_3 + 21NH_4NO_3 + 12H_2O$$

10.17 Convert the following into balanced chemical equations:

(*a*) $NCl_3 + H_2O \longrightarrow NH_3 + HOCl$ (*b*) $PCl_3 + H_2O \longrightarrow H_3PO_3 + HCl$
(*c*) $SbCl_3 + H_2O \longrightarrow Sb(O)Cl + HCl$

▌ (*a*) $NCl_3 + 3H_2O \longrightarrow NH_3 + 3HOCl$ (*b*) $PCl_3 + 3H_2O \longrightarrow H_3PO_3 + 3HCl$
(*c*) $SbCl_3 + H_2O \longrightarrow Sb(O)Cl + 2HCl$

10.18 Complete and balance the following equations using Prob. 10.17 and the periodic table as aids, if necessary:

(*a*) $AsCl_3 + H_2O \longrightarrow H_3AsO_3 +$ (*b*) $BiCl_3 + H_2O \longrightarrow$

▌ (*a*) $AsCl_3 + 3H_2O \longrightarrow H_3AsO_3 + 3HCl$ (*b*) $BiCl_3 + H_2O \longrightarrow Bi(O)Cl + 2HCl$

The reactions of $AsCl_3$ and $BiCl_3$ are analogous to those of the chlorides of the elements immediately above them in the periodic table.

10.19 Balance

(*a*) $H_2SO_4 + NaOH \longrightarrow NaHSO_4 + H_2O$
(*b*) $H_2SO_4 + NaOH \longrightarrow Na_2SO_4 + H_2O$

▮ (*a*) $H_2SO_4 + NaOH \longrightarrow NaHSO_4 + H_2O$
(*b*) $H_2SO_4 + 2NaOH \longrightarrow Na_2SO_4 + 2H_2O$

10.20 Balance

(*a*) $C_2H_6O + O_2 \longrightarrow CO_2 + H_2O$ (*b*) $CoBr_3 + Na_2CO_3 \longrightarrow Co_2(CO_3)_3 + NaBr$

▮ (*a*) $C_2H_6O + 3O_2 \longrightarrow 2CO_2 + 3H_2O$ (*b*) $2CoBr_3 + 3Na_2CO_3 \longrightarrow Co_2(CO_3)_3 + 6NaBr$

10.2 PREDICTION OF PRODUCTS

10.21 Classify each of the following substances as acid, base, acid anhydride, or basic anhydride: (*a*) H_2SO_3 (*b*) NH_3 (*c*) LiOH (*d*) Li_2O (*e*) Cl_2O_3 (*f*) BaO (*g*) CO_2 (*h*) CrO

▮ (*a*) Acid (*b*) basic anhydride (*c*) base (*d*) basic anhydride
(*e*) acid anhydride (*f*) basic anhydride (*g*) acid anhydride (*h*) basic anhydride

10.22 Classify simple inorganic chemical reactions into four types (leaving a fifth type for more complex reactions to be discussed later).

▮ 1. Combination reactions, e.g., $2Na + Cl_2 \longrightarrow 2NaCl$
2. Decomposition reactions, e.g., $2HgO \longrightarrow 2Hg + O_2$
3. Replacement reactions, e.g., $2NaI + Cl_2 \longrightarrow 2NaCl + I_2$
4. Double replacement reactions, e.g., $AgNO_3 + NaCl \longrightarrow AgCl + NaNO_3$
5. More complex reactions

10.23 Which type of simple chemical reaction (Prob. 10.22) depends on the relative activities of two metals or two nonmetals?

▮ The replacement reactions

10.24 Which type of simple chemical reaction (Prob. 10.22) depends on the solubilities of the reactants and/or products?

▮ Double replacement reactions

10.25 Name five ions or classes of ions whose compounds are virtually all soluble in water.

▮ Acetates, nitrates, chlorates, alkali metal ions, and ammonium ion

10.26 Which metal chlorides are insoluble in water?

▮ AgCl, $PbCl_2$, and Hg_2Cl_2

10.27 Select the compounds from the list below which are insoluble in water.

HCl NH_3 $NaClO_3$ $BaSO_4$ $AgNO_3$ $PbCl_2$ Cu_2O $CuSO_4$ $Pb(C_2H_3O_2)_2$ AgBr

▮ $BaSO_4$ $PbCl_2$ Cu_2O AgBr

10.28 Give the formulas of three different bromide salts that are essentially insoluble in cold water.

▮ AgBr, $PbBr_2$, Hg_2Br_2 (similar to the corresponding chlorides)

10.29 Indicate whether each of the following statements is true for the element in its uncombined (elemental) state only, in its compounds only, or both. (*a*) The halogens occur in diatomic molecules—two like atoms together. (*b*) The metallic elements conduct electricity well. (*c*) Sulfur occurs naturally as a mixture of isotopes. (*d*) The sodium ion is always monopositive. (*e*) Nitrogen atoms are capable of forming covalent bonds.

▮ (*a*) Element only (*b*) element only (*c*) both (*d*) compounds only (*e*) both

Write complete and balanced equations for each of the following reactions. If there is no reaction, write "no reaction." State the type of reaction, as described in Prob. 10.22.

10.30 $H_2O_2 \xrightarrow{\text{heat}}$

▮ $2H_2O_2 \xrightarrow{\text{heat}} 2H_2O + O_2$ decomposition

10.31 $H_2 + O_2 \xrightarrow{\text{heat}}$

▮ $2H_2 + O_2 \xrightarrow{\text{heat}} 2H_2O$ combination

10.32 Sodium plus chlorine

▮ $2Na + Cl_2 \longrightarrow 2NaCl$ combination

10.33 $FeCl_2 + AgNO_3 \longrightarrow$

▮ $FeCl_2 + 2AgNO_3 \longrightarrow Fe(NO_3)_2 + 2AgCl$ double replacement

10.34 $C_2H_6O + O_2 \longrightarrow$

▮ $C_2H_6O + 3O_2 \longrightarrow 2CO_2 + 3H_2O$ (combustion)

10.35 $Cu + Fe(NO_3)_2 \longrightarrow$

▮ $Cu + Fe(NO_3)_2 \longrightarrow$ no reaction

10.36 $Fe_2(SO_4)_3 + BaCl_2 \longrightarrow$

▮ $Fe_2(SO_4)_3 + 3BaCl_2 \longrightarrow 2FeCl_3 + 3BaSO_4$ double replacement

10.37 $CaCO_3 + HCl \longrightarrow$

▮ $CaCO_3 + 2HCl \longrightarrow CaCl_2 + H_2O + CO_2$ double replacement followed by decomposition

10.38 $H_2SO_4 + NaOH$ (excess) $\longrightarrow$

▮ $H_2SO_4 + 2NaOH$ (excess) $\longrightarrow Na_2SO_4 + 2H_2O$ double replacement

10.39 $H_2SO_4 + NaOH$ (limited)

▮ $H_2SO_4 + NaOH$ (limited) $\longrightarrow NaHSO_4 + H_2O$ double replacement

10.40 $Ba(OH)_2 + HCl$

▮ $Ba(OH)_2 + 2HCl \longrightarrow BaCl_2 + 2H_2O$ double replacement

10.41 $Na + H_2O \longrightarrow$

▮ $2Na + 2H_2O \longrightarrow 2NaOH + H_2$ replacement

10.42 $CaO + CO_2 \longrightarrow$

▮ $CaO + CO_2 \longrightarrow CaCO_3$ combination

10.43 $BaSO_4 + CuCl_2 \longrightarrow$

▮ $BaSO_4 + CuCl_2 \longrightarrow$ no reaction

10.44 $H_2O \xrightarrow[\text{NaCl}]{\text{electrolysis}}$

▮ $2H_2O \xrightarrow[\text{NaCl}]{\text{electrolysis}} 2H_2 + O_2$ decomposition

10.45 $Zn + H_2SO_4 \longrightarrow$

▮ $Zn + H_2SO_4 \longrightarrow ZnSO_4 + H_2$ replacement

10.46 $C + O_2$ (excess) ⟶

▮ $C + O_2$ (excess) ⟶ CO_2 combination

10.47 $C + O_2$ (limited) ⟶

▮ $2C + O_2$ (limited) ⟶ $2CO$ combination

10.48 $Fe + Cl_2 \longrightarrow$

▮ $2Fe + 3Cl_2 \longrightarrow 2FeCl_3$ combination

10.49 $Fe + HCl \longrightarrow$

▮ $Fe + 2HCl \longrightarrow FeCl_2 + H_2$ replacement

Write complete and balanced equations for each of the following reactions. If there is no reaction, write "no reaction."

10.50 $Zn + FeCl_2 \longrightarrow$

▮ $Zn + FeCl_2 \longrightarrow ZnCl_2 + Fe$

10.51 $F_2 + NaCl \longrightarrow$

▮ $F_2 + 2NaCl \longrightarrow 2NaF + Cl_2$

10.52 $NaCl + AgNO_3 \longrightarrow$

▮ $NaCl + AgNO_3 \longrightarrow AgCl + NaNO_3$

10.53 $AgCl + NaNO_3 \longrightarrow$

▮ $AgCl + NaNO_3 \longrightarrow$ no reaction

10.54 $H_3PO_4 + NaOH$ (limited) ⟶

▮ $H_3PO_4 + NaOH$ (limited) ⟶ $NaH_2PO_4 + H_2O$

10.55 $KHSO_4 + KOH \longrightarrow$

▮ $KHSO_4 + KOH \longrightarrow K_2SO_4 + H_2O$

10.56 $SO_2 + H_2O \longrightarrow$

▮ $SO_2 + H_2O \longrightarrow H_2SO_3$

10.57 $P_2O_5 + H_2O \longrightarrow$

▮ $P_2O_5 + 3H_2O \longrightarrow 2H_3PO_4$

10.58 $BaO + H_2O \longrightarrow$

▮ $BaO + H_2O \longrightarrow Ba(OH)_2$

10.59 $ZnO + H_2O \longrightarrow$

▮ $ZnO + H_2O \longrightarrow Zn(OH)_2$

10.60 $CrO + SO_2 \longrightarrow$

▮ $CrO + SO_2 \longrightarrow CrSO_3$

10.61 $K + Cl_2 \longrightarrow$

▮ $2K + Cl_2 \longrightarrow 2KCl$

10.62 $AgNO_3 + Cu \longrightarrow$

$$2AgNO_3 + Cu \longrightarrow Cu(NO_3)_2 + 2Ag$$

10.63 $NaI + Cl_2 \longrightarrow$

$$2NaI + Cl_2 \longrightarrow 2NaCl + I_2$$

10.64 $HgO \xrightarrow{\text{heat}}$

$$2HgO \xrightarrow{\text{heat}} 2Hg + O_2$$

10.65 $AgNO_3 + CuCl_2 \longrightarrow$

$$2AgNO_3 + CuCl_2 \longrightarrow 2AgCl + Cu(NO_3)_2$$

10.66 $HCl + NaOH \longrightarrow$

$$HCl + NaOH \longrightarrow NaCl + H_2O$$

10.67 $HCl + Ba(OH)_2 \longrightarrow$

$$2HCl + Ba(OH)_2 \longrightarrow BaCl_2 + 2H_2O$$

10.68 $Zn + HCl \longrightarrow$

$$Zn + 2HCl \longrightarrow H_2 + ZnCl_2$$

10.69 $CaCO_3 \xrightarrow{\text{heat}}$

$$CaCO_3 \xrightarrow{\text{heat}} CaO + CO_2$$

10.70 $KClO_3 \xrightarrow{\text{heat}}$

$$2KClO_3 \xrightarrow{\text{heat}} 2KCl + 3O_2$$

10.71 $H_2O + Cl_2O_7 \longrightarrow$

$$H_2O + Cl_2O_7 \longrightarrow 2HClO_4$$

10.72 $Ba(OH)_2 + H_2SO_4 \longrightarrow$

$$Ba(OH)_2 + H_2SO_4 \longrightarrow BaSO_4 + 2H_2O$$

10.73 $BaCl_2 + K_2SO_4 \longrightarrow$

$$BaCl_2 + K_2SO_4 \longrightarrow BaSO_4 + 2KCl$$

10.74 H_2SO_4 (excess) $+ Zn(OH)_2 \longrightarrow$

$$H_2SO_4 + Zn(OH)_2 \longrightarrow ZnSO_4 + 2H_2O$$

10.75 $Cl_2 + NaI \longrightarrow$

$$Cl_2 + 2NaI \longrightarrow 2NaCl + I_2$$

10.76 $Cl_2 + Al \longrightarrow$

$$3Cl_2 + 2Al \longrightarrow 2AlCl_3$$

10.77 $CO + O_2 \longrightarrow$

$$2CO + O_2 \longrightarrow 2CO_2$$

10.78 $H_2 + P \longrightarrow$

$$3H_2 + 2P \longrightarrow 2PH_3$$

10.79 $C + Cl_2 \longrightarrow$

▮ $$C + 2Cl_2 \longrightarrow CCl_4$$

10.80 $C_6H_{12}O_6 + O_2$ (excess) $\longrightarrow$

▮ $$C_6H_{12}O_6 + 6O_2 \longrightarrow 6CO_2 + 6H_2O$$

10.81 $CH_4 + O_2$ (limited) $\longrightarrow$

▮ $$2CH_4 + 3O_2 \text{ (limited)} \longrightarrow 2CO + 4H_2O$$

10.82 $Fe + Cu(NO_3)_2 \longrightarrow$

▮ $$Fe + Cu(NO_3)_2 \longrightarrow Fe(NO_3)_2 + Cu$$

10.83 $HCl + Cu \longrightarrow$

▮ $$HCl + Cu \longrightarrow \text{ no reaction}$$

10.84 $HCl + K_2CO_3 \longrightarrow$

▮ $$2HCl + K_2CO_3 \longrightarrow H_2O + CO_2 + 2KCl$$

10.85 $H_2SO_4 + MgO \longrightarrow$

▮ $$H_2SO_4 + MgO \longrightarrow MgSO_4 + H_2O$$

10.86 $NH_3 + CO_2 + H_2O \longrightarrow$

▮ $$2NH_3 + CO_2 + H_2O \longrightarrow (NH_4)_2CO_3$$

10.87 $C_6H_6 + O_2$ (excess) $\longrightarrow$

▮ $$2C_6H_6 + 15O_2 \text{ (excess)} \longrightarrow 12CO_2 + 6H_2O$$

10.88 $NH_3 + HCl \longrightarrow$

▮ $$NH_3 + HCl \longrightarrow NH_4Cl$$

10.89 $NaHCO_3 + HCl \longrightarrow$

▮ $$NaHCO_3 + HCl \longrightarrow NaCl + H_2O + CO_2$$

10.90 $Na_2CO_3 + HCl \longrightarrow$

▮ $$Na_2CO_3 + 2HCl \longrightarrow 2NaCl + H_2O + CO_2$$

10.91 $H_2CO_3 \longrightarrow$

▮ $$H_2CO_3 \longrightarrow H_2O + CO_2$$

10.92 Complete and balance equations for the following reactions.

(*a*) $CaO + H_2O \longrightarrow$ (*b*) $SO_2 + H_2O \longrightarrow$ (*c*) $Mg(s) + N_2(g) \longrightarrow$
(*d*) $Cu + HCl(aq) \longrightarrow$ (*e*) $Zn + H_2SO_4(aq) \longrightarrow$ (*f*) $Pb + CuSO_4(aq) \longrightarrow$

▮ (*a*) $CaO + H_2O \longrightarrow Ca(OH)_2$ (*b*) $SO_2 + H_2O \longrightarrow H_2SO_3$
(*c*) $3Mg + N_2 \longrightarrow Mg_3N_2$ (*d*) No reaction
(*e*) $Zn + H_2SO_4 \longrightarrow ZnSO_4 + H_2$ (*f*) $Pb + CuSO_4 \longrightarrow PbSO_4 + Cu$

10.93 Complete the balance equations for the following reactions:

(*a*) $H_2O_2 \longrightarrow$ (*b*) $C + H_2O \xrightarrow{600\,^\circ C}$

(*c*) $O_2 \xrightarrow{\text{uv light}}$ (*d*) $Al + NaOH \longrightarrow$

▌ (*a*) $2H_2O_2 \longrightarrow 2H_2O + O_2$ (*b*) $C + H_2O \longrightarrow CO + H_2$
(*c*) $3O_2 \longrightarrow 2O_3$ (ozone) (*d*) $6H_2O + 2Al + 2NaOH \longrightarrow 2NaAl(OH)_4 + 3H_2$

(Al is amphoteric and very active, so it can displace hydrogen from basic solutions.)

10.94 Predict the products: $NH_4Cl + NaOH \longrightarrow$

▌ $$NH_4Cl + NaOH \longrightarrow NaCl + NH_3 + H_2O$$

(Un-ionized products are formed from ionic reactants; one of the products is unstable and decomposes to NH_3 and H_2O.)

10.95 Predict the products: $NaC_2H_3O_2 + HCl \longrightarrow$

▌ $$NaC_2H_3O_2 + HCl \longrightarrow NaCl + HC_2H_3O_2$$

(Un-ionized product formed from ionic reactants)

10.96 Complete and balance the following equations.
(*a*) $Li + O_2 \longrightarrow$ (*b*) $Mg + HCl(aq) \longrightarrow$ (*c*) $C_6H_6 + O_2$ (excess) $\longrightarrow$
(*d*) $HCl(aq) + H_2SO_4(aq) \longrightarrow$ (*e*) $K + H_2O \longrightarrow$

▌ (*a*) $4Li + O_2 \longrightarrow 2Li_2O$ (*b*) $Mg + 2HCl \longrightarrow MgCl_2 + H_2$
(*c*) $2C_6H_6 + 15O_2 \longrightarrow 12CO_2 + 6H_2O$ or $C_6H_6 + 7\frac{1}{2}O_2 \longrightarrow 6CO_2 + 3H_2O$
(*d*) no reaction (*e*) $2K + 2H_2O \longrightarrow 2KOH + H_2$

10.97 What are the products of the reaction of calcium and water?

▌ $$Ca + 2H_2O \longrightarrow Ca(OH)_2 + H_2$$

10.98 Complete and balance:

(*a*) $HClO_3 + Ba(OH)_2 \longrightarrow$ (*b*) $Mg + HBr \longrightarrow$ (*c*) $F_2 + KI \longrightarrow$

▌ (*a*) $2HClO_3 + Ba(OH)_2 \longrightarrow Ba(ClO_3)_2 + 2H_2O$
(*b*) $Mg + 2HBr \longrightarrow MgBr_2 + H_2$
(*c*) $F_2 + 2KI \longrightarrow 2KF + I_2$

10.99 Complete and balance

(*a*) $Zn(ClO_3)_2(aq) + NaOH(aq) \longrightarrow$ (*b*) $La(OH)_3 + HClO_3 \longrightarrow$ (*c*) $C_2H_6O + O_2$ (excess) $\longrightarrow$

▌ (*a*) $Zn(ClO_3)_2 + 2NaOH \longrightarrow Zn(OH)_2 + 2NaClO_3$
(*b*) $La(OH)_3 + 3HClO_3 \longrightarrow La(ClO_3)_3 + 3H_2O$
(*c*) $C_2H_6O + 3O_2 \longrightarrow 2CO_2 + 3H_2O$

10.100 Predict the products of the following combinations. Balance each equation.

(*a*) $Na + Cl_2 \longrightarrow$ (*b*) $BaCl_2 + AgNO_3 \longrightarrow$
(*c*) $Zn + CuCl_2 \longrightarrow$ (*d*) $C_2H_5OH + O_2$ (excess) $\longrightarrow$

▌ (*a*) $2Na + Cl_2 \longrightarrow 2NaCl$ (*b*) $BaCl_2 + 2AgNO_3 \longrightarrow Ba(NO_3)_2 + 2AgCl$
(*c*) $Zn + CuCl_2 \longrightarrow Cu + ZnCl_2$ (*d*) $C_2H_5OH + 3O_2$ (excess) $\longrightarrow 3H_2O + 2CO_2$

10.101 Complete and balance the following equations:

(*a*) $Mg + Cu(NO_3)_2 \longrightarrow$ (*b*) $HBrO_3 + Ba(OH)_2 \longrightarrow$

▌ (*a*) $Mg + Cu(NO_3)_2 \longrightarrow Mg(NO_3)_2 + Cu$ (*b*) $2HBrO_3 + Ba(OH)_2 \longrightarrow Ba(BrO_3)_2 + 2H_2O$

10.102 Complete and balance the following equations.

(*a*) $Mg + FeCl_3 \longrightarrow$ (*b*) $HClO_3 + Ba(OH)_2 \longrightarrow$
(*c*) $C_6H_{12}O_2 + O_2$ (excess) $\longrightarrow$ (*d*) $Al + Br_2 \longrightarrow$

■ (*a*) $3Mg + 2FeCl_3 \longrightarrow 3MgCl_2 + 2Fe$ (*b*) $2HClO_3 + Ba(OH)_2 \longrightarrow Ba(ClO_3)_2 + 2H_2O$
(*c*) $C_6H_{12}O_2 + 8O_2$ (excess) $\longrightarrow 6CO_2 + 6H_2O$ (*d*) $2Al + 3Br_2 \longrightarrow 2AlBr_3$

10.3 NET IONIC EQUATIONS

10.103 What species that appear in complete equations are omitted from net ionic equations?

■ Only ions in solution that appear unchanged on both sides of the equation are omitted in the net ionic equation. If an ion changes in any way, for example, changes into part of a solid or part of a covalent compound, it appears as the ion on one side.

Write (*a*) ionic equations and (*b*) net ionic equations for each of the following complete equations.

10.104 $K_2SO_4(aq) + BaCl_2(aq) \longrightarrow BaSO_4(s) + 2KCl(aq)$

■ (*a*) $2K^+ + SO_4^{2-} + Ba^{2+} + 2Cl^- \longrightarrow BaSO_4 + 2K^+ + 2Cl^-$ (*b*) $SO_4^{2-} + Ba^{2+} \longrightarrow BaSO_4$

10.105 $2HNO_3(aq) + Ca(HCO_3)_2(aq) \longrightarrow Ca(NO_3)_2(aq) + 2H_2O + 2CO_2(g)$

■ (*a*) $2H^+ + 2NO_3^- + Ca^{2+} + 2HCO_3^- \longrightarrow Ca^{2+} + 2NO_3^- + 2H_2O + 2CO_2$
(*b*) $H^+ + HCO_3^- \longrightarrow CO_2 + H_2O$

10.106 The statement is often made in elementary texts that all nitrates, acetates, and chlorates are soluble, as well as all sodium and potassium salts. Using a table of solubilities, such as may be found in the *Handbook of Chemistry and Physics* (CRC Press, Boca Raton, Florida), find at least one exception to this statement. Is the example a common chemical?

■ Such rarely encountered compounds as sodium aluminum orthosilicate and sodium pyroantimonate are insoluble.

10.107 Aqueous copper(II) nitrate reacts with potassium iodide to yield solid copper(I) iodide, potassium nitrate, and iodine. Write a balanced net ionic equation for the reaction.

■
$$2Cu^{2+} + 4I^- \longrightarrow 2CuI + I_2$$

10.108 Write net ionic equations for the processes which occur when solutions of the following electrolytes are mixed: (*a*) $AgClO_3(aq)$ and $Na_2S(aq)$ (*b*) $(NH_4)_3PO_4(aq)$ and $HgSO_4(aq)$.

■ (*a*) Ag_2S is insoluble while $NaClO_3$ is soluble.

$$2Ag^+ + \cancel{2ClO_3^-} + \cancel{2Na^+} + S^{2-} \longrightarrow Ag_2S(s) + \cancel{2Na^+} + \cancel{2ClO_3^-}$$

Net equation:
$$2Ag^+ + S^{2-} \longrightarrow Ag_2S(s)$$

(*b*) $Hg_3(PO_4)_2$ is insoluble while $(NH_4)_2SO_4$ is soluble.

$$\cancel{6NH_4^+} + 2PO_4^{3-} + 3Hg^{2+} + \cancel{3SO_4^{2-}} \longrightarrow \cancel{6NH_4^+} + \cancel{3SO_4^{2-}} + Hg_3(PO_4)_2(s)$$

Net equation:
$$3Hg^{2+} + 2PO_4^{3-} \longrightarrow Hg_3(PO_4)_2(s)$$

To ascertain that a particular net ionic reaction actually occurs it is necessary to perform the experiment or to predict the behavior of the ions theoretically.

10.109 Assuming that no chemical reaction takes place, state which of the following mixtures would be expected to be completely soluble if a sample containing 1 g of each was shaken in 100 mL of water (*a*) NaCl, AgCl, and $PbCl_2$ (*b*) $Ba(ClO_3)_2$, NaCl, and KNO_3 (*c*) $Ba(ClO_3)_2$, $BaCO_3$, and $BaCl_2$

■ (*b*) [In (*a*), the AgCl and $PbCl_2$ are insoluble; in (*c*), the $BaCO_3$ is insoluble.]

10.110 Temporarily hard water, but not permanently hard water, can be softened by boiling. Explain.

■
$$Ca(HCO_3)_2(aq) \longrightarrow CaCO_3(s) + CO_2 + H_2O$$

10.111 Write net ionic equations corresponding to the following reactions:

(*a*) $Na_2CO_3(aq) + Ca(NO_3)_2(aq) \longrightarrow 2NaNO_3(aq) + CaCO_3(s)$
(*b*) $NaOH(aq) + HNO_3(aq) \longrightarrow NaNO_3(aq) + H_2O(l)$
(*c*) Find the number of mmol of NaOH required to react with 35.0 mmol of H^+ in a nitric acid solution.

▮ (*a*) $CO_3^{2-} + Ca^{2+} \longrightarrow CaCO_3$ (*b*) $OH^- + H^+ \longrightarrow H_2O$

(*c*) Since the H^+ reacts with OH^- in a 1:1 mole ratio [part (*b*)], and since there is 1 mol of OH^- in each mole of NaOH, it takes 35.0 mmol of NaOH. [The unit millimole (0.001 mol) is a very useful unit for dilute solution calculations.]

10.112 Write *balanced* net ionic equations for the following:

(*a*) $La_2(CO_3)_3(s) + HCl \longrightarrow LaCl_3 + CO_2 + H_2O$

(*b*) $5FeCl_2 + KMnO_4 + 8HCl \longrightarrow 5FeCl_3 + MnCl_2 + KCl + 4H_2O$ (all aqueous)

▮ (*a*) $La_2(CO_3)_3 + 6H^+ \longrightarrow 2La^{3+} + 3CO_2 + 3H_2O$

(*b*) $5Fe^{2+} + MnO_4^- + 8H^+ \longrightarrow 5Fe^{3+} + Mn^{2+} + 4H_2O$

10.113 Write balanced net ionic equations corresponding to the following (unbalanced) equations:

(*a*) $Ba(NO_3)_2(aq) + H_2SO_4(aq) \longrightarrow BaSO_4(s) + HNO_3(aq)$

(*b*) $HClO_3 + Fe(OH)_2(s) \longrightarrow Fe(ClO_3)_2 + H_2O$

▮ (*a*) $Ba^{2+} + SO_4^{2-} \longrightarrow BaSO_4$ (*b*) $2H^+ + Fe(OH)_2(s) \longrightarrow Fe^{2+} + 2H_2O$

Write net ionic equations for the following reactions.

10.114 $Fe_2(SO_4)_3(aq) + Fe(s) \longrightarrow 3FeSO_4(aq)$

▮ $2Fe^{3+} + Fe \longrightarrow 3Fe^{2+}$

10.115 $NaCl + AgNO_3 \longrightarrow AgCl + NaNO_3$

▮ $Cl^- + Ag^+ \longrightarrow AgCl$

10.116 $HCl + NaHCO_3 \longrightarrow NaCl + CO_2 + H_2O$

▮ $H^+ + HCO_3^- \longrightarrow CO_2 + H_2O$

10.117 $H_2SO_4 + BaCl_2 \longrightarrow BaSO_4 + 2HCl$

▮ $SO_4^{2-} + Ba^{2+} \longrightarrow BaSO_4$

10.118 $NaOH + NH_4Cl \longrightarrow NH_3 + H_2O + NaCl$

▮ $OH^- + NH_4^+ \longrightarrow NH_3 + H_2O$

10.119 $HC_2H_3O_2 + NaOH \longrightarrow NaC_2H_3O_2 + H_2O$

▮ $HC_2H_3O_2 + OH^- \longrightarrow C_2H_3O_2^- + H_2O$

10.120 $Ba(OH)_2 + 2HCl \longrightarrow BaCl_2 + 2H_2O$

▮ $OH^- + H^+ \longrightarrow H_2O$

10.121 $Fe + 2FeCl_3 \longrightarrow 3FeCl_2$

▮ $Fe + 2Fe^{3+} \longrightarrow 3Fe^{2+}$

10.122 $Cu(OH)_2 + 2HClO_3 \longrightarrow Cu(ClO_3)_2 + 2H_2O$

▮ $Cu(OH)_2(s) + 2H^+ \longrightarrow Cu^{2+} + 2H_2O$

10.123 $H_3PO_4 + 2NaOH \longrightarrow Na_2HPO_4 + 2H_2O$

▮ $H_3PO_4 + 2OH^- \longrightarrow HPO_4^{2-} + 2H_2O$

10.124 $CuCl_2 + H_2S \longrightarrow CuS + 2HCl$

▮ $$Cu^{2+} + H_2S \longrightarrow CuS + 2H^+$$

10.125 $ZnS + 2HCl \longrightarrow H_2S + ZnCl_2$

▮ $$ZnS + 2H^+ \longrightarrow H_2S + Zn^{2+}$$

10.126 $AlCl_3 + 4NaOH \longrightarrow NaAl(OH)_4 + 3NaCl$

▮ $$Al^{3+} + 4OH^- \longrightarrow Al(OH)_4^-$$

10.127 $CaCO_3 + CO_2 + H_2O \longrightarrow Ca(HCO_3)_2$

▮ $$CaCO_3 + CO_2 + H_2O \longrightarrow Ca^{2+} + 2HCO_3^-$$

10.128 $NaHCO_3 + NaOH \longrightarrow Na_2CO_3 + H_2O$

▮ $$HCO_3^- + OH^- \longrightarrow CO_3^{2-} + H_2O$$

10.129 $2AgNO_3 + H_2S \longrightarrow Ag_2S + 2HNO_3$

▮ $$2Ag^+ + H_2S \longrightarrow Ag_2S + 2H^+$$

(Concentrated nitric acid will oxidize S^{2-}, but very dilute HNO_3 will not.)

10.130 $Zn + 2HCl \longrightarrow ZnCl_2 + H_2$

▮ $$Zn + 2H^+ \longrightarrow Zn^{2+} + H_2$$

10.131 $Zn + CuCl_2 \longrightarrow ZnCl_2 + Cu$

▮ $$Zn + Cu^{2+} \longrightarrow Zn^{2+} + Cu$$

10.132 $Zn + HgCl_2 \longrightarrow ZnCl_2 + Hg$

▮ $$Zn + Hg^{2+} \longrightarrow Zn^{2+} + Hg$$

10.133 $HClO_3 + NaOH \longrightarrow NaClO_3 + H_2O$

▮ $$H^+ + OH^- \longrightarrow H_2O$$

10.134 $HCl + NaOH \longrightarrow NaCl + H_2O$

▮ $$H^+ + OH^- \longrightarrow H_2O$$

10.135 $HClO_4 + NaOH \longrightarrow NaClO_4 + H_2O$

▮ $$H^+ + OH^- \longrightarrow H_2O$$

10.136 $HBrO_3 + NaOH \longrightarrow NaBrO_3 + H_2O$

▮ $$H^+ + OH^- \longrightarrow H_2O$$

10.137 $HClO_3 + KOH \longrightarrow KClO_3 + H_2O$

▮ $$H^+ + OH^- \longrightarrow H_2O$$

10.138 $HClO_3 + LiOH \longrightarrow LiClO_3 + H_2O$

▮ $$H^+ + OH^- \longrightarrow H_2O$$

10.139 $HClO_3 + RbOH \longrightarrow RbClO_3 + H_2O$

▮ $$H^+ + OH^- \longrightarrow H_2O$$

10.140 $16HCl + 2KMnO_4 \longrightarrow 2MnCl_2 + 8H_2O + 5Cl_2 + 2KCl$

▮ $$10Cl^- + 16H^+ + 2MnO_4^- \longrightarrow 2Mn^{2+} + 8H_2O + 5Cl_2$$

10.141 $2FeCl_2 + 2HCl + H_2O_2 \longrightarrow 2FeCl_3 + 2H_2O$

■ $$2Fe^{2+} + 2H^+ + H_2O_2 \longrightarrow 2Fe^{3+} + 2H_2O$$

10.142 $Cu + 4HNO_3 \longrightarrow 2NO_2 + 2H_2O + Cu(NO_3)_2$

■ $$Cu + 4H^+ + 2NO_3^- \longrightarrow 2NO_2 + 2H_2O + Cu^{2+}$$

10.143 $3Cu + 8HNO_3 \longrightarrow 2NO + 4H_2O + 3Cu(NO_3)_2$

■ $$3Cu + 8H^+ + 2NO_3^- \longrightarrow 2NO + 4H_2O + 3Cu^{2+}$$

10.144 $5Na_2C_2O_4 + 2KMnO_4 + 8H_2SO_4 \longrightarrow 10CO_2 + 2MnSO_4 + 8H_2O + K_2SO_4 + 5Na_2SO_4$

■ $$5C_2O_4^{2-} + 16H^+ + 2MnO_4^- \longrightarrow 10CO_2 + 2Mn^{2+} + 8H_2O$$

10.145 $CdS + I_2 \longrightarrow CdI_2 + S$

■ $$CdS + I_2 \longrightarrow Cd^{2+} + 2I^- + S$$

10.146 $2MnO + 5PbO_2 + 10HNO_3 \longrightarrow 2HMnO_4 + 5Pb(NO_3)_2 + 4H_2O$

■ $$2MnO + 5PbO_2 + 8H^+ \longrightarrow 2MnO_4^- + 5Pb^{2+} + 4H_2O$$

10.147 $3Na_2HAsO_3 + NaBrO_3 + 6HCl \longrightarrow NaBr + 3H_3AsO_4 + 6NaCl$

■ $$3HAsO_3^{2-} + BrO_3^- + 6H^+ \longrightarrow Br^- + 3H_3AsO_4$$

10.148 $5U(SO_4)_2 + 2H_2O + 2KMnO_4 \longrightarrow 2MnSO_4 + 5UO_2SO_4 + 2H_2SO_4 + K_2SO_4$

■ $$5U^{4+} + 2H_2O + 2MnO_4^- \longrightarrow 2Mn^{2+} + 5UO_2^{2+} + 4H^+$$

10.149 $I_2 + 2Na_2S_2O_3 \longrightarrow Na_2S_4O_6 + 2NaI$

■ $$I_2 + 2S_2O_3^{2-} \longrightarrow S_4O_6^{2-} + 2I^-$$

10.150 $4KI + 4HCl + Ca(OCl)_2 \longrightarrow CaCl_2 + 2H_2O + 2I_2 + 4KCl$

■ $$2I^- + 2H^+ + OCl^- \longrightarrow Cl^- + H_2O + I_2$$

10.151 $2NaOCl + 2NaOH + Bi_2O_3 \longrightarrow 2NaBiO_3 + H_2O + 2NaCl$

■ $$2OCl^- + 2OH^- + Bi_2O_3 \longrightarrow 2BiO_3^- + H_2O + 2Cl^-$$

10.152 $10KOH + 6K_3Fe(CN)_6 + Cr_2O_3 \longrightarrow 6K_4Fe(CN)_6 + 2K_2CrO_4 + 5H_2O$

■ $$10OH^- + 6Fe(CN)_6^{3-} + Cr_2O_3 \longrightarrow 6Fe(CN)_6^{4-} + 2CrO_4^{2-} + 5H_2O$$

10.153 $(NH_4)_2S_2O_8 + 2H_2O + MnSO_4 \longrightarrow MnO_2 + 2H_2SO_4 + (NH_4)_2SO_4$

■ $$S_2O_8^{2-} + 2H_2O + Mn^{2+} \longrightarrow MnO_2 + 4H^+ + 2SO_4^{2-}$$

10.154 $14HCl + K_2Cr_2O_7 + 3SnCl_2 \longrightarrow 2CrCl_3 + 3SnCl_4 + 7H_2O + 2KCl$

■ $$14H^+ + Cr_2O_7^{2-} + 3Sn^{2+} + 12Cl^- \longrightarrow 2Cr^{3+} + 3SnCl_4 + 7H_2O$$

10.155 $2NaOH + 2CoCl_2 + 2H_2O + Na_2O_2(s) \longrightarrow 2Co(OH)_3 + 4NaCl$

■ $$2OH^- + 2Co^{2+} + 2H_2O + Na_2O_2(s) \longrightarrow 2Co(OH)_3 + 2Na^+$$

10.156 $H_2O + 7KCN + 2Cu(NH_3)_4Cl_2 \longrightarrow 2K_2Cu(CN)_3 + 6NH_3 + KCNO + 2NH_4Cl + 2KCl$

■ $$H_2O + 7CN^- + 2Cu(NH_3)_4^{2+} \longrightarrow 2Cu(CN)_3^{2-} + 6NH_3 + CNO^- + 2NH_4^+$$

10.157 $4Ag + 8KCN + 2H_2O + O_2 \longrightarrow 4KAg(CN)_2 + 4KOH$

■ $$4Ag + 8CN^- + 2H_2O + O_2 \longrightarrow 4Ag(CN)_2^- + 4OH^-$$

10.158 $4HCl + 3WO_3 + SnCl_2 \longrightarrow H_2SnCl_6 + W_3O_8 + H_2O$

▮ $$2H^+ + 3WO_3 + 6Cl^- + Sn^{2+} \longrightarrow SnCl_6^{2-} + W_3O_8 + H_2O$$

10.159 $2HCl + 7KNO_2 + CoCl_2 \longrightarrow K_3Co(NO_2)_6 + NO + H_2O + 4KCl$

▮ $$2H^+ + 7NO_2^- + Co^{2+} \longrightarrow Co(NO_2)_6^{3-} + NO + H_2O$$

10.160 $2HCl + FeCl_2 + V(OH)_4Cl \longrightarrow VOCl_2 + 3H_2O + FeCl_3$

▮ $$2H^+ + Fe^{2+} + V(OH)_4^+ \longrightarrow VO^{2+} + 3H_2O + Fe^{3+}$$

10.161 $2HI + 2HNO_2 \longrightarrow 2NO + 2H_2O + I_2$

▮ $$2I^- + 4H^+ + 2NO_2^- \longrightarrow 2NO + 2H_2O + I_2$$

10.162 $4Au + 16KCN + 6H_2O + 3O_2 \longrightarrow 4KAu(CN)_4 + 12KOH$

▮ $$4Au + 16CN^- + 6H_2O + 3O_2 \longrightarrow 4Au(CN)_4^- + 12OH^-$$

10.163 $4KOH + 4KMnO_4 \longrightarrow 4K_2MnO_4 + O_2 + 2H_2O$

▮ $$4OH^- + 4MnO_4^- \longrightarrow 4MnO_4^{2-} + O_2 + 2H_2O$$

10.164 In Prob. 10.117, $BaSO_4$ was written as a unit. Is $BaSO_4$ ionic?

▮ Yes, $BaSO_4$ is ionic; it is written as a complete compound because it does not exist as ions in solution, but as ions in a solid.

Write one or more complete equations for each of the following net ionic equations.

10.165 $NH_3 + H^+ \longrightarrow NH_4^+$

▮ $NH_3 + HCl \longrightarrow NH_4Cl$ or $NH_3 + HNO_3 \longrightarrow NH_4NO_3$ or $2NH_3 + H_2SO_4 \longrightarrow (NH_4)_2SO_4$

or ammonia plus any other strong acid to yield the ammonium salt.

10.166 $Co^{2+} + S^{2-} \longrightarrow CoS$

▮ $$CoCl_2 + K_2S \longrightarrow CoS + 2KCl$$

or

$$Co(ClO_3)_2 + BaS \longrightarrow CoS + Ba(ClO_3)_2$$

or

$$CoSO_4 + (NH_4)_2S \longrightarrow CoS + (NH_4)_2SO_4$$

or any other soluble cobalt(II) salt with any soluble sulfide.

10.167 $CO_2 + 2OH^- \longrightarrow CO_3^{2-} + H_2O$

▮ $$CO_2 + 2NaOH \longrightarrow Na_2CO_3 + H_2O$$

or CO_2 plus any soluble hydroxide to give the corresponding soluble carbonate, but not

$$CO_2 + Ba(OH)_2 \longrightarrow BaCO_3(s) + H_2O$$

10.168 $CO_2 + OH^- \longrightarrow HCO_3^-$

▮ $CO_2 + NaOH \longrightarrow NaHCO_3$ or $2CO_2 + Ba(OH)_2 \longrightarrow Ba(HCO_3)_2$

but not

$$2CO_2 + Mg(OH)_2 \longrightarrow Mg(HCO_3)_2$$

because the $Mg(OH)_2$ is a solid.

Write net ionic equations for the following.

10.169 $NaC_2H_3O_2 + HCl \longrightarrow HC_2H_3O_2 + NaCl$

▮ $C_2H_3O_2^- + H^+ \longrightarrow HC_2H_3O_2$

10.170 $NH_4Cl + NaOH \longrightarrow NaCl + NH_3 + H_2O$

▮ $NH_4^+ + OH^- \longrightarrow NH_3 + H_2O$

10.171 (*a*) $Mg + HCl(aq) \longrightarrow$ (*b*) $Fe(OH)_3(s) + HNO_3(aq) \longrightarrow$

▮ (*a*) $Mg + 2H^+ \longrightarrow Mg^{2+} + H_2$ (*b*) $Fe(OH)_3 + 3H^+ \longrightarrow Fe^{3+} + 3H_2O$

10.172 (*a*) $CaCO_3(s) + HCl(aq) \longrightarrow CO_2(g) + H_2O + CaCl_2(aq)$
(*b*) $Pb(NO_3)_2(aq) + NaI(aq) \longrightarrow$ [Lead(II) iodide is insoluble in water.]

▮ (*a*) $CaCO_3 + 2H^+ \longrightarrow CO_2 + H_2O + Ca^{2+}$ (*b*) $Pb^{2+} + 2I^- \longrightarrow PbI_2(s)$

10.173 Write one or two overall equations corresponding to each of the following net ionic equations:
(*a*) $5Fe^{2+} + MnO_4^- + 8H^+ \longrightarrow 5Fe^{3+} + Mn^{2+} + 4H_2O$ (*b*) $2I^- + Cl_2 \longrightarrow 2Cl^- + I_2$

▮ (*a*) $10FeSO_4 + 2KMnO_4 + 8H_2SO_4 \longrightarrow 5Fe_2(SO_4)_3 + 2MnSO_4 + 8H_2O + K_2SO_4$
(*b*) $2KI + Cl_2 \longrightarrow 2KCl + I_2$

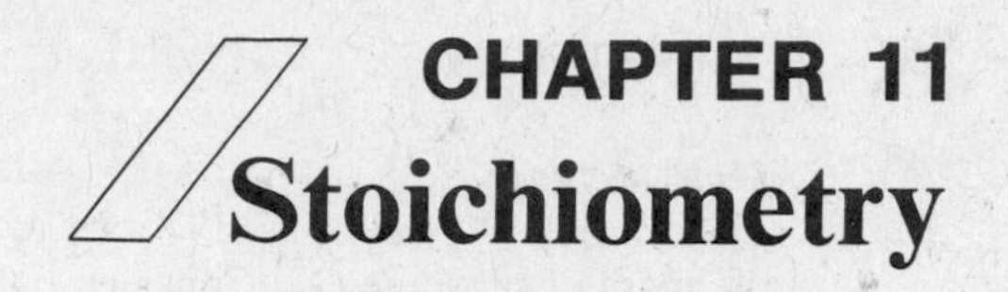

CHAPTER 11
Stoichiometry

11.1 QUANTITIES IN CHEMICAL REACTIONS

11.1 Calculate the number of mol of ammonia, NH_3, required to produce 2.50 mol of $Cu(NH_3)_4SO_4$ according to the equation: $CuSO_4 + 4NH_3 \rightarrow Cu(NH_3)_4SO_4$.

▮

$$(2.50 \text{ mol } Cu(NH_3)_4SO_4)\left(\frac{4 \text{ mol } NH_3}{\text{mol } Cu(NH_3)_4SO_4}\right) = 10.0 \text{ mol } NH_3$$

11.2 Calculate the number of mol of $Ca(HCO_3)_2$ required to prepare 1.50 mol of CO_2 according to the equation

$$Ca(HCO_3)_2 + 2HCl \longrightarrow CaCl_2 + 2CO_2 + 2H_2O$$

▮

$$(1.50 \text{ mol } CO_2)\left(\frac{1 \text{ mol } Ca(HCO_3)_2}{2 \text{ mol } CO_2}\right) = 0.750 \text{ mol } Ca(HCO_3)_2$$

11.3 Calculate the mass of $BaCO_3$ produced when excess CO_2 is bubbled through a solution containing 0.205 mol of $Ba(OH)_2$.

▮

$$Ba(OH)_2 + CO_2 \longrightarrow BaCO_3 + H_2O$$

$$(0.205 \text{ mol } Ba(OH)_2)\left(\frac{1 \text{ mol } BaCO_3}{\text{mol } Ba(OH)_2}\right)\left(\frac{197.4 \text{ g } BaCO_3}{\text{mol } BaCO_3}\right) = 40.5 \text{ g}$$

11.4 The equation for the preparation of phosphorus in an electric furnace is

$$2Ca_3(PO_4)_2 + 6SiO_2 + 10C \longrightarrow 6CaSiO_3 + 10CO + P_4$$

Determine **(*a*)** the number of mol of phosphorus formed for each mol of $Ca_3(PO_4)_2$ used, **(*b*)** the number of g of phosphorus formed per mol of $Ca_3(PO_4)_2$ used, **(*c*)** the number of g of phosphorus formed per g of $Ca_3(PO_4)_2$ used, **(*d*)** the number of lb of phosphorus formed per lb of $Ca_3(PO_4)_2$ used, **(*e*)** the number of tons of phosphorus formed per ton of $Ca_3(PO_4)_2$ used, **(*f*)** the number of mol each of SiO_2 and C required per mol of $Ca_3(PO_4)_2$ used.

▮ **(*a*)** From the equation, 1 mol P_4 is obtained for each 2 mol $Ca_3(PO_4)_2$ used, or, $\frac{1}{2}$ mol P_4 per mol $Ca_3(PO_4)_2$. **(*b*)** Molecular weight of P_4 is 124. Then $\frac{1}{2}$ mol $P_4 = \frac{1}{2} \times 124 = 62$ g P_4. **(*c*)** One mol of $Ca_3(PO_4)_2$ (310 g) yields $\frac{1}{2}$ mol P_4 (62 g). Then 1.0 g $Ca_3(PO_4)_2$ gives

$$\frac{1.0 \text{ g } Ca_3(PO_4)_2}{[310 \text{ g } Ca_3(PO_4)_2]/(62 \text{ g } P_4)} = 0.20 \text{ g } P_4$$

(*d*) 0.20 lb **(*e*)** 0.20 ton **(*f*)** From the equation, 1.0 mol $Ca_3(PO_4)_2$ requires 3 mol SiO_2 and 5 mol C.

11.5 Caustic soda, NaOH, can be prepared commercially by the reaction of Na_2CO_3 with slaked lime, $Ca(OH)_2$. How many g of NaOH can be obtained by treating 1.000 kg of Na_2CO_3 with $Ca(OH)_2$?

▮ First write the balanced equation for the reaction.

$$\underset{(1 \text{ mol} = 106.0 \text{ g})}{Na_2CO_3} + Ca(OH)_2 \longrightarrow \underset{(2 \text{ mol} = 2(40.0 \text{ g}) = 80.0 \text{ g})}{2NaOH} + CaCO_3$$

The mass ratio 106.0/80.0 is all that is needed to solve the problem. We show four methods, all equivalent, of handling the arithmetic.

First Method

$$106.0 \text{ g } Na_2CO_3 \text{ gives } 80.0 \text{ g NaOH} \quad \text{so} \quad 1 \text{ g } Na_2CO_3 \text{ gives } \frac{80.0}{106.0} \text{ g NaOH}$$

and

$$1000 \text{ g } Na_2CO_3 \text{ gives } 1000 \times \frac{80.0}{106.0} = 755 \text{ g NaOH}$$

Mole Method

The symbol $n(X)$ will be used to refer to the number of moles of a substance whose formula is X, and $m(X)$ will denote the mass of substance X. Consider 1000 g Na_2CO_3.

$$n(Na_2CO_3) = \frac{1000 \text{ g}}{106.0 \text{ g/mol}} = 9.433 \text{ mol } Na_2CO_3$$

From the coefficients in the balanced equation, $n(NaOH) = 2n(Na_2CO_3) = 2(9.433) = 18.87$ mol NaOH.

$$m(NaOH) = (18.87 \text{ mol NaOH})(40.0 \text{ g NaOH/mol NaOH}) = 755 \text{ g NaOH}$$

Proportion Method

Let x = number of g of NaOH obtained from 1000 g of Na_2CO_3. It is known that 106.0 g of Na_2CO_3 gives 80.0 g of NaOH; then, by proportion,

$$\frac{106.0 \text{ g } Na_2CO_3}{80.0 \text{ g NaOH}} = \frac{1000 \text{ g } Na_2CO_3}{x} \qquad x = (1000 \text{ g } Na_2CO_3)\left(\frac{80.0 \text{ g NaOH}}{106.0 \text{ g } Na_2CO_3}\right) = 755 \text{ g NaOH}$$

Note. It should be evident that 1000 *lb* of Na_2CO_3 will give 755 *lb* of NaOH, and that 1000 *tons* of Na_2CO_3 will give 755 *tons* of NaOH.

Factor-Label Method

As in the previous method, x = mass of NaOH obtained. Now x is equated to the 1000 g Na_2CO_3, and the right side of the equation is multiplied by successive conversion factors until it has the desired units g NaOH.

$$x = (1000 \text{ g } Na_2CO_3)\left(\frac{1 \text{ mol } Na_2CO_3}{106.0 \text{ g } Na_2CO_3}\right)\left(\frac{2 \text{ mol NaOH}}{1 \text{ mol } Na_2CO_3}\right)\left(\frac{40.0 \text{ g NaOH}}{1 \text{ mol NaOH}}\right) = 755 \text{ g NaOH}$$

11.6 The equation for the reaction of sucrose (sugar) with oxygen is

$$C_{12}H_{22}O_{11} + 12O_2 \longrightarrow 12CO_2 + 11H_2O$$

How many g of CO_2 is produced per g of sucrose used? How many mol of oxygen gas is needed to react with 1.00 g of sucrose?

▮ Since the equation states that 12 mol of CO_2 is produced per mol of sucrose which reacts, the number of mol of sucrose in 1.00 g of sucrose must be calculated.

$$\frac{1.00 \text{ g sucrose}}{342 \text{ g sucrose/mol}} = 0.00292 \text{ mol sucrose}$$

$$(0.00292 \text{ mol sucrose})\left(\frac{12 \text{ mol } CO_2}{1 \text{ mol sucrose}}\right) = 0.0350 \text{ mol } CO_2 \qquad (0.0350 \text{ mol } CO_2)\left(\frac{44.0 \text{ g } CO_2}{\text{mol } CO_2}\right) = 1.54 \text{ g } CO_2$$

According to the balanced chemical equation, for every mol of CO_2 produced 1 mol of O_2 is needed, and in this case 0.0350 mol of O_2 will be used up.

11.7 How many g of $CaCl_2$ does it take to produce 14.3 g of AgCl when treated with excess $AgNO_3$? $Ca(NO_3)_2$ is the other product.

▮
$$CaCl_2 + 2AgNO_3 \longrightarrow 2AgCl + Ca(NO_3)_2$$

$$(14.3 \text{ g AgCl})\left(\frac{1 \text{ mol AgCl}}{142.4 \text{ g AgCl}}\right)\left(\frac{1 \text{ mol } CaCl_2}{2 \text{ mol AgCl}}\right)\left(\frac{111 \text{ g } CaCl_2}{\text{mol } CaCl_2}\right) = 5.57 \text{ g } CaCl_2$$

11.8 A 0.6000 mol sample of Cu_2S is roasted in excess oxygen to yield copper metal and sulfur dioxide. Calculate the mass of copper metal produced.

▮
$$Cu_2S + O_2 \longrightarrow 2Cu + SO_2 \qquad (0.6000 \text{ mol } Cu_2S)\left(\frac{2 \text{ mol Cu}}{\text{mol } Cu_2S}\right)\left(\frac{63.55 \text{ g Cu}}{\text{mol Cu}}\right) = 76.26 \text{ g Cu}$$

11.9 ***(a)*** Write a balanced chemical equation for the reaction of $ZnCl_2$ with excess NaOH to produce $Na_2Zn(OH)_4$, sodium zincate. ***(b)*** What mass of sodium zincate can be produced from 2.00 g of $ZnCl_2$ with excess NaOH by this reaction?

▮ (*a*) $ZnCl_2 + 4NaOH \longrightarrow Na_2Zn(OH)_4 + 2NaCl$

(*b*) $$(2.00 \text{ g } ZnCl_2)\left(\frac{1 \text{ mol } ZnCl_2}{136 \text{ g } ZnCl_2}\right)\left(\frac{1 \text{ mol } Na_2Zn(OH)_4}{\text{mol } ZnCl_2}\right)\left(\frac{179 \text{ g } Na_2Zn(OH)_4}{\text{mol } Na_2Zn(OH)_4}\right) = 2.63 \text{ g}$$

11.10 Calculate the mass of CaO which will react with 6.92 g of HCl to form H_2O, Ca^{2+}, and Cl^-.

▮ $CaO + 2HCl \longrightarrow Ca^{2+} + 2Cl^- + H_2O$ $$(6.92 \text{ g HCl})\left(\frac{1 \text{ mol HCl}}{36.5 \text{ g HCl}}\right)\left(\frac{1 \text{ mol CaO}}{2 \text{ mol HCl}}\right)\left(\frac{56.0 \text{ g CaO}}{\text{mol CaO}}\right) = 5.31 \text{ g}$$

11.11 Consider the combustion of amyl alcohol, $C_5H_{11}OH$. $2C_5H_{11}OH + 15O_2 \rightarrow 10CO_2 + 12H_2O$. (*a*) How many mol of O_2 is needed for the combustion of 1.0 mol of amyl alcohol? (*b*) How many mol of H_2O is formed for each mol of O_2 consumed? (*c*) How many g of CO_2 are produced for each mol of amyl alcohol burned? (*d*) How many g of CO_2 is produced for each g of amyl alcohol burned? (*e*) How many tons of CO_2 are produced for each ton of amyl alcohol burned?

▮

(*a*) $$(1.0 \text{ mol } C_5H_{11}OH)\left(\frac{15 \text{ mol } O_2}{2 \text{ mol } C_5H_{11}OH}\right) = 7.5 \text{ mol } O_2$$

(*b*) $$(1.0 \text{ mol } O_2)\left(\frac{12 \text{ mol } H_2O}{15 \text{ mol } O_2}\right) = 0.80 \text{ mol } H_2O$$

(*c*) $$(1.0 \text{ mol } C_5H_{11}OH)\left(\frac{10 \text{ mol } CO_2}{2 \text{ mol } C_5H_{11}OH}\right)\left(\frac{44.0 \text{ g } CO_2}{\text{mol } CO_2}\right) = 220 \text{ g } CO_2$$

(*d*) $$(1.0 \text{ g } C_5H_{11}OH)\left(\frac{1 \text{ mol } C_5H_{11}OH}{88.0 \text{ g } C_5H_{11}OH}\right)\underbrace{\left(\frac{220 \text{ g } CO_2}{\text{mol } C_5H_{11}OH}\right)}_{\text{from part } (c)} = 2.5 \text{ g } CO_2$$

(*e*) 2.5 tons CO_2

11.12 (*a*) Calculate the mass of $KClO_3$ necessary to produce 1.23 g of O_2. (*b*) What mass of KCl is produced along with this quantity of oxygen?

▮ (*a*) $2KClO_3 \longrightarrow 2KCl + 3O_2$

$$(1.23 \text{ g } O_2)\left(\frac{1 \text{ mol } O_2}{32.0 \text{ g } O_2}\right)\left(\frac{2 \text{ mol } KClO_3}{3 \text{ mol } O_2}\right)\left(\frac{122.6 \text{ g } KClO_3}{\text{mol } KClO_3}\right) = 3.14 \text{ g } KClO_3$$

(*b*) $(3.14 \text{ g } KClO_3) - (1.23 \text{ g } O_2) = 1.91 \text{ g KCl}$

11.13 Calculate the number of mol of calcium chloride needed to react with excess silver nitrate to produce 6.60 g of AgCl.

▮ $CaCl_2 + 2AgNO_3 \longrightarrow 2AgCl + Ca(NO_3)_2$ $$(6.60 \text{ g AgCl})\left(\frac{1 \text{ mol AgCl}}{143 \text{ g AgCl}}\right)\left(\frac{1 \text{ mol } CaCl_2}{2 \text{ mol AgCl}}\right) = 0.0231 \text{ mol } CaCl_2$$

11.14 Chloropicrin, CCl_3NO_2, can be made cheaply for use as an insecticide by a process which utilizes the reaction $CH_3NO_2 + 3Cl_2 \rightarrow CCl_3NO_2 + 3HCl$. How much nitromethane, CH_3NO_2, is needed to form 300 g of chloropicrin?

▮ $$(300 \text{ g } CCl_3NO_2)\left(\frac{1 \text{ mol } CCl_3NO_2}{164.5 \text{ g } CCl_3NO_2}\right)\left(\frac{1 \text{ mol } CH_3NO_2}{1 \text{ mol } CCl_3NO_2}\right)\left(\frac{61.0 \text{ g } CH_3NO_2}{\text{mol } CH_3NO_2}\right) = 111 \text{ g } CH_3NO_2$$

11.15 Ethyl alcohol (C_2H_5OH) is made by the fermentation of glucose ($C_6H_{12}O_6$), as indicated by the equation $C_6H_{12}O_6 \rightarrow 2C_2H_5OH + 2CO_2$. How many metric tons of alcohol can be made from 3.00 metric tons of glucose?

▮ $$(3.00 \times 10^6 \text{ g } C_6H_{12}O_6)\left(\frac{1 \text{ mol } C_6H_{12}O_6}{180.0 \text{ g } C_6H_{12}O_6}\right)\left(\frac{2 \text{ mol } C_2H_5OH}{\text{mol } C_6H_{12}O_6}\right)\left(\frac{46.0 \text{ g } C_2H_5OH}{\text{mol } C_2H_5OH}\right) = 1.53 \times 10^6 \text{ g } C_2H_5OH$$
$$= 1.53 \text{ ton } C_2H_5OH$$

11.16 In a rocket motor fueled with butane, C_4H_{10}, how many kg of liquid oxygen should be provided with each kg of butane to provide for complete combustion? $2C_4H_{10} + 13O_2 \rightarrow 8CO_2 + 10H_2O$

▮ $$(1000 \text{ g } C_4H_{10})\left(\frac{1 \text{ mol } C_4H_{10}}{58.0 \text{ g } C_4H_{10}}\right)\left(\frac{13 \text{ mol } O_2}{2 \text{ mol } C_4H_{10}}\right)\left(\frac{32.0 \text{ g } O_2}{\text{mol } O_2}\right)\left(\frac{1 \text{ kg } O_2}{10^3 \text{ g } O_2}\right) = 3.59 \text{ kg } O_2$$

11.17 What mass of KI is needed to produce 69.6 g of K_2SO_4 by the reaction $8KI + 5H_2SO_4 \rightarrow 4K_2SO_4 + 4I_2 + H_2S + 4H_2O$?

▮
$$(69.6\text{ g }K_2SO_4)\left(\frac{1\text{ mol }K_2SO_4}{174\text{ g }K_2SO_4}\right)\left(\frac{8\text{ mol KI}}{4\text{ mol }K_2SO_4}\right)\left(\frac{166\text{ g KI}}{\text{mol KI}}\right) = 133\text{ g KI}$$

11.18 How much iron(III) oxide will be produced by the complete oxidation of 200 g of iron? The reaction is $4Fe + 3O_2 \rightarrow 2Fe_2O_3$.

▮
$$(200\text{ g Fe})\left(\frac{1\text{ mol Fe}}{55.85\text{ g Fe}}\right)\left(\frac{2\text{ mol }Fe_2O_3}{4\text{ mol Fe}}\right)\left(\frac{159.7\text{ g }Fe_2O_3}{\text{mol }Fe_2O_3}\right) = 286\text{ g }Fe_2O_3$$

11.19 (*a*) How many lb of ZnO will be formed when 1.00 lb of zinc blende, ZnS, is strongly heated in air? The reaction is $2ZnS + 3O_2 \rightarrow 2ZnO + 2SO_2$. (*b*) How many tons of ZnO will be formed from 1.00 ton of ZnS? (*c*) How many kg of ZnO will be formed from 1.00 kg of ZnS?

▮
$$(1.00\text{ g ZnS})\left(\frac{1\text{ mol ZnS}}{97.4\text{ g ZnS}}\right)\left(\frac{2\text{ mol ZnO}}{2\text{ mol ZnS}}\right)\left(\frac{81.4\text{ g ZnO}}{\text{mol ZnO}}\right) = 0.836\text{ g ZnO}$$

The ratio of lb of ZnO to lb of ZnS is the same as the ratio of tons ZnO to tons ZnS, or kg to kg, or any other mass unit. (*a*) 0.836 lb ZnO (*b*) 0.836 ton (*c*) 0.836 kg

11.20 How much $KClO_3$ must be heated to obtain 2.50 g of oxygen?

▮
$$2KClO_3 \longrightarrow 2KCl + 3O_2 \qquad (2.50\text{ g }O_2)\left(\frac{1\text{ mol }O_2}{32.0\text{ g }O_2}\right)\left(\frac{2\text{ mol }KClO_3}{3\text{ mol }O_2}\right)\left(\frac{122\text{ g }KClO_3}{\text{mol }KClO_3}\right) = 6.35\text{ g }KClO_3$$

11.21 Iodine can be made by the reaction $2NaIO_3 + 5NaHSO_3 \rightarrow 3NaHSO_4 + 2Na_2SO_4 + H_2O + I_2$. To produce each kg of iodine, how much $NaIO_3$ and how much $NaHSO_3$ must be used?

▮
$$(1000\text{ g }I_2)\left(\frac{1\text{ mol }I_2}{253.8\text{ g }I_2}\right) = 3.94\text{ mol }I_2$$

$$(3.94\text{ mol }I_2)\left(\frac{2\text{ mol }NaIO_3}{\text{mol }I_2}\right)\left(\frac{197.9\text{ g }NaIO_3}{\text{mol }NaIO_3}\right) = 1.56 \times 10^3\text{ g} = 1.56\text{ kg }NaIO_3$$

$$(3.94\text{ mol }I_2)\left(\frac{5\text{ mol }NaHSO_3}{\text{mol }I_2}\right)\left(\frac{104\text{ g }NaHSO_3}{\text{mol }NaHSO_3}\right) = 2.05 \times 10^3\text{ g} = 2.05\text{ kg }NaHSO_3$$

11.22 A portable hydrogen generator utilizes the reaction $CaH_2 + 2H_2O \rightarrow Ca(OH)_2 + 2H_2$. How many g of H_2 can be produced by a 70 g cartridge of CaH_2?

▮
$$(70\text{ g }CaH_2)\left(\frac{1\text{ mol }CaH_2}{42\text{ g }CaH_2}\right)\left(\frac{2\text{ mol }H_2}{\text{mol }CaH_2}\right)\left(\frac{2.0\text{ g }H_2}{\text{mol }H_2}\right) = 6.7\text{ g }H_2$$

11.23 Calculate the number of g of SO_2 which can be prepared by the treatment of 100 g of Na_2SO_3 with HCl. Write a balanced chemical equation for the reaction, including the two other (familiar) products.

▮
$$Na_2SO_3 + 2HCl \longrightarrow SO_2 + 2NaCl + H_2O$$

H_2SO_3 decomposes to $SO_2 + H_2O$ with a little energy, which is provided by the reaction.

$$(100\text{ g }Na_2SO_3)\left(\frac{1\text{ mol }Na_2SO_3}{126\text{ g }Na_2SO_3}\right)\left(\frac{1\text{ mol }SO_2}{1\text{ mol }Na_2SO_3}\right)\left(\frac{64.1\text{ g }SO_2}{\text{mol }SO_2}\right) = 50.9\text{ g }SO_2$$

11.24 Cu_2S reacts upon heating in oxygen to produce copper metal and sulfur dioxide. (*a*) Write a balanced chemical equation for the reaction. (*b*) How many g of copper can be obtained from 500 g of Cu_2S by this process?

▮
(*a*) $Cu_2S + O_2 \longrightarrow 2Cu + SO_2$ (*b*) $(500\text{ g }Cu_2S)\left(\frac{1\text{ mol }Cu_2S}{159\text{ g }Cu_2S}\right)\left(\frac{2\text{ mol Cu}}{\text{mol }Cu_2S}\right)\left(\frac{63.5\text{ g Cu}}{\text{mol Cu}}\right) = 399\text{ g Cu}$

11.25 Calculate the number of g of CO_2 which can be produced by burning 90.0 g of ethane, C_2H_6, in excess oxygen.

▮
$$2C_2H_6 + 7O_2 \longrightarrow 4CO_2 + 6H_2O \qquad (90.0\text{ g }C_2H_6)\left(\frac{1\text{ mol }C_2H_6}{30.0\text{ g }C_2H_6}\right)\left(\frac{4\text{ mol }CO_2}{2\text{ mol }C_2H_6}\right)\left(\frac{44.0\text{ g }CO_2}{\text{mol }CO_2}\right) = 264\text{ g }CO_2$$

11.26 What mass of AgCl can be obtained from 100 g of $[Ag(NH_3)_2]Cl$ by means of the reaction

$$[Ag(NH_3)_2]Cl + 2HNO_3 \rightarrow AgCl + 2NH_4NO_3?$$

▌ $$(100\text{ g }Ag(NH_3)_2Cl)\left(\frac{1\text{ mol }Ag(NH_3)_2Cl}{177.38\text{ g }Ag(NH_3)_2Cl}\right)\left(\frac{1\text{ mol AgCl}}{\text{mol }Ag(NH_3)_2Cl}\right)\left(\frac{143.3\text{ g AgCl}}{\text{mol AgCl}}\right) = 80.8\text{ g AgCl}$$

11.27 CaC_2 is made in an electric furnace by the reaction $CaO + 3C \rightarrow CaC_2 + CO$. How much CaO is to be added to the furnace charge for each 40 tons of CaC_2 produced?

▌ $$(40\text{ g }CaC_2)\left(\frac{1\text{ mol }CaC_2}{64\text{ g }CaC_2}\right)\left(\frac{1\text{ mol CaO}}{\text{mol }CaC_2}\right)\left(\frac{56\text{ g CaO}}{\text{mol CaO}}\right) = 35\text{ g CaO}$$

Thus, the answer is 35 tons CaO.

11.28 In the *Mond process* for purifying nickel, the volatile nickel carbonyl, $Ni(CO)_4$, is produced by the reaction $Ni + 4CO \rightarrow Ni(CO)_4$. How much CO is used up in volatilizing 2.00 kg of nickel?

▌ $$(2.00 \times 10^3\text{ g Ni})\left(\frac{1\text{ mol Ni}}{58.7\text{ g Ni}}\right)\left(\frac{4\text{ mol CO}}{\text{mol Ni}}\right)\left(\frac{28.0\text{ g CO}}{\text{mol CO}}\right) = 3.82 \times 10^3\text{ g} = 3.82\text{ kg CO}$$

11.29 How much 83.4% pure salt cake (Na_2SO_4) could be produced from 250 kg of 94.5% pure salt in the reaction $2NaCl + H_2SO_4 \rightarrow Na_2SO_4 + 2HCl$?

▌ $$(250 \times 10^3\text{ g mixture})\left(\frac{94.5\text{ g NaCl}}{100\text{ g mixture}}\right)\left(\frac{1\text{ mol NaCl}}{58.5\text{ g NaCl}}\right)\left(\frac{1\text{ mol }Na_2SO_4}{2\text{ mol NaCl}}\right)\left(\frac{142\text{ g }Na_2SO_4}{\text{mol }Na_2SO_4}\right) = 287 \times 10^3\text{ g}$$

$$= 287\text{ kg }Na_2SO_4$$

$$(287 \times 10^3\text{ g }Na_2SO_4)\left(\frac{100\text{ g new mixture}}{83.4\text{ g }Na_2SO_4}\right) = 344 \times 10^3\text{ g} = 344\text{ kg mixture}$$

11.30 How many kg of H_2SO_4 can be prepared from 3.00 kg of cuprite, Cu_2S, if each atom of S in Cu_2S is converted into 1 molecule of H_2SO_4?

▌ $$(3.00\text{ kg }Cu_2S)\left(\frac{10^3\text{ g}}{\text{kg}}\right)\left(\frac{1\text{ mol }Cu_2S}{159.1\text{ g }Cu_2S}\right)\left(\frac{1\text{ mol S}}{\text{mol }Cu_2S}\right)\left(\frac{1\text{ mol }H_2SO_4}{\text{mol S}}\right)\left(\frac{98.0\text{ g }H_2SO_4}{\text{mol }H_2SO_4}\right) = 1850\text{ g }H_2SO_4$$

$$= 1.85\text{ kg }H_2SO_4$$

11.31 A process designed to remove organic sulfur from coal prior to combustion involves the reactions

$$X{-}S{-}Y + 2NaOH \longrightarrow X{-}O{-}Y + Na_2S + H_2O$$
$$CaCO_3 \longrightarrow CaO + CO_2$$
$$Na_2S + CO_2 + H_2O \longrightarrow Na_2CO_3 + H_2S$$
$$CaO + H_2O \longrightarrow Ca(OH)_2$$
$$Na_2CO_3 + Ca(OH)_2 \longrightarrow CaCO_3 + 2NaOH$$

In the processing of 200 metric tons of a coal having a 1.0% sulfur content, how much limestone ($CaCO_3$) must be decomposed to provide enough $Ca(OH)_2$ to regenerate the NaOH used in the original leaching step?

▌ $$(200 \times 10^6\text{ g coal})\left(\frac{1.0\text{ g S}}{100\text{ g coal}}\right)\left(\frac{1\text{ mol S}}{32.0\text{ g S}}\right)\left(\frac{1\text{ mol }CaCO_3}{\text{mol S}}\right)\left(\frac{100\text{ g }CaCO_3}{\text{mol }CaCO_3}\right) = 6.25 \times 10^6\text{ g }CaCO_3$$

$$= 6.25\text{ metric ton }CaCO_3$$

11.32 Polyethylene can be produced from calcium carbide according to the following sequence of reactions:

$$CaC_2 + H_2O \longrightarrow CaO + HC{\equiv}CH \qquad HC{\equiv}CH + H_2 \longrightarrow H_2C{=}CH_2 \qquad nH_2C{=}CH_2 \longrightarrow (CH_2CH_2)_n$$

Calculate the mass of polyethylene which can be produced from 20.0 kg of CaC_2.

▌ $$(20.0\text{ kg }CaC_2)\left(\frac{10^3\text{ g}}{\text{kg}}\right)\left(\frac{1\text{ mol }CaC_2}{64.10\text{ g}}\right)\left(\frac{1\text{ mol }C_2H_4}{\text{mol }CaC_2}\right)\left(\frac{28.05\text{ g }C_2H_4}{\text{mol }C_2H_4}\right) = 8.75\text{ kg }C_2H_4$$

The mass of polyethylene is the same as the mass of ethylene, 8.75 kg, as required by the law of conservation of mass.

11.33 Commercial sodium "hydrosulfite" is 90.1% pure $Na_2S_2O_4$. How much of the commercial product could be made by using 100 metric tons of zinc with a sufficient supply of the other reactants? The reactions are

$$Zn + 2SO_2 \longrightarrow ZnS_2O_4 \qquad ZnS_2O_4 + Na_2CO_3 \longrightarrow ZnCO_3 + Na_2S_2O_4$$

▮ $$(100 \times 10^6 \text{ g Zn})\left(\frac{1 \text{ mol Zn}}{65.4 \text{ g Zn}}\right)\left(\frac{1 \text{ mol } Na_2S_2O_4}{\text{mol Zn}}\right)\left(\frac{174 \text{ g } Na_2S_2O_4}{\text{mol } Na_2S_2O_4}\right) = 266 \times 10^6 \text{ g } Na_2S_2O_4$$

$$(266 \times 10^6 \text{ g } Na_2S_2O_4)\left(\frac{100 \text{ g commercial}}{90.1 \text{ g } Na_2S_2O_4}\right) = 295 \times 10^6 \text{ g commercial product} = 295 \text{ metric tons}$$

11.34 Fluorocarbon polymers can be made by fluorinating polyethylene according to the reaction $(CH_2)_n + 4nCoF_3 \rightarrow (CF_2)_n + 2nHF + 4nCoF_2$, where n is a large integer. The CoF_3 can be regenerated by the reaction $2CoF_2 + F_2 \rightarrow 2CoF_3$. ***(a)*** If the HF formed in the first reaction cannot be reused, how many kg of fluorine are consumed per kg of fluorocarbon produced, $(CF_2)_n$? ***(b)*** If the HF can be recovered and electrolyzed to hydrogen and fluorine, and if this fluorine is used for regenerating CoF_3, what is the net consumption of fluorine per kg of fluorocarbon?

▮ ***(a)*** $$[1.00 \times 10^3 \text{ g } (CF_2)_n]\left(\frac{1 \text{ mol } (CF_2)_n}{50n \text{ g } (CF_2)_n}\right)\left(\frac{4n \text{ mol } CoF_3}{\text{mol } (CF_2)_n}\right)\left(\frac{1 \text{ mol } F_2}{2 \text{ mol } CoF_3}\right)\left(\frac{38 \text{ g } F_2}{\text{mol } F_2}\right) = 1.52 \times 10^3 \text{ g}$$

$$= 1.52 \text{ kg } F_2$$

Note that n cancels. ***(b)*** If $4nCoF_3$ yields $4nCoF_2$, $4n$F atoms are consumed. Saving $2n$F atoms by recovery of the F from HF reduces consumption by half. Hence, $1.52/2 = 0.76$ kg of F_2 is required.

11.35 The "roasting" of 100.0 g of a copper ore yielded 75.4 g of 89.5% pure copper. If the ore is composed of Cu_2S and CuS with 11.0% inert impurity, calculate the percent of Cu_2S in the ore. The equations are

$$Cu_2S + O_2 \longrightarrow 2\,Cu + SO_2 \qquad \text{and} \qquad CuS + O_2 \longrightarrow Cu + SO_2$$

▮ $$(75.4 \text{ g product})\left(\frac{89.5 \text{ g Cu}}{100 \text{ g product}}\right)\left(\frac{1 \text{ mol Cu}}{63.54 \text{ g Cu}}\right) = 1.06 \text{ mol Cu}$$

In the sample, 89.0 g of which is composed of the sulfides, let x = mass of Cu_2S; then $89.0 - x$ = mass of CuS. Equating the number of moles of Cu produced:

$$\frac{2x}{159.15} + \frac{89.0 - x}{95.61} = 1.06 \qquad x = 62 \text{ g } Cu_2S \qquad 89.0 - x = 27 \text{ g CuS}$$

Hence 62% of the sample is Cu_2S.

11.36 ***(a)*** What mass of CO would be produced by the reaction of 16.0 g of O_2 with excess carbon according to the equation $2C + O_2 \rightarrow 2CO$? ***(b)*** If the atomic weight of carbon were set at 50.00 u, what would be the atomic weight of oxygen? The molecular weight of CO? ***(c)*** Using the atomic weights and molecular weight of part (*b*), calculate the mass of CO that would be produced by the reaction of 16.0 g of O_2 with excess carbon. ***(d)*** How does this exercise demonstrate the arbitrary nature of the atomic weight scale?

▮ ***(a)*** $$(16.0 \text{ g } O_2)\left(\frac{1 \text{ mol } O_2}{32.0 \text{ g } O_2}\right)\left(\frac{2 \text{ mol CO}}{\text{mol } O_2}\right) = 1.0 \text{ mol CO} \qquad (1.0 \text{ mol CO})\left(\frac{28.0 \text{ g CO}}{\text{mol CO}}\right) = 28.0 \text{ g CO}$$

(b) Oxygen is 16.0/12.0 times as massive as carbon, no matter what the atomic weight scale. The atomic and molecular weights are therefore: For carbon: 50.0. For oxygen: $(50.0)\left(\frac{16.0}{12.0}\right) = 66.7$. For CO: $50.0 + 66.7 = 116.7$.

(c) Under the new scale,

$$(16.0 \text{ g } O_2)\left(\frac{1 \text{ mol } O_2}{133.4 \text{ g } O_2}\right)\left(\frac{2 \text{ mol CO}}{\text{mol } O_2}\right)\left(\frac{116.7 \text{ g CO}}{\text{mol CO}}\right) = 28.0 \text{ g CO}$$

(d) The answers in parts (*a*) and (*c*) are the same, which indicates that the choice of the basis of the atomic weight scale is arbitrary.

11.37 Carbon disulfide, CS_2, can be made from by-product SO_2. The overall reaction is $5C + 2SO_2 \rightarrow CS_2 + 4CO$. How much CS_2 can be produced from 540 kg of waste SO_2 with excess coke, if the SO_2 conversion is 82.0%?

$$(540 \times 10^3 \text{ g } SO_2)\left(\frac{1 \text{ mol } SO_2}{64.0 \text{ g } SO_2}\right)\left(\frac{1 \text{ mol } CS_2}{2 \text{ mol } SO_2}\right)\left(\frac{76.0 \text{ g } CS_2}{\text{mol } CS_2}\right) = 321 \times 10^3 \text{ g } CS_2$$

Theoretical yield is 321 kg CS_2. (321 kg)(0.820) = 263 kg actually produced.

11.38 A 55.0 g sample of impure zinc reacts with exactly 129 cm³ of hydrochloric acid which has a density of 1.18 g/cm³ and contains 35.0% HCl by mass. What is the percent of metallic zinc in the sample? Assume that the impurity is inert to HCl.

$$(129 \text{ cm}^3 \text{ solution})\left(\frac{1.18 \text{ g solution}}{\text{cm}^3 \text{ solution}}\right)\left(\frac{35.0 \text{ g HCl}}{100 \text{ g solution}}\right)\left(\frac{1 \text{ mol HCl}}{36.5 \text{ g HCl}}\right)\left(\frac{1 \text{ mol Zn}}{2 \text{ mol HCl}}\right)\left(\frac{65.4 \text{ g Zn}}{\text{mol Zn}}\right) = 47.7 \text{ g Zn}$$

$$\left(\frac{47.7 \text{ g Zn}}{55.0 \text{ g sample}}\right)(100\%) = 86.7\% \text{ Zn}$$

11.39 Find the number of mol of chloride ion needed to react with sufficient silver nitrate to make 10.0 g of AgCl. What mass of $CaCl_2$ is required to provide this number of mol of Cl^-?

The net ionic equation is $Ag^+ + Cl^- \rightarrow AgCl(s)$. The number of mol of AgCl, which is equal to the number of mol of Cl^-, is given by

$$\text{mol AgCl desired} = \frac{10.0 \text{ g}}{143.4 \text{ g/mol}} = 0.0697 \text{ mol}$$

Hence 0.0697 mol of Cl^- is required. Since $CaCl_2$ contains 2 mol of chloride ion per mol of substance,

$$\text{mol } CaCl_2 \text{ required} = \frac{0.0697 \text{ mol } Cl^-}{2 \text{ mol } Cl^-/\text{mol } CaCl_2} = 0.0348 \text{ mol } CaCl_2$$

and

$$(0.0348 \text{ mol } CaCl_2)\left(\frac{111 \text{ g } CaCl_2}{\text{mol } CaCl_2}\right) = 3.86 \text{ g } CaCl_2 \text{ required}$$

11.40 A 1.2048 g sample of impure Na_2CO_3 is dissolved and allowed to react with a solution of $CaCl_2$. The resulting $CaCO_3$, after precipitation, filtration, and drying, was found to weigh 1.0362 g. Assuming that the impurities do not contribute to the weight of the precipitate, calculate the percent purity of the Na_2CO_3.

The equation for the reaction is $Na_2CO_3 + CaCl_2 \rightarrow CaCO_3 + 2NaCl$. First find the amount of $CaCO_3$:

$$n(CaCO_3) = \frac{1.0362 \text{ g } CaCO_3}{100.09 \text{ g } CaCO_3/\text{mol}} = 0.010353 \text{ mol}$$

From the coefficients in the balanced equation,

$$n(Na_2CO_3) = n(CaCO_3) = 0.010353 \text{ mol } Na_2CO_3$$

Now calculate the mass of pure Na_2CO_3 in the sample.

$$m(Na_2CO_3) = (0.010353 \text{ mol})(105.99 \text{ g } Na_2CO_3/\text{mol}) = 1.0973 \text{ g } Na_2CO_3$$

The percent purity is obtained by dividing the mass of Na_2CO_3 by the mass of the sample and multiplying by 100%.

$$\% \text{ purity} = \left(\frac{1.0973 \text{ g}}{1.2048 \text{ g}}\right)(100\%) = 91.08\%$$

11.41 (*a*) How much bismuth nitrate, $Bi(NO_3)_3 \cdot 5H_2O$, would be formed from a solution of 15.0 g of bismuth in nitric acid? $Bi + 4HNO_3 + 3H_2O \rightarrow Bi(NO_3)_3 \cdot 5H_2O + NO$. (*b*) How much 30.0% nitric acid (containing 30.0% HNO_3 by mass) is required to react with this amount of bismuth?

$$(15.0 \text{ g Bi})\left(\frac{1 \text{ mol Bi}}{209 \text{ g Bi}}\right) = 0.0718 \text{ mol Bi}$$

(*a*) $$(0.0718 \text{ mol Bi})\left(\frac{1 \text{ mol } Bi(NO_3)_3 \cdot 5H_2O}{\text{mol Bi}}\right)\left(\frac{485 \text{ g } Bi(NO_3)_3 \cdot 5H_2O}{\text{mol } Bi(NO_3)_3 \cdot 5H_2O}\right) = 34.8 \text{ g compound}$$

(*b*) $$(0.0718 \text{ mol Bi})\left(\frac{4 \text{ mol } HNO_3}{\text{mol Bi}}\right)\left(\frac{63.0 \text{ g } HNO_3}{\text{mol } HNO_3}\right)\left(\frac{100 \text{ g solution}}{30.0 \text{ g } HNO_3}\right) = 60.3 \text{ g solution}$$

11.42 One of the reactions used in the petroleum industry for improving the octane rating of fuels is $C_7H_{14} \rightarrow C_7H_8 + 3H_2$. The two hydrocarbons appearing in this equation are liquids; the hydrogen formed is a gas. What is the percent reduction in liquid weight accompanying the completion of the above reaction?

▮ $$\text{Ratio} = \frac{(1 \text{ mol } C_7H_8)(92.0 \text{ g/mol})}{(1 \text{ mol } C_7H_{14})(98.0 \text{ g/mol})} = 0.939 \qquad \% \text{ reduction} = \left(\frac{1.000 - 0.939}{1.000}\right)(100\%) = 6.1\%$$

11.43 $KClO_4$ may be made by means of the following series of reactions:

$$Cl_2 + 2KOH \longrightarrow KCl + KClO + H_2O$$
$$3KClO \longrightarrow 2KCl + KClO_3$$
$$4KClO_3 \longrightarrow 3KClO_4 + KCl$$

How much Cl_2 is needed to prepare 200 g $KClO_4$ by the above sequence?

▮ The mole method and the factor-label method are the simplest routes to the solution of this problem.

Mole Method

$$n(KClO) = n(Cl_2) \qquad n(KClO_3) = \tfrac{1}{3}n(KClO) = \tfrac{1}{3}n(Cl_2)$$

$$n(KClO_4) = \tfrac{3}{4}n(KClO_3) = (\tfrac{3}{4})(\tfrac{1}{3})n(Cl_2) = \tfrac{1}{4}n(Cl_2) \qquad n(KClO_4) = \frac{200 \text{ g } KClO_4}{139 \text{ g } KClO_4/\text{mol } KClO_4} = 1.44 \text{ mol } KClO_4$$

$$n(Cl_2) = 4 \times 1.44 = 5.76 \text{ mol } Cl_2 \qquad m(Cl_2) = (5.76 \text{ mol } Cl_2)(71.0 \text{ g } Cl_2/\text{mol } Cl_2) = 409 \text{ g } Cl_2$$

Factor-Label Method

$$x \text{ g } Cl_2 = \left(\frac{200 \text{ g } KClO_4}{139 \text{ g } KClO_4/\text{mol } KClO_4}\right)\left(\frac{4 \text{ mol } KClO_3}{3 \text{ mol } KClO_4}\right)\left(\frac{3 \text{ mol } KClO}{1 \text{ mol } KClO_3}\right)\left(\frac{1 \text{ mol } Cl_2}{1 \text{ mol } KClO}\right)\left(\frac{71.0 \text{ g } Cl_2}{1 \text{ mol } Cl_2}\right) = 409 \text{ g } Cl_2$$

11.44 A 10.20 mg sample of an organic compound containing carbon, hydrogen, and oxygen only was burned in excess oxygen, yielding 23.10 mg of CO_2 and 4.72 mg of H_2O. Calculate the empirical formula of the compound.

▮ The masses of C and H in the compound are calculated from the masses of the products:

$$(23.10 \text{ mg } CO_2)\left(\frac{1 \text{ mmol } CO_2}{44.0 \text{ mg } CO_2}\right)\left(\frac{1 \text{ mmol C}}{\text{mmol } CO_2}\right) = 0.525 \text{ mmol C} \qquad (0.525 \text{ mmol C})\left(\frac{12.0 \text{ mg C}}{\text{mmol C}}\right) = 6.30 \text{ mg C}$$

$$(4.72 \text{ mg } H_2O)\left(\frac{1 \text{ mmol } H_2O}{18.0 \text{ mg } H_2O}\right)\left(\frac{2 \text{ mmol H}}{\text{mmol } H_2O}\right) = 0.524 \text{ mmol H} \qquad (0.524 \text{ mmol H})\left(\frac{1.008 \text{ mg H}}{\text{mmol H}}\right) = 0.528 \text{ mg H}$$

The mass of oxygen in the unknown compound is determined by difference:

$$(10.20 \text{ mg}) - (6.30 \text{ mg}) - (0.528 \text{ mg}) = 3.37 \text{ mg O} \qquad (3.37 \text{ mg O})\left(\frac{1 \text{ mmol O}}{16.0 \text{ mg O}}\right) = 0.211 \text{ mmol O}$$

The mole ratio is (0.525 mmol C):(0.524 mmol H):(0.211 mmol O) or (5 mol C):(5 mol H):(2 mol O). The empirical formula is $C_5H_5O_2$.

11.45 Calculate the amount of lime (CaO) that can be prepared by heating 200 kg of limestone that is 95.0% pure $CaCO_3$.

▮ The amount of pure $CaCO_3$ in 200 kg limestone is $0.950 \times 200 \text{ kg} = 190 \text{ kg } CaCO_3$; formula weights of $CaCO_3$ and CaO are 100 and 56.1. The balanced equation for the reaction is

$$\underset{(1 \text{ mol} = 100 \text{ g})}{CaCO_3} \longrightarrow \underset{(1 \text{ mol} = 56.1 \text{ g})}{CaO} + CO_2$$

First Method

$$100 \text{ g } CaCO_3 \text{ gives } 56.1 \text{ g CaO} \qquad 1 \text{ g } CaCO_3 \text{ gives } \frac{56.1}{100} \text{ g CaO, or } 0.561 \text{ g CaO}$$

Then 1 kg $CaCO_3$ gives 0.561 kg CaO and 190 kg $CaCO_3$ gives $190(0.561 \text{ kg CaO}) = 107 \text{ kg CaO}$

Mole Method

$$n(\text{CaCO}_3) = \frac{190 \times 10^3 \text{ g CaCO}_3}{100 \text{ g CaCO}_3/\text{mol CaCO}_3} = 1.90 \times 10^3 \text{ mol CaCO}_3$$

$$n(\text{CaO}) = n(\text{CaCO}_3) = 1.90 \times 10^3 \text{ mol CaO}$$

$$m(\text{CaO}) = (1.90 \times 10^3 \text{ mol CaO})(56.1 \text{ g CaO/mol CaO}) = 107 \times 10^3 \text{ g CaO} = 107 \text{ kg CaO}$$

Factor-Label Method

$$\text{Amount of CaO} = (190 \times 10^3 \text{ g CaCO}_3)\left(\frac{1 \text{ mol CaCO}_3}{100 \text{ g CaCO}_3}\right)\left(\frac{1 \text{ mol CaO}}{1 \text{ mol CaCO}_3}\right)\left(\frac{56.1 \text{ g CaO}}{1 \text{ mol CaO}}\right)$$

$$= 107 \times 10^3 \text{ g CaO} = 107 \text{ kg CaO}$$

11.46 Most commercial hydrochloric acid is prepared by heating NaCl with concentrated H_2SO_4. How much sulfuric acid containing 90.0% H_2SO_4 by weight is needed for the production of 2000 kg of concentrated hydrochloric acid containing 42.0% HCl by weight?

▮ (1) Amount of pure HCl in 2000 kg of 42.0% acid is (0.420)(2000 kg) = 840 kg. (2) Determine the amount of H_2SO_4 required to produce 840 kg HCl.

$$\underset{}{2\text{NaCl}} + \underset{(1 \text{ mol} = 98.1 \text{ g})}{\text{H}_2\text{SO}_4} \longrightarrow \text{Na}_2\text{SO}_4 + \underset{(2 \text{ mol} = 2(36.46) = 72.92 \text{ g})}{2\text{HCl}}$$

From the equation,

$$72.92 \text{ g HCl requires } 98.1 \text{ g H}_2\text{SO}_4, \qquad 1 \text{ g HCl requires } \frac{98.1}{72.92} \text{ g H}_2\text{SO}_4$$

$$1 \text{ kg HCl requires } \frac{98.1}{72.92} \text{ kg H}_2\text{SO}_4 \quad \text{and} \quad 840 \text{ kg HCl requires } (840)\left(\frac{98.1}{72.92} \text{ kg}\right) = 1130 \text{ kg H}_2\text{SO}_4.$$

(3) Finally, determine the amount of sulfuric acid solution containing 90.0% H_2SO_4 that can be made from 1130 kg of pure H_2SO_4. Since 0.900 kg of pure H_2SO_4 makes 1.00 kg of 90.0% solution, 1130 kg of pure H_2SO_4 will make

$$\frac{1130 \text{ kg H}_2\text{SO}_4}{0.900 \text{ kg H}_2\text{SO}_4/\text{kg soln}} = 1260 \text{ kg soln}$$

11.47 A 1.00 g total sample of a mixture of $CaCO_3(s)$ (FW = 100) and glass beads liberates 0.220 g of CO_2 (MW = 44.0) upon treatment with excess HCl (MW = 36.5). (Glass does not react with HCl.) ***(a)*** Calculate the weight of $CaCO_3$ in the original mixture. ***(b)*** Calculate the weight percent of calcium in the original mixture.

▮

$$\text{CaCO}_3 + 2\text{HCl} \longrightarrow \text{CO}_2 + \text{H}_2\text{O} + \text{CaCl}_2$$

$$(0.220 \text{ g CO}_2)\left(\frac{1 \text{ mol CO}_2}{44.0 \text{ g CO}_2}\right)\left(\frac{1 \text{ mol CaCO}_3}{\text{mol CO}_2}\right) = (0.00500 \text{ mol CaCO}_3)\left(\frac{100 \text{ g CaCO}_3}{\text{mol CaCO}_3}\right) = 0.500 \text{ g CaCO}_3$$

$$(0.00500 \text{ mol CaCO}_3)\left(\frac{1 \text{ mol Ca}}{\text{mol CaCO}_3}\right)\left(\frac{40.0 \text{ g Ca}}{\text{mol Ca}}\right) = 0.200 \text{ g Ca} \qquad \left(\frac{0.200 \text{ g Ca}}{1.00 \text{ g total}}\right)(100\%) = 20.0\% \text{ Ca}$$

11.48 A particular 100-octane aviation gasoline used 1.00 cm³ of tetraethyllead, $(C_2H_5)_4Pb$, of density 1.66 g/cm³, per liter of product. This compound is made as follows: $4C_2H_5Cl + 4NaPb \rightarrow (C_2H_5)_4Pb + 4NaCl + 3Pb$. How many g of ethyl chloride, C_2H_5Cl, is needed to make enough tetraethyllead for 1.00 L of gasoline?

▮ The mass of 1.00 cm³ $(C_2H_5)_4Pb$ is (1.00 cm³)(1.66 g/cm³) = 1.66 g; this is the amount needed per liter. In terms of moles,

$$\text{Number of mol } (\text{C}_2\text{H}_5)_4\text{Pb needed} = \frac{1.66 \text{ g}}{323 \text{ g/mol}} = 0.00514 \text{ mol}$$

The chemical equation shows that 1 mol $(C_2H_5)_4Pb$ requires 4 mol C_2H_5Cl. Hence 4(0.00514) = 0.0206 mol C_2H_5Cl is needed.

$$m(\text{C}_2\text{H}_5\text{Cl}) = (0.0206 \text{ mol})(64.5 \text{ g/mol}) = 1.33 \text{ g C}_2\text{H}_5\text{Cl}$$

11.49 A carbonate, MCO_3, of an unknown metal is treated with HCl to produce CO_2. Upon treating 1.00 g of MCO_3 with excess HCl, 0.522 g of CO_2 is produced. Find the weight percent of the metal in MCO_3.

▮
$$MCO_3 + 2HCl \longrightarrow MCl_2 + H_2O + CO_2$$

$$(0.522\text{ g } CO_2)\left(\frac{1\text{ mol } CO_2}{44.0\text{ g } CO_2}\right)\left(\frac{1\text{ mol } MCO_3}{\text{mol } CO_2}\right) = 0.01186\text{ mol } MCO_3 \qquad \frac{1.00\text{ g } MCO_3}{0.01186\text{ mol } MCO_3} = 84.3\text{ g/mol } MCO_3$$

$$(84.3\text{ g/mol } MCO_3) - (60.0\text{ g/mol } CO_3^{2-}) = 24.3\text{ g/mol } M^{2+} \qquad \left(\frac{24.3\text{ g/mol } M^{2+}}{84.3\text{ g/mol } MCO_3}\right)(100\%) = 28.8\%$$

11.50 How many kg of pure H_2SO_4 could be obtained from 2.00 kg of pure iron pyrites (FeS_2) according to the following reactions?

$$4FeS_2 + 11O_2 \longrightarrow 2Fe_2O_3 + 8SO_2 \qquad 2SO_2 + O_2 \longrightarrow 2SO_3 \qquad SO_3 + H_2O \longrightarrow H_2SO_4$$

▮ First it should be noted that there is no by-product loss or other permanent loss of sulfur, so that it is not even necessary to balance the equations or to use them further. Each incoming atom of sulfur produces an outgoing molecule of H_2SO_4. (One formula unit of FeS_2 contains 2 atoms of S, and one molecule of H_2SO_4 contains 1 atom of S.) Hence

$$1\text{ mol } FeS_2 \longrightarrow 2\text{ mol } H_2SO_4$$

$$n(FeS_2) = \frac{2000\text{ g}}{120\text{ g/mol}} = 16.7\text{ mol } FeS_2 \qquad n(H_2SO_4) = 2n(FeS_2) = 2 \times 16.7 = 33.4\text{ mol } H_2SO_4$$

$$m(H_2SO_4) = (33.4\text{ mol})(98.0\text{ g/mol}) = 3270\text{ g} = 3.27\text{ kg } H_2SO_4$$

Another Method
1 mol FeS_2 (120 g FeS_2) yields 2 mol H_2SO_4 ($2 \times 98 = 196$ g H_2SO_4). Thus

$$120\text{ kg } FeS_2 \text{ yields } 196\text{ kg } H_2SO_4 \qquad \text{and} \qquad 1\text{ kg } FeS_2 \text{ yields } \frac{196}{120} = 1.63\text{ kg } H_2SO_4$$

$$2\text{ kg } FeS_2 \text{ yields } 2 \times 1.63 = 3.26\text{ kg } H_2SO_4$$

11.51 When 2.86 g of a mixture of 1-butene, C_4H_8, and butane, C_4H_{10}, was burned in excess oxygen, 8.80 g of CO_2 and 4.14 g of H_2O were obtained. Calculate the percentage by mass of butane in the original mixture.

▮
$$(8.80\text{ g } CO_2)\left(\frac{1\text{ mol } CO_2}{44.0\text{ g } CO_2}\right) = 0.200\text{ mol } CO_2 \qquad (4.14\text{ g } H_2O)\left(\frac{1\text{ mol } H_2O}{18.0\text{ g } H_2O}\right) = 0.230\text{ mol } H_2O$$

The mixture contains 0.200 mol of carbon atoms and 0.460 mol of hydrogen atoms. Let x = number of moles of C_4H_8 and y = number of moles of C_4H_{10}. Then

$$4(x + y) = 0.200$$
$$8x + 10y = 0.460$$
$$8x = 0.400 - 8y \qquad \text{(doubling the first equation)}$$
$$(0.400 - 8y) + 10y = 0.460 \qquad \text{(substituting the last equation into the second)}$$
$$2y = 0.060$$
$$y = 0.030\text{ mol } C_4H_{10} \qquad x = 0.020\text{ mol } C_4H_8$$
$$(0.030\text{ mol } C_4H_{10})\left(\frac{58\text{ g}}{\text{mol}}\right) = 1.74\text{ g } C_4H_{10}$$
$$(0.020\text{ mol } C_4H_8)\left(\frac{56\text{ g}}{\text{mol}}\right) = 1.12\text{ g } C_4H_8$$
$$\text{total} = 2.86\text{ g}$$

$$\left(\frac{1.12\text{ g } C_4H_8}{2.86\text{ g total}}\right)(100\%) = 39.2\%\ C_4H_8;\ \text{hence } 60.8\%\ C_4H_{10}$$

11.52 In one process for waterproofing, a fabric is exposed to $(CH_3)_2SiCl_2$ vapor. The vapor reacts with hydroxyl groups on the surface of the fabric or with traces of water to form the waterproofing film $[(CH_3)_2SiO]_n$, by the reaction

$$n(CH_3)_2SiCl_2 + 2nOH^- \longrightarrow 2nCl^- + nH_2O + [(CH_3)_2SiO]_n$$

where n stands for a large integer. The waterproofing film is deposited on the fabric layer upon layer. Each layer is 6.0 Å thick [the thickness of the $(CH_3)_2SiO$ group]. How much $(CH_3)_2SiCl_2$ is needed to waterproof one side of a piece of fabric, 1.00 m by 3.00 m, with a film 300 layers thick? The density of the film is 1.0 g/cm³.

▮

$$\text{mass of film} = (\text{volume of film})(\text{density of film})$$
$$= (\text{area of film})(\text{thickness of film})(\text{density of film})$$
$$= (100 \text{ cm} \times 300 \text{ cm})(300 \times 6.0 \text{ Å})(10^{-8} \text{ cm/Å})(1.0 \text{ g/cm}^3) = 0.54 \text{ g}$$

The equation involves n moles each of $(CH_3)_2SiCl_2$ and $(CH_3)_2SiO$. Therefore 1 mol (74 g) of $(CH_3)_2SiO$ in the film requires 1 mol (129 g) of $(CH_3)_2SiCl_2$. Then 0.54 g of $(CH_3)_2SiO$ requires (0.54 g)(129/74) = 0.94 g $(CH_3)_2SiCl_2$.

11.53 What is the percent free SO_3 in an oleum (considered as a solution of SO_3 in H_2SO_4) that is labeled "109% H_2SO_4"? Such a designation refers to the total mass of pure H_2SO_4, 109 g, that would be present after dilution of 100 g of the oleum, when all free SO_3 would combine with water to form H_2SO_4.

▮ 9 g H_2O will combine with all the free SO_3 in 100 g of the oleum to give a total of 109 g H_2SO_4. The equation $H_2O + SO_3 \rightarrow H_2SO_4$ indicates that 1 mol H_2O (18 g) combines with 1 mol SO_3 (80 g). Then 9 g H_2O combines with

$$(\tfrac{9}{18})(80 \text{ g}) = 40 \text{ g } SO_3$$

Thus 100 g of the oleum contains 40 g SO_3, or the percent free SO_3 in the oleum is 40%.

11.54 The plastics industry uses large amounts of phthalic anhydride, $C_8H_4O_3$, made by the controlled oxidation of naphthalene. $2C_{10}H_8 + 9O_2 \rightarrow 2C_8H_4O_3 + 4CO_2 + 4H_2O$. Since some of the naphthalene is oxidized to other products, only 70.0% of the maximum yield predicted by the equation is actually obtained. How many phthalic anhydride would be produced in practice by the oxidation of 200 lb of $C_{10}H_8$?

▮

$$(200 \text{ g } C_{10}H_8)\left(\frac{1 \text{ mol } C_{10}H_8}{128 \text{ g } C_{10}H_8}\right)\left(\frac{2 \text{ mol } C_8H_4O_3}{2 \text{ mol } C_{10}H_8}\right)\left(\frac{148 \text{ g } C_8H_4O_3}{\text{mol } C_8H_4O_3}\right) = 231 \text{ g}$$

$$(231 \text{ g theoretical})\left(\frac{70.0 \text{ g actual}}{100 \text{ g theoretical}}\right) = 162 \text{ g} \qquad 162 \text{ lb } C_8H_4O_3 \text{ produced}$$

11.55 The empirical formula of a commerical ion-exchange resin is $C_8H_7SO_3Na$. The resin can be used to soften water according to the reaction $Ca^{2+} + 2C_8H_7SO_3Na \rightarrow (C_8H_7SO_3)_2Ca + 2Na^+$. What would be the maximum uptake of Ca^{2+} by the resin, expressed in mol/g resin?

▮

$$\frac{1 \text{ mol } Ca^{2+}}{(2 \text{ mol resin})(206 \text{ g resin/mol resin})} = 0.00246 \text{ mol } Ca^{2+}/\text{g resin}$$

11.56 One gram (dry weight) of green algae was able to absorb 5.5×10^{-3} mol CO_2 per hour by photosynthesis. If the fixed carbon atoms were all stored after photosynthesis as starch, $(C_6H_{10}O_5)_n$, how long would it take for the algae to double their own weight? Neglect the increase in photosynthetic rate due to the increasing amount of living matter.

▮ The part of the equation which is important to the calculation is $6nCO_2 \rightarrow (C_6H_{10}O_5)_n$. 1.00 g of $(C_6H_{10}O_5)_n$ must be produced to double the weight. Per hour,

$$[1.00 \text{ g } (C_6H_{10}O_5)_n]\left(\frac{1 \text{ mol starch}}{162n \text{ g starch}}\right)\left(\frac{6n \text{ mol } CO_2}{\text{mol starch}}\right) = 0.0370 \text{ mol } CO_2 \qquad \text{Note that the } n\text{'s cancel.}$$

$$(0.0370 \text{ mol } CO_2)\left(\frac{1 \text{ h}}{5.5 \times 10^{-3} \text{ mol } CO_2}\right) = 6.7 \text{ h}$$

11.57 The chemical formula of the chelating agent Versene is $C_2H_4N_2(C_2H_2O_2Na)_4$. If each mol of this compound could bind 1 mol of Ca^{2+}, what would be the rating of pure Versene, expressed as mg $CaCO_3$ bound per g of chelating agent? Here the Ca^{2+} is expressed in terms of the amount of $CaCO_3$ it could form.

▮

$$\text{For each mol of compound,} \quad (1 \text{ mol } Ca^{2+})\left(\frac{1 \text{ mol } CaCO_3}{\text{mol } Ca^{2+}}\right)\left(\frac{100 \text{ g } CaCO_3}{\text{mol } CaCO_3}\right)\left(\frac{10^3 \text{ mg}}{\text{g}}\right) = 1.00 \times 10^5 \text{ mg}$$

$$(1 \text{ mol compound})\left(\frac{380 \text{ g compound}}{\text{mol compound}}\right) = 380 \text{ g compound}$$

$$\frac{1.00 \times 10^5 \text{ mg } CaCO_3}{380 \text{ g compound}} = 263 \text{ mg } CaCO_3/\text{g compound}$$

11.58 A mixture of NaCl and KCl weighed 5.4892 g. The sample was dissolved in water and treated with an excess of silver nitrate in solution. The resulting AgCl weighed 12.7052 g. What was the percentage NaCl in the mixture?

The two parallel reactions are $NaCl + AgNO_3 \rightarrow AgCl + NaNO_3$ and $KCl + AgNO_3 \rightarrow AgCl + KNO_3$. Here the conservation of Cl atoms requires that the number of moles of AgCl formed equal the *sum* of the numbers of moles of NaCl and KCl.

$$n(\text{AgCl}) = \frac{12.7052 \text{ g AgCl}}{143.321 \text{ g AgCl/mol}} = 0.088649 \text{ mol}$$

Then $n(\text{NaCl}) + n(\text{KCl}) = 0.088649$ mol, or

$$\frac{m(\text{NaCl})}{58.443 \text{ g/mol}} + \frac{m(\text{KCl})}{74.551 \text{ g/mol}} = 0.088649 \text{ mol} \qquad (1)$$

A second equation for the unknown masses is provided by the data:

$$m(\text{NaCl}) + m(\text{KCl}) = 5.4892 \text{ g} \qquad (2)$$

Eliminating $m(\text{KCl})$ between (1) and (2) and solving for $m(\text{NaCl})$, we obtain $m(\text{NaCl}) = 4.0624$ g. Then

$$\left(\frac{4.0624 \text{ g}}{5.4892 \text{ g}}\right)(100\%) = 74.01\% \text{ NaCl}$$

11.59 A mixture of NaCl and NaBr weighing 3.5084 g was dissolved and treated with enough $AgNO_3$ to precipitate all of the chloride and bromide as AgCl and AgBr. The washed precipitate was treated with KCN to solubilize the silver, and the resulting solution was electrolyzed. The equations are:

$$NaCl + AgNO_3 \longrightarrow AgCl + NaNO_3 \qquad NaBr + AgNO_3 \longrightarrow AgBr + NaNO_3$$
$$AgCl + 2KCN \longrightarrow KAg(CN)_2 + KCl \qquad AgBr + 2KCN \longrightarrow KAg(CN)_2 + KBr$$
$$4KAg(CN)_2 + 4KOH \longrightarrow 4Ag + 8KCN + O_2 + 2H_2O$$

After the final step was complete, the deposit of metallic silver weighed 5.5028 g. What was the composition of the initial mixture?

Let x = mass NaCl; then $3.5084 - x$ = mass NaBr

$$\text{mol NaCl} = \frac{x}{22.990 + 35.453} \qquad \text{mol NaBr} = \frac{3.5084 - x}{22.990 + 79.904}$$

$$\text{mol Ag} = (5.5028)\left(\frac{1 \text{ mol Ag}}{107.868 \text{ g Ag}}\right) = 0.051014 \text{ mol Ag} = \text{mol NaCl} + \text{mol NaBr}$$

$$0.051014 = \frac{x}{58.443} + \frac{3.5084 - x}{102.894} \qquad x = 2.2886 \text{ g}$$

$$\%\text{NaCl} = \left(\frac{2.2886 \text{ g}}{3.5084 \text{ g}}\right)(100\%) = 65.23\% \text{ NaCl} \qquad \%\text{NaBr} = 100.00\% - 65.23\% = 34.77\% \text{ NaBr}$$

11.60 A mixture of $NaHCO_3$ and Na_2CO_3 weighed 1.0235 g. The dissolved mixture was reacted with excess $Ba(OH)_2$ to form 2.1028 g $BaCO_3$, by the reactions

$$Na_2CO_3 + Ba(OH)_2 \longrightarrow BaCO_3 + 2NaOH \qquad NaHCO_3 + Ba(OH)_2 \longrightarrow BaCO_3 + NaOH + H_2O$$

What was the percentage $NaHCO_3$ in the original mixture?

Let x = mass $NaHCO_3$; then $1.0235 - x$ = mass Na_2CO_3 and

$$\text{mol NaHCO}_3 = \frac{x}{84.01} \qquad \text{mol Na}_2\text{CO}_3 = \frac{1.0235 - x}{105.99}$$

$$\text{mol BaCO}_3 = \underset{\text{(from 2nd reaction)}}{\text{mol NaHCO}_3} + \underset{\text{(from 1st reaction)}}{\text{mol Na}_2\text{CO}_3}$$

$$\frac{2.1028}{197.34} = \frac{x}{84.01} + \frac{1.0235 - x}{105.99}$$

$$x = 0.4052 \text{ g NaHCO}_3 \qquad \left(\frac{0.4052}{1.0235}\right)(100\%) = 39.6\% \text{ NaHCO}_3$$

11.61 (*a*) Balance the following equations:

$$Na_2CO_3 + HCl \longrightarrow NaCl + CO_2 + H_2O \qquad NaHCO_3 + HCl \longrightarrow NaCl + CO_2 + H_2O$$

(*b*) A sample contains a mixture of $NaHCO_3$ and Na_2CO_3. HCl is added to 15.0 g of the sample, yielding 11.0 g of NaCl. What percent of the sample is Na_2CO_3?

▮ (*a*) $Na_2CO_3 + 2HCl \longrightarrow 2NaCl + CO_2 + H_2O$

$NaHCO_3 + HCl \longrightarrow NaCl + CO_2 + H_2O$

(*b*) Let x = number of g of Na_2CO_3. Then $15.0 - x$ = number of g of $NaHCO_3$.

$$(11.0\text{ g NaCl})\left(\frac{1\text{ mol}}{58.5\text{ g}}\right) = 0.188\text{ mol NaCl produced}$$

The NaCl is produced by the reaction of $(x/106)$ mol of Na_2CO_3 and $(15.0 - x)/84.0$ mol of $NaHCO_3$. Each mol of Na_2CO_3 produces 2 mol of NaCl.

$$\frac{2x}{106} + \frac{15.0 - x}{84.0} = 0.188$$

$$0.01887x + 0.1786 - 0.01190x = 0.188$$

$$x = 1.35\text{ g }Na_2CO_3 \qquad 15.0 - x = 13.65\text{ g }NaHCO_3$$

$$\left(\frac{1.35\text{ g }Na_2CO_3}{15.0\text{ g total}}\right)(100\%) = 9.00\%\ Na_2CO_3$$

11.62 Four grams of a mixture of calcium carbonate and sand is treated with an excess of hydrochloric acid, and 0.880 g of CO_2 is produced. What is the percent of $CaCO_3$ in the original mixture?

▮

$$\frac{0.880\text{ g }CO_2}{44.0\text{ g }CO_2/\text{mol}} = 0.0200\text{ mol }CO_2 \qquad (0.0200\text{ mol }CO_2)\left(\frac{1\text{ mol }CaCO_3}{1\text{ mol }CO_2}\right) = 0.0200\text{ mol }CaCO_3$$

$$(0.0200\text{ mol }CaCO_3)\left(\frac{100\text{ g }CaCO_3}{\text{mol }CaCO_3}\right) = 2.00\text{ g }CaCO_3$$

$$\%\ CaCO_3\text{ in mixture} = \frac{2.00\text{ g }CaCO_3}{4.00\text{ g mixture}} \times 100\% = 50.0\%$$

11.63 A 5.00 g sample of a natural gas, consisting of methane, CH_4, and ethylene, C_2H_4, was burned in excess oxygen, yielding 14.5 g of CO_2 and some H_2O as products. What percent of the sample was ethylene?

▮ The equations for the reactions are

$$C_2H_4 + 3O_2 \longrightarrow 2CO_2 + 2H_2O \qquad CH_4 + 2O_2 \longrightarrow CO_2 + 2H_2O$$

The total number of mol of carbon dioxide formed can be calculated from the observed mass of carbon dioxide:

$$\frac{14.5\text{ g }CO_2}{44.0\text{ g }CO_2/\text{mol }CO_2} = 0.330\text{ mol }CO_2$$

The total number of mol of CO_2 can be expressed in terms of the quantities of C_2H_4 and CH_4 which have reacted. Let x equal the number of g of ethylene in the original mixture. Then $5.00 - x$ will be the number of g of methane.

$$\text{mol }CO_2\text{ from }C_2H_4 = \left(\frac{x\text{ g }C_2H_4}{28.0\text{ g }C_2H_4/\text{mol}}\right)\left(\frac{2\text{ mol }CO_2}{\text{mol }C_2H_4}\right) = \frac{2x}{28.0}\text{ mol }CO_2$$

$$\text{mol }CO_2\text{ from }CH_4 = \left(\frac{(5.00 - x)\text{ g }CH_4}{16.0\text{ g }CH_4/\text{mol}}\right)\left(\frac{1\text{ mol }CO_2}{\text{mol }CH_4}\right) = \frac{5.00 - x}{16.0}\text{ mol }CO_2$$

The sum of these expressions must equal the number of moles calculated above. Hence

$$\frac{2x}{28.0}\text{ mol} + \frac{5.00 - x}{16.0}\text{ mol} = 0.330\text{ mol}$$

Solving for x, $(0.0714x) + (0.312 - 0.0625x) = 0.330$, so $x = 2.02\text{ g }C_2H_4$. Then $5.00 - x = 2.98\text{ g }CH_4$ and

$$\%\ C_2H_4 = \left(\frac{2.02\text{ g }C_2H_4}{5.00\text{ g total}}\right)(100\%) = 40.4\%$$

11.2 LIMITING QUANTITIES

11.64 How can you recognize a limiting quantities problem?

▮ The quantities of two (or more) reactants are given.

11.65 For the reaction $Ba(OH)_2 + 2HClO_3 \rightarrow Ba(ClO_3)_2 + 2H_2O$, calculate the number of mol of H_2O formed when 0.100 mol of $Ba(OH)_2$ is treated with 0.250 mol of $HClO_3$.

▮ 0.100 mol $Ba(OH)_2$ takes 0.200 mol $HClO_3$. $Ba(OH)_2$ is the limiting quantity and produces 0.200 mol H_2O in this reaction.

11.66 When copper is heated with an excess of sulfur, Cu_2S is formed. How many g of Cu_2S could be produced if 100 g of copper is heated with 50 g of sulfur?

▮
$$(100\text{ g Cu})\left(\frac{1\text{ mol Cu}}{63.5\text{ g Cu}}\right)\left(\frac{1\text{ mol Cu}_2\text{S}}{2\text{ mol Cu}}\right)\left(\frac{159\text{ g Cu}_2\text{S}}{\text{mol Cu}_2\text{S}}\right) = 125\text{ g Cu}_2\text{S}$$

Since the problem states that S is in excess, the 50 g value is not used in the calculation.

11.67 Calculate the mass of carbon tetrachloride which can be produced by the reaction of 10.0 g of carbon with 100.0 g of chlorine. Determine the mass of excess reagent left unreacted.

▮ First the number of moles of each reactant is determined:

$$(10.0\text{ g C})\left(\frac{1\text{ mol C}}{12.0\text{ g C}}\right) = 0.833\text{ mol C} \qquad (100.0\text{ g Cl}_2)\left(\frac{1\text{ mol Cl}_2}{70.9\text{ g Cl}_2}\right) = 1.41\text{ mol Cl}_2$$

From the balanced chemical equation $C + 2Cl_2 \longrightarrow CCl_4$, it is seen that each mol of C requires 2 mol of Cl_2.

$$(0.833\text{ mol C})\left(\frac{2\text{ mol Cl}_2}{1\text{ mol C}}\right) = 1.67\text{ mol Cl}_2\text{ required}$$

However, only 1.41 mol of Cl_2 is available; hence the carbon is in excess.

$$(1.41\text{ mol Cl}_2)\left(\frac{1\text{ mol C}}{2\text{ mol Cl}_2}\right) = 0.705\text{ mol C required}$$

The excess C is $0.833 - 0.705 = 0.128$ mol C. In this case Cl_2 is the ***limiting reagent***, which determines the quantity of CCl_4 which can be produced:

$$(1.41\text{ mol Cl}_2)\left(\frac{1\text{ mol CCl}_4}{2\text{ mol Cl}_2}\right) = 0.705\text{ mol CCl}_4 \qquad (0.705\text{ mol CCl}_4)\left(\frac{153.8\text{ g CCl}_4}{\text{mol CCl}_4}\right) = 108\text{ g CCl}_4$$

The mass of unreacted carbon is determined from the number of moles in excess:

$$(0.128\text{ mol C})\left(\frac{12.0\text{ g C}}{\text{mol C}}\right) = 1.54\text{ g C unreacted}$$

11.68 For the reaction $4Fe + 3O_2 \rightarrow 2Fe_2O_3$, 4.80 g of oxygen is used to burn 0.150 mol of iron. What mass of Fe_2O_3 will be produced? What mass of Fe will be left over at the end of the reaction? What mass of O_2 will be left over at the end of the reaction?

▮
$$(4.80\text{ g O}_2)\left(\frac{1\text{ mol O}_2}{32.0\text{ g O}_2}\right) = (0.150\text{ mol O}_2\text{ present})\left(\frac{4\text{ mol Fe}}{3\text{ mol O}_2}\right) = 0.200\text{ mol Fe required}$$

Hence Fe is present in limiting quantity, and no Fe will remain after the reaction.

$$(0.150\text{ mol Fe})\left(\frac{3\text{ mol O}_2}{4\text{ mol Fe}}\right)\left(\frac{32.0\text{ g O}_2}{\text{mol O}_2}\right) = 3.60\text{ g O}_2\text{ required}$$

$$(4.80\text{ g O}_2\text{ present}) - (3.60\text{ g O}_2\text{ required}) = 1.20\text{ g O}_2\text{ in excess}$$

$$(0.150\text{ mol Fe})\left(\frac{2\text{ mol Fe}_2\text{O}_3}{4\text{ mol Fe}}\right)\left(\frac{159.7\text{ g Fe}_2\text{O}_3}{\text{mol Fe}_2\text{O}_3}\right) = 12.0\text{ g Fe}_2\text{O}_3\text{ produced}$$

11.69 What mass of solid is produced by treatment of aqueous solutions containing 2.00 g of $AgNO_3$ and 4.00 g of KBr, respectively?

▮

$$Ag^+ + Br^- \longrightarrow AgBr$$

$$(2.00\text{ g AgNO}_3)\left(\frac{1\text{ mol AgNO}_3}{169.9\text{ g AgNO}_3}\right) = 0.0118\text{ mol AgNO}_3 \qquad (4.00\text{ g KBr})\left(\frac{1\text{ mol KBr}}{119.0\text{ g KBr}}\right) = 0.0336\text{ mol KBr}$$

$AgNO_3$ is the limiting reagent.

$$(0.0118\text{ mol AgNO}_3)\left(\frac{1\text{ mol AgBr}}{\text{mol AgNO}_3}\right)\left(\frac{187.8\text{ g AgBr}}{\text{mol AgBr}}\right) = 2.21\text{ g}$$

11.70 A mixture containing 100 g H_2 and 100 g O_2 is ignited so that water is formed according to the reaction $2H_2 + O_2 \rightarrow 2H_2O$. How much water is formed?

▮ It is necessary first to determine which, if any, substance is in excess. The mole method is the simplest method for this type of problem.

$$n(H_2) = \frac{100\text{ g}}{2.02\text{ g/mol}} = 49.5\text{ mol H}_2 \qquad n(O_2) = \frac{100\text{ g}}{32.0\text{ g/mol}} = 3.13\text{ mol O}_2$$

If all the hydrogen were to be used, $\frac{1}{2}(49.5) = 24.8$ mol O_2 would be required. Obviously the hydrogen cannot all be used. Since O_2 is present in limiting amount, the calculations must be based on the amount of O_2. Counting only those moles which participate in the reaction,

$$n(H_2O) = 2n(O_2) = 2 \times 3.13 = 6.26\text{ mol H}_2O \qquad m(H_2O) = (6.26\text{ mol})(18.0\text{ g/mol}) = 113\text{ g H}_2O$$

The amount of H_2 consumed is $(6.26\text{ mol})(2.02\text{ g/mol}) = 12.6$ g. The reaction mixture will contain, in addition to the 113 g H_2O, 87.4 g of unreacted H_2.

11.71 ***(a)*** What mass of P_4O_{10} will be obtained from the reaction of 1.33 g of P_4 and 5.07 g of O_2? ***(b)*** What mass of P_4O_6 will be obtained from the reaction of 4.07 g of P_4 and 2.01 g of O_2?

▮ ***(a)*** Present are

$$(1.33\text{ g P}_4)\left(\frac{1\text{ mol P}_4}{123.9\text{ g P}_4}\right) = 0.0107\text{ mol P}_4 \quad \text{and} \quad (5.07\text{ g O}_2)\left(\frac{1\text{ mol O}_2}{32.0\text{ g O}_2}\right) = 0.158\text{ mol O}_2$$

The 0.0107 mol of P_4 requires, from the balanced chemical equation,

$$(0.0107\text{ mol P}_4)\left(\frac{5\text{ mol O}_2}{\text{mol P}_4}\right) = 0.0535\text{ mol O}_2\text{ required}$$

Hence there is more oxygen present than required, and P_4 is in limiting quantity.

$$(0.0107\text{ mol P}_4)\left(\frac{1\text{ mol P}_4O_{10}}{\text{mol P}_4}\right)\left(\frac{283.9\text{ g P}_4O_{10}}{\text{mol P}_4O_{10}}\right) = 3.05\text{ g} \qquad \textbf{(b)}\ 4.60\text{ g P}_4O_6$$

11.72 When solutions containing 2.00 g of Na_2SO_4 and 3.00 g of $BaCl_2$ are mixed, what mass of $BaSO_4$ is produced?

▮

$$Na_2SO_4 + BaCl_2 \longrightarrow BaSO_4(s) + 2NaCl$$

$$(2.00\text{ g Na}_2SO_4)\left(\frac{1\text{ mol Na}_2SO_4}{142\text{ g Na}_2SO_4}\right) = 0.0141\text{ mol Na}_2SO_4 \qquad (3.00\text{ g BaCl}_2)\left(\frac{1\text{ mol BaCl}_2}{208.2\text{ g BaCl}_2}\right) = 0.0144\text{ mol BaCl}_2$$

There are present 0.0141 mol Na_2SO_4 and 0.0144 mol $BaCl_2$. Since they react in a 1:1 mole ratio, as shown in the balanced chemical equation, the $BaCl_2$ is in excess, and the rest of the calculation is based on the 0.0141 mol Na_2SO_4, the limiting quantity.

$$(0.0141\text{ mol Na}_2SO_4)\left(\frac{1\text{ mol BaSO}_4}{1\text{ mol Na}_2SO_4}\right)\left(\frac{233.4\text{ g BaSO}_4}{\text{mol BaSO}_4}\right) = 3.29\text{ g BaSO}_4$$

11.73 How much carbon monoxide is produced from the reaction of 1.00 kg of octane, C_8H_{18}, and 1.00 kg of oxygen?

▮

$$(1.00\text{ kg C}_8H_{18})\left(\frac{1000\text{ g}}{\text{kg}}\right)\left(\frac{1\text{ mol C}_8H_{18}}{114\text{ g C}_8H_{18}}\right) = 8.77\text{ mol C}_8H_{18}\text{ present}$$

$$(1.00\text{ kg O}_2)\left(\frac{1000\text{ g}}{\text{kg}}\right)\left(\frac{1\text{ mol O}_2}{32.0\text{ g O}_2}\right) = 31.25\text{ mol O}_2\text{ present}$$

$$2C_8H_{18} + 17O_2 \longrightarrow 16CO + 18H_2O$$

$$(8.77 \text{ mol } C_8H_{18})\left(\frac{17 \text{ mol } O_2}{2 \text{ mol } C_8H_{18}}\right) = 74.5 \text{ mol } O_2 \text{ required.}$$ Hence O_2 is in limiting quantity.

$$(31.25 \text{ mol } O_2)\left(\frac{16 \text{ mol CO}}{17 \text{ mol } O_2}\right)\left(\frac{28.0 \text{ g CO}}{\text{mol CO}}\right) = 824 \text{ g CO}$$

11.74 What masses of P_4O_6 and P_4O_{10} will be produced by the combustion of 2.00 g of P_4 in 2.00 g of oxygen, leaving no P_4 or O_2?

▮

$$P_4 + 5O_2 \longrightarrow P_4O_{10} \qquad P_4 + 3O_2 \longrightarrow P_4O_6$$

$$(2.00 \text{ g } P_4)\left(\frac{1 \text{ mol } P_4}{123.88 \text{ g}}\right) = 0.0161 \text{ mol } P_4 \qquad (2.00 \text{ g } O_2)\left(\frac{1 \text{ mol } O_2}{32.0 \text{ g}}\right) = 0.0625 \text{ mol } O_2$$

There is not enough oxygen to produce P_4O_{10} exclusively, which would have required 5(0.0161 mol) = 0.0805 mol O_2. Thus

$$(0.0161 \text{ mol } P_4)\left(\frac{3 \text{ mol } O_2}{\text{mol } P_4}\right) = 0.0483 \text{ mol } O_2 \text{ used to make } P_4O_6.$$

As yet, 0.0142 mol of O_2 is unreacted. This oxygen can then react according to the equation $P_4O_6 + 2O_2 \rightarrow P_4O_{10}$. Then

$$(0.0142 \text{ mol } O_2)\left(\frac{1 \text{ mol } P_4O_{10}}{2 \text{ mol } O_2}\right) = 0.00710 \text{ mol } P_4O_{10}$$

Of the 0.0161 mol of P_4O_6 originally formed, 0.00710 mol is converted to P_4O_{10}, leaving 0.0090 mol.

$$(0.00710 \text{ mol } P_4O_{10})\left(\frac{284 \text{ g}}{\text{mol}}\right) = 2.02 \text{ g } P_4O_{10} \qquad (0.0090 \text{ mol } P_4O_6)\left(\frac{220 \text{ g}}{\text{mol}}\right) = 1.98 \text{ g } P_4O_6$$

11.75 A mixture of 1.00 ton of CS_2 and 2.00 tons of Cl_2 is passed through a hot reaction tube, where the reaction $CS_2 + 3Cl_2 \rightarrow CCl_4 + S_2Cl_2$ takes place. ***(a)*** How much CCl_4 can be made by complete reaction of the limiting starting material? ***(b)*** Which starting material is in excess, and how much of it remains unreacted?

▮ The ratio of tons is the same as the ratio of g, so solve the problem using g:

$$(1.00 \text{ g } CS_2)\left(\frac{1 \text{ mol } CS_2}{76.0 \text{ g } CS_2}\right)\left(\frac{3 \text{ mol } Cl_2}{\text{mol } CS_2}\right)\left(\frac{71.0 \text{ g } Cl_2}{\text{mol } Cl_2}\right) = 2.80 \text{ g } Cl_2 \text{ needed}$$

Since there is 2.00 g Cl_2 present, Cl_2 is the limiting quantity.

$$(2.00 \text{ g } Cl_2)\left(\frac{1 \text{ mol } Cl_2}{71.0 \text{ g } Cl_2}\right)\left(\frac{1 \text{ mol } CCl_4}{3 \text{ mol } Cl_2}\right)\left(\frac{154 \text{ g } CCl_4}{\text{mol } CCl_4}\right) = 1.45 \text{ g } CCl_4 \qquad \textbf{(a)}\ 1.45 \text{ ton}$$

$$(2.00 \text{ g } Cl_2)\left(\frac{1 \text{ mol } Cl_2}{71.0 \text{ g } Cl_2}\right)\left(\frac{1 \text{ mol } CS_2}{3 \text{ mol } Cl_2}\right)\left(\frac{76.0 \text{ g } CS_2}{\text{mol } CS_2}\right) = 0.714 \text{ g } CS_2 \text{ used up}$$

$$(1.00 \text{ g } CS_2 \text{ present}) - (0.714 \text{ g used}) = 0.286 \text{ g } CS_2 \text{ excess} \qquad \textbf{(b)}\ 0.286 \text{ ton}$$

11.76 The reaction $2Al + 3MnO \rightarrow Al_2O_3 + 3Mn$ proceeds until the limiting substance is all consumed. A mixture containing 110 g Al and 200 g MnO was heated to initiate the reaction. Which initial substance remained in excess, and by how much?

▮

$$(110 \text{ g Al})\left(\frac{1 \text{ mol Al}}{27.0 \text{ g Al}}\right) = 4.07 \text{ mol Al present} \qquad (200 \text{ g MnO})\left(\frac{1 \text{ mol MnO}}{70.9 \text{ g MnO}}\right) = 2.82 \text{ mol MnO present}$$

Since the ratio of Al/MnO present exceeds that required by the equation (2/3), the Al is present in excess.

$$(2.82 \text{ mol MnO})\left(\frac{2 \text{ mol Al}}{3 \text{ mol MnO}}\right)\left(\frac{27.0 \text{ g Al}}{\text{mol Al}}\right) = 50.8 \text{ g Al required} \qquad (100.0 \text{ g}) - (50.8 \text{ g}) = 49.2 \text{ g Al excess}$$

11.77 Calculate the number of g of NaCl formed by the reaction of 100 g of Na_2CO_3 with 100 g of HCl. Also calculate the number of g of the excess reagent which remains unreacted.

$$(100\text{ g }Na_2CO_3)\left(\frac{1\text{ mol }Na_2CO_3}{106\text{ g }Na_2CO_3}\right) = 0.943\text{ mol }Na_2CO_3\text{ present}$$

$$(100\text{ g HCl})\left(\frac{1\text{ mol HCl}}{36.5\text{ g HCl}}\right) = 2.74\text{ mol HCl present}$$

$$Na_2CO_3 + 2HCl \longrightarrow 2NaCl + CO_2 + H_2O$$

$$(0.943\text{ mol }Na_2CO_3)\left(\frac{2\text{ mol HCl}}{\text{mol }Na_2CO_3}\right) = 1.89\text{ mol HCl required}$$

The HCl is in excess. The quantity of NaCl is therefore based on the quantity of Na_2CO_3 present—the limiting quantity:

$$(0.943\text{ mol }Na_2CO_3)\left(\frac{2\text{ mol NaCl}}{1\text{ mol }Na_2CO_3}\right)\left(\frac{58.5\text{ g NaCl}}{\text{mol NaCl}}\right) = 110\text{ g NaCl}$$

The excess HCl is (2.74 mol) − (1.89 mol) = 0.85 mol.

$$(0.85\text{ mol HCl})\left(\frac{36.5\text{ g HCl}}{1\text{ mol HCl}}\right) = 31\text{ g HCl in excess}$$

11.78 How much potassium chloride is produced from the reaction of 2.00 g of K and 3.00 g of Cl_2?

▮

$$(2.00\text{ g K})\left(\frac{1\text{ mol K}}{39.1\text{ g K}}\right) = 0.512\text{ mol K present} \qquad (3.00\text{ g }Cl_2)\left(\frac{1\text{ mol }Cl_2}{70.9\text{ g }Cl_2}\right) = 0.0423\text{ mol }Cl_2\text{ present}$$

$$2K + Cl_2 \longrightarrow 2KCl$$

Required for 0.0512 mol K is

$$(0.0512\text{ mol K})\left(\frac{1\text{ mol }Cl_2}{2\text{ mol K}}\right) = 0.0256\text{ mol }Cl_2. \qquad \text{Hence } Cl_2 \text{ is in excess.}$$

$$(0.0512\text{ mol K})\left(\frac{1\text{ mol KCl}}{\text{mol K}}\right)\left(\frac{74.5\text{ g KCl}}{\text{mol KCl}}\right) = 3.81\text{ g KCl}$$

11.79 When 20 g of lithium reacts with 30 g of oxygen, calculate the number of g of Li_2O formed. Which is in excess, Li or O_2, and by how many grams?

▮

$$4Li + O_2 \longrightarrow 2Li_2O$$

$$(20.0\text{ g Li})\left(\frac{1\text{ mol Li}}{6.94\text{ g Li}}\right) = 2.88\text{ mol Li} \qquad (30.0\text{ g }O_2)\left(\frac{1\text{ mol }O_2}{32.0\text{ g }O_2}\right) = 0.938\text{ mol }O_2\text{ present}$$

$$(2.88\text{ mol Li})\left(\frac{1\text{ mol }O_2}{4\text{ mol Li}}\right) = 0.720\text{ mol }O_2\text{ required, hence }O_2\text{ is in excess.}$$

$$\text{Excess }O_2 = (0.938\text{ mol}) - (0.720\text{ mol}) = (0.218\text{ mol})\left(\frac{32.0\text{ g}}{\text{mol}}\right) = 7.00\text{ g }O_2\text{ excess}$$

$$(2.88\text{ mol Li})\left(\frac{1\text{ mol }Li_2O}{2\text{ mol Li}}\right)\left(\frac{29.9\text{ g }Li_2O}{\text{mol }Li_2O}\right) = 43.0\text{ g }Li_2O$$

11.80 How many mL of 0.200 M NaOH will completely neutralize 100 mL of 0.250 M H_2SO_4?

▮

$$(100\text{ mL }H_2SO_4)\left(\frac{0.250\text{ mmol }H_2SO_4}{\text{mL }H_2SO_4}\right)\left(\frac{2\text{ mmol NaOH}}{\text{mmol }H_2SO_4}\right)\left(\frac{1\text{ mL NaOH}}{0.200\text{ mmol NaOH}}\right) = 250\text{ mL NaOH}$$

11.81 The insecticide DDT is made by the reaction

$$CCl_3CHO\text{ (chloral)} + 2C_6H_5Cl\text{ (chlorobenzene)} \longrightarrow (ClC_6H_4)_2CHCCl_3\text{ (DDT)} + H_2O$$

If 100 lb of chloral were treated with 100 lb of chlorobenzene, how much DDT would be formed? Assume the reaction goes to completion without side reactions or losses.

▮

$$(100\text{ g }CCl_3CHO)\left(\frac{1\text{ mol }CCl_3CHO}{147.5\text{ g }CCl_3CHO}\right)\left(\frac{2\text{ mol }C_6H_5Cl}{\text{mol }CCl_3CHO}\right)\left(\frac{112.5\text{ g }C_6H_5Cl}{\text{mol }C_6H_5Cl}\right) = 153\text{ g }C_6H_5Cl\text{ needed}$$

Since there is only 100 g C_6H_5Cl present, C_6H_5Cl is in limiting quantity.

$$(100\text{ g }C_6H_5Cl)\left(\frac{1\text{ mol }C_6H_5Cl}{112.5\text{ g }C_6H_5Cl}\right)\left(\frac{1\text{ mol DDT}}{2\text{ mol }C_6H_5Cl}\right)\left(\frac{353.7\text{ g DDT}}{\text{mol DDT}}\right) = 157\text{ g DDT}$$

157 lb DDT is produced.

11.82 The reduction of Cr_2O_3 by Al proceeds quantitatively on ignition of a suitable fuse. The reaction is $2Al + Cr_2O_3 \rightarrow Al_2O_3 + 2Cr$. ***(a)*** How much metallic chromium can be made by bringing to reaction temperature a mixture of 5.0 kg Al and 20.0 kg Cr_2O_3? ***(b)*** Which reactant remains at the completion of the reaction, and how much?

▮

$$\frac{(5.0\times 10^3\text{ g Al})(1\text{ mol Al}/27\text{ g Al})}{(20.0\times 10^3\text{ g }Cr_2O_3)(1\text{ mol }Cr_2O_3/152\text{ g }Cr_2O_3)} = 1.41\text{ mol Al/mol }Cr_2O_3\text{ present}$$

Since 2 mol Al/mol Cr_2O_3 is required, Al is the limiting quantity.

(a) $(5.0\times 10^3\text{ g Al})\left(\frac{1\text{ mol Al}}{27.0\text{ g Al}}\right)\left(\frac{2\text{mol Cr}}{2\text{ mol Al}}\right)\left(\frac{52.0\text{ g Cr}}{\text{mol Cr}}\right) = 9.6\times 10^3\text{ g Cr} = 9.6\text{ kg Cr}$

(b) $(5.0\times 10^3\text{ g Al})\left(\frac{1\text{ mol Al}}{27.0\text{ g Al}}\right)\left(\frac{1\text{ mol }Cr_2O_3}{2\text{ mol Al}}\right)\left(\frac{152\text{ g }Cr_2O_3}{\text{mol }Cr_2O_3}\right) = 14.1\text{ kg }Cr_2O_3\text{ needed}$

$(20.0\text{ kg }Cr_2O_3\text{ present}) - (14.1\text{ kg }Cr_2O_3\text{ used}) = 5.9\text{ kg }Cr_2O_3\text{ excess}$

11.83 Silver may be removed from solutions of its salts by reaction with metallic zinc according to the reaction $Zn + 2Ag^+ \rightarrow Zn^{2+} + 2Ag$. A 50 g piece of zinc was thrown into a 100 L vat containing 3.0 g Ag^+/L. ***(a)*** Which reactant was completely consumed? ***(b)*** How much of the other substance remained unreacted?

▮ $\left(\frac{3.0\text{ g}}{\text{L}}\right)(100\text{ L}) = (300\text{ g }Ag^+)\left(\frac{1\text{ mol }Ag^+}{107.9\text{ g }Ag^+}\right) = 2.8\text{ mol }Ag^+$ $\quad (50\text{ g Zn})\left(\frac{1\text{ mol Zn}}{65.4\text{ g Zn}}\right) = 0.76\text{ mol Zn}$

(a) Since the mole ratio Ag^+/Zn present (2.8/0.76) exceeds the mole ratio required (2/1) in the reaction, the Ag^+ is in excess; the Zn is completely consumed.

(b) $(0.76\text{ mol Zn})\left(\frac{2\text{ mol }Ag^+}{\text{mol Zn}}\right) = 1.5\text{ mol }Ag^+\text{ used}$

$(2.8\text{ mol}) - (1.5\text{ mol}) = (1.3\text{ mol }Ag^+\text{ excess})\left(\frac{107.9\text{ g }Ag^+}{\text{mol }Ag^+}\right) = 140\text{ g }Ag^+\text{ excess}$

11.84 Carbon reacts with oxygen to yield carbon monoxide or carbon dioxide, depending on the quantity of oxygen available per mol of carbon. Calculate the number of mol of each product produced when 100 g of oxygen reacts with ***(a)*** 10.0 g of carbon ***(b)*** 100 g of carbon ***(c)*** 60.0 g of carbon.

▮

$$(100\text{ g }O_2)\left(\frac{1\text{ mol }O_2}{32.0\text{ g }O_2}\right) = 3.125\text{ mol }O_2$$

(a) $(10.0\text{ g C})\left(\frac{1\text{ mol C}}{12.0\text{ g C}}\right) = 0.833\text{ mol C}$

There is excess oxygen even for the reaction in which carbon dioxide is produced.

$$(0.833\text{ mol C})\left(\frac{1\text{ mol }CO_2}{\text{mol C}}\right) = 0.833\text{ mol }CO_2$$

(b) 100 g of carbon is 8.33 mol of carbon. There is excess carbon even for the reaction in which carbon monoxide is produced.

$$(3.125\text{ mol }O_2)\left(\frac{2\text{ mol CO}}{\text{mol }O_2}\right) = 6.25\text{ mol CO}$$

(c) $(60.0\text{ g C})\left(\frac{1\text{ mol C}}{12.0\text{ g C}}\right) = 5.00\text{ mol C}$

There is excess oxygen for the formation of CO, but insufficient for the formation of CO_2. Hence, some of each of these products is formed. One may solve by simultaneous equations or assume that first all of the carbon is converted to CO and then the excess O_2 oxidizes some of the CO to CO_2.

5.00 mol of carbon requires 2.50 mol of O_2 to produce 5.00 mol of CO, leaving 0.625 mol of O_2 in excess. This quantity of O_2 reacts with 1.25 mol of CO in the reaction $CO + \frac{1}{2}O_2 \rightarrow CO_2$ to produce 1.25 mol of CO_2 and leaves 3.75 mol of CO unreacted.

11.3 SOLUTE CONCENTRATIONS, PHYSICAL UNITS

11.85 How would you prepare 60 mL of an aqueous solution containing 0.030 g $AgNO_3$ per mL?

▮ Since each mL of solution is to contain 0.030 g $AgNO_3$, 60 mL of solution should contain

$$(0.030\ \text{g/mL})(60\ \text{mL}) = 1.8\ \text{g}\ AgNO_3$$

Thus, dissolve 1.8 g of $AgNO_3$ in about 50 mL of water. Then add sufficient water to make the volume exactly 60 mL. Stir thoroughly to ensure uniformity. (Note that 60 mL is to be the volume of the final solution, not the volume of the water used to make up the solution.)

11.86 How many g of a 5.0% by weight NaCl solution are necessary to yield 3.2 g NaCl?

▮ A 5.0% NaCl solution contains 5.0 g NaCl in 100 g of solution. Then:

$$1\ \text{g NaCl is contained in}\ \frac{100}{5.0}\ \text{g solution}$$

and

$$3.2\ \text{g NaCl is contained in}\ (3.2)\left(\frac{100}{5.0}\ \text{g solution}\right) = 64\ \text{g solution}$$

Or, by proportion, letting x = required number of g of solution,

$$\frac{5.0\ \text{g NaCl}}{100\ \text{g solution}} = \frac{3.2\ \text{g NaCl}}{x} \qquad \text{therefore} \qquad x = 64\ \text{g solution}$$

Or, by the use of a quantitative factor, $(3.2\ \text{g NaCl})\left(\frac{100\ \text{g solution}}{5.0\ \text{g NaCl}}\right) = 64\ \text{g solution}$

11.87 Calculate the mass of anhydrous HCl in 5.00 mL concentrated hydrochloric acid (density 1.19 g/mL) containing 37.23% HCl by weight.

▮ The mass of 5.00 mL solution is $(5.00\ \text{mL})(1.19\ \text{g/mL}) = 5.95\ \text{g}$. The solution contains 37.23% HCl by weight. Hence the mass of HCl in 5.95 g solution is $(0.3723)(5.95\ \text{g}) = 2.22$ g anhydrous HCl

11.88 Calculate the volume of concentrated sulfuric acid (density 1.84 g/mL), containing 98.0% H_2SO_4 by weight, that would contain 40.0 g pure H_2SO_4.

▮ One mL of solution has a mass of 1.84 g and contains $(0.98)(1.84\ \text{g}) = 1.80$ g pure H_2SO_4. Then 40.0 g H_2SO_4 is contained in $(40.0/1.80)(1\ \text{mL solution}) = 22.2$ mL solution

11.89 To what extent must a solution of concentration 40 mg $AgNO_3$ per mL be diluted to yield one of concentration 16 mg $AgNO_3$ per mL?

▮ Let V be the volume to which 1 mL of the original solution must be diluted to yield a solution of concentration 16 mg $AgNO_3$ per mL. Because the amount of solute does not change with dilution,

$$\text{volume}_1 \times \text{concentration}_1 = \text{volume}_2 \times \text{concentration}_2 \qquad \text{or} \qquad (1\ \text{mL})(40\ \text{mg/mL}) = V(16\ \text{mg/mL})$$

Solving, $V = 2.5$ mL. Each mL of the original solution must be diluted to a volume of 2.5 mL.

Another Method

The amount of solute per mL of diluted solution will be 16/40 as much as in the original solution. Hence $(40/16)\ \text{mL} = 2.5\ \text{mL}$ of the diluted solution will contain as much solute as 1 mL of the original solution. Note that 2.5 is not the number of mL of water to be added but the final volume of the solution after water has been added to 1 mL of the original solution. The dilution formula always gives answers in terms of the *total* volume of solution. If we can assume that there is no volume shrinkage or expansion on dilution, the amount of water to be added in this problem is 1.5 mL per mL of original solution.

11.90 A procedure calls for 100 cm^3 of 20.0% H_2SO_4, density 1.14 g/cm^3. How much concentrated acid, of density 1.84 g/cm^3 and containing 98.0% H_2SO_4 by weight, must be diluted with water to prepare 100 cm^3 acid of the required strength?

▌ The concentrations must first be changed from a mass basis to a volumetric basis, so that the dilution equation will apply.

$$\text{mass of } H_2SO_4 \text{ per cm}^3 \text{ of 20.0\% acid} = (0.200)(1.14\ \text{g/cm}^3) = 0.228\ \text{g/cm}^3$$
$$\text{mass of } H_2SO_4 \text{ per cm}^3 \text{ of 98.0\% acid} = (0.980)(1.84\ \text{g/cm}^3) = 1.80\ \text{g/cm}^3$$

Let V be the volume of 98.0% acid required for 100 cm³ of 20.0% acid. Then

$$\text{volume}_1 \times \text{concentration}_1 = \text{volume}_2 \times \text{concentration}_2 \qquad \text{or} \qquad (100\ \text{cm}^3)(0.228\ \text{g/cm}^3) = (V)(1.80\ \text{g/cm}^3)$$

Solving, $V = 12.7\ \text{cm}^3$ of the concentrated acid.

11.91 How much NH_4Cl is required to prepare 100 mL of a solution containing 70 mg NH_4Cl per mL?

▌
$$(100\ \text{mL})\left(\frac{70\ \text{mg}}{\text{mL}}\right)\left(\frac{10^{-3}\ \text{g}}{\text{mg}}\right) = 7.0\ \text{g}$$

11.92 How many g of concentrated hydrochloric acid, containing 37.9% HCl by weight, will contain 5.00 g HCl?

▌
$$(5.00\ \text{g HCl})\left(\frac{100\ \text{g solution}}{37.9\ \text{g HCl}}\right) = 13.2\ \text{g solution}$$

11.93 To prepare 100.0 g of a 19.7% by weight solution of NaOH, how many g each of NaOH and H_2O are needed?

▌
$$(100.0\ \text{g solution})\left(\frac{19.7\ \text{g NaOH}}{100\ \text{g solution}}\right) = 19.7\ \text{g NaOH} \qquad (100.0\ \text{g}) - (19.7\ \text{g}) = 80.3\ \text{g water}$$

11.94 Calculate the volume occupied by 100 g of sodium hydroxide solution of density 1.20 g/mL.

▌
$$(100\ \text{g solution})\left(\frac{1\ \text{mL}}{1.20\ \text{g}}\right) = 83.3\ \text{mL}$$

11.95 What volume of dilute nitric acid, of density 1.11 g/mL and 19% HNO_3 by weight, contains 10 g HNO_3?

▌
$$(10\ \text{g } HNO_3)\left(\frac{100\ \text{g solution}}{19\ \text{g } HNO_3}\right)\left(\frac{1\ \text{mL solution}}{1.11\ \text{g solution}}\right) = 47\ \text{mL}$$

11.96 Ammonia gas is passed into water, yielding a solution of density of 0.93 g/cm³ and containing 18.6% NH_3 by weight. What is the mass of NH_3 per cm³ of solution?

▌
$$\left(\frac{0.93\ \text{g}}{\text{cm}^3\ \text{solution}}\right)\left(\frac{18.6\ \text{g } NH_3}{100\ \text{g solution}}\right) = 0.17\ \text{g/cm}^3$$

11.97 Hydrogen chloride gas is passed into water, yielding a solution of density 1.12 g/mL and containing 30.5% HCl by weight. What is the concentration of the HCl solution?

▌
$$\left(\frac{30.5\ \text{g HCl}}{100\ \text{g solution}}\right)\left(\frac{1.12\ \text{g solution}}{\text{mL solution}}\right) = 0.342\ \text{g HCl/mL solution}$$

11.98 Given 100 cm³ of pure water at 4 °C, what volume of a solution of hydrochloric acid, density 1.175 g/cm³ and containing 34.4% HCl by weight, could be prepared?

▌
$$34.4\%\ \text{HCl in water is} \quad \frac{34.4\ \text{g HCl}}{100\ \text{g total}} \quad \text{or} \quad \frac{34.4\ \text{g HCl}}{65.6\ \text{g water}} \quad \text{or} \quad \frac{65.6\ \text{water}}{100\ \text{g total}}$$

and
$$(100\ \text{cm}^3\ \text{water})\left(\frac{1.000\ \text{g water}}{\text{cm}^3\ \text{water}}\right)\left(\frac{100\ \text{g solution}}{65.6\ \text{g water}}\right)\left(\frac{1\ \text{cm}^3\ \text{solution}}{1.175\ \text{g solution}}\right) = 130\ \text{cm}^3$$

11.99 An excellent solution for cleaning grease stains from cloth or leather consists of the following: carbon tetrachloride 80% (by volume), ligroin 16%, amyl alcohol 4%. How many mL of each should be taken to make up 75 mL of solution? (Assume no volume change on mixing.)

▌ $(75\ \text{mL})(0.80) = 60\ \text{mL } CCl_4 \qquad (75\ \text{mL})(0.16) = 12\ \text{mL ligroin} \qquad (75\ \text{mL})(0.04) = 3\ \text{mL amyl alcohol}$

11.100 A liter of milk weighs 1.032 kg. The butterfat it contains to the extent of 4.0% by volume has a density of 865 kg/m³. What is the density of the fat-free "skimmed" milk?

▌ On the basis of 1.0 m³ of milk:

$$\text{Mass of butterfat} = (0.040\ \text{m}^3)(865\ \text{kg/m}^3) = 35\ \text{kg}$$

$$\text{Mass of skimmed milk} = \left(\frac{1.032\ \text{kg}}{\text{L}}\right)10^3\ \text{L} - (35\ \text{kg}) = 997\ \text{kg} \qquad \text{Volume} = 0.96\ \text{m}^3$$

$$\text{Density} = \frac{997\ \text{kg}}{0.96\ \text{m}^3} = 1040\ \text{kg/m}^3$$

11.101 To make a benzene-soluble cement, melt 49 g rosin in an iron pan and add 28 g each shellac and beeswax. How much of each component should be taken to make 75 kg cement?

▌

$$49\ \text{kg rosin} + 28\ \text{kg shellac} + 28\ \text{kg beeswax} = 105\ \text{kg total}$$

$$(75\ \text{kg total})\left(\frac{49\ \text{kg rosin}}{105\ \text{kg total}}\right) = 35\ \text{g rosin} \qquad (75\ \text{kg total})\left(\frac{28\ \text{kg shellac}}{105\ \text{kg total}}\right) = 20\ \text{kg shellac}$$

The same mass of beeswax as shellac is required.

11.102 A solution contains 75 mg NaCl per mL. To what extent must it be diluted to give a solution of concentration 15 mg NaCl per mL of solution?

▌ The concentration is reduced to one-fifth (without change in the number of g of solute), hence the volume must be increased fivefold. Dilute it to 5.0 times its former volume.

11.103 How many mL of a solution of concentration 100 mg Co^{2+} per ML is needed to prepare 1.5 L of solution of concentration 20 mg Co^{2+} per mL?

▌

$$V_2 = \frac{C_1V_1}{C_2} = \frac{(1500\ \text{mL})(20\ \text{mg/mL})}{100\ \text{mg/mL}} = 300\ \text{mL}$$

11.104 What volume of 95.0% alcohol by weight (density 0.809 g/cm³) must be used to prepare 150 cm³ of 30.0% alcohol by weight (density 0.957 g/cm³)?

▌

$$(150\ \text{cm}^3\ \text{solution})\left(\frac{0.957\ \text{g solution}}{\text{cm}^3\ \text{solution}}\right)\left(\frac{30.0\ \text{g alcohol}}{100\ \text{g solution}}\right) = 43.1\ \text{g alcohol}$$

$$(43.1\ \text{g alcohol})\left(\frac{100\ \text{g solution}}{95.0\ \text{g alcohol}}\right)\left(\frac{1\ \text{cm}^3\ \text{solution}}{0.809\ \text{g solution}}\right) = 56.1\ \text{cm}^3$$

11.105 How much $NaNO_3$ must be weighed out to make 50.0 mL of an aqueous solution containing 70.0 mg Na^+ per mL?

▌

$$\text{Mass of Na}^+\ \text{in 50.0 mL solution} = (50.0\ \text{mL})(70.0\ \text{mg/mL}) = 3500\ \text{mg} = 3.50\ \text{g Na}^+$$

Formula weight of $NaNO_3$ is 85.0; atomic weight of Na is 23.0. Then:

$$23.0\ \text{g Na}^+\ \text{is contained in 85.0 g NaNO}_3 \qquad 1\ \text{g Na}^+\ \text{is contained in}\ \frac{85.0}{23.0}\ \text{g NaNO}_3$$

and

$$3.50\ \text{g Na}^+\ \text{is contained in}\ (3.50)\left(\frac{85.0}{23.0}\ \text{g}\right) = 12.9\ \text{g NaNO}_3$$

Or, by direct use of quantitative factors,

$$x\ \text{g NaNO}_3 = (50.0\ \text{mL solution})\left(\frac{70.0\ \text{mg Na}^+}{1\ \text{mL solution}}\right)\left(\frac{85.0\ \text{g NaNO}_3}{23.0\ \text{g Na}^+}\right)\left(\frac{1\ \text{g}}{1000\ \text{mg}}\right) = 12.9\ \text{g NaNO}_3$$

11.106 Calculate the mass of $Al_2(SO_4)_3 \cdot 18H_2O$ needed to make up 50.0 mL of an aqueous solution of concentration 40.0 mg Al^{3+} per mL.

▮ Mass of Al^{3+} in 50.0 mL of solution = (50.0 mL)(40.0 mg/mL) = 2000 mg = 2.00 g Al^{3+}

Atomic weight of Al is 27 g/mol; formula weight of $Al_2(SO_4)_3 \cdot 18H_2O$ is 666 g/mol.

$$x \text{ g } Al_2(SO_4)_3 \cdot 18H_2O = (2.00 \text{ g } Al^{3+})\left(\frac{666 \text{ g } Al_2(SO_4)_3 \cdot 18H_2O}{54.0 \text{ g } Al^{3+}}\right) = 24.7 \text{ g } Al_2(SO_4)_3 \cdot 18H_2O$$

A solution of identical composition could be prepared from the appropriate amount of anhydrous $Al_2(SO_4)_3$. To compute the appropriate amount in this case, the formula weight of $Al_2(SO_4)_3$, 342, would be used instead of 666. The experimenter would find that more water is needed to make up the 50.0 mL of solution from the anhydrous salt than from the hydrate. In general, hydrated salts differ from the anhydrous salts only in the crystalline state. In solution, the water of hydration and the water of the solvent become indistinguishable from each other.

11.107 Describe how you would prepare 50 g of a 12.0% $BaCl_2$ solution, starting with $BaCl_2 \cdot 2H_2O$ and pure water.

▮ A 12.0% $BaCl_2$ solution contains 12.0 g $BaCl_2$ per 100 g of solution, or 6.0 g $BaCl_2$ in 50 g of solution. Formula weight of $BaCl_2$ is 208 g/mol; of $BaCl_2 \cdot 2H_2O$, 244 g/mol. Therefore,

$$208 \text{ g } BaCl_2 \text{ is contained in } 244 \text{ g } BaCl_2 \cdot 2H_2O \qquad 1 \text{ g } BaCl_2 \text{ is contained in } \frac{244}{208} \text{ g } BaCl_2 \cdot 2H_2O$$

and

$$6.0 \text{ g } BaCl_2 \text{ is contained in } (6.0)\left(\frac{244}{208} \text{ g}\right) = 7.0 \text{ g } BaCl_2 \cdot 2H_2O$$

The desired solution is prepared by dissolving 7.0 g of $BaCl_2 \cdot 2H_2O$ in 43 g (43 mL) of water.

11.108 How much $BaCl_2$ would be needed to make 250 mL of a solution having the same concentration of Cl^- as one containing 3.78 g NaCl per 100 mL?

▮

$$\left(\frac{3.78 \text{ g NaCl}}{100 \text{ mL}}\right)\left(\frac{1 \text{ mol } Cl^-}{58.5 \text{ g NaCl}}\right)\left(\frac{1 \text{ mol } BaCl_2}{2 \text{ mol } Cl^-}\right)\left(\frac{208 \text{ g } BaCl_2}{\text{mol } BaCl_2}\right) = 0.0672 \text{ g } BaCl_2/\text{mL}$$

$$(250 \text{ mL})\left(\frac{0.0672 \text{ g}}{\text{mL}}\right) = 16.8 \text{ g } BaCl_2$$

11.109 Exactly 4.00 g of a solution of sulfuric acid was diluted with water and excess $BaCl_2$ was added. The washed and dried precipitate of $BaSO_4$ weighed 4.08 g. Find the percent H_2SO_4 in the original acid solution.

▮ First determine the mass of H_2SO_4 required to precipitate 4.08 g $BaSO_4$ by the reaction $H_2SO_4 + BaCl_2 \rightarrow 2HCl + BaSO_4$. The equation shows that 1 mol $BaSO_4$ (233.4 g) requires 1 mol H_2SO_4 (98.08 g). Therefore, 4.08 g $BaSO_4$ requires

$$\left(\frac{4.08 \text{ g } BaSO_4}{233.4 \text{ g } BaSO_4}\right)(98.08 \text{ g } H_2SO_4) = 1.72 \text{ g } H_2SO_4$$

and

$$\text{fraction } H_2SO_4 \text{ by weight} = \frac{\text{mass of } H_2SO_4}{\text{mass of solution}} = \frac{1.72 \text{ g}}{4.00 \text{ g}} = 0.430 = 43.0\%$$

11.110 How much $CrCl_3 \cdot 6H_2O$ is needed to prepare 1.00 L of solution containing 20.0 g Cr^{3+} per L?

▮

$$(1.00 \text{ L})\left(\frac{20.0 \text{ g } Cr^{3+}}{\text{L}}\right)\left(\frac{1 \text{ mol } Cr^{3+}}{52.0 \text{ g } Cr^{3+}}\right)\left(\frac{1 \text{ mol compound}}{\text{mol } Cr^{3+}}\right)\left(\frac{266.4 \text{ g compound}}{\text{mol compound}}\right) = 102 \text{ g compound}$$

11.111 How many g of Na_2CO_3 are needed to prepare 500 cm^3 of a solution containing 10.0 mg CO_3^{2-} per cm^3?

▮

$$(500 \text{ cm}^3)\left(\frac{10.0 \text{ mg } CO_3^{2-}}{\text{cm}^3}\right)\left(\frac{1 \text{ g}}{10^3 \text{ mg}}\right)\left(\frac{1 \text{ mol } CO_3^{2-}}{60.0 \text{ g } CO_3^{2-}}\right)\left(\frac{1 \text{ mol } Na_2CO_3}{\text{mol } CO_3^{2-}}\right)\left(\frac{106 \text{ g } Na_2CO_3}{\text{mol } Na_2CO_3}\right) = 8.83 \text{ g } Na_2CO_3$$

11.112 How many mL of a solution containing 40.0 g $CaCl_2$ per L are needed to react with 0.642 g pure Na_2CO_3? $CaCO_3$ is formed in the reaction.

▮

$$Na_2CO_3 + CaCl_2 \longrightarrow CaCO_3 + 2NaCl$$

$$(0.642 \text{ g } Na_2CO_3)\left(\frac{1 \text{ mol } Na_2CO_3}{106 \text{ g } Na_2CO_3}\right)\left(\frac{1 \text{ mol } CaCl_2}{\text{mol } Na_2CO_3}\right)\left(\frac{111 \text{ g } CaCl_2}{\text{mol } CaCl_2}\right)\left(\frac{1 \text{ L}}{40.0 \text{ g}}\right)\left(\frac{10^3 \text{ mL}}{\text{L}}\right) = 16.8 \text{ mL}$$

11.113 How much $CaCl_2 \cdot 6H_2O$ and water must be weighed out to make 100 g of a solution that is 5.0% $CaCl_2$?

▌

$$(100 \text{ g solution})\left(\frac{5.0 \text{ g CaCl}_2}{100 \text{ g solution}}\right)\left(\frac{1 \text{ mol CaCl}_2}{111 \text{ g CaCl}_2}\right)\left(\frac{1 \text{ mol hydrate}}{\text{mol CaCl}_2}\right)\left(\frac{219 \text{ g hydrate}}{\text{mol hydrate}}\right) = 9.9 \text{ g CaCl}_2 \cdot 6\text{H}_2\text{O}$$

$$(100 \text{ g}) - (9.9 \text{ g}) = 90 \text{ g water}$$

11.4 MOLARITY

11.114 What volume of 1.71 M NaCl solution contains 0.20 mol NaCl?

▌

$$\text{volume} = \frac{0.20 \text{ mol NaCl}}{1.71 \text{ mol/L}} = 0.117 \text{ L} = 117 \text{ mL}$$

11.115 What volume of 3.0 M NaOH (formula weight = 40 g/mol) can be prepared with 84.0 g NaOH?

▌

$$(84.0 \text{ g NaOH})\left(\frac{1 \text{ mol NaOH}}{40.0 \text{ g NaOH}}\right)\left(\frac{1.0 \text{ L}}{3.0 \text{ mol NaOH}}\right) = 0.70 \text{ L}$$

11.116 What is the molarity of NaOH in a solution which contains 24.0 g NaOH dissolved in 300 mL of solution?

▌

$$(24.0 \text{ g NaOH})\left(\frac{1 \text{ mol NaOH}}{40.0 \text{ g NaOH}}\right) = 0.600 \text{ mol NaOH} \qquad M = \frac{0.600 \text{ mol}}{0.300 \text{ L}} = 2.00 \text{ M}$$

11.117 Calculate the volume of 2.50 M sugar solution which contains 0.400 mol sugar.

▌

$$(0.400 \text{ mol})\left(\frac{1 \text{ L}}{2.50 \text{ mol}}\right) = 0.160 \text{ L}$$

11.118 How many mL of water must be added to 200 mL of 0.65 M HCl to dilute the solution to 0.20 M?

▌

$$(200 \text{ mL HCl})\left(\frac{0.650 \text{ mmol HCl}}{\text{mL}}\right) = 130 \text{ mmol HCl} \qquad V_{\text{final}} = \frac{130 \text{ mmol}}{0.200 \text{ mmol/mL}} = 650 \text{ mL}$$

Approximately (650 mL) − (200 mL) = 450 mL must be added.

11.119 How much 1.00 M HCl should be mixed with what volume of 0.250 M HCl in order to prepare 1.00 L of 0.500 M HCl?

▌ Let x = L of 0.25 M HCl; then $1.00 - x$ = L of 1.00 M HCl, and

$$x(0.25) + (1.0 - x)(1.00) = (1.00)(0.500)$$

Thus $x = 0.667 \text{ L} = 667$ mL of 0.25 M HCl and $1.00 - x = 0.333 \text{ L} = 333$ mL of 1.00 M HCl.

11.120 What is the molar concentration of a solution containing 16.0 g CH_3OH in 200 cm^3 solution?

▌ The molecular weight of CH_3OH is 32.0 g/mol.

$$M = \text{molar concentration} = \frac{n(\text{solute})}{\text{volume of solution in L}} = \frac{\dfrac{16.0 \text{ g}}{32.0 \text{ g/mol}}}{0.200 \text{ L}} = 2.50 \text{ mol/L} = 2.50 \text{ M}$$

11.121 What volume of 0.30 M Na_2SO_4 solution is required to prepare 2.0 L of a solution 0.40 M in Na^+?

▌

$$(2.0 \text{ L})\left(\frac{0.40 \text{ mol Na}^+}{\text{L}}\right)\left(\frac{1 \text{ mol Na}_2\text{SO}_4}{2 \text{ mol Na}^+}\right)\left(\frac{1 \text{ L}}{0.30 \text{ mol Na}_2\text{SO}_4}\right) = 1.3 \text{ L}$$

11.122 To what extent must a 0.500 M $BaCl_2$ solution be diluted to yield one of concentration 20.0 mg Ba^{2+} per mL?

▌ The original solution contains 0.500 mol $BaCl_2$ or Ba^{2+} per L. The mass of Ba^{2+} in 0.500 mol is

$$(0.500 \text{ mol})(137.3 \text{ g/mol}) = 68.6 \text{ g Ba}^{2+}$$

Thus 0.500 M $BaCl_2$ contains 68.6 g Ba^{2+} per L, or 68.6 mg Ba^{2+} per mL.

The problem now is to find the extent to which a 68.6 mg Ba^{2+} per mL solution must be diluted to yield one of concentration 20.0 mg Ba^{2+} per mL.

$$\text{volume}_1 \times \text{concentration}_1 = \text{volume}_2 \times \text{concentration}_2 \qquad \text{or} \qquad (1\ \text{mL})(68.6\ \text{mg/mL}) = (V)(20.0\ \text{mg/mL})$$

Solving, $V = 3.43$ mL. Each mL of 0.500 M $BaCl_2$ must be diluted with water to a volume of 3.43 mL.

11.123 Calculate the final concentration of HNO_3 if 0.20 mol HNO_3 is added to a beaker containing 2.0 L of 1.1 M HNO_3 and enough pure water is added to give a final volume of 3.0 L.

$$n = (0.20\ \text{mol}) + (2.0\ \text{L})\left(\frac{1.1\ \text{mol}}{\text{L}}\right) = 2.4\ \text{mol} \qquad M = (2.4\ \text{mol})/(3.0\ \text{L}) = 0.80\ \text{M}$$

11.124 How many mL of 2.00 M $Pb(NO_3)_2$ contain 600 mg Pb^{2+}?

One liter of 1.00 M $Pb(NO_3)_2$ contains 1.00 mol Pb^{2+}. Then 1.00 L of 2.00 M contains 2.00 mol Pb^{2+}, or 414 g Pb^{2+} per L, or 414 mg Pb^{2+} per mL. Hence 600 mg Pb^{2+} is contained in

$$\frac{600\ \text{mg}}{414\ \text{mg/mL}} = 1.45\ \text{mL of 2.00 M } Pb(NO_3)_2.$$

11.125 How many g solute is required to prepare 1.000 L of 1.000 M $Pb(NO_3)_2$? What is the molar concentration of the solution with respect to each of the ions?

A 1.000 M solution contains 1.000 mol solute in 1.000 L solution. The formula weight of $Pb(NO_3)_2$ is 331.2; hence 331.2 g $Pb(NO_3)_2$ is needed. A 1 M solution of $Pb(NO_3)_2$ is 1 M with respect to Pb^{2+} and 2 M with respect to NO_3^-.

11.126 Determine the molar concentration of each of the following solutions: ***(a)*** 18.0 g $AgNO_3$ per L solution, ***(b)*** 12.00 g $AlCl_3 \cdot 6H_2O$ per L solution. Formula weight of $AgNO_3$ is 169.9 g/mol; of $AlCl_3 \cdot 6H_2O$, 241.4 g/mol.

(a) $\dfrac{18.0\ \text{g/L}}{169.9\ \text{g/mol}} = 0.106\ \text{mol/L} = 0.106\ \text{M}$ ***(b)*** $\dfrac{12.00\ \text{g/L}}{241.4\ \text{g/mol}} = 0.0497\ \text{mol/L} = 0.0497\ \text{M}$

11.127 What is the molar concentration of a solution containing 37.5 g $Ba(MnO_4)_2$ per L, and what is the molar concentration with respect to each type of ion?

$$\left(\frac{37.5\ \text{g } Ba(MnO_4)_2}{\text{L}}\right)\left(\frac{1\ \text{mol } Ba(MnO_4)_2}{375\ \text{g } Ba(MnO_4)_2}\right) = 0.100\ \text{mol/L} = 0.100\ \text{M}$$

$$\left(\frac{0.100\ \text{mol } Ba(MnO_4)_2}{\text{L}}\right)\left(\frac{1\ \text{mol } Ba^{2+}}{\text{mol } Ba(MnO_4)_2}\right) = 0.100\ \text{mol/L} = 0.100\ \text{M } Ba^{2+}$$

$$\left(\frac{0.100\ \text{mol } Ba(MnO_4)_2}{\text{L}}\right)\left(\frac{2\ \text{mol } MnO_4^-}{\text{mol } Ba(MnO_4)_2}\right) = 0.200\ \text{mol } MnO_4^-/\text{L} = 0.200\ \text{M } MnO_4^-$$

11.128 How many g solute are required to prepare 1.00 L of 1.00 M $CaCl_2 \cdot 6H_2O$?

$$(1.00\ \text{L})\left(\frac{1.00\ \text{mol}}{\text{L}}\right)\left(\frac{219\ \text{g } CaCl_2 \cdot 6H_2O}{\text{mol}}\right) = 219\ \text{g}$$

11.129 Exactly 100 g NaCl is dissolved in sufficient water to give 1500 cm^3 solution. What is the molar concentration?

$$\left(\frac{100\ \text{g NaCl}}{1.500\ \text{L}}\right)\left(\frac{1\ \text{mol NaCl}}{58.5\ \text{g NaCl}}\right) = 1.14\ \text{M}$$

11.130 Determine the molar concentration of each of the following solutions: ***(a)*** 166 g KI per L solution, ***(b)*** 33.0 g $(NH_4)_2SO_4$ in 200 mL solution, ***(c)*** 12.5 g $CuSO_4 \cdot 5H_2O$ in 100 mL solution, ***(d)*** 10.0 mg Al^{3+} per mL solution.

(a) $$\left(\frac{166\ \text{g KI}}{\text{L}}\right)\left(\frac{1\ \text{mol KI}}{166\ \text{g KI}}\right) = \frac{1.00\ \text{mol}}{\text{L}} = 1.00\ \text{M}$$

(b) $$\left(\frac{33.0\ \text{g } (NH_4)_2SO_4}{0.200\ \text{L}}\right)\left(\frac{1\ \text{mol } (NH_4)_2SO_4}{132\ \text{g } (NH_4)_2SO_4}\right) = 1.25\ \text{M}$$

(c) $$\left(\frac{12.5\ \text{g } CuSO_4 \cdot 5H_2O}{0.100\ \text{L}}\right)\left(\frac{1\ \text{mol } CuSO_4 \cdot 5H_2O}{250\ \text{g } CuSO_4 \cdot 5H_2O}\right) = 0.500\ \text{M}$$

(d) $$\left(\frac{10.0\ \text{mg } Al^{3+}}{\text{mL}}\right)\left(\frac{1\ \text{mmol } Al^{3+}}{27.0\ \text{mg } Al^{3+}}\right) = 0.370\ \text{mmol/mL} = 0.370\ \text{M}$$

11.131 What volume of 0.200 M $Ni(NO_3)_2 \cdot 6H_2O$ contains 500 mg Ni^{2+}?

▮

$$(500 \text{ mg } Ni^{2+})\left(\frac{1 \text{ mmol}}{58.7 \text{ mg}}\right)\left(\frac{1 \text{ mL}}{0.200 \text{ mmol}}\right) = 42.6 \text{ mL}$$

11.132 Which two of the following solutions contain approximately equal hydrogen ion concentrations?

(1) 50 mL 0.10M HCl + 25 mL H_2O (3) 50 mL 0.10M H_2SO_4 + 25 mL H_2O
(2) 50 mL 0.10M HCl + 50 mL H_2O (4) 25 mL 0.10M H_2SO_4 + 50 mL H_2O

▮ (1) and (4). (1) contains 5.0 mmol H^+ in 75 mL; (2) contains 5.0 mmol H^+ in 100 mL; (3) contains 10 mmol H^+ in 75 mL; (4) contains 5.0 mmol in 75 mL (assuming complete ionization of the H_2SO_4).

11.133 Calculate the final concentration of solute when 2.0 L of 3.0 M sugar solution and 3.0 L of 2.5 M sugar solution are mixed and then diluted to 10.0 L with water.

▮

Solution 1: (2.0 L)(3.0 M) = 6.0 mol sugar
Solution 2: (3.0 L)(2.5 M) = 7.5 mol sugar
Combined: 13.5 mol sugar

Final concentration: $\frac{13.5 \text{ mol}}{10.0 \text{ L}} = 1.35 \text{ M}$

11.134 If 40.00 mL of 1.600 M HCl and 60.00 mL of 2.000 M NaOH are mixed, what are the molar concentrations of Na^+, Cl^-, and OH^- in the resulting solution? Assume a total volume of 100.00 mL.

▮

(40.00 mL)(1.600 M HCl) yields 64.00 mmol H^+ and 64.00 mmol Cl^-
(60.00 mL)(2.000 M NaOH) yields 120.0 mmol OH^- and 120.0 mmol Na^+

$$H^+ + OH^- \longrightarrow H_2O$$

The 64.00 mmol of H^+ reacts with 64.00 mmol of OH^-, leaving 56.0 mmol of OH^- in the solution.

$$\frac{56.0 \text{ mmol } OH^-}{100.0 \text{ mL}} = 0.560 \text{ M } OH^- \qquad \frac{64.00 \text{ mmol } Cl^-}{100.0 \text{ mL}} = 0.640 \text{ M } Cl^- \qquad \frac{120.0 \text{ mmol } Na^+}{100 \text{ mL}} = 1.200 \text{ M } Na^+$$

11.135 What volume of 0.300 M H_2SO_4 is required to exactly neutralize 200 mL of 0.500 M NaOH?

▮

$$2NaOH + H_2SO_4 \longrightarrow Na_2SO_4 + 2H_2O$$

$$(200 \text{ mL NaOH})\left(\frac{0.500 \text{ mmol NaOH}}{\text{mL}}\right)\left(\frac{1 \text{ mmol } H_2SO_4}{2 \text{ mmol NaOH}}\right)\left(\frac{1 \text{ mL } H_2SO_4}{0.300 \text{ mmol } H_2SO_4}\right) = 167 \text{ mL } H_2SO_4$$

11.136 Calculate the molarity of the original H_3PO_4 solution if 20.0 mL of H_3PO_4 solution is required to completely neutralize 40.0 mL of 0.0500 M $Ba(OH)_2$ solution.

▮

$$[40.0 \text{ mL } Ba(OH)_2]\left(\frac{0.0500 \text{ mmol } Ba(OH)_2}{\text{mL}}\right)\left(\frac{2 \text{ mmol } H_3PO_4}{3 \text{ mmol } Ba(OH)_2}\right) = 1.33 \text{ mmol } H_3PO_4$$

$$\frac{1.33 \text{ mmol } H_3PO_4}{20.0 \text{ mL } H_3PO_4} = 0.0667 \text{ M } H_3PO_4$$

11.137 What concentration of NaCl finally results from the mixing of 2.00 L of 4.00 M NaCl with 3.00 L of 1.50 M NaCl plus sufficient water to dilute the solution to 10.0 L?

▮

(2.00 L)(4.00 M) = 8.00 mol
(3.00 L)(1.50 M) = 4.50 mol
12.50 mol

$$\frac{12.50 \text{ mol}}{10.0 \text{ L}} = 1.25 \text{ M}$$

11.138 What volume of 0.50 M $BaCl_2$ will contain 3.0 mol of chloride ion?

▮

$$(3.0 \text{ mol } Cl^-)\left(\frac{1 \text{ mol } BaCl_2}{2 \text{ mol } Cl^-}\right)\left(\frac{1 \text{ L } BaCl_2}{0.50 \text{ mol } BaCl_2}\right) = 3.0 \text{ L}$$

11.139 How would one prepare exactly 300 mL of 5.00 M HCl solution by diluting sufficient 12.00 M stock solution?

▮ Required: $(300\text{ mL})(5.00\text{ M}) = 1500\text{ mmol}$

Stock solution used: $(1500\text{ mmol})\left(\frac{1\text{ mL}}{12.00\text{ mmol}}\right) = 125\text{ mL}$

Dilute 125 mL of 12.00 M stock solution with sufficient water to make 300 mL of solution.

11.140 How many mol of chloride ion, Cl^-, is present in 50.0 mL of 0.200 M $BaCl_2$?

▮ $(0.0500\text{ L})\left(\frac{0.200\text{ mol }BaCl_2}{\text{L}}\right) = 0.0100\text{ mol }BaCl_2 \qquad (0.0100\text{ mol }BaCl_2)\left(\frac{2\text{ mol }Cl^-}{\text{mol }BaCl_2}\right) = 0.0200\text{ mol }Cl^-$

The 0.200 M $BaCl_2$ solution is a solution of 0.200 M Ba^{2+} and 0.400 M Cl^-.

11.141 An aqueous solution of HCl contains 28% HCl by weight and has a density of 1.20 g/mL. Find the molarity of the solution.

▮ One liter of solution has a mass of 1200 g.

$$\left(\frac{1200\text{ g solution}}{\text{L}}\right)\left(\frac{28\text{ g HCl}}{100\text{ g solution}}\right)\left(\frac{1\text{ mol HCl}}{36.5\text{ g HCl}}\right) = \frac{9.2\text{ mol HCl}}{\text{L}} = 9.2\text{ M HCl}$$

11.142 Calculate the molarity of a solution prepared by adding 100.0 g NaCl to sufficient water to make 1.00 L solution.

▮ $$\frac{100.0\text{ g NaCl}}{58.5\text{ g/mol}} = 1.71\text{ mol NaCl} \qquad \frac{1.71\text{ mol NaCl}}{1.00\text{ L solution}} = 1.71\text{ M}$$

11.143 How many mL of a 0.100 M solution must be added to water to make 2.00 L of 0.0250 M solution?

▮ $$(2.00\text{ L})\left(\frac{0.0250\text{ mol}}{\text{L}}\right) = 0.0500\text{ mol required} \qquad \frac{0.0500\text{ mol required}}{0.100\text{ mol/L}} = 0.500\text{ L} = 500\text{ mL}$$

Hence 500 mL of 0.100 M solution must be diluted to 2.00 L.

11.144 If 20.0 mL of 1.00 M $CaCl_2$ and 60.0 mL of 0.200 M $CaCl_2$ are mixed, what will be the molarity of the final solution?

▮ The total number of mmol of $CaCl_2$ is the sum of the number of mmol in the two solutions:

$$(20.0\text{ mL})\left(\frac{1.00\text{ mmol}}{\text{mL}}\right) + (60.0\text{ mL})\left(\frac{0.200\text{ mmol}}{\text{mL}}\right) = 20.0\text{ mmol} + 12.0\text{ mmol} = 32.0\text{ mmol}$$

The total volume of the final solution is 80.0 mL. Its molarity is therefore $\frac{32.0\text{ mmol}}{80.0\text{ mL}} = 0.400\text{ M}$

11.145 How much $(NH_4)_2SO_4$ is required to prepare 400 mL of M/4 solution? (The notation M/4 is sometimes used in place of $\frac{1}{4}$M.)

▮ Formula weight of $(NH_4)_2SO_4$ is 132.1 g/mol. One liter of M/4 solution contains $\frac{1}{4}(132.1\text{ g}) = 33.02\text{ g}$ $(NH_4)_2SO_4$. Then 400 mL of M/4 solution requires $(0.400\text{ L})(33.02\text{ g/L}) = 13.21\text{ g }(NH_4)_2SO_4$.

Another Method

$$\text{Mass} = (\text{molar concentration}) \times (\text{formula weight}) \times (\text{volume of solution})$$
$$= (\tfrac{1}{4}\text{ mol/L})(132.1\text{ g/mol})(0.400\text{ L}) = 13.21\text{ g }(NH_4)_2SO_4$$

11.146 Calculate the mass of chloride ion in 1.00 L of each of the following solutions: ***(a)*** 10.0% NaCl (density 1.07 g/mL), ***(b)*** 10.0% KCl (density 1.06 g/mL), ***(c)*** 1.00 M NaCl, ***(d)*** 1.00 M KCl. ***(e)*** Calculate the molarities of the first two solutions.

▮

(a) $(1000\text{ mL})\left(\frac{1.07\text{ g}}{\text{mL}}\right)\left(\frac{10.0\text{ g NaCl}}{100\text{ g solution}}\right)\left(\frac{35.5\text{ g }Cl^-}{58.5\text{ g NaCl}}\right) = 64.9\text{ g }Cl^-$

(b) $(1000\text{ mL})\left(\frac{1.06\text{ g}}{\text{mL}}\right)\left(\frac{10.0\text{ g KCl}}{100\text{ g solution}}\right)\left(\frac{35.5\text{ g }Cl^-}{74.6\text{ g KCl}}\right) = 50.4\text{ g }Cl^-$ ***(c)*** and ***(d)*** 35.5 g Cl^-

(e) $\frac{64.9\text{ g }Cl^-/\text{L}}{35.5\text{ g/mol}} = 1.83\text{ M }Cl^- \qquad \frac{50.4\text{ g }Cl^-/\text{L}}{35.5\text{ g/mol}} = 1.42\text{ M }Cl^-$

11.147 What is the concentration of each ion in solution when 10.0 g $CaCl_2$ and 20.0 g NaCl are diluted with water to 500 mL?

▌ $$(10.0\text{ g CaCl}_2)\left(\frac{1\text{ mol CaCl}_2}{111\text{ g CaCl}_2}\right) = 0.0901\text{ mol CaCl}_2 \qquad (20.0\text{ g NaCl})\left(\frac{1\text{ mol NaCl}}{58.5\text{ g NaCl}}\right) = 0.342\text{ mol NaCl}$$

0.0901 mol $CaCl_2$ is 0.0901 mol Ca^{2+} + 0.180 mol Cl^-			
0.342 mol NaCl is		0.342 mol Cl^-	+ 0.342 mol Na^+
Moles of solute:	0.0901 mol Ca^{2+}	0.522 mol Cl^-	0.342 mol Na^+
Concentrations:	$\frac{0.0901\text{ mol}}{0.500\text{ L}}$	$\frac{0.522\text{ mol}}{0.500\text{ L}}$	$\frac{0.342\text{ mol}}{0.500\text{ L}}$
	0.180 M Ca^{2+}	1.04 M Cl^-	0.684 M Na^+

11.148 Calculate the final concentration of solute after 250 mL of 5.0 M antifreeze solution and 500 mL of 4.0 M antifreeze solution are combined and diluted to 2.0 L with water.

▌ $$(0.250\text{ L})(5.0\text{ M}) = 1.2\text{ mol}$$
$$(0.500\text{ L})(4.0\text{ M}) = 2.0\text{ mol}$$
$$3.2\text{ mol} \qquad \frac{3.2\text{ mol}}{2.0\text{ L}} = 1.6\text{ M}$$

11.149 Calculate the concentration of all ions in solution when 3.0 L of 4.0 M NaCl and 4.0 L of 2.0 M $CoCl_2$ are combined and diluted to 10.0 L.

▌ $$(3.0\text{ L})(4.0\text{ M}) = 12\text{ mol NaCl} \qquad (4.0\text{ L})(2.0\text{ M}) = 8.0\text{ mol CoCl}_2$$

12 mol NaCl is	12 mol Na^+ + 12 mol Cl^-		
8.0 mol $CoCl_2$ is		16 mol Cl^-	+ 8.0 mol Co^{2+}
Moles solute:	12 mol Na^+	28 mol Cl^-	8.0 mol Co^{2+}
Concentrations:	$\frac{12\text{ mol}}{10.0\text{ L}}$	$\frac{28\text{ mol}}{10.0\text{ L}}$	$\frac{8.0\text{ mol}}{10.0\text{ L}}$
	1.2 M Na^+	2.8 M Cl^-	0.80 M Co^{2+}

11.150 Determine the molar concentration of each ionic species in solution after each of the following operations: ***(a)*** 200 mL of 2.0 M NaCl is diluted to 500 mL ***(b)*** 200 mL of 2.0 M $BaCl_2$ is diluted to 500 mL ***(c)*** 200 mL of 3.00 M NaCl is added to 300 mL of 4.0 M NaCl ***(d)*** 200 mL of 2.0 M $BaCl_2$ is added to 400 mL of 3.0 M $BaCl_2$ and 400 mL of water ***(e)*** 300 mL of 3.0 M NaCl is added to 200 mL of 4.0 M $BaCl_2$ ***(f)*** 400 mL of 2.00 M HCl is added to 150 mL of 4.00 M NaOH ***(g)*** 100 mL of 2.0 M HCl and 200 mL of 1.5 M NaOH are added to 150 mL of 4.0 M NaCl and 50 mL of water.

▌ ***(a)*** $$(200\text{ mL})\left(\frac{2.0\text{ mmol}}{\text{mL}}\right) = 400\text{ mmol}$$

Each mmol of NaCl consists of 1 mmol of Na^+ and 1 mmol of Cl^-; there is 400 mmol of each ion present.

$$\frac{400\text{ mmol Na}^+}{500\text{ mL}} = 0.800\text{ M Na}^+ \qquad \frac{400\text{ mmol Cl}^-}{500\text{ mL}} = 0.800\text{ M Cl}^-$$

(b) The 400 mmol of $BaCl_2$ (see part *(a)*) contains 400 mmol of Ba^{2+} ions and 800 mmol of Cl^- ions, since there are 2 mmol of Cl^- per mmol of $BaCl_2$.

$$\frac{400\text{ mmol Ba}^{2+}}{500\text{ mL}} = 0.800\text{ M Ba}^{2+} \qquad \frac{800\text{ mmol Cl}^-}{500\text{ mL}} = 1.60\text{ M Cl}^-$$

(c) Assuming that the total volume will be 500 mL (a very good approximation):

$$(200\text{ mL})\left(\frac{3.00\text{ mmol NaCl}}{\text{mL}}\right) = 600\text{ mmol NaCl from 1st solution}$$

$$(300\text{ mL})\left(\frac{4.0\text{ mmol NaCl}}{\text{mL}}\right) = 1200\text{ mmol NaCl from 2nd solution}$$

There is 1800 mmol NaCl, consisting of 1800 mmol Na^+ ions and 1800 mmol Cl^- ions. The concentrations are

$$\frac{1800 \text{ mmol } Na^+}{500 \text{ mL}} = 3.6 \text{ M } Na^+ \text{ and the same concentration of } Cl^-.$$

(*d*) The numbers of moles of each type of ion may be calculated as above. The water contains no solute and hence affects only the total volume of the final solution. The Ba^{2+} concentration is calculated in one step:

$$\frac{(200 \text{ mL})(2.0 \text{ M } Ba^{2+}) + (400 \text{ mL})(3.0 \text{ M } Ba^{2+})}{1000 \text{ mL}} = 1.6 \text{ M } Ba^{2+}$$

The Cl^- concentration is twice as great, since there are two Cl^- ions per Ba^{2+} ion.

$$(1.6 \text{ M } Ba^{2+})\left(\frac{2 \text{ mol } Cl^-}{\text{mol } Ba^{2+}}\right) = 3.2 \text{ M } Cl^-$$

(*e*) $(300 \text{ mL})(3.0 \text{ M NaCl}) = 900 \text{ mmol NaCl}$ $\qquad (200 \text{ mL})(4.0 \text{ M } BaCl_2) = 800 \text{ mmol } BaCl_2$

There are present 900 mmol Na^+, 800 mmol Ba^{2+}, and 2500 mmol Cl^-, of which 900 mmol came from the NaCl and 1600 mmol came from the $BaCl_2$.

$$\frac{900 \text{ mmol } Na^+}{500 \text{ mL}} = 1.8 \text{ M } Na^+ \qquad \frac{2500 \text{ mmol } Cl^-}{500 \text{ mL}} = 5.0 \text{ M } Cl^- \qquad \frac{800 \text{ mmol } Ba^{2+}}{500 \text{ mL}} = 1.6 \text{ M } Ba^{2+}$$

(*f*) When these reagents are mixed, a chemical reaction occurs.

400 mL of 2.00 M HCl yields 800 mmol H^+ and 800 mmol Cl^-

150 mL of 4.00 M NaOH yields 600 mmol OH^- and 600 mmol Na^+

$$H^+ + OH^- \longrightarrow H_2O$$

The 600 mmol OH^- neutralizes 600 mmol H^+, leaving 200 mmol H^+ in solution.

$$\frac{800 \text{ mmol } Cl^-}{550 \text{ mL}} = 1.45 \text{ M } Cl^- \qquad \frac{600 \text{ mmol } Na^+}{550 \text{ mL}} = 1.09 \text{ M } Na^+$$

$$\frac{200 \text{ mmol } H^+ \text{ remaining}}{550 \text{ mL}} = 0.364 \text{ M } H^+$$

(*g*) Again a chemical reaction is expected.

100 mL of 2.0 M HCl yields 200 mmol H^+ and 200 mmol Cl^-

200 mL of 1.5 M NaOH yields 300 mmol OH^- and 300 mmol Na^+

150 mL of 4.0 M NaCl yields 600 mmol Cl^- and 600 mmol Na^+

$$H^+ + OH^- \longrightarrow H_2O$$

$$\frac{100 \text{ mmol } OH^- \text{ remaining}}{500 \text{ mL}} = 0.20 \text{ M } OH^- \qquad \frac{800 \text{ mmol } Cl^- \text{ total}}{500 \text{ mL}} = 1.6 \text{ M } Cl^-$$

$$\frac{900 \text{ mmol } Na^+ \text{ total}}{500 \text{ mL}} = 1.8 \text{ M } Na^+$$

11.151 How would one prepare exactly 3.00 L of 1.00 M NaOH by mixing portions of stock solutions of 2.50 M NaOH and 0.400 M NaOH?

▮ Needed: $\qquad (3.00 \text{ L})\left(\frac{1.00 \text{ mol}}{\text{L}}\right) = 3.00 \text{ mol}$

Let x = number of L of 2.50 M NaOH; then $3.00 - x$ = number of L of 0.400 M NaOH. The number of moles of solute from the more concentrated solution is $2.50x$; that from the less concentrated solution is $(0.400)(3.00 - x)$. The total number of moles is 3.00.

$$(2.50x) + (0.400)(3.00 - x) = 3.00$$

$$x = 0.857 \text{ L of } 2.50 \text{ M NaOH} \qquad 3.00 - x = 2.14 \text{ L of } 0.400 \text{ M NaOH}$$

11.152 Calculate the concentration of each type of ion which remains in solution when each of the following sets of solutions is mixed: (*a*) 100 mL of 0.50 M NaCl + 50 mL of 0.25 M KCl (*b*) 100 mL of 0.50 M NaCl + 50 mL of 0.25 M $AgNO_3$ (*c*) 100 mL of 0.50 M NaCl + 50 mL containing 1.00 mmol NaCl + 100 mL of water.

▮ (*a*) 100 mL of 0.50 M NaCl yields 50 mmol Na^+ and 50 mmol Cl^-
50 mL of 0.25 M KCl yields 12.5 mmol K^+ and 12.5 mmol Cl^-

$$\frac{50 \text{ mmol Na}^+}{150 \text{ mL}} = 0.33 \text{ M Na}^+ \qquad \frac{12.5 \text{ mmol K}^+}{150 \text{ mL}} = 0.083 \text{ M K}^+ \qquad \frac{62.5 \text{ mmol Cl}^-}{150 \text{ mL}} = 0.42 \text{ M Cl}^-$$

(*b*) $Ag^+ + Cl^- \longrightarrow AgCl$, 0.33 M Na^+ (same as in part (*a*)).

$$\frac{37.5 \text{ mmol Cl}^- \text{ remaining in solution}}{150 \text{ mL}} = 0.25 \text{ M Cl}^- \qquad \frac{12.5 \text{ mmol NO}_3^-}{150 \text{ mL}} = 0.083 \text{ M NO}_3^-$$

(*c*) $\dfrac{51 \text{ mmol Na}^+}{250 \text{ mL}} = 0.20 \text{ M Na}^+$ and an equal concentration of Cl^-.

11.153 What volume of 96.0% H_2SO_4 solution (density 1.83 g/mL) is required to prepare 2.00 L of 3.00 M H_2SO_4 solution?

▮ The 96.0% H_2SO_4 means 96.0 parts (e.g., grams) of H_2SO_4 per 100 parts (g) of solution.

$$(2.00 \text{ L})(3.00 \text{ mol/L}) = 6.00 \text{ mol}$$

$$(6.00 \text{ mol})\left(\frac{98.0 \text{ g H}_2\text{SO}_4}{\text{mol H}_2\text{SO}_4}\right)\left(\frac{100 \text{ g solution}}{96.0 \text{ g H}_2\text{SO}_4}\right)\left(\frac{1 \text{ mL}}{1.83 \text{ g solution}}\right) = 335 \text{ mL}$$

11.154 How many mL of 3.000 M HCl should be added to react completely with 12.35 g $NaHCO_3$? The reaction is $HCl + NaHCO_3 \rightarrow NaCl + CO_2 + H_2O$.

▮

$$(12.35 \text{ g NaHCO}_3)\left(\frac{1 \text{ mol NaHCO}_3}{84.01 \text{ g NaHCO}_3}\right)\left(\frac{1 \text{ mol HCl}}{1 \text{ mol NaHCO}_3}\right)\left(\frac{1000 \text{ mL}}{3.000 \text{ mol HCl}}\right) = 49.00 \text{ mL}$$

11.155 $Zn + H_2SO_4 \rightarrow ZnSO_4 + H_2$ What volume of 3.00 M H_2SO_4 is required to react with 10.0 g of zinc?

▮

$$\left(\frac{10.0 \text{ g Zn}}{65.4 \text{ g Zn/mol Zn}}\right)\left(\frac{1 \text{ mol H}_2\text{SO}_4}{1 \text{ mol Zn}}\right) = 0.153 \text{ mol H}_2\text{SO}_4$$

$$\frac{0.153 \text{ mol H}_2\text{SO}_4}{3.00 \text{ mol/L}} = 0.0510 \text{ L} = 51.0 \text{ mL}$$

11.156 What mass of $(NH_4)_2CO_3$ (FW = 96.0 g/mol) (*a*) contains 0.40 mol NH_4^+? (*b*) contains 6.02×10^{23} hydrogen atoms? (*c*) will produce 6.0 mol CO_2 when treated with sufficient acid? (*d*) is required to prepare 200 mL of 0.10 M ammonium carbonate solution?

▮

(*a*) $(0.40 \text{ mol NH}_4^+)\left(\dfrac{1 \text{ mol (NH}_4)_2\text{CO}_3}{2 \text{ mol NH}_4^+}\right)\left(\dfrac{96.0 \text{ g (NH}_4)_2\text{CO}_3}{\text{mol (NH}_4)_2\text{CO}_3}\right) = 19.2 \text{ g}$

(*b*) $(1 \text{ mol H atoms})\left(\dfrac{1 \text{ mol (NH}_4)_2\text{CO}_3}{8 \text{ mol H atoms}}\right)\left(\dfrac{96.0 \text{ g (NH}_4)_2\text{CO}_3}{\text{mol (NH}_4)_2\text{CO}_3}\right) = 12.0 \text{ g}$

(*c*) $(6.0 \text{ mol CO}_2)\left(\dfrac{1 \text{ mol (NH}_4)_2\text{CO}_3}{\text{mol CO}_2}\right)\left(\dfrac{96.0 \text{ g (NH}_4)_2\text{CO}_3}{\text{mol (NH}_4)_2\text{CO}_3}\right) = 576 \text{ g}$

(*d*) $(0.200 \text{ L})(0.10 \text{ M}) = (0.020 \text{ mol})\left(\dfrac{96.0 \text{ g}}{\text{mol}}\right) = 1.92 \text{ g}$

11.157 Calculate the mass of $Ba(ClO_3)_2$ formed from 0.100 L of 3.00 M $HClO_3$ solution plus excess $Ba(OH)_2$.

▮

$$Ba(OH)_2 + 2HClO_3 \longrightarrow Ba(ClO_3)_2 + 2H_2O$$

$$(0.100 \text{ L})(3.00 \text{ mol/L}) = 0.300 \text{ mol HClO}_3$$

$$(0.300 \text{ mol HClO}_3)\left(\frac{1 \text{ mol Ba(ClO}_3)_2}{2 \text{ mol HClO}_3}\right)\left(\frac{304 \text{ g Ba(ClO}_3)_2}{\text{mol Ba(ClO}_3)_2}\right) = 45.6 \text{ g Ba(ClO}_3)_2$$

11.158 How many mL of 0.5000 M $KMnO_4$ solution will react completely with 20.00 g of $K_2C_2O_4 \cdot H_2O$ according to the following equation?

$$16H^+ + 2MnO_4^- + 5C_2O_4^{2-} \longrightarrow 10CO_2 + 2Mn^{2+} + 8H_2O$$

$$(20.00\text{ g } K_2C_2O_4\cdot H_2O)\left(\frac{1\text{ mol } K_2C_2O_4\cdot H_2O}{184.2\text{ g}}\right)\left(\frac{2\text{ mol } MnO_4^-}{5\text{ mol } K_2C_2O_4\cdot H_2O}\right)\left(\frac{1\text{ L}}{0.5000\text{ mol } MnO_4^-}\right)\left(\frac{10^3\text{ mL}}{\text{L}}\right)$$
$$= 86.9\text{ mL}$$

11.159 Calculate the molarity of each ion in solution after 2.00 L of 3.00 M $AgNO_3$ is mixed with 3.00 L of 1.00 M $BaCl_2$.

$$BaCl_2 + 2AgNO_3 \longrightarrow Ba(NO_3)_2 + 2AgCl \qquad \text{(balanced)}$$

$$(2.00\text{ L})(3.00\text{ M}) = 6.00\text{ mol } AgNO_3 \qquad (3.00\text{ L})(1.00\text{ M}) = 3.00\text{ mol } BaCl_2$$

6.00 mol $AgNO_3$ is 6.00 mol Ag^+ + 6.00 mol NO_3^-

3.00 mol $BaCl_2$ is 6.00 mol Cl^- + 3.00 mol Ba^{2+}

Yields: 6.00 mol AgCl(*s*) $\quad \dfrac{6.00\text{ mol } NO_3^-}{5.00\text{ L}} \quad \dfrac{3.00\text{ mol } Ba^{2+}}{5.00\text{ L}}$

$1.20\text{ M } NO_3^- \qquad 0.600\text{ M } Ba^{2+}$

11.160 Calculate the concentrations of all species remaining in solution after treatment of 50.0 mL of 0.300 M HCl with 50.0 mL of 0.400 M NH_3.

(50.0 mL)(0.300 M) yields 15.0 mmol H^+ and 15.0 mmol Cl^-

(50.0 mL)(0.400 M) yields 20.0 mmol NH_3

$$NH_3 + H^+ \longrightarrow NH_4^+$$

The reaction yields 15.0 mmol of NH_4^+, leaving 5.0 mmol NH_3 unreacted. The 15.0 mmol of Cl^- is unchanged.

$$\frac{15.0\text{ mmol } NH_4^+}{100\text{ mL}} = 0.150\text{ M } NH_4^+ \qquad \frac{15.0\text{ mmol } Cl^-}{100\text{ mL}} = 0.150\text{ M } Cl^- \qquad \frac{5.0\text{ mmol } NH_3}{100\text{ mL}} = 0.050\text{ M } NH_3$$

11.161 Calculate the concentration of an HCl solution if 2.50 mL of the solution took 4.50 mL of 3.00 M NaOH to neutralize.

$$HCl + NaOH \longrightarrow NaCl + H_2O$$

$$(4.50\times 10^{-3}\text{ L NaOH})\left(\frac{3.00\text{ mol NaOH}}{\text{L NaOH}}\right)\left(\frac{1\text{ mol HCl}}{1\text{ mol NaOH}}\right) = 13.5\times 10^{-3}\text{ mol HCl}$$

$$\frac{13.5\times 10^{-3}\text{ mol HCl}}{2.50\times 10^{-3}\text{ L HCl}} = 5.40\text{ M HCl}$$

11.162 How many mL of 0.0250 M HBr are required to neutralize 25.0 mL of 0.0200 M $Ba(OH)_2$? (Assume complete neutralization.)

$$2HBr + Ba(OH)_2 \longrightarrow BaBr_2 + 2H_2O \qquad (25.0\text{ mL})(0.0200\text{ M}) = 0.500\text{ mmol } Ba(OH)_2$$

This quantity of $Ba(OH)_2$ contains 1.00 mmol of OH^-, which will neutralize 1.00 mmol of HBr.

$$(1.00\text{ mmol})\left(\frac{1\text{ mL}}{0.0250\text{ mmol}}\right) = 40.0\text{ mL}$$

11.163 How many mL of 3.00 M HCl should be added to react completely with 16.8 g of $NaHCO_3$? $HCl + NaHCO_3 \rightarrow NaCl + CO_2 + H_2O$.

$$(16.8\text{ g } NaHCO_3)\left(\frac{1\text{ mol } NaHCO_3}{84.0\text{ g } NaHCO_3}\right)\left(\frac{1\text{ mol HCl}}{\text{mol } NaHCO_3}\right)\left(\frac{1\text{ L HCl}}{3.00\text{ mol HCl}}\right) = 6.67\times 10^{-2}\text{ L} = 66.7\text{ mL}$$

11.164 How many mL of 0.600 M NaOH will be needed to react with 138.0 mL of 4.00 M HCl?

$$NaOH + HCl \longrightarrow NaCl + H_2O$$

$$(138.0\text{ mL})\left(\frac{4.00\text{ mmol}}{\text{mL}}\right) = 552\text{ mmol HCl} \qquad (552\text{ mmol HCl})\left(\frac{1\text{ mmol NaOH}}{\text{mmol HCl}}\right) = 552\text{ mmol NaOH}$$

$$(552\text{ mmol NaOH})\left(\frac{1\text{ mL NaOH}}{0.600\text{ mmol}}\right) = 920\text{ mL NaOH}$$

11.165 Calculate the concentration of $Ba(OH)_2$ solution if it takes 42.00 mL of 4.000 M HCl to neutralize 31.50 mL of the base. Write a balanced chemical equation for the reaction.

▌

$$Ba(OH)_2 + 2HCl \longrightarrow BaCl_2 + 2H_2O$$

$$(42.00\ \text{mL})\left(\frac{1\ \text{L}}{1000\ \text{mL}}\right)\left(\frac{4.000\ \text{mol}}{\text{L}}\right) = 0.1680\ \text{mol HCl}$$

$$(0.1680\ \text{mol HCl})\left(\frac{1\ \text{mol Ba(OH)}_2}{2\ \text{mol HCl}}\right) = 0.08400\ \text{mol Ba(OH)}_2$$

$$\frac{0.08400\ \text{mol Ba(OH)}_2}{0.03150\ \text{L}} = 2.667\ \text{M Ba(OH)}_2$$

11.166 When 150.0 mL of 2.000 M NaOH was added to 100.0 mL of a sulfuric acid solution, it required 43.0 mL of 0.5000 M HCl to neutralize the excess base. What was the original concentration of H_2SO_4?

▌

$$(150.0\ \text{mL})(2.000\ \text{M NaOH}) = 300.0\ \text{mmol NaOH added}$$

$$(43.0\ \text{mL})(0.5000\ \text{M HCl}) = 21.5\ \text{mmol HCl needed}$$

Thus 278.5 mmol of NaOH reacted with the H_2SO_4.

$$H_2SO_4 + 2NaOH \longrightarrow Na_2SO_4 + 2H_2O$$

$$(278.5\ \text{mmol NaOH})\left(\frac{1\ \text{mmol H}_2\text{SO}_4}{2\ \text{mmol NaOH}}\right) = 139.2\ \text{mmol H}_2\text{SO}_4 \qquad \frac{139.2\ \text{mmol}}{100.0\ \text{mL}} = 1.392\ \text{M H}_2\text{SO}_4$$

11.167 Calculate the molarity of each type of ion remaining in solution after 20.0 mL of 6.00 M HCl is mixed with 50.0 mL of 2.00 M $Ba(OH)_2$ and 30.0 mL of water.

▌

$$20.0\ \text{mL of 6.00 M HCl yields 120 mmol H}^+ \text{ and 120 mmol Cl}^-$$

$$50.0\ \text{mL of 2.00 M Ba(OH)}_2 \text{ yields 200 mmol OH}^- \text{ and 100 mmol Ba}^{2+}$$

$$H^+ + OH^- \longrightarrow H_2O$$

$$\frac{80\ \text{mmol OH}^-\ \text{excess}}{100\ \text{mL}} = 0.80\ \text{M OH}^- \qquad \frac{120\ \text{mmol Cl}^-}{100\ \text{mL}} = 1.20\ \text{M Cl}^- \qquad \frac{100\ \text{mmol Ba}^{2+}}{100\ \text{mL}} = 1.00\ \text{M Ba}^{2+}$$

11.168 What volume of 0.20 M H_2SO_4 is required to produce 34.0 g H_2S by the reaction $8KI + 5H_2SO_4 \rightarrow 4K_2SO_4 + 4I_2 + H_2S + 4H_2O$?

▌

$$(34.0\ \text{g H}_2\text{S})\left(\frac{1\ \text{mol H}_2\text{S}}{34.0\ \text{g H}_2\text{S}}\right)\left(\frac{5\ \text{mol H}_2\text{SO}_4}{\text{mol H}_2\text{S}}\right)\left(\frac{1\ \text{L}}{0.20\ \text{mol H}_2\text{SO}_4}\right) = 25.0\ \text{L}$$

11.169 What volume of 4.40 M H_2SO_4 solution is needed to react exactly with 100 g Al?

▌ The balanced equation for the reaction is $2Al + 3H_2SO_4 \rightarrow Al_2(SO_4)_3 + 3H_2$.

Mole Method

$$\text{Number of moles of Al in 100 g Al} = \frac{100\ \text{g}}{27.0\ \text{g/mol}} = 3.70\ \text{mol}$$

$$\text{Number of moles of H}_2\text{SO}_4 \text{ required for 3.70 mol Al} = \tfrac{3}{2}(3.70) = 5.55\ \text{mol}$$

$$\text{Volume of 4.40 M H}_2\text{SO}_4 \text{ containing 5.55 mol} = \frac{5.55\ \text{mol}}{4.40\ \text{mol/L}} = 1.26\ \text{L}$$

Factor-Label Method

$$x\ \text{L solution} = \left(\frac{100\ \text{g Al}}{27.0\ \text{g Al/mol Al}}\right)\left(\frac{3\ \text{mol H}_2\text{SO}_4}{2\ \text{mol Al}}\right)\left(\frac{1\ \text{L solution}}{4.40\ \text{mol H}_2\text{SO}_4}\right) = 1.26\ \text{L solution}$$

11.170 In standardizing a solution of $AgNO_3$ it was found that 50.0 mL was required to precipitate all the chloride ion contained in 36.0 mL of 0.520 M NaCl. How many g of Ag could be obtained from 100 mL of the $AgNO_3$ solution?

▌ In the precipitation of AgCl, equimolar amounts of Ag^+ and Cl^- are needed; therefore equal numbers of moles of $AgNO_3$ and NaCl must have been used.

$$n(\text{AgNO}_3) = n(\text{NaCl}) = (0.0360\ \text{L})(0.520\ \text{mol/L}) = 0.01872\ \text{mol}$$

Then 50.0 mL of the $AgNO_3$ solution contains 0.01872 mol $AgNO_3$, or 0.01872 mol Ag, and so 100 mL of solution contains

$$\left(\frac{100\text{ mL}}{50.0\text{ mL}}\right)(0.01872\text{ mol Ag})(107.9\text{ g Ag/mol Ag}) = 4.04\text{ g Ag}$$

11.171 Exactly 40.0 mL of 0.225 M $AgNO_3$ was required to react exactly with 55.0 mL of a solution of NaCN, according to the equation $Ag^+ + 2CN^- \rightarrow Ag(CN)_2^-$. What is the molar concentration of the NaCN solution?

▮ $n(AgNO_3) = (0.0400\text{ L})(0.225\text{ mol/L}) = 0.00900\text{ mol}$ $\quad n(NaCN) = 2 \times n(AgNO_3) = 0.0180\text{ mol}$

Then 25.0 mL of the NaCN solution contains 0.0180 mol NaCN, so that $M = \dfrac{0.0180\text{ mol}}{0.055\text{ L}} = 0.327\text{ M}$

11.172 How many mL of 0.25 M $BaCl_2$ is required to precipitate all the sulfate ion from 10 mL of a solution containing 100 g of Na_2SO_4 per liter?

▮
$$BaCl_2 + Na_2SO_4 \longrightarrow BaSO_4 + 2NaCl$$

$$(0.010\text{ L})\left(\frac{100\text{ g }Na_2SO_4}{\text{L}}\right)\left(\frac{1\text{ mol }Na_2SO_4}{142\text{ g }Na_2SO_4}\right)\left(\frac{1\text{ mol }BaCl_2}{1\text{ mol }Na_2SO_4}\right)\left(\frac{1\text{ L}}{0.25\text{ mol }BaCl_2}\right) = 28 \times 10^{-3}\text{ L} = 28\text{ mL}$$

11.173 A 40.0 mL sample of Na_2SO_4 solution is treated with an excess of $BaCl_2$. If the mass of the precipitated $BaSO_4$ is 1.756 g, what is the molar concentration of the Na_2SO_4 solution?

▮
$$BaCl_2 + Na_2SO_4 \longrightarrow BaSO_4 + 2NaCl$$

$$(1.756\text{ g }BaSO_4)\left(\frac{1\text{ mol }BaSO_4}{233\text{ g }BaSO_4}\right)\left(\frac{1\text{ mol }Na_2SO_4}{1\text{ mol }BaSO_4}\right) = 7.54 \times 10^{-3}\text{ mol} \qquad \frac{7.54 \times 10^{-3}\text{ mol}}{0.0400\text{ L}} = 0.188\text{ M }Na_2SO_4$$

11.174 What was the thorium content of a sample that required 35.0 mL of 0.0300 M $H_2C_2O_4$ to precipitate $Th(C_2O_4)_2$?

▮
$$Th^{4+} + 2C_2O_4^{2-} \longrightarrow Th(C_2O_4)_2$$

$$(35.0\text{ mL})\left(\frac{0.0300\text{ mmol }H_2C_2O_4}{\text{mL}}\right)\left(\frac{1\text{ mmol }Th^{4+}}{2\text{ mmol }H_2C_2O_4}\right)\left(\frac{232\text{ mg}}{\text{mmol }Th^{4+}}\right) = 122\text{ mg }Th^{4+}$$

11.175 What molar concentration of $K_4Fe(CN)_6$ should be used so that 43.0 mL of the solution titrates 150.0 mg Zn (dissolved) by forming $K_2Zn_3[Fe(CN)_6]_2$?

▮
$$(150\text{ mg Zn})\left(\frac{1\text{ mmol Zn}}{65.4\text{ mg Zn}}\right)\left(\frac{2\text{ mmol }Fe(CN)_6^{4-}}{3\text{ mmol Zn}}\right) = 1.53\text{ mmol }Fe(CN)_6^{4-} \qquad \frac{1.53\text{ mmol}}{43.0\text{ mL}} = 0.0356\text{ M}$$

11.176 How many g of copper will be replaced from 2.00 L of 1.50 M $CuSO_4$ solution by 40.0 g aluminum?

▮ This is a limiting quantities problem.

$$(2.00\text{ L})\left(\frac{1.50\text{ mol }CuSO_4}{\text{L}}\right) = 3.00\text{ mol }CuSO_4 \qquad (40.0\text{ g Al})\left(\frac{1\text{ mol Al}}{27.0\text{ g Al}}\right) = 1.48\text{ mol Al}$$

$$2Al + 3CuSO_4 \longrightarrow Al_2(SO_4)_3 + 3Cu$$

The copper is in excess (3 mol $CuSO_4$ would react with 2 mol Al if it were present).

$$(1.48\text{ mol Al})\left(\frac{3\text{ mol Cu}}{2\text{ mol Al}}\right)\left(\frac{63.5\text{ g Cu}}{\text{mol Cu}}\right) = 141\text{ g Cu}$$

11.177 Calculate the mass of CuS produced and the concentration of H^+ ion produced by bubbling excess H_2S into 1.00 L of 0.10 M $CuCl_2$ solution. The equation is $Cu^{2+} + H_2S(g) \rightarrow CuS(s) + 2H^+$.

▮
$$(1.00\text{ L})(0.10\text{ M}) = 0.10\text{ mol }Cu^{2+}$$

$$(0.10\text{ mol }Cu^{2+})\left(\frac{1\text{ mol CuS}}{1\text{ mol }Cu^{2+}}\right)\left(\frac{95.6\text{ g CuS}}{\text{mol CuS}}\right) = 9.6\text{ g CuS} \qquad (0.10\text{ mol }Cu^{2+})\left(\frac{2\text{ mol }H^+}{1\text{ mol }Cu^{2+}}\right) = 0.20\text{ mol }H^+$$

$$\frac{0.20\text{ mol }H^+}{1.00\text{ L}} = 0.20\text{ M }H^+$$ (Assuming no volume change of the liquid.)

11.178 When 50.00 mL of a nitric acid solution was titrated with 0.334 M NaOH, it required 42.80 mL of the base to achieve the equivalence point. What is the molarity of the nitric acid solution? What mass of HNO_3 was dissolved in 90.00 mL of solution?

▮

$$(42.80\ \text{mL})(0.334\ \text{mmol/mL}) = 14.3\ \text{mmol NaOH used}$$

$$H^+ + OH^- \longrightarrow H_2O$$

14.3 mmol NaOH reacts with 14.3 mmol HNO_3. There was 14.3 mmol HNO_3 in the 50.00 mL.

$$\frac{14.3\ \text{mmol HNO}_3}{50.00\ \text{mL}} = 0.286\ \text{M HNO}_3 \qquad (0.286\ \text{mmol/mL})(90.00\ \text{mL})(63.0\ \text{mg/mmol}) = 1620\ \text{mg} = 1.62\ \text{g}$$

11.179 The acidic substance in vinegar is acetic acid, $HC_2H_3O_2$. When 6.00 g of a certain vinegar was titrated with 0.100 M NaOH, 40.11 mL of base had to be added to reach the equivalence point. What percent by mass of this sample of vinegar is acetic acid?

▮ $HC_2H_3O_2 + NaOH \longrightarrow NaC_2H_3O_2 + H_2O \qquad (40.11\ \text{mL})(0.100\ \text{mmol/mL}) = 4.01\ \text{mmol NaOH}$

The same number of mmol of acid must have been present, as required by the balanced chemical equation.

$$(4.01\ \text{mmol HC}_2\text{H}_3\text{O}_2)(60.0\ \text{mg/mmol}) = 241\ \text{mg HC}_2\text{H}_3\text{O}_2 \qquad \left(\frac{241\ \text{mg HC}_2\text{H}_3\text{O}_2}{6000\ \text{mg vinegar}}\right)(100\%) = 4.01\%\ \text{HC}_2\text{H}_3\text{O}_2$$

11.180 Two drops of phenolphthalein solution was added to 40.00 mL of an HCl solution, and this solution was titrated with 0.1000 M NaOH. When 30.00 mL of base had been added, part of the solution turned pink, but the color disappeared upon mixing the solution. Addition of NaOH solution was continued dropwise until a one drop addition produced a lasting pink color. At this point the volume of base added was 32.56 mL. What was the concentration of the HCl solution?

▮ The reaction is $HCl + NaOH \rightarrow NaCl + H_2O$. The number of mmol of added base is

$$(32.56\ \text{mL})(0.1000\ \text{mmol/mL}) = 3.256\ \text{mmol NaOH}$$

Thus 3.256 mmol of HCl was originally present in the 40.00 mL of solution.

$$\frac{3.256\ \text{mmol}}{40.00\ \text{mL}} = \frac{0.08140\ \text{mmol}}{\text{mL}} = 0.08140\ \text{M}$$ The concentration of the HCl solution was 0.08140 M.

11.181 (*a*) Balance the following equation, which represents the combustion of pyrites, FeS_2, a pollution-causing impurity in some coals: $FeS_2 + O_2 \rightarrow SO_2 + FeO$ (*b*) What volume of 6.0 M NaOH would be required to react with the SO_2 produced from 1.0 metric ton (10^3 kg) of coal containing 0.050% by mass of pyrites impurity?

▮ (*a*) $2FeS_2 + 5O_2 \longrightarrow 4SO_2 + 2FeO$

(*b*) $$(1.0 \times 10^6\ \text{g coal})\left(\frac{0.050\ \text{g FeS}_2}{100\ \text{g coal}}\right)\left(\frac{1\ \text{mol FeS}_2}{120\ \text{g FeS}_2}\right)\left(\frac{4\ \text{mol SO}_2}{2\ \text{mol FeS}_2}\right) = 8.32\ \text{mol SO}_2$$

$$2NaOH + SO_2 \longrightarrow Na_2SO_3 + H_2O \qquad (8.32\ \text{mol SO}_2)\left(\frac{2\ \text{mol NaOH}}{\text{mol SO}_2}\right)\left(\frac{1\ \text{L}}{6.0\ \text{mol NaOH}}\right) = 2.8\ \text{L}$$

11.82 What volume of 3.00 M HNO_3 can react completely with 15.0 g of a brass (90.0% Cu, 10.0% Zn) according to the following equations?

$$Cu + 4H^+(aq) + 2NO_3^-(aq) \longrightarrow 2NO_2(g) + Cu^{2+} + 2H_2O$$
$$4Zn + 10H^+(aq) + NO_3^-(aq) \longrightarrow NH_4^+ + 4Zn^{2+} + 3H_2O$$

What volume of NO_2 gas at 25 °C and 1.00 atm pressure would be produced?

▮

$$(13.5\ \text{g Cu})\left(\frac{1\ \text{mol Cu}}{63.5\ \text{g}}\right)\left(\frac{4\ \text{mol HNO}_3}{\text{mol Cu}}\right) = 0.850\ \text{mol HNO}_3$$

$$(1.5\ \text{g Zn})\left(\frac{1\ \text{mol Zn}}{65.37\ \text{g}}\right)\left(\frac{10\ \text{mol HNO}_3}{4\ \text{mol Zn}}\right) = 0.057\ \text{mol HNO}_3$$

$$\text{Total quantity of HNO}_3 = 0.907\ \text{mol HNO}_3$$

$$(0.907 \text{ mol HNO}_3)\left(\frac{1 \text{ L}}{3.00 \text{ mol}}\right) = 0.302 \text{ L} = 302 \text{ mL}$$

$$(13.5 \text{ g Cu})\left(\frac{1 \text{ mol Cu}}{63.5 \text{ g}}\right)\left(\frac{2 \text{ mol NO}_2}{\text{mol Cu}}\right) = 0.425 \text{ mol NO}_2$$

$$V = \frac{nRT}{P} = \frac{(0.425 \text{ mol})(0.0821 \text{ L·atm/mol·K})(298 \text{ K})}{1.00 \text{ atm}} = 10.4 \text{ L}$$

11.183 Calculate the percent of BaO in 29.0 g of a mixture of BaO and CaO which just reacts with 100.8 mL of 6.00 M HCl.

▮

$$\text{BaO} + 2\text{HCl} \longrightarrow \text{BaCl}_2 + \text{H}_2\text{O} \qquad \text{CaO} + 2\text{HCl} \longrightarrow \text{CaCl}_2 + \text{H}_2\text{O}$$

$$\text{HCl used} = (100.8 \text{ mL})(6.00 \text{ M}) = 605 \text{ mmol}$$

Let x = number of g BaO; then $29.0 - x$ = number of g CaO

$$\frac{x}{153.4} = \text{number of mol BaO} \qquad \text{and} \qquad \frac{29.0 - x}{56.08} = \text{number of mol CaO}$$

Since 2 mol of HCl is needed for each mol of metal oxide, the number of mol of HCl used is

$$\frac{2x}{153.4} + \frac{2(29.0 - x)}{56.08} = 0.605 \qquad 0.01304x + 1.034 - 0.03566x = 0.605$$

$$x = 19.0 \text{ g BaO} \qquad 29.0 - x = 10.0 \text{ g CaO}$$

$$\left(\frac{19.0 \text{ g BaO}}{29.0 \text{ g total}}\right)(100\%) = 65.5\% \text{ BaO}$$

11.184 Determine the percent BaO in a 10.0 g mixture of BaO and CaO which requires 100 mL of 2.50 M HCl to react with it completely.

▮

$$2\text{HCl} + \text{BaO} \longrightarrow \text{BaCl}_2 + \text{H}_2\text{O} \qquad 2\text{HCl} + \text{CaO} \longrightarrow \text{CaCl}_2 + \text{H}_2\text{O}$$

The 0.250 mol HCl will react with 0.125 mol metal oxides.

Let x = number of mol of BaO; then $0.125 - x$ = number of mol CaO.

$$(56.0)(0.125 - x) + 153x = 10.0 \qquad \text{So} \qquad x = 0.0309 \text{ mol BaO}$$

$$(0.0309 \text{ mol})(153 \text{ g/mol}) = 4.73 \text{ g BaO} \qquad 47.3\% \text{ BaO}$$

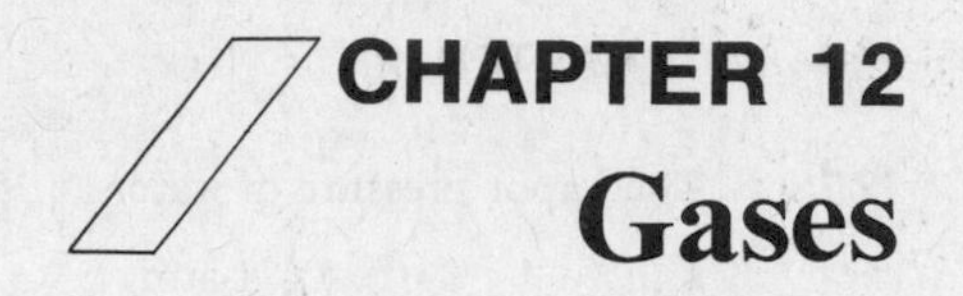

Gases

12.1 UNITS OF PRESSURE AND TEMPERATURE

***Warning*: In all gas-law equations temperature must be expressed in kelvins (K).**

12.1 Express the following gas pressures in atm: (*a*) 300 cmHg (*b*) 28 lb/in.^{2} (*c*) 380 torr (*d*) 0.760 torr (*e*) 7.60 torr.

▌ (*a*) $(300\text{ cmHg})\left(\dfrac{1\text{ atm}}{76.0\text{ cmHg}}\right) = 3.95\text{ atm}$ (*b*) $\left(\dfrac{28\text{ lb}}{\text{in.}^2}\right)\left(\dfrac{1\text{ atm}}{14.7\text{ lb/in.}^2}\right) = 1.9\text{ atm}$

(*c*) $(380\text{ torr})\left(\dfrac{1\text{ atm}}{760\text{ torr}}\right) = 0.500\text{ atm}$ (*d*) $(0.760\text{ torr})\left(\dfrac{1\text{ atm}}{760\text{ torr}}\right) = 0.00100\text{ atm}$

(*e*) $(7.60\text{ torr})\left(\dfrac{1\text{ atm}}{760\text{ torr}}\right) = 0.0100\text{ atm}$

12.2 How high a column of air would be necessary to cause the barometer to read 76 cmHg if the atmosphere were of uniform density 1.2 kg/m^3? The density of mercury is 13.6×10^3 kg/m^3.

▌

$$\text{Pressure of Hg} = \text{pressure of air}$$

$$(\text{Height of Hg}) \times (\text{density of Hg}) \times (g) = (\text{height of air}) \times (\text{density of air}) \times (g)$$

$$(0.76\text{ m})(13\,600\text{ kg/m}^3) = (h)(1.2\text{ kg/m}^3) \qquad h = \frac{(0.76\text{ m})(13\,600\text{ kg/m}^3)}{1.2\text{ kg/m}^3} = 8.6\text{ km}$$

(Incidentally, the atmosphere is not uniform in density as one ascends.)

12.3 Calculate the difference in pressure between the top and bottom of a vessel 76.00 cm deep when filled at 25°C with (*a*) water (*b*) mercury. Density of mercury at 25 °C is 13.53 g/cm^3; of water, 0.997 g/cm^3.

▌ (*a*) Pressure = height × density × g = $(0.7600\text{ m})(997\text{ kg/m}^3)(9.81\text{ m/s}^2) = 7.43 \times 10^3\text{ Pa} = 7.43\text{ kPa}$

(*b*) Pressure = $(0.7600\text{ m})(13\,530\text{ kg/m}^3)(9.81\text{ m/s}^2) = 100.9 \times 10^3\text{ Pa} = 100.9\text{ kPa}$

12.4 The Rankine scale is an absolute temperature scale which has degrees the same size as on the Fahrenheit scale. What is the freezing point of water in °R?

▌ The size of a kelvin is 180/100 times as great as the size of each °R, since there is a 180° difference between the freezing point and normal boiling point of water on the Rankine scale, and a 100 K difference.

$$(273\text{ K})\left(\frac{180\ ^\circ\text{R}}{100\text{ K}}\right) = 491\ ^\circ\text{R}$$

12.5 Calculate the mass of mercury in a uniform column 760 mm high and 1.00 cm^2 in cross-sectional area. Calculate the mass of mercury in a column of equal height but with 2.00 cm^2 cross-sectional area. Compare the pressures at the bases of the two columns. (The density of mercury is 13.6 g/cm^3.)

▌

$$760\text{ mm} = 76.0\text{ cm}$$

$$V = \text{area} \times \text{height} = (1.00\text{ cm}^2)(76.0\text{ cm}) = 76.0\text{ cm}^3 \qquad (76.0\text{ cm}^3)\left(\frac{13.6\text{ g}}{\text{cm}^3}\right) = 1030\text{ g} = 1.03\text{ kg}$$

In a 2.00 cm^2 column, there would be twice the volume, hence twice the mass of mercury. That mass would rest on twice the area and so exert the same pressure. The pressure does not depend on the cross section of the tube, but only on the vertical height of the mercury.

12.6 Express the standard atmosphere in (*a*) bars (*b*) pounds force per square inch.
1 atm = 101 325 Pa = 101 325 N/m^2 1 bar = 10^5 Pa 1 lbf = 4.448 N

▌

$$1\text{ atm} = (101\,325\text{ Pa})\left(\frac{1\text{ bar}}{10^5\text{ Pa}}\right) = 1.013\,25\text{ bar}$$

$$1\text{ atm} = \left(\frac{101\,325\text{ N}}{\text{m}^2}\right)\left(\frac{1\text{ lbf}}{4.448\text{ N}}\right)\left(\frac{1\text{ m}}{100\text{ cm}}\right)^2\left(\frac{2.54\text{ cm}}{1\text{ in.}}\right)^2 = \frac{14.70\text{ lbf}}{\text{in.}^2}$$

12.7 The vapor pressure of water at 25 °C is 23.8 torr. Express this in (*a*) atm (*b*) kPa.

▮ (*a*) $(23.8\ \text{torr})\left(\frac{1\ \text{atm}}{760\ \text{torr}}\right) = 0.0313\ \text{atm}$ (*b*) $(0.0313\ \text{atm})\left(\frac{101.325\ \text{kPa}}{\text{atm}}\right) = 3.17\ \text{kPa}$

12.8 Camphor has been found to undergo a crystalline modification at a temperature of 148 °C and a pressure of $3.09 \times 10^9\ \text{N/m}^2$. What is the transition pressure in atm?

▮
$$3.09 \times 10^9\ \text{N/m}^2 = (3.09 \times 10^9\ \text{Pa})\left(\frac{1\ \text{atm}}{101\,325\ \text{Pa}}\right) = 3.05 \times 10^4\ \text{atm}$$

12.9 An abrasive, borazon, is made by heating ordinary boron nitride to 3000 °F at one million lb/in.^{2}. Express the experimental conditions in °C and atm.

▮ $\left(\frac{1.00 \times 10^6\ \text{lbf}}{\text{in.}^2}\right)\left(\frac{1\ \text{atm}}{14.7\ \text{lbf/in.}^2}\right) = 6.80 \times 10^4\ \text{atm}$ $\quad ^\circ\text{C} = (\frac{5}{9})(^\circ\text{F} - 32^\circ) = (\frac{5}{9})(3000^\circ - 32^\circ) = 1650\ ^\circ\text{C}$

12.10 In a satellite flyby of Mercury in 1974, the planet's atmospheric pressure was observed to be 2×10^{-9} mbar. What fraction is this of the earth's atmospheric pressure?

▮
$$(2 \times 10^{-9}\ \text{mbar})\left(\frac{10^{-3}\ \text{bar}}{\text{mbar}}\right)\left(\frac{10^5\ \text{Pa}}{\text{bar}}\right)\left(\frac{1\ \text{atm}}{101\,325\ \text{Pa}}\right) = 2 \times 10^{-12}\ \text{atm}$$

12.11 (*a*) Show how the value of R in L·atm/K·mol can be derived from the value in SI units. (*b*) Express R in calories.

▮ (*a*)
$$R = 8.314\ \text{J/K}\cdot\text{mol} = (8.314\ \text{N}\cdot\text{m/K}\cdot\text{mol})\left(\frac{1\ \text{atm}}{1.013 \times 10^5\ \text{N/m}^2}\right)\left(\frac{10^3\ \text{dm}^3}{1\ \text{m}^3}\right)\left(\frac{1\ \text{L}}{\text{dm}^3}\right)$$
$$= 0.0821\ \text{L}\cdot\text{atm/K}\cdot\text{mol}$$

(*b*)
$$R = (8.314\ \text{J/K}\cdot\text{mol})\left(\frac{1\ \text{cal}}{4.184\ \text{J}}\right) = 1.987\ \text{cal/K}\cdot\text{mol}$$

12.2 BOYLE'S LAW, CHARLES' LAW, COMBINED GAS LAW

12.12 A 10.0 cm column of air is trapped by a column of Hg 8.00 cm long in a capillary tube of uniform bore when the tube is held horizontally in a room at 0.9400 atm pressure (Fig. 12.1). What will be the length of the air column when the tube is held (*a*) vertically with the open end up (*b*) vertically with the open end down (*c*) at a 45° angle from vertical with the open end up?

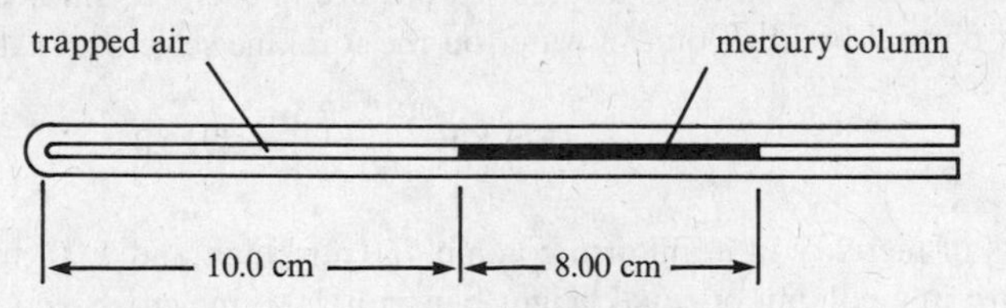

Fig. 12.1

▮ A convenient pressure unit here is the mmHg = torr. Thus, the given barometric pressure is

$$(0.9400\ \text{atm})\left(\frac{760.0\ \text{torr}}{1\ \text{atm}}\right) = 714.4\ \text{torr}$$

The volume of trapped air is equal to its length l times the (constant) cross-sectional area A. At constant temperature for a given mass of gas,

$$P_1V_1 = P_2V_2 = P_1l_1A = P_2l_2A \qquad P_1l_1 = P_2l_2$$

(*a*) The weight of mercury represents an increase in the pressure on the gas of 8.00 cmHg = 80.0 mmHg = 80.0 torr.

$$P_2 = (714.4\ \text{torr}) + (80.0\ \text{torr}) = 794.4\ \text{torr}$$
$$l_2 = \frac{P_1l_1}{P_2} = \frac{(714.4\ \text{torr})(10.0\ \text{cm})}{794.4\ \text{torr}} = 8.99\ \text{cm}$$

(*b*) The weight of the mercury decreases the pressure by 80.0 torr.

$$P_2 = 634.4 \text{ torr} \qquad l_2 = \frac{(714.4 \text{ torr})(10.0 \text{ cm})}{634.4 \text{ torr}} = 11.3 \text{ cm}$$

(*c*) The weight of the mercury is borne partially by the gas and partially by the glass. The vertical height of the mercury is a measure of the additional pressure on the gas. A 45° right triangle has sides in the ratio $1:1:\sqrt{2}$. Thus the vertical height $= (80.0 \text{ mm})/\sqrt{2} = 56.6$ mm. The pressure is

$$(714.4 \text{ torr}) + (56.6 \text{ torr}) = 771.0 \text{ torr} \qquad l_2 = \frac{(714.4 \text{ torr})(10.0 \text{ cm})}{771.0 \text{ torr}} = 9.27 \text{ cm}$$

12.13 A 1.00 L sample of gas at 760 torr is compressed to 0.800 L at constant temperature. Calculate the final pressure of the gas.

▮
$$PV = k = (760 \text{ torr})(1.00 \text{ L}) = 760 \text{ torr·L}$$

But after compression, $PV = k$, $P(0.800 \text{ L}) = 760 \text{ torr·L}$ thus $P = 950$ torr

12.14 Calculate the volume which 4.00 L of gas at 0 °C will occupy at 100 °C and the same pressure.

▮
$$V_1 = k'T_1 \qquad 4.00 \text{ L} = k'(0° + 273°) = k'(273 \text{ K}) \qquad k' = 0.01465 \text{ L/K}$$

New conditions: $V_2 = k'T_2 \qquad V_2 = (0.01465 \text{ L/K})[(100 + 273) \text{ K}] = 5.46 \text{ L}$

Note that temperatures must be expressed in kelvins. This calculation may be done by combining the separate steps, as follows:

$$\frac{V_1}{T_1} = k' = \frac{V_2}{T_2} \qquad V_2 = \frac{V_1 T_2}{T_1} = \frac{(4.00 \text{ L})(373 \text{ K})}{273 \text{ K}} = 5.46 \text{ L}$$

12.15 A sample of gas occupies 2.00 L at 760 torr. Calculate the volume it will occupy at 1.25 atm and the same temperature.

▮

	1	2
P	1.00 atm 760 torr	1.25 atm
V	2.00 L	V_2

$$P_1V_1 = P_2V_2 \qquad V_2 = \frac{P_1V_1}{P_2} = \frac{(1.00 \text{ atm})(2.00 \text{ L})}{1.25 \text{ atm}} = 1.60 \text{ L}$$

12.16 Calculate the temperature at which a 2.00 L sample of gas at 27 °C would occupy 3.00 L if its pressure were changed from 1.00 atm to 800 torr.

▮

	1	2
P	1.00 atm	(800/760) atm
V	2.00 L	3.00 L
T	300 K	T_2

$$\frac{P_1V_1}{T_1} = \frac{P_2V_2}{T_2} \qquad \frac{T_1}{P_1V_1} = \frac{T_2}{P_2V_2}$$

$$T_2 = \frac{T_1P_2V_2}{P_1V_1} = \frac{(300 \text{ K})[(800/760) \text{ atm}](3.00 \text{ L})}{(1.00 \text{ atm})(2.00 \text{ L})} = 474 \text{ K}$$

12.17 A 4.00 L sample of gas at 30 °C and 1.00 atm is changed to 0 °C and 800 torr. What is its new volume?

▮

	1	2
P	1.00 atm	(800/760) atm
V	4.00 L	V_2
T	303 K	273 K

$$V_2 = \frac{P_1V_1T_2}{P_2T_1} = \frac{(1.00 \text{ atm})(4.00 \text{ L})(273 \text{ K})}{[(800/760) \text{ atm}](303 \text{ K})} = 3.42 \text{ L}$$

12.18 A sample of gas at 1.20 atm and 27 °C is heated at constant pressure to 57 °C. Its final volume is 4.75 L. What was its original volume?

▮

	1	2
P	1.20 atm	1.20 atm
V	V_1	4.75 L
T	300 K	330 K

From Charles' law:

$$V_1 = \frac{V_2T_1}{T_2} = \frac{(4.75 \text{ L})(300 \text{ K})}{330 \text{ K}} = 4.32 \text{ L}$$

12.19 At what temperature does a sample of gas occupy 4.00 L at 1.11 atm, if it initially occupied 2.22 L at 1.00 atm and 60 °C?

▮

	1	2
P	1.00 atm	1.11 atm
V	2.22 L	4.00 L
T	333 K	T_2

$$\frac{P_1V_1}{T_1} = \frac{P_2V_2}{T_2} \qquad \frac{T_1}{P_1V_1} = \frac{T_2}{P_2V_2}$$

$$T_2 = \frac{T_1P_2V_2}{P_1V_1} = \frac{(333\text{ K})(1.11\text{ atm})(4.00\text{ L})}{(1.00\text{ atm})(2.22\text{ L})} = 666\text{ K}$$

12.20 A sample of gas at 0 °C and 1.00 atm pressure occupies 2.50 L. What change in temperature is necessary to adjust the pressure of that gas to 1.50 atm after it has been transferred to a 2.00 L container?

▮

$$\frac{P_1V_1}{T_1} = \frac{P_2V_2}{T_2} \qquad T_2 = \frac{P_2V_2T_1}{P_1V_1} = \frac{(1.50\text{ atm})(2.00\text{ L})(0° + 273°)}{(1.00\text{ atm})(2.50\text{ L})} = 328\text{ K}$$

Hence, an increase in temperature of 55 K, from 273 K to 328 K, is required.

12.21 Consider several samples of gas undergoing change as described in the table below. Determine the missing value for each sample.

	initial conditions			final conditions		
	pressure	**volume**	**temperature**	**pressure**	**volume**	**temperature**
(*a*)	760 torr	1.00 L	25 °C	760 torr	____	200 °C
(*b*)	1.00 atm	500 mL	127 °C	____	200 mL	127 °C
(*c*)	1.23 atm	700 mL	250 K	650 torr	____	200 °C
(*d*)	600 lb/in.^{2}	3.25 L	300 K	____	1.75 L	100 °C
(*e*)	____	1.00 L	300 K	700 torr	3000 mL	400 K
(*f*)	800 torr	7.50 L	300 K	____	2.25 L	350 K

▮ (*a*) P is constant. Hence, $V_2 = V_1T_2/T_1 = (1.00\text{ L})(473\text{ K})/(298\text{ K}) = 1.59\text{ L}$

(*b*) T is constant. Hence, $P_2 = P_1V_1/V_2 = (1.00\text{ atm})(500\text{ mL})/(200\text{ mL}) = 2.50\text{ atm}$

(*c*) $$V_2 = \frac{P_1V_1T_2}{T_1P_2} = \frac{(1.23\text{ atm})(700\text{ mL})(473\text{ K})}{(250\text{ K})(\frac{650}{760}\text{ atm})} = 1900\text{ mL} = 1.90\text{ L}$$

(*d*) $$P_2 = \frac{P_1V_1T_2}{T_1V_2} = \frac{(600\text{ lb/in.}^2)(3.25\text{ L})(373\text{ K})}{(300\text{ K})(1.75\text{ L})} = 1390\text{ lb/in.}^2$$

(*e*) $$P_1 = \frac{P_2V_2T_1}{T_2V_1} = \frac{(700\text{ torr})(3.00\text{ L})(300\text{ K})}{(400\text{ K})(1.00\text{ L})} = 1580\text{ torr}$$

(*f*) $$P_2 = \frac{P_1V_1T_2}{T_1V_2} = \frac{(800\text{ torr})(7.50\text{ L})(350\text{ K})}{(300\text{ K})(2.25\text{ L})} = 3110\text{ torr}$$

12.22 A mass of oxygen occupies 7.00 L under a pressure of 740 torr. Determine the volume of the same mass of gas at standard pressure, the temperature remaining constant.

▮ Standard pressure is 760 torr. Boyle's law gives

$$P_1V_1 = P_2V_2 \qquad \text{or} \qquad V_2 = \frac{P_1}{P_2}V_1 = \left(\frac{740\text{ torr}}{760\text{ torr}}\right)(7.00\text{ L}) = 6.82\text{ L}$$

12.23 An ideal gas initially at 710 torr and 30.0 °C occupies 2600 mL. Calculate the final temperature if the conditions are changed to a pressure of 1.20 atm and a volume of 3.20 L.

▮ Both values of P as well as those for V must be expressed in the same units for use in the combined gas law.

$$T_2 = \frac{P_2V_2T_1}{P_1V_1} = \frac{(1.20\text{ atm})(760\text{ torr/atm})(3.20\text{ L})(303\text{ K})}{(710\text{ torr})(2.60\text{ L})} = 479\text{ K} = 206\text{ °C}$$

12.24 A mass of neon occupies 400 mL at 100 °C. Find its volume at 0 °C, the pressure remaining constant.

▮ Charles' law gives

$$\frac{V_1}{T_1} = \frac{V_2}{T_2} \qquad \text{or} \qquad V_2 = \frac{T_2}{T_1} V_1 = \frac{(0 + 273)\ \text{K}}{(100 + 273)\ \text{K}} (400\ \text{mL}) = 293\ \text{mL}$$

12.25 A steel tank contains carbon dioxide at 27 °C and is at a pressure of 11.0 atm. Determine the internal gas pressure when the tank and its contents are heated to 100 °C.

▮ Gay-Lussac's law gives

$$\frac{P_1}{T_1} = \frac{P_2}{T_2} \qquad \text{or} \qquad P_2 = \frac{T_2}{T_1} P_1 = \left(\frac{(100 + 273)\ \text{K}}{(27 + 273)\ \text{K}}\right)(11.0\ \text{atm}) = 13.7\ \text{atm}$$

12.26 When 2.00 L of gas at 760 torr and 27 °C is expanded to 3.00 L at 77 °C, what will the final pressure be?

▮

$$P_2 = \frac{P_1 V_1 T_2}{T_1 V_2} = \frac{(760\ \text{torr})(2.00\ \text{L})(350\ \text{K})}{(3.00\ \text{L})(300\ \text{K})} = 591\ \text{torr}$$

12.27 Given 20.0 L of ammonia at 15 °C and 760 torr. Determine its volume at 50 °C and 800 torr.

▮ The combined gas law gives (15 °C = 288 K, 50 °C = 323 K):

$$\frac{P_1 V_1}{T_1} = \frac{P_2 V_2}{T_2} \qquad \text{or} \qquad V_2 = V_1\left(\frac{T_2}{T_1}\right)\left(\frac{P_1}{P_2}\right) = (20.0\ \text{L})\left(\frac{323\ \text{K}}{288\ \text{K}}\right)\left(\frac{760\ \text{torr}}{800\ \text{torr}}\right) = 21.3\ \text{L}$$

12.28 The volume of a quantity of sulfur dioxide at 18 °C and 1500 torr is 4.3 ft³. Calculate its volume at STP.

▮ STP means 0 °C and 760 torr. Converting to absolute temperatures and applying the combined gas law,

$$V_2 = V_1\left(\frac{T_2}{T_1}\right)\left(\frac{P_1}{P_2}\right) = (4.3\ \text{ft}^3)\left(\frac{273\ \text{K}}{291\ \text{K}}\right)\left(\frac{1500\ \text{torr}}{760\ \text{torr}}\right) = 8.0\ \text{ft}^3$$

12.29 To how many atm pressure must 2.00 L of gas measured at 1.00 atm and −20 °C be subjected to be compressed to 0.500 L when the temperature is 40 °C?

▮ The combined gas law gives

$$\frac{P_1 V_1}{T_1} = \frac{P_2 V_2}{T_2} \qquad \text{or} \qquad P_2 = P_1\left(\frac{V_1}{V_2}\right)\left(\frac{T_2}{T_1}\right) = (1.00\ \text{atm})\left(\frac{2.00\ \text{L}}{0.500\ \text{L}}\right)\left(\frac{313\ \text{K}}{253\ \text{K}}\right) = 4.95\ \text{atm}$$

12.30 A mass of oxygen occupies 40.0 ft³ at 758 torr. Compute its volume at 700 torr, temperature remaining constant.

▮

$$P_1 V_1 = P_2 V_2 \qquad V_2 = \frac{V_1 P_1}{P_2} = (40.0\ \text{ft}^3)\left(\frac{758\ \text{torr}}{700\ \text{torr}}\right) = 43.3\ \text{ft}^3$$

12.31 Ten liters of hydrogen under 7.0 atm pressure is contained in a cylinder which has a movable piston. The piston is moved in until the same mass of gas occupies 4.0 L at the same temperature. Find the pressure in the cylinder.

▮

$$P_1 V_1 = P_2 V_2 \qquad P_2 = \frac{P_1 V_1}{V_2} = \frac{(7.0\ \text{atm})(10\ \text{L})}{4.0\ \text{L}} = 18\ \text{atm}$$

12.32 A given mass of chlorine occupies 25.0 L at 20 °C. Determine its volume at 45 °C, pressure remaining constant.

▮

$$\frac{V_1}{T_1} = \frac{V_2}{T_2} \qquad V_2 = \frac{V_1 T_2}{T_1} = \frac{(25.0\ \text{L})(45^\circ + 273^\circ)}{20^\circ + 273^\circ} = 27.1\ \text{L}$$

12.33 A quantity of hydrogen is confined in a chamber of constant volume. When the chamber is immersed in a bath of melting ice, the pressure of the gas is 1000 torr. ***(a)*** What is the Celsius temperature when the pressure manometer indicates an absolute pressure of 400 torr? ***(b)*** What pressure will be indicated when the chamber is brought to 100 °C?

▮

$$\frac{P_1}{T_1} = \frac{P_2}{T_2}$$ (Gay-Lussac's law, or derive from the combined gas law with $V_1 = V_2$.)

Note: Melting ice implies 0 °C.

$$(a)\quad T_2 = \frac{T_1P_2}{P_1} = \frac{(0° + 273°)(400\text{ torr})}{1000\text{ torr}} = 109\text{ K} = -164\text{ °C}$$

$$(b)\quad P_2 = \frac{P_1T_2}{T_1} = \frac{(1000\text{ torr})(100° + 273°)}{0° + 273°} = 1.37 \times 10^3\text{ torr}$$

12.34 Given 1000 ft^3 of helium at 15 °C and 763 torr, calculate the volume at −6 °C and 620 torr.

▮
$$\frac{P_1V_1}{T_1} = \frac{P_2V_2}{T_2} \qquad V_2 = \frac{P_1V_1T_2}{T_1P_2} = \frac{(763\text{ torr})(1000\text{ ft}^3)(267\text{ K})}{(288\text{ K})(620\text{ torr})} = 1140\text{ ft}^3$$

12.35 A mass of gas at 80 °C and 785 torr occupies 350 mL. What volume will the gas occupy at STP?

▮
$$\frac{P_1V_1}{T_1} = \frac{P_2V_2}{T_2} \qquad V_2 = \frac{P_1V_1T_2}{T_1P_2} = \frac{(785\text{ torr})(350\text{ mL})(273\text{ K})}{(353\text{ K})(760\text{ torr})} = 280\text{ mL}$$

12.36 If a mass of gas occupies 3.00 L at STP, what volume will it occupy at 300 °C and 25 atm?

▮
$$\frac{P_1V_1}{T_1} = \frac{P_2V_2}{T_2} \qquad V_2 = \frac{P_1V_1T_2}{T_1P_2} = \frac{(3.00\text{ atm})(1.00\text{ L})(573\text{ K})}{(273\text{ K})(25\text{ atm})} = 0.25\text{ L}$$

12.37 If a gas occupies 15.7 ft^3 at 60 °F and 14.7 lbf/in.^{2}, what volume would it occupy at 100°F and 35 lbf/in.^{2}?

▮
$$60\text{ °F} = \tfrac{5}{9}(60° - 32°) = 15.6\text{ °C} = 289\text{ K} \qquad 100\text{ °F} = \tfrac{5}{9}(100° - 32°) = 37.8\text{ °C} = 311\text{ K}$$

$$\frac{P_1V_1}{T_1} = \frac{P_2V_2}{T_2} \qquad V_2 = \frac{P_1V_1T_2}{T_1P_2} = \frac{(14.7\text{ lbf/in.}^2)(15.7\text{ ft}^3)(311\text{ K})}{(289\text{ K})(35\text{ lbf/in.}^2)} = 7.1\text{ ft}^3$$

12.38 A mass of gas occupies 0.825 L at −30 °C and 556 Pa. What is the pressure if the volume becomes 2.00 L and the temperature 20 °C?

▮
$$\frac{P_1V_1}{T_1} = \frac{P_2V_2}{T_2} \qquad P_2 = \frac{P_1V_1T_2}{T_1V_2} = \frac{(556\text{ Pa})(0.825\text{ L})(293\text{ K})}{(243\text{ K})(2.00\text{ L})} = 277\text{ Pa}$$

12.39 Calculate the final Celsius temperature required to change 10.0 L of helium at 100 K and 0.150 atm to 30.0 L at 0.200 atm.

▮
$$\frac{P_1V_1}{T_1} = \frac{P_2V_2}{T_2} \qquad T_2 = \frac{T_1P_2V_2}{P_1V_1} = \frac{(100\text{ K})(0.200\text{ atm})(30.0\text{ L})}{(0.150\text{ atm})(10.0\text{ L})} = 400\text{ K} = 127\text{ °C}$$

12.40 One mole of a gas occupies 22.4 L at STP (*a*) What pressure would be required to compress 1.00 mol of oxygen into a 5.00 L container held at 100 °C? (*b*) What maximum Celsius temperature would be permitted if this amount of oxygen were held in 5.00 L at a pressure not exceeding 3.00 atm? (*c*) What capacity would be required to hold this same amount if the conditions were fixed at 100 °C and 3.00 atm?

▮
$$\frac{P_1V_1}{T_1} = \frac{P_2V_2}{T_2}$$

$$(a)\quad P_2 = \frac{P_1V_1T_2}{T_1V_2} = \frac{(1.00\text{ atm})(22.4\text{ L})(373\text{ K})}{(273\text{ K})(5.00\text{ L})} = 6.12\text{ atm}$$

$$(b)\quad T_2 = \frac{T_1P_2V_2}{P_1V_1} = \frac{(273\text{ K})(3.00\text{ atm})(5.00\text{ L})}{(1.00\text{ atm})(22.4\text{ L})} = 183\text{ K} = -90\text{ °C}$$

$$(c)\quad V_2 = \frac{P_1V_1T_2}{T_1P_2} = \frac{(1.00\text{ atm})(22.4\text{ L})(373\text{ K})}{(273\text{ K})(3.00\text{ atm})} = 10.2\text{ L}$$

12.41 A certain container holds 3.87 g neon at STP. What mass of neon will it hold at 100 °C and 10.0 atm?

▮ The 3.87 g neon under the new conditions will occupy

$$V_2 = \frac{P_1V_1T_2}{T_1P_2} = \frac{(1.00\text{ atm})(V_1)(373\text{ K})}{(273\text{ K})(10.0\text{ atm})} = 0.137\ V_1 \text{ or } 13.7\% \text{ of the original volume}$$

Thus the container will have room for $(3.87\text{ g})\left(\frac{V_1}{0.137\,V_1}\right) = 28.2\text{ g}$

Note: A more direct way to solve this problem is by using the ideal gas law (Sec. 12.3) twice.

12.42 At a certain altitude in the upper atmosphere, the temperature is estimated to be −100 °C and the density just 10^{-9} that of the earth's atmosphere at STP. Assuming a uniform atmospheric composition, what is the pressure in torr at this altitude?

▌

$$P_2 = \frac{P_1V_1T_2}{T_1V_2} = \frac{P_1T_2}{T_1}\left(\frac{d_2}{d_1}\right) = \frac{(760\text{ torr})(173\text{ K})}{(273\text{ K})} \times 10^{-9} = 4.8 \times 10^{-7}\text{ torr}$$

12.43 The density of helium is 0.1784 kg/m³ at STP. If a given mass of helium at STP is allowed to expand to 1.400 times its initial volume by changing the temperature and pressure, compute its resultant density.

▌ The density of a gas varies inversely with the volume.

$$\text{Resultant density} = (0.1784\text{ kg/m}^3)\left(\frac{1}{1.400}\right) = 0.1274\text{ kg/m}^3 \qquad \text{Note that}\quad 1\text{ kg/m}^3 = 1\text{ g/L.}$$

12.44 The density of oxygen is 1.43 g/L at STP. Determine the density of oxygen at 17 °C and 800 torr.

▌ The combined gas law shows that the density of an ideal gas varies inversely with the absolute temperature and directly with the pressure.

$$d_2 = d_1\left(\frac{T_1}{T_2}\right)\left(\frac{P_2}{P_1}\right) = (1.43\text{ g/L})\left(\frac{273\text{ K}}{290\text{ K}}\right)\left(\frac{800\text{ torr}}{760\text{ torr}}\right) = 1.42\text{ g/L}$$

12.45 If the density of a certain gas at 30 °C and 768 torr is 1.35 kg/m³, find its density at STP.

▌ Since the mass does not change, the density is inversely proportional to the volume:

$$\frac{d_{STP}}{d_2} = \frac{V_2}{V_{STP}} \qquad \frac{P_{STP}V_{STP}}{T_{STP}} = \frac{P_2V_2}{T_2}$$

$$\frac{P_{STP}T_2}{P_2T_{STP}} = \frac{V_2}{V_{STP}} = \frac{(760\text{ torr})(303\text{ K})}{(768\text{ torr})(273\text{ K})} = 1.10 \qquad d_{STP} = 1.10d_2 = (1.10)(1.35\text{ kg/m}^3) = 1.48\text{ kg/m}^3$$

12.46 At the top of a mountain the thermometer reads 0 °C and the barometer reads 710 mmHg. At the bottom of the mountain the temperature is 30 °C and the pressure is 760 mmHg. Compare the density of the air at the top with that at the bottom.

▌ Density is inversely proportional to volume:

$$\frac{d_{top}}{d_{bottom}} = \frac{V_b}{V_t} = \frac{T_bP_t}{T_tP_b} = \frac{(303\text{ K})(710\text{ mmHg})}{(273\text{ K})(760\text{ mmHg})} = 1.04 \qquad \text{The density ratio is 1.04 to 1.}$$

12.47 At 0 °C the density of nitrogen at 1.00 atm is 1.25 kg/m³. The nitrogen which occupied 1500 cm³ at STP was compressed at 0 °C to 575 atm, and the gas volume was observed to be 3.92 cm³, in violation of Boyle's law. What was the final density of this nonideal gas?

▌ Since the mass of the nitrogen is constant,

$$\text{Final density} = \frac{\text{initial mass}}{\text{final volume}} = \frac{(1.25\text{ kg/m}^3)(1500 \times 10^{-6}\text{ m}^3)}{3.92 \times 10^{-6}\text{ m}^3} = 478\text{ kg/m}^3$$

12.48 A volume of 95 cm³ of nitrous oxide at 27 °C is collected over mercury in a graduated tube, the level of mercury inside the tube being 40 mm above the outside mercury level when the barometer reads 750 torr. **(*a*)** Compute the volume of the same mass of gas at STP. **(*b*)** What volume would the same mass of gas occupy at 40 °C, the barometric pressure being 745 torr and the level of mercury inside the tube 25 mm below that outside?

▌ Remember that 1 torr = 1 mmHg.

(*a*) $P_1 = (750\text{ torr}) - (40\text{ torr}) = 710\text{ torr}$ $\qquad V_2 = \frac{P_1V_1T_2}{T_1P_2} = \frac{(710\text{ torr})(95\text{ cm}^3)(273\text{ K})}{(300\text{ K})(760\text{ torr})} = 81\text{ cm}^3$

(*b*) $P_2 = (745 \text{ torr}) + (25 \text{ torr}) = 770 \text{ torr}$ $\quad V_2 = \dfrac{(710 \text{ torr})(95 \text{ cm}^3)(313 \text{ K})}{(300 \text{ K})(770 \text{ torr})} = 91 \text{ cm}^3$

12.49 A quantity of gas is collected in a graduated tube over the mercury. The volume of gas at 20 °C is 50.0 mL, and the level of the mercury in the tube is 100 mm above the outside mercury level. The barometer reads 750 torr. Find the volume at STP.

▮ Since the level of mercury inside the tube is 200 mm higher than outside, the pressure of the gas is $100 \text{ mmHg} = 100 \text{ torr}$ less than the atmospheric pressure of 750 torr.

$$\text{Volume of STP} = (50.0 \text{ mL})\left(\frac{273 \text{ K}}{293 \text{ K}}\right)\left(\frac{(750 - 100) \text{ torr}}{760 \text{ torr}}\right) = 39.8 \text{ mL}$$

12.50 The respiration of a suspension of yeast cells was measured by observing the decrease in pressure of gas above the cell suspension. The apparatus was arranged so that the gas was confined to a constant volume, 16.0 cm^3, and the entire pressure change was caused by uptake of oxygen by the cells. The pressure was measured in a manometer, the fluid of which had a density of 1.034 g/cm^3. The entire apparatus was immersed in a thermostat at 37 °C. In a 30 min observation period the fluid in the open side of the manometer dropped 37 mm. Neglecting the solubility of oxygen in the yeast suspension, compute the rate of oxygen consumption by the cells in mm^3 of O_2 (STP) per hour.

▮ Convert the observed pressure drop to mmHg: $(37 \text{ mm})\left(\dfrac{1.034 \text{ g/cm}^3}{13.6 \text{ g/cm}^3}\right) = 2.8 \text{ mmHg in 30 min}$, or 5.6 mmHg drop per hour. With 5.6 mmHg as the oxygen partial pressure, change to STP:

$$V_2 = \frac{P_1V_1T_2}{T_1P_2} = \frac{(5.6 \text{ mmHg})(16.0 \text{ cm}^3)(273 \text{ K})}{(310 \text{ K})(760 \text{ mmHg})} = 0.10 \text{ cm}^3 = 1.0 \times 10^2 \text{ mm}^3$$

12.51 A gas at a pressure of 5.0 atm is heated from 0° to 546 °C and simultaneously compressed to one-third of its original volume. What will be the final pressure, in atm?

▮

$$P_2 = \frac{P_1V_1T_2}{T_1V_2} = \frac{(5.0 \text{ atm})(V_1)(546 + 273) \text{ K}}{(273 \text{ K})(\frac{1}{3}V_1)} = 45 \text{ atm}$$

12.52 To what temperature must a neon gas sample be heated to double its pressure if the initial volume of gas at 75 °C is decreased by 15.0%?

▮

initial conditions	final conditions
P_1	$2P_1$
V_1	$0.850V_1$
75 °C = 348 K	T_2

$$\frac{P_1V_1}{T_1} = \frac{P_2V_2}{T_2} = \frac{(2P_1)(0.850V_1)}{T_2}$$

$$T_2 = \frac{(2P_1)(0.850V_1)(348 \text{ K})}{P_1V_1} = 592 \text{ K}$$

12.53 Obtain the combined gas law from Charles' law and Gay-Lussac's law.

▮ Charles' law says that V/T has a constant value for a given mass of gas at a given pressure. Thus $V/T = f(m, P)$, where f is an undetermined function. Similarly, Gay-Lussac's law may be written $P/T = g(m, V)$, where g is another undetermined function. Multiply the first equation by P and the second by V, to obtain two expressions, each equal to PV/T and thus equal to each other:

$$Pf(m, P) = Vg(m, V)$$

In this last equation, when P changes, the right side does not change; therefore the left side cannot change either and must be a function of m alone. Likewise, when V changes, the left side does not change; therefore the right side is independent of V, and is a function of m alone. Hence

$$Pf(m, P) = Vg(m, V) = h(m) \qquad \text{and} \qquad PV/T = h(m)$$

which is the combined gas law. [The function $h(m)$ is a constant times the number of moles of gas present.]

12.54 S. Cannizzaro (1858) was able to demonstrate that hydrogen gas consists of molecules with an even number of hydrogen atoms, based on the assumption that Avogadro's hypothesis is true. Avogadro's hypothesis states that equal volumes of gas under the same conditions of temperature and pressure contain equal numbers of molecules. Using the data from the following table, show precisely how he demonstrated this fact.

gases, 100 °C	density, g/L	percent hydrogen
Hydrogen	0.0659	100.0
Hydrogen chloride	1.19	2.76
Water vapor	0.589	11.2
Ammonia	0.557	17.7
Methane	0.524	25.1

▌ The number of hydrogen per L of gas, obtained by multiplying the density by the percent hydrogen, is converted to a fraction. The results are

	hydrogen per liter, g	ratio
Hydrogen	0.0659	2
Hydrogen chloride	0.0328	1
Water vapor	0.0659	2
Ammonia	0.0986	3
Methane	0.1315	4

The ratio of the number of g hydrogen per liter from one substance to another is shown in the last column. If equal volumes of all the gases contain equal numbers of molecules, and the volume of hydrogen gas has twice the number of hydrogen atoms as the same volume of hydrogen chloride (since it has twice the mass of hydrogen), then each hydrogen molecule must have twice the number of hydrogen atoms as each hydrogen chloride molecule. Thus the number of hydrogen atoms in a hydrogen molecule must be even.

12.3 MOLES OF GAS AND IDEAL GAS LAW

12.55 Which of the following, if any, contains the greatest number of N atoms? **(*a*)** 22.4 L nitrogen gas at STP (0 °C and 1 atm) **(*b*)** 500 mL of 2.00 M NH_3 **(*c*)** 1.00 mol NH_4Cl **(*d*)** 6.02×10^{23} molecules of NO_2.

▌ **(*a*)** because 1 mol N_2 has 2 mol N atoms, **(*b*)** 1 mol of NH_3 has 1 mol N atoms, **(*c*)** 1 mol of NH_4Cl has 1 mol N atoms, **(*d*)** 1 mol of NO_2 has 1 mol N atoms.

12.56 Compute the weight (mass) of 6.00 L of ammonia gas, NH_3, at STP.

▌

$$n = \frac{6.00\ \text{L}}{22.4\ \text{L/mol (STP)}} = (0.268\ \text{mol})\left(\frac{17.0\ \text{g}}{\text{mol}}\right) = 4.55\ \text{g}$$

12.57 Determine the volume occupied by 4.0 g oxygen at STP.

▌

$$\text{Volume of 4.0 g } O_2 \text{ at STP} = \text{number of moles in 4.0 g } O_2 \times \text{standard molar volume}$$
$$= \left(\frac{4.0\ \text{g}}{32\ \text{g/mol}}\right)(22.4\ \text{L/mol}) = 2.8\ \text{L}$$

12.58 What volume of hydrogen will combine with 22 L chlorine to form hydrogen chloride? What volume of hydrogen chloride will be formed? Assume the same temperature and pressure for all gases.

▌ The balanced equation for this reaction is

$$\underset{\text{(1 molecule)}}{H_2(g)} + \underset{\text{(1 molecule)}}{Cl_2(g)} \longrightarrow \underset{\text{(2 molecules)}}{2HCl(g)}$$

The equation shows that 1 molecule H_2 reacts with 1 molecule Cl_2 to form 2 molecules HCl. But, by Avogadro's hypothesis, equal numbers of molecules of *gases* under the same conditions of temperature and pressure occupy equal volumes. Therefore the equation also indicates that 1 volume of H_2 reacts with 1 volume of Cl_2 to form 2 volumes of HCl(*g*). Thus 22 L of H_2 combines with 22 L of Cl_2 to form $2 \times 22\ \text{L} = 44\ \text{L}$ of HCl.

12.59 What volume of hydrogen will unite with 16 ft³ of nitrogen to form ammonia? What volume of ammonia will be produced, all at the same temperature and pressure?

$$N_2(g) + 3H_2(g) \longrightarrow 2NH_3(g)$$

By the principles stated in Problem 12.58, 16 ft^3 N_2 reacts with $3 \times 16 = 48$ ft^3 H_2 to form $2 \times 16 = 32$ ft^3 NH_3.

12.60 Calculate the volume of 0.3000 mol of a gas at 60 °C and 0.821 atm.

$$V = \frac{nRT}{P} = \frac{(0.3000 \text{ mol})(0.0821 \text{ L·atm/mol·K})(333 \text{ K})}{0.821 \text{ atm}} = 9.99 \text{ L}$$

12.61 What weight of hydrogen at STP could be contained in a vessel that holds 4.8 g oxygen at STP?

The same number of moles of H_2 as O_2 can be held in the same volume at the same temperature and pressure.

$$n(H_2) = n(O_2) = \frac{4.8 \text{ g}}{32.0 \text{ g/mol}} = 0.15 \text{ mol} \qquad (0.15 \text{ mol } H_2)\left(\frac{2.00 \text{ g}}{\text{mol}}\right) = 0.30 \text{ g } H_2$$

12.62 Calculate the volume of 8.40 g N_2 at $t = 100$ °C and $P = 800$ torr.

$$(8.40 \text{ g } N_2)\left(\frac{1 \text{ mol } N_2}{28.0 \text{ g } N_2}\right) = 0.300 \text{ mol } N_2$$

$$V = \frac{nRT}{P} = \frac{(0.300 \text{ mol})(0.0821 \text{ L·atm/mol·K})(373 \text{ K})}{(800/760) \text{ atm}} = 8.73 \text{ L}$$

12.63 In order to economize on the oxygen supply in space ships, it has been suggested that the oxygen in exhaled CO_2 be converted to water by a reduction with hydrogen. The CO_2 output per astronaut has been estimated as 1.00 kg per 24 h day. An experimental catalytic converter reduces CO_2 at a rate of 600 mL (STP) per min. What fraction of the time would such a converter have to operate in order to keep up with the CO_2 output of one astronaut?

In one day,

$$(1000 \text{ g})\left(\frac{1 \text{ mol}}{44.0 \text{ g}}\right) = 22.7 \text{ mol } CO_2 \text{ produced}$$

$$\left(\frac{0.600 \text{ L}}{\text{min}}\right)\left(\frac{1 \text{ mol}}{22.4 \text{ L (STP)}}\right)\left(\frac{60 \text{ min}}{\text{h}}\right)\left(\frac{24 \text{ h}}{\text{day}}\right) = 38.6 \text{ mol/day capability} \qquad \left(\frac{22.7 \text{ mol}}{38.6 \text{ mol}}\right)(100\%) = 58.8\%$$

12.64 A natural gas sample contains 84% (by volume) CH_4, 10% C_2H_6, 3% C_3H_8, and 3% N_2. If a series of catalytic reactions could be used for converting all the carbon atoms of the gas into butadiene, C_4H_6, with 100% efficiency, how much butadiene could be prepared from 100 g of the natural gas?

Molecular weights: $CH_4 = 16.0$, $C_2H_6 = 30.0$, $C_3H_8 = 44.0$, $N_2 = 28.0$, $C_4H_6 = 54.0$. The volume percent of a gas mixture is the same as the mole percent.

$$\begin{aligned} 100 \text{ mol mixture} &= 84 \text{ mol } CH_4 + 10 \text{ mol } C_2H_6 + 3 \text{ mol } C_3H_8 + 3 \text{ mol } N_2 \\ &= 84(16.0 \text{ g } CH_4) + 10(30.0 \text{ g } C_2H_6) + 3(44.0 \text{ g } C_3H_8) + 3(28.0 \text{ g } N_2) = 1860 \text{ g natural gas} \end{aligned}$$

The number of mol C in 100 mol mixture is $84(1) + 10(2) + 3(3) = 113$ mol C. Since 4 mol C gives 1 mol (54.0 g) C_4H_6, 113 mol C gives

$$(\tfrac{113}{4} \text{ mol})(54.0 \text{ g/mol}) = 1530 \text{ g } C_4H_6$$

Then 1860 g natural gas yields 1530 g C_4H_6 and 100 g natural gas yields $\left(\frac{100}{1860}\right)(1530 \text{ g}) = 82.0 \text{ g } C_4H_6$

12.65 What volume will 1.216 g of SO_2 gas occupy at 18 °C and 755 torr?

$$(1.216 \text{ g } SO_2)\left(\frac{1 \text{ mol } SO_2}{64.06 \text{ g } SO_2}\right) = 0.0190 \text{ mol } SO_2$$

$$V = \frac{nRT}{P} = \frac{(0.0190 \text{ mol})(0.0821 \text{ L·atm/mol·K})(291 \text{ K})}{(755/760) \text{ atm}} = 0.457 \text{ L}$$

12.66 Compute the volume of 14 g of nitrous oxide, N_2O, at STP

$$(14 \text{ g})\left(\frac{1 \text{ mol}}{44 \text{ g}}\right) = 0.32 \text{ mol} \qquad V = \frac{nRT}{P} = \frac{(0.32 \text{ mol})(0.0821 \text{ L·atm/mol·K})(273 \text{ K})}{1.00 \text{ atm}} = 7.2 \text{ L}$$

12.67 What is the volume of 18.0 g of pure water at 1.00 atm and 4 °C?

▌ $$(18.0\text{ g})\left(\frac{1.00\text{ mL}}{1.00\text{ g}}\right) = 18.0\text{ mL}$$

Do not use the ideal gas law for this problem, because water at 4 °C is not a gas.

12.68 Assuming the same pressure in each case, calculate the mass of hydrogen required to inflate a balloon to a certain volume V at 100 °C if 3.5 g helium is required to inflate the balloon to half the volume, $0.50V$, at 25 °C.

▌ When quantities are not known explicitly but only in relationship to other quantities, dividing one equation by another often eliminates many unknowns. For example,

$$P(\text{H}_2)V(\text{H}_2) = n(\text{H}_2)RT(\text{H}_2) \qquad \text{Dividing:} \quad \frac{P(\text{H}_2)V(\text{H}_2)}{P(\text{He})V(\text{He})} = \frac{n(\text{H}_2)RT(\text{H}_2)}{n(\text{He})RT(\text{He})}$$
$$P(\text{He})V(\text{He}) = n(\text{He})RT(\text{He})$$

But in this exercise $P(\text{H}_2) = P(\text{He})$, $\frac{1}{2}V(\text{H}_2) = V(\text{He})$, $T(\text{H}_2) = 373\text{ K}$, and $T(\text{He}) = 298\text{ K}$.

Therefore,
$$\frac{V(\text{H}_2)}{\frac{1}{2}V(\text{H}_2)} = \frac{n(\text{H}_2)(373\text{ K})}{n(\text{He})(298\text{ K})}$$

$$n(\text{H}_2) = \frac{2(298)}{373}n(\text{He}) = \left(\frac{2(298)}{373}\right)\left(\frac{3.5\text{ g}}{4.0\text{ g/mol}}\right) = 1.4\text{ mol H}_2 \qquad (1.4\text{ mol H}_2)(2.0\text{ g/mol}) = 2.8\text{ g H}_2$$

12.69 Liquefied natural gas (LNG) is mainly methane (MW = 16.0). A 10.0 m³ tank is constructed to store LNG at −164 °C and 1.0 atm pressure, under which conditions its density is 415 kg/m³. Calculate the volume of a storage tank capable of holding the same mass of LNG as a gas at 20 °C and 1.0 atm pressure.

▌ In liquid form:
$$(10.0\text{ m}^3)\left(\frac{415\text{ kg}}{\text{m}^3}\right)\left(\frac{1000\text{ mol}}{16.0\text{ kg}}\right) = 2.59 \times 10^5\text{ mol}$$

In gaseous form:

$$V = \frac{nRT}{P} = \frac{(2.59 \times 10^5\text{ mol})(0.0821\text{ L·atm/mol·K})(293\text{ K})}{1.00\text{ atm}} = 6.24 \times 10^6\text{ L} = 6240\text{ m}^3$$

12.70 An electronic vacuum tube was sealed off during manufacture at a pressure of 1.8×10^{-5} torr at 27 °C. Its volume is 100 cm³. Compute the number of gas molecules remaining in the tube.

▌ $$n = \frac{PV}{RT} = \frac{\{[(1.8 \times 10^{-5})/(760)]\text{ atm}\}(0.100\text{ L})}{(0.0821\text{ L·atm/mol·K})(300\text{ K})} = 9.6 \times 10^{-11}\text{ mol}$$

$$(9.6 \times 10^{-11}\text{ mol})(6.02 \times 10^{23}\text{ molecules/mol}) = 5.8 \times 10^{13}\text{ molecules}$$

12.71 Calculate the number of mol of gas in (*a*) Problem 12.17 (*b*) Problem 12.18.

▌ (*a*) $$n = \frac{(1.00\text{ atm})(4.00\text{ L})}{(0.0821\text{ L·atm/mol·K})(303\text{ K})} = 0.161\text{ mol}$$

(*b*) $$n = \frac{(1.20\text{ atm})(4.75\text{ L})}{(0.0821\text{ L·atm/mol·K})(330\text{ K})} = 0.210\text{ mol}$$

12.72 What volume would 25.0 g argon occupy at 90 °C and 735 torr?

▌ with
$$n = \frac{m}{\text{MW}}, \qquad V = \frac{mRT}{(\text{MW})P} = \frac{(25.0\text{ g})(0.0821\text{ L·atm/K·mol})(363\text{ K})}{(39.9\text{ g/mol})[(735/760)\text{ atm}]} = 19.3\text{ L}$$

Another Method

At STP this much argon would occupy $V_1 = \left(\frac{25.0}{39.9}\text{ mol}\right)(22.4\text{ L/mol}) = 14.0\text{ L}$

This problem can be completed by the usual procedure for converting from STP to an arbitrary set of conditions.

$$V_2 = V_1\left(\frac{T_2}{T_1}\right)\left(\frac{P_1}{P_2}\right) = (14.0\text{ L})\left(\frac{363\text{ K}}{273\text{ K}}\right)\left(\frac{760\text{ torr}}{735\text{ torr}}\right) = 19.3\text{ L}$$

12.73 An iron meteorite was analyzed for its isotopic argon content. The amount of ^{36}Ar was 0.200 mm³ (STP) per kg of meteorite. If each ^{36}Ar atom had been formed by a single cosmic event, how many such events must there have been per kg of meteorite?

▮
$$n = \frac{PV}{RT} = \frac{(1.00\ \text{atm})(0.200 \times 10^{-6}\ \text{L})}{(0.0821\ \text{L·atm/mol·K})(273\ \text{K})} = 8.92 \times 10^{-9}\ \text{mol}$$

$$(8.92 \times 10^{-9}\ \text{mol})\left(\frac{6.02 \times 10^{23}\ \text{atoms}}{\text{mol}}\right) = 5.4 \times 10^{15}\ \text{atoms} \qquad \text{Hence, } 5.4 \times 10^{15}\ \text{events}$$

12.74 Mercury diffusion pumps may be used in the laboratory to produce a high vacuum. Cold traps are generally placed between the pump and the system to be evacuated. These cause the condensation of mercury vapor and prevent mercury from diffusing back into the system. The minimum pressure of mercury that can exist in the system is the vapor pressure of mercury at the temperature of the cold trap. Calculate the number of mercury atoms per unit volume in a cold trap maintained at −120 °C. The vapor pressure of mercury at this temperature is 10^{-16} torr.

▮
$$\text{Moles per liter} = \frac{n}{V} = \frac{P}{RT} = \frac{(10^{-16}/760)\ \text{atm}}{(0.0821\ \text{L·atm/K·mol})(153\ \text{K})} = 1.0 \times 10^{-20}\ \text{mol/L}$$

$$\text{Atoms per liter} = (1.0 \times 10^{-20}\ \text{mol/L})(6.0 \times 10^{23}\ \text{atoms/mol}) = 6000\ \text{atoms/L} \quad \text{or} \quad 6\ \text{atoms/cm}^3$$

12.75 Determine the density of H_2S gas at 27 °C and 2.00 atm.

▮
$$PV = nRT \qquad n = m/\text{MW} \qquad PV = mRT/\text{MW}$$

$$d = \frac{m}{V} = \frac{P(\text{MW})}{RT} = \frac{(2.00\ \text{atm})(34.0\ \text{g/mol})}{(0.0821\ \text{L·atm/mol·K})(300\ \text{K})} = 2.76\ \text{g/L}$$

12.76 Compute the approximate density of methane, CH_4, at 20 °C and 6.00 atm.

▮
$$d = \frac{P(\text{MW})}{RT} = \frac{(16.0\ \text{g/mol})(6.00\ \text{atm})}{(0.0821\ \text{L·atm/K·mol})(293\ \text{K})} = 3.99\ \text{g/L}$$

Another Method

$$\text{Density} = \frac{\text{mass of 1 mol}}{\text{volume of 1 mol}} = \frac{16.0\ \text{g}}{(22.4\ \text{L})\left(\dfrac{1.00\ \text{atm}}{6.00\ \text{atm}}\right)\left(\dfrac{293\ \text{K}}{273\ \text{K}}\right)} = 3.99\ \text{g/L}$$

12.77 A gas cylinder contains 370 g oxygen gas at 30.0 atm pressure and 25 °C. What mass of oxygen would escape if first the cylinder were heated to 75 °C and then the valve were held open until the gas pressure was 1.00 atm, the temperature being maintained at 75 °C?

▮ The volume is found from the initial conditions:

$$n = (370\ \text{g})\left(\frac{1\ \text{mol}}{32.0\ \text{g}}\right) = 11.6\ \text{mol}$$

$$V = \frac{nRT}{P} = \frac{(11.6\ \text{mol})(0.0821\ \text{L·atm/mol·K})(298\ \text{K})}{30.0\ \text{atm}} = 9.43\ \text{L}$$

The final number of moles is
$$n = \frac{PV}{RT} = \frac{(1.00\ \text{atm})(9.43\ \text{L})}{(0.0821\ \text{L·atm/mol·K})(348\ \text{K})} = 0.330\ \text{mol}$$

$$(0.330\ \text{mol})\left(\frac{32.0\ \text{g}}{\text{mol}}\right) = 10.6\ \text{g} \qquad (370\ \text{g initial}) - (10.6\ \text{g final}) = 359\ \text{g escaped}$$

12.78 A refrigeration tank holding 5.00 L Freon gas ($C_2Cl_2F_4$) at 25 °C and 3.00 atm pressure developed a leak. When the leak was discovered and repaired, the tank had lost 76.0 g of the gas. What was the pressure of the gas remaining in the tank at 25 °C?

▮ The mass of gas initially present is calculated:

$$n = \frac{PV}{RT} = \frac{(3.00\ \text{atm})(5.00\ \text{L})}{(0.0821\ \text{L·atm/mol·K})(298\ \text{K})} = 0.613\ \text{mol} \qquad (0.613\ \text{mol})\left(\frac{171\ \text{g}}{\text{mol}}\right) = 105\ \text{g originally present}$$

The quantity remaining and the final pressure are then calculated:

$$(105 \text{ g}) - (76 \text{ g}) = 29 \text{ g remaining} \qquad (29 \text{ g})\left(\frac{1 \text{ mol}}{171 \text{ g}}\right) = 0.17 \text{ mol}$$

$$P = \frac{nRT}{V} = \frac{(0.17 \text{ mol})(0.0821 \text{ L·atm/mol·K})(298 \text{ K})}{5.00 \text{ L}} = 0.83 \text{ atm}$$

12.79 A quantity of hydrogen gas occupies a volume of 30.0 mL at a certain temperature and pressure. What volume would half this mass of hydrogen occupy at triple the absolute temperature if the pressure were one-ninth that of the original gas?

▮

	initial	**final**
V	30.0 mL	V_2
P	P_1	$P_2 = \frac{1}{9}P_1$
T	T_1	$T_2 = 3T_1$
n	n_1	$n_2 = \frac{1}{2}n_1$

$$\frac{P_1V_1}{P_2V_2} = \frac{n_1RT_1}{n_2RT_2} \qquad \frac{P_1(30.0 \text{ mL})}{(\frac{1}{9}P_1)(V_2)} = \frac{n_1T_1}{(\frac{1}{2}n_1)(3T_1)} \qquad V_2 = (\tfrac{1}{2})(3)(9)(30.0 \text{ mL}) = 405 \text{ mL}$$

12.80 A 10.0 L cylinder of oxygen at 4.00 atm pressure and 17 °C developed a leak. When the leak was repaired, 2.50 atm of oxygen remained in the cylinder, still at 17 °C. How many mol of gas escaped?

▮

$$\text{Moles escaped} = \text{original moles} - \text{final moles} = \frac{(4.00)(10.0)}{(0.0821)(290)} - \frac{(2.50)(10.0)}{(0.0821)(290)} = 0.630 \text{ mol}$$

12.81 One of the important attributes of hydrogen as a vehicular fuel is its compactness. Compare the numbers of hydrogen atoms per m^3 available in **(*a*)** hydrogen gas under a pressure of 14.0 MPa at 300 K; **(*b*)** liquid hydrogen at 20 K at a density of 70.0 kg/m^3; **(*c*)** the solid compound $DyCo_3H_5$, having a density of 8200 kg/m^3 at 300 K, all of whose hydrogen may be made available for combustion.

▮ **(*a*)** 1.00 atm = 100.9 kPa [from Problem 12.3(*b*); a more precise value is 101.3 kPa]

$$P = 14.0 \text{ MPa} = (14.0 \times 10^3 \text{ kPa})\left(\frac{1 \text{ atm}}{100.9 \text{ kPa}}\right) = 139 \text{ atm}$$

$$T = 300 \text{ K} \qquad V = 1.00 \text{ m}^3 = 1000 \text{ L}$$

$$n = \frac{PV}{RT} = \frac{(139 \text{ atm})(1000 \text{ L})}{(0.0821 \text{ L·atm/mol·K})(300 \text{ K})} = 5.63 \times 10^3 \text{ mol}$$

$$(5.63 \times 10^3 \text{ mol})\left(\frac{6.02 \times 10^{23} \text{ molecules H}_2}{\text{mol H}_2}\right)\left(\frac{2 \text{ atoms}}{\text{molecule}}\right) = 6.8 \times 10^{27} \text{ atoms}$$

(*b*)
$$(70.0 \text{ kg})\left(\frac{1000 \text{ g}}{\text{kg}}\right)\left(\frac{1 \text{ mol H}_2}{2.0 \text{ g H}_2}\right)\left(\frac{6.02 \times 10^{23} \text{ molecules H}_2}{\text{mol H}_2}\right)\left(\frac{2 \text{ atoms H}}{\text{molecule H}_2}\right) = 4.2 \times 10^{28} \text{ atoms H}$$

(*c*)
$$(8200 \text{ kg DyCo}_3\text{H}_5)\left(\frac{1000 \text{ g}}{\text{kg}}\right)\left(\frac{1 \text{ mol compd}}{344.2 \text{ g}}\right)\left(\frac{5 \text{ mol H}}{\text{mol compd}}\right)\left(\frac{6.02 \times 10^{23} \text{ H atoms}}{\text{mol H}}\right) = 7.2 \times 10^{28} \text{ atoms}$$

12.82 A volume of 105 mL of pure water at 4 °C is saturated with NH_3 gas, yielding a solution of density 0.900 g/mL and containing 30.0% NH_3 by weight. Find the volume of the ammonia solution resulting, and the volume of the ammonia gas at 5 °C and 775 torr which was used to saturate the water.

▮

$$(105 \text{ mL water})\left(\frac{1.000 \text{ g water}}{\text{mL water}}\right)\left(\frac{100 \text{ g solution}}{70.0 \text{ g water}}\right)\left(\frac{1 \text{ mL solution}}{0.900 \text{ g solution}}\right) = 167 \text{ mL}$$

$$(105 \text{ mL water})\left(\frac{1.000 \text{ g water}}{\text{mL water}}\right)\left(\frac{30.0 \text{ g NH}_3}{70.0 \text{ g water}}\right)\left(\frac{1 \text{ mol NH}_3}{17.0 \text{ g NH}_3}\right) = 2.65 \text{ mol NH}_3$$

$$V = \frac{nRT}{P} = \frac{(2.65 \text{ mol})(0.0821 \text{ L·atm/mol·K})(278 \text{ K})}{(775/760) \text{ atm}} = 59.3 \text{ L}$$

12.83 An empty steel gas tank with valve weighs 125 lb. Its capacity is 1.50 ft³. When the tank is filled with oxygen to 2000 lbf/in.² at 25 °C, what percent of the total weight of the full tank is O_2? Assume the validity of the ideal gas law.

▮

$$(125\ \text{lb})\left(\frac{1\ \text{kg}}{2.20\ \text{lb}}\right) = 56.8\ \text{kg} \qquad V = (1.50\ \text{ft}^3)\left(\frac{(12.0)(2.54)\ \text{cm}}{\text{ft}}\right)^3\left(\frac{1\ \text{L}}{1000\ \text{cm}^3}\right) = 42.5\ \text{L}$$

$$P = (2000\ \text{lbf/in.}^2)\left(\frac{1\ \text{atm}}{14.7\ \text{lbf/in.}^2}\right) = 136\ \text{atm} \qquad n = \frac{PV}{RT} = \frac{(136\ \text{atm})(42.5\ \text{L})}{(0.0821\ \text{L}\cdot\text{atm/mol}\cdot\text{K})(298\ \text{K})} = 236\ \text{mol}$$

$$(236\ \text{mol})\left(\frac{0.032\ \text{kg}}{\text{mol}}\right)\left(\frac{2.2\ \text{lb}}{\text{kg}}\right) = 16.6\ \text{lb}\ O_2 \qquad \left(\frac{16.6\ \text{lb}}{(125\ \text{lb}) + (16.6\ \text{lb})}\right)(100\%) = 11.7\%$$

12.84 Pure oxygen gas is not necessarily the most compact source of oxygen for confined fuel systems because of the weight of the cylinder necessary to confine the gas. Other compact sources are hydrogen peroxide and lithium peroxide. The oxygen-yielding reactions are:

$$2H_2O_2 \longrightarrow 2H_2O + O_2 \qquad 2Li_2O_2 \longrightarrow 2Li_2O + O_2$$

Rate (*a*) 65% (by weight) H_2O_2 and (*b*) pure Li_2O_2 in terms of % of total weight which is "available" oxygen. Neglect the weights of the containers. Compare with Problem 12.83.

▮ (*a*) 1.00 kg sample contains 650 g H_2O_2.

$$(650\ \text{g}\ H_2O_2)\left(\frac{1\ \text{mol}\ H_2O_2}{34.0\ \text{g}\ H_2O_2}\right)\left(\frac{1\ \text{mol}\ O_2}{2\ \text{mol}\ H_2O_2}\right)\left(\frac{32.0\ \text{g}\ O_2}{\text{mol}\ O_2}\right) = 0.31 \times 10^3\ \text{g}\ O_2 = 0.31\ \text{kg}\ O_2$$

31% O_2 available

(*b*)
$$(1.00 \times 10^3\ \text{g}\ Li_2O_2)\left(\frac{1\ \text{mol}\ Li_2O_2}{45.88\ \text{g}\ Li_2O_2}\right)\left(\frac{1\ \text{mol}\ O_2}{2\ \text{mol}\ Li_2O_2}\right)\left(\frac{32.0\ \text{g}\ O_2}{\text{mol}\ O_2}\right) = 0.35 \times 10^3\ \text{g}\ O_2 = 0.35\ \text{kg}\ O_2$$

35% O_2 available

12.85 What percent of a sample of nitrogen must be allowed to escape if its temperature, pressure, and volume are to be changed from 220 °C, 3.00 atm, and 1.65 L to 110 °C, 0.700 atm, and 1.00 L, respectively?

▮

$$n_1 = \frac{(3.00\ \text{atm})(1.65\ \text{L})}{R(493\ \text{K})} \qquad n_2 = \frac{(0.700\ \text{atm})(1.00\ \text{L})}{R(383\ \text{K})}$$

$$\text{Fraction remaining} = \frac{n_2}{n_1} = \frac{(0.700\ \text{atm})(1.00\ \text{L})/R(383\ \text{K})}{(3.00\ \text{atm})(1.65\ \text{L})/R(493\ \text{K})} = 0.182$$

$$\text{Fraction escaped} = 1.000 - 0.182 = 0.818 \qquad \text{Percent escaped} = 81.8\%$$

12.4 DALTON'S LAW

12.86 Show that the volume ratio of two gases, before mixing and at the same temperature and pressure, is equal to the pressure ratio after the two gases are mixed.

▮ Before mixing, T and P are the same for the two gases:
$$\frac{V_1}{V_2} = \frac{n_1RT/P}{n_2RT/P} = \frac{n_1}{n_2}$$

After mixing, T and V are the same:
$$\frac{P_1}{P_2} = \frac{n_1RT/V}{n_2RT/V} = \frac{n_1}{n_2}$$

From this proof, it is seen that the terms "volume percent" and "mole percent" are equivalent, and that the term "partial pressure percent" could also be used synonymously.

12.87 A mixture of gases at 760 torr contains 55.0% nitrogen, 25.0% oxygen, and 20.0% carbon dioxide by volume. What is the partial pressure of each gas in torr?

▮ A fundamental property of gases is that each component of a gas mixture occupies the entire volume of the mixture. The percent composition by volume refers to the volumes of the separate gases *before* mixing. Thus 55.0 volumes of nitrogen, 25.0 volumes of oxygen, and 20.0 volumes of carbon dioxide, each at 760 torr, are mixed to give 100.0 volumes of mixture at 760 torr. As the volume of each component is increased by mixing, its pressure must (Boyle's law) decrease proportionately.

$$\text{Partial pressure of } N_2 = (760\ \text{torr})(55.0/100) = 418\ \text{torr}$$
$$\text{Partial pressure of } O_2 = (760\ \text{torr})(25.0/100) = 190\ \text{torr}$$
$$\text{Partial pressure of } CO_2 = (760\ \text{torr})(20.0/100) = 152\ \text{torr}$$

Check: Total pressure = sum of partial pressures = 760 torr

12.88 A 1.00 L vessel containing 1.00 g H_2 gas at 27.0 °C is connected to a 2.00 L vessel containing 88.0 g CO_2 gas, also at 27.0 °C. When the gases are completely mixed, what are the partial pressures and total pressure (in atm)?

▮

$$(1.00\ \text{g}\ H_2)\left(\frac{1\ \text{mol}\ H_2}{2.00\ \text{g}\ H_2}\right) = 0.500\ \text{mol}\ H_2 \qquad (88.0\ \text{g}\ CO_2)\left(\frac{1\ \text{mol}\ CO_2}{44.0\ \text{g}\ CO_2}\right) = 2.00\ \text{mol}\ CO_2$$

$$P(H_2) = \frac{n(H_2O)RT}{V} = \frac{(0.500\ \text{mol})(0.0821\ \text{L·atm/mol·K})(300\ \text{K})}{3.00\ \text{L}} = 4.10\ \text{atm}$$

$$P(CO_2) = \frac{(2.00\ \text{mol})(0.0821\ \text{L·atm/mol·K})(300\ \text{K})}{3.00\ \text{L}} = 16.4\ \text{atm} \qquad P(\text{total}) = (4.10\ \text{atm}) + (16.4\ \text{atm}) = 20.5\ \text{atm}$$

12.89 In a gaseous mixture at 20 °C the partial pressures of the components are: hydrogen, 150 torr; carbon dioxide, 200 torr; methane, 320 torr; ethylene, 105 torr. What are the total pressure of the mixture and the volume percent of hydrogen?

▮

$$\text{Total pressure of mixture} = \text{sum of partial pressures} = 150 + 200 + 320 + 105 = 775\ \text{torr}$$

$$\text{Volume fraction of } H_2 = \frac{\text{partial pressure of } H_2}{\text{total pressure of mixture}} = \frac{150\ \text{torr}}{775\ \text{torr}} = 0.194 = 19.4\%$$

12.90 A 200 mL flask contained oxygen at 220 torr, and a 300 mL flask contained nitrogen at 100 torr. The two flasks were then connected so that each gas filled their combined volume. Assuming no change in temperature, what was the partial pressure of each gas in the final mixture and what was the total pressure?

▮ The final total volume was 500 mL.

Oxygen:
$$P_{\text{final}} = P_{\text{initial}}\left(\frac{V_{\text{initial}}}{V_{\text{final}}}\right) = (220\ \text{torr})\left(\frac{200}{500}\right) = 88\ \text{torr}$$

Nitrogen:
$$P_f = P_i\left(\frac{V_i}{V_f}\right) = (100\ \text{torr})\left(\frac{300}{500}\right) = 60\ \text{torr}$$

$$\text{Total pressure} = (88\ \text{torr}) + (60\ \text{torr}) = 148\ \text{torr}$$

12.91 If 3.00 L oxygen gas is collected over water at 27 °C when the barometric pressure is 787 torr, what is the volume of dry gas at 0 °C and 1.00 atm, assuming ideal behavior? Vapor pressure of water at 27 °C is 27 torr.

▮

$$P(O_2) = (787\ \text{torr}) - (27\ \text{torr}) = 760\ \text{torr} = 1.00\ \text{atm} \qquad n(O_2) = \frac{PV}{RT} = \frac{(1.00)(3.00)}{(0.0821)(300)} = 0.122\ \text{mol}$$

$$V = \frac{nRT}{P} = \frac{(0.122\ \text{mol})(0.0821\ \text{L·atm/mol·K})(273\ \text{K})}{1.00\ \text{atm}} = 2.73\ \text{L}$$

12.92 Exactly 100 mL of oxygen is collected over water at 23 °C and 800 torr. Compute the standard volume of the dry oxygen. Vapor pressure of water at 23 °C is 21.1 torr.

▮ The gas collected is actually a mixture of oxygen and water vapor. The partial pressure of water vapor in the mixture at 23 °C is 21.1 torr. Hence

$$\text{Pressure of dry oxygen} = (\text{total pressure}) - (\text{vapor pressure of water}) = (800\ \text{torr}) - (21\ \text{torr}) = 779\ \text{torr}$$

Thus, for the dry oxygen, $V_1 = 100\ \text{mL}$, $T_1 = 23 + 273 = 296\ \text{K}$, $P_1 = 779\ \text{torr}$. Converting to STP,

$$V_2 = V_1\left(\frac{T_2}{T_1}\right)\left(\frac{P_1}{P_2}\right) = (100\ \text{mL})\left(\frac{273\ \text{K}}{296\ \text{K}}\right)\left(\frac{779\ \text{torr}}{760\ \text{torr}}\right) = 94.5\ \text{mL}$$

12.93 In a basal metabolism measurement timed at 6.00 min, a patient exhaled 52.5 L of air, measured over water at 20 °C. The vapor pressure of water at 20 °C is 17.5 torr. The barometric pressure was 750 torr. The exhaled air analyzed 16.75 vol % oxygen, and the inhaled air 20.32 vol % oxygen, both on a dry basis. Neglecting any solubility of the gases in water and any difference in the total volumes of inhaled and exhaled air, calculate the rate of oxygen consumption by the patient in mL (STP) per minute.

▮

$$\text{Volume of dry air at STP} = (52.5\text{ L})\left(\frac{273\text{ K}}{293\text{ K}}\right)\left[\frac{(750 - 17.5)\text{ torr}}{760\text{ torr}}\right] = 47.1\text{ L}$$

$$\text{Rate of oxygen consumption} = \frac{\text{volume of oxygen (STP) consumed}}{\text{time in which this volume was consumed}}$$

$$= \frac{(0.2032 - 0.1675)(47.1\text{ L})}{6.00\text{ min}} = 0.280\text{ L/min} = 280\text{ mL/min}$$

12.94 Exactly 0.550 L nitrogen is collected over water at 25 °C and 755 torr. The gas is saturated with water vapor. Compute the volume of the nitrogen in the dry condition at STP. Vapor pressure of water at 25 °C is 23.8 torr.

▮

$$P(N_2) = P(\text{total}) - P(H_2O) = (755\text{ torr}) - (23.8\text{ torr}) = 731\text{ torr}$$

$$V_2 = \frac{P_1V_1T_2}{T_1P_2} = \frac{(0.550\text{ L})(731\text{ torr})(273\text{ K})}{(298\text{ K})(760\text{ torr})} = 0.485\text{ L}$$

12.95 A dry gas occupied 150 mL at STP. If this same mass of gas were collected over water at 23 °C and a total gas pressure of 745 torr, what volume would it occupy? The vapor pressure of water at 23 °C is 21 torr.

▮ The final pressure of the dry gas must be used. (Imagine expanding the dry gas to V_2 and then adding the water vapor, which increases the pressure but not the volume.)

$$P(\text{gas}) = P(\text{total}) - P(H_2O) = (745\text{ torr}) - (21\text{ torr}) = 724\text{ torr}$$

$$\frac{P_1V_1}{T_1} = \frac{P_2V_2}{T_2} \qquad V_2 = \frac{P_1V_1T_2}{T_1P_2} = \frac{(760\text{ torr})(150\text{ mL})(296\text{ K})}{(273\text{ K})(724\text{ torr})} = 171\text{ mL}$$

12.96 A mixture of N_2, NO, and NO_2 was analyzed by selective absorption of the oxides of nitrogen. The initial volume of the mixture was 2.74 cm^3. After treatment with water, which absorbed the NO_2, the volume was 2.02 cm^3. A ferrous sulfate solution was then shaken with the residual gas to absorb the NO, after which the volume was 0.25 cm^3. All volumes were measured at barometric pressure. Neglecting water vapor, what was the volume percent of each gas in the original mixture?

▮

$$\%\,N_2 = \left(\frac{0.25\text{ cm}^3}{2.74\text{ cm}^3}\right)(100\%) = 9.1\%\,N_2 \qquad \%\,\text{NO} = \left(\frac{(2.02 - 0.25)\text{ cm}^3}{2.74\text{ cm}^3}\right)(100\%) = 64.6\%\,\text{NO}$$

$$\%\,NO_2 = \left(\frac{(2.74 - 2.02)\text{ cm}^3}{2.74\text{ cm}^3}\right)(100\%) = 26.3\%\,NO_2$$

12.97 A 250 mL flask contained krypton at 500 torr. A 450 mL flask contained helium at 950 torr. The contents of the two flasks were mixed by opening a stopcock connecting them. Assuming that all operations were carried out at a uniform constant temperature, calculate the final total pressure and the volume percent of each gas in the mixture. Neglect the volume of the stopcock.

▮

$$P_2(\text{Kr}) = P_1(\text{Kr})\left(\frac{V_1}{V_2}\right) = (500\text{ torr})\left(\frac{250\text{ mL}}{700\text{ mL}}\right) = 179\text{ torr} \qquad P_2(\text{He})\left(\frac{V_1}{V_2}\right) = (950\text{ torr})\left(\frac{450\text{ mL}}{700\text{ mL}}\right) = 611\text{ torr}$$

$$\text{Total pressure} = 790\text{ torr} \qquad \text{Vol }\% = \left(\frac{179\text{ torr}}{790\text{ torr}}\right)(100\%) = 22.7\%\,\text{Kr} \quad \text{and} \quad 77.3\%\,\text{He}$$

12.98 The vapor pressure of water at 80 °C is 355 torr. A 100 mL vessel contained water-saturated oxygen at 80 °C, the total gas pressure being 760 torr. The contents of the vessel were pumped into a 50.0 mL vessel at the same temperature. What were the partial pressures of oxygen and of water vapor, and what was the total pressure in the final equilibrated state? Neglect the volume of any water which might condense.

▮ The partial pressure of water remains the same, 355 torr; the excess water condenses. The partial pressure of oxygen doubles, $2(760 - 355) = 810$ torr. The total pressure is just the sum, 1165 torr.

12.99 Show that for a mixture of gases $P(\text{total})V = n(\text{total})RT$.

▮ The total number of moles of gas is given by

$$n(\text{total}) = n_1 + n_2 + \cdots = \frac{P_1V_1}{RT_1} + \frac{P_2V_2}{RT_2} + \cdots$$

Since all the gases in the mixture are in the same container, they all occupy the same volume and they all have the same temperature:

$$n(\text{total}) = \frac{P_1V}{RT} + \frac{P_2V}{RT} + \cdots = \frac{V}{RT}(P_1 + P_2 + \cdots)$$

Thus $n(\text{total})RT = V(P_1 + P_2 + \cdots) = VP(\text{total})$

12.100 Oxygen is collected over water at 25 °C in a 2.00 L vessel at a total barometric pressure of 765 torr. Calculate the number of moles of oxygen collected.

▮ The vapor pressure of water at 25 °C is 24 torr. Thus the pressure due to the oxygen is (765 torr) − (24 torr) = 741 torr. The number of moles of oxygen may be calculated:

$$n = \frac{PV}{RT} = \frac{(\frac{741}{760}\text{ atm})(2.00\text{ L})}{(0.0821\text{ L}\cdot\text{atm/mol}\cdot\text{K})(298\text{ K})} = 0.0797\text{ mol}$$

12.101 If 2.0 g N_2, 0.40 g H_2, and 9.0 g O_2 are put into a 1.00 L container at 27 °C, what is the total pressure in the container?

▮ See Problem 12.99.

$$n(\text{total}) = \left(\frac{2.0\text{ g}}{28.0\text{ g/mol}}\right) + \left(\frac{0.40\text{ g}}{2.0\text{ g/mol}}\right) + \left(\frac{9.0\text{ g}}{32.0\text{ g/mol}}\right) = 0.55\text{ mol}$$

$$P(\text{total}) = \frac{(0.55\text{ mol})(0.0821\text{ L}\cdot\text{atm/mol}\cdot\text{K})(300\text{ K})}{1.00\text{ L}} = 14\text{ atm}$$

12.102 A 15.0 L vessel containing 5.65 g N_2 is connected by means of a valve to a 6.00 L vessel containing 5.00 g oxygen. After the valve is opened and the gases are allowed to mix, what will be the partial pressure of each gas and the total pressure at 27 °C?

▮ The total volume after mixing is 21.0 L

$$(5.65\text{ g N}_2)\left(\frac{1\text{ mol N}_2}{28.0\text{ g}}\right) = 0.202\text{ mol N}_2 \qquad (5.00\text{ g O}_2)\left(\frac{1\text{ mol O}_2}{32.0\text{ g}}\right) = 0.156\text{ mol O}_2$$

$$P(\text{N}_2) = \frac{nRT}{V} = \frac{(0.202\text{ mol})(0.0821\text{ L}\cdot\text{atm/mol}\cdot\text{K})(300\text{ K})}{21.0\text{ L}} = 0.237\text{ atm}$$

$$P(\text{O}_2) = \frac{(0.156\text{ mol})(0.0821\text{ L}\cdot\text{atm/mol}\cdot\text{K})(300\text{ K})}{21.0\text{ L}} = 0.183\text{ atm}$$

$$P(\text{total}) = P(\text{N}_2) + P(\text{O}_2) = (0.237\text{ atm}) + (0.183\text{ atm}) = 0.420\text{ atm}$$

12.103 Into a 5.00 L container at 18 °C are placed 0.200 mol H_2, 20.0 g CO_2, and 14.00 g O_2. Calculate the total pressure in the container and the partial pressure of each gas.

▮

$$P(\text{H}_2) = \frac{nRT}{V} = \frac{(0.200\text{ mol})(0.0821\text{ L}\cdot\text{atm/mol}\cdot\text{K})(291\text{ K})}{5.00\text{ L}} = 0.956\text{ atm}$$

$$n(\text{CO}_2) = (20.0\text{ g})\left(\frac{1\text{ mol}}{44.0\text{ g}}\right) = 0.4545\text{ mol CO}_2$$

$$P(\text{CO}_2) = \frac{(0.4545\text{ mol})(0.0821\text{ L}\cdot\text{atm/mol}\cdot\text{K})(291\text{ K})}{5.00\text{ L}} = 2.17\text{ atm}$$

$$n(\text{O}_2) = (14.00\text{ g})\left(\frac{1\text{ mol}}{32.00\text{ g}}\right) = 0.4375\text{ mol O}_2$$

$$P(\text{O}_2) = \frac{(0.4375\text{ mol})(0.0821\text{ L}\cdot\text{atm/mol}\cdot\text{K})(291\text{ K})}{5.00\text{ L}} = 2.09\text{ atm}$$

$$P(\text{total}) = (0.956\text{ atm}) + (2.17\text{ atm}) + (2.09\text{ atm}) = 5.22\text{ atm}$$

12.104 How many g of oxygen is contained in 10.5 L of oxygen measured over water at 25 °C and 740 torr? Vapor pressure of water at 25 °C is 24 torr.

▮ $$P(O_2) = P(\text{total}) - P(\text{water}) = (740 \text{ torr}) - (24 \text{ torr}) = 716 \text{ torr}$$

$$n = \frac{PV}{RT} = \frac{(\frac{716}{760} \text{ atm})(10.5 \text{ L})}{(0.0821 \text{ L·atm/mol·K})(298 \text{ K})} = 0.404 \text{ mol} \qquad (0.404 \text{ mol})\left(\frac{32.0 \text{ g}}{\text{mol}}\right) = 12.9 \text{ g}$$

12.105 If 0.10 mol N_2, 0.30 mol H_2, and 0.20 mol He are placed in a 10.0 L container at 25 °C, ***(a)*** what are the volume, temperature, and pressure of each gas? ***(b)*** What is the total pressure? ***(c)*** If 0.20 mol of NH_3 and 0.20 mol of He are placed in a similar container at the same temperature, what is the total pressure in the container? ***(d)*** Would a reaction converting some N_2 and H_2 into NH_3 raise or lower the total pressure in the first container?

▮ ***(a)*** $V = 10.0$ L for each; $T = 298$ K for each.

$$P(N_2) = \frac{nRT}{V} = \frac{(0.10 \text{ mol})(0.0821 \text{ L·atm/mol·K})(298 \text{ K})}{10.0 \text{ L}} = 0.24 \text{ atm}$$

In the same manner, $P(H_2) = 0.73$ atm and $P(\text{He}) = 0.49$ atm

(b) $P(\text{total}) = P(N_2) + P(H_2) + P(\text{He}) = 1.46$ atm,

(c) In the same manner, $P'(\text{total}) = 0.98$ atm ***(d)*** The pressure would be lowered.

12.106 Relative humidity is defined as the ratio of the partial pressure of water in air at a given temperature to the vapor pressure of water at that temperature. Calculate the mass of water per liter of air at ***(a)*** 20 °C and 45% relative humidity, ***(b)*** 0 °C and 95% relative humidity. ***(c)*** Discuss whether temperature or relative humidity has the greater effect on the mass of water vapor in the air.

▮ ***(a)*** The pressure of water in the air is 45% of 17.5 torr (the vapor pressure of water at 20 °C).

$$P(H_2O) = 0.45(17.5 \text{ torr}) = 7.9 \text{ torr} = 0.0104 \text{ atm}$$

$$n = \frac{PV}{RT} = \frac{(0.0104 \text{ atm})(1.00 \text{ L})}{(0.0821 \text{ L·atm/mol·K})(293 \text{ K})} = 4.3 \times 10^{-4} \text{ mol}$$

$$(4.3 \times 10^{-4} \text{ mol})\left(\frac{18 \text{ g}}{\text{mol}}\right) = 7.8 \times 10^{-3} \text{ g} = 7.8 \text{ mg}$$

(b) $P(H_2O) = (0.95)(4.58 \text{ torr}) = 4.4$ torr. The same procedure as in part *(a)* yields the result 4.6 mg.

(c) Despite the higher percent humidity in part *(b)*, there is less water in the air because of the much lower vapor pressure of water at 0 °C. Both temperature and relative humidity play important roles in determining the mass of water in the air.

12.107 How much water vapor is contained in a cubic room 4.0 m along an edge if the relative humidity is 50% and the temperature is 27 °C? The vapor pressure of water at 27 °C is 26.7 torr. (The relative humidity expresses the partial pressure of water as a percent of the water vapor pressure.)

▮ $$V = (4.0 \text{ m})^3 = 64 \text{ m}^3 = 64 \times 10^3 \text{ L} \qquad P = (0.50)[P(\text{vapor})] = (0.50)(26.7 \text{ torr}) = 13.4 \text{ torr}$$

$$n = \frac{PV}{RT} = \frac{[(13.4/760) \text{ atm}](64 \times 10^3 \text{ L})}{(0.0821 \text{ L·atm/mol·K})(300 \text{ K})} = 45.8 \text{ mol} \qquad (45.8 \text{ mol})\left(\frac{18.0 \text{ g}}{\text{mol}}\right)\left(\frac{1 \text{ kg}}{10^3 \text{ g}}\right) = 0.82 \text{ kg}$$

12.5 MOLECULAR WEIGHTS OF GASES

12.108 Avogadro's law states that equal volumes of gases under the same conditions of temperature and pressure contain equal numbers of molecules. Develop from that law the fact that at a given temperature and pressure, the density of a pure gas is proportional to its molecular weight.

▮ $$PV = nRT = \frac{m}{\text{MW}} RT$$ where m is the mass of the gas and MW is the molecular weight.

Rearranging algebraically yields $$\frac{m}{V} = \frac{P}{RT}(\text{MW})$$

Since m/V is the density of the gas, and under the conditions stated in the problem T and P are constant, the equation can be restated Density = const. × MW which is a mathematical statement of direct proportionality.

12.109 Determine the approximate molecular weight of a gas if 560 mL weighs 1.80 g at STP.

▮ $$PV = \left(\frac{m}{\text{MW}}\right) RT \qquad \text{MW} = \frac{mRT}{PV} = \frac{(1.80 \text{ g})(0.0821 \text{ L·atm/K·mol})(273 \text{ K})}{(1.00 \text{ atm})(0.560 \text{ L})} = 72.0 \text{ g/mol}$$

12.110 At 18 °C and 765 torr, 2.29 L of a gas weighs 3.71 g. Calculate the approximate molecular weight of the gas.

▮ Converting the data to atm and K, $P = \frac{765}{760}$ atm $T = (18 + 273)$ K $= 291$ K

Then
$$MW = \frac{mRT}{PV} = \frac{(3.71\ \text{g})(0.0821\ \text{L·atm/K·mol})(291\ \text{K})}{(\frac{765}{760}\ \text{atm})(2.29\ \text{L})} = 38.5\ \text{g/mol}$$

12.111 If the density of carbon monoxide is 3.17 g/L at −20 °C and 2.35 atm, what is its approximate molecular weight?

▮
$$MW = \frac{dRT}{P} = \frac{(3.17\ \text{g/L})(0.0821\ \text{L·atm/K·mol})(253\ \text{K})}{2.35\ \text{atm}} = 28.0\ \text{g/mol}$$

12.112 At 750 torr and 27 °C, 0.60 g of a certain gas occupies 0.50 L. Calculate its molecular weight.

▮
$$PV = nRT \qquad (\tfrac{750}{760}\ \text{atm})(0.50\ \text{L}) = n\left(0.0821\ \frac{\text{L·atm}}{\text{mol·K}}\right)(300\ \text{K}) \qquad n = 2.0 \times 10^{-2}\ \text{mol}$$

$$MW = \frac{\text{mass}}{\text{number of moles}} = \frac{0.60\ \text{g}}{2.0 \times 10^{-2}\ \text{mol}} = 30\ \text{g/mol}$$

12.113 Calculate the MW of a gas which has a density of 1.48 g/L at 100 °C and 600 torr.

▮ Per liter:
$$n = \frac{PV}{RT} = \frac{(\frac{600}{760}\ \text{atm})(1.00\ \text{L})}{(0.0821\ \text{L·atm/mol·K})(373\ \text{K})} = 0.0258\ \text{mol}$$

$$MW = \frac{1.48\ \text{g/L}}{0.0258\ \text{mol/L}} = 57.4\ \text{g/mol}$$

12.114 If 300 mL of a gas weighs 0.368 g at STP, what is its molecular weight?

▮
$$n = \frac{PV}{RT} = \frac{(1.00\ \text{atm})(0.300\ \text{L})}{(0.0821\ \text{L·atm/mol·K})(273\ \text{K})} = 13.4\ \text{mmol}$$

$$MW = \frac{368\ \text{mg}}{13.4\ \text{mmol}} = 27.5\ \text{mg/mmol} = 27.5\ \text{g/mol}$$

12.115 At 27 °C, 2.40 g of a certain gas occupies a volume of 9.84 L at a pressure of 172 torr. Calculate the MW of the gas.

▮
$$n = \frac{PV}{RT} = \frac{[(172/760)\ \text{atm}](9.84\ \text{L})}{(0.0821\ \text{L·atm/mol·K})(300\ \text{K})} = 0.0904\ \text{mol} \qquad \frac{2.40\ \text{g}}{0.0904\ \text{mol}} = 26.5\ \text{g/mol}$$

12.116 A 2.00 g sample of gas at 27 °C and 1.00 atm occupies 1.00 L. What is the MW of the gas?

▮
$$n = \frac{PV}{RT} = \frac{(1.00\ \text{atm})(1.00\ \text{L})}{(0.0821\ \text{L·atm/mol·K})(300\ \text{K})} = 0.0406\ \text{mol} \qquad MW = \frac{2.00\ \text{g}}{0.0406\ \text{mol}} = 49.3\ \text{g/mol}$$

12.117 A 1.525 g mass of a volatile liquid is vaporized, giving 500 cm³ of vapor when measured over water at 30 °C and 770 torr. The vapor pressure of water at 30 °C is 32 torr. What is the MW of the substance?

▮ For the vapor,
$$P = (770\ \text{torr}) - (32\ \text{torr}) = (738\ \text{torr})\left(\frac{1\ \text{atm}}{760\ \text{torr}}\right) = 0.971\ \text{atm}$$

$$n = \frac{PV}{RT} = \frac{(0.971\ \text{atm})(0.500\ \text{L})}{(0.0821\ \text{L·atm/mol·K})(303\ \text{K})} = 0.0195\ \text{mol} \qquad MW = \frac{1.525\ \text{g}}{0.0195\ \text{mol}} = 78.2\ \text{g/mol}$$

12.118 Argon gas liberated from crushed meteorites does not have the same isotopic composition as atmospheric argon. The gas density of a particular sample of meteoritic Ar was found to be 1.481 g/L at 27 °C and 740 torr. What is the average atomic weight of this sample of argon?

▮ One liter of sample contains
$$n = \frac{PV}{RT} = \frac{[\frac{740}{760}\ \text{atm}](1.00\ \text{L})}{(0.0821\ \text{L·atm/mol·K})(300\ \text{K})} = 0.0395\ \text{mol} \qquad AW = \frac{1.481\ \text{g/L}}{0.0395\ \text{mol/L}} = 37.5\ \text{g/mol}$$

12.119 The density of an unknown gas at 0.821 atm and 27 °C is 2.50 g/L. What is the MW of this gas? (Assume ideal behavior.)

In 1.00 L: $n = \frac{PV}{RT} = \frac{(0.821\ \text{atm})(1.00\ \text{L})}{(0.0821\ \text{L·atm/mol·K})(300\ \text{K})} = 0.0333\ \text{mol} \qquad \text{MW} = \frac{2.50\ \text{g/L}}{0.0333\ \text{mol/L}} = 75.0\ \text{g/mol}$

12.120 Find the MW of a gas if 75.0 g of this gas exerts a pressure of 2.00 atm in a 30.0 L vessel at 500 K.

$$n = \frac{PV}{RT} = \frac{(2.00\ \text{atm})(30.0\ \text{L})}{(0.0821\ \text{L·atm/mol·K})(500\ \text{K})} = 1.46\ \text{mol} \qquad \text{MW} = \frac{75.0\ \text{g}}{1.46\ \text{mol}} = 51.3\ \text{g/mol}$$

12.121 The density of NO was very carefully determined to be 0.2579 kg/m^3 at a temperature and pressure at which oxygen's measured density was 0.2749 kg/m^3. On the basis of this information and the known atomic weight of oxygen, calculate the atomic weight of nitrogen.

Density and *molecular* weight are proportional (Problem 12.108); hence

$$\frac{d_1}{\text{MW}_1} = \frac{d_2}{\text{MW}_2} \qquad \frac{0.2579\ \text{g/L}}{\text{MW}_1} = \frac{0.2749\ \text{g/L}}{32.00\ \text{g/mol}} \qquad \text{MW}_1 = (32.00\ \text{g/mol})\left(\frac{0.2579}{0.2749}\right) = 30.02\ \text{g/mol NO}$$

$$\text{AW} = (30.02\ \text{g/mol}) - (16.00\ \text{g/mol}) = 14.02\ \text{g/mol} = 14.02\ \text{u}$$

12.122 (*a*) Assuming that air is 79 mol % N_2, 20% O_2, and 1% Ar, compute the average MW of air. (*b*) What is the density of air at 25 °C and 1 atm?

(*a*) $\text{MW} = (0.79)(28.0) + (0.20)(32.0) + (0.01)(39.9) = 28.9\ \text{g/mol}$

(*b*) $d = \frac{(\text{MW})P}{RT} = \frac{(28.9\ \text{g/mol})(1.00\ \text{atm})}{(0.0821\ \text{L·atm/mol·K})(298\ \text{K})} = 1.18\ \text{g/L}$

12.123 Devise a method for determining the molecular weight of a low-boiling liquid.

Heat the sample to vaporize the liquid at a known temperature and pressure in a vessel of known mass and volume, displacing all other gases and making sure that all the liquid phase has evaporated. Cool the vessel, allowing the gas to condense to liquid. Weigh the vessel and its contents. The data now available are P, V, T, and m. From the first three of these, the number of moles can be calculated. Then the MW is equal to m/n.

12.124 Find the MW of a gas whose density at 40 °C and 758 torr is 1.386 kg/m^3.

$$\text{MW} = \frac{mRT}{PV} = \left(\frac{m}{V}\right)\left(\frac{RT}{P}\right) = d\left(\frac{RT}{P}\right)$$

$$\left(\frac{1.386\ \text{kg}}{\text{m}^3}\right)\left(\frac{1000\ \text{g}}{\text{kg}}\right)\left(\frac{1\ \text{m}}{10\ \text{dm}}\right)^3 = 1.386\ \text{g/dm}^3 = 1.386\ \text{g/L}$$

$$\text{MW} = \frac{(1.386\ \text{g/L})(0.0821\ \text{L·atm/mol·K})(313\ \text{K})}{(758/760)\ \text{atm}} = 35.7\ \text{g/mol}$$

12.125 A gaseous compound has an empirical formula CH_3. Its density is 2.06 g/L at the same temperature and pressure at which oxygen has a density of 2.21 g/L. Calculate the molecular formula of the gas.

Rearrange $PV = nRT$ to the form $n/V = P/RT$. Since P and T are the same for the oxygen and the unknown gas (X), their numbers of mol/L are also equal. Per liter:

$$n(O_2) = n(\text{X}) \qquad \frac{2.21\ \text{g}}{32.0\ \text{g/mol}} = \frac{2.06\ \text{g}}{\text{MW(X)}}$$

$$\text{MW(X)} = \left(\frac{2.06}{2.21}\right)\left(\frac{32.0\ \text{g}}{\text{mol}}\right) = 29.8\ \text{g/mol}$$

The empirical formula, CH_3, has a formula weight of 15 g/mol; so there must be $29.8/15 \approx 2$ formula units in a molecule of X. The molecular formula is $(CH_3)_2$ or C_2H_6.

12.126 A gas is composed of 30.4% N and 69.6% O. Its density is 11.1 g/L at −20 °C and 2.50 atm. What are the empirical and molecular formulas of the gas?

$$(30.4\ \text{g N})\left(\frac{1\ \text{mol N}}{14.0\ \text{g N}}\right) = 2.17\ \text{mol N} \qquad (69.6\ \text{g O})\left(\frac{1\ \text{mol O}}{16.0\ \text{g O}}\right) = 4.35\ \text{mol O}$$

The mole ratio is 1 mol N to 2 mol O, and the empirical formula is NO_2. That formula has a formula weight of 46. The number of mol in 1 L of the gas is

$$n = \frac{PV}{RT} = \frac{(2.50 \text{ atm})(1.00 \text{ L})}{(0.0821 \text{ L·atm/mol·K})(253 \text{ K})} = 0.120 \text{ mol}$$

The molecular weight is given by $\dfrac{11.1 \text{ g/L}}{0.120 \text{ mol/L}} = 92.5 \text{ g/mol}$.

There are $\dfrac{92.5 \text{ g/mol}}{46 \text{ g/formula unit}} \approx 2$ formula units per mol. The molecular formula is therefore N_2O_4.

12.127 An organic compound had the following analysis: C, 55.8% by weight, H, 7.03%; O, 37.2%. A 1.500 g sample was vaporized and was found to occupy 530 cm^3 at 100 °C and 740 torr. What is the molecular formula of the compound?

▮ The approximate MW, calculated from the gas density data, is 89.0 g/mol. The empirical formula, calculated from the percentage composition data, is C_2H_3O, and the empirical formula weight is 43.0. The exact MW must therefore be $2 \times 43.0 = 86.0$, since this is the only integral multiple of 43.0 which is reasonably close to the approximate MW of 89. The molecule must therefore contain twice the number of atoms in the empirical formula, and the molecular formula must be $C_4H_6O_2$.

Another Method

Instead of calculating the empirical formula, we can use the composition data to calculate the number of mol of atoms of each element in 89 g of compound.

$$n(\text{C}) = \frac{(0.558)(89 \text{ g})}{12.0 \text{ g/mol}} = 4.1 \qquad n(\text{H}) = \frac{(0.0703)(89 \text{ g})}{1.01 \text{ g/mol}} = 6.2 \qquad n(\text{O}) = \frac{(0.372)(89 \text{ g})}{16.0 \text{ g/mol}} = 2.1$$

These numbers approximate the numbers of atoms in a molecule, the small deviations from integral values resulting from the approximate nature of the MW. The molecular formula, $C_4H_6O_2$, is obtained without going through the intermediate evaluation of an empirical formula.

12.128 A gaseous compound is composed of 85.7% by mass carbon and 14.3% by mass hydrogen. Its density is 2.28 g/L at 300 K and 1.00 atm pressure. Determine the molecular formula of the compound.

▮ In 100 g of compound:

$$(85.7 \text{ g C})\left(\frac{1 \text{ mol C}}{12.0 \text{ g C}}\right) = 7.14 \text{ mol C} \qquad (14.3 \text{ g H})\left(\frac{1 \text{ mol H}}{1.008 \text{ g H}}\right) = 14.2 \text{ mol H}$$

The empirical formula is CH_2 (FW = 14 g/empirical formula unit). Per liter:

$$n = \frac{PV}{RT} = \frac{(1.00 \text{ atm})(1.00 \text{ L})}{(0.0821 \text{ L·atm/mol·K})(300 \text{ K})} = 0.0406 \text{ mol}$$

$$\frac{2.28 \text{ g/L}}{0.0406 \text{ mol/L}} = 56.2 \text{ g/mol, or 4 empirical formula units.}$$

Hence the molecular formula is C_4H_8.

12.129 Exactly 250 cm^3 of a gas at STP weighs 0.291 g. The composition of the gas is as follows: C, 92.24%; H, 7.76%. Derive its molecular formula.

▮

$$(92.24 \text{ g C})\left(\frac{1 \text{ mol C}}{12.01 \text{ g C}}\right) = 7.680 \text{ mol C} \qquad (7.76 \text{ g H})\left(\frac{1 \text{ mol H}}{1.008 \text{ g H}}\right) = 7.70 \text{ mol H}$$

Empirical formula: CH

$$n = \frac{PV}{RT} = \frac{(1.00 \text{ atm})(0.250 \text{ L})}{(0.0821 \text{ L·atm/mol·K})(273 \text{ K})} = 0.0112 \text{ mol} \qquad \text{MW} = \frac{0.291 \text{ g}}{0.0112 \text{ mol}} = 26.0 \text{ g/mol}$$

$$\text{Empirical FW} = 13.0 \qquad \frac{\text{MW}}{\text{FW}} = \frac{26.0}{13.0} = 2.0$$

The molecular formula is $(CH)_2$ or C_2H_2.

12.130 A hydrocarbon has the following composition: C = 82.66%; H, 17.34%. The density of the vapor is 0.2308 g/L at 30 °C and 75 torr. Determine its MW and its molecular formula.

▌

$$(82.66\text{ g C})\left(\frac{1\text{ mol C}}{12.0\text{ g C}}\right) = 6.89\text{ mol C} \qquad (17.34\text{ g H})\left(\frac{1\text{ mol H}}{1.00\text{ g H}}\right) = 17.3\text{ mol H}$$

$$\frac{6.89}{6.89} = 1 \qquad \frac{17.3}{6.89} = 2.5$$

Multiplying by 2 yields integers. The ratio is 2 mol C to 5 mol H, giving the empirical formula C_2H_5.

$$\text{MW} = \frac{dRT}{P} = \frac{(0.2308\text{ g/L})(0.0821\text{ L·atm/mol·K})(303\text{ K})}{(75/760)\text{ atm}} = 58.2\text{ g/mol}$$

$$\frac{58.2\text{ g/mol}}{29.0\text{ g/form. unit}} = \frac{2\text{ formula units}}{\text{mol}} \qquad \text{The molecular formula is } C_4H_{10}.$$

12.131 One of the methods for estimating the temperature of the center of the sun is based on the ideal gas law. If the center is assumed to consist of gases whose average MW is 2.0, and if the density and pressure are 1.4×10^3 kg/m³ and 1.3×10^9 atm, respectively, calculate the temperature.

▌

$$T = \frac{(\text{MW})PV}{mR} = \frac{(2.0\text{ g/mol})(1.3 \times 10^9\text{ atm})}{(0.0821\text{ L·atm/mol·K})(1.4 \times 10^3\text{ g/L})} = 2.3 \times 10^7\text{ K}$$

12.132 An empty flask open to the air weighed 24.173 g. The flask was filled with the vapor of an organic liquid and was sealed off at the barometric pressure and at 100 °C. At room temperature the flask then weighed 25.002 g. The flask was then opened and filled with water at room temperature, after which it weighed 176 g. The barometric reading was 725 mmHg. All weighings were done at the temperature of the room, 25 °C. What is the MW of the organic vapor? Allow for the buoyancy of the air in the weighing of the sealed-off flask, using the result of Problem 12.122(*b*).

▌

$$m(H_2O) = (176\text{ g}) - (24.173\text{ g}) = 152\text{ g} \qquad V = (152\text{ g})\left(\frac{1\text{ mL}}{0.997\text{ g}}\right) = 152\text{ mL at }25\text{ °C}$$

$$m(\text{air}) = (0.152\text{ L})(1.18\text{ g/L}) = 0.179\text{ g}$$

$$m(\text{vapor}) = m(\text{flask + vapor}) - m(\text{flask}) = (25.002\text{ g}) - [(24.173\text{ g}) - (0.179\text{ g})] = 1.008\text{ g}$$

$$\text{MW} = \frac{mRT}{PV} = \frac{(1.008\text{ g})(0.0821\text{ L·atm/mol·K})(373\text{ K})}{[(725/760)\text{ atm}](0.152\text{ L})} = 213\text{ g/mol}$$

12.133 Three volatile compounds of a certain element have gaseous densities calculated back to STP as follows: 6.75, 9.56, and 10.08 kg/m³. The three compounds contain 96.0%, 33.9%, and 96.4% of the element in question, respectively. What is the most probable atomic weight of the element?

▌ The molecular weights are calculated as

$$(6.75\text{ g/L})[22.4\text{ L (STP)/mol}] = 151\text{ g/mol}$$
$$(9.56\text{ g/L})[22.4\text{ L (STP)/mol}] = 214\text{ g/mol}$$
$$(10.08\text{ g/L})[22.4\text{ L (STP)/mol}] = 226\text{ g/mol}$$

Multiplying by the percentage of each element gives the number of g of that element per mol of compound.

$$151 \times \frac{96.0}{100.0} = 145\text{ g} \qquad \frac{145\text{ g}}{72.5\text{ g}} = 2$$

$$214 \times \frac{33.9}{100.0} = 72.5\text{ g} \qquad \frac{72.5\text{ g}}{72.5\text{ g}} = 1$$

$$226 \times \frac{96.4}{100.0} = 218\text{ g} \qquad \frac{218\text{ g}}{72.5\text{ g}} = 3$$

Dividing by the largest common multiple yields the number of atoms per molecule or a multiple thereof. Presumably there is one atom of element per molecule in the second compound, two atoms per molecule in the first compound, and three atoms per molecule in the third. The AW is 72.5. If the AW were 36.2, there would be 4, 2, and 6 atoms per molecule, which is possible since only these data are available. In fact, $72.5/n$ for any small integer n is a possible atomic weight.

12.134 It is found that in 11.2 L at STP of any gaseous compound of X there is never less than 15.5 g of X. Also, this volume of the vapor of X itself at STP weighs 62 g. What conclusions may be drawn from these data with reference to the atomic and molecular weights of X?

▮ In 22.4 L there is never less than $2 \times 15.5\text{ g} = 31.0\text{ g}$ of X. As one atom is the smallest amount of an element that can exist in a molecule, 1 mol of atoms is the smallest amount that can exist in 1 mol of a compound (22.4 L for gaseous compounds at STP). Hence the approximate atomic weight of X is 31.0.

Also, 22.4 L of X vapor weighs $2 \times 62\text{ g} = 124\text{ g}$. Hence the approximate molecular weight of X is 124, and the molecule contains $\frac{124}{31} = 4$ atoms (X_4).

12.6 REACTIONS INVOLVING GASES

12.135 When sufficient energy is added to a mixture of N_2 and O_2, the gases react to form oxides of nitrogen. Name two sources—one natural and one man-made—which contribute to air pollution from nitrogen oxides.

▮ Possible answers include lightning, automobile engines, sparking electric motors, and airplane engines.

12.136 What volume of oxygen, measured at STP (0 °C and 1 atm), can be prepared by heating 2.16 g HgO?

▮

$$2HgO \longrightarrow 2Hg + O_2 \qquad (2.16\text{ g HgO})\left(\frac{1\text{ mol HgO}}{216\text{ g HgO}}\right)\left(\frac{1\text{ mol }O_2}{2\text{ mol HgO}}\right) = 0.00500\text{ mol }O_2$$

$$V = \frac{nRT}{P} = \frac{(0.00500\text{ mol})(0.0821\text{ L·atm/mol·K})(273\text{ K})}{1.00\text{ atm}} = 0.112\text{ L}$$

12.137 Calcium metal reacts with hydrochloric acid to yield hydrogen and calcium chloride. Write a balanced chemical equation for the reaction. Determine the volume of hydrogen gas at 1.00 atm pressure and 18 °C produced from the reaction of 12.2 g of calcium with excess HCl.

▮

$$Ca + 2HCl \longrightarrow CaCl_2 + H_2 \qquad (12.2\text{ g Ca})\left(\frac{1\text{ mol Ca}}{40.08\text{ g Ca}}\right)\left(\frac{1\text{ mol }H_2}{1\text{ mol Ca}}\right) = 0.304\text{ mol }H_2$$

$$V = \frac{nRT}{P} = \frac{(0.304\text{ mol})(0.0821\text{ L·atm/mol·K})(291\text{ K})}{1.00\text{ atm}} = 7.27\text{ L}$$

12.138 Calculate the volume of CO_2 at 25 °C and 765 mmHg required to produce 100 g of $Ca(HCO_3)_2$ by its reaction with $Ca(OH)_2$.

▮

$$2CO_2 + Ca(OH)_2 \longrightarrow Ca(HCO_3)_2$$

$$(100\text{ g }Ca(HCO_3)_2)\left(\frac{1\text{ mol }Ca(HCO_3)_2}{162\text{ g }Ca(HCO_3)_2}\right)\left(\frac{2\text{ mol }CO_2}{1\text{ mol }Ca(HCO_3)_2}\right) = 1.23\text{ mol }CO_2$$

$$V = \frac{nRT}{P} = \frac{(1.23\text{ mol})(0.0821\text{ L·atm/mol·K})(298\text{ K})}{(765/760)\text{ atm}} = 29.9\text{ L}$$

12.139 Calculate the volume of O_2 which can be prepared at 60 °C and 760 torr by the decomposition of 20.0 g H_2O_2 to H_2O and O_2.

▮

$$2H_2O_2 \longrightarrow 2H_2O + O_2 \qquad (20.0\text{ g }H_2O_2)\left(\frac{1\text{ mol }H_2O_2}{34.0\text{ g }H_2O_2}\right)\left(\frac{1\text{ mol }O_2}{2\text{ mol }H_2O_2}\right) = 0.294\text{ mol }O_2$$

$$V = \frac{nRT}{P} = \frac{(0.294\text{ mol }O_2)(0.0821\text{ L·atm/mol·K})(333\text{ K})}{1.00\text{ atm}} = 8.04\text{ L}$$

12.140 ***(a)*** Determine the volume of oxygen gas at 27 °C and 0.821 atm produced by decomposition of 2.44 g $KClO_3$. KCl is the other product. ***(b)*** How many atoms of oxygen are there in this quantity of product?

▮ ***(a)*** $2KClO_3 \longrightarrow 2KCl + 3O_2$

$$(2.44\text{ g }KClO_3)\left(\frac{1\text{ mol }KClO_3}{122\text{ g }KClO_3}\right)\left(\frac{3\text{ mol }O_2}{2\text{ mol }KClO_3}\right) = 0.0300\text{ mol }O_2$$

$$V = \frac{nRT}{P} = \frac{(0.0300\text{ mol})(0.0821\text{ L·atm/mol·K})(300\text{ K})}{0.821\text{ atm}} = 0.900\text{ L}$$

(b)

$$(0.0300\text{ mol }O_2)\left(\frac{6.02 \times 10^{23}\text{ molecules }O_2}{\text{mol }O_2}\right)\left(\frac{2\text{ atoms O}}{\text{molecule }O_2}\right) = 3.60 \times 10^{22}\text{ O atoms}$$

12.141 How many L of oxygen, at standard conditions, can be obtained from 110 g of potassium chlorate?

▮
$$\underset{(2\ \text{mol})}{2KClO_3(s)} \longrightarrow 2KCl + \underset{(3\ \text{mol})}{3O_2(g)}$$

Molar Method
The equation shows that 2 mol $KClO_3$ gives 3 mol O_2.

$$n(KClO_3) = \frac{110\ \text{g}}{122.6\ \text{g/mol}} = 0.897\ \text{mol}\ KClO_3 \qquad n(O_2) = (3/2)n(KClO_3) = (3/2)(0.897) = 1.35\ \text{mol}\ O_2$$

$$\text{Volume of 1.35 mol } O_2 \text{ at STP} = (1.35\ \text{mol}\ O_2)(22.4\ \text{L/mol}) = 30.2\ \text{L}\ O_2$$

Another Method
The equation shows that 2 mol $KClO_3$ (2 × 122.6 = 245.2 g) gives 3 molar volumes O_2 (3 × 22.4 L = 67.2 L). Then

$$245.2\ \text{g}\ KClO_3 \text{ gives } 67.2\ \text{L}\ O_2 \text{ at STP} \qquad 1\ \text{g}\ KClO_3 \text{ gives } \frac{67.2}{245.2}\ \text{L}\ O_2 \text{ at STP}$$

and
$$110\ \text{g}\ KClO_3 \text{ gives } 110 \times \frac{67.2}{245.2}\ \text{L} = 30.1\ \text{L}\ O_2 \text{ at STP}$$

12.142 What volume of oxygen, at 18 °C and 750 torr, can be obtained from 110 g $KClO_3$?

▮ This problem is identical with Problem 12.141, except that the 30.2 L of O_2 at 0 °C and 760 torr must be converted to liters of O_2 at 18 °C and 750 torr.

$$\text{Volume at 18 °C and 750 torr} = (30.2\ \text{L})\left(\frac{(273+18)\text{K}}{273\ \text{K}}\right)\left(\frac{760\ \text{torr}}{750\ \text{torr}}\right) = 32.6\ \text{L}$$

Another Method

$$V = \frac{nRT}{P} = \frac{(1.35\ \text{mol})(0.0821\ \text{L·atm/K·mol})(291\ \text{K})}{(750/760)\ \text{atm}} = 32.7\ \text{L}$$

12.143 What volume of $H_2(g)$ at 27 °C and 680 torr is produced by the reaction of 15.0 g Al metal with excess HCl(*aq*) to produce $AlCl_3$ and H_2?

▮
$$2Al + 6HCl \longrightarrow 2AlCl_3 + 3H_2 \qquad (15.0\ \text{g Al})\left(\frac{1\ \text{mol Al}}{27.0\ \text{g Al}}\right)\left(\frac{3\ \text{mol}\ H_2}{2\ \text{mol Al}}\right) = 0.833\ \text{mol}\ H_2$$

$$V = \frac{nRT}{P} = \frac{(0.833\ \text{mol})(0.0821\ \text{L·atm/mol·K})(300\ \text{K})}{(680/760)\ \text{atm}} = 22.9\ \text{L}$$

12.144 What mass of $C_6H_{13}Br$ will be produced by a reaction giving 65% yield if 12.5 mL liquid C_6H_{12} (density = 0.673 g/mL) is treated with 2.70 L HBr(*g*) at STP?

▮
$$(12.5\ \text{mL})\left(\frac{0.673\ \text{g}}{\text{mL}}\right)\left(\frac{1\ \text{mol}\ C_6H_{12}}{84.0\ \text{g}\ C_6H_{12}}\right) = 0.100\ \text{mol}\ C_6H_{12}$$

$$n = \frac{PV}{RT} = \frac{(1.00\ \text{atm})(2.70\ \text{L})}{(0.0821\ \text{L·atm/mol·K})(273\ \text{K})} = 0.120\ \text{mol HBr}$$

$$C_6H_{12} + HBr \longrightarrow C_6H_{13}Br$$

The C_6H_{12} is in limiting quantity, and 0.100 mol of $C_6H_{13}Br$ can be prepared.

$$(0.100\ \text{mol}\ C_6H_{13}Br)\left(\frac{165\ \text{g}\ C_6H_{13}Br}{1\ \text{mol}\ C_6H_{13}Br}\right) = 16.5\ \text{g}\ C_6H_{13}Br$$

Theoretically, 16.5 g of the compound can be produced. But there is only a 65% yield. The actual number of g which will be produced is therefore

$$(16.5\ \text{g calculated})\left(\frac{65\ \text{g produced}}{100\ \text{g calculated}}\right) = 10.7\ \text{g produced}$$

12.145 How many g of zinc must be dissolved in sulfuric acid in order to obtain 600 cm³ of hydrogen at 20 °C and 770 torr?

$$\underset{(1\ \text{mol})}{\text{Zn}(s)} + \text{H}_2\text{SO}_4 \longrightarrow \text{ZnSO}_4 + \underset{(1\ \text{mol})}{\text{H}_2(g)}$$

The number of mol of H_2 may be found by either of the following methods:

$$n(\text{H}_2) = \frac{PV}{RT} = \frac{[(770/760)\ \text{atm}](0.600\ \text{L})}{(0.0821\ \text{L}\cdot\text{atm/K}\cdot\text{mol})(293\ \text{K})} = 0.0253\ \text{mol H}_2$$

or

$$V_{\text{STP}} = (600\ \text{cm}^3)\left(\frac{273\ \text{K}}{(273+20)\ \text{K}}\right)\left(\frac{770\ \text{torr}}{760\ \text{torr}}\right) = 566\ \text{cm}^3 = 0.566\ \text{L}$$

$$n(\text{H}_2) = \frac{0.566\ \text{L}}{22.4\ \text{L/mol}} = 0.0253\ \text{mol H}_2$$

The equation shows that 1 mol H_2 requires 1 mol Zn. Then 0.0253 mol H_2 requires 0.0253 mol Zn, the mass of which is $m(\text{Zn}) = (0.0253\ \text{mol})(65.4\ \text{g/mol}) = 1.65\ \text{g Zn}$

12.146 How many g of $KClO_3$ is needed to prepare 1.8 L of oxygen which is collected over water at 22 °C and 760 torr? Vapor pressure of water at 22 °C is 19.8 torr.

$$P(\text{O}_2) = (760\ \text{torr}) - (19.8\ \text{torr}) = 740\ \text{torr} \qquad n(\text{O}_2) = \frac{PV}{RT} = \frac{[(740/760)\ \text{atm}](1.8\ \text{L})}{(0.0821\ \text{L}\cdot\text{atm/mol}\cdot\text{K})(295\ \text{K})} = 0.0724\ \text{mol O}_2$$

$$2\text{KClO}_3 \longrightarrow 2\text{KCl} + 3\text{O}_2 \qquad (0.0724\ \text{mol O}_2)\left(\frac{2\ \text{mol KClO}_3}{3\ \text{mol O}_2}\right)\left(\frac{122\ \text{g KClO}_3}{\text{mol KClO}_3}\right) = 5.9\ \text{g KClO}_3$$

12.147 What volume of CO_2 at 350 °C and 1.00 atm pressure would be produced from the complete thermal decomposition of 1.00 kg of $MgCO_3$?

$$\frac{1000\ \text{g MgCO}_3}{84.3\ \text{g/mol}} = 11.9\ \text{MgCO}_3$$

$MgCO_3 \rightarrow MgO + CO_2$ Since 1 mol of CO_2 is produced per mole of $MgCO_3$ decomposed, 11.9 mol of CO_2 is generated in this case.

$$PV = nRT \qquad (1.00\ \text{atm})V = (11.9\ \text{mol})\left(0.0821\ \frac{\text{L}\cdot\text{atm}}{\text{mol}\cdot\text{K}}\right)[(350+273)\text{K}] \qquad V = 609\ \text{L}$$

12.148 What volume of O_2 at STP is required for the complete combustion of 1.00 mol of carbon disulfide, CS_2? What volumes of CO_2 and SO_2 at STP are produced?

$$\underset{(1\ \text{mol})}{\text{CS}_2(l)} + \underset{(3\ \text{mol})}{3\text{O}_2(g)} \longrightarrow \underset{(1\ \text{mol})}{\text{CO}_2(g)} + \underset{(2\ \text{mol})}{2\text{SO}_2(g)}$$

The equation shows that 1 mol CS_2 reacts with 3 mol O_2 to form 1 mol CO_2 and 2 mol SO_2. One mole of a *gas* at STP occupies 22.4 L. Therefore:

$$\text{Volume of 3.00 mol O}_2\ \text{at STP} = 3.00 \times 22.4 = 67.2\ \text{L O}_2$$
$$\text{Volume of 1.00 mol CO}_2\ \text{at STP} = 1.00 \times 22.4 = 22.4\ \text{L CO}_2$$
$$\text{Volume of 2.00 mol SO}_2\ \text{at STP} = 2.00 \times 22.4 = 44.8\ \text{L SO}_2$$

12.149 All the oxygen in $KClO_3$ can be converted to O_2 by heating in the presence of a catalyst. **(*a*)** What volume of oxygen measured at 20 °C and 0.996 atm pressure can be prepared from 450 g $KClO_3$? **(*b*)** If the oxygen were collected over water at 20 °C and 0.996 atm barometric pressure, what would be the volume of gas collected?

$$2\text{KClO}_3 \longrightarrow 2\text{KCl} + 3\text{O}_2$$

(*a*) $(450\ \text{g KClO}_3)\left(\frac{1\ \text{mol KClO}_3}{122.6\ \text{g}}\right)\left(\frac{3\ \text{mol O}_2}{2\ \text{mol KClO}_3}\right) = 5.51\ \text{mol O}_2$

$$V = \frac{nRT}{P} = \frac{(5.51\ \text{mol})(0.0821\ \text{L}\cdot\text{atm/mol}\cdot\text{K})(293\ \text{K})}{0.996\ \text{atm}} = 133\ \text{L}$$

(*b*) At 20 °C the vapor pressure of water is 17.5 torr = 0.0230 atm; hence the partial pressure of O_2 is 0.996 − 0.0230 = 0.973 atm. Then, for the O_2,

$$V = \frac{nRT}{P} = \frac{(5.51\ \text{mol})(0.0821\ \text{L}\cdot\text{atm/mol}\cdot\text{K})(293\ \text{K})}{0.973\ \text{atm}} = 136\ \text{L}$$

12.150 Calculate the change in pressure when 1.04 mol NO and 20.0 g O_2 in a 20.0 L vessel originally at 27 °C react to produce the maximum quantity of NO_2 possible according to the equation $2NO + O_2 \longrightarrow 2NO_2$.

▮ Before the reaction there are 1.04 mol of NO and $(20.0\text{ g }O_2)\left(\frac{1\text{ mol }O_2}{32.0\text{ g }O_2}\right) = 0.625\text{ mol }O_2$.

$$n(\text{total}) = 1.665\text{ mol} \qquad P = \frac{nRT}{V} = \frac{(1.665\text{ mol})(0.0821\text{ L}\cdot\text{atm/mol}\cdot\text{K})(300\text{ K})}{20.0\text{ L}} = 2.05\text{ atm}$$

After the reaction, $2NO + O_2 \longrightarrow 2NO_2$, there would be present 1.04 mol of NO_2, and 0.105 mol of oxygen would remain unreacted. The total number of moles of gas is 1.145 mol. The pressure is

$$P = \frac{nRT}{V} = \frac{(1.145\text{ mol})(0.0821\text{ L}\cdot\text{atm/mol}\cdot\text{K})(300\text{ K})}{20.0\text{ L}} = 1.41\text{ atm}$$

There is a 0.64 atm pressure reduction, which results from the reaction of a greater number of moles of gas to yield a lesser number of moles of gas.

12.151 Ethane gas, C_2H_6, burns in air as indicated by the equation $2C_2H_6 + 7O_2 \longrightarrow 4CO_2 + 6H_2O$. Determine the number of (*a*) mol CO_2 and H_2O formed when 1 mol C_2H_6 is burned, (*b*) liters of O_2 required to burn 1.0 L of C_2H_6, (*c*) liters of CO_2 formed when 25 L of C_2H_6 is burned, (*d*) liters (STP) CO_2 formed when 1 mol C_2H_6 is burned, (*e*) mol CO_2 formed when 25 L (STP) C_2H_6 is burned, (*f*) g CO_2 formed when 25 L (STP) C_2H_6 is burned.

▮ (*a*) $(1\text{ mol }C_2H_6)\left(\frac{4\text{ mol }CO_2}{2\text{ mol }C_2H_6}\right) = 2\text{ mol }CO_2 \qquad (1\text{ mol }C_2H_6)\left(\frac{6\text{ mol }H_2O}{2\text{ mol }C_2H_6}\right) = 3\text{ mol }H_2O$

(*b*) The volume ratio of gases at the same temperature and pressure is equal to the mole ratio.

$$(1.0\text{ L }C_2H_6)\left(\frac{7\text{ L }O_2}{2\text{ L }C_2H_6}\right) = 3.5\text{ L }O_2$$

(*c*) $(25\text{ L }C_2H_6)\left(\frac{4\text{ L }CO_2}{2\text{ L }C_2H_6}\right) = 50\text{ L }CO_2$

(*d*) $(2.00\text{ mol }CO_2)\left(\frac{22.4\text{ L}}{\text{mol}}\right) = 44.8\text{ L}$ [from part (*a*)]

(*e*) $(25\text{ L (STP) }C_2H_6)\left(\frac{1\text{ mol }C_2H_6}{22.4\text{ L (STP)}}\right)\left(\frac{4\text{ mol }CO_2}{2\text{ mol }C_2H_6}\right) = 2.23\text{ mol }CO_2$

(*f*) $(2.23\text{ mol }CO_2)\left(\frac{44.0\text{ g}}{\text{mol}}\right) = 98.2\text{ g }CO_2$

12.152 What volume of hydrochloric acid solution, of density 1.18 g/cm³ and containing 35.0% by weight HCl, must be allowed to react with zinc in order to liberate 5.00 g of hydrogen?

▮

$$Zn + 2HCl \longrightarrow H_2 + ZnCl_2$$

$$(5.00\text{ g }H_2)\left(\frac{1\text{ mol }H_2}{2.016\text{ g }H_2}\right)\left(\frac{2\text{ mol HCl}}{1\text{ mol }H_2}\right)\left(\frac{36.5\text{ g HCl}}{\text{mol HCl}}\right) = 181\text{ g HCl}$$

$$(181\text{ g HCl})\left(\frac{100\text{ g solution}}{35.0\text{ g HCl}}\right)\left(\frac{1\text{ cm}^3}{1.18\text{ g solution}}\right) = 438\text{ cm}^3$$

12.153 Fifty grams of aluminum is to be treated with a 10% excess of H_2SO_4. The chemical equation for the reaction is $2Al + 3H_2SO_4 \rightarrow Al_2(SO_4)_3 + 3H_2$. (*a*) What volume of concentrated sulfuric acid, of density 1.80 g/cm³ and containing 96.5% H_2SO_4 by weight, must be taken? (*b*) What volume of hydrogen would be collected over water at 20 °C and 785 torr? Vapor pressure of water at 20 °C is 17.5 torr.

▮ (*a*) $(50.0\text{ g Al})\left(\frac{1\text{ mol Al}}{27.0\text{ g Al}}\right)\left(\frac{3\text{ mol }H_2SO_4}{2\text{ mol Al}}\right)\left(\frac{98.0\text{ g }H_2SO_4}{\text{mol }H_2SO_4}\right) = 272\text{ g }H_2SO_4$

$$(272\text{ g }H_2SO_4)\left(\frac{100\text{ g solution}}{96.5\text{ g }H_2SO_4}\right)\left(\frac{1\text{ cm}^3\text{ solution}}{1.80\text{ g solution}}\right) = 157\text{ cm}^3\text{ solution}$$

10% excess = $(1.10)(157\text{ cm}^3) = 173\text{ cm}^3$

(*b*) $$(50.0 \text{ g Al})\left(\frac{1 \text{ mol Al}}{27.0 \text{ g Al}}\right)\left(\frac{3 \text{ mol H}_2}{2 \text{ mol Al}}\right) = 2.78 \text{ mol H}_2$$

$$P(\text{H}_2) = P(\text{total}) - P(\text{H}_2\text{O}) = (785 \text{ torr}) - (17.5 \text{ torr}) = 768 \text{ torr}$$

$$V = \frac{nRT}{P} = \frac{(2.78 \text{ mol})(0.0821 \text{ L·atm/mol·K})(293 \text{ K})}{(768/760) \text{ atm}} = 66.2 \text{ L}$$

12.154 What volume of H_2S gas measured at 1.00 atm and 0 °C is produced when 16.6 g of KI reacts with excess H_2SO_4?

$$8\text{KI} + 5\text{H}_2\text{SO}_4 \longrightarrow 4\text{K}_2\text{SO}_4 + 4\text{I}_2 + \text{H}_2\text{S} + 4\text{H}_2\text{O}$$

▮ $$(16.6 \text{ g KI})\left(\frac{1 \text{ mol KI}}{166 \text{ g KI}}\right)\left(\frac{1 \text{ mol H}_2\text{S}}{8 \text{ mol KI}}\right)\left(\frac{22.4 \text{ L (STP)}}{\text{mol H}_2\text{S}}\right) = 0.280 \text{ L} = 280 \text{ mL}$$

12.155 After 11.2 g of carbon reacts with oxygen originally occupying 21.2 L at 18 °C and 750 torr, the cooled gases are passed through 3.00 L of 2.50 M NaOH solution. Determine the concentration of NaOH remaining in solution which is not converted to Na_2CO_3. *Note*: CO does not react with NaOH under these conditions.

▮ $$(11.2 \text{ g C})\left(\frac{1 \text{ mol C}}{12.01 \text{ g C}}\right) = 0.933 \text{ mol C}$$

$$n(\text{O}_2) = \frac{PV}{RT} = \frac{[(750/760) \text{ atm}](21.2 \text{ L})}{(0.0821 \text{ L·atm/mol·K})(291 \text{ K})} = 0.876 \text{ mol O}_2$$

$$\text{C} + \text{O}_2 \longrightarrow \text{CO}_2 \qquad \text{C} + \tfrac{1}{2}\text{O}_2 \longrightarrow \text{CO}$$

Let x = number of mol of CO_2 formed and y = number of mol of CO formed

$$x + y = 0.933 \qquad x + 0.5y = 0.876 \qquad \text{yielding} \qquad y = 0.114 \text{ mol CO} \qquad x = 0.819 \text{ mol CO}_2$$

$$2\text{NaOH} + \text{CO}_2 \longrightarrow \text{Na}_2\text{CO}_3$$

$$(3.00 \text{ L})(2.50 \text{ mol/L}) = 7.50 \text{ mol NaOH present initially}$$

$$(0.819 \text{ mol CO}_2)\left(\frac{2 \text{ mol NaOH}}{\text{mol CO}_2}\right) = 1.64 \text{ mol NaOH reacting}$$

Hence there is 5.86 mol of NaOH remaining.

$$\frac{5.86 \text{ mol NaOH}}{3.00 \text{ L}} = 1.95 \text{ M NaOH}$$

12.156 Two gases in adjoining vessels were brought into contact by opening a stopcock between them. The one vessel measured 0.250 L and contained NO at 800 torr and 220 K; the other measured 0.100 L and contained O_2 at 600 torr and 220 K. The reaction to form $N_2O_4(s)$ exhausts the limiting reactant completely. **(*a*)** Neglecting the vapor pressure of N_2O_4, what is the pressure and composition of the gas remaining at 220 K after completion of the reaction? **(*b*)** What weight of N_2O_4 is formed?

▮ $$n(\text{NO}) = \frac{PV}{RT} = \frac{[(800/760) \text{ atm}](0.250 \text{ L})}{(0.0821 \text{ L·atm/mol·K})(220 \text{ K})} = 0.0146 \text{ mol NO}$$

$$n(\text{O}_2) = \frac{PV}{RT} = \frac{[(600/760) \text{ atm}](0.100 \text{ L})}{(0.0821 \text{ L·atm/mol·K})(220 \text{ K})} = 0.00437 \text{ mol O}_2$$

$$2\text{NO} + \text{O}_2 \longrightarrow \text{N}_2\text{O}_4(s) \qquad (0.00437 \text{ mol O}_2)\left(\frac{2 \text{ mol NO}}{1 \text{ mol O}_2}\right) = 0.00874 \text{ mol NO used up}$$

(*a*) NO is in excess: $(0.0146 \text{ mol}) - (0.00874 \text{ mol}) = 0.0059 \text{ mol excess NO}$

$$P = \frac{nRT}{V} = \frac{(0.0059 \text{ mol})(0.0821 \text{ L·atm/mol·K})(220 \text{ K})}{0.350 \text{ L}} = 0.304 \text{ atm}$$

$$(0.304 \text{ atm})\left(\frac{760 \text{ torr}}{\text{atm}}\right) = 231 \text{ torr}$$

(*b*) $$(0.00437 \text{ mol O}_2)\left(\frac{1 \text{ mol N}_2\text{O}_4}{1 \text{ mol O}_2}\right) = (0.00437 \text{ mol N}_2\text{O}_4)\left(\frac{92.0 \text{ g}}{\text{mol}}\right) = 0.402 \text{ g N}_2\text{O}_4$$

12.157 Given the equation $2CO(g) + O_2(g) \rightarrow 2CO_2(g)$, what volume of CO_2 at 1.00 atm and 0 °C can be produced from 112 g CO and 48 g O_2? What reagent is present in excess?

▮ $(112\text{ g})\left(\dfrac{1\text{ mol}}{28.0\text{ g}}\right) = 4.00\text{ mol CO present}$ $\quad (48\text{ g } O_2)\left(\dfrac{1\text{ mol}}{32.0\text{ g}}\right) = 1.5\text{ mol } O_2\text{ present}$

O_2 is in limiting quantity; CO is in excess.

$$(1.5\text{ mol } O_2)\left(\frac{2\text{ mol } CO_2}{\text{mol } O_2}\right) = 3.0\text{ mol } CO_2 \qquad V = \frac{nRT}{P} = \frac{(3.0\text{ mol})(0.0821\text{ L}\cdot\text{atm/mol}\cdot\text{K})(273\text{ K})}{1.00\text{ atm}} = 67.2\text{ L}$$

12.158 A 0.750 g sample of solid benzoic acid, $C_7H_6O_2$, was placed in a 0.500 L pressurized reaction vessel filled with O_2 at 10.0 atm pressure and 25 °C. To the extent of the availability of oxygen, the benzoic acid was burned completely to water and CO_2. What were the final mole fractions (x) of CO_2 and H_2O vapor in the resulting gas mixture brought to the initial temperature? The vapor pressure of water at 25 °C is 23.8 torr. Neglect both the volume occupied by nongaseous substances and the solubility of CO_2 in H_2O. (The water pressure in the gas phase cannot exceed the vapor pressure of water, so most of the water is condensed to the liquid.)

▮

$$n(O_2) = \frac{PV}{RT} = \frac{(0.500\text{ L})(10.0\text{ atm})}{(0.0821\text{ L}\cdot\text{atm/mol}\cdot\text{K})(298\text{ K})} = 0.204\text{ mol } O_2$$

$$n(\text{acid}) = (0.750\text{ g})\left(\frac{1\text{ mol}}{122\text{ g}}\right) = 0.00615\text{ mol acid}$$

$$2C_7H_6O_2 + 15O_2 \longrightarrow 14CO_2 + 6H_2O$$

$$(0.00615\text{ mol acid})\left(\frac{15\text{ mol } O_2}{2\text{ mol acid}}\right) = 0.0461\text{ mol } O_2\text{ required}$$

$$(0.204\text{ mol}) - (0.0461\text{ mol}) = 0.158\text{ mol } O_2\text{ in excess}$$

$$(0.00615\text{ mol acid})\left(\frac{14\text{ mol } CO_2}{2\text{ mol acid}}\right) = 0.0430\text{ mol } CO_2\text{ produced}$$

$$(0.00615\text{ mol acid})\left(\frac{6\text{ mol } H_2O}{2\text{ mol acid}}\right) = 0.0184\text{ mol } H_2O\text{ produced}$$

If all the water remained in the vapor phase,

$$P(H_2O) = \frac{(0.0184\text{ mol } H_2O)(0.0821\text{ L}\cdot\text{atm/mol}\cdot\text{K})(298\text{ K})}{0.500\text{ L}} = 0.903\text{ atm}$$

Since this pressure greatly exceeds P_{vap}, most of the water condenses to liquid. The number of moles of water in the vapor phase depends on the vapor pressure at 25 °C.

$$P(H_2O) = (23.8\text{ torr})\left(\frac{1\text{ atm}}{760\text{ torr}}\right) = 0.0313\text{ atm}$$

$$n = \frac{PV}{RT} = \frac{(0.0313\text{ atm})(0.500\text{ L})}{(0.0821\text{ L}\cdot\text{atm/mol}\cdot\text{K})(298\text{ K})} = 6.4 \times 10^{-4}\text{ mol}$$

Thus in the vapor phase there is

$$(0.158\text{ mol } O_2\text{ excess}) + (0.0430\text{ mol } CO_2) + [6.4 \times 10^{-4}\text{ mol } H_2O(g)] = 0.202\text{ mol total}$$

$$x(CO_2) = \frac{0.0430}{0.202} = 0.213 \qquad x(H_2O) = \frac{6.4 \times 10^{-4}}{0.202} = 0.0032$$

12.159 Chemical absorbers can be used to remove exhaled CO_2 of space travelers in short space flights. Li_2O is one of the most efficient in terms of absorbing capacity per unit weight. If the reaction is $Li_2O + CO_2 \rightarrow Li_2CO_3$, what is the absorption efficiency of pure Li_2O in L CO_2 (STP) per kg?

▮

$$(1.00\text{ kg } Li_2O)\left(\frac{1000\text{ g}}{\text{kg}}\right)\left(\frac{1\text{ mol } Li_2O}{29.88\text{ g } Li_2O}\right)\left(\frac{1\text{ mol } CO_2}{1\text{ mol } Li_2O}\right)\left(\frac{22.4\text{ L (STP)}}{\text{mol } CO_2}\right) = 750\text{ L}$$

12.160 In an auto engine with no pollution controls, about 5% of the fuel (assume 100% octane, C_8H_{18}) is unburned. Calculate the relative masses of CO and C_8H_{18} emitted in the exhaust gas. Calculate the relative volumes of these gases.

▮

$$2C_8H_{18} + 17O_2 \longrightarrow 16CO + 18H_2O$$

Per mole of fuel, 0.05 mol is unreacted, and $(0.95\ \text{mol})\left(\frac{16\ \text{mol CO}}{2\ \text{mol } C_8H_{18}}\right) = 7.6\ \text{mol}$ of CO is produced. Since the temperatures and pressures of the two gases are the same, the volume ratio equals the mole ratio:

$$\frac{7.6\ \text{mol CO}}{0.05\ \text{mol } C_8H_{18}} = 1.5 \times 10^2\ \text{mol CO/mol } C_8H_{18}$$

The mass ratio is $\frac{(7.6\ \text{mol CO})(28.0\ \text{g/mol})}{(0.05\ \text{mol } C_8H_{18})(114\ \text{g/mol})} = 37\ \text{g CO/g } C_8H_{18}$

To one significant figure, there is about 40 times the mass of CO than of C_8H_{18}.

12.161 Into a 3.00 L container at 25 °C are placed 1.23 mol of O_2 and 2.73 mol of C. ***(a)*** What is the initial pressure? ***(b)*** If the carbon and oxygen react as completely as possible to form CO, what will the final pressure in the container be at 25 °C?

▮ ***(a)*** Only the oxygen is gaseous; the carbon exerts no pressure.

$$P = \frac{nRT}{V} = \frac{(1.23\ \text{mol})(0.0821\ \text{L·atm/mol·K})(298\ \text{K})}{3.00\ \text{L}} = 10.0\ \text{atm}$$

(b) $C + \frac{1}{2}O_2 \rightarrow CO$. Two moles of CO is produced for every mol O_2 used up, yielding 2.46 mol CO. The final pressure is

$$P = \frac{(2.46\ \text{mol})(0.0821\ \text{L·atm/mol·K})(298\ \text{K})}{3.00\ \text{L}} = 20.0\ \text{atm}$$

12.162 Calcium carbide, CaC_2, reacts with water to produce acetylene, C_2H_2, and calcium hydroxide, $Ca(OH)_2$. Calculate the volume of $C_2H_2(g)$ at 25 °C and 0.950 atm produced from the reaction of 128 g CaC_2 with 45.0 g water.

▮

$$CaC_2 + 2H_2O \longrightarrow Ca(OH)_2 + C_2H_2$$

$$(128\ \text{g } CaC_2)\left(\frac{1\ \text{mol } CaC_2}{64.0\ \text{g}}\right) = 2.00\ \text{mol } CaC_2 \qquad (45.0\ \text{g } H_2O)\left(\frac{1\ \text{mol } H_2O}{18.0\ \text{g}}\right) = 2.50\ \text{mol } H_2O$$

H_2O is the limiting reagent.

$$(2.50\ \text{mol } H_2O)\left(\frac{1\ \text{mol } C_2H_2}{2\ \text{mol } H_2O}\right) = 1.25\ \text{mol } C_2H_2$$

$$V = \frac{nRT}{P} = \frac{(1.25\ \text{mol})(0.0821\ \text{L·atm/mol·K})(298\ \text{K})}{0.950\ \text{atm}} = 32.2\ \text{L}$$

12.163 A reaction mixture for the combustion of SO_2 was prepared by opening a stopcock connecting two separate chambers, one having a volume of 2.125 L filled at 0.750 atm with SO_2 and the other having a 1.500 L volume filled at 0.500 atm with O_2; both gases were at 80 °C. ***(a)*** What were the mole fraction of SO_2 in the mixture, the total pressure, and the partial pressures? ***(b)*** If the mixture was passed over a catalyst that promoted the formation of SO_3 and then was returned to the original two connected vessels, what were the mole fractions in the final mixture, and what was the final total pressure? Assume that the conversion of SO_2 is complete to the extent of the availability of O_2.

▮ ***(a)***

$$n(SO_2) = \frac{PV}{RT} = \frac{(0.750\ \text{atm})(2.125\ \text{L})}{(0.0821\ \text{L·atm/K·mol})(353\ \text{K})} = 0.0550\ \text{mol}$$

$$n(O_2) = \frac{PV}{RT} = \frac{(0.500\ \text{atm})(1.500\ \text{L})}{(0.0821\ \text{L·atm/K·mol})(353\ \text{K})} = 0.0259\ \text{mol}$$

Each mole fraction is evaluated by dividing the number of mol of the component by the total number of mol in the mixture.

$$x(SO_2) = \frac{n(SO_2)}{n(SO_2) + n(O_2)} = \frac{0.0550\ \text{mol}}{(0.0550 + 0.0259)\ \text{mol}} = \frac{0.0550}{0.0809} = 0.680 \qquad x(O_2) = \frac{0.0259}{0.0809} = 0.320$$

Note that the sum of the mole fractions is unity.

The total pressure can best be evaluated by using total volume $(2.125\ \text{L} + 1.500\ \text{L} = 3.625\ \text{L})$ and total number of moles (0.0809 mol).

$$P(\text{total}) = \frac{nRT}{V} = \frac{(0.0809\ \text{mol})(0.0821\ \text{L·atm/K·mol})(353\ \text{K})}{(3.625\ \text{L})} = 0.647\ \text{atm}$$

Partial pressures in a mixture are easily computed by multiplying the respective mole fractions by the total pressure. This statement can be proved as follows:

$$\frac{P(\mathrm{X})}{P(\mathrm{tot})} = \frac{n(\mathrm{X})\,RT/V}{nRT/V} = \frac{n(\mathrm{X})}{n} = x(\mathrm{X})$$

Thus,

$$P(SO_2) = x(SO_2) \times p(\text{total}) = 0.680 \times 0.647 \text{ atm} = 0.440 \text{ atm}$$

$$P(O_2) = x(O_2) \times p(\text{total}) = 0.320 \times 0.647 \text{ atm} = 0.207 \text{ atm}$$

The sum of the partial pressures must of course equal the total pressure.

(*b*) The chemical equation for the reaction is $2SO_2 + O_2 \rightarrow 2SO_3$. All substances are gases under the experimental conditions. The number of mol O_2 required for stoichiometric conversion is half the number of mol SO_2; but the number of mol O_2 available, 0.0259, was less than half of 0.0550. Therefore, all the O_2 was used up and an excess of SO_2 remained. For the quantities of gases participating or formed in the reaction,

$$n(O_2) = 0.0259 \text{ mol} \qquad n(SO_2) = 2 \times n(O_2) = 2 \times 0.0259 \text{ mol} = 0.0518 \text{ mol}$$

The product would be $n(SO_3) = n(SO_2) = 0.0518$ mol, so the residue would be $(0.0550 - 0.0518)$ mol $= 0.0032$ mol SO_2. After the reaction is completed,

$$x(SO_3) = \frac{n(SO_3)}{n(SO_3) + n(SO_2)} = \frac{0.0518 \text{ mol}}{(0.0518 + 0.0032) \text{ mol}} = 0.942 \qquad x(SO_2) = \frac{n(SO_2)}{n(SO_3) + n(SO_2)} = \frac{0.0032 \text{ mol}}{0.0550 \text{ mol}} = 0.058$$

$$P(\text{total}) = \frac{n(\text{total})\,RT}{V} = \frac{(0.0550 \text{ mol})(0.0821 \text{ L·atm/K·mol})(353 \text{ K})}{3.625 \text{ L}} = 0.440 \text{ atm}$$

Note that the final *total* pressure is the same as the initial *partial* pressure of SO_2. This happens because the only gases present at the end are SO_2 and SO_3. Because each mol SO_3 was made at the expense of 1 mol SO_2, the sum of the numbers of moles of the two gases had to equal the initial number of moles of SO_2. The final total pressure is less than the initial total pressure because the consumption of the oxygen caused a decrease in the total number of moles of gas.

12.164 It took exactly 3.00 mL of 6.00 M HCl to convert 1.200 g of a mixture of $NaHCO_3$ and Na_2CO_3 to NaCl, CO_2, and H_2O. Calculate the volume of CO_2 liberated at 25 °C and 760 torr.

▮ $$NaHCO_3 + HCl \longrightarrow NaCl + H_2O + CO_2 \qquad Na_2CO_3 + 2HCl \longrightarrow 2NaCl + H_2O + CO_2$$

$$(3.00 \text{ mL})(6.00 \text{ M}) = 18.0 \text{ mmol HCl used}$$

Let x = number of g $NaHCO_3$, then $1.200 - x$ = number of g Na_2CO_3. There are thus present $(x/84.0)$ mol of $NaHCO_3$, which reacts with $(x/84.0)$ mol of HCl, and $(1.200 - x)/106$ mol of Na_2CO_3, which reacts with $2(1.200 - x)/106$ mol of HCl, since they react in a 2:1 mole ratio.

$$\frac{x}{84.0} + 2\left(\frac{1.200 - x}{106}\right) = 0.0180 \qquad 0.0119x + 0.0226 - 0.0189x = 0.0180$$

$$x = 0.66 \text{ g } NaHCO_3 \qquad 1.200 - x = 0.54 \text{ g } Na_2CO_3$$

$$(0.66 \text{ g } NaHCO_3)\left(\frac{1 \text{ mol } NaHCO_3}{84.0 \text{ g}}\right)\left(\frac{1 \text{ mol } CO_2}{\text{mol } NaHCO_3}\right) = 7.9 \times 10^{-3} \text{ mol } CO_2$$

$$(0.54 \text{ g } Na_2CO_3)\left(\frac{1 \text{ mol } Na_2CO_3}{106 \text{ g } Na_2CO_3}\right)\left(\frac{1 \text{ mol } CO_2}{\text{mol } Na_2CO_3}\right) = 5.1 \times 10^{-3} \text{ mol } CO_2$$

$$\text{Total } CO_2 = 13.0 \times 10^{-3} \text{ mol } CO_2$$

$$V = \frac{nRT}{P} = \frac{(13.0 \times 10^{-3} \text{ mol})(0.0821 \text{ L·atm/mol·K})(298 \text{ K})}{1.00 \text{ atm}} = 0.318 \text{ L}$$

12.165 A 50 mL sample of a hydrogen-oxygen mixture was placed in a gas buret at 18 °C and confined at barometric pressure. A spark was passed through the sample so that the formation of water could go to completion. The resulting pure gas had a volume of 10 mL at barometric pressure. What was the initial mole fraction of hydrogen in the mixture (*a*) if the residual gas after sparking was hydrogen (*b*) if the residual gas was oxygen?

▮ If they had been at the same temperature and pressure but in separate containers, H_2 and O_2 would react in a 2:1 volume ratio. The 40 mL used up could thus be considered 13.3 mL O_2 and 26.7 mL H_2.

(*a*) Total volume = (13.3 mL O_2) + (26.7 mL H_2) + (10.0 mL H_2 in excess) = 50.0 mL

$$\% H_2 = \left(\frac{36.7 \text{ mL } H_2}{50.0 \text{ mL total}}\right)(100\%) = 73.4\% H_2 \qquad x(H_2) = 0.734$$

(*b*) Total volume = (13.3 mL O_2) + (10.0 mL O_2 in excess) + (26.7 mL H_2) = 50.0 mL

$$\% H_2 = \left(\frac{26.7 \text{ mL } H_2}{50.0 \text{ mL total}}\right)(100\%) = 53.4\% H_2 \qquad x(H_2) = 0.534$$

12.166 Only gases remain after 18.0 g carbon is treated with 5.00 L O_2 at 18 °C and 5.00 atm pressure. Determine the concentration of $NaHCO_3$ and of Na_2CO_3 produced by the reaction of the CO_2 with 0.500 L of 2.00 M NaOH. *Note*: CO does not react with NaOH under these conditions.

▮

$$n(O_2) = \frac{PV}{RT} = \frac{(5.00 \text{ atm})(5.00 \text{ L})}{(0.0821 \text{ L·atm/mol·K})(291 \text{ K})} = 1.05 \text{ mol } O_2 \qquad (18.0 \text{ g C})\left(\frac{1 \text{ mol C}}{12.0 \text{ g C}}\right) = 1.50 \text{ mol C}$$

Let x = number of moles of CO produced and y = number of moles of CO_2 produced.

$$x + y = 1.50 \qquad 0.5x + y = 1.05 \qquad \text{therefore} \qquad x = 0.90 \text{ mol CO} \qquad y = 0.60 \text{ mol } CO_2$$

$$NaOH + CO_2 \longrightarrow NaHCO_3 \qquad 2NaOH + CO_2 \longrightarrow Na_2CO_3 + H_2O$$

With 0.60 mol CO_2 and 1.00 mol NaOH, let w = number of moles of $NaHCO_3$ and z = number of moles of Na_2CO_3.

$$z + w = 0.60 \qquad 2z + w = 1.00 \qquad \text{therefore} \qquad z = 0.40 \text{ mol } Na_2CO_3 \qquad w = 0.20 \text{ mol } NaHCO_3$$

$$\frac{0.40 \text{ mol } Na_2CO_3}{0.500 \text{ L}} = 0.80 \text{ M } Na_2CO_3 \qquad \frac{0.20 \text{ mol } NaHCO_3}{0.500 \text{ L}} = 0.40 \text{ M } NaHCO_3$$

12.167 Only gases remain after 18.0 g carbon is treated with 25.0 L air at 17 °C and 5.00 atm pressure. (Assume 19.0% by volume O_2, 80% N_2, and 1.0% CO_2.) Determine the concentrations of $NaHCO_3$ and Na_2CO_3 produced by the reaction of the mixture of gases with 0.750 L of 1.00 M NaOH.

▮

$$n(O_2) = 0.190n(\text{air}) = \frac{(0.190)(5.00 \text{ atm})(25.0 \text{ L})}{(0.0821 \text{ L·atm/mol·K})(290 \text{ K})} = 0.998 \text{ mol } O_2$$

$$n(CO_2) = 0.010n(\text{air}) = 0.0525 \text{ mol } CO_2$$

The 1.50 mol C and 0.998 mol O_2 produce 0.496 mol CO_2 and 1.004 mol CO (see method, Problem 12.166). The 0.496 mol CO_2 produced from the C and O_2 plus the 0.0525 mol originally present make 0.548 mol CO_2 total. The 0.750 mol NaOH with the 0.548 mol CO_2 produces 0.346 mol $NaHCO_3$ and 0.202 mol Na_2CO_3 (see Problem 12.166).

$$\frac{0.346 \text{ mol } NaHCO_3}{0.750 \text{ L}} = 0.461 \text{ M } NaHCO_3 \qquad \frac{0.202 \text{ mol } Na_2CO_3}{0.750 \text{ L}} = 0.269 \text{ M } Na_2CO_3$$

12.168 When 0.75 mol solid "A_4" and 2 mol gaseous O_2 are heated in a sealed vessel (bomb), completely using up the reactants and producing only one compound, it is found that when the temperature is reduced to the initial temperature, the contents of the vessel exhibit a pressure equal to half the original pressure. What conclusions can be drawn from these data about the product of the reaction?

▮ From the reactant ratio, it is observed that the ratio of atoms of A to O must be 3:4. Empirical formula: A_3O_4. Since the product exerts considerable pressure, it must be a gas. That the pressure is equal to half that of the oxygen initially present implies that the number of molecules of product is half the number of molecules of oxygen initially present. Hence the number of oxygen atoms per molecule must be 4. The molecular formula is therefore A_3O_4.

12.169 The industrial production of NH_3 by use of a natural gas feedstock can be represented by the following simplified set of reactions.

$$CH_4 + H_2O \longrightarrow CO + 3H_2 \qquad (1)$$

$$2CH_4 + O_2 \longrightarrow 2CO + 4H_2 \qquad (2)$$

$$CO + H_2O \longrightarrow CO_2 + H_2 \qquad (3)$$

$$N_2 + 3H_2 \longrightarrow 2NH_3 \qquad (4)$$

By assuming (i) that only the above reactions take place, plus the chemical absorption of CO_2, (ii) that natural gas consists just of CH_4, (iii) that air consists of 0.80 mol fraction N_2 and 0.20 mol fraction O_2, and (iv) that the ratio of conversion of CH_4 by processes (*1*) and (*2*) is controlled through admitting oxygen for reaction (*2*) by adding just enough air so that the mol ratio of N_2 to H_2 is exactly 1:3, consider the overall efficiency of a process in which 1200 m^3 (STP) of natural gas yields 1.00 metric ton of NH_3. ***(a)*** How many mol NH_3 would be formed from each mol natural gas if there were complete conversion of the natural gas subject to the stated assumptions? ***(b)*** What percentage of the maximum yield calculated in (*a*) is the actual yield?

▮ ***(a)*** The mole ratio of N_2 to H_2 must be 1:3. Each 4.0 mol N_2 is accompanied by 1.0 mol O_2. Each mole of CO produced in reactions (*1*) and (*2*) eventually produces 1 mol H_2 in reaction (*3*). Each 1.0 mol O_2 thus generates 6.0 mol H_2 [reaction (*2*)]. Therefore, 6.0 mol more of H_2 is required to get a 4.0:12.0 mol ratio. Thus for 4.0 mol N_2 we need: 1.5 mol CH_4 from reaction (*1*), producing 6.0 mol H_2; 2.0 mol CH_4 from reaction (*2*), producing 6.0 mol H_2; 3.5 mol CH_4 total, producing 12.0 mol H_2, which reacts with 4.0 mol N_2 to yield 8.0 mol NH_3. For each mol CH_4, 8.0/3.5 = 2.29 mol of NH_3 is produced.

(b) $$n = \frac{(1200\ \text{m}^3)(10^3\ \text{L/m}^3)(1.00\ \text{atm})}{(0.0821\ \text{L·atm/mol·K})(273\ \text{K})} = 53.5 \times 10^3\ \text{mol CH}_4$$

Theoretical yield: $$(53.5 \times 10^3\ \text{mol CH}_4)\left(\frac{2.29\ \text{mol NH}_3}{\text{mol CH}_4}\right) = 122.5 \times 10^3\ \text{mol NH}_3$$

$$(1.00 \times 10^6\ \text{g NH}_3)\left(\frac{1\ \text{mol NH}_3}{17.0\ \text{g NH}_3}\right) = 58.8 \times 10^3\ \text{mol NH}_3\ \text{produced} \qquad \left(\frac{58.8 \times 10^3}{122.5 \times 10^3}\right)(100\%) = 48.0\%\ \text{yield}$$

12.170 A mixture of Na_2CO_3 and $NaHCO_3$ has a mass of 22.0 g. Treatment with excess HCl solution liberates 6.00 L CO_2 at 25 °C and 0.947 atm pressure. Determine the percent Na_2CO_3 in the mixture.

▮ $$\text{Na}_2\text{CO}_3 + 2\text{HCl} \longrightarrow \text{CO}_2 + \text{H}_2\text{O} + 2\text{NaCl} \qquad \text{NaHCO}_3 + \text{HCl} \longrightarrow \text{CO}_2 + \text{H}_2\text{O} + \text{NaCl}$$

The number of mol CO_2 produced is equal to the number of mol Na_2CO_3 plus $NaHCO_3$ initially present.

$$n = \frac{PV}{RT} = \frac{(0.947\ \text{atm})(6.00\ \text{L})}{(0.0821\ \text{L·atm/mol·K})(298\ \text{K})} = 0.232\ \text{mol}$$

Let x = number of g Na_2CO_3, then $22.0 - x$ = number of g $NaHCO_3$. The number of mol of each reactant is equal to the mass divided by the formula weight. Hence

$$\frac{x}{106.0} + \frac{22.0 - x}{84.01} = 0.232 \qquad x = 12.1\ \text{g Na}_2\text{CO}_3 \qquad 22.0 - x = 9.9\ \text{g NaHCO}_3$$

$$\left(\frac{12.1\ \text{g Na}_2\text{CO}_3}{22.0\ \text{g total}}\right)(100\%) = 55.0\%\ \text{Na}_2\text{CO}_3$$

12.171 A mixture of $Al_2(CO_3)_3$ and sand, containing 35.0% $Al_2(CO_3)_3$, when treated with excess HCl gave 3.96 g CO_2 gas. Sand does not react with HCl. $Al_2(CO_3)_3$ reacts as follows: $Al_2(CO_3)_3 + 6HCl \rightarrow 2AlCl_3 + 3H_2O + 3CO_2$. ***(a)*** Find the mass in g of $Al_2(CO_3)_3$. ***(b)*** Find the mass in g of HCl used up. ***(c)*** Find the mass in g of $AlCl_3$ formed. ***(d)*** Find the volume at 27 °C and 2.00 atm of CO_2 gas formed. ***(e)*** Find the mass in g of the mixture.

▮ ***(a)*** $$(3.96\ \text{g})\left(\frac{1\ \text{mol}}{44.0\ \text{g}}\right) = 0.0900\ \text{mol CO}_2$$

$$(0.0900\ \text{mol CO}_2)\left(\frac{1\ \text{mol Al}_2(\text{CO}_3)_3}{3\ \text{mol CO}_2}\right) = 0.0300\ \text{mol Al}_2(\text{CO}_3)_3$$

$$(0.0300\ \text{mol})\left(\frac{234\ \text{g}}{\text{mol}}\right) = 7.02\ \text{g Al}_2(\text{CO}_3)_3$$

(b) $$(0.0900\ \text{mol CO}_2)\left(\frac{6\ \text{mol HCl}}{3\ \text{mol CO}_2}\right)\left(\frac{36.5\ \text{g}}{\text{mol HCl}}\right) = 6.57\ \text{g HCl}$$

(c) $$(0.0900\ \text{mol CO}_2)\left(\frac{2\ \text{mol AlCl}_3}{3\ \text{mol CO}_2}\right)\left(\frac{133.5\ \text{g}}{\text{mol AlCl}_3}\right) = 8.01\ \text{g AlCl}_3$$

(d) $$V = \frac{nRT}{P} = \frac{(0.0900\ \text{mol})(0.0821\ \text{L·atm/mol·K})(300\ \text{K})}{2.00\ \text{atm}} = 1.11\ \text{L}$$

(e) $$(7.02\ \text{g Al}_2(\text{CO}_3)_3)\left(\frac{100\ \text{g mixture}}{35.0\ \text{g Al}_2(\text{CO}_3)_3}\right) = 20.1\ \text{g mixture}$$

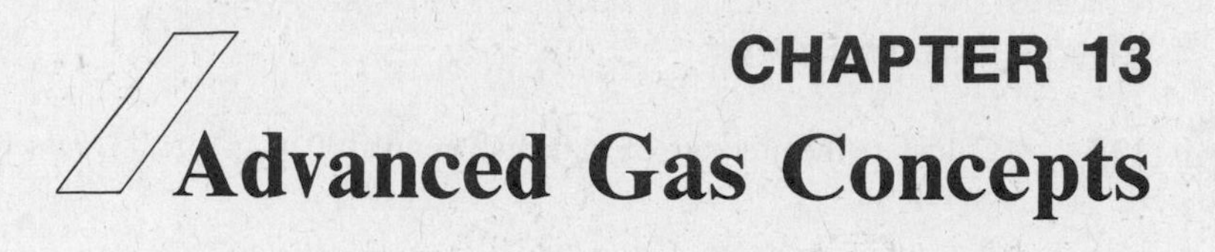

CHAPTER 13 Advanced Gas Concepts

Note: Before working the problems of this chapter, you should familiarize yourself with Table 13.1, below.

13.1 VAN DER WAALS EQUATION

13.1 Identify the postulate of the kinetic molecular theory most closely identified with the van der Waals constant *a*, and explain the relationship. Repeat the procedure for the constant *b*.

▌ The constant *a* reflects the intermolecular attractions in real gases. The kinetic molecular theory postulates no intermolecular attractions, which is a good approximation in many cases. The constant *b* reflects the actual volume of the molecules themselves. The kinetic molecular theory postulates that the volume of the molecules is zero (or negligible), also sometimes a good approximation.

13.2 In the van der Waals equation

$$\left(P + \frac{n^2a}{V^2}\right)(V - nb) = nRT$$

the constant *a* accounts for what property of molecules of real gases?

▌ Intermolecular attraction

13.3 Using the van der Waals equation, calculate the pressure exerted by 10.0 mol carbon dioxide in a 2.00 L vessel at 47 °C. Repeat the calculation using the equation of state for an ideal gas. Compare these results with the experimentally observed pressure of 82 atm.

▌

$$\left(P + \frac{n^2a}{V^2}\right)(V - nb) = nRT$$

From Table 13.1, $a = 3.59\ \text{L}^2\cdot\text{atm/mol}$ and $b = 0.0427\ \text{L/mol}$.

$$\left(P + \frac{100 \times 3.59}{4.00}\ \text{atm}\right)\left[(2.00\ \text{L}) - (10.0\ \text{mol})\left(0.0427\ \frac{\text{L}}{\text{mol}}\right)\right] = 10.0(0.0821)(320)\ \text{L}\cdot\text{atm}$$

$$P = 77\ \text{atm}$$

From the ideal gas equation: $PV = nRT$ $\quad (2.00\ \text{L})P = (10.0)(0.0821)(320)\ \text{L}\cdot\text{atm}$ $\quad P = 131\ \text{atm}$

Since the total pressure is so high, one would not expect the gas to behave ideally. The van der Waals equation gives a result much closer to the experimental value.

TABLE 13.1 Van der Waals Constants

gas	*a* $\text{L}^2\cdot\text{atm/mol}^2$	*b* L/mol	gas	*a* $\text{L}^2\cdot\text{atm/mol}^2$	*b* L/mol
He	0.0341	0.0237	C_2H_4	4.47	0.0571
Ne	0.2107	0.0171	CO_2	3.59	0.0427
H_2	0.244	0.0266	NH_3	4.17	0.0371
N_2	1.39	0.0391	H_2O	5.46	0.0305
CO	1.49	0.0399	Hg	8.09	0.0170
O_2	1.36	0.0318			

13.4 Calculate the pressure of 0.60 mol NH_3 gas in a 3.00 L vessel at 25 °C (*a*) with the ideal gas law (*b*) with the van der Waals equation.

▌ (*a*) $$P = \frac{nRT}{V} = \frac{(0.60\ \text{mol})(0.0821\ \text{L}\cdot\text{atm/mol}\cdot\text{K})(298\ \text{K})}{3.00\ \text{L}} = 4.9\ \text{atm}$$

(*b*) $$P = \frac{nRT}{V - nb} - \frac{n^2a}{V^2}$$

$$= \frac{(0.60\ \text{mol})(0.0821\ \text{L}\cdot\text{atm/mol}\cdot\text{K})(298\ \text{K})}{(3.00\ \text{L}) - (0.60\ \text{mol})(0.0371\ \text{L/mol})} - \frac{(0.60\ \text{mol})^2(4.17\ \text{L}^2\cdot\text{atm/mol}^2)}{(3.00\ \text{L})^2} = 4.8\ \text{atm}$$

13.5 Calculate the pressure of 15.00 mol neon at 30 °C in a 12.0 L vessel using (*a*) the ideal gas law (*b*) the van der Waals equation.

▮ (*a*) $P = \dfrac{(15.00\ \text{mol})(0.0821\ \text{L·atm/mol·K})(303\ \text{K})}{12.0\ \text{L}} = 31.1\ \text{atm}$

(*b*) $P = \dfrac{nRT}{V - nb} - \dfrac{n^2a}{V^2}$

$$= \frac{(15.00\ \text{mol})(0.0821\ \text{L·atm/mol·K})(303\ \text{K})}{(12.0\ \text{L}) - (15.00\ \text{mol})(0.0171\ \text{L/mol})} - \frac{(15.00\ \text{mol})^2(0.2107\ \text{L}^2\text{·atm/mol}^2)}{(12.0\ \text{L})^2} = 31.4\ \text{atm}$$

13.6 (*a*) Under what sets of experimental conditions is the van der Waals equation more applicable than the ideal gas law equation? Calculate the pressure of 12.0 mol CO at 25 °C in a 10.0 L vessel using (*b*) the ideal gas law (*c*) the van der Waals equation. Calculate the pressure of 0.120 mol CO at 25 °C in a 10.0 L vessel using (*d*) the ideal gas law (*e*) the van der Waals equation.

▮ (*a*) The van der Waals equation is more applicable than the ideal gas equation (but still by no means exact) at high pressure and low temperature, where gases exhibit nonideal behavior. (Note that the van der Waals equation, like the ideal gas equation, is more precise at low pressures than at high pressures.)

(*b*) $P = \dfrac{(12.0\ \text{mol})(0.0821\ \text{L·atm/mol·K})(298\ \text{K})}{10.0\ \text{L}} = 29.4\ \text{atm}$

(*c*) $P = \dfrac{(12.0\ \text{mol})(0.0821\ \text{L·atm/mol·K})(298\ \text{K})}{(10.0\ \text{L}) - (12.0\ \text{mol})(0.0399\ \text{L/mol})} - \dfrac{(12.00\ \text{mol})^2(1.49\ \text{L}^2\text{·atm/mol}^2)}{(10.0\ \text{L})^2} = 28.7\ \text{atm}$

(*d*) $P = \dfrac{(0.120\ \text{mol})(0.0821\ \text{L·atm/mol·K})(298\ \text{K})}{10.0\ \text{L}} = 0.294\ \text{atm}$

(*e*) $P = \dfrac{(0.120\ \text{mol})(0.0821\ \text{L·atm/mol·K})(298\ \text{K})}{(10.0\ \text{L}) - (0.120\ \text{mol})(0.0399\ \text{L/mol})} - \dfrac{(0.120\ \text{mol})^2(1.49\ \text{L}^2\text{·atm/mol}^2)}{(10.0\ \text{L})^2} = 0.294\ \text{atm}$

13.7 Using the van der Waals equation, calculate the pressure of 10.0 mol NH_3 gas in a 10.0 L vessel at 0 °C.

$$\left(P + n^2\frac{a}{V^2}\right)(V - nb) = nRT \qquad a = 4.2\ \text{L}^2\text{·atm/mol}^2 \qquad b = 0.037\ \text{L/mol}$$

▮

$$\left[P + (10.0\ \text{mol})^2\left(\frac{4.2\ \text{L}^2\text{·atm/mol}^2}{(10.0\ \text{L})^2}\right)\right][(10.0\ \text{L}) - (10.0\ \text{mol})(0.037\ \text{L/mol})] = nRT$$

$$(P + 4.2\ \text{atm})(9.6\ \text{L}) = (10.0\ \text{mol})(0.0821\ \text{L·atm/mol·K})(273\ \text{K}) \qquad P = 19.1\ \text{atm}$$

13.8 What volume would 3.00 mol oxygen occupy at 50.0 atm pressure and 100 °C according to (*a*) the ideal gas law (*b*) the van der Waals equation? [*Hint*: Solve by successive approximations, using the volume obtained in part (*a*) in the pressure correction term to calculate a better answer in part (*b*). Use that answer to compute a still better answer, continuing until the results of two successive steps differ by a negligible amount.]

▮ (*a*) $V = nRT/P = 1.84\ \text{L}$

(*b*) $V = \dfrac{nRT}{P + n^2a/V^2} + nb$

$$= \frac{(3.00\ \text{mol})(0.0821\ \text{L·atm/mol·K})(373\ \text{K})}{(50.0\ \text{atm}) + (3.00\ \text{mol})^2(1.36\ \text{L}^2\text{·atm/mol}^2)/(1.84\ \text{L})^2} + (3.00\ \text{mol})(0.0318\ \text{L/mol}) = 1.81\ \text{L}$$

Putting this new value for V into the denominator of the first term on the right of the equation yields a value of $V = 1.81$ L. When the two successive values come out the same, the process is complete. (To check, use the value 1.81 L to calculate a value for P.)

13.9 According to the van der Waals equation, how many mol ammonia will occupy 7.00 L at 20.0 atm and 100 °C? (See the hint in Problem 13.8.)

▮

$$n = \frac{PV}{RT} = \frac{(20.0\ \text{atm})(7.00\ \text{L})}{(0.0821\ \text{L·atm/mol·K})(373\ \text{K})} = 4.57\ \text{mol}$$

Inserting this value into the van der Waals equation in the form

$$n = \left(P + \frac{n^2a}{V^2}\right)\left(\frac{V - nb}{RT}\right) = \frac{\left[\left(20.0 + \frac{(4.57)^2(4.17)}{(7.00)^2}\right)\text{atm}\right]\{[7.00 - (4.57)(0.0371)]\text{ L}\}}{(0.0821)(373)\text{ L}\cdot\text{atm/mol}} = 4.86\text{ mol}$$

Substituting the 4.86 mol for n in two places on the right side of the same equation and solving yields a value of 4.90 mol. Substituting 4.90 mol then yields a value of 4.91 mol. Substituting this value again yields a value of 4.91 mol, which is the solution.

13.10 For a given number of moles of gas, show that the van der Waals equation predicts greater deviation from ideal behavior ***(a)*** at high pressure rather than low pressure at a given temperature, ***(b)*** at low temperature rather than high temperature at a given pressure.

▌

$$\left(P + \frac{n^2a}{V^2}\right)(V - nb) = nRT \quad \text{compared to} \quad PV = nRT$$

(a) At high pressure, the volume is low. Hence the n^2a/V^2 term is more important than at low pressure and high volume. (The square term makes the change in pressure more important for n^2a/V^2 than for P.) The constant nb term is also more important when subtracted from the smaller V. ***(b)*** At low temperatures and a given pressure, V must be low, and the effect is as in part (*a*).

13.11 Assuming ideal behavior, calculate the volume per mol of gaseous water at 1.00 atm pressure and 100 °C. Also calculate the volume of 1.00 mol liquid water at 100 °C (density = 0.958 g/mL). Assuming that this value is the total volume of the molecules themselves, with negligible space between molecules, calculate the percent of the total volume of gaseous water at 100 °C which is "free volume." Compare the volume of 1.00 mol water molecules at 100 °C to the value of the van der Waals constant b for water (Table 13.1). Suggest a reason for the difference.

▌ For the gas: $$V = \frac{nRT}{P} = \frac{(1.00\text{ mol})(0.0821\text{ L}\cdot\text{atm/mol}\cdot\text{K})(373\text{ K})}{1.00\text{ atm}} = 30.6\text{ L}$$

For the liquid: $V = \text{mass/density} = (18.0\text{ g})/(0.958\text{ g/mL}) = 18.8\text{ mL}$

The percentage of the volume occupied is $$\left(\frac{18.8\text{ mL}}{30{,}600\text{ mL}}\right)(100\%) = 0.0614\%$$

The percentage of free volume is $(100.0000\%) - (0.0614\%) = 99.9386\%$

13.12 Compare the values of the van der Waals constants for NH_3 and N_2. Explain why the value of a is larger for NH_3 but that of b is larger for N_2.

▌ The constant a is related to intermolecular forces; those for NH_3 are much higher because of hydrogen bonding. The constant b is related to molecular volume. The hydrogen atoms of NH_3 take up practically no volume, and ammonia is little over half the volume of N_2.

13.13 According to the ideal gas law, at constant volume the pressure of a gas becomes zero at 0 K and is independent of the volume. Solve the van der Waals equation explicitly for pressure. Is the variation of pressure of a real gas with temperature at constant volume independent of the volume? Using the data of Table 13.1, determine the absolute temperature at which 1.0 mol helium gas in a volume of 1.0 L would have zero pressure. Repeat the calculation for ethylene, C_2H_4. Compare these temperatures with the normal boiling points of He and C_2H_4.

▌

$$P = \frac{nRT}{V - nb} - \frac{n^2a}{V^2}$$

The variation is not independent of volume, since the correction factors are more or less important depending on the value of the volume.

For He: $$0 = \frac{(1.0\text{ mol})(0.0821\text{ L}\cdot\text{atm/mol}\cdot\text{K})(T)}{(1.0\text{ L}) - (1.0\text{ mol})(0.0237\text{ L/mol})} - \frac{(1.0\text{ mol})^2(0.0341\text{ L}^2\cdot\text{atm/mol}^2)}{(1.0\text{ L})^2}$$

$$\frac{(0.0821\text{ atm/K})T}{0.98} = 0.0341\text{ atm} \qquad T = 0.41\text{ K} \quad (\text{normal boiling point} = 4\text{ K})$$

For C_2H_4: $$\frac{(0.0821\text{ atm/K})T}{0.94} = 4.47\text{ atm} \qquad T = 51\text{ K} \quad (\text{normal boiling point} = 169\text{ K})$$

13.14 Expand the van der Waals equation (for 1 mol gas) into a cubic equation with respect to the volume. At the critical temperature T_c, and the critical pressure P_c, the volume of 1 mol is the critical volume V_c. From this, obtain expressions for a and b in terms of critical data.

▮

$$\left(P + \frac{a}{V^2}\right)(V - b) = RT \qquad PV + \frac{a}{V} - \frac{ab}{V^2} - bP = RT \qquad V^3 - \frac{bP + RT}{P}V^2 + \frac{a}{P}V - \frac{ab}{P} = 0$$

When $P = P_c$ and $T = T_c$, this cubic equation must have the unique solution $V = V_c$. Hence the equation must be identical to $(V - V_c)^3 = V^3 - 3V^2V_c + 3VV_c^2 - V_c^3 = 0$.

Equating the coefficients of V^2 in the two forms:

$$\frac{bP_c + RT_c}{P_c} = 3V_c \qquad b = \frac{3V_cP_c - RT_c}{P_c} = 3V_c - \frac{RT_c}{P_c}$$

Equating the coefficients of V:

$$\frac{a}{P_c} = 3V_c^2 \qquad a = 3V_c^2P_c$$

Equating the coefficients of V^0:

$$V_c^3 = ab/P_c \qquad ab = V_c^3P_c \qquad a = 3V_c^2P_c \quad \text{(from above)} \qquad b = \tfrac{1}{3}V_c$$

13.15 Expand the van der Waals equation (for n mol gas) into a cubic equation in V. Solve this equation to calculate the volume occupied by 3.00 mol oxygen at 50.0 atm pressure and 100 °C. Compare the result with that obtained by successive approximations in Problem 13.8.

▮

$$\left(P + \frac{n^2a}{V^2}\right)(V - nb) = nRT$$

$$PV - Pnb + \frac{n^2a}{V} - \frac{n^3ab}{V^2} = nRT \qquad V^3 - \left(\frac{nRT}{P} + nb\right)V^2 + \left(\frac{n^2a}{P}\right)V - \frac{n^3ab}{P} = 0$$

An equation in the form $y^3 + py^2 + qy + r = 0$ is converted into the equation $x^3 + vx + w = 0$ by the substitution $y = x - p/3$ where $v = \frac{1}{3}(3q - p^2)$ and $w = \frac{1}{27}(2p^3 - 9pq + 27r)$. Using the further substitutions

$$A = \sqrt[3]{-\frac{w}{2} + \sqrt{\frac{w^2}{4} + \frac{v^3}{27}}}$$

and

$$B = \sqrt[3]{-\frac{w}{2} - \sqrt{\frac{w^2}{4} + \frac{v^3}{27}}}$$

three possible solutions are obtained:

$$x = A + B \qquad x = -\frac{A + B}{2} + \sqrt{-3}\left(\frac{A - B}{2}\right) \qquad x = -\frac{A + B}{2} - \sqrt{-3}\left(\frac{A - B}{2}\right)$$

For the given data,

$$p = -\left(\frac{nRT}{P} + nb\right) = -\left(\frac{(3.00)(0.0821)(373)}{50.0} + (3.00)(0.0318)\right) = -1.93$$

$$q = \frac{n^2a}{P} = \frac{(3.00)^2(1.36)}{50.0} = 0.245$$

$$r = -\frac{n^3ab}{P} = -\frac{(3.00)^3(1.36)(0.0318)}{50.0} = -0.02335$$

Then

$$v = \tfrac{1}{3}(3(0.245) - (-1.93)^2) = -0.997$$

$$w = \tfrac{1}{27}[2(-1.93)^3 - 9(-1.93)(0.245) + 27(-0.02335)] = -0.398$$

$$A = \sqrt[3]{-\left(\frac{-0.398}{2}\right) + \sqrt{\frac{(0.398)^2}{4} + \frac{(-0.997)^3}{27}}} = 0.632$$

$$B = \sqrt[3]{0.199 - 0.0539} = 0.525$$

$$x = A + B = 0.632 + 0.525 = 1.157 \quad \text{(the other solutions for } x \text{ are imaginary)}$$

$$y = 1.157 - (-1.93/3) = 1.80 \text{ L}$$

This value is very close to that found in Problem 13.8. Indeed, it is not worth the extra effort to obtain an "exact" solution with the cubic equation, since the van der Waals equation itself is far from exact.

13.2 BASICS OF KINETIC MOLECULAR THEORY

13.16 Which postulate(s) of the kinetic molecular theory can be used to justify (*a*) Dalton's law of partial pressures (*b*) Graham's law of effusion?

▮ (*a*) The facts that the molecules of each gas are in constant random motion and exert no forces on each other are necessary to have pressures independent of the presence of other gases. (*b*) The average translational kinetic energy is directly proportional to absolute temperature.

$$m_1u_1^2 = m_2u_2^2 \qquad \frac{u_2}{u_1} = \sqrt{\frac{m_1}{m_2}} = \sqrt{\frac{M_1}{M_2}}$$

13.17 What is the relationship between k, the Boltzmann constant, and R, the ideal gas law constant?

▮
$$R = N_Ak = (6.02 \times 10^{23})k$$

13.18 If separate samples of argon, neon, nitrogen, and ammonia, all at the same initial temperature and pressure, are expanded adiabatically to double their original volumes, which would require the greatest quantity of heat to restore the original temperature?

▮ NH_3, which has the highest intermolecular forces—hydrogen bonding—would take the greatest quantity of energy to overcome those forces. Hence, it would take the most heat to restore the original temperature.

13.19 Explain why spark plugs are not necessary in a diesel engine.

▮ The compression of the gases heats them above the ignition temperature of the fuel.

13.20 Show that change in momentum per unit time has the same units as mass times acceleration.

▮ Change in momentum, $\Delta(mv)$, has the units kg·m/s; so $\Delta(mv)/\Delta t$ has units

$$\frac{\text{kg·m/s}}{\text{s}} = \text{kg·(m/s}^2)$$

13.21 When a tire is pumped up rapidly, its temperature rises. Would this effect be expected if air were an ideal gas? Explain.

▮ Even if air were ideal, which it is not, the compression would cause a temperature rise because energy is expended on the gas in the process. The energy is converted to kinetic energy of the molecules: heat.

13.22 When CO_2 under high pressure is released from a fire extinguisher, particles of solid CO_2 are formed, despite the low sublimation temperature of CO_2 at 1.0 atm pressure (−77 °C). Explain this phenomenon.

▮ The gas does work pushing back the atmosphere. The energy comes from the kinetic energy of the molecules, lowering the average kinetic energy—the temperature.

13.23 Explain why Boyle's law cannot be used to calculate the volume of a real gas which is changed from its initial state to its final state by an adiabatic expansion.

▮ The temperature is lowered in this experiment.

13.24 Prove that for a vector velocity **u**, $u^2 = u_x^2 + u_y^2 + u_z^2$; that is, the square of the speed equals the sum of the squares of the speeds along the three coordinate axes.

▮ Apply the Pythagorean theorem twice in Fig. 13.1:

$$u_x^2 + u_y^2 = h^2 \qquad h^2 + u_z^2 = u^2 \quad \text{so} \quad u_x^2 + u_y^2 + u_z^2 = u^2$$

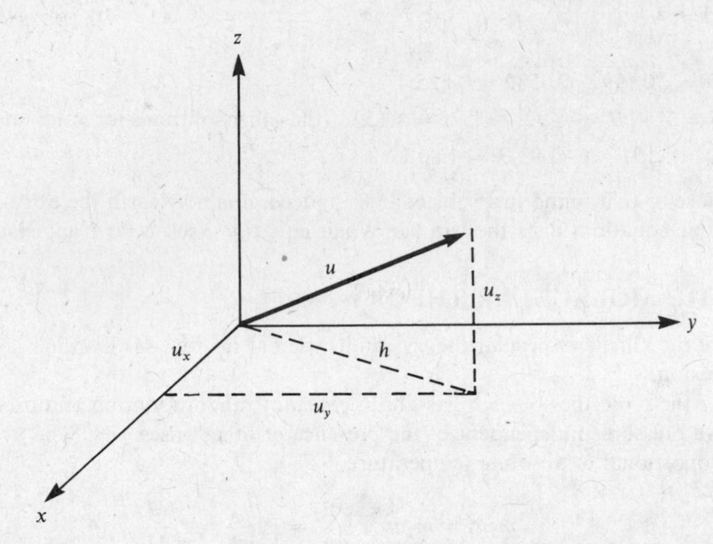

Fig. 13.1

13.25 What is the numerical value of N/n, where N is the number of molecules in a given sample of gas and n is the number of moles of the gas?

▮ $$\frac{(N \text{ molecules})/(\text{volume } V)}{(n \text{ mol})/(\text{volume } V)} = \text{number of molecules/mol} = 6.02 \times 10^{23} \equiv N_A \quad \text{(Avogadro's number)}$$

13.26 Predict from the kinetic molecular theory what the effect will be on the pressure of a gas inside a cubic box of sides l of reducing the size so that each side measures $l/2$. Assume no change in temperature.

▮ The volume of the box will be one-eighth the original volume: $(l/2)^3 = V/8$ (Fig. 13.2). With no change in temperature, the molecules will retain the original average kinetic energy and thus the original mean square speed: $\overline{\text{KE}} = \frac{3}{2}kT = \frac{1}{2}m\overline{u^2}$. The molecules will strike wall A (Fig 13.2) twice as often, since the distance in the x direction that they have to go is only half as great. Also, the area of wall A will be reduced to one-fourth of the original area, and the pressure (force per unit area) will be increased on that account by a factor of 4. Thus the pressure due to collisions on wall A will be increased by a factor of 2 because the molecules hit the wall more often and by a factor of 4 because the wall is smaller, for a total increase of a factor of 8. This result is exactly that predicted by Boyle's law.

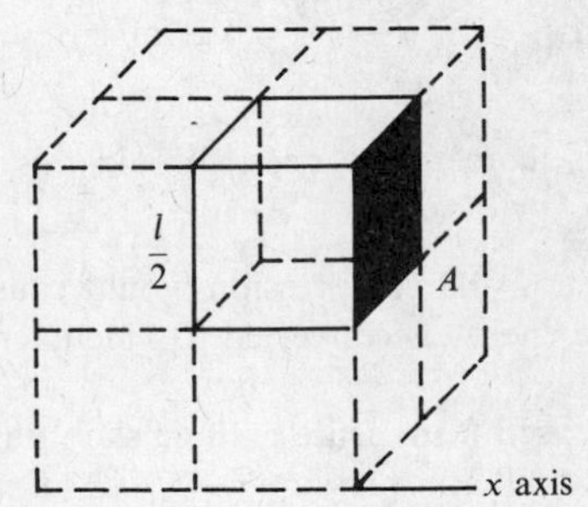

Fig. 13.2

13.27 Consider a sample of gas in a fixed-volume container. From the arguments of the kinetic molecular theory, show that quadrupling the absolute temperature causes a quadrupling in pressure.

▮ Quadrupling the absolute temperature causes a quadrupling in the average kinetic energy of the molecules. This increase in average KE causes an increase of the root mean square speed by a factor of 2:

$$\overline{\text{KE}}_1 = \tfrac{1}{2}mu_1^2 \qquad \overline{\text{KE}}_2 = \tfrac{1}{2}mu_2^2 = 4\overline{\text{KE}}_1 = \tfrac{1}{2}m(2u_1)^2$$

Thus the molecules are traveling twice as fast at the higher temperature, and therefore they hit the walls twice as often. In addition, because they are traveling twice as fast, their momentum is doubled, and their change in momentum at the wall also doubles—they hit the wall twice as hard each time. The combined effect of hitting the wall twice as often with twice the change in momentum each time they hit causes a fourfold increase in pressure, in agreement with Charles' law.

13.28 Compare the average velocity of the molecules in a sample of air 25 °C to its bulk velocity in a 10 mi/h wind.

▮ In a breeze, all the molecules have a translational kinetic energy in the direction of the breeze superimposed on their random motions. In random motion the number of molecules and their average velocities in any given direction are equal to those in any other direction. In a breeze, more molecules will be moving in one direction and with a greater average velocity than in any other direction. (The difference in average velocities in the various directions will be very small, however.)

13.29 Which one(s) of the following represent vector quantities: ***(a)*** force ***(b)*** momentum ***(c)*** energy ***(d)*** speed ***(e)*** velocity ***(f)*** temperature ***(g)*** pressure.

▮ Vector quantities have direction as well as magnitude. Thus ***(a)*** force, ***(b)*** momentum, ***(e)*** velocity are vector quantities.

13.30 What factors influence the magnitude of the temperature drop on expansion of a pure real gas into a vacuum?

▮ The intermolecular forces (which affect the quantity of heat lost) and the heat capacity (which together with the heat lost affects the temperature drop).

13.31 The absolute zero of temperature may be determined experimentally in several ways. List the various experimental techniques which might be used for this purpose.

▮ The absolute temperature may be determined from gas data, $(PV = nRT)$, free energy variation with temperature $(\Delta G = \Delta H - T\,\Delta S)$, electrochemical potential $[\varepsilon = \varepsilon^\circ - (RT/nF)(\ln Q)]$, and $\overline{\text{KE}} = \frac{3}{2}\,kT$.

13.32 A given quantity of real gas is expanded adiabatically into a vacuum, from an initial volume of 1 L to a final volume of 2 L, and the temperature of the gas decreases by 1.0 °C. If half the given quantity of the same gas were expanded adiabatically into a vacuum from 1 to 2 L, would the same decrease in temperature be expected? Explain.

▮ Since the gas is at a lower pressure with half the number of moles, there is less intermolecular attraction to overcome, so the energy loss from the kinetic energy is lower. The temperature drop is less.

13.33 Which of the postulates of the kinetic molecular theory are only approximations when applied to real gases?

▮ The approximations are that there are no intermolecular forces between molecules and that the molecules have negligible sizes.

13.3 KINETIC ENERGIES OF GAS MOLECULES

13.34 For the following set of speeds, calculate ***(a)*** the average speed ***(b)*** the mean square speed ***(c)*** the root mean square speed ***(d)*** the most probable speed.

10 m/s   15 m/s   20 m/s   25 m/s   25 m/s   30 m/s

▮

(a) With all speeds in m/s: $$\frac{10 + 15 + 20 + 25 + 25 + 30}{6} = 21 \text{ m/s}$$

(b) $$\frac{(10)^2 + (15)^2 + (20)^2 + (25)^2 + (25)^2 + (30)^2}{6} = 479 \text{ m}^2/\text{s}^2$$

(c) $\sqrt{479 \text{ m}^2/\text{s}^2} = 22$ m/s   ***(d)*** 25 m/s. (That speed is the one most often found—here, twice.)

13.35 Calculate the kinetic energy of gas molecules at 0 °C.

▮ $$\overline{\text{KE}} = \tfrac{3}{2}kT = (1.50)(1.381 \times 10^{-23} \text{ J/molecule}\cdot\text{K})(273 \text{ K}) = 5.66 \times 10^{-21} \text{ J/molecule}$$

Note that in the calculation of average kinetic energy, the species of the gas molecules is unimportant.

13.36 Calculate the root mean square speed, u, of a hydrogen molecule at 0 °C.

▮ The root mean square speed is obtained by using the equation relating the average KE of gas molecules with the absolute temperature of the gas:

$$\overline{\text{KE}} = \tfrac{1}{2}mu^2 = \tfrac{3}{2}kT \quad \text{where} \quad k = \frac{R}{N_A} \quad \text{(see Problem 13.17)}$$

To obtain u in cm/s, it is necessary to use the mass of the molecule in g equal to the molecular weight M divided by Avogadro's number N_A:

$$\frac{1}{2}\frac{M}{N}u^2 = \frac{3}{2}\frac{R}{N}T$$

$$u = \sqrt{\frac{3RT}{M}} = \sqrt{\frac{3(8.31 \times 10^7 \text{ erg/mol·K})(273 \text{ K})}{2.01 \text{ g/mol}}}$$

$$= \sqrt{3.39 \times 10^{10} \text{ cm}^2/\text{s}^2} = 1.84 \times 10^5 \text{ cm/s} = 1.84 \text{ km/s} = 4140 \text{ mi/h}$$

13.37 Calculate the ratio of ***(a)*** rms speed to most probable speed ***(b)*** average speed to most probable speed.

▮ ***(a)*** $$\frac{u_{\text{rms}}}{u_{\text{mp}}} = \frac{(3kT/m)^{1/2}}{(2kT/m)^{1/2}} = (\tfrac{3}{2})^{1/2} = 1.22$$

(b) $$\frac{u_{\text{avg}}}{u_{\text{mp}}} = \frac{(8kT/\pi m)^{1/2}}{(2kT/m)^{1/2}} = \left(\frac{4}{\pi}\right)^{1/2} = 1.13$$

13.38 At what temperature would the most probable speed of CO_2 molecules be twice that at 50 °C

▮ $$\left(\frac{2kT}{m}\right)^{1/2} = 2\left(\frac{2k(323 \text{ K})}{m}\right)^{1/2} \qquad T = 4(323 \text{ K}) = 1292 \text{ K} = 1019 \text{ °C}$$

13.39 Calculate the root mean square velocity of H_2 at 100 °C.

▮ $$u_{\text{rms}} = \left(\frac{3kT}{m}\right)^{1/2} = \left[\frac{3(1.381 \times 10^{-23} \text{ J/K})(373 \text{ K})(1 \text{ m}^2\text{·kg/s}^2\text{·J})}{(2.016 \text{ u})(1.66 \times 10^{-27} \text{ kg/u})}\right]^{1/2}$$

$$= (4.62 \times 10^6 \text{ m}^2/\text{s}^2)^{1/2} = 2.15 \times 10^3 \text{ m/s} = 2.15 \text{ km/s}$$

13.40 What is the ratio of the average molecular kinetic energy of UF_6 to that of H_2, both at 300 K?

▮ The average KE depends only on T, not on the substance. The ratio is thus 1:1.

13.41 What is the kinetic energy, in kcal, of a mole of CO_2 at 450 K?

▮ $$\overline{\text{KE}} = \tfrac{3}{2}kT = \tfrac{3}{2}(1.381 \times 10^{-23} \text{ J/K})(450 \text{ K}) = 9.32 \times 10^{-21} \text{ J} = 9.32 \times 10^{-24} \text{ kJ}$$

Per mole, $$(9.32 \times 10^{-24} \text{ kJ})(6.02 \times 10^{23}) = (5.61 \text{ kJ})\left(\frac{1 \text{ kcal}}{4.184 \text{ kJ}}\right) = 1.34 \text{ kcal}$$

13.42 At what temperature would N_2 molecules have the same average speed as He atoms at 330 K?

▮ $$u_{\text{avg}}(\text{He}) = \left(\frac{8kT}{\pi m}\right)^{1/2} = \left(\frac{8k(330 \text{ K})}{\pi(4.0 \text{ u})}\right)^{1/2} \qquad u_{\text{avg}}(\text{N}_2) = \left(\frac{8kT}{\pi(28.0 \text{ u})}\right)^{1/2}$$

Since the average speeds are the same, the right-hand sides of these equations are equal, as are their squares:

$$\frac{8k(330 \text{ K})}{\pi(4.0 \text{ u})} = \frac{8kT}{\pi(28.0 \text{ u})} \qquad T = \left(\frac{28.0}{4.0}\right)(330 \text{ K}) = 2310 \text{ K}$$

13.43 At what temperature will helium atoms have the same root mean square speed as hydrogen molecules have at 300 K?

▮ $$\overline{\text{KE}} = \tfrac{3}{2}kT = \tfrac{1}{2}mu^2$$

If u is the same for He and H_2, u^2 is also the same

$$u^2 = \frac{3kT}{m} = \frac{3k(300 \text{ K})}{2 \text{ u}} = \frac{3kT_{\text{He}}}{4 \text{ u}} \qquad T_{\text{He}} = 2(300 \text{ K}) = 600 \text{ K}$$

13.44 At what temperature will hydrogen molecules have the same root mean square speed as nitrogen molecules have at 35 °C?

▌ Both gases have the same root mean square speed, u:

$$u = \sqrt{\frac{3RT}{M}} = \sqrt{\frac{3R(308\ \text{K})}{28.0\ \text{g/mol}}} = \sqrt{\frac{3RT}{2.00\ \text{g/mol}}} \qquad \sqrt{\frac{308\ \text{K}}{28.0}} = \sqrt{\frac{T}{2.00}} \qquad T = 22.0\ \text{K}$$

13.45 Calculate the root mean square speed and the average kinetic energy of oxygen molecules at 18 °C.

▌ Working in SI units,

$$u = \sqrt{\frac{3RT}{M}} = \sqrt{\frac{3(8.31\ \text{J/mol}\cdot\text{K})(291\ \text{K})}{32.0 \times 10^{-3}\ \text{kg/mol}}} = 4.76 \times 10^2\ \text{m/s}$$

$$\overline{\text{KE}} = \tfrac{3}{2}kT = \tfrac{3}{2}\left(\frac{8.31\ \text{J/mol}\cdot\text{K}}{6.02 \times 10^{23}\ \text{molecules/mol}}\right)(291\ \text{K}) = 6.03 \times 10^{-21}\ \text{J/molecule}$$

$$\text{Check:}\quad \tfrac{1}{2}mu^2 = \tfrac{1}{2}[(32\ \text{u})(1.66 \times 10^{-27}\ \text{kg/u})](4.76 \times 10^2\ \text{m/s})^2 = 6.02 \times 10^{-21}\ \text{kg}\cdot\text{m}^2/\text{s}^2 = 6.02 \times 10^{-21}\ \text{J}$$

13.46 (*a*) Show that the average of the squares of the following set of numbers is different from the square of the average of the numbers: 5, 10, 15, 20. (*b*) Which of the results is larger? (*c*) For molecules having a given mass, why does the root mean square speed, rather than the average speed, have greater physical significance?

▌ (*a*)

$$\frac{(5)^2 + (10)^2 + (15)^2 + (20)^2}{4} = 187.5 \qquad \left(\frac{5 + 10 + 15 + 20}{4}\right)^2 = 156.25$$

(*b*) The average of the squares is larger, since the process of squaring increases the importance of the larger numbers more than it decreases that of the smaller. (This is true for any set of numbers.) (*c*) Root mean square speed is directly related to kinetic energy.

13.47 The total kinetic energy of molecules of a given gaseous substance has contributions from translation, molecular vibration, and molecular rotation. (*a*) Which one(s) of these components is (are) proportional to the absolute temperature of the gas? (*b*) Which one(s) is (are) related to the specific heat of the substance? (*c*) Which one(s) is (are) related to the infrared absorptions of the molecules?

▌ (*a*) Translational, (*b*) all three, (*c*) rotational and vibrational.

13.48 At what temperature will hydrogen molecules have the same KE as nitrogen molecules have at 35 °C?

▌ 35 °C. (The KE depends only on the absolute temperature, not on the identity of the gas. Compare Problem 13.44.)

13.49 Show that the ideal gas law can be written $P = \frac{2}{3}\varepsilon$, where ε is the kinetic energy per unit volume.

▌

$$PV = nRT = \frac{N}{N_A}RT = N\frac{R}{N_A}T = NkT = \tfrac{2}{3}N(\tfrac{3}{2}kT) = \tfrac{2}{3}(N \times \overline{\text{KE}}) = \tfrac{2}{3}(\text{total KE})$$

$$\text{so}\quad P = \tfrac{2}{3}\left(\frac{\text{total KE}}{V}\right) = \tfrac{2}{3}\varepsilon$$

13.50 Distinguish between the total kinetic energy of a molecule and its translational kinetic energy. For what type of gas molecules are they the same?

▌ The total KE of a molecule includes its translation KE plus its rotational and vibrational KEs. Only for molecules in which the latter two are zero—monatomic molecules—is the total energy equal to the translational kinetic energy.

13.51 At what temperature would the average translational kinetic energy of gaseous hydrogen molecules equal the energy required to dissociate the molecules into atoms (104 kcal/mol)?

▌

$$\overline{\text{KE}} = \tfrac{3}{2}kT$$

For 1 mol:

$$N_A\,\overline{\text{KE}} = \tfrac{3}{2}RT = \tfrac{3}{2}(1.987\ \text{cal/mol}\cdot\text{K})T = 1.04 \times 10^5\ \text{cal/mol} \qquad T = \frac{2(1.04 \times 10^5\ \text{cal/mol})}{3(1.987\ \text{cal/mol}\cdot\text{K})} = 34\,900\ \text{K}$$

13.52 Does a molecule of a gas sample which has a kinetic energy equal to the average KE of the gas sample also have a speed equal to the average speed of the molecules of the sample? Explain.

▮ No; the molecule with average KE has average u^2. But the average u^2 is not the square of the average u (see Problem 13.46).

13.53 Explain in terms of atomic and molecular motions why the heat capacity of CH_3OCH_3, 66.5 J/mol·K, is so much greater than that of He, 20.9 J/mol·K.

▮ CH_3OCH_3 has its rotational and vibrational energy increased by an increase in temperature. The extra energy required is reflected by the greater heat capacity.

13.54 Carbon dioxide and helium are stored in separate but identical containers initially at the same temperature and pressure. Using data from Table 13.1, determine **(*a*)** which gas (if either) would require the greater number of calories to reheat to the original temperature if the volume of each were doubled by sudden expansion into a vacuum; **(*b*)** which gas, if either, would do more work on the surroundings if the original volume of each gas were doubled by expanding against a constant pressure of 1 atm at constant temperature; **(*c*)** which gas would show the greater rate of pressure decrease if each gas were allowed to effuse from its container through a very small orifice. **(*d*)** Molecules of which gas should travel the greater average distance between collisions? **(*e*)** If a given quantity of heat were added to each gas, which would experience the greater rise in temperature?

▮ **(*a*)** CO_2, which lost more heat to overcome intermolecular forces. **(*b*)** He, which has more energy to push back the atmosphere rather than having to overcome the intermolecular forces. **(*c*)** He, which has a lower molecular weight and thus a higher rate of effusion. **(*d*)** He, which is smaller. **(*e*)** He, for which all the energy would go into translation.

13.55 A study of automobile speeds on a highway yielded the following data for a 1 h period. What are **(*a*)** the average speed, **(*b*)** the root mean square speed, and **(*c*)** the most probable speed of this group of cars?

speed (s)	number of cars (n)	speed (s)	number of cars (n)
47	2	56	210
48	7	57	192
49	20	58	144
50	45	59	112
51	102	60	63
52	150	61	26
53	205	62	11
54	220	63	4
55	250	66	2

▮ **(*a*)** $\bar{v}$ = avg speed = $(\sum n_i s_i)$/total number = 97 253/1765 = 55.10

(*b*) u = rms speed = $\sqrt{(\sum n_i s_i^2)/\text{total number}} = \sqrt{5\,372\,657/1765} = 55.17$

(The equation $\bar{v} = (\sqrt{8/3\pi})u$ is valid for molecules but not for automobiles.)

(*c*) Most probable speed = speed of highest number of cars = 55

13.56 Explain why you would suspect that some data were missing if the second pair of columns of the data in Problem 13.55 had been accidentally omitted.

▮ One expects a "bell-shaped" curve of speeds, or some similar variation. One should suspect a set of data in which the most probable value is the highest value.

13.4 GRAHAM'S LAW

13.57 What is the ratio of the rate of effusion of neon gas to that of helium gas at the same temperature and pressure?

▮
$$\frac{r(\text{Ne})}{r(\text{He})} = \sqrt{\frac{M(\text{He})}{M(\text{Ne})}} = \sqrt{\frac{4}{20}} = \frac{1}{\sqrt{5}}$$

13.58 Compute the relative rates of effusion of H_2 and CO through a fine pinhole.

▮
$$\frac{r(H_2)}{r(CO)} = \sqrt{\frac{M(CO)}{M(H_2)}} = \sqrt{\frac{28.0}{2.00}} = 3.74$$

13.59 Calculate the ratio of rates of effusion of H_2 and O_2, both at 0 °C and 1 atm pressure.

▮
$$\frac{r(H_2)}{r(O_2)} = \sqrt{\frac{32.0}{2.0}} = 4.0$$
H_2 effuses four times as fast as O_2 under these conditions.

13.60 A certain saturated hydrocarbon effuses about half as fast as methane. What is the molecular formula of this hydrocarbon?

▮
$$\frac{\text{Rate}}{\text{Rate of } CH_4} = \sqrt{\frac{16\text{ g/mol}}{M}} \approx \frac{1}{2} \qquad M \approx 64\text{ g/mol}$$

Butane, C_4H_{10}, has a molecular weight of 58 g/mol.

13.61 Uranium isotopes have been separated by taking advantage of the different rates of effusion of the two isotopic forms of UF_6. One form contains uranium of atomic weight 238, and the other of atomic weight 235. What are the relative rates of effusion of these two molecules?

▮
$$M_{\text{heavy}} = 238 + 6(19) = 352 \qquad M_{\text{light}} = 235 + 6(19) = 349$$
$$r_h/r_l = \sqrt{349/352} = 0.996$$

13.62 The pressure in a vessel that contained pure oxygen dropped from 2000 torr to 1500 torr in 55 min as the oxygen leaked through a small hole into a vacuum. When the same vessel was filled with another gas, the pressure dropped from 2000 torr to 1500 torr in 85 min. What is the molecular weight of the second gas?

▮
$$\frac{r_1}{r_2} = \frac{1/55\text{ min}}{1/85\text{ min}} = \sqrt{\frac{M}{32.0}}$$
$$M = 76\text{ g/mol}$$

Note that the *rates* are inversely proportional to the *times* of effusion.

13.63 A large cylinder of helium filled at 2000 lbf/in.^{2} had a small thin orifice through which helium escaped into an evacuated space at the rate of 6.4 mmol/h. How long would it take for 10 mmol of CO to leak through a similar orifice if the CO were confined at the same pressure?

▮
$$\frac{r(\text{He})}{r(\text{CO})} = \sqrt{\frac{M(\text{CO})}{M(\text{He})}} = \sqrt{\frac{28.0}{4.00}} = \sqrt{7.00} = 2.65$$

Helium escapes at a rate 2.65 times as fast as CO.

$$r(\text{CO}) = \frac{r(\text{He})}{2.65} = \left(\frac{6.4\text{ mmol}}{\text{h}}\right)\left(\frac{1}{2.65}\right) = 2.4\text{ mmol CO/h} \qquad (10\text{ mmol CO})\left(\frac{1\text{ h}}{2.4\text{ mmol CO}}\right) = 4.2\text{ h}$$

13.64 At 1200 °C, the following equilibrium is established between chlorine atoms and chlorine molecules: $Cl_2 \rightleftharpoons 2Cl$. The composition of the equilibrium mixture may be determined by measuring the rate of effusion of the mixture through a pinhole. It is found that at 1200 °C and 1.8 torr the mixture effuses 1.16 times as fast as krypton effuses under the same conditions. Calculate the fraction of chlorine molecules dissociated into atoms.

▮
$$\frac{r}{r(\text{Kr})} = \sqrt{\frac{M(\text{Kr})}{\overline{M}}} = 1.16 \qquad \overline{M} = M(\text{Kr})/(1.16)^2 = (83.80)/(1.16)^2 = 62.28$$

Let $2x$ = mol Cl atoms present per mol Cl_2 originally present. Then $1 - x$ = mol Cl_2 molecules present per mol Cl_2 originally present. $1 + x$ = mol Cl plus Cl_2 molecules per mol Cl_2 originally present. The average molecular weight is then

$$\overline{M} = \frac{(2x)(35.5) + (1 - x)(71.0)}{1 + x} = 62.28 \qquad \frac{71.0}{1 + x} = 62.28$$

$x = 0.140$; i.e., 14.0% of the molecular chlorine dissociated.

13.65 A space capsule is filled with neon gas at 1.00 atm and 290 K. The gas effuses through a pinhole into outer space at such a rate that the pressure drops by 0.30 torr/s. (*a*) If the capsule were filled with ammonia at the same temperature and pressure, what would be the rate of pressure drop? (*b*) If the capsule were filled with 30.0 mol %

helium, 20.0 mol % oxygen, and 50.0 mol % nitrogen at a total pressure of 1.00 atm and a temperature of 290 K, what would be the corresponding rate of pressure drop?

▮ The rate of pressure drop is directly proportional to the rate of effusion.

(*a*) $$\frac{r(\text{Ne})}{r(\text{NH}_3)} = \sqrt{\frac{M(\text{NH}_3)}{M(\text{Ne})}} = \sqrt{\frac{17.0}{20.2}} = 0.917 \qquad r(\text{NH}_3) = r(\text{Ne})/0.917 = \frac{0.30\ \text{torr/s}}{0.917} = 0.33\ \text{torr/s}$$

(*b*) The average molecular weight of the gas is $0.300(4.0) + 0.200(32.0) + 0.500(28.0) = 21.6$

$$\text{Rate} = \left(\sqrt{\frac{20.2}{21.6}}\right) r(\text{Ne}) = 0.29\ \text{torr/s}$$

13.66 At the start of an experiment, one end of a U tube of 6 mm glass tubing is immersed in concentrated ammonia solution and the other end is immersed in concentrated hydrochloric acid solution. At the point in the tube where vapors of NH_3 and HCl meet, a white cloud of $NH_4Cl(s)$ forms. At what fraction of the distance along the tube from the ammonia solution does the white cloud *first* form?

▮ The ratio of rates of diffusion is given by $$\frac{r(\text{HCl})}{r(\text{NH}_3)} = \sqrt{\frac{17.0}{36.5}} = 0.682$$

The HCl travels 68.2% as far as the NH_3 in a given time. If x is the distance traveled by the NH_3, $0.682x$ is the distance traveled by the HCl. The total length of the tube is $1.682x$, and the fraction traveled by the NH_3 is

$$\frac{x}{1.682x} = 0.595$$

13.67 A porous cup filled with hydrogen gas at atmospheric pressure is connected to a glass tube which has one end immersed in water, as shown in Fig. 13.3. Explain why the water rises in the glass tube.

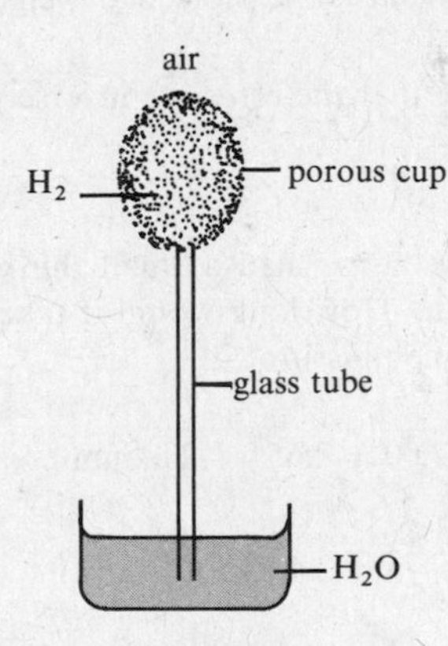

Fig. 13.3

▮ The water would rise in the tube because the H_2 effuses out of the porous cup faster than the air effuses in. Thus the pressure in the cup drops, and the water is pushed up the tube by the air pressure.

13.68 The separation factor, f, of a process of isotopic separation is defined as the ratio of the relative *concentration* of a given species after processing to its relative *concentration* before processing:

$$f = \frac{n'_1/n'_2}{n_1/n_2}$$

where n_1 and n_2 are the concentrations of two species before processing. For a single-stage diffusion process, the maximum separation factor is merely the ratio of the rates of diffusion of the isotopic molecules as determined from Graham's law. Naturally occurring uranium consists of 99.3% by mass of ^{238}U and 0.7% of ^{235}U. (***a***) Calculate the overall separation factor necessary to achieve a product containing 99.7% ^{235}U starting with natural uranium. (***b***) What is the theoretical separation factor for a single diffusion step using isotopic molecules of UF_6? (***c***) How many ideal diffusion steps would be required to produce 99.7% pure ^{235}U? (***d***) Natural hydrogen consists of 99.8% ^{1}H and 0.02% ^{2}H by mass. In the manner outlined above, calculate the number of ideal diffusion steps required to produce 99% pure ^{1}H^{2}H from ordinary gaseous hydrogen.

▮ (*a*) $$f = \frac{0.997/0.003}{0.007/0.993} = 4.7 \times 10^4$$

(b) The molecular weight of $^{238}UF_6$ is 352; that of $^{235}UF_6$ is 349.

$$f' = \frac{r(^{235}U)}{r(^{238}U)} = \sqrt{\frac{352}{349}} = 1.0043$$

(c) $(1.0043)^n = 4.7 \times 10^4 \qquad n(\log 1.0043) = (\log 4.7) + 4 \qquad n = \frac{4.67}{0.00186} = 2.5 \times 10^3$

(d) The percentage of $^2H_2$ is insignificant, and the percentage of H_2 which is $^1H^2H$ is 0.02%. If no bonds are broken, the enrichment product is 99% $^1H^2H$.

$$f = \frac{0.99/0.01}{0.0002/0.9998} = 5.0 \times 10^5 \qquad f' = \sqrt{\frac{3.00}{2.00}} = 1.22$$

$$n = \frac{(\log 5.0) + 5}{\log 1.22} = \frac{5.7}{0.086} = 66$$

13.69 Problem 13.68(*d*) suggests that ^2H in hydrogen gas prepared from naturally occurring sources of hydrogen is more likely to be in the form of $^1H^2H$ than in the form $^2H_2$. Hence, to obtain pure ^2H by a diffusion method, isotopic *compounds* of hydrogen are used. What characteristics would be desirable in such a compound? What compound(s) can you suggest?

▮ The desirable features are low molecular weight, volatility, combination with an element having one predominant isotope, and the incorporation of only one hydrogen atom. HF might be thought of as a likely example, because of its relatively low molecular weight and the fact that fluorine is practically 100% ^{19}F. However, the hydrogen bonding makes the effective molecular weight, even in the gas phase, about 80 at 1 atm pressure, and each "molecule" would contain about 4 H atoms. Moreover, the H atoms would be exchanged among the "molecules." HF could be used at very low pressure to minimize hydrogen bonding.

CHAPTER 14
Solids and Liquids

Note: Before working the problems of this chapter, you should familiarize yourself with Tables 14.1, p. 246 and 14.2, p. 262.

14.1 CRYSTAL STRUCTURE

14.1 Explain why an end-centered unit cell cannot be cubic. What is the highest possible symmetry for this type of unit cell?

▌ A cubic unit cell must have the six faces the same. The highest symmetry possible for an end-centered unit cell is tetragonal.

14.2 Explain why uncharged atoms or molecules never crystallize in simple cubic structures.

▌ Uncharged atoms and molecules pack more efficiently in closest-packed structures.

14.3 Explain why a hexagonal closest-packed structure and a cubic closest-packed structure for a given element would be expected to have the same density.

▌ The two structures have the same coordination number, hence the same packing fraction.

14.4 Metallic gold crystallizes in the face-centered cubic lattice. The length of the cubic unit cell (Fig. 14.1) is $a = 4.070$ Å. **(*a*)** What is the closest distance between gold atoms? **(*b*)** How many "nearest neighbors" does each gold atom have at the distance calculated in (*a*)? **(*c*)** What is the density of gold? **(*d*)** Prove that the *packing factor* for gold, the fraction of the total volume occupied by the atoms themselves, is 0.74.

▌ **(*a*)** Consider a corner gold atom in Fig. 14.1. The closest distance to another corner atom is *a*. The distance to an atom at the center of a face is one-half the diagonal of that face, i.e.,

$$\frac{1}{2}(a\sqrt{2}) = \frac{a}{\sqrt{2}} < a$$

Thus the closest distance between atoms is $\dfrac{4.070}{\sqrt{2}} = 2.878$ Å

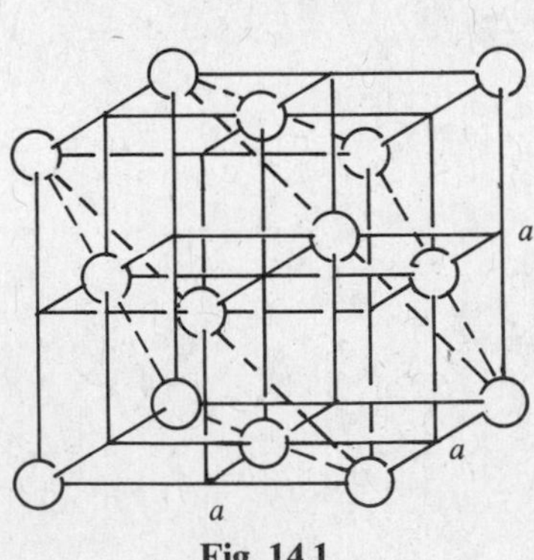

Fig. 14.1

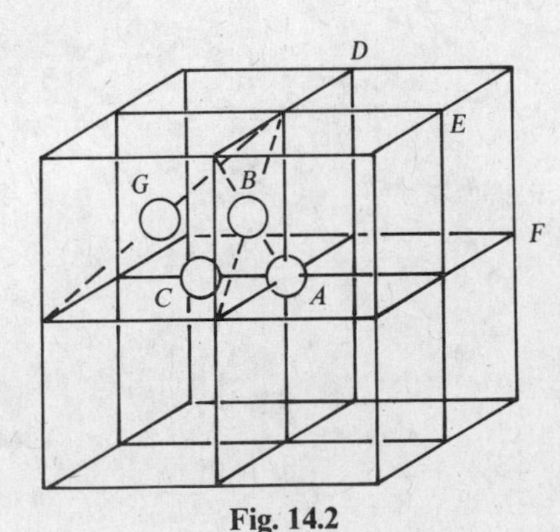

Fig. 14.2

(*b*) The problem is to find how many face centers are equidistant from a corner atom. Point *A* in Fig. 14.2 may be taken as the reference corner atom. In that same figure, *B* is one of the face-center points at the nearest distance to *A*. In plane *ABD* in the figure there are three other points equally close to *A*: the centers of the squares in the upper right, lower left, and lower right quadrants of the plane, measured around *A*. Plane *ACE*, parallel to the plane of the paper, also has points in the centers of each of the squares in the four quadrants around *A*. Also, plane *ACF*, perpendicular to the plane of the paper, has points in the centers of each of the squares in the four quadrants around *A*. Thus there are 12 nearest neighbors in all, the number expected for a close-packed structure. The same result would have been obtained by counting the nearest neighbors around *B*, a face-centered point.

(*c*) For the face-centered cubic structure, with 8 corners and 6 face centers,

$$\text{Mass per unit cell} = \tfrac{1}{8}(8m) + \tfrac{1}{2}(6m) = 4m$$

where m is the mass of a single gold atom, 197.0 u. Converting to grams,

$$m = (197.0\ \text{u})\left(\frac{1\ \text{g}}{6.023 \times 10^{23}\ \text{u}}\right) = 3.27 \times 10^{-22}\ \text{g} \qquad \text{and} \qquad \text{Density} = \frac{4m}{a^3} = \frac{4(3.27 \times 10^{-22}\ \text{g})}{(4.070 \times 10^{-8}\ \text{cm})^3} = 19.4\ \text{g/cm}^3$$

The reverse of this type of calculation can be used for a precise determination of Avogadro's number, provided the lattice dimension, the density, and the atomic weight are known precisely.

(*d*) Since the atoms at closest distance are in contact in a close-packed structure, the closest distance between centers calculated in (*a*), $a/\sqrt{2}$, must equal the sum of the radii of the two spherical atoms, $2r$. Thus, $r = a/2^{3/2}$. From (*c*), there are 4 gold atoms per unit cell. Then,

$$\text{Volume of 4 gold atoms} = 4\left(\frac{4}{3}\pi r^3\right) = 4\left(\frac{4}{3}\pi\right)\left(\frac{a}{2^{3/2}}\right)^3 = \frac{\pi a^3}{3\sqrt{2}}$$

and

$$\text{Packing fraction} = \frac{\text{volume of 4 gold atoms}}{\text{volume of unit cell}} = \frac{\pi a^3/3\sqrt{2}}{a^3} = \frac{\pi}{3\sqrt{2}} = 0.7405$$

Note that the parameter a for the gold unit cell cancels, so the result holds for any cubic close-packed structure.

14.5 Show that the tetrahedral and octahedral holes in gold are appropriately named. Find the closest distance between an impurity atom and a gold atom if the impurity atom occupies a tetrahedral hole. An octahedral hole. How many holes of each type are there per gold atom?

▌ Examine Fig. 14.1, and imagine a hole in the center of the upper left front minicube. This hole is equidistant from the four occupied corners of the minicube, the common distance being half a body diagonal of the minicube, or

$$\frac{1}{2}\sqrt{\left(\frac{a}{2}\right)^2 + \left(\frac{a}{2}\right)^2 + \left(\frac{a}{2}\right)^2} = \frac{a\sqrt{3}}{4}$$

These four occupied corners define a regular tetrahedron (see Problem 7.76), and the center of the tetrahedron is the point equidistant from the corners, which we showed is the location of the hole. This justifies the name "tetrahedral hole." As the unit cell contains 8 tetrahedral holes (one in each minicube) and 4 gold atoms (Problem 14.4), there are $8 \div 4 = 2$ tetrahedral holes per gold atom.

Now consider the hole at the center of the unit cell of Fig. 14.1. This hole is equidistant from the centers of all six faces of the unit cell, all of which are the nearest occupied sites to the hole. These six points are the vertices of an eight-faced figure; the faces are congruent equilateral triangles (whose edges are face diagonals of the minicubes). Such a figure is a regular octahedron, and the hole is at its center; hence the name "octahedral hole." The distance between the hole and a nearest-neighbor atom is $a/2$. A similar proof can be made for an octahedral hole on the center of an edge of the unit cell in Fig. 14.1 if we note that the actual crystal lattice consists of a three-dimensional stack of unit cells, as in Fig. 14.2. Each such edge-center hole is shared by 4 unit cells, so that the number of octahedral holes per unit cell is $1 + \frac{1}{4}(12) = 4$. The ratio of octahedral holes to gold atoms is thus 4 to 4, or 1.

Note the competing advantages of tetrahedral and octahedral holes for housing impurities or second components of an alloy. If the crystal forces, whatever their nature, depend mostly on interactions between nearest neighbors, the octahedral hole has an advantage in having more nearest neighbors to interact with (6, as opposed to 4). However, the tetrahedral hole has a shorter nearest-neighbor distance ($a\sqrt{3}/4 = 0.433a$, as opposed to $0.500a$), giving it the advantage of a greater potential to interact with any one host atom.

14.6 CsCl crystallizes in a cubic structure that has a Cl^- at each corner and a Cs^+ at the center of the unit cell. Use the ionic radii listed in Table 14.1 to predict the lattice constant a and compare this value with the value of a calculated from the observed density of CsCl, 3.97 g/cm^3.

▌ Figure 14.3*a* shows a schematic view of the unit cell, where the open circles represent Cl^- and the filled circle Cs^+. The circles are made small with respect to the unit cell length a to show more clearly the locations of the various ions. Figure 14.3*b* is a more realistic representation of the right triangle ABC, showing anion-cation-anion contact along the body diagonal AC.

TABLE 14.1

ion	ionic radius	ion	ionic radius
Li^+	0.60 Å	Cd^{2+}	0.97 Å
Na^+	0.95 Å	Ni^{2+}	0.69 Å
K^+	1.33 Å	Al^{3+}	0.50 Å
Cs^+	1.69 Å	H^-	2.08 Å
Ag^+	1.26 Å	F^-	1.36 Å
Mg^{2+}	0.65 Å	Cl^-	1.81 Å
Ca^{2+}	0.99 Å	Br^-	1.95 Å
Sr^{2+}	1.13 Å	I^-	2.16 Å
Ba^{2+}	1.35 Å	O^{2-}	1.40 Å
Zn^{2+}	0.74 Å	S^{2-}	1.84 Å

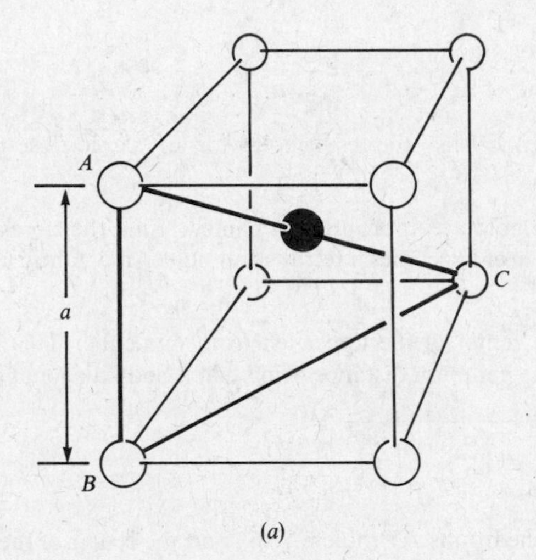

(*a*)

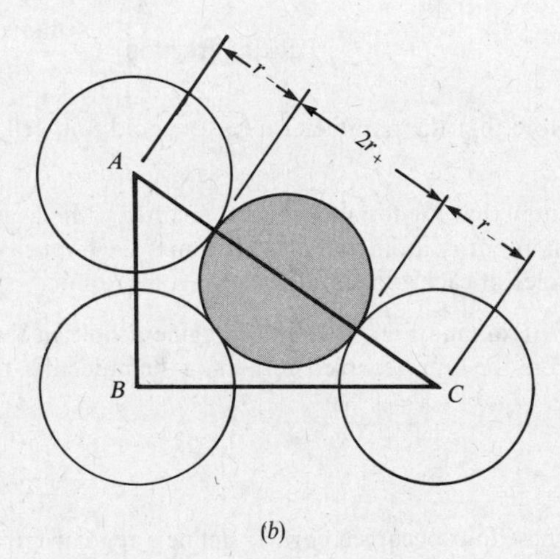

(*b*)

Fig. 14.3

Let us assume that the closest Cs^+-to-Cl^- distance is the sum of the ionic radii of Cs^+ and Cl^-, or $1.69 + 1.81 = 3.50$ Å. This distance is one-half the cube diagonal, or $a\sqrt{3}/2$. Then

$$\frac{a\sqrt{3}}{2} = 3.50 \text{ Å} \qquad \text{so} \qquad a = \frac{2(3.50 \text{ Å})}{\sqrt{3}} = 4.04 \text{ Å}$$

The density can be used to compute a if we count the number of ions of each type per unit cell. The number of assigned Cl^- ions per unit cell is one-eighth the number of corner Cl^- ions, or $\frac{1}{8}(8) = 1$. The only Cs^+ in the unit cell is the center Cs^+, so the assigned number of cesium ions is also 1. (This type of assignment of ions or atoms in a compound must always agree with the empirical formula of the compound, as the 1:1 ratio in this problem does.) The assigned mass per unit cell is thus that of 1 formula unit of CsCl,

$$\left(\frac{132.9 + 35.4}{6.02 \times 10^{23}}\right)\text{g} = 2.797 \times 10^{-22} \text{ g}$$

$$\text{Volume of unit cell} = a^3 = \frac{\text{mass}}{\text{density}} = \frac{2.797 \times 10^{-22} \text{ g}}{3.97 \text{ g/cm}^3} = 70.4 \times 10^{-24} \text{ cm}^3$$

and therefore $a = \sqrt[3]{70.4 \times 10^{-24} \text{ cm}^3} = 4.13 \times 10^{-8} \text{ cm} = 4.13$ Å. This value, based on the experimental density, is to be considered the more reliable since it is based on a measured property of CsCl, while the ionic radii are based on averages over many different compounds.

14.7 The cesium chloride structure is shown in Fig. 14.3. To what system does cesium chloride belong? Is the unit cell compound or simple?

▎ The cesium chloride unit cell is a simple cubic unit cell. The center of the unit cell is occupied by an ion of opposite charge from the ions at the cell corners and is not the site of a lattice point.

14.8 Potassium crystallizes in a body-centered cubic lattice (Fig. 14.4), with a unit cell length $a = 5.20$ Å. **(a)** What is the distance between nearest neighbors? **(b)** What is the distance between next-nearest neighbors? **(c)** How many nearest neighbors does each K atom have? **(d)** How many next-nearest neighbors does each K have? **(e)** What is the calculated density of crystalline K?

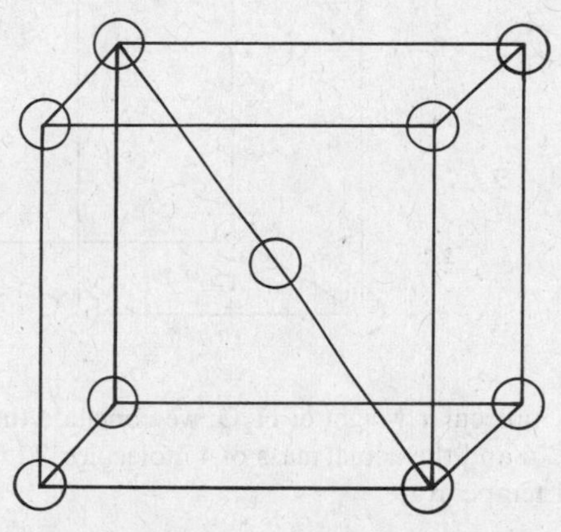

Fig. 14.4

▮ **(a)** The body diagonal is $\sqrt{3}(5.20\ \text{Å}) = 9.01$ Å. The nearest neighbors along that diagonal are half that distance apart, 4.50 Å. **(b)** 5.20 Å, along the cell edge. **(c)** 8 (For example, the body center is next to 8 corners.) **(d)** 6 (The corner has neighbors along each cell edge: up, down, in, out, left, right.) **(e)** Two K atoms per unit cell: one at the body center and one-eighth at each of eight corners. Their mass:

$$2\left(\frac{39.1\ \text{g}}{6.02 \times 10^{23}}\right) = 1.30 \times 10^{-22}\ \text{g}$$

The volume is
$$\left[(5.20\ \text{Å})\left(\frac{10^{-8}\ \text{cm}}{1\ \text{Å}}\right)\right]^3 = 1.41 \times 10^{-22}\ \text{cm}^3$$

The density is
$$\frac{1.30 \times 10^{-22}\ \text{g}}{1.41 \times 10^{-22}\ \text{cm}^3} = 0.925\ \text{g/cm}^3$$

14.9 The CsCl structure (Fig. 14.3) is observed in alkali halides only when the radius of the cation is sufficiently large to keep its eight nearest-neighbor anions from touching. What minimum value for the ratio of cation to anion radii, r_+/r_-, is needed to prevent this contact?

▮ In the CsCl structure, the nearest cation-anion distance occurs along the diagonal of the unit cell cube, while the nearest anion-anion distance occurs along a unit cell edge. This relationship is shown in Fig. 14.3*b*. In the figure, $\overline{AB} = a$, $\overline{BC} = a\sqrt{2}$, and $\overline{AC} = a\sqrt{3}$. If we assume anion-cation contact along $\overline{AC}$, then $\overline{AC} = 2(r_+ + r_-) = a\sqrt{3}$. In the limiting case, where anions touch along the edge of the unit cell, $2r_- = a$. Dividing the former equation by the latter,

$$\frac{r_+}{r_-} + 1 = \sqrt{3} \qquad \text{or} \qquad \frac{r_+}{r_-} = \sqrt{3} - 1 = 0.732$$

If the ratio were less than this critical value, anions would touch (increasing the repulsive forces) and cation and anion would be separated (decreasing the attractive forces). Both effects would tend to make the structure unstable.

14.10 **(a)** Calculate the number of CsCl formula units in a unit cell (Fig. 14.3). **(b)** What is the coordination number of each type of ion?

▮ **(a)** One. (One Cl atom at the center, and one $(8 \times \frac{1}{8})$ Cs atom at the corners.) **(b)** Eight.

14.11 Ice crystallizes in a hexagonal lattice. At the low temperature at which the structure was determined, the lattice constants were $a = 4.53$ Å and $c = 7.41$ Å (Fig. 14.5). How many H_2O molecules are contained in a unit cell?

▮ The volume V of the unit cell in Fig. 14.5 is

$$V = (\text{area of rhombus base}) \times (\text{height } c) = (a^2 \sin 60°)c = (4.53\ \text{Å})^2(0.866)(7.41\ \text{Å}) = 132\ \text{Å}^3 = 132 \times 10^{-24}\ \text{cm}^3$$

Although the density of ice at the experimental temperature is not stated, it could not be very different from the value at 0 °C, 0.92 g/cm³.

$$\text{Mass of unit cell} = V \times \text{density} = (132 \times 10^{-24}\ \text{cm}^3)(0.92\ \text{g/cm}^3)(6.02 \times 10^{23}\ \text{u/g}) = 73\ \text{u}$$

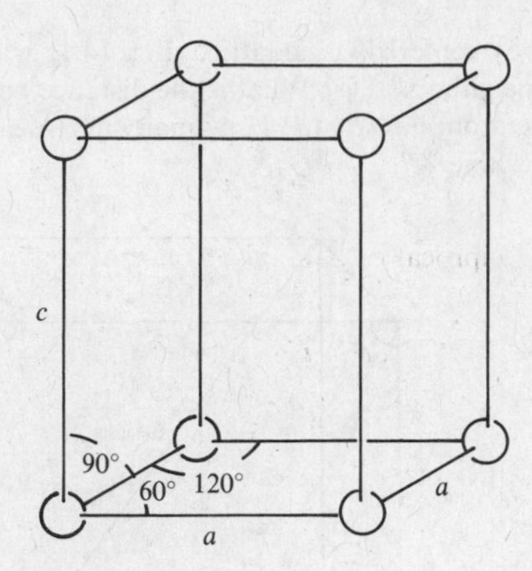

Fig. 14.5

This is close to 4 times the molecular weight of H_2O; we conclude that there are 4 molecules of H_2O per unit cell. The discrepancy between 73 u and the actual mass of 4 molecules, 72 u, is undoubtedly due to the uncertainty in the density at the experimental temperature.

14.12 $BaTiO_3$ crystallizes in the perovskite structure. This structure may be described as a cubic lattice, with barium ions occupying the corners of the unit cell, oxide ions occupying the face centers, and titanium ions occupying the centers of the unit cells. ***(a)*** If titanium is described as occupying holes in the Ba-O lattice, what type of hole does it occupy? ***(b)*** What fraction of the holes of this type does it occupy? ***(c)*** Can you suggest a reason why it occupies certain holes of this type but not the other holes of the same type?

▮ ***(a)*** These are octahedral holes. ***(b)*** The octahedral holes at the centers of unit cells constitute just one-fourth of all the octahedral holes in a face-centered cubic lattice. (See Problem 14.5.) ***(c)*** An octahedral hole at the center of a unit cell, occupied by a titanium ion, has 6 nearest-neighbor oxide ions. The other octahedral holes, located at the centers of the edges of the unit cell, have 6 nearest neighbors each, as is the case with any octahedral hole, but 2 of the 6 neighbors are barium ions (at the unit-cell corners terminating the given edge) and 4 are oxide ions. The proximity of two cations, Ba^{2+} and Ti^{4+}, would be electrostatically unfavorable.

14.13 The Cl_2 molecules in crystalline chlorine are arranged in herringbone arrays within sheets, held together by covalent bonds. The shortest Cl-to-Cl distance within these sheets is 1.98 Å. Adjacent sheets are held together by van der Waals forces, and the closest approach between Cl atoms on neighboring sheets is 3.69 Å. In solid HCl observed at low temperatures, the H—Cl bond length is 1.25 Å and the shortest Cl-to-Cl distance is 3.69 Å. ***(a)*** Estimate the van der Waals radius of Cl. ***(b)*** How do you account for the shortest distance between Cl atoms in HCl?

▮ ***(a)*** The closest distance between centers of Cl atoms in adjacent sheets may be taken as twice the van der Waals radius. The van der Waals radius would then be half of 3.69 Å, or 1.85 Å. A general rule of thumb for estimating a van der Waals radius for an atom is to add 0.8 Å to its covalent radius; this method would have given 1.8 Å (see Table 7.1).

(b) Because of the polar character of HCl, a stable crystal arrangement requires that the hydrogen of one HCl molecule be close to the chlorine of the neighboring molecule. Thus, the shortest Cl-to-Cl distance in HCl is really the distance across the array Cl—H ···· Cl. If the dotted line represented only van der Waals attraction, the Cl-to-Cl distance should equal the H—Cl bond length plus the van der Waals radii of hydrogen and chlorine. To estimate the van der Waals radius of H, we first approximate its covalent radius by subtracting the covalent radius of Cl, 0.99 Å, from the H—Cl bond length; we get 0.26 Å. (The covalent radius of H is much more variable from compound to compound than the radii of other atoms, but figures in the 0.3 to 0.4 Å range are common.) The estimated van der Waals radius of H is then $0.3 + 0.8 = 1.1$ Å. Finally, the total estimated Cl-to-Cl distance would then equal

$$r(\text{HCl}) + r_{\text{v.d.W.}}(\text{H}) + r_{\text{v.d.W.}}(\text{Cl}) = 1.25 + 1.1 + 1.85 = 4.2\ \text{Å}$$

The fact that the actual Cl-to-Cl distance is much less than this shows that the interaction between HCl molecules is stronger than van der Waals. The shortening must be attributed to the hydrogen bonding between the H of one molecule and the Cl of another.

14.14 If the density of crystalline CsCl is 3.988 g/cm³, calculate the volume effectively occupied by a single CsCl ion pair in the crystal.

▮

$$(1\ \text{CsCl unit})\left(\frac{168.4\ \text{g}}{6.02 \times 10^{23}\ \text{pairs}}\right)\left(\frac{1\ \text{cm}^3}{3.988\ \text{g}}\right) = 7.014 \times 10^{-23}\ \text{cm}^3$$

14.15 Refer to Problem 14.14. Calculate the smallest Cs-to-Cs internuclear distance, equal to the length of the side of a cube corresponding to the volume of one CsCl ion pair.

▌ There are $\dfrac{1}{7.014 \times 10^{-23}} = 1.426 \times 10^{22}$ units/cm³ $\quad \sqrt[3]{1.426 \times 10^{22}} = 2.42 \times 10^{7}$ units/cm along the edge

The edge of each cube is the reciprocal of the number of units per centimeter:

$$\frac{1}{2.42 \times 10^{7} \text{ units/cm}} = 4.125 \times 10^{-8} \text{ cm} = 4.125 \text{ Å}$$

14.16 Refer to Problems 14.14 and 14.15. Calculate the smallest Cs-to-Cl internuclear distance in the crystal, assuming each Cs^+ ion to be located in the center of a cube with Cl^- ions at each corner of the cube.

▌ The Cs-to-Cl distance, which is the distance from the corner to the center of a cube of edge 4.125 Å is

$$\frac{\sqrt{3}}{2}(4.125 \text{ Å}) = 3.572 \text{ Å}$$

14.17 What sort of electromagnetic radiation has wavelengths comparable to the dimensions found in Problems 14.15 and 14.16?

▌ X-rays

14.18 The hexagonal close-packed lattice can be represented by Fig. 14.5, if $c = a\sqrt{8/3} = 1.633a$. There is an atom at each corner of the unit cell and another atom which can be located by moving one-third the distance along the diagonal of the rhombus base, starting at the lower left-hand corner and moving perpendicularly upward by $c/2$ (Fig. 14.6). Mg crystallizes in this lattice and has a density of 1.74 g/cm³. ***(a)*** What is the volume of the unit cell? ***(b)*** What is a? ***(c)*** What is the distance between nearest neighbors? ***(d)*** How many nearest neighbors does each atom have?

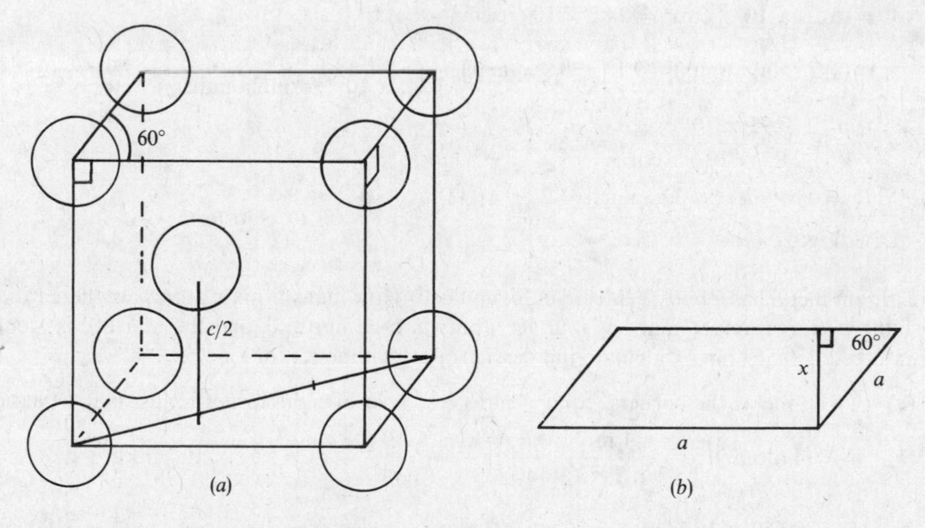

Fig. 14.6

▌ ***(a)*** The volume is calculated from the density. Per unit cell:

$$\text{Mass} = (2 \text{ atoms})\left(\frac{24.3 \text{ g}}{6.02 \times 10^{23} \text{ atoms}}\right) = 8.07 \times 10^{-23} \text{ g}$$

$$\text{Volume} = (8.07 \times 10^{-23} \text{ g})\left(\frac{(1.00 \times 10^{8} \text{ Å})^3}{1.74 \text{ g}}\right) = 46.4 \text{ Å}^3$$

(b) The base of the cell (Fig. 14.6*b*) has an area calculated as follows:

$$\sin 60° = x/a \qquad x = a(\sin 60°) = a(0.8660) \qquad \text{Area} = ax = 0.8660a^2$$

The volume is $\quad (0.8660a^2)c = (0.8660a^2)(1.633a) = 1.414a^3 = 46.4 \text{ Å}^3. \qquad a = 3.20 \text{ Å}$

(c) 3.20 Å. Nearest neighbors are along the base edge. ***(d)*** 12. The 12 are most easily viewed as 6 in the same plane, 3 above, and 3 below. The plane of the six atoms around the central atom is not shown in Fig. 14.6.

14.19 The NaCl lattice has the cubic unit cell shown in Fig. 14.7. KBr crystallizes in this lattice. **(a)** How many K^+ ions and how many Br^- ions are in each unit cell? **(b)** Assuming the additivity of ionic radii, what is a? **(c)** Calculate the density of a perfect KBr crystal. **(d)** What minimum value of r_+/r_- is needed to prevent anion-anion contact in this structure?

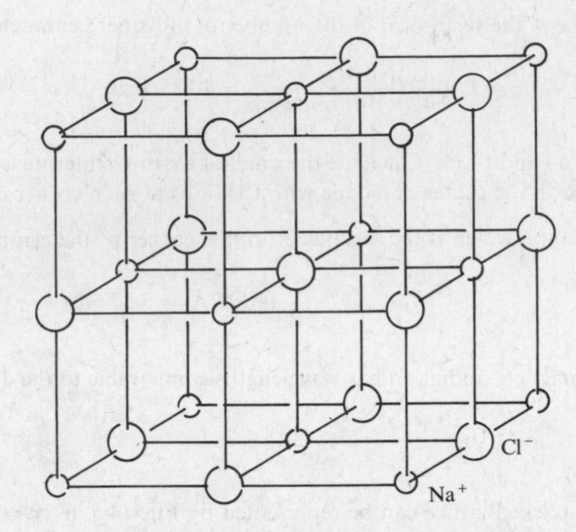

Fig. 14.7

▮ **(a)** Four of each: $\frac{1}{4}$ Br along each of 12 edges plus one at the center; $\frac{1}{8}$ K at each of eight corners, plus $\frac{1}{2}$ K at each of six face centers.

(b) $a = 2(\text{K}^+ \text{ radius} + \text{Br}^- \text{ radius}) = 2(1.33\text{ Å} + 1.95\text{ Å}) = 6.56\text{ Å}$ (See Table 14.1.)

(c) $V = (6.56 \times 10^{-8}\text{ cm})^3 = 2.82 \times 10^{-22}\text{ cm}^3$

$$m = (4 \text{ formula units})[(39.1 + 79.9)\text{ g/mol}]\left(\frac{1 \text{ mol}}{6.02 \times 10^{23} \text{ formula units}}\right) = 7.91 \times 10^{-22}\text{ g}$$

$$d = \frac{m}{V} = \frac{7.91 \times 10^{-22}\text{ g}}{2.82 \times 10^{-22}\text{ cm}^3} = 2.80\text{ g/cm}^3$$

(d) Side $= a_+ + a_-$ Diagonal $= 2a_- = (\sqrt{2})(a_+ + a_-)$

Dividing by $\sqrt{2}$, $(\sqrt{2})a_- = a_+ + a_-$ $(\sqrt{2} - 1)a_- = a_+$ $a_+/a_- = 0.414$

14.20 **(a)** Lithium metal has a body-centered cubic unit cell. How many lithium atoms are there in a unit cell? **(b)** Krypton crystallizes in a structure that has four Kr atoms in each unit cell, and the unit cell is a cube. The edge length of the unit cell is 0.559 nm. Calculate the density of crystalline Kr, in kg/m³.

▮ **(a)** Two (one at the corners, equivalent to $\frac{1}{8}$ at each of eight corners, plus one at the body center).

(b) $$m = (4 \text{ atoms})\left(\frac{1 \text{ mol}}{6.02 \times 10^{23} \text{ atoms}}\right)\left(\frac{83.8 \times 10^{-3}\text{ kg}}{\text{mol}}\right) = 5.57 \times 10^{-25}\text{ kg}$$

$$V = (0.559 \times 10^{-9}\text{ m})^3 = 1.75 \times 10^{-28}\text{ m}^3 \qquad d = \frac{m}{V} = \frac{5.57 \times 10^{-25}\text{ kg}}{1.75 \times 10^{-28}\text{ m}^3} = 3190\text{ kg/m}^3$$

14.21 A solid has a structure in which W atoms are located at the cube corners of the unit cell, O atoms are located at the cube edges, and Na atoms at the cube centers. What type of lattice is represented by this compound? What is the formula for the compound?

▮ The lattice type is simple cubic. The lattice points are the locations of the W atoms, and the locations of the other atoms are not lattice points because their chemical environment is different. The formula includes one W atom ($\frac{1}{8}$ at each of 8 corners), three O atoms ($\frac{1}{4}$ at each of 12 cell edges), and one Na atom (at the cube center). The formula is therefore $NaWO_3$.

14.22 How many unit cells are there **(a)** in a 1.00 g cube-shaped ideal crystal of NaCl? **(b)** along each edge of that crystal?

▮ (*a*) $(1.00\text{ g})\left(\frac{1\text{ mol}}{58.5\text{ g}}\right)\left(\frac{6.02\times10^{23}\text{ formula units}}{\text{mol}}\right)\left(\frac{1\text{ unit cell}}{4\text{ formula units}}\right) = 2.57\times10^{21}\text{ unit cells}$

(*b*) Number per edge $= \sqrt[3]{2.57\times10^{21}} = 1.37\times10^{7}$

14.23 What type of unit cell best describes the NaCl crystal lattice diagramed in Fig. 14.7.

▮ Either Na^+ or Cl^- ions may be chosen as the lattice points. Choosing Cl^- ions, one sees that the simplest cubic arrangement contains a chloride ion on the center of each face as well as at the corners of the unit cell; hence the unit cell is face-centered cubic.

14.24 Determine the number of formula units of NaCl in the unit cell (Fig. 14.7).

▮ Since the cell is face-centered cubic (Problem 14.23), there are: At the 8 corners, $8\times\frac{1}{8} = 1\ Cl^-$ ion; at the 6 faces, $6\times\frac{1}{2} = 3\ Cl^-$ ions; along the 12 edges, $12\times\frac{1}{4} = 3\ Na^+$ ions; and at the cube center, 1 Na^+ ion; for a total of 4 Cl^- and 4 Na^+ ions. Hence, the unit cell contains four NaCl units.

14.25 Perform the following exercise: (i) Arrange 15 balls in a triangle, as in preparing for a billiards game. (ii) Place a second layer of balls in the depressions of the first layer, then a third layer (one ball) in a depression directly over the center ball of the first layer. (iii) On the slanting side of the pyramid created, identify a square. Find the portion of the face-centered cube of which the top ball forms one corner and identify as many other corners of the smallest such cube as are visible. How can a *cubic* closest-packed arrangement start out with a *hexagonal* layer?

▮ The atoms are in an arrangement that is square from one viewpoint but hexagonal from another.

14.26 Arrange uniform balls in a rectangle or square. Pile a second layer of identical balls in the holes of the first layer, and add as many identical balls to the third layer as will fit. Remove balls until you can identify a hexagonal pattern of balls slanting up the three layers. Explain how a cubic pattern can generate a closest-packed arrangement. (Compare Problem 14.25.)

▮ Although the illustrations in a text are adequate for some people, there is no substitute for actually making the model. The arrangement that is a hexagonal pattern when viewed from one direction is a cubic pattern when viewed from a different angle.

14.27 How many octahedral sites per sphere are in a cubic closest-packed (face-centered cubic) structure?

▮ One.

14.28 MgS and CaS both crystallize in the NaCl-type lattice (Fig. 14.7). From ionic radii listed in Table 14.1, what conclusion can you draw about anion-cation contact in these crystals?

▮ There can be either anion-anion contact or anion-cation contact, with a geometric ratio of 0.414 [see Problem 14.19(*d*)] for either.

$$r(\text{Mg})/r(\text{S}) = (0.65\text{ Å})/(1.84\text{ Å}) = 0.35 \qquad r(\text{Ca})/r(\text{S}) = (0.99\text{ Å})/(1.84\text{ Å}) = 0.54$$

Ca^{2+} and S^{2-} can be in contact, but Mg^{2+} and S^{2-} cannot. In MgS, if Mg^{2+} and S^{2-} were in contact there would not be enough room for the sulfide ions along the diagonal of a square constituting one-quarter of a unit cell face.

14.29 Each rubidium halide crystallizing in the NaCl-type lattice has a unit cell length 0.30 Å greater than that for the corresponding potassium salt of the same halogen. What is the ionic radius of Rb^+ computed from these data?

▮
$$r(\text{Rb}^+) = r(\text{K}^+) + \tfrac{1}{2}(0.30\text{ Å}) = 1.33\text{ Å} + 0.15\text{ Å} = 1.48\text{ Å}$$

14.30 Iron crystallizes in several modifications. At about 910 °C, the body-centered cubic α form undergoes a transition to the face-centered cubic γ form. Assuming that the distance between nearest neighbors is the same in the two forms at the transition temperature, calculate the ratio of the density of γ iron to that of α iron at the transition temperature.

▮ Body-centered cubic: Body diagonal $= 4r(\text{Fe}) = \sqrt{3}\,a$ thus $a = (4/\sqrt{3})r(\text{Fe})$ (See Fig. 14.8.) Two atoms per unit cell (one at center; $\frac{1}{8}$ at each of 8 corners).

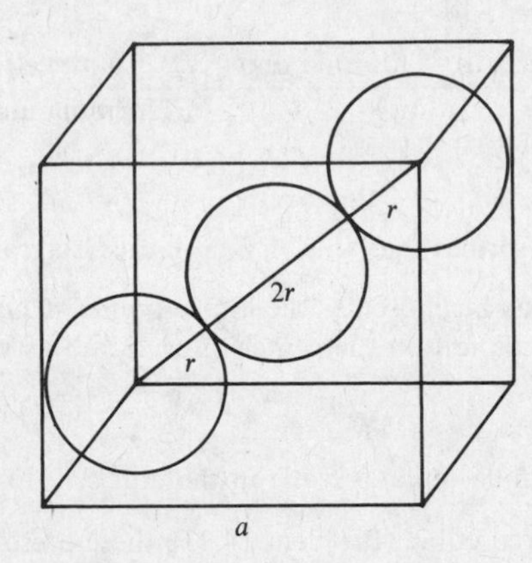

Fig. 14.8

Face-centered cubic: Face diagonal $= 4r(\text{Fe}) = \sqrt{2}\,a'$ $\quad a' = (2\sqrt{2})r(\text{Fe})$ Four atoms per unit cell ($\frac{1}{8}$ at each of 8 corners; $\frac{1}{2}$ at each of six faces)

$$\text{Density ratio} = \frac{4/[2\sqrt{2}r(\text{Fe})]^3}{2/[(4/\sqrt{3})r(\text{Fe})]^3} = \frac{4/(2^{9/2})}{2/(4^3/3^{3/2})} = \frac{4\sqrt{2}}{3\sqrt{3}} = 1.09$$

14.31 A compound alloy of gold and copper crystallizes in a cubic lattice in which the gold atoms occupy the lattice points at the corners of a cube and the copper atoms occupy the centers of each of the cube faces. What is the formula of this compound?

▮ $\frac{1}{8} \times 8 = 1$ Au atom/unit cell $\quad \frac{1}{2} \times 6 = 3$ Cu atoms/unit cell $\quad$ The empirical formula is $AuCu_3$.

14.32 The ZnS zinc blende structure is cubic. The unit cell may be described as a face-centered sulfide ion sublattice with zinc ions in the centers of alternating minicubes made by partitioning the main cube into 8 equal parts (Fig. 14.9). **(a)** How many nearest neighbors does each Zn^{2+} have? **(b)** How many nearest neighbors does each S^{2-} have? **(c)** What angle is made by the lines connecting any Zn^{2+} to any two of its nearest neighbors? **(d)** What minimum r_+/r_- ratio is needed to avoid anion-anion contact, if closest cation-anion pairs are assumed to touch?

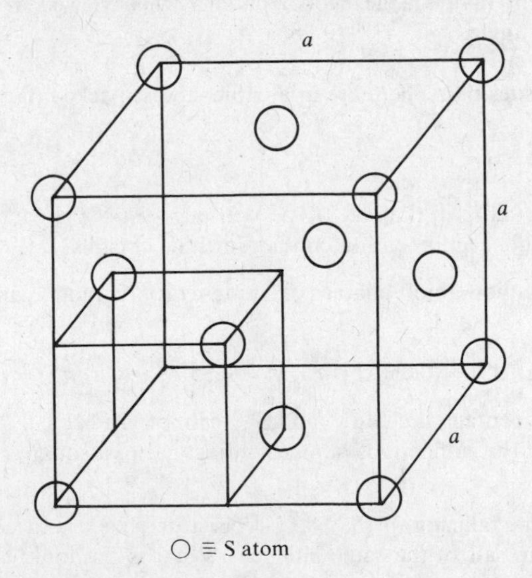

○ ≡ S atom

Fig. 14.9

▮ **(a)** 4 **(b)** 4 (Of 8 little boxes around each S atom, 4 are filled with Zn atoms and 4 are empty.) **(c)** Tetrahedral angle, 109°28′

(d) Body diagonal of little cube $= 2r(\text{S}) + 2r(\text{Zn}) = \sqrt{3}(a/2)$

Face diagonal of little cube $= 2r(\text{S}) = \sqrt{2}(a/2)$

$$\frac{a}{2} = \sqrt{2}r(\text{S}) = \frac{2}{\sqrt{3}}[r(\text{S}) + r(\text{Zn})] \qquad r(\text{S}) = \sqrt{\frac{2}{3}}[r(\text{S}) + r(\text{Zn})]$$

$$\left(1 - \sqrt{\frac{2}{3}}\right)[r(\text{S})] = \sqrt{\frac{2}{3}}[r(\text{Zn})] \qquad r(\text{Zn})/r(\text{S}) = \frac{1 - \sqrt{2/3}}{\sqrt{2/3}} = 0.225$$

14.33 Why does ZnS not crystallize in the NaCl structure? (*Hint*: Refer to Problem 14.19.)

▌ The r_+/r_- ratio is 0.402, too low to avoid anion-anion contact in the NaCl structure.

14.34 Silver iodide crystallizes in the cubic close-packed zinc blende structure. Assuming that the iodide ions occupy the lattice points, what fraction of the tetrahedral sites is occupied by silver ions?

▌ 50%. (There are two tetrahedral sites per I^- ion.)

14.35 Compute the packing factor for spheres occupying **(*a*)** a body-centered cubic and **(*b*)** a simple cubic structure, where closest neighbors in both cases are in contact.

▌ **(*a*)** 2 atoms per unit cell (one atom at center, $\frac{1}{8}$ at each of 8 corners)

$$\text{Body diagonal} = \sqrt{3}a = 4r \quad \text{(Fig. 14.8)} \qquad \text{Volume of atoms} = 2\left[\frac{4\pi}{3}\left(\frac{\sqrt{3}}{4}a\right)^3\right] = 0.680a^3$$

$$\text{Volume of unit cell} = a^3 \qquad \text{Packing factor} = (0.680a^3)/a^3 = 0.680$$

(*b*) 1 atom per unit cell

$$r = a/2 \qquad \text{Volume of atom:} \quad V = \frac{4\pi}{3}\left(\frac{a}{2}\right)^3 = 0.524a^3$$

$$\text{Volume of unit cell} = a^3 \qquad \text{Packing factor} = (0.524a^3)/a^3 = 0.524$$

14.36 Titanium crystallizes in a face-centered cubic lattice. It reacts with carbon or hydrogen interstitially, by allowing atoms of these elements to occupy holes in the host lattice. Hydrogen occupies tetrahedral holes, but carbon occupies octahedral holes. **(*a*)** Predict the formulas of titanium hydride and titanium carbide formed by saturating the titanium lattice with either "foreign" element. **(*b*)** What is the maximum ratio of "foreign" atom radius to host atom radius that can be tolerated in a tetrahedral hole without causing a strain in the host lattice? **(*c*)** What is the maximum allowable radius ratio in an octahedral hole? **(*d*)** Account for the fact that hydrogen occupies tetrahedral holes while carbon occupies octahedral holes.

▌ **(*a*)** Hydride: There are 4 atoms per unit cell, and 8 tetrahedral sites per unit cell, for a 2:1 ratio of tetrahedral sites to atoms. The formula would be TiH_2. Carbide: There are four octahedral sites per unit cell, and the atomic ratio would be 1:1. The formula would be TiC. **(*b*)** 0.225 [see Problem 14.32(*d*).] **(*c*)** 0.414 **(*d*)** The hydrogen atom is small enough to fit into a tetrahedral hole, but the carbon atom is not.

14.37 Iron crystallizes in two body-centered cubic lattices, the α form below 910 °C and the δ form above 1400 °C, and in a face-centered cubic γ form between these two temperatures. Of these three crystalline phases, only the γ form can dissolve appreciable amounts of carbon. **(*a*)** What is the symmetry of a hole centered in a face of a body-centered cubic unit cell, and what is the maximum ratio of a foreign particle radius occupying such a hole to the host ion radius? **(*b*)** What is the symmetry of a hole whose coordinates are $(0, a/2, a/4)$ in a body-centered cubic unit cell, and what is the maximum ratio of a foreign particle radius occupying such a hole to the host ion radius? **(*c*)** Why does only the γ form dissolve appreciable amounts of carbon? (Compare with the results of Problem 14.36.)

▌ In body-centered cubic, the atom at the body center is in contact with those at the corners; the corner atoms do not touch each other. The body diagonal is $\sqrt{3}a = 4r$. Thus $a = (4/\sqrt{3})r$. **(*a*)** From body center to body center is a and also $2r + 2r_h$ [Fig. 14.10*a*], where r_h is the radius of hole ×.

$$a = 2r + 2r_h = \frac{4}{\sqrt{3}}r \qquad r + r_h = \frac{2}{\sqrt{3}}r \qquad \frac{r_h}{r} = \left(\frac{2}{\sqrt{3}} - 1\right) = 0.155$$

The face diagonals are longer $(\sqrt{2}a)$ than the distance between body centers, so the geometry of the hole is a shortened octahedron, called a distorted octahedral. **(*b*)** Consider the body-centered atoms (A and B) and the two corner atoms at the base of the face joining the two unit cells (C and D in Fig. 14.10*b*). The distance from the center of the line joining the top atoms to that joining the bottom atoms is $a/2$; the bottom atoms are a half unit cell below the top ones. The hole h is located halfway down that line, as required by symmetry. The triangle of interest therefore has one side $a/2$ and one side $a/4$ (Fig. 14.10*c*). The hypotenuse is $\sqrt{5/16}\,a$.

$$r + r_h = \sqrt{\tfrac{5}{16}}a = \left(\sqrt{\tfrac{5}{16}}\right)\left(\frac{4r}{\sqrt{3}}\right) = \sqrt{\tfrac{5}{3}}\,r \qquad r_h = \left(\sqrt{\tfrac{5}{3}} - 1\right)r \qquad \frac{r_h}{r} = \sqrt{\tfrac{5}{3}} - 1 = 0.291$$

(*c*) Carbon is too big to occupy either type of hole in a body-centered structure but can fit within an octahedral hole of a face-centered structure.

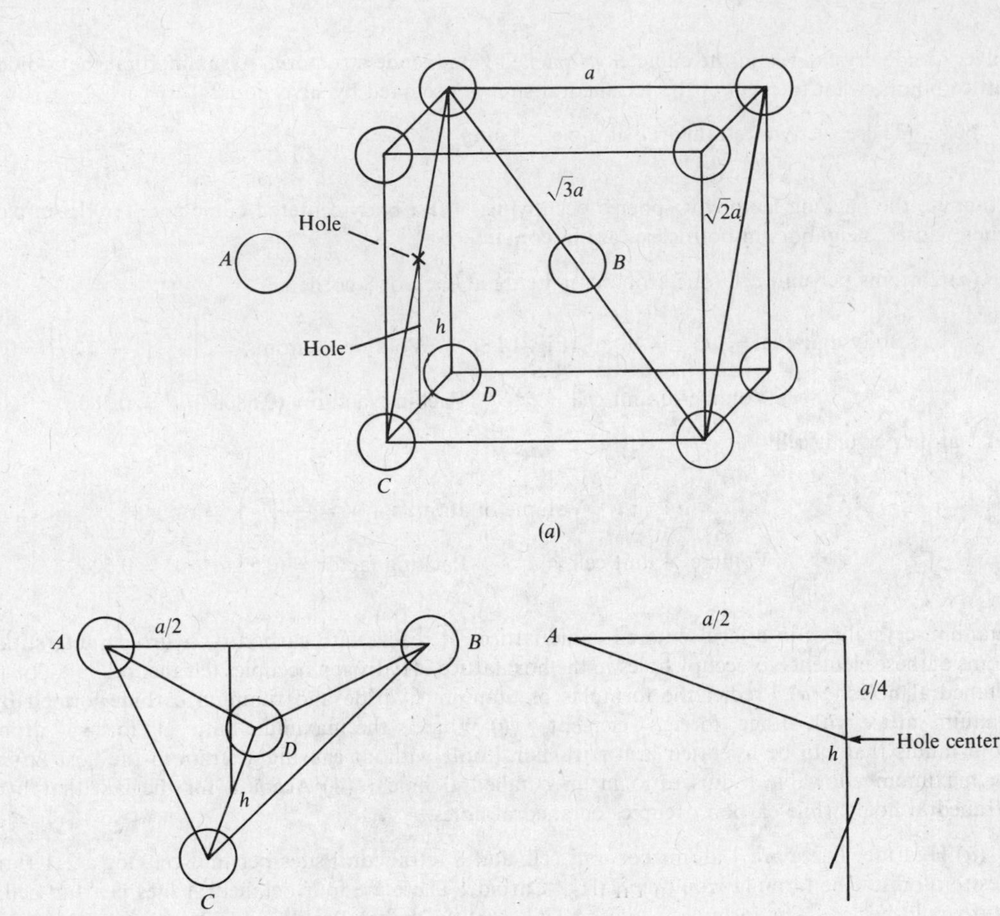

Fig. 14.10

14.38 Calculate the number of formula units in each of the following types of unit cells: (*a*) MgO in a rock salt type unit cell (*b*) ZnS in a zinc blende structure (*c*) platinum in a face-centered cubic unit cell.

▮ (*a*) 4 (the same as NaCl) (*b*) 4 (*c*) 4 (1 at the corners, 3 at the face centers)

14.39 A mineral having the formula AB_2 crystallizes in the cubic close-packed lattice, with the A atoms occupying the lattice points. What is the coordination number of the A atoms? of the B atoms? What fraction of the tetrahedral sites is occupied by B atoms?

▮ 8, 4, 100%

14.40 The intermetallic compound LiAg crystallizes in a cubic lattice in which both lithium and silver atoms have coordination numbers of 8. To what crystal class does the unit cell belong?

▮ Simple cubic (CsCl) structure, Fig. 14.3

14.41 Locate the 4 threefold axes, the 3 fourfold axes, and the 9 mirror planes in a cube.

▮ An *n*-fold axis of symmetry is a line about which a figure can be rotated 360°/*n* to yield a figure indistinguishable from the original. The four 3-fold axes of a cube are its four diagonals. There are three 4-fold axes, through the centers of opposite faces. The nine mirror planes cut opposite faces in half in two ways: parallel to edges (3 possibilities) or through face diagonals (6 possibilities).

14.42 Predict the coordination number of the cation in crystals of each of the following compounds:

	cation radius	anion radius
(*a*) MgO	0.65 Å	1.40 Å
(*b*) MgS	0.65 Å	1.84 Å
(*c*) CsCl	1.69 Å	1.81 Å

▮

	radius ratio	coordination number
(*a*)	0.46	6
(*b*)	0.35	4
(*c*)	0.934	8

14.43 The metal ion–halide ion distances in several alkali metal halides are given in the table. Suggest why there is such a small difference in internuclear distance between LiI and NaI compared to that between LiCl and NaCl.

	distance, nm	
	I^-	Cl^-
Li^+	0.310	0.249
Na^+	0.317	0.279
K^+	0.352	0.314
Rb^+	0.367	0.329

▮ Iodide ion–iodide ion contact in LiI makes closer approach impossible.

14.44 Calculate the ionic radius of the fluoride ion using only the following interionic distances and the radius of the iodide ion, 2.19 Å. (Do not use tabulated values.) RbI, 3.67 Å; RbBr, 3.44 Å; KBr, 3.29 Å; KCl, 3.14 Å; NaCl, 2.79 Å; NaF, 2.31 Å.

▮ All distances are in angstrom units:

		I^-	2.19
RbI	3.67	Rb^+	$3.67 - 2.19 = 1.48$
RbBr	3.44	Br^-	$3.44 - 1.48 = 1.96$
KBr	3.29	K^+	$3.29 - 1.96 = 1.33$
KCl	3.14	Cl^-	$3.14 - 1.33 = 1.81$
NaCl	2.79	Na^+	$2.79 - 1.81 = 0.98$
NaF	2.31	F^-	$2.31 - 0.98 = 1.33$

14.45 A salt MY crystallizes in the cesium chloride structure. The anions at the corners touch each other and the cation in the center. What is the radius ratio, r_+/r_-, for this structure?

▮ The ions shown in Fig. 14.11*a* are shown in a different perspective in Fig. 14.11*b*. The important distances are marked in Fig. 14.11*c*. The distance between opposite corners of the face of a cube with sides $2r_-$ is $2\sqrt{2}\,r_-$. The distance between the centers of touching anions is $2r_-$. Therefore the distance between opposite anions is equal to $\sqrt{12}\,r_-$, which is also equal to $2r_- + 2r_+$ (see Fig. 14.11*b*).

$$\sqrt{12}\,r_- = 2r_- + 2r_+ \qquad 3.464r_- = 2r_- + 2r_+ \qquad 1.464r_- = 2r_+ \qquad \frac{r_+}{r_-} = 0.732$$

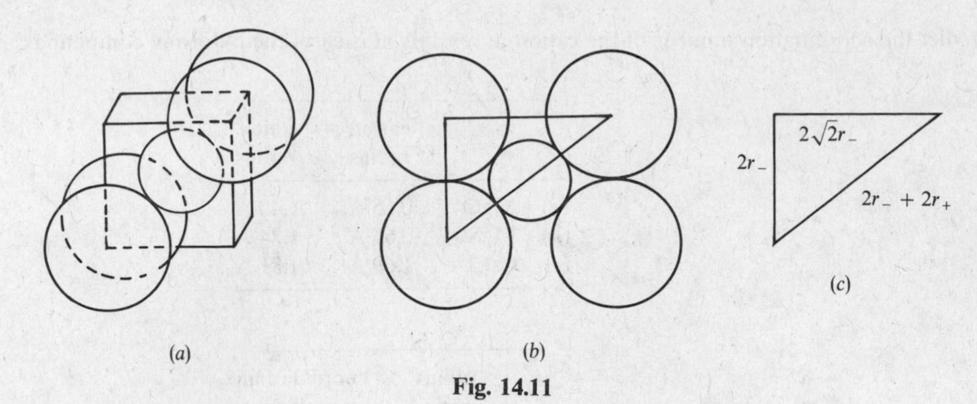

Fig. 14.11

14.46 Estimate ionic radii for M^+, R^+, Q^-, and T^- from the following internuclear distances in NaCl-type crystals:

salt	anion-anion distance, Å	cation-anion distance, Å
MT	2.40	1.70
MQ	1.63	1.15
RT	2.66	1.88
RQ	2.09	1.48

▮ Q^- is smaller than T^-. The differences in the internuclear distances between MT and RT indicate that the cation must not be touching all the anions in MT (smaller cation, larger anion; see Fig. 14.12*a*). There must be anion-anion contact in this salt. Therefore, the anion-anion distance is equal to $2r_-$. The remaining radius ratios are suggested in Fig. 14.12*b*. T^- has a radius 1.20 Å; hence R^+ has a radius 1.88 Å − 1.20 Å = 0.68 Å. By similar reasoning, Q^- has a radius 1.48 Å − 0.68 Å = 0.80 Å, and M^+ has a radius 1.15 Å − 0.80 Å = 0.35 Å. Note that the M^+ radius cannot be calculated from the M^+–T^- distance, since there is no cation-anion contact in MT.

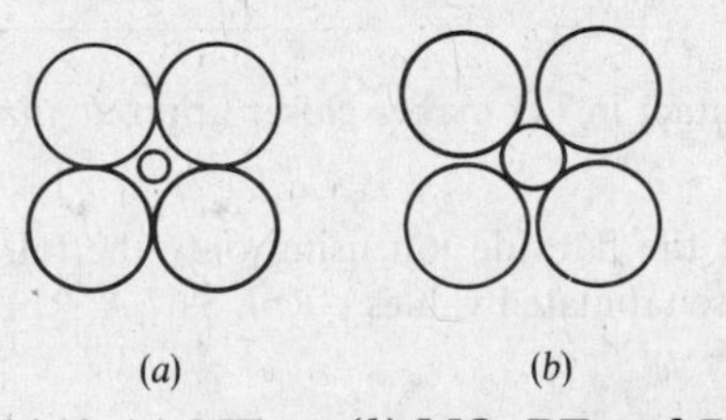

Fig. 14.12 (*a*) MT (*b*) MQ, RT, and RQ

14.47 Suggest the probable structures of the unit cells of each of the following:

	cationic radius	anionic radius
(*a*) RbBr	1.48 Å	1.95 Å
(*b*) MgTe	0.65 Å	2.21 Å
(*c*) MgO	0.65 Å	1.40 Å
(*d*) BaO	1.35 Å	1.40 Å

▮

radius ratio	structure
(*a*) 0.759	CsCl (actually, NaCl)
(*b*) 0.29	Zinc blende
(*c*) 0.46	Rock salt
(*d*) 0.964	CsCl

14.48 Calculate the value of Avogadro's number from the internuclear distance of adjacent ions in NaCl, 0.282 nm, and the density of solid NaCl, $2.17 \times 10^3\ \text{kg/m}^3$.

▌ One mole of NaCl has a mass of 58.5×10^{-3} kg and therefore has a volume

$$(58.5 \times 10^{-3}\ \text{kg})\left(\frac{1\ \text{m}^3}{2.17 \times 10^3\ \text{kg}}\right) = 2.70 \times 10^{-5}\ \text{m}^3$$

A unit cell, containing 4 NaCl formula units, has sides of length 0.564 nm. The volume of a unit cell is

$$(0.564 \times 10^{-9}\ \text{m})^3 = 1.79 \times 10^{-28}\ \text{m}^3$$

The number of unit cells per mole is then

$$\frac{2.70 \times 10^{-5}\ \text{m}^3}{1.79 \times 10^{-28}\ \text{m}^3} = 1.51 \times 10^{23}$$

Since there are 4 NaCl formula units per unit cell, there are $N_A = (4)(1.51 \times 10^{23}) = 6.04 \times 10^{23}$ formula units per mole.

14.49 Ionic radii have been estimated (by Linus Pauling) by assuming that the internuclear distances between isoelectronic ions is inversely proportional to the effective nuclear charges of the ions. Effective nuclear charges are obtained by subtracting an amount reflecting the shielding of the inner electrons from the actual nuclear charges. For neon-type ions, Na^+ and F^-, for example, the factor is 4.15*e*. On this basis, calculate the ionic radii of Na^+ and F^- from the experimentally determined internuclear distance in NaF, 2.31 Å.

▌ Let $x = Na^+$ radius; then $2.31 - x = F^-$ radius.

$$\frac{x}{2.31 - x} = \frac{(9 - 4.15)e}{(11 - 4.15)e} = 0.708 \qquad x = (0.708)(2.31 - x) = 1.64 - 0.708x$$

$$x = 0.96\ \text{Å} \qquad \text{and} \qquad 2.31 - x = 1.35\ \text{Å}$$

The accepted values are 0.95 Å and 1.36 Å, respectively.

14.50 Show that the face-centered cubic lattice of NaCl (Fig. 14.7) also contains a smaller, tetragonal, unit cell. How many NaCl formula units are there in one such cell? Outline on a figure of the lattice a still smaller unit cell, containing only one NaCl formula unit per unit cell. Calculate the dimensions and angles of this simple unit cell.

▌ There are two normal unit cells shown in Fig. 14.13*a*, with a smaller unit cell drawn in. The small tetragonal unit cell contains two formula units of NaCl. The slanting unit cell lengths are easily shown to be $l/\sqrt{2}$, where l is the length of the face-centered cubic unit cell. Thus the volume of the tetragonal cell is one-half that of the face-centered cubic cell. In Fig. 14.13*b* is shown an even smaller unit cell, containing only one formula unit of NaCl. The unit cell length is again $l/\sqrt{2}$. The angle between the two edges on the front face (heavier lines) is 90°. This unit cell extends only $l/2$ toward the back, and therefore the volume of the cell is $l/\sqrt{2} \times l/\sqrt{2} \times l/2 = V/4$; the cubic unit cell volume is V.

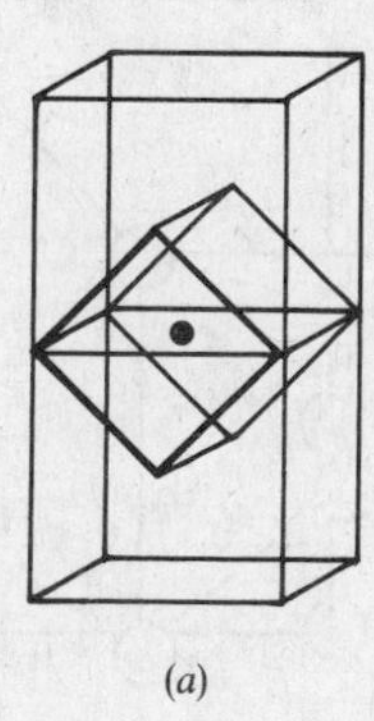
(*a*)

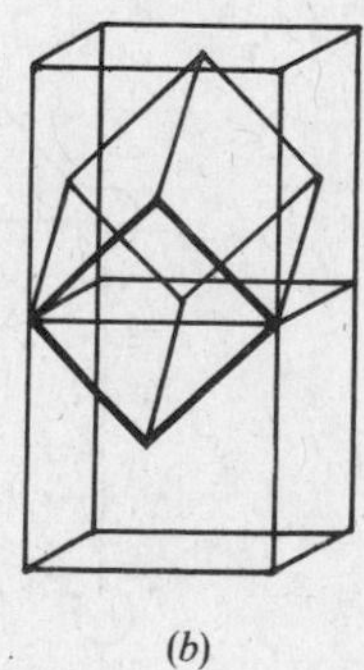
(*b*)

Fig. 14.13

14.51 The length of a unit cell in the nickel crystal is 0.352 nm. Diffraction of x-rays of 0.154 nm wavelength from a nickel crystal occurs at 22.2°, 25.9°, and 38.2°. Show that these data are consistent with a face-centered cubic crystal structure.

Using Bragg's law, $n\lambda = 2d \sin \theta$, and assuming that the diffractions are first order ($n = 1$), the distances are calculated to be 0.204 nm, 0.176 nm, and 0.124 nm. The distance 0.176 nm is seen to be half the unit cell length, corresponding to the perpendicular distance between layers of atoms in one face of the unit cell and the center (see Fig. 14.14*a*). The distance 0.204 nm turns out to be $1/\sqrt{3}$ times the unit cell length, corresponding to the perpendicular distance between a corner of the unit cell and the plane of the three adjacent corners (see Fig. 14.14*c*). The distance 0.124 nm corresponds to $\sqrt{2}/4$ times the unit cell dimension—one-fourth of the distance from one edge of the cell to the opposite edge (see Fig. 14.14*b*). The plane intersects the face-centered atoms of two sides. The plane including the edge atoms is equivalent to the plane across the center of the unit cell.

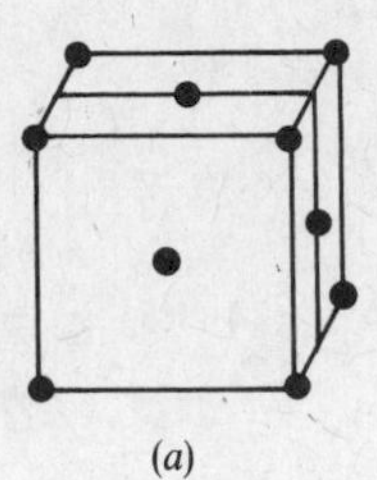
(*a*)

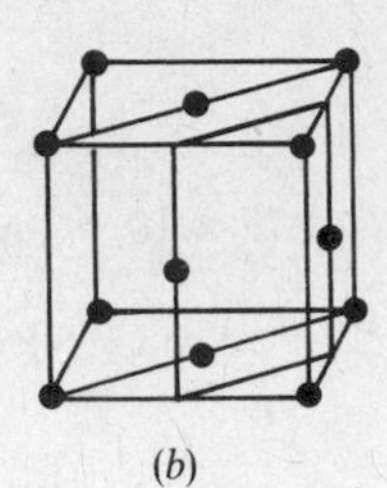
(*b*)

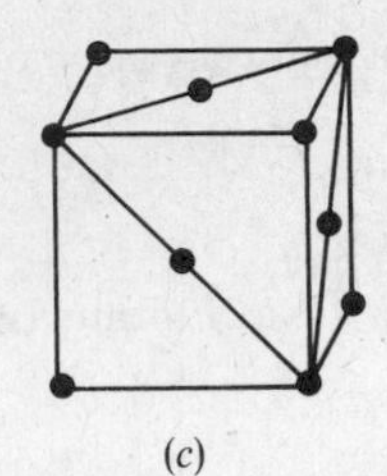
(*c*)

Fig. 14.14

14.2 CRYSTAL ENERGIES

14.52 The melting point of quartz, one of the crystalline forms of SiO_2, is 1610 °C, and the sublimation point of CO_2 is −79 °C. How similar do you expect the crystal structures of these two substances to be?

The big difference in melting points suggests a difference in type of crystal binding. The intermolecular forces in solid CO_2 must be very low to be overcome by a low-temperature sublimation. CO_2 is actually a molecular lattice held together only by the weak van der Waals forces between discrete CO_2 molecules. SiO_2, on the other hand, is a covalent lattice, with a three-dimensional network of bonds; each silicon atom is bonded tetrahedrally to 4 oxygens, and each oxygen is bonded to 2 silicons.

14.53 Contrast the visible changes which occur while heat is added to **(*a*)** an ice cube **(*b*)** a bar of chocolate. **(*c*)** Which type of behavior is characteristic of an amorphous solid? **(*d*)** List three other common amorphous materials.

(*a*) As heat is added, some ice melts completely; the rest stays completely solid. **(*b*)** Chocolate softens throughout, gradually. **(*c*)** The chocolate type. **(*d*)** Examples include glass, wax, butter, plastic, molasses, rubber.

14.54 In the hexagonal ice structure (Fig. 14.15), each oxygen is coordinated tetrahedrally with 4 other oxygens, with an intervening hydrogen between adjoining oxygens. ΔH of sublimation of ice at 0 °C is 51.0 kJ/mol H_2O. It has been estimated by comparison with non-hydrogen-bonded solids having intermolecular van der Waals forces similar to those in ice that the ΔH of sublimation would be only 15.5 kJ/mol H_2O if ice were not hydrogen-bonded. From these data, estimate the strength of the hydrogen bond in ice.

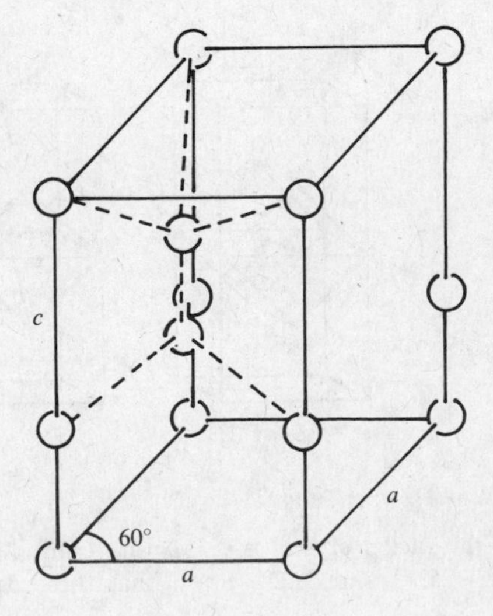

Fig. 14.15

The excess of ΔH of sublimation above that of a non-hydrogen-bonded solid can be attributed to hydrogen bonds.

$$\Delta H(\text{excess}) = 51.0 - 15.5 = 35.5\ \text{kJ/mol}\ H_2O$$

Each H_2O is hydrogen bonded to 4 other H_2O molecules, through O—H—O linkages (indicated in Fig. 14.15 only for the two interior molecules). Each such hydrogen-bonded linkage is shared by two H_2O molecules (to which the two oxygen atoms belong). Thus, on the average, each H_2O can be assigned 4 halves, or 2 hydrogen bonds.

$$\Delta H(\text{hydrogen bond}) = \frac{35.5\ \text{kJ/mol}\ H_2O}{2\ \text{mol H bonds/mol}\ H_2O} = 17.8\ \text{kJ/mol hydrogen bond}$$

Observe from Fig. 14.15 that $8(\frac{1}{8}) + 4(\frac{1}{4}) + 2 = 4$ H_2O molecules are uniquely assigned to the unit cell. This is in agreement with the result of Problem 14.53.

14.55 Use the Born-Haber cycle and the following data to calculate the electron affinity of chlorine: $\Delta H_f(\text{RbCl}) = -102.9$ kcal/mol; IP = 95 kcal/mol; $\Delta H_{\text{sub}}(\text{Rb}) = +20.5$ kcal/mol; $D(Cl_2) = +54$ kcal/mol; $U(\text{RbCl}) = -166$ kcal/mol.

All values are in kilocalories:

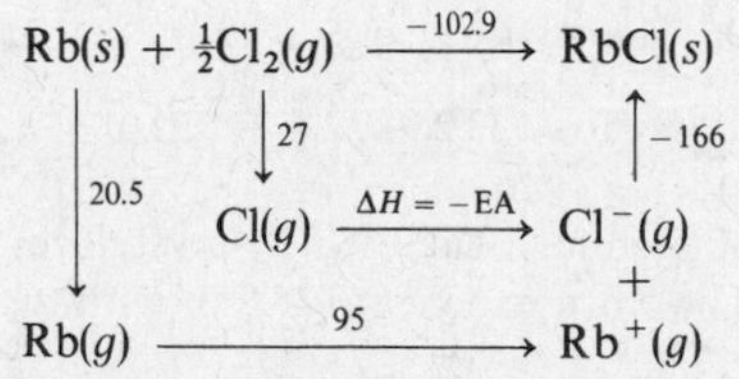

$$\text{Rb}(s) + \tfrac{1}{2}Cl_2(g) \xrightarrow{-102.9} \text{RbCl}(s)$$

- $\tfrac{1}{2}Cl_2(g) \xrightarrow{27} \text{Cl}(g)$
- $\text{Rb}(s) \xrightarrow{20.5} \text{Rb}(g)$
- $\text{Cl}(g) \xrightarrow{\Delta H = -\text{EA}} \text{Cl}^-(g)$
- $\text{Rb}(g) \xrightarrow{95} \text{Rb}^+(g)$
- $\text{Cl}^-(g) + \text{Rb}^+(g) \xrightarrow{-166} \text{RbCl}(s)$

$$\Delta H = -\text{EA} = -102.9 - (20.5 + 27 + 95 - 166) = -79\ \text{kcal/mol}$$
$$= (-79\ \text{kcal/mol})(4.184\ \text{kJ/kcal}) = -330\ \text{kJ/mol}$$

The positive value of the electron affinity indicates that energy is *released* when an electron is added to a gaseous chlorine atom to form a gaseous chloride ion.

14.56 With reference to Fig. 14.16, how do you account for the differences in melting point between (*a*) and (*b*), between (*c*) and (*d*), and between these two differences?

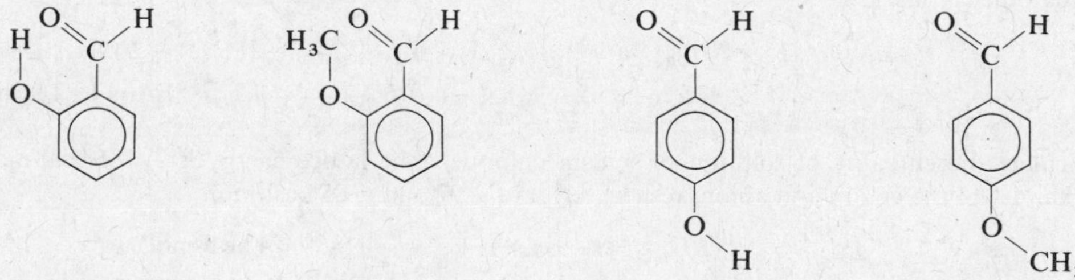

o-Hydroxybenzaldehyde	*o*-Methoxybenzaldehyde	*p*-Hydroxybenzaldehyde	*p*-Methoxybenzaldehyde
mp = 266 °C	mp = 309 °C	mp = 388 °C	mp = 273 °C
(*a*)	(*b*)	(*c*)	(*d*)

Fig. 14.16

The crystal forces in (*b*) and (*d*) are largely van der Waals. Compounds (*a*) and (*c*), containing the polar —OH group, are capable of hydrogen bonding. In the case of (*c*), the hydrogen bonding is from the —OH of one molecule to the doubly bonded oxygen of the neighboring molecule; and the resulting *inter*molecular attraction leads to an increase in melting point as compared with (*d*), the non-hydrogen-bonded control. In the case of (*a*), the molecular structure allows *intra*molecular hydrogen bonding from the —OH group of each molecule to the doubly bonded oxygen of the same molecule; in the absence of strong *inter*molecular hydrogen bonding, the difference in melting point as compared with the reference substance (*b*) should be small, related perhaps to differences in the crystal structure or to the van der Waals forces, which should be slightly larger for (*b*) than for (*a*) on account of the slightly increased molecular weight.

14.57 Explain why the Madelung constant does not depend on the charges of the ions in a crystal.

▮ The Madelung constant depends only on the geometry of the crystal. The energy of the crystal lattice depends on the charges on the ions, but the magnitudes of the charges are explicitly included in such equations as

$$U_0 = -1.75N_AZ_1Z_2e^2/r_0$$

and the coefficient, the Madelung constant, is not affected by the magnitude of the charges.

14.58 Calculate the lattice energy of magnesium sulfide using the following energies (all in kcal/mol): $\Delta H_f(\text{MgS}) = -82.2$; $\Delta H_{\text{sub}}(\text{Mg}) = 36.5$; $\text{IP}_1 + \text{IP}_2 = 520.6$; $\Delta H_{\text{atom}} = 133.2$; $\text{EA}_1 + \text{EA}_2 = -72.4$.

$$\text{Mg}(s) + \tfrac{1}{8}\text{S}_8(s) \xrightarrow{-82.2} \text{MgS}(s)$$

$$\text{S}_8(s) \xrightarrow{133.2} \text{S}(g) \xrightarrow{-72.4} \text{S}^{2-}(g)$$

$$\text{Mg}(s) \xrightarrow{36.5} \text{Mg}(g) \xrightarrow{520.6} \text{Mg}^{2+}(g)$$

$$\text{S}^{2-}(g) + \text{Mg}^{2+}(g) \xrightarrow{U} \text{MgS}(s)$$

All quantities are given in kcal. The lattice energy is thus

$$U = (-82.2) - (36.5 + 133.2 - 72.4 + 520.6) = -700.1 \text{ kcal/mol} = -2929 \text{ kJ/mol}$$

14.59 Calculate the lattice energy of cesium iodide, which crystallizes with the cesium chloride structure and has an interionic distance of 3.95 Å. The Born exponent for CsI, with two Xe type ions, is 12.

▮ Refer to Problem 14.62.

$$U_0 = -\left(\frac{(1)^2(6.02 \times 10^{23})(1.76)(4.80 \times 10^{-10})^2}{3.95 \times 10^{-8}}\right)\left(1 - \frac{1}{12}\right) = -5.67 \times 10^{12} \text{ erg/mol} = -135 \text{ kcal/mol}$$

14.60 Calculate the electron affinity of iodine, given the following data (all values in kcal/mol):

$$U(\text{NaI}) = -165.4, \quad \Delta H_f(\text{NaI}) = -64.8, \quad \Delta H_{\text{sub}}(\text{Na}) = 25.9,$$
$$\text{IP(Na)} = 118.4, \quad \tfrac{1}{2}(\Delta H_{\text{sub}}(\text{I}_2)) + \Delta H_{\text{diss}}(\text{I}_2) = 25.5$$

▮ The negative of the electron affinity of iodine is the enthalpy of formation of NaI minus the enthalpies of each of the other terms:

$$-\text{EA} = \Delta H_f - \Delta H_{\text{sub}}(\text{Na}) - \tfrac{1}{2}(\Delta H_{\text{sub}}(\text{I}_2)) + \Delta H_{\text{diss}}(\text{I}_2) - \text{IP(Na)} - U(\text{NaI})$$
$$= -64.8 - 25.9 - 25.5 - 118.4 - (-165.4) = -69.2 \qquad \text{EA} = 69.2 \text{ kcal/mol}$$

14.61 Calculate the enthalpy of solution of sodium chloride. The lattice energy is -186 kcal/mol, and the solvation enthalpies of the cation and anion, respectively, are -97 and -85 kcal/mol.

▮
$$\Delta H_{\text{soln}} = (-97) + (-85) - (-186) = 4 \text{ kcal/mol}$$

14.62 Calculate the proton affinity of ammonia, $\text{NH}_3(g) + \text{H}^+(g) \rightarrow \text{NH}_4^+(g)$, given the following data: NH_4F crystallizes in a ZnS (wurtzite) structure (Madelung constant 1.641); the Born exponent is 8; the ammonium ion to fluoride ion distance is 2.63 Å; the enthalpy of formation of NH_4F is -112 kcal/mol; the enthalpy of formation of ammonia gas is -280 kcal/mol. Use other data from tables as needed.

▮ The lattice energy per mol NH_4F formula units, U_0, is calculated from the Born exponent n' and the Madelung constant A for a wurtzite structure:

$$U_0 = -\frac{z_1z_2N_AAe^2}{r_0}\left(1 - \frac{1}{n'}\right) = -\frac{(1)(1)(6.023 \times 10^{23})(1.641)(4.80 \times 10^{-10} \text{ esu})^2}{2.63 \times 10^{-8} \text{ cm}}(1 - \tfrac{1}{8})$$
$$= -7.58 \times 10^{12} \text{ ergs} = -7.58 \times 10^5 \text{ J} = -758 \text{ kJ} = -181 \text{ kcal}$$

The difference between ΔH_f values of NH_4F and NH_3 yields:

$$\tfrac{1}{2}\text{N}_2(g) + 2\text{H}_2(g) + \tfrac{1}{2}\text{F}_2(g) \longrightarrow \text{NH}_4\text{F}(s) \qquad \Delta H_f = -112$$
$$\tfrac{1}{2}\text{N}_2(g) + \tfrac{3}{2}\text{H}_2(g) \longrightarrow \text{NH}_3(g) \qquad \Delta H_f = -280$$
$$\text{NH}_3(g) + \tfrac{1}{2}\text{H}_2(g) + \tfrac{1}{2}\text{F}_2(g) \longrightarrow \text{NH}_4\text{F}(s) \qquad \Delta H_f = +168$$

To replace the elements in the last equation by their ions, one must subtract from this equation those representing the dissociation energies of H_2 and F_2 as well as the ionization potential of hydrogen and the electron affinity of fluorine, yielding

$$\tfrac{1}{2}H_2(g) \longrightarrow H(g) \qquad D = 52.1 \text{ kcal}$$
$$H(g) \longrightarrow H^+(g) \qquad (13.527 \text{ eV})(23.06 \text{ kcal/eV}) = 311.9 \text{ kcal}$$
$$\tfrac{1}{2}F_2(g) \longrightarrow F(g) \qquad D = 18.9 \text{ kcal}$$
$$F(g) \longrightarrow F^-(g) \qquad \Delta H = -\text{EA} = -79.6 \text{ kcal}$$
$$NH_3(g) + H^+(g) + F^-(g) \longrightarrow NH_4F(s) \qquad \Delta H = -135 \text{ kcal}$$

Subtraction of the lattice energy equation yields the desired proton affinity:

$$NH_4^+(g) + F^-(g) \longrightarrow NH_4F(s) \qquad U_0 = -181 \text{ kcal}$$
$$NH_3(g) + H^+(g) \longrightarrow NH_4^+(g) \qquad \text{PA} = \quad 46 \text{ kcal}$$

14.63 In solid ammonia, each NH_3 molecule has six other NH_3 molecules as nearest neighbors. ΔH of sublimation of NH_3 at the melting point is 30.8 kJ/mol, and the estimated ΔH of sublimation in the absence of hydrogen bonding is 14.4 kJ/mol. What is the strength of a hydrogen bond in solid ammonia?

▮ Total strength of all hydrogen bonds = 30.8 − 14.4 = 16.4 kJ/mol. There are 6 nearest neighbors, but each hydrogen bond involves 2 molecules. Thus we divide the total strength by $\frac{6}{2} = 3$, obtaining 5.5 kJ/mol.

14.64 Calculate the lattice energy in terms of e^2/r of a "crystal" consisting of 14 sodium ions and 13 chloride ions arranged in one isolated unit cell, with Na^+ ions at the corners.

▮ The unit cell of interest is pictured in Fig. 14.17. The task is to determine the number of interactions of the various types. For example, there are 54 cation-anion neighbors at a distance *r*. (There are 12 in each of three planes, pictured in Fig. 14.18*a*, plus 18 connecting the three planes.) There are 36 cation-cation next-nearest neighbors (four in each of nine planes, as shown in Fig. 14.18*b*. Continuing in this manner, one develops the data in Table 14.2 (c ≡ cation, a ≡ anion).

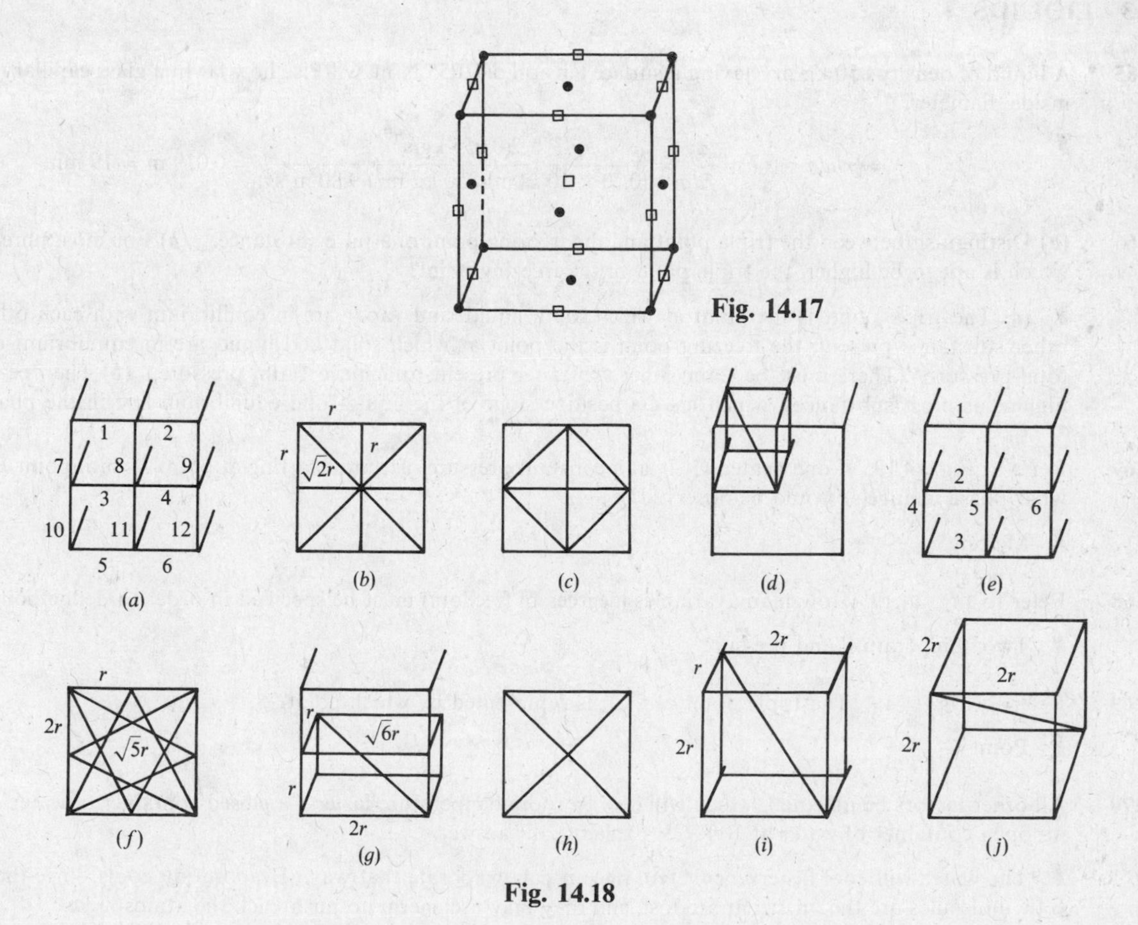

Fig. 14.17

Fig. 14.18

TABLE 14.2

Fig. 14.18	type	distance	number of units	interactions per unit	interactions total	energy contribution
(*a*)	c–a	r	3 planes; 18 others	12; 1	54	$-54e^2/r$
(*b*)	c–c	$\sqrt{2}r$	9 planes	4	36	$+36e^2/\sqrt{2}r$
(*c*)	a–a	$\sqrt{2}r$	9 planes	4	36	$+36e^2/\sqrt{2}r$
(*d*)	c–a	$\sqrt{3}r$	8 cubes	4	32	$-32e^2/\sqrt{3}r$
(*e*)	c–c; a–a	$\sqrt{2}r$	3 planes; 9 others	6; 1	27	$+27e^2/2r$
(*f*)	c–a	$\sqrt{5}r$	9 planes	8	72	$-72e^2/\sqrt{5}r$
(*g*)	c–c; a–a	$\sqrt{6}r$	12 rect. prisms	4	48	$+48e^2/\sqrt{6}r$
(*h*)	c–c	$\sqrt{8}r$	9 planes	2	18	$+18e^2/\sqrt{8}r$
(*i*)	c–a	$\sqrt{9}r$	6 prisms	4	24	$-24e^2/3r$
(*j*)	c–c	$\sqrt{12}r$	1 cube	4	4	$+4e^2/\sqrt{12}r$

The total lattice energy for this isolated unit cell is the sum of these energy values:

$$U_0 = \left(-54 + \frac{36}{\sqrt{2}} + \frac{36}{\sqrt{2}} - \frac{32}{\sqrt{3}} + \frac{27}{2} - \frac{72}{\sqrt{5}} + \frac{48}{\sqrt{6}} + \frac{18}{\sqrt{8}} - \frac{24}{3} + \frac{4}{\sqrt{12}}\right)\frac{e^2}{r}$$

$$= (-54 + 25.46 + 25.46 - 18.48 + 13.5 - 32.20 + 19.60 + 6.36 - 8 + 1.15)e^2/r = -21.15e^2/r$$

14.3 LIQUIDS

14.65 A liquid of density 850 kg/m^3 having a surface tension of 0.055 N/m, will rise how far in a glass capillary of 1.40 mm inside diameter?

▮ $$\gamma = \tfrac{1}{2}rhdg \qquad h = \frac{2\gamma}{rdg} = \frac{2(0.055\ \text{kg/s}^2)}{(0.70 \times 10^{-3}\ \text{m})(850\ \text{kg/m}^3)(9.80\ \text{m/s}^2)} = 0.019\ \text{m} = 19\ \text{mm}$$

14.66 (*a*) Distinguish between the triple point and the freezing point of a pure substance. (*b*) For most pure substances, which is apt to be higher, the triple point or the freezing point?

▮ (*a*) The triple point is the point at which solid, liquid, and vapor are in equilibrium with each other, with no other substance present; the freezing point is the point at which solid and liquid are in equilibrium under 1 atm total pressure. (There must be some other substance present to achieve 1 atm pressure.) (*b*) The freezing point is higher for most substances, which have a positive slope of the liquid-solid equilibrium line in the phase diagram.

14.67 Refer to Fig. 14.19. If one heated CCl_4 at a constant pressure of 1 atm starting at point *D* until point *E* is reached, what phase change(s) would be observed?

▮ Melting

14.68 Refer to Fig. 14.19. How many variables (degrees of freedom) must be specified in order to define point *E*?

▮ Two: temperature and pressure

14.69 Refer to Fig. 14.19. The triple point of CCl_4 is represented by which point?

▮ Point *C*

14.70 All other factors being equal, which will cool to room temperature faster—a closed container of water at 100 °C or an open container of water at 100 °C? Explain your answer.

▮ The water will cool faster uncovered, since it can evaporate that way. Evaporation cools, since the most energetic molecules are the ones that are lost, and they may use energy to push back the atmosphere.

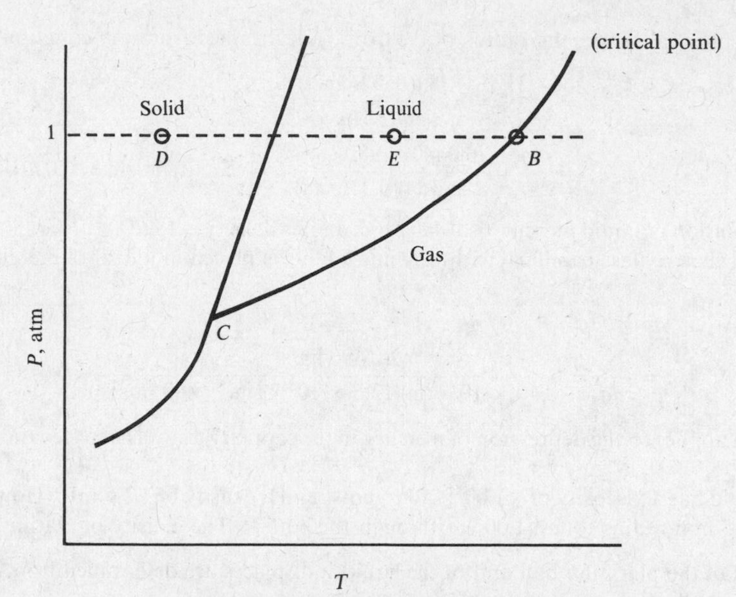

Fig. 14.19 Phase Diagram of CCl_4.

14.71 An astronaut in an orbiting spaceship spilled a few drops of his drink, and the liquid floated around the cabin. In what geometric shape was each drop most likely to be found? Explain.

▌ The drops would be spherical—the shape with the smallest surface area per unit volume, in response to surface tension.

14.72 The enthalpy of fusion of H_2O at 0 °C is 1.435 kcal/mol. Calculate ΔS for the following process at 0 °C: $H_2O(l) \rightleftharpoons H_2O(s)$.

▌

$$\text{Since} \quad \Delta G = 0, \quad \Delta S = \frac{\Delta H}{T} = \frac{-1435 \text{ cal/mol}}{273 \text{ K}} = -5.26 \text{ cal/mol·K}$$

14.73 When a liquid in a thermostated chamber is vaporized at constant temperature by the very slow withdrawal of a piston, the vapor is continuously in equilibrium with the liquid. The heat absorbed from the bath in the process is equal to what thermodynamic property of the liquid? Write an expression for calculating the value of $\Delta G°$ for this isothermal process at constant pressure.

▌ The heat is absorbed at constant pressure, with no work other than expansion. Hence the heat absorbed is equal to ΔH_{vap}. Since the liquid and vapor are in equilibrium, $\Delta G = 0$.

14.74 As heat is removed from a liquid which tends to supercool, its temperature drops below the freezing point and then rises suddenly. What is the source of the heat which causes the temperature rise?

▌ The enthalpy of fusion

14.75 Explain why water would completely fill a fine capillary tube which is open at both ends when one end is immersed in water.

▌ The surface tension pulls the water into the capillary. In a fine capillary, the surface tension is great enough to overcome the attraction of gravity on the water.

14.76 In a measurement of surface tension by the falling drop method, 5 drops of a liquid of density 0.797 g/mL weighed 0.220 g. Calculate the surface tension of the liquid.

▌

The mass of the average drop is $\frac{0.220 \text{ g}}{5} = 0.0440 \text{ g}$, and its volume is

$$\frac{0.0440 \text{ g}}{0.797 \text{ g/mL}} = 0.0552 \text{ mL}$$

Assuming a spherical shape, the radius of the drop is determined from the equation $V = \frac{4}{3}\pi r^3$:

$$r = \sqrt[3]{\frac{3V}{4\pi}} = \sqrt[3]{\frac{3(0.0552\ \text{cm}^3)}{4(3.14)}} = 0.236\ \text{cm}$$

$$\gamma = \frac{mg}{2\pi r} = \frac{(0.0440\ \text{g})(980\ \text{cm/s}^2)}{2(3.14)(0.236\ \text{cm})} = 29.1\ \text{dyne/cm} = 0.0291\ \text{N/m}$$

14.77 The surface tension of liquid mercury is 0.490 N/m, and its density is $13.6 \times 10^3\ \text{kg/m}^3$. How far will the mercury level be *depressed* when a glass capillary with 0.40 mm radius is placed in a dish of mercury?

▮

$$\gamma = \tfrac{1}{2}rhdg$$

$$h = \frac{2\gamma}{rdg} = \frac{2(0.490\ \text{kg/s}^2)}{(0.40 \times 10^{-3}\ \text{m})(13.6 \times 10^3\ \text{kg/m}^3)(9.80\ \text{m/s}^2)} = 0.0183\ \text{m} = 18.3\ \text{mm}$$

The equation applies to the depression of mercury in the capillary as well as to elevations of other liquids in capillaries.

14.78 A certain liquid has a viscosity of 1.00×10^4 poise and a density of 3.2 g/mL. How long will it take for a platinum ball with a 2.5 mm radius to fall 1.00 cm through the liquid? The density of platinum is $21.4\ \text{g/cm}^3$.

▮ The mass of the platinum ball and of the liquid it displaces are determined from the respective densities and the volume of the ball. Volume is given by

$$V = \tfrac{4}{3}\pi r^3 = \tfrac{4}{3}(3.14)(0.25\ \text{cm})^3 = 0.0654\ \text{cm}^3$$

$$m(\text{Pt}) = (0.0654\ \text{cm}^3)(21.4\ \text{g/cm}^3) = 1.40\ \text{g} \qquad m(\text{liq}) = (0.0654\ \text{cm}^3)(3.2\ \text{g/cm}^3) = 0.21\ \text{g}$$

$$v = \frac{(m - m_0)g}{6\pi r\eta} = \frac{(1.40\ \text{g} - 0.21\ \text{g})(980\ \text{cm/s}^2)}{6(3.14)(0.25\ \text{cm})(1.00 \times 10^4\ \text{poise})} = 2.47 \times 10^{-2}\ \text{cm/s}$$

$$t = \frac{d}{v} = \frac{1.00\ \text{cm}}{2.47 \times 10^{-2}\ \text{cm/s}} = 40.5\ \text{s}$$

14.79 (*a*) On the phase diagram for water, Fig. 14.20, what feature represents the equilibrium of solid water and water vapor? (*b*) What phase change(s) occur(s) when a sample at point E is heated at constant pressure until point F is reached? (*c*) What is the temperature at which line $\overline{BC}$ intersects the 1 atm line called?

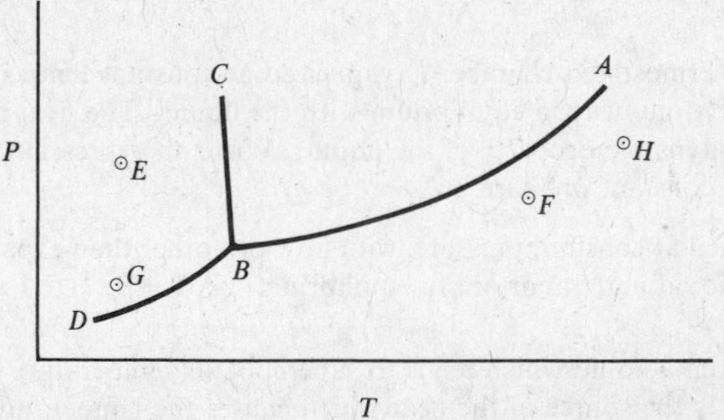

Fig. 14.20

▮ (*a*) Line $\overline{DB}$ represents the solid-gas equilibrium. (*b*) When a sample at point E is heated, melting and later vaporization occurs until the temperature reaches that represented by point F. (*c*) The freezing point is the temperature at which line $\overline{BC}$ crosses the $P = 1$ atm line.

14.80 Knowing that the density of ice is less than that of water, explain why the slope of the solid-liquid equilibrium line in the phase diagram of water (Fig. 14.20) is in accord with Le Châtelier's principle.

▮

$$H_2O(l) \rightleftharpoons H_2O(s)$$

As the pressure increases at constant temperature, this equilibrium shifts to the liquid phase—the more dense phase. Thus a movement straight up on the phase diagram, increasing the pressure at constant temperature, causes melting as one crosses the solid-liquid equilibrium line. The slope is correct.

14.81 Substance X has its triple point at 18 °C and 0.5 atm; its normal melting point is 20 °C, and its normal boiling point is 300 °C. Sketch the phase diagram for X.

▌ See Fig. 14.21.

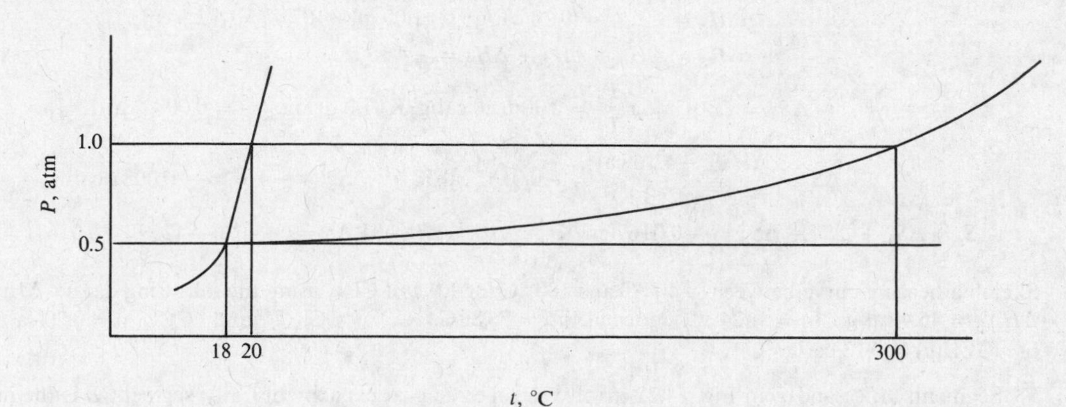

Fig. 14.21 (not drawn to scale)

14.82 Use the Clausius-Clapeyron equation to estimate the boiling point of water at 24 torr pressure. The average ΔH_{vap} over the temperature range is 10.12 kcal/mol.

▌ Since the normal boiling point of water is known to be 100 °C and P_{vap} at that temperature is 1.00 atm, the Clausius-Clapeyron equation may be applied to determine the temperature at which the vapor pressure is 24 torr:

$$\text{At } 100\ °\text{C (373 K)}: \quad P_{vap} = 760 \text{ torr} \qquad \text{At } T: \quad P_{vap} = 24 \text{ torr}$$

$$\log\frac{760}{24} = \frac{10.12 \times 10^3}{(2.30)(1.99)}\left(\frac{1}{T} - \frac{1}{373}\right) \qquad T = 298 \text{ K} = 25\ °\text{C}$$

To make water boil at 25 °C, it must be placed in a closed container from which air is pumped until the pressure of the gas phase is 24 torr. If all the water is to be boiled at this temperature, continuous pumping must be applied so that the gas pressure is maintained at 24 torr.

14.83 Between 20 and 80 °C, the enthalpy of vaporization of benzene is 7800 cal/mol. At 26 °C, the vapor pressure of benzene is 100 torr. Calculate its vapor pressure at 60 °C.

▌

$$\log\frac{100 \text{ torr}}{P_{60°}} = \frac{7800 \text{ cal/mol}}{(2.30)(1.99 \text{ cal/mol}\cdot\text{K})}\left(\frac{1}{333 \text{ K}} - \frac{1}{299 \text{ K}}\right) = -0.58 \qquad P_{60°} = 380 \text{ torr}$$

14.84 The vapor pressure of ethyl acetate is 300 torr at 51 °C, and its enthalpy of vaporization is 9.0 kcal/mol. Estimate the normal boiling point of ethyl acetate.

▌

$$\log\frac{P_2}{P_1} = \frac{\Delta H°}{2.30R}\left(\frac{1}{T_1} - \frac{1}{T_2}\right) \qquad \log\frac{760}{300} = \left(\frac{9.0 \times 10^3}{(2.30)(1.987)}\right)\left(\frac{1}{324} - \frac{1}{T_2}\right)$$

$$\frac{1}{324} - \frac{1}{T_2} = 2.052 \times 10^{-4} \qquad T_2 = 350 \text{ K (2 significant figures)}$$

14.85 Using the data $\Delta H_{fus}^{-24°} = 4.16$ cal/g and $C_p(s) = C_p(l) = 0.20$ cal/g·°C, calculate the entropy change and the free energy change for the freezing of 1.00 g supercooled $CCl_4(l)$ at −34 °C to solid at that same temperature.

▌

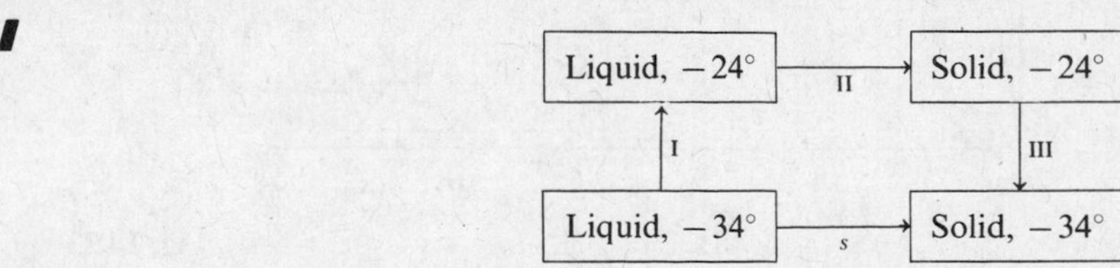

The freezing process at −34 °C, denoted *s* on the diagram, can be accomplished by the sum of processes I, II, and III. The enthalpy and entropy changes are thus the sums of these properties for processes I, II, and III.

$$\Delta H_1 = C_p\,\Delta t = (0.20\ \text{cal/g}\cdot{}^\circ\text{C})(1.00\ \text{g})(+10\ {}^\circ\text{C}) = 2.0\ \text{cal}$$
$$\Delta H_2 = (-4.16\ \text{cal/g})(1.00\ \text{g}) = -4.16\ \text{cal}$$
$$\Delta H_3 = C_p\,\Delta t = (0.20\ \text{cal/g}\cdot{}^\circ\text{C})(1.00\ \text{g})(-10\ {}^\circ\text{C}) = -2.0\ \text{cal}$$
$$\Delta H_8 = \Delta H_1 + \Delta H_2 + \Delta H_3 = -4.16\ \text{cal}$$
$$\Delta S_1 = 2.30 C_p \log\frac{T_2}{T_1} = 2.30(0.20\ \text{cal/g}\cdot\text{K})(1.0\ \text{g})\log\frac{249}{239} = 0.0082\ \text{cal/K}$$
$$\Delta S_2 = \frac{\Delta H}{T} = \frac{-4.16\ \text{cal}}{249\ \text{K}} = -0.0167\ \text{cal/K} \qquad \Delta S_3 = -\Delta S_1 = -0.0082\ \text{cal/K}$$
$$\Delta S_s = \Delta S_1 + \Delta S_2 + \Delta S_3 = -0.0167\ \text{cal/K} \qquad \Delta G = \Delta H - T\,\Delta S = (-4.16) - (239)(-0.0167) = -0.17\ \text{cal}$$

14.86 Sketch a heating curve between -40 °C and 100 °C for 1.0 g of CCl_4 using the following data: $\Delta H_{fus} = 4.16$ cal/g, $\Delta H_{vap} = 46.4$ cal/g, fp $= -24$ °C, normal bp $= 76.8$ °C, C_p of $CCl_4(l)$ and $CCl_4(s) = 0.20$ cal/g·°C, and C_p of $CCl_4(g) = 0.13$ cal/g·°C.

▌ Segments *A*, *C*, and *E* in Fig. 14.22 involve the specific heat capacity of CCl_4; segment *B* is the melting process, involving ΔH_{fus}; and part *D* is the vaporization process, involving ΔH_{vap}. The number of calories for each individual step is given below the corresponding line segment. The total number of calories added to that point is shown on the horizontal axis.

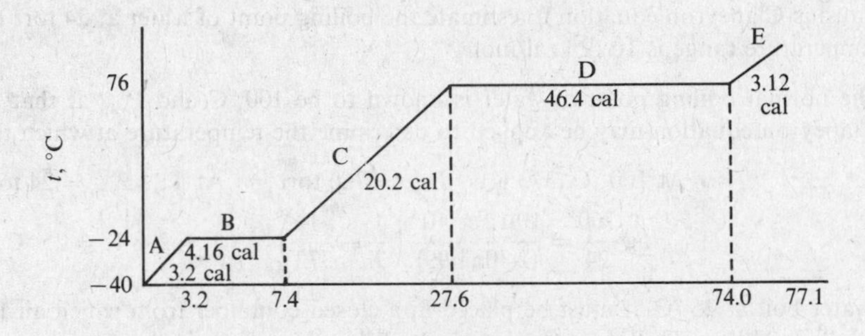

Fig. 14.22 (not drawn to scale)

14.87 The densities of ethyl alcohol liquid and vapor are tabulated below. By means of a graph, estimate the critical temperature of ethyl alcohol.

temperature, °C	density, g/mL vapor	density, g/mL liquid	temperature, °C	density, g/mL vapor	density, g/mL liquid
100	0.004	0.716	220	0.085	0.496
150	0.019	0.649	240	0.172	0.383
200	0.051	0.557			

▌ From Fig. 14.23, the critical temperature is approximately 245 °C.

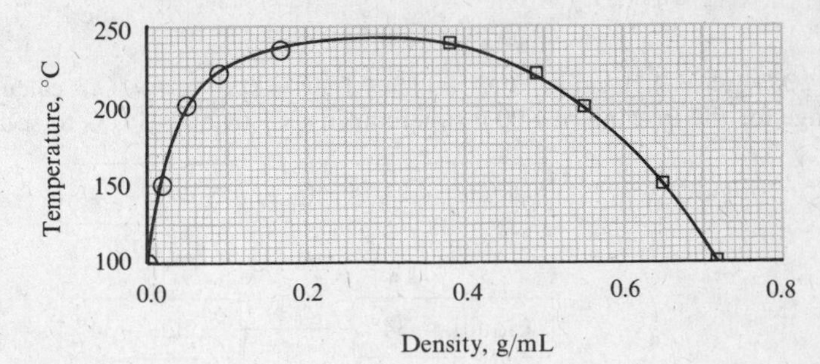

Fig. 14.23

14.88 The enthalpy change of the reaction $H^+ + OH^- \rightarrow H_2O$ is practically independent of temperature. Calculate the pH of pure water at 100 °C.

▮ $$\Delta G° = \Delta H° - T\,\Delta S° = (-13\,600 \text{ cal}) - (373 \text{ K})(19.23 \text{ cal/mol·K}) = -20\,770 \text{ cal/mol}$$

Assuming that the entropy change at 100 °C is the same as that at 25 °C, and taking $S°$ values from Tables 18.5 and 18.6, one calculates $\Delta S°$ as follows:

$$\Delta S° = 16.73 - (-2.5) = 19.23 \text{ cal/mol·K}$$

$$-\Delta G° = (2.30RT)(\log K) \qquad \log K = \frac{-\Delta G°}{2.30RT} = \frac{20\,770 \text{ cal/mol}}{(2.30)(1.987 \text{ cal/mol·K})(373 \text{ K})} = +12.2$$

$$K = 1.5 \times 10^{12} \qquad K_w = 1/K = 6.7 \times 10^{-13} \qquad [H_3O^+] = \sqrt{K_w} = 8.1 \times 10^{-7} \qquad \text{pH} = 6.1$$

14.89 As supercooled water freezes spontaneously, its temperature rises to 0 °C. What is ΔH for the spontaneous process: $H_2O(l)(-10\ °C) \rightarrow H_2O(s)(0\ °C)$?

▮ ΔH is zero. No energy is transferred to or from the system. The energy liberated in the freezing process warms the system to 0 °C.

14.90 By means of calculus, derive the expression for the temperature variation of entropy at constant pressure: $\Delta S = 2.30C_p \log (T_2/T_1)$.
What assumption about C_p is made in this derivation?

▮ $$\int dS = \int \frac{dH}{T} = \int_1^2 \frac{C_p\,dT}{T}$$

Assuming C_p to be constant over the range of temperatures involved,

$$dS = C_p \frac{dT}{T} \qquad \text{and} \qquad \Delta S = C_p \ln (T_2/T_1) = (2.30C_p) \log (T_2/T_1)$$

CHAPTER 15

Oxidation and Reduction

15.1 OXIDATION NUMBER; OXIDIZING AND REDUCING AGENTS

15.1 What is the oxidation number of sulfur in each of the following cases? (*a*) S^{2-} (*b*) H_2SO_4 (*c*) $S_2O_3^{2-}$ (*d*) CS_2 (*e*) S_8 (*f*) $Na_2S_4O_6$ (*g*) S_2Cl_2.

▌ (*a*) The oxidation number is -2, which is equal to the charge on the monatomic ion. (*b*) In H_2SO_4, the oxidation numbers must total zero. With 2 hydrogen at $+1$, 4 oxygen at -2, sulfur at x,

$$2(+1) + x + 4(-2) = 0 \qquad x = +6 \qquad \text{The oxidation number of sulfur is } +6$$

(*c*) In $S_2O_3^{2-}$, the oxidation numbers must total -2. With 3 oxygen at -2 and 2 sulfur at x,

$$2x + 3(-2) = -2 \qquad x = +2 \qquad \text{The oxidation number of sulfur is } +2.$$

(*d*) From just the rules stated above, any assignment of oxidation numbers which gives a net charge of zero could be made. However, noting that sulfur is in the same periodic group with oxygen, and by analogy with CO_2, the sulfur is usually assigned an oxidation number of -2. Then, 2 sulfur at $-2 = -4$, carbon at $+4 = +4$, and the net charge = 0. (*e*) The oxidation number of any free element is zero. (*f*) The oxidation numbers of alkali and alkaline earth metals in their compounds are uniformly $+1$ and $+2$, respectively. The oxidation number of sulfur can be established on this basis: 2 sodium at $+1$, 4 sulfur at x, 6 oxygen at -2, or

$$2(+1) + 4x + 6(-2) = 0 \qquad x = 2.5$$

(*g*) The oxidation number of chlorine is -1, as in all halogen compounds with elements other than oxygen or another halogen.

$$2x + 2(-1) = 0 \qquad x = +1$$

Therefore the oxidation number of sulfur must be $+1$.

15.2 In $H_2AsO_4^-$, what is the oxidation number of (*a*) hydrogen (*b*) oxygen (*c*) arsenic?

▌ (*a*) $+1$ (*b*) -2 (*c*) $+5$

15.3 State the oxidation number of the underlined element in each of the following: (*a*) $\underline{P}_2O_7^{4-}$ (*b*) $\underline{C}_3O_2$ (*c*) $\underline{Mn}O_4^-$ (*d*) $\underline{Mn}O_4^{2-}$ (*e*) $\underline{V}O_2^+$ (*f*) $\underline{U}O_2^{2+}$ (*g*) $\underline{Cl}O_3^-$

▌ The oxidation number of the underlined element is represented by x.

(*a*) $2x + 7(-2) = -4$ $\quad x = +5$ $\qquad$ (*b*) $3x + 2(-2) = 0$ $\quad x = +\frac{4}{3}$
(*c*) $x + 4(-2) = -1$ $\quad x = +7$ $\qquad$ (*d*) $x + 4(-2) = -2$ $\quad x = +6$
(*e*) $x + 2(-2) = +1$ $\quad x = +5$ $\qquad$ (*f*) $+6$ $\quad$ (*g*) $+5$

15.4 What is the oxidation state of hydrogen in each of the following? (*a*) HCl (*b*) H^+ (*c*) NaH (*d*) H_2 (*e*) $LiAlH_4$

▌ Hydrogen has a $+1$ oxidation state in all its compounds except those with group IA, IIA, or IIIA metals. It has 0 oxidation state as an element. (*a*) $+1$ (*b*) $+1$ (*c*) -1 (*d*) 0 (*e*) -1

15.5 Which of the following is the anhydride of HNO_3? (*a*) N_2O (*b*) NO (*c*) NO_2 (*d*) N_2O_4 (*e*) N_2O_5

▌

$$N_2O_5 + H_2O \longrightarrow 2HNO_3$$

N_2O_5 is the anhydride. The easiest way to tell is to check the oxidation number of nitrogen. HNO_3 and N_2O_5 both have nitrogen in the $+5$ oxidation state.

15.6 Calculate the oxidation number of the underlined element in each of the following: (*a*) $\underline{S}O_3^{2-}$ (*b*) $H_2\underline{C}O$ (*c*) $Na_2\underline{Cr}_2O_7$ (*d*) $\underline{O}_3$

▌ (*a*) $+4$ (*b*) 0 (*c*) $+6$ (*d*) 0

15.7 (*a*) What is the oxidation state of hydrogen in $LiAlH_4$? (*b*) Write an equation for the reaction of hydrogen in that oxidation state with water.

▌ (*a*) -1 (*b*) $H^- + H_2O \longrightarrow H_2 + OH^-$

15.8 What is the oxidation state of nitrogen in each of the following? (*a*) NH_3 (*b*) HN_3 (*c*) N_2H_4 (*d*) NO_2 (*e*) N_2O_4 (*f*) NO_2^- (*g*) NH_2OH (*h*) NO (*i*) HNO_3 (*j*) N_2O (*k*) HCN

▌ (*a*) -3 (H has an oxidation state of $+1$.) (*b*) $-\frac{1}{3}$ (Oxidation states need not be integral.) (*c*) -2 (*d*) $+4$ (*e*) $+4$ (*f*) $+3$ [$x + 2(-2) = -1$, the charge on the ion.] (*g*) -1 (*h*) $+2$ (*i*) $+5$ (*j*) $+1$ (*k*) -3 (C is $+2$ in CN^- and HCN, analogous to C in CO, but this value is somewhat arbitrary.)

15.9 Which one(s) of the following involve redox reactions? (*a*) Burning of gasoline (*b*) evaporation of water (*c*) human respiration (*d*) preparation of metals from their ores (*e*) production by lightning of nitrogen oxides from nitrogen and oxygen in the atmosphere (*f*) production by lightning of ozone (O_3) from O_2 (*g*) reaction of H_2SO_4 with NaOH

▌ *a, c, d, e.* [*b* is a physical change; *f* is the change of an element in one form (oxidation state = 0) to another form (oxidation state = 0); *g* is the reaction of an acid with a base.]

15.10 In the process of drying dishes with a dishtowel, what may be considered the wetting agent? the drying agent? What happens to the wetting agent? to the drying agent? In an oxidation-reduction reaction, what happens to the oxidizing agent? to the reducing agent?

▌ The wetting agent is the dish; the drying agent is the towel. (The drying agent gets wet during the process, since the water must go into it.) In a redox reaction, the oxidizing agent is reduced; the reducing agent is oxidized. (The electrons are transferred in a manner similar to the water in the drying analogy.)

15.11 Determine the oxidation state of chlorine in (*a*) ClO^- (*b*) ClO_2^- (*c*) ClO_3^- (*d*) ClO_4^-. (*e*) Name each of these ions.

▌

		oxidation state	(*e*) name
(*a*)	ClO^-	$+1$	Hypochlorite ion
(*b*)	ClO_2^-	$+3$	Chlorite ion
(*c*)	ClO_3^-	$+5$	Chlorate ion
(*d*)	ClO_4^-	$+7$	Perchlorate ion

15.12 Determine the oxidation number of (*a*) N in HN_3 (*b*) O in KO_2. (*c*) Draw electron dot diagrams for both of these species.

▌ (*a*) $-\frac{1}{3}$ (*b*) $-\frac{1}{2}$ (*c*) H:N̤̈:N:::N: K⁺ :Ö:Ö:⁻

15.13 Given that in these compounds the oxidation state of hydrogen is $+1$, of oxygen -2, and of fluorine -1, determine the oxidation states of the other elements. (*a*) PH_3 (*b*) H_2S (*c*) CrF_3 (*d*) H_2SO_4 (*e*) H_2SO_3 (*f*) Al_2O_3.

▌ (*a*) 3H represent an oxidation-state sum of $+3$ ($+1$ for each of the three hydrogens). Then the oxidation state of P must be -3, since the algebraic sum of the oxidation states of all atoms in a compound must equal zero. (*b*) Oxidation-state sum of 2H is $+2$; then oxidation state of S is -2. (*c*) Oxidation-state sum of 3F is -3; then oxidation state of Cr is $+3$. (*d*) Oxidation-state sum of 2H is $+2$ and of 4O is -8, or a total of -6; then oxidation state of S is $+6$. The sulfate ion, SO_4^{2-}, has a charge of -2, since the total oxidation state of two H atoms is $+2$. (*e*) Oxidation-state sum of 2H is $+2$ and of 3O is -6, or a total of -4; then oxidation state of S is $+4$. The sulfite ion, SO_3^{2-}, has an ionic charge of -2, since the total oxidation state of 2H is $+2$. (*f*) Oxidation-state sum of 3O is -6; then oxidation-state sum of 2Al is $+6$, and oxidation state of Al is $\frac{1}{2}(+6) = +3$.

15.14 Determine the oxidation state of the underlined element in (*a*) $K_4\underline{P}_2O_7$ (*b*) $Na\underline{Au}Cl_4$ (*c*) $Rb_4Na[H\underline{V}_{10}O_{28}]$ (*d*) $\underline{I}Cl$ (*e*) $Ba_2\underline{Xe}O_6$ (*f*) $\underline{O}F_2$ (*g*) $Ca(\underline{Cl}O_2)_2$.

▌ (*a*) $+5$ (*b*) $+3$ (*c*) $+5$ (*d*) $+1$ (*e*) $+8$ (*f*) $+2$ (*g*) $+3$

15.15 What is the oxidation number of Cl in $Ba(ClO_3)_2$?

▌ The ClO_3^- ion has Cl in the $+5$ oxidation state.

15.16 What are the maximum and minimum oxidation states of nitrogen in its compounds?

▌ +5 (equal to the group number) and −3 (equal to the group number minus 8)

15.17 State the oxidation number of the underlined element: (*a*) $\underline{Mn}O_4^-$ (*b*) $\underline{S}O_2$ (*c*) $\underline{Se}O_3^{2-}$ (*d*) $K_2\underline{S}$.

▌ (*a*) +7 (*b*) +4 (*c*) +4 (*d*) −2

15.18 State the maximum positive oxidation state for each of the following elements in any of their compounds (all of their compounds): (*a*) Ti (*b*) Zn (*c*) C (*d*) Cl.

▌ (*a*) +4 (*b*) +2 (*c*) +4 (*d*) +7

15.19 Considering oxidation states, predict the formulas of the compounds which are likely to be produced by the combination of (*a*) Cs and I (*b*) S and F.

▌ (*a*) CsI (*b*) SF_6 (The highest oxidation number of sulfur is +6; the negative oxidation number of fluorine is −1. Hence the compound is SF_6. Others could be SF_4 and SF_2.)

15.20 Name the four most important oxidation states exhibited by sulfur.

▌ +6, +4, 0, −2

15.21 What is the oxidation state of the underlined element in each of the following: (*a*) $\underline{Cr}_2O_7^{2-}$ (*b*) $Na_2\underline{S}_2$

▌ (*a*) +6 (*b*) −1 (in S_2^{2-} ion)

15.22 Predict the highest and lowest possible oxidation states of each of the following elements: (*a*) Ta (*b*) Te (*c*) Tc (*d*) Ti (*e*) Tl.

▌ The maximum oxidation state is the group number. The minimum oxidation state for the metals is zero; for the nonmetals it is equal to the group number minus 8. (*a*) +5, 0 (*b*) +6, −2 (*c*) +7, 0 (*d*) +4, 0 (*e*) +3, 0.

15.23 Which of the following are examples of disproportionation reactions? What criteria determine whether a reaction is a disproportionation?

(*a*) $Ag(NH_3)_2^+ + 2H^+ \longrightarrow Ag^+ + 2NH_4^+$

(*b*) $Cl_2 + 2OH^- \longrightarrow ClO^- + Cl^- + H_2O$

(*c*) $CaCO_3 \longrightarrow CaO + CO_2$

(*d*) $2HgO \longrightarrow 2Hg + O_2$

(*e*) $Cu_2O + 2H^+ \longrightarrow Cu + Cu^{2+} + H_2O$

(*f*) $CuS + O_2 \longrightarrow Cu + SO_2$

(*g*) $2HCuCl_2 \xrightarrow[\text{dilute with } H_2O]{} Cu + Cu^{2+} + 4Cl^- + 2H^+$

▌ Equations *b*, *e*, and *g* represent disproportionations, in that in each reaction the oxidizing agent and the reducing agent are the same.

15.24 In the following reaction, identify the species oxidized, the species reduced, the oxidizing agent, and the reducing agent.

$$Fe^{2+} + 2H^+ + NO_3^- \longrightarrow Fe^{3+} + NO_2 + H_2O$$

▌ The Fe^{2+} is oxidized to Fe^{3+} by the nitrate ion, which is the oxidizing agent. Nitrogen is reduced from nitrogen(V) in the nitrate to nitrogen(IV) in NO_2, by the reducing agent Fe^{2+}.

15.25 State the oxidation number of the underlined element(s) in each of the following: (*a*) $H_2\underline{S}O_4$ (*b*) $\underline{V}(\underline{Br}O_2)_2$ (*c*) $K\underline{Cl}O_3$ (*d*) $\underline{Br}O_3^-$

▌ (*a*) 6 (*b*) V^{II}, Br^{III} (deduced from BrO_2^-) (*c*) 5 (*d*) 5

15.26 Name the compounds in each of the following sets: (*a*) $FeCl_2$, $FeCl_3$ (*b*) VCl_2, VCl_3, VCl_4 (*c*) UCl_3 (*d*) Cu_2O, CuO (*e*) HNO_2, HNO_3

▌ (*a*) Iron(II) chloride and iron(III) chloride. (*b*) Vanadium(II) chloride, vanadium(III) chloride, and vanadium(IV) chloride. The use of roman numerals is specific and easy; one does not have to remember that VCl_2 is hypovanadous chloride, VCl_3 is vanadous chloride, and VCl_4 is vanadic chloride. (*c*) Uranium(III) chloride. (*d*) Copper(I) oxide and copper(II) oxide. (*e*) Nitrous acid and nitric acid are familiar examples of the nomenclature of acids, for which the Stock system, using oxidation numbers, is not commonly used.

15.27 Write the formulas for sulfide ion, sulfite ion, sulfate ion, and thiosulfate ion (in which one oxygen atom of sulfate is replaced by a sulfur atom). Compare the oxidation numbers of sulfur in these species. Use this exercise to distinguish between the concepts of oxidation number and charge. Is there any direct relationship between the oxidation number of sulfur and the charge on the ion?

▮ The formulas are S^{2-}, SO_3^{2-}, SO_4^{2-}, $S_2O_3^{2-}$, respectively. The oxidation numbers of sulfur are -2, $+4$, $+6$, and $+2$, respectively. Since the charge on each species is $2-$, it is apparent that charge and oxidation number are not the same. The charge is the sum of the oxidation numbers of all the atoms in a species and refers to the species as a whole. Oxidation number refers to atoms individually. There is no direct relationship between the oxidation number of a single atom and the charge on the ion as a whole.

15.28 Both VO_2^+ and VO^{2+} are known as the "vanadyl" ion. (*a*) Determine the oxidation number of vanadium in each. (*b*) Which of the following names corresponds to which ion? oxovanadium(IV) ion, dioxovanadium(V) ion.

▮ (*a*) and (*b*) VO_2^+ contains V^V and is named dioxovanadium(V) ion; VO^{2+} contains V^{IV} and is named oxovanadium(IV) ion.

15.29 Hydrogen peroxide, H_2O_2, may act as an oxidizing agent or a reducing agent. Explain why this behavior is possible. Write an equation for the disproportionation of H_2O_2.

▮ The oxygen exists in H_2O_2 in an intermediate oxidation state; it can be either oxidized or reduced depending on the reagents used. It can disproportionate according to the equation $2H_2O_2 \rightarrow 2H_2O + O_2$

15.30 Which of the following are (*a*) very good oxidizing agents? (*b*) very good reducing agents? (*c*) neither? MnO_4^-, I^-, Cl^-, Ce^{4+}, $Cr_2O_7^{2-}$, Na, Na^+, CrO_4^{2-}, HNO_3, Fe^{2+}, F_2, F^-

▮ (*a*) MnO_4^-, Ce^{4+}, $Cr_2O_7^{2-}$, CrO_4^{2-}, HNO_3, F_2 (*b*) I^-, Na, Fe^{2+} (*c*) Cl^-, Na^+, F^-

15.31 Which of the following equations represent oxidation-reduction reactions? Identify each oxidizing agent and each reducing agent.

(*a*) $K + O_2 \longrightarrow KO_2$ (*b*) $H_2O_2 + KOH \longrightarrow KHO_2 + H_2O$

(*c*) $Ca(HCO_3)_2 \xrightarrow[\text{heat}]{} CaCO_3 + CO_2 + H_2O$ (*d*) $Cr_2O_7^{2-} + 2OH^- \longrightarrow 2CrO_4^{2-} + H_2O$

(*e*) $H_2O_2 \longrightarrow H_2O + \frac{1}{2}O_2$

▮ (*a*) The oxidizing agent is O_2; the reducing agent is K. (*b*), (*c*), and (*d*) are not redox reactions. (*e*) The H_2O_2, which disproportionates, is both oxidizing and reducing agent.

15.32 Predict the products of each of the following. If no reaction occurs, write "nr."

(*a*) $Na + ZnCl_2 \longrightarrow$ (*b*) $F_2 + NaCl \longrightarrow$ (*c*) $I_2 + Fe \longrightarrow$ (*d*) $I^- + Fe^{2+} \longrightarrow$

▮ (*a*) $2Na + ZnCl_2 \longrightarrow Zn + 2NaCl$ (*b*) $F_2 + 2NaCl \longrightarrow 2NaF + Cl_2$

(*c*) $I_2 + Fe \longrightarrow Fe^{2+} + 2I^-$ (*d*) $I^- + Fe^{2+} \longrightarrow nr$

Fe is a better reducing agent than I^-. Thus, reaction (*c*) goes, but reaction (*d*) does not.

15.2 BALANCING REDOX EQUATIONS

There are very many ways to balance redox equations systematically. Choose one (or two) methods, learn them, and use them exclusively. Since various instructors and textbooks use different methods, problems are solved by several different methods here.

15.33 If the problem doesn't state "acid solution" or "basic solution" explicitly, how can you tell which type of solution should be assumed?

▮ If an acid or base is one of the reactants or products, the solution is the same. If ammonia is present, the solution is basic; if ammonium ion is present, it is acidic. If metals which would form insoluble hydroxides are shown in their ionic form, the solution is acidic.

15.34 State whether each of the following equations should be balanced in acid or basic solution:

(*a*) $HNO_3 + Fe^{2+} \longrightarrow Fe^{3+} + NO_2$ (*b*) $NH_3 + MnO_4^- \longrightarrow MnO_2 + NO_2$

(*c*) $Fe^{3+} + H_2O_2 \longrightarrow Fe^{2+} + O_2$ (*d*) $Cr(OH)_2 + I_2 \longrightarrow Cr(OH)_3 + 2I^-$

▌ (*a*) acidic (*b*) basic (*c*) acidic (*d*) basic

15.35 Balance the oxidation-reduction equation $H^+NO_3^- + H_2S \longrightarrow NO + S + H_2O$.

▌ *Ion-Electron Method*

(1) The skeleton equation is given above.

(2) The oxidizing agent is nitrate ion, NO_3^-, since it contains the element (N), which undergoes a decrease in oxidation state. The reducing agent is H_2S, since it contains the element (S), which undergoes an increase in oxidation state. (S^{2-} might have been selected as the reducing agent, but H_2S is preferable because of the very slight degree of ionization of the acid in nitric acid solution. Only a very small fraction exists as S^{2-}.)

(3) The partial skeleton equation for the reduction half-reaction: $NO_3^- \longrightarrow NO$

(4) The partial skeleton equation for the oxidation half-reaction: $H_2S \longrightarrow S$

(5) (*a*) In the first partial equation, $2H_2O$ must be added to the right side in order to balance the oxygen atoms. Then $4H^+$ must be added to the left side to balance the H atoms.

$$4H^+ + NO_3^- \longrightarrow NO + 2H_2O$$

(*b*) The atoms in the second partial equation can be balanced by adding $2H^+$ to the right side.

$$H_2S \longrightarrow S + 2H^+$$

(6) (*a*) In equation 5(*a*), the net charge on the left is $+4 - 1 = +3$, and on the right it is 0. Therefore 3 electrons must be added to the left side:

$$4H^+ + NO_3^- + 3e^- \longrightarrow NO + 2H_2O$$

(*b*) In equation 5(*b*), the net charge on the left is 0, and on the right it is +2. Hence 2 electrons must be added to the right side.

$$H_2S \longrightarrow S + 2H^+ + 2e^-$$

(7) Equation 6(*a*) must be multiplied by 2, and equation 6(*b*) by 3.

(*a*) $8H^+ + 2NO_3^- + 6e^- \longrightarrow 2NO + 4H_2O$ (*b*) $3H_2S \longrightarrow 3S + 6H^+ + 6e^-$

(8) Addition of equations 7(*a*) and 7(*b*) results in

$$8H^+ + 2NO_3^- + 3H_2S + 6e^- \longrightarrow 2NO + 4H_2O + 3S + 6H^+ + 6e^-$$

Since $6H^+$ and $6e^-$ are common to both sides, they may be canceled.

$$2H^+ + 2NO_3^- + 3H_2S \longrightarrow 2NO + 4H_2O + 3S$$

This form of the equation shows all the ions and compounds in the proper form.

(9) If we want to know how much HNO_3 is required, we merely combine H^+ with NO_3^-.

$$2HNO_3 + 3H_2S \longrightarrow 2NO + 4H_2O + 3S$$

Oxidation-State Method

(1) Note that the oxidation state of N changes from +5 in NO_3^- to +2 in NO.

(2) The oxidation state of S changes from −2 in H_2S to 0 in S.

(3) Electron balance *diagrams* can be written as follows. (These are not *equations*.)

$$N^{V} + 3e^- \longrightarrow N^{II} \quad (1)$$

$$S^{(-2)} \longrightarrow S + 2e^- \quad (2)$$

(4) In order that the number of electrons lost shall equal the number gained, we must multiply diagram (*1*) by 2, and (*2*) by 3.

$$2N^{V} + 6e^- \longrightarrow 2N^{II} \qquad 3S^{(-2)} \longrightarrow 3S^0 + 6e^-$$

Hence the coefficients of $H^+NO_3^-$ and of NO are 2, and those of H_2S and S are 3. Part of the skeleton equation can now be filled in.

$$2H^+NO_3^- + 3H_2S \longrightarrow 2NO + 3S$$

(5) The 8 atoms of H on the left (2 from $H^+NO_3^-$ plus 6 from H_2S) must form $4H_2O$ on the right. The final and complete equation is

$$2H^+NO_3^- + 3H_2S \longrightarrow 2NO + 3S + 4H_2O$$

Note that the oxygen atoms were balanced automatically, without special attention.

The first partly filled skeleton equation could have been written in terms of NO_3^- rather than $H^+NO_3^-$. For variety in the subsequent solved problems, the neutral-compound notation will be used in the oxidation-state method, and the ionic notation in the ion-electron method.

Alternatively, a diagram using the actual chemical species involved can be used in place of steps (3) and (4). The change in oxidation state for each atom times the number of atoms gives the number of electrons involved.

$$(-2 \rightarrow 0) = +2$$

$$HNO_3 + H_2S \longrightarrow NO + S$$

$$(+5 \rightarrow +2) = -3$$

Yielding: $2HNO_3 + 3H_2S \longrightarrow 2NO + 2S$ (still incomplete)

15.36 Balance the following oxidation-reduction equation:

$$K^+MnO_4^- + K^+Cl^- + (H^+)_2SO_4^{2-} \longrightarrow Mn^{2+}SO_4^{2-} + (K^+)_2SO_4^{2-} + H_2O + Cl_2$$

Ion-Electron Method

The skeleton partial equations may be written

(*a*) $MnO_4^- \longrightarrow Mn^{2+}$ (*b*) $Cl^- \longrightarrow Cl_2$

Partial (*a*) requires $4H_2O$ on the right to balance the oxygen atoms, then $8H^+$ on the left to balance the hydrogen. Partial (*b*) balances in routine fashion.

(*a*) $8H^+ + MnO_4^- \longrightarrow Mn^{2+} + 4H_2O$ (*b*) $2Cl^- \longrightarrow Cl_2$

The net charge on the left of partial (*a*) is $+8 - 1 = +7$, and on the right it is $+2$; therefore 5 electrons must be added to the left. In partial (*b*) the net charge on the left is -2, and on the right it is 0; therefore 2 electrons must be added to the right.

(*a*) $8H^+ + MnO_4^- + 5e^- \longrightarrow Mn^{2+} + 4H_2O$ (*b*) $2Cl^- \longrightarrow Cl_2 + 2e^-$

The multiplying factors are seen to be 2 and 5, respectively.

$$16H^+ + 2MnO_4^- + 10e^- \longrightarrow 2Mn^{2+} + 8H_2O$$
$$10Cl^- \longrightarrow 5Cl_2 + 10e^-$$
$$\overline{16H^+ + 2MnO_4^- + 10Cl^- \longrightarrow 2Mn^{2+} + 8H_2O + 5Cl_2}$$

Since MnO_4^- was added as $KMnO_4$, $2MnO_4^-$ introduces $2K^+$ to the left side of the equation; and since K^+ does not react, the same number will appear on the right side. Since Cl^- was added as KCl, the $10Cl^-$ introduces $10K^+$ to each side of the equation. Since H^+ was added as H_2SO_4, $16H^+$ introduces $8SO_4^{2-}$ to each side of the equation. Then

$$16H^+ + 8SO_4^{2-} + 2K^+ + 2MnO_4^- + 10K^+ + 10Cl^- \longrightarrow 2Mn^{2+} + 8H_2O + 5Cl_2 + 12K^+ + 8SO_4^{2-}$$

Pairs of ions may be grouped on the left side to show that the chemical reagents were H_2SO_4, $KMnO_4$, and KCl. Pairs of ions may be grouped on the right to show that evaporation of the solution, after reaction has taken place, would cause the crystallization of $MnSO_4$ and K_2SO_4.

$$8(H^+)_2SO_4^{2-} + 2K^+MnO_4^- + 10K^+Cl^- \longrightarrow 2Mn^{2+}SO_4^{2-} + 6(K^+)_2SO_4^{2-} + 5Cl_2 + 8H_2O$$

Oxidation-State Method

Mn undergoes a change in oxidation state from $+7$ in MnO_4^- to $+2$ in Mn^{2+}. Cl undergoes a change in oxidation state from -1 in Cl^- to 0 in Cl_2. The electron balance diagrams are

$$Mn^{VII} + 5e^- \longrightarrow Mn^{II} \quad (1)$$

$$2Cl^{(-1)} \longrightarrow 2Cl^0 + 2e^- \quad (2)$$

Diagram (2) was written in terms of two Cl atoms because these atoms occur in pairs in the product Cl_2. The multiplying factors are 2 and 5, just as in the previous method.

$$2Mn^{VII} + 10e^- \longrightarrow 2Mn^{II}$$
$$10Cl^{(-1)} \longrightarrow 10Cl^0 + 10e^-$$

Hence the coefficients of $K^+MnO_4^-$ and $Mn^{2+}SO_4^{2-}$ are 2, of K^+Cl^- is 10, of Cl_2 is 5 $(\frac{1}{2} \times 10)$.

$$2K^+MnO_4^- + 10K^+Cl^- \longrightarrow 2Mn^{2+}SO_4^{2-} + 5Cl_2 \quad \text{(incomplete)}$$

So far no provision has been made for the H_2O, $(H^+)_2SO_4^{2-}$, and $(K^+)_2SO_4^{2-}$. The 8 atoms of oxygen from $2K^+MnO_4^-$ form $8H_2O$. For $8H_2O$ we need 16 atoms of hydrogen, which can be furnished by $8(H^+)_2SO_4^{2-}$. The 12 atoms of K $(10K^+Cl^- + 2K^+MnO_4^-)$ yield $6(K^+)_2SO_4^{2-}$. Note that all the oxygen in the oxidizing agent is converted to water. The sulfate radical retains its identity throughout the reaction.

$$2K^+MnO_4^- + 10K^+Cl^- + 8(H^+)_2SO_4^{2-} \longrightarrow 2Mn^{2+}SO_4^{2-} + 5Cl_2 + 6(K^+)_2SO_4^{2-} + 8H_2O$$

15.37 Balance the following oxidation-reduction equation:

$$(K^+)_2Cr_2O_7^{2-} + H^+Cl^- \longrightarrow K^+Cl^- + Cr^{3+}(Cl^-)_3 + H_2O + Cl_2$$

▮ *Ion-Electron Method*

The balancing of the partial equation for the oxidizing agent proceeds as follows.

$$Cr_2O_7^{2-} \longrightarrow Cr^{3+} \qquad Cr_2O_7^{2-} \longrightarrow 2Cr^{3+} \qquad Cr_2O_7^{2-} \longrightarrow 2Cr^{3+} + 7H_2O$$
$$14H^+ + Cr_2O_7^{2-} \longrightarrow 2Cr^{3+} + 7H_2O$$
$$14H^+ + Cr_2O_7^{2-} + 6e^- \longrightarrow 2Cr^{3+} + 7H_2O \quad \text{(balanced)}$$

The balancing of the partial equation for the reducing agent is as follows.

$$Cl^- \longrightarrow Cl_2 \qquad 2Cl^- \longrightarrow Cl_2 \qquad 2Cl^- \longrightarrow Cl_2 + 2e^- \quad \text{(balanced)}$$

The overall equation is

$$1 \times [14H^+ + Cr_2O_7^{2-} + 6e^- \longrightarrow 2Cr^{3+} + 7H_2O]$$
$$3 \times [\qquad 2Cl^- \longrightarrow Cl_2 + 2e^- \qquad]$$
$$\overline{14H^+ + Cr_2O_7^{2-} + 6Cl^- \longrightarrow 2Cr^{3+} + 7H_2O + 3Cl_2}$$

The $14H^+$ was added as $14H^+Cl^-$, and 6 of the 14 chloride ions were oxidized. To each side of the equation 8 more Cl^- can be added to represent those Cl^- which were not oxidized. Similarly, $2K^+$ may be added to each side to show that $Cr_2O_7^{2-}$ came from $(K^+)_2Cr_2O_7^{2-}$.

$$14H^+ + 6Cl^- + 8Cl^- + 2K^+ + Cr_2O_7^{2-} \longrightarrow 2Cr^{3+} + 2K^+ + 8Cl^- + 3Cl_2 + 7H_2O$$
$$14H^+Cl^- + (K^+)_2Cr_2O_7^{2-} \longrightarrow 2Cr^{3+}(Cl^-)_3 + 2K^+Cl^- + 3Cl_2 + 7H_2O$$

Oxidation-State Method

The electron balance diagrams are written in terms of 2 atoms of Cr and 2 atoms of Cl because of the appearance of pairs of atoms of these kinds in $(K^+)_2Cr_2O_7^{2-}$ and Cl_2.

For the oxidizing agent: $\quad 2Cr^{VI} \longrightarrow 2Cr^{III} \qquad 2Cr^{VI} + 6e^- \longrightarrow 2Cr^{III}$

For the reducing agent: $\quad 2Cl^{(-1)} \longrightarrow 2Cl^0 \qquad 2Cl^{(-1)} \longrightarrow 2Cl^0 + 2e^-$

Multiplying and adding:

$$1 \times [2Cr^{VI} + 6e^- \longrightarrow 2Cr^{III} \qquad]$$
$$3 \times [\qquad 2Cl^{(-1)} \longrightarrow 2Cl^0 + 2e^-]$$
$$\overline{2Cr^{VI} + 6Cl^{(-1)} \longrightarrow 2Cr^{III} + 6Cl^0}$$

Hence $\quad (K^+)_2Cr_2O_7^{2-} + 6H^+Cl^- \longrightarrow 2Cr^{3+}(Cl^-)_3 + 3Cl_2 \quad \text{(incomplete)}$

The equation is still unbalanced, as no provision has been made for the K^+Cl^-, H_2O, or the H^+Cl^- that acts as an acid (as opposed to the H^+Cl^- that acts as reducing agent).

By inspection, the 7 atoms of oxygen in $(K^+)_2Cr_2O_7^{2-}$ form $7H_2O$. For $7H_2O$ we need 14 atoms of H, which can be furnished by $14H^+Cl^-$. Since 6 of the chloride ions were oxidized to Cl_2, the remaining 8 $(14 - 6)$ should appear on the right as K^+Cl^- or $Cr^{3+}(Cl^-)_3$. Moreover, one $(K^+)_2Cr_2O_7^{2-}$ yields $2K^+Cl^-$. Thus the coefficient of H^+Cl^- is 14, of H_2O is 7, and of K^+Cl^- is 2.

$$(K^+)_2Cr_2O_7^{2-} + 14H^+Cl^- \longrightarrow 2Cr^{3+}(Cl^-)_3 + 3Cl_2 + 7H_2O + 2K^+Cl^- \quad \text{(balanced)}$$

Note that here again all the oxygen in the oxidizing agent is converted to water.

15.38 Balance the following oxidation-reduction equation by the oxidation-state method.

$$FeS_2 + O_2 \longrightarrow Fe_2O_3 + SO_2$$

▮ The two special features of this problem are that both the iron and sulfur in FeS_2 undergo a change in oxidation state and that the reduction product of oxygen gas occurs in combination with both iron and sulfur. The electron balance diagram for oxidation must contain Fe and S atoms in the ratio of 1:2, since this is the ratio in which they are oxidized. The two diagrams, with their multiplying factors, are

$$4 \times [Fe^{II} + 2S^{(-1)} \longrightarrow Fe^{III} + 2S^{IV} + 11e^-]$$
$$11 \times [\ 2O^0 + 4e^- \longrightarrow 2O^{(-2)}\ ]$$
$$4Fe^{II} + 8S^{(-1)} + 22O^0 \longrightarrow 4Fe^{III} + 8S^{IV} + 22O^{(-2)}$$

Hence $$4FeS_2 + 11O_2 \longrightarrow 2Fe_2O_3 + 8SO_2 \qquad \text{(balanced)}$$

15.39 Balance the following oxidation-reduction equation by the ion-electron method.

$$Zn + Na^+NO_3^- + Na^+OH^- \longrightarrow (Na^+)_2ZnO_2^{2-} + NH_3 + H_2O$$

▮ The skeleton partial for the oxidizing agent is $NO_3^- \rightarrow NH_3$. In alkaline solutions, each excess oxygen atom is balanced by adding one H_2O to the same side of the equation and $2OH^-$ to the opposite side. Each excess hydrogen atom is balanced by adding one OH^- to the same side and one H_2O to the opposite side. In this example we must add $3H_2O$ to the left and $6OH^-$ to the right to balance the excess oxygen of the NO_3^-. Also, we must add $3OH^-$ to the right and $3H_2O$ to the left to balance the excess hydrogen of the NH_3. In all, $9OH^-$ must be added to the right and $6H_2O$ to the left.

$$NO_3^- + 6H_2O \longrightarrow NH_3 + 9OH^- \qquad NO_3^- + 6H_2O + 8e^- \longrightarrow NH_3 + 9OH^- \quad \text{(balanced)}$$

The balancing of the partial equation for the reducing agent is as follows.

$$Zn \longrightarrow ZnO_2^{2-} \qquad 4OH^- + Zn \longrightarrow ZnO_2^{2-} + 2H_2O$$
$$4OH^- + Zn \longrightarrow ZnO_2^{2-} + 2H_2O + 2e^- \quad \text{(balanced)}$$

The overall equation is

$$1 \times [NO_3^- + 6H_2O + 8e^- \longrightarrow NH_3 + 9OH^-\ ]$$
$$4 \times [\ 4OH^- + Zn \longrightarrow ZnO_2^{2-} + 2H_2O + 2e^-]$$
$$NO_3^- + 6H_2O + 4Zn + 16OH^- \longrightarrow NH_3 + 9OH^- + 4ZnO_2^{2-} + 8H_2O$$

Canceling, $$NO_3^- + 4Zn + 7OH^- \longrightarrow NH_3 + 4ZnO_2^{2-} + 2H_2O$$

Adding ions and combining, $$Na^+NO_3^- + 4Zn + 7Na^+OH^- \longrightarrow NH_3 + 4(Na^+)_2ZnO_2^{2-} + 2H_2O$$

15.40 Balance the following equation by the ion-electron method.

$$HgS + H^+Cl^- + H^+NO_3^- \longrightarrow (H^+)_2HgCl_4^{2-} + NO + S + H_2O$$

▮ The partial equation for the oxidizing agent is balanced by the straightforward procedure.

$$4H^+ + NO_3^- + 3e^- \longrightarrow NO + 2H_2O$$

The skeleton for the reducing agent is $HgS \rightarrow S$. The skeleton contains both reduced and oxidized forms of sulfur, the only element undergoing a change in oxidation state. The imbalance is not one of hydrogen or oxygen atoms but of mercury. According to the overall equation, the form in which mercury exists among the products is the ion $HgCl_4^{2-}$. If such an ion is added to the right to balance the mercury, then chloride ions must be added to the left to balance the chlorine. In general, it is always allowable to add ions necessary for complex formation when such an addition does not require the introduction of new oxidation states.

$$HgS + 4Cl^- \longrightarrow S + HgCl_4^{2-} + 2e^- \qquad \text{(balanced)}$$

The overall equation is

$$2 \times [4H^+ + NO_3^- + 3e^- \longrightarrow NO + 2H_2O\ ]$$
$$3 \times [\ HgS + 4Cl^- \longrightarrow S + HgCl_4^{2-} + 2e^-]$$
$$8H^+ + 2NO_3^- + 3HgS + 12Cl^- \longrightarrow 2NO + 4H_2O + 3S + 3HgCl_4^{2-}$$

Adding ions and combining, $$3HgS + 2H^+NO_3^- + 12H^+Cl^- \longrightarrow 3S + 3(H^+)_2HgCl_4^{2-} + 2NO + 4H_2O$$

15.41 Complete and balance the following skeleton equation for a reaction in acid solution.

$$H_2O_2 + MnO_4^- + \qquad \longrightarrow$$

▮ The products must be deduced from chemical experience. MnO_4^- contains manganese in its highest oxidation state. Hence, if reaction occurs at all, MnO_4^- must be reduced. Since the solution is acid, the reduction product will be Mn^{2+}. The H_2O_2 must therefore act as a reducing agent in this reaction, and its only possible oxidation product is O_2. The usual procedure may be followed from this point, leading to the following solution.

$$2 \times [MnO_4^- + 8H^+ + 5e^- \longrightarrow Mn^{2+} + 4H_2O\ ]$$
$$5 \times [\qquad H_2O_2 \longrightarrow O_2 + 2H^+ + 2e^-]$$
$$2MnO_4^- + 16H^+ + 5H_2O_2 \longrightarrow 2Mn^{2+} + 8H_2O + 5O_2 + 10H^+$$

After canceling,
$$2MnO_4^- + 6H^+ + 5H_2O_2 \longrightarrow 2Mn^{2+} + 8H_2O + 5O_2$$

The above equation is as definite a statement as can be made within the given overall skeleton. If we are to write an equation in terms of neutral substances, we are free to decide which salt of MnO_4^- and which acid to use. If we use $K^+MnO_4^-$ and $(H^+)_2SO_4^{2-}$, we obtain

$$2K^+MnO_4^- + 3(H^+)_2SO_4^{2-} + 5H_2O_2 \longrightarrow 2Mn^{2+}SO_4^{2-} + 5O_2 + (K^+)_2SO_4^{2-} + 8H_2O$$

15.42 Compare the ion-electron method of balancing equations to the oxidation-state method.

▮ Both methods lead to the correct form of the balanced equation. The ion-electron method has two advantages:

(1) It differentiates those components of a system which react from those which do not. In Problem 15.41, any permanaganate and any strong acid could have been used, not necessarily $K^+MnO_4^-$ and $(H^+)_2SO_4^{2-}$. (The use of complete formulas for neutral substances is helpful only for the calculation of mass relations. Since permanganate ion cannot be weighed on a balance, it is necessary to choose one particular permanganate, like $K^+MnO_4^-$, and weigh an amount of $K^+MnO_4^-$ that will give the correct weight of MnO_4^-.)

(2) The half-reactions given by the partial equations of the ion-electron method can actually be made to occur independently. Many oxidation-reduction reactions can be carried out as galvanic cell processes for producing an electrical potential. This can be done by placing the reducing agent and oxidizing agent in separate vessels and making electrical connections between the two. It has been found that the reaction taking place in each beaker corresponds exactly to a partial equation written according to the rules for the ion-electron method.

Some chemists prefer to use the ion-electron method for oxidation-reduction reactions carried out in dilute aqueous solutions, where free ions have more or less independent existence, and to use the oxidation-state method for oxidation-reduction reactions between solid chemicals or for reactions in concentrated acid media.

Write a net ionic equation and/or a complete equation for each of the reactions in Problems 15.43–15.62.

15.43 $CuS + H^+NO_3^-(\textit{dilute}) \longrightarrow Cu^{2+}(NO_3^-)_2 + S + H_2O + NO$

▮

$$(5 \rightarrow 2) = -3$$
$$CuS + H^+ + NO_3^- \longrightarrow Cu(NO_3)_2 + S + H_2O + NO$$
$$(-2 \rightarrow 0) = +2$$

Balance the elements which change oxidation state:

$$3CuS + H^+ + 2NO_3^- \longrightarrow Cu(NO_3)_2 + 3S + H_2O + 2NO$$

Balance the other elements. Note that $6NO_3^-$ ions which do not change oxidation state must be added to the two which do.

$$3CuS + 8H^+ + 8NO_3^- \longrightarrow 3Cu(NO_3)_2 + 3S + 4H_2O + 2NO$$

15.44 $K^+MnO_4^- + H^+Cl^- \longrightarrow K^+Cl^- + Mn^{2+}(Cl^-)_2 + H_2O + Cl_2$

▮

$$MnO_4^- + Cl^- \longrightarrow Mn^{2+} + Cl_2$$

The half-reactions:

$$MnO_4^- \longrightarrow Mn^{2+} \qquad 2Cl^- \longrightarrow Cl_2$$
$$MnO_4^- \longrightarrow Mn^{2+} + 4H_2O \qquad 2Cl^- \longrightarrow Cl_2 + 2e^-$$
$$8H^+ + MnO_4^- \longrightarrow Mn^{2+} + 4H_2O$$
$$5e^- + 8H^+ + MnO_4^- \longrightarrow Mn^{2+} + 4H_2O$$
$$10e^- + 16H^+ + 2MnO_4^- \longrightarrow 2Mn^{2+} + 8H_2O \qquad 10Cl^- \longrightarrow 5Cl_2 + 10e^-$$

Combining the half-reactions:

$$10Cl^- + 16H^+ + 2MnO_4^- \longrightarrow 2Mn^{2+} + 8H_2O + 5Cl_2$$

or

$$16HCl + 2KMnO_4 \longrightarrow 2MnCl_2 + 8H_2O + 5Cl_2 + 2KCl$$

15.45 $Fe^{2+}(Cl^-)_2 + H_2O_2 + H^+Cl^- \longrightarrow Fe^{3+}(Cl^-)_3 + H_2O$

▮

$$Fe^{2+} + H_2O_2 \longrightarrow Fe^{3+} + H_2O$$

The half-reactions:

$$Fe^{2+} \longrightarrow Fe^{3+} + e^- \qquad 2H^+ + H_2O_2 \longrightarrow 2H_2O$$
$$2Fe^{2+} \longrightarrow 2Fe^{3+} + 2e^- \qquad 2H^+ + H_2O_2 + 2e^- \longrightarrow 2H_2O$$

Combining half-reactions:

$$2Fe^{2+} + 2H^+ + H_2O_2 \longrightarrow 2Fe^{3+} + 2H_2O$$

or

$$2FeCl_2 + 2HCl + H_2O_2 \longrightarrow 2FeCl_3 + 2H_2O$$

15.46 $As_2S_5 + H^+NO_3^-(conc) \longrightarrow H_3AsO_4 + H^+HSO_4^- + H_2O + NO_2$

▮

$$(5 \rightarrow 4) = -1$$
$$As_2S_5 + HNO_3 \longrightarrow H_3AsO_4 + 5H_2SO_4 + H_2O + NO_2$$
$$5(-2 \rightarrow +6) = 40$$

Balance the elements which change oxidation state:

$$As_2S_5 + 40HNO_3 \longrightarrow H_3AsO_4 + 5H_2SO_4 + H_2O + 40NO_2$$

Balance the rest:

$$As_2S_5 + 40HNO_3 \longrightarrow 2H_3AsO_4 + 5H_2SO_4 + 12H_2O + 40NO_2$$

15.47 $Cu + H^+NO_3^-(conc) \longrightarrow Cu^{2+}(NO_3^-)_2 + H_2O + NO_2$

▮

$$Cu \longrightarrow Cu^{2+} + 2e^- \qquad NO_3^- \longrightarrow NO_2$$
$$NO_3^- \longrightarrow NO_2 + H_2O$$
$$2H^+ + NO_3^- \longrightarrow NO_2 + H_2O$$
$$e^- + 2H^+ + NO_3^- \longrightarrow NO_2 + H_2O$$
$$2e^- + 4H^+ + 2NO_3^- \longrightarrow 2NO_2 + 2H_2O$$

Combining half-reactions: $Cu + 4H^+ + 2NO_3^- \longrightarrow 2NO_2 + 2H_2O + Cu^{2+}$

or $Cu + 4HNO_3 \longrightarrow 2NO_2 + 2H_2O + Cu(NO_3)_2$

15.48 $Cu + H^+NO_3^-(dil) \longrightarrow Cu^{2+}(NO_3^-)_2 + H_2O + NO$

▮

$$Cu \longrightarrow Cu^{2+} + 2e^- \qquad NO_3^- \longrightarrow NO$$
$$NO_3^- \longrightarrow NO + 2H_2O$$
$$4H^+ + NO_3^- \longrightarrow NO + 2H_2O$$
$$3e^- + 4H^+ + NO_3^- \longrightarrow NO + 2H_2O$$
$$3Cu \longrightarrow 3Cu^{2+} + 6e^- \qquad 6e^- + 8H^+ + 2NO_3^- \longrightarrow 2NO + 4H_2O$$

Combining half-reactions: $3Cu + 8H^+ + 2NO_3^- \longrightarrow 2NO + 4H_2O + 3Cu^{2+}$

or $3Cu + 8HNO_3 \longrightarrow 2NO + 4H_2O + 3Cu(NO_3)_2$

15.49 $Zn + H^+NO_3^-(dil) \longrightarrow Zn^{2+}(NO_3^-)_2 + H_2O + NH_4^+NO_3^-$

▮

$$\overset{(+5 \to -3) = -8}{\mathrm{Zn + HNO_3 \longrightarrow Zn(NO_3)_2 + H_2O + NH_4NO_3}}$$
$$(0 \to +2) = +2$$

Balance the atoms which change oxidation state:

$$4\mathrm{Zn} + \mathrm{HNO_3} \longrightarrow 4\mathrm{Zn(NO_3)_2} + \mathrm{H_2O} + \mathrm{NH_4NO_3}$$

Balance the rest: $$4\mathrm{Zn} + 10\mathrm{HNO_3} \longrightarrow 4\mathrm{Zn(NO_3)_2} + 3\mathrm{H_2O} + \mathrm{NH_4NO_3}$$

15.50 $(Na^+)_2C_2O_4^{2-} + K^+MnO_4^- + (H^+)_2SO_4^{2-} \longrightarrow (K^+)_2SO_4^{2-} + (Na^+)_2SO_4^{2-} + H_2O + Mn^{2+}SO_4^{2-} + CO_2$

▮

$$C_2O_4^{2-} \longrightarrow CO_2 \qquad MnO_4^- \longrightarrow Mn^{2+}$$
$$C_2O_4^{2-} \longrightarrow 2CO_2 + 2e^- \qquad MnO_4^- \longrightarrow Mn^{2+} + 4H_2O$$
$$8H^+ + MnO_4^- \longrightarrow Mn^{2+} + 4H_2O$$
$$5e^- + 8H^+ + MnO_4^- \longrightarrow Mn^{2+} + 4H_2O$$
$$5C_2O_4^{2-} \longrightarrow 10CO_2 + 10e^- \qquad 10e^- + 16H^+ + 2MnO_4^- \longrightarrow 2Mn^{2+} + 8H_2O$$
$$5C_2O_4^{2-} + 16H^+ + 2MnO_4^- \longrightarrow 10CO_2 + 2Mn^{2+} + 8H_2O$$

or $$5Na_2C_2O_4 + 2KMnO_4 + 8H_2SO_4 \longrightarrow 10CO_2 + 2MnSO_4 + 8H_2O + K_2SO_4 + 5Na_2SO_4$$

15.51 $CdS + I_2 + H^+Cl^- \longrightarrow Cd^{2+}(Cl^-)_2 + H^+I^- + S$

▮

$$2(0 \to -1) = -2$$
$$CdS + I_2 \longrightarrow Cd^{2+} + 2I^- + S$$
$$(-2 \to 0) = +2$$

$$CdS + I_2 \longrightarrow Cd^{2+} + S + 2I^- \quad \text{or} \quad CdS + I_2 \longrightarrow CdI_2 + S$$

15.52 $MnO + PbO_2 + H^+NO_3^- \longrightarrow H^+MnO_4^- + Pb^{2+}(NO_3^-)_2 + H_2O$

▮

$$(2 \to 7) = +5$$
$$MnO + PbO_2 + H^+ + NO_3^- \longrightarrow MnO_4^- + Pb^{2+} + H_2O$$
$$(4 \to 2) = -2$$

$$2MnO + 5PbO_2 \longrightarrow 2MnO_4^- + 5Pb^{2+}$$
$$2MnO + 5PbO_2 \longrightarrow 2MnO_4^- + 5Pb^{2+} + 4H_2O$$
$$2MnO + 5PbO_2 + 8H^+ \longrightarrow 2MnO_4^- + 5Pb^{2+} + 4H_2O$$

or $$2MnO + 5PbO_2 + 10HNO_3 \longrightarrow 2HMnO_4 + 5Pb(NO_3)_2 + 4H_2O$$

15.53 $Cr^{3+}(I^-)_3 + K^+OH^- + Cl_2 \longrightarrow (K^+)_2CrO_4^{2-} + K^+IO_4^- + K^+Cl^- + H_2O$
(Note that both the chromium and the iodide are oxidized in this reaction.)

▮

$$2(3 \to 6) + 6(-1 \to +7) = 54$$
$$2CrI_3 + KOH + Cl_2 \longrightarrow Cr_2O_7^{2-} + 6IO_4^- + H_2O + 2Cl^-$$
$$2(0 \to -1) = -2$$

$$2CrI_3 + 27Cl_2 \longrightarrow Cr_2O_7^{2-} + 6IO_4^- + 54Cl^-$$
$$2CrI_3 + 62OH^- + 27Cl_2 \longrightarrow Cr_2O_7^{2-} + 6IO_4^- + 31H_2O + 54Cl^-$$

or $$2CrI_3 + 62KOH + 27Cl_2 \longrightarrow K_2Cr_2O_7 + 6KIO_4 + 31H_2O + 54KCl$$

15.54 $(Na^+)_2HAsO_3^{2-} + K^+BrO_3^- + H^+Cl^- \longrightarrow Na^+Cl^- + K^+Br^- + H_3AsO_4$

▮

$$Na_2HAsO_3 + NaBrO_3 + HCl \longrightarrow NaCl + NaBr + H_3AsO_4$$

$$HAsO_3^{2-} \longrightarrow H_3AsO_4 \qquad BrO_3^- \longrightarrow Br^-$$
$$HAsO_3^{2-} + H_2O \longrightarrow H_3AsO_4 + 2e^- \qquad BrO_3^- + 6H^+ \longrightarrow Br^- + 3H_2O$$
$$3HAsO_3^{2-} + 3H_2O \longrightarrow 3H_3AsO_4 + 6e^- \qquad BrO_3^- + 6H^+ + 6e^- \longrightarrow Br^- + 3H_2O$$

Combining half-reactions:

$$3HAsO_3^{2-} + BrO_3^- + 6H^+ + 3H_2O \longrightarrow Br^- + 3H_3AsO_4 + 3H_2O$$
$$3HAsO_3^{2-} + BrO_3^- + 6H^+ \longrightarrow Br^- + 3H_3AsO_4$$

or $$3Na_2HAsO_3 + NaBrO_3 + 6HCl \longrightarrow NaBr + 3H_3AsO_4 + 6NaCl$$

15.55 $(Na^+)_2TeO_3^{2-} + Na^+I^- + H^+Cl^- \longrightarrow Na^+Cl^- + Te + H_2O + I_2$

▮

$$2(-1 \rightarrow 0) = +2$$
$$Na_2TeO_3 + 2\,NaI + HCl \longrightarrow NaCl + Te + H_2O + I_2$$
$$(+4 \rightarrow 0) = -4$$

$$Na_2TeO_3 + 4NaI \longrightarrow Te + 2I_2 \qquad Na_2TeO_3 + 4NaI \longrightarrow 6NaCl + Te + 2I_2$$
$$Na_2TeO_3 + 4NaI + 6HCl \longrightarrow 6NaCl + Te + 3H_2O + 2I_2$$

15.56 $U^{4+}(SO_4^{2-})_2 + K^+MnO_4^- + H_2O \longrightarrow (H^+)_2SO_4^{2-} + (K^+)_2SO_4^{2-} + Mn^{2+}SO_4^{2-} + UO_2^{2+}SO_4^{2-}$

▮

$$U^{4+} + MnO_4^- + H_2O \longrightarrow Mn^{2+} + UO_2^{2+}$$
$$U^{4+} + 2H_2O \longrightarrow UO_2^{2+}$$
$$U^{4+} + 2H_2O \longrightarrow UO_2^{2+} + 4H^+ + 2e^-$$

$$8H^+ + MnO_4^- \longrightarrow Mn^{2+} + 4H_2O$$
$$5e^- + 8H^+ + MnO_4^- \longrightarrow Mn^{2+} + 4H_2O$$
$$10e^- + 16H^+ + 2MnO_4^- \longrightarrow 2Mn^{2+} + 8H_2O$$

$$5U^{4+} + 10H_2O \longrightarrow 5UO_2^{2+} + 20H^+ + 10e^-$$

Combining half-reactions:

$$5U^{4+} + 16H^+ + 10H_2O + 2MnO_4^- \longrightarrow 2Mn^{2+} + 8H_2O + 5UO_2^{2+} + 20H^+$$
$$5U^{4+} + 2H_2O + 2MnO_4^- \longrightarrow 2Mn^{2+} + 5UO_2^{2+} + 4H^+$$

or $$5U(SO_4)_2 + 2H_2O + 2KMnO_4 \longrightarrow 2MnSO_4 + 5UO_2SO_4 + 2H_2SO_4 + K_2SO_4$$

15.57 $I_2 + (Na^+)_2S_2O_3^{2-} \longrightarrow (Na^+)_2S_4O_6^{2-} + Na^+I^-$

▮

$$I_2 + S_2O_3^{2-} \longrightarrow S_4O_6^{2-} + I^-$$
$$2e^- + I_2 \longrightarrow 2I^- \qquad 2S_2O_3^{2-} \longrightarrow S_4O_6^{2-} + 2e^-$$
$$I_2 + 2S_2O_3^{2-} \longrightarrow S_4O_6^{2-} + 2I^-$$

or $$I_2 + 2Na_2S_2O_3 \longrightarrow Na_2S_4O_6 + 2NaI$$

15.58 $Ca^{2+}(OCl^-)_2 + K^+I^- + H^+Cl^- \longrightarrow I_2 + Ca^{2+}(Cl^-)_2 + H_2O + K^+Cl^-$

▮

$$OCl^- + I^- + H^+ \longrightarrow I_2 + Cl^- + H_2O$$
$$OCl^- \longrightarrow Cl^-$$
$$2H^+ + OCl^- \longrightarrow Cl^- + H_2O$$
$$2e^- + 2H^+ + OCl^- \longrightarrow Cl^- + H_2O \qquad 2I^- \longrightarrow I_2 + 2e^-$$

Combining: $$2I^- + 2H^+ + OCl^- \longrightarrow Cl^- + H_2O + I_2$$

or $$4KI + 4HCl + Ca(OCl)_2 \longrightarrow CaCl_2 + 2H_2O + 2I_2 + 4KCl$$

15.59 $Bi_2O_3 + Na^+OH^- + Na^+OCl^- \longrightarrow Na^+BiO_3^- + Na^+Cl^- + H_2O$

▮

$$Bi_2O_3 + OH^- + OCl^- \longrightarrow BiO_3^- + Cl^- + H_2O$$

$$Bi_2O_3 \longrightarrow 2BiO_3^-$$
$$6OH^- + Bi_2O_3 \longrightarrow 2BiO_3^- + 3H_2O$$
$$6OH^- + Bi_2O_3 \longrightarrow 2BiO_3^- + 3H_2O + 4e^-$$

$$OCl^- \longrightarrow Cl^-$$
$$H_2O + OCl^- \longrightarrow Cl^- + 2OH^-$$
$$2e^- + H_2O + OCl^- \longrightarrow Cl^- + 2OH^-$$
$$4e^- + 2H_2O + 2OCl^- \longrightarrow 2Cl^- + 4OH^-$$

Combining half-reactions:

$$2H_2O + 2OCl^- + 6OH^- + Bi_2O_3 \longrightarrow 2BiO_3^- + 3H_2O + 2Cl^- + 4OH^-$$
$$2OCl^- + 2OH^- + Bi_2O_3 \longrightarrow 2BiO_3^- + H_2O + 2Cl^-$$

or
$$2NaOCl + 2NaOH + Bi_2O_3 \longrightarrow 2NaBiO_3 + H_2O + 2NaCl$$

15.60 $(K^+)_3Fe(CN)_6^{3-} + Cr_2O_3 + K^+OH^- \longrightarrow (K^+)_4Fe(CN)_6^{4-} + (K^+)_2CrO_4^{2-} + H_2O$

▌

$$2(+3 \rightarrow +6) = +6$$
$$Fe(CN)_6^{3-} + Cr_2O_3 + OH^- \longrightarrow Fe(CN)_6^{4-} + 2CrO_4^{2-} + H_2O$$
$$(+3 \rightarrow +2) = -1$$

$$6Fe(CN)_6^{3-} + Cr_2O_3 \longrightarrow 6Fe(CN)_6^{4-} + 2CrO_4^{2-}$$
$$10OH^- + 6Fe(CN)_6^{3-} + Cr_2O_3 \longrightarrow 6Fe(CN)_6^{4-} + 2CrO_4^{2-} + 5H_2O$$

or
$$10KOH + 6K_3Fe(CN)_6 + Cr_2O_3 \longrightarrow 6K_4Fe(CN)_6 + 2K_2CrO_4 + 5H_2O$$

15.61 $H^+NO_3^- + H^+I^- \longrightarrow NO + I_2 + H_2O$

▌

$$H^+ + NO_3^- + I^- \longrightarrow NO + I_2 + H_2O$$

$$NO_3^- \longrightarrow NO + 2H_2O$$
$$4H^+ + NO_3^- \longrightarrow NO + 2H_2O$$
$$3e^- + 4H^+ + NO_3^- \longrightarrow NO + 2H_2O \qquad 2I^- \longrightarrow I_2 + 2e^-$$
$$6e^- + 8H^+ + 2NO_3^- \longrightarrow 2NO + 4H_2O \qquad 6I^- \longrightarrow 3I_2 + 6e^-$$

Combining half-reactions: $6I^- + 8H^+ + 2NO_3^- \longrightarrow 2NO + 4H_2O + 3I_2$

or $6HI + 2HNO_3 \longrightarrow 2NO + 4H_2O + 3I_2$

15.62 $Mn^{2+}SO_4^{2-} + (NH_4^+)_2S_2O_8^{2-} + H_2O \longrightarrow MnO_2 + (H^+)_2SO_4^{2-} + (NH_4^+)_2SO_4^{2-}$

▌

$$Mn^{2+} + S_2O_8^{2-} + H_2O \longrightarrow MnO_2 + SO_4^{2-}$$

$$Mn^{2+} \longrightarrow MnO_2 \qquad 2e^- + S_2O_8^{2-} \longrightarrow 2SO_4^{2-}$$
$$2H_2O + Mn^{2+} \longrightarrow MnO_2 + 4H^+$$
$$2H_2O + Mn^{2+} \longrightarrow MnO_2 + 4H^+ + 2e^-$$

Combining half-reactions:

$$S_2O_8^{2-} + 2H_2O + Mn^{2+} \longrightarrow MnO_2 + 4H^+ + 2SO_4^{2-}$$

or
$$(NH_4)_2S_2O_8 + 2H_2O + MnSO_4 \longrightarrow MnO_2 + 2H_2SO_4 + (NH_4)_2SO_4$$

Balance the equations in Problems 15.63–15.65 by the oxidation-state method.

15.63 $NH_3 + O_2 \longrightarrow NO + H_2O$

▌

$$(-3 \rightarrow +2) = +5$$
$$NH_3 + O_2 \longrightarrow NO + 2\,H_2O$$
$$2(0 \rightarrow -2) = -4$$

$$4NH_3 + 5O_2 \longrightarrow 4NO + 6H_2O$$

15.64 $CuO + NH_3 \longrightarrow N_2 + H_2O + Cu$

▌

$$(+2 \rightarrow 0) = -2$$
$$CuO + 2NH_3 \longrightarrow N_2 + H_2O + Cu$$
$$2(-3 \rightarrow 0) = -6$$

$$3CuO + 2NH_3 \longrightarrow N_2 + 3Cu$$
$$3CuO + 2NH_3 \longrightarrow N_2 + 3H_2O + 3Cu$$

15.65 $(K^+)_2Cr_2O_7^{2-} + Sn^{2+}(Cl^-)_2 + H^+Cl^- \longrightarrow Cr^{3+}(Cl^-)_3 + SnCl_4 + K^+Cl^- + H_2O$

▮

$$(+2 \rightarrow +4) = +2$$

$$Cr_2O_7^{2-} + Sn^{2+} + H^+ \longrightarrow 2\,Cr^{3+} + SnCl_4 + H_2O$$

$$2(+6 \rightarrow +3) = -6$$

$$Cr_2O_7^{2-} + 3Sn^{2+} \longrightarrow 2Cr^{3+} + SnCl_4$$

$$14H^+ + Cr_2O_7^{2-} + 3Sn^{2+} + 12Cl^- \longrightarrow 2Cr^{3+} + 3SnCl_4 + 7H_2O$$

or

$$14HCl + K_2Cr_2O_7 + 3SnCl_2 \longrightarrow 2CrCl_3 + 3SnCl_4 + 7H_2O + 2KCl$$

Balance the oxidation-reduction equations of Problems 15.66–15.82.

15.66 $NaHSO_4 + Al + NaOH \longrightarrow Na_2S + Al_2O_3 + H_2O$

▮

$$2Al \longrightarrow Al_2O_3 + 6e^- \qquad 8e^- + HSO_4^- \longrightarrow S^{2-}$$

$$2Al + 6OH^- \longrightarrow Al_2O_3 + 3H_2O + 6e^- \qquad 8e^- + HSO_4^- + 3H_2O \longrightarrow S^{2-} + 7OH^-$$

$$8Al + 24OH^- \longrightarrow 4Al_2O_3 + 12H_2O + 24e^- \qquad 24e^- + 3HSO_4^- + 9H_2O \longrightarrow 3S^{2-} + 21OH^-$$

Combining:

$$8Al + 3OH^- + 3HSO_4^- \longrightarrow 4Al_2O_3 + 3H_2O + 3S^{2-}$$

or

$$8Al + 3NaOH + 3NaHSO_4 \longrightarrow 4Al_2O_3 + 3H_2O + 3Na_2S$$

15.67 $K_2Cr_2O_7 + HCl \longrightarrow CrCl_3 + Cl_2 + H_2O + KCl$

▮

$$2(+6 \rightarrow +3) = -6$$

$$K_2Cr_2O_7 + 2HCl \longrightarrow 2CrCl_3 + Cl_2 + H_2O + KCl$$

$$2(-1 \rightarrow 0) = +2$$

$$K_2Cr_2O_7 + 6HCl \longrightarrow 2CrCl_3 + 3Cl_2 + H_2O + KCl$$

$$K_2Cr_2O_7 + 14HCl \longrightarrow 2CrCl_3 + 3Cl_2 + 7H_2O + 2KCl$$

15.68 $Co^{2+}(Cl^-)_2 + Na_2O_2(s) + Na^+OH^- + H_2O \longrightarrow Co(OH)_3 + Na^+Cl^-$

▮

$$CoCl_2 + Na_2O_2 + NaOH + H_2O \longrightarrow Co(OH)_3 + NaCl$$

$$Co^{2+} \longrightarrow Co(OH)_3 \qquad Na_2O_2 \longrightarrow 2OH^-$$

$$3OH^- + Co^{2+} \longrightarrow Co(OH)_3 + e^- \qquad Na_2O_2 \longrightarrow 2OH^- + 2Na^+$$

$$6OH^- + 2Co^{2+} \longrightarrow 2Co(OH)_3 + 2e^- \qquad 2e^- + 2H_2O + Na_2O_2 \longrightarrow 2OH^- + 2Na^+$$

Combining half-reactions:

$$6OH^- + 2Co^{2+} + 2H_2O + Na_2O_2 \longrightarrow 2Co(OH)_3 + 4OH^- + 2Na^+$$

$$2OH^- + 2Co^{2+} + 2H_2O + Na_2O_2 \longrightarrow 2Co(OH)_3 + 2Na^+$$

or

$$2NaOH + 2CoCl_2 + 2H_2O + Na_2O_2(s) \longrightarrow 2Co(OH)_3 + 4NaCl$$

15.69 $Cu(NH_3)_4^{2+}(Cl^-)_2 + K^+CN^- + H_2O \longrightarrow NH_3 + NH_4^+Cl^- + (K^+)_2Cu(CN)_3^{2-} + K^+CNO^- + K^+Cl^-$

▮

$$Cu(NH_3)_4^{2+} + CN^- + H_2O \longrightarrow Cu(CN)_3^{2-} + CNO^-$$

$$e^- + 3CN^- + Cu(NH_3)_4^{2+} \longrightarrow Cu(CN)_3^{2-} + 4NH_3$$

$$CN^- \longrightarrow CNO^-$$

$$H_2O + 2NH_3 + CN^- \longrightarrow CNO^- + 2NH_4^+ + 2e^-$$

$$2e^- + 6CN^- + 2Cu(NH_3)_4^{2+} \longrightarrow 2Cu(CN)_3^{2-} + 8NH_3$$

Combining half-reactions:

$$H_2O + 7CN^- + 2Cu(NH_3)_4^{2+} \longrightarrow 2Cu(CN)_3^{2-} + 6NH_3 + CNO^- + 2NH_4^+$$

or

$$H_2O + 7KCN + 2Cu(NH_3)_4Cl_2 \longrightarrow 2K_2Cu(CN)_3 + 6NH_3 + KCNO + 2NH_4Cl + 2KCl$$

15.70 $Sb_2O_3 + K^+IO_3^- + H^+Cl^- + H_2O \longrightarrow HSb(OH)_6 + K^+Cl^- + ICl$

▮

$$(+5 \to +1) = -4$$

$$Sb_2O_3 + KIO_3 + HCl + H_2O \longrightarrow 2HSb(OH)_6 + KCl + ICl$$

$$2(+3 \to +5) = +4$$

$$Sb_2O_3 + KIO_3 \longrightarrow 2HSb(OH)_6 + ICl$$

$$Sb_2O_3 + KIO_3 + 2HCl + 6H_2O \longrightarrow 2HSb(OH)_6 + KCl + ICl$$

15.71 $Ag + K^+CN^- + O_2 + H_2O \longrightarrow K^+Ag(CN)_2^- + K^+OH^-$

▮

$$Ag + CN^- + O_2 + H_2O \longrightarrow Ag(CN)_2^- + OH^-$$

$$4e^- + 2H_2O + O_2 \longrightarrow 4OH^-$$

$$Ag + 2CN^- \longrightarrow Ag(CN)_2^- + e^-$$

$$4Ag + 8CN^- \longrightarrow 4Ag(CN)_2^- + 4e^-$$

Combining half-reactions:

$$4Ag + 8CN^- + 2H_2O + O_2 \longrightarrow 4Ag(CN)_2^- + 4OH^-$$

or

$$4Ag + 8KCN + 2H_2O + O_2 \longrightarrow 4KAg(CN)_2 + 4KOH$$

15.72 $WO_3 + Sn^{2+}(Cl^-)_2 + H^+Cl^- \longrightarrow W_3O_8 + (H^+)_2SnCl_6^{2-} + H_2O$

▮

$$WO_3 + Sn^{2+} + H^+ \longrightarrow W_3O_8 + SnCl_6^{2-} + H_2O$$

$$3WO_3 \longrightarrow W_3O_8 \qquad Sn^{2+} \longrightarrow SnCl_6^{2-}$$

$$2e^- + 2H^+ + 3WO_3 \longrightarrow W_3O_8 + H_2O \qquad 6Cl^- + Sn^{2+} \longrightarrow SnCl_6^{2-} + 2e^-$$

Combining half-reactions:

$$2H^+ + 3WO_3 + 6Cl^- + Sn^{2+} \longrightarrow SnCl_6^{2-} + W_3O_8 + H_2O$$

or

$$4HCl + 3WO_3 + SnCl_2 \longrightarrow H_2SnCl_6 + W_3O_8 + H_2O$$

15.73 $Co^{2+}(Cl^-)_2 + K^+NO_2^- + H^+Cl^- \longrightarrow (K^+)_3Co(NO_2)_6^{3-} + NO + K^+Cl^- + H_2O$

▮

$$Co^{2+} + NO_2^- \longrightarrow Co(NO_2)_6^{3-} + NO + H_2O$$

$$Co^{2+} \longrightarrow Co(NO_2)_6^{3-} \qquad NO_2^- \longrightarrow NO + H_2O$$

$$6NO_2^- + Co^{2+} \longrightarrow Co(NO_2)_6^{3-} + e^- \qquad e^- + 2H^+ + NO_2^- \longrightarrow NO + H_2O$$

Combining: $2H^+ + 7NO_2^- + Co^{2+} \longrightarrow Co(NO_2)_6^{3-} + NO + H_2O$

or $2HCl + 7KNO_2 + CoCl_2 \longrightarrow K_3Co(NO_2)_6 + NO + H_2O + 4KCl$

15.74 $V(OH)_4^+Cl^- + Fe^{2+}(Cl^-)_2 + H^+Cl^- \longrightarrow VO^{2+}(Cl^-)_2 + Fe^{3+}(Cl^-)_3 + H_2O$

▮

$$V(OH)_4^+ + Fe^{2+} + H^+ \longrightarrow VO^{2+} + Fe^{3+} + H_2O$$

$$V(OH)_4^+ \longrightarrow VO^{2+} \qquad Fe^{2+} \longrightarrow Fe^{3+} + e^-$$

$$2H^+ + e^- + V(OH)_4^+ \longrightarrow VO^{2+} + 3H_2O$$

Combining: $2H^+ + Fe^{2+} + V(OH)_4^+ \longrightarrow VO^{2+} + 3H_2O + Fe^{3+}$

or $2HCl + FeCl_2 + V(OH)_4Cl \longrightarrow VOCl_2 + 3H_2O + FeCl_3$

15.75 $KClO_3 + H_2SO_4 \longrightarrow KHSO_4 + O_2 + ClO_2 + H_2O$

▮

$$3(-2 \to 0) = +6$$

$$KClO_3 + H_2SO_4 \longrightarrow KHSO_4 + O_2 + ClO_2 + H_2O$$

$$(+5 \to +4) = -1$$

$$4KClO_3 \longrightarrow O_2 + 4ClO_2$$

$$4H_2SO_4 + 4KClO_3 \longrightarrow 4KHSO_4 + O_2 + 4ClO_2 + 2H_2O$$

15.76 $Sn + HNO_3 \longrightarrow SnO_2 + NO_2 + H_2O$

▮

$$(0 \rightarrow +4) = +4$$
$$Sn + HNO_3 \longrightarrow SnO_2 + NO_2 + H_2O$$
$$(+5 \rightarrow +4) = -1$$

$$Sn + 4HNO_3 \longrightarrow SnO_2 + 4NO_2 + 2H_2O$$

15.77 $I_2 + HNO_3 \longrightarrow HIO_3 + NO_2 + H_2O$

▮

$$2(0 \rightarrow 5) = +10$$
$$I_2 + HNO_3 \longrightarrow 2HIO_3 + NO_2 + H_2O$$
$$(+5 \rightarrow +4) = -1$$

$$I_2 + 10HNO_3 \longrightarrow 2HIO_3 + 10NO_2 \qquad I_2 + 10HNO_3 \longrightarrow 2HIO_3 + 10NO_2 + 4H_2O$$

15.78 $KI + H_2SO_4 \longrightarrow K_2SO_4 + I_2 + H_2S + H_2O$

▮

$$2(-1 \rightarrow 0) = +2$$
$$2KI + H_2SO_4 \longrightarrow K_2SO_4 + I_2 + H_2S + H_2O$$
$$(+6 \rightarrow -2) = -8$$

$$8KI + H_2SO_4 \longrightarrow 4I_2 + H_2S \qquad 8KI + 5H_2SO_4 \longrightarrow 4K_2SO_4 + 4I_2 + H_2S + 4H_2O$$

15.79 $KBr + H_2SO_4 \longrightarrow K_2SO_4 + Br_2 + SO_2 + H_2O$

▮

$$2(-1 \rightarrow 0) = +2$$
$$2KBr + H_2SO_4 \longrightarrow K_2SO_4 + Br_2 + SO_2 + H_2O$$
$$(+6 \rightarrow +4) = -2$$

$$2KBr + H_2SO_4 \longrightarrow Br_2 + SO_2 \qquad 2KBr + 2H_2SO_4 \longrightarrow K_2SO_4 + Br_2 + SO_2 + 2H_2O$$

15.80 $Cr_2O_3 + Na_2CO_3 + KNO_3 \longrightarrow Na_2CrO_4 + CO_2 + KNO_2$

▮

$$(+5 \rightarrow +3) = -2$$
$$Cr_2O_3 + Na_2CO_3 + KNO_3 \longrightarrow 2Na_2CrO_4 + CO_2 + KNO_2$$
$$2(+3 \rightarrow +6) = +6$$

$$Cr_2O_3 + 3KNO_3 \longrightarrow 2Na_2CrO_4 + 3KNO_2$$
$$Cr_2O_3 + 2Na_2CO_3 + 3KNO_3 \longrightarrow 2Na_2CrO_4 + 2CO_2 + 3KNO_2$$

15.81 $P_2H_4 \longrightarrow PH_3 + P_4H_2$

▮

$$2e^- + 2H^+ + P_2H_4 \longrightarrow 2PH_3 \qquad 2P_2H_4 \longrightarrow P_4H_2 + 6H^+ + 6e^-$$
$$6e^- + 6H^+ + 3P_2H_4 \longrightarrow 6PH_3$$

Combining and dividing by 4:

$$5P_2H_4 \longrightarrow 6PH_3 + P_4H_2$$

15.82 $Ca_3(PO_4)_2 + SiO_2 + C \longrightarrow CaSiO_3 + P_4 + CO$

▮

$$(0 \rightarrow +2) = +2$$
$$Ca_3(PO_4)_2 + SiO_2 + C \longrightarrow CaSiO_3 + P_4 + CO$$
$$4(+5 \rightarrow 0) = -20$$

$$2Ca_3(PO_4)_2 + 10C \longrightarrow P_4 + 10CO$$
$$2Ca_3(PO_4)_2 + 6SiO_2 + 10C \longrightarrow 6CaSiO_3 + P_4 + 10CO$$

Complete and balance the skeleton equations of Problems 15.83–15.91 by the ion-electron method.

15.83 $I^- + NO_2^- \longrightarrow I_2 + NO$ (acid solution)

▮

$$I^- + NO_2^- \longrightarrow I_2 + NO$$

$$2I^- \longrightarrow I_2 + 2e^- \qquad NO_2^- \longrightarrow NO$$

$$NO_2^- \longrightarrow NO + H_2O$$

$$e^- + 2H^+ + NO_2^- \longrightarrow NO + H_2O$$

$$2e^- + 4H^+ + 2NO_2^- \longrightarrow 2NO + 2H_2O$$

Combining: $$2I^- + 4H^+ + 2NO_2^- \longrightarrow 2NO + 2H_2O + I_2$$

or $$2HI + 2HNO_2 \longrightarrow 2NO + 2H_2O + I_2$$

15.84 $Au + CN^- + O_2 \longrightarrow Au(CN)_4^-$ (aqueous solution)

▮ [*Warning*: Keep CN^- out of acid solution!]

$$Au + 4CN^- \longrightarrow Au(CN)_4^- + 3e^- \qquad 4e^- + 2H_2O + O_2 \longrightarrow 4OH^-$$

$$4Au + 16CN^- \longrightarrow 4Au(CN)_4^- + 12e^- \qquad 12e^- + 6H_2O + 3O_2 \longrightarrow 12OH^-$$

Combining: $$4Au + 16CN^- + 6H_2O + 3O_2 \longrightarrow 4Au(CN)_4^- + 12OH^-$$

or $$4Au + 16KCN + 6H_2O + 3O_2 \longrightarrow 4KAu(CN)_4 + 12KOH$$

15.85 $MnO_4^- \longrightarrow MnO_4^{2-} + O_2$ (alkaline solution)

▮

$$e^- + MnO_4^- \longrightarrow MnO_4^{2-} \qquad 4OH^- \longrightarrow O_2 + 2H_2O + 4e^-$$

$$4e^- + 4MnO_4^- \longrightarrow 4MnO_4^{2-}$$

$$4OH^- + 4MnO_4^- \longrightarrow 4MnO_4^{2-} + O_2 + 2H_2O$$

or $$4KOH + 4KMnO_4 \longrightarrow 4K_2MnO_4 + O_2 + 2H_2O$$

15.86 $P \longrightarrow PH_3 + H_2PO_2^-$ (alkaline solution)

▮

$$P \longrightarrow PH_3 \qquad P \longrightarrow H_2PO_2^-$$

$$3e^- + 3H_2O + P \longrightarrow PH_3 + 3OH^- \qquad 2OH^- + P \longrightarrow H_2PO_2^- + e^-$$

$$6OH^- + 3P \longrightarrow 3H_2PO_2^- + 3e^-$$

Combining: $$3H_2O + 4P + 3OH^- \longrightarrow PH_3 + 3H_2PO_2^-$$

15.87 $Zn + As_2O_3 \longrightarrow AsH_3$ (acid solution)

▮

$$Zn \longrightarrow Zn^{2+} + 2e^- \qquad 12e^- + 12H^+ + As_2O_3 \longrightarrow 2AsH_3 + 3H_2O$$

$$6Zn \longrightarrow 6Zn^{2+} + 12e^-$$

$$6Zn + 12H^+ \longrightarrow As_2O_3 \longrightarrow 2AsH_3 + 3H_2O + 6Zn^{2+}$$

15.88 $Zn + ReO_4^- \longrightarrow Re^-$ (acid solution)

▮

$$Zn \longrightarrow Zn^{2+} + 2e^- \qquad 8e^- + 8H^+ + ReO_4^- \longrightarrow Re^- + 4H_2O$$

$$4Zn \longrightarrow 4Zn^{2+} + 8e^-$$

$$4Zn + 8H^+ + ReO_4^- \longrightarrow Re^- + 4H_2O + 4Zn^{2+}$$

15.89 $ClO_2 + O_2^{2-} \longrightarrow ClO_2^-$ (alkaline solution)

▮

$$e^- + ClO_2 \longrightarrow ClO_2^-$$

Since ClO_2 is reduced, the oxygen must be oxidized: $O_2^{2-} \longrightarrow O_2 + 2e^-$ $\qquad 2e^- + 2ClO_2 \longrightarrow 2ClO_2^-$

$$O_2^{2-} + 2ClO_2 \longrightarrow 2ClO_2^- + O_2$$

15.90 $Cl_2 + IO_3^- \longrightarrow IO_4^-$ (alkaline solution)

▮ Since iodine is oxidized, Cl_2 must be reduced.

$$2e^- + Cl_2 \longrightarrow 2Cl^- \qquad 2OH^- + IO_3^- \longrightarrow IO_4^- + H_2O + 2e^-$$

$$Cl_2 + 2OH^- + IO_3^- \longrightarrow IO_4^- + H_2O + 2Cl^-$$

15.91 $V \longrightarrow HV_6O_{17}^{3-} + H_2$ (alkaline solution)

▮ $6V \longrightarrow HV_6O_{17}^{3-}$

$33OH^- + 6V \longrightarrow HV_6O_{17}^{3-} + 16H_2O$ $\qquad 2e^- + 2H_2O \longrightarrow H_2 + 2OH^-$

$33OH^- + 6V \longrightarrow HV_6O_{17}^{3-} + 16H_2O + 30e^-$ $\qquad 30e^- + 30H_2O \longrightarrow 15H_2 + 30OH^-$

Combining: $14H_2O + 3OH^- + 6V \longrightarrow HV_6O_{17}^{3-} + 15H_2$

15.92 In Problem 15.41, what volume of O_2 at STP is produced for each g H_2O_2 consumed?

▮
$$(1.00 \text{ g } H_2O_2)\left(\frac{1 \text{ mol } H_2O_2}{34.0 \text{ g } H_2O_2}\right)\left(\frac{5 \text{ mol } O_2}{5 \text{ mol } H_2O_2}\right) = 0.0294 \text{ mol}$$
$$V = \frac{nRT}{P} = \frac{(0.0294 \text{ mol } O_2)(0.0821 \text{ L·atm/mol·K})(273 \text{ K})}{1.00 \text{ atm}} = 0.659 \text{ L}$$

15.93 How much $KMnO_4$ is needed to oxidize 100 g $Na_2C_2O_4$?

▮ See the equation for the reaction in Problem 15.50.
$$(100 \text{ g } Na_2C_2O_4)\left(\frac{1 \text{ mol } Na_2C_2O_4}{134 \text{ g } Na_2C_2O_4}\right)\left(\frac{2 \text{ mol } KMnO_4}{5 \text{ mol } Na_2C_2O_4}\right)\left(\frac{158 \text{ g } KMnO_4}{\text{mol } KMnO_4}\right) = 47 \text{ g}$$

15.94 When NaBr is treated with concentrated H_2SO_4, SO_2, HBr, and Br_2 are produced; when NaCl is treated with concentrated H_2SO_4, HCl is produced but no Cl_2 or SO_2 is produced. ***(a)*** Write balanced chemical equations for all these reactions. ***(b)*** On the basis of the facts given above, predict which of the following reactions will occur:

(i) $Br_2 + 2NaCl \longrightarrow Cl_2 + 2NaBr$ (ii) $Cl_2 + 2NaBr \longrightarrow Br_2 + 2NaCl$

▮ ***(a)*** $Br^- + H_2SO_4(conc) \longrightarrow HBr + HSO_4^-$

$2Br^- + 3H_2SO_4(conc) \longrightarrow Br_2 + SO_2 + 2H_2O + 2HSO_4^-$

$Cl^- + H_2SO_4(conc) \longrightarrow HCl + HSO_4^-$

(b) From the equations above, it is seen that Br^- is more easily oxidized than Cl^-. Hence Cl_2 will oxidize Br^-, but not vice versa. (ii) $2Br^- + Cl_2 \longrightarrow Br_2 + 2Cl^-$

15.95 Complete and balance the following:

(a) $Sn^{2+} + H^+ + NO_3^- \longrightarrow Sn^{4+} + NO$ ***(b)*** $Cu + Ag^+ \longrightarrow Ag + Cu^{2+}$

(c) $Cr_2O_7^{2-} + I^- \longrightarrow I_2 + Cr^{3+}$ ***(d)*** $Zn + H^+ + NO_3^- \longrightarrow Zn^{2+} + NH_4^+$

(e) $Br_2 + I^- \longrightarrow I_2 + Br^-$ ***(f)*** $ClO^- + I^- \longrightarrow I_2 + Cl^-$

(g) $S_2O_3^{2-} + Ag^+ \longrightarrow Ag + S_4O_6^{2-}$

▮ ***(a)*** $Sn^{2+} \longrightarrow Sn^{4+}$ $\qquad NO_3^- \longrightarrow NO$

$Sn^{2+} \longrightarrow Sn^{4+} + 2e^-$ $\qquad 3e^- + NO_3^- \longrightarrow NO$

$4H^+ + 3e^- + NO_3^- \longrightarrow NO$

$4H^+ + 3e^- + NO_3^- \longrightarrow NO + 2H_2O$

$3Sn^{2+} \longrightarrow 3Sn^{4+} + 6e^-$ $\qquad 8H^+ + 6e^- + 2NO_3^- \longrightarrow 2NO + 4H_2O$

$3Sn^{2+} + 8H^+ + 2NO_3^- \longrightarrow 3Sn^{4+} + 2NO + 4H_2O$

(b) $Cu + 2Ag^+ \longrightarrow 2Ag + Cu^{2+}$ The charge must be balanced.

(c) $I^- \longrightarrow I_2$ $\qquad Cr_2O_7^{2-} \longrightarrow Cr^{3+}$

$2I^- \longrightarrow I_2$ $\qquad Cr_2O_7^{2-} \longrightarrow 2Cr^{3+}$

$2I^- \longrightarrow I_2 + 2e^-$ $\qquad 6e^- + Cr_2O_7^{2-} \longrightarrow 2Cr^{3+}$

$14H^+ + 6e^- + Cr_2O_7^{2-} \longrightarrow 2Cr^{3+}$

$6I^- \longrightarrow 3I_2 + 6e^-$ $\qquad 14H^+ + 6e^- + Cr_2O_7^{2-} \longrightarrow 2Cr^{3+} + 7H_2O$

$6I^- + 14H^+ + Cr_2O_7^{2-} \longrightarrow 2Cr^{3+} + 3I_2 + 7H_2O$

(d) $4Zn + 10H^+ + NO_3^- \longrightarrow 4Zn^{2+} + NH_4^+ + 3H_2O$

(e) $Br_2 + 2I^- \longrightarrow I_2 + 2Br^-$ ***(f)*** $2H^+ + ClO^- + 2I^- \longrightarrow I_2 + Cl^- + H_2O$

(g) $2S_2O_3^{2-} + 2Ag^+ \longrightarrow 2Ag + S_4O_6^{2-}$

15.96 Complete and balance the following equation in basic solution: $Br_2 + IO_3^- \longrightarrow Br^- + IO_4^-$

$$2OH^- + IO_3^- \longrightarrow IO_4^- + H_2O + 2e^- \qquad 2e^- + Br_2 \longrightarrow 2Br^-$$
$$2OH^- + IO_3^- + Br_2 \longrightarrow 2Br^- + IO_4^- + H_2O$$

15.97 Complete and balance the following equations:

(*a*) $CrO + H_2O \longrightarrow Cr(OH)_2$ (*b*) $Cr_2O_3 + HCl \longrightarrow CrCl_3 + H_2O$
(*c*) $Cr_2O_3 + NaOH \longrightarrow NaCr(OH)_4$ (*d*) $CrO_3 + NaOH \longrightarrow Na_2CrO_4 + H_2O$
(*e*) Does the oxidation number of chromium in its oxides have any effect on the tendency of the oxide to act as an acid anhydride or a basic anhydride?

(*a*) $CrO + H_2O \longrightarrow Cr(OH)_2$ (*b*) $Cr_2O_3 + 6HCl \longrightarrow 2CrCl_3 + 3H_2O$
(*c*) $Cr_2O_3 + 2NaOH + 3H_2O \longrightarrow 2NaCr(OH)_4$ (*d*) $CrO_3 + 2NaOH \longrightarrow Na_2CrO_4 + H_2O$
(*e*) The higher the oxidation state, the more acidic the oxide. CrO is basic. Cr_2O_3 is amphoteric; it reacts with acid or base. CrO_3 reacts with base; it is acidic.

15.98 Complete and balance the following equations:

(*a*) $MnO_4^- + Sn^{2+} \longrightarrow Mn^{2+} + Sn^{4+}$ (*b*) $Mn^{2+} + Sn^{4+} \longrightarrow MnO_4^- + Sn^{2+}$
(*c*) Is it possible to predict from the equation alone whether a reaction will go as the equation is written? (*d*) Could a prediction in the above case be made if the following reactions are known to occur?

$$16H^+ + 2MnO_4^- + 10Cl^- \longrightarrow 5Cl_2 + 2Mn^{2+} + 8H_2O \qquad Sn + 2Cl_2 \longrightarrow SnCl_4$$

(*a*) $2MnO_4^- + 16H^+ + 5Sn^{2+} \longrightarrow 2Mn^{2+} + 8H_2O + 5Sn^{4+}$
(*b*) $2Mn^{2+} + 8H_2O + 5Sn^{4+} \longrightarrow 2MnO_4^- + 16H^+ + 5Sn^{2+}$
(*c*) Since the second equation is merely the reverse of the first, and both are balanced, the possibility of balancing an equation cannot imply that the reaction will go as written. (*d*) The addition equations show that MnO_4^- can oxidize Cl^- to Cl_2 and that Cl_2 can oxidize Sn to Sn^{IV}. Thus MnO_4^- is a better oxidizing agent than Sn^{4+}, and reaction (*a*) will proceed spontaneously.

15.99 Complete and balance the following equations:

(*a*) $S_2O_8^{2-} + I^- \longrightarrow I_3^- + SO_2$ (*b*) $[CrCl_6]^{3-} + Zn \longrightarrow [ZnCl_4]^{2-} + Cr^{2+}$
(*c*) $CH_3CHO + Cr_2O_7^{2-} \longrightarrow Cr^{3+} + CH_3CO_2H$ (*d*) $H_2SO_4(conc) + Br^- \longrightarrow Br_2 + SO_2$
(*e*) $Br_2 + OH^- \longrightarrow Br^- + BrO_3^-$ (*f*) $MnO_4^{2-} + H^+ \longrightarrow MnO_4^- + Mn^{2+}$
(*g*) $AuCl_4^- + Sn^{2+} \longrightarrow Sn^{4+} + AuCl$ (*h*) $Au(CN)_2^- + Zn \longrightarrow Zn(CN)_4^{2-} + Au$
(*i*) $NO_2 + H_2O \longrightarrow NO + NO_3^-$ (*j*) $CrO_4^{2-} + Cu_2O \longrightarrow Cu(OH)_2 + Cr(OH)_4^-$

(*a*) $8H^+ + S_2O_8^{2-} + 9I^- \longrightarrow 3I_3^- + 2SO_2 + 4H_2O$
(*b*)
$$CrCl_6^{3-} \longrightarrow Cr^{2+} \qquad Zn \longrightarrow ZnCl_4^{2-}$$
$$e^- + CrCl_6^{3-} \longrightarrow Cr^{2+} \qquad Zn \longrightarrow ZnCl_4^{2-} + 2e^-$$
$$e^- + CrCl_6^{3-} \longrightarrow Cr^{2+} + 6Cl^- \qquad 4Cl^- + Zn \longrightarrow ZnCl_4^{2-} + 2e^-$$
$$2e^- + 2CrCl_6^{3-} \longrightarrow 2Cr^{2+} + 12Cl^-$$
$$2CrCl_6^{3-} + Zn \longrightarrow ZnCl_4^{2-} + 8Cl^- + 2Cr^{2+}$$
(*c*) $3CH_3CHO + Cr_2O_7^{2-} + 8H^+ \longrightarrow 3CH_3CO_2H + 4H_2O + 2Cr^{3+}$
(Regard the CH_3 group as R, which does not change in oxidation state.)
(*d*) $3H_2SO_4(conc) + 2Br^- \longrightarrow Br_2 + SO_2 + 2H_2O + 2HSO_4^-$
(*e*)
$$Br_2 \longrightarrow Br^- \qquad Br_2 \longrightarrow BrO_3^-$$
$$Br_2 \longrightarrow 2Br^- \qquad Br_2 \longrightarrow 2BrO_3^-$$
$$2e^- + Br_2 \longrightarrow 2Br^- \qquad Br_2 \longrightarrow 2BrO_3^- + 10e^-$$
$$10e^- + 5Br_2 \longrightarrow 10Br^- \qquad 12OH^- + Br_2 \longrightarrow 2BrO_3^- + 10e^-$$
$$12OH^- + Br_2 \longrightarrow 2BrO_3^- + 10e^- + 6H_2O$$
$$12OH^- + 6Br_2 \longrightarrow 10Br^- + 2BrO_3^- + 6H_2O$$
$$6OH^- + 3Br_2 \longrightarrow 5Br^- + BrO_3^- + 3H_2O$$

(*f*) $MnO_4^{2-} \longrightarrow MnO_4^-$

$MnO_4^{2-} \longrightarrow MnO_4^- + e^-$ $MnO_4^{2-} \longrightarrow Mn^{2+}$

$4MnO_4^{2-} \longrightarrow 4MnO_4^- + 4e^-$ $4e^- + MnO_4^{2-} \longrightarrow Mn^{2+}$

$8H^+ + 4e^- + MnO_4^{2-} \longrightarrow Mn^{2+} + 4H_2O$

$8H^+ + 5MnO_4^{2-} \longrightarrow 4MnO_4^- + Mn^{2+} + 4H_2O$

(*g*) $AuCl_4^- + Sn^{2+} \longrightarrow Sn^{4+} + AuCl + 3Cl^-$

(*h*) $2Au(CN)_2^- + Zn \longrightarrow Zn(CN)_4^{2-} + 2Au$

(*i*) $3NO_2 + H_2O \longrightarrow NO + 2NO_3^- + 2H^+$

(*j*) $2CrO_4^{2-} + 3Cu_2O + 11H_2O \longrightarrow 6Cu(OH)_2 + 2Cr(OH)_4^- + 2OH^-$

15.100 Write balanced equations for the following reactions. In each case assume that the second reagent is in large excess.

(*a*) $H_3PO_4 + NaOH \longrightarrow$ (*b*) $NaOH + H_3PO_4 \longrightarrow$

(*c*) $H_3PO_4 + Ba(OH)_2 \longrightarrow$ (*d*) $CrO_4^{2-} + H^+ \longrightarrow Cr_2O_7^{2-}$

▮ (*a*) $H_3PO_4 + 3NaOH \longrightarrow Na_3PO_4 + 3H_2O$ (*b*) $NaOH + H_3PO_4 \longrightarrow NaH_2PO_4 + H_2O$

(*c*) $2H_3PO_4 + 3Ba(OH)_2 \longrightarrow Ba_3(PO_4)_2 + 6H_2O$ (*d*) $2CrO_4^{2-} + 2H^+ \longrightarrow Cr_2O_7^{2-} + H_2O$

15.101 Complete and balance the following:

(*a*) $Br^- + BrO_3^- + H^+ \longrightarrow$ (*b*) $Cu + Cu^{2+} + OH^- \longrightarrow$

(*c*) $P_4 + OH^- \longrightarrow PH_3 + PHO_3^{2-}$

▮ Reactions (*a*) and (*b*) are each the opposite of a disproportionation reaction, in acidic and basic solution, respectively. [Compare Problems 15.99(*e*) and 15.23(*e*).]

(*a*) $5Br^- + BrO_3^- + 6H^+ \longrightarrow 3Br_2 + 3H_2O$ (*b*) $Cu + Cu^{2+} + 2OH^- \longrightarrow Cu_2O + H_2O$

(*c*) $P_4 + 4OH^- + 2H_2O \longrightarrow 2PH_3 + 2PHO_3^{2-}$

15.102 Complete and balance the following equation, in basic solution,

$$Hg_2(CN)_2 + Ce^{4+} \longrightarrow CO_3^{2-} + NO_3^- + Hg(OH)_2 + Ce^{3+}$$

(*a*) by considering the carbon in $Hg_2(CN)_2$ to be in the −4 oxidation state and the nitrogen to be in the +3 oxidation state (*b*) by considering C as +4 and N as +5 (*c*) by considering Hg to be +2 and carbon −4. (*d*) Explain why the same result is obtained regardless of the choice of oxidation states.

▮ $Hg_2(CN)_2 \longrightarrow CO_3^{2-} + NO_3^- + Hg(OH)_2$ $Hg_2(CN)_2 \longrightarrow 2CO_3^{2-} + 2NO_3^- + 2Hg(OH)_2$

(*a*) If C has an oxidation state of −4 and N is +3, then Hg is +1. In the products, the oxidation numbers are +4, +5, and +2, respectively. The total increase in oxidation number is 22. (*b*) If C is +4 and N is +5, the Hg must be −9. Then the only element oxidized is Hg, and the total increase in oxidation number is again 22. (The value for the oxidation number of Hg, −9, is impossible of course, but it still gives the same change in oxidation number.) (*c*) If Hg is +2 and C is −4, then N must be +2. The oxidation still involves a +22 change in oxidation number. (*d*) No matter which set of oxidation numbers is used, as long as they total zero—the charge on the $Hg_2(CN)_2$—the same number of electrons must be added to the half-reaction.

$$Hg_2(CN)_2 \longrightarrow 2CO_3^{2-} + 2NO_3^- + 2Hg(OH)_2 + 22e^-$$

$$28OH^- + Hg_2(CN)_2 \longrightarrow 2CO_3^{2-} + 2NO_3^- + 2Hg(OH)_2 + 22e^-$$

$$28OH^- + Hg_2(CN)_2 \longrightarrow 2CO_3^{2-} + 2NO_3^- + 2Hg(OH)_2 + 22e^- + 12H_2O$$

$$22Ce^{4+} + 28OH^- + Hg_2(CN)_2 \longrightarrow 2CO_3^{2-} + 2NO_3^- + 2Hg(OH)_2 + 12H_2O + 22Ce^{3+}$$

15.103 Complete and balance the following, and give the oxidation state of the indicated element in the *missing* product:

(*a*) Oxygen in: $H_2O_2 + I_2 \longrightarrow I^- +$ (*b*) Oxygen in: $H_2O_2 + Sn^{2+} \longrightarrow Sn^{4+} +$

(*c*) Manganese in: $MnO_4^{2-} + H^+ \longrightarrow Mn^{2+} +$ (*d*) Nitrogen in: $NO_2 + H_2O \longrightarrow NO +$

▮ (*a*) Since the iodine is reduced, the H_2O_2 must be oxidized—to O_2, in which the oxygen has an oxidation state of zero.

$$H_2O_2 + I_2 \longrightarrow 2I^- + O_2 + 2H^+$$

(*b*) The Sn^{2+} is oxidized; the H_2O_2 must be reduced—to H_2O, in which the oxygen has an oxidation state of -2.

$$2H^+ + H_2O_2 + Sn^{2+} \longrightarrow Sn^{4+} + 2H_2O$$

(*c*) The MnO_4^{2-} has been reduced to Mn^{2+}; it must also be oxidized—to Mn^{VII}—since H^+ is already in its maximum oxidation state.

$$5MnO_4^{2-} + 8H^+ \longrightarrow Mn^{2+} + 4MnO_4^- + 4H_2O$$

(*d*) NO_2 disproportionates to NO and NO_3^-, in which nitrogen has an oxidation state of $+5$.

$$3NO_2 + H_2O \longrightarrow NO + 2NO_3^- + 2H^+$$

15.104 Complete and balance the following:

(*a*) $MnO_4^- + Cr^{3+} \longrightarrow Cr_2O_7^{2-} + Mn^{2+}$ (*b*) $CrO_4^{2-} + Fe(OH)_2 \longrightarrow Fe(OH)_3 + Cr(OH)_4^-$

(*c*) $AuCl_4^- + Zn \longrightarrow Au + Zn^{2+} +$ (*d*) $Zn + OH^- \longrightarrow Zn(OH)_4^{2-} +$

▮ (*a*) $6MnO_4^- + 10Cr^{3+} + 11H_2O \longrightarrow 5Cr_2O_7^{2-} + 6Mn^{2+} + 22H^+$

(*b*)

$$CrO_4^{2-} \longrightarrow Cr(OH)_4^- \qquad Fe(OH)_2 \longrightarrow Fe(OH)_3$$
$$3e^- + CrO_4^{2-} \longrightarrow Cr(OH)_4^- \qquad Fe(OH)_2 \longrightarrow Fe(OH)_3 + e^-$$
$$3e^- + CrO_4^{2-} \longrightarrow Cr(OH)_4^- + 4OH^- \qquad OH^- + Fe(OH)_2 \longrightarrow Fe(OH)_3 + e^-$$
$$3e^- + 4H_2O + CrO_4^{2-} \longrightarrow Cr(OH)_4^- + 4OH^- \qquad 3OH^- + 3Fe(OH)_2 \longrightarrow 3Fe(OH)_3 + 3e^-$$
$$4H_2O + CrO_4^{2-} + 3Fe(OH)_2 \longrightarrow Cr(OH)_4^- + OH^- + 3Fe(OH)_3$$

(*c*) $2AuCl_4^- + 3Zn \longrightarrow 2Au + 3Zn^{2+} + 8Cl^-$

(*d*) $Zn + 2OH^- + 2H_2O \longrightarrow Zn(OH)_4^{2-} + H_2$

15.105 Write an equation describing the oxidation of ammonia by oxygen to NO in basic solution. In the commercial Ostwald process, for the production of nitric acid, this reaction is carried out directly in the gaseous state. Explain why the same equation describes the direct reaction and the reaction in basic solution.

▮

$$O_2 \longrightarrow 2OH^- \qquad NH_3 \longrightarrow NO$$
$$4e^- + O_2 \longrightarrow 2OH^- \qquad NH_3 \longrightarrow NO + 5e^-$$
$$4e^- + O_2 \longrightarrow 2OH^- + 2OH^- \qquad 5OH^- + NH_3 \longrightarrow NO + 5e^-$$
$$4e^- + 2H_2O + O_2 \longrightarrow 4OH^- \qquad 5OH^- + NH_3 \longrightarrow NO + 5e^- + 4H_2O$$
$$4NH_3 + 5O_2 \longrightarrow 4NO + 6H_2O$$

As long as the same products are produced from a given set of reactants, the equation will be the same no matter what the medium.

15.106 Complete and balance the following equations:

(*a*) $Fe(CN)_6^{4-} + H^+ + MnO_4^- \longrightarrow Fe^{3+} + CO_2 + NO_3^- + Mn^{2+}$

(*b*) $Cu_3P + Cr_2O_7^{2-} \longrightarrow Cu^{2+} + H_3PO_4 + Cr^{3+}$ (*c*) $H_2SO_4 + I^- \longrightarrow I_2 + H_2S$

▮ (*a*) $Fe(CN)_6^{4-} \longrightarrow Fe^{3+} + 6CO_2 + 6NO_3^-$

The total change in oxidation number is 61 (see Problem 15.102).

$$Fe(CN)_6^{4-} \longrightarrow Fe^{3+} + 6CO_2 + 6NO_3^- + 61e^-$$
$$Fe(CN)_6^{4-} \longrightarrow Fe^{3+} + 6CO_2 + 6NO_3^- + 61e^- + 60H^+$$
$$30H_2O + Fe(CN)_6^{4-} \longrightarrow Fe^{3+} + 6CO_2 + 6NO_3^- + 61e^- + 60H^+$$
$$MnO_4^- + 5e^- + 8H^+ \longrightarrow Mn^{2+} + 4H_2O$$
$$5Fe(CN)_6^{4-} + 188H^+ + 61MnO_4^- \longrightarrow 5Fe^{3+} + 30CO_2 + 30NO_3^- + 61Mn^{2+} + 94H_2O$$

(*b*)

$$Cu_3P \longrightarrow 3Cu^{2+} + H_3PO_4$$
$$Cu_3P \longrightarrow 3Cu^{2+} + H_3PO_4 + 11e^-$$
$$4H_2O + Cu_3P \longrightarrow 3Cu^{2+} + H_3PO_4 + 11e^- + 5H^+$$
$$6e^- + 14H^+ + Cr_2O_7^{2-} \longrightarrow 2Cr^{3+} + 7H_2O$$
$$6Cu_3P + 124H^+ + 11Cr_2O_7^{2-} \longrightarrow 18Cu^{2+} + 6H_3PO_4 + 22Cr^{3+} + 53H_2O$$

(*c*) $H_2SO_4 + 2I^- + 2H^+ \longrightarrow I_2 + SO_2 + 2H_2O$

15.107 Write a balanced chemical equation for the oxidation of phosphorus(III) sulfide by nitric acid. The products include NO and SO_2.

▮

$$3P_4S_6 + 44H^+ + 44NO_3^- \longrightarrow 44NO + 12H_3PO_4 + 18SO_2 + 4H_2O$$

15.108 Cyanide ion is oxidized by powerful oxidizing agents to NO_3^- and CO_2 or CO_3^{2-}, depending on the acidity of the reaction mixture. Nitric acid, a powerful oxidizing agent, is reduced by moderate reducing agents to NO plus other products. Write a complete and balanced equation for the reaction of nitric acid with potassium cyanide. If this reaction were actually carried out, what safety precautions would be necessary?

▮

$$15H_2O + 3CN^- \longrightarrow 3CO_2 + 3NO_3^- + 30e^- + 30H^+$$
$$30e^- + 40H^+ + 10NO_3^- \longrightarrow 10NO + 20H_2O$$
$$3CN^- + 10H^+ + 7NO_3^- \longrightarrow 5H_2O + 3CO_2 + 10NO$$

Acid (H^+) and cyanide ion react to yield HCN, a deadly gas, unless the oxidizing properties of the acid cause the oxidation of the cyanide before the HCN can escape.

15.109 If all reactants and products are known, the equation for any reaction may be balanced by the following procedure.

(1) Using alphabetic or other symbols, write arbitrary coefficients in front of each formula in the equation.

(2) On the basis that the number of atoms of each element must be the same on both sides of the equation, derive the relative magnitudes of the coefficients.

(3) Assign one coefficient a numerical value and find the values of the other coefficients as indicated by their relative magnitudes. For example, the alphabetic coefficients in the unbalanced equation

$$aKMnO_4 + bKCl + cH_2SO_4 \longrightarrow dCl_2 + eMnSO_4 + fH_2O + gK_2SO_4$$

must be related as follow: $a = e$, $2b = d$, $c = f = e + g = 4a$, $a + b = 2g$. If a is assigned a value of 10, then $e = 10$, $c = f = 40$, $g = 30$, $b = 50$, $d = 25$. Dividing each of these values by 5 to obtain the smallest integral ratios yields $a{:}b{:}c{:}d{:}e{:}f{:}g = 2{:}10{:}8{:}5{:}2{:}8{:}6$. Hence the balanced equation is

$$2KMnO_4 + 10KCl + 8H_2SO_4 \longrightarrow 5Cl_2 + 2MnSO_4 + 8H_2O + 6K_2SO_4$$

Use this method to balance the following equations:

(a) $KIO_3 + KI + H_2SO_4 \longrightarrow KI_3 + K_2SO_4 + H_2O$
(b) $Pb(N_3)_2 + Co(MnO_4)_3 \longrightarrow CoO + MnO_2 + Pb_3O_4 + NO$
(c) $KOH + K_4Fe(CN)_6 + Ce(NO_3)_4 \longrightarrow Fe(OH)_3 + Ce(OH)_3 + K_2CO_3 + KNO_3 + H_2O$

▮ *(a)* $aKIO_3 + bKI + cH_2SO_4 \longrightarrow dKI_3 + eK_2SO_4 + fH_2O$

$a + b = d + 2e$ (balancing the K atoms)
$a + b = 3d$ (balancing the I atoms)
$c = e = f$ (balancing the H and the S atoms)
$3a + 4c = 4e + f$ (balancing the O atoms)

Combination of the last two equations yields $3a = f$. Let $a = 1$. Then $c = e = f = 3$. Subtraction of the first two equations yields $d = e = 3$. Hence $b = 3(3) - 1 = 8$. The complete equation:

$$KIO_3 + 8KI + 3H_2SO_4 \longrightarrow 3KI_3 + 3K_2SO_4 + 3H_2O$$

(b) $30Pb(N_3)_2 + 44Co(MnO_4)_3 \longrightarrow 132MnO_2 + 44CoO + 180NO + 10Pb_3O_4$
(c) $258KOH + K_4Fe(CN)_6 + 61Ce(NO)_4 \longrightarrow 61Ce(OH)_3 + Fe(OH)_3 + 36H_2O + 6K_2CO_3 + 250KNO_3$

15.110 Complete and balance the following equation:

$$Bi(OH)_3 + SnO_2^{2-} \longrightarrow Bi + SnO_3^{2-} \quad \text{(basic solution)}$$

▮

$$3e^- + Bi(OH)_3 \longrightarrow Bi + 3OH^- \qquad 2OH^- + SnO_2^{2-} \longrightarrow SnO_3^{2-} + H_2O + 2e^-$$
$$2Bi(OH)_3 + 3SnO_2^{2-} \longrightarrow 3SnO_3^{2-} + 3H_2O + 2Bi$$

15.111 Balance: $MnO_2 + H_2O_2 \longrightarrow MnO_4^- + H_2O$

▮

$$4OH^- + MnO_2 \longrightarrow MnO_4^- + 2H_2O + 3e^- \qquad 2e^- + H_2O_2 \longrightarrow 2OH^-$$
$$8OH^- + 2MnO_2 \longrightarrow 2MnO_4^- + 4H_2O + 6e^- \qquad 6e^- + 3H_2O_2 \longrightarrow 6OH^-$$
$$2OH^- + 2MnO_2 + 3H_2O_2 \longrightarrow 2MnO_4^- + 4H_2O$$

15.112 Complete and balance the following equations:

(*a*) $Fe(CN)_6^{3-} + Cr_2O_7^{2-} \longrightarrow Fe^{3+} + Cr^{3+} + CO_2 + NO_3^-$

(*b*) $Cr(OH)_4^- + MnO_4^- \longrightarrow MnO_2 + CrO_4^{2-}$

▮ (*a*)

$$Fe(CN)_6^{3-} + 30H_2O \longrightarrow Fe^{3+} + 6CO_2 + 6NO_3^- + 60e^- + 60H^+$$
$$140H^+ + 60e^- + 10Cr_2O_7^{2-} \longrightarrow 20Cr^{3+} + 70H_2O$$
$$\overline{Fe(CN)_6^{3-} + 80H^+ + 10Cr_2O_7^{2-} \longrightarrow Fe^{3+} + 6CO_2 + 6NO_3^- + 20Cr^{3+} + 40H_2O}$$

(*b*)

$$4OH^- + Cr(OH)_4^- \longrightarrow CrO_4^{2-} + 3e^- + 4H_2O$$
$$2H_2O + 3e^- + MnO_4^- \longrightarrow MnO_2 + 4OH^-$$
$$\overline{MnO_4^- + Cr(OH)_4^- \longrightarrow CrO_4^{2-} + 2H_2O + MnO_2}$$

15.113 Complete and balance the following equations:

(*a*) $MnO_4^- + Cl^- \longrightarrow Cl_2 + Mn^{2+}$ (*b*) $Mn(OH)_2 + MnO_4^- \longrightarrow MnO_4^{2-}$

▮ (*a*)

$$5e^- + MnO_4^- + 8H^+ \longrightarrow Mn^{2+} + 4H_2O$$
$$2Cl^- \longrightarrow Cl_2 + 2e^-$$
$$\overline{2MnO_4^- + 16H^+ + 10Cl^- \longrightarrow 2Mn^{2+} + 8H_2O + 5Cl_2}$$

(*b*)

$$Mn(OH)_2 + 6OH^- \longrightarrow MnO_4^{2-} + 4e^- + 4H_2O$$
$$4e^- + 4MnO_4^- \longrightarrow 4MnO_4^{2-}$$
$$\overline{4MnO_4^- + Mn(OH)_2 + 6OH^- \longrightarrow 5MnO_4^{2-} + 4H_2O}$$

15.114 H_2SO_4 acts as an oxidizing agent, a dehydrating agent, and an acid. Select equations from the following which illustrate each type of behavior:

(*a*) $C_6H_{12}O_6 \xrightarrow{H_2SO_4(conc)} 6C + 6H_2O$

(*b*) $5H_2SO_4(conc) + 4Zn \longrightarrow H_2S + 4Zn^{2+} + 4SO_4^{2-} + 4H_2O$

(*c*) $H_2SO_4(dil) + Zn \longrightarrow Zn^{2+} + H_2 + SO_4^{2-}$

(*d*) $H_2SO_4(dil) + ZnCO_3 \longrightarrow Zn^{2+} + CO_2 + SO_4^{2-} + H_2O$

▮ (*a*) Dehydrating agent (it removes H_2O from $C_6H_{12}O_6$) (*b*) Oxidizing agent (it is reduced to H_2S) (*c*) Acid and oxidizing agent (all strong acids can liberate H_2 with an active metal like zinc, unless another part of the molecule is more easily reduced) (*d*) Acid.

15.3 CALCULATIONS INVOLVING REDOX EQUATIONS

15.115 (*a*) Balance the following oxidation-reduction equation in acidic solution:

$$MnO_4^- + C_2O_4^{2-} \longrightarrow Mn^{2+} + CO_2$$

(*b*) Find the volume of a 0.200 M solution of MnO_4^- which will react with 50.0 mL of 0.100 M solution of $C_2O_4^{2-}$.

▮ (*a*)

$$MnO_4^- + C_2O_4^{2-} \longrightarrow Mn^{2+} + CO_2$$
$$8H^+ + 5e^- + MnO_4^- \longrightarrow Mn^{2+} + 4H_2O$$
$$C_2O_4^{2-} \longrightarrow 2CO_2 + 2e^-$$
$$\overline{16H^+ + 2MnO_4^- + 5C_2O_4^{2-} \longrightarrow 2Mn^{2+} + 8H_2O + 10CO_2}$$

(*b*) $$(50.0\text{ mL})\left(\frac{0.100\text{ mmol } C_2O_4^{2-}}{\text{mL}}\right)\left(\frac{2\text{ mmol } MnO_4^-}{5\text{ mmol } C_2O_4^{2-}}\right) = 2.00\text{ mmol } MnO_4^- \qquad \frac{2.00\text{ mmol}}{0.200\text{ M}} = 10.0\text{ mL}$$

15.116 What mass of N_2H_4 can be oxidized to N_2 by 24.0 g K_2CrO_4, which is reduced to $Cr(OH)_4^-$?

▮ The balanced equation is $4H_2O + 3N_2H_4 + 4CrO_4^{2-} \rightarrow 3N_2 + 4Cr(OH)_4^- + 4OH^-$

$$(24.0\text{ g } K_2CrO_4)\left(\frac{1\text{ mol } K_2CrO_4}{194.2\text{ g } K_2CrO_4}\right)\left(\frac{3\text{ mol } N_2H_4}{4\text{ mol } CrO_4^{2-}}\right) = (0.0927\text{ mol } N_2H_4)\left(\frac{32.0\text{ g}}{\text{mol}}\right) = 2.97\text{ g}$$

15.117 How many mol $FeCl_3$ can be prepared by the reaction of 10.0 g $KMnO_4$, 1.07 mol $FeCl_2$, and 500 mL of 3.00 M HCl? $MnCl_2$ is the reduction product.

▮

$$MnO_4^- + 5Fe^{2+} + 8H^+ \longrightarrow Mn^{2+} + 5Fe^{3+} + 4H_2O$$

$$(10.0 \text{ g } KMnO_4)\left(\frac{1 \text{ mol}}{158.0 \text{ g}}\right) = 0.0633 \text{ mol } MnO_4^-$$

$$(0.0633 \text{ mol } MnO_4^-)\left(\frac{5 \text{ mol } Fe^{2+}}{\text{mol } MnO_4^-}\right) = 0.316 \text{ mol } Fe^{2+} \text{ reacts}$$

$$(0.0633 \text{ mol } MnO_4^-)\left(\frac{8 \text{ mol } H^+}{\text{mol } MnO_4^-}\right) = 0.506 \text{ mol } H^+ \text{ reacts}$$

The 1.07 mol Fe^{2+} and $(0.500 \text{ L})\left(\frac{3.00 \text{ mol}}{\text{L}}\right) = 1.50 \text{ mol } H^+$ are present in excess. MnO_4^- is in limiting quantity; 0.316 mol of Fe^{3+} is produced.

15.118 H_2O_2 is reduced rapidly by Sn^{2+}, the products being Sn^{4+} and water. H_2O_2 decomposes slowly at room temperature to yield O_2 and water. Calculate the volume of O_2 produced at 20 °C and 1.00 atm when 200 g of 10.0% by mass H_2O_2 in water is treated with 100.0 mL of 2.00 M Sn^{2+} and then the mixture is allowed to stand until no further reaction occurs.

▮

$$2H^+ + H_2O_2 + Sn^{2+} \longrightarrow Sn^{4+} + 2H_2O$$

$$(100.0 \text{ mL})(2.00 \text{ M}) = (200 \text{ mmol } Sn^{2+})\left(\frac{1 \text{ mmol } H_2O_2}{\text{mmol } Sn^{2+}}\right) = 200 \text{ mmol } H_2O_2$$

$$(200 \text{ g sample})\left(\frac{10.0 \text{ g } H_2O_2}{100 \text{ g sample}}\right)\left(\frac{1 \text{ mol } H_2O_2}{34.0 \text{ g } H_2O_2}\right) = 0.588 \text{ mol } H_2O_2$$

$$H_2O_2 \text{ initially present} - H_2O_2 \text{ reduced by } Sn^{2+} = H_2O_2 \text{ remaining}$$

$$(588 \text{ mmol}) - (200 \text{ mmol}) = 388 \text{ mmol}$$

$$2H_2O_2 \longrightarrow 2H_2O + O_2 \qquad (388 \text{ mmol } H_2O_2)\left(\frac{1 \text{ mmol } O_2}{2 \text{ mmol } H_2O_2}\right) = 194 \text{ mmol } O_2$$

$$V = \frac{nRT}{P} = \frac{(0.194 \text{ mol})(0.0821 \text{ L}\cdot\text{atm/mol}\cdot\text{K})(293 \text{ K})}{1.00 \text{ atm}} = 4.67 \text{ L}$$

15.119 It requires 40.05 mL of 1.000 M Ce^{4+} to titrate 20.00 mL of 1.000 M Sn^{2+} to Sn^{4+}. What is the oxidation state of the cerium in the reduction product?

▮

$$(40.05 \text{ mL})\left(\frac{1.00 \text{ mmol}}{\text{mL}}\right) = 40.0 \text{ mmol } Ce^{4+} \qquad (20.00 \text{ mL})\left(\frac{1.00 \text{ mmol}}{\text{mL}}\right) = 20.0 \text{ mmol } Sn^{2+}$$

$$(20.00 \text{ mmol } Sn^{2+})\left(\frac{2 \text{ mmol } e^-}{\text{mmol } Sn^{2+}}\right) = 40.0 \text{ mmol } e^-$$

The numbers of moles of Ce^{4+} and e^- are equal; hence 1 mol of electrons is involved in the reduction of each mole of Ce^{4+}: $Ce^{4+} + e^- \rightarrow Ce^{n+}$ The change in oxidation number of cerium must be equal to the number of moles of electrons; Ce^{3+} is the product.

15.120 A volume of 12.53 mL of 0.05093 M selenium dioxide, SeO_2, reacted with exactly 25.52 mL of 0.1000 M $CrSO_4$. In the reaction, Cr^{2+} was oxidized to Cr^{3+}. To what oxidation state was the selenium converted by the reaction?

▮

$$SeO_2\text{:} \quad (12.53 \text{ mL})\left(\frac{0.05093 \text{ mmol}}{\text{mL}}\right) = 0.6382 \text{ mmol } SeO_2$$

$$CrSO_4\text{:} \quad (25.52 \text{ mL})\left(\frac{0.1000 \text{ mmol}}{\text{mL}}\right) = 2.552 \text{ mmol } Cr^{2+}$$

$$Cr^{2+} \longrightarrow Cr^{3+} + e^- \qquad (2.552 \text{ mmol } Cr^{2+})\left(\frac{1 \text{ mmol } e^-}{\text{mmol } Cr^{2+}}\right) = 2.552 \text{ mmol } e^-$$

$$\frac{2.552 \text{ mmol } e^-}{0.6382 \text{ mmol } SeO_2} = 3.999 \text{ mmol } e^-/\text{mmol } SeO_2$$

The reduction of 1 mol SeO_2 requires 4 mol electrons; hence Se^{IV} is reduced to Se^0.

15.121 Determine which reagent is in excess and by how much if 100.0 g P_4O_6 is treated with 100.0 g $KMnO_4$ in HCl solution to form H_3PO_4 and $MnCl_2$.

▮

$$(100.0\text{ g KMnO}_4)\left(\frac{1\text{ mol KMnO}_4}{158.0\text{ g KMnO}_4}\right) = 0.6329\text{ mol KMnO}_4$$

$$(100.0\text{ g P}_4\text{O}_6)\left(\frac{1\text{ mol P}_4\text{O}_6}{219.9\text{ g P}_4\text{O}_6}\right) = 0.4548\text{ mol P}_4\text{O}_6$$

$$24H^+ + 18H_2O + 8MnO_4^- + 5P_4O_6 \longrightarrow 20H_3PO_4 + 8Mn^{2+}$$

$$(0.6329\text{ mol MnO}_4^-)\left(\frac{5\text{ mol P}_4\text{O}_6}{8\text{ mol MnO}_4^-}\right) = 0.3956\text{ mol P}_4\text{O}_6\text{ required}$$

Hence P_4O_6 is in excess by $0.4548 - 0.3956 = 0.0592$ mol.

$$(0.0592\text{ mol})\left(\frac{219.9\text{ g}}{\text{mol}}\right) = 13.0\text{ g}$$

15.122 If 10.00 g V_2O_5 is dissolved in acid and reduced to V^{2+} by treatment with zinc metal, how many mol I_2 could be reduced by the resulting V^{2+} solution, as it is reoxidized to V^{IV}?

▮

$$(10.00\text{ g V}_2\text{O}_5)\left(\frac{1\text{ mol V}_2\text{O}_5}{181.9\text{ g V}_2\text{O}_5}\right) = 0.05498\text{ mol V}_2\text{O}_5$$

$$6e^- + 10H^+ + V_2O_5 \longrightarrow 2V^{2+} + 5H_2O$$

$$(0.05498\text{ mol V}_2\text{O}_5)\left(\frac{2\text{ mol V}^{2+}}{\text{mol V}_2\text{O}_5}\right) = 0.1100\text{ mol V}^{2+}$$

$$H_2O + V^{2+} + I_2 \longrightarrow 2I^- + VO^{2+} + 2H^+$$

$$(0.1100\text{ mol V}^{2+})\left(\frac{1\text{ mol I}_2}{\text{mol V}^{2+}}\right) = 0.1100\text{ mol I}_2$$

15.123 What mass of $K_2Cr_2O_7$ is required to produce from excess oxalic acid, $H_2C_2O_4$, 5.00 L CO_2 at 75 °C and 1.07 atm pressure? The reduction product of $Cr_2O_7^{2-}$ is Cr^{3+}.

▮

$$8H^+ + Cr_2O_7^{2-} + 3H_2C_2O_4 \longrightarrow 6CO_2 + 2Cr^{3+} + 7H_2O$$

$$n = \frac{PV}{RT} = \frac{(1.07\text{ atm})(5.00\text{ L})}{(0.0821\text{ L·atm/mol·K})(348\text{ K})} = 0.187\text{ mol CO}_2$$

$$(0.187\text{ mol CO}_2)\left(\frac{1\text{ mol Cr}_2\text{O}_7^{2-}}{6\text{ mol CO}_2}\right) = 0.0312\text{ mol Cr}_2\text{O}_7^{2-}$$

$$(0.0312\text{ mol K}_2\text{Cr}_2\text{O}_7)\left(\frac{294.2\text{ g K}_2\text{Cr}_2\text{O}_7}{\text{mol K}_2\text{Cr}_2\text{O}_7}\right) = 9.18\text{ g}$$

15.124 A 10.00 g mixture of Cu_2S and CuS was treated with 200.0 mL of 0.7500 M MnO_4^- in acid solution, producing SO_2, Cu^{2+}, and Mn^{2+}. The SO_2 was boiled off, and the excess MnO_4^- was titrated with 175.0 mL of 1.000 M Fe^{2+} solution. Write balanced chemical equations for all the reactions. Calculate the percent CuS in the original mixture.

▮

$$(200.0\text{ mL})(0.7500\text{ mmol MnO}_4^-/\text{mL}) = 150.0\text{ mmol MnO}_4^-\text{ present}$$

$$8H^+ + MnO_4^- + 5Fe^{2+} \longrightarrow 5Fe^{3+} + Mn^{2+} + 4H_2O$$

$$(175.0\text{ mL})(1.000\text{ M Fe}^{2+})\left(\frac{1\text{ mmol MnO}_4^-}{5\text{ mmol Fe}^{2+}}\right) = 35.0\text{ mmol MnO}_4^-\text{ excess}$$

Hence, 115.0 mmol of MnO_4^- reacted.

$$44H^+ + 8MnO_4^- + 5Cu_2S \longrightarrow 10Cu^{2+} + 5SO_2 + 8Mn^{2+} + 22H_2O$$

$$28H^+ + 6MnO_4^- + 5CuS \longrightarrow 5Cu^{2+} + 5SO_2 + 6Mn^{2+} + 14H_2O$$

Let x = number of g CuS, then $10.0 - x$ = number of g Cu_2S. Then there are $(x/95.606)$ mol CuS and $(10.0 - x)/159.15$ mol Cu_2S.

$$\left(\frac{x}{95.606}\text{ mol CuS}\right)\left(\frac{6\text{ mol MnO}_4^-}{5\text{ mol CuS}}\right) = 0.01255x\text{ mol MnO}_4^-\text{ reacted with CuS}$$

$$\left(\frac{10.0 - x}{159.15}\text{ mol Cu}_2\text{S}\right)\left(\frac{8\text{ mol MnO}_4^-}{5\text{ mol Cu}_2\text{S}}\right) = 0.1005 - 0.01005x\text{ mol MnO}_4^-\text{ reacted with Cu}_2\text{S}$$

$$(0.1005 - 0.01005x) + 0.01255x = 0.115$$

$$x = 5.79\text{ g CuS} \qquad 10.0 - x = 4.21\text{ g Cu}_2\text{S} \qquad 57.9\%\text{ CuS}$$

15.125 Consider the following balanced equations:

$$H_2O_2 + Sn^{4+} \longrightarrow Sn^{2+} + O_2 + 2H^+$$
$$3H_2O_2 + Sn^{4+} \longrightarrow Sn^{2+} + 2O_2 + 2H_2O + 2H^+$$

Does the fact that these two equations are complete and balanced mean that the reaction between H_2O_2 and Sn^{4+} can lead to different sets of products depending on the mole ratio of H_2O_2 to Sn^{4+}, or is some other explanation possible? If necessary, write another equation to support your answer.

▌ The second equation is a combination of the first equation and $2H_2O_2 \rightarrow O_2 + 2H_2O$. Two reactions should not be written in one equation.

15.126 A 6.000 g sample contained Fe_3O_4, Fe_2O_3, and inert materials. It was treated with an excess of aqueous KI in acid, which reduced all the iron to Fe^{2+}. The resulting solution was diluted to 50.00 mL, and a 10.00 mL sample of it was taken. The liberated iodine in the small sample was titrated with 5.500 mL of 1.00 M $Na_2S_2O_3$ solution, yielding $S_4O_6^{2-}$. The iodine from another 25.00 mL sample was extracted, after which the Fe^{2+} was titrated with 3.20 mL of 1.000 M MnO_4^- in H_2SO_4 solution. Calculate the percentages of Fe_3O_4 and of Fe_2O_3 in the original mixture.

▌

$$Fe_3O_4 + 8H^+ + 2I^- \longrightarrow 3Fe^{2+} + I_2 + 4H_2O$$
$$Fe_2O_3 + 6H^+ + 2I^- \longrightarrow 2Fe^{2+} + I_2 + 3H_2O$$

In a 10.00 mL sample, the iodine oxidized

$$(5.500\text{ mL})(1.000\text{ M } Na_2S_2O_3) = 5.500\text{ mmol } Na_2S_2O_3$$

There was thus

$$(5.500\text{ mmol } S_2O_3^{2-})\left(\frac{1\text{ mmol } I_2}{2\text{ mmol } S_2O_3^{2-}}\right) = 2.750\text{ mmol } I_2$$

In the 50.00 mL sample, there must have been 13.75 mmol I_2.

In a 25.00 mL sample, the total iron in the sample was reduced.

$$(3.20\text{ mL})(1.000\text{ M } MnO_4^-) = 3.20\text{ mmol } MnO_4^-$$
$$MnO_4^- + 8H^+ + 5Fe^{2+} \longrightarrow 5Fe^{3+} + Mn^{2+} + 4H_2O$$
$$(3.20\text{ mmol } MnO_4^-)\left(\frac{5\text{ mmol } Fe^{2+}}{1\text{ mmol } MnO_4^-}\right) = 16.0\text{ mmol } Fe^{2+}$$

Since there is 16.0 mmol of Fe^{2+} in the 25.00 mL portion, there must have been 32.0 mmol in the entire 50.00 mL sample.

Let x = number of mmol Fe_3O_4 and y = number of mmol Fe_2O_3. Then

$$3x + 2y = 32.0\text{ mmol}$$
$$x + y = 13.75\text{ mmol} \qquad \text{(Since } Fe_2O_3 \text{ and } Fe_3O_4 \text{ each produce 1 mol of } I_2 \text{ per mole.)}$$
$$x = 4.50\text{ mmol } Fe_3O_4 = 1.04\text{ g } Fe_3O_4, \text{ or } 17.4\%$$
$$y = 9.25\text{ mmol } Fe_2O_3 = 1.48\text{ g } Fe_2O_3, \text{ or } 24.6\%$$

15.127 In Problem 15.36, how much Cl_2 could be produced by the reaction of 100 g of $KMnO_4$?

▌ This mass problem is solved like all other mass problems if the complete balanced equation is used; it shows that 2 mol $KMnO_4$ can produce 5 mol Cl_2. The problem can also be solved directly from the balanced ionic equation showing that 2 MnO_4^- ions yield 5 molecules of Cl_2. We need only the chemical equivalence of $2MnO_4^-$ with $2KMnO_4$. In either case, the solution is

$$x\text{ g } Cl_2 = (100\text{ g } KMnO_4)\left(\frac{1\text{ mol } KMnO_4}{158\text{ g } KMnO_4}\right)\left(\frac{5\text{ mol } Cl_2}{2\text{ mol } KMnO_4}\right)\left(\frac{70.9\text{ g } Cl_2}{1\text{ mol } Cl_2}\right) = 112\text{ g } Cl_2$$

15.128 It requires 40.00 mL of 0.5000 M Ce^{4+} to titrate 10.00 mL of 1.000 M Sn^{2+} to Sn^{4+}. What is the oxidation state of cerium in the reduction product?

▌ $(40.00\text{ mL})(0.5000\text{ mmol/mL}) = 20.00\text{ mmol } Ce^{4+}$ $\qquad (10.00\text{ mL})(1.000\text{ mmol/mL}) = 10.00\text{ mmol } Sn^{2+}$

$$Sn^{2+} \longrightarrow Sn^{4+} + 2e^-$$

$$(10.00\text{ mmol } Sn^{2+})\left(\frac{2\text{ mmol } e^-}{\text{mmol } Sn^{2+}}\right) = 20.00\text{ mmol } e^- \qquad \frac{20.00\text{ mmol } e^-}{20.00\text{ mmol } Ce^{4+}} = \frac{1\text{ mol } e^-}{\text{mol } Ce^{4+}}$$

$$Ce^{4+} + e^- \longrightarrow Ce^{3+}$$

15.129 Exactly 40 mL of an acidified solution of 0.4000 M iron(II) ion is titrated with potassium permanganate solution. After addition of 32.00 mL $KMnO_4$, one additional drop turns the iron solution purple. Calculate the concentration of the permanganate solution.

▮ The reaction which occurs is $8H^+ + 5Fe^{2+} + MnO_4^- \longrightarrow Mn^{2+} + 5Fe^{3+} + 4H_2O$

The number of mmol iron(II) ion present is (40.00 mL)(0.4000 mmol/mL) = 16.00 mmol Fe^{2+}. Hence the number of mmol MnO_4^- is

$$(16.00 \text{ mmol Fe}^{2+})\left(\frac{1 \text{ mmol MnO}_4^-}{5 \text{ mmol Fe}^{2+}}\right) = 3.200 \text{ mmol MnO}_4^-$$

The concentration is then $\dfrac{3.200 \text{ mmol}}{32.00 \text{ mL}} = 0.1000 \text{ M MnO}_4^-$

CHAPTER 16

Other Concentration Units

16.1 NORMALITY IN ACID-BASE REACTIONS

16.1 What is the difference between the definition of an equivalent in an acid-base reaction and an equivalent in an oxidation-reduction reaction?

▌ An equivalent in an acid-base reaction is that amount of a substance which reacts with or liberates 1 mol of hydrogen ions; an equivalent in a redox reaction is that amount of substance which reacts with or liberates 1 mol of electrons.

16.2 Where reagent 1 and reagent 2 are involved in a chemical reaction, explain why it is correct to calculate $V_1N_1 = V_2N_2$ but it may not be correct to calculate $V_1M_1 = V_2M_2$.

▌ The first equation reduces to $\text{eq}_1 = \text{eq}_2$, which is true for any two reagents in a chemical reaction. The second reduces to $n_1 = n_2$, which is true only if the coefficients of the reagents are the same in the balanced equation. (The second equation is true for a dilution problem or for any problem in which we are sure that the numbers of moles are the same.)

16.3 What volume of a 0.232 N solution contains (*a*) 3.17 meq of solute (*b*) 6.5 eq of solute?

▌ (*a*) $(3.17\text{ meq})\left(\dfrac{1\text{ mL}}{0.232\text{ meq}}\right) = 13.7\text{ mL}$ (*b*) $(6.5\text{ eq})\left(\dfrac{1\text{ L}}{0.232\text{ eq}}\right) = 28.0\text{ L}$

16.4 Calculate the normality of each of the following solutions: (*a*) 7.88 g of HNO_3 per L solution (*b*) 26.5 g of Na_2CO_3 per L solution (if acidified to form CO_2).

▌ (*a*) Equivalent weight of HNO_3 = formula weight = 63.02 g/eq

$$N = \text{normality} = \frac{(7.88\text{ g})/(63.02\text{ g/eq})}{\text{L}} = 0.1251\text{ eq/L} = 0.1251\text{ N}$$

(*b*) Equivalent weight of $Na_2CO_3 = (\tfrac{1}{2})(\text{formula weight}) = (\tfrac{1}{2})(106.0) = 53.0$

$$N = \frac{(26.5\text{ g})/(53.0\text{ g/eq})}{1.00\text{ L}} = 0.500\text{ N}$$

16.5 How many (*a*) equivalents and (*b*) meq of solute are present in 60 mL of 4.0 N solution?

▌ (*a*) number of equivalents = (number of liters) × (normality) = (0.060 L)(4.0 eq/L) = 0.24 eq

(*b*) (0.24 eq)(1000 meq/eq) = 240 meq

Another Method

$$\text{Number of meq} = (\text{number of mL}) \times (\text{normality}) = (60\text{ mL})(4.0\text{ meq/mL}) = 240\text{ meq}$$

16.6 How many equivalents of solute are contained in (*a*) 1 L of 2 N solution, (*b*) 1 L of 0.5 N solution, (*c*) 0.5 L of 0.2 N solution?

▌ A 1 N solution contains 1 eq of solute in 1 L of solution. (*a*) 1 L of 2 N solution contains 2 eq of solute. (*b*) 1 L of 0.5 N solution contains 0.5 eq of solute. (*c*) 0.5 L of 0.2 N solution contains (0.5 L)(0.2 eq/L) = 0.1 eq of solute.

16.7 What is the normality of 0.300 M H_3PO_3 when it undergoes the following reaction?

$$H_3PO_3 + 2OH^- \longrightarrow HPO_3^{2-} + 2H_2O$$

▌ The equation shows that there is 2 eq per mol.

$$\left(\frac{0.300\text{ mol}}{\text{L}}\right)\left(\frac{2\text{ eq}}{\text{mol}}\right) = 0.600\text{ N}$$

16.8 Calculate the approximate volume of water that must be added to 250 mL of 1.25 N solution to make it 0.500 N (neglecting volume changes).

▌ $$V_2 = \frac{N_1 V_1}{N_2} = \frac{(1.25\ \text{N})(250\ \text{mL})}{0.500\ \text{N}} = 625\ \text{mL}$$

The volume increase is $V_2 - V_1 = 625\ \text{mL} - 250\ \text{mL} = 375\ \text{mL}$ (When liquids are mixed, the final volume is not necessarily the sum of the volumes of the original liquids. For dilute aqueous solutions being mixed with each other or with water, however, the volumes are very nearly additive.)

16.9 What volumes of 12.0 N and 3.00 N HCl must be mixed to give 1.00 L of 6.00 N HCl?

▌ $$(1.00\ \text{L})\left(\frac{6.00\ \text{eq}}{\text{L}}\right) = 6.00\ \text{eq}$$

Let x = number of L of concentrated solution. Then $1.00 - x$ = number of L of dilute solution.

$$(x)\left(\frac{12.0\ \text{eq}}{\text{L}}\right) + (1.00 - x)\left(\frac{3.00\ \text{eq}}{\text{L}}\right) = 6.00\ \text{eq}$$

$$x = 0.333\ \text{L of conc. solution} \qquad 1 - x = 0.667\ \text{L of dil. solution}$$

16.10 What volumes of N/2 and of N/10 HCl must be mixed to give 2.0 L of N/5 HCl?

▌ Let x = volume of N/2 required; then $2.0\ \text{L} - x$ = volume of N/10 required.

$$\text{Number of eq of N/5} = (\text{number of eq of N/2}) + (\text{number of eq of N/10})$$

$$(2.0\ \text{L})(\tfrac{1}{5}\ \text{N}) = (x)(\tfrac{1}{2}\ \text{N}) + (2.0\ \text{L} - x)(\tfrac{1}{10}\ \text{N})$$

Solving, $x = 0.50$ L. Thus 0.50 L of N/2 and 1.5 L of N/10 are required.

16.11 Determine the normality of an H_3PO_4 solution, 40.0 mL of which neutralized 120 mL of 0.531 N NaOH.

▌ The solutions react exactly with each other. Therefore

$$(\text{Volume } H_3PO_4) \times (\text{normality } H_3PO_4) = (\text{volume NaOH}) \times (\text{normality NaOH})$$

$$(40.0\ \text{mL})(\text{normality } H_3PO_4) = (120\ \text{mL})(0.531\ \text{N}) \qquad \text{Normality } H_3PO_4 = \frac{(120\ \text{mL})(0.531\ \text{N})}{40.0\ \text{mL}} = 1.59\ \text{N}$$

Note: In this problem we do not have to know whether 1, 2, or 3 hydrogens of H_3PO_4 were replaced. The normality was determined by reaction of the acid with a base of known concentration. Therefore the acid will have the same normality, 1.59 N, in reactions with any strong base under similar conditions. In order to know the molar concentration of the acid, however, it would be necessary to know the number of hydrogens replaced in the reaction.

In a case like this, where a substance can have several equivalent weights, the normality determined by one type of reaction is not necessarily the normality in other types of reaction.

16.12 How many mL of 6.0 N NaOH is required to neutralize 30 mL of 4.0 N HCl?

▌ $$(\text{Volume HCl}) \times (\text{normality HCl}) = (\text{volume NaOH}) \times (\text{normality NaOH})$$

$$(30\ \text{mL})(4.0\ \text{N}) = (\text{volume NaOH})(6.0\ \text{N}) \qquad \text{Volume NaOH} = \frac{(30\ \text{mL})(4.0\ \text{N})}{6.0\ \text{N}} = 20\ \text{mL}$$

16.13 ***(a)*** What volume of 5.00 N H_2SO_4 is required to neutralize a solution containing 2.50 g NaOH? ***(b)*** How many g pure H_2SO_4 are required?

▌ ***(a)*** One equivalent of H_2SO_4 reacts completely with 1 eq of NaOH. The equivalent weight of NaOH is the formula weight, 40.0 g/mol.

$$\text{Number of equivalents in 2.50 g NaOH} = \frac{2.50\ \text{g}}{40.0\ \text{g/eq}} = 0.0625\ \text{eq NaOH}$$

Therefore, 0.0625 eq H_2SO_4 is required.

$$(0.0625\ \text{eq})\left(\frac{1\ \text{L}}{5.00\ \text{eq}}\right) = 0.0125\ \text{L} = 12.5\ \text{mL of 5.00 N solution}$$

(b) Equivalent weight of $H_2SO_4 = (\tfrac{1}{2})(\text{formula weight}) = \tfrac{1}{2}(98.08) = 49.04$ g/eq

Mass of H_2SO_4 required $= (0.0625\ \text{eq})(49.04\ \text{g/eq}) = 3.07$ g

16.14 A 0.250 g sample of a solid acid was dissolved in water and exactly neutralized by 40.0 mL of 0.125 N base. What is the equivalent weight of the acid?

$$\text{Number of meq base} = (40.0\text{ mL})(0.125\text{ meq/mL}) = 5.00\text{ meq}$$

$$\text{Number of meq acid} = \text{number of meq base} = 5.00\text{ meq}$$

$$\text{Equivalent weight of acid} = \frac{250\text{ mg}}{5.00\text{ meq}} = 50.0\text{ mg/meq} = 50.0\text{ g/eq}$$

16.15 A 48.4 mL sample of HCl solution requires 1.240 g of pure $CaCO_3$ for complete neutralization. Calculate the normality of the acid.

Each CO_3^{2-} ion requires two H^+ for neutralization: $CO_3^{2-} + 2H^+ \rightarrow CO_2 + H_2O$. Thus, the equivalent weight of $CaCO_3$ is one-half the formula weight, or 50.05 g/eq.

$$\text{Number of equivalents in 1.240 g } CaCO_3 = \frac{1.240\text{ g}}{50.05\text{ g/eq}} = 0.0248\text{ eq } CaCO_3$$

Hence, 48.4 mL of acid solution contains 0.0248 eq HCl, and 1.000 L contains

$$\frac{1000}{48.4}(0.0248) = 0.512\text{ eq HCl} \qquad \text{The acid is 0.512 N.}$$

16.16 Exactly 50.0 mL of Na_2CO_3 solution is equivalent to 56.3 mL of 0.102 N HCl in an acid-base neutralization. How many g $CaCO_3$ would be precipitated if an excess of $CaCl_2$ solution were added to 100 mL of this Na_2CO_3 solution?

In 56.3 mL of the acid there is (56.3 mL)(0.102 meq/mL) = 5.74 meq HCl. Hence 50.0 mL Na_2CO_3 solution contains 5.74 meq Na_2CO_3, and 100 mL contains 2(5.74) = 11.48 meq Na_2CO_3. Since each CO_3^{2-} neutralizes two H^+, each meq of Na_2CO_3 has $\frac{1}{2}$ mmol of CO_3^{2-}.

$$\text{Number of mmol } CO_3^{2-} = \tfrac{1}{2}(11.48) = 5.74\text{ mmol}$$

The number of mmol $CaCO_3$ in the precipitate equals the number of mmol CO_3^{2-} available. Since the formula weight of $CaCO_3$ is 100.1 g/mol,

$$\text{Mass of 5.74 mmol } CaCO_3 = (5.74\text{ mmol})(100.1\text{ mg/mmol}) = 575\text{ mg} = 0.575\text{ g}$$

16.17 How many cm^3 of concentrated sulfuric acid, of density 1.84 g/cm^3 and containing 98.0% H_2SO_4 by weight, should be taken to make (*a*) 1.00 L of normal solution (*b*) 1.00 L of 3.00 N solution (*c*) 200 cm^3 of 0.500 N solution? Assume complete neutralization.

$$\text{Equivalent weight of } H_2SO_4 = (\tfrac{1}{2})(\text{formula weight}) = \tfrac{1}{2}(98.1) = 49.0\text{ g/eq}$$

The H_2SO_4 content of 1.00 L of the concentrated acid is (0.980)(1000 cm^3)(1.84 g/cm^3) = 1800 g H_2SO_4. The normality of the concentrated acid is

$$\frac{1800\text{ g } H_2SO_4/\text{L}}{49.0\text{ g } H_2SO_4/\text{eq}} = 36.7\text{ eq/L}$$

The dilution formula, $V_{conc} \times N_{conc} = V_{dil} \times N_{dil}$, can now be applied to each case.

(*a*) $$V_{conc} = \frac{(1.00\text{ L})(1.00\text{ N})}{36.7\text{ N}} = 0.0272\text{ L} = 27.2\text{ cm}^3\text{ conc. acid}$$

(*b*) $$V_{conc} = \frac{(1.00\text{ L})(3.00\text{ N})}{36.7\text{ N}} = 0.0817\text{ L} = 81.7\text{ cm}^3\text{ conc. acid}$$

(*c*) $$V_{conc} = \frac{(200\text{ cm}^3)(0.500\text{ N})}{36.7\text{ N}} = 2.72\text{ cm}^3\text{ conc. acid}$$

16.18 Compute the volume of concentrated H_2SO_4 (density 1.835 g/mL, 93.2% H_2SO_4 by weight) required to make up 500 mL of 3.00 N acid for complete neutralization.

For $H_2SO_4 + 2OH^- \longrightarrow 2H_2O + SO_4^{2-}$, H_2SO_4 has 2 eq per mol.

$$(0.500\text{ L})\left(\frac{3.00\text{ eq}}{\text{L}}\right)\left(\frac{49.0\text{ g}}{\text{eq}}\right)\left(\frac{100\text{ g solution}}{93.2\text{ g } H_2SO_4}\right)\left(\frac{1\text{ mL}}{1.835\text{ g solution}}\right) = 43.0\text{ mL}$$

16.19 Compute the volume of conc. HCl (density 1.19 g/cm³, 38% HCl by weight) required to make up 18 L of N/50 acid.

▮
$$(18\text{ L})\left(\frac{1\text{ eq}}{50\text{ L}}\right)\left(\frac{36.5\text{ g HCl}}{\text{eq HCl}}\right)\left(\frac{100\text{ g solution}}{38\text{ g HCl}}\right)\left(\frac{1\text{ cm}^3}{1.19\text{ g solution}}\right) = 29\text{ cm}^3$$

16.20 How many kg of wet NaOH containing 12% water is required to prepare 60 L of 0.50 N solution?

▮ One liter of 0.50 N NaOH contains (0.50)(40 g) = 20 g = 0.020 kg NaOH. Then 60 L of 0.50 N NaOH contains (60)(0.020 kg) = 1.20 kg NaOH. The wet NaOH contains 100% − 12% = 88% pure NaOH. Then 88 kg pure NaOH is contained in 100 kg wet NaOH, 1 kg pure NaOH is contained in $\frac{100}{88}$ kg wet NaOH, and

$$1.20\text{ kg pure NaOH is contained in }(1.20)(\tfrac{100}{88}\text{ kg}) = 1.36\text{ kg wet NaOH}$$

Or, by proportion, letting x = mass of wet NaOH that gives 1.20 kg pure NaOH,

$$\frac{100\text{ kg wet}}{88\text{ kg pure}} = \frac{x}{1.20\text{ kg pure}} \qquad \text{or} \qquad x = 1.36\text{ kg wet NaOH}$$

16.21 A 10.0 g sample of "gas liquor" is boiled with an excess of NaOH, and the resulting ammonia is passed into 60 cm³ of 0.90 N H_2SO_4. Exactly 10.0 cm³ of 0.40 N NaOH is required to neutralize the excess sulfuric acid (not neutralized by the NH_3). Determine the percent ammonia in the "gas liquor" examined.

▮
$$\begin{aligned}\text{Number of meq NH}_3\text{ in 10.0 g of gas liquor} &= (\text{number of meq H}_2\text{SO}_4) - (\text{number of meq NaOH})\\ &= (60\text{ cm}^3)(0.90\text{ meq/cm}^3) - (10.0\text{ cm}^3)(0.40\text{ meq/cm}^3)\\ &= 50\text{ meq NH}_3\end{aligned}$$

In neutralization experiments, the equivalent weight of NH_3 is numerically the same as the molecular weight, 17.0 g/eq, in accord with the equation $NH_3 + H^+ \longrightarrow NH_4^+$. Then the mass of NH_3 in the sample is

$$(50\text{ meq})(17.0\text{ mg/meq}) = 850\text{ mg} = 0.85\text{ g} \qquad \text{and} \qquad \text{Fraction NH}_3\text{ in sample} = \frac{0.85\text{ g}}{10.0\text{ g}} = 0.085 = 8.5\%$$

16.22 A 40.8 mL sample of an acid is equivalent to 50.0 mL of a Na_2CO_3 solution, 25.0 mL of which is equivalent to 23.8 mL of a 0.102 N HCl. What is the normality of the first acid?

▮
$$\text{Volume}_1 \times \text{normality}_1 = \text{volume}_2 \times \text{normality}_2$$
$$(40.8\text{ mL})N_1 = \left[\frac{50.0}{25.0}(23.8\text{ mL})\right](0.102\text{ N}) \qquad N_1 = 0.119\text{ N}$$

16.23 A 50.0 mL sample of NaOH solution requires 27.8 mL of 0.100 N acid in titration. What is its normality? How many mg NaOH are in each mL?

▮
$$N_1V_1 = N_2V_2 \qquad N_1 = \frac{(0.100\text{ N})(27.8\text{ mL})}{50.0\text{ mL}} = 0.0556\text{ N}$$
$$\left(\frac{0.0556\text{ eq}}{\text{L}}\right)\left(\frac{40.0\text{ g}}{\text{eq}}\right)\left(\frac{1\text{ L}}{1000\text{ mL}}\right) = 2.22\times10^{-3}\text{ g/mL} = 2.22\text{ mg/mL}$$

16.24 In standardizing HCl, 22.5 mL was required to neutralize 25.0 mL of 0.100 N Na_2CO_3 solution. What is the normality of the HCl solution? How much water must be added to 200 mL of it to make it 0.100 N? Neglect volume changes.

▮
$$N_1 = \frac{N_2V_2}{V_1} = \frac{(0.100\text{ N})(25.0\text{ mL})}{22.5\text{ mL}} = 0.111\text{ N} \qquad V_2 = \frac{(200\text{ mL})(0.111\text{ N})}{0.100\text{ N}} = 222\text{ mL}$$

To get 222 mL from a solution originally 200 mL, add 22 mL.

16.25 What volume of 0.0224 N adipic acid solution would be used in the titration of 1.022 cm³ of 0.0317 N $Ba(OH)_2$?

▮
$$V_1 = \frac{N_2V_2}{N_1} = \frac{(0.0317\text{ N})(1.022\text{ cm}^3)}{0.0224\text{ N}} = 1.45\text{ cm}^3$$

16.26 Exactly 21.0 mL of 0.80 N acid was required to neutralize completely 1.12 g of an impure sample of calcium oxide. What is the purity of the CaO?

▌

$$(21.0\text{ mL})(0.800\text{ N}) = 16.8\text{ meq H}^+$$

$$2H^+ + CaO \longrightarrow Ca^{2+} + H_2O$$

$$(16.8\text{ meq CaO})\left(\frac{28.0\text{ mg}}{\text{meq}}\right) = 470\text{ mg} \qquad \left(\frac{470\text{ mg}}{1120\text{ mg}}\right)(100\%) = 42.0\%\text{ CaO}$$

16.27 By the *Kjeldahl method*, the nitrogen contained in a foodstuff is converted into ammonia. If the ammonia from 5.0 g of a foodstuff is just sufficient to neutralize 20 mL of 0.100 N acid, calculate the percent nitrogen in the foodstuff.

▌

$$(20\text{ mL})(0.100\text{ N}) = (2.0\text{ meq})\left(\frac{17\text{ mg}}{\text{meq}}\right) = 34\text{ mg NH}_3$$

$$(34\text{ mg NH}_3)\left(\frac{14\text{ mg N}}{17\text{ mg NH}_3}\right) = 28\text{ mg N} \qquad \left(\frac{28\text{ mg N}}{5000\text{ mg sample}}\right)(100\%) = 0.56\%\text{ N}$$

16.28 What is the purity of concentrated H_2SO_4 (density 1.800 g/mL) if 5.00 mL is neutralized by 84.6 mL of 2.000 N NaOH?

▌

$$(84.6\text{ mL})(2.000\text{ N}) = (169.2\text{ meq H}_2\text{SO}_4)\left(\frac{49.0\text{ mg}}{\text{meq}}\right) = 8.29 \times 10^3\text{ mg} = 8.29\text{ g H}_2\text{SO}_4$$

$$(5.00\text{ mL})\left(\frac{1.800\text{ g}}{\text{mL}}\right) = 9.00\text{ g solution} \qquad \left(\frac{8.29\text{ g}}{9.00\text{ g solution}}\right)(100\%) = 92.1\%$$

16.29 A 10.0 mL portion of $(NH_4)_2SO_4$ solution was treated with excess NaOH. The NH_3 gas evolved was absorbed in 50.00 mL of 0.1000 N HCl. To neutralize the remaining HCl, 21.50 mL of 0.0980 N NaOH was required. What is the molar concentration of the $(NH_4)_2SO_4$? How many g $(NH_4)_2SO_4$ are in 1.00 L solution?

▌

$$(50.00\text{ mL})(0.1000\text{ N HCl}) = 5.000\text{ meq HCl}$$

$$(21.50\text{ mL})(0.0980\text{ N NaOH}) = 2.107\text{ meq NaOH for excess HCl}$$

$$\text{Net:} \qquad 2.893\text{ meq HCl used, therefore } 2.893\text{ meq NH}_3$$

$$(NH_4)_2SO_4 + 2OH^- \longrightarrow 2NH_3 + SO_4^{2-} + 2H_2O$$

$$(2.893\text{ meq NH}_3)\left(\frac{1\text{ meq (NH}_4)_2\text{SO}_4}{1\text{ meq NH}_3}\right)\left(\frac{1\text{ mmol (NH}_4)_2\text{SO}_4}{2\text{ meq (NH}_4)_2\text{SO}_4}\right) = 1.45\text{ mmol (NH}_4)_2\text{SO}_4$$

$$\frac{1.45\text{ mmol}}{10.0\text{ mL}} = 0.145\text{ M} \qquad (1.00\text{ L})\left(\frac{0.145\text{ mol}}{\text{L}}\right)\left(\frac{132\text{ g}}{\text{mol}}\right) = 19.1\text{ g/L}$$

16.30 Exactly 50.0 mL of a solution of Na_2CO_3 was titrated with 65.8 mL of 3.00 N HCl.

$$CO_3^{2-} + 2H^+ \longrightarrow CO_2 + H_2O$$

If the density of the Na_2CO_3 solution is 1.25 g/mL, what percent Na_2CO_3 by weight does it contain?

▌

$$(65.8\text{ mL})\left(\frac{3.00\text{ meq}}{\text{mL}}\right) = 197\text{ meq HCl}, \qquad \text{which yields } 197\text{ meq Na}_2\text{CO}_3$$

$$(197\text{ meq Na}_2\text{CO}_3)\left(\frac{53.0\text{ mg Na}_2\text{CO}_3}{\text{meq Na}_2\text{CO}_3}\right) = 10\,500\text{ mg} = 10.5\text{ g Na}_2\text{CO}_3$$

$$(50.0\text{ mL})\left(\frac{1.25\text{ g}}{\text{mL}}\right) = 62.5\text{ g solution} \qquad \left(\frac{10.5\text{ g Na}_2\text{CO}_3}{62.5\text{ g solution}}\right)(100\%) = 16.8\%\text{ Na}_2\text{CO}_3$$

16.31 What is the equivalent weight of an acid 1.243 g of which required 31.72 cm^3 of 0.1923 N standard base for neutralization?

▌

$$(31.72\text{ cm}^3)\left(\frac{0.1923\text{ meq}}{\text{cm}^3}\right) = 6.100\text{ meq base}, \qquad \text{hence } 6.100\text{ meq acid}$$

$$\frac{1.243\text{ g}}{0.006100\text{ eq}} = 203.8\text{ g/eq}$$

16.32 The molecular weight of an organic acid was determined by the following study of its barium salt. 4.290 g of the salt was converted to the free acid by reaction with 21.64 mL of 0.954 N H_2SO_4. The barium salt was known to contain 2 mol water of hydration per mol Ba^{2+}, and the acid was known to be monoprotic (monobasic). What is the molecular weight of the anhydrous acid?

▮ We will represent the acid HA. Then $BaA_2 + H_2SO_4 \rightarrow 2HA + BaSO_4$

$$(21.64\text{ mL})\left(\frac{0.954\text{ meq}}{\text{mL}}\right) = 20.64\text{ meq HA}, \qquad \text{yielding } 20.64\text{ meq } BaA_2$$

$$(20.64\text{ meq } BaA_2)\left(\frac{1\text{ mmol}}{2\text{ meq}}\right) = 10.32\text{ mmol } BaA_2 \qquad \frac{4.290\text{ g } BaA_2\cdot 2H_2O}{10.32\times 10^{-3}\text{ mol } BaA_2\cdot 2H_2O} = 415.7\text{ g/mol}$$

Two A^- anions thus weigh $415.7 - 137.3 - 2(18.0) = 242.4$ g/mol. HA has a molecular weight of $(242.4/2) + 1 = 122.2$ g/mol.

16.2 NORMALITY IN REDOX REACTIONS

16.33 What is the normality of a MnO_4^- solution if 32.00 mL of the solution is required to titrate 40.00 mL of 0.4000 N Fe^{2+}. Compare this result with the molarity of MnO_4^- calculated in Problem 15.129, and explain the relationship of the concentration units.

▮ $$[N\,(MnO_4^-)](32.00\text{ mL}) = (0.4000\text{ N})(40.00\text{ mL}) \qquad N\,(MnO_4^-) = 0.5000\text{ N}$$

The concentration may be expressed as 0.1000 M or 0.5000 N. As seen by the balanced equation in Problem 15.129, there are 5 eq/mol; hence the normality is 5 times the molarity.

16.34 A solution of $KMnO_4$ is reduced to MnO_2. The normality of the solution is 1.752. What is its molarity?

▮ $Mn^{VII} \longrightarrow Mn^{IV}$ involves three electrons per Mn atom. $$\left(\frac{1.752\text{ eq}}{\text{L}}\right)\left(\frac{1\text{ mol}}{3\text{ eq}}\right) = 0.584\text{ mol/L} = 0.584\text{ M}$$

16.35 What is the equivalent weight of H_2SO_4 when it is reduced to SO_2?

▮ $$2H^+ + H_2SO_4 + 2e^- \longrightarrow SO_2 + 2H_2O \qquad \left(\frac{98.0\text{ g } H_2SO_4}{\text{mol } H_2SO_4}\right)\left(\frac{1\text{ mol } H_2SO_4}{2\text{ mol } e^-}\right) = \frac{49.0\text{ g } H_2SO_4}{\text{eq}}$$

16.36 What is the normality of a solution of sulfuric acid made by dissolving 2 mol H_2SO_4 in sufficient water to make 1 L solution if ***(a)*** the solution is to be completely neutralized with NaOH ***(b)*** the H_2SO_4 in the solution is to be reduced to H_2S?

▮ ***(a)*** $H_2SO_4 + 2NaOH \longrightarrow Na_2SO_4 + 2H_2O$

$$(2\text{ mol } H_2SO_4)\left(\frac{2\text{ mol } OH^-}{\text{mol } H_2SO_4}\right) = 4.0\text{ eq } H_2SO_4 \qquad \frac{4.0\text{ eq}}{\text{L}} = 4.0\text{ N}$$

(b) $H_2SO_4 + 8H^+ + 8e^- \longrightarrow H_2S + 4H_2O$

$$(2\text{ mol } H_2SO_4)\left(\frac{8\text{ mol } e^-}{\text{mol } H_2SO_4}\right) = 16\text{ mol } e^- = 16\text{ eq } H_2SO_4 \qquad \frac{16\text{ eq } H_2SO_4}{\text{L}} = 16\text{ N}$$

16.37 In a reaction, $Cr_2O_7^{2-}$ was reduced to Cr^{3+}. What is the concentration of a 0.1 M solution of $Cr_2O_7^{2-}$ expressed in equivalents per liter?

▮ $$6e^- + 14H^+ + Cr_2O_7^{2-} \longrightarrow 2Cr^{3+} + 7H_2O$$

Since 1 mol of $Cr_2O_7^{2-}$ reacts with 6 mol e^-, it is 6 eq.

$$\left(\frac{0.1\text{ mol}}{\text{L}}\right)\left(\frac{6\text{ eq}}{\text{mol}}\right) = \frac{0.6\text{ eq}}{\text{L}} = 0.6\text{ N}$$

16.38 How many equivalents are there per mol of H_2S in its oxidation to SO_2?

▮ $$2H_2O + H_2S \longrightarrow SO_2 + 6H^+ + 6e^-$$

There are 6 eq/mol, corresponding to 6 mol e^-/mol H_2S.

16.39 The following is a completely balanced equation: $3Sn + 12HCl + 4HNO_3 \rightarrow 3SnCl_4 + 4NO + 8H_2O$. In this reaction, what fractions of the formula weights are the equivalent weights of Sn and HNO_3, respectively?

▮ The half-reactions are

$$4Cl^- + Sn \longrightarrow SnCl_4 + 4e^- \qquad 3e^- + 4H^+ + NO_3^- \longrightarrow NO + 2H_2O$$

In the oxidation, there are 4 eq Sn/mol Sn; the equivalent weight is $\frac{1}{4}$ the formula weight. In the reduction, there are 3 eq HNO_3/mol HNO_3; the equivalent weight is $\frac{1}{3}$ the formula weight.

16.40 Consider the following equations:

$$Cr_2O_7^{2-} + 14H^+ + 6e^- \longrightarrow 2Cr^{3+} + 7H_2O$$
$$Cr_2O_7^{2-} + 3SO_3^{2-} + 8H^+ \longrightarrow 2Cr^{3+} + 3SO_4^{2-} + 4H_2O$$

(a) How many mol dichromate ion are represented in each equation? ***(b)*** How many eq dichromate ion are represented in each? ***(c)*** What is the normality of a 0.200 M solution of dichromate ion when used in the second reaction?

▌ ***(a)*** and ***(b)*** There are 1 mol and 6 eq of $Cr_2O_7^{2-}$ represented in each equation, despite the fact that the 6 mol of electrons are not shown explicitly in equation (b). ***(c)*** 0.200 M = (0.200 mol/L)(6 eq/mol) = 1.20 N

16.41 Given the unbalanced equation

$$K^+MnO_4^- + K^+I^- + (H^+)_2SO_4^{2-} \longrightarrow (K^+)_2SO_4^{2-} + Mn^{2+}SO_4^{2-} + I_2 + H_2O$$

(a) How many g $KMnO_4$ are needed to make 500 mL 0.250 N solution? ***(b)*** How many g KI are needed to make 25.0 mL 0.360 N solution?

▌ ***(a)*** In this oxidation-reduction reaction, the oxidation state of Mn changes from +7 in MnO_4^- to +2 in Mn^{2+}. Hence

$$\text{Equivalent weight of } KMnO_4 = \frac{\text{formula weight}}{\text{oxidation state change}} = \frac{158}{5} = 31.6 \text{ g/eq}$$

Then 0.500 L of 0.250 N requires (0.500 L)(0.250 eq/L)(31.6 g/eq) = 3.95 g $KMnO_4$.

(b) The oxidation state of I changes from −1 in I^- to 0 in I_2. Hence

$$\text{equivalent weight of KI} = \text{formula weight} = 166 \text{ g/eq}$$

Then 0.0250 L of 0.360 N requires (0.0250 L)(0.360 eq/L)(166 g/eq) = 1.49 g KI

16.42 Given the unbalanced equation

$$K^+MnO_4^- + Mn^{2+}SO_4^{2-} + H_2O \longrightarrow MnO_2 + (K^+)_2SO_4^{2-} + (H^+)_2SO_4^{2-}$$

How many g $KMnO_4$ is needed to make 500 mL of 0.250 N solution?

▌ In this oxidation-reduction reaction, the oxidation state of Mn changes from +7 in MnO_4^- to +4 in MnO_2. Hence

$$\text{Equivalent weight of } KMnO_4 = \frac{\text{formula weight}}{\text{oxidation state change}} = \frac{158}{3} = 52.7 \text{ g/eq}$$

Then 0.500 L of 0.250 N requires (0.500 L)(0.250 eq/L)(52.7 g/eq) = 6.59 g $KMnO_4$. Compare this with Problem 16.41(*a*).

16.43 How many g solute is required to prepare 1.000 L of 1.000 N solution of ***(a)*** LiOH ***(b)*** Br_2 (as oxidizing agent) ***(c)*** H_3PO_4 (for a reaction in which three H atoms are replaced)?

▌ A normal solution contains 1 eq solute in 1 L solution. Formula weight of LiOH is 23.95; of Br_2, 159.82; of H_3PO_4, 97.99. ***(a)*** One liter of 1.000 N LiOH requires 23.95 g LiOH. ***(b)*** Note from the partial equation $Br_2 + 2e^- \longrightarrow 2Br^-$ that two electrons react per Br_2. Thus, the equivalent weight of Br_2 is *half* its molecular weight, and 1.000 L of 1.000 N Br_2 requires (159.82/2) g = 79.91 g Br_2. ***(c)*** One liter of 1.000 N H_3PO_4 requires (97.99/3) g = 32.66 g H_3PO_4.

16.44 What mass of phosphoric acid, H_3PO_4, is required to make 550 mL of 0.400 N solution ***(a)*** assuming complete neutralization of the acid ***(b)*** assuming reduction to HPO_3^{2-}?

▌ ***(a)*** $H_3PO_4 + 3OH^- \qquad PO_4^{3-} + 3H_2O$

An equivalent is defined for a nonredox reaction, in part, as the quantity of reagent which can react with 1 mol of OH^-. The 1 mol of H_3PO_4, reacting with 3 mol of OH^-, thus represents 3 eq.

$$3 \text{ eq } H_3PO_4 = 1 \text{ mol } H_3PO_4$$

$$(550 \text{ mL})\left(\frac{1 \text{ L}}{1000 \text{ mL}}\right)\left(\frac{0.400 \text{ eq}}{\text{L}}\right) = 0.220 \text{ eq} \qquad (0.220 \text{ eq})\left(\frac{1 \text{ mol } H_3PO_4}{3 \text{ eq } H_3PO_4}\right)\left(\frac{98.0 \text{ g}}{\text{mol}}\right) = 7.19 \text{ g}$$

(b) $H_3PO_4 + 2e^- \longrightarrow HPO_3^{2-} + H_2O$

For a redox reaction, 1 eq is the quantity of substance which reacts with 1 mol e^-. Thus there are 2 eq in 1 mol H_3PO_4 when it undergoes this reaction.

$$(0.220\ \text{eq})\left(\frac{1\ \text{mol}}{2\ \text{eq}}\right)\left(\frac{98.0\ \text{g}}{\text{mol}}\right) = 10.8\ \text{g}$$

16.45 Determine the mass of $KMnO_4$ required to make 80.0 mL of N/8 $KMnO_4$ when the latter acts as an oxidizing agent in acid solution and Mn^{2+} is a product of the reaction.

■

$$8H^+ + 5e^- + MnO_4^- \longrightarrow Mn^{2+} + 4H_2O$$

$$(0.0800\ \text{L})\left(\frac{1\ \text{eq}}{8.00\ \text{L}}\right)\left(\frac{1\ \text{mol KMnO}_4}{5\ \text{eq KMnO}_4}\right)\left(\frac{158\ \text{g KMnO}_4}{\text{mol KMnO}_4}\right) = 0.316\ \text{g}$$

16.46 Given the unbalanced equation $Cr_2O_7^{2-} + Fe^{2+} + H^+ \rightarrow Cr^{3+} + Fe^{3+} + H_2O$, ***(a)*** What is the normality of a $K_2Cr_2O_7$ solution 35.0 mL of which contains 3.87 g of $K_2Cr_2O_7$? ***(b)*** What is the normality of the $FeSO_4$ solution 750 mL of which contains 96.3 g of $FeSO_4$?

■ ***(a)*** $Cr_2O_7^{2-} + 6e^- + 14H^+ \longrightarrow Cr^{3+} + 7H_2O$

$$\left(\frac{3.87\ \text{g K}_2\text{Cr}_2\text{O}_7}{35.0\ \text{mL}}\right)\left(\frac{1\ \text{mol K}_2\text{Cr}_2\text{O}_7}{294\ \text{g K}_2\text{Cr}_2\text{O}_7}\right)\left(\frac{6\ \text{eq K}_2\text{Cr}_2\text{O}_7}{\text{mol K}_2\text{Cr}_2\text{O}_7}\right)\left(\frac{1000\ \text{mL}}{\text{L}}\right) = 2.26\ \text{N}$$

(b) $Fe^{2+} \longrightarrow Fe^{3+} + e^-$

$$\frac{96.3\ \text{g FeSO}_4}{750\ \text{mL}}\left(\frac{1\ \text{mol FeSO}_4}{152\ \text{g FeSO}_4}\right)\left(\frac{1\ \text{eq FeSO}_4}{\text{mol FeSO}_4}\right)\left(\frac{1000\ \text{mL}}{\text{L}}\right) = 0.845\ \text{N}$$

16.47 Calculate the mass of oxalic acid, $H_2C_2O_4$, which can be oxidized to CO_2 by 100.0 mL of an MnO_4^- solution, 10.0 mL of which is capable of oxidizing 50.0 mL of 1.00 N I^- to I_2.

■ $H_2C_2O_4 \longrightarrow 2CO_2 + 2H^+ + 2e^-$

$$MnO_4^- + 8H^+ + 5e^- \longrightarrow Mn^{2+} + 4H_2O$$

$$2MnO_4^- + 16H^+ + 10e^- \longrightarrow 2Mn^{2+} + 8H_2O$$

$5H_2C_2O_4 \longrightarrow 10CO_2 + 10H^+ + 10e^-$

$$5H_2C_2O_4 + 2MnO_4^- + 6H^+ \longrightarrow 10CO_2 + 2Mn^{2+} + 8H_2O$$

$$(50.0\ \text{mL})(1.00\ \text{N}) = 50.0\ \text{meq I}^-$$

The 50.0 meq I^- is equivalent to 50.0 meq MnO_4^-. The balanced equation for the reaction of I^- with MnO_4^- is not required, since 1 eq of any substance reacts with 1 eq of any reactant.

$$\frac{50.0\ \text{meq MnO}_4^-}{10.0\ \text{mL}} = 5.00\ \text{N MnO}_4^-$$

$$(100\ \text{mL})(5.00\ \text{N}) = 500\ \text{meq MnO}_4^-, \qquad \text{therefore 500 meq H}_2\text{C}_2\text{O}_4$$

From the half-reaction above:

$$(500\ \text{meq H}_2\text{C}_2\text{O}_4)\left(\frac{1\ \text{mmol H}_2\text{C}_2\text{O}_4}{2\ \text{meq H}_2\text{C}_2\text{O}_4}\right) = 250\ \text{mmol} \qquad (0.250\ \text{mol})\left(\frac{90.0\ \text{g}}{\text{mol}}\right) = 22.5\ \text{g}$$

16.48 What mass of $Na_2S_2O_3 \cdot 5H_2O$ is needed to make up 500 cm^3 of 0.200 N solution for the reaction $2S_2O_3^{2-} + I_2 \longrightarrow S_4O_6^{2-} + 2I^-$?

■

$$(0.500\ \text{L})\left(\frac{0.200\ \text{eq}}{\text{L}}\right)\left(\frac{2\ \text{mol S}_2\text{O}_3^{2-}}{2\ \text{eq S}_2\text{O}_3^{2-}}\right)\left(\frac{248\ \text{g Na}_2\text{S}_2\text{O}_3\cdot 5\text{H}_2\text{O}}{\text{mol S}_2\text{O}_3^{2-}}\right) = 24.8\ \text{g}$$

16.49 Calculate the number of g $FeSO_4$ that will be oxidized by 24.0 mL of 0.250 N $KMnO_4$ in a solution acidified with sulfuric acid. The unbalanced equation for the reaction is

$$MnO_4^- + Fe^{2+} + H^+ \longrightarrow Fe^{3+} + Mn^{2+} + H_2O$$

and the normality of the $KMnO_4$ is with respect to this reaction.

▮ It is not necessary to balance the complete equation. All that need be known is that the iron changes in oxidation state from +2 in Fe^{2+} to +3 in Fe^{3+}. Then

$$\text{Equiv wt of FeSO}_4 = \frac{\text{formula weight}}{\text{oxidation state change}} = \frac{152}{1} = 152 \text{ g/eq}$$

The same result may be found from the balanced partial equation $Fe^{2+} \rightarrow Fe^{3+} + e^-$.

$$\text{Equiv wt of FeSO}_4 = \frac{\text{formula weight}}{\text{number of electrons transferred}} = \frac{152}{1} = 152 \text{ g/eq}$$

Let x = required mass of $FeSO_4$.

$$\text{Number of eq KMnO}_4 = \text{number of eq FeSO}_4$$

$$(\text{Volume KMnO}_4) \times (\text{normality KMnO}_4) = \frac{\text{mass of FeSO}_4}{\text{equivalent weight of FeSO}_4}$$

$$(0.0240 \text{ L})(0.250 \text{ eq/L}) = \frac{x}{152 \text{ g/eq}} \qquad \text{Thus, } x = 0.912 \text{ g FeSO}_4.$$

16.50 What volume of 0.1000 N $FeSO_4$ is required to reduce 4.000 g $KMnO_4$ in a solution acidified with sulfuric acid?

▮ The normality of the $FeSO_4$ is with respect to the oxidation-reduction reaction given in Problem 16.49. In this reaction the Mn changes in oxidation state from +7 in MnO_4^- to +2 in Mn^{2+}. The net change is 5. Or, from the balanced partial equation

$$MnO_4^- + 8H^+ + 5e^- \longrightarrow Mn^{2+} + 4H_2O$$

it can be seen that the electron transfer is 5 for each MnO_4^-. The equivalent weight of $KMnO_4$ in this reaction is then

$$(\tfrac{1}{5})(\text{formula weight}) = \tfrac{1}{5}(158.0) = 31.60 \text{ g/eq}$$

$$\text{Number of eq FeSO}_4 = \text{number of eq KMnO}_4$$

$$(\text{Volume FeSO}_4)(0.1000 \text{ eq/L}) = \frac{4.000 \text{ g}}{31.60 \text{ g/eq}} \qquad \text{Volume FeSO}_4 = 1.266 \text{ L}$$

16.51 Exactly 400 mL of an acid solution, when acted upon by an excess of zinc, evolved 2.430 L of H_2 gas measured over water at 21 °C and 747.5 torr. What is the normality of the acid? Vapor pressure of water at 21 °C is 18.6 torr.

▮

$$2HCl + Zn \longrightarrow H_2 + ZnCl_2$$

$$n = \frac{PV}{RT} = \frac{(747.5 \text{ torr} - 18.6 \text{ torr})(1 \text{ atm}/760 \text{ torr})(2.430 \text{ L})}{(0.0821 \text{ L·atm/mol·K})(294 \text{ K})} = 0.0966 \text{ mol H}_2$$

$$(0.0966 \text{ mol H}_2)\left(\frac{2 \text{ eq HCl}}{\text{mol H}_2}\right) = 0.193 \text{ eq HCl} \qquad \frac{0.193 \text{ eq}}{0.400 \text{ L}} = 0.483 \text{ N}$$

16.52 Determine the equivalent weight of bromine in each of the following: ***(a)*** The reduction half-reaction for the disproportionation of Br_2 in base ***(b)*** the oxidation half-reaction, which yields bromate ion ***(c)*** the overall reaction ***(d)*** What is the relationship between the answer to (c) and the answers to (a) and (b)? Explain fully.

▮ ***(a)*** $2e^- + Br_2 \longrightarrow 2Br^- \qquad 10e^- + 5Br_2 \longrightarrow 10Br^-$

(b) $12OH^- + Br_2 \longrightarrow 2BrO_3^- + 6H_2O + 10e^-$

(c) $12OH^- + 6Br_2 \longrightarrow 10Br^- + 2BrO_3^- + 6H_2O$

Equivalent weights of Br_2: in *(a)* 79.9 g/mol (2 eq/mol), in *(b)* 16.0 g/mol (10 eq/mol), in *(c)* 96.0 g/mol (10 eq/6 mol). ***(d)*** The electrons produced by the oxidation of some of the Br_2 are the same ones used up in the reduction of the rest of the Br_2. The equivalent weight of the Br_2 in the overall reaction is the sum of that of the two half-reactions (for a species which disproportionates).

16.53 What volume of 3.00 N H_2SO_4 is needed to liberate 185 L of hydrogen gas at STP when treated with an excess of zinc?

▮

$$(185 \text{ L H}_2(\text{STP}))\left(\frac{1 \text{ mol H}_2}{22.4 \text{ L H}_2(\text{STP})}\right)\left(\frac{2 \text{ eq H}_2\text{SO}_4}{\text{mol H}_2}\right)\left(\frac{1 \text{ L}}{3.00 \text{ eq}}\right) = 5.51 \text{ L}$$

16.54 How many L hydrogen at STP would be replaced from 500 mL of 3.78 N HCl by 125 g zinc?

▮

$$2HCl + Zn \longrightarrow H_2 + ZnCl_2$$

$$(125\text{ g Zn})\left(\frac{1\text{ mol Zn}}{65.4\text{ g Zn}}\right)\left(\frac{2\text{ eq Zn}}{\text{mol Zn}}\right) = 3.82\text{ eq Zn} \qquad (0.500\text{ L})\left(\frac{3.78\text{ eq HCl}}{\text{L}}\right) = 1.89\text{ eq HCl}$$

HCl is in limiting quantity, so 1.89 eq of H_2 is produced.

$$(1.89\text{ eq }H_2)\left(\frac{1\text{ mol }H_2}{2\text{ eq }H_2}\right) = 0.945\text{ mol }H_2 \qquad V = (0.945\text{ mol})\left(\frac{22.4\text{ L(STP)}}{\text{mol}}\right) = 21.2\text{ L}$$

16.55 A ferrous sulfate solution was standardized by titration. A 25.00 mL portion of the solution required 42.08 mL of 0.08000 N ceric sulfate for complete oxidation. What is the normality of the ferrous sulfate?

▮

$$N_1 = \frac{N_2 V_2}{V_1} = \frac{(42.08\text{ mL})(0.08000\text{ N }Ce^{4+})}{25.00\text{ mL}} = 0.1347\text{ N}$$

16.56 How many mL of 0.0257 N KIO_3 would be needed to reach the end point in the oxidation of 34.2 mL of 0.0416 N hydrazine in hydrochloric acid solution?

▮

$$V_1 = \frac{N_2 V_2}{N_1} = \frac{(34.2\text{ mL})(0.0416\text{ N})}{0.0257\text{ N}} = 55.4\text{ mL}$$

16.57 How many g $FeCl_2$ will be oxidized by 28 mL of 0.25 N $K_2Cr_2O_7$ in HCl solution? The unbalanced equation is $Fe^{2+} + Cr_2O_7^{2-} + H^+ \longrightarrow Fe^{3+} + Cr^{3+} + H_2O$.

▮ The half-reaction of interest is $Fe^{2+} \longrightarrow Fe^{3+} + e^-$.

$$(28\text{ mL})(0.25\text{ N}) = 7.0\text{ meq }K_2Cr_2O_7\text{, which reacts with 7.0 meq }Fe^{2+}$$

$$(7.0\text{ meq }Fe^{2+})\left(\frac{1\text{ mmol }Fe^{2+}}{\text{meq }Fe^{2+}}\right)\left(\frac{55.8\text{ mg}}{\text{mmol}}\right) = 390\text{ mg} = 0.39\text{ g}$$

16.58 What mass of MnO_2 is reduced by 35 mL of 0.16 N oxalic acid, $H_2C_2O_4$, in sulfuric acid solution? The unbalanced equation is $MnO_2 + H^+ + H_2C_2O_4 \longrightarrow CO_2 + H_2O + Mn^{2+}$.

▮ The half-reaction of interest is $MnO_2 + 4H^+ + 2e^- \longrightarrow Mn^{2+} + 2H_2O$

$$(35\text{ mL})(0.16\text{ N}) = 5.6\text{ meq} \qquad (5.6\text{ meq }MnO_2)\left(\frac{1\text{ mmol}}{2\text{ meq}}\right)\left(\frac{86.9\text{ mg}}{\text{mmol}}\right) = 240\text{ mg} = 0.24\text{ g}$$

16.59 How many g $KMnO_4$ is required to oxidize 2.40 g $FeSO_4$ in a solution acidified with sulfuric acid? What is the equivalent weight of $KMnO_4$ in this reaction?

▮

$$8H^+ + MnO_4^- + 5Fe^{2+} \longrightarrow 5Fe^{3+} + Mn^{2+} + 4H_2O$$

$$\left(\frac{158.0\text{ g}}{\text{mol}}\right)\left(\frac{1\text{ mol}}{5\text{ eq}}\right) = 31.60\text{ g/eq}$$

$$(2.40\text{ g }FeSO_4)\left(\frac{1\text{ mol }FeSO_4}{151.8\text{ g }FeSO_4}\right)\left(\frac{5\text{ eq }FeSO_4}{5\text{ mol }FeSO_4}\right)\left(\frac{1\text{ eq }KMnO_4}{\text{eq }FeSO_4}\right)\left(\frac{31.60\text{ g }KMnO_4}{\text{eq }KMnO_4}\right) = 0.500\text{ g }KMnO_4$$

16.60 Find the equivalent weight of $KMnO_4$ in the reaction $Mn^{2+} + MnO_4^- + H_2O \rightarrow MnO_2 + H^+$ (unbalanced). How many g $MnSO_4$ is oxidized by 1.25 g $KMnO_4$?

▮

$$4H^+ + 3e^- + MnO_4^- \longrightarrow MnO_2 + 2H_2O$$

$$(158.0\text{ g/mol})\left(\frac{1\text{ mol}}{3\text{ eq}}\right) = 52.7\text{ g/eq}$$

$$2H_2O + Mn^{2+} \longrightarrow MnO_2 + 4H^+ + 2e^-$$

$$(1.25\text{ g }KMnO_4)\left(\frac{1\text{ eq }KMnO_4}{52.7\text{ g }KMnO_4}\right)\left(\frac{1\text{ eq }MnSO_4}{1\text{ eq }KMnO_4}\right)\left(\frac{1\text{ mol }MnSO_4}{2\text{ eq }MnSO_4}\right)\left(\frac{151\text{ g }MnSO_4}{\text{mol }MnSO_4}\right) = 1.79\text{ g }MnSO_4$$

16.61 (*a*) What volume of 0.400 N $K_2Cr_2O_7$ is required to liberate the chlorine from 1.20 g of NaCl in a solution acidified with H_2SO_4?

$$Cr_2O_7^{2-} + Cl^- + H^+ \longrightarrow Cr^{3+} + Cl_2 + H_2O \qquad \text{(unbalanced)}$$

(*b*) How many g $K_2Cr_2O_7$ is required? (*c*) How many g chlorine is liberated?

▮ (*a*) $2Cl^- \longrightarrow Cl_2 + 2e^-$

$$(1.20\text{ g NaCl})\left(\frac{1000\text{ mmol NaCl}}{58.5\text{ g NaCl}}\right)\left(\frac{2\text{ meq}}{2\text{ mmol}}\right) = 20.5\text{ meq} \qquad (20.5\text{ meq})\left(\frac{1\text{ mL}}{0.400\text{ meq}}\right) = 51.2\text{ mL}$$

(*b*) $6e^- + 14H^+ + Cr_2O_7^{2-} \longrightarrow 2Cr^{3+} + 7H_2O$

$$(20.5\text{ meq})\left(\frac{1\text{ mmol}}{6\text{ meq}}\right)\left(\frac{294\text{ mg}}{\text{mmol}}\right) = 1.01 \times 10^3\text{ mg} = 1.01\text{ g}$$

(*c*) $$(20.5\text{ meq }Cl_2)\left(\frac{1\text{ mmol }Cl_2}{2\text{ meq }Cl_2}\right)\left(\frac{71.0\text{ mg }Cl_2}{\text{mmol }Cl_2}\right) = 728\text{ mg }Cl_2 = 0.728\text{ g }Cl_2$$

16.62 If 25.0 mL of an iodine solution is equivalent to 0.125 g of $K_2Cr_2O_7$, to what volume should 1.000 L be diluted to make the solution one tenth normal?

▮

$$(0.125\text{ g }K_2Cr_2O_7)\left(\frac{1\text{ mol }K_2Cr_2O_7}{294\text{ g }K_2Cr_2O_7}\right)\left(\frac{6\text{ eq}}{\text{mol}}\right) = 2.55 \times 10^{-3}\text{ eq} \qquad \text{Hence 2.55 meq }I_2\text{ reacts.}$$

The aliquot (small sample) is $\dfrac{2.55\text{ meq }I_2}{25.0\text{ mL}} = 0.102\text{ N}$. The large sample is thus 0.102 N also.

$$V_2 = \frac{N_1V_1}{N_2} = \frac{(1.000\text{ L})(0.102\text{ N})}{0.100\text{ N}} = 1.020\text{ L}$$

16.63 How many g $KMnO_4$ should be taken to make up 250 mL of a solution of such concentration that 1 mL is equivalent to 5.00 mg iron in $FeSO_4$?

▮ Note that it is 5.00 mg Fe, not 5.00 mg $FeSO_4$, which is stated in the problem.

$$8H^+ + MnO_4^- + 5Fe^{2+} \longrightarrow 5Fe^{3+} + Mn^{2+} + 4H_2O$$

$$(5.00\text{ mg }Fe^{3+})\left(\frac{1\text{ meq }Fe^{3+}}{55.85\text{ mg }Fe^{3+}}\right) = 0.0895\text{ meq }Fe^{3+} \qquad \text{Hence 0.0895 meq }KMnO_4\text{ in 1 mL.}$$

$$(250\text{ mL})\left(\frac{0.0895\text{ meq}}{\text{cm}^3}\right) = 22.4\text{ meq in sample} \qquad (22.4\text{ meq})\left(\frac{1\text{ mmol}}{5\text{ meq}}\right)\left(\frac{158\text{ mg}}{\text{mmol}}\right) = 707\text{ mg} = 0.707\text{ g}$$

16.64 How many g iodine is present in a solution which requires 40.0 mL of 0.112 N $Na_2S_2O_3$ to react with it? The reaction is $S_2O_3^{2-} + I_2 \longrightarrow S_4O_6^{2-} + I^-$ (unbalanced).

▮

$$2S_2O_3^{2-} + I_2 \longrightarrow S_4O_6^{2-} + 2I^-$$

$$(40.0\text{ mL})\left(\frac{0.112\text{ meq}}{\text{mL}}\right) = 4.48\text{ meq }Na_2S_2O_3$$

Hence $$(4.48\text{ meq }I_2)\left(\frac{1\text{ mmol }I_2}{2\text{ meq }I_2}\right)\left(\frac{254\text{ mg }I_2}{\text{mmol }I_2}\right) = 569\text{ mg }I_2 = 0.569\text{ g }I_2$$

16.65 To how many mg iron (Fe^{2+}) is 1.00 mL of 0.1055 N $K_2Cr_2O_7$ equivalent?

▮

$$(1.00\text{ mL})\left(\frac{0.1055\text{ meq}}{\text{mL}}\right) = 0.1055\text{ meq }K_2Cr_2O_7$$

$$Fe^{2+} \longrightarrow Fe^{3+} + e^- \qquad (0.1055\text{ meq }Fe^{2+})\left(\frac{55.8\text{ mg}}{\text{meq}}\right) = 5.89\text{ mg }Fe^{2+}$$

16.66 Reducing sugars are sometimes characterized by a number R_{Cu}, which is defined as the number of mg of copper reduced by 1 g of the sugar, in which the half-reaction for the copper is

$$Cu^{2+} + OH^- \longrightarrow Cu_2O + H_2O \qquad \text{(unbalanced)}$$

It is sometimes more convenient to determine the reducing power of a carbohydrate by an indirect method. In this method 43.2 mg of the carbohydrate was oxidized by an excess of $K_3Fe(CN)_6$. The $Fe(CN)_6^{4-}$ formed in this reaction required 5.29 cm³ of 0.0345 N $Ce(SO_4)_2$ for reoxidation to $Fe(CN)_6^{3-}$ [the normality of the cerium(IV) sulfate solution is given with respect to the reduction of Ce^{4+} to Ce^{3+}]. Determine the R_{Cu} value for the sample. (*Hint*: The number of meq of Cu in a direct oxidation is the same as the number of meq of Ce^{4+} in the indirect method.)

■

$$2e^- + 2Cu^{2+} + 2OH^- \longrightarrow Cu_2O + H_2O$$

$$(5.29 \text{ cm}^3 \text{ Ce}^{4+})\left(\frac{0.0345 \text{ meq}}{\text{cm}^3}\right) = 0.183 \text{ meq Ce}^{4+} \text{ used}$$

Hence 0.183 meq iron(II) formed, 0.183 meq carbohydrate oxidized, 0.183 meq Cu^{2+} reduced. Thus 43.2 mg of sugar was reduced by

$$(0.183 \text{ meq Cu}^{2+})\left(\frac{63.5 \text{ mg Cu}^{2+}}{\text{meq}}\right) = 11.6 \text{ mg Cu}^{2+}$$

For each g of sugar:

$$(1000 \text{ mg sugar})\left(\frac{11.6 \text{ mg Cu}^{2+}}{43.2 \text{ mg sugar}}\right) = 269 \text{ mg Cu}^{2+}$$

$$R_{Cu} = 269$$

16.67 An acid solution of a $KReO_4$ sample containing 26.83 mg of combined rhenium was reduced by passage through a column of granulated zinc. The effluent solution, including the washings from the column, was then titrated with 0.1000 N $KMnO_4$; 11.45 mL of the standard permanganate was required for the reoxidation of all the rhenium to the perrhenate ion, ReO_4^-. Assuming that rhenium was the only element reduced, what is the oxidation state to which rhenium was reduced by the zinc column?

■

$$(11.45 \text{ mL})\left(\frac{0.1000 \text{ meq MnO}_4^-}{\text{mL}}\right) = 1.145 \text{ meq MnO}_4^- \qquad (26.83 \text{ mg Re})\left(\frac{1 \text{ mmol Re}}{186.2 \text{ mg Re}}\right) = 0.1441 \text{ mmol Re}$$

$$\frac{1.145 \text{ meq Re}}{0.1441 \text{ mmol Re}} = 7.95 \text{ meq/mmol}$$

Hence there was an 8-electron reduction, leaving Re in the −1 oxidation state (Re^-).

16.68 The iodide content of a solution was determined by titration with cerium(IV) sulfate in the presence of HCl, in which I^- is converted to ICl. A 250 mL sample of the solution required 20.0 mL of 0.0500 N Ce^{4+} solution. What is the iodide concentration in the original solution, in g/L?

■

$$(20.0 \text{ mL})\left(\frac{0.0500 \text{ meq}}{\text{mL}}\right) = 1.00 \text{ meq Ce}^{4+} \qquad I^- + Cl^- \longrightarrow ICl + 2e^-$$

$$\left(\frac{1.00 \text{ meq I}^-}{250 \text{ mL}}\right)\left(\frac{1 \text{ mmol I}^-}{2 \text{ meq I}^-}\right)\left(\frac{127 \text{ mg I}^-}{\text{mmol I}^-}\right) = 0.254 \text{ mg/mL} = 0.254 \text{ g/L}$$

16.69 A 0.518 g sample of limestone is dissolved, and then the calcium is precipitated as calcium oxalate, CaC_2O_4. After filtering and washing the precipitate, it requires 40.0 mL of 0.250 N $KMnO_4$ solution acidified with sulfuric acid to titrate it. What is the percent CaO in the limestone? The unbalanced equation for the titration is

$$MnO_4^- + CaC_2O_4 + (H^+)_2SO_4^{2-} \longrightarrow CaSO_4 + Mn^{2+} + CO_2 + H_2O.$$

■

$$(40.0 \text{ mL})\left(\frac{0.250 \text{ meq}}{\text{mL}}\right) = 10.0 \text{ meq MnO}_4^-, \qquad \text{hence } 10.0 \text{ meq C}_2\text{O}_4^{2-}$$

$$C_2O_4^{2-} \longrightarrow 2CO_2 + 2e^- \qquad (10.0 \text{ meq C}_2\text{O}_4^{2-})\left(\frac{1 \text{ mmol}}{2 \text{ meq}}\right) = 5.00 \text{ mmol C}_2\text{O}_4^{2-}$$

$$5.00 \text{ mmol CaO} = (5.00 \text{ mmol CaO})\left(\frac{56.0 \text{ mg CaO}}{\text{mmol CaO}}\right) = 280 \text{ mg CaO}$$

$$\left(\frac{280 \text{ mg CaO}}{518 \text{ mg sample}}\right)(100\%) = 54.1\% \text{ CaO}$$

16.3 MOLE FRACTION AND MOLALITY

Caution: Be sure to distinguish carefully between molarity and molality. There is only a one-letter difference in the spelling of the words, and the difference in the symbols is the difference between M and m, but:

Molarity is the number of moles of solute per *liter* of *solution*.

Molality is the number of moles of solute per *kilogram* of *solvent*.

16.70 Calculate the mole fraction of water in a mixture consisting of 9.0 g water, 120 g acetic acid, and 115 g ethyl alcohol.

▮ The molecular weights of water, acetic acid, and ethyl alcohol are 18, 60, and 46 g/mol, respectively. The mixture thus contains 0.50 mol water, 2.0 mol acetic acid, and 2.5 mol ethyl alcohol, for a total of 5 mol:

$$x(H_2O) = \frac{0.50}{5.0} = 0.10$$

In other words, one out of each 10 molecules in the solution is a water molecule.

16.71 What is the mole fraction of H_2 in a gaseous mixture containing 1.0 g H_2, 8.0 g O_2, and 16 g CH_4?

▮

$$(1.0\text{ g } H_2)\left(\frac{1\text{ mol } H_2}{2.0\text{ g } H_2}\right) = 0.50\text{ mol } H_2 \qquad (8.0\text{ g } O_2)\left(\frac{1\text{ mol } O_2}{32.0\text{ g } O_2}\right) = 0.25\text{ mol } O_2$$

$$(16\text{ g } CH_4)\left(\frac{1\text{ mol } CH_4}{16.0\text{ g } CH_4}\right) = 1.0\text{ mol } CH_4$$

$$x(H_2) = \frac{0.50\text{ mol } H_2}{1.75\text{ mol total}} = 0.29$$

16.72 A solution contains 116 g acetone (CH_3COCH_3), 138 g ethyl alcohol (C_2H_5OH), and 126 g water. Determine the mole fraction of each.

▮

$$(116\text{ g } C_3H_6O)\left(\frac{1\text{ mol } C_3H_6O}{58.0\text{ g } C_3H_6O}\right) = 2.00\text{ mol } C_3H_6O$$

$$(138\text{ g } C_2H_5OH)\left(\frac{1\text{ mol } C_2H_5OH}{46.0\text{ g } C_2H_5OH}\right) = 3.00\text{ mol } C_2H_5OH \qquad (126\text{ g } H_2O)\left(\frac{1\text{ mol } H_2O}{18.0\text{ g } H_2O}\right) = 7.00\text{ mol } H_2O$$

$$\text{Total} = 2.00 + 3.00 + 7.00 = 12.00\text{ mol}$$

$$x(C_3H_6O) = \frac{2.00}{12.00} = 0.167 \qquad x(C_2H_5OH) = \frac{3.00}{12.00} = 0.250 \qquad x(H_2O) = \frac{7.00}{12.00} = 0.583$$

16.73 A solution contains 18.0 g glucose (MW = 180), 24 g acetic acid (MW = 60), and 81 g water (MW = 18). What is the mole fraction of acetic acid in the solution?

▮

$$\frac{18.0\text{ g glucose}}{180\text{ g/mol}} = 0.100\text{ mol glucose} \qquad \frac{24\text{ g acetic acid}}{60\text{ g/mol}} = 0.40\text{ mol acetic acid}$$

$$\frac{81\text{ g water}}{18\text{ g/mol}} = 4.5\text{ mol water} \qquad x(\text{acetic acid}) = \frac{0.40}{4.5 + 0.4 + 0.1} = 0.080$$

16.74 Determine the mole fractions of both substances in a solution containing 36.0 g water and 46.0 g glycerin, $C_3H_5(OH)_3$.

▮ The molecular weight of $C_3H_5(OH)_3$ is 92.0; of H_2O, 18.0.

$$n(\text{glycerin}) = \frac{46.0\text{ g}}{92.0\text{ g/mol}} = 0.500\text{ mol} \qquad n(\text{water}) = \frac{36.0\text{ g}}{18.0\text{ g/mol}} = 2.00\text{ mol}$$

$$\text{Total number of moles} = 0.50 + 2.00 = 2.50\text{ mol}$$

$$x(\text{glycerin}) = \text{mole fraction of glycerin} = \frac{n(\text{glycerin})}{\text{total number of moles}} = \frac{0.50}{2.50} = 0.20$$

$$x(\text{water}) = \text{mole fraction of water} = \frac{n(\text{water})}{\text{total number of moles}} = \frac{2.00}{2.50} = 0.80$$

Check: Sum of mole fractions = 0.20 + 0.80 = 1.00.

16.75 The density of a 2.0 M solution of acetic acid (MW = 60) in water is 1.02 g/mL. Calculate the mole fraction of acetic acid.

▮ Per liter of solution: (2.0 mol acid)(60.0 g/mol) = 120 g acid

$$(1.000\text{ L})(1.02\text{ kg/L}) = 1.02\text{ kg solution} = 1020\text{ g solution}$$

$$(1020\text{ g solution}) - (120\text{ g acid}) = 900\text{ g water}$$

$$(900\text{ g } H_2O)\left(\frac{1\text{ mol } H_2O}{18.0\text{ g } H_2O}\right) = 50\text{ mol } H_2O \qquad x(\text{acid}) = \frac{2.0\text{ mol}}{(50\text{ mol}) + (2.0\text{ mol})} = 0.038$$

16.76 A solution contains 10.0 g acetic acid, CH_3COOH, in 125 g water. What is the concentration of the solution expressed as **(a)** mole fractions of CH_3COOH and H_2O **(b)** molality?

▮ **(a)** $(10.0\ \text{g}\ HC_2H_3O_2)\left(\frac{1\ \text{mol}\ HC_2H_3O_2}{60.0\ \text{g}\ HC_2H_3O_2}\right) = 0.167\ \text{mol}\ HC_2H_3O_2$

$$(125\ \text{g}\ H_2O)\left(\frac{1\ \text{mol}\ H_2O}{18.0\ \text{g}\ H_2O}\right) = 6.94\ \text{mol}\ H_2O$$

$$x(\text{acid}) = \frac{0.167}{0.167 + 6.94} = 0.0235 \qquad x(\text{water}) = 1.0000 - 0.0235 = 0.9765$$

(b) $\frac{0.167\ \text{mol}\ HC_2H_3O_2}{0.125\ \text{kg}\ H_2O} = 1.34\ \text{m}$

16.77 Calculate the molalities and the mole fractions of acetic acid in two solutions prepared by dissolving 120 g acetic acid **(a)** in 100 g water **(b)** in 100 g ethyl alcohol.

▮ **(a)** In water:

$$\text{Molality} = \frac{2.00\ \text{mol acetic acid}}{0.10\ \text{kg water}} = 20\ \text{m} \qquad x(\text{acetic acid}) = \frac{2.00\ \text{mol}}{(2.00 + 5.55)\ \text{mol}} = 0.265$$

(b) In ethyl alcohol:

$$\text{Molality} = \frac{2.00\ \text{mol acetic acid}}{0.10\ \text{kg ethyl alcohol}} = 20\ \text{m} \qquad x(\text{acetic acid}) = \frac{2.00\ \text{mol}}{(2.00 + 2.17)\ \text{mol}} = 0.480$$

16.78 The density of a 2.03 M solution of acetic acid in water is 1.017 g/mL. Calculate the molality of the solution.

▮ One liter of the solution, containing 2.03 mol solute, has a mass of 1017 g. Thus the solution contains $2.03\ \text{mol} \times 60.0\ \text{g/mol} = 122\ \text{g}$ acetic acid and also $1017 - 122 = 895$ g water. The molality is

$$\frac{2.03\ \text{mol solute}}{0.895\ \text{kg water}} = 2.27\ \text{m}$$

16.79 What mass of ammonium chloride is dissolved in 100 g water in each of the following solutions? **(a)** 1.10 m NH_4Cl solution **(b)** A solution which is 75% water by mass **(c)** A solution with a mole fraction of 0.15 NH_4Cl.

▮ **(a)** $(100\ \text{g}\ H_2O)\left(\frac{1.10\ \text{mol}}{1000\ \text{g}\ H_2O}\right)\left(\frac{53.5\ \text{g}}{\text{mol}}\right) = 5.88\ \text{g}$ **(b)** $(100\ \text{g}\ H_2O)\left(\frac{25.0\ \text{g}\ NH_4Cl}{75.0\ \text{g}\ H_2O}\right) = 33.3\ \text{g}$

(c) $(100\ \text{g}\ H_2O)\left(\frac{1\ \text{mol}\ H_2O}{18.0\ \text{g}\ H_2O}\right) = 5.55\ \text{mol}\ H_2O$ $(5.55\ \text{mol}\ H_2O)\left(\frac{0.15\ \text{mol}\ NH_4Cl}{0.85\ \text{mol}\ H_2O}\right)\left(\frac{53.5\ \text{g}}{\text{mol}\ NH_4Cl}\right) = 52.5\ \text{g}$

16.80 What is the molality of a solution which contains 20.0 g cane sugar, $C_{12}H_{22}O_{11}$, dissolved in 125 g water?

▮ The molecular weight of $C_{12}H_{22}O_{11}$ is 342.

$$m = \text{molality} = \frac{n(\text{solute})}{\text{mass of solvent in kg}} = \frac{(20.0\ \text{g})/(342\ \text{g/mol})}{0.125\ \text{kg}} = 0.468\ \text{mol/kg}$$

16.81 The molality of a solution of ethyl alcohol, C_2H_5OH, in water is 1.54 mol/kg. How many g alcohol is dissolved in 2.50 kg water?

▮ The molecular weight of C_2H_5OH is 46.1 g/mol. Since the molality is 1.54, 1 kg water dissolves 1.54 mol alcohol. Then 2.50 kg water dissolves $(2.50)(1.54) = 3.85$ mol alcohol, and

$$\text{Mass of alcohol} = (3.85\ \text{mol})(46.1\ \text{g/mol}) = 177\ \text{g alcohol}$$

16.82 Calculate the **(a)** molar concentration and **(b)** molality of a sulfuric acid solution of density 1.198 g/cm³, containing 27.0% H_2SO_4 by weight.

▮ **(a)** Each cm³ of acid solution has a mass of 1.198 g and contains $(0.270)(1.198) = 0.324\ \text{g}\ H_2SO_4$. Since the molecular weight of H_2SO_4 is 98.1,

$$M = \frac{n(H_2SO_4)}{\text{volume of solution in L}} = \frac{(0.324\ \text{g})/(98.1\ \text{g/mol})}{(1\ \text{cm}^3)(10^{-3}\ \text{L/cm}^3)} = 3.30\ \text{mol/L} = 3.30\ \text{M}$$

(*b*) From (*a*), there is 324 g, or 3.30 mol, of solute per liter of solution. The amount of water in 1 L of solution is 1198 g − 324 g = 874 g H_2O. Hence

$$m = \frac{n(\text{solute})}{\text{mass of solvent in kg}} = \frac{3.30\ \text{mol}\ H_2SO_4}{0.874\ \text{kg}\ H_2O} = 3.78\ \text{mol/kg} = 3.78\ \text{m}$$

16.83 Calculate the molality of a solution containing (*a*) 0.65 mol glucose, $C_6H_{12}O_6$, in 250 g water (*b*) 45 g glucose in 1.00 kg water (*c*) 18 g glucose in 200 g water.

▮ (*a*) $\frac{0.65\ \text{mol}}{0.250\ \text{kg}} = 2.6\ \text{m}$ (*b*) $\left(\frac{45\ \text{g glucose}}{\text{kg water}}\right)\left(\frac{1\ \text{mol glucose}}{180\ \text{g glucose}}\right) = 0.25\ \text{m}$

(*c*) $\left(\frac{18\ \text{g glucose}}{200\ \text{g water}}\right)\left(\frac{1\ \text{mol glucose}}{180\ \text{g glucose}}\right)\left(\frac{10^3\ \text{g}}{\text{kg}}\right) = 0.50\ \text{m}$

16.84 How many g $CaCl_2$ should be added to 300 mL water to make up a 2.46 m solution?

▮ Assuming that water has a density of 1.00 g/mL,

$$(300\ \text{mL water})\left(\frac{1.00\ \text{g water}}{\text{mL water}}\right)\left(\frac{1\ \text{kg}}{10^3\ \text{g}}\right)\left(\frac{2.46\ \text{mol}}{\text{kg}}\right)\left(\frac{111\ \text{g}\ CaCl_2}{\text{mol}}\right) = 81.9\ \text{g}\ CaCl_2$$

16.85 A solution contains 57.5 mL ethyl alcohol (C_2H_5OH) and 600 mL benzene (C_6H_6). How many g alcohol are in 1000 g benzene? What is the molality of the solution? Density of C_2H_5OH is 0.800 g/mL; of C_6H_6, 0.900 g/mL.

▮
$$(57.5\ \text{mL}\ C_2H_5OH)\left(\frac{0.800\ \text{g}}{\text{mL}}\right) = 46.0\ \text{g}\ C_2H_5OH \qquad (600\ \text{mL}\ C_6H_6)\left(\frac{0.900\ \text{g}}{\text{mL}}\right)\left(\frac{1\ \text{kg}}{1000\ \text{g}}\right) = 0.540\ \text{kg}\ C_6H_6$$

$$\frac{46.0\ \text{g}\ C_2H_5OH}{0.540\ \text{kg}\ C_6H_6} = \frac{85.2\ \text{g}\ C_2H_5OH}{\text{kg}\ C_6H_6} \qquad \left(\frac{85.2\ \text{g}\ C_2H_5OH}{\text{kg}}\right)\left(\frac{1\ \text{mol}\ C_2H_5OH}{46.0\ \text{g}\ C_2H_5OH}\right) = 1.85\ \text{mol/kg} = 1.85\ \text{m}$$

16.86 What is the mole fraction of the solute in a 1.00 m aqueous solution?

▮ There is 1.00 mol solute in 1.00 kg H_2O.

$$(1.00\ \text{kg})\left(\frac{10^3\ \text{g}}{\text{kg}}\right)\left(\frac{1\ \text{mol}\ H_2O}{18.0\ \text{g}\ H_2O}\right) = 55.6\ \text{mol}\ H_2O \qquad x = \frac{1.00}{1.00 + 55.6} = 0.0177$$

16.87 An aqueous solution labeled 35.0% $HClO_4$ had a density of 1.251 g/cm^3. What are the molar concentration and molality of the solution?

▮ Consider 1.000 L of solution: Its mass is 1.251 kg = 1251 g. It contains

$$(1251\ \text{g solution})\left(\frac{35.0\ \text{g}\ HClO_4}{100.0\ \text{g solution}}\right)\left(\frac{1\ \text{mol}\ HClO_4}{100.5\ \text{g}\ HClO_4}\right) = 4.36\ \text{mol}\ HClO_4$$

$$\frac{4.36\ \text{mol}\ HClO_4}{1.251\ \text{kg}\ (0.650)} = 5.36\ \text{m} \qquad 4.36\ \text{mol/L} = 4.36\ \text{M}$$

16.88 A sucrose solution was prepared by dissolving 0.0135 kg $C_{12}H_{22}O_{11}$ in enough water to make exactly 0.1000 L solution, which was then found to have a density of 1.050 kg/L. Compute the molar concentration and molality of the solution.

▮
$$(0.0135\ \text{kg}\ C_{12}H_{22}O_{11})\left(\frac{1\ \text{mol}\ C_{12}H_{22}O_{11}}{342\ \text{g}\ C_{12}H_{22}O_{11}}\right)\left(\frac{1000\ \text{g}}{\text{kg}}\right) = 0.0395\ \text{mol}\ C_{12}H_{22}O_{11}$$

$$(0.1000\ \text{L})\left(\frac{1.050\ \text{kg}}{\text{L}}\right) = 0.1050\ \text{kg solution} \qquad (0.1050\ \text{kg}) - (0.0135\ \text{kg}) = 0.0915\ \text{kg}\ H_2O$$

$$\frac{0.0395\ \text{mol}}{0.100\ \text{L}} = 0.395\ \text{M} \qquad \frac{0.0395\ \text{mol}}{0.0915\ \text{kg}} = 0.432\ \text{m}$$

16.89 Determine the volume of dilute nitric acid (density 1.11 g/mL, 19.0% HNO_3 by weight) that can be prepared by diluting with water 50 mL of the concentrated acid (density 1.42 g/mL, 69.8% HNO_3 by weight). Calculate the molar concentrations and molalities of the concentrated and dilute acids.

▌ $$(50.0 \text{ mL solution})\left(\frac{1.42 \text{ g solution}}{\text{mL solution}}\right)\left(\frac{69.8 \text{ g HNO}_3}{100 \text{ g solution}}\right) = 49.6 \text{ g HNO}_3$$

$$(49.6 \text{ g HNO}_3)\left(\frac{100 \text{ g solution}}{19.0 \text{ g HNO}_3}\right)\left(\frac{1 \text{ mL}}{1.11 \text{ g}}\right) = 235 \text{ mL} \qquad (49.6 \text{ g HNO}_3)\left(\frac{1 \text{ mol HNO}_3}{63.0 \text{ g HNO}_3}\right) = 0.787 \text{ mol HNO}_3$$

$$\frac{0.787 \text{ mol}}{0.0500 \text{ L}} = 15.7 \text{ M} \qquad \frac{0.787 \text{ mol}}{0.235 \text{ L}} = 3.35 \text{ M} \qquad (50.0 \text{ mL})\left(\frac{1.42 \text{ g}}{\text{mL}}\right) = 71.0 \text{ g solution}$$

$$\text{Mass of H}_2\text{O} = 71.0 \text{ g} - 49.6 \text{ g} = 21.4 \text{ g H}_2\text{O} \qquad \frac{0.787 \text{ mol}}{0.0214 \text{ kg}} = 36.8 \text{ m}$$

$$(235 \text{ mL})\left(\frac{1.11 \text{ g}}{\text{mL}}\right) = 261 \text{ g solution} \qquad \text{Mass of H}_2\text{O} = 261 \text{ g} - 49.6 \text{ g} = 211 \text{ g H}_2\text{O}$$

$$\frac{0.787 \text{ mol}}{0.211 \text{ kg}} = 3.73 \text{ m}$$

16.90 Calculate the molarity, molality, and mole fraction of ethyl alcohol in a solution of total volume 95 mL prepared by adding 50 mL ethyl alcohol (density = 0.789 g/mL) to 50 mL water (density = 1.00 g/mL). Calculate the molality of water in alcohol.

▌ $$(50 \text{ mL})\left(\frac{0.789 \text{ g}}{\text{mL}}\right)\left(\frac{1 \text{ mol}}{46.0 \text{ g}}\right) = 0.86 \text{ mol C}_2\text{H}_5\text{OH} \qquad (50 \text{ mL})\left(\frac{1.0 \text{ g}}{\text{mL}}\right)\left(\frac{1 \text{ mol}}{18 \text{ g}}\right) = 2.8 \text{ mol H}_2\text{O}$$

$$\text{Molality} = \frac{0.86 \text{ mol C}_2\text{H}_5\text{OH}}{0.050 \text{ kg H}_2\text{O}} = 17 \text{ m}$$

$$\text{Mole fraction} = \frac{0.86 \text{ mol}}{0.86 \text{ mol} + 2.8 \text{ mol}} = 0.23 \qquad \text{Molarity} = \frac{0.86 \text{ mol}}{0.095 \text{ L}} = 9.1 \text{ M}$$

$$\text{Molality of water} = \frac{2.8 \text{ mol H}_2\text{O}}{(0.050 \text{ L})(0.789 \text{ kg/L}) \text{ C}_2\text{H}_5\text{OH}} = 71 \text{ m}$$

16.91 For a solute of molecular weight W, show that the molar concentration M and molality m of the solution are related by

$$M\left(\frac{W}{1000} + \frac{1}{m}\right) = d$$

where d is the solution density in g/mL. (*Hint*: Show that each mL of solution contains $MW/1000$ g of solute and M/m g of solvent.) Use this relation to check the answers to Problems 16.87 and 16.88.

▌ The number of g solute per mL solution is given by

$$\left(\frac{M \text{ mol}}{\text{L}}\right)\left(\frac{1 \text{ L}}{10^3 \text{ mL}}\right)\left(\frac{W \text{ g}}{\text{mol}}\right) = \frac{MW}{1000} \text{ g solute/mL soln}$$

The number of g solvent per mL solution is given by

$$\left(\frac{M \text{ mol solute}}{\text{L}}\right)\left(\frac{\text{kg solvent}}{m \text{ mol solute}}\right)\left(\frac{\text{L}}{10^3 \text{ mL}}\right)\left(\frac{10^3 \text{ g}}{\text{kg}}\right) = \frac{M}{m} \text{ g solvent/mL soln}$$

$$\text{Density} = \frac{\text{total g}}{\text{mL}} = \frac{MW}{1000} + \frac{M}{m} = M\left(\frac{W}{1000} + \frac{1}{m}\right)$$

16.92 Show algebraically that the sum of the mole fractions of all components of a solution must equal 1.00.

▌ In a solution of $A + B + C + \cdots$,

$$x(\text{A}) = \frac{a \text{ mol A}}{(a \text{ mol A}) + (b \text{ mol B}) + (c \text{ mol C}) + \cdots}$$

Similarly for the mole fractions of B and C, etc., all the fractions having the same denominator. The total of all the mole fractions is thus

$$\frac{(a \text{ mol A}) + (b \text{ mol B}) + (c \text{ mol C}) + \cdots}{(a \text{ mol A}) + (b \text{ mol B}) + (c \text{ mol C}) + \cdots} = 1$$

16.93 The density of 10.0% by mass KCl solution in water is 1.06 g/mL. Calculate the molarity, molality, and mole fraction of KCl in this solution.

▌ In each 100.0 g of solution are

$$(10.0\text{ g KCl})\left(\frac{1\text{ mol}}{74.5\text{ g}}\right) = 0.134\text{ mol KCl} \quad \text{and} \quad (90.0\text{ g }H_2O)\left(\frac{1\text{ mol}}{18.0\text{ g}}\right) = 5.00\text{ mol }H_2O$$

$$(100\text{ g solution})\left(\frac{1\text{ mL}}{1.06\text{ g}}\right) = 94.3\text{ mL} = 0.0943\text{ L}$$

$$\text{Molarity} = \frac{0.134\text{ mol KCl}}{0.0943\text{ L}} = 1.42\text{ M} \qquad \text{Molality} = \frac{0.134\text{ mol KCl}}{0.0900\text{ kg }H_2O} = 1.49\text{ m}$$

$$\text{Mole fraction} = \frac{0.134\text{ mol}}{(0.134\text{ mol}) + (5.00\text{ mol})} = 0.0261$$

CHAPTER 17

Properties of Solutions

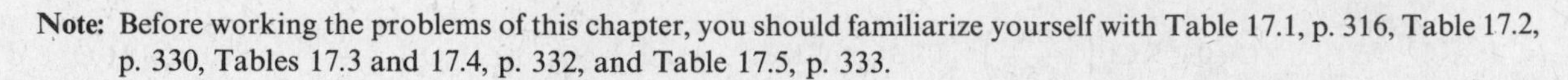

Note: Before working the problems of this chapter, you should familiarize yourself with Table 17.1, p. 316, Table 17.2, p. 330, Tables 17.3 and 17.4, p. 332, and Table 17.5, p. 333.

17.1 RAOULT'S LAW AND VAPOR PRESSURE LOWERING

17.1 At 40 °C the vapor pressure, in torr, of methyl alcohol–ethyl alcohol solutions is represented by

$$P = 119x + 135$$

where x is the mole fraction of methyl alcohol. What are the vapor pressures of the pure components at this temperature?

▌

$$\text{If} \quad x = 0, \qquad P = 135 \text{ torr}$$
$$\text{If} \quad x = 1, \qquad P = 119 + 135 = 254 \text{ torr}$$

17.2 Ethylene bromide, $C_2H_4Br_2$, and 1,2-dibromopropane, $C_3H_6Br_2$, form a series of ideal solutions over the whole range of composition. At 85 °C the vapor pressures of these two pure liquids are 173 and 127 torr, respectively. (*a*) If 10.0 g of ethylene bromide is dissolved in 80.0 g of 1,2-dibromopropane, calculate the partial pressure of each component and the total pressure of the solution at 85 °C. (*b*) Calculate the mole fraction of ethylene bromide in the vapor in equilibrium with the above solution. (*c*) What would be the mole fraction of ethylene bromide in a solution at 85 °C equilibrated with a 50:50 mole mixture in the vapor?

▌ (*a*)

$$(10.0 \text{ g } C_2H_4Br_2)\left(\frac{1 \text{ mol } C_2H_4Br_2}{188 \text{ g } C_2H_4Br_2}\right) = 0.0532 \text{ mol } C_2H_4Br_2$$

$$(80.0 \text{ g } C_3H_6Br_2)\left(\frac{1 \text{ mol } C_3H_6Br_2}{202 \text{ g } C_3H_6Br_2}\right) = 0.396 \text{ mol } C_3H_6Br_2$$

$$x(C_2H_4Br_2) = \frac{0.0532}{0.0532 + 0.396} = 0.118$$

$$P(C_2H_4Br_2) = P°x(C_2H_4Br_2) = (173 \text{ torr})(0.118) = 20.4 \text{ torr}$$

$$P(C_3H_6Br_2) = P°x(C_3H_6Br_2) = (127 \text{ torr})(0.882) = 112 \text{ torr}$$

$$P(\text{total}) = 132 \text{ torr}$$

(*b*)

$$x(C_2H_4Br_2) = \frac{20.4 \text{ torr}}{132 \text{ torr}} = 0.155$$

(*c*)

$$P(C_2H_4Br_2) = P(C_3H_6Br_2) = 173x(C_2H_4Br_2) = 127(1 - x(C_2H_4Br_2))$$

$$300x = 127$$

$$x = 0.423$$

17.3 The vapor pressure of pure liquid solvent A is 0.80 atm. When a nonvolatile substance B is added to the solvent, its vapor pressure drops to 0.60 atm. What is the mole fraction of component B in the solution?

▌

$$P = xP° \qquad x(A) = \frac{P}{P°} = \frac{0.60 \text{ atm}}{0.80 \text{ atm}} = 0.75 \qquad x(B) = 1 - x(A) = 1 - 0.75 = 0.25$$

17.4 The vapor pressure of pure water at 26 °C is 25.21 torr. What is the vapor pressure of a solution which contains 20.0 g glucose, $C_6H_{12}O_6$, in 70 g water?

▌

$$\frac{20.0 \text{ g glucose}}{180 \text{ g/mol}} = 0.111 \text{ mol glucose} \qquad \frac{70.0 \text{ g } H_2O}{18.0 \text{ g/mol}} = 3.89 \text{ mol } H_2O$$

$$x(H_2O) = \frac{3.89}{0.111 + 3.89} = 0.972$$

$$P = P°x(H_2O) = (25.21 \text{ torr})(0.972) = 24.5 \text{ torr}$$

17.5 The vapor pressure of pure water at 25 °C is 23.76 torr. The vapor pressure of a solution containing 5.40 g of a nonvolatile substance in 90.0 g water is 23.32 torr. Compute the molecular weight of the solute.

$$P = P°x$$

$$x = \frac{23.32 \text{ torr}}{23.76 \text{ torr}} = 0.981 = \frac{5.00}{5.00 + z}$$

$$(5.00 + z)(0.981) = 5.00$$

$$z = 0.0968 \text{ mol}$$

$$\frac{5.40 \text{ g}}{0.0968 \text{ mol}} = 55.8 \text{ g/mol}$$

17.6 At 20 °C the vapor pressure of methyl alcohol (CH_3OH) is 94 torr and the vapor pressure of ethyl alcohol (C_2H_5OH) is 44 torr. Being closely related, these compounds form a two-component system which adheres quite closely to Raoult's law throughout the entire range of concentrations. If 20 g of CH_3OH is mixed with 100 g of C_2H_5OH, determine the partial pressure exerted by each and the total pressure of the solution. Calculate the composition of the vapor above the solution by applying Dalton's law.

In an ideal solution of two liquids, there is no distinction between solute and solvent, and Raoult's law holds for each component of such solutions. Hence, when two liquids are mixed to give an ideal solution, the partial pressure of each liquid is equal to its vapor pressure multiplied by its mole fraction in the solution. The molecular weights of CH_3OH and C_2H_5OH are 32 and 46, so

$$\text{Partial pressure of } CH_3OH = (94 \text{ torr})\left[\frac{\frac{20}{32} \text{ mol } CH_3OH}{\frac{20}{32} \text{ mol } CH_3OH + \frac{100}{46} \text{ mol } C_2H_5OH}\right] = (94 \text{ torr})(0.22) = 21 \text{ torr}$$

$$\text{Partial pressure of } C_2H_5OH = (44 \text{ torr})\left[\frac{\frac{100}{46} \text{ mol } C_2H_5OH}{\frac{20}{32} \text{ mol } CH_3OH + \frac{100}{46} \text{ mol } C_2H_5OH}\right] = (44 \text{ torr})(0.78) = 34 \text{ torr}$$

The total pressure of the gaseous mixture is the sum of the partial pressures of all the components (Dalton's law): Total pressure of solution = (21 + 34) torr = 55 torr. Dalton's law also indicates that the mole fraction of any component of a gaseous mixture is equal to its pressure fraction, i.e., its partial pressure divided by the total pressure.

$$\text{Mole fraction of } CH_3OH \text{ in vapor} = \frac{\text{partial pressure of } CH_3OH}{\text{total pressure}} = \frac{21 \text{ torr}}{55 \text{ torr}} = 0.38$$

$$\text{Mole fraction of } C_2H_5OH \text{ in vapor} = \frac{\text{partial pressure of } C_2H_5OH}{\text{total pressure}} = \frac{34 \text{ torr}}{55 \text{ torr}} = 0.62$$

Since the mole fraction for (ideal) gases is the same as the volume fraction, we may also say that the vapor consists of 38% CH_3OH by volume. Note that the vapor is relatively richer in the more volatile component, methyl alcohol (mole fraction, 0.38), than is the liquid (mole fraction of CH_3OH, 0.22).

17.7 At 30 °C, pure benzene (molecular weight 78.1 g/mol) has a vapor pressure of 121.8 torr. Dissolving 15.0 g of a nonvolatile solute in 250 g of benzene produced a solution having a vapor pressure of 120.2 torr. Determine the approximate molecular weight of the solute.

Let W be the molecular weight of the solute.

$$\text{Number of moles of benzene in 250 g} = \frac{250 \text{ g}}{78.1 \text{ g/mol}} = 3.20 \text{ mol benzene}$$

$$\text{Number of moles of solute in 15.0 g} = \frac{15.0}{W} \text{ mol solute}$$

Substituting in the relation vp solution = (vp pure solvent)(mole fraction solvent),

$$120.2 \text{ torr} = (121.8 \text{ torr})\left[\frac{3.20 \text{ mol}}{(15.0/W) \text{ mol} + 3.20 \text{ mol}}\right] \quad \text{or} \quad 120.2 = (121.8)\left(\frac{3.20W}{15.0 + 3.20W}\right)$$

Solving, $W = 350$. Note that the accuracy of the calculation is limited by the term $121.8 - 120.2$ that appears in the expansion. The answer is significant only to 1 part in 16.

17.8 The vapor pressure of water at 28 °C is 28.35 torr. Compute the vapor pressure at 28 °C of a solution containing 68 g of cane sugar, $C_{12}H_{22}O_{11}$, in 1000 g of water.

$$\text{Mol of } C_{12}H_{22}O_{11} \text{ in 68 g} = \frac{68 \text{ g}}{342 \text{ g/mol}} = 0.20 \text{ mol } C_{12}H_{22}O_{11}$$

$$\text{Mol of } H_2O \text{ in 1000 g} = \frac{1000 \text{ g}}{18.02 \text{ g/mol}} = 55.49 \text{ mol } H_2O$$

$$\text{Total mol} = (0.20 + 55.49) \text{ mol} = 55.69 \text{ mol}$$

$$\text{Mole fraction } C_{12}H_{22}O_{11} = \frac{0.20}{55.69} = 0.0036 \qquad \text{Mole fraction } H_2O = \frac{55.49}{55.69} = 0.9964$$

First method

Vapor pressure of solution = (vp of pure solvent)(mole fraction of solvent) = (28.35 torr)(0.9964) = 28.25 torr

Second method

Vapor pressure depression = ΔP = (vp of pure solvent)(mole fraction of solute) = (28.35 torr)(0.0036) = 0.10 torr

Vapor pressure of solution = (28.35 − 0.10) torr = 28.25 torr

17.9 Calculate the mole fraction of toluene in the vapor phase which is in equilibrium with a solution of benzene and toluene having a mole fraction of toluene 0.500. The vapor pressure of pure benzene is 119 torr; that of toluene is 37.0 torr at the same temperature.

Assuming ideal behavior,

$$P_b = x_b P_b^\circ = 0.500(119 \text{ torr}) = 59.5 \text{ torr}$$
$$P_t = x_t P_t^\circ = 0.500(37.0 \text{ torr}) = 18.5 \text{ torr}$$
$$P_{\text{tot}} = 78.0 \text{ torr}$$

$$\text{Mole fraction toluene} = \frac{18.5 \text{ torr}}{78.0 \text{ torr}} = 0.237$$

17.10 What is the composition of the vapor which is in equilibrium at 30 °C with a benzene-toluene solution with a mole fraction of benzene of 0.400? with a mole fraction of benzene of 0.6000? $P_b^\circ = 119$ torr and $P_t^\circ = 37.0$ torr

$P_b = (0.400)(119 \text{ torr}) = 47.6$ torr $\quad P_t = (0.600)(37.0 \text{ torr}) = 22.2$ torr $\quad P_{\text{tot}} = (47.6 + 22.2) \text{ torr} = 69.8$ torr

The composition of the vapor is determined by applying Dalton's law of partial pressures:

$$x_b = \frac{P_b}{P_{\text{tot}}} = \frac{47.6 \text{ torr}}{69.8 \text{ torr}} = 0.682 \qquad x_t = \frac{22.2 \text{ torr}}{69.8 \text{ torr}} = 0.318 = 1.000 - 0.682$$

Similarly, for the case of the solution in which the mole fraction of toluene is 0.400,

$$x_b = \frac{71.5}{86.3} = 0.829 \qquad x_t = \frac{14.8}{86.3} = 0.171$$

17.11 At 30 °C, the vapor pressure of pure benzene, C_6H_6, is 119 torr, while that of toluene, $C_6H_5CH_3$, is 37 torr. Assuming ideal behavior, plot a vapor pressure–composition diagram for solutions of the two at 30 °C. Include on the diagram the compositions of vapor in equilibrium with the solutions.

See Fig. 17.1. The vapor pressure of each component is calculated at several different points, and a smooth curve is drawn through them to represent the pressure-composition curve of the vapor. (The corresponding curve

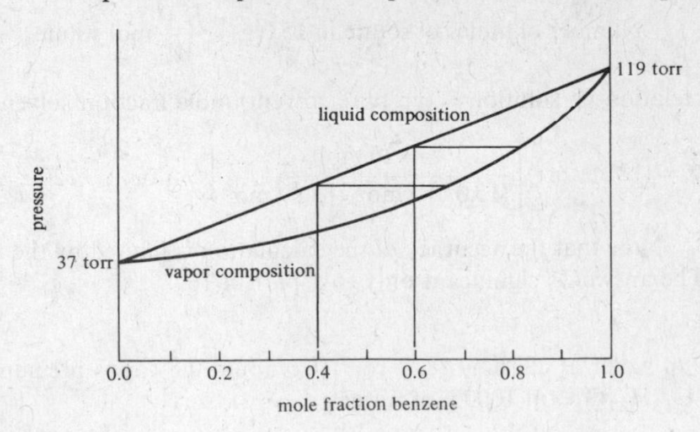

Fig. 17.1

for the liquid phase in an ideal solution is a straight line connecting the vapor pressure of the pure components.) An example is given:

At
$$x_b = 0.25 \qquad P_b = (0.25)(119 \text{ torr}) = 30 \text{ torr}$$
$$x_t = 0.75 \qquad P_t = (0.75)(37 \text{ torr}) = 28 \text{ torr}$$
$$P_{tot} = 58 \text{ torr}$$

In the vapor phase,
$$x_b = \frac{30 \text{ torr}}{58 \text{ torr}} = 0.52$$

17.12 At 50 °C the vapor pressure of pure CS_2 is 854 torr. A solution of 2.0 g of sulfur in 100 g of CS_2 has a vapor pressure 848.9 torr. Determine the formula of the sulfur molecule.

▮ The mole fraction of CS_2 is determined from Raoult's law:

$$x = \frac{P}{P^\circ} = \frac{848.9 \text{ torr}}{854 \text{ torr}} = 0.994$$

The mole fraction of the sulfur molecules, S_i, is thus 0.006. The molality of S_i is

$$\frac{0.006 \text{ mol}}{(0.994 \text{ mol } CS_2)(76 \text{ g/mol})(1 \text{ kg}/10^3 \text{ g})} = 0.079 \text{ m}$$

Hence 0.079 mol S_i/kg CS_2 is 20 g S_i/kg CS_2; hence the molecular weight of S_i is

$$\frac{20 \text{ g}}{0.079 \text{ mol}} = 250 \text{ g/mol}, \qquad \text{so} \qquad x = \frac{250}{32} = 8 \qquad \text{The formula is } S_8.$$

17.13 Calculate the vapor pressure lowering of a 0.100 m aqueous solution of nonelectrolyte at 75 °C.

▮ The vapor pressure of pure water, P_A°, at 75 °C is determined as follows:

$$\log \frac{P_2}{P_1} = \frac{\Delta H^\circ}{2.30}\left(\frac{1}{T_1} - \frac{1}{T_2}\right) = \frac{9720 \text{ cal}}{2.30(1.987 \text{ cal/mol}\cdot\text{K})}\left(\frac{1}{348} - \frac{1}{373}\right) = 0.410$$
$$P_2/P_1 = 2.57 \qquad P_1 = 296 \text{ torr}$$

The vapor pressure lowering due to 0.100 m solute is

$$\Delta P = mM_A P_A^\circ = \left(\frac{0.100 \text{ mol}}{1000 \text{ g}}\right)\left(\frac{18.0 \text{ g}}{\text{mol}}\right)(296 \text{ torr}) = 0.533 \text{ torr}$$

17.14 Calculate the composition of the vapor in equilibrium with an ideal solution of ethylbenzene ($P_e^\circ = 10.0$ torr) and methylbenzene ($P_m^\circ = 37.0$ torr) in which the mole fraction of ethylbenzene in the liquid is 0.35. Calculate the total vapor pressure of the solution.

▮
$$P_e = x_e P_e^\circ = (0.35)(10.0 \text{ torr}) = 3.5 \text{ torr} \qquad P_t = x_t P_t^\circ = (0.65)(37.0 \text{ torr}) = 24 \text{ torr}$$

The number of moles of each substance in the vapor phase is proportional to its pressure in the vapor phase. The mole fraction in the vapor phase is thus

$$x_e = \frac{3.5 \text{ torr}}{(3.5 + 24) \text{ torr}} = 0.13 \qquad P_{\text{tot}} = (3.5 + 24) \text{ torr} = 28 \text{ torr}$$

(It may be seen that the mole fraction of the more volatile component has been increased in the vapor phase.)

17.15 The vapor pressure of pure benzene, C_6H_6, at 50 °C is 268 torr. How many mol of nonvolatile solute per mol of benzene is required to prepare a solution of benzene having a vapor pressure of 167.0 torr at 50 °C?

▮
$$P_b = x_b P_b^\circ \qquad x_b = P_b/P_b^\circ = 167.0 \text{ torr}/268 \text{ torr} = 0.623 \qquad x_u = 1.000 - 0.623 = 0.377$$

There is 0.377 mol of solute per 0.623 mol of benzene:

$$\frac{0.377 \text{ mol solute}}{0.623 \text{ mol benzene}} = 0.605 \text{ mol solute/mol benzene}$$

17.16 At 25 °C, the vapor pressure of methyl alcohol, CH_3OH, is 96.0 torr. What is the mole fraction of CH_3OH in a solution in which the (partial) vapor pressure of CH_3OH is 23.0 torr at 25 °C?

■
$$x = \frac{P}{P^\circ} = \frac{23.0\text{ torr}}{96.0\text{ torr}} = 0.240$$

17.17 At 25 °C, the vapor pressure of pure benzene is 100 torr, while that of pure ethyl alcohol is 44 torr. Assuming ideal behavior, calculate the vapor pressure at 25 °C of a solution which contains 10.0 g of each substance.

■
$$(10.0\text{ g }C_6H_6)\left(\frac{1\text{ mol}}{78.0\text{ g}}\right) = 0.128\text{ mol }C_6H_6 \qquad (10.0\text{ g }C_2H_5OH)\left(\frac{1\text{ mol}}{46.0\text{ g}}\right) = 0.217\text{ mol }C_2H_5OH$$

$$x_b = \frac{0.128\text{ mol}}{(0.128 + 0.217)\text{ mol}} = 0.371 \qquad x_a = 1.000 - 0.371 = 0.629$$

$$P_b = x_b P_b^\circ = (0.371)(100\text{ torr}) = 37.1\text{ torr} \qquad P_a = x_a P_a^\circ = (0.629)(44\text{ torr}) = 28\text{ torr}$$

$$P_{\text{tot}} = 65\text{ torr}$$

17.18 Estimate the lowering of the vapor pressure due to the solute in a 1.0 m aqueous solution at 100 °C.

■ The normal boiling point of water is 100 °C; hence

$$P^\circ(H_2O) = 1.00\text{ atm} \qquad M(H_2O) = 18.0\text{ g/mol}$$

$$\Delta P = mM_A P_A^\circ = \left(\frac{1.0\text{ mol}}{1000\text{ g}}\right)\left(\frac{18.0\text{ g}}{\text{mol}}\right)(1.0\text{ atm}) = 0.018\text{ atm} = 14\text{ torr}$$

17.2 FREEZING POINT DEPRESSION AND BOILING POINT ELEVATION

TABLE 17.1 Boiling Point Elevation and Freezing Point Depression Data

solvent	normal boiling point, °C	K_b, °C/m	freezing point, °C	K_f, °C/m
Acetic acid	118.9	3.1	16.6	3.9
Benzene	80.1	2.53	5.5	5.12
Chloroform	61.2	3.63	—	—
Naphthalene	—	—	80.22	6.85
Water	100.0	0.512	0.00	1.86

In the problems that follow, temperature on the Celsius scale will be denoted t, with T denoting (as previously) absolute temperature. Recall that the two scales utilize the same size of a degree so that for differences in temperatures: 1 °C = 1 K.

17.19 Calculate the freezing point of the solution containing 15.6 g of solute/kg of benzene for each of the following solutes:

(*a*) CH_3 CH_3 (*b*) CH_3 CH_3 CH_3 CH_3

■ (*a*) $(15.6\text{ g})\left(\frac{1\text{ mol}}{156\text{ g}}\right) = 0.100\text{ mol}$

Since the solute is dissolved in 1 kg of solvent, the molality is 0.100 m.

$$\Delta t = (0.100\text{ m})(4.9\text{ °C/m}) = 0.49\text{ °C} = 5.5^\circ - t$$
$$t = (5.5 - 0.49)\text{ °C} = 5.0\text{ °C}$$

(*b*) $(15.6\text{ g})\left(\frac{1\text{ mol}}{234\text{ g}}\right) = 0.0667\text{ mol}$

In 1.00 kg of benzene:

$$\Delta t = (0.0667\text{ m})(4.9\text{ °C/m}) = 0.33\text{ °C} = 5.5^\circ - t$$
$$t = (5.5 - 0.33)\text{ °C} = 5.2\text{ °C}$$

17.20 What is the freezing point of a 10% (by weight) solution of CH_3OH in water?

▮ 1.00 kg contains 0.100 kg CH_3OH (and 0.900 kg H_2O).

$$(100\text{ g }CH_3OH)\left(\frac{1\text{ mol}}{32.0\text{ g}}\right) = 3.12\text{ mol} \qquad \frac{3.12\text{ mol}}{0.900\text{ kg}} = 3.47\text{ m}$$

$$\Delta t = Km = (1.86\text{ °C/m})(3.47\text{ m}) = 6.45\text{ °C}$$

$$\text{fp} = -6.45\text{ °C}$$

17.21 When 10.6 g of a nonvolatile substance is dissolved in 740 g of ether, its boiling point is raised 0.284 °C. What is the molecular weight of the substance? Molal boiling-point constant for ether is 2.11 °C·kg/mol.

▮

$$\Delta t = Km = 0.284\text{ °C}$$

$$m = \frac{0.284\text{ °C}}{2.11\text{ °C/m}} = 0.135\text{ m} = \frac{0.135\text{ mol}}{\text{kg}}$$

$$\frac{10.6\text{ g}}{0.740\text{ kg ether}} = \frac{14.3\text{ g}}{\text{kg ether}} \qquad \frac{14.3\text{ g/kg}}{0.135\text{ mol/kg}} = 106\text{ g/mol}$$

17.22 The freezing point of a sample of naphthalene was found to be 80.6 °C. When 0.512 g of a substance is dissolved in 7.03 g naphthalene, the solution has a freezing point of 75.2 °C. What is the molecular weight of the solute? The molal freezing point constant of naphthalene is 6.80 °C·kg/mol.

▮

$$\Delta t_f = (80.6 - 75.2)\text{ °C} = 5.4\text{ °C} = K_f m$$

$$m = \frac{5.4\text{ °C}}{6.80\text{ °C/m}} = 0.79\text{ m} = \frac{0.79\text{ mol}}{\text{kg solvent}}$$

$$\frac{0.512\text{ g}}{0.00703\text{ kg}} = \frac{72.8\text{ g}}{\text{kg}} \qquad \frac{72.8\text{ g/kg}}{0.79\text{ mol/kg}} = 92\text{ g/mol}$$

17.23 If glycerin, $C_3H_5(OH)_3$, and methyl alcohol, CH_3OH, sell at the same price per pound, which would be cheaper for preparing an antifreeze solution for the radiator of an automobile?

▮ A given mass of methyl alcohol, because it has a lower molecular weight, contains more moles than the same mass of glycerin. More moles of solute means higher molality and thus lower freezing point. It would take less methyl alcohol to protect a radiator to a given freezing point. (Practically, however, methyl alcohol vaporizes too readily to be used in modern, high-temperature auto radiators.)

17.24 How much ethyl alcohol, C_2H_5OH, must be added to 1.00 L of water so that the solution will not freeze at −4 °F?

▮

$$-4\text{ °F} = \tfrac{5}{9}(-4 - 32)\text{ °C} = -20\text{ °C}$$

$$\Delta t = 20\text{ °C} = Km = (1.86\text{ °C/m})(m)$$

$$m = \frac{20\text{ °C}}{1.86\text{ °C/m}} = 10.7\text{ m} \qquad (10.7\text{ mol})\left(\frac{46.0\text{ g}}{\text{mol}}\right) = 495\text{ g}$$

17.25 If the radiator of an automobile contains 12 L of water, how much would the freezing point be lowered by the addition of 5 kg of Prestone (glycol, $C_2H_4(OH)_2$)? How many kg of Zerone (methyl alcohol, CH_3OH) would be required to produce the same result? Assume 100% purity.

▮

$$(5.0\text{ kg }C_2H_4(OH)_2)\left(\frac{10^3\text{ g}}{\text{kg}}\right)\left(\frac{1\text{ mol}}{62.0\text{ g}}\right) = 80.6\text{ mol} \qquad \frac{80.6\text{ mol}}{12\text{ kg}} = 6.7\text{ m}$$

$$\Delta t = (1.86\text{ °C/m})(6.7\text{ m}) = 12\text{ °C} \qquad \text{fp} = -12\text{ °C}$$

With methyl alcohol, to get the same freezing point depression we would need the same number of moles:

$$(80.6\text{ mol})\left(\frac{32\text{ g}}{\text{mol}}\right)\left(\frac{1\text{ kg}}{10^3\text{ g}}\right) = 2.6\text{ kg}$$

17.26 A solution containing 4.50 g of a nonelectrolyte dissolved in 125 g of water freezes at −0.372 °C. Calculate the approximate molecular weight of the solute.

▌ *First method*
First compute the molality from the freezing point equation.

$$\Delta t_f = K_f m$$
$$0.372\ °\text{C} = (1.86\ °\text{C}\cdot\text{kg/mol})m$$
$$m = \frac{0.372\ °\text{C}}{1.86\ °\text{C}\cdot\text{kg/mol}} = 0.200\ \text{mol/kg}$$

From the definition of molality compute the number of moles of solute in the sample.

$$n(\text{solute}) = (0.200\ \text{mol solute/kg solvent})(0.125\ \text{kg solvent}) = 0.025\ \text{mol solute}$$

Then
$$\text{Molecular weight} = \frac{4.50\ \text{g solute}}{0.025\ \text{mol solute}} = 180\ \text{g/mol}$$

Second method
For the given concentration,

$$1\ \text{g}\ H_2O\ \text{contains}\ \frac{4.50}{125}\ \text{g solute} \qquad 1000\ \text{g}\ H_2O\ \text{contains}\ 1000\left(\frac{4.50}{125}\right) = 36.0\ \text{g solute}$$

Now, since 1.86 °C lowering is produced by 1 mol solute in 1000 g H_2O, 0.372 °C lowering is produced by

$$\frac{0.372}{1.86} = 0.200\ \text{mol solute}$$

in 1000 g H_2O. Thus 0.200 mol solute is equivalent to 36.0 g solute, and

$$\text{Molecular weight} = \frac{36.0\ \text{g}}{0.200\ \text{mol}} = 180\ \text{g/mol}$$

17.27 C_6H_6 freezes at 5.5 °C. At what temperature will a solution of 10.0 g of C_4H_{10} in 200 g of C_6H_6 freeze? The molal freezing point depression constant of C_6H_6 is 5.12 °C/m.

▌
$$(10.0\ \text{g}\ C_4H_{10})\left(\frac{1\ \text{mol}\ C_4H_{10}}{58.0\ \text{g}\ C_4H_{10}}\right) = 0.172\ \text{mol}\ C_4H_{10} \qquad \frac{0.172\ \text{mol}\ C_4H_{10}}{0.200\ \text{kg}\ C_6H_6} = 0.860\ \text{m}$$
$$\Delta t = (5.12\ °\text{C/m})(0.860\ \text{m}) = 4.40\ °\text{C}$$
$$t = (5.5 - 4.4)\ °\text{C} = 1.1\ °\text{C}$$

17.28 An aqueous solution boils at 100.50 °C. What is the freezing point of the solution?

▌
$$m = \Delta t_b/K_b = \Delta t_f/K_f = (0.50\ °\text{C})/(0.51\ °\text{C/m}) = \Delta t_f/(1.86\ °\text{C/m})$$
$$\Delta t_f = 1.82° \quad \text{from which } t = -1.82\ °\text{C}$$

17.29 Explain by use of a phase diagram why the normal boiling point of water is raised by addition of a nonvolatile solute while the freezing point is lowered. Label all pertinent points explicitly.

▌ See Fig. 17.2. The vapor pressure is lowered so that the temperature at which it reaches 1 atm is higher (point *A*). The intersection of the curve with the solid-vapor curve (which is not affected by solute) is at a lower temperature (point *B*).

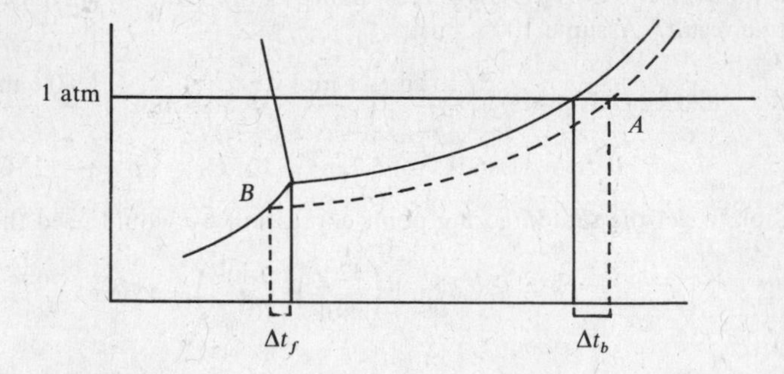

Fig. 17.2

17.30 The freezing point of pure camphor is 178.4 °C, and its molal freezing point constant, K_f, is 40.0 °C·kg/mol. Find the freezing point of a solution containing 1.50 g of a compound of molecular weight 125 in 35.0 g of camphor.

▮ The first step is to find the molality (m) of the solution.

$$m = \frac{\text{mol of solute}}{\text{kg of solvent}} = \frac{(1.50/125)\text{ mol solute}}{(35/1000)\text{ kg solvent}} = 0.343\text{ mol/kg}$$

$$\text{Lowering of freezing point} = \Delta t_f = K_f m = (40.0\text{ °C·kg/mol})(0.343\text{ mol/kg}) = 13.7\text{ °C}$$

$$\text{Freezing point of solution} = (\text{freezing point of pure solvent}) - \Delta t_f = (178.4 - 13.7)\text{ °C} = 164.7\text{ °C}$$

17.31 Bromoform has a normal freezing point of 7.80 °C and its $K_f = 14.4$ °C/m. A solution of 2.58 g of an unknown in 100 g of bromoform freezes at 5.43 °C. What is the molecular weight of the unknown?

▮

$$\Delta t = (7.80 - 5.43)\text{ °C} = 2.37\text{ °C} \qquad m = \frac{\Delta t}{K_f} = \frac{2.37\text{ °C}}{14.4\text{ °C/m}} = 0.165\text{ m} \qquad \frac{2.58\text{ g}/0.100\text{ kg}}{0.165\text{ mol/kg}} = 156\text{ g/mol}$$

17.32 When 30.0 g of a nonvolatile solute having the empirical formula CH_2O is dissolved in 800 g of water, the solution freezes at -1.16 °C. What is the molecular formula of the solute? $K_f = 1.86$ °C/m.

▮ The molality is given by

$$m = \frac{\Delta t}{K_f} = \frac{1.16\text{ °C}}{1.86\text{ °C/m}} = 0.624\text{ m} \qquad \frac{30.0\text{ g}}{0.800\text{ kg solvent}} = \frac{37.5\text{ g}}{\text{kg solvent}}$$

Hence 37.5 g is equivalent to 0.624 mol.

$$\frac{37.5\text{ g}}{0.624\text{ mol}} = 60.1\text{ g/mol}$$

The empirical formula weight is 30 g/eq. formula unit. There must be two units per molecule; the formula is $C_2H_4O_2$.

17.33 The freezing point of a solution containing 2.40 g of a compound in 60.0 g of benzene is 0.10 °C lower than that of pure benzene. What is the molecular weight of the compound? (K_f is 5.12 °C/m for benzene.)

▮

$$m = \frac{\Delta t}{K_f} = \frac{0.10\text{ °C}}{5.12\text{ °C/m}} = 0.0195\text{ m} \qquad \left(\frac{2.40\text{ g}}{60.0\text{ g benzene}}\right)\left(\frac{10^3\text{ g benzene}}{\text{kg benzene}}\right) = 40.0\text{ g/kg benzene}$$

Thus there is 40.0 g of solute and 0.0195 mol of solute per kg of benzene.

$$\frac{40.0\text{ g}}{0.0195\text{ mol}} = 2050\text{ g/mol}$$

17.34 The freezing point of a solution containing 4.80 g of a compound in 60.0 g of benzene is 4.50 °C. What is the molecular weight of the compound?

▮

$$\Delta t_f = (5.5 - 4.5)\text{ °C} = 1.0\text{ °C}$$

$$m = \frac{\Delta t_f}{K_f} = \frac{1.0\text{ °C}}{5.1\text{ °C/m}} = 0.20\text{ m} = 0.20\text{ mol/kg} \qquad M_{\text{compound}} = \frac{4.8\text{ g}/60.0\text{ g benzene}}{0.20\text{ mol}/1000\text{ g benzene}} = 400\text{ g/mol}$$

17.35 An aqueous solution containing 288 g of a nonvolatile compound having the stoichiometric composition $C_nH_{2n}O_n$ in 90.0 g of water boils at 101.24 °C at 1.00 atm pressure. What is the molecular formula of the compound?

▮

$$\Delta t_b = (101.24 - 100.00)\text{ °C} = 1.24\text{ °C}$$

$$m = \frac{\Delta t_b}{K_b} = \frac{1.24\text{ °C}}{0.512\text{ °C/m}} = 2.42\text{ m} \qquad M = \frac{288\text{ g}/90.0\text{ g water}}{2.42\text{ mol}/1000\text{ g water}} = 1320\text{ g/mol}$$

The weight of a molecule is 1320 u. If the value of n were 1, corresponding to CH_2O, the weight would be 30 u. Hence n must be $1320/30 = 44$, and the molecular formula is $C_{44}H_{88}O_{44}$.

17.36 Calculate the molecular weight of a substance which forms a 7.0% by mass solution in water which freezes at -0.89 °C.

▮

$$\text{Molality} = \frac{\Delta t}{K_f} = \frac{0.89°}{1.86°/\text{m}} = 0.48\text{ m}$$

The solution is 0.48 mol/kg of solvent, and also 70 g/kg of solvent. Thus 70 g is 0.48 mol. The molecular weight is

$$\frac{70\text{ g}}{0.48\text{ mol}} = 150\text{ g/mol}$$

17.37 Calculate (*a*) the freezing point of a solution of 0.0100 mol of glucose dissolved in 100 g of water, and (*b*) the freezing point of a 0.100 m solution of naphthalene in benzene.

▮ (*a*) $\text{Concentration} = \dfrac{0.0100\text{ mol}}{0.100\text{ kg}} = 0.100\text{ m}$

$$\Delta t_f = (1.86\text{ °C/m})(0.100\text{ m}) = 0.186\text{ °C}$$

Since the freezing point of pure water is 0.000 °C, the freezing point of the solution will be −0.186 °C.

(*b*) $\Delta t_f = (5.12\text{ °C/m})(0.100\text{ m}) = 0.512\text{ °C}$

The freezing point is lowered by 0.512 °C, from 5.5 to 5.0 °C.

17.38 A solution containing 3.50 g of solute X in 50.0 g of water has a volume of 52.5 mL and a freezing point of −0.86 °C. (*a*) Calculate the molality, mole fraction, and molarity of X. (*b*) Calculate the molecular weight of X.

▮ (*a*) $\text{Molality} = \dfrac{\Delta t}{K_f} = \dfrac{0.86\text{ °C}}{1.86\text{ °C/m}} = 0.46\text{ m}$

Thus there is 0.46 mol of X/kg of water, or 0.46 mol of X per 55.5 mol of water. The mole fraction is therefore

$$x(\text{X}) = \frac{0.46\text{ mol}}{(0.46 + 55.5)\text{ mol}} = 0.0082$$

The solution has a density of

$$\frac{(50.0\text{ g }H_2O) + (3.50\text{ g X})}{52.5\text{ mL}} = 1.02\text{ g/mL}$$

Per kg of water there is

$$\left(\frac{3.50\text{ g X}}{50.0\text{ g }H_2O}\right)(1000\text{ g }H_2O) = 70.0\text{ g X}$$

The total mass of the solution is thus 1070 g; its total volume is

$$(1070\text{ g})\left(\frac{1\text{ mL}}{1.02\text{ g}}\right) = 1050\text{ mL} = 1.05\text{ L}$$

Then

$$\text{molarity} = \frac{0.46\text{ mol}}{1.05\text{ L}} = 0.44\text{ M}$$

(*b*) $\dfrac{70.0\text{ g}}{0.46\text{ mol}} = 152\text{ g/mol}$

17.39 Camphor, $C_{10}H_{16}O$, which has a freezing point of 174 °C, has a freezing point depression constant of 40.0 °C/m. Explain the usefulness and the limitations of camphor as a solvent for determination of molecular weights. For what kind(s) of solutes would camphor be especially useful?

▮ The freezing point depression constant is so large that an accurate freezing point depression can be obtained on a solution with very few moles dissolved per kilogram of camphor. Thus, high molecular weight materials give measurable freezing point depressions in this solvent. There is a problem in getting many materials to dissolve in camphor, however.

17.40 When 36.0 g of a solute having the empirical formula CH_2O is dissolved in 1.20 kg of water, the solution freezes at −0.93 °C. What is the molecular formula of the solute?

▮

$$m = \frac{\Delta t}{K_f} = \frac{0.93°}{1.86°/\text{m}} = 0.50\text{ m}$$

Per kg of solvent, there is 0.50 mol of solute, which has a mass of 36.0 g:

$$\frac{36.0\text{ g}}{1.20\text{ kg}} = 30.0\text{ g/kg} \qquad \frac{30.0\text{ g}}{0.50\text{ mol}} = 60\text{ g/mol}$$

The empirical formula weight is 30 g/empirical formula unit; there must be 60/30 = 2 empirical formula units per molecule. The formula is thus $(CH_2O)_2$ or $C_2H_4O_2$.

17.41 Calculate the freezing point of 0.200 m solutions of fructose, $C_6H_{12}O_6$, (*a*) in water (*b*) in acetic acid

▮ (*a*) $\Delta t = K_f m = (1.86\ °C/m)(0.200\ m) = 0.372\ °C \qquad t = (0.000 - 0.372)\ °C = -0.372\ °C$

(*b*) $\Delta t = K_f m = (3.9\ °C/m)(0.200\ m) = 0.78\ °C \qquad t = (16.6 - 0.78)\ °C = 15.8\ °C$

17.42 A solution of 10.0 g of a nonionic solute in 100 g of benzene freezes at 4.2 °C. Calculate the molecular weight of the solute.

▮ The freezing point depression, Δt, is $(5.5 - 4.2)\ °C = 1.3\ °C$

$$m = \frac{\Delta t}{K} = \frac{1.3\ °C}{5.12\ °C/m} = 0.25\ m$$

The solution contains 100 g of solute/kg of benzene. It contains 0.25 mol solute/kg of benzene, as determined from the freezing point depression. Thus the molecular weight is

$$\frac{100\ g}{0.25\ mol} = 400\ g/mol$$

17.43 Distinguish (*a*) between the boiling point of a liquid and the normal boiling point of the liquid, (*b*) between freezing point and freezing point depression.

▮ (*a*) The boiling point is the temperature at which the vapor pressure of the liquid equals the surrounding pressure. The normal boiling point is the temperature at which the vapor pressure is 1 atm. (*b*) The freezing point is the temperature at which solid first crystallizes from solution; the freezing point depression is the difference between that temperature and the freezing point of the pure solvent. For example, a certain benzene solution starts to freeze at 4.5 °C; its freezing point depression is $(5.5 - 4.5)\ °C = 1.0\ °C$.

17.44 A solution containing 6.35 g of a nonelectrolyte dissolved in 500 g of water freezes at −0.465 °C. Determine the molecular weight of the solute.

▮

$$\Delta t = Km \qquad m = \frac{\Delta t}{K} = \frac{0.465\ °C}{1.86\ °C/m} = 0.250\ m = \frac{0.250\ mol}{kg\ solvent}$$

$$\frac{6.35\ g\ solute}{500\ g\ solvent} = \frac{12.7\ g\ solute}{kg\ solvent} \qquad MW = \frac{12.7\ g}{0.250\ mol} = 50.8\ g/mol$$

17.45 A solution containing 3.24 g of a nonvolatile nonelectrolyte and 200 g of water boils at 100.130 °C at 1 atm. What is the molecular weight of the solute?

▮

$$\Delta t = Km \qquad m = \frac{\Delta t}{K} = \frac{(100.130 - 100.000)\ °C}{0.513\ °C/m} = 0.253\ m = \frac{0.253\ mol}{kg\ solvent}$$

$$\frac{3.24\ g\ solute}{200\ g\ solvent} = \frac{16.2\ g\ solute}{kg\ solvent} \qquad \frac{16.2\ g}{0.253\ mol} = 64.0\ g/mol$$

17.46 Calculate the freezing point and the boiling point at 1 atm of a solution containing 30.0 g cane sugar (molecular weight 342 g/mol) and 150 g water.

▮

$$(30.0\ g)\left(\frac{1\ mol}{342\ g}\right) = 0.0877\ mol \qquad \frac{0.0877\ mol}{0.150\ kg} = 0.585\ m$$

$$\Delta t_f = K_f m = (1.86\ °C/m)(0.585\ m) = 1.09\ °C \qquad fp = (0 - 1.09)\ °C = -1.09\ °C$$

$$\Delta t = K_b m = (0.513\ °C/m)(0.585\ m) = 0.300\ °C \qquad bp = (100.000 + 0.300)\ °C = 100.300\ °C$$

17.47 A solution was made up by dissolving 3.75 g of a pure hydrocarbon in 95.0 g of acetone. The boiling point of pure acetone was observed to be 55.95 °C, and of the solution, 56.50 °C. If the molal boiling point constant of acetone is 1.71 °C·kg/mol, what is the approximate molecular weight of the hydrocarbon?

▮ *First method*

Compute the molality (m) from the boiling-point equation.

$$\Delta t_b = K_b m \qquad (56.50 - 55.95)\ °C = 1.71m$$

Solving, $m = 0.322$ mol solute/kg solvent. Now find the number of moles of solute in the weighed sample.

$$n(\text{solute}) = \left(0.322\,\frac{\text{mol solute}}{\text{kg solvent}}\right)(0.0950\text{ kg solvent}) = 0.0306\text{ mol solute}$$

Then
$$\text{MW} = \frac{3.75\text{ g solute}}{0.0306\text{ mol solute}} = 123\text{ g/mol}$$

Second method
For the given concentration,

$$1\text{ g acetone contains }\frac{3.75}{95.0}\text{ g solute} \qquad 1000\text{ g acetone contains }1000\left(\frac{3.75}{95.0}\right) = 39.5\text{ g solute}$$

Since 1.71 °C elevation is produced by 1 mol solute per 1000 g acetone, 0.55 °C elevation is produced by

$$\frac{0.55}{1.71} = 0.322\text{ mol solute}$$

per 1000 g acetone. Thus 0.322 mol solute is equivalent to 39.5 g solute, and

$$\text{MW} = \frac{39.5\text{ g}}{0.322\text{ mol}} = 123\text{ g/mol}$$

17.48 The molecular weight of an organic compound is 58.0 g/mol. Compute the boiling point of a solution containing 24.0 g of the solute and 600 g of water, when the barometric pressure is such that pure water boils at 99.725 °C.

■

$$\text{Molality} = m = \frac{n(\text{solute})}{\text{number of kg solvent}} = \frac{(24.0/58.0)\text{ mol solute}}{0.600\text{ kg solvent}} = 0.690\text{ mol/kg}$$

$$\text{Elevation of boiling point} = \Delta t_b = K_b m = (0.513\text{ °C}\cdot\text{kg/mol})(0.690\text{ mol/kg}) = 0.354\text{ °C}$$

$$\text{Boiling point of solution} = (\text{boiling point of water}) + \Delta t_b = (99.725 + 0.354)\text{ °C} = 100.079\text{ °C}$$

17.49 A certain solution of benzoic acid in benzene has a freezing point of 3.1 °C and a normal boiling point of 82.6 °C. Explain these observations, and suggest structures for the solute particles at the two temperatures.

■

$$m = \frac{\Delta t}{K_f} = \frac{(5.5 - 3.1)\text{ °C}}{5.12\text{ °C/m}} = 0.47\text{ m} \qquad m = \frac{\Delta t}{K_b} = \frac{(82.6 - 80.1)\text{ °C}}{2.53\text{ °C/m}} = 0.99\text{ m}$$

It is evident that the number of moles of solute particles at the boiling point is twice that at the freezing point of the solution. In fact, benzoic acid dimerizes (forms double molecules) by hydrogen bonding at the lower temperature, but the dimers are broken up at the elevated temperature.

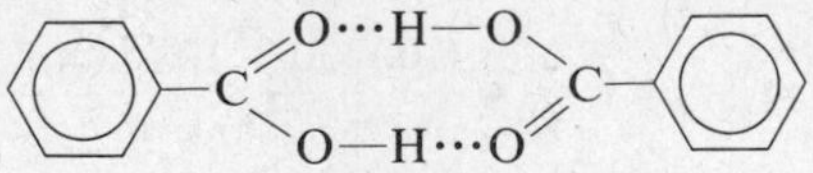

17.50 An aqueous solution of a nonvolatile solute boils at 100.17 °C. At what temperature will this solution freeze?

■ The molality is calculated from the boiling point elevation:

$$m = \frac{\Delta t}{K_b} = \frac{0.17\text{ °C}}{0.512\text{ °C}} = 0.33\text{ m}$$

The freezing point depression is calculated from that molality:

$$\Delta t = K_f m = (1.86\text{ °C/m})(0.33\text{ m}) = 0.62\text{ °C}$$

The freezing point is

$$t = (0.00 - 0.62)\text{ °C} = -0.62\text{ °C}$$

17.51 A certain nonvolatile nonelectrolyte contains 40.0% carbon, 6.7% hydrogen, and 53.3% oxygen. An aqueous solution containing 5.00% by mass of the solute boils at 100.15°C. Determine the molecular formula of the compound.

■

$$(40.0\text{ g C})\left(\frac{1\text{ mol C}}{12.0\text{ g C}}\right) = 3.33\text{ mol C} \qquad (6.7\text{ g H})\left(\frac{1\text{ mol H}}{1.0\text{ g H}}\right) = 6.7\text{ mol H}$$

$$(53.3\text{ g O})\left(\frac{1\text{ mol O}}{16.0\text{ g O}}\right) = 3.33\text{ mol O}$$

The empirical formula is CH_2O, which has a formula weight of 30 g/empirical formula unit

$$m = \frac{\Delta t}{K} = \frac{0.15\ °C}{0.512\ °C/m} = 0.29\ m$$

$$\frac{50.0\ g/0.950\ kg\ H_2O}{0.29\ mol/kg\ H_2O} = 180\ g/mol \qquad \text{(2 significant figures)}$$

$$\frac{180}{30} = 6\ \text{units}$$

The molecular formula is $C_6H_{12}O_6$.

17.52 Pure benzene freezes at 5.45 °C. A solution containing 7.24 g of $C_2H_2Cl_4$ in 115.3 g of benzene was observed to freeze at 3.55 °C. What is the molal freezing point constant of benzene?

▌

$$\Delta t = (5.45 - 3.55)\ °C = 1.90\ °C$$

$$(7.24\ g\ C_2H_2Cl_4)\left(\frac{1\ mol}{168\ g}\right) = 0.0431\ mol$$

$$\frac{0.0431\ mol\ C_2H_2Cl_4}{0.1153\ kg\ C_6H_6} = 0.374\ m \qquad K_f = \frac{\Delta t}{m} = \frac{1.90\ °C}{0.374\ m} = 5.08\ °C/m$$

17.3 OSMOTIC PRESSURE

17.53 The osmotic pressure of blood is 7.65 atm at 37 °C. How much glucose should be used per L for an intravenous injection that is to have the same osmotic pressure as blood?

▌

$$n = \frac{\pi V}{RT} = \frac{(7.65\ atm)(1.00\ L)}{(0.0821\ L\cdot atm/mol\cdot K)(310\ K)} = 0.301\ mol \qquad (0.301\ mol)\left(\frac{180\ g\ C_6H_{12}O_6}{mol}\right) = 54.2\ g$$

17.54 What is the osmotic pressure at 0 °C of an aqueous solution containing 46.0 g of glycerin ($C_3H_8O_3$) per L?

▌

$$\pi V = nRT$$

$$(46.0\ g)\left(\frac{1\ mol}{92.0\ g}\right) = 0.500\ mol$$

$$\pi = \frac{(0.500\ mol)(0.0821\ L\cdot atm/mol\cdot K)(273\ K)}{1.00\ L} = 11.2\ atm$$

17.55 An aqueous solution of urea had a freezing point of −0.52 °C. Predict the osmotic pressure of the same solution at 37 °C. Assume that the molar concentration and the molality are numerically equal.

▌ The concentration of the solution is not specified, but the effective molality may be inferred from the freezing point lowering.

$$m = \frac{\Delta t_f}{K_f} = \frac{0.52\ °C}{1.86\ °C\cdot kg/mol} = 0.280\ mol/kg$$

The assumption that the molality and molar concentration are equal is not very bad for dilute aqueous solutions. (The relation found in Prob. 16.91 shows that $M \approx m$ when $d \approx 1\ g/cm^3$ and $W \ll 1000/m$. Urea has a molecular weight of 60 g/mol.) Then 0.280 mol/L may be used for the molar concentration in the osmotic pressure equation.

$$\pi = MRT = (0.280\ mol/L)(0.082\ L\cdot atm/K\cdot mol)(310\ K) = 7.1\ atm$$

17.56 The osmotic pressure of a solution of a synthetic polyisobutylene in benzene was determined at 25 °C. A sample containing 0.20 g of solute/100 cm³ of solution developed a rise of 2.4 mm at osmotic equilibrium. The density of the solution was 0.88 g/cm³. What is the molecular weight of the polyisobutylene?

▌ The osmotic pressure is equal to that of a column of the solution 2.4 mm high.

$$\pi = (\text{height})(\text{density})(g) = (2.4 \times 10^{-3}\ m)(0.88 \times 10^3\ kg/m^3)(9.81\ m/s^2) = 20.7\ Pa$$

The molar concentration can now be determined from the osmotic pressure equation.

$$M = \frac{\pi}{RT} = \frac{20.7\ N/m^2}{(8.314\ J/K\cdot mol)(298\ K)} = 8.3 \times 10^{-3}\ mol/m^3 = 8.3 \times 10^{-6}\ mol/L$$

The solution contained 0.20 g solute/100 mL solution, or 2.0 g/L, and has been found to contain 8.3×10^{-6} mol/L. Then

$$\text{MW} = \frac{2.0\ \text{g}}{8.3 \times 10^{-6}\ \text{mol}} = 2.4 \times 10^5\ \text{g/mol}$$

17.57 What would be the osmotic pressure at 17 °C of an aqueous solution containing 1.75 g of sucrose ($C_{12}H_{22}O_{11}$) per 150 cm^3 of solution?

▮

$$\text{Molar concentration} = M = \frac{1.75\ \text{g/(342 g/mol)}}{0.150\ \text{L}} = 0.0341\ \text{mol/L}$$

$$\text{Osmotic pressure} = \pi = MRT = (0.0341\ \text{mol/L})(0.0821\ \text{L·atm/K·mol})(290\ \text{K}) = 0.81\ \text{atm}$$

17.58 Calculate the osmotic pressure of a 0.100 M solution of a nonelectrolyte at 0 °C.

▮

$$\pi = \frac{n}{V}RT = \left(\frac{0.100\ \text{mol}}{\text{L}}\right)\left(\frac{0.0821\ \text{L·atm}}{\text{mol·K}}\right)(273\ \text{K}) = 2.24\ \text{atm}$$

17.59 What is the molecular weight, M_A, of a solute, A, if the osmotic pressure of a solution containing 10.0 g/L is 10.0 torr at 27 °C?

▮

$$\pi V = nRT = \frac{m}{M_A}RT \qquad M_A = \frac{mRT}{\pi V} = \frac{(10.0\ \text{g})(0.0821\ \text{L·atm/mol·K})(300\ \text{K})}{[(10.0/760)\ \text{atm}](1.00\ \text{L})} = 18\,700\ \text{g/mol}$$

17.60 A 250-mL water solution containing 48.0 g of sucrose, $C_{12}H_{22}O_{11}$, at 300 K is separated from pure water by means of a semipermeable membrane. What pressure must be applied above the solution in order to just prevent osmosis?

▮

$$n = \frac{48.0\ \text{g}}{342\ \text{g/mol}} = 0.140\ \text{mol}$$

$$\pi = \frac{nRT}{V} = \frac{(0.140\ \text{mol})(0.0821\ \text{L·atm/mol·K})(300\ \text{K})}{0.250\ \text{L}} = 13.8\ \text{atm}$$

17.61 Calculate the osmotic pressure of an aqueous solution which contains 4.00 g of glucose, $C_6H_{12}O_6$, in 250 mL of solution at 25 °C.

▮

$$(4.00\ \text{g})\left(\frac{1\ \text{mol}}{180\ \text{g}}\right) = 0.0222\ \text{mol}\ C_6H_{12}O_6$$

$$\pi = \frac{nRT}{V} = \frac{(0.0222\ \text{mol})(0.0821\ \text{L·atm/mol·K})(298\ \text{K})}{0.250\ \text{L}} = 2.17\ \text{atm}$$

17.62 A solution of crab hemocyanin, a pigmented protein extracted from crabs, was prepared by dissolving 0.750 g in 125 cm^3 of an aqueous medium. At 4 °C an osmotic pressure rise of 2.6 mm of the solution was observed. The solution had a density of 1.00 g/cm^3. Determine the molecular weight of the protein.

▮

$$n = \frac{\pi V}{RT} = \frac{[2.6\ \text{mm}/(13.6 \times 760\ \text{mm/atm})](0.125\ \text{L})}{(0.0821\ \text{L·atm/mol·K})(277\ \text{K})} = 1.38 \times 10^{-6}\ \text{mol}$$

$$\frac{0.750\ \text{g}}{1.38 \times 10^{-6}\ \text{mol}} = 5.4 \times 10^5\ \text{g/mol}$$

17.4 OTHER PROPERTIES OF SOLUTIONS

17.63 In the process of **recrystallization**, an impure solid sample is dissolved in a minimum quantity of hot solvent; the solution is filtered and allowed to cool, whereupon purer solute separates. ***(a)*** Explain why this process removes both soluble and insoluble impurities from the solid sample. ***(b)*** Which compound, KCl or KNO_3, would be more apt to be easily purified by recrystallization? (*Hint*: see Fig. 17.3.)

▮ ***(a)*** The impurities which are less soluble than the desired product are separated by filtration of the hot solution. As the solution is cooled, the sample crystallizes and some of the impurities remain in solution. The sample has been purified somewhat. Repetition of the procedure may be used to further purify the sample. ***(b)*** KNO_3 is more easily purified by this method because its solubility changes with temperature so much more (see Fig. 17.3).

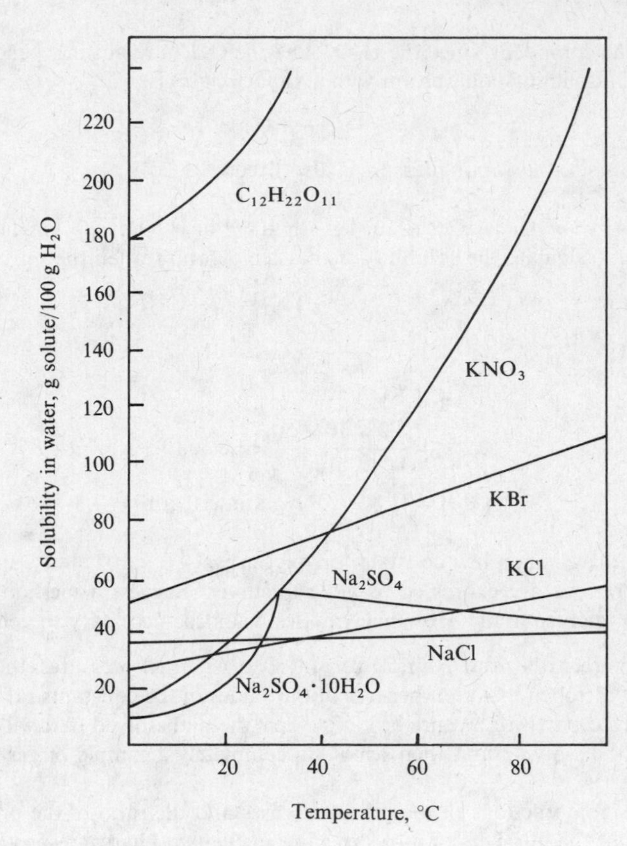

Fig. 17.3

17.64 Solid $Ce_2(SO_4)_3$ has a positive enthalpy of solution—it dissolves in water to a greater extent at lower temperature than at higher. Suggest how this substance can be purified by recrystallization. (*Hint*: see Prob. 17.63.)

▌ The substance is dissolved to make a saturated solution at low temperature; the solution is filtered and then heated. The $Ce_2(SO_4)_3$ crystallizes out at the higher temperature, whereas the impurities tend to become even more soluble at the elevated temperature. The purified $Ce_2(SO_4)_3$ is filtered from the hot solution.

17.65 A clear solution is heated in an open vessel, whereupon a solid separates from the hot solution. Given two possible explanations for this behavior. Devise an experiment which would distinguish between these two possibilities.

▌ The material could have a positive enthalpy of solution or the solvent may merely have evaporated.

17.66 Explain why the melting point of a substance gives an indication of the purity of the substance. Explain why one usually recrystallizes an unknown product until successive recrystallized samples show no increase in melting point.

▌ Impurities cause a freezing point (melting point) depression. The more impurities, the lower the freezing point. If one purifies an unknown until the melting point stops rising, one can usually assume that all of the impurities have been removed.

17.67 Determine the number of degrees of freedom in a system containing a solution of NaCl in water, ice, and solid NaCl, all in equilibrium with each other.

▌ There are 4 phases—solution, solid water, solid salt, and vapor. There are 2 components. Hence

$$p + f = c + 2$$
$$f = 2 + 2 - 4 = 0$$

The number of degrees of freedom is zero. The temperature, the vapor pressure of the water, and the concentration of the saturated solution are all fixed.

17.68 Pure water contains H_3O^+ and OH^- in addition to H_2O. How many degrees of freedom are there in pure water at 50 °C and 1 atm pressure?

▮ There is only one compoent, since the H_3O^+ and the OH^- are not independently variable. Thus the number of degrees of freedom (assuming equilibrium with a vapor phase) is

$$f = c + 2 - p = 1 + 2 - 2 = 1$$

If the temperature is set, the vapor pressure is also fixed.

17.69 The solubility of N_2 in water is 2.2×10^{-4} g in 100 g of H_2O at 20 °C when the pressure of nitrogen over the solution is 1.2 atm. Calculate the solubility at that temperature when the nitrogen pressure is 10 atm.

▮ By Henry's law,

$$[N_2] = KP(N_2)$$

Per 100 g of water,

$$2.2 \times 10^{-4}\ \text{g} = K(1.2\ \text{atm})$$

$$K = \frac{2.2 \times 10^{-4}\ \text{g}}{1.2\ \text{atm}} = 1.8 \times 10^{-4}\ \text{g/atm}$$

$$[N_2] = (1.8 \times 10^{-4}\ \text{g/atm})(10\ \text{atm}) = 1.8 \times 10^{-3}\ \text{g} = 1.8\ \text{mg}$$

17.70 In water at 20 °C, the Henry's law constant for oxygen is 4.6×10^4 atm, and for nitrogen it is 8.2×10^4 atm, where the concentrations are expressed as mole fractions. Suggest a method based on these data for preparing 99 mole % pure oxygen from air. How many cycles would be necessary to achieve this result?

▮ Air is 79 mole percent N_2 and 20 mole percent O_2. At equal pressures, the concentration of oxygen would be about twice that of nitrogen in water, because of the values of the constants. If a sample of air is partially dissolved in water, the dissolved part will be enriched in oxygen; the undissolved part will be enriched in nitrogen. Successive treatments partially dissolving and then removing completely a sample of gas will produce gas successively richer in oxygen.

The ratio of the mole fractions of the dissolved gases and the ratio of the pressures of the undissolved gases are equal to the ratios of the numbers of moles of gases in the two phases, respectively. Henry's law can be expressed, using the data given in the exercise, in the forms

$$x(N_2)(8.2 \times 10^4\ \text{atm}) = P(N_2) \qquad x(O_2)(4.6 \times 10^4\ \text{atm}) = P(O_2)$$

$$\frac{P(N_2)}{P(O_2)} = \frac{8.2}{4.6}\left[\frac{N(N_2)}{N(O_2)}\right]$$

The initial dissolved gas will have a mole fraction ratio

$$\frac{x(N_2)}{x(O_2)} = \frac{79}{20} \times \frac{4.6}{8.2} = 2.22$$

All the gas is removed under vacuum from the water, and the new gas phase is partially redissolved, whereupon the $P(N_2)/P(O_2)$ ratio is 2.22/1. The ratio of mole fractions is

$$\frac{x(N_2)}{x(O_2)} = 2.22 \times \frac{4.6}{8.2}$$

It is evident that each repetition of the procedure results in multiplication of the pressure ratio by the factor 4.6/8.2. For the n times that the procedure must be repeated to obtain 90% pure oxygen, the simple equation may be formulated

$$\frac{79}{20} \times \left(\frac{4.6}{8.2}\right)^n = \frac{10}{90} \qquad \left(\frac{4.6}{8.2}\right)^n = \frac{10}{90} \times \frac{20}{79}$$

$$n \log \frac{4.6}{8.2} = n(-0.2510) = \log 0.0281 = -1.551$$

$$n = 6.2$$

It will take 7 cycles to achieve over 90% pure O_2.

17.71 At 20 °C and a total pressure of 760 torr, 1 L of water dissolves 0.043 g of pure oxygen or 0.019 g of pure nitrogen. Assuming that dry air is composed of 20% oxygen and 80% nitrogen (by volume), determine the masses of oxygen and nitrogen dissolved by 1 L of water at 20 °C exposed to air at a total pressure of 760 torr.

▮ The *solubility* of a gas, i.e., the concentration of the dissolved gas in the presence of excess gas, may be expressed as

$$\text{Solubility of Y} = K_H(\text{Y}) \times P(\text{Y})$$

In words, the solubility of a gas dissolved from a gaseous mixture (air in the present case) is directly proportional to the partial pressure of the gas; the proportionality constant, K_H, is called the *Henry's law constant*. (Some authors define the Henry's law constant as the reciprocal of the K_H used here.) To evaluate K_H from the data, note that when pure oxygen is equilibrated with water at a total pressure of 760 torr,

$$P(O_2) = (760 \text{ torr}) - (\text{vapor pressure of water})$$

Then, from the data,

$$K_H(O_2) = \frac{\text{solubility of } O_2}{P(O_2)} = \frac{0.043 \text{ g/L}}{760 \text{ torr} - \text{vp}} \qquad K_H(N_2) = \frac{\text{solubility of } N_2}{P(N_2)} = \frac{0.019 \text{ g/L}}{760 \text{ torr} - \text{vp}}$$

When water is exposed to air at a total pressure of 760 torr,

$$P(O_2) = (0.20)(760 \text{ torr} - \text{vp}) \qquad P(N_2) = (0.80)(760 \text{ torr} - \text{vp})$$

Hence

$$\text{Solubility of } O_2 \text{ from air} = K_H(O_2) \times P(O_2) = \left(\frac{0.043 \text{ g/L}}{760 \text{ torr} - \text{vp}}\right)(0.20)(760 \text{ torr} - \text{vp}) = 0.0086 \text{ g/L}$$

Similarly, the solubility of N_2 from air is $(0.80)(0.019 \text{ g/L}) = 0.015$ g/L.

17.72 A gaseous mixture of hydrogen and oxygen contains 70% hydrogen and 30% oxygen by volume. If the gas mixture at a pressure of 2.5 atm (excluding the vapor pressure of water) is allowed to saturate water at 20 °C, the water is found to contain 31.5 mL (STP) of hydrogen/L. Find the solubility of hydrogen (reduced to STP) at 20 °C and 1 atm partial pressure of hydrogen.

▮ Since the volume of a gas at STP depends only on the mass, the volume of the dissolved gas (reduced to STP) is proportional to the partial pressure of the gas.

$$\text{Partial pressure of hydrogen} = (0.70)(2.5 \text{ atm}) = 1.75 \text{ atm}$$

$$\text{Solubility of } H_2 \text{ at 20 °C and 1 atm} = \left(\frac{1.00 \text{ atm}}{1.75 \text{ atm}}\right)(31.5 \text{ mL/L}) = 18.0 \text{ mL (STP)/L}$$

17.73 If 29 mg of N_2 dissolves in 1 L of water at 0 °C and 760 torr N_2 pressure, how much N_2 will dissolve in 1 L of water at 0 °C and 5.00 atm N_2 pressure?

▮ According to Henry's Law,

$$[N_2] = KP(N_2)$$

This equation holds for both sets of conditions. Dividing the equation for one set of conditions by that for the other yields

$$\frac{[N_2]}{29 \text{ mg/L}} = \frac{K(5.00 \text{ atm})}{K(1.00 \text{ atm})} = 5.00 \qquad [N_2] = 5.00(29 \text{ mg/L}) = 145 \text{ mg/L}$$

17.74 At 20 °C and 1.00 atm partial pressure of hydrogen, 18 mL of hydrogen, measured at STP, dissolves in 1 L of water. If water at 20 °C is exposed to a gaseous mixture having a total pressure of 1400 torr (excluding the vapor pressure of water) and containing 68.5% H_2 by volume, find the volume of H_2, measured at STP, which will dissolve in 1 L of water.

▮

$$P(H_2) = (1400 \text{ torr})(0.685) = (959 \text{ torr})\left(\frac{1 \text{ atm}}{760 \text{ torr}}\right) = 1.26 \text{ atm}$$

$$\frac{V}{18 \text{ mL}} = \frac{1.26 \text{ atm}}{1.00 \text{ atm}}$$

$$V = 23 \text{ mL}$$

17.75 A liter of CO_2 gas at 15 °C and 1.00 atm dissolves in 1.00 L of water at the same temperature when the pressure of CO_2 is 1.00 atm. Compute the molar concentration of CO_2 in a solution over which the partial pressure of CO_2 is 150 torr at this temperature.

▮

$$n = \frac{PV}{RT} = \frac{(1.00 \text{ atm})(1.00 \text{ L})}{(0.0821 \text{ L·atm/mol·K})(288 \text{ K})} = 0.0423 \text{ mol}$$

The concentration at 1.00 atm partial pressure is 0.0423 M. At $\frac{150}{760}$ atm partial pressure, the concentration is $0.0423(\frac{150}{760}) = 8.35 \times 10^{-3}$ M = 8.35 mM.

17.76 (*a*) An aqueous solution of iodine, of volume 25 mL and containing 2 mg of iodine, is shaken with 5 mL of CCl_4, and the CCl_4 is allowed to separate. Given that the solubility of iodine per unit volume is 85 times greater in CCl_4 than in water at the temperature of the experiment and that both saturated solutions may be considered to be "dilute," calculate the quantity of iodine remaining in the water layer. (*b*) If a second extraction is made of the water layer using another 5 mL of CCl_4, calculate the quantity of iodine remaining after the second extraction.

▮ (*a*) Let

$$x = \text{mg of iodine in } H_2O \text{ layer at equilibrium}$$
$$2 - x = \text{mg of iodine in } CCl_4 \text{ layer at equilibrium}$$

The concentration of iodine in the water layer will be $x/25$ (mg/mL of water), and the concentration of iodine in the CCl_4 layer will be $(2 - x)/5$ (mg/mL of CCl_4). Hence

$$\frac{\text{Conc. } I_2 \text{ in } CCl_4}{\text{Conc. } I_2 \text{ in } H_2O} = \frac{85}{1} \quad \text{or} \quad \frac{(2-x)/5}{x/25} = \frac{85}{1} \quad \text{or} \quad \frac{2-x}{x} = 17$$

Solving, $x = 0.11$ mg iodine.

(*b*) Let

$$y = \text{mg of iodine in } H_2O \text{ layer after second extraction}$$
$$0.11 - y = \text{mg of iodine in } CCl_4 \text{ layer after second extraction}$$

The concentration of iodine in the water layer will be $y/25$, and the concentration in the CCl_4 layer will be $(0.11 - y)/5$. Hence

$$\frac{\text{Conc. } I_2 \text{ in } CCl_4}{\text{Conc. } I_2 \text{ in } H_2O} = \frac{85}{1} \quad \text{or} \quad \frac{(0.11-y)/5}{y/25} = \frac{85}{1} \quad \text{or} \quad \frac{0.11-y}{y} = 17$$

Solving, $y = 0.0061$ mg iodine.

17.77 (*a*) The solubility of iodine per unit volume is 200 times greater in ether than in water at a particular temperature. If an aqueous solution of iodine, 30 mL in volume and containing 2.0 mg of iodine, is shaken with 30 mL of ether and the ether is allowed to separate, what quantity of iodine remains in the water layer? (*b*) What quantity of iodine remains in the water layer if only 3 mL of ether is used? (*c*) How much iodine is left in the water layer if the extraction in (*b*) is followed by a second extraction, again using 3 mL of ether? (*d*) Which method is more efficient, a single large washing or repeated small washings?

▮ (*a*) Let $x = \text{mg } I_2$ in water at equilibrium. Then $2.0 - x = \text{mg } I_2$ in ether at equilibrium.

$$\frac{2.0 - x}{30} = 200\left(\frac{x}{30}\right)$$
$$2.0 = 201x$$
$$x = 0.010 \text{ mg } I_2$$

(*b*)
$$\frac{2.0 - x}{3.0} = 200\left(\frac{x}{30}\right)$$
$$20 - 10x = 200x$$
$$x = 0.095 \text{ mg } I_2$$

(*c*) Let $y = \text{mg } I_2$ in water after second equilibrium.

$$\frac{0.095 - y}{3.0} = 200\left(\frac{y}{30}\right)$$
$$y = 0.0045 \text{ mg } I_2$$

(*d*) Repeated extractions with small volumes are more effective. When 6 mL of ether are used 3 mL at a time, the I_2 concentration in water is lower than the concentration found after 30 mL of ether is used all at once [part (*a*)].

17.78 The ratio of the solubility of stearic acid per unit volume of *n*-heptane to that in 97.5% acetic acid is 4.95. How many extractions of a 10 mL solution of stearic acid in 97.5% acetic with successive 10 mL portions of *n*-heptane are needed to reduce the residual stearic acid content of the acetic acid layer to less than 0.5% of its original value?

▌ The ratio in each extraction is 1.00/4.95. The ratio of the concentration in the acetic acid to the total concentration is

$$\frac{1.00}{1.00 + 4.95} = \frac{1.00}{5.95}$$

But n repeated extractions leaves a concentration of 0.005:

$$\left(\frac{1.00}{5.95}\right)^n = 0.005$$

$$-n \log 5.95 = \log 0.005$$

$$n = 2.97$$

Three extractions will suffice.

17.79 One method of purifying penicillin is by extraction. The distribution coefficient for penicillin G between isopropyl ether and an aqueous phosphate medium is 0.34 (lower solubility in the ether). The corresponding ratio for penicillin F is 0.68. A preparation of penicillin G has penicillin F as a 10.0% impurity. ***(a)*** If an aqueous phosphate solution of this preparation is extracted with an equal volume of isopropyl ether, what will be the percent recovery of the initial G in the residual aqueous-phase product after one extraction, and what will be the percent impurity in this product? ***(b)*** Compute the same two quantities for the product remaining in the aqueous phase after a second extraction with an equal volume of ether.

▌ Let C = original penicillin G concentration. Then $0.100C$ = original penicillin F concentration.

(a) After one extraction, let x mol/L of G remain in the phosphate.

$$\frac{\text{G concentration in ether}}{\text{G concentration in phosphate}} = 0.34 = \frac{C - x}{x}$$

$$C - x = 0.34\,x$$

$$\frac{x}{C} = \frac{1}{1.34} = 0.75 = 75\%\ \text{recovery}$$

Let y mol/L of F remain in the phosphate.

$$\frac{\text{F concentration in ether}}{\text{F concentration in phosphate}} = 0.68 = \frac{0.100C - y}{y}$$

$$0.68y = 0.100C - y$$

$$y = \frac{0.100C}{1.68} = 0.060C$$

$$\frac{y}{x} = \frac{0.060C}{0.75C} \times 100 = 8.0\%\ \text{impurity}$$

(b) Two extractions yield:

$$\left(\frac{1}{1.34}\right)^2 = 0.56\% = 56\%\ \text{recovery of G}$$

$$\left(\frac{1}{1.68}\right)^2 = 0.35 = 35\%\ \text{recovery of F}$$

$$35\%\ \text{of}\ 0.100C = 0.035C \qquad \text{Impurity} = \frac{0.035C}{0.56C} \times 100 = 6.4\%\ \text{impurity}$$

17.80 The molecular weight of a newly synthesized organic compound was determined by the method of *isothermal distillation.* In this procedure two solutions, each in an open calibrated vial, are placed side by side in a closed chamber. One of the solutions contained 9.3 mg of the new compound, the other 13.2 mg of azobenzene (MW = 182). Both were dissolved in portions of the same solvent. During a period of 3 days of equilibration, solvent distilled from one vial into the other until the same partial pressure of solvent was reached in the two vials. At this point the distillation of solvent stopped. Neither of the solutes distilled at all. The volumes of the two solutions at equilibrium were then read on the calibration marks of the vials. The solution containing the new compound occupied 1.72 mL and the azobenzene solution occupied 1.02 mL. What is the molecular weight of the new compound? The mass of solvent in solution may be assumed to be proportional to the volume of the solution.

▮
$$P = P^\circ x_1 = P^\circ x_2$$

Since the partial pressures of the two solutions in the same solvent are the same, their mole fractions must also be the same.

$$\frac{n_{\text{azo}}}{n_{\text{azo}} + (n_{\text{solvent}})_1} = \frac{n_X}{n_X + (n_{\text{solvent}})_2}$$

The subscripts for the solvent terms refer to the solution in which the solvent is found. Since the masses of the solutes are so low, the denominators may be approximated as the masses of the solvent alone, which is proportional to the number of moles of solvent alone.

$$\frac{n_{\text{azo}}}{(n_{\text{solvent}})_1} = \frac{n_X}{(n_{\text{solvent}})_2}$$

$$\frac{n_{\text{azo}}}{n_X} = \frac{(n_{\text{solvent}})_1}{(n_{\text{solvent}})_2} = \frac{(m_{\text{solvent}})_1}{(m_{\text{solvent}})_2} = \frac{(V_{\text{solvent}})_1}{(V_{\text{solvent}})_2} = \frac{1.02\text{ mL}}{1.72\text{ mL}}$$

The mole ratio of solutes is equal to

$$\frac{m_{\text{azo}}/\text{MW}_{\text{azo}}}{m_X/\text{MW}_X} = \frac{1.02}{1.72}$$

$$\text{MW}_X = \frac{m_X}{m_{\text{azo}}}(\text{MW}_{\text{azo}})\left(\frac{1.02}{1.72}\right) = \left(\frac{9.3\text{ mg X}}{13.2\text{ mg azo}}\right)(182\text{ g/mol})\left(\frac{1.02}{1.72}\right) = 76.0\text{ g/mol}$$

17.81 Describe how and explain why a standard solution (of precisely known concentration) of HCl can be prepared from an aqueous solution of HCl without chemical analysis.

▮ The HCl-H_2O eutectic has a definite composition at any given pressure. Distillation of the mixture will sooner or later yield a constant boiling mixture, which has a composition which has been determined previously to an accuracy sufficient for analytical determinations (to at least 4 significant figures).

17.82 Using Table 17.2, devise a method for preparing practically pure ethyl alcohol from the alcohol-water azeotrope.

▮ Add benzene, distill the three-component azeotrope until all the water has been removed; then distill the benzene from the alcohol.

TABLE 17.2 Azeotropic Mixtures

components		normal boiling points, (°C)	normal boiling point of azeo-trope, (°C)	mole fractions
ethyl alcohol	C_2H_5OH	78.5	78.2	0.91
water	H_2O	100.0		0.09
acetone	CH_3COCH_3	56.5	39.3	0.40
carbon disulfide	CS_2	46.3		0.60
ethyl alcohol	C_2H_5OH	78.5	65.0	0.385
carbon tetrachloride	CCl_4	76.8		0.615
benzene	C_6H_6	80.1		0.559
ethyl alcohol	C_2H_5OH	78.5	64.6	0.218
water	H_2O	100.0		0.223
hydrogen chloride	HCl	−83.7	108.6	0.11
water	H_2O	100.0		0.89

17.83 How many degrees of freedom are there for a system containing ice, aqueous NaCl solution, and solid NaCl in a bottle under a nitrogen atmosphere?

▮
$p = 4$ (ice, NaCl(s), solution, gas); $c = 3$ (H_2O, NaCl, N_2)

$f = c + 2 - p = 3 + 2 - 4 = 1$ There is 1 degree of freedom.

17.84 Explain on a molecular basis why benzene and toluene form nearly ideal solutions with each other.

▮ They have very similar structures, differing only by the replacement of a hydrogen atom by a methyl group.

17.5 SOLUTIONS OF STRONG ELECTROLYTES

17.85 Of the following 0.10 m aqueous solution, which one will exhibit the largest freezing point depression?

(*a*) KCl (*b*) $C_6H_{12}O_6$ (*c*) K_2SO_4 (*d*) $Al_2(SO_4)_3$ (*e*) NaCl

▮ KCl has 2 ions per formula unit. $C_6H_{12}O_6$ has 1 molecule per formula unit. K_2SO_4 has 3 ions per formula unit. $Al_2(SO_4)_3$ has 5 ions per formula unit. NaCl has 2 ions per formula unit. $Al_2(SO_4)_3$ thus has the greatest freezing point depression.

17.86 Arrange the following aqueous solutions in order of increasing freezing points (that is, lowest first): (*a*) 0.10 m C_2H_5OH (*b*) 0.10 m $Ba_3(PO_4)_2$ (*c*) 0.10 m Na_2SO_4 (*d*) 0.10 m KCl (*e*) 0.10 m Li_3PO_4

▮ $b < e < c < d < a$. $Ba_3(PO_4)_2$ has 5 mol of ions per mol of formula units; Li_3PO_4 has 4; Na_2SO_4 has 3; and KCl has 2. C_2H_5OH is not ionic; it has 1 mol of molecules per mol of formula units.

17.87 A 0.100 m solution of $NaClO_3$ freezes at −0.3433 °C. What would you predict for the boiling point of this aqueous solution at 1 atm pressure? At 0.00100 m concentration of this same salt, the electrical interferences between the ions no longer exist, because the ions are, on the average, too far apart from each other. Predict the freezing point of this more dilute solution.

▮

(*a*) $$m_{\text{effective}} = \frac{\Delta t_f}{K_f} = \frac{0.3433\ ^\circ\text{C}}{1.86\ ^\circ\text{C/m}} = 0.185\ \text{m}$$

$$\Delta t_b = K_b m_{\text{effective}} = (0.513\ ^\circ\text{C/m})(0.185\ \text{m}) = 0.095\ ^\circ\text{C}$$

$$t = (100.000 + 0.095)\ ^\circ\text{C} = 100.095\ ^\circ\text{C}$$

(*b*) 0.00100 m $NaClO_3$ at such low concentrations is 0.00100 m Na^+ plus 0.00100 m ClO_3^-, a 0.00200 m solution:

$$\Delta t = (1.86\ ^\circ\text{C/m})(0.00200\ \text{m}) = 0.00372\ ^\circ\text{C} \qquad t = -0.00372\ ^\circ\text{C}$$

17.88 Estimate the equivalent conductivity at infinite dilution for $HClO_4$, using the data of Tables 17.3 and 17.4.

TABLE 17.3 Equivalent Conductivity, Λ, of Some Electrolytes in Water at 25 °C (cm²/Ω·eq)

	eq/L				infinitely dilute,
	1.00	0.10	0.01	0.001	Λ_0^a
NaCl		106.74	118.51	123.74	126.45
KCl	111.9	129.0	141.3	147.0	149.9
HCl	332.8	391.3	412.0	421.4	426.2
$HC_2H_3O_2$		5.2	16.3	49.2	(390.7)
NH_3		3.6	11.3	34.0	(271.0)
$NaC_2H_3O_2$		72.8	83.8	88.5	91.0
NH_4Cl		128.8	141.3	146.8	149.7

[a] Extrapolated (or calculated) values.

TABLE 17.4 Illustration of Kohlrausch's Rule: Λ_0 at 25 °C (cm²/Ω·eq)

	anion		
cation	Cl^-	I^-	ClO_4^-
K^+	149.9	150.4	140.0
Na^+	126.5	126.9	117.5
difference	23.4	23.5	22.5

▮ $$\Lambda_0(HClO_4) = \Lambda_0(HCl) + \Lambda_0(NaClO_4) - \Lambda_0(NaCl) = 426.2 + 117.5 - 126.5 = 417.2\ \text{cm}^2/\Omega\cdot\text{eq}$$

17.89 Show how the equivalent conductivity at infinite dilution for acetic acid was calculated from other data in Table 17.3.

▮ $$\Lambda_0(HC_2H_3O_2) = \Lambda_0(NaC_2H_3O_2) + \Lambda_0(HCl) - \Lambda_0(NaCl) = 91.0 + 426.2 - 126.5 = 390.7\ \text{cm}^2/\Omega\cdot\text{eq}$$

17.90 The equivalent conductivity of a solution containing 2.54 g of $CuSO_4$/L is 91.0 cm^2/Ω·eq. **(*a*)** Calculate the specific conductivity of the solution. **(*b*)** What is the resistance of a cm^3 of this solution when placed between two electrodes 1.00 cm apart, each having an area of 1.00 cm^2?

▮ **(*a*)** $\Lambda = \dfrac{\kappa}{C}$, where $C = N/1000 = \text{eq/cm}^3$

$$\left(\frac{2.54\ \text{g}\ CuSO_4}{\text{L}}\right)\left(\frac{1\ \text{mol}}{159.5\ \text{g}}\right)\left(\frac{2\ \text{eq}}{\text{mol}}\right) = 0.0318\ \text{N}\ CuSO_4$$

$$\kappa = \left(\frac{91.0\ \text{cm}^2}{\Omega\cdot\text{eq}}\right)\left(\frac{0.0318\ \text{eq}}{1000\ \text{cm}^3}\right) = 2.89 \times 10^{-3}\ \Omega^{-1}\cdot\text{cm}^{-1}$$

(*b*) $$R = \frac{1}{\kappa}\frac{l}{A} = \left(\frac{1}{2.89 \times 10^{-3}\ \Omega^{-1}\cdot\text{cm}^{-1}}\right)\left(\frac{1.00\ \text{cm}}{1.00\ \text{cm}^2}\right) = 346\ \Omega$$

17.91 The equivalent conductivity of a 0.0100 M aqueous solution of ammonia is 10 cm^2/Ω·eq. The equivalent conductivity of ammonia at infinite dilution is calculated to be 238 cm^2/Ω·eq. Calculate the percent ionization of ammonia in 0.0100 M aqueous solution.

▮ $$\frac{10\ \text{cm}^2/\Omega\cdot\text{eq}}{238\ \text{cm}^2/\Omega\cdot\text{eq}} \times 100\% = 4.2\%\ \text{ionized}$$

17.92 The freezing point of a solution composed of 10.0 g of KCl in 100 g of water is −4.5 °C. Calculate the van't Hoff factor, *i*, for this solution.

▮ $$\Delta t = iKm$$

$$m = \frac{100\ \text{g/kg}\ H_2O}{74.55\ \text{g/mol}} = 1.34\ \text{m}$$

$$i = \frac{\Delta t}{Km} = \frac{4.5\ °\text{C}}{(1.86\ °\text{C/m})(1.34\ \text{m})} = 1.8$$

17.93 For each of the following statements about the nature of aqueous solutions of strong electrolytes **(*a*)** give the experimental observations which tend to support the statement and **(*b*)** suggest experimental methods for proving or disproving the statement.

(i) Ions are formed when charged electrodes are placed in a solution of a strong electrolyte.
(ii) Ions are formed when strong electrolytes are dissolved in water.
(iii) Not only is it unnecessary to dissolve an electrolyte in water in order for ions to exist, but in solution there is considerable interionic attraction.

▮ (i) For strong electrolytes, the ions are present even before immersion of charged electrodes in the solution. Early experimenters could not prove the presence of ions except by electrolysis. Freezing point depression methods prove that ions are present even in the absence of electrodes. (ii) Only for strong acids does the solution process produce ions. Most strong electrolytes, 100% ionized in solution, are ionic even in the pure state. The conductivity of molten salts proves that the existence of ions does not depend on the solution process for salts. (The absence of conductivity in such strong acids as pure liquid HI shows that these compounds are covalent when pure and form ions by reaction with the solvent.) (iii) The conductivity of molten salts proves the first statement; the values of *i* lower than the number of ions per formula unit proves the second. Statement iii is the present theory of strong electrolytes.

17.94 Calculate the value of α for 0.010 m aqueous acetic acid from the data of Table 17.5 and again from the data of Table 17.3.

▮ From Table 17.5: $$\alpha_{\text{acetic acid}} = \frac{1.043 - 1}{2.00 - 1} = 0.043$$

From Table 17.3: $$\alpha_{\text{acetic acid}} = \frac{16.3}{391} = 0.042$$

TABLE 17.5 Some Representative Values of the van't Hoff Factor, i

	molality				infinite
	0.10	0.01	0.001	0.00001	dilution[a]
Surose	1.01	1.00	1.00	1.00	1.00
$HC_2H_3O_2$	1.013	1.043	1.15	1.75	(2.00)
HCl	1.89	1.94	1.98		2.00
KCl	1.85	1.94	1.98		2.00
$MgSO_4$	1.21	1.53	1.82		2.00
K_2SO_4	2.32	2.70	2.84		3.00

[a] Extrapolated (or calculated) values.

17.95 Using the data of Table 17.5, calculate the value of K_a for a 0.010 M solution of acetic acid. As shown above, the value of α for acetic acid in 0.010 M solution is 0.043:

$$K_a = \frac{(0.043)^2(0.010)}{0.957} = 1.9 \times 10^{-5}$$

17.96 Estimate the equivalent conductivity of Na_2SO_4 in 0.00100 N solutions from the following equivalent conductivities in 0.00100 N solutions: NaCl, 123.7 cm²/Ω·eq; KCl, 147.0 cm²/Ω·eq; K_2SO_4, 152.1 cm²/Ω·eq.

$$\Lambda(K_2SO_4) + \Lambda(NaCl) - \Lambda(KCl) = \Lambda(Na_2SO_4)$$
$$152.1 + 123.7 - 147.0 = 128.8\ \text{cm}^2/\Omega\cdot\text{eq}$$

(Note that for *equivalent* conductivities, the values for K_2SO_4 and Na_2SO_4 are *not* halved.)

17.97 Estimate the equivalent conductivity of a 0.0100 M acetic acid solution from its conductivity at infinite dilution, 390.7 cm²/Ω·eq at 25 °C, and the value of its dissociation constant, $K_a = 1.8 \times 10^{-5}$.

$$K_a = \frac{[H_3O^+][C_2H_3O_2^-]}{[HC_2H_3O_2]} = 1.8 \times 10^{-5}$$

Let $x = [H_3O^+]$

then $x = [C_2H_3O_2^-]$ and $0.0100 - x = [HC_2H_3O_2]$

$$K_a = \frac{x^2}{0.0100 - x} = 1.8 \times 10^{-5} \simeq \frac{x^2}{0.0100}$$
$$x^2 = 1.8 \times 10^{-7}$$
$$x = 4.2 \times 10^{-4}$$
$$\frac{\Lambda}{\Lambda_0} = \frac{4.2 \times 10^{-4}}{1.0 \times 10^{-2}} = \frac{\Lambda}{390.7} = 4.2 \times 10^{-2}$$
$$\Lambda = (390.7)(4.2 \times 10^{-2}) = 16\ \text{cm}^2/\Omega\cdot\text{eq}$$

17.98 Chloroacetic acid, a monoprotic acid has a K_a of 1.36×10^{-3}. Compute the freezing point of a 0.10 M solution of this acid. Assume that the stoichiometric molar concentration and molality are the same in this case.

$$HC_2H_2ClO_2 + H_2O \rightleftharpoons C_2H_2ClO_2^- + H_3O^+ \qquad K_a = \frac{[H_3O^+][C_2H_2ClO_2^-]}{[HC_2H_2ClO_2]} = 1.36 \times 10^{-3}$$

$$\frac{x^2}{0.10 - x} = 1.36 \times 10^{-3}$$
$$x^2 = 1.36 \times 10^{-4} - (1.36 \times 10^{-3})x$$
$$x^2 + (1.36 \times 10^{-3})x - 1.36 \times 10^{-4} = 0$$
$$x = \frac{-1.36 \times 10^{-3} + \sqrt{(1.36 \times 10^{-3})^2 - 4(-1.36 \times 10^{-4})}}{2} = 0.0110\ \text{M}$$

$$[C_2H_2ClO_2^-] = [H_3O^+] = 1.10 \times 10^{-2} \qquad [HC_2H_2ClO_2] = 0.10 - 1.10 \times 10^{-2} = 0.09\ \text{M}$$

$$\text{Total solute particles} = [C_2H_2ClO_2H] + [C_2H_2ClO_2^-] + [H_3O^+] = 0.112\ \text{M} \approx 0.112\ \text{m}$$

$$\Delta t = K_f m = (-1.86\ ^\circ\text{C/m})(0.112\ \text{m}) = -0.21\ ^\circ\text{C}$$

17.99 A 0.025 M solution of monobasic acid had a freezing point of −0.060 °C. What are K_a and pK_a for the acid?

▌ $$m = \Delta t/K_b = -(0.060\ °C)/(-1.86\ °C/m) = 3.2 \times 10^{-2}\ m \approx 3.2 \times 10^{-2}\ M \text{ of total particles}$$

Each molecule which ionizes increases the number of solute particles by one.

$$HA \rightleftharpoons H^+ + A^-$$

The increase in the number of particles is therefore equal to the number of each type of ion. As a result, there is $(0.032 - 0.025)\ M = 0.007\ M\ H^+$, $0.007\ M\ A^-$, and 0.018 M HA in the solution:

$$K_a = \frac{(0.007)^2}{0.018} = 3 \times 10^{-3} \qquad pK_a = 2.5$$

17.100 A 0.100 M solution of an acid (density = 1.010 g/cm³) is 4.5% ionized. Compute the freezing point of the solution. The molecular weight of the acid is 300.

▌ We must first determine the molality of the solution, i.e., the number of moles of acid dissolved in 1 kg of water.

$$\text{Mass of 1.000 L of solution} = (1000\ cm^3)(1.010\ g/cm^3) = 1010\ g$$

$$\text{Mass of solute in 1.000 L solution} = (0.100\ mol)(300\ g/mol) = 30\ g$$

$$\text{Mass of water in 1.000 L of solution} = (1010 - 30)\ g = 980\ g$$

$$\text{Molality of solution} = \frac{0.100\ \text{mol acid}}{0.980\ \text{kg water}} = 0.102\ mol/kg$$

If the acid were not ionized at all, the freezing point lowering would be

$$1.86 \times 0.102 = 0.190\ °C$$

Because of ionization, the total number of dissolved particles is greater than 0.102 mol/kg of solvent. The freezing point depression is determined by the total number of dissolved particles, regardless of whether they are charged or uncharged.

Let α = fraction ionized. For every mol of acid added to the solution, there will be $(1 - \alpha)$ mol of un-ionized acid at equilibrium, α mol of H^+, and α mol of anion base conjugate to the acid, or a total of $(1 + \alpha)$ mol of dissolved particles. Hence the molality with respect to all dissolved particles is $(1 + \alpha)$ times the molality computed without regard to ionization.

$$\text{Freezing point depression} = (1 + \alpha)(0.190\ °C) = (1.045)(0.190\ °C) = 0.199\ °C$$

The freezing point of the solution is −0.199 °C.

17.101 Explain why 0.100 m NaCl in water does *not* have a freezing point equal to **(*a*)** −0.183 °C **(*b*)** −0.366 °C

▌ **(*a*)** There are 2 mol of ions per mol of NaCl. **(*b*)** There is interionic attraction, which reduces somewhat the independence of the ions and therefore their effect on the freezing point. (The value of *i* is somewhat below 2.)

17.102 State one method by which one could distinguish between the compounds $[Co(NH_3)_5SO_4]Br$ and $[Co(NH_3)_5Br]SO_4$.

▌ The compounds can be distinguished by their conductivities [as well as by their infrared spectra (coordinated sulfate is not tetrahedral in its vibrational motions), x-ray diffraction, and chemical methods].

CHAPTER 18
Thermodynamics

Notes on thermochemical calculations. In some books, ΔE is defined as $q - w$, where w is the work *produced by* the system. In other books, $\Delta E = q + w$, where w is the work *done on* the system. (In both cases, q is heat *into* the system.) Be sure to follow the convention used in your course.

In all calculations, but especially in thermodynamics calculations, the student must be diligent about the units and the signs. Tables of data are often used in which the entries have units different from the units stated in the problem. Signs are sometimes deduced from the wording of the problem (if heat was removed from a system, then q is taken as negative).

In the problems that follow, c is used for specific heat capacity, and C for molar heat capacity.

18.1 HEAT, INTERNAL ENERGY, ENTHALPY

18.1 Identify the kind of energy change usually associated with each of the following: **(*a*)** toaster **(*b*)** radio **(*c*)** automobile engine **(*d*)** automobile battery **(*e*)** automobile generator **(*f*)** model airplane engine **(*g*)** steam engine **(*h*)** friction **(*i*)** ski jump **(*j*)** fluorescent lamp **(*k*)** photoelectric cell **(*l*)** automobile starter motor **(*m*)** furnace

▮ **(*a*)** Electric energy to heat **(*b*)** electric energy to sound **(*c*)** chemical energy to mechanical energy **(*d*)** chemical energy to electric energy and vice versa **(*e*)** kinetic energy to electric energy **(*f*)** chemical energy to kinetic energy (then to potential energy **(*g*)** heat to kinetic energy **(*h*)** mechanical energy to heat **(*i*)** potential energy to kinetic energy **(*j*)** electric energy to light **(*k*)** light to electric energy **(*l*)** electric energy to kinetic energy **(*m*)** chemical energy to heat

18.2 Distinguish between a change of state and a phase change. Give an example of each.

▮ *State* sometimes refers to phase, as in the liquid state. A change of state in that sense is a phase change. However, in thermodynamics, systems exist in *states*, as for example initial and final states, which describe perhaps more than one phase. A change in state may involve 0, 1, or more phase changes.

18.3 Contrast the meaning of the word *molar* in the term *molar heat capacity* with that in the term *molar solution*.

▮ The word *molar* in molar heat capacity means *per mole*. In molar solution, the word means the number of moles of solute per liter of solution.

18.4 In a certain process, 678 J of heat is absorbed by a system while 294 J of work is done on the system. What is the change in internal energy for the process?

▮
$$q = +678\ \text{J}$$

Since work is done on the system, w has a positive value:

$$w = +294\ \text{J} \qquad \Delta E = q + w = 678\ \text{J} + 294\ \text{J} = 972\ \text{J}$$

18.5 If 1500 cal of heat is added to a system while the system does work equivalent to 2500 cal by expanding against the surrounding atmosphere, what is the value of ΔE for the system?

▮
$$\Delta E = q + w = [1500 + (-2500)]\ \text{cal} = -1000\ \text{cal}$$

When the first law of thermodynamics is expressed as above, work done *by* a system is negative.

18.6 In a certain process, 500 J of work is done on a system which gives off 200 J of heat. What is the value of ΔE for the process?

▮
$$\Delta E = q + w = [(-200) + 500]\ \text{J} = 300\ \text{J}$$

18.7 Starting with the definition $H \equiv E + PV$, prove that the enthalpy change for a process with no work other than expansion is equal to the heat added at constant pressure.

▮
$$\Delta H = \Delta E + \Delta(PV) = \Delta E + P\,\Delta V \qquad \text{(for } P = \text{const.)}$$
$$\Delta E = q + w = q - P\,\Delta V \qquad \text{(for no other work besides expansion)}$$

so
$$\Delta H = (q - P\,\Delta V) + P\,\Delta V = q$$

18.8 A sample of 0.20 mol of a gas at 44 °C and 1.5 atm pressure is cooled to 27 °C and compressed to 3.0 atm. Calculate ΔV. Suppose the original sample of gas were heated at constant volume until its pressure was 3.0 atm and then cooled at constant pressure to 27 °C. What would ΔV have been? Is volume a state function? Is your answer limited to this sample of gas?

▮

$$V_1 = \frac{(0.20\ \text{mol})R(317\ \text{K})}{1.5\ \text{atm}} \qquad V_2 = \frac{(0.20\ \text{mol})R(300\ \text{K})}{3.0\ \text{atm}}$$

$$\Delta V = V_2 - V_1 = (0.20\ \text{mol})(0.0821\ \text{L·atm/mol·K})\left(\frac{300}{3.0} - \frac{317}{1.5}\right) = -1.83\ \text{L}$$

Since the initial and final states are the same in each process, the change in volume is the same. Volume is a state function.

18.9 Given the following information,

$$A + B \longrightarrow C + D \qquad \Delta H^\circ = -10.0\ \text{kcal}$$

$$C + D \longrightarrow E \qquad \Delta H^\circ = 15.0\ \text{kcal}$$

calculate ΔH° for each of the following reactions:

(*a*) $C + D \longrightarrow A + B$ (*b*) $2C + 2D \longrightarrow 2A + 2B$ (*c*) $A + B \longrightarrow E$

▮ (*a*) Since the reaction is the reverse of the first given, the sign of ΔH is opposite also, $+10.0$ kcal. (*b*) Since there are twice the number of mol of each reactant, the value of ΔH will be doubled also.

$$\Delta H = 20.0\ \text{kcal}$$

(*c*) Adding the two equations given, and canceling the substances which appear on both sides, yields the equation given in (*c*). The values of ΔH are added also, yielding $\Delta H = +5.0$ kcal.

18.10 Show that the product of pressure times volume, PV, has the dimensions of energy.

▮ The product PV has dimensions

$$\frac{\text{Force}}{\text{Area}} \times \text{volume} = \text{force} \times \text{length} = \text{work} = \text{energy}$$

(cf. Prob. 18.7).

18.11 A system is changed from an initial state to a final state by a manner such that $\Delta H = q$. If the change from the initial state to the final state were made by a different path, would ΔH be the same as that for the first path? would q?

▮ ΔH is the same, since it is a state function, but q most probably will be different for the different paths.

18.12 If 1.000 kcal of heat is added to 1.200 L of oxygen in a cylinder at constant pressure of 1.00 atm, the volume increases to 1.500 L. Calculate ΔE for the process.

▮ By Prob. 18.7, $\Delta H = q = 1.000$ kcal and

$$\Delta E = \Delta H - P\,\Delta V = (1.000\ \text{kcal}) - (1.00\ \text{atm})\left(\frac{101.3\ \text{kN/m}^2}{\text{atm}}\right)(0.300 \times 10^{-3}\ \text{m}^3)\left(\frac{1\ \text{kcal}}{4.184\ \text{kJ}}\right) = 0.993\ \text{kcal}$$

18.13 In the equation for expansion work against the atmosphere, $w = -P\,\Delta V$, which one(s) of the following does P represent? (i) the pressure of the system, (ii) the pressure of the surroundings, (iii) some constant pressure.

▮ P represents the pressure of the surroundings, but it must be a constant for the equation $w = -P\,\Delta V$ to be valid [otherwise, $w = -\Delta(PV)$]. Also, the pressure of the system must be equal to that of the surroundings or the system will adjust to make them equal. All three are thus equal to P.

18.14 The reaction of cyanamide, $NH_2CN(s)$, with oxygen was run in a bomb calorimeter, and ΔE was found to be -742.7 kJ/mol of $NH_2CN(s)$ at 298 K. Calculate ΔH_{298} for the reaction.

$$NH_2CN(s) + \tfrac{3}{2}O_2 \longrightarrow N_2 + CO_2 + H_2O(l)$$

▌ The number of moles of gaseous reactant (O_2) is $\frac{3}{2}$; the number of moles of gaseous products ($N_2 + CO_2$) is 2. Therefore

$$\Delta H = \Delta E + \Delta(PV) = \Delta E + (\Delta n)RT$$

$$\Delta n = 2 - \tfrac{3}{2} = \tfrac{1}{2} \qquad \Delta H = -742.7\ \text{kJ} + (0.500\ \text{mol})(8.314\ \text{J/mol·K})(298\ \text{K}) = -742.7\ \text{kJ} + 1240\ \text{J} = -741.5\ \text{kJ}$$

18.15 A lead bullet weighing 18.0 g and traveling at 500 m/s is embedded in a wooden block weighing 1.00 kg. If both the block and the bullet were initially at 25.0 °C, what is the final temperature of the block containing the bullet? Assume no heat loss to the surroundings. (Heat capacity of wood, 0.500 kcal/kg·K; of lead, 0.030 kcal/kg·K.)

▌
$$\text{KE} = \tfrac{1}{2}mv^2 = \tfrac{1}{2}(18.0 \times 10^{-3}\ \text{kg})(500\ \text{m/s})^2 = 2.25 \times 10^3\ \text{J} = (2.25\ \text{kJ})\left(\frac{1\ \text{kcal}}{4.184\ \text{kJ}}\right) = 0.538\ \text{kcal} = q$$

But
$$q = \Delta t[(18.0 \times 10^{-3}\ \text{kg})(0.030\ \text{kcal/kg·K}) + (1.00\ \text{kg})(0.500\ \text{kcal/kg·K})] = \Delta t(0.500\ \text{kcal/K})$$

Hence
$$\Delta t = \frac{0.538\ \text{kcal}}{0.500\ \text{kcal/K}} = 1.08\ \text{K} = 1.08\ \text{°C}$$

$$t_{\text{final}} = (25.0 + 1.08)\ \text{°C} = 26.1\ \text{°C}$$

18.16 The most exothermic "ordinary" chemical reaction for a given mass of reactants is

$$2\text{H} \longrightarrow \text{H}_2 \qquad \Delta E = -103\ \text{kcal}$$

Calculate the theoretical decrease in mass after the combination of 2.0 mol of hydrogen atoms to form 1.0 mol of hydrogen molecules.

▌
$$E = mc^2$$

The change in mass, Δm, corresponding to a change in energy, ΔE, is given by

$$\Delta m = \frac{\Delta E}{c^2} = \frac{(-103\ \text{kcal})(4184\ \text{J/kcal})}{(3.0 \times 10^8\ \text{m/s})^2} = -4.79 \times 10^{-12}\ \text{kg}$$

18.2 HEAT CAPACITY AND CALORIMETRY

Note: Before working the problems in this section, you should familiarize yourself with Table 18.1 on p. 338.

18.17 Determine the heat capacity of water in kJ/kg·K.

▌
$$\left(\frac{1.000\ \text{kcal}}{\text{kg·K}}\right)\left(\frac{4.184\ \text{kJ}}{\text{kcal}}\right) = 4.184\ \text{kJ/kg·K}$$

18.18 The enthalpy of fusion per g and the corresponding molecular weights are given for several substances:

	enthalpy of fusion, cal/g	**molecular weight, g/mol**
(a)	80	18
(b)	45	20
(c)	90	30
(d)	45	60
(e)	45	30

Identify the two substances having the same molar enthalpies of fusion.

▌ (*c*) and (*d*) have equal values of ΔH_{fus}.

(*a*) (80 cal/g)(18 g/mol) = 1440 cal/mol **(*b*)** (45 cal/g)(20 g/mol) = 900 cal/mol

(*c*) (90 cal/g)(30 g/mol) = 2700 cal/mol **(*d*)** (45 cal/g)(60 g/mol) = 2700 cal/mol

(*e*) (45 cal/g)(30 g/mol) = 1350 cal/mol

18.19 How many calories are required to heat each of the following from 15 to 65 °C: **(*a*)** 1.0 g water **(*b*)** 5.0 g Pyrex glass **(*c*)** 20 g platinum? Specific heat capacity of Pyrex glass, 0.20 cal/g·K; of platinum, 0.032 cal/g·K.

▌
$$q = mc\,\Delta t$$

(a) Heat = (1.0 g)(1.00 cal/g·°C)[(65 − 15) °C] = 50 cal **(b)** Heat = (5.0 g)(0.20 cal/g·°C)(50 °C) = 50 cal

(c) Heat = (20 g)(0.032 cal/g·°C)(50 °C) = 32 cal

18.20 The specific heat of aluminum is 0.214 cal/g·°C. Calculate the heat necessary to raise the temperature of 40.0 g of aluminum from 20.0 to 32.3 °C.

$$\text{Heat} = (\text{mass})(\text{specific heat})(\text{temperature change}) = (40.0\ \text{g})(0.214\ \text{cal/g}\cdot{}^\circ\text{C})(12.3\ {}^\circ\text{C}) = 105\ \text{cal}$$

18.21 The specific heat of silver is 0.0565 cal/g·°C. Assuming no loss of heat to the surroundings or to the container, calculate the final temperature when 100 g of silver at 40.0 °C is immersed in 60.0 g of water at 10.0 °C.

Let t be the final temperature.

$$\text{Heat} = (\text{mass})(\text{specific heat})(\text{temperature change})$$
$$\text{Heat loss by silver} = (100\ \text{g})(0.0565\ \text{cal/g}\cdot{}^\circ\text{C})[(40.0 - t)\ {}^\circ\text{C}]$$
$$\text{Heat gain by water} = (60.0\ \text{g})(1.00\ \text{cal/g}\cdot{}^\circ\text{C})[(t - 10.0)\ {}^\circ\text{C}]$$

Since the heat lost by the silver is gained by the water, the two products are equal:

$$(100\ \text{g})(0.0565\ \text{cal/g}\cdot{}^\circ\text{C})[(40.0 - t)\ {}^\circ\text{C} = (60.0\ \text{g})(1.00\ \text{cal/g}\cdot{}^\circ\text{C})[(t - 10.0)\ {}^\circ\text{C}]$$
$$t = 12.6\ {}^\circ\text{C}$$

18.22 Calculate the number of kcal necessary to raise the temperature of 60.0 g of aluminum from 35 to 55 °C.

The molar heat capacity of Al is 5.8 cal/mol·°C (Table 18.1).

$$q = mc\,\Delta t = (60.0\ \text{g})\left[\left(\frac{5.8\ \text{cal}}{\text{mol}\cdot{}^\circ\text{C}}\right)\left(\frac{1\ \text{mol}}{27.0\ \text{g}}\right)\right](20\ {}^\circ\text{C}) = 258\ \text{cal} = 0.26\ \text{kcal}$$

TABLE 18.1 Mean Molar Heat Capacities at 1.00 atm and 298 K

metallic elements	C_p cal/mol·°C	C_p J/mol·K	other elements and compounds	C_p cal/mol·°C	C_p J/mol·K
Ag	6.1	26	C	2.04	8.5
Al	5.8	24	H_2	6.90	28.9
Au	6.07	25.4	N_2	6.94	29.0
Bi	6.1	26	O_2	7.05	29.5
Cd	6.2	26	Al_2O_3	18.96	79.33
Cr	5.6	23	CH_4	8.60	36.0
Cu	5.85	24.5	C_2H_6	12.71	53.18
Fe	5.9	25	CO	6.97	29.2
Pb	6.4	27	CO_2	8.96	37.5
Sn	6.4	27	Fe_2O_3	24.91	104.2
Zn	6.06	25.4	HBr	6.58	27.5
			$H_2O(g)$	5.92	24.8
			$H_2O(l)$	18.00	75.3
			$H_2O(s)$	8.8	36.8
			NH_3	8.63	36.1
			SnO_2	13.53	56.61

18.23 Determine the final temperature of a system after 100.0 g of zinc at 95 °C is immersed into 50.0 g of water at 15 °C.

The absolute value of the heat lost by the zinc is equal to the absolute value of the heat gained by the water.

$$|\text{Heat lost}| = |\text{heat gained}|$$

$$100\ \text{g}\left(\frac{6.06\ \text{cal}}{\text{mol}\cdot{}^\circ\text{C}}\right)\left(\frac{1\ \text{mol}}{65.37\ \text{g}}\right)(95^\circ - t) = 50.0\ \text{g}\left(\frac{1.00\ \text{cal}}{\text{g}\cdot{}^\circ\text{C}}\right)(t - 15\ {}^\circ\text{C})$$
$$9.27(95^\circ - t) = 50t - 750^\circ$$
$$t = 28\ {}^\circ\text{C}$$

18.24 If 2.00 kJ of energy raises the temperature of a calorimeter containing 45.0 g of water from 23.0 to 32.0 °C, what is the water equivalent of the calorimeter?

▌ If the calorimeter were constructed of water, the "total mass" of it plus its contents would be given by

$$\text{Heat} = (\text{"total mass"})(\text{heat capacity})(\Delta t)$$

$$\text{"Total mass"} = \frac{\text{heat}}{(c)(\Delta t)} = \frac{2.00\ \text{kJ}}{(4.184\ \text{kJ/kg}\cdot\text{K})(9.0\ \text{K})} = 0.053\ \text{kg} = 53\ \text{g}$$

The actual mass of water contained is 45.0 g: hence the "effective mass" of the calorimeter in heat absorbing capacity—its water equivalent—is

$$(53 - 45.0)\ \text{g} = 8\ \text{g}$$

18.25 Calculate the enthalpy of condensation of 1.00 g of water at 100 °C, if $\Delta H_{\text{condensation}} = -9.72$ kcal/mol.

▌

$$(1.00\ \text{g})\left(\frac{1\ \text{mol}}{18.0\ \text{g}}\right)\left(\frac{-9720\ \text{cal}}{\text{mol}}\right) = -540\ \text{cal}$$

18.26 Calculate the enthalpy of sublimation of 100 g of carbon dioxide at 183 K, if $\Delta H_{\text{sub}} = 3.87$ kcal/mol.

▌

$$(100\ \text{g})\left(\frac{1\ \text{mol}}{44.0\ \text{g}}\right)\left(\frac{3.87\ \text{kcal}}{\text{mol}}\right) = 8.80\ \text{kcal}$$

18.27 A 45.0 g sample of an alloy was heated to 90.0 °C and then dropped into a beaker containing 82.0 g of water at 23.50 °C. The temperature of the water rose to a final 26.25 °C. What is the specific heat capacity of the alloy?

▌

$$q = mc(\Delta t) = (82.0\ \text{g})(1.00\ \text{cal/g}\cdot{}^\circ\text{C})(2.75\ {}^\circ\text{C}) = 225.5\ \text{cal} = \text{heat removed from alloy}$$

$$c = \frac{q}{m(\Delta t)} = \frac{225.5\ \text{cal}}{(45.0\ \text{g})[(90.0 - 26.25)\ {}^\circ\text{C}]} = 0.0786\ \text{cal/g}\cdot{}^\circ\text{C}$$

18.28 If the specific heat capacity of a substance is h cal/g·K, what is its specific heat in Btu/lb·°F?

▌

$$\left(h\frac{\text{cal}}{\text{g}\cdot\text{K}}\right)\left(\frac{1\ \text{Btu}}{252\ \text{cal}}\right)\left(\frac{454\ \text{g}}{\text{lb}}\right)\left(\frac{5\ \text{K}}{9\ {}^\circ\text{F}}\right) = 1.00\ h\ \text{Btu/lb}\cdot{}^\circ\text{F}$$

Temperature *differences* in °C are the same as those in kelvins.

18.29 What is the enthalpy change when 1.00 g of water is frozen at 0 °C? $\Delta H_{\text{fus}} = 1.435$ kcal/mol.

▌ Freezing is just the reverse of melting; hence the enthalpy change must be −1.435 kcal/mol.

$$\Delta H = \frac{-1435\ \text{cal/mol}}{18.0\ \text{g/mol}} = -79.7\ \text{cal/g}$$

Therefore, 79.7 cal of heat is liberated when 1.00 g of water is frozen at 0 °C and 1.00 atm pressure.

18.30 A 25.0 g sample of an alloy was heated to 100.0 °C and dropped into a beaker containing 90 g of water at 25.32 °C. The temperature of the water rose to a final value of 27.18 °C. Neglecting heat losses to the room and the heat capacity of the beaker itself, what is the specific heat of the alloy?

▌

$$\text{Heat lost by alloy} = \text{heat absorbed by water}$$

$$(25.0\ \text{g})(c)[(100.0 - 27.2)\ \text{K}] = (90\ \text{g})(1.00\ \text{cal/g}\cdot\text{K})[(27.18 - 25.32)\ \text{K}]$$

from which $c = 0.092$ cal/g·K.

18.31 Outline a practical method by which the specific heat of a solid material may be determined in the laboratory.

▌ Heat the weighed material to a specific temperature in a water bath, rapidly dry it, and transfer it into a weighed quantity of cold water at a measured temperature, taking care to avoid splattering. Measure the final temperature of the system, and calculate as in Prob. 18.23.

18.32 When 10.000 kJ of electric energy is added to a calorimeter containing 400 g of water, the temperature of the calorimeter and contents is raised 5.00 °C. Calculate the water equivalent of the calorimeter.

▌

$$(10.000\ \text{kJ})\left(\frac{1.000\ \text{kcal}}{4.184\ \text{kJ}}\right) = 2.390\ \text{kcal}$$

The mass of water which would be raised 5.00 °C by this quantity of energy is given by

$$\frac{2390\text{ cal}}{(1.00\text{ cal/g}\cdot{}^\circ\text{C})(5.00\text{ }^\circ\text{C})} = 478\text{ g}$$

Since there is only 400 g of water present, the calorimeter must absorb the rest of the heat. The calorimeter is equivalent in heat-absorbing capacity to 78 g of water.

18.33 Calculate the heat evolved when 0.75 mol of molten aluminum at its melting point of 658 °C is solidified and cooled to 25 °C. The enthalpy of fusion of aluminum is 76.8 cal/g.

▌ Heat is evolved (*a*) when the liquid solidifies and (*b*) when it cools. (Heat lost by the aluminum is negative.)

$$\text{Heat}(a) = (0.75\text{ mol})\left(\frac{27.0\text{ g}}{\text{mol}}\right)\left(\frac{-76.8\text{ cal}}{\text{g}}\right) = -1555\text{ cal}$$

$$\Delta t = t_2 - t_1 = (25 - 658)\,^\circ\text{C} = -633\text{ }^\circ\text{C}$$

$$\text{Heat}(b) = (0.75\text{ mol})\left(\frac{5.8\text{ cal}}{\text{mol}\cdot{}^\circ\text{C}}\right)(-633\text{ }^\circ\text{C}) = -2754\text{ cal}$$

$$\text{Total heat evolved} = \text{heat}(a) + \text{heat}(b) = [(-1555) + (-2754)]\text{ cal} = -4300\text{ cal} = -4.3\text{ kcal}$$

18.34 The melting point of a certain substance is 70 °C, its normal boiling point is 450 °C, its enthalpy of fusion is 30.0 cal/g, its enthalpy of vaporization is 45.0 cal/g, and its specific heat is 0.215 cal/g·K. Calculate the heat required to convert 100.0 g of the substance from the solid state at 70 °C to vapor at 450 °C.

▌

$$\Delta H_{\text{fus}} = (100.0\text{ g})\left(\frac{30.0\text{ cal}}{\text{g}}\right) = 3000\text{ cal}$$

$$\Delta H_{\text{heating}} = (100.0\text{ g})\left(\frac{0.215\text{ cal}}{\text{g}\cdot\text{K}}\right)[(450 - 70)\text{ K}] = 8170\text{ cal}$$

$$\Delta H_{\text{vap}} = (100.0\text{ g})\left(\frac{45.0\text{ cal}}{\text{g}}\right) = 4500\text{ cal}$$

$$\Delta H_{\text{total}} = \Delta H_{\text{fus}} + \Delta H_{\text{heating}} + \Delta H_{\text{vap}} = 15.7\text{ kcal}$$

18.35 How much heat is required to raise the temperature of 10.0 g of $H_2O(s)$ from −10 °C to $H_2O(l)$ at 10 °C?

▌ The added heat causes the following changes:

$$\underset{-10^\circ}{\text{solid}} \xrightarrow[1]{} \underset{0^\circ}{\text{solid}} \xrightarrow[2]{} \underset{0^\circ}{\text{liquid}} \xrightarrow[3]{} \underset{+10^\circ}{\text{liquid}}$$

$$\Delta H_1 = m(c)\,\Delta t = 10\text{ g }(0.50\text{ cal/g}\cdot{}^\circ\text{C})(10\text{ }^\circ\text{C}) = 50\text{ cal}$$

$$\Delta H_2 = m(\Delta H_{\text{fus}}) = 10\text{ g }(80\text{ cal/g}) = 800\text{ cal}$$

$$\Delta H_3 = m(c)\,\Delta t = 10\text{ g }(1.0\text{ cal/g}\cdot{}^\circ\text{C})(10\text{ }^\circ\text{C}) = 100\text{ cal}$$

$$\Delta H_{\text{total}} = 950\text{ cal}$$

18.36 How much heat is necessary to convert 100 g of ice at 0 °C to water vapor (steam) at 100 °C?

$$\Delta H_{\text{fus}} = 80\text{ kcal/kg} \qquad \Delta H_{\text{vap}} = 540\text{ kcal/kg} \qquad \text{Heat capacity} = 1.00\text{ kcal/kg}\cdot\text{K}$$

▌ The heat may be considered absorbed in three steps: fusion, heating, and vaporization. The total heat required is the sum of all three.

$$\Delta H_{\text{fus}} = 0.100\text{ kg}(80\text{ kcal/kg}) = 8.0\text{ kcal}$$

$$\Delta H_{\text{warming}} = 0.100\text{ kg}(1.00\text{ kcal/kg}\cdot\text{K})(100\text{ K}) = 10.0\text{ kcal}$$

$$\Delta H_{\text{vap}} = 0.100\text{ kg}(540\text{ kcal/kg}) = 54.0\text{ kcal}$$

$$\Delta H_{\text{total}} = 72.0\text{ kcal}$$

18.37 A 0.300-kg piece of thallium metal at a temperature of 100 °C, when added to a certain mass of water at 10.0 °C, raises the temperature of the water to 25.0 °C. Find the mass of the water. Specific heat of thallium is 0.0333 cal/g·°C and that of water is 1.00 cal/g·°C.

$$\text{Heat lost by thallium} = 300\text{ g}(0.0333\text{ cal/g}\cdot{}^\circ\text{C})(75\ {}^\circ\text{C}) = 750\text{ cal}$$
$$\text{Heat gained by water} = 750\text{ cal} = m(1.00\text{ cal/g}\cdot{}^\circ\text{C})(15\ {}^\circ\text{C})$$
$$m = 50\text{ g}$$

18.38 When 20 g of Al is heated from 90 to 115 °C, 105 cal is required. What is the value of C_p (in cal/mol·°C) for Al?

$$\text{Heat} = nC_p\,\Delta t = (\tfrac{20}{27}\text{ mol})C_p(115 - 90)\ {}^\circ\text{C} = 105\text{ cal} \qquad C_p = 5.7\text{ cal/mol}\cdot{}^\circ\text{C}$$

18.39 How much heat is given up when 20.0 g of steam at 100 °C is condensed and cooled to 20 °C?

$$\Delta H(\text{condensation}) = -(\text{mass})(\text{heat of vaporization}) = -(20.0\text{ g})(540\text{ cal/g}) = -10.8\text{ kcal}$$
$$\Delta H(\text{cooling}) = mc\,\Delta T = (1.00\text{ cal/g}\cdot\text{K})(20.0\text{ g})[(20 - 100)\text{ K}] = -1.6\text{ kcal}$$
$$\Delta H(\text{total}) = -10.8 - 1.6 = -12.4\text{ kcal.}$$

The amount of heat given up is 12.4 kcal.

18.40 How much heat is required to convert 40.0 g of ice $(c = 0.5\text{ cal/g}\cdot\text{K})$ at -10 °C to steam $(c = 0.33\text{ cal/g}\cdot\text{K})$ at 120 °C?

$$\Delta H(\text{ice}) = mc\,\Delta T = (0.5\text{ cal/g}\cdot\text{K})(40.0\text{ g})(10\text{ K}) = 0.2\text{ kcal}$$
$$\Delta H(\text{melting}) = m(\text{heat of fusion}) = (40.0\text{ g})(80\text{ cal/g}) = 3.2\text{ kcal}$$
$$\Delta H(\text{water}) = mc\,\Delta T = (1.00\text{ cal/g}\cdot\text{K})(40.0\text{ g})(100\text{ K}) = 4.0\text{ kcal}$$
$$\Delta H(\text{vap}) = m(\text{heat of vaporization}) = (40.0\text{ g})(540\text{ cal/g}) = 21.6\text{ kcal}$$
$$\Delta H(\text{steam}) = mc\,\Delta T = (0.33\text{ cal/g}\cdot\text{K})(40.0\text{ g})(20\text{ K}) = 0.3\text{ kcal}$$
$$\Delta H(\text{total}) = (0.2 + 3.2 + 4.0 + 21.6 + 0.3)\text{kcal} = 29.3\text{ kcal}$$

18.41 ***(a)*** How many calories are required to heat 100 g of copper $(c = 0.092\text{ cal/g}\cdot\text{K})$ from 10 to 100 °C? ***(b)*** The same quantity of heat as in *(a)* is added to 100 g of aluminum $(c = 0.217\text{ cal/g}\cdot\text{K})$ at 10 °C. Which gets hotter, the copper or aluminum?

(a) $\Delta H = mc\,\Delta T = (0.092\text{ cal/g}\cdot\text{K})(100\text{ g})[(100 - 10)\text{ K}] = 830\text{ cal}$

(b) Since the specific heat capacity of copper is less than that of aluminum, less heat is required to raise the temperature of a mass of copper by 1 K than is required for an equal mass of aluminum. Hence the copper gets hotter.

18.42 Determine the resulting temperature, t, when 150 g of ice at 0 °C is mixed with 300 g of water at 50 °C.

(1) Consider the heat absorbed by the ice and by the water from it.

$$\Delta H(\text{fusion}) = (80.0\text{ cal/g})(150\text{ g}) = 1.20 \times 10^4\text{ cal}$$
$$\Delta H(\text{heating}) = mc\,\Delta T = (1.00\text{ cal/g}\cdot\text{K})(150\text{ g})[(t - 0)\text{ K}]$$

(2) Now consider the heat lost by the hot water.

$$\Delta H = mc\,\Delta T = (1.00\text{ cal/g}\cdot\text{K})(300\text{ g})[(t - 50)\text{ K}]$$

where $t < 50$.

(3) The sum of the ΔH's must equal 0 since heat is assumed not to leak into or out of the total system treated in (1) and (2).

$$1.20 \times 10^4 + 150t + 300(t - 50) = 0$$

from which $t = 6.7$ °C.

18.43 Determine the resulting temperature when 1.00 kg of ice at 0 °C is mixed with 9.00 kg of water at 50 °C. Heat of fusion of ice is 80.0 cal/g.

$$(1.00\text{ kg ice})\left(\frac{80.0\text{ kcal}}{\text{kg}}\right) = 80.0\text{ kcal} \qquad \text{to melt the ice to water at 0 °C}$$
$$q = mc\,\Delta t$$
$$80.0 \times 10^3\text{ cal} = (9.00 \times 10^3\text{ g})\left(\frac{1.00\text{ cal}}{\text{g}\cdot{}^\circ\text{C}}\right)\Delta t$$
$$\Delta t = 8.89\ {}^\circ\text{C}$$

The temperature drop due to the melting gets the temperature down to (50 − 8.89) °C = 41 °C. The heat gained by the melted water is equal to the heat lost by the water from 41°:

$$(1.00 \times 10^3\ \text{g})c(t - 0°) = (9.00 \times 10^3\ \text{g})c(41° - t)$$

$$t = 37\ °\text{C}$$

18.44 How much heat is required to change 10 g of ice at 0 °C to steam at 100 °C? Heat of vaporization of water at 100 °C is 540 cal/g.

$$\Delta H_{\text{total}} = \Delta H_{\text{fus}} + \Delta H_{\text{heating}} + \Delta H_{\text{vap}}$$

$$= 10\ \text{g}\left[\frac{80\ \text{cal}}{\text{g}} + \left(\frac{1.00\ \text{cal}}{\text{g}\cdot°\text{C}}\right)(100\ °\text{C}) + \frac{540\ \text{cal}}{\text{g}}\right] = 7200\ \text{cal} = 7.2\ \text{kcal}$$

18.45 Calculate the final temperature of the system after 10.0 g of ice at −10.0 °C is treated with 2.00 g of water vapor at 115.0 °C and 1.00 atm pressure. $\Delta H_{\text{vap}} = 9.72$ kcal/mol.

▮ The heat lost by the water vapor will be gained by the ice. Consideration must be given to the heat lost by (1) cooling the vapor to the condensation point, (2) condensing it to liquid, and (3) cooling the liquid. Similarly, the heat transferred to the ice will be used in (1) warming the solid to the melting point, (2) melting it, and (3) warming the liquid water. The individual steps may be done alternately to ensure that enough heat is available for the various steps. Warming the solid to 0 °C takes

$$(10.0\ \text{g})\left(\frac{1\ \text{mol}}{18.0\ \text{g}}\right) = (0.555\ \text{mol})\left(\frac{8.8\ \text{cal}}{\text{mol}\cdot°\text{C}}\right)(10\ °\text{C}) = 49\ \text{cal}$$

Cooling the vapor to 100 °C yields

$$(2.00\ \text{g})\left(\frac{1\ \text{mol}}{18.0\ \text{g}}\right) = (0.111\ \text{mol})\left(\frac{5.92\ \text{cal}}{\text{mol}\cdot°\text{C}}\right)(-15\ °\text{C}) = -9.87\ \text{cal}$$

Since more heat is needed, the condensation of the water yields

$$(0.111\ \text{mol})\left(\frac{-9720\ \text{cal}}{\text{mol}}\right) = -1080\ \text{cal}$$

This may be used to melt the ice:

$$(0.555\ \text{mol})\left(\frac{1435\ \text{cal}}{\text{mol}}\right) = 800\ \text{cal}$$

with some heat left over to warm the water. So far the heat transferred may be summarized

lost	gained
−9.87 cal	49 cal
−1080 cal	800 cal
−1090 cal	849 cal

The calculation is completed from the knowledge that 10.0 g of liquid water at 0 °C and 2.00 g of liquid water at 100 °C plus 1090 849 = 240 cal (2 sig. fig.) of heat are to be mixed.

$$(10.0\ \text{g})(1.00\ \text{cal/g}\cdot°\text{C})(t - 0.0°) = (2.00\ \text{g})(1.00\ \text{cal/g}\cdot°\text{C})(100° - t) + 240\ \text{cal}$$

$$10.0t = 200 - 2.00t + 240$$

$$12.0t = 440$$

$$t = 37\ °\text{C}$$

18.46 Calculate the enthalpy change accompanying the freezing of 1.00 mol of water at −10.0 °C to ice at −10.0 °C. $\Delta H_{\text{fus}} = 1.435$ kcal/mol at 0 °C.

▮

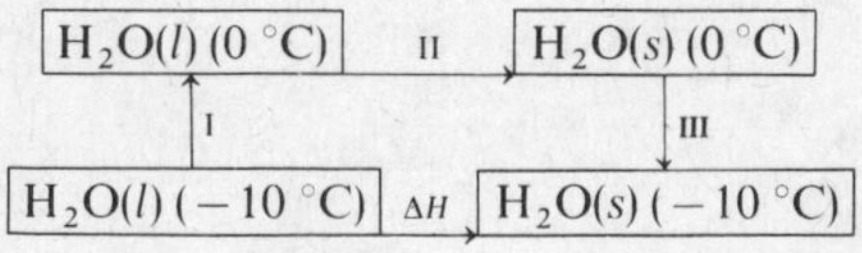

$H_2O(l)$ (0 °C)	—II→	$H_2O(s)$ (0 °C)
↑ I		↓ III
$H_2O(l)$ (−10 °C)	—ΔH→	$H_2O(s)$ (−10 °C)

$$\Delta H_{II} = -1435 \text{ cal} \qquad \Delta H_I = (18.0 \text{ g})(1 \text{ cal/g·K})(10 \text{ K}) = 180 \text{ cal}$$

$$\Delta H_{III} = (1.00 \text{ mol})(8.8 \text{ cal/mol·K})(-10 \text{ K}) = -88 \text{ cal} \qquad \Delta H = \Delta H_I + \Delta H_{II} + \Delta H_{III} = -1343 \text{ cal} = -1.343 \text{ kcal}$$

18.47 A steam boiler is made of steel and weighs 900 kg. The boiler contains 400 kg of water. Assuming that 70% of the heat is delivered to boiler and water, how much heat is required to raise the temperature of the whole from 10 to 100 °C? Heat capacity of steel is 0.11 kcal/kg·K.

▮

$$\Delta H(\text{heating}) = [mc(\text{boiler}) + mc(\text{water})]\,\Delta T$$
$$= [(0.11)(900) \text{ kcal/K} + (1.00)(400) \text{ kcal/K}](90 \text{ K}) = 44\,900 \text{ kcal}$$
$$\text{Input required} = \frac{44\,900 \text{ kcal}}{0.70} = 64\,000 \text{ kcal}$$

18.48 The heat of combustion of ethane gas, C_2H_6, is 368 kcal/mol. Assuming that 60.0% of the heat is useful, how many m^3 of ethane measured at STP must be burned to supply enough heat to convert 50.0 kg of water at 10 °C to steam at 100 °C?

▮

$$\text{Heat per g} = (90 \text{ cal to heat to } 100°\text{C}) + (540 \text{ cal to vaporize}) = 630 \text{ cal total}$$
$$(50.0 \times 10^3 \text{ g } H_2O)\left(\frac{630 \text{ cal}}{\text{g}}\right) = 31\,500 \times 10^3 \text{ cal} = 31\,500 \text{ kcal needed}$$
$$(31\,500 \text{ kcal needed})\left(\frac{100 \text{ kcal total}}{60.0 \text{ kcal needed}}\right) = 52\,500 \text{ kcal total} \qquad (52\,500 \text{ kcal})\left(\frac{1 \text{ mol}}{368 \text{ kcal}}\right) = 143 \text{ mol } C_2H_6$$
$$V = \frac{nRT}{P} = \frac{(143 \text{ mol})(0.0821 \text{ L·atm/mol·K})(273 \text{ K})}{(1.00 \text{ atm})} = 3210 \text{ L} = 3.21 \text{ m}^3$$

18.49 When 1.0 kg of anthracite coal is burned about 7300 kcal is evolved. What amount of coal is required to heat 4.0 kg of water from room temperature (20 °C) to the boiling point (at 1 atm pressure), assuming that all the heat is available?

▮

$$\Delta H(\text{heating}) = mc\,\Delta T = (1.00 \text{ kcal/kg·K})(4.0 \text{ kg})[(100 - 20) \text{ K}] = 320 \text{ kcal}$$
$$\text{Amount of coal required} = \frac{320 \text{ kcal}}{7300 \text{ kcal/kg}} = 0.044 \text{ kg} = 44 \text{ g}$$

18.50 Calculate the final temperature of the system after a 100-g piece of lead at 45.0 °C is added to a mixture of 2.00 g of ice in 10.0 g of water at 0 °C.

▮ The heat lost by the lead will first melt ice at 0 °C, and only then, if sufficient, warm the total quantity of water. To melt the ice would take

$$(2.00 \text{ g})\left(\frac{1 \text{ mol}}{18.0 \text{ g}}\right)\left(\frac{1435 \text{ cal}}{\text{mol}}\right) = 160 \text{ cal}$$

But even if the lead were cooled to 0 °C, there is only

$$(100 \text{ g})\left(\frac{1 \text{ mol}}{207.2 \text{ g}}\right)\left(\frac{6.4 \text{ cal}}{\text{mol·°C}}\right)(45 \text{ °C}) = 139 \text{ cal}$$

Since there is insufficient heat to melt all the ice, some will remain frozen, and the final temperature will be 0 °C.

18.51 The combustion of 5.00 g of coke raised the temperature of 1.00 kg of water from 10 to 47 °C. Calculate the heat value of coke in kcal/g.

▮

$$q = mc\,\Delta t = (1.00 \times 10^3 \text{ g})(1.00 \text{ cal/g·°C})(47 - 10) \text{ °C} = 37 \times 10^3 \text{ cal} = 37 \text{ kcal}$$
$$\frac{37 \text{ kcal}}{5.00 \text{ g coke}} = 7.4 \text{ kcal/g coke}$$

18.52 Assuming that 50% of the heat is useful, how many kg of water at 15.0 °C can be heated to 95.0 °C by burning 200 L of methane, CH_4, measured at STP? The heat of combustion of methane is 211 kcal/mol.

▮

$$n = \frac{PV}{RT} = \frac{(1.00 \text{ atm})(200 \text{ L})}{(0.0821 \text{ L·atm/mol·K})(273 \text{ K})} = 8.92 \text{ mol } CH_4 \qquad (8.92 \text{ mol } CH_4)\left(\frac{211 \text{ kcal}}{\text{mol } CH_4}\right) = 1880 \text{ kcal}$$
$$\text{Useful heat} = (1880 \text{ kcal})(0.50) = 940 \text{ kcal}$$
$$m = \frac{q}{c\,\Delta t} = \frac{940 \text{ kcal}}{(1.00 \text{ kcal/kg·K})(80.0 \text{ K})} = 11.8 \text{ kg}$$

18.53 Exactly 3.00 g of carbon was burned to CO_2 in a copper calorimeter. The mass of the calorimeter is 1500 g and the mass of water in the calorimeter is 2000 g. The initial temperature was 20.0 °C and the final temperature 31.3 °C. Calculate the heat value of carbon in cal/g. Specific heat capacity of copper is 0.092 cal/g·K.

▮
$$q(\text{calorimeter}) = [mc(\text{Cu}) + mc(\text{H}_2\text{O})]\,\Delta T$$
$$= [(0.092\ \text{cal/g·K})(1500\ \text{g}) + (1.00\ \text{cal/g·K})(2000\ \text{g})][(31.3\text{–}20.0)\ \text{K}] = 2.4 \times 10^4\ \text{cal}$$
$$\text{Heat value of carbon} = \frac{2.4 \times 10^4\ \text{cal}}{3.00\ \text{g}} = 8.0 \times 10^3\ \text{cal/g}$$

18.54 When 156 g of Al at 50 °C is added to 100 g of water at 20 °C, the final temperature becomes 30 °C. Find the specific heat of Al.

▮
$$(156\ \text{g Al})c[(50 - 30)\ \text{K}] = (100\ \text{g H}_2\text{O})(4.18\ \text{J/g·K})[(30 - 20)\ \text{K}]$$
$$c = 1.3\ \text{J/g·K}$$

18.55 An insulated container holds 30.0 g of water (specific heat 4.184 J/g·K) and 10.0 g of Al (specific heat 0.890 J/g·K), all at 35.0 °C.

(*a*) How much heat must be removed to cool the contents (water + Al) to 0.0 °C without freezing the water?
(*b*) What is the smallest mass of ice at −5.0 °C that must be used in part (*a*) to cool the contents to 0.0 °C?

Specific heat of ice 2.092 J/g·K ΔH_{fus} of ice 6.02 kJ/mol

▮ (*a*) Heat = (30.0 g H_2O)(4.184 J/g·K)(35.0 K) + (10.0 g Al)(0.890 J/g·K)(35.0 K) = 4700 J

(*b*)
$$\text{Heat} = m\left((2.092\ \text{J/g·K})(5.0\ \text{K}) + \frac{6020\ \text{J/mol}}{18.0\ \text{g/mol}}\right) = 4700\ \text{J} \qquad m = \frac{4700\ \text{J}}{345\ \text{J/g}} = 13.6\ \text{g ice}$$

18.56 How much energy is needed to raise 50 g of $H_2O(l)$ at 50 °C to $H_2O(g)$ at 200 °C?

ΔH_{vap} at 100 °C = 40.7 kJ/mol

Heat capacity of liquid water = 4.18 kJ/kg·K Heat capacity of water vapor = 1.84 kJ/kg·K

▮
$$\Delta H = \Delta H_{\text{liquid}} + \Delta H_{\text{vap}} + \Delta H_{\text{gas}}$$
$$= (50 \times 10^{-3}\ \text{kg})\left[\left(\frac{4.184\ \text{kJ}}{\text{kg·K}}\right)(50\ \text{K}) + \left(\frac{1\ \text{mol}}{18.0 \times 10^{-3}\ \text{kg}}\right)\left(\frac{40.7\ \text{kJ}}{\text{mol}}\right) + \left(\frac{1.84\ \text{kJ}}{\text{kg·K}}\right)(100\ \text{K})\right] = 133\ \text{kJ}$$

18.57 Assuming no loss of heat to the surroundings, determine the final temperature of the water produced from 2.00 mol of H_2 and 1.00 mol of O_2 initially at 25 °C which are allowed to react to form $H_2O(g)$. ($\Delta H_f = -57.79$ kcal/mol).

▮ The reaction yields 115.58 kcal, which is then added to the 2.00 mol of $H_2O(g)$.

$$115.58 \times 10^3\ \text{cal} = (2.00\ \text{mol})\left(\frac{5.92\ \text{cal}}{\text{mol·°C}}\right)\Delta t$$
$$t_f - t_i = \Delta t = 9762\ \text{°C}$$
$$t_f = (9762 + 25)\ \text{°C} = 9790\ \text{°C} \qquad \text{(3 sig. fig.)}$$

18.3 LAW OF DULONG AND PETIT

18.58 Estimate the specific heat of platinum.

▮ According to the law of Dulong and Petit, the molar heat capacity of Pt (or any other metallic element) is 6 cal/mol·°C. The specific heat is therefore given by

$$\frac{6\ \text{cal}}{\text{mol·°C}}\left(\frac{1\ \text{mol}}{195\ \text{g}}\right) = 0.03\ \text{cal/g·°C}$$

18.59 A 40.0 g sample of a metal at 50.0 °C is immersed in 100.0 g of water at 10.0 °C. The final temperature of the system is 13.0 °C. What is the specific heat of the metal? What is the approximate atomic weight of the metal?

▮
$$\text{Heat lost} = -\text{heat gained}$$
$$(40.0\ \text{g})c(37.0\ \text{°C}) = (100\ \text{g})(1.0\ \text{cal/g·°C})(3.0\ \text{°C})$$
$$c = 0.20\ \text{cal/g·°C}$$

According to the law of Dulong and Petit, the metal has a molar heat capacity ≈ 6 cal/mol·°C. Hence the atomic weight is approximately

$$\frac{6\ \text{cal/mol}\cdot°\text{C}}{0.20\ \text{cal/g}\cdot°\text{C}} = 30\ \text{g/mol} = 30\ \text{u}$$

18.60 Estimate the final temperature of a system after 1.6 mol of an unknown crystalline metal at 60.0 °C was immersed in 100 g of water at 20.0 °C.

▌ From the law of Dulong and Petit, $C = 6\ \text{cal/mol}\cdot°\text{C}$

$$(1.6\ \text{mol})(6\ \text{cal/mol}\cdot°\text{C})(60.0 - t) = 100\ \text{g}(1.0\ \text{cal/g}\cdot°\text{C})(t - 20.0)$$

$$t = 24\ °\text{C}$$

18.61 A metallic element whose specific heat is 0.11 cal/g·°C forms an oxide containing 22.27% oxygen. Identify the element.

▌ The approximate atomic weight is determined from the law of Dulong and Petit:

$$\frac{6\ \text{cal/mol}\cdot°\text{C}}{0.11\ \text{cal/g}\cdot°\text{C}} = 55\ \text{g/mol}$$

In a 100 g sample:

$$(22.27\ \text{g O})\left(\frac{1\ \text{mol O}}{16.00\ \text{g O}}\right) = 1.392\ \text{mol O} \qquad (77.73\ \text{g M})\left(\frac{1\ \text{mol M}}{55\ \text{g M}}\right) = 1.4\ \text{mol M}$$

The empirical formula is MO. Then, if there is 1.392 mol of O present, there must be exactly 1.392 mol of M present also.

$$\frac{77.73\ \text{g}}{1.392\ \text{mol}} = 55.84\ \text{g/mol} = 55.84\ \text{u}$$

The metal most closely resembles Fe.

18.62 Two 10 g bars of different metals are heated to 60 °C and then immersed in identical, insulated containers each containing 200 g of water at 20 °C. Will the metal with higher or lower atomic weight cause a greater temperature rise in the water?

▌ From the law of Dulong and Petit, the molar heat capacities are known to be approximately equal. Hence

$$c_1(\text{AW}_1) = c_2(\text{AW}_2)$$

The metal with the lower atomic weight will have a higher specific heat, and will cause a greater temperature rise.

18.63 A certain metallic element formed an oxide which contained 89.70% M. The specific heat capacity of M was found to be 0.0305 cal/g·K. What is the exact atomic weight of the metal?

▌ The heat capacity information allows the calculation of an *approximate* atomic weight. By itself, the chemical analysis allows the calculation of an *exact* combining weight but not an atomic weight, because the formula of the oxide is not known. Somehow the two pieces of information must be combined.

From the law of Dulong and Petit,

$$\text{AW} \approx \frac{6\ \text{cal/mol}\cdot\text{K}}{0.0305\ \text{cal/g}\cdot\text{K}} = 200\ \text{g/mol}$$

In dealing with the chemical analysis of the oxide, let us calculate the mass of M combining with 1 mol of oxygen atoms.

$$m(\text{M})/\text{mol O} = \frac{89.70\ \text{g M}}{\dfrac{10.30\ \text{g O}}{16.00\ \text{g O/mol O}}} = 139.34\ \text{g M/mol O}$$

Next set up a table of likely formulas for the oxide by analogy with some of the known simpler metal oxides.

possible formula	M_2O	MO	M_2O_3	MO_2	M_2O_5	MO_3
$n(\text{O})/n(\text{M})$	$\frac{1}{2}$	1	$\frac{3}{2}$	2	$\frac{5}{2}$	3
$139.34\ n(\text{O})/n(\text{M}) = \text{AW}$	69.67	139.34	209.01	278.68	348.35	418.02

The third row is the number of g of M per mol of M, i.e., the exact atomic weight of M. Only the entry for M_2O_3 is close to the approximate value given by the law of Dulong and Petit. Hence the atomic weight of M is 209.01.

18.64 An element was found to have a specific heat capacity of 0.0276 cal/g·K. If 114.79 g of a chloride of this element contained 79.34 g of the metallic element, what is the exact atomic weight of this element?

▮

$$\text{Approximate atomic weight} = \frac{6\ \text{cal/g}\cdot\text{K}}{0.0276\ \text{cal/g}\cdot\text{K}} = 220\ \text{g/mol}$$

$$m(\text{chlorine}) = (114.79\ \text{g total}) - (79.34\ \text{g metal}) = 35.45\ \text{g Cl}$$

$$(35.45\ \text{g Cl})\left(\frac{1\ \text{mol Cl}}{35.45\text{g Cl}}\right) = 1.000\ \text{mol Cl} \qquad 79.34\ \text{g} = \text{equivalent weight of M}$$

$$\frac{220\ \text{g/mol}}{79.34\ \text{g/eq}} = 3\ \text{eq/mol} \qquad \text{(round to nearest integer)}$$

$$\text{AW} = (79.34\ \text{g/eq})(3\ \text{eq/mol}) = 238.0\ \text{g/mol}$$

18.65 The specific heat capacity of a solid element M is 0.0442 cal/g·K. A sulfate of this element was purified and found to contain 42.2% of the element by weight. ***(a)*** What is the exact atomic weight from this determination? ***(b)*** What is the formula of the sulfate?

▮

$$\text{Approximate atomic weight} = \frac{6\ \text{cal/mol}\cdot\text{K}}{0.0442\ \text{cal/g}\cdot\text{K}} = 140\ \text{g/mol}$$

$$(42.2\ \text{g M})\left(\frac{1\ \text{mol M}}{140\ \text{g M}}\right) = 0.30\ \text{mol M} \qquad (57.8\ \text{g } SO_4^{2-})\left(\frac{1\ \text{mol } SO_4^{2-}}{96.0\ \text{g } SO_4^{2-}}\right) = 0.602\ \text{mol } SO_4^{2-}$$

The compound is $M(SO_4)_2$. The exact fraction of a mol is $0.602/2 = 0.301$ mol M = 42.2 g M

$$\frac{42.2\ \text{g}}{0.301\ \text{mol}} = 140.2\ \text{g/mol}$$

The metal ion is Ce^{4+}, and the compound is $Ce(SO_4)_2$.

18.66 A "new" element, El, forms a compound with chlorine which contains 1.455 g Cl/g El. The specific heat of pure El is found to be 0.050 cal/g·K. Estimate the atomic weight of El, and deduce the formula of its compound with chlorine.

▮ According to the law of Dulong and Petit,

$$\text{Atomic weight} \approx \frac{6\ \text{cal/mol}\cdot\text{K}}{0.050\ \text{cal/g}\cdot\text{K}} = 120\ \text{g/mol}$$

Hence for 1 mol of El atoms, there must be

$$(120\ \text{g El})\left(\frac{1.455\ \text{g Cl}}{\text{g El}}\right) = 175\ \text{g Cl} \qquad (175\ \text{g Cl})\left(\frac{1\ \text{mol Cl}}{35.5\ \text{g Cl}}\right) = 5\ \text{mol Cl}$$

There are 5 mol Cl atoms/mol El atoms; the formula is $ElCl_5$. These results suggest that the "new" element is antimony, which has an atomic weight of 121.75 g/mol and forms a compound having the formula $SbCl_5$.

18.4 ENTHALPY CHANGE

Note: Before working the problem in this section, familiarize yourself with Table 18.2, p. 349.

18.67 An early-model Concorde supersonic airplane consumed 4700 gal of aviation fuel per hour at cruising speed. The density of the fuel is 6.65 lb/gal and ΔH of combustion is $-10\,500$ kcal/kg. Express the power consumption in megawatts ($1\ \text{MW} = 10^6\ \text{W} = 10^6\ \text{J/s}$) during supersonic cruising.

▮

$$\frac{(4700\ \text{gal})\left(\frac{6.65\ \text{lb}}{\text{gal}}\right)\left(\frac{1\ \text{kg}}{2.20\ \text{lb}}\right)\left(\frac{10\,500\ \text{kcal}}{\text{kg}}\right)\left(\frac{4.18\ \text{kJ}}{\text{kcal}}\right)\left(\frac{1\ \text{MW}}{10^3\ \text{kW}}\right)}{\text{h}\left(\frac{3600\ \text{s}}{\text{h}}\right)} = 173\ \text{MW}$$

18.68 Two solutions, initially at 25.08 °C, were mixed in an insulated bottle. One contained 400 mL of 0.200 M weak monoprotic acid solution. The other contained 100 mL of a solution having 0.800 mol NaOH/L. After mixing, the temperature rose to 26.25 °C. How much heat is evolved in the neutralization of 1 mol of the acid? Assume that the densities of all solutions are 1.00 g/cm³ and that their specific heat capacities are all 4.2 J/g·K. (These assumptions are in error by several percent, but the subsequent errors in the final result partly cancel each other.)

▮

$$\text{Heat} = mc\,\Delta t = (500\text{ g})(4.2\text{ J/g}\cdot\text{K})[(26.25 - 25.08)\text{ K}] = 2.5 \times 10^3\text{ J} = 2.5\text{ kJ}$$

$$(400\text{ mL})\left(\frac{0.200\text{ mol}}{1000\text{ mL}}\right) = 0.0800\text{ mol acid} \qquad (100\text{ mL})\left(\frac{0.800\text{ mol}}{1000\text{ mL}}\right) = 0.0800\text{ mol base}$$

$$\frac{2.5\text{ kJ}}{0.0800\text{ mol}} = 31\text{ kJ/mol}$$

18.69 The thermochemical equation for the combustion of ethylene gas, C_2H_4, is

$$C_2H_4(g) + 3O_2(g) \longrightarrow 2CO_2(g) + 2H_2O(l) \qquad \Delta H^\circ = -337\text{ kcal}$$

Assuming 70.0% efficiency, how many kg of water at 20 °C can be converted into steam at 100 °C by burning 1.00 m³ of C_2H_4 gas measured at STP?

▮

$$n(C_2H_4) = \frac{(1.00\text{ m}^3)(1000\text{ L/m}^3)}{22.4\text{ L/mol}} = 44.6\text{ mol}$$

$$\Delta H(1\text{ m}^3) = n(C_2H_4) \times \Delta H(1\text{ mol}) = (44.6\text{ mol})(-337\text{ kcal/mol}) = -1.50 \times 10^4\text{ kcal}$$

The useful heat is then $(0.700)(1.50 \times 10^4\text{ kcal}) = 1.05 \times 10^4\text{ kcal}$.
For the overall process, consider two stages:

$$H_2O(l, 20\text{ °C}) \longrightarrow H_2O(l, 100\text{ °C}) \qquad \Delta H = (1.00\text{ kcal/kg}\cdot\text{K})(80\text{ K}) = 80\text{ kcal/kg}$$

$$H_2O(l, 100\text{ °C}) \longrightarrow H_2O(g, 100\text{ °C}) \qquad \Delta H = 540\text{ kcal/kg}$$

$$\Delta H(\text{total}) = 620\text{ kcal/kg}$$

The mass of water converted is then equal to the amount of heat available divided by the heat requirement per kg.

$$m(H_2O) = \frac{1.05 \times 10^4\text{ kcal}}{620\text{ kcal/kg}} = 16.9\text{ kg}$$

18.70 Use the data below to calculate ΔH for the reaction $N_2O_4 + 3CO \longrightarrow N_2O + 3CO_2$

compound	ΔH_f°, kJ/mol
$CO(g)$	−110
$CO_2(g)$	−393
$N_2O(g)$	81
$N_2O_4(g)$	9.7

▮ $\Delta H = \Delta H_f^\circ(N_2O) + 3\Delta H_f^\circ(CO_2) - \Delta H_f^\circ(N_2O_4) - 3\Delta H_f^\circ(CO) = [81 + 3(-393) - 9.7 - 3(-110)]\text{ kJ} = -778\text{ kJ}$

18.71 The heat of combustion of carbon to CO_2 is −393.5 kJ/mol. Calculate the heat released upon formation of 35.2 g of CO_2 from carbon and oxygen gas.

▮

$$(35.2\text{ g }CO_2)\left(\frac{1\text{ mol }CO_2}{44.0\text{ g }CO_2}\right)\left(\frac{-393.5\text{ kJ}}{\text{mol }CO_2}\right) = -315\text{ kJ}$$

18.72 Calculate ΔH_f° of $C_6H_{12}O_6(s)$ from the following data:

ΔH_{comb} of $C_6H_{12}O_6(s) = -2816$ kJ/mol ΔH_f° of $CO_2(g) = -393.5$ kJ/mol ΔH_f° of $H_2O(l) = -285.9$ kJ/mol

▮

$$C_6H_{12}O_6 + 6O_2 \longrightarrow 6CO_2 + 6H_2O$$

$$\Delta H = 6\Delta H_f^\circ(CO_2) + 6\Delta H_f^\circ(H_2O) - \Delta H_f(C_6H_{12}O_6)$$

$$\Delta H_f(C_6H_{12}O_6) = [6(-393.5) + 6(-285.9) - (-2816)]\text{ kJ} = -1260\text{ kJ}$$

Since 1 mol of $C_6H_{12}O_6$ was involved, this is −1260 kJ/mol.

18.73 Write equations for the two reactions corresponding to the following ΔH_f° values. Combine these equations to give that for the reaction

$$2NO_2(g) \longrightarrow N_2O_4(g)$$

Calculate the value of ΔH for this reaction, and state whether the reaction is endothermic or exothermic.

$$\Delta H_f^\circ \text{ for } NO_2(g) = 33.84 \text{ kJ/mol} \qquad \Delta H_f^\circ \text{ for } N_2O_4(g) = 9.66 \text{ kJ/mol}$$

▌ (*1*) $\frac{1}{2}N_2 + O_2 \longrightarrow NO_2 \qquad \Delta H = 33.84 \text{ kJ}$

(*2*) $N_2 + 2O_2 \longrightarrow N_2O_4 \qquad \Delta H = 9.66 \text{ kJ}$

(*2*) − 2(*1*) $2NO_2 \longrightarrow N_2O_4 \qquad \Delta H = -58.02 \text{ kJ (exothermic reaction)}$

18.74 Consider the reaction

$$2C_2H_6(g) + 7O_2(g) \longrightarrow 4CO_2(g) + 6H_2O(l) \qquad \Delta H = -3119 \text{ kJ}$$

Calculate ΔH_f° for $C_2H_6(g)$.

▌
$$\Delta H = 4\Delta H_f^\circ(CO_2) + 6\Delta H_f^\circ(H_2O) - 2\Delta H_f^\circ(C_2H_6) - 7\Delta H_f^\circ(O_2)$$

$$\Delta H_f^\circ(C_2H_6) = \frac{[4(-393.5) + 6(-285.8) - (-3119)] \text{ kJ}}{2 \text{ mol}} = -84.9 \text{ kJ/mol}$$

18.75 Propane gas is used as a fuel for household heating. Assuming no heat loss, calculate the number of cubic feet ($1 \text{ ft}^3 = 28.3 \text{ L}$) of propane gas, measured at 25 °C and 1.00 atm pressure, which would be necessary to heat 50 gal (1 gal = 3.785 L) of water from 60 to 150 °F. $\Delta H_{comb} = -526.3$ kcal/mol.

▌
$$(50 \text{ gal})\left(\frac{3.785 \text{ L}}{\text{gal}}\right)\left(\frac{10^3 \text{ ml}}{\text{L}}\right)\left(\frac{1.00 \text{ cal}}{\text{mL}}\right)\left(\frac{1 \text{ kcal}}{10^3 \text{ cal}}\right)\left(\frac{1 \text{ mol } C_3H_8}{526.3 \text{ kcal}}\right) = 0.36 \text{ mol } C_3H_8$$

$$V = \frac{nRT}{P} = \frac{(0.36 \text{ mol})(0.0821 \text{ L·atm/mol·K})(298 \text{ K})}{1.00 \text{ atm}} = (8.8 \text{ L})\left(\frac{1 \text{ ft}^3}{28.3 \text{ L}}\right) = 0.31 \text{ ft}^3$$

18.76 A 1.250-g sample of benzoic acid, $C_7H_6O_2$, was placed in a combustion bomb. The bomb was filled with an excess of oxygen at high pressure, sealed, and immersed in a pail of water which served as a calorimeter. The heat capacity of the entire apparatus was found to be 2422 cal/K, including the bomb, the pail, a thermometer, and the water. The oxidation of the benzoic acid was triggered by passing an electric spark through the sample. After complete combustion of the sample, the thermometer immersed in the water registered a temperature 3.256 K greater than before the combustion. What is ΔE_{comb} per mol of benzoic acid combusted in a bomb-type calorimeter? Assume that no correction need be made for the sparking process.

▌
$$q(\text{acid}) = -q(\text{calorimeter}) = -(2422 \text{ cal/K})(3.256 \text{ K}) = -7.89 \text{ kcal}$$

$$\Delta E_{comb} = \frac{q(\text{acid})}{n} = \frac{-7.89 \text{ kcal}}{(1.250 \text{ g})/(122.1 \text{ g/mol})} = -771 \text{ kcal/mol}$$

18.77 (*a*) What do (or what could) we call the following processes?

(i) $H_2O(s) \longrightarrow H_2O(l)$ (ii) $CH_4 + 2O_2 \longrightarrow CO_2 + 2H_2O$

(iii) $CO_2(s) \longrightarrow CO_2(g)$ (iv) $2Na + Cl_2 \longrightarrow 2NaCl$

(*b*) Name the enthalpy changes associated with the processes in part (*a*). (*c*) In there anything inherently different about the enthalpy changes named in (*b*), despite their different names?

▌ (*a*) The processes are called fusion (or melting), combustion, sublimation, and formation, respectively. (*b*) The associated enthalpy changes are the enthalpies of fusion, combustion, sublimation, and formation, respectively. (*c*) Despite the fact that they have different names, these and other enthalpy changes are merely enthalpy changes for different processes; they are not inherently different.

18.78 Calculate the standard enthalpy change at 298 K for the reaction

$$CH_4(g, 1 \text{ atm}) + 4CuO(s) \longrightarrow CO_2(g, 1 \text{ atm}) + 2H_2O(l) + 4Cu(s)$$

▌ The sum of the enthalpies of formation of the products (from Table 18.2) is

$$[-94.05 + 2(-68.32) + 4(0)] \text{ kcal} = -230.7 \text{ kcal}$$

TABLE 18.2 Standard Enthalpies of Formation, ΔH_f°, at 298 K

	kcal/mol*		kcal/mol*		kcal/mol*
$AgBr(s)$	−23.8	$Cr_2O_3(s)$	−272.7	$NH_4Cl(s)$	−75.38
$AgCl(s)$	−30.36	$CuO(s)$	−37.6	$NH_3(g)$	−11.04
$AgI(s)$	−14.9	$Cu_2S(s)$	−19.0	$N_2H_4(g)$	12.05
$Al_2O_3(s)$	−399.1	$CuSO_4(s)$	−184.03	$N_2O(g)$	19.49
$Au(OH)_3(s)$	−100	$Fe_2O_3(s)$	−196.5	$NO(g)$	21.60
$BF_3(g)$	−265.4	$FeS(s)$	−22.8	$NO_2(g)$	8.09
$B_2H_6(g)$	7.5	$HBr(g)$	−8.66	$N_2O_4(g)$	2.19
$B_2O_3(s)$	−305.3	$HCl(g)$	−22.02	$N_2O_4(l)$	−4.66
$BaCO_3(s)$	−290.8	$HF(g)$	−64.2	$NaCl(s)$	−98.23
$BaO(s)$	−133.5	$HI(g)$	6.20	$NaHCO_3(s)$	−169.8
$BaSO_4(s)$	−350.2	$HNO_3(l)$	−41.61	$Na_2CO_3(s)$	−341.8
$Br(g)$	26.73	$H_2O(g)$	−57.79	$O_3(g)$	34.1
$C(s;$ diamond$)$	0.45	$H_2O(l)$	−68.32	$PCl_3(l)$	−76.4
$CF_4(g)$	−221	$H_2O_2(l)$	−44.88	$PCl_3(g)$	−68.6
$CH_4(g)$	−17.89	$H_2S(g)$	−4.815	$PCl_5(g)$	−89.6
$CH_3OH(g)$	−47.96	$H_2S(aq)$	−9.5	$PH_3(g)$	5.5
$C_9H_{20}(l)$	−65.85	$H_2SO_4(l)$	−193.91	$P_2O_5(s)$	−365.83
$(CH_3)_2N_2H_2(l)$	13.3	$HgO(s)$	−21.7	$POCl_3(g)$	−133.48
$C(NO_2)_4(l)$	8.8	$I_2(g)$	14.92	$SO_2(g)$	−70.96
$CO(g)$	−26.41	$KCl(s)$	−104.18	$SO_3(g)$	−94.45
$CO_2(g)$	−94.05	$KClO_3(s)$	−93.5	$SiO_2(s)$	−209.9
$CaC_2(s)$	−14.2	$KClO_4(s)$	−102.8	$SiCl_4(g)$	−145.7
$CaCO_3(s)$	−288.5	$LiAlH_4(s)$	−24	$SiF_4(g)$	−361.29
$CaCl_2(s)$	−190.0	$LiBH_4(s)$	−45	$WO_3(s)$	−201.5
$Ca(OH)_2(s)$	−235.80	$Li_2O(s)$	−143.1	$ZnO(s)$	−83.2
$CaO(s)$	−151.9	$NF_3(g)$	−30.4	$ZnS(s)$	−48.5
$ClF_3(l)$	−45.3				

* To convert to kJ/mol, multiply tabular entries by 4.184.

The related sum for the reactants is

$$[(-17.89) + 4(-37.6)]\ \text{kcal} = -168.3\ \text{kcal}$$

Then

$$\Delta H_f^\circ = -230.7 - (-168.3) = -62.4\ \text{kcal}$$

18.79 At 25 °C and 1 atm, which one(s) of the following, if any, has a nonzero ΔH_f°? ***(a)*** Fe ***(b)*** O ***(c)*** C(*s*) ***(d)*** Ne

▮ ***(b)*** ΔH_f° is 0 for elements in their standard states—Fe, C, and Ne. O, because it is not in its standard state, has a nonzero ΔH_f°. (O_2 is the standard state.)

18.80 Calculate the enthalpy of combustion of $C_2H_4(g)$.

▮

$$C_2H_4 + 3O_2 \longrightarrow 2CO_2 + 2H_2O(l)$$

For $C_2H_4(g)$ $\Delta H_f^\circ = 12.4$ kcal/mol. For $CO_2(g)$ $\Delta H_f^\circ = -94.0$ kcal/mol. For $H_2O(l)$ $\Delta H_f^\circ = -68.3$ kcal/mol.

$$\Delta H = 2\Delta H_f^\circ(CO_2) + 2\Delta H_f^\circ(H_2O) - \Delta H_f^\circ(C_2H_4) = 2(-94.0) + 2(-68.3) - (12.4) = -337\ \text{kcal}$$

18.81 $\Delta H = -84.4$ kcal for the following reaction at 298 K.

$$F_2 + 2HCl \longrightarrow 2HF + Cl_2$$

All substances are gaseous. $\Delta H_f^\circ(HF) = -64.2$ kcal/mol. What is $\Delta H_f^\circ(HCl)$, in kcal/mol?

▮ $\Delta H = 2\Delta H_f^\circ(HF) - 2\Delta H_f^\circ(HCl) = -84.4$ kcal $\quad \Delta H_f^\circ(HCl) = \dfrac{[2(-64.2) - (-84.4)]\ \text{kcal}}{2\ \text{mol}} = -22.0$ kcal/mol

18.82 The enthalpy of transition of crystalline boron to amorphous boron at 1500 °C is 0.40 kcal/mol. What is the enthalpy change accompanying the conversion of 50.0 g of crystalline boron at that temperature?

▌
$$(50.0 \text{ g B})\left(\frac{1 \text{ mol B}}{10.81 \text{ g B}}\right)\left(\frac{0.40 \text{ kcal}}{\text{mol B}}\right) = 1.9 \text{ kcal}$$

18.83 The enthalpy change for which of the following processes represents the enthalpy of formation of AgCl? Explain.

(*a*) $Ag^+(aq) + Cl^-(aq) \longrightarrow AgCl(s)$ (*b*) $Ag(s) + \frac{1}{2}Cl_2(g) \longrightarrow AgCl(s)$
(*c*) $AgCl(s) \longrightarrow Ag(s) + \frac{1}{2}Cl_2(g)$ (*d*) $Ag(s) + AuCl(s) \longrightarrow Au(s) + AgCl(s)$

▌ (*b*). (Formation of the compound from its elements in their standard states.)

18.84 Which one(s) of the following equations have enthalpy changes equal to (*a*) $\Delta H_f^\circ(CO_2)$? (*b*) $\Delta H_{comb}(C)$? (*c*) $\Delta H_{comb}(CO)$? (*d*) $\Delta H_f^\circ(CO)$?

(i) $C + O_2 \longrightarrow CO_2$ (ii) $C + \frac{1}{2}O_2 \longrightarrow CO$ (iii) $CO + \frac{1}{2}O_2 \longrightarrow CO_2$

▌ (*a*) i (In iii, CO_2 is not formed from its elements.) (*b*) i (In ii, the C does not react as much as possible with O_2.) (*c*) iii (*d*) ii

18.85 The enthalpy of combustion of glucose, $C_6H_{12}O_6$, is −673 kcal/mol at 298 K. Calculate the standard enthalpy of formation of glucose. Sketch an enthalpy diagram for the process.

▌
$$C_6H_{12}O_6(s) + 6O_2(g) \longrightarrow 6CO_2(g) + 6H_2O(l)$$

Since
$$\Delta H_{reaction} = \sum \Delta H_f^\circ(\text{products}) - \sum \Delta H_f^\circ(\text{reactants})$$
$$\Delta H_{comb} = [6\Delta H_f^\circ(CO_2) + 6\Delta H_f^\circ(H_2O)] - [\Delta H_f^\circ(C_6H_{12}O_6) + 6\Delta H_f^\circ(O_2)]$$
$$-673 \text{ kcal} = 6(-94.05) + 6(-68.32) - \Delta H_f^\circ(C_6H_{12}O_6) - 6(0.0)$$
$$\Delta H_f^\circ(C_6H_{12}O_6) = -301 \text{ kcal/mol}$$

The enthalpy diagram, Fig. 18.1, emphasizes that the enthalpy change for the conversion of 6 mol of carbon, 6 mol of hydrogen molecules, and 9 mol of oxygen molecules to 6 mol of carbon dioxide and 6 mol of water is the same whether or not one first "prepares" a mol of glucose and then allows that to react with the rest of the oxygen to give the same products.

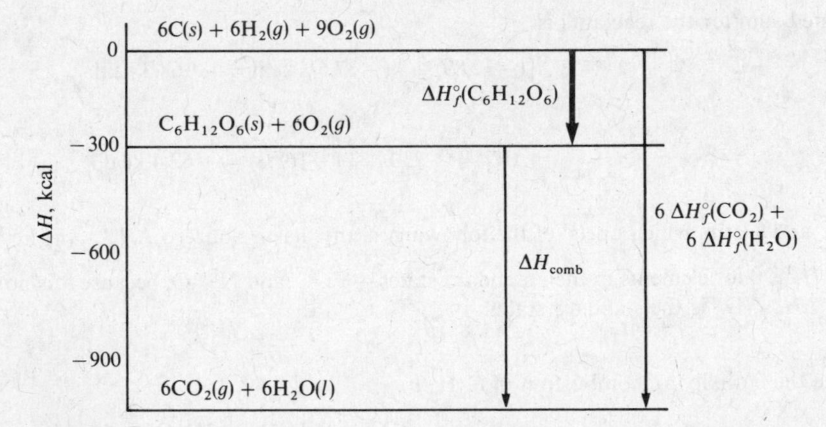

Fig. 18.1

18.86 In Fig. 18.1, why is the state $6C(s) + 6H_2(g) + 9O_2(g)$ assigned a value of 0 kcal? Would it make a difference in the magnitudes of the calculated enthalpy changes if this state were assigned some other value? Explain.

▌ The elements in their standard states are all defined as having ΔH_f° values of 0. Even if the value for the elements were defined otherwise, the values of the enthalpies of reaction would not change. The other states would merely change in accordance with the changed definitions, maintaining the same ΔH.

18.87 Determine the enthalpy change for the reaction of 96.0 g of sulfur trioxide with excess barium oxide to yield barium sulfate at 298 K.

▌

$$SO_3 + BaO \longrightarrow BaSO_4$$

$$\Delta H = \Delta H_f^\circ(BaSO_4) - \Delta H_f^\circ(BaO) - \Delta H_f^\circ(SO_3) = [(-350.2) - (-133.5) - (-94.45)] \text{ kcal/mol} = -122.25 \text{ kcal/mol}$$

$$(96.0 \text{ g})\left(\frac{1 \text{ mol } SO_3}{80.0 \text{ g}}\right)\left(\frac{-122.25 \text{ kcal}}{\text{mol}}\right) = -146.7 \text{ kcal}$$

18.88 Calculate ΔH° for reduction of iron(III) oxide by aluminum (thermite reaction) at 25 °C.

▌ We must start with a balanced equation for the reaction. We may then write in parentheses under each formula the enthalpy of formation, taken from Table 18.2, and multiply each enthalpy by the corresponding number of moles in the balanced equation. Remember that ΔH_f° for any element in its standard state is 0, by definition.

	$2Al$	$+\ Fe_2O_3$	$\longrightarrow 2Fe$	$+\ Al_2O_3$
$n(\Delta H_f^\circ)$	2(0)	1(−196.5)	2(0)	1(−399.1)

Then ΔH° of the reaction is given by

$$\Delta H^\circ = (\text{sum of } n\,\Delta H_f^\circ \text{ of products}) - (\text{sum of } n\,\Delta H_f^\circ \text{ of reactants}) = -399.1 - (-196.5) = -202.6 \text{ kcal}$$

This is ΔH° for the reduction of 1 mol Fe_2O_3. In SI units, $\Delta H^\circ = -848$ kJ.

18.89 ΔH_f° for N(g) (not the standard state of nitrogen) has been determined as 472.7 kJ/mol, and for O(g) as 249.2 kJ/mol. What is ΔH° for the hypothetical upper-atmosphere reaction

$$N(g) + O(g) \longrightarrow NO(g)$$

both in kJ and in kcal?

▌

	$N(g)$	$+\ O(g)$	$\longrightarrow NO(g)$
ΔH_f°, kJ	472.7	249.2	89.7

where ΔH_f° for NO(g) has been computed from Table 18.2. Then

$$\Delta H^\circ = 89.7 - (472.7 + 249.2) = -632.2 \text{ kJ} = \frac{-632.2 \text{ kJ}}{4.184 \text{ kJ/kcal}} = -151.1 \text{ kcal}$$

18.90 Calculate the mass of mercury which can be liberated from HgO at 25 °C by the treatment of excess HgO with 10.0 kcal of heat.

▌

$$HgO \longrightarrow Hg + \tfrac{1}{2}O_2$$

$$\Delta H = -\Delta H_f^\circ(HgO) = 21.7 \text{ kcal/mol HgO} \qquad \text{(from Table 18.2)}$$

$$(10.0 \text{ kcal})\left(\frac{1 \text{ mol Hg}}{21.7 \text{ kcal}}\right)\left(\frac{200.59 \text{ g}}{\text{mol Hg}}\right) = 92.4 \text{ g Hg}$$

18.91 Aluminum metal is a very effective reagent for reducing oxides to their elements. An example is the thermite reaction:

$$Fe_2O_3(s) + 2Al(s) \longrightarrow Al_2O_3(s) + 2Fe(s)$$

Calculate the standard enthalpy change when 1.00 mol of each of the following oxides is reduced by aluminum at 298 K: ***(a)*** Fe_2O_3 ***(b)*** SiO_2 ***(c)*** CuO ***(d)*** CaO

▌ ***(a)*** $\Delta H = \Delta H_f^\circ(Al_2O_3) - \Delta H_f^\circ(Fe_2O_3) = [(-399.1) - (-196.5)] \text{ kcal} = -202.6 \text{ kcal}$

(b) $4Al + 3SiO_2 \longrightarrow 2Al_2O_3 + 3Si \qquad \Delta H = 2\Delta H_f^\circ(Al_2O_3) - 3\Delta H_f^\circ(SiO_2)$

$(2 \text{ mol})(-399.1 \text{ kcal/mol}) - (3 \text{ mol})(-209.9 \text{ kcal/mol}) = -168.5 \text{ kcal}$

(c) $3CuO + 2Al \longrightarrow 3Cu + Al_2O_3$

$\Delta H = \Delta H_f^\circ(Al_2O_3) - 3\Delta H_f^\circ(CuO) = (-399.1 \text{ kcal}) - (3 \text{ mol})(-37.6 \text{ kcal/mol}) = -286.3 \text{ kcal}$

(d) 56.6 kcal

18.92 Calculate the heat produced when 1.00 gal of octane, C_8H_{18}, reacts with oxygen to form carbon monoxide and water vapor at 25 °C. (The density of octane is 0.7025 g/mL; 1.00 gal = 3.785 L.) $\Delta H_{comb}(C_8H_{18}) = -1302.7$ kcal

▌

$$(3785 \text{ mL})\left(\frac{0.7025 \text{ g}}{\text{mL}}\right)\left(\frac{1 \text{ mol}}{114.2 \text{ g}}\right) = 23.28 \text{ mol } C_8H_{18} \qquad C_8H_{18} + \tfrac{17}{2}O_2 \longrightarrow 8CO + 9H_2O$$

Per mol of C_8H_{18}:

$\Delta H_f^\circ(C_8H_{18})$ must be determined from its ΔH_{comb}: $\quad C_8H_{18} + \frac{25}{2}O_2 \longrightarrow 8CO_2 + 9H_2O$

$$\Delta H_{comb} = -1302.7 \text{ kcal} = 8\Delta H_f^\circ(CO_2) + 9\Delta H_f^\circ(H_2O) - \Delta H_f^\circ(C_8H_{18})$$
$$\Delta H_f^\circ(C_8H_{18}) = +1302.7 + 8(-94.05) + 9(-68.32) = -64.6 \text{ kcal}$$
$$\Delta H = 8\Delta H_f^\circ(CO) + 9\Delta H_f^\circ(H_2O) - \Delta H_f^\circ(C_8H_{18}) = [8(-26.41) + 9(-57.79) - (-64.6)] \text{ kcal}$$
$$= -666.8 \text{ kcal} \qquad (\text{per mol of } C_8H_{18})$$
$$\Delta H = (-666.8 \text{ kcal/mol})(23.28 \text{ mol}) = -15\,520 \text{ kcal}$$

18.93 Calculate the enthalpy of decomposition of $CaCO_3$ into CaO and CO_2.

▮

$$\begin{array}{lccc} & CaCO_3(s) & CaO(s) + & CO_2(g) \\ \Delta H_f^\circ & -288.5 & -151.9 & -94.0 \end{array}$$
$$\Delta H^\circ = (-151.9 - 94.0) - (-288.5) = 42.6 \text{ kcal}$$

This is the enthalpy change for the decomposition of 1 mol $CaCO_3$. A positive value signifies an endothermic reaction.

18.94 The heat evolved on combustion of acetylene gas, C_2H_2, at 25 °C is 310.5 kcal/mol. Determine the enthalpy of formation of acetylene gas.

▮ The complete combustion of an organic compound involves the formation of CO_2 and H_2O.

$$\begin{array}{lcccc} & C_2H_2(g) + & \frac{5}{2}O_2(g) \longrightarrow & 2CO_2(g) + & H_2O(l) \qquad \Delta H^\circ = -310.5 \text{ kcal} \\ n(\Delta H_f^\circ) & x & 0 & 2(-94.0) & -68.3 \end{array}$$

Thus
$$-310.5 = [2(-94.0) + (-68.3)] - x$$

Solving, $x = \Delta H_f^\circ(C_2H_2) = +54.2$ kcal/mol.

18.95 How much heat will be required to make 1.000 kg of CaC_2 from CaO(s) and C(s)?

▮

$$\begin{array}{lcccc} & CaO(s) + & 3C(s) \longrightarrow & CaC_2(s) + & CO(g) \\ \Delta H_f^\circ & -151.9 & 0 & -14.2 & -26.4 \end{array}$$
$$\Delta H^\circ = -(14.2 + 26.4) - (-151.9) = +111.3 \text{ kcal}$$

This is the heat required to make 1.000 mol CaC_2; 1.000 kg CaC_2 will require

$$\left(\frac{1000 \text{ g } CaC_2}{64.10 \text{ g } CaC_2/\text{mol}}\right)(111.3 \text{ kcal/mol}) = 1736 \text{ kcal}$$

18.96 What is the heat of sublimation of solid iodine at 25 °C?

▮ The enthalpy of sublimation is merely the enthalpy of formation of $I_2(g)$.

$$\Delta H_{sub} = 14.92 \text{ kcal/mol} = 62.4 \text{ kJ/mol}$$

18.97 The standard enthalpy of formation of H(g) has been determined to be 218.0 kJ/mol. Calculate ΔH° in kJ for

$$H(g) + Br(g) \qquad HBr(g)$$

▮ $\Delta H^\circ = \Delta H_f^\circ(HBr) - \Delta H_f^\circ(H) - \Delta H_f^\circ(Br) = [(-36.38) - (218.0) - (111.84)] \text{ kJ} = -366.2 \text{ kJ}$

18.98 Repeat Prob. 18.97 for the reaction $H(g) + Br_2(l) \qquad HBr(g) + Br(g)$

▮ $\Delta H^\circ = \Delta H_f^\circ(HBr) + \Delta H_f^\circ(Br) - \Delta H_f^\circ(H) - \Delta H_f^\circ(Br_2) = [(-36.38) + (111.84) - (218.0) - 0] \text{ kJ} = -142.5 \text{ kJ}$

18.99 Given

$$N_2(g) + 3H_2(g) \longrightarrow 2NH_3(g) \qquad \Delta H^\circ = -22.0 \text{ kcal}$$

What is the standard enthalpy of formation of NH_3 gas?

▮

$$\frac{-22.0 \text{ kcal}}{2 \text{ mol}} = -11.0 \text{ kcal/mol}$$

18.100 Determine ΔH° of decomposition of 1.000 mol of solid $KClO_3$ into solid KCl and gaseous oxygen.

▮ $$KClO_3 \longrightarrow KCl + \tfrac{3}{2}O_2 \qquad \Delta H^\circ = [(-436.7) - (-391.2)] \text{ kJ} = -45.5 \text{ kJ}$$

18.101 Find the heat evolved in slaking 1.00 kg of quicklime (CaO) according to the reaction

$$CaO(s) + H_2O(l) \longrightarrow Ca(OH)_2(s)$$

▮ $$\Delta H^\circ = \Delta H_f^\circ(Ca(OH)_2) - \Delta H_f^\circ(CaO) - \Delta H_f^\circ(H_2O)$$
$$= [(-986.2) - (-634.3) - (-285.83)] \text{ kJ/mol} = -66.1 \text{ kJ/mol}$$
$$(1000 \text{ g CaO})\left(\frac{1 \text{ mol CaO}}{56.0 \text{ g CaO}}\right)\left(\frac{-66.1 \text{ kJ}}{\text{mol CaO}}\right) = -1180 \text{ kJ}$$

18.102 The heat liberated on complete combustion of 1.00 mol of CH_4 gas to $CO_2(g)$ and $H_2O(l)$ is 890 kJ. Determine the enthalpy of formation of 1.00 mol of CH_4 gas.

▮ $$CH_4 + 2O_2 \longrightarrow CO_2 + 2H_2O$$
$$\Delta H^\circ = \Delta H_f^\circ(CO_2) + 2\Delta H_f^\circ(H_2O) - \Delta H_f^\circ(CH_4)$$
$$\Delta H_f^\circ(CH_4) = [(-393.51) + 2(-285.83) - (-890)] \text{ kJ} = -75 \text{ kJ}$$

18.103 The heat evolved on combustion of 1.00 g of starch, $(C_6H_{10}O_5)_x$, into $CO_2(g)$ and $H_2O(l)$ is 4.18 kcal. Compute the standard enthalpy of formation of 1.00 g of starch.

▮ $$xO_2 + (C_6H_{10}O_5)_x \longrightarrow 6xCO_2 + 5xH_2O$$

The problem may be solved using the empirical formula, $C_6H_{10}O_5$.

$$\Delta H_f^\circ(\text{starch}) = 6x\,\Delta H_f^\circ(CO_2) + 5x\,\Delta H_f^\circ(H_2O) - \Delta H^\circ$$
$$= 6x\left(\frac{-94.05 \text{ kcal}}{\text{mol}}\right) + 5x\left(\frac{-68.32 \text{ kcal}}{\text{mol}}\right) - \left(\frac{-4.18 \text{ kcal}}{\text{g}}\right)\left(\frac{162x \text{ g}}{\text{mol}}\right) = -229x \text{ kcal}$$

Per g of starch:

$$\frac{-229x \text{ kcal}}{\text{mol}}\left(\frac{1 \text{ mol}}{162x \text{ g}}\right) = -1.41 \text{ kcal/g}$$

18.104 Calculate ΔH at 25 °C for the reaction of 1.000 mol of carbon with excess carbon dioxide to produce carbon monoxide.

▮ $$C + CO_2 \longrightarrow 2CO$$

Since the enthalpy of formation of an element is 0:

$$\Delta H = 2\,\Delta H_f^\circ(CO) - \Delta H_f^\circ(CO_2) = [2(-26.41) - (-94.05)] \text{ kcal} = +41.23 \text{ kcal}$$

18.105 The heat evolved on combustion into $CO_2(g)$ and $H_2O(l)$ of 1.000 mol C_2H_6 is 368.4 kcal, and of 1.000 mol C_2H_4 is 337.2 kcal. Calculate ΔH of the following reaction: $C_2H_4 + H_2(g) \rightarrow C_2H_6$.

▮
(1)	$C_2H_6 + \frac{7}{2}O_2 \longrightarrow 2CO_2 + 3H_2O$	$\Delta H = -368.4$ kcal
(2)	$C_2H_4 + 3O_2 \longrightarrow 2CO_2 + 2H_2O$	$\Delta H = -337.2$ kcal
(3)	$H_2 + \frac{1}{2}O_2 \longrightarrow H_2O$	$\Delta H = -\ 68.32$ kcal
(2) − (1) + (3)	$C_2H_4 + H_2 \longrightarrow C_2H_6$	$\Delta H = -\ 37.1$ kcal

18.106 The thermochemical equation for the dissociation of hydrogen gas into atoms may be written:

$$H_2 \longrightarrow 2H \qquad \Delta H = 436 \text{ kJ}$$

What is the ratio of the energy yield on combustion of hydrogen atoms to steam to the yield on combustion of an equal mass of hydrogen molecules to steam?

▮
(1)	$H_2 + \frac{1}{2}O_2 \longrightarrow H_2O(g)$	$\Delta H = -241.81$ kJ
(2)	$H_2 \longrightarrow 2H$	$\Delta H = 436$ kJ
(1) − (2)	$2H + \frac{1}{2}O_2 \longrightarrow H_2O(g)$	$\Delta H = -678$ kJ

The ratio of the last reaction to the first, the combustion of H atoms to H_2 molecules, is

$$\frac{-678 \text{ kJ}}{-241.81 \text{ kJ}} = 2.80$$

18.107 The commercial production of water gas utilizes the reaction $C + H_2O(g) \rightarrow H_2 + CO$. The required heat for this endothermic reaction may be supplied by adding a limited amount of air and burning some carbon to carbon dioxide. How many g of carbon must be burned to CO_2 to provide enough heat for the water gas conversion of 100 g of carbon? Neglect all heat losses to the environment.

▌

$$C + H_2O(g) \longrightarrow H_2 + CO$$

$$\Delta H = \Delta H_f^\circ(CO) - \Delta H_f^\circ(H_2O) = [(-110.53) - (-241.81)] \text{ kJ/mol} = 131.28 \text{ kJ/mol}$$

$$(100 \text{ g C})\left(\frac{1 \text{ mol C}}{12.0 \text{ g C}}\right)\left(\frac{131.28 \text{ kJ}}{\text{mol}}\right) = 1094 \text{ kJ}$$

To provide that energy,

$$(1094 \text{ kJ})\left(\frac{1 \text{ mol C}}{393.51 \text{ kJ}}\right)\left(\frac{12.0 \text{ g C}}{\text{mol C}}\right) = 33.4 \text{ g C}$$

18.108 The reversible reaction

$$Na_2SO_4 \cdot 10H_2O \longrightarrow Na_2SO_4 + 10H_2O \qquad \Delta H = +18.8 \text{ kcal}$$

goes completely to the right at temperatures above 32.4 °C, and remains completely on the left below this temperature. This system has been used in some solar houses for heating at night with the energy absorbed from the sun's radiation during the day. How many cubic feet of fuel gas could be saved per night by the reversal of the dehydration of a fixed charge of 100 lb $Na_2SO_4 \cdot 10H_2O$? Assume that the fuel value of the gas is 2000 Btu/ft³.

▌

$$(100 \text{ lb } Na_2SO_4 \cdot 10H_2O)\left(\frac{1.00 \text{ kg}}{2.20 \text{ lb}}\right)\left(\frac{10^3 \text{ g}}{\text{kg}}\right)\left(\frac{1 \text{ mol}}{322 \text{ g}}\right)\left(\frac{18.8 \text{ kcal}}{\text{mol}}\right) = 2650 \text{ kcal}$$

$$(2650 \text{ kcal})\left(\frac{1 \text{ Btu}}{0.252 \text{ kcal}}\right)\left(\frac{1 \text{ ft}^3}{2000 \text{ Btu}}\right) = 5.3 \text{ ft}^3$$

18.109 Using data from Tables 18.1 and 18.2, calculate the enthalpy change for the reaction between carbon and oxygen at 400 °C to form 1.00 mol of carbon dioxide at 400 °C.

▌ The enthalpy of reaction at 673 K may be obtained from the corresponding enthalpy change at 298 K and the heat capacity of each reactant and product. All the enthalpy changes are identified in Fig. 18.2.

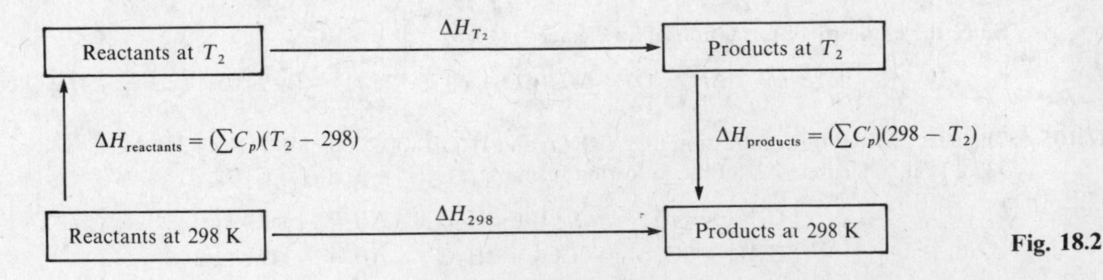

Fig. 18.2

$$\Delta H_{673} = \Delta H_{298} - \Delta H_{\text{products}} - \Delta H_{\text{reactants}} \qquad \Delta H_{298} = -94.05 \text{ kcal}$$

$$\Delta H_{\text{products}} = C_p(CO_2)(298 - 673) = (8.96 \text{ cal/mol}\cdot\text{K})(-375 \text{ K})(1.00 \text{ mol}) = -3360 \text{ cal} = -3.36 \text{ kcal}$$

$$\Delta H_{\text{reactants}} = [C_p(O_2) + C_p(C)](673 - 298) = [(7.05 + 2.04) \text{ cal/mol}\cdot\text{K}](375 \text{ K})(1.00 \text{ mol}) = +3410 \text{ cal} = +3.41 \text{ kcal}$$

$$\Delta H_{673} = (-94.05 + 3.36 - 3.41) \text{ kcal} = -94.10 \text{ kcal}$$

18.110 An important criterion for the desirability of fuel reactions for rockets is the fuel value in kJ/g of reactant. Compute this quantity for each of the following reactions. Assume that the reactants are present in stoichiometric proportions.

(*a*) $N_2H_4(l) + 2H_2O_2(l) \longrightarrow N_2(g) + 4H_2O(g)$

(*b*) $2LiBH_4(s) + KClO_4(s) \longrightarrow Li_2O(s) + B_2O_3(s) + KCl(s) + 4H_2(g)$

(*c*) $6LiAlH_4(s) + 2C(NO_2)_4(l) \longrightarrow 3Al_2O_3(s) + 3Li_2O(s) + 2CO_2(g) + 4N_2(g) + 12H_2(g)$

(*d*) $4HNO_3(l) + 5N_2H_4(l) \longrightarrow 7N_2(g) + 12H_2O(g)$

(*e*) $7N_2O_4(l) + C_9H_{20}(l) \longrightarrow 9CO_2(g) + 10H_2O(g) + 7N_2(g)$

(*f*) $4ClF_3(l) + (CH_3)_2N_2H_2(l) \longrightarrow 2CF_4(g) + N_2(g) + 4HCl(g) + 4HF(g)$

▮ (*a*) $N_2H_4 + 2H_2O_2 \longrightarrow N_2 + 4H_2O$

$$\Delta H^\circ = 4\Delta H_f^\circ(H_2O) - \Delta H_f^\circ(N_2H_4) - 2\Delta H_f^\circ(H_2O_2) = [4(-241.81) - (50.63) - 2(-187.8)]\text{ kJ}$$
$$= -642.27\text{ kJ}$$

$$(1\text{ mol }N_2H_4)\left(\frac{32.0\text{ g}}{\text{mol}}\right) + (2\text{ mol }H_2O_2)\left(\frac{34.0\text{ g}}{\text{mol}}\right) = 100.0\text{ g total} \qquad \frac{-642.27\text{ kJ}}{100.0\text{ g}} = -6.42\text{ kJ/g}$$

(*b*)
$$\Delta H^\circ = \Delta H_f^\circ(Li_2O) + \Delta H_f^\circ(B_2O_3) + \Delta H_f^\circ(KCl) - 2\Delta H_f^\circ(LiBH_4) - \Delta H_f^\circ(KClO_4)$$
$$= [(-598.7) + (-1272.8) + (-436.7) - 2(-188) - (-430.1)]\text{ kJ} = -1502\text{ kJ}$$

$$\text{Mass} = (2\text{ mol LiBH}_4)\left(\frac{21.78\text{ g}}{\text{mol}}\right) + (1\text{ mol KClO}_4)\left(\frac{138.6\text{ g}}{\text{mol}}\right) = 182.2\text{ g} \qquad \frac{-1502\text{ kJ}}{182.2\text{ g}} = -8.24\text{ kJ/g}$$

(*c*)
$$\Delta H^\circ = 3\Delta H_f^\circ(Al_2O_3) + 3\Delta H_f^\circ(Li_2O) + 2\Delta H_f^\circ(CO_2) - 6\Delta H_f^\circ(LiAlH_4) - 2\Delta H_f^\circ(C(NO_2)_4)$$
$$= [3(-1675) + 3(-598.7) + 2(-393.51) - 6(-100) - 2(36.8)]\text{ kJ} = -7082\text{ kJ}$$

$$\text{Mass} = (6\text{ mol LiAlH}_4)\left(\frac{37.9\text{ g}}{\text{mol}}\right) + (2\text{ mol C(NO}_2)_4)\left(\frac{196\text{ g}}{\text{mol}}\right) = 619\text{ g} \qquad \frac{-7082\text{ kJ}}{619\text{ g}} = -11.4\text{ kJ/g}$$

(*d*)
$$\Delta H^\circ = 12\Delta H_f^\circ(H_2O) - 4\Delta H_f^\circ(HNO_3) - 5\Delta H_f^\circ(N_2H_4)$$
$$= [12(-241.81) - 4(-174.10) - 5(50.63)]\text{ kJ} = -2458.5\text{ kJ}$$

$$\text{Mass} = (4\text{ mol HNO}_3)\left(\frac{63.0\text{ g}}{\text{mol}}\right) + (5\text{ mol N}_2\text{H}_4)\left(\frac{32.0\text{ g}}{\text{mol}}\right) = 412\text{ g} \qquad \frac{-2458.5\text{ kJ}}{412\text{ g}} = -5.97\text{ kJ/g}$$

(*e*)
$$\Delta H^\circ = 9\Delta H_f^\circ(CO_2) + 10\Delta H_f^\circ(H_2O) - 7\Delta H_f^\circ(N_2O_4) - \Delta H_f^\circ(C_9H_{20})$$
$$= [9(-393.51) + 10(-241.81) - 7(-19.50) - (-275.5)]\text{ kJ} = -5547.7\text{ kJ}$$

$$\text{Mass} = (7\text{ mol N}_2\text{O}_4)\left(\frac{92.0\text{ g}}{\text{mol}}\right) + (1\text{ mol C}_9\text{H}_{20})\left(\frac{128\text{ g}}{\text{mol}}\right) = 772\text{ g} \qquad \frac{-5547.7\text{ kJ}}{772\text{ g}} = -7.19\text{ kJ/g}$$

(*f*)
$$\Delta H^\circ = 2\Delta H_f^\circ(CF_4) + 4\Delta H_f^\circ(HCl) + 4\Delta H_f^\circ(HF) - 4\Delta H_f^\circ(ClF_3) - \Delta H_f^\circ((CH_3)_2N_2H_2)$$
$$= [2(-925) + 4(-92.31) + 4(-271.1) - 4(-190) - (55.6)]\text{ kJ} = -2600\text{ kJ}$$

$$\text{Mass} = (4\text{ mol ClF}_3)\left(\frac{92.5\text{ g}}{\text{mol}}\right) + (1\text{ mol (CH}_3)_2\text{N}_2\text{H}_2)\left(\frac{60.0\text{ g}}{\text{mol}}\right) = 430\text{ g} \qquad \frac{-2600\text{ kJ}}{430\text{ g}} = -6.05\text{ kJ/g}$$

18.111 Calculate the enthalpy of combustion of 100.0 g of CO at 125 °C.

▮

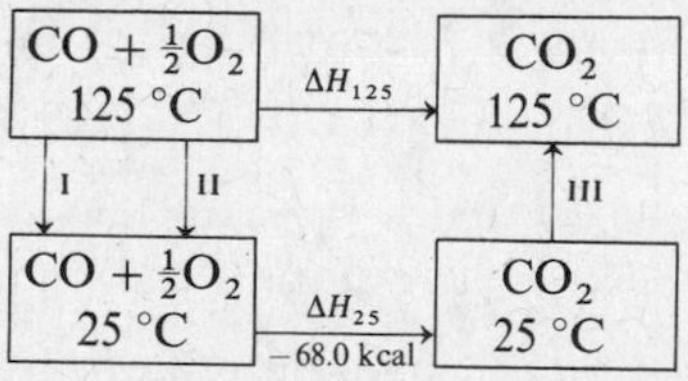

$CO + \frac{1}{2}O_2$, 125 °C $\xrightarrow{\Delta H_{125}}$ CO_2, 125 °C

I, II ↓ (to $CO + \frac{1}{2}O_2$, 25 °C) III ↑ (from CO_2, 25 °C)

$CO + \frac{1}{2}O_2$, 25 °C $\xrightarrow[-68.0\text{ kcal}]{\Delta H_{25}}$ CO_2, 25 °C

Fig. 18.3

$$\Delta H_{125} = \Delta H_{\text{I}} + \Delta H_{\text{II}} + \Delta H_{25} + \Delta H_{\text{III}}$$

Using Fig. 18.3, we have, per mol of CO:

$$\Delta H_{\text{I}} = (6.97\text{ cal/mol}\cdot\text{K})(-100\text{ K}) = -697\text{ cal}$$
$$\Delta H_{\text{II}} = (0.5\text{ mol})(7.05\text{ cal/mol}\cdot\text{K})(-100\text{ K}) = -352\text{ cal}$$
$$\Delta H_{\text{III}} = (8.96\text{ cal/mol}\cdot\text{K})(+100\text{ K}) = +896\text{ cal}$$
$$\Delta H_{25} = -68.0\text{ kcal} = -68\,000\text{ cal} \qquad \Delta H_{125} = -68\,153\text{ cal} = -68.2\text{ kcal}$$
$$(100\text{ g CO})\left(\frac{1\text{ mol}}{28.0\text{ g}}\right)\left(\frac{-68.2\text{ kcal}}{\text{mol CO}}\right) = -244\text{ kcal}$$

18.112 The Solvay process for the industrial production of Na_2CO_3 involves the following reactions:

$$CaCO_3 \longrightarrow CaO + CO_2$$
$$2CO_2 + 2NaCl + 2H_2O + 2NH_3 \longrightarrow 2NaHCO_3 + 2NH_4Cl$$
$$2NaHCO_3 \xrightarrow[\text{heat}]{} Na_2CO_3 + CO_2 + H_2O$$
$$CaO + H_2O \longrightarrow Ca(OH)_2$$
$$Ca(OH)_2 + 2NH_4Cl \longrightarrow CaCl_2 + 2NH_3 + 2H_2O$$

(*a*) Calculate ΔH for each step in the process. (*b*) Determine the overall ΔH for the process. (*c*) Write an equation for the net reaction. (*d*) Calculate ΔH for the net reaction, and compare this result to that of part (*b*).

▮ (*a*) ΔH for each reaction is ΔH_f° of products minus reactants, all in kcal:

$$\Delta H_1^\circ = (-94.05) + (-151.9) - (-288.5) = +42.55$$
$$\Delta H_2^\circ = 2(-169.8) + 2(-75.38) - 2(-94.05) - 2(-98.23) - 2(-68.32) - 2(-11.04) = 52.92$$
$$\Delta H_3^\circ = (-94.05) + (-68.32) + (-341.8) - 2(-169.8) = -164.57$$
$$\Delta H_4^\circ = (-235.80) - (-151.9) - (-68.32) = -15.58$$
$$\Delta H_5^\circ = (-190.0) + 2(-11.04) + 2(-68.32) - (-235.80) - 2(-75.38) = +37.84$$

(*b*) $\Delta H_{\text{total}}^\circ = \Delta H_1^\circ + \Delta H_2^\circ + \Delta H_3^\circ + \Delta H_4^\circ + \Delta H_5^\circ = 42.55 + 52.92 - 164.57 - 15.58 + 37.84 = -46.84\text{ kcal}$

(*c*) $2NaCl + CaCO_3 \longrightarrow CaCl_2 + Na_2CO_3$

(*d*) $\Delta H^\circ = \Delta H_f^\circ(CaCl_2) + \Delta H_f^\circ(Na_2CO_3) - 2\Delta H_f^\circ(NaCl) - \Delta H_f^\circ(CaCO_3)$
$= (-190.0) + (-341.8) - 2(-98.23) - (-288.5) = -46.84\text{ kcal}$

The result is the same as in part (*b*).

18.113 The contact process for the commercial production of sulfuric acid involves the following set of reactions:

$$S_8 + 8O_2 \longrightarrow 8SO_2$$
$$2SO_2 + O_2 \xrightarrow{V_2O_5} 2SO_3$$
$$SO_3 + H_2O \longrightarrow H_2SO_4$$

(*a*) Calculate ΔH for each step. (*b*) For which step is heat evolution most likely to be a problem? (*c*) Determine the ΔH of the overall reaction per mol of H_2SO_4.

▮ (*a*) $\Delta H_1 = 8\Delta H_f^\circ(SO_2) = (8\text{ mol})(-70.96\text{ kcal/mol}) = -567.68\text{ kcal}$

$\Delta H_2 = 2\Delta H_f^\circ(SO_3) - 2\Delta H_f^\circ(SO_2) = (2\text{ mol})(-94.45\text{ kcal/mol}) - (2\text{ mol})(-70.96\text{ kcal/mol})$
$= -46.98\text{ kcal}$

$\Delta H_3 = \Delta H_f^\circ(H_2SO_4) - \Delta H_f^\circ(SO_3) - \Delta H_f^\circ(H_2O) = [(-193.91) - (-94.45) - (-68.32)]\text{ kcal} = -31.14\text{ kcal}$

(*b*) Step 3, for which there is great evolution of heat in the presence of liquids.

(*c*) $\Delta H_{\text{overall}} = \frac{1}{8}\Delta H_1 + \frac{1}{2}\Delta H_2 + \Delta H_3 = [(-70.96) + (-23.49) + (-31.14)]\text{ kcal} = -125.59\text{ kcal}$

also equal to $\Delta H_f^\circ(H_2SO_4) - \Delta H_f^\circ(H_2O) = [(-193.91) - (-68.32)]\text{ kcal} = -125.59\text{ kcal}$

18.114 What is the heat of vaporization of water per g at 25 °C and 1 atm?

▮ We can write the thermochemical equation for the process:

$$H_2O(l) \longrightarrow H_2O(g)$$

ΔH° can be evaluated by subtracting ΔH_f° of reactants from ΔH_f° of products, as tabulated in Table 18.2:

$$\Delta H^\circ = \Delta H_f^\circ\text{ (products)} - \Delta H_f^\circ\text{ (reactants)} = -57.80 - (-68.32) = 10.52\text{ kcal}$$

The enthalpy of vaporization per gram is

$$\frac{10.52 \times 10^3\text{ cal/mol}}{18.02\text{ g/mol}} = 584\text{ cal/g}$$

Note that the heat of vaporization at 25 °C is greater than the value (540 cal/g) at 100 °C.

18.115 Calculate the enthalpy change at 25 °C for the reaction of 50.0 g of CO with oxygen to form CO_2.

▮

$$CO + \tfrac{1}{2}O_2 \longrightarrow CO_2$$

$$\Delta H = \Delta H_f^\circ(CO_2) - \Delta H_f(CO) = [-94.05 - (-26.41)] \text{ kcal/mol} = -67.64 \text{ kcal/mol}$$

$$(50.0 \text{ g CO})\left(\frac{1 \text{ mol CO}}{28.0 \text{ g CO}}\right)\left(\frac{-67.64 \text{ kcal}}{\text{mol}}\right) = -121 \text{ kcal}$$

Alternatively, ΔH_{comb} could have been used.

18.116 Calculate the enthalpy change for the reaction of 1.00 mol of carbon dioxide with 1.00 mol of carbon at 600 °C to produce carbon monoxide at 600 °C.

▮ From Fig. 18.4:

$$\Delta H_{600} = \Delta H_I + \Delta H_{II} + \Delta H_{25} + \Delta H_{III}$$

$$\Delta H_{25} = 2\Delta H_f^\circ(CO) - \Delta H_f^\circ(CO_2) = [2(-26.41) - (-94.05)] \text{ kcal} = +41.23 \text{ kcal}$$

$$\Delta H_I + \Delta H_{II} = (1.00 \text{ mol})[(2.04 + 8.96) \text{ cal/mol·°C}](-575 \text{ °C}) = -6325 \text{ cal} = -6.325 \text{ kcal}$$

$$\Delta H_{III} = (2.00 \text{ mol})(6.97 \text{ cal/mol·°C})(+575 \text{ °C}) = 8016 \text{ cal} = 8.016 \text{ kcal}$$

$$\Delta H_{600} = [(-6.325) + (41.23) + (8.016)] \text{ kcal} = +42.92 \text{ kcal}$$

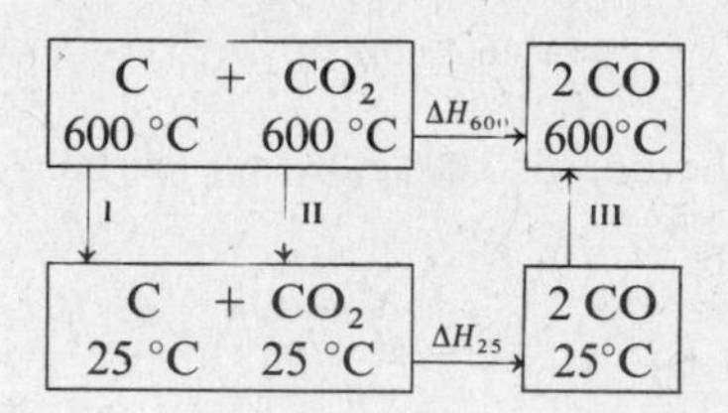

Fig. 18.4

18.117 What quantity of heat is yielded to the surroundings when 0.100 mol of C_8H_{18} at 25 °C is completely burned at constant pressure in 1.25 mol of oxygen gas at 25 °C, yielding CO_2 and $H_2O(g)$ at 300 °C?

▮

$$\Delta H = 0.100\,\Delta H_{comb} = (0.100 \text{ mol})(-1302.7 \text{ kcal/mol}) = -130.27 \text{ kcal}$$

130.27 kcal of energy is available, to (1) raise the temperature of the CO_2 and H_2O produced, plus (2) escape to the surroundings. The quantity which is required to heat the products is found as follows:

$$C_8H_{18} + \tfrac{25}{2}O_2 \longrightarrow 8CO_2 + 9H_2O$$

From 0.100 mol of C_8H_{18}, 0.800 mol of CO_2 and 0.900 mol of H_2O are formed.

$$CO_2(25 \text{ °C}) \xrightarrow{1} CO_2(300 \text{ °C})$$

$$H_2O(l)(25 \text{ °C}) \xrightarrow{2} H_2O(l)(100 \text{ °C}) \xrightarrow{3} H_2O(g)(100 \text{ °C}) \xrightarrow{4} H_2O(g)(300 \text{ °C})$$

$$\Delta H_1 = (0.800 \text{ mol})(8.96 \text{ cal/mol·°C})(275 \text{ °C}) = 1971 \text{ cal}$$

$$\Delta H_2 = (0.900 \text{ mol})(18.0 \text{ cal/mol·°C})(75 \text{ °C}) = 1215 \text{ cal}$$

$$\Delta H_3 = (0.900 \text{ mol})(9720 \text{ cal/mol}) = 8748 \text{ cal}$$

$$\Delta H_4 = (0.900 \text{ mol})(5.92 \text{ cal/mol·°C})(200 \text{ °C}) = 1066 \text{ cal}$$

$$\Delta H_{total} = 13\,000 \text{ cal} = 13.00 \text{ kcal}$$

$$\text{Released to surroundings} = [(130.27) - (13.00)] \text{ kcal}$$
$$= -117.27 \text{ kcal}$$

18.118 A sample of 0.100 mmol of H_2 and 0.0500 mmol of O_2 at 25 °C in a sealed bomb is ignited by an electric spark. Calculate the final temperature of the (gaseous) water produced. Ignore the energy of the spark and any heat loss to the surroundings. *Note*: Molar heat capacity at constant volume, C_v, is given by $C_v = C_p - R$.

▮

$$\Delta H = (0.100 \text{ mol } H_2O \text{ formed})(-57.79 \text{ kcal/mol}) = -5.779 \text{ kcal}$$

In the closed system, the heat evolved is equal to $-\Delta E$. ($\Delta H = q$ only for constant pressure systems.)

$$\Delta E = \Delta H - \Delta(PV) = \Delta H - (\Delta n)RT$$

where Δn is the change in the number of mol of gases:

$$\Delta n = 0.100 - 0.100 - 0.050 = -0.050 \text{ mol}$$
$$\Delta E = -5779 \text{ cal} - (-0.050 \text{ mol})(1.99 \text{ cal/mol·K})(298 \text{ K})$$
$$q = \Delta E = -5749 \text{ cal}$$

The heat evolved from the chemical reaction (−5749 cal) is added to the water formed (+5749 cal).

$$q = nC\,\Delta t \qquad C_v = C_p - 1.987 = 5.92 - 1.987 = 3.93 \text{ cal/mol·K}$$
$$\Delta t = \frac{q}{nC} = \frac{+5749 \text{ cal}}{(0.100 \text{ mol})(3.93 \text{ cal/mol·K})} = 14\,600 \text{ K}$$
$$t_{\text{final}} = 14\,600\ ^\circ\text{C} \qquad \text{(to 3 significant figures)}$$

18.119 When 12.0 g of carbon reacted with a limited quantity of oxygen, 57.5 kcal of heat was produced. Calculate the number of moles of CO and the number of moles of CO_2 produced.

$$C + O_2 \longrightarrow CO_2 \qquad \Delta H_f = -94.05 \text{ kcal}$$
$$C + \tfrac{1}{2}O_2 \longrightarrow CO \qquad \Delta H_f = -26.41 \text{ kcal}$$

▮ Let x = number of moles of CO_2 produced; then $1 - x$ = number of moles of CO produced

$$x(-94.05) + (1 - x)(-26.41) = -57.5$$
$$x = 0.460 \text{ mol } CO_2 \qquad 1 - x = 0.540 \text{ mol CO}$$

18.120 Calculate the enthalpy of combustion of 3.000 g of coke (carbon) at 100 °C.

▮

$$\begin{array}{ccc} C + O_2 & \xrightarrow{100\ ^\circ C} & CO_2 \\ \downarrow \text{I} \quad \downarrow \text{II} & & \uparrow \text{III} \\ C + O_2 & \xrightarrow{25\ ^\circ C} & CO_2 \end{array}$$

Per mol of C:

$$\Delta H_\text{I} + \Delta H_\text{II} + \Delta H_\text{III} = (1.00 \text{ mol})[(2.04 + 7.05 - 8.96) \text{ cal/mol·}^\circ\text{C}](-75\ ^\circ\text{C}) = -10 \text{ cal}$$
$$\Delta H_{100} = \Delta H_{25} + \Delta H_\text{I} + \Delta H_\text{II} + \Delta H_\text{III} = [(-94.05) - (0.010)] \text{ kcal} = -94.06 \text{ kcal}$$

For 3.000 g of C:

$$(3.000 \text{ g})\left(\frac{1 \text{ mol C}}{12.01 \text{ g}}\right)\left(\frac{-94.06 \text{ kcal}}{\text{mol}}\right) = -23.50 \text{ kcal}$$

18.121 In certain areas where coal is cheap, artificial gas is produced for household use by the "water gas" reaction:

$$C(s) + H_2O(g) \xrightarrow[600\ ^\circ C]{} H_2(g) + CO(g)$$

Assuming that coke is 100% carbon, calculate the maximum heat obtainable at 298 K from the combustion of 1.00 kg of coke, and compare this value to the maximum heat obtainable at 298 K from burning the water gas produced from 1.00 kg of coke.

▮

$$(1000 \text{ g})\left(\frac{1 \text{ mol C}}{12.0 \text{ g}}\right)\left(\frac{-94.05 \text{ kcal}}{\text{mol C}}\right) = -7840 \text{ kcal from coke}$$
$$(1000 \text{ g})\left(\frac{1 \text{ mol C}}{12.0 \text{ g}}\right)\left(\frac{1 \text{ mol CO} + 1 \text{ mol } H_2}{\text{mol C}}\right) = 83.3 \text{ mol CO} + 83.3 \text{ mol } H_2$$
$$(83.3 \text{ mol})\left(\frac{-68.32 \text{ kcal}}{\text{mol } H_2} + \frac{-68.0 \text{ kcal}}{\text{mol CO}}\right) = -11\,360 \text{ kcal from water gas}$$

More energy is obtainable from water gas, but of course energy was added to the $C + H_2O$ to get the water gas in the first place.

18.122 From ΔH_{comb} values for methane, ethane, and propane, −210.8, −368.4, and −526.3 kcal/mol, respectively, estimate the increase in ΔH_{comb} per added CH_2 group in a hydrocarbon. Predict ΔH_{comb} for octane on this basis, and compare this value to the usually accepted value, −1302.7 kcal.

▌ The difference in ΔH_{comb} between CH_4 and C_2H_6 is 157.6 kcal; that between C_2H_6 and C_3H_8 is 157.9 kcal. Assuming that 157.75 kcal, the average, is added for each of the next five CH_2 units added, the calculated ΔH_{comb} is

$$\Delta H_{comb}(C_8H_{18}) = \Delta H_{comb}(C_3H_8) + 5(-157.75 \text{ kcal}) = -1310 \text{ kcal}$$

This estimate is near the tabulated value, -1302.7 kcal.

18.123 When 12.0 g of carbon reacted with oxygen to form CO and CO_2 at 25 °C and constant pressure, 75.0 kcal of heat was liberated and no carbon remained. Calculate the mass of oxygen which reacted.

▌

$$C + O_2 \longrightarrow CO_2 \qquad \Delta H = -94.05 \text{ kcal/mol}$$
$$C + \tfrac{1}{2}O_2 \longrightarrow CO \qquad \Delta H = -26.41 \text{ kcal/mol}$$

Since there is 1.00 mol of carbon present, a total of 1.00 mol of CO plus CO_2 is formed. Let n = number of moles of CO_2 formed: then $1.00 - n$ = number of moles of CO formed.

$$n(-94.05 \text{ kcal/mol}) + (1.00 - n)(-26.41 \text{ kcal/mol}) = -75.0 \text{ kcal}$$
$$-67.64n = -48.59 \qquad n = 0.718 \text{ mol } CO_2 \qquad 1.00 - n = 0.282 \text{ mol CO}$$

It takes 0.718 mol of O_2 to form CO_2 plus (0.282/2) mol of O_2 to form CO, for a total of 0.859 mol of O_2.

$$(0.859 \text{ mol } O_2)\left(\frac{32.0 \text{ g}}{\text{mol}}\right) = 27.5 \text{ g } O_2$$

18.124 Using the data of Table 18.2 and the ΔH_{comb} values given, calculate the enthalpy of formation of each of the following: (*a*) propane, $\Delta H_{comb} = -526.3$ kcal/mol (*b*) carbon disulfide, $\Delta H_{comb} = -246.6$ kcal/mol (*c*) naphthalene, $\Delta H_{comb} = -1232.5$ kcal/mol

▌ (*a*) $C_3H_8 + 5O_2 \longrightarrow 3CO_2 + 4H_2O$

Per mol of C_3H_8:

$$\Delta H_{comb} = 3\Delta H_f^\circ(CO_2) + 4\Delta H_f^\circ(H_2O) - \Delta H_f^\circ(C_3H_8)$$
$$\Delta H_f = 3(-94.05) + 4(-68.32) - (-526.3) = -29.1 \text{ kcal}$$

Hence:

$$\Delta H_f^\circ = -29.1 \text{ kcal/mol}$$

(*b*) $+81.6$ kcal/mol (*c*) $+18.7$ kcal/mol

18.125 A 1.00-L sample of a mixture of methane gas and oxygen, measured at 25 °C and 740 torr, was allowed to react at constant pressure in a calorimeter which, together with its contents, had a heat capacity of 1260 cal/K. The complete combustion of the methane to carbon dioxide and water caused a temperature rise in the calorimeter of 0.667 K. What was the mole percent of CH_4 in the original mixture?

▌

$$\text{Heat generated} = (0.667 \text{ K})(1260 \text{ cal/K}) = 840 \text{ cal}$$
$$CH_4 + 2O_2 \longrightarrow CO_2 + 2H_2O \qquad \Delta H = -210.8 \text{ kcal}$$
$$(840 \text{ cal})\left(\frac{1 \text{ mol } CH_4}{210\,800 \text{ cal}}\right) = 3.98 \times 10^{-3} \text{ mol } CH_4$$
$$n_{total} = \frac{PV}{RT} = \frac{(\frac{740}{760} \text{ atm})(1.00 \text{ L})}{(0.0821 \text{ L}\cdot\text{atm/mol}\cdot\text{K})(298 \text{ K})} = 0.0398 \text{ mol}$$
$$\text{Mole percent } CH_4 = \frac{3.98 \times 10^{-3} \text{ mol } CH_4}{0.0398 \text{ mol total}} \times 100\% = 10.0 \text{ mole percent } CH_4$$

(The problem stated that all the CH_4 was burned; 840 cal also could have been generated if the O_2 were in limited supply, but then CO would have been produced.)

18.126 List the types of data (e.g., ΔH_f° for NaCl at 298 K) which are necessary to determine the quantity of (electric) energy required to decompose 1.00 mol of molten NaCl to its elements at its melting point.

▌ Figure 18.5 shows that the following are required: ΔH_f°(NaCl), ΔH_{fus}(NaCl), mp(NaCl), C_p(NaCl), ΔH_{fus}(Na), mp(Na), C_p(Na(*l*)), C_p(Na(*s*)), $C_p(Cl_2)$

$NaCl(l)$ mp $\longrightarrow$ $Na(l) + \frac{1}{2}Cl_2(g)$

ΔH_{fus} $\qquad C_p$

$Na(l)$ mp

$NaCl(s)$ mp $\qquad \Delta H_{fus} \qquad C_p$

C_p $\qquad Na(s)$ mp

C_p

$NaCl(s) \xrightarrow{\Delta H_f^\circ} Na(s) + \frac{1}{2}Cl_2(g)$ **Fig. 18.5**

18.127 When 0.100 mol of C_8H_{18} at 25 °C is completely burned at constant pressure in some oxygen gas at 25 °C, yielding as products gaseous H_2O, CO, and CO_2 at 300 °C, the process yielded 90.20 kcal of heat to the surroundings. Calculate the work done and the number of moles of CO and of CO_2 produced.

I

$$C_8H_{18} + \tfrac{25}{2}O_2 \xrightarrow{\Delta H_{comb}} 8CO_2 + 9H_2O(l)$$

$$\begin{aligned}\Delta H_f^\circ(C_8H_{18}) &= 8\Delta H_f^\circ(CO_2) + 9\Delta H_f^\circ(H_2O(l)) - \Delta H_{comb}\\ &= [8(-94.05) + 9(-68.32) - (-1302.7)]\text{ kcal}\\ &= -64.6\text{ kcal}\end{aligned}$$

The reactions under consideration are

$$C_8H_{18} + \tfrac{25}{2}O_2 \xrightarrow{1} 8CO_2 + 9H_2O(g) \qquad C_8H_{18} + \tfrac{17}{2}O_2 \xrightarrow{2} 8CO + 9H_2O(g)$$

$$\Delta H_1 = [8(-94.05) + 9(-57.79) - (-64.6)]\text{ kcal} = -1207.9\text{ kcal/mol }C_8H_{18}$$

$$\Delta H_2 = [8(-26.41) + 9(-57.79) - (-64.6)]\text{ kcal} = -666.8\text{ kcal/mol }C_8H_{18}$$

Let x = moles of CO produced; $0.800 - x$ = moles of CO_2 produced. Heat generated by the reaction (in kcal):

$$\frac{666.8x}{8} + \frac{(0.800 - x)}{8}(1207.9) = 120.79 - 67.64x$$

Heat utilized (in kcal):

$$\begin{aligned}90.20 + 275[x(6.97) + (0.800 - x)(8.96) + (0.900)(5.92)] \times 10^{-3} &= 90.20 + 275(12.5 - 1.99x) \times 10^{-3}\\ &= 90.20 + 3.44 - 0.547x = 93.64 - 0.547x\end{aligned}$$

Since the heat generated must equal the heat utilized

$$\begin{aligned}120.79 - 67.64x &= 93.64 - 0.547x\\ 27.15 &= 67.09x\\ x &= 0.405\text{ mol CO}\\ 0.800 - x &= 0.395\text{ mol }CO_2\end{aligned}$$

The number of moles of O_2 initially present was

$$0.395 + 0.405/2 = 0.5975\text{ mol}$$

$$w = -P\,\Delta V$$

At pressure P

$$V_1 = \frac{(0.6975\text{ mol})(1.99\text{ cal/mol·K})(298\text{ K})}{P} = \frac{414}{P} \qquad V_2 = \frac{(1.700\text{ mol})(1.99\text{ cal/mol·K})(573\text{ K})}{P} = \frac{1938}{P}$$

$$\Delta V = \frac{1524}{P} \qquad P\,\Delta V = 1524\text{ cal} \qquad w = -1.52\text{ kcal}$$

18.128 Only gases remain after 15.50 g of carbon is treated with 25.0 L of air at 25 °C and 5.50 atm pressure. (Assume 19.0% by volume oxygen, 80.0% nitrogen, 1.0% carbon dioxide.) Determine the heat evolved under constant pressure.

I

$$(15.50\text{ g C})\left(\frac{1\text{ mol C}}{12.0\text{ g C}}\right) = 1.292\text{ mol C}$$

$$n(O_2) = \frac{PV}{RT} = \frac{(5.50\text{ atm})(0.190)(25.0\text{ L})}{(0.0821\text{ L·atm/mol·K})(298\text{ K})} = 1.068\text{ mol }O_2$$

$$C + O_2 \longrightarrow CO_2 \qquad \text{Let } x = \text{number of moles }CO_2$$

$$C + \tfrac{1}{2}O_2 \longrightarrow CO \qquad 1.292 - x = \text{number of moles CO}$$

The number of moles of O atoms is the same before and after the reaction:

$$2x + (1.292 - x) = 2(1.068)$$
$$x = 0.844 \text{ mol } CO_2$$
$$1.292 - x = 0.448 \text{ mol CO}$$
$$(0.844 \text{ mol } CO_2)(-94.05 \text{ kcal/mol } CO_2) = -79.4 \text{ kcal}$$
$$(0.448 \text{ mol CO})(-26.41 \text{ kcal/mol CO}) = -11.8 \text{ kcal}$$
$$-91.2 \text{ kcal total}$$

18.129 By means of integral calculus, calculate the enthalpy change accompanying the reaction of 1.00 mol of carbon with water in the water gas reaction at 600 °C. The heat capacities (cal/K) of hydrogen, carbon monoxide, water vapor, and carbon are given in terms of absolute temperature as follows:

$$H_2: \quad C_p = 6.95 - (0.0001999)T + (4.8 \times 10^{-7})T^2$$
$$CO: \quad C_p = 6.42 + (0.001665)T - (1.96 \times 10^{-7})T^2$$
$$H_2O: \quad C_p = 7.256 + (2.298 \times 10^{-3})T + (2.83 \times 10^{-7})T^2$$
$$C: \quad C_p = 2.04$$

▌

$$\begin{array}{ccc} C + H_2O & \xrightarrow{25^\circ} & CO + H_2 \\ \uparrow \text{I} \quad \uparrow \text{II} & & \downarrow \text{III} \quad \downarrow \text{IV} \\ C + H_2O & \xrightarrow{600^\circ} & CO + H_2 \end{array}$$

$$\Delta H_{\text{I}} = \int_{25}^{600} C_p\, dT = C_p\, \Delta t = (2.04 \text{ cal/K})(575 \text{ K}) = 1170 \text{ cal} = 1.17 \text{ kcal}$$

$$\Delta H_{\text{II}} = \int_{25}^{600} [7.256 + (2.298 \times 10^{-3})T + (2.83 \times 10^{-7})T^2]\, dT$$

$$= (7.256)(575) + (2.298 \times 10^{-3})\left(\frac{873^2 - 298^2}{2}\right) + (2.83 \times 10^{-7})\left(\frac{873^3 - 298^3}{3}\right) = 5000 \text{ cal} = 5.00 \text{ kcal}$$

$$\Delta H_{\text{III}} = 6.42(-575) + (0.001665)\left(\frac{298^2 - 873^2}{2}\right) + 1.96 \times 10^{-7}\left(\frac{298^3 - 873^3}{3}\right) = -4290 \text{ cal} = -4.29 \text{ kcal}$$

$$\Delta H_{\text{IV}} = 6.95(-575) + (0.0001999)\left(\frac{298^2 - 873^2}{2}\right) + (4.8 \times 10^{-7})\left(\frac{298^3 - 873^3}{3}\right) = -4170 \text{ cal} = -4.17 \text{ kcal}$$

$$\Delta H_{25} = \Delta H_f^\circ(CO) - \Delta H_f^\circ(H_2O) = (-26.41 \text{ kcal}) - (-57.79 \text{ kcal}) = 31.38 \text{ kcal}$$

$$\Delta H_{600} = \Delta H_{25} + \Delta H_{\text{I}} + \Delta H_{\text{II}} + \Delta H_{\text{III}} + \Delta H_{\text{IV}} = 31.38 + 1.17 + 5.00 - 4.29 - 4.17 = 29.09 \text{ kcal}$$

18.5 ENTHALPIES OF IONS IN SOLUTION

18.130 Calculate ΔH_f° for chloride ion from the following data:

$$\tfrac{1}{2}H_2(g) + \tfrac{1}{2}Cl_2(g) \longrightarrow HCl(g) \qquad \Delta H_f^\circ = -22.1 \text{ kcal}$$
$$HCl(g) + nH_2O \longrightarrow H^+(aq) + Cl^-(aq) \qquad \Delta H_{298} = -17.9 \text{ kcal}$$

▌ (Here nH_2O signifies a large excess of water molecules, and $H^+(aq)$ and $Cl^-(aq)$ represent these ions in the presence of the large excess of water.) Adding the two equations,

$$\tfrac{1}{2}H_2(g) + \tfrac{1}{2}Cl_2(g) + nH_2O \longrightarrow H^+(aq) + Cl^-(aq) \qquad \Delta H_{298} = -40.0 \text{ kcal}$$

Remembering that the enthalpy of reaction is the difference between the enthalpies of formation of the products and those of the reactants and considering that the enthalpy of the water on the two sides of the equation cancels out:

$$-40.0 \text{ kcal} = \Delta H_f^\circ(H^+) + \Delta H_f^\circ(Cl^-) - \tfrac{1}{2}[\Delta H_f^\circ(H_2) + \Delta H_f^\circ(Cl_2)]$$
$$= 0 + \Delta H_f^\circ(Cl^-) - 0 - 0$$
$$\Delta H_f^\circ(Cl^-) = -40.0 \text{ kcal}$$

18.131 Calculate the enthalpy change of the reaction

$$2H^+(aq) + CO_3^{2-}(aq) \longrightarrow CO_2(g) + H_2O(l)$$

TABLE 18.3 Standard Enthalpies of Formation of Aqueous Ions at Unit Activity and 298 K

	ΔH_f°			ΔH_f°	
ion	kcal/mol	kJ/mol	ion	kcal/mol	kJ/mol
H^+	0.00	0.00	OH^-	−54.96	−229.99
Ag^+	25.31	105.9	F^-	−78.66	−329.1
K^+	−60.04	−251.2	Cl^-	−40.00	−167.08
Ca^{2+}	−129.77	−542.96	Br^-	−28.9	−121
Mg^{2+}	−110.41	−461.96	I^-	−13.4	−56.1
Ba^{2+}	−128.67	−538.36	HS^-	−4.22	−17.7
Cu^{2+}	15.39	64.4	S^{2-}	10.0	41.8
Zn^{2+}	−36.34	−153.89	CO_3^{2-}	−161.63	−676.26
Hg^{2+}	41.56	173.9	SO_4^{2-}	−216.9	−907.5
Fe^{2+}	−21.0	−87.9			

$$\Delta H^\circ = \Delta H_f^\circ(CO_2) + \Delta H_f^\circ(H_2O) - \Delta H^\circ(CO_3^{2-}) - 2\Delta H_f^\circ(H^+)$$
$$= -94.0 + (-68.3) - (-161.6) - 2(0) = -0.7 \text{ kcal}$$

18.132 Calculate ΔH for the reactions:

(*a*) $Ag_2O + C \longrightarrow CO + 2Ag$ (*b*) $BaCl_2(aq) + H_2SO_4(aq) \longrightarrow BaSO_4(s) + 2HCl(aq)$

(*a*) $\Delta H = \Delta H_f^\circ(CO) - \Delta H_f^\circ(Ag_2O) = -26.41 - (-7.31) = -19.10$ kcal

(*b*) $\Delta H = \Delta H_f^\circ(BaSO_4) - \Delta H_f^\circ(Ba^{2+}) - \Delta H_f^\circ(SO_4^{2-}) = [-350.2 - (-128.67) - (-216.9)]$ kcal $= -4.6$ kcal

18.133 Calculate the enthalpy change for each of the following reactions at 25 °C:

(*a*) $Ba^{2+}(aq) + CO_3^{2-}(aq) \longrightarrow BaCO_3(s)$ (*b*) $Ag^+(aq) + Br^-(aq) \longrightarrow AgBr(s)$

ΔH is the difference between the ΔH_f° (products) and ΔH_f°(reactants), including this case in which the reactants happen to be ions in solution. The data are obtained from Tables 18.2 and 18.3.

(*a*) $\Delta H = [(-290.8) - (-128.67) - (-161.63)]$ kcal $= -0.5$ kcal

(*b*) $\Delta H = [(-23.8) - (+25.31) - (-28.9)]$ kcal $= -20.2$ kcal

18.134 Calculate ΔH for the process at 25 °C of dissolving 1.00 mol of KCl in a large excess of water:

$$KCl(s) \longrightarrow K^+(aq) + Cl^-(aq)$$

Does this process represent an ionization reaction? Explain.

Data from Tables 18.2 and 18.3,

$$\Delta H = \Delta H_f^\circ(K^+(aq)) + \Delta H_f^\circ(Cl^-(aq)) - \Delta H_f^\circ(KCl) = [(-60.04) + (-40.0) - (-104.18)] \text{ kcal} = +4.1 \text{ kcal}$$

The "reaction" represents a solution process, not an ionization. The KCl solid is ionic before being dissolved as well as after.

18.135 Calculate ΔH for each of the following reactions:

(*a*) $BaCO_3(s) + 2HCl(aq) \longrightarrow BaCl_2(aq) + CO_2(g) + H_2O(l)$

(*b*) $AgNO_3(aq) + NaCl(aq) \longrightarrow NaNO_3(aq) + AgCl(s)$

(*c*) $HNO_3(aq) + NaOH(aq) \longrightarrow NaNO_3(aq) + H_2O(l)$

(*d*) $HCl(aq) + KOH(aq) \longrightarrow KCl(aq) + H_2O(l)$

(*e*) $LiOH(aq) + HClO_3(aq) \longrightarrow LiClO_3(aq) + H_2O(l)$

The values of ΔH are obtained most easily from the net ionic equations. ΔH_f° values for aqueous salts are not equal to ΔH_f° values for the pure solid.

(*a*) $\Delta H = \Delta H_f^\circ(\mathrm{Ba^{2+}}) + \Delta H_f^\circ(\mathrm{CO_2}) + \Delta H_f^\circ(\mathrm{H_2O}) - \Delta H_f^\circ(\mathrm{BaCO_3}) - 2\Delta H_f^\circ(\mathrm{H^+})$

$= [(-128.67) + (-94.05) + (-68.32) - (-290.8) - 2(0)]\ \text{kcal} = -0.2\ \text{kcal}$

(*b*) $\Delta H = [(-30.36) - (25.31) - (-40.0)]\ \text{kcal} = -15.7\ \text{kcal}$

(*c*), (*d*), and (*e*) $\mathrm{H^+ + OH^- \longrightarrow H_2O}$

$$\Delta H = [(-68.32) - (0) - (-54.96)]\ \text{kcal} = -13.36\ \text{kcal}$$

18.136 How many kJ of heat will be evolved in making 22.4 L at STP (1.00 mol) of H_2S from FeS and dilute hydrochloric acid?

■

$$\begin{array}{lcccc} & \mathrm{FeS}(s) + & \mathrm{2H^+}(aq) \longrightarrow & \mathrm{Fe^{2+}}(aq) + & \mathrm{H_2S}(g) \\ \Delta H_f^\circ & -95.4 & 0 & -87.9 & -20.6 \end{array}$$

Since HCl and $FeCl_2$ are strong electrolytes, only their essential ions need be written.

$$\Delta H^\circ = -(87.9 + 20.6) + 95.4 = -13.1\ \text{kJ/mol}\ \mathrm{H_2S}$$

18.137 Is the process of dissolving H_2S gas in water endothermic or exothermic? To what extent?

■

$$\mathrm{H_2S}(g) \longrightarrow \mathrm{H_2S}(aq)$$

$$\Delta H^\circ = \Delta H_f^\circ(aq) - \Delta H_f^\circ(g) = [-9.5 - (-4.8)]\ \text{kcal} = -4.7\ \text{kcal}$$

The process is exothermic.

18.138 How much heat is released on dissolving 1.000 mol of HCl(*g*) in a large amount of water? (*Hint*: HCl is completely ionized in dilute solution.)

■

$$\mathrm{HCl}(g) \longrightarrow \mathrm{H_3O^+ + Cl^-}$$

$$\Delta H^\circ = \Delta H_f^\circ(\mathrm{H_3O^+}) + \Delta H_f^\circ(\mathrm{Cl^-}) - \Delta H_f^\circ(\mathrm{HCl}) = [0.00 + (-40.00) - (-22.06)]\ \text{kcal/mol} = -17.94\ \text{kcal/mol HCl}$$

18.139 In an ice calorimeter, a chemical reaction is allowed to occur in thermal contact with an ice-water mixture at 0 °C. Any heat liberated by the reaction is used to melt some ice; the volume change of the ice-water mixture indicates the amount of melting. When solutions containing 1.00 mmol each of $AgNO_3$ and NaCl were mixed in such a calorimeter, both solutions having been precooled to 0 °C, 0.20 g of ice melted. Assuming complete reaction in this experiment, what is ΔH for the reaction $\mathrm{Ag^+ + Cl^- \rightarrow AgCl}$?

■

$$(0.20\ \text{g ice})\left(\frac{-80\ \text{cal}}{\text{g ice}}\right) = -16\ \text{cal} \qquad \text{liberated for 1 mmol AgCl formed}$$

$$-16\ \text{cal/mmol} = -16\ \text{kcal/mol}$$

18.140 The heat released on neutralization of CsOH with all strong acids is 13.4 kcal/mol. The heat released on neutralization of CsOH with HF (weak acid) is 16.4 kcal/mol. Calculate ΔH° of ionization of HF in water.

■

(1)	$\mathrm{OH^- + H^+ \longrightarrow H_2O}$	$\Delta H^\circ = -13.4$ kcal
(2)	$\mathrm{OH^- + HF \longrightarrow H_2O + F^-}$	$\Delta H^\circ = -16.4$ kcal
(2) – (1)	$\mathrm{HF \quad H^+ + F^-}$	$\Delta H^\circ = -3.0$ kcal

18.141 Calculate ΔH° for the reaction $\mathrm{CuSO_4}(aq) + \mathrm{Zn}(s) \rightarrow \mathrm{ZnSO_4}(aq) + \mathrm{Cu}(s)$.

■ The net ionic equation is $\mathrm{Zn + Cu^{2+} \rightarrow Cu + Zn^{2+}}$

$$\Delta H^\circ = \Delta H_f^\circ(\mathrm{Zn^{2+}}) - \Delta H_f^\circ(\mathrm{Cu^{2+}}) = [(-153.89) - (+64.4)]\ \text{kJ} = -218.3\ \text{kJ}$$

18.142 The amount of heat evolved in dissolving $CuSO_4$ is 17.9 kcal/mol. What is ΔH_f° for $SO_4^{2-}(aq)$?

■

$$\mathrm{CuSO_4}(s) \longrightarrow \mathrm{Cu^{2+}}(aq) + \mathrm{SO_4^{2-}}(aq)$$

$$\Delta H^\circ = \Delta H_f^\circ(\mathrm{Cu^{2+}}) + \Delta H_f^\circ(\mathrm{SO_4^{2-}}) - \Delta H_f^\circ(\mathrm{CuSO_4})$$

$$-17.9\ \text{kcal/mol} = (+15.4\ \text{kcal/mol}) + \Delta H_f^\circ(\mathrm{SO_4^{2-}}) - (-184.03\ \text{kcal/mol})$$

$$\Delta H_f^\circ(\mathrm{SO_4^{2-}}) = -217.3\ \text{kcal/mol}$$

18.143 The heat of solution of $CuSO_4 \cdot 5H_2O$ in a large amount of water is 1.3 kcal/mol (endothermic). Calculate the heat of reaction for

$$CuSO_4(s) + 5H_2O(l) \longrightarrow CuSO_4 \cdot 5H_2O(s)$$

Use data from Prob. 18.142.

▮

$$CuSO_4 + 5H_2O \longrightarrow CuSO_4 \cdot 5H_2O$$
$$\Delta H^\circ = \Delta H_f^\circ(CuSO_4 \cdot 5H_2O) - \Delta H_f^\circ(CuSO_4) - 5\Delta H_f^\circ(H_2O)$$

Per mol of salt:

(1)	$CuSO_4 \cdot 5H_2O \longrightarrow Cu^{2+}(aq) + SO_4^{2-}(aq)$	$\Delta H^\circ = +1.3$ kcal
(2)	$CuSO_4 \longrightarrow Cu^{2+}(aq) + SO_4^{2-}(aq)$	$\Delta H^\circ = -17.9$ kcal
(2) − (1)	$CuSO_4 + 5H_2O \longrightarrow CuSO_4 \cdot 5H_2O$	$\Delta H^\circ = -19.2$ kcal

The reaction is exothermic.

18.144 Calculate the enthalpy of neutralization of a strong acid by a strong base in water. The heat liberated on neutralization of HCN (weak acid) by NaOH is 2.9 kcal/mol. How many kcal are absorbed in ionizing 1 mol of HCN in water?

▮ The basic equation for neutralization is as follows:

$$H^+(aq) + OH^-(aq) \longrightarrow H_2O(l)$$
$$\Delta H_f^\circ \qquad 0 \qquad -55.0 \text{ kcal} \qquad -68.3 \text{ kcal}$$

Thus $\Delta H^\circ = -68.3 - (-55.0) = -13.3$ kcal.

The neutralization of HCN(*aq*) by NaOH(*aq*) may be thought of as the sum of two processes, ionization of HCN(*aq*) and neutralization of $H^+(aq)$ with $OH^-(aq)$. [Since NaOH is a strong base, NaOH(*aq*) implies complete ionization, and a separate thermochemical equation for the ionization need not be written.] We may therefore construct the following thermochemical cycle.

	$HCN(aq) \longrightarrow H^+(aq) + CN^-(aq)$	$\Delta H^\circ(\text{ionization}) = x$
	$H^+(aq) + OH^-(aq) \longrightarrow H_2O(l)$	$\Delta H^\circ = -13.3$ kcal
Sum:	$HCN(aq) + OH^-(aq) \longrightarrow H_2O(l) + CN^-(aq)$	$\Delta H^\circ(\text{experimental}) = -2.9$ kcal

ΔH°(experimental) is negative because heat is liberated on neutralization.

From the principle of additivity,

$$x + (-13.3) = -2.9 \qquad \text{thus} \qquad x = 10.4 \text{ kcal}$$

The ionization process is endothermic to the extent of 10.4 kcal/mol.

18.145 The solution of $CaCl_2 \cdot 6H_2O$ in a large volume of water is endothermic to the extent of 3.5 kcal/mol. For the reaction

$$CaCl_2(s) + 6H_2O(l) \longrightarrow CaCl_2 \cdot 6H_2O(s)$$

$\Delta H = -23.2$ kcal. What is the heat of solution of $CaCl_2$ (anhydrous) in a large volume of water?

▮

	$CaCl_2 \cdot 6H_2O \longrightarrow Ca^{2+}(aq) + 2Cl^-(aq)$	$\Delta H^\circ = 3.5$ kcal
	$CaCl_2 + 6H_2O \longrightarrow CaCl_2 \cdot 6H_2O$	$\Delta H^\circ = -23.2$ kcal
Adding:	$CaCl_2 \xrightarrow{H_2O} Ca^{2+}(aq) + 2Cl^-(aq)$	$\Delta H^\circ = -19.7$ kcal

18.146 Calculate the enthalpy changes, ΔH_{298}°, for the following reactions in aqueous solution:

$$HS^- \longrightarrow H^+ + S^{2-}$$
$$OH^- + HS^- \longrightarrow S^{2-} + H_2O$$

Combine these two results to obtain the enthalpy change for the reaction

$$H^+ + OH^- \longrightarrow H_2O$$

Compare this result to that derived directly from data in Table 18.3.

▮ $$HS^- \longrightarrow H^+ + S^{2-} \qquad \Delta H = 10.0 + 0 - (-4.22) = 14.22 \text{ kcal}$$
$$OH^- + HS^- \longrightarrow S^{2-} + H_2O \qquad \Delta H = -68.32 + 10.0 - (-54.96) - (-4.22) = +0.86 \text{ kcal}$$

Subtracting the first equation (and its ΔH value) from the second (and its ΔH value) yields

$$OH^- + H^+ \longrightarrow H_2O \qquad \Delta H = 0.86 - 14.22 = -13.36 \text{ kcal}$$

The same value is obtained directly from ΔH_f° values from Tables 18.2 and 18.3.

18.6 FREE ENERGY CHANGE AND ENTROPY

Note: Before working the problems in this section, familiarize yourself with Tables 18.4 (p. 366), 18.5 (p. 367), and 18.6 (p. 369).

18.147 Molar entropies are quoted in J/mol·K; molar heat capacities are quoted in J/mol·K or J/mol·°C. Explain.

▮ Dimensionally,

$$\text{Entropy} = \frac{\text{heat}}{\text{absolute temperature}} \qquad \text{Heat capacity} = \frac{\text{heat}}{\text{temperature change}}$$

Since a temperature *change* is the same on the Kelvin and Celsius scales, either unit may be used in writing heat capacity.

18.148 For the reaction at 298 K,

$$2A + B \longrightarrow C$$

$\Delta H = 100$ kcal and $\Delta S = 0.050$ kcal/K. Assuming ΔH and ΔS to be constant over the temperature range, at what temperature will the reaction become spontaneous?

▮ The reaction will just be spontaneous when $\Delta G = 0$.

$$\Delta G = \Delta H - T\,\Delta S = 0 \qquad \Delta H = T\,\Delta S \qquad T = \frac{\Delta H}{\Delta S} = \frac{100 \text{ kcal}}{0.050 \text{ kcal/K}} = 2000 \text{ K}$$

18.149 For the reaction at 298 K:

$$A(g) + B(g) \longrightarrow E(g)$$
$$\Delta E = -3.00 \text{ kcal} \qquad \Delta S = -10.0 \text{ cal/K}$$

Calculate ΔG. Predict whether the reaction may occur spontaneously, as written.

▮ $$\Delta H = \Delta E + \Delta(PV) = \Delta E + \Delta(nRT) = \Delta E + (\Delta n)RT$$
$$= (-3000 \text{ cal}) + (-1 \text{ mol})(1.987 \text{ cal/mol·K})(298 \text{ K}) = -3592 \text{ cal}$$
$$\Delta G = \Delta H - T\,\Delta S = -3592 - 298(-10.0) = -612 \text{ cal}$$

Since ΔG is negative, the reaction will be spontaneous as written.

18.150 For the reaction $2Cl(g) \rightarrow Cl_2(g)$, what are the signs of ΔH and ΔS?

▮ ΔH is negative (the bond energy is released) and ΔS is negative (there is less randomness among the molecules than among the atoms).

18.151 A reaction has a value of $\Delta H = -40$ kcal at 400 K. Above 400 K, the reaction is spontaneous; below that temperature, it is not. Calculate ΔG and ΔS at 400 K.

▮ $$\Delta G = \Delta H - T\,\Delta S = 0 \quad \text{at 400 K}$$
$$\Delta S = \frac{\Delta H}{T} = \frac{-40\,000 \text{ cal/K}}{400 \text{ K}} = -100 \text{ cal/K}$$

18.152 Using data from Table 18.4, calculate ΔG° at 298 K for the reaction

$$CH_4(g) + 2O_2 \rightarrow CO_2(g) + 2H_2O(l)$$

▮ $$\Delta G^\circ = [\Delta G_f^\circ(CO_2) + 2\Delta G_f^\circ(H_2O)] - [\Delta G_f^\circ(CH_4) + 2\Delta G_f^\circ(O_2)]$$
$$= [-94.3 + (-113.4)] \text{ kcal} - [(-12.14) + 0] \text{ kcal} = -195.6 \text{ kcal}$$

TABLE 18.4 Standard Free Energies of Formation, ΔG_f°, at 298 K

	kcal/mol*		kcal/mol*		kcal/mol*
$Ag_2O(s)$	−2.679	$CH_3Cl(g)$	−19.6	$H_2O(g)$	−54.63
$Al_2O_3(s)$	−376.8	$C_6H_{12}O_6(s;$ glucose$)$	−215	$H_2S(g)$	−7.89
$BaO(s)$	−126.3	$CCl_4(l)$	−33.3	$NH_3(g)$	−3.97
$BaSO_4(s)$	−350.2	$CO(g)$	−32.780	$NO_2(g)$	12.4
$CH_4(g)$	−12.14	$CO_2(g)$	−94.3	$N_2O_4(g)$	23.38
$C_2H_4(g)$	16.28	$ClO_2(g)$	25.9	$PCl_3(l)$	−65.11
$C_2H_6(g)$	−7.86	$Cl_2O(g)$	23.4	$PCl_3(g)$	−64.01
$C_6H_6(l)$	29.8	$CaO(s)$	−144.4	$PCl_5(g)$	−72.90
$CH_3OH(l)$	−39.76	$CaCO_3(s)$	−269.8	$SiO_2(s)$	−192.4
$CH_3OH(g)$	−38.72	$HCl(g)$	−22.7	$Sn(s;$ gray$)$	0.03
$C_2H_5OH(g)$	−40.29	$H_2O(l)$	−56.7		

* To convert to kJ/mol, multiply tabular entries by 4.184.

18.153 Without consulting entropy tables, predict the sign of ΔS for each of the following processes.

(*a*) $O_2(g) \longrightarrow 2O(g)$ (*b*) $N_2(g) + 3H_2(g) \longrightarrow 2NH_3(g)$
(*c*) $C(s) + H_2O(g) \longrightarrow CO(g) + H_2(g)$ (*d*) $Br_2(l) \longrightarrow Br_2(g)$
(*e*) $N_2(g, 10 \text{ atm}) \longrightarrow N_2(g, 1 \text{ atm})$ (*f*) Desalination of seawater.
(*g*) Devitrification of glass. (*h*) Hard boiling of an egg.
(*i*) $C(s, \textit{graphite}) \longrightarrow C(s, \textit{diamond})$

I (*a*) Positive. There is an increase in the number of gas molecules. (*b*) Negative. There is a decrease in the number of gas molecules. (*c*) Positive. There is an increase in the number of gas molecules. (*d*) Positive. S is always greater for a gas than for its corresponding liquid. (*e*) Positive. Entropy increases on expansion. (*f*) Negative. Desalination is the opposite of solution; a solute must be removed from a solution. (*g*) Negative. Devitrification is the onset of crystallization in a supercooled liquid. (*h*) Positive. The fundamental process in the "boiling" of an egg is not a literal boiling, in the sense of vaporization, but a denaturation of the egg protein. A protein is a large molecule which exists in a particular configuration in the so-called native state but may occupy a large number of almost random configurations in the denatured state, resulting from rotations around the bonds. The increase in the number of possible configurations is analogous to the melting process. (*i*) Negative. Diamond, being a harder solid, would be expected to have more restricted atomic motions within the crytal.

18.154 At its melting point, 0 °C, the enthalpy of fusion of water is 1.435 kcal/mol. What is the molar entropy change for the melting of ice at 0 °C?

I At 0 °C, $\Delta G = 0$ and $\Delta S = \dfrac{\Delta H}{T} = \dfrac{1435 \text{ cal/mol}}{273 \text{ K}} = 5.26 \text{ cal/mol}\cdot\text{K}$

18.155 Calculate ΔS for the vaporization of water at 100 °C. $\Delta H(\text{vap}) = 40.7$ kJ/mol.

I Recall that $q = \Delta H$ at constant pressure. Since the vaporization at 100 °C is reversible, $\Delta G = 0$ and

$$\Delta S = \frac{q_{\text{rev}}}{T} = \frac{4.07 \times 10^4 \text{ J/mol}}{373.1 \text{ K}} = 109.1 \text{ J/K}\cdot\text{mol}$$

18.156 What is the standard free energy change for the melting of 3.0 mol of water at 0 °C? Determine the entropy change for this process. Is the entropy greater for the liquid or the solid? $\Delta H_{\text{fus}} = 1.435$ kcal/mol.

I ΔG° is 0 for this reversible process.

$$\Delta G = \Delta H - T\,\Delta S = 0 \qquad \Delta S = \frac{\Delta H}{T} = \frac{(+1435 \text{ cal/mol})}{273 \text{ K}} = 5.26 \text{ cal/mol}\cdot\text{K}$$

$$(5.26 \text{ cal/mol}\cdot\text{K})(3.0 \text{ mol}) = 16 \text{ cal/K}$$

The entropy is greater for the liquid, as seen by the positive sign of ΔS and as expected for the melting of a solid.

18.157 After comparing data in Table 18.2 and the answer to Prob. 18.153(*i*), how do you account for the fact that ΔH and ΔS for the phase transition from diamond to graphite are not related by the same equation that applied in Prob. 18.155?

From the table, the formation of diamond from graphite (the standard state of carbon) is accompanied by a *positive* ΔH of 0.45 kcal/mol at 25 °C. From Prob. 18.153(*i*), ΔS for the same process is *negative*. Since 25 °C is not the transition temperature, the process is not a reversible one at 25 °C.

18.158 Using data from Tables 18.2 and 18.4, calculate ΔS°_{298} for the reaction of 100 g of nitrogen with oxygen according to the equation

$$N_2(g) + 2O_2(g) \longrightarrow 2NO_2(g)$$

Per mol of N_2 used:

$$\Delta H^\circ = 2\Delta H^\circ_f = (2\text{ mol})(8.09\text{ kcal/mol}) = 16.2\text{ kcal}$$
$$\Delta G^\circ = 2\Delta G^\circ_f = (2\text{ mol})(12.4\text{ kcal/mol}) = 24.8\text{ kcal}$$
$$\Delta S^\circ = (\Delta H^\circ - \Delta G^\circ)/T = (16.2\text{ kcal} - 24.8\text{ kcal})/(298\text{ K}) = -29\text{ cal/K}$$

Per 100 g of N_2:

$$(100\text{ g }N_2)\left(\frac{1\text{ mol }N_2}{28.0\text{ g }N_2}\right) = 3.57\text{ mol }N_2$$
$$\Delta S^\circ = (3.57\text{ mol }N_2)(-29\text{ cal/K}\cdot\text{mol }N_2) = -100\text{ cal/K}$$

18.159 Calculate ΔH°_f for $C_2H_5OH(g)$ at 25 °C.

For the special process in which a substance in its standard state is formed from its elements in their standard states,

$$\Delta G^\circ_f = \Delta H^\circ_f - T\,\Delta S^\circ_f$$

ΔH°_f can be computed by using data from Tables 18.4 and 18.5.

TABLE 18.5 Absolute Entropies, S°, at 298 K

	J/mol·K*		J/mol·K*		J/mol·K*
Ag(*s*)	28	Ca(*s*)	41.6	$N_2(g)$	191
AgBr(*s*)	107	CaO(*s*)	40	$NH_3(g)$	192.5
AgCl(*s*)	96.2	$Ca(OH)_2(s)$	76.1	$NH_4Cl(s)$	94.6
$Ag_2O(s)$	121.3	$Cl_2(g)$	222.96	NO(*g*)	210
Al(*s*)	28.3	$Cl_2O(g)$	266.10	$NO_2(g)$	239.95
$AlCl_3(s)$	167	Cu(*s*)	33.3	$N_2O_4(g)$	304.18
$Al_2O_3(s)$	51.00	CuO(*s*)	43.5	Na(*s*)	51.0
Ar(*g*)	154.7	$Cu_2O(s)$	101	NaCl(*s*)	72.38
B(*s*)	6.7	$F_2(g)$	203	Ne(*g*)	146.2
BaO(*s*)	70.3	Fe(*s*)	27.2	$O_2(g)$	205.04
$BaSO_4(s)$	132	FeO(*s*)	54.0	$PCl_3(l)$	217.1
$Br_2(l)$	152.23	$Fe_2O_3(s)$	90.0	$PCl_3(g)$	311.7
$Br_2(g)$	245.35	$H_2(g)$	130.57	S(*s*)	31.9
C(*s*; graphite)	5.74	HBr(*g*)	198	$SO_2(g)$	249
$CH_4(g)$	186	HCl(*g*)	187	$SO_3(s)$	52.3
$C_2H_2(g)$	209.2	HF(*g*)	174	$SO_3(l)$	95.6
$C_2H_4(g)$	219.5	HI(*g*)	206	$SO_3(g)$	256.2
$C_2H_6(g)$	229	$H_2O(l)$	69.950	Si(*s*)	18.9
$C_3H_8(g)$	270	$H_2O(g)$	188.724	$SiO_2(s)$	42
$C_4H_{10}(g)$	310.0	$H_2S(g)$	205	Sn(*s*; gray)	44.1
$C_6H_6(l)$	173	He(*g*)	126.1	Sn(*s*; white)	51.5
$CH_3OH(l)$	126.8	Hg(*l*)	76.02	Xe(*g*)	169.6
$C_2H_5OH(g)$	274.2	$I_2(s)$	116	Zn(*s*)	41.6
$CCl_4(l)$	214	Kr(*g*)	164.0	ZnO(*s*)	43.9
CO(*g*)	197.56	Mg(*s*)	32.5		
$CO_2(g)$	213.68	$MgCO_3(s)$	65.7		

* To convert to cal/mol·K, divide tabular entries by 4.184.

Write the balanced formation equation for $C_2H_5OH(g)$, and write the $nS°$ value under each substance.

$$\begin{array}{lcccc} & 2C(s) + & 3H_2(g) + & \frac{1}{2}O_2(g) \longrightarrow & C_2H_5OH(g) \\ nS° & 2(5.74) & 3(130.57) & \frac{1}{2}(205.04) & 274.2 \end{array}$$

For the process,

$$\Delta S° = 274.2 - 2(5.74) - 3(130.57) - \tfrac{1}{2}(205.04) = -231.5 \text{ J/K}$$

then

$$\Delta S_f° = \frac{-231.5 \text{ J/K}}{1 \text{ mol } C_2H_5OH} = -231.5 \text{ J/K·mol}$$

Now,

$$\Delta H_f° = \Delta G_f° + T\,\Delta S_f° = -168.57 \text{ kJ/mol} + (298.1 \text{ K})(-0.2315 \text{ kJ/K·mol}) = -237.60 \text{ kJ/mol}$$

Note again that $\Delta G_f°$ for a substance is listed as a single entry in Table 18.4, but $\Delta S_f°$ must be computed by taking the difference of the tabulated absolute entropies of the substance and its constituent elements from Table 18.5.

18.160 (*a*) What is $\Delta G°$ at 25 °C for the following reaction?

$$H_2(g) + CO_2(g) \rightleftharpoons H_2O(g) + CO(g)$$

(*b*) What is ΔG at 25 °C under conditions where the partial pressures of H_2, CO_2, H_2O, and CO are 10.00, 20.00, 0.02000, and 0.01000 atm, respectively?

▮ First tabulate the $(n\,\Delta G_f°)$ value under each substance (in this case, $n = 1$ mol for each substance).

$$\begin{array}{lcccc} & H_2(g) + & CO_2(g) \rightleftharpoons & H_2O(g) + & CO(g) \\ n\,\Delta G_f° & 0 & -394.37 & -228.58 & -137.15 \text{ kJ} \end{array}$$

(*a*) The computation of $\Delta G°$ is analogous to that of $\Delta H°$ (see Prob. 18.88).

$$\Delta G° = (-228.58 - 137.15) - (0 - 394.37) = 28.64 \text{ kJ}$$

(*b*) $\Delta G = \Delta G° + 2.303RT \log Q$

$$= (28.64 \text{ kJ}) + (8.314 \times 10^{-3} \text{ kJ/K})(298.1 \text{ K})\left[2.303 \log \frac{P(H_2O)P(CO)}{P(H_2)P(CO_2)}\right]$$

$$= \left[28.64 + 5.708 \log \frac{(0.02000)(0.01000)}{(10.00)(20.00)}\right] \text{kJ} = (28.64 - 34.25) \text{ kJ} = -5.61 \text{ kJ}$$

Note that the reaction, although not possible under standard conditions, becomes possible $(\Delta G < 0)$ under this set of experimental conditions. In this case, Q was evaluated in terms of partial pressures, since $\Delta G°$ was defined in terms of a standard state of 1 atm for each substance. In general, Q must be expressed in terms of the same concentration measure as used to define the standard states.

18.161 Calculate the absolute entropy of $CH_3OH(g)$ at 25 °C.

▮ Although there is no $S°$ entry in Table 18.5 for $CH_3OH(g)$, the $\Delta G_f°$ value for this substance listed in Table 18.4, the $\Delta H_f°$ value listed in Table 18.2, and the $S°$ values for the constituent elements may be combined to yield the desired value.

$$\Delta S_f° = \frac{\Delta H_f° - \Delta G_f°}{T} = \frac{(-200.7 + 162.0) \text{ kJ/mol}}{298.1 \text{ K}} = -129.8 \text{ J/K·mol}$$

From the equation for the formation of 1 mol of $CH_3OH(g)$ under standard conditions,

$$C(s) + 2H_2(g) + \tfrac{1}{2}O_2(g) \longrightarrow CH_3OH(g)$$

we can write

$$-129.8 \text{ J/K} = (1 \text{ mol})[S°(CH_3OH)] - (1 \text{ mol})[S°(C)] - (2 \text{ mol})[S°(H_2)] - (\tfrac{1}{2} \text{ mol})[S°(O_2)]$$
$$= (1 \text{ mol})[S°(CH_3OH)] - [5.7 + 2(130.6) + \tfrac{1}{2}(205.0)] \text{ J/K}$$

Solving, $S°(CH_3OH) = 239.6$ J/K·mol.

TABLE 18.6 Standard Entropies of Aqueous Ions at 298 K

	$S°$			$S°$	
	cal/mol·K	J/mol·K		cal/mol·K	J/mol·K
H^+	0.0	0.0	OH^-	−2.5	−10.5
Na^+	14.4	60.2	F^-	3.0	12.6
K^+	24.5	102.5	Cl^-	13.2	55.2
Ag^+	17.7	74.1	Br^-	19.3	80.8
Ba^{2+}	3.0	12.6	I^-	26.1	109.2
Ca^{2+}	−13.2	−55.2	HS^-	14.6	61.1
Cu^{2+}	−23.6	−98.7	S^{2-}	5.3	22.2
Zn^{2+}	−25.4	−106.3	SO_4^{2-}	4.1	17.2
			CO_3^{2-}	−12.7	−53.1

18.162 The enthalpy change for a certain reaction at 298 K is −15.0 kcal/mol. The entropy change under these conditions is −7.2 cal/mol·K. Calculate the free energy change for the reaction, and predict whether the reaction may occur spontaneously.

$$\Delta G = \Delta H - T\,\Delta S = (-15\,000 \text{ cal/mol}) - (298 \text{ K})(-7.2 \text{ cal/mol·K}) = -12.9 \text{ kcal/mol}$$

Since the value of ΔG is negative, the reaction may proceed spontaneously.

18.163 For the reaction

$$2\text{A}(g) + \text{B}(g) \longrightarrow 2\text{D}(g)$$

$\Delta E°_{298} = -2.50$ kcal and $\Delta S°_{298} = -10.5$ cal/K. Calculate $\Delta G°_{298}$ for the reaction, and predict whether the reaction may occur spontaneously.

$$\Delta G° = \Delta H° - T\,\Delta S° = \Delta E° + \Delta(PV) - T\,\Delta S°$$

Assuming ideal gas behavior,

$$\Delta G° = \Delta E° + (\Delta n)RT - T\,\Delta S°$$

Using the value of $R = 1.987$ cal/mol·K, and the fact that 2 mol of gas (D) is produced from 3 mol (2A + B),

$$(\Delta n)RT = (-1 \text{ mol})(1.987 \text{ cal/mol·K})(298 \text{ K}) = -592 \text{ cal}$$
$$\Delta G° = -2500 \text{ cal} + (-592 \text{ cal}) - (298 \text{ K})(-10.5 \text{ cal/K}) = -3.09 \text{ kcal} + 3.13 \text{ kcal} = 0.04 \text{ kcal}$$

Since the value of $\Delta G°$ is positive, the indicated reaction cannot be spontaneous.

18.164 Estimate the normal boiling point of PCl_3.

The normal boiling point is the temperature at which $\Delta G°$ for the following reaction is 0.

$$PCl_3(l) \rightleftharpoons PCl_3(g)$$

The reaction is not spontaneous at 25 °C, where, according to Table 18-4,

$$\Delta G° = (1 \text{ mol})[\Delta G_f°(PCl_3, g)] - (1 \text{ mol})[\Delta G_f°(PCl_3, l)] = -267.8 + 272.4 = 4.6 \text{ kJ}$$

If we assume that $\Delta H°$ and $\Delta S°$ are both independent of temperature, then the temperature dependence of $\Delta G°$ is given by the factor T. If $\Delta H°$ and $\Delta S°$ are known from the data at 25 °C, T can be calculated to satisfy the condition that $\Delta G°$ equals 0.

$$\Delta G° = \Delta H° - T\,\Delta S° = 0 \qquad \text{or} \qquad T(\text{op}) = \Delta H°/\Delta S°$$

Now

$$\Delta S° = (1 \text{ mol})[S°(g)] - (1 \text{ mol})[S°(l)] = 311.7 - 217.1 = 94.6 \text{ J/K}$$

and, from Table 18.2,

$$\Delta H° = (1 \text{ mol})[\Delta H_f°(g)] - (1 \text{ mol})[\Delta H_f°(l)] = -287.0 - (-319.7) = 32.7 \text{ kJ}$$

Then

$$T(\text{bp}) = \frac{\Delta H^\circ}{\Delta S^\circ} = \frac{32.8 \times 10^3\ \text{J}}{94.6\ \text{J/K}} = 347\ \text{K}$$

or 74 °C. The observed boiling point, 75 °C, is very close to the estimated 74 °C.

The validity of this type of estimate is no better than the assumption of constancy of ΔS° and ΔH°. In general, the smaller the temperature range over which the extrapolation must be made, the greater the accuracy of the prediction.

18.165 For the reaction at 25 °C

$$X_2O_4(l) \longrightarrow 2XO_2(g)$$

$\Delta E = 2.1$ kcal and $\Delta S = 20$ cal/K. **(a)** Calculate ΔG for the reaction. **(b)** Is the reaction spontaneous as written?

▮ **(a)** $\Delta G = \Delta H - T\,\Delta S = \Delta E + \Delta(PV) - T\,\Delta S$

Since liquid does not exert significant pressure in a system with gas present, and there is an increase of 2 mol of gas,

$$\Delta(PV) = (\Delta n)RT = 2RT$$

$$\Delta G = \Delta E + 2RT - T\,\Delta S = (2.1\ \text{kcal}) + 2(1.99\ \text{cal/mol}\cdot\text{K})(298\ \text{K}) - (298\ \text{K})(20\ \text{cal/K})$$
$$= [(2.1) + (1.19) - (5.96)]\ \text{kcal} = -2.7\ \text{kcal}$$

(b) Since the value of ΔG is negative, the reaction can be spontaneous.

18.166 Explain in terms of thermodynamic properties why heating to a high temperature causes decomposition of $CaCO_3$ to CaO and CO_2.

▮

$$\Delta G = \Delta H - T\,\Delta S \qquad CaCO_3 \longrightarrow CaO + CO_2$$

Increasing T makes the second term relatively more important. Since ΔS is positive for the decomposition reaction, the increasing magnitude of $T\,\Delta S$ causes ΔG to become negative at high temperature, signaling the spontaneous decomposition of $CaCO_3$.

18.167 Calculate ΔS_f° at 25 °C for $PCl_5(g)$.

▮

$$\Delta H_f^\circ = -374.9\ \text{kJ/mol} \qquad \Delta G_f^\circ = -305.0\ \text{kJ/mol}$$

$$\Delta G^\circ = \Delta H^\circ - T\,\Delta S^\circ \qquad \Delta S_f^\circ = \frac{\Delta H_f^\circ - \Delta G_f^\circ}{T} = \frac{[(-374.9) - (-305.0)]\ \text{kJ/mol}}{298\ \text{K}} = -235\ \text{J/mol}\cdot\text{K}$$

118.168 Calculate S° at 25 °C for $PCl_5(g)$.

▮ Note that S° and ΔS_f° are not the same, since the S° values for the elements at 25 °C are nonzero. Therefore, consider the reaction (for which data are available).

$$PCl_3(g) + Cl_2(g) \longrightarrow PCl_5(g)$$

$$\Delta H^\circ = \Delta H_f^\circ(PCl_5) - \Delta H_f^\circ(PCl_3) = [(-374.9) - (-287.0)]\ \text{kJ/mol} = -87.9\ \text{kJ/mol}$$
$$\Delta G^\circ = \Delta G_f^\circ(PCl_5) - \Delta G_f^\circ(PCl_3) = [(-305.0) - (-267.8)]\ \text{kJ/mol} = -37.2\ \text{kJ/mol}$$
$$\Delta S^\circ = \frac{\Delta H^\circ - \Delta G^\circ}{T} = S^\circ(PCl_5) - S^\circ(PCl_3) - S^\circ(Cl_2)$$
$$= \frac{-50.7\ \text{kJ/mol}}{298\ \text{K}} = S^\circ(PCl_5) - 311.7\ \text{J/mol}\cdot\text{K} - 223.0\ \text{J/mol}\cdot\text{K}$$
$$S^\circ(PCl_5) = [(-170) + (311.7) + (223.0)]\ \text{J/mol}\cdot\text{K} = 364.7\ \text{J/mol}\cdot\text{K}$$

18.169 Consider the production of water gas:

$$C(s,\ graphite) + H_2O(g) \rightleftharpoons CO(g) + H_2(g)$$

(a) What is ΔG° for this reaction at 25 °C? **(b)** Estimate the temperature at which $\Delta G^\circ = 0$ for the reaction.

▮ **(a)** $\Delta G^\circ = \Delta G_f^\circ(CO) - \Delta G_f^\circ(H_2O) = [(-137.15) - (-228.58)]\ \text{kJ/mol} = 91.43\ \text{kJ/mol}$

(b) At $\Delta G^\circ = 0$, $\Delta H^\circ = T\,\Delta S^\circ$

Assuming that ΔH° is constant over the temperature range in question,

$$\Delta H^\circ = [(-110.53) - (-241.81)] \text{ kJ/mol} = +131.28 \text{ kJ/mol}$$

$$\Delta S^\circ = S^\circ(CO) + S^\circ(H_2) - S^\circ(C) - S^\circ(H_2O) = (197.56 + 130.57 - 5.74 - 188.724) \text{ J/mol·K} = 133.67 \text{ J/mol·K}$$

$$T = \frac{\Delta H^\circ}{\Delta S^\circ} = \frac{131.28 \times 10^3 \text{ J/mol}}{133.67 \text{ J/mol·K}} = 982 \text{ K}$$

The extrapolation for this estimate is extended over such a large temperature range that an appreciable error might be expected. However, the actual experimental value, 947 K, is not far from the estimate.

18.170 Calculate ΔH_f° for $Cl_2O(g)$ at 25 °C.

▌

$$Cl_2 + \tfrac{1}{2}O_2 \longrightarrow Cl_2O$$

$$\Delta G^\circ = \Delta G_f^\circ = 97.9 \text{ kJ/mol}$$

$$\Delta S^\circ = S^\circ(Cl_2O) - S^\circ(Cl_2) - \tfrac{1}{2}S^\circ(O_2) = [266.10 - 222.96 - \tfrac{1}{2}(205.04)] \text{ J/mol·K} = -59.38 \text{ J/mol·K}$$

$$\Delta H^\circ = \Delta G^\circ + T\,\Delta S^\circ = 97\,900 \text{ J/mol} + (298 \text{ K})(-59.38 \text{ J/mol·K}) = 80\,200 \text{ J/mol} = 80.2 \text{ kJ/mol}$$

18.171 Predict the phase-transition temperature for the conversion of gray to white tin, using the tabulated thermodynamic data.

▌

$$\text{Sn(gray)} \longrightarrow \text{Sn(white)}$$

$$\Delta G^\circ = 0 \qquad \Delta H^\circ = T\,\Delta S^\circ$$

At 25 °C,

$$\Delta H^\circ = \Delta G^\circ + T\,\Delta S^\circ = (-120 \text{ J/mol}) + (298 \text{ K})[(51.5 - 44.1) \text{ J/mol·K}] = 2.1 \text{ kJ/mol}$$

Assuming that ΔH° and ΔS° are constant over the temperature range:

$$T = \frac{\Delta H^\circ}{\Delta S^\circ} = \frac{2100 \text{ J/mol}}{7.4 \text{ J/mol·K}} = 280 \text{ K} \qquad \text{(2 significant figures)}$$

The observed transition temperature is 286 K; the calculated temperature is reasonably close.

18.172 ΔG_f° for the formation of $HI(g)$ from its gaseous elements is -10.10 kJ/mol at 500 K. When the partial pressure of HI is 10.0 atm, and of I_2 0.001 atm, what must the partial pressure of hydrogen be at this temperature to reduce the magnitude of ΔG for the reaction to 0?

▌

$$\tfrac{1}{2}H_2 + \tfrac{1}{2}I_2 \longrightarrow HI \qquad \Delta G^\circ = -10.10 \text{ kJ at 500 K}$$

At $\Delta G = 0$, the reaction is at equilibrium.

$$-\Delta G^\circ = RT \ln K = 2.303RT \log K$$

$$\log K = \frac{-\Delta G^\circ}{2.303RT} = \frac{+10.10 \times 10^3 \text{ J}}{(2.303)(8.31 \text{ J/K})(500 \text{ K})} = 1.055$$

$$K = 11.36 = \frac{P(HI)}{[P(H_2)]^{1/2}[P(I_2)]^{1/2}} = \frac{10.0}{10^{-3/2}[P(H_2)]^{1/2}}$$

[Note the omission of the unit mol^{-1} from the gas constant; this is necessary to make log K and K pure numbers.] Squaring and rearranging yields

$$P(H_2) = \frac{100}{(10^{-3})(11.36)^2} = 775 \text{ atm}$$

18.173 Calculate ΔS° for the combusion of 1 mol of glucose at 25 °C from the enthalpy of combustion at that temperature, -673 kcal/mol, and data from Table 18.4.

▌

$$C_6H_{12}O_6 + 6O_2 \longrightarrow 6CO_2 + 6H_2O(l)$$

$$\Delta G = \Delta H - T\,\Delta S$$

$$\Delta G_{comb} = (6 \text{ mol})(-94.3 \text{ kcal/mol}) + (6 \text{ mol})(-56.7 \text{ kcal/mol}) - (-215 \text{ kcal}) = -691.0 \text{ kcal}$$

$$\Delta S = (\Delta H - \Delta G)/T = [(-673) - (-691.0)] \text{ kcal/298 K} = 60.4 \text{ cal/K}$$

18.174 Using data from Table 18.4, determine which one(s) of the following reactions is (are) feasible at 298 K with all species in their standard states:

(*a*) $Cl_2O \longrightarrow Cl_2 + \frac{1}{2}O_2$ (*b*) $ClO_2 \longrightarrow \frac{1}{2}Cl_2 + O_2$
(*c*) $Cl_2O + \frac{3}{2}O_2 \longrightarrow 2ClO_2$ (*d*) $CCl_4 + 5O_2 \longrightarrow CO_2 + 4ClO_2$

▮ (*a*) $\Delta G° = \Delta G_f°(Cl_2) + \frac{1}{2}\Delta G_f°(O_2) - \Delta G_f°(Cl_2O) = -23.4$ kcal

Since $\Delta G°$ has a negative value, the reaction may proceed.

(*b*) $\Delta G°$ has a negative value, -29.5 kcal, and the reaction may proceed spontaneously,

(*c*) $\Delta G° = 2\Delta G_f°(ClO_2) - \Delta G_f°(Cl_2O) = 2(29.5) - (23.4) = +35.6$ kcal

$\Delta G°$ is positive, hence this reaction cannot be spontaneous.

(*d*) $\Delta G° = [(-94.3) + 4(29.5) - (-33.3)]$ kcal $= +57.0$ kcal

$\Delta G°$ is positive; the reaction is not spontaneous.

18.175 Under what conditions could the decomposition of $Ag_2O(s)$ into $Ag(s)$ and $O_2(g)$ proceed spontaneously at 25 °C?

▮
$$Ag_2O \longrightarrow 2Ag + \tfrac{1}{2}O_2 \qquad \Delta G° = -\Delta G_f° = +11.21 \text{ kJ}$$

For spontaneous decomposition, ΔG must be at least slightly negative.

$$\Delta G = \Delta G° + 2.303RT \log [P(O_2)]^{1/2}$$
$$0 = +11\,210 \text{ J} + 2.303(8.31 \text{ J/K})(298 \text{ K}) \log [P(O_2)]^{1/2}$$

Solving, $P(O_2) = 0.000\,116$ atm $= 0.089$ torr

18.176 It is possible for heat to flow from a body at a lower temperature to one at a higher temperature. Explain under what conditions such a process might occur.

▮ With the expenditure of energy, such as in a refrigerator, one can get heat to flow to a higher temperature body.

18.177 Calculate the enthalpy changes for the reactions

$$SiO_2(s) + 4HF(g) \longrightarrow SiF_4(g) + 2H_2O(g)$$
$$SiO_2(s) + 4HCl(g) \longrightarrow SiCl_4(g) + 2H_2O(g)$$

Explain why hydrofluoric acid attacks glass, whereas hydrochloric acid does not.

▮ Per mole of SiO_2,

$$\Delta H = -361.29 + 2(-57.79) - (-209.9) - 4(-64.2) = -10.17 \text{ kcal}$$
$$\Delta H = -145.7 + 2(-57.79) - (-209.9) - 4(-22.02) = 36.7 \text{ kcal}$$

The reaction of HF, with comparable ΔS, has a much lower ΔG value at any given temperature, and thus HF is much more apt to attack glass (containing SiO_2).

18.178 Given the following reactions with their enthalpy changes,

$$N_2(g) + 2O_2(g) \longrightarrow 2NO_2(g) \qquad \Delta H_{298} = 16.18 \text{ kcal}$$
$$N_2(g) + 2O_2(g) \longrightarrow N_2O_4(g) \qquad \Delta H_{298} = 2.31 \text{ kcal}$$

calculate the enthalpy of dimerization of NO_2. Is N_2O_4 apt to be stable with respect to NO_2 at 25 °C? Explain your answer.

▮ Subtraction of the first equation from the second yields

$$2NO_2 \longrightarrow N_2O_4 \qquad \Delta H_{298} = -13.87 \text{ kcal}$$

The value of $\Delta G = \Delta H - T\,\Delta S$ is apt to remain negative at low temperature but become positive at high temperature. The reaction should be spontaneous as written only at low temperature.

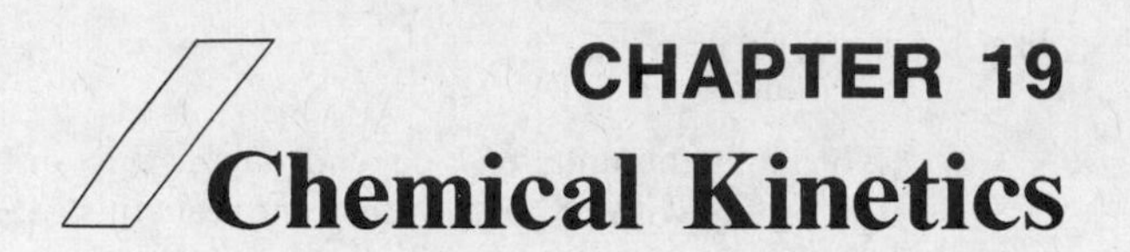

CHAPTER 19
Chemical Kinetics

Note: Before working the problems in this chapter, you should familiarize yourself with Table 19.1, p. 378.

19.1 RATE LAWS

19.1 Distinguish explicitly between the rate of a chemical reaction and the rate constant for the reaction.

▌ Rate $= k[A]^n[B]^m$ The rate is a variable, proportional to the concentrations and to the rate constant. The rate constant, k, is a number which depends on the temperature only and not on the concentrations.

19.2 A reaction between A and B is second order. Write three different rate law expressions which might possibly apply to the reaction.

▌ Those with integral orders are: Rate $= k[A][B]$ Rate $= k[A]^2$ Rate $= k[B]^2$

19.3 In a catalytic experiment involving the Haber process, $N_2 + 3H_2 \rightarrow 2NH_3$, the rate of reaction was measured as

$$\text{Rate} = \frac{\Delta[NH_3]}{\Delta t} = 2.0 \times 10^{-4}\ \text{mol}\cdot\text{L}^{-1}\cdot\text{s}^{-1}$$

If there were no side reactions, what was the rate of reaction expressed in terms of (*a*) N_2 (*b*) H_2?

▌ (*a*) From the coefficients in the balanced equation. $\Delta n(N_2) = -\frac{1}{2}\Delta n(NH_3)$. Therefore,

$$-\frac{\Delta[N_2]}{\Delta t} = \frac{1}{2}\frac{\Delta[NH_3]}{\Delta t} = 1.0 \times 10^{-4}\ \text{mol}\cdot\text{L}^{-1}\cdot\text{s}^{-1}$$

(*b*) Similarly,

$$-\frac{\Delta[H_2]}{\Delta t} = \frac{3}{2}\frac{\Delta[NH_3]}{\Delta t} = 3.0 \times 10^{-4}\ \text{mol}\cdot\text{L}^{-1}\cdot\text{s}^{-1}$$

19.4 What are the units of the rate constant, k, for (*a*) a zero order reaction, (*b*) a first order reaction, (*c*) a second order reaction, (*d*) a third order reaction, (*e*) a half order reaction, when concentrations are expressed in mol/L and time in s?

▌ In each case, we write out the full rate equation and find the units of k that will satisfy the equation.

(*a*) $-\frac{\Delta[A]}{\Delta t} = k$ Units of k = units of $\frac{[A]}{t} = \frac{\text{mol/L}}{\text{s}} = \text{mol}\cdot\text{L}^{-1}\cdot\text{s}^{-1}$

Note that the units of $\Delta[A]$, the change in concentration, are the same as the units of $[A]$ itself; similarly for Δt.

(*b*) $-\frac{\Delta[A]}{\Delta t} = k[A]$ or $k = -\frac{1}{[A]}\frac{\Delta[A]}{\Delta t}$

$$\text{Units of } k = \left(\frac{1}{\text{mol/L}}\right)\left(\frac{\text{mol/L}}{\text{s}}\right) = \text{s}^{-1}$$

First order reactions are the only reactions for which k has the same numerical value, regardless of the units used for expressing the concentrations of the reactants or products.

(*c*) $-\frac{\Delta[A]}{\Delta t} = k[A]^2$ | $-\frac{\Delta[A]}{\Delta t} = k[A][B]$

$k = \frac{1}{[A]^2}\frac{\Delta[A]}{\Delta t}$ | $k = -\frac{1}{[A][B]}\frac{\Delta[A]}{\Delta t}$

$$\text{Units of } k = \left(\frac{1}{(\text{mol/L})^2}\right)\frac{\text{mol/L}}{\text{s}} = \text{L}\cdot\text{mol}^{-1}\cdot\text{s}^{-1}$$

Note that the units of k depend on the *total* order of the reaction, not on the way the total order is composed of the orders with respect to different reactants.

(*d*) $-\frac{\Delta[A]}{\Delta t} = k[A]^3$ or $k = -\frac{1}{[A]^3}\frac{\Delta[A]}{\Delta t}$

$$\text{Units of } k = \left(\frac{1}{(\text{mol/L})^3}\right)\left(\frac{\text{mol/L}}{\text{s}}\right) = \text{L}^2\cdot\text{mol}^{-2}\cdot\text{s}^{-1}$$

(*e*) $-\frac{\Delta[A]}{\Delta t} = k[A]^{1/2}$ or $k = -\frac{1}{[A]^{1/2}}\frac{\Delta[A]}{\Delta t}$

$$\text{Units of } k = \left(\frac{1}{(\text{mol/L})^{1/2}}\right)\left(\frac{\text{mol/L}}{\text{s}}\right) = \text{mol}^{1/2}\cdot\text{L}^{-1/2}\cdot\text{s}^{-1}$$

19.5 Which of the following will react fastest (produce most product in a given time) and which will react at the highest *rate*? **(*a*)** 1 mol of A and 1 mol of B in a 1-L vessel **(*b*)** 2 mol of A and 2 mol of B in a 2-L vessel **(*c*)** 0.2 mol of A and 0.2 mol of B in a 0.1-L vessel.

▮ Fastest, (*b*). Since there is more reactant present, there will be more product produced per unit time. The highest rate, the greatest change in *concentration* per unit time, is (*c*), simply because the vessel with the greatest concentration will also experience the greatest change in concentration per unit time.

19.6 For the reaction

$$3BrO^- \longrightarrow BrO_3^- + 2Br^-$$

in alkaline aqueous solution, the value of the second order (in BrO^-) rate constant at 80 °C in the rate law for $-\Delta[BrO^-]/\Delta t$ was found to be 0.056 $\text{L}\cdot\text{mol}^{-1}\cdot\text{s}^{-1}$. What is the rate constant when the rate law is written for **(*a*)** $\Delta[BrO_3^-]/\Delta t$ **(*b*)** $\Delta[Br^-]/\Delta t$?

▮ **(*a*)** BrO_3^- appears at a rate one-third that of disappearance of BrO^-.

$$\frac{0.056\ \text{L}\cdot\text{mol}^{-1}\cdot\text{s}^{-1}}{3} = 0.019\ \text{L}\cdot\text{mol}^{-1}\cdot\text{s}^{-1}$$

(*b*) Two-thirds of the BrO^- rate:

$$\frac{2(0.056\ \text{L}\cdot\text{mol}^{-1}\cdot\text{s}^{-1})}{3} = 0.037\ \text{L}\cdot\text{mol}^{-1}\cdot\text{s}^{-1}$$

19.7 Five 5.00-mL samples of 1.0 M reagent B were poured into five vessels containing 5.00-mL samples of A, each having the concentration tabulated below. The temperature was 25 °C, and all other conditions were constant. The initial rates are also tabulated. Another set of experiments was performed in which the concentration of B was varied; these results are also tabulated. What is the order of the reaction with respect to A and with respect to B? What is the value of the rate constant?

conc A, M	conc B, M	initial rate, M/s	conc A, M	conc B, M	initial rate, M/s
1.0	1.0	1.2×10^{-2}	1.0	1.0	1.2×10^{-2}
2.0	1.0	2.3×10^{-2}	1.0	2.0	4.8×10^{-2}
4.0	1.0	4.9×10^{-2}	1.0	4.0	1.9×10^{-1}
8.0	1.0	9.6×10^{-2}	1.0	8.0	7.6×10^{-1}
16.0	1.0	1.9×10^{-1}	1.0	16.0	3.0

▮ It is apparent from the data on the left that, all other factors being held constant, the initial rate of the reaction is proportional to the initial concentration of A. The reaction is first order with respect to A. Note that no conclusions about the order with respect to B can be drawn from the data on the left-hand side of the table. The data on the right show that the initial rate is proportional to the square of the initial concentration of B, and therefore the reaction is second order with respect to B. The reaction is third order overall. The numerical value of the rate constant, k, may be obtained by substituting data for each experiment into the rate law expression:

$$\text{Rate} = k[A]^m[B]^n = k[A][B]^2$$

Thus for the data in the first line on the left,

$$1.2 \times 10^{-2}\ \text{M/s} = k(1.0\ \text{M})(1.0\ \text{M})^2 \quad \text{so} \quad k = 1.2 \times 10^{-2}\ \text{M}^{-2}\cdot\text{s}^{-1}$$

19.8 A reaction between substances A and B is represented stoichiometrically by $A + B \rightarrow C$. Observations on the rate of this reaction are obtained in three separate experiments as follows:

	initial concentrations			
	$[A]_0$, M	$[B]_0$, M	duration of experiment, Δt, h	final concentration, $[A]_f$, M
(*1*)	0.1000	1.0	0.50	0.0975
(*2*)	0.1000	2.0	0.50	0.0900
(*3*)	0.0500	1.0	2.00	0.0450

What is the order with respect to each reactant, and what is the value of the rate constant?

▮ First, let us tabulate the rate of the reaction for each experiment, noting that $\Delta[A] = [A]_f - [A]_0$ is in each case small enough in magnitude to allow the rate to be expressed in terms of changes over the entire time of the experiment.

	$[A]_0$, M	$[B]_0$, M	$\Delta[A]$, M	Δt, h	$-\dfrac{\Delta[A]}{\Delta t}$, $M\cdot h^{-1}$
(*1*)	0.1000	1.0	−0.0025	0.50	0.0050
(*2*)	0.1000	2.0	−0.0100	0.50	0.0200
(*3*)	0.0500	1.0	−0.0050	2.00	0.0025

In comparing experiments (*1*) and (*2*), we note that [A] is the same in both, but [B] is twice as great in (*2*). Since the rate in (*2*) is 4 times that in (*1*), the reaction must be second order in B.

In comparing experiments (*1*) and (*3*), we note that [B] is the same in both, but [A] is twice as great in (*1*). Since the rate in (*1*) is twice that in (*3*), the reaction must be first order in A.

The rate equation may be written accordingly as

$$\text{Rate} = \frac{-\Delta[A]}{\Delta t} = k[A][B]^2$$

and k may be evaluated from any of the experimental points; we take (*1*) as an example, using an average value of [A].

$$k = \frac{-\Delta[A]/\Delta t}{[A][B]^2} = \frac{0.0050\ \text{M}\cdot\text{h}^{-1}}{(0.099\ \text{M})(1.0\,\text{M})^2} = 0.051\ \text{L}^2\cdot\text{mol}^{-2}\cdot\text{h}^{-1} = 1.4 \times 10^{-5}\ \text{L}^2\cdot\text{mol}^{-2}\cdot\text{s}^{-1}.$$

19.9 The following reaction was studied at 25 °C in benzene solution containing 0.1 M pyridine:

$$\underset{A}{CH_3OH} + \underset{B}{(C_6H_5)_3CCl} \longrightarrow \underset{C}{CH_3OC(C_6H_5)_3} + HCl$$

The following sets of data were observed.

initial concentrations				
$[A]_0$, M	$[B]_0$, M	$[C]_0$, M	Δt, min	Final [C], M
0.10	0.05	0.0000	25	0.0033
0.10	0.10	0.0000	15.0	0.0039
0.20	0.10	0.0000	7.5	0.0077

What rate law is consistent with the above data and what is the best average value for the rate constant, expressed in seconds and molar concentration units?

▌ Average initial rate ($M^{-1}\cdot min^{-1}$)

$$0.0033/25 = 0.000132 \qquad 0.0039/15 = 0.00026 \qquad 0.0077/7.5 = 0.00103$$

When [B] is doubled, the rate is doubled, hence the reaction is first order in B. When [A] is doubled, the rate is quadrupled, hence the reaction is second order in A.

$$\text{Rate} = k[A]^2[B]$$

	(1)	(2)	(3)
$k = \dfrac{\text{rate}}{[A]^2[B]}$	$\dfrac{0.000132}{(0.10)^2(0.05)}$	$\dfrac{0.00026}{(0.10)^2(0.10)}$	$\dfrac{0.00103}{(0.20)^2(0.10)}$
$k =$	0.264	0.26	0.258

The average of these rates is 0.26 $L^2\cdot mol^{-2}\cdot min^{-1}$ or 4.3×10^{-3} $L^2\cdot mol^{-2}\cdot s^{-1}$.

19.10 The reaction $2NO(g) + Cl_2(g) \rightleftharpoons 2NOCl$ was studied at −10 °C, and the following data were obtained:

	initial concentration, mol/L		
Run	**NO**	**Cl_2**	**initial rate of formation of NOCl, mol/L·min**
1	0.10	0.10	0.18
2	0.10	0.20	0.35
3	0.20	0.20	1.45

What is the order of reaction with respect to NO and with respect to Cl_2?

▌ When the Cl_2 concentration is doubled, holding the NO concentration constant (compare runs 1 and 2), the initial rate doubles. Hence the reaction is first order with respect to Cl_2. When the NO concentration is doubled (compare runs 2 and 3), the initial rate quadruples. Hence the reaction is second order with respect to NO.

19.11 For the data of Prob. 19.10, what is the numerical value of the rate constant at −10 °C?

▌ $$\text{Rate} = k[NO]^2[Cl_2] = k(0.10)(0.10)^2 = 0.18\ \text{mol/L}\cdot\text{min} \qquad k = 180\ \text{L/mol}\cdot\text{min}$$

19.12 The hydrolysis of methyl acetate in alkaline solution,

$$CH_3COOCH_3 + OH^- \longrightarrow CH_3COO^- + CH_3OH$$

followed rate = $k[CH_3OOCH_3][OH^-]$, with k equal to 0.137 $L\cdot mol^{-1}\cdot s^{-1}$ at 25 °C. A reaction mixture was prepared to have initial concentrations of methyl acetate and OH^- of 0.050 M each. How long would it take for 5.0% of the methyl acetate to be hydrolyzed at 25 °C?

▌ $$\text{Initial rate} = (0.137\ M^{-1}\cdot s^{-1})(0.050\ M)^2 = 3.4 \times 10^{-4}\ M\cdot s^{-1} \qquad \frac{0.050(0.050\ M)}{3.4 \times 10^{-4}\ M\cdot s^{-1}} = 7.4\ s$$

(This answer assumes that the initial rate does not drop appreciably as some of the reactants are used up.)

19.13 The reaction $A + 2B \rightarrow C + 2D$ is run three times. In the second run, the initial concentration of A is double that in the first run, and the initial rate of the reaction is double that of the first run. In the third run, the initial concentration of each reactant is double the respective concentrations in the first run, and the initial rate is double that of the first run. What is the order of the reaction with respect to each reactant?

▌ From the first statement it may be deduced that the reaction is first order with respect to A. Since doubling both concentrations causes only a doubling of the rate, which is caused by the doubling of the A concentration, the doubling of the B concentration has had no effect on the rate; the rate is zero order with respect to B.

19.14 A viral preparation was inactivated in a chemical bath. The inactivation process was found to be first order in virus concentration, and at the beginning of the experiment 2.0% of the virus was found to be inactivated per minute. Evaluate k for the inactivation process.

From the first order rate law,

$$k = -\frac{\Delta[A]}{[A]}\frac{1}{\Delta t}$$

It is seen that only the *fractional* change in concentration, $-\Delta[A]/[A]$, is needed; namely, 0.020 when $\Delta t = 1$ min = 60 s. This form of the equation may be used for the *initial* rate when the value of [A] is not changing appreciably; that condition is certainly met when only 2% is inactivated in the first minute.

$$k = \frac{0.020}{60\ \text{s}} = 3.3 \times 10^{-4}\ \text{s}^{-1}$$

19.15 The rate of the following reaction in aqueous solution, in which the initial concentration of the complex compound is 0.100 *M*, is to be studied:

$$[Co(NH_3)_5Cl]^{2+} + H_2O \longrightarrow [Co(NH_3)_5(H_2O)]^{3+} + Cl^-$$

Explain why under these conditions it is possible to determine the order of the reaction with respect to $[Co(NH_3)_5Cl]^{2+}$, but not with respect to water. Support your explanation with an appropriate calculation.

This reaction is pseudo first order. No matter how complete the reaction of the complex ion, the concentration of the water changes very little, and therefore it is impossible to determine the order with respect to water. Quantitatively, the concentration of water is about 55.5 M. If the reaction went to completion, the water concentration could not fall below 55.4 M. Less than 0.2% change is insufficient to allow a determination of how the water concentration affects the rate of reaction.

19.16 A certain reaction, $A + B \rightarrow C$, is first order with respect to each reactant, with $k = 1.0 \times 10^{-2}\ \text{L·mol}^{-1}\text{·s}^{-1}$. Calculate the concentration of A remaining after 100 s if the initial concentration of each reactant was 0.100 M.

Since the concentrations of the two reactants start equal and remain equal throughout the reaction, the reaction can be treated as a simple second order reaction.

$$\frac{1}{[A]} = kt + \frac{1}{[A_0]} = \frac{1.0 \times 10^{-2}\ \text{L}}{\text{mol·s}}(100\ \text{s}) + \frac{1\ \text{L}}{0.100\ \text{mol}} = 11\ \text{L/mol}$$

$$[A] = 0.091\ \text{M}$$

19.17 A certain reaction, $A + B \rightarrow$ products, is first order with respect to each reactant, with $k = 5.0 \times 10^{-3}\ \text{M}^{-1}\text{·s}^{-1}$. Calculate the concentration of A remaining after 100 s if the initial concentration of A was 0.100 M and that of B was 6.00 M. State any approximation made in obtaining your result.

Since the concentration of B is 6.0 M, and throughout the reaction the B concentration cannot possibly fall below 5.9 M, it is useful to treat the B concentration as a constant. The rate expression can be written in the form

$$\text{Rate} = 6.00k[A]$$

Treating $6.00k$ as a new constant rate, k', one can use the first order rate law equation in the form

$$\log [A] = \log [A_0] - 6.00kt/2.30$$

$$\log [A] = -1.00 - \frac{(3.0 \times 10^{-2})(100)}{2.30} = -1.00 - 1.304 = -2.30$$

$$[A] = 5.0 \times 10^{-3}\ \text{M}$$

19.18 For a reaction in which 1 mol of reactant produces 1 mol of product, show that each of the expressions on the left-hand side of Table 19.1 becomes identical to the corresponding expression on the right.

For a reaction $A \rightarrow X$, the concentration of X at any time is equal to the mol/L of A which has reacted; that is, $[X] = [A_0] - [A]$. Rearranging yields

$$[A] = [A_0] - [X]$$

Substitution of $[A_0] - [X]$ for [A] in the equations in terms of [A] in the table yields the equations in terms of [X].

TABLE 19.1 Concentration as a Function of Time*

order (n) with respect to A	rate equation in terms of reactant concentration	rate equation in terms of product concentration	units of k
0	$[A_0] - [A] = kt$	$[X] = kt$	$mol \cdot L^{-1} \cdot s^{-1}$
1	$\log [A_0] - \log [A] = \frac{kt}{2.30}$ or $2.30 \log \frac{[A]}{[A_0]} = -kt$	$\log [A_0] - \log([A_0] - [X]) = \frac{kt}{2.30}$ or $2.30 \log \frac{[A_0] - [X]}{[A_0]} = -kt$	s^{-1}
2	$\frac{1}{[A]} - \frac{1}{[A_0]} = kt$	$\frac{1}{[A_0] - [X]} - \frac{1}{[A_0]} = kt$	$L \cdot mol^{-1} \cdot s^{-1}$
3	$\frac{1}{[A]^2} - \frac{1}{[A_0]^2} = 2kt$	$\frac{1}{([A_0] - [X])^2} - \frac{1}{[A_0]^2} = 2kt$	$L^2 \cdot mol^{-2} \cdot s^{-1}$

* Rate = $k[A]^n$.
$[A]$ = concentration of a reactant at a given time, t(s).
$[A_0]$ = initial concentration of the reactant.
$[X]$ = concentration of a product at a given time, t(s).

19.19 At 326 °C, 1,3-butadiene dimerizes according to the equation

$$2C_4H_6(g) \longrightarrow C_8H_{12}(g)$$

In a given experiment, the initial pressure of C_4H_6 was 632.0 torr at 326 °C. Determine the order of the reaction and the value of the rate constant from the accompanying data.

time, min	total pressure, torr	time, min	total pressure, torr
0.00	632.0	24.55	546.8
3.25	618.5	42.50	509.3
12.18	584.2	68.05	474.6

▌ The total pressure is due to both C_4H_6 and C_8H_{12}. Each molecule of C_8H_{12} formed is produced from two molecules of C_4H_6. Therefore the total number of molecules is reduced by one each time one molecule of C_8H_{12} is formed. And since the pressure of each gas is proportional to the number of its molecules, the total pressure in the system is reduced as much as the partial pressure of the C_8H_{12} formed. The reduction in total pressure is equal to the original total pressure, P_0, minus the total pressure at any time, P_{tot}.

$$P(C_8H_{12}) = P_0 - P_{tot}$$

From Dalton's law of partial pressures

$$P(C_4H_6) = P_{tot} - P(C_8H_{12})$$

The pressures may be tabulated as follows:

t, min	P_{tot}, torr	$P(C_8H_{12})$	$P(C_4H_6)$	$\log P(C_4H_6)$	$1/P(C_4H_6)$, torr^{-1}
0.00	632.0	0.0	632.0	2.801	1.58×10^{-3}
3.25	618.5	13.5	605.0	2.782	1.65×10^{-3}
12.18	584.2	47.8	536.4	2.730	1.86×10^{-3}
24.55	546.8	85.2	461.6	2.664	2.17×10^{-3}
42.50	509.3	122.7	386.6	2.587	2.59×10^{-3}
68.05	474.6	157.4	317.2	2.502	3.15×10^{-3}

When $\log P$ is plotted versus time (Fig. 19.1(a)), a straight line does not result. But the plot of $1/P$ versus t (Fig. 19.1(b)) produces a straight line; hence the reaction is second order. The rate constant is evaluated by rearrangement of the equation of Table 19.1 to the form

$$\frac{1}{P} = kt + \frac{1}{P_0}$$

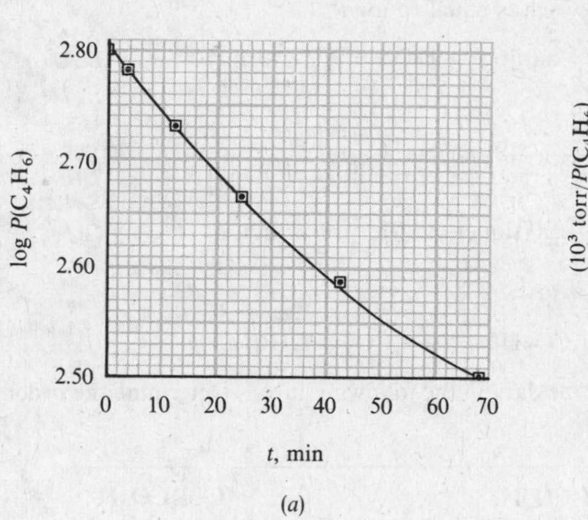

(a)

(b)

Fig. 19.1

analogous to $y = mx + b$. The slope is therefore equal to k.

$$k = \frac{(3.00 - 1.58) \times 10^{-3}\ \text{torr}^{-1}}{(60.0 - 0.00)\ \text{min}} = 2.37 \times 10^{-5}\ \text{torr}^{-1}\cdot\text{min}^{-1}$$

19.20 Determine the order of the reaction and the value of the rate constant for the decomposition of N_2O_5 from the following data:

t, min	$P(N_2O_5)$, torr	t, min	$P(N_2O_5)$, torr
1.83	0.826	79.50	0.372
16.33	0.708	109.66	0.277
38.50	0.562	149.80	0.184
61.17	0.449		

▮

t, min	P, torr	log P
1.83	0.826	−0.0830
16.33	0.708	−0.150
38.50	0.562	−0.250
61.17	0.449	−0.348
79.50	0.372	−0.429
109.66	0.277	−0.558
149.80	0.184	−0.735

The reaction is first order, since the log P versus t plot (Fig. 19.2) yields a straight line, corresponding to $y = mx + b$.

$$\log P = -\frac{k}{2.30}t + \log P_0$$

Fig. 19.2

From the graph the y intercept is -0.800, which is equal to $\log P_0$.

$$P_0 = \text{antilog}(-0.800) = 0.158 \text{ torr}$$

The slope of the line is

$$\frac{(-0.800) - (-0.100)}{164.0 - 4.5} = -4.39 \times 10^{-3} \text{ min}^{-1}$$

and $$k = -2.30(\text{slope}) = 1.01 \times 10^{-2} \text{ min}^{-1}$$

19.21 The reaction

$$CCl_3CO_2H(aq) \longrightarrow CO_2 + CHCl_3$$

was found to proceed at 70 °C according to the data in the following table. Determine the order of the reaction and the value of the rate constant.

time, h	CCl_3CO_2H concentration, M	time, h	CCl_3CO_2H concentration, M
0.00	0.10000	3.00	0.08314
1.00	0.09403	4.00	0.07817
2.00	0.08842	5.00	0.07351

■

t, h	conc, M	log conc
0.000	0.10000	−1.000
1.000	0.09403	−1.027
2.000	0.08842	−1.053
3.000	0.08314	−1.080
4.000	0.07817	−1.107
5.000	0.07351	−1.134

Since log conc is proportional to time (Fig. 19.3), the reaction is first order. (See Prob. 19.20.) The rate constant is given by

$$k = -2.30(\text{slope}) = -2.30\left(\frac{1.14 - 1.00}{5.25 - 0.00}\right) = -0.061 \text{ h}^{-1}$$

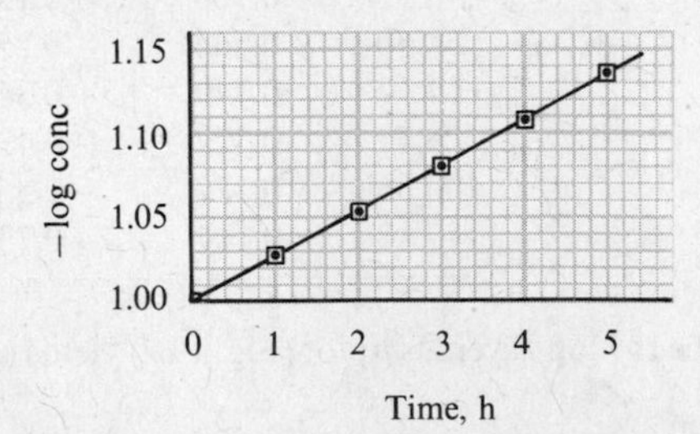

Fig. 19.3

19.22 For the reaction of reagent A in solution to give products, the following data were obtained:

t, min	[A]	*t*, min	[A]
0.0	0.583	18.0	0.170
9.0	0.343	20.0	0.133
12.0	0.257	24.0	0.118
14.0	0.223	30.0	0.079

Determine the order of the reaction, and calculate the value of the rate constant.

▌

t, min	[A]	log [A]
0.0	0.583	−0.234
9.0	0.343	−0.465
12.0	0.257	−0.590
14.0	0.223	−0.652
18.0	0.170	−0.770
20.0	0.133	−0.876
24.0	0.118	−0.928
30.0	0.079	−1.10

Figure 19.4 shows that the reaction is first order. Note that the intercept is approximately −0.200, as drawn from the majority of the points. It is not necessarily equal to the value of log [A] at $t = 0$.

$$\text{Slope} = -\frac{k}{2.30} = -0.029 \qquad k = (2.30)(0.029) = 0.067\ \text{min}^{-1}$$

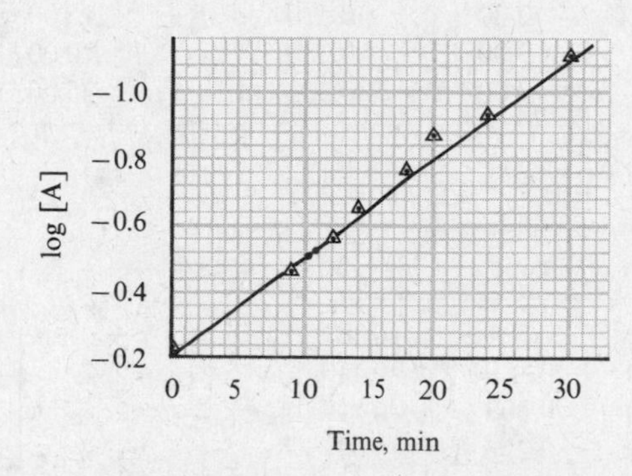

Fig. 19.4

19.23 For the reaction $A \rightarrow C + D$, the initial concentration of A is 0.010 M. After 100 s, the concentration of A is 0.0010 M. The rate constant of the reaction has the numerical magnitude of 9.0. What is the order of the reaction?

▌ The data may be tested to see if they fit any of the equations of Table 19.1:

order	equation from table	trial data
1st	$\log \frac{[A]}{[A_0]} = -\frac{kt}{2.30}$	$\log \frac{10^{-3}}{10^{-2}} \neq -\frac{9.0(100)}{2.30}$
2nd	$\frac{1}{[A]} - \frac{1}{[A_0]} = kt$	$\frac{1}{10^{-3}} - \frac{1}{10^{-2}} = 9.0(100)$
3rd	$\frac{1}{[A]^2} - \frac{1}{[A_0]^2} = 2kt$	$\frac{1}{10^{-6}} - \frac{1}{10^{-4}} \neq 9.0(100)(2)$
0	$[A] - [A_0] = kt$	$10^{-3} - 10^{-2} \neq 9.0(100)$

The reaction may be seen to be second order, for that equation is the only one the data fit.

19.24 From the accompanying data for the reaction $A \xrightarrow{\text{catalyst}}$ products, calculate the value of k.

time, min	[A]	time, min	[A]
0.0	0.100	2.0	0.080
1.0	0.090	3.0	0.070

t, min	[A]	$[A]-[A_0]$
0.0	0.100	0.000
1.0	0.090	−0.010
2.0	0.080	−0.020
3.0	0.070	−0.030

Inspection of the data is sufficient to see that $[A]-[A_0]$ is proportional to t, and therefore the reaction is zero order, with $k = 0.010$ M/min.

19.25 Graph the following data from an experimental determination of the rate of a reaction. From the slope of the line, calculate the order of the reaction and the value of the rate constant.

time, min	[reactant]	time, min	[reactant]
0.00	0.500	15.00	0.236
5.00	0.389	20.00	0.184
10.00	0.303	25.00	0.143

▮

t, min	conc	log conc
0.00	0.500	−0.301
5.00	0.389	−0.410
10.00	0.303	−0.519
15.00	0.236	−0.627
20.00	0.184	−0.735
25.00	0.143	−0.845

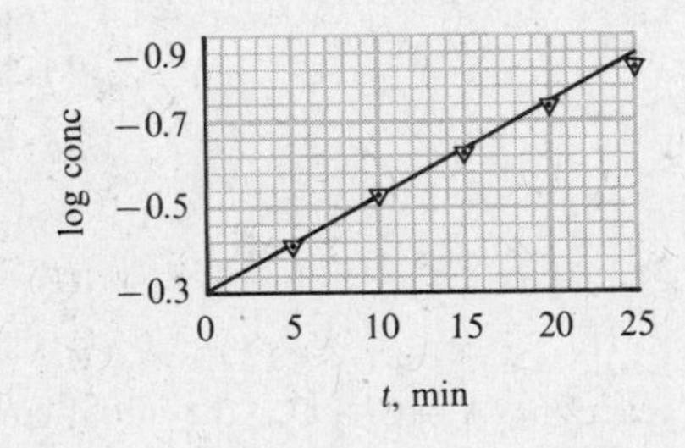

Fig. 19.5

From Fig. 19.5,

$$\text{Slope} = -\frac{0.845 - 0.301}{25.00 - 0.00} = -0.0218 \text{ min}^{-1}$$

$$k = -2.30(\text{slope}) = -2.30(-0.0218 \text{ min}^{-1}) = 0.0500 \text{ min}^{-1}$$

The reaction is first order.

19.26 At 25 °C, the second order rate constant for the reaction $I^- + ClO^- \rightarrow IO^- + Cl^-$ is 0.0606 $M^{-1}\cdot s^{-1}$. If a solution is initially 3.50×10^{-3} M with respect to each reactant, what will be the concentration of each species present after 300 s?

▮ Since the concentrations of the two reactants are equal at the start and remain so throughout the entire reaction, the reaction can be treated as a simple second order reaction.

$$k = 0.0606 \text{ M}^{-1}\cdot\text{s}^{-1} \qquad [B_0] = [A_0] = 3.50 \times 10^{-3} \text{ M}$$

$$\frac{1}{[A]} - \frac{1}{[A_0]} = kt$$

$$\frac{1}{[A]} - \frac{1}{3.5 \times 10^{-3}} = (0.0606)(300) = 18.18$$

$$\frac{1}{[A]} = 18.18 + 285.7 = 303.9$$

$$[A] = 3.29 \times 10^{-3} \text{ M} = [B]$$

19.27 Substance A reacts according to a first order rate law with $k = 5.0 \times 10^{-5}\ \text{s}^{-1}$. **(a)** If the initial concentration of A is 1.00 M, what is the initial rate? **(b)** What is the rate after 1.00 h?

▮ *(a)* rate $= k[\text{A}] = (5.0 \times 10^{-5}\ \text{s}^{-1})(1.00\ \text{M}) = 5.0 \times 10^{-5}\ \text{M/s}$

(b) $$2.30 \log \frac{[\text{A}]}{[\text{A}_0]} = -kt = -(5.0 \times 10^{-5}\ \text{s}^{-1})(3600\ \text{s}) = -0.18$$

$$\log \frac{[\text{A}]}{[\text{A}_0]} = -0.0783$$

$$[\text{A}] = 0.835\ \text{M}$$

$$\text{Rate} = k[\text{A}] = (5.0 \times 10^{-5}\ \text{s}^{-1})(0.835\ \text{M}) = 4.2 \times 10^{-5}\ \text{M/s}$$

19.28 For the nonequilibrium process $\text{A} + \text{B} \rightarrow \text{products}$, the rate is first order with respect to A and second order with respect to B. If 1.0 mol each of A and B were introduced into a 1.0-L vessel, and the initial rate were $1.0 \times 10^{-2}\ \text{mol/L·s}$ calculate the rate when half the reactants have been turned into products.

▮
$$\text{Rate} = k[\text{A}][\text{B}]^2$$

$$1.0 \times 10^{-2}\ \text{mol/L·s} = k(1.0\ \text{mol/L})^3 \qquad k = 1.0 \times 10^{-2}\ \text{L}^2/\text{mol}^2\text{·s}$$

$$\text{Rate}_2 = (1.0 \times 10^{-2}\ \text{L}^2/\text{mol}^2\text{·s})(0.50\ \text{mol/L})^3 = 1.2 \times 10^{-3}\ \text{mol/L·s}$$

19.29 At 25 °C, the second order rate constant for the reaction $\text{I}^- + \text{ClO}^- \rightarrow \text{IO}^- + \text{Cl}^-$ is $0.0606\ \text{L·mol}^{-1}\text{·s}^{-1}$. If a solution is initially 1.0 M in I^- and 5.0×10^{-4} M in ClO^-, what additional information is required to determine what the concentration of ClO^- will be after 300 s?

▮ It is necessary to know the rate law expression, which, from the statement of the problem, could be any of the following (among others):

$$\text{Rate} = k[\text{I}^-][\text{ClO}^-] \qquad \text{or} \qquad \text{Rate} = k[\text{I}^-]^2 \qquad \text{or} \qquad \text{Rate} = k[\text{ClO}^-]^2$$

19.30 The approach to the following equilibrium was observed kinetically from both directions:

$$\text{PtCl}_4^{2-} + \text{H}_2\text{O} \rightleftharpoons \text{Pt(H}_2\text{O)Cl}_3^- + \text{Cl}^-$$

At 25 °C, it was found that at 0.3 ionic strength:

$$-\frac{\Delta[\text{PtCl}_4^{2-}]}{\Delta t} = (3.9 \times 10^{-5}\ \text{s}^{-1})[\text{PtCl}_4^{2-}] - (2.1 \times 10^{-3}\ \text{L·mol}^{-1}\text{·s}^{-1})[\text{Pt(H}_2\text{O)Cl}_3^-][\text{Cl}^-]$$

What is the value of K'_4 for the complexation of the fourth Cl^- by Pt(II) (the apparent equilibrium constant for the reverse of the reaction as written above) at 0.3 ionic strength?

▮ At equilibrium,
$$\frac{\Delta[\text{PtCl}_4^{2-}]}{\Delta t} = 0$$

Hence, $(3.9 \times 10^{-5}\ \text{s}^{-1})[\text{PtCl}_4^{2-}] = (2.1 \times 10^{-3}\ \text{L·mol}^{-1}\text{·s}^{-1})[\text{Pt(H}_2\text{O)Cl}_3^-][\text{Cl}^-]$

$$K'_4 = \frac{[\text{PtCl}_4^{2-}]}{[\text{Pt(H}_2\text{O)Cl}_3^-][\text{Cl}^-]} = \frac{2.1 \times 10^{-3}}{3.9 \times 10^{-5}} = 54$$

19.31 The esterification of acetic anhydride by ethyl alcohol can be represented by the following balanced equation:

$$\underset{\text{A}}{(\text{CH}_3\text{CO})_2\text{O}} + \underset{\text{B}}{\text{C}_2\text{H}_5\text{OH}} \longrightarrow \text{CH}_3\text{COOC}_2\text{H}_5 + \text{CH}_3\text{COOH}$$

When the reaction is carried out in dilute hexane solution, the rate may be represented by $k[\text{A}][\text{B}]$. When ethyl alcohol (B) is the solvent, the rate may be represented by $k[\text{A}]$. (The values of k are not the same in the two cases.) Explain the difference in the apparent order of the reaction.

▮ When a solvent is also a reactant, its concentration is so large compared with the extent of reaction that it does not change. (Compare the convention we used, in studying the thermodynamics of aqueous solutions, of viewing water as always being in its standard state in all dilute solutions.) Thus, the dependence of the rate on the concentration of ethyl alcohol cannot be determined unless ethyl alcohol becomes a solute in some other solvent, so that its concentration can be varied.

19.32 The complexation of Fe^{2+} with the chelating agent dipyridyl (abbreviated dipy) has been studied kinetically in both the forward and reverse directions. For the complexation reaction

$$Fe^{2+} + 3\ dipy \longrightarrow Fe(dipy)_3^{2+}$$

the rate of formation of the complex at 25 °C is given by

$$\text{Rate} = (1.45 \times 10^{13}\ L^3\cdot mol^{-3}\cdot s^{-1})[Fe^{2+}][dipy]^3$$

and for the reverse of the above reaction, the rate of disappearance of the complex is

$$(1.22 \times 10^{-4}\ s^{-1})[Fe(dipy)_3^{2+}]$$

What is K_s, the stability constant for the complex?

Not all reactions can be studied conveniently in both directions. When this is possible, we know that at dynamic equilibrium the rate of formation of the complex must equal the rate of decomposition, since concentrations of the various species stay fixed.

$$\text{Rate of forward reation} = \text{rate of reverse reaction}$$

$$(1.45 \times 10^{13}\ L^3\cdot mol^{-3}\cdot s^{-1})[Fe^{2+}][dipy]^3 = (1.22 \times 10^{-4}\ s^{-1})[Fe(dipy)_3^{2+}]$$

Solving,

$$K_s = \frac{[Fe(dipy)_3^{2+}]}{[Fe^{2+}][dipy]^3} = \frac{1.45 \times 10^{13}}{1.22 \times 10^{-4}} = 1.19 \times 10^{17}$$

(In the equilibrium constant equation, as opposed to rate equations, [X] is dimensionless, being the concentration relative to the standard state of 1 mol/L.)

19.33 Bicyclohexane was found to undergo two parallel first order rearrangements. At 730 K, the first order rate constant for the formation of cyclohexene was measured as $1.26 \times 10^{-4}\ s^{-1}$, and for the formation of methylcyclopentene the rate constant was $3.8 \times 10^{-5}\ s^{-1}$. What was the percentage distribution of the rearrangement products?

$$\frac{1.26 \times 10^{-4}\ s^{-1}}{(1.26 \times 10^{-4}\ s^{-1}) + (3.8 \times 10^{-5}\ s^{-1})} = 77\%\ \text{cyclohexene}$$

Hence 23% methylcyclopentene

19.34 Using calculus, derive the expressions of Table 19.1 from the corresponding rate law expressions.

The rate of a reaction is equal to $-d[A]/dt$. Therefore

zero order	first order	second order	third order
$-\frac{d[A]}{dt} = k$	$-\frac{d[A]}{dt} = k[A]$	$-\frac{d[A]}{dt} = k[A]^2$	$-\frac{d[A]}{dt} = k[A]^3$
$d[A] = -k\,dt$	$-\frac{d[A]}{[A]} = k\,dt$	$\frac{d[A]}{[A]^2} = -k\,dt$	$\frac{d[A]}{[A]^3} = -k\,dt$
$[A] - [A_0] = -kt$	$\ln[A] - \ln[A_0] = -kt$	$-\frac{1}{[A]} + \frac{1}{[A_0]} = -kt$	$\frac{1}{-2[A]^2} + \frac{1}{-2[A_0]^2} = -kt$
$[A_0] - [A] = kt$	$2.30 \log\frac{[A]}{[A_0]} = -kt$	$\frac{1}{[A]} - \frac{1}{[A_0]} = kt$	$\frac{1}{[A]^2} - \frac{1}{[A_0]^2} = 2kt$

19.35 A certain reaction, $A + B \rightarrow C$, is first order with respect to each reactant, with $k = 1.0 \times 10^{-3}\ M^{-1}\cdot s^{-1}$. Using calculus, calculate the concentration of A remaining after 100 s if the initial concentration of A was 0.100 M and that of B was 0.200 M.

$$\frac{d[C]}{dt} = k[A][B]$$

Let $[C] = x$.

then $[A] = 0.100 - x$ and $[B] = 0.200 - x$

$$\frac{dx}{(0.100 - x)(0.200 - x)} = k\,dt$$

This equation can be integrated by the method of partial fractions. Alternatively, a table of integrals may be consulted; one finds the formula

$$\frac{1}{ad - bc}\left(\ln\frac{c + dx}{a + bx} - \ln\frac{c}{a}\right) = kt$$

Substituting and solving:

$$\frac{1}{(-0.100 + 0.200)\ \text{M}}\left(\ln\frac{0.200 - x}{0.100 - x} - \ln\frac{0.200}{0.100}\right) = (1.0 \times 10^{-3}\ \text{M}^{-1}\cdot\text{s}^{-1})(100\ \text{s})$$

$$10.0 \ln\left(\frac{0.200 - x}{0.100 - x} - \ln 2.00\right) = 0.10$$

$$\ln\frac{0.200 - x}{0.100 - x} - 0.693 = 0.010$$

$$\ln\frac{0.200 - x}{0.100 - x} = 0.703$$

$$\frac{0.200 - x}{0.100 - x} = 2.02$$

$$x = 0.002\ \text{M}$$

19.2 HALF-LIFE

19.36 A first order reaction in aqueous solution was too fast to be detected by a procedure that could have followed a reaction having a half-life of at least 2.0 ns. What is the minimum value of k for this reaction?

▮

$$k = \frac{0.693}{t_{1/2}} = \frac{0.693}{2.0 \times 10^{-9}\ \text{s}} = 3.5 \times 10^{8}\ \text{s}^{-1}$$

19.37 The half-life of a first order reaction is 2.50 h. Calculate the value of the rate constant in s^{-1}.

▮

$$k = \frac{0.693}{t_{1/2}} = \frac{0.693}{2.50\ \text{h}}\left(\frac{1\ \text{h}}{3600\ \text{s}}\right) = 7.70 \times 10^{-5}\ \text{s}^{-1}$$

19.38 What is the half-life of a first order reaction for which $k = 7.1 \times 10^{-5}\ \text{s}^{-1}$?

▮

$$t_{1/2} = \frac{0.693}{k} = \frac{0.693}{7.1 \times 10^{-5}\ \text{s}^{-1}} = 9.8 \times 10^{3}\ \text{s} = 2.7\ \text{h}$$

19.39 Sucrose decomposes in acid solution into glucose and fructose according to a first order rate law, with a half-life of 3.33 h at 25 °C. What fraction of a sample of sucrose remains after 9.00 h?

▮ At any time, t, the fraction of sucrose remaining is $[\text{A}]/[\text{A}_0]$.

$$2.30 \log\frac{[\text{A}]}{[\text{A}_0]} = -kt = -\left(\frac{0.693}{t_{1/2}}\right)t = -\left(\frac{0.693}{3.33\ \text{h}}\right)(9.00\ \text{h})$$

$$\log\frac{[\text{A}]}{[\text{A}_0]} = -0.814$$

$$\frac{[\text{A}]}{[\text{A}_0]} = 0.153$$

$$\text{fraction remaining} = 0.153$$

19.40 Gaseous cyclobutane isomerizes to butadiene in a first order process which has a k value at 153 °C of $3.3 \times 10^{-4}\ \text{s}^{-1}$. How many minutes would it take for the isomerization to proceed 40% to completion at this temperature?

▮

$$t_{1/2} - \frac{0.693}{k} = \frac{0.693}{3.3 \times 10^{-4}\ \text{s}^{-1}} = 2.1 \times 10^{3}\ \text{s}$$

$$-2.303 \log\frac{C}{C_0} = -\left(\frac{0.693}{t_{1/2}}\right)t$$

40% to completion means 60% unreacted:

$$2.303 \log (0.60) = -(3.3 \times 10^{-4}\ \text{s}^{-1})t \qquad t = (1.55 \times 10^3\ \text{s})\left(\frac{1\ \text{min}}{60\ \text{s}}\right) = 26\ \text{min}$$

19.41 Azomethane, $(CH_3)_2N_2$, decomposes with a first order rate according to the equation

$$(CH_3)_2N_2(g) \longrightarrow N_2(g) + C_2H_6(g)$$

The following data were obtained for the decomposition in a 200-mL flask at 300 °C:

time, min	total pressure, torr
0	36.2
15	42.4
30	46.5
48	53.1
75	59.3

Calculate the rate constant and the half-life for this reaction.

▌ It may be seen that for each mole of azomethane which decomposes, 2 mol of gaseous products are obtained. Since in a constant volume at constant temperature the pressure of a gas is directly proportional to the number of moles of gas, the total pressure in the flask will increase as the reaction proceeds. The pressure increase is equal to the partial pressure of N_2 (or C_2H_6). Since the reaction is first order, the rate equation in terms of pressure (concentration) of N_2 is as follows (Table 19.1):

$$\log \frac{P_0 - P_{N_2}}{P_0} = -\frac{k}{2.30}t$$

A plot of the logarithm term against time, t, gives a straight line with a slope equal to $-k/2.30$. In this case, the slope is -5.67×10^{-3}/min, and therefore $k = 1.30 \times 10^{-2}$/min. The half-life is given by $0.693/k = t_{1/2}$; therefore $t_{1/2} = 53.3$ min.

19.42 The decomposition of SO_2Cl_2 to SO_2 and Cl_2 is first order. When 0.10 mol of $SO_2Cl_2(g)$ is heated at 600 K in a 1.0-L vessel, the following data are obtained:

time, h	pressure, atm	time, h	pressure, atm
0.0	4.91	4.0	7.31
1.0	5.58	8.0	8.54
2.0	6.32	16.0	9.50

(*a*) Calculate the rate constant for the decomposition of SO_2Cl_2 at 600 K. (*b*) What is the half-life of the reaction? (*c*) What would be the pressure in the vessel after 20 min?

▌

$$SO_2Cl_2 \longrightarrow SO_2 + Cl_2$$

$$P(SO_2) = P_{tot} - P_0 \qquad P(SO_2Cl_2) = P_{tot} - 2P(SO_2)$$

t, h	P_{tot}	$P(SO_2)$	$P(SO_2Cl_2)$	$\log P(SO_2Cl_2)$
0.0	4.91	0.00	4.91	0.691
1.0	5.58	0.67	4.24	0.627
2.0	6.32	1.41	3.50	0.544
4.0	7.31	2.40	2.51	0.400
8.0	8.54	3.63	1.28	0.107
16.0	9.50	4.59	0.32	−0.495

From Fig. 19.6,

$$\text{Slope} = \frac{-0.50 - 0.70}{16.0 - 0.0} = -0.075$$

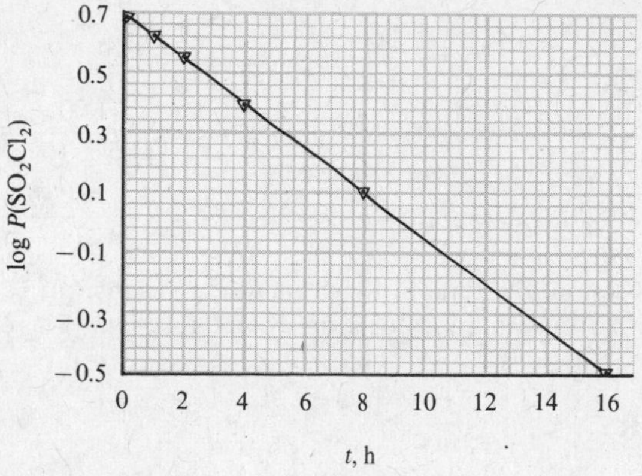

Fig. 19.6

(*a*) $k = -2.30(\text{slope}) = -2.30(-0.075\ \text{h}^{-1}) = 0.17\ \text{h}^{-1}$

(*b*) $$t_{1/2} = \frac{0.693}{k} = \frac{0.693}{0.17\ \text{h}^{-1}} = 4.1\ \text{h}$$

(*c*) $$\log \frac{P}{P_0} = -\frac{(0.17\ \text{h}^{-1})(0.33\ \text{h})}{2.30} = -0.024$$

$$P = P_0(0.945) = 4.6\ \text{atm}$$

$$P_{\text{tot}} = (4.91\ \text{atm}) + (4.91 - 4.6)\ \text{atm} = 5.2\ \text{atm}$$

After 20.0 h: $$\log \frac{P}{P_0} = -\frac{(0.17\ \text{h}^{-1})(20.0\ \text{h})}{2.30} = -1.478$$

$$\frac{P}{P_0} = 0.0332; \qquad P = 0.0332(4.91) = 0.16\ \text{atm}$$

$$P_{\text{tot}} = (4.91\ \text{atm}) + (4.91 - 0.16)\ \text{atm} = 9.66\ \text{atm}$$

19.43 For the process described in Prob. 19.14, how much time would be required for the virus to become (*a*) 50% inactivated (*b*) 75% inactivated?

▌ The method used in Prob. 19.14 cannot be used here because [A] changes appreciably over the course of the reaction.

(*a*) The time for 50% reaction is the half-life,

$$t_{1/2} = \frac{0.693}{k} = \frac{0.693}{3.3 \times 10^{-4}\ \text{s}^{-1}} = 2.1 \times 10^3\ \text{s} = 35\ \text{min}$$

(*b*) If 75% of the virus is inactivated, the fraction remaining $[\text{A}]/[\text{A}]_0$, is 0.25.

$$t = -\frac{2.303 \log \dfrac{[\text{A}]}{[\text{A}]_0}}{k} = -\frac{2.303 \log 0.25}{3.3 \times 10^{-4}\ \text{s}^{-1}} = 4.2 \times 10^3\ \text{s} = 70\ \text{min}$$

An alternate solution is to apply the half-life concept twice. Since it takes 35 min for half of the virus to become inactivated regardless of the initial concentration, the time for the virus to be reduced from 50% to 25% full strength will be another half-life. The total time for reduction to quarter strength is thus two half-lives, or 70 min.

Similarly, the total time required to reduce the initial activity to 1/8 is three half-lives; to 1/16, four half-lives; and so on. This method can be used only for first order reactions.

19.44 If in the fermentation of sugar in an enzymatic solution that is initially 0.12 M the concentration of the sugar is reduced to 0.06 M in 10 h and to 0.03 M in 20 h, what is the order of the reaction and what is the rate constant?

▌ This problem is analogous to Prob. 19.43. Since doubling the time doubles the fractional reduction of the reactant concentration, the reaction must be first order. The rate constant may be evaluated from the half-life equation:

$$k = \frac{0.693}{t_{1/2}} = \frac{0.693}{10\ \text{h}} = 6.9 \times 10^{-2}\ \text{h}^{-1} = \frac{6.9 \times 10^{-2}\ \text{h}^{-1}}{3.6 \times 10^3\ \text{s}\cdot\text{h}^{-1}} = 1.9 \times 10^{-5}\ \text{s}^{-1}$$

19.45 The rate constant for a certain second order reaction is $8.00 \times 10^{-5}\ \text{M}^{-1}\cdot\text{min}^{-1}$. How long will it take a 1.00 M solution to be reduced to 0.500 M in reactant? How long will it take from that point until the solution is 0.250 M in reactant? Explain why the term *half-life* is not used often for second order reactions.

$$\frac{1}{[A]} - \frac{1}{[A_0]} = kt$$

$$2.00 - 1.00 = (8.00 \times 10^{-5})t \qquad t = \frac{1.00}{8.00 \times 10^{-5}} = 1.25 \times 10^4 \text{ min}$$

$$4.00 - 2.00 = (8.00 \times 10^{-5})t \qquad t = \frac{2.00}{8.00 \times 10^{-5}} = 2.50 \times 10^4 \text{ min}$$

For a second order reaction, the time in which half the reactant will be used up is dependent on the initial concentration, in contrast to the situation for a first order reaction. Therefore, the concept of half-life is much less useful for second order reactions.

19.3 COLLISION THEORY

19.46 The number of collisions of molecules per unit volume per unit time is given by $Z = 2(n')^2\sigma^2\sqrt{\pi RT/M}$. Verify that Z carries the proper units (e.g., $\text{cm}^{-3}\cdot\text{s}^{-1}$).

Let n' = number of molecules per cm^3

$$Z \propto (n')^2\sigma^2\sqrt{RT/M}$$

$$\left(\frac{1}{\text{cm}^3}\right)^2 (\text{cm})^2 \sqrt{\left(\frac{\text{erg}}{\text{K}\cdot\text{mol}}\right)(\text{K})\bigg/\frac{\text{g}}{\text{mol}}} = \frac{1}{\text{cm}^4}\sqrt{\frac{\text{erg}}{\text{g}}} = \frac{1}{\text{cm}^4}\sqrt{\left(\frac{\text{cm}}{\text{s}}\right)^2} = \text{cm}^{-3}\cdot\text{s}^{-1}$$

19.47 Calculate the number of collisions per second per cm^3 at 0 °C in hydrogen gas at a pressure of 1.0 atm.

To apply the formula of Prob. 19.46, it is first necessary to determine the number of molecules per cubic centimeter:

$$n' = Nn = N\frac{PV}{RT} = \frac{(6.02 \times 10^{23} \text{ molecules/mol})(1.00 \text{ atm})(0.00100 \text{ L})}{(0.0821 \text{ L}\cdot\text{atm/mol}\cdot\text{K})(273 \text{ K})} = 2.69 \times 10^{19} \text{ molecules}$$

Then σ for hydrogen is calculated from the van der Waals constant, $b = 26.6 \text{ cm}^3/\text{mol}$ (from Table 13.1).

$$V_M = \frac{b}{4} = \frac{26.6 \text{ cm}^3/\text{mol}}{4} = 6.65 \text{ cm}^3/\text{mol}$$

$$\frac{\text{Volume}}{\text{Molecule}} = \frac{V_M}{N} = \frac{6.65 \text{ cm}^3/\text{mol}}{6.02 \times 10^{23} \text{ molecules/mol}} = \frac{1.10 \times 10^{-23} \text{ cm}^3}{\text{molecule}}$$

Since $V = \frac{4}{3}\pi r^3$ for any sphere, and the radius of the molecule is $\sigma/2$,

$$V = \frac{4}{3}\pi r^3 = \frac{4}{3}\pi\left(\frac{\sigma}{2}\right)^3 = \frac{\pi\sigma^3}{6} \qquad \sigma = \sqrt[3]{\frac{6V}{\pi}} = \sqrt[3]{\frac{6(1.10 \times 10^{-23}) \text{ cm}^3}{3.14}} = 2.76 \times 10^{-8} \text{ cm}$$

$$Z = 2(2.69 \times 10^{19})^2(2.76 \times 10^{-8})^2\sqrt{\frac{3.14(8.31 \times 10^7)(273)}{2.01}} = 2.08 \times 10^{29} \text{ collisions cm}^{-3}\cdot\text{s}^{-1}$$

19.48 Three sets of railroad tracks are to accommodate trains which are 14 feet wide. What must be the minimum distance between the *centers* of the two outermost tracks? Explain why the effective collision diameter of a molecule is 2σ.

28 ft. The molecule must have clearance of σ on each side.

19.49 A gaseous molecule A can undergo a unimolecular decomposition into C if it acquires a critical amount of energy. Such a suitably energized molecule of A, designated as A*, can be formed by a collision between two ordinary A molecules. Competing with the unimolecular decomposition of A* into C is the bimolecular deactivation of A* by collision with an ordinary A molecule. ***(a)*** Write a balanced equation and the rate law for each of the above steps. ***(b)*** Making the assumption that A* disappears by all processes at the same rate at which it is formed, what would be the rate law for the formation of C, in terms of [A] and constants of individual steps only? ***(c)*** What limiting order in A would the formation of C exhibit at low pressures of A, and what limiting order at high pressures?

(a) (*1*) Activation: $A + A \longrightarrow A^* + A \qquad \frac{\Delta[A^*]}{\Delta t} = k_1[A]^2$

(*2*) Deactivation: $A^* + A \longrightarrow A + A \qquad -\frac{\Delta[A^*]}{\Delta t} = k_2[A^*][A]$

(*3*) Reaction: $A^* \longrightarrow C \qquad -\frac{\Delta[A^*]}{\Delta t} = k_3[A^*]$

(*b*) Note that A* appears in each of the three separate steps. The net change in [A*] can be evaluated by summing over all three steps.

$$\left(\frac{\Delta[\mathrm{A}^*]}{\Delta t}\right)_{\text{net}} = k_1[\mathrm{A}]^2 - k_2[\mathrm{A}^*][\mathrm{A}] - k_3[\mathrm{A}^*]$$

From the assumption that the net rate of change of [A*] is 0, the right-hand side can be equated to 0.

$$k_1[\mathrm{A}]^2 - k_2[\mathrm{A}^*][\mathrm{A}] - k_3[\mathrm{A}^*] = 0 \qquad \text{or} \qquad [\mathrm{A}^*] = \frac{k_1[\mathrm{A}]^2}{k_3 + k_2[\mathrm{A}]}$$

Inserting this value into the rate law for step (*3*), and recognizing that $-\Delta[\mathrm{A}^*] = \Delta[\mathrm{C}]$ for this step,

$$\frac{\Delta[\mathrm{C}]}{\Delta t} = \frac{k_3 k_1[\mathrm{A}]^2}{k_3 + k_2[\mathrm{A}]} \tag{4}$$

Thus, the formation of C follows a complex kinetic formulation, not represented by a simple order.

(*c*) At very low pressures (i.e., small [A]), the second term in the denominator on the right side of (*4*) becomes negligible in comparison with the first.

$$\text{Low-pressure limit:} \qquad \frac{\Delta[\mathrm{C}]}{\Delta t} = \frac{k_3 k_1[\mathrm{A}]^2}{k_3} = k_1[\mathrm{A}]^2$$

The order appears to be second, with k_1 the second order rate constant.

At very high pressures, on the other hand, the first term of the denominator becomes negligible in comparison with the second.

$$\text{High-pressure limit:} \qquad \frac{\Delta[\mathrm{C}]}{\Delta t} = \frac{k_3 k_1[\mathrm{A}]^2}{k_2[\mathrm{A}]} = \frac{k_3 k_1}{k_2}[\mathrm{A}]$$

Now the order appears to be first, with k_3k_1/k_2 as the apparent rate constant.

This problem illustrates the concept of the rate-limiting step. In the familiar example of a bucket brigade, in which a water bucket passes sequentially from one person to the next until it finally reaches the reservoir that is to be filled, the overall rate of transfer of water cannot be greater than the rate of the slowest step. Here the activation step (*1*) becomes the slowest step when [A] is low and the reaction step (*3*) becomes the slowest step when [A] is high. Step (*1*), depending on the square of [A], is more sensitive to pressure than step (*3*).

19.50 In terms of reaction kinetics, explain why each of the following speeds up a chemical reaction: (*a*) catalyst (*b*) increase in temperature (*c*) increase in concentration.

▌ (*a*) A catalyst lowers the activation energy; therefore more molecules in the sample have sufficient energy to react. (*b*) A temperature increase causes a greater fraction of the molecules to have an energy at least equal to the activation energy. (*c*) A concentration increase causes a greater number of collisions per second, therefore a greater number of effective collisions per second.

19.51 The reaction rate of a normal reaction approximately doubles with an increase of 10 °C. Explain in terms of collision theory why this doubling occurs.

▌ The number of molecules exceeding the activation energy approximately doubles by this change in temperature (see Fig. 19.7).

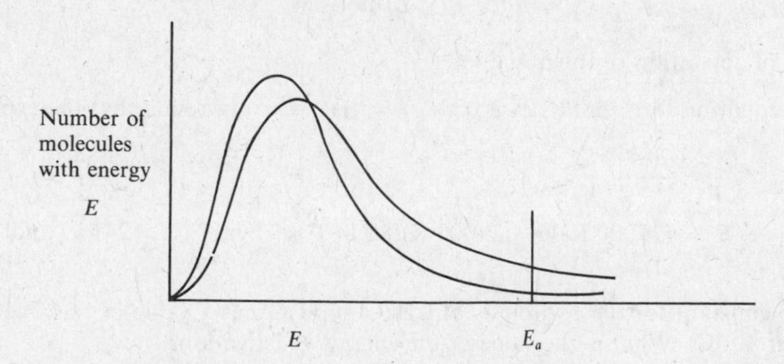

Fig. 19.7

19.52 According to Fig. 19.8, the enthalpy change of the reaction $A + B \rightarrow M + N$ is represented by what value?

y

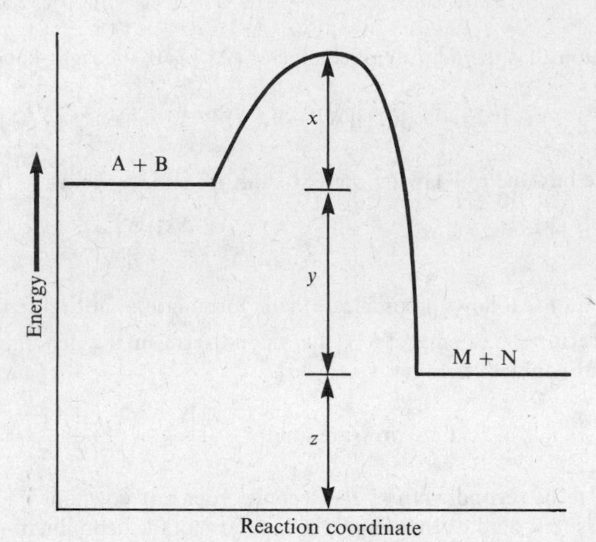

Fig. 19.8

19.53 According to Fig. 19.8, the activation energy of the reaction $M + N \rightarrow A + B$ is represented by what value?

$x + y$

19.54 Refer to Fig. 19.8. In the presence of a catalyst, the activation energy for the conversion $A + B \rightarrow M + N$ is somewhat less than x. Explain.

The catalyst speeds the reaction, which means that more molecules have sufficient energy to react. Thus the activation energy must have been lowered, since the molecules are no more energetic.

19.55 List the ways an activated complex differs from an ordinary molecule.

An activated complex by definition is unstable and has no independent existence. ΔH_f° is probably positive. The bond orders are apt to be unusual, and the bond lengths and angles are likely to differ from those of analogous stable molecules.

19.56 Explain why transition states cannot be isolated as independent chemical species.

Transition states have maximum energies. No matter how their bonds vibrate, the resulting molecules will have lower energies and therefore be more stable.

19.57 The decomposition of N_2O into N_2 and O in the presence of gaseous argon follows second order kinetics, with

$$k = (5.0 \times 10^{11}\ \mathrm{L \cdot mol^{-1} \cdot s^{-1}})e^{-29\,000\ \mathrm{K}/T}$$

What is the energy of activation of this reaction?

Comparing the equation for k in this case with $k = Ae^{-E_a/RT}$, we note that the exponent of e is $-E_a/RT$.

$$\frac{E_a}{RT} = \frac{29\,000\ \mathrm{K}}{T}$$

$$E_a = (29\,000\ \mathrm{K})R = (29\,000\ \mathrm{K})(8.314\ \mathrm{J \cdot K^{-1} \cdot mol^{-1}}) = 241\ \mathrm{kJ \cdot mol^{-1}}$$

19.58 The first order rate constant for the hydrolysis of CH_3Cl in H_2O has a value of $3.32 \times 10^{-10}\ \mathrm{s^{-1}}$ at 25° C and $3.13 \times 10^{-9}\ \mathrm{s^{-1}}$ at 40 °C. What is the value of the energy of activation?

$$E_a = 2.303\,R\left(\frac{T_1T_2}{T_2 - T_1}\right)\left(\log\frac{k_2}{k_1}\right)$$

$$= (2.303)(8.314\ \text{J}\cdot\text{K}^{-1}\cdot\text{mol}^{-1})\left(\frac{298 \times 313}{313 - 298}\ \text{K}\right)\left(\log\frac{3.13 \times 10^{-9}}{3.32 \times 10^{-10}}\right)$$

$$= (119\ \text{kJ}\cdot\text{mol}^{-1})(\log 9.4) = 115\ \text{kJ}\cdot\text{mol}^{-1}$$

19.59 A second order reaction whose rate constant at 800 °C was found to be $5.0 \times 10^{-3}\ \text{L}\cdot\text{mol}^{-1}\cdot\text{s}^{-1}$ has an activation energy of $45\ \text{kJ}\cdot\text{mol}^{-1}$. What is the value of the rate constant at 875 °C?

$$\log\frac{k_2}{k_1} = \frac{E_a(T_2 - T_1)}{2.303\,RT_1T_2} = \frac{(4.5 \times 10^4\ \text{J}\cdot\text{mol}^{-1})[(875 - 800)\ \text{K}]}{(2.303)(8.314\ \text{J}\cdot\text{K}^{-1}\cdot\text{mol}^{-1})(1073\ \text{K})(1148\ \text{K})} = 0.1431$$

$$\frac{k_2}{k_1} = 1.39$$

$$k_2 = (1.39)(5.0 \times 10^{-3}\ \text{L}\cdot\text{mol}^{-1}\cdot\text{s}^{-1}) = 7.0 \times 10^{-3}\ \text{L}\cdot\text{mol}^{-1}\cdot\text{s}^{-1}$$

19.60 The rate constant for the first order decomposition of ethylene oxide into CH_4 and CO may be described by

$$\log k(\text{in s}^{-1}) = 14.34 - \frac{1.25 \times 10^4\ \text{K}}{T}$$

(*a*) What is the energy of activation for this reaction? (*b*) What is the value of k at 670 K?

$$k = Ae^{-E_a/RT}$$

(*a*) $\log k = \log A - \dfrac{E_a}{2.303RT} \qquad \dfrac{E_a}{2.303RT} = \dfrac{1.25 \times 10^4\ \text{K}}{T}$

$$E_a = (1.25 \times 10^4\ \text{K})(2.303)(8.31\ \text{J/mol}\cdot\text{K}) = 239 \times 10^3\ \text{J/mol} = 239\ \text{kJ/mol}$$

(*b*) $\log k = 14.34 - \dfrac{1.25 \times 10^4\ \text{K}}{670\ \text{K}} = -4.32 \qquad k = 4.8 \times 10^{-5}\ \text{s}^{-1}$

19.61 The first order gaseous decomposition of N_2O_4 into NO_2 has a k value of $4.5 \times 10^3\ \text{s}^{-1}$ at 1 °C and an energy of activation of $58\ \text{kJ}\cdot\text{mol}^{-1}$. At what temperature would k be $1.00 \times 10^4\ \text{s}^{-1}$?

$$\log k = \log A - \frac{E_a}{2.303RT}$$

$$\log(4.5 \times 10^3) = \log A - \frac{58 \times 10^3\ \text{J/mol}}{2.303(8.31\ \text{J/mol}\cdot\text{K})(274\ \text{K})}$$

$$\log A = 14.71$$

$$\log(1.00 \times 10^4) = 14.71 - \frac{58 \times 10^3}{2.303(8.31)T}$$

$$4.00 = 14.71 - \frac{3031}{T}$$

$$T = 283\ \text{K}$$

19.62 Biochemists often define Q_{10} for a reaction as the ratio of the rate constant at 37 °C to the rate constant at 27 °C. What must be the energy of activation for a reaction that has a Q_{10} of 2.5?

$$\log k = \log A - \frac{E_a}{2.303(8.31\ \text{J/mol}\cdot\text{K})T}$$

$$\log k_{37} - \log k_{27} = \log\frac{k_{37}}{k_{27}} = -\frac{E_a}{2.303R}\left(\frac{1}{310} - \frac{1}{300}\right)$$

$$\log 2.5 = +\frac{E_a}{2.303(8.31)}\left(\frac{1}{300} - \frac{1}{310}\right) \qquad E_a = 71 \times 10^3\ \text{J/mol} = 71\ \text{kJ/mol}$$

19.63 In gaseous reactions important for the understanding of the upper atmosphere, H_2O and O react bimolecularly to form two OH radicals. ΔH for this reaction is 72 kJ at 500 K and E_a is $77\ \text{kJ}\cdot\text{mol}^{-1}$. Estimate E_a for the bimolecular recombination of two OH radicals to form H_2O and O.

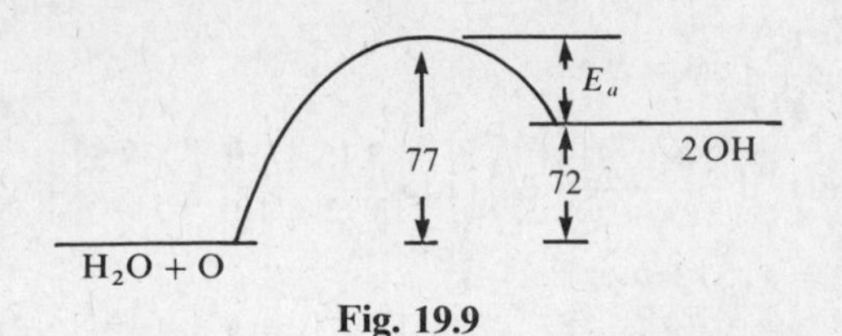

Fig. 19.9

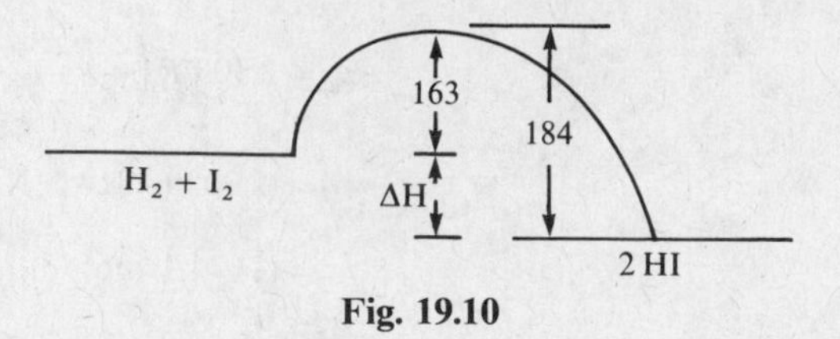

Fig. 19.10

See Fig. 19.9.

$$E_{a(\text{rev})} = 77 - 72 = 5\ \text{kJ}\cdot\text{mol}^{-1}$$

19.64 H_2 and I_2 react bimolecularly in the gas phase to form HI, and HI in turn decomposes bimolecularly into H_2 and I_2. The energies of activation for these two reactions were observed to be 163 and 184 kJ·mol^{-1}, respectively, over the same temperature ranges near 100 °C. What do you predict from these data for ΔH of the gaseous reaction $H_2 + I_2 \rightleftharpoons 2HI$ at 100 °C?

See Fig. 19.10.

$$\Delta H = 163 - 184 = -21\ \text{kJ}$$

19.65 From the following data, estimate the activation energy for the reaction

$$H_2 + I_2 \longrightarrow 2\ \text{HI}$$

T, K	$1/T$, K^{-1}	log *k*
769	1.3×10^{-3}	2.9
667	1.5×10^{-3}	1.1

$$\text{Slope} = \frac{(2.9 - 1.1)}{(1.3 - 1.5) \times 10^{-3}\ \text{K}^{-1}} = -9 \times 10^3\ \text{K}$$

$$E_a = -2.30R(\text{slope}) = -2.30(1.99\ \text{cal/mol}\cdot\text{K})(-9 \times 10^3\ \text{K}) = 4 \times 10^4\ \text{cal/mol} = 40\ \text{kcal/mol}$$

19.66 Distinguish between an unstable intermediate and a transition state or activated complex.

An unstable intermediate is an actual chemical species (which perhaps can be stabilized under different reaction conditions). It has the normal bond orders for its atoms. It represents a minimum on the potential energy curve (Fig. 19.11), albeit a small minimum. The activated complex is the postulated species which has maximum energy during the conversion from reactants to products. No matter which way the bond lengths and strengths vary, stabilization results. The bond orders of the atoms of an activated complex are sometimes unusual.

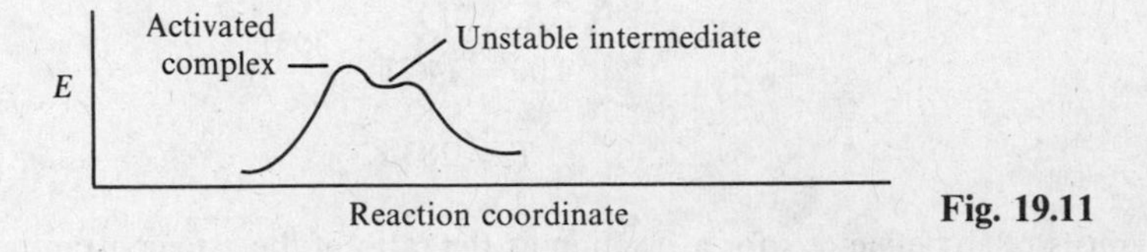

Fig. 19.11

19.67 Calculate the activation energy for a reaction which doubles in rate when the temperature is raised from 18 to 28 °C.

With energies measured in cal,

$$\log k = \log A - \frac{E_a}{2.30R}\left(\frac{1}{T}\right) \qquad \log k_1 - \log k_2 = -\frac{E_a}{2.30R}\left(\frac{1}{T_1} - \frac{1}{T_2}\right)$$

$$\log\frac{k_1}{k_2} = \log\tfrac{1}{2} = -\frac{E_a}{2.30(1.987)}\left(\frac{1}{291} - \frac{1}{301}\right) = -2.50 \times 10^{-5} E_a$$

$$E_a = 1.20 \times 10^4\ \text{cal/mol} = 12.0\ \text{kcal/mol}$$

19.68 The activation energy of the reaction A + B → products is 24.6 kcal/mol. At 40 °C the products are formed at the rate of 0.133 M/min. What will be the rate of product formation at 80 °C?

$$\log k_{40} - \log k_{80} = -\frac{24\,600 \text{ cal/mol}}{2.30(1.987 \text{ cal/mol·K})}\left[\left(\frac{1}{313} - \frac{1}{353}\right)\text{K}^{-1}\right] = -1.94$$

$$\frac{k_{40}}{k_{80}} = 0.0115 \qquad k_{80} = \frac{0.133 \text{ M/min}}{0.0115} = 11.6 \text{ M/min}$$

19.69 The trans → cis isomerization of 1,2-dichloroethylene proceeds with an energy of activation of 55.3 kcal·mol^{-1}. ΔH associated with the reaction is 1.0 kcal. What do you predict is the value of E_a for the reverse cis → trans isomerization?

$$E_{a(\text{rev})} = E_{a(\text{for})} - \Delta H = 55.3 - 1.0 = 54.3 \text{ kcal·mol}^{-1}$$

19.4 REACTION MECHANISMS

19.70 What is the rate law for the single-step reaction A + B → 2C?

Since this is a single-step reaction, the order with respect to each reactant is equal to the molecularity of the reactant.

$$\text{Rate} = k[\text{A}][\text{B}]$$

19.71 If a rate law has the form: $\text{Rate} = k[\text{A}][\text{B}]^{3/2}$ can the reaction be an elementary process? Explain.

No. An elementary process would have a rate law with orders equal to its molecularities, which are necessarily integers.

19.72 The initial rate of the reaction

$$\text{A} + \text{B} \longrightarrow \text{C} + \text{D}$$

is doubled if the initial concentration of A is doubled, but is quadrupled if the initial concentration of B is doubled. **(*a*)** What is the order with respect to each of the reactants? **(*b*)** What is the overall order of the reaction? **(*c*)** Could this possibly be a single-step reaction? Explain.

(*a*) The rate is proportional to the concentration of A, and so is first order with respect to A. It is proportional to the *square* of the concentration of B, and so is second order with respect to B. **(*b*)** The reaction is third order (1 + 2) overall. **(*c*)** The reaction could not be a single-step reaction, in which the molecularity is equal to the order. For a single-step reaction having this equation, the rate would be first order with respect to each reactant.

19.73 Write the equation which represents the rate law for each of the following elementary processes:

$$\text{A} + \text{B} \longrightarrow \text{C} + \text{D}$$
$$2\text{A} \longrightarrow \text{E} + \text{F}$$

The first elementary process is first order with respect to A and with respect to B, and therefore second order overall:

$$\text{Rate} = k[\text{A}][\text{B}]$$

Although the second elementary process involves only one reactant, A, it a bimolecular process, and therefore is second order:

$$\text{Rate} = k[\text{A}]^2$$

19.74 What is meant by an "accepted" mechanism? Is there any such thing as a proved mechanism?

An accepted mechanism is one which fits all the data available. No mechanism can be proved, since it is always possible that further data will necessitate a change.

19.75 Explain why a catalyst which could accelerate the reactions

$$2NO_2 + 4CO \longrightarrow N_2 + 4CO_2$$
$$2NO + 2CO \longrightarrow N_2 + 2CO_2$$

would be of great value.

Such a catalyst would reduce pollution from both carbon monoxide and oxides of nitrogen, producing harmless CO_2 and N_2.

19.76 A possible mechanism for the reaction $2H_2 + 2NO \longrightarrow N_2 + 2H_2O$ is

$$2NO \rightleftharpoons N_2O_2$$
$$N_2O_2 + H_2 \longrightarrow N_2O + H_2O$$
$$N_2O + H_2 \longrightarrow N_2 + H_2O$$

If the second step is rate determining, what is the rate law for this reaction?

▮ $$\text{Rate} = k_2[N_2O_2][H_2] \qquad K_1 = \frac{[N_2O_2]}{[NO]^2} \qquad \text{hence } [N_2O_2] = K_1[NO]^2$$
$$\text{Rate} = k_2K_1[NO]^2[H_2] = k[NO]^2[H_2]$$

19.77 For the reaction $2NO + Br_2 \rightarrow 2NOBr$, the following mechanism has been suggested:

$$NO + Br_2 \longrightarrow NOBr_2 \quad \text{(fast)}$$
$$NOBr_2 + NO \longrightarrow 2NOBr \quad \text{(slow)}$$

Determine a rate law consistent with this mechanism.

▮ $$K_1 = \frac{[NOBr_2]}{[NO][Br_2]} \qquad \text{hence } [NOBr_2] = K_1[NO][Br_2]$$
$$\text{Rate} = k_2[NOBr_2][NO] = k_2(K_1[NO][Br_2])[NO] = k[NO]^2[Br_2]$$

19.78 For the hydrolysis of methyl formate, $HCOOCH_3$, in acid solutions,

$$\text{Rate} = k[HCOOCH_3][H^+]$$

The balanced equation is: $HCOOCH_3 + H_2O \rightarrow HCOOH + CH_3OH$. Why does $[H^+]$ appear in the rate law, when it does not appear in the balanced equation?

▮ H^+ is a catalyst for the reaction. It is actually a reactant in an early intermediate stage of the reaction and is then released back into the solution at a later stage.

19.79 Deduce rate law expressions for the conversion of H_2 and I_2 to HI at 400 °C corresponding to each of the following mechanisms:

(*a*) $H_2 + I_2 \longrightarrow 2HI$ (one step)

(*b*) $I_2 \rightleftharpoons 2I$
$2I + H_2 \longrightarrow 2HI$ (slow)

(*c*) $I_2 \rightleftharpoons 2I$
$I + H_2 \rightleftharpoons IH_2$
$IH_2 + I \longrightarrow 2HI$ (slow)

(*d*) Can the observed rate law expression rate $= k[H_2][I_2]$ distinguish among these mechanisms? (*e*) If it is known that ultraviolet light causes the reaction of H_2 and I_2 to proceed at 200 °C with the same rate law expression, which of these mechanisms becomes most improbable? Are any of these mechanisms proved?

▮ (*a*) $\text{Rate} = k[H_2][I_2]$

(*b*) $K_1 = \dfrac{[I]^2}{[I_2]} \qquad \text{Rate} = k_2[I]^2[H_2] = k_2K_1[I_2][H_2] = k[H_2][I_2]$

(*c*) $K_1 = [I]^2/[I_2] \qquad K_2 = [IH_2]/[I][H_2]$
$\text{Rate} = k_3[IH_2][I] = k_3(K_2[I][H_2])[I] = k_3K_2[H_2](K_1[I_2]) = k[H_2][I_2]$

(*d*) All these mechanisms suggest the same rate law expression; hence the rate law expression cannot distinguish among them.
(*e*) Ultraviolet light tends to produce atoms from molecules. Hence more I atoms are expected, even at 200 °C, in the presence of ultraviolet light. These free atoms would not be expected to affect the reaction if mechanism (*a*) were the correct mechanism, and hence these new data suggest that mechanism (*a*) is incorrect.

19.80 Write a stoichiometric equation for the reaction whose mechanism is detailed below. Determine the value of the equilibrium constant for the first step. Write a rate law equation for the overall reaction in terms of its initial reactants.

$$A_2 \rightleftharpoons 2A \qquad k_1 = 10^{10}\ \text{s}^{-1} \quad \text{(forward)}$$
$$k_{-1} = 10^{10}\ \text{M}^{-1}\,\text{s}^{-1} \quad \text{(reverse)}$$
$$A + C \longrightarrow AC \qquad k_2 = 10^{-4}\ \text{M}^{-1}\,\text{s}^{-1}$$

▌ The second step is the slow step, as seen by the values of the rate constants. The overall reaction, obtained by doubling the second step reaction and adding the first step, is

$$A_2 + 2C \longrightarrow 2AC$$

The equilibrium constant expression for the first step is used to calculate an expression for [A]:

$$K_1 = \frac{[A]^2}{[A_2]} = \frac{k_1}{k_{-1}} = \frac{10^{10}}{10^{10}} = 1 \qquad [A] = \sqrt{K_1[A_2]}$$

$$\text{Rate} = k_2[A][C] = k_2\sqrt{K_1[A_2]}\,[C] = k_2[C][A_2]^{1/2} \qquad (K_1 = 1)$$

19.81 For the reaction $A + B \rightarrow C + D$, the rate law equation is $\text{Rate} = k_1[A] + k_2[A][B]$, where k_1 and k_2 represent two different constants. The products of this reaction are formed by two different mechanisms. What may be said about the relative magnitudes of the rates of the two individual mechanisms?

▌ The rates of formation of products must be of the same order of magnitude. If they were very different, the rate of the overall reaction would reflect only the faster mechanism.

19.82 Write a rate law expression for the overall reaction between A and B to yield C and D by the two mechanisms proposed below. What will be the initial rate of conversion of A and B in a solution containing 0.10 M of each? At what concentration(s) of A and/or B will the inherent rates be equal?

I. $A + B \longrightarrow AB^* \longrightarrow C + D \qquad k_1' = 1 \times 10^{-5}\ M^{-1}\cdot s^{-1}$

II. $\begin{cases} A \longrightarrow A^* \longrightarrow E & k_1 = 1 \times 10^{-4}\ s^{-1} \\ E + B \longrightarrow C + D & k_2 = 1 \times 10^{10}\ M^{-1}\cdot s^{-1} \end{cases}$

▌

$$\text{Rate}_I = k_1'[A][B] = (1 \times 10^{-5}\ M^{-1}\cdot s^{-1})(0.10\ M)^2 = 1 \times 10^{-7}\ M\cdot s^{-1}$$

$$\text{Rate}_{II} = k_1[A] = (1 \times 10^{-4}\ s^{-1})(0.10\ M) = 1 \times 10^{-5}\ M\cdot s^{-1}$$

$$k_1'[A][B] = k_1[A]$$

$$[B] = \frac{k_1}{k_1'} = \frac{(1 \times 10^{-4}\ s^{-1})}{(1 \times 10^{-5}\ M^{-1}\cdot s^{-1})} = 10\ M$$

19.83 *trans*-1,2-Dideuterocyclopropane (A) undergoes a first order decomposition. The observed rate constant at a certain temperature, measured in terms of the disappearance of A, was $1.52 \times 10^{-4}\ s^{-1}$. Analysis of the reaction products showed that the reaction followed two parallel paths, one leading to dideuteropropene (B) and the other to *cis*-1,2-dideuterocyclopropane (C). B was found to constitute 11.2% of the reaction product, independent of the extent of reaction. What is the order of the reaction for each of the paths, and what is the value of the rate constant for the formation of each of the products?

▌ Since the percent distribution of B and C is always the same, B and C must be formed at a fixed ratio of rates. Since A disappears by a first order process, B and C must each be formed by a first order process. We can then write

$$\frac{\Delta[B]}{\Delta t} = k_B[A] \qquad \frac{\Delta[C]}{\Delta t} = k_C[A]$$

We now want an equation to represent the rate of formation of all products—that is, B and C combined—which should equal the rate of disappearance of A.

$$\frac{\Delta([B]+[C])}{\Delta t} = \frac{\Delta[B]}{\Delta t} + \frac{\Delta[C]}{\Delta t} = (k_B + k_C)[A] = k_A[A] \qquad \text{or} \qquad k_A = k_B + k_C$$

In other words, the rate constant for the disappearance of A is equal to the sum of the rate constants for the formation of B and of C. Also, since B is formed at a rate equal to 11.2% of the rate of disappearance of A,

$$\frac{\Delta[A]}{\Delta t} = k_B[B] = (0.112)\left(-\frac{\Delta[A]}{\Delta t}\right) = (0.112)k_A[A]$$

or $$k_B = (0.112)k_A = (0.112)(1.52 \times 10^{-4}\ s^{-1}) = 1.7 \times 10^{-5}\ s^{-1}$$

Then $$k_C = k_A - k_B = 1.52 \times 10^{-4} - 1.7 \times 10^{-5} = 1.35 \times 10^{-4}\ s^{-1}$$

19.84 In a certain polluted atmosphere containing O_3 at a steady-state concentration of 2.0×10^{-8} mol/L, the hourly production of O_3 by all sources was estimated as 7.2×10^{-15} mol/L. If the only mechanism for destruction of O_3 is the second order reaction $2O_3 \rightarrow 3O_2$, calculate the rate constant for the destruction reaction, defined by the rate law for $-\Delta[O_3]/\Delta t$.

▮ At the steady state, the rate of destruction of O_3 must equal the rate of its generation, 7.2×10^{-15} mol·L^{-1}·h^{-1}. From the second order rate law,

$$k = -\frac{1}{[O_3]^2}\frac{\Delta[O_3]}{\Delta t} = \frac{1}{(2.0 \times 10^{-8}\ \text{mol/L})^2}\ \frac{7.2 \times 10^{-15}\ \text{mol}\cdot\text{L}^{-1}\cdot\text{h}^{-1}}{3.6 \times 10^3\ \text{s}\cdot\text{h}^{-1}} = 5 \times 10^{-3}\ \text{L}\cdot\text{mol}^{-1}\cdot\text{s}^{-1}$$

19.85 Predict the form of the rate law for the reaction $2A + B \rightarrow$ products, if the first step is the reversible dimerization of A, followed by reaction of A_2 with B in a bimolecular rate-controlling step. Assume that the equilibrium concentration of A_2 is very small compared with [A].

▮

$$2A \overset{1}{\rightleftharpoons} A_2 \qquad K_1 = \frac{[A_2]}{[A]^2}$$

$$A_2 + B \xrightarrow{2} \text{products} \qquad \text{Rate} = k_2[A_2][B] = k_2(K_1[A]^2)[B] = k[A]^2[B]$$

Note that capital K refers to an equilibrium constant, while lower case k refers to a rate constant.

19.86 The rate law expression for the reaction $2NO + O_2 \rightarrow 2NO_2$ is

$$\text{Rate} = k[NO][O_2]$$

Write a detailed mechanism for the reaction in accord with these data. Be sure to indicate the rate determining step.

▮ One possibility: $NO + O_2 \rightarrow NO_3$ (slow)

$$NO_3 + NO \longrightarrow 2NO_2 \quad \text{(fast)}$$

19.87 For a reaction $A + B \rightarrow C + D$, the rate law is $\text{Rate} = k_1[A][B] - k_2[A]$. What can be said about the nature of this reaction from this rate law? (*Hint*: See Prob. 19.81).

▮ The reaction evidently involves at least two steps of approximately the same rate, one of which includes the reaction of A and B to produce products and the other of which involves the *production* of A and B at a rate proportional to [A] and to k_2.

19.88 For the reaction $A + B \rightarrow C + D$, the rate law is $\text{Rate} = k_1[A][B] - k_2[C][D]$. What can be said about the nature of this reaction? (*Hint*: See Prob. 19.81.)

▮ This reaction also must involve two steps of approximately the same rate. One involves the reaction of A and B, the other reaction of C and D. A possible mechanism could be a one-step reaction of A and B to produce C and D with a one-step reverse reaction in which C and D produce A and B. The rate of the forward reaction must be somewhat higher than that of the reverse reaction, or equilibrium would be established with very little product produced.

19.89 The following reaction was observed in aqueous solution: $2Cu^{2+} + 6CN^- \rightarrow 2Cu(CN)_2^- + (CN)_2$ and the rate law was found to be of the form $k[Cu^{2+}]^2[CN^-]^6$. If the first step is the rapid development of the complexation equilibrium to form the relatively unstable (with respect to reversal of the complexation step) $Cu(CN)_3^-$, what rate-limiting step could account for the observed kinetic data?

▮

$$Cu^{2+} + 3CN^- \rightleftharpoons Cu(CN)_3^-$$

$$2Cu(CN)_3^- \longrightarrow 2Cu(CN)_2^- + (CN)_2 \quad \text{(slow step)}$$

19.90 Above 500 °C, the reaction $NO_2 + CO \rightarrow CO_2 + NO$ obeys the rate law: $\text{Rate} = k[NO_2][CO]$. Below 500 °C, the rate law for this reaction is $\text{Rate} = k'[NO_2]^2$. Suggest mechanisms for each of these cases.

▮ Possible mechanisms:

Above 500 °C: $NO_2 + CO \longrightarrow CO_2 + NO$ (single step)

Below 500 °C: $2NO_2 \longrightarrow NO + NO_3$ (slow)

$NO_3 + CO \longrightarrow CO_2 + NO_2$ (fast)

19.91 For the reaction $2Fe^{2+} + H_2O_2 \rightarrow 2Fe^{3+} + 2OH^-$ the rate of formation of Fe^{3+} is given by $\text{Rate} = k[Fe^{2+}][H_2O_2]$. Suggest a mechanism for the reaction, and indicate the probable rate-determining step.

▮ A possible mechanism:

$$Fe^{2+} + H_2O_2 \longrightarrow Fe^{3+} + OH^- + OH \quad \text{(slow)}$$

$$Fe^{2+} + OH \longrightarrow Fe^{3+} + OH^- \quad \text{(fast)}$$

19.92 The conversion of the D optical isomer of gaseous

$$C_2H_5-\underset{\displaystyle CH_3}{\overset{\displaystyle I}{CH}}$$

into the L isomer in the presence of iodine vapor follows the law Rate $= kP(A)P(I_2)^{1/2}$, where A represents the D isomer. (Partial pressures are a legitimate form for expression of concentrations in rate laws.) Suggest a mechanism that could account for the fractional order.

▮ I_2 can undergo a slight dissociation into I atoms. This system easily comes to equilibrium.

$$I_2 \rightleftharpoons 2I \qquad K_p = \frac{P(I)^2}{P(I_2)}$$

The relative partial pressure of iodine atoms can be evaluated in terms of this equilibrium as

$$P(I) = K_p^{1/2} \times P(I_2)^{1/2}$$

The numerical values of the *actual* partial pressures of I and I_2 stand in this same relationship, since the actual pressure is the relative pressure times the standard-state pressure, 1 atm.

If an intermediate stage of the reaction involves the addition of an iodine atom to A, followed by a subsequent loss of the iodine atom that was initially a part of the A molecule, and if the addition of I to A is bimolecular (with a rate constant k_2) and rate-determining, then:

$$\begin{aligned}\text{Observed rate} &= \text{rate of I addition}\\ &= k_2P(A)P(I) = k_2P(A) \times K_p^{1/2} \times P(I_2)^{1/2} = (k_2K_p^{1/2})P(A)P(I_2)^{1/2}\end{aligned}$$

The numerical value of the term in parentheses, $(k_2K_p^{1/2})$, is the numerical value of the apparent rate constant for the overall $\frac{3}{2}$-order reaction.

The above mechanism is plausible and is consistent with the observations. We cannot be sure from the kinetic data alone, however, that there is not some other mechanism which is also consistent with the observations. Other types of experiments are needed to confirm a mechanism based on rate data.

19.93 The hydrolysis of $(i\text{-}C_3H_7O)_2POF$ was studied at different acidities. The apparent first order rate constant, k, at a particular temperature was found to depend on pH but not on the nature or concentration of the buffer used to regulate the pH. k was fairly constant from pH 4 to pH 7, but rose from this constant value with decreasing pH below 4 or with increasing pH above 7. What is the nature of the phenomenon responsible for this behavior?

▮ The reaction is catalyzed by both H^+ and OH^-.

19.94 It has been found that the rates of reaction of a ketone in mildly basic solution are identical for the following three reactions: ***(a)*** reaction with Br_2 leading to the substitution of an H on the ketone by a Br, ***(b)*** conversion of the D isomer of the ketone into an equimolar mixture of the D and L isomers, ***(c)*** isotopic exchange of a hydrogen atom on the carbon next to the C=O group of the ketone by a deuterium atom in the solvent. The rate of each of these reactions is equal to $k[\text{ketone}][OH^-]$ and is independent of $[Br_2]$. What can be concluded about the mechanism from these observations?

▮ The rate-determining step for all three reactions must be the preliminary reaction of the ketone with OH^-, probably leading to the conjugate base of the ketone. The conjugate base subsequently reacts very rapidly with ***(a)*** Br_2 ***(b)*** some acid in the medium or ***(c)*** the deuterated solvent.

19.95 Devise a mechanism for the following decomposition which will fit the given rate law:

$$COCl_2 \longrightarrow CO + Cl_2 \qquad \text{Rate} = k[COCl_2][Cl_2]^{1/2}$$

▮ One possible mechanism:

$$\begin{array}{llll} COCl_2 \overset{1}{\rightleftharpoons} COCl + Cl & \text{(fast)} & \text{initiation} \\ Cl + COCl_2 \overset{2}{\longrightarrow} COCl + Cl_2 & \text{(slow)} & \rbrace\ \text{chain} \\ COCl \overset{3}{\longrightarrow} CO + Cl & \text{(fast)} & \\ 2Cl \overset{4}{\rightleftharpoons} Cl_2 & \text{(fast)} & \text{termination} \end{array}$$

The key to figuring out such a mechanism is to recognize from the half power in the rate expression that a Cl atom is involved in the rate-determining step, and that there must be an equilibrium between the chlorine atoms and molecules. The overall stoichiometry is governed by the chain steps, which occur many times more often than the other steps.

19.96 The rate law expression for the reaction of H_2 and Br_2 is

$$\text{Rate} = \frac{k[H_2][Br_2]^{1/2}}{1 + [HBr]/k'[Br_2]}$$

Derive a mechanism to fit this rate law. In your derivation, assume that all the steps have rates on the same order of magnitude and that a "steady state" is established—that is, after a short time of concentration buildup, any free atoms react as rapidly as they are formed. Thus throughout most of the reaction, the concentrations of free atoms do not change with time. Show specifically that your mechanism yields a rate law expression which corresponds to the experimental rate law expression.

▮

$$Br_2 \underset{-1}{\overset{1}{\rightleftharpoons}} 2Br$$

$$Br + H_2 \underset{-2}{\overset{2}{\rightleftharpoons}} HBr + H$$

$$H + Br_2 \overset{3}{\longrightarrow} HBr + Br$$

The rate of formation of HBr is given by

$$\text{Rate} = k_2[Br][H_2] + k_3[H][Br_2] - k_{-2}[H][HBr] = k_2[Br][H_2] + [H](k_3[Br_2] - k_{-2}[HBr])$$

The change in concentration of the bromine and hydrogen atoms per unit time, derived from the equations for the elementary processes, is 0. Hence, for Br atoms:

$$2k_1[Br_2] - k_2[Br][H_2] + k_3[H][Br_2] + k_{-2}[HBr][H] - k_{-1}[Br]^2 = 0$$

and for H atoms:

$$k_2[Br][H_2] - k_3[H][Br_2] - k_{-2}[HBr][H] = 0$$

Adding these two equations yields:

$$2k_1[Br_2] - k_{-1}[Br]^2 = 0$$

$[Br] = \sqrt{2k_1/k_{-1}}[Br_2]^{1/2} = a[Br_2]^{1/2}$, where a is a new constant. Substitution of the value of [Br] into the equation for H atoms yields:

$$k_2a[Br_2]^{1/2}[H_2] - k_3[H][Br_2] - k_{-2}[H][HBr] = 0 \qquad [H] = \frac{k_2a[Br_2]^{1/2}[H_2]}{k_3[Br_2] + k_{-2}[HBr]}$$

Substitution of the values of [H] and [Br] into the rate of formation of HBr equation, above, yields

$$\text{Rate} = k_2[H_2]a[Br_2]^{1/2} + \left(\frac{k_2a[Br_2]^{1/2}[H_2]}{k_3[Br_2] + k_{-2}[HBr]}\right)(k_3[Br_2] - k_{-2}[HBr])$$

The first term and the numerator of the second term of this equation are the same. Collecting terms yields

$$\text{Rate} = k_2[H_2]a[Br_2]^{1/2}\left(1 + \frac{k_3[Br_2] - k_{-2}[HBr]}{k_3[Br_2] + k_{-2}[HBr]}\right)$$

$$= k_2[H_2]a[Br_2]^{1/2}\left(\frac{k_3[Br_2] + k_{-2}[HBr] + k_3[Br_2] - k_{-2}[HBr]}{k_3[Br_2] + k_{-2}[HBr]}\right)$$

Elimination of like terms in the numerator followed by division of both numerator and denominator by $[Br_2]$ yields

$$\text{Rate} = \frac{k_2a(2k_3)[H_2][Br_2]^{1/2}}{k_3 + k_{-2}[HBr]/[Br_2]} = \frac{2k_2a[H_2][Br_2]^{1/2}}{1 + (k_{-2}/k_3)([HBr]/[Br_2])} = \frac{k[H_2][Br_2]^{1/2}}{1 + k'[HBr]/[Br_2]}$$

Replacement of the products of constants by two new constants—k and k'—yields an equation in the exact form of the experimental rate law expression.

CHAPTER 20
Equilibrium

20.1 LE CHATELIER'S PRINCIPLE

20.1 What effect will the addition of more nitrogen have on the following equilibrium, observed in a vessel at constant volume?

$$N_2(g) + 3H_2(g) \rightleftharpoons 2NH_3(g)$$

The equilibrium must shift to the right to use up some of the added nitrogen, thus lowering its concentration and thereby reducing the stress. A new equilibrium is established. Note that the equilibrium does not shift back to the left because of the additional ammonia formed, since that was produced by the reaction itself. Note also that Le Châtelier's principle alone does not tell *how much* the equilibrium will be shifted. No matter how large the quantity of nitrogen added, some hydrogen will be present at the new equilibrium state.

In the application of Le Châtelier's principle, heat energy may be treated as one of the reactants or products of the reaction. A more complete description of the reaction between nitrogen and hydrogen includes the heat produced:

$$N_2(g) + 3H_2(g) \rightleftharpoons 2NH_3(g) + 22 \text{ kcal}$$

20.2 What is the effect of an increase in temperature on the equilibrium of Prob. 20.1, at constant pressure?

Heat must be added to raise the temperature of the system. Therefore in order to use up some of the added heat, the equilibrium must shift to the left.

20.3

$$N_2 + 3H_2 \rightleftharpoons 2NH_3 + 22 \text{ kcal}$$

What is the effect on this equilibrium of halving the volume, thus initially doubling the total pressure?

The reaction of a total of 4 mol of hydrogen and nitrogen would produce 2 mol of ammonia. In a given volume, fewer moles of gas exert less pressure; thus, as required by Le Châtelier's principle, if additional pressure is applied, the equilibrium will shift to the right, in the direction of fewer moles of gas.

20.4 What is the effect of halving the pressure by doubling the volume on the following system at 500 °C?

$$H_2(g) + I_2(g) \rightleftharpoons 2HI(g)$$

Since the equation states that 2 mol of gaseous reactant(s) would produce 2 mol of gaseous product(s), changing the total pressure will not shift the equilibrium at all.

20.5 What is the effect of reducing the volume on the system described below?

$$2C(s) + O_2(g) \rightleftharpoons 2CO(g)$$

The equation states that the reaction of 1 mol of gaseous reactant (and 2 mol of solid reactant) produces 2 mol of gaseous product. Reducing the volume affects the gases much more than it does the solid, and so this equilibrium will be shifted to the left by the increased pressure caused by the reduction in volume.

20.6 The balanced chemical equation

$$2CO + O_2 \rightleftharpoons 2CO_2$$

signifies which one(s) of the following? **(*a*)** One can add to a vessel only 2 mol of CO for each mol of O_2 added. **(*b*)** No matter how much of these two reagents are added to a vessel, only 1 mol of O_2 will react, and it will react with 2 mol of CO. **(*c*)** When they react, CO reacts with O_2 in a 2:1 mole ratio. **(*d*)** When 2 mol of CO and 1 mol of O_2 are placed in a vessel, they will react to give 2 mol of CO_2.

The only correct answer is (*c*). (If the reaction were one which went to completion, answer (*d*) would also have been correct.)

20.7 What is the effect on the following equilibrium if each of the indicated stresses is applied?

$$\tfrac{1}{2}N_2 + O_2 \rightleftharpoons NO_2 + \text{heat}$$

(*a*) increase in N_2 concentration (*b*) decrease in temperature (*c*) increase in volume (*d*) decrease in O_2 concentration (*e*) addition of a catalyst

▌ The equilibrium will shift to the (*a*) right (*b*) right (*c*) left (*d*) left (*e*) no shift will occur

20.8 In the Haber process for the production of ammonia from N_2 and H_2, a total pressure of 200 atm is used. Explain why such a high pressure is desirable.

▌ The high pressure tends to shift the equilibrium toward greater production of ammonia.

20.9 $N_2 + O_2 \rightleftharpoons 2NO$. State the effect upon the reaction equilibrium of (*a*) increased temperature (*b*) decreased pressure (*c*) higher concentration of O_2 (*d*) lower concentration of N_2 (*e*) higher concentration of NO (*f*) presence of a catalyst

▌ (*a*) $\Delta H^\circ = 2\Delta H_f^\circ(\text{NO}) = 2(+89.75\ \text{kJ})$

The reaction is endothermic, so adding heat to raise the temperature favors the forward reaction. (*b*) Since there are the same number of moles *of gases* on the two sides of the equation, neither is favored. (*c*) Forward reaction is favored by higher concentration of reactant. (*d*) Reverse reaction is favored by lower concentration of reactant. (*e*) Reverse reaction is favored by higher concentration of product. (*f*) No change. A catalyst does not affect the position of the equilibrium.

20.10 For the gaseous equilibrium at high temperature

$$PCl_5(g) \rightleftharpoons PCl_3(g) + Cl_2(g)$$

explain the effect upon the material distribution of (*a*) increased temperature (*b*) increased pressure (*c*) higher concentration of Cl_2 (*d*) higher concentration of PCl_5 (*e*) presence of a catalyst

▌ (*a*) When the temperature of a system in equilibrium is raised (by addition of heat), the equilibrium point is displaced in the direction which absorbs heat. Table 18.2 can be used to determine that the *forward* reaction as written is endothermic.

$$\Delta H^\circ = [\Delta H_f^\circ(PCl_3)] + [\Delta H_f^\circ(Cl_2)] - [\Delta H_f^\circ(PCl_5)] = -287.0 + 0 - (-374.9) = 87.9\ \text{kJ}$$

Hence increasing the temperature will cause more PCl_5 to dissociate.

(*b*) When the pressure of a system in equilibrium is increased, the equilibrium point is displaced in the direction of the smaller volume. One volume each of PCl_3 and Cl_2, a total of 2 gas volumes, form 1 volume of PCl_5. Hence a pressure increase will promote the reaction between PCl_3 and Cl_2 to form more PCl_5.

(*c*) Increasing the concentration of any component will displace the equilibrium in the direction which tends to lower the concentration of the component added. Increasing the concentration of Cl_2 will result in the consumption of more PCl_3 and the formation of more PCl_5, and this action will tend to offset the increased concentration of Cl_2.

(*d*) Increasing the concentration of PCl_5 will result in the formation of more PCl_3 and Cl_2.

(*e*) A catalyst accelerates both forward and backward reactions equally. It speeds up the approach to equilibrium, but does not favor reaction in either direction over the other.

20.11 Phosphorus pentachloride (PCl_5) dissociates according to the equation

$$PCl_5(g) \rightleftharpoons PCl_3(g) + Cl_2(g)$$

(*a*) What effect will an increase in the total pressure caused by a decrease in volume have on the equilibrium? Explain your answer. (*b*) If the reaction is endothermic, how will an increase in temperature affect the quantity of Cl_2 produced? Justify your answer.

▌ (*a*) The equilibrium will shift left, to decrease the number of moles of gas and therefore lessen the increase in pressure. (*b*) The equilibrium will shift to the right. It will shift to use up some of the heat which was added to raise the temperature. More Cl_2 will be produced.

20.12 Predict the effects upon the following reaction equilibria of increased temperature and of increased pressure.

1. $CO(g) + H_2O(g) \rightleftharpoons CO_2(g) + H_2(g)$

2. $2SO_2(g) + O_2(g) \rightleftharpoons 2SO_3(g)$

3. $N_2O_4(g) \rightleftharpoons 2NO_2(g)$

4. $H_2O(g) \rightleftharpoons H_2(g) + \frac{1}{2}O_2(g)$

5. $2O_3(g) \rightleftharpoons 3O_2(g)$
6. $CO(g) + 2H_2(g) \rightleftharpoons CH_3OH(g)$
7. $CaCO_3(s) \rightleftharpoons CaO(s) + CO_2(g)$
8. $C(s) + H_2O(g) \rightleftharpoons H_2(g) + CO(g)$
9. $4HCl(g) + O_2(g) \rightleftharpoons 2H_2O(g) + 2Cl_2(g)$
10. $C(s, diamond) \rightleftharpoons C(s, graphite)$ (This equilibrium can exist only under very special conditions.) Densities of diamond and graphite are 3.5 and 2.3 g/cm^3, respectively.

▮ See Table 20.1. Exothermic reactions will be shifted left by adding heat to increase the temperature; endothermic reactions will be shifted right. Adding pressure (by reducing the volume) causes a shift toward fewer moles of *gas*; reducing the pressure causes a shift toward more moles of *gas*. If no gases are involved in the reaction, higher pressure favors shift toward higher density solid or liquid.

TABLE 20.1

reaction type	effect of increased T	side with fewer moles of gas	effect of increased P
1. Exothermic	shift left	neither	no shift
2. Exothermic	shift left	right	shift right
3. Endothermic	shift right	right	shift right
4. Endothermic	shift right	left	shift left
5. Exothermic	shift left	left	shift left
6. Exothermic	shift left	right	shift right
7. Endothermic	shift right	left	shift left
8. Endothermic	shift right	left	shift left
9. Exothermic	shift left	right	shift right
10. Exothermic	shift left	(diamond has greater density)	shift left

20.13 What conditions would you suggest for the manufacture of ammonia by the Haber process?

$$N_2(g) + 3H_2(g) \rightleftharpoons 2NH_3(g) \qquad \Delta H = -22 \text{ kcal}$$

▮ From the sign of ΔH, we see that the forward reaction gives off heat (exothermic). Therefore the equilibrium formation of NH_3 is favored by as low a temperature as practicable. In a reaction like this, the choice of temperature requires a compromise between equilibrium considerations and rate considerations. The lower the temperature, the higher the equilibrium yield of NH_3. On the other hand, the rate at which the system will reach equilibrium is lower, the lower the temperature.

The forward reaction is accompanied by a decrease in mol, since 4 mol of initial gases yield 2 mol of NH_3. Therefore increased pressure will give a larger proportion of NH_3 in the equilibrium mixture.

A catalyst should be employed to speed up the approach to equilibrium.

20.14 A system at equilibrium is described by the equation

$$\text{Heat} + SO_2Cl_2 \rightleftharpoons SO_2 + Cl_2$$

Why does the temperature of the system increase when Cl_2 is added to the equilibrium mixture at constant volume?

▮ The Cl_2 causes a shift to the left, producing heat and thus raising the temperature.

20.15 Predict the effect on each of the following equilibria of decreasing the volume of the system:

(*a*) $\text{Heat} + MgCO_3(s) \rightleftharpoons MgO(s) + CO_2(g)$ **(*b*)** $2C(s) + O_2(g) \rightleftharpoons 2CO(g) + \text{heat}$

▮ Decreasing the volume increases the partial pressure of the *gases*. The equilibria will shift to decrease the number of mol of *gases*: **(*a*)** to the left **(*b*)** to the left.

20.16 Determine if possible what shift each of the following combinations of stresses would cause in the following equilibrium system:

$$2CO + O_2 \rightleftharpoons 2CO_2 + \text{heat}$$

(*a*) Addition of CO and removal of CO_2 at constant volume (*b*) increase in temperature and decrease in volume (*c*) addition of O_2 and decrease in volume (*d*) addition of a catalyst and decrease in temperature (*e*) addition of CO and increase in temperature at constant volume

▮ Each of the stresses is considered individually. If both stresses would cause the same direction of shift, the shift is determinable. If the two stresses would cause shifts in opposing directions, no deduction is possible since Le Chatelier's principle is not quantitative. (*a*) Right (*b*) undeterminable (*c*) right (*d*) right (*e*) undeterminable

20.2 EQUILIBRIUM CONSTANTS

20.17 Write an equilibrium constant expression for each of the following reactions. What relationship do the constants have to one another?

(*a*) $H_2(g) + I_2(g) \underset{500\,°C}{\rightleftharpoons} 2HI(g)$ (*b*) $2HI(g) \underset{500\,°C}{\rightleftharpoons} H_2(g) + I_2(g)$

▮ The equilibrium constant expressions for the two reactions are reciprocals of each other:

(*a*) $K = \dfrac{[HI]^2}{[H_2][I_2]}$ (*b*) $K = \dfrac{[H_2][I_2]}{[HI]^2}$

20.18 Write equilibrium constant expressions for the following reactions:

(*a*) $NO_2(g) + SO_2(g) \rightleftharpoons NO(g) + SO_3(g)$
(*b*) $2SO_2(g) + O_2(g) \rightleftharpoons 2SO_3(g)$
(*c*) $2SO_2(g) + O_2(g) \rightleftharpoons 2SO_3(g)$ + heat
(*d*) $Ca(HCO_3)_2(s) \rightleftharpoons CaO(s) + 2CO_2(g) + H_2O(g)$
(*e*) $CO(g) + Cl_2(g) \rightleftharpoons COCl_2(g)$

▮ (*a*) $K = \dfrac{[NO][SO_3]}{[NO_2][SO_2]}$ (*b*) and (*c*) $K = \dfrac{[SO_3]^2}{[SO_2]^2[O_2]}$

(*d*) $K = [CO_2]^2[H_2O]$ (*e*) $K = \dfrac{[COCl_2]}{[CO][Cl_2]}$

20.19 Calculate the value of the equilibrium constant for the reaction

$$A + B \rightleftharpoons C + D$$

if at equilibrium there are 1.0 mol of A, 2.0 mol of B, 6.0 mol of C, and 20 mol of D in a 1.0-L vessel.

▮
$$K = \frac{[C][D]}{[A][B]} = \frac{(6.0\text{ mol/L})(20\text{ mol/L})}{(1.0\text{ mol/L})(2.0\text{ mol/L})} = 60$$

20.20 At a certain temperature, the value of the equilibrium constant corresponding to the equation

$$N_2 + 2O_2 \rightleftharpoons 2NO_2$$

is 100. Write the equilibrium constant expressions for each of the following reactions, and calculate the value of the equilibrium constant for each:

(*a*) $2NO_2 \rightleftharpoons N_2 + 2O_2$ (*b*) $NO_2 \rightleftharpoons \frac{1}{2}N_2 + O_2$

▮ (*a*) Since the reaction is the reverse of the one stated in the exercise, the equilibrium constant is the reciprocal:

$$K = \frac{[N_2][O_2]^2}{[NO_2]^2} = 0.010$$

(*b*) Since the reaction has half the number of mol of each substance as that in part (*a*), the value of K is the square root of that in part (*a*).

$$K = \frac{[N_2]^{1/2}[O_2]}{[NO_2]} = \sqrt{0.010} = 0.10$$

20.21 Calculate the value of the equilibrium constant for the reaction below if there are present at equilibrium 5.0 mol of N_2, 7.0 mol of O_2, and 0.10 mol of NO_2 in a 1.5-L vessel at a certain temperature.

$$N_2 + 2O_2 \rightleftharpoons 2NO_2 + \text{heat}$$

If the temperature is increased, would the value of the equilibrium constant for the reaction increase, decrease, or remain unchanged?

▮ $$[N_2] = 5.0 \text{ mol}/1.5 \text{ L} = 3.3 \text{ mol/L} \qquad [O_2] = 7.0 \text{ mol}/1.5 \text{ L} = 4.7 \text{ mol/L}$$

$$[NO_2] = 0.10 \text{ mol}/1.5 \text{ L} = 0.067 \text{ mol/L}$$

$$K = \frac{[NO_2]^2}{[N_2][O_2]^2} = \frac{(0.067)^2}{(3.3)(4.7)^2} = 6.1 \times 10^{-5}$$

According to Le Chatelier's principle, an increase in temperature would cause this equilibrium to shift to the left. Thus the value of K should decrease, since the concentration of product would be less and those of the reactants would be greater.

20.22 For the reaction

$$E + F \rightleftharpoons G + H$$

one starts with 6.0 mol of E and 7.0 mol of F in a 1.0-L vessel. When equilibrium is attained, 4.5 mol of G is formed. Calculate the value of the equilibrium constant for the reaction.

▮ As shown by the balanced chemical equation, if 4.5 mol of G has been produced, then 4.5 mol of H has also been produced and 4.5 mol each of E and F have been consumed. The quantities of reactants present at equilibrium are equal to those initially present minus what has been used up. The quantities of products present at equilibrium are equal to those initially present plus what has been produced. It is often helpful to tabulate these quantities as follows:

	initially present	**consumed by the reaction**	**produced by the reaction**	**present at equilibrium**
E	6.0	4.5		1.5
F	7.0	4.5		2.5
G	0.0		4.5	4.5
H	0.0		4.5	4.5

The concentrations in the two middle columns of such a table always have values related by the balanced chemical equation.

$$K = \frac{[G][H]}{[E][F]} = \frac{(4.5)(4.5)}{(1.5)(2.5)} = 5.4$$

20.23 Determine the value of the equilibrium constant for the reaction

$$A + B \rightleftharpoons 2C$$

if 1.0 mol of A, 1.4 mol of B, and 0.50 mol of C are placed in a 1.0-L vessel and allowed to come to equilibrium. The final concentration of C is 0.75 mol/L.

▮ $$K = \frac{[C]^2}{[A][B]}$$

The sequence of construction of the following table is shown by the circled numbers.

	initial	**produced**	**used up**	**equilibrium**
[A]	1.0 ①		0.125 ④	0.875 ⑤
[B]	1.4 ①		0.125 ④	1.275 ⑤
[C]	0.50 ①	0.25 ③		0.75 ②

$$K = \frac{(0.75)^2}{(0.875)(1.275)} = 0.50$$

20.24 At a certain temperature 1.00 mol of $PCl_3(g)$ and 2.00 mol of $Cl_2(g)$ were placed in a 3.00-L container. When equilibrium was established, only 0.700 mol of PCl_3 remained. Calculate the value of the equilibrium constant for the reaction

$$PCl_3 + Cl_2 \rightleftharpoons PCl_5(g)$$

	initial	produced	used up	equilibrium
$[PCl_3]$	0.333		0.100	0.233
$[Cl_2]$	0.667		0.100	0.567
$[PCl_5]$	0.000	0.100		0.100

$$K = \frac{[PCl_5]}{[PCl_3][Cl_2]} = \frac{(0.100)}{(0.233)(0.567)} = 0.757$$

20.25 Determine the value of the equilibrium constant for the reaction

$$A + 2B \rightleftharpoons 2C$$

if 1.0 mol of A and 1.5 mol of B are placed in a 2.0-L vessel and allowed to come to equilibrium. The equilibrium concentration of C is 0.35 mol/L.

$$K = \frac{[C]^2}{[A][B]^2}$$

The sequence of construction of the following table is shown by the circled numbers.

	initial	produced	used up	equilibrium
[A]	0.50 ①		0.175 ④	0.325 ⑤
[B]	0.75 ①		0.35 ④	0.40 ⑤
[C]	0.00 ①	0.35 ③		0.35 ②

$$K = \frac{(0.35)^2}{(0.325)(0.40)^2} = 2.4$$

20.26 For the reaction of XO with O_2 to form XO_2, the equilibrium constant at 398 K is 1.0×10^{-4} L/mol. If 1.0 mol of XO and 2.0 mol of O_2 are placed in a 1.0-L vessel and allowed to come to equilibrium, what will be the equilibrium concentration of each of the species?

$$2XO + O_2 \rightleftharpoons 2XO_2 \qquad K = \frac{[XO_2]^2}{[XO]^2[O_2]} = 1.0 \times 10^{-4}\ \text{L/mol}$$

[The equation $XO + \frac{1}{2}O_2 \rightleftharpoons XO_2$ would have an equilibrium constant with units $(\text{L/mol})^{1/2}$.]
The sequence of construction of the concentration table is shown by the circled numbers.

	initial	produced	used up	equilibrium
$[XO]$	1.0 ①		$2x$ ④	$1.0 - 2x \simeq 1.0$ ⑤
$[O_2]$	2.0 ①		x ④	$2.0 - x \simeq 2.0$ ⑤
$[XO_2]$	0.0 ①	$2x$ ③		$2x$ ②

① The initial concentrations are given in the statement of the problem; the $[XO_2]$ by implication. ② Letting the equilibrium concentration of XO_2 equal $2x$ allows the deduction ③ that $2x$ mol/L was produced by the reaction. ④ The concentrations of XO and O_2 used up, $2x$ and x, are in the ratio of the coefficients in the balanced chemical equation. ⑤ Knowledge of the initial concentrations and how much is used up in the reaction allows calculation of the equilibrium concentrations. The latter are approximated as 1.0 and 2.0 mol/L, subject to later checking. These values are substituted in the equilibrium constant expression:

$$K = \frac{(2x)^2}{(1.0)^2(2.0)} = 1.0 \times 10^{-4}$$

$x = 7.1 \times 10^{-3}$ (The approximations made, e.g. $1.0 - 0.014 \simeq 1.0$, are within 1.4% of the true value, which is acceptable.)

$$[XO_2] = 2x = 1.4 \times 10^{-2} \text{ mol/L} \qquad [XO] = 1.0 - 0.014 \simeq 1.0 \text{ mol/L} \qquad [O_2] = 2.0 - 0.0071 \simeq 2.0 \text{ mol/L}$$

20.27 Determine the equilibrium concentration of each of the species which react according to the equation

$$A + B \rightleftharpoons C + 2D$$

if the value of the equilibrium constant is 1.8×10^{-6} after 1.0 mol of C and 1.0 mol of D are placed in a 1.0 L vessel and allowed to come to equilibrium.

▮

$$K = \frac{[C][D]^2}{[A][B]} = 1.8 \times 10^{-6}$$

The sequence of construction of the following table is shown by the circled numbers.

	initial	produced	used up	equilibrium
[A]	0 ①	$\frac{1.0 - x}{2}$ ④		$0.50 - 0.50x \simeq 0.50$ ⑤
[B]	0 ①	$\frac{1.0 - x}{2}$ ④		$0.50 - 0.50x \simeq 0.50$ ⑤
[C]	1.0 ①		$\frac{1.0 - x}{2}$ ④	$0.50 + 0.50x \simeq 0.50$ ⑤
[D]	1.0 ①		$1.0 - x$ ③	x ②

② Since the D will be used up faster than C and since the equilibrium lies far to the left $(K \ll 1)$, let the equilibrium D concentration equal x. ③ Since initially it was 1.0 and finally it is x, $1.0 - x$ must be used up. ④ By the ratio in the balanced chemical equation, the concentrations of the other substances produced or used up can be calculated. ⑤ These are added (A and B, which are produced) or subtracted (C, which is used up) to determine the equilibrium concentrations.

$$K = \frac{0.50x^2}{(0.50)(0.50)} = 1.8 \times 10^{-6} \qquad x^2 = 9.0 \times 10^{-7}$$

$$x = 9.5 \times 10^{-4} \text{ mol/L} = [D] \qquad [C] = [B] = [A] = 0.50 \text{ mol/L}$$

20.28 At 90 °C, the following equilibrium is established:

$$H_2(g) + S(s) \rightleftharpoons H_2S(g) \qquad K = 6.8 \times 10^{-2}$$

If 0.20 mol of hydrogen and 1.0 mol of sulfur are heated to 90 °C in a 1.0 L vessel, what will be the partial pressure of H_2S at equilibrium?

▮

$$K = \frac{[H_2S]}{[H_2]} = 6.8 \times 10^{-2}$$

	initial	produced	used up	equilibrium
$[H_2]$	0.20		x	$0.20 - x \simeq 0.20$
$[H_2S]$	0.00	x		x

$$K = \frac{x}{0.20} = 6.8 \times 10^{-2} \qquad x = 1.4 \times 10^{-2} \text{ mol/L}$$

$$P(H_2S) = \left(\frac{n}{V}\right)RT = (1.4 \times 10^{-2} \text{ mol/L})(0.0821 \text{ L·atm/mol·K})(363 \text{ K}) = 0.42 \text{ atm}$$

20.29 Determine the value of the equilibrium constant for the reaction

$$A + 2B \rightleftharpoons 2C$$

if 1.0 mol of A, 2.0 mol of B, and 3.0 mol of C are placed in a 1.0-L vessel and allowed to come to equilibrium. The final concentration of C is 1.4 mol/L.

▌

$$K = \frac{[C]^2}{[A][B]^2}$$

	initial	produced	used up	equilibrium
[A]	1.0	0.80		1.8
[B]	2.0	1.6		3.6
[C]	3.0		1.6	1.4

$$K = \frac{(1.4)^2}{(1.8)(3.6)^2} = 8.4 \times 10^{-2}$$

20.30 When 3.00 mol of A and 1.00 mol of B are mixed in a 1.00 L vessel, the following reaction takes place:

$$A(g) + B(g) \rightleftharpoons 2C(g)$$

The equilibrium mixture contains 0.50 mol of C. What is the value of the equilibrium constant for the reaction?

▌

$$K = \frac{[C]^2}{[A][B]}$$

	initial	produced	used up	equilibrium
[A]	3.00		0.25	2.75
[B]	1.00		0.25	0.75
[C]	0.00	0.50		0.50

$$K = \frac{(0.50)^2}{(0.75)(2.75)} = 0.12$$

20.31 For the reaction $A + 2B \rightleftharpoons C + D$, the equilibrium constant is 1.0×10^8. ***(a)*** Calculate the equilibrium concentration of A if 1.0 mol of A and 3.0 mol of B are placed in a 1.0-L vessel and allowed to come to equilibrium. ***(b)*** If 1.0 mol of C and 3.0 mol of D were placed in a 1.0-L vessel, calculate the equilibrium concentration of B.

▌

$$K = \frac{[C][D]}{[A][B]^2} = 1.0 \times 10^8$$

(a)

	initial	produced	used up	equilibrium
[A]	1.0		$1.0 - x$	x
[B]	3.0		$2.0 - 2x$	$1.0 + 2x \simeq 1.0$
[C]	0.0	$1.0 - x$		$1.0 - x \simeq 1.0$
[D]	0.0	$1.0 - x$		$1.0 - x \simeq 1.0$

$$K = \frac{(1.0)(1.0)}{x(1.0)^2} = 1.0 \times 10^8 \qquad x = [A] = 1.0 \times 10^{-8} \text{ mol/L}$$

(b)

	initial	produced	used up	equilibrium
[A]	0.0	x		x
[B]	0.0	$2x$		$2x$
[C]	1.0		x	$1.0 - x \simeq 1.0$
[D]	3.0		x	$3.0 - x \simeq 3.0$

$$K = \frac{(3.0)(1.0)}{x(2x)^2} = 1.0 \times 10^8 \qquad 4x^3 = 3.0 \times 10^{-8}$$

$$x = \sqrt[3]{7.5 \times 10^{-9}} = 2.0 \times 10^{-3} = [A] \qquad [B] = 2x = 4.0 \times 10^{-3} \text{ mol/L}$$

20.32 At a certain temperature, the equilibrium constant for the reaction of CO with O_2 to produce CO_2 is 5.0×10^3 L/mol. Calculate [CO] at equilibrium if 1.0 mol each of CO and O_2 are placed in a 2.0-L vessel and allowed to come to equilibrium.

▌

$$2CO + O_2 \longrightarrow 2CO_2 \qquad K = \frac{[CO_2]^2}{[CO]^2[O_2]} = 5.0 \times 10^3$$

	initial	**produced**	**used up**	**equilibrium**
[CO]	0.50		0.50 − x	x
[O₂]	0.50		(0.50 − x)/2	0.25 + x/2 ≈ 0.25
[CO₂]	0.00	0.50 − x		0.50 − x ≈ 0.50

$$K = \frac{(0.50)^2}{x^2(0.25)} = 5.0 \times 10^3 \qquad x^2 = 0.20 \times 10^{-3} = 2.0 \times 10^{-4} \qquad x = 1.4 \times 10^{-2} \text{ mol/L} = [CO]$$

The approximations $(0.25 + x/2 \approx 0.25$ and $0.50 - x \approx 0.50)$ are within 3% and are acceptable.

20.33 A theoretically computed equilibrium constant for the polymerization of formaldehyde, HCHO, to glucose, $C_6H_{12}O_6$, in aqueous solution is 6×10^{22}.

$$6HCHO \longrightarrow C_6H_{12}O_6$$

If a 1.0 M solution of glucose were to reach dissociation equilibrium with respect to the above equation, what would be the concentration of formaldehyde in the solution?

▌

$$6HCHO \longrightarrow C_6H_{12}O_6 \qquad K = \frac{[C_6H_{12}O_6]}{[HCHO]^6} = \frac{1.00}{[HCHO]^6} = 6 \times 10^{22}$$

$$[HCHO]^6 = \frac{1.00}{6 \times 10^{22}} \qquad [HCHO] = 1.6 \times 10^{-4} \text{ M}$$

20.34 At 46 °C, K_p for the reaction

$$N_2O_4(g) \rightleftharpoons 2NO_2(g)$$

is 0.66 atm. Compute the percent dissociation of N_2O_4 at 46 °C and a total pressure of 380 torr. What are the partial pressures of N_2O_4 and NO_2 at equilibrium?

▌

$$K_p = \frac{P(NO_2)^2}{P(N_2O_4)} = 0.66 \qquad P(NO_2) + P(N_2O_4) = 380 \text{ torr} = 0.500 \text{ atm}$$

$$\frac{P^2}{0.500 - P} = 0.66$$

$$P^2 + 0.66P - 0.33 = 0$$

$$P(NO_2) = 0.332 \text{ atm} \qquad P(N_2O_4) = 0.168 \text{ atm}$$

Since each mol of N_2O_4 which dissociates produces 2 mol of NO_2, the percent dissociated is given by

$$\frac{0.5P(NO_2)}{P°(N_2O_4)} = \frac{0.5(0.332 \text{ atm})}{[0.5(0.332) + 0.168] \text{ atm}} \times 100\% = 50\%$$

20.35 Determine the equilibrium concentration of NH_3 after 2.0 mol of N_2 and 3.0 mol of H_2 are placed in a 1.0-L vessel and are allowed to achieve equilibrium.

$K = 0.444$ for the formation of ammonia. Set up; do not solve.

■

$$N_2 + 3H_2 \rightleftharpoons 2NH_3 \qquad K = \frac{[NH_3]^2}{[N_2][H_2]^3} = 0.444$$

	initial	produced	used up	equilibrium
$[NH_3]$	0	$2x$		$2x$
$[N_2]$	2.0		x	$2.0 - x$
$[H_2]$	3.0		$3x$	$3.0 - 3x$

$$\frac{(2x)^2}{(2.0 - x)(3.0 - 3x)^3} = 0.444$$

One cannot neglect x in this case.

20.36 Calculate the equilibrium concentration of T after 1.0 mol of G, 2.0 mol of J, and 0.50 mol of D are put into a 1.0-L vessel and allowed to come to equilibrium.

$$2G + J \rightleftharpoons D + 2T \qquad K = 1.5 \times 10^{-3}$$

■

$$K = \frac{[D][T]^2}{[G]^2[J]} = 1.5 \times 10^{-3}$$

	initial	produced	used up	equilibrium
[G]	1.0		$2x$	$1.0 - 2x \cong 1.0$
[J]	2.0		x	$2.0 - x \cong 2.0$
[D]	0.50	x		$0.50 + x \cong 0.50$
[T]	0	$2x$		$2x$

$$K = \frac{(0.50)(2x)^2}{(1.0)^2(2.0)} = x^2 = 1.5 \times 10^{-3} \qquad x = 0.039 \qquad 2x = [T] = 0.078$$

20.37 Given the reaction

$$2NO_2 \rightleftharpoons N_2O_4$$

Initially, 1.0 mol of NO_2 and 1.0 mol of N_2O_4 are placed in a 1.0-L vessel. At equilibrium, 0.75 mol of N_2O_4 is present in the flask. What is the value of K for the reaction?

■

	initial	used	produced	equilibrium
$[NO_2]$	1.0		0.50	1.50
$[N_2O_4]$	1.0	0.25		0.75

$$K = \frac{[N_2O_4]}{[NO_2]^2} = \frac{(0.75)}{(1.50)^2} = 0.33$$

20.38 $PCl_5(g) \rightleftharpoons PCl_3(g) + Cl_2(g)$. Calculate the moles of Cl_2 produced at equilibrium when 1.00 mol of PCl_5 is heated at 250 °C in a vessel having a capacity of 10.0 L. At 250 °C, $K = 0.041$ for this dissociation.

■

$$K = \frac{[PCl_3][Cl_2]}{[PCl_5]} = 0.041$$

	initial	produced	used up	equilibrium
$[PCl_3]$		x		x
$[Cl_2]$		x		x
$[PCl_5]$	0.100		x	$0.100 - x$

$$K = \frac{x^2}{(0.100 - x)} = 0.041 \qquad x = \frac{-0.041 + \sqrt{(0.041)^2 + 4(0.0041)}}{2} = 0.047 \text{ mol/L}$$

$$(0.047 \text{ mol/L})(10.0 \text{ L}) = 0.47 \text{ mol}$$

20.39 $H_2(g) + I_2(g) \rightleftharpoons 2HI(g)$. When 46.0 g of I_2 and 1.00 g of H_2 are heated to equilibrium at 470 °C, the equilibrium mixture contains 1.9 g I_2. ***(a)*** How many mol of each gas are present in the equilibrium mixture? ***(b)*** Compute the equilibrium constant.

Since the same number of moles of gas appears on both sides of the equation, the equilibrium constant expression can be stated as a ratio of moles instead of concentrations.

$$(46.0 \text{ g } I_2)\left(\frac{1 \text{ mol } I_2}{254 \text{ g } I_2}\right) = 0.181 \text{ mol } I_2 \text{ initially present} \qquad (1.00 \text{ g } H_2)\left(\frac{1 \text{ mol } H_2}{2.00 \text{ g } H_2}\right) = 0.500 \text{ mol } H_2$$

$$(1.9 \text{ g } I_2)\left(\frac{1 \text{ mol } I_2}{254 \text{ g } I_2}\right) = 0.0075 \text{ mol } I_2 \text{ at equilibrium}$$

	initial	**produced**	**used up**	**equilibrium**
$[I_2]$	0.181		0.174	0.0075
$[H_2]$	0.500		0.174	0.326
$[HI]$	0	2(0.174)		0.348

$$K = \frac{[HI]^2}{[I_2][H_2]} = \frac{(0.348)^2}{(0.326)(0.0075)} = 50$$

20.40 Exactly 1 mol each of H_2 and I_2 are heated in a 30-L evacuated chamber to 470 °C. Using the value of K from Prob. 20.39, determine ***(a)*** how many mol of I_2 remain unreacted when equilibrium is established, ***(b)*** the total pressure in the chamber, ***(c)*** the partial pressures of I_2 and of HI in the equilibrium mixture. ***(d)*** Now if one additional mol of H_2 is introduced into this equilibrium system, how many mol of the original iodine will remain unreacted?

(a) $$K = \frac{[HI]^2}{[H_2][I_2]} = \frac{(2x)^2}{(1.00 - x)^2} = 50$$

The easy way to solve for x is to take the square root of this equation:

$$2x = 7.1(1.00 - x) \qquad x = 0.78 \text{ mol } I_2 \text{ reacted} \qquad 1.00 - x = 0.22 \text{ mol } I_2 \text{ remaining}$$

(b) The number of moles of gas does not change as the reaction proceeds at 470 °C; hence 2.00 mol of gas remains at equilibrium.

$$P = \frac{nRT}{V} = \frac{(2.00 \text{ mol})(0.0821 \text{ L}\cdot\text{atm/mol}\cdot\text{K})(743 \text{ K})}{30 \text{ L}} = 4.1 \text{ atm}$$

(c) $$P(I_2) = \frac{(0.22 \text{ mol})(0.0821 \text{ L}\cdot\text{atm/mol}\cdot\text{K})(743 \text{ K})}{30 \text{ L}} = 0.45 \text{ atm}$$

$$P(HI) = \frac{(1.56 \text{ mol})(0.0821 \text{ L}\cdot\text{atm/mol}\cdot\text{K})(743 \text{ K})}{30 \text{ L}} = 3.17 \text{ atm}$$

(d) $$K = \frac{(2x)^2}{(1.00 - x)(2.00 - x)} = 50$$

$$4x^2 = (2.00 - 3.00x + x^2)50 \qquad 46x^2 - 150x + 100 = 0$$

$$x = \frac{+150 \pm \sqrt{(-150)^2 - 400(46)}}{92} = 2.2 \text{ mol (impossible) or } 0.93 \text{ mol}$$

1.00 mol initial − 0.93 mol reacted = 0.07 mol unreacted

20.41 A saturated solution of iodine in water contains 0.330 g I_2/L. More than this can dissolve in a KI solution because of the following equilibrium:

$$I_2(aq) + I^- \rightleftharpoons I_3^-$$

A 0.100 M KI solution (0.100 M I^-) actually dissolves 12.5 g of iodine/L, most of which is converted to I_3^-. Assuming that the concentration of I_2 in all saturated solutions is the same, calculate the equilibrium constant for the above reaction. What is the effect of adding water to a clear saturated solution of I_2 in the KI solution?

▮

$$(0.330 \text{ g } I_2)\left(\frac{1 \text{ mol } I_2}{254 \text{ g } I_2}\right) = 1.30 \times 10^{-3} \text{ mol} \qquad (12.5 \text{ g } I_2)\left(\frac{1 \text{ mol } I_2}{254 \text{ g } I_2}\right) = 0.0492 \text{ mol}$$

$$K = \frac{[I_3^-]}{[I_2][I^-]}$$

At equilibrium,

$$[I_2] = 1.30 \times 10^{-3} \text{ M} \qquad [I_3^-] = (0.0492 \text{ M}) - (0.00130 \text{ M}) = 0.0479 \text{ M}$$

$$[I^-] = (0.100 \text{ M}) - (0.0479 \text{ M}) = 0.0521 \text{ M}$$

$$K = \frac{0.0479}{(1.30 \times 10^{-3})(0.0521)} = 707$$

Reverse reaction is favored.

20.42 The equilibrium

$$p\text{-Xyloquinone} + \text{methylene white} \rightleftharpoons p\text{-xylohydroquinone} + \text{methylene blue}$$

may be studied conveniently by observing the difference in color between methylene blue and methylene white. One mmol of methylene blue was added to 1.00 L of solution that was 0.24 M in *p*-xylohydroquinone and 0.0120 M in *p*-xyloquinone. It was then found that 4.0% of the added methylene blue was reduced to methylene white. What is the equilibrium constant for the above reaction? The equation is balanced with 1 molecule of each of the 4 substances.

▮

$$K = \frac{[\text{XHQ}][\text{blue}]}{[\text{XQ}][\text{white}]} = \frac{(0.24)(0.96 \times 10^{-3})}{(0.012)(0.040 \times 10^{-3})} = 480$$

20.43 The equilibrium constant for the reaction H_3BO_3 + glycerin ⇌ (H_3BO_3-glycerin) is 0.90. How much glycerin should be added per liter of 0.10 M H_3BO_3 solution so that 60% of the H_3BO_3 is converted to the boric acid–glycerin complex?

▮

$$K = \frac{[\text{complex}]}{[H_3BO_3][\text{glycerin}]} = 0.90 \qquad \frac{[\text{complex}]}{[H_3BO_3]} = \frac{60}{40} = 1.5$$

$$K = \frac{1.5}{[\text{glycerin}]} = 0.90 \qquad [\text{glycerin}] = 1.5/0.90 = 1.7 \text{ M}$$

20.44 Sulfide ion in alkaline solution reacts with solid sulfur to form polysulfide ions having formulas S_2^{2-}, S_3^{2-}, S_4^{2-}, and so on. The equilibrium constant for the formation of S_2^{2-} is 12 and for the formation of S_3^{2-} is 130, both from S and S^{2-}. What is the equilibrium constant for the formation of S_3^{2-} from S_2^{2-} and S?

▮ To avoid confusion, let us designate the equilibrium constants for the various reactions by subscripts. Also, we note that only the ion concentrations appear in the equilibrium constant equations, because solid sulfur, S, is always in its standard state.

$$S + S^{2-} \rightleftharpoons S_2^{2-} \qquad K_1 = [S_2^{2-}]/[S^{2-}] = 12$$

$$2S + S^{2-} \rightleftharpoons S_3^{2-} \qquad K_2 = [S_3^{2-}]/[S^{2-}] = 130$$

$$S + S_2^{2-} \rightleftharpoons S_3^{2-} \qquad K_3 = [S_3^{2-}]/[S_2^{2-}]$$

The desired constant, K_3, expresses the equilibrium ratio of S_3^{2-} to S_2^{2-} concentrations in a solution equilibrated with solid sulfur. Such a solution must also contain sulfide ion, S^{2-}, resulting from the dissociation of S_2^{2-} by the reverse of the first equation. Since all four species, S, S^{2-}, S_2^{2-}, and S_3^{2-}, are present, all the equilibria represented above must be satisfied. The three equilibrium ratios are not all independent, because

$$\frac{[S_3^{2-}]}{[S_2^{2-}]} = \frac{[S_3^{2-}]/[S^{2-}]}{[S_2^{2-}]/[S^{2-}]} \qquad \text{or} \qquad K_3 = \frac{K_2}{K_1} = \frac{130}{12} = 11$$

The result, $K_2 = K_1K_3$, is a general one for any case where one chemical equation (the second in this case) can be written as the sum of two other equations (the first and third in this case).

20.3 K_p

Note that for a mixture of ideal gases, *percent by volume* is another way of saying *mole percent.*

20.45 $K = 49$ for the following reaction at 127 °C:

$$A(g) + B(s) \rightleftharpoons C(g) + D(g)$$

Calculate K_p for the reaction.

▮
$$K = \frac{[C][D]}{[A]} = \frac{(P_C/RT)(P_D/RT)}{P_A/RT} = \frac{K_p}{RT} = 49 \text{ mol/L}$$
$$K_p = (49 \text{ mol/L})(0.0821 \text{ L·atm/mol·K})(400 \text{ K}) = 1.6 \times 10^3 \text{ atm}$$

[This calculation is generalized in Prob. 20.47.]

20.46 (*a*) What effect would the introduction of He gas have on the partial pressure of each gas in a system containing N_2, H_2, and NH_3 at equilibrium? (*b*) What effect, if any, will the added He have on the position of the equilibrium?

▮ (*a*) Introduction of He(*g*) has no effect on the partial pressures of the gases already present, and so (*b*) it does not affect the position of the equilibrium, which is governed by

$$K_p = \frac{P(NH_3)^2}{P(N_2)P(H_2)^3}$$

20.47 Derive the relationship between K and K_p for a reaction in which gases are involved

$$K_p = K(RT)^{\Delta n}$$

where Δn is the difference in the number of moles of gases between products and reactants.

▮ For the reaction

$$a\,A(g) + b\,B(s) \longrightarrow c\,C(g) + d\,D(s) \qquad K = \frac{[C]^c[D]^d}{[A]^a[B]^b}$$

The activities of the solids are each equal to 1. The concentrations of the gases, in mol/L, are given by

$$\frac{n}{V} = \frac{P}{RT} \qquad K = \frac{(P_C/RT)^c}{(P_A/RT)^a} = K_p(RT)^{a-c} = K_p(RT)^{-\Delta n} \qquad K_p = K(RT)^{\Delta n}$$

Δn is the change in the number of moles of gaseous compounds involved in the reaction.

20.48 In a 10.0-L evacuated chamber, 0.500 mol H_2 and 0.500 mol I_2 react at 448 °C.

$$H_2(g) + I_2(g) \rightleftharpoons 2HI(g)$$

At the given temperature, and for a standard state of 1 mol/L, $K = 50$ for the reaction. (*a*) What is the value of K_p? (*b*) What is the total pressure in the chamber? (*c*) How many moles of the iodine remain unreacted at equilibrium? (*d*) What is the partial pressure of each component in the equilibrium mixture?

▮ (*a*) The same number of mol of gas occurs on both sides of the equation. Hence this reaction is not affected by a volume change or a pressure change, and $K_p = K = 50$. This can be proved in detail as follows.

Since the volume of the chamber is 10 L,

$$K = \frac{[HI]^2}{[H_2][I_2]} = \frac{\left[\dfrac{n(HI)}{10}\right]^2}{\left[\dfrac{n(H_2)}{10}\right]\left[\dfrac{n(I_2)}{10}\right]} = \frac{n(HI)^2}{n(H_2)n(I_2)}$$

We may evaluate K_p by noting that the partial pressure of any component in a gas mixture is equal to the total pressure times the mole fraction of that component (see Prob. 12.163). Thus, taking all pressures relative to 1 atm,

$$K_p = \frac{P(HI)^2}{P(H_2)P(I_2)} = \frac{\left\{\dfrac{n(HI)}{n(\text{total})} \times P(\text{total})\right\}^2}{\left\{\dfrac{n(H_2)}{n(\text{total})} \times P(\text{total})\right\}\left\{\dfrac{n(I_2)}{n(\text{total})} \times P(\text{total})\right\}} = \frac{n(HI)^2}{n(H_2)n(I_2)} = K$$

(**b**) Before the reaction sets in, the total number of mol of gas is $0.5 + 0.5 = 1$. During the reaction, there is no change in the total number of mol. (For every mol of H_2 that reacts, 1 mol of I_2 will react and 2 mol of HI will be formed.) Then, the total pressure will remain the same, because it depends only on the total number of mol of gas. The total pressure can be computed from the ideal gas law:

$$P(\text{total}) = \frac{n(\text{total})RT}{V}$$

$$= \frac{(1.00\ \text{mol})(0.0821\ \text{L·atm/K·mol})(721\ \text{K})}{10\ \text{L}} = 5.9\ \text{atm}$$

(**c**) Let x = number of moles of iodine reacting.

	H_2	+	I_2	$\rightleftharpoons$ 2HI
n at start:	0.5		0.5	0
Change by reaction:	$-x$		$-x$	$2x$
n at equilibrium:	$0.5 - x$		$0.5 - x$	$2x$

Note the 2:1 ratio of mol of HI formed to mol of H_2 reacted. This ratio is demanded by the coefficients in the balanced chemical equation. Regardless of how complete or incomplete the reaction may be, 2 mol of HI is always formed for every 1 mol of H_2 that reacts.

From the expression for K found in (*a*),

$$K = 50 = \frac{(2x)^2}{(0.5 - x)(0.5 - x)} \quad \text{or} \quad \sqrt{50} = 7.1 = \frac{2x}{0.5 - x}$$

Then $2x = 7.1(0.5 - x)$. Solving, $x = 0.39$. Thus, 0.39 mol I_2 reacts, leaving $0.5 - 0.39 = 0.11$ mol I_2 unreacted at equilibrium.

(**d**) $P(I_2) = \frac{n(I_2)}{n(\text{total})} \times (\text{total pressure}) = \left(\frac{0.11}{1}\right)(5.9\ \text{atm}) = 0.65\ \text{atm}$

$P(H_2) = P(I_2) = 0.65\ \text{atm}$

$P(\text{HI}) = (\text{total pressure}) - [P(H_2) + P(I_2)] = 5.9 - 1.3 = 4.6\ \text{atm}$

or $\quad P(\text{HI}) = \frac{n(\text{HI})}{n(\text{total})} \times (\text{total pressure}) = \left(\frac{0.78}{1}\right)(5.9\ \text{atm}) = 4.6\ \text{atm}$

20.49 For the reaction

$$CaCO_3(s) \rightleftharpoons CaO(s) + CO_2(g)$$

$K_p = 1.16$ atm at 800 °C. If 20.0 g of $CaCO_3$ was put into a 10.0-L container and heated to 800 °C, what percent of the $CaCO_3$ would remain unreacted at equilibrium?

▌ $K_p = P(CO_2) = 1.16\ \text{atm} \qquad n(CO_2) = \frac{PV}{RT} = \frac{(1.16\ \text{atm})(10.0\ \text{L})}{(0.0821\ \text{L·atm/mol·K})(1073\ \text{K})} = 0.132\ \text{mol}$

Moles $CaCO_3$ initially present $= \frac{20.0\ \text{g}}{100\ \text{g/mol}} = 0.200\ \text{mol} \qquad \frac{0.132\ \text{mol}\ CO_2}{0.200\ \text{mol}\ CaCO_3} \times 100\% = 66.0\%$ decomposed

Hence 34.0% remains undecomposed.

20.50 For the reaction at 800 K

$$3H_2(g) + N_2(g) \rightleftharpoons 2NH_3(g)$$

determine the relationship between the values of K and K_p.

▌ By Prob. 20.47,

$$K_p = K/(RT)^2 = K/(0.0821)^2(800)^2 = 2.32 \times 10^{-4}\ K$$

20.51 At 27 °C and 1.00 atm, N_2O_4 is 20.0% dissociated into NO_2. Find (**a**) K_p and (**b**) the percent dissociation at 27 °C and a total pressure of 0.10 atm. (**c**) What is the extent of dissociation in a 69-g sample of N_2O_4 confined in a 20-L vessel at 27 °C?

▮ (*a*) When 1.00 mol of N_2O_4 gas dissociates completely, 2.00 mol of NO_2 gas is formed. Since this problem does not specify a particular size of reaction vessel or a particular weight of sample, we are free to choose 1.00 mol (92 g) as the starting amount of N_2O_4. For the given total pressure of 1.00 atm, the tabular computation of the partial pressures proceeds as follows:

	$N_2O_4(g)$	$\rightleftharpoons$	$2NO_2(g)$
n at start:	1.00		0
Change by reaction:	−0.20		+0.40
n at equilibrium:	1.00 − 0.20 = 0.80		0.40
Mole fraction:	$\frac{0.80}{0.80 + 0.40} = 0.67$		$\frac{0.40}{0.80 + 0.40} = 0.33$
Partial pressure = (mole frac.)(1 atm):	0.67 atm		0.33 atm

Then, in terms of the numerical values of the partial pressures,

$$K_p = \frac{P(NO_2)^2}{P(N_2O_4)} = \frac{(0.33)^2}{0.67} = 0.17$$

(*b*) Let α = fraction of N_2O_4 dissociated at equilibrium at 0.100 atm total pressure.

	N_2O_4	$\rightleftharpoons$	$2NO_2$
n at start:	1		0
Change by reaction:	$-\alpha$		2α
n at equilibrium:	$1 - \alpha$		2α
Mole fraction:	$\frac{1-\alpha}{1+\alpha}$		$\frac{2\alpha}{1+\alpha}$
Partial pressure:	$\left(\frac{1-\alpha}{1+\alpha} \times 0.10\right)$ atm		$\left(\frac{2\alpha}{1+\alpha} \times 0.10\right)$ atm

From (*a*), $K_p = 0.17$. Thus,

$$0.17 = K_p = \frac{P(NO_2)^2}{P(N_2O_4)} = \frac{\left(\frac{2\alpha}{1+\alpha} \times 0.10\right)^2}{\frac{1-\alpha}{1+\alpha} \times 0.10} = \frac{0.40\alpha^2}{1-\alpha^2}$$

or $0.4\alpha^2 = 0.17(1 - \alpha^2)$. Solving, $\alpha = 0.55 = 55\%$ dissociated at 27 °C and 0.10 atm.

Note that a larger fraction of the N_2O_4 is dissociated at 0.10 atm than at 1.0 atm. This is in agreement with Le Chatelier's principle: decreasing the pressure should favor the side with the greater number of moles of gas molecules (NO_2).

(*c*) If the sample were all N_2O_4, it would contain

$$\frac{69\text{ g}}{92\text{ g/mol}} = 0.75\text{ mol}$$

Let α be the fractional dissociation. The tabular analysis follows:

	N_2O_4	$\rightleftharpoons$	$2NO_2$
n at start:	0.75		0
Change by reaction:	-0.75α		$+2(0.75\alpha)$
n at equilibrium	$0.75(1 - \alpha)$		1.50α

Because the total gas pressure is unknown, it is simplest to obtain the partial pressures directly from Dalton's law (Chap. 5).

$$P(N_2O_4) = \frac{n(N_2O_4)RT}{V} = \frac{[0.75(1-\alpha)\text{ mol}](0.082\text{ L·atm/K·mol})(300\text{ K})}{20\text{ L}} = 0.92(1-\alpha)\text{ atm}$$

$$P(NO_2) = \frac{n(NO_2)RT}{V} = \frac{1.50\alpha}{0.75(1-\alpha)}[0.92(1-\alpha)\text{ atm}] = 1.84\alpha\text{ atm}$$

Then
$$0.17 = K_p = \frac{P(NO_2)^2}{P(N_2O_4)} = \frac{(1.84\alpha)^2}{0.92(1-\alpha)} = \frac{3.68\alpha^2}{1-\alpha}$$

or
$$3.68\alpha^2 + 0.17\alpha - 0.17 = 0$$

Solving,
$$\alpha = \frac{-0.17 \pm \sqrt{(0.17)^2 + 4(0.17)(3.68)}}{2(3.68)} = \frac{-0.17 \pm 1.59}{7.36} = -0.24 \text{ or } +0.19$$

The negative root must obviously be discarded. The extent of dissociation is 19%.

20.52 Pure PCl_5 is introduced into an evacuated chamber and comes to equilibrium (see Prob. 20.38) at 250 °C and 2.00 atm. The equilibrium gas contains 40.7% chlorine by volume.

(a) 1. What are the partial pressures of the gaseous components at equilibrium?
2. Calculate K_p at 250 °C for the reaction as written in Prob. 20.38.

(b) If the gas mixture is expanded to 0.200 atm at 250 °C, calculate:
1. the percentage of PCl_5 that would be dissociated at equilibrium
2. the percent by volume of Cl_2 in the equilibrium gas mixture
3. the partial pressure of Cl_2 in the equilibrium mixture

▮
$$PCl_5 \longrightarrow PCl_3 + Cl_2$$

(a) 1. The gas contains 40.7% Cl_2 and 40.7% PCl_3, since they are produced in a 1:1 mole ratio. The PCl_5 percentage is $100.0 - 2(40.7) = 18.6\%$.

$$P(Cl_2) = P(PCl_3) = (2.00\text{ atm})(0.407) = 0.814\text{ atm} \qquad P(PCl_5) = (2.00\text{ atm})(0.186) = 0.372\text{ atm}$$

2. $$K_p = \frac{P(Cl_2)P(PCl_3)}{P(PCl_5)} = \frac{(0.814\text{ atm})^2}{0.372\text{ atm}} = 1.78\text{ atm}$$

(b) $$P(Cl_2) + P(PCl_3) + P(PCl_5) = 0.200\text{ atm} = 2P(PCl_3) + \frac{P(PCl_3)^2}{1.78\text{ atm}}$$

$$0.562P(PCl_3)^2 + 2P(PCl_3) - 0.200 = 0$$

$$P(PCl_3) = 0.097\text{ atm} = P(Cl_2) \qquad P(PCl_5) = \frac{(0.097)^2}{1.78} = 5.29 \times 10^{-3}\text{ atm}$$

$$\text{Fraction dissociated} = \frac{P(PCl_3)}{P(PCl_3) + P(PCl_5)} = \frac{0.097}{0.097 + 0.00529} = 0.948$$

$$\text{Percent dissociated} = 0.948 \times 100\% = 94.8\%$$

$$\text{Mole percent } Cl_2 = \left(\frac{0.097}{0.097 + 0.097 + 0.00529}\right) \times 100\% = 48.7\%\ Cl_2$$

20.53 At 500 °C, the equilibrium constant of 3.90×10^{-3}/atm is found for the reaction

$$\tfrac{1}{2}N_2 + \tfrac{3}{2}H_2 \rightleftharpoons NH_3$$

If sufficient ammonia were introduced into an evacuated container at 500 °C to give a pressure of 1.00 atm before any decomposition occurred, what would be the partial pressures of N_2, H_2, and NH_3 at equilibrium?

▮
$$K_p = \frac{P(NH_3)}{P(N_2)^{1/2}P(H_2)^{3/2}} = 3.90 \times 10^{-3}$$

	initial	produced	used up	equilibrium
$P(NH_3)$	1.00		$1.00 - x$	x
$P(N_2)$	0.00	$(1.00 - x)/2$		$0.500 - x/2 \simeq 0.500$
$P(H_2)$	0.00	$3(1.00 - x)/2$		$1.50 - 3x/2 \simeq 1.50$

$$K_p = \frac{x}{(0.500)^{1/2}(1.50)^{3/2}} = 3.90 \times 10^{-3}$$

$$x = 5.07 \times 10^{-3} \text{ atm} = P(NH_3) \qquad P(N_2) = 0.498 \text{ atm} \qquad P(H_2) = 1.50 \text{ atm}$$

20.54 Ammonium hydrogen sulfide dissociates as follows:

$$NH_4HS(s) \rightleftharpoons H_2S(g) + NH_3(g)$$

If solid NH_4HS is placed in an evacuated flask at a certain temperature, it will dissociate until the total gas pressure is 500 torr. ***(a)*** Calculate the value of the equilibrium constant for the dissociation reaction. ***(b)*** Additional NH_3 is introduced into the equilibrium mixture without change in temperature until the partial pressure of ammonia is 700 torr. What is the partial pressure of H_2S under these conditions? What is the total pressure in the flask?

▮ The equilibrium constant expression contains the pressures of the gases only:

$$K_p = P(H_2S)P(NH_3)$$

(a) Since the gases are produced in a 1:1 ratio by the equation given, $P(H_2S) = P(NH_3)$. The total pressure is 500 torr; each partial pressure must be $250 \text{ torr} = \frac{250}{760} \text{ atm} = 0.329 \text{ atm}$.

$$K_p = (0.329 \text{ atm})^2 = 0.108 \text{ atm}^2$$

(b) Introduction of additional NH_3 will cause the equilibrium to shift to the left, using up some H_2S. But K_p remains constant.

$$K_p = P(H_2S)P(NH_3) = 0.108 \text{ atm}^2$$

$$P(H_2S) = \frac{K_p}{P(NH_3)} = \frac{0.108 \text{ atm}^2}{700/760 \text{ atm}} = 0.117 \text{ atm} = 89.3 \text{ torr} \qquad P_{tot} = P(H_2S) + P(NH_3) = 789 \text{ torr}$$

20.55 At 817 °C, K_p for the reaction between pure CO_2 and excess hot graphite to form $2CO(g)$ is 10. ***(a)*** What is the analysis of the gases at equilibrium at 817 °C and a total pressure of 4.0 atm? What is the partial pressure of CO_2 at equilibrium? ***(b)*** At what total pressure will the gas mixture analyze 6% CO_2 by volume?

▮ ***(a)*** Consider 1.0 mol CO_2 at the start. Let α be the fraction of CO_2 that converts.

	$CO_2(gas)$	$+ C(solid) \rightleftharpoons$	$2CO(gas)$
n at start:	1		0
Change by reaction:	$-\alpha$		2α
n at equilibrium:	$1-\alpha$		2α
Partial pressure:	$\left(\frac{1-\alpha}{1+\alpha} \times 4.0\right)$ atm		$\left(\frac{2\alpha}{1+\alpha} \times 4.0\right)$ atm

Since pure solids are themselves standard states, graphite is not included in the equilibrium constant, as long as it is present in excess.

$$K_p = \frac{P(CO)^2}{P(CO_2)} \quad \text{or} \quad 10 = \frac{\left(\frac{2\alpha}{1+\alpha} \times 4.0\right)^2}{\frac{1-\alpha}{1+\alpha} \times 4.0} = \frac{16\alpha^2}{1-\alpha^2} \quad \text{or} \quad 26\alpha^2 = 10$$

Solving, $\alpha = \sqrt{10/26} = 0.62$. Then, by substitution,

$$\text{Mole fraction } CO_2 \text{ at equilibrium} = \frac{1-\alpha}{1+\alpha} = \frac{0.38}{1.62} = 23\% \quad \text{(by volume)}$$

$$\text{Mole fraction CO at equilibrium} = \frac{2\alpha}{1+\alpha} = 77\% \quad \text{(by volume)}$$

$$P(CO_2) = (0.23 \times 4.0) \text{ atm} = 0.92 \text{ atm}$$

(b) Let the total pressure be a atm, or simply a, relative to the 1 atm standard state. We have

$$\% \, CO_2 = 6.0\% \text{ by volume} \qquad \% \, CO = 100\% - 6.0\% = 94\% \text{ by volume}$$

Thus, $P(CO_2) = 0.060a$, $P(CO) = 0.94a$, and

$$K_p = 10 = \frac{(0.94a)^2}{0.060a} \qquad \text{from which} \quad a = 0.68$$

We could have predicted, without solving the problem numerically, that the pressure for 6.0% CO_2 would be less than 4.0 atm, since decreasing the pressure favors an increase in the number of gas molecules. The equilibrium gases in (*b*) have more CO (side with the larger number of gas molecules), 94%, than the equilibrium gases in (*a*), 77% CO.

20.56 Would 1.00% CO_2 in the air be sufficient to prevent any loss in weight when Ag_2CO_3 is dried at 120 °C?

$$Ag_2CO_3(s) \rightleftharpoons Ag_2O(s) + CO_2(g) \qquad K_p = 0.0095 \quad \text{at } 120°C$$

How low would the partial pressure of CO_2 have to be to promote this reaction at 120 °C?

▮ Since the solids are in their standard states, they do not appear in the expression for the equilibrium constant. Hence

$$K_p = P(CO_2) = 0.0095$$

from which the partial pressure of CO_2 is 0.0095 atm. This is its partial pressure in equilibrium with the two solids, Ag_2CO_3 and Ag_2O. If some Ag_2CO_3 were placed in a *closed* vessel, a small amount would decompose into CO_2 and Ag_2O until the partial pressure of CO_2 reached 0.0095 atm. The decomposition would then stop, as the system would have reached equilibrium. If a partial pressure of CO_2 greater than this value (e.g., 1%, or 0.01 atm) is applied to the system, the above reaction will be reversed, all of the Ag_2O will be converted to Ag_2CO_3, and there will no longer be an equilibrium of the three components. In other words, there would be no loss in weight of Ag_2CO_3.

If, on the other hand, the partial pressure of CO_2 near the system is less than 0.0095 atm, some Ag_2CO_3 will dissociate according to the above equation. The amount of decomposition will depend on how much the surrounding air is stirred. If the air remains quiet, a small amount of decomposition will suffice to build up the partial pressure to the equilibrium value. If the air is changed rapidly, the CO_2 is removed from contact with the Ag_2CO_3, and more Ag_2CO_3 will have to decompose to again make up the equilibrium partial pressure of CO_2.

Regardless of the amount of stirring, Ag_2CO_3 would eventually decompose completely in an *open* vessel at a temperature (greater than 120 °C) at which the equilibrium partial pressure of CO_2 is 1 atm. At such a temperature the CO_2 from the decomposition would push the air away from the sample, and the falling CO_2 pressure near the sample resulting from the outward movement of gas would have to be compensated for by progressive decomposition.

20.57 For the reaction

$$SnO_2(s) + 2H_2(g) \rightleftharpoons 2H_2O(g) + Sn(l)$$

calculate K_p **(a)** at 900 K, where the equilibrium steam-hydrogen mixture was 45% H_2 by volume; **(b)** at 1100 K, where the equilibrium steam-hydrogen mixture was 24% H_2 by volume. **(c)** Would you recommend higher or lower temperatures for more efficient reduction of tin?

▮
$$K_p = P(H_2O)^2/P(H_2)^2$$

(a) $K_p = \left(\frac{0.55}{0.45}\right)^2 = 1.5$ **(b)** $K_p = \left(\frac{0.76}{0.24}\right)^2 = 10$

(c) The higher temperature yields greater conversion (and also a higher rate of conversion).

20.58 Equilibrium constants are given (in atm) for the following reactions at 0 °C:

$$SrCl_2 \cdot 6H_2O(s) \rightleftharpoons SrCl_2 \cdot 2H_2O(s) + 4H_2O(g) \qquad K_p = 6.89 \times 10^{-12}$$
$$Na_2HPO_4 \cdot 12H_2O(s) \rightleftharpoons Na_2HPO_4 \cdot 7H_2O(s) + 5H_2O(g) \qquad K_p = 5.25 \times 10^{-13}$$
$$Na_2SO_4 \cdot 10H_2O(s) \rightleftharpoons Na_2SO_4(s) + 10H_2O(g) \qquad K_p = 4.08 \times 10^{-25}$$

The vapor pressure of water at 0 °C is 4.58 torr. **(a)** Calculate the pressure of water vapor in equilibrium at 0 °C with each of $SrCl_2 \cdot 6H_2O$, $Na_2HPO_4 \cdot 12H_2O$, and $Na_2SO_4 \cdot 10H_2O$. Express the pressures in torr. **(b)** Which is the most effective drying agent at 0 °C: $SrCl_2 \cdot 2H_2O$, $Na_2HPO_4 \cdot 7H_2O$, or Na_2SO_4? **(c)** At what relative humidities will $Na_2SO_4 \cdot 10H_2O$ be efflorescent when exposed to the air at 0 °C? **(d)** At what relative humidities will Na_2SO_4 be deliquescent (i.e. absorb moisture) when exposed to the air at 0 °C?

▮ (*a*) $K_p = P(H_2O)^4 = 6.89 \times 10^{-12}$ $P(H_2O) = 1.62 \times 10^{-3}$ atm = 1.23 torr

$K_p = P(H_2O)^5 = 5.25 \times 10^{-13}$ $P(H_2O) = 3.50 \times 10^{-3}$ atm = 2.66 torr

$K_p = P(H_2O)^{10} = 4.08 \times 10^{-25}$ $P(H_2O) = 3.64 \times 10^{-3}$ atm = 2.77 torr

(*b*) $SrCl_2 \cdot 2H_2O$, since it has the lowest vapor pressure of water in equilibrium with it.

(*c*) The hydrate will lose water (effloresce) below 2.77 torr; that is, below $\frac{2.77 \text{ torr}}{4.58 \text{ torr}} \times 100\% = 60.5\%$ relative humidity.

(*d*) The dehydrated salt will absorb water above 60.5% relative humidity.

20.59 The moisture content of a gas is often expressed in terms of the *dew point*. The dew point is the temperature to which the gas must be cooled before the gas becomes saturated with water vapor. At this temperature, water or ice (depending on the temperature) will be deposited on a solid surface.

The efficiency of $CaCl_2$ as a drying agent was measured by a dew point experiment. Air at 0 °C was allowed to pass slowly over large trays containing $CaCl_2$. The air was then passed through a glass vessel through which a copper rod was sealed. The rod was cooled by immersing the emergent part of it in a dry ice bath. The temperature of the rod inside the glass vessel was measured by a thermocouple. As the rod was cooled slowly the temperature at which the first crystals of frost were deposited was observed to be −43 °C. The vapor pressure of ice at this temperature is 0.07 torr. Assuming that the $CaCl_2$ owes its desiccating properties to the formation of $CaCl_2 \cdot 2H_2O$, calculate K_p at that temperature for the reaction

$$CaCl_2 \cdot 2H_2O(s) \rightleftharpoons CaCl_2(s) + 2H_2O(g)$$

▮ At −43 °C, $P(H_2O) = 0.07$ torr

$$K_p = P(H_2O)^2 = \left(\frac{0.07}{760} \text{ atm}\right)^2 = 8 \times 10^{-9} \text{ atm}^2$$

20.60 The equilibrium constant for the reaction

$$CO(g) + H_2O(g) \rightleftharpoons CO_2(g) + H_2(g)$$

at 986 °C is 0.63. A mixture of 1.00 mol of water vapor and 3.00 mol of CO is allowed to come to equilibrium at a total pressure of 2.00 atm. (*a*) How many mol of H_2 are present at equilibrium? (*b*) What are the partial pressures of the gases in the equilibrium mixture?

▮
$$K_p = \frac{P(CO_2)P(H_2)}{P(CO)P(H_2O)} = 0.63$$

Since the number of mol of gas does not change during the process, the pressure does not change either. The 4.00 mol of gas exerts a total pressure of 2.00 atm, hence in this volume each mol of gas exerts 0.500 atm.

	initial	produced	used up	equilibrium
$P(CO)$	3.00/2.00		x	$1.50 - x$
$P(H_2O)$	1.00/2.00		x	$0.500 - x$
$P(CO_2)$	0	x		x
$P(H_2)$	0	x		x

$$\frac{x^2}{(1.50 - x)(0.500 - x)} = 0.63 \qquad 0.37x^2 + 1.26x - 0.472 = 0$$

$$x = 0.34 \text{ atm} = P(CO_2) = P(H_2)$$

$$P(H_2O) = 0.500 - x = 0.16 \text{ atm} \qquad P(CO) = 1.50 - x = 1.16 \text{ atm}$$

Since each mol of gas exerts 0.500 atm, the number of mol of hydrogen is 2.00(0.34) = 0.68 mol.

20.61 $2NOBr(g) \rightleftharpoons 2NO(g) + Br_2(g)$. If nitrosyl bromide (NOBr) is 34% dissociated at 25 °C and a total pressure of 0.25 atm, calculate K_p for the dissociation at this temperature.

▮
$$P(NO) = 2P(Br_2) \qquad \text{(from the balanced equation)}$$

Because it is 34% dissociated,

$$P(\text{NOBr}) = 0.66P^\circ(\text{NOBr}) \qquad [P^\circ(\text{NOBr}) \text{ represents the original pressure}]$$

$$P(\text{NO}) = 0.34P^\circ(\text{NOBr}) \qquad P(\text{Br}_2) = 0.17P^\circ(\text{NOBr})$$

$$P(\text{NOBr}) + P(\text{NO}) + P(\text{Br}_2) = 0.25 \text{ atm}$$

$$0.66P^\circ + 0.34P^\circ + 0.17P^\circ = 0.25 \text{ atm}$$

$$1.17P^\circ = 0.25 \text{ atm}$$

$$P^\circ(\text{NOBr}) = \frac{0.25 \text{ atm}}{1.17} = 0.214 \text{ atm}$$

$$P(\text{NOBr}) = (0.66)(0.214 \text{ atm}) = 0.14 \text{ atm} \qquad P(\text{NO}) = (0.34)(0.214 \text{ atm}) = 0.073 \text{ atm}$$

$$P(\text{Br}_2) = 0.17(0.214 \text{ atm}) = 0.036 \text{ atm}$$

$$K_p = \frac{P(\text{NO})^2 P(\text{Br}_2)}{P(\text{NOBr})^2} = 0.0098 \times 10^{-3}$$

20.62 Under what conditions will $CuSO_4 \cdot 5H_2O$ be *efflorescent* at 25 °C? How good a drying agent is $CuSO_4 \cdot 3H_2O$ at the same temperature? For the reaction

$$CuSO_4 \cdot 5H_2O(s) \rightleftharpoons CuSO_4 \cdot 3H_2O(s) + 2H_2O(g)$$

K_p at 25 °C is 1.086×10^{-4} atm²; the vapor pressure of water at 25 °C is 23.8 torr.

▌ An efflorescent salt is one that loses water to the atmosphere. This will occur if the water vapor pressure in equilibrium with the salt is greater than the water vapor pressure in the atmosphere. The mechanism by which $CuSO_4 \cdot 5H_2O$ could be efflorescent is that the salt would lose 2 molecules of H_2O and simultaneously form 1 formula unit of $CuSO_4 \cdot 3H_2O$ for each unit of the original salt that dissociates. Then the above equilibrium equation would apply, since all three components would be present.

Since $CuSO \cdot 5H_2O$ and $CuSO_4 \cdot 3H_2O$ are both solids,

$$K_p = P(H_2O)^2 = 1.086 \times 10^{-4}$$

where $P(H_2O)$ is the partial pressure of water vapor (relative to the standard state pressure of 1 atm) in equilibrium with the two solids. Solving,

$$P(H_2O) = 1.042 \times 10^{-2} \text{ atm} = (1.042 \times 10^{-2} \text{ atm})(760 \text{ torr/atm}) = 7.92 \text{ torr}$$

Since $P(H_2O)$ is less than the vapor pressure of water at the same temperature, $CuSO_4 \cdot 5H_2O$ will not always effloresce. It will effloresce only on a dry day, when the partial pressure of moisture in the air is less than 7.92 torr. This will occur when the relative humidity is less than

$$\frac{7.92 \text{ torr}}{23.8 \text{ torr}} = 0.333 = 33.3\%$$

$CuSO_4 \cdot 3H_2O$ could act as a drying agent by reacting with 2 molecules of H_2O to form $CuSO_4 \cdot 5H_2O$. The same equilibrium would be set up as above, and the vapor pressure of water would be fixed at 7.92 torr. In other words, $CuSO_4 \cdot 3H_2O$ can reduce the moisture content of any confined volume of gas to 7.92 torr. It cannot reduce the moisture content below this value.

To find out the conditions under which $CuSO_4 \cdot 3H_2O$ would be efflorescent, we would have to know the equilibrium constant for another reaction, which shows the product of dehydration of $CuSO_4 \cdot 3H_2O$. This reaction is

$$CuSO_4 \cdot 3H_2O(s) \rightleftharpoons CuSO_4 \cdot H_2O(s) + 2H_2O(g)$$

20.4 THERMODYNAMICS OF EQUILIBRIUM

20.63 If K is not numerically equal to K_p, how can both of the following equations be valid?

$$\Delta G^\circ = -2.30RT \log K \qquad \Delta G^\circ = -2.30RT \log K_p$$

▌ Two different values of ΔG° will be obtained, but one refers to the standard state in which all reactants and products are 1 M and the other refers to the standard state where all reactants and products are at 1 atm.

20.64 Calculate ΔG and ΔG° for the reaction at equilibrium at 27 °C

$$A + B \rightleftharpoons C + D$$

for which $K = 10$.

▮ $\Delta G = 0$ for any system at equilibrium.

$$\Delta G^\circ = -2.30RT \log K = -2.30(8.31 \text{ J/K})(300 \text{ K})(1.0) = -5.73 \text{ kJ}$$

Energies of reaction are not usually specified per mol, so we drop that unit from R when calculating them.

20.65 Which of the following statements is correct for a reversible process in a state of equilibrium?

(*a*) $\Delta G = -2.30RT \log K$ (*b*) $\Delta G = 2.30RT \log K$

(*c*) $\Delta G^\circ = -2.30RT \log K$ (*d*) $\Delta G^\circ = 2.30RT \log K$

▮ (*c*). At equilibrium, $\Delta G = 0$. The larger ΔG°, the less likely the reaction is spontaneous, and the smaller the value of the equilibrium constant.

20.66 At 490 °C, the value of the equilibrium constant, K_p, is 45.9 for the reaction

$$H_2(g) + I_2(g) \rightleftharpoons 2HI(g)$$

Calculate the value of ΔG° for the reaction at that temperature.

▮ $$\Delta G^\circ = -2.30RT \log K_p = -2.30(1.987 \text{ cal/K})(763 \text{ K})(\log 45.9) = -5.79 \times 10^3 \text{ cal} = -5.79 \text{ kcal}$$

20.67 For the reaction at 298 K:

$$A(g) + B(g) \rightleftharpoons D(g) + C(g)$$

$$\Delta H^\circ = -29.8 \text{ kcal} \quad \text{and} \quad \Delta S^\circ = -0.100 \text{ kcal/K}$$

What is the value of the equilibrium constant for the reaction?

▮ $$\Delta G^\circ = \Delta H^\circ - T\,\Delta S^\circ = (-29.8 \text{ kcal}) - (298 \text{ K})(-0.100 \text{ kcal/K}) = 0.0$$

$$\Delta G^\circ = -2.30RT \log K = 0.0$$

$$\log K = 0.0$$

$$K = 1.0$$

20.68 Calculate the equilibrium concentration ratio of C to A if 2.0 mol each of A and B were allowed to come to equilibrium at 300 K.

$$A + B \rightleftharpoons C + D \qquad \Delta G^\circ = 460 \text{ cal}$$

▮ $$-\Delta G^\circ = 2.30RT \log K$$

Adjusting the units of R, we have

$$\log K = \frac{-460 \text{ cal}}{(2.30)(1.99 \text{ cal/K})(300 \text{ K})} = -0.336 \qquad K = 0.461 = \frac{[C][D]}{[A][B]} = \frac{[C]^2}{[A]^2}$$

$$\frac{[C]}{[A]} = \sqrt{0.461} = 0.679$$

20.69 Predict whether sulfur dioxide will reduce copper(II) oxide at 298 K.

$$SO_2(g) + CuO(s) \longrightarrow Cu(s) + SO_3(g)$$

▮ The enthalpy change for the above reaction, calculated from the values of enthalpies of formation from Table 18.2, is 14.1 kcal. Similarly, from the data of Table 18.5, the entropy change is calculated to be -0.60 cal/K.

$$\Delta G^\circ = \Delta H^\circ - T\,\Delta S^\circ = 14.1 \text{ kcal} - (298 \text{ K})(-0.60 \times 10^{-3} \text{ kcal/K}) = 14.3 \text{ kcal}$$

The change in free energy is calculated to be positive, and the reaction will not occur.

20.70 Calculate the standard entropy change and the free energy change when 1.00 mol of water is formed from its elements at 25 °C.

▮ $$H_2(g) + \tfrac{1}{2}O_2(g) \longrightarrow H_2O(l)$$

$$\Delta S_f^\circ = S^\circ(H_2O) - S^\circ(H_2) - \tfrac{1}{2}S^\circ(O_2) = 16.73 - 31.21 - 24.50 = -38.98 \text{ cal/mol}\cdot\text{K}$$

$\Delta G^\circ = \Delta H^\circ - T\,\Delta S^\circ$. Using ΔH_f° from Table 18.2 yields

$$\Delta G_f^\circ = -68\,320 - 298(-38.98) = -56\,700 \text{ cal/mol} = -56.7 \text{ kcal/mol}$$

20.71 When 1.00 mol of pure ethyl alcohol is mixed with 1.00 mol of acetic acid at room temperature, the equilibrium mixture contains $\frac{2}{3}$ mol each of ester and water. **(a)** What is the equilibrium constant? **(b)** What is $\Delta G°$ for the reaction? **(c)** How many mol of ester are formed at equilibrium when 3.00 mol of alcohol is mixed with 1.00 mol of acid? All substances are liquids at the reaction temperature.

▌ (*a*)

	alcohol	acid	ester	water
	$C_2H_5OH(l)$ +	$CH_3COOH(l) \rightleftharpoons$	$CH_3COOC_2H_5(l)$ +	$H_2O(l)$
(1) n at start:	1	1	0	0
(2) Change by reaction:	$-\frac{2}{3}$	$-\frac{2}{3}$	$+\frac{2}{3}$	$+\frac{2}{3}$
(3) n at equilibrium:	$1-\frac{2}{3}=\frac{1}{3}$	$1-\frac{2}{3}=\frac{1}{3}$	$\frac{2}{3}$	$\frac{2}{3}$

This tabular representation is a convenient way of doing the bookkeeping for equilibrium problems. Under each substance in the balanced equation an entry is made on three lines: (1) the amount of starting material; (2) the change (plus or minus) in the number of mol due to the attainment of equilibrium; and (3) the equilibrium amount, which is the algebraic sum of entries (1) and (2). The entries in line (2) must be in the same ratio to each other as the coefficients in the balanced chemical equation. The equilibrium constant can be found from the entries in line (3).

Let V = L of mixture and choose, as usual, 1 mol/L as the standard-state concentration. Then

$$K = \frac{[\text{ester}][\text{water}]}{[\text{alcohol}][\text{acid}]} = \frac{\left(\frac{2/3}{V}\right)\left(\frac{2/3}{V}\right)}{\left(\frac{1/3}{V}\right)\left(\frac{1/3}{V}\right)} = 4$$

Note that the concentration of water is not so large compared with the other reaction components that it remains constant under all reaction conditions; the concentration of water must therefore appear in the expression for K, along with the concentrations of the reactants and of the other product.

(b) $\Delta G° = -2.303RT \log K = -(8.314\ \text{J/K})(298.1\ \text{K})[(2.303)(\log 4)] = -3440\ \text{J} = -3.44\ \text{kJ}$

Because K is independent of the choice of standard-state concentration (the sum of the exponents in the numerator equals the sum of the exponents in the denominator), so also is $\Delta G°$.

(c) Let x = moles of alcohol reacting.

	C_2H_5OH +	$CH_3COOH \rightleftharpoons$	$CH_3COOC_2H_5$ +	H_2O
n at start:	3.00	1.00	0	0
Change by reaction:	$-x$	$-x$	$+x$	$+x$
n at equilibrium:	$3.00 - x$	$1.00 - x$	x	x

$$K = 4 = \frac{\left(\frac{x}{V}\right)\left(\frac{x}{V}\right)}{\left(\frac{3.00-x}{V}\right)\left(\frac{1.00-x}{V}\right)} = \frac{x^2}{3 - 4x + x^2}$$

Then $x^2 = 4(3 - 4x + x^2)$ or $3x^2 - 16x + 12 = 0$. Solving by the quadratic formula,

$$x = \frac{-b \pm \sqrt{b^2 - 4ac}}{2a} = \frac{+16 \pm \sqrt{(16)^2 - 4(3)(12)}}{2(3)} = \frac{16 \pm 10.6}{6} = 4.4 \quad \text{or} \quad 0.90$$

Of the two roots to the quadratic equation, only one has physical meaning. The applicable root can generally be selected very easily. In this problem, we started with 3.00 mol of alcohol and 1.00 mol of acid. We can see from the reaction equation that we cannot form more than 1 mol of ester, even if we use up all the acid. Therefore the correct root of the quadratic is 0.90. Consequently, 0.9 mol of ester is formed at equilibrium.

Note that the number of mol of ester formed is greater than the number of moles of ester in equilibrium in (*a*) (0.90 compared with 2/3). This result was to be expected because of the increased concentration of alcohol, one of the reactants. A further addition of alcohol would lead to an even greater yield of ester, but in no case could the amount of ester formed exceed 1 mol, since 1 mol would represent 100% conversion of the acid.

In practice, the actual selection of the excess concentration to be used may depend on economic factors. If alcohol is cheap compared to acid and ester, a large excess of alcohol might be used to assure a high percent conversion of the acid. If, on the other hand, alcohol costs more per mol than acid, it would be more sensible to use an excess of acid and aim for a high percentage conversion of the alcohol.

20.72 In an experiment at 490 °C, the following equilibrium composition was obtained for the reaction

$$H_2(g) + I_2(g) \rightleftharpoons 2HI(g) + \text{heat}$$

$$[H_2] = 8.62 \times 10^{-4}\ \text{mol/L} \qquad [I_2] = 2.63 \times 10^{-3}\ \text{mol/L} \qquad [HI] = 1.02 \times 10^{-2}\ \text{mol/L}$$

(*a*) Evaluate the equilibrium constant. (*b*) Calculate $\Delta G°$. (*c*) Predict the effect of increasing the volume on the equilibrium. (*d*) Predict the effect of increasing the temperature on the equilibrium.

▌ (*a*) $K = \dfrac{[HI]^2}{[H_2][I_2]} = \dfrac{(1.02 \times 10^{-2})^2}{(8.62 \times 10^{-4})(2.63 \times 10^{-3})} = 45.9$

(*b*) $\Delta G° = -2.30RT \log K = -(2.30)(1.99\ \text{cal/K})(763\ \text{K})(\log 45.9) = -5.80 \times 10^3\ \text{cal} = -5.80\ \text{kcal}$

(*c*) There should be no effect. Increasing the volume decreases the partial pressure of each component. But since there are equal numbers of mol of gas on each side of the equation, there will be no shift of the equilibrium.

(*d*) The equilibrium will shift left.

20.73 Calculate $\Delta E°$ for the reaction $2A(g) + B(g) \longrightarrow A_2B(g)$ for which $\Delta S° = 5.0\ \text{J/K}$, $K = 1.0 \times 10^{-10}$, and $T = 300\ \text{K}$.

▌ The change in the number of moles of gas, Δn, is -2.

$$\Delta G° = -2.30RT \log K = -2.30(8.31\ \text{J/K})(300\ \text{K})(-10) = 57.3\ \text{kJ}$$

$$\Delta H° = \Delta G° + T\,\Delta S° = 57.3 \times 10^3\ \text{J} + (298\ \text{K})(5.0\ \text{J/K}) = 58.8\ \text{kJ}$$

Assuming ideal gas behavior, $\Delta(PV) = (\Delta n)RT$, and

$$\Delta E° = \Delta H° - \Delta(PV) = \Delta H° - (\Delta n)RT = (58.8\ \text{kJ}) - (-2)(8.314 \times 10^{-3}\ \text{kJ/K})(300\ \text{K}) = 63.8\ \text{kJ}$$

20.74 Calculate the value of the equilibrium constant for the following reaction at 300 K and constant pressure:

$$A(g) + B(g) \rightleftharpoons C(g) + D(g) + E(g)$$

using the ideal gas law and the following data:

$$\Delta E° = -90.0\ \text{kcal} \qquad \Delta S° = 100\ \text{cal/K}$$

▌ $\Delta n = 1$

$$\Delta H° = \Delta E° + \Delta(PV) = \Delta E° + (\Delta n)RT = \Delta E° + RT = -90\,000 + 2.0(300) = -89\,400\ \text{cal}$$

$$\Delta G° = \Delta H° - T\,\Delta S° = -89\,400 - (300)(100) = -119\,400\ \text{cal}$$

$$-\Delta G° = 2.30RT \log K$$

$$\log K = \frac{+119\,400}{(2.30)(2.00)(300)} = 86.5 \qquad K = 3 \times 10^{86}$$

20.75 Assuming that the heat capacities of $H_2(g)$, $N_2(g)$, and $NH_3(g)$ do not vary with temperature and further assuming that ΔS is independent of temperature for the reaction

$$3H_2(g) + N_2(g) \longrightarrow 2NH_3(g)$$

estimate the minimum temperature at which this reaction will occur spontaneously with all reactants at unit activity.

▌ At the minimum temperature at which the reaction occurs spontaneously, $\Delta G = \Delta H - T\,\Delta S = 0$. (Any higher temperature would make ΔG negative; any lower would make it positive.) Then $\Delta H = T\,\Delta S$ and $T = \Delta H/\Delta S$. From Tables 18.2 and 18.5,

$$\Delta H = 2\Delta H_f°(NH_3) = 2(-11.04) = -22.08\ \text{kcal} = -92.38\ \text{kJ}$$

$$\Delta S = 2S°(NH_3) - 3S°(H_2) - S°(N_2) = 2(192.5) - 3(130.57) - 191 = -197.7\ \text{J/K}$$

$$T = \frac{-92.38\ \text{kJ}}{-0.1977\ \text{kJ/K}} = 467\ \text{K}$$

20.76 For the reaction

$$C_2H_6(g) \rightleftharpoons C_2H_4(g) + H_2(g)$$

$\Delta G°$ is 5.35 kcal at 900 K. Calculate the mole percent of hydrogen present at equilibrium if pure C_2H_6 is passed over a suitable dehydrogenation catalyst at 900 K and 1.00 atm pressure.

$$\log K = -\frac{\Delta G^\circ}{2.30RT} = -\frac{5350}{(2.30)(1.99)(900)} = -1.30 \qquad K_p = 5.0 \times 10^{-2} = \frac{P(H_2)P(C_2H_4)}{P(C_2H_6)}$$

	initial	produced	used up	equilibrium
$P(C_2H_6)$	1.00		x	$1.00 - x$
$P(C_2H_4)$		x		x
$P(H_2)$		x		x

$$K_p = \frac{x^2}{1.00 - x} = 5.0 \times 10^{-2} \qquad x^2 + 0.050x - 0.050 = 0$$

$$x = \frac{-0.050 \pm \sqrt{(0.050)^2 + 4(0.050)}}{2} = 0.20 \text{ atm}$$

$$P(H_2) = P(C_2H_4) = 0.20 \text{ atm} \qquad P(C_2H_6) = 0.80 \text{ atm} \qquad P_{tot} = 1.20 \text{ atm}$$

$$\text{Mole percent } H_2 = \frac{0.20}{1.20} \times 100\% = 17 \text{ mole percent } H_2$$

20.77 Given the reaction

$$CaCO_3(s) \rightleftharpoons CaO(s) + CO_2(g)$$

estimate the pressure of CO_2 in equilibrium with a mixture of $CaCO_3$ and CaO at 298 K.

▮ From Table 18.4:

$$\Delta G^\circ = \Delta G_f^\circ(CaO) + \Delta G_f^\circ(CO_2) - \Delta G_f^\circ(CaCO_3) = -144.4 + (-94.3) - (-269.8) = 31.1 \text{ kcal}$$

$$\Delta G^\circ = -2.30RT \log K$$

$$\log K = -\frac{\Delta G^\circ}{2.30RT} = \frac{-31\,100}{2.30(1.99)(298)} = -22.8 \qquad K = 1.6 \times 10^{-23} \text{ atm} = P(CO_2)$$

20.78 Using data from Table 18.5, $\Delta H_{comb}(C_6H_6) = -782.3$ kcal/mol and $\Delta H_{comb}(C_2H_2) = -312.0$ kcal/mol, ascertain which of the following reactions is feasible at 298 K with both reactants in their standard states:

$$C_6H_6 \longrightarrow 3C_2H_2 \qquad 3C_2H_2 \longrightarrow C_6H_6$$

▮
$$C_6H_6 + \tfrac{15}{2}O_2 \longrightarrow 6CO_2 + 3H_2O \qquad \Delta H_{comb} = -782.3 \text{ kcal}$$

$$3C_2H_2 + \tfrac{15}{2}O_2 \longrightarrow 6CO_2 + 3H_2O \qquad \Delta H_{comb} = 3(-312.0) \text{ kcal}$$

Subtracting the chemical equations and enthalpy change values yields

$$C_6H_6 \longrightarrow 3C_2H_2 \qquad \Delta H = +153.7 \text{ kcal}$$

$$\Delta S = 3S^\circ(C_2H_2) - S^\circ(C_6H_6) = 3(49.99) - 41.3 = 108.7 \text{ cal/K}$$

$$\Delta G = \Delta H - T\,\Delta S = (153\,700 \text{ cal}) - (298 \text{ K})(108.7 \text{ cal/K}) = 121\,300 \text{ cal}$$

Since the value of ΔG is positive, this reaction cannot be spontaneous. The opposite reaction might go (in the presence of suitable catalysts and conditions).

20.79 Calculate the standard free energy change of the following reaction. Determine the value of its equilibrium constant at 298 K.

$$CH_4 + 2O_2 \rightleftharpoons CO_2 + 2H_2O(g)$$

▮ ΔG° cannot be calculated from the data of Table 18.4 since the ΔG_f° for gaseous water is not included. It must be calculated from ΔH° and S° data (Tables 18.2 and 18.5).

$$\Delta H^\circ = \Delta H_f^\circ(CO_2) + 2\Delta H_f^\circ(H_2O(g)) - \Delta H_f^\circ(CH_4) = (-94.05) + 2(-57.79) - (-17.89) = -191.7 \text{ kcal}$$

$$\Delta S^\circ = S^\circ(CO_2) + 2S^\circ(H_2O(g)) - S^\circ(CH_4) - 2S^\circ(O_2)$$

$$= (213.68) + 2(188.72) - (186) - 2(205) = -5 \text{ J/K} \approx -1 \text{ cal/K}$$

$$\Delta G^\circ = \Delta H^\circ - T\,\Delta S^\circ = (-191\,700 \text{ cal}) - (298 \text{ K})(-1 \text{ cal/K}) = -191\,400 \text{ cal}$$

$$\log K = \frac{+191\,400}{2.30(1.99)(298)} = 140 \qquad K = 10^{140}$$

20.80 Using appropriate thermodynamic data, calculate the equilibrium constant for the reaction at 25 °C

$$CH_4 + Cl_2 \rightleftharpoons CH_3Cl + HCl$$

▌ ΔG_f° values of the compounds can be obtained from Table 18.4, while that of Cl_2 is 0.

$$\Delta G^\circ = \Delta G_f^\circ(HCl) + \Delta G_f^\circ(CH_3Cl) - \Delta G_f^\circ(CH_4) = (-22.7) + (-19.6) - (-12.14) = -30.2 \text{ kcal}$$

$$\Delta G^\circ = -2.30RT \log K$$

$$\log K = -\frac{\Delta G^\circ}{2.30RT} = -\left(\frac{-30\,200 \text{ cal}}{2.30(1.99 \text{ cal/K})(298 \text{ K})}\right) = 22.1$$

$$K = 1.3 \times 10^{22}$$

20.81 Calculate the value of the equilibrium constant for the reaction, at 25 °C.

$$2H_2 + O_2 \rightleftharpoons 2H_2O(l)$$

from the data of Tables 18.2 and 18.5.

▌
$$\Delta H^\circ = 2\Delta H_f^\circ = 2(-68.32) = -136.64 \text{ kcal}$$

$$\Delta S^\circ = 2S^\circ(H_2O) - 2S^\circ(H_2) - S^\circ(O_2) = 2(69.95) - 2(130.57) - (205.04) = -326.3 \text{ J/K} = -78.0 \text{ cal/K}$$

$$\Delta G^\circ = \Delta H^\circ - T\Delta S^\circ = (-136\,640 \text{ cal}) - (298 \text{ K})(-78.0 \text{ cal/K}) = -113\,400 \text{ cal}$$

$$\Delta G^\circ = -2.30RT \log K$$

$$\log K = -\frac{\Delta G^\circ}{2.303RT} = -\frac{-113\,400 \text{ cal}}{2.303(1.987 \text{ cal/K})(298.2 \text{ K})} = 83.10$$

$$K = 1.3 \times 10^{83}$$

20.82 From the data of Tables 18.2 and 18.5, calculate ΔG_f° of 1.00 mol of $NO_2(g)$ at 298 K. Is NO_2 stable at 298 K with respect to decomposition into its elements?

▌
$$\tfrac{1}{2}N_2 + O_2 \longrightarrow NO_2$$

From Table 18.2

$$\Delta H_f^\circ = 8.09 \text{ kcal}$$

From Table 18.5

$$\Delta S_f^\circ = S^\circ(NO_2) - \tfrac{1}{2}S^\circ(N_2) - S^\circ(O_2) = 57.5 - \tfrac{1}{2}(45.7) - (49.0) = -14.4 \text{ cal/K}$$

$$\Delta G_f^\circ = \Delta H_f^\circ - T\Delta S_f^\circ = (8090 \text{ cal}) - (298 \text{ K})(-14.4 \text{ cal/K}) = 12\,380 \text{ cal}$$

Since ΔG_f° has a positive value, ΔG° for the decomposition would have a negative value, and the decomposition could occur spontaneously.

20.83 Using data from Tables 18.2 and 18.5, calculate ΔG_{298}° for the reaction of 50.0 g of nitrogen with oxygen according to the equation

$$N_2(g) + O_2(g) \longrightarrow 2NO(g)$$

Can NO decompose into its elements at 298 K?

▌ For convenience, the calculation is started for the reaction of 1.000 mol of N_2. From Table 18.2

$$\Delta H^\circ = 2\Delta H_f^\circ(NO) = (2 \text{ mol})(21.60 \text{ kcal/mol}) = 43.20 \text{ kcal}$$

From Table 18.5

$$\Delta S^\circ = 2S^\circ(NO) - S^\circ(N_2) - S^\circ(O_2) = 2(50.3) - (45.7) - (49.0) = 5.9 \text{ cal/K}$$

$$\Delta G^\circ = \Delta H^\circ - T\Delta S^\circ = (43\,200 \text{ cal}) - (298 \text{ K})(5.9 \text{ cal/K}) = 41\,400 \text{ cal} = 41.4 \text{ kcal}$$

For 50.0 g of N_2:

$$\Delta G^\circ = (50.0 \text{ g } N_2)\left(\frac{1 \text{ mol } N_2}{28.0 \text{ g}}\right)\left(\frac{41.4 \text{ kcal}}{\text{mol } N_2}\right) = 73.9 \text{ kcal}$$

Since ΔG° for the combination reaction has a positive value, the decomposition can proceed spontaneously.

20.84 For the reaction

$$CO(g) + H_2O(g) \rightleftharpoons CO_2(g) + H_2(g)$$

the equilibrium constant at 1250 K is 0.63. ***(a)*** Calculate $\Delta G°$ for the reaction at 1250 K. ***(b)*** At 1250 K, does $CO(g)$ react with $H_2O(g)$ to form $CO_2(g)$ and $H_2(g)$ in a system in which all four are present at 1.0 atm, or does the reverse reaction occur?

▮ ***(a)*** $\Delta G° = -2.30RT \log K = -2.30(8.31\text{ J/K})(1250\text{ K})(\log 0.63) = +4.79\text{ kJ}$

(b) Since $\Delta G° > 0$, this reaction does not occur, but the reverse reaction does.

20.85 Consider the gaseous reaction

$$2NO_2 \rightleftharpoons N_2O_4$$

(a) Calculate $\Delta G°$ and K_p for this reaction at 25 °C. ***(b)*** Calculate $\Delta G°$ and K_p for the reverse reaction:

$$N_2O_4 \rightleftharpoons 2NO_2$$

(c) Calculate $\Delta G°$ and K_p for the forward reaction written with different coefficients:

$$NO_2 \rightleftharpoons \tfrac{1}{2}N_2O_4$$

▮ ***(a)*** $\Delta G° = [\Delta G_f°(N_2O_4)] - 2[\Delta G_f°(NO_2)] = 97.82 - 2(51.30) = -4.78\text{ kJ} = -RT \ln K_p$

$$\log K_p = \frac{-\Delta G°}{2.303RT} = \frac{4.78 \times 10^3\text{ J}}{(2.303)(8.314\text{ J/K})(298.1\text{ K})} = 0.837 \qquad K_p = 6.87$$

(b) $\Delta G° = 2[\Delta G_f°(NO_2)] - [\Delta G_f°(N_2O_4)] = 2(51.30) - 97.82 = 4.78\text{ kJ}$

$$\log K_p = \frac{-\Delta G°}{2.303RT} = \frac{-4.78 \times 10^3}{(2.303)(8.314)(298.1)} = -0.837 \qquad K_p = 0.146$$

Parts (*a*) and (*b*) illustrate the general requirement that ΔG of a reverse reaction is the negative of ΔG for the forward reaction and K for a reverse reaction is the reciprocal of K for the forward reaction.

(c) $\Delta G° = \tfrac{1}{2}[\Delta G_f°(N_2O_4)] - [\Delta G_f°(NO_2)] = \tfrac{1}{2}(97.82) - 51.30 = -2.39\text{ kJ}$

$$\log K_p = \frac{-\Delta G°}{2.303RT} = \frac{2.39 \times 10^3}{(2.303)(8.314)(298.1)} = 0.418 \qquad K_p = 2.62$$

Parts (*a*) and (*c*) illustrate the general result that ΔG for a reaction with halved coefficients is half the ΔG value for the standard coefficients and that K for the reaction with halved coefficients is the one-half power of the K for the standard reaction. Note that the value of the ratio

$$\frac{P(N_2O_4)}{P(NO_2)^2}$$

at equilibrium must be independent of the way we write the balanced equation. Any equilibrium constant for the reaction must involve the two partial pressures in exactly this way. For part (*a*), K is equal to this ratio; for part (*c*), K is equal to the square root of this ratio.

20.86 In the preparation of quicklime from limestone, the reaction is

$$CaCO_3(s) \rightleftharpoons CaO(s) + CO_2(g)$$

Experiments carried out between 850 and 950 °C led to a set of K_p values fitting an empirical equation

$$\log K_p = 7.282 - \frac{8500}{T}$$

where T is the absolute temperature. If the reaction is carried out in quiet air, what temperature would be predicted from this equation for the complete decomposition of the limestone?

▮ If the $CaCO_3$ can decompose to yield CO_2 at 1.00 atm (to push back the air), it will do so.

$$K_p = P(CO_2) = 1.00 \qquad \log K_p = 0.00 = 7.282 - \frac{8500}{T} \qquad T = 1167\text{ K} = 894\text{ °C}$$

20.87 When α-D-glucose is dissolved in water it undergoes a partial conversion to β-D-glucose, a sugar of the same molecular weight but of slightly different physical properties. This conversion, called *mutarotation*, stops when 63.6% of the glucose is in the β form. Assuming that equilibrium has been attained, calculate K and ΔG° for the reaction α-D-glucose ⇌ β-D-glucose at the experimental temperature.

▌

$$\text{Alpha form} \rightleftharpoons \text{beta form} \qquad K = \frac{63.6}{36.4} = 1.75$$

$$\Delta G^\circ = -2.303RT \log \text{K} = -(2.303)(8.31\ \text{J/K})(298\ \text{K}) \log 1.75 = -1.38\ \text{kJ}$$

20.88 A quantity of PCl_5 was heated in a 12-L vessel at 250 °C.

$$PCl_5(g) \rightleftharpoons PCl_3(g) + Cl_2(g)$$

At equilibrium the vessel contains 0.21 mol PCl_5, 0.32 mol PCl_3, and 0.32 mol Cl_2. **(*a*)** Compute the equilibrium constant K for the dissociation of PCl_5 at 250 °C when concentrations are referred to the standard state of 1 mol/L. **(*b*)** What is ΔG° for the reaction?

▌

$$(a) \quad K = \frac{[PCl_3][Cl_2]}{[PCl_5]} = \frac{\left(\frac{0.32}{12}\right)\left(\frac{0.32}{12}\right)}{\left(\frac{0.21}{12}\right)} = 0.041$$

$$(b) \quad \Delta G^\circ = -RT \ln K = -(8.314\ \text{J/K})(523\ \text{K})[(2.303)(\log 0.041)] = +13\,900\ \text{J} = +13.9\ \text{kJ}$$

20.89 NO and Br_2 at initial partial pressures of 98.4 and 41.3 torr, respectively, were allowed to react at 300 K. At equilibrium the total pressure was 110.5 torr. Calculate the value of the equilibrium constant and the standard free energy change at 300 K for the reaction $2NO(g) + Br_2(g) \rightleftharpoons 2NOBr(g)$.

▌

	initial	**produced**	**used up**	**equilibrium**
$P(NO)$	98.4		x	$98.4 - x$
$P(Br_2)$	41.3		$x/2$	$41.3 - x/2$
$P(NOBr)$		x		x

The total pressure at equilibrium of all three gases is 110.5 torr.

$$(98.4 - x) + (41.3 - x/2) + x = 110.5 \qquad x = 58.4\ \text{torr} = P(NOBr)$$

$$P(NOBr) = 58.4\ \text{torr} = 7.68 \times 10^{-2}\ \text{atm} \qquad P(NO) = 98.4 - x = 40.0\ \text{torr} = 5.26 \times 10^{-2}\ \text{atm}$$

$$P(Br_2) = 41.3 - x/2 = 12.1\ \text{torr} = 1.59 \times 10^{-2}\ \text{atm}$$

The pressures must be expressed in atm since ΔG° is defined with reagents in their standard states, and standard state for gases is expressed as 1.00 atm.

$$K_p = \frac{P(NOBr)^2}{P(NO)^2 P(Br_2)} = \frac{(7.68 \times 10^{-2})^2}{(5.26 \times 10^{-2})^2(1.59 \times 10^{-2})} = 134$$

$$\Delta G^\circ = -2.30RT \log K = -2.30(1.99)(300)(\log 134) = -2.92\ \text{kcal} = -12.2\ \text{kJ}$$

20.90 **(*a*)** Show that

$$2.30 \log K = -\frac{\Delta H^\circ}{RT} + \frac{\Delta S^\circ}{R}$$

(*b*) Given the data below for the system $H_2(g) + \frac{1}{2}S_2(g) \rightleftharpoons H_2S(g)$, and assuming that ΔH° and ΔS° are constant, calculate the value of ΔH° in the temperature range 1000 to 1700 K.

***T*, K**	**log K_p**	***T*, K**	**log K_p**
1023	2.025	1473	0.643
1218	1.305	1667	0.257
1362	0.902		

▌ (*a*) $-\Delta G° = 2.30RT \log K = -\Delta H° + T\,\Delta S°$ $2.30 \log K = -\dfrac{\Delta H°}{RT} + \dfrac{\Delta S°}{R}$

(*b*) The equation in part (*a*) is in the form of a straight line $(y = mx + b)$. If one plots $\log K$ on the y axis and $1/T$ on the x axis, the slope m equals $-\Delta H°/2.30R$ and the intercept b equals $\Delta S°/R$. (The fact that the data yield a straight line when plotted this way supports the assumptions that $\Delta H°$ and $\Delta S°$ are constant throughout this temperature range.)

T	$\log K_p$	$(1/T) \times 10^3$
1023	2.025	0.9775
1218	1.305	0.8210
1362	0.902	0.7342
1473	0.643	0.6789
1667	0.257	0.6000

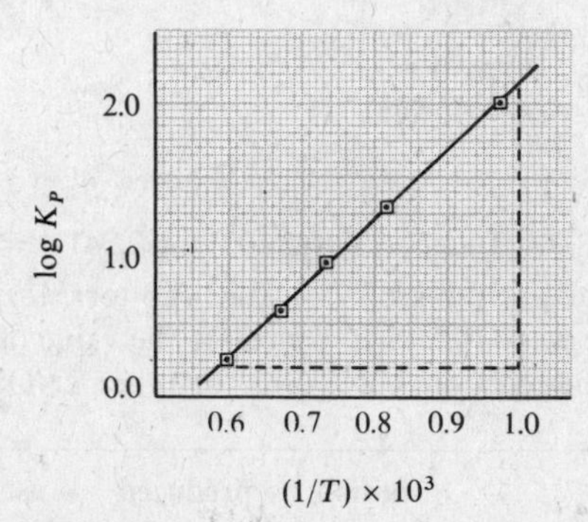

Fig. 20.1

From Fig. 20.1,

$$\text{Slope} = \frac{2.13 - 0.20}{(1.00 - 0.590) \times 10^{-3}} = 4.7 \times 10^3 \qquad -\frac{\Delta H°}{2.30R} = 4.7 \times 10^3$$

$$\Delta H° = -(4.7 \times 10^3)(2.30)(1.99) = -2.2 \times 10^4 \text{ cal} = -22 \text{ kcal}$$

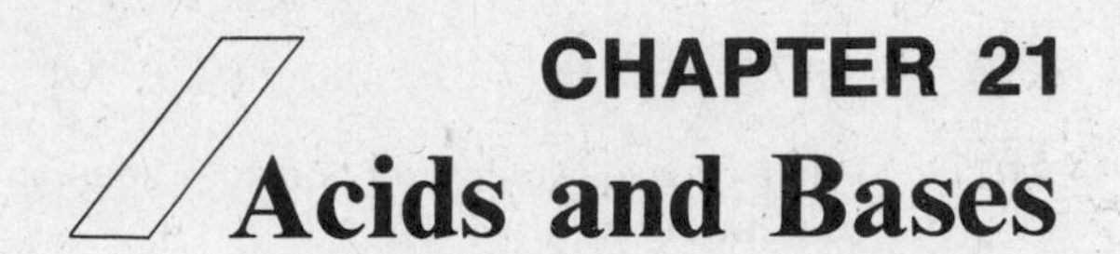

CHAPTER 21
Acids and Bases

21.1 ACID-BASE THEORY

Note: H^+ is often used as an abbreviation for H_3O^+ in aqueous solutions. In equations where this abbreviation is used, less water is included. Thus, the following equations both represent the same reaction in aqueous solution:

$$H_2O + HCl \rightleftharpoons Cl^- + H_3O^+$$
$$HCl \rightleftharpoons Cl^- + H^+$$

Note that not every reagent which is added to a solution will react in a manner such as given in Chap. 11. Some reagents might merely affect the position of an equilibrium.

21.1 When a strong acid dissolves in water, there are ions in the solution afterward. When a soluble metal hydroxide dissolves in water, there are ions in solution afterward. Besides the difference in the identity of the ions, what is the basic difference in these processes?

▌ The dissolving process causes the formation of the ions in the case of the strong acids; the ions are present from the beginning in the case of the metal hydroxides.

21.2 (*a*) Distinguish between acid strength and acid concentration. (*b*) Distinguish between a weak base and an insoluble base.

▌ (*a*) The concentration of acid in a solution is determined by how many mol of acid is dissolved per L of solution; its strength is determined by how completely it ionizes. Both of these factors affect the hydronium ion concentration. (*b*) A weak base furnishes little hydroxide ion to a solution because it ionizes only partially; an insoluble base furnishes little hydroxide ion because it dissolves only slightly.

21.3 Draw an electron dot diagram to illustrate that "ammonium hydroxide" could not exist as a *weak* base. Explain the weakly basic properties of a solution of ammonia in water.

▌

$$\left[\begin{array}{c} \text{H} \\ \text{H}:\ddot{\text{N}}:\text{H} \\ \text{H} \end{array}\right]^+ \quad :\ddot{\underset{\cdot\cdot}{\text{O}}}:\text{H}^-$$

The diagram shows that there is no covalent bonding of the OH^- ion to the NH_4^+ ion. Since solutions of ammonia in water do *not* act as strong bases, this representation must be incorrect. The fact that ammonia acts as a weak base is explained by the equilibrium

$$NH_3 + H_2O \rightleftharpoons NH_4^+ + OH^-$$

21.4 Which of the following reagents are strong electrolytes? (*a*) NH_3 (*b*) NH_4Cl (*c*) $HC_2H_3O_2$ (*d*) $NaC_2H_3O_2$ (*e*) HCl (*f*) NaCl

▌ (*b*), (*d*), (*e*), and (*f*). All salts (*b*, *d*, and *f*), soluble hydroxides, and the acids (*e*) HCl, $HClO_3$, $HClO_4$, HBr, HI, HNO_3, and H_2SO_4 (first proton) are strong electrolytes.

21.5 Write the equilibrium equation for a solution containing acetic acid, $HC_2H_3O_2$, and sodium acetate, $NaC_2H_3O_2$.

▌

$$HC_2H_3O_2 + H_2O \rightleftharpoons C_2H_3O_2^- + H_3O^+$$

Note that the added sodium acetate merely shifted this equilibrium to the left.

21.6 What effect does sodium acetate have on a solution of acetic acid, according to Le Chatelier's principle?

▌ The sodium acetate shifts the ionization reaction to the left. Note that the sodium acetate does not react with the acetic acid to produce other products.

21.7 What reagent would you add to a solution to increase the (*a*) acetate ion concentration? (*b*) ammonium ion concentration?

▌ (*a*) Sodium acetate, or any soluble acetate, but not acetic acid. (*b*) Ammonium chloride, or any soluble ammonium salt, but not ammonia (or ammonium hydroxide—a poorer name for ammonia in water solution).

21.8 Write equilibrium constant expressions for the following reactions:
(*a*) $HC_2H_3O_2 + H_2O \rightleftharpoons H_3O^+ + C_2H_3O_2^-$
(*b*) $HC_2H_3O_2 + CH_3OH \underset{\text{no solvent}}{\rightleftharpoons} H_2O + C_2H_3O_2CH_3$
(*c*) In which of these expressions may $[H_2O]$ be omitted?

▌ (*a*) $K_a = \dfrac{[H_3O^+][C_2H_3O_2^-]}{[HC_2H_3O_2]}$ (*b*) $K = \dfrac{[H_2O][C_2H_3O_2CH_3]}{[HC_2H_3O_2][CH_3OH]}$

(*c*) Note that only in dilute aqueous solution can the (constant) concentration of water be incorporated into the equilibrium constant.

21.9 Which of the reagents listed below could be added to water to make 0.10 M solutions of each of the following ions? (*a*) NH_4^+ (*b*) $C_2H_3O_2^-$ (*c*) Cl^-

(i) NH_3 (ii) NH_4Cl (iii) $HC_2H_3O_2$ (iv) $NaC_2H_3O_2$ (v) HCl (vi) NaCl

▌ (*a*) ii. NH_4Cl will furnish 0.10 M NH_4^+ to a solution. (NH_3, a weak electrolyte, does not dissociate sufficiently to provide 0.10 M NH_4^+.)
(*b*) iv. $NaC_2H_3O_2$. ($HC_2H_3O_2$, a weak electrolyte, does not dissociate sufficiently.)
(*c*) v and vi. HCl and NaCl. Both of these are strong electrolytes, which are completely ionic.

21.10 Sodium chloride, hydrogen chloride gas, and water are put into a vessel. Sodium acetate, acetic acid, and water are put into a second vessel. (*a*) Write a balanced chemical equation for the reaction which occurs in the first vessel. (*b*) Does the added NaCl influence the reaction in any way? (*c*) Is it necessary for every substance which is placed in a vessel to be a reactant in any reaction which might occur? (*d*) Write a balanced equation for the equilibrium reaction which occurs in the second vessel. (*e*) Does the added sodium acetate influence the reaction in any way? (*f*) Are all the substances in the second vessel reactants?

▌ (*a*) $HCl + H_2O \longrightarrow H_3O^+ + Cl^-$ (*b*) no (*c*) no (*d*) $HC_2H_3O_2 + H_2O \longrightarrow H_3O^+ + C_2H_3O_2^-$
(*e*) The acetate ion represses the ionization of the acetic acid (*f*) no

21.11 (*a*) Which of the following 0.10 M solutions are acidic, which are basic, and which are neutral? (*b*) Arrange them in order of increasing pH. (i) NH_4Cl (ii) NaOH (iii) $HC_2H_3O_2$ (iv) NaCl (v) $NH_3 + NH_4Cl$ (vi) NH_3 (vii) HCl

▌

acid			neutral	basic		
HCl	$HC_2H_3O_2$	NH_4Cl	NaCl	($NH_3 + NH_4Cl$)	NH_3	NaOH

Increasing pH $\longrightarrow$

NH_4Cl is acidic because of the hydrolysis of NH_4^+. The buffer solution is less basic than ammonia alone because the NH_4^+ represses the ionization of NH_3 in the buffer.

21.12 For each of the following salts, state which ion (if either) hydrolyzes more in water at 25 °C. (*a*) $NaC_2H_3O_2$ (*b*) NH_4Cl (*c*) MgS (*d*) NaCl (*e*) $NH_4C_2H_3O_2$

▌ (*a*) $C_2H_3O_2^-$ (*b*) NH_4^+ (*c*) S^{2-} (*d*) neither (*e*) both hydrolyze to the same extent, since the values for K_a for acetic acid and K_b for ammonia happen to be equal.

21.13 What is the conjugate acid of HS^-?

▌ H_2S

21.14 Write complete net ionic equations for the following processes. Which combinations of reactants will react less than 2% of the theoretically possible extent? Which one(s) will react until more than 98% of the limiting quantity is used up?

(*a*) $HC_2H_3O_2 + H_2O$ (*b*) $C_2H_3O_2^- + H_2O$ (*c*) $C_2H_3O_2^- + H_3O^+$
(*d*) $NaOH + HC_2H_3O_2$ (*e*) $NaC_2H_3O_2 + HCl(aq)$ (*f*) $HCl(g) + H_2O$

(*g*) $Cl^- + H_3O^+$ (*h*) $Cl^- + H_2O$ (*i*) $NH_4^+ + NaOH$
(*j*) $NH_4^+ + OH^-$ (*k*) $NH_3 + H_2O$ (*l*) $NH_3 + H_3O^+$
(*m*) $NH_3 + HCl(aq)$ (*n*) $Na^+ + OH^-$

(*a*) $HC_2H_3O_2 + H_2O \longrightarrow H_3O^+ + C_2H_3O_2^-$ < 2%
(*b*) $C_2H_3O_2^- + H_2O \longrightarrow HC_2H_3O_2 + OH^-$ < 2%
(*c*) $C_2H_3O_2^- + H_3O^+ \longrightarrow HC_2H_3O_2 + H_2O$ > 98%
(*d*) $OH^- + HC_2H_3O_2 \longrightarrow C_2H_3O_2^- + H_2O$ > 98%
(*e*) $C_2H_3O_2^- + H_3O^+ \longrightarrow HC_2H_3O_2 + H_2O$ > 98%
(*f*) $HCl + H_2O \longrightarrow H_3O^+ + Cl^-$ > 98%
(*g*) $Cl^- + H_3O^+ \longrightarrow$ n.r. < 2%
(*h*) $Cl^- + H_2O \longrightarrow$ n.r. < 2%
(*i*) $NH_4^+ + OH^- \longrightarrow NH_3 + H_2O$ > 98%
(*j*) $NH_4^+ + OH^- \longrightarrow NH_3 + H_2O$ > 98%
(*k*) $NH_3 + H_2O \longrightarrow NH_4^+ + OH^-$ < 2%
(*l*) $NH_3 + H_3O^+ \longrightarrow NH_4^+ + H_2O$ > 98%
(*m*) $NH_3 + H_3O^+ \longrightarrow NH_4^+ + H_2O$ > 98%
(*n*) $Na^+ + OH^- \longrightarrow$ n.r.

21.15 Write the formulas for the conjugate bases of the following acids: (*a*) HCN (*b*) HCO_3^- (*c*) $N_2H_5^+$ (*d*) C_2H_5OH

In each case the conjugate base is derived from the acid by the loss of a proton. (*a*) CN^- (*b*) CO_3^{2-} (*c*) N_2H_4 (*d*) $C_2H_5O^-$

21.16 Write the formulas for the conjugate acids of the following bases: (*a*) $HC_2H_3O_2$ (*b*) HCO_3^- (*c*) C_5H_5N (*d*) $N_2H_5^+$

In each case the conjugate acid is formed from the base by the addition of a proton.

(*a*) $H_2C_2H_3O_2^+$. This is a species which might form in liquid acetic acid upon addition of a strong acid. (*b*) H_2CO_3. Note that HCO_3^- can serve as both an acid [Prob. 21.15(*b*)] and a base. (*c*) $C_5H_5NH^+$. (*d*) $N_2H_6^{2+}$. Note that bases, like acids, can be polyfunctional. The second proton is accepted by N_2H_4, however, only with great difficulty.

21.17 Liquid NH_3, like water, is an amphiprotic solvent. Write the equation for the autoionization of NH_3.

$$2NH_3 \rightleftharpoons NH_4^+ + NH_2^-$$

21.18 Aniline, $C_6H_5NH_2$, is a weak organic base in aqueous solutions. Suggest a solvent in which aniline would become a strong base.

A solvent is needed which has appreciably stronger acid properties than water; one such solvent is liquid acetic acid.

21.19 The self-ionization constant for pure formic acid, $K = [HCOOH_2^+][HCOO^-]$, has been estimated as 10^{-6} at room temperature. What percentage of formic acid molecules in pure formic acid, HCOOH, are converted to formate ion? The density of formic acid is 1.22 g/cm³.

$$\left(\frac{1.22\text{ g}}{\text{cm}^3}\right)\left(\frac{10^3\text{ cm}^3}{\text{L}}\right)\left(\frac{1\text{ mol HCOOH}}{46.0\text{ g HCOOH}}\right) = \frac{26.5\text{ mol}}{\text{L}}$$

$$2HCOOH \rightleftharpoons HCOO^- + HCOOH_2^+ \qquad K = 10^{-6} = [HCOO^-][HCOOH_2^+]$$

Since each ion is produced in equal numbers of moles,

$$[HCOO^-] = [HCOOH_2^+] = 10^{-3}\text{ M}$$

$$\%\text{ dissociated to formate ion} = \left(\frac{10^{-3}}{26.5}\right) \times 100\% = 0.004\%$$

21.20 Would each of the following ions in solution tend to produce an acidic, a basic, or a neutral solution? (*a*) $C_2H_3O_2^-$ (*b*) Na^+ (*c*) SO_3^{2-} (*d*) F^- (*e*) NH_4^+

▌ (*a*), (*c*), and (*d*) basic. Each is the conjugate of a weak acid. (*b*) neutral. It is the conjugate of a strong base. (*e*) acidic. It is the conjugate of a weak base.

21.21 Write an equation showing the dissociation of HX, and label all acids and bases.

▌

$$HX + H_2O \rightleftharpoons H_3O^+ + X^-$$

acid (HX), base (H_2O), acid (H_3O^+), base (X^-)

H_2O and H_3O^+: conjugates

HX and X^-: conjugates

21.22 NH_4ClO_4 and $HClO_4 \cdot H_2O$ both crystallize in the same orthorhombic structure, having unit-cell volumes of 395 and 370 Å^3, respectively. How do you account for the similarity in crystal structure and crystal dimension?

▌ Both substances are ionic substances with lattice sites occupied by cations and anions. In "perchloric acid monohydrate," the cation is H_3O^+ and there are no H_2O *molecules* of hydration. The cations in the two crystals, H_3O^+ and NH_4^+, should occupy nearly equal amounts of space because they are isoelectronic (have the same number of electrons).

The data in this problem have been cited as one of the proofs for the existence of the hydronium ion, H_3O^+.

21.23 (*a*) Give an explanation for the decreasing acid strength in the series $HClO_4$, $HClO_3$, $HClO_2$. (*b*) What would be the relative basic strengths of ClO_4^-, ClO_3^-, ClO_2^-? (*c*) Keeping in mind the discussion of (*a*), how do you account for the fact that there is very little difference in acid strength in the series H_3PO_4, H_3PO_3, H_3PO_2?

▌ (*a*) The Lewis structures are:

```
     ..               ..
    :O:              :O:
     |                |
 ..      ..       ..      ..       ..  ..  ..
:O—Cl—O—H        :O—Cl—O—H        :O—Cl—O—H
 ..  |   ..       ..      ..       ..  ..  ..
    :O:
     ..
```

Since oxygen is more electronegative than chlorine, each terminal oxygen tends to withdraw electrons from the chlorine and in turn from the O—H bond, thus increasing the tendency for the proton to dissociate. In general, the greater the number of terminal oxygens in an oxy-acid, the stronger the acid.

(*b*) A conjugate acid-base pair stand in complementary relationship to each other: the stronger an acid, the weaker its conjugate base. Thus the order of decreasing basic strength is ClO_2^-, ClO_3^-, ClO_4^-, the reverse of the order of decreasing strength of the conjugate acids.

(*c*) The hydrogens in these acids are not all bonded to oxygens. From the Lewis structures,

```
       ..                 ..               ..
      :O:                :O:              :O:
       |                  |                |
   ..      ..         ..      ..               ..
H—O—P—O—H          H—O—P—O—H          H—P—O—H
   ..  |   ..         ..  |   ..           |   ..
      :O:                 H                H
       |
       H
```

it is seen that the number of terminal oxygen atoms is 1 in all three acids. The electronegativities of P and H are almost the same. Thus, no great differences in acidity are expected.

21.24 Which one(s) of the following reactions is (are) Lewis acid-base reactions?

$$NH_3 + BF_3 \longrightarrow H_3N{:}BF_3 \qquad Ca + S \longrightarrow Ca^{2+} + S^{2-}$$

▌ Only the first reaction is defined as a Lewis acid-base reaction. Lewis acid-base reactions involve "donation of an electron pair from base to acid." This donation, however, must result in a coordinate covalent bond.

21.25 (*a*) Is a solution of $AlCl_3$ in water acidic, basic, or neutral? (*b*) Write a net ionic equation to support your answer.

▌ (*a*) Acidic

(*b*) $Al(H_2O)_6^{3+} \rightleftharpoons [Al(H_2O)_5OH]^{2+} + H^+$ or $Al(H_2O)_6^{3+} + H_2O \rightleftharpoons [Al(H_2O)_5OH]^{2+} + H_3O^+$

21.26 In dilute benzene solutions, equimolar additions of $(C_4H_9)_3N$ and HCl produce a substance with a dipole moment. In the same solvent, equimolar additions of $(C_4H_9)_3N$ and SO_3 produce a substance having an almost identical dipole moment. What is the nature of the polar substances formed and what is the unifying feature of HCl and SO_3?

▌ Both HCl and SO_3 are Lewis acids and can react with the amine base to form polar substances which could presumably undergo ionic dissociation in a solvent sufficiently more polar than benzene. The reactions may be represented as follows ($R = C_4H_9$):

$$R_3N{:} + SO_3 \longrightarrow R_3N{-}SO_3$$

Even if a double bond structure were written for SO_3 in which the sulfur atom already had an octet, sulfur is a third-period element which can expand its electron complement in the valence shell to exceed the octet. In any event, sulfur expands its number of valence electrons by attaching to the lone pair on the nitrogen. The N—S bond will be polar because of the big difference in electronegativity between N and S.

$$R_3N{:} + H{-}\ddot{Cl}{:} \longrightarrow R_3N{-}H{\cdots}\ddot{Cl}{:}$$

Here the proton of the HCl attaches to the lone pair on the N. The connection between the N—H and Cl is designated by ---, symbolizing an electron pair on the Cl connected to nitrogen by a hydrogen bond.

21.27 How do you account for the formation of $S_2O_3^{2-}$ from SO_3^{2-} and S in terms of Lewis acid theory?

▌ A sulfur atom is electron deficient and can thus be regarded as an acid. SO_3^{2-}, for which an octet structure can be written, is the base.

$$:\ddot{S} + [SO_3]^{2-} \longrightarrow [S{-}SO_3]^{2-}$$

21.28 A certain reaction is catalyzed by acids, and the catalytic activity for 0.1 M solutions of the acids in water was found to decrease in the order HCl, HCOOH, $HC_2H_3O_2$. The same reaction takes place in anhydrous ammonia, but the three acids all have the same catalytic effect in 0.1 M solutions. Explain.

▌ The order of catalytic activity in water is the same as the order of acidity. In anhydrous ammonia all three acids are strong.

21.29 The amino acid glycine exists predominantly in the form $^+NH_3CH_2COO^-$. Write formulas for ***(a)*** the conjugate base and ***(b)*** the conjugate acid of glycine.

▌ *(a)* $NH_2CH_2COO^-$ *(b)* $^+NH_3CH_2COOH$

21.30 In the reaction of BeF_2 with $2F^-$ to form BeF_4^{2-}, which reactant is the Lewis acid and which is the Lewis base?

▌ BeF_2 is the acid and F^- is the base.

21.31 Liquid ammonia ionizes to a slight extent. At -50 °C, its ion product is $K_{NH_3} = [NH_4^+][NH_2^-] = 10^{-30}$. How many amide ions, NH_2^-, are present per mm³ of pure liquid ammonia?

▌

$$2NH_3 \rightleftharpoons NH_4^+ + NH_2^-$$

$$K = x^2 = 10^{-30} \qquad \text{thus} \qquad x = 10^{-15}\ \text{M} = [NH_2^-]$$

$$\left(\frac{10^{-15}\ \text{mol}}{\text{L}}\right)\left(\frac{1\ \text{L}}{10^6\ \text{mm}^3}\right)\left(\frac{6\times10^{23}\ \text{ions}}{\text{mol}}\right) = 600\ \text{ions/mm}^3$$

21.32 Which is the strongest acid of the following?

(*a*) NH_3 (*b*) $HClO$ (*c*) $HClO_2$ (*d*) $HClO_3$ (*e*) $C_2H_3O_2^-$

▮ $HClO_3$ is strongest. NH_3 and $C_2H_3O_2^-$ are bases. In $HClO_3$, because of the extra oxygen atoms, electrons are pulled away from the H—O bond more than the corresponding bonds in HClO and $HClO_2$.

21.2 IONIZATION CONSTANTS

Note: Before working the following problems, familiarize yourself with Table 21.1, p. 434

21.33 When 0.100 mol of ammonia, NH_3, is dissolved in sufficient water to make 1.00 L of solution, the solution is found to have a hydroxide ion concentration of 1.34×10^{-3} M. Calculate K_b for ammonia.

▮

$$NH_3 + H_2O \rightleftharpoons NH_4^+ + OH^- \qquad K_b = \frac{[NH_4^+][OH^-]}{[NH_3]} \qquad [OH^-] = 1.34 \times 10^{-3}\text{ M}$$

Ionization of the base produces equal concentrations of hydroxide and ammonium ions:

$$[NH_4^+] = [OH^-] = 1.34 \times 10^{-3}\text{ M}$$

Also

$$[NH_3] = 0.100 - (1.34 \times 10^{-3}) = 0.099\text{ M} \qquad K_b = \frac{(1.34 \times 10^{-3})^2}{(0.099)} = 1.81 \times 10^{-5}$$

21.34 Calculate the hydronium ion concentration of a solution containing 0.200 mol of $HC_2H_3O_2$ in 1.00 L of solution. $K_a = 1.80 \times 10^{-5}$. What is the percent ionization of the acid?

▮

$$HC_2H_3O_2 + H_2O \rightleftharpoons H_3O^+ + C_2H_3O_2^- \qquad K_a = \frac{[H_3O^+][C_2H_3O_2^-]}{[HC_2H_3O_2]} = 1.80 \times 10^{-5}$$

At equilibrium, let $[H_3O^+] = x$ then $[C_2H_3O_2^-] = x$ and $[HC_2H_3O_2] = 0.200 - x$

$$\frac{x^2}{0.200 - x} = 1.80 \times 10^{-5} \qquad \text{or} \qquad x^2 + (1.80 \times 10^{-5})x - (3.60 \times 10^{-6}) = 0$$

Application of the quadratic formula yields

$$x = \frac{-1.80 \times 10^{-5} + \sqrt{(3.24 \times 10^{-10}) + (1.44 \times 10^{-5})}}{2} = 1.89 \times 10^{-3}\text{ M} = [H_3O^+]$$

The percent ionization of the acid is the numher of mol which ionized, divided by the number of mol of acid originally present, times 100. The moles which ionized are represented by the acetate ion, $C_2H_3O_2^-$. Since the ion and the nonionized acid are both in the same volume of solution, their concentrations can be used instead of mol.

$$\%\text{ ionized} = \frac{[C_2H_3O_2^-]}{[HC_2H_3O_2]_{\text{original}}} \times 100\% = \frac{1.89 \times 10^{-3}}{0.200} \times 100\% = 0.945\%$$

The acid is less than 1% ionized.

21.35 At 25 °C, a 0.0100 M ammonia solution is 4.1% ionized. Calculate (*a*) the concentration of the OH^- and NH_4^+ ions, (*b*) the concentration of molecular ammonia, (*c*) the ionization constant of aqueous ammonia, (*d*) $[OH^-]$ after 0.0090 mol of NH_4Cl is added to 1.00 L of the above solution, (*e*) $[OH^-]$ of a solution prepared by dissolving 0.010 mol of NH_3 and 0.0050 mol of HCl per L.

▮

$$NH_3 + H_2O \rightleftharpoons NH_4^+ + OH^-$$

The label on a solution refers to the stoichiometric or weight composition and does not indicate the concentration of any particular component of an ionic equilibrium. Thus the designation 0.0100 M NH_3 means that the solution might have been made by dissolving 0.0100 mol ammonia in enough water to make 1.00 L of solution. It does not mean that the concentration of un-ionized ammonia in solution, $[NH_3]$, is 0.0100. Rather, the sum of the ionized and the un-ionized ammonia is equal to 0.0100 mol/L.

(*a*) $[NH_4^+] = [OH^-] = (0.041)(0.0100) = 0.00041$ (*b*) $[NH_3] = 0.0100 - 0.00041 = 0.0096$

(*c*) $K_b = \dfrac{[NH_4^+][OH^-]}{[NH_3]} = \dfrac{(0.00041)(0.00041)}{0.0096} = 1.8 \times 10^{-5}$

(*d*) Since the base is so slightly ionized, we may assume that (1) the $[NH_4^+]$ is completely derived from the NH_4Cl and that (2) the $[NH_3]$ at equilibrium is the same as the stoichiometric molar concentration of the base. Then

$$K_b = \frac{[NH_4^+][OH^-]}{[NH_3]}$$

gives $$[OH^-] = \frac{K_b[NH_3]}{[NH_4^+]} = \frac{(1.8 \times 10^{-5})(0.0100)}{0.0090} = 2.0 \times 10^{-5}$$

The addition of NH_4Cl represses the ionization of NH_3, thus reducing greatly the $[OH^-]$ of the solution; this phenomenon is called the *common-ion effect*, and is in accord with Le Chatelier's Principle.

Assumptions made to simplify the solution of a problem should always be verified after the problem has been solved. In this case we assumed that practically all the NH_4^+ came from the added NH_4Cl. In addition to this, there is a small amount of NH_4^+ resulting from the dissociation of the NH_3. From the chemical equation given in the statement of the problem, the amount of NH_4^+ coming from the NH_3 must equal the amount of OH^-, which we now know is 2.0×10^{-5} M. The correct total $[NH_4^+]$ is then the sum of the contributions of NH_4Cl and NH_3, $0.0090 + (2.0 \times 10^{-5})$. This sum is indeed equal to 0.0090, well within our 10% allowance. We have thus justified our assumption.

The reader may wonder why we must make assumptions at all. This problem could have been solved by a more complete analysis as follows, without the assumption.

	NH_3	+ H_2O ⇌	NH_4^+	+ OH^-
M at start from NH_3:	0.0100		0	0
M at start from NH_4^+:	0		0.0090	0
Change by reaction:	$-x$		$+x$	$+x$
M at equilibrium:	$0.0100 - x$		$0.0090 + x$	x

$$[OH^-] = x = \frac{K_b[NH_3]}{[NH_4^+]} = \frac{(1.8 \times 10^{-5})(0.0100 - x)}{0.0090 + x}$$

This is a quadratic equation in x and can be solved by the usual methods. In the solution,

$$x = \frac{-b \pm \sqrt{b^2 - 4ac}}{2a}$$

the acceptable root requires the same sign for the square root as the sign of b. The value of the square root, however, is so close to the value of b that the square root must be evaluated to 4 significant figures in order that x be known to even 2 figures. There are mathematical methods of solving the problem by approximations without having to evaluate the square root to 4 significant figures, but these mathematical approximations are essentially equivalent to the *chemical* approximations we had originally made.

(*e*) Since HCl is a strong acid, the 0.0050 mol of HCl will react completely with 0.0050 mol of NH_3 to form 0.0050 mol NH_4^+. Of the original 0.0100 mol of NH_3, only half will remain as un-ionized ammonia.

$$[OH^-] = \frac{K_b[NH_3]}{[NH_4^+]} = \frac{(1.8 \times 10^{-5})(0.0050)}{0.0050} = 1.8 \times 10^{-5}$$

Check of assumption: The amount of NH_4^+ contributed by protonation of NH_3 must be equal to the amount of OH^-, or 1.8×10^{-5} mol/L. This is indeed small compared with the 0.0050 mol/L formed by neutralization of NH_3 with HCl.

21.36 Calculate the pH of a 0.10 M NH_3 solution.

▮

$$NH_3 + H_2O \rightleftharpoons NH_4^+ + OH^-$$

$$K_b = \frac{[NH_4^+][OH^-]}{[NH_3]} = \frac{x^2}{0.10} = 1.8 \times 10^{-5} \qquad x^2 = 1.8 \times 10^{-6}$$

$$x = 1.35 \times 10^{-3} = [OH^-] \qquad pOH = 2.87 \qquad pH = 11.13$$

21.37 Calculate the hydronium ion concentration and the percent dissociation of an acid with $K_a = 1.0 \times 10^{-6}$ in (*a*) 0.10 M solution, (*b*) 0.0010 M solution. (*c*) Explain precisely the difference between the terms *higher hydronium ion concentration* and *greater percent dissociation*, as applied to weak acid solutions.

▮

$$HA + H_2O \rightleftarrows H_3O^+ + A^- \qquad K_a = \frac{[H_3O^+][A^-]}{[HA]} = 1.0 \times 10^{-6}$$

(a)

	initial	produced	used up	equilibrium
$[H_3O^+]$	0	x		x
$[A^-]$	0	x		x
$[HA]$	0.10		x	$0.10 - x \simeq 0.10$

$$\frac{x^2}{0.10} = 1.0 \times 10^{-6} \quad \text{or} \quad x^2 = 1.0 \times 10^{-7} = 10 \times 10^{-8} \quad \text{thus} \quad x = 3.2 \times 10^{-4}$$

$$\% \text{ ionization} = \frac{x}{0.10} \times 100\% = \frac{3.2 \times 10^{-4}}{0.10} \times 100\% = 0.32\%$$

(b) $$\frac{y^2}{0.0010} = 1.0 \times 10^{-6} \quad \text{or} \quad y^2 = 1.0 \times 10^{-9} = 10 \times 10^{-10} \quad \text{thus} \quad y = 3.2 \times 10^{-5}$$

$$\% \text{ ionization} = \frac{y}{0.0010} \times 100\% = \frac{3.2 \times 10^{-5}}{0.0010} \times 100\% = 3.2\%$$

(c) Despite the lower hydronium ion concentration, the weak acid is ionized to a greater percent in part (*b*), where it is more dilute.

21.38 What is the H_3O^+ concentration of a solution of 0.600 M $HC_2H_3O_2$?

▮

$$HC_2H_3O_2 + H_2O \longrightarrow C_2H_3O_2^- + H_3O^+$$

$$K_a = 1.8 \times 10^{-5} \quad \text{(from Table 21.1)}$$

$$\frac{[C_2H_3O_2^-][H_3O^+]}{[HC_2H_3O_2]} = 1.8 \times 10^{-5} = \frac{x^2}{0.600} \quad \text{thus} \quad x = 3.3 \times 10^{-3} = [H_3O^+]$$

TABLE 21.1 Values of K_a and K_b at 25 °C

		K_1	K_2	K_3
Acids				
Acetic	$HC_2H_3O_2$	1.8×10^{-5}		
Benzoic	$HC_7H_5O_2$	6.3×10^{-5}		
Boric	H_3BO_3	7.3×10^{-10}		
Carbonic	H_2CO_3	4.5×10^{-7}	4.7×10^{-11}	
Chloroacetic	$HC_2H_2ClO_2$	1.36×10^{-3}		
Cyanic	$HOCN$	3.3×10^{-4}		
Dichloroacetic	$HC_2HCl_2O_2$	5.0×10^{-2}		
Formic	$HCOOH$ or $HCHO_2$	1.8×10^{-4}		
Hydrocyanic	HCN	6.2×10^{-10}		
Hydrofluoric	HF	6.7×10^{-4}		
Hydrogen sulfate ion	HSO_4^-		1.02×10^{-2}	
Hydrogen sulfide	H_2S	1.0×10^{-7}	1.0×10^{-14}	
Nitrous	HNO_2	4.5×10^{-4}		
Oxalic	$H_2C_2O_4$	5.6×10^{-2}	5.4×10^{-5}	
Phenol	HC_6H_5O	1.05×10^{-10}		
Phosphoric	H_3PO_4	7.1×10^{-3}	6.3×10^{-8}	4.5×10^{-13}
Phthalic	$H_2C_8H_4O_4$	1.1×10^{-3}	3.9×10^{-6}	
Bases				
Ammonia	NH_3	1.8×10^{-5}		
Ethylenediamine	$NH_2C_2H_4NH_2$	8.5×10^{-5}	7.1×10^{-8}	
Pyridine	C_5H_5N	1.52×10^{-9}		

21.39 Consider the reaction $A^- + H_3O^+ \rightleftharpoons HA + H_2O$. The K_a value for the acid HA is 1.0×10^{-6}. What is the value of the K for this reaction?

▮ The reaction is the reverse of the ionization reaction of HA; hence the equilibrium constant is the reciprocal of K_a.

$$K = \frac{[HA]}{[A^-][H_3O^+]} = \frac{1}{K_a} = \frac{1}{1.0 \times 10^{-6}} = 1.0 \times 10^6$$

21.40 Write equilibrium constant expressions for the following equations. Show how they are related.

(*a*) $Na_2CO_3 + HCl \rightleftharpoons NaHCO_3 + NaCl$ (*b*) $2Na^+ + CO_3^{2-} + H^+ + Cl^- \rightleftharpoons 2Na^+ + HCO_3^- + Cl^-$
(*c*) $CO_3^{2-} + H^+ \rightleftharpoons HCO_3^-$

▮ (*a*) $K = \dfrac{[NaHCO_3][NaCl]}{[Na_2CO_3][HCl]}$ (*b*) $K = \dfrac{[Na^+]^2[HCO_3^-][Cl^-]}{[Na^+]^2[CO_3^{2-}][H^+][Cl^-]} = \dfrac{[HCO_3^-]}{[CO_3^{2-}][H^+]}$

(*c*) $K = \dfrac{[HCO_3^-]}{[CO_3^{2-}][H^+]}$

The values of the constants in (*b*) and (*c*), of course, are equal. The constant of part (*a*) is inapplicable, since it incorporates [NaCl] when NaCl is not involved in the actual reaction (which takes place between the ions).

21.41 What is the OH^- concentration of a 0.20 M solution of NH_3? ($K_b = 1.8 \times 10^{-5}$.)

▮
$$NH_3 + H_2O \longrightarrow NH_4^+ + OH^-$$

$$K_b = \frac{[NH_4^+][OH^-]}{[NH_3]} = 1.8 \times 10^{-5} = \frac{x^2}{0.20} \qquad \text{thus} \qquad x = [OH^-] = 1.9 \times 10^{-3}\ \text{M}$$

21.42 Calculate the molar concentration of an acetic acid solution which is 2.0% ionized.

▮
$$HC_2H_3O_2 \rightleftharpoons H^+ + C_2H_3O_2^-$$

Let x = molar concentration of acetic acid solution. Then

$$[H^+] = [C_2H_3O_2^-] = 0.020x \qquad \text{and} \qquad [HC_2H_3O_2] = x - 0.020x \approx x$$

the approximation being well within 10%.

$$\frac{[H^+][C_2H_3O_2^-]}{[HC_2H_3O_2]} = K_a \qquad \text{or} \qquad \frac{(0.020x)(0.020x)}{x} = 1.8 \times 10^{-5}$$

Solving, $(0.020)^2x = 1.8 \times 10^{-5}$ or $x = 0.045$

21.43 Calculate the percent ionization of a 1.00 M solution of hydrocyanic acid, HCN.

▮
$$HCN \rightleftharpoons H^+ + CN^-$$

Since H^+ and CN^- are present in the solution only as a result of the ionization, their concentrations must be equal. (The contribution of the ionization of water to $[H^+]$ can be safely neglected.)

Let $x = [H^+] = [CN^-]$. Then $[HCN] = 1.00 - x$. Let us assume that x will be very small compared with 1.00, so that $[HCN] = 1.00$, within the allowed 10% error. Then

$$K_a = \frac{[H^+][CN^-]}{[HCN]} \qquad \text{or} \qquad 6.2 \times 10^{-10} = \frac{x^2}{1.00} \qquad \text{thus} \qquad x = 2.5 \times 10^{-5}.$$

$$\%\ \text{ionization} = \frac{\text{ionized HCN}}{\text{total HCN}} \times 100 = \frac{2.5 \times 10^{-5}\ \text{mol/L}}{1.00\ \text{mol/L}} \times 10^2 = 0.0025\%$$

Check of assumption: $x(=2.5 \times 10^{-5})$ is indeed very small compared with 1.00. (Or, as long as the percent ionization is less than 10%, the assumption is justified.)

21.44 In 1.00 L of 0.100 M solution, 0.00135 mol of ammonia dissociates. Calculate the value of the equilibrium constant.

▮
$$NH_3 + H_2O \rightleftharpoons NH_4^+ + OH^-$$

$$K_b = \frac{[NH_4^+][OH^-]}{[NH_3]} = \frac{(0.00135)^2}{(0.100 - 0.00135)} = 1.84 \times 10^{-5}$$

21.45 The $[H^+]$ in a 0.072 M solution of benzoic acid is 2.1×10^{-3}. Compute K_a for the acid from this information.

$$HC_7H_5O_2 \rightleftharpoons H^+ + C_7H_5O_2^-$$

▮ Since the hydrogen ion and benzoate ion come only from the ionization of the acid, their concentrations must be equal. (We have neglected the contribution of the ionization of water to $[H^+]$.)

$$[H^+] = [C_7H_5O_2^-] = 2.1 \times 10^{-3} \qquad [HC_7H_5O_2] = 0.072 - (2.1 \times 10^{-3}) = 0.070$$

$$K_a = \frac{[H^+][C_7H_5O_2^-]}{[HC_7H_5O_2]} = \frac{(2.1 \times 10^{-3})^2}{0.070} = 6.3 \times 10^{-5}$$

21.46 The ionization constant of formic acid, HCO_2H, is 1.80×10^{-4}. What is the percent ionization of a 0.00100 M solution of this acid?

▮ Let $x = [H^+] = [HCO_2^-]$; then $[HCO_2H] = 0.00100 - x$. Assume, as in Prob. 21.43, that the percent ionization is less than 10% and that the formic acid concentration, $0.00100 - x$, may be approximated as 0.00100. Then

$$K_a = \frac{[H^+][HCO_2^-]}{[HCO_2H]} = \frac{x^2}{0.00100} = 1.80 \times 10^{-4} \qquad \text{or} \qquad x = 4.2 \times 10^{-4}$$

In checking the assumption, we see that x is not negligible compared with 0.00100. Therefore the assumption and the solution based on it must be rejected and the full quadratic form of the equation must be solved.

$$\frac{x^2}{0.00100 - x} = 1.80 \times 10^{-4} \qquad \text{thus} \qquad x = 3.4 \times 10^{-4}$$

(We reject the negative root, -5.2×10^{-4}.)

$$\%\ \text{ionization} = \frac{\text{ionized } HCO_2H}{\text{total } HCO_2H} \times 100 = \frac{3.4 \times 10^{-4}}{0.00100} \times 100 = 34\%$$

This exact solution (34% ionization) shows that the solution based on the original assumption was in error by about 20%.

21.47 What concentration of acetic acid is needed to give a $[H^+]$ of 3.5×10^{-4} M?

▮ Let x = mol of acetic acid/L.

$$[H^+] = [C_2H_3O_2^-] = 3.5 \times 10^{-4} \qquad [HC_2H_3O_2] = x - 3.5 \times 10^{-4}$$

$$\frac{[H^+][C_2H_3O_2^-]}{[HC_2H_3O_2]} = \frac{(3.5 \times 10^{-4})^2}{x - 3.5 \times 10^{-4}} = 1.8 \times 10^{-5} \qquad \text{or} \qquad x = 7.2 \times 10^{-3}$$

21.48 The K_a of an unknown acid, HX, was determined to be 1.0×10^{-6}. Calculate the $[H_3O^+]$ and $[X^-]$ in a 0.10 M solution of HX.

▮

$$HX + H_2O \longrightarrow H_3O^+ + X^-$$

$$K_a = \frac{[H_3O^+][X^-]}{[HX]} = \frac{x^2}{0.10 - x} \approx \frac{x^2}{0.10} = 1.0 \times 10^{-6} \qquad \text{thus} \qquad x = [H_3O^+] = [X^-] = 3.2 \times 10^{-4}$$

21.49 What is the CN^- ion concentration in a 0.010 M solution of HCN?

▮

$$HCN + H_2O \longrightarrow H_3O^+ + CN^-$$

$$K_a = \frac{[CN^-][H_3O^+]}{[HCN]} = \frac{x^2}{0.010} = 6.2 \times 10^{-10} \qquad \text{thus} \qquad x = 2.5 \times 10^{-6}\ \text{M} = [CN^-]$$

21.50 Calculate the $[H^+]$ and the $[OH^-]$ in 0.100 M $HC_2H_3O_2$ which is 1.31% ionized.

▮

$$[H^+] = (0.0131)(0.100) = 1.31 \times 10^{-3} \qquad [OH^-] = \frac{1.00 \times 10^{-14}}{[H^+]} = \frac{1.00 \times 10^{-14}}{1.31 \times 10^{-3}} = 7.6 \times 10^{-12}$$

$[H^+]$ is computed as if the $HC_2H_3O_2$ were the only contributor, whereas $[OH^-]$ is based on the ionization of water. If water ionizes to supply OH^-, it must supply an equal amount of H^+ at the same time. Implied in this solution is the assumption that water's contribution to $[H^+]$, 7.6×10^{-12} mol/L, is negligible compared with that of the $HC_2H_3O_2$. (This assumption will be valid in all but the most dilute or weakest acid solutions.) In computing $[OH^-]$, however, water is the only source and it therefore cannot be overlooked.

21.51 Determine the $[OH^-]$ and the $[H^+]$ in 0.0100 M ammonia solution which is 4.1% ionized.

▮

$$[OH^-] = (0.041)(0.0100) = 4.1 \times 10^{-4} \qquad [H^+] = \frac{1.00 \times 10^{-14}}{[OH^-]} = \frac{1.00 \times 10^{-14}}{4.1 \times 10^{-4}} = 2.4 \times 10^{-11}$$

Here we have made the assumption that the contribution of water to $[OH^-]$ (equal to $[H^+]$, or 2.4×10^{-11}) is negligible compared with that of NH_3. K_w is used to compute $[H^+]$, since water is the only supplier of H^+. In general, $[H^+]$ for acidic solutions can usually be computed without regard to the water equilibrium; then K_w is used to compute $[OH^-]$. Conversely, $[OH^-]$ for basic solutions can usually be computed without regard to the water equilibrium; then K_w is used to compute $[H^+]$.

21.52 Calculate the ionization constant of formic acid, $HCHO_2$, which ionizes 4.2% in 0.10 M solution.

▮

$$HCHO_2 + H_2O \rightleftharpoons CHO_2^- + H_3O^+ \qquad K_a = \frac{[H_3O^+][CHO_2^-]}{[HCHO_2]} = \frac{(4.2 \times 10^{-3})^2}{0.10 - 0.0042} = 1.8 \times 10^{-4}$$

21.53 A solution of acetic acid is 1.0% ionized. Determine the molar concentration of acetic acid and the $[H^+]$ of the solution.

▮

$$HC_2H_3O_2 + H_2O \rightleftharpoons C_2H_3O_2^- + H_3O^+ \qquad K_a = \frac{[H_3O^+][C_2H_3O_2^-]}{[HC_2H_3O_2]} = 1.8 \times 10^{-5}$$

Since $$\frac{[C_2H_3O_2^-]}{[HC_2H_3O_2]} = 0.010 \qquad [H_3O^+](0.010) = 1.8 \times 10^{-5} \quad \text{and} \quad [H_3O^+] = 1.8 \times 10^{-3}\ M$$

$$[HC_2H_3O_2]_0 = 100[H_3O^+] = 1.8 \times 10^{-1}\ M$$

21.54 Determine (*a*) the degree of ionization and (*b*) the $[OH^-]$ of a 0.080 M solution of NH_3.

▮

$$NH_3 + H_2O \rightleftharpoons NH_4^+ + OH^- \qquad K_b = \frac{[NH_4^+][OH^-]}{[NH_3]} = 1.8 \times 10^{-5}$$

At equilibrium, $[NH_4^+] = [OH^-]$ since both were produced by the ionization reaction. Let $[NH_4^+] = [OH^-] = x$

$$\frac{x^2}{0.080} = 1.8 \times 10^{-5} \qquad \text{thus} \qquad x = 1.2 \times 10^{-3}$$

$$\%\ \text{ionized} = \left(\frac{1.2 \times 10^{-3}}{0.080}\right) \times 100\% = 1.5\%$$

21.55 Fluoroacetic acid has a K_a of 2.6×10^{-3}. What concentration of the acid is needed so that $[H^+]$ is 2.0×10^{-3}?

▮

$$HC_2H_2FO_2 + H_2O \rightleftharpoons C_2H_2FO_2^- + H_3O^+$$

$$K_a = \frac{[H_3O^+][C_2H_2FO_2^-]}{[HC_2H_2FO_2]} = \frac{(2.0 \times 10^{-3})^2}{[HC_2H_2FO_2]} = 2.6 \times 10^{-3}$$

Thus $$[HC_2H_2FO_2] = \frac{(2.0 \times 10^{-3})^2}{2.6 \times 10^{-3}} = 1.5 \times 10^{-3}\ M \text{ remaining in solution.}$$

$$\text{Total concentration} = (2.0 \times 10^{-3}) + (1.5 \times 10^{-3}) = 3.5 \times 10^{-3}\ M$$

21.56 What molar concentration of NH_3 provides a $[OH^-]$ of 1.5×10^{-3}?

▮

$$NH_3 + H_2O \rightleftharpoons NH_4^+ + OH^-$$

$$K_b = \frac{[NH_4^+][OH^-]}{[NH_3]} = \frac{(1.5 \times 10^{-3})(1.5 \times 10^{-3})}{[NH_3]} = 1.8 \times 10^{-5}$$

$$\text{Equilibrium } [NH_3] = \frac{(1.5 \times 10^{-3})^2}{(1.8 \times 10^{-5})} = 0.125\ M \qquad \text{and} \qquad \text{Total } [NH_3] = (1.5 \times 10^{-3}) + 0.125 = 0.13\ M$$

21.57 What is $[NH_4^+]$ in a solution that is 0.0200 M NH_3 and 0.0100 M KOH?

▮

$$NH_3 + H_2O \rightleftharpoons NH_4^+ + OH^-$$

$$K_b = \frac{[NH_4^+][OH^-]}{[NH_3]} = \frac{[NH_4^+](0.0100)}{(0.0200)} = 1.8 \times 10^{-5} \qquad \text{so} \qquad [NH_4^+] = 3.6 \times 10^{-5}$$

21.58 What is $[HCOO^-]$ in a solution that is both 0.015 M HCOOH and 0.020 M HCl?

$$HCOOH + H_2O \rightleftharpoons HCOO^- + H_3O^+$$

The HCl reacts completely with water to yield 0.020 M H_3O^+.

$$K_a = \frac{[H_3O^+][HCOO^-]}{[HCOOH]} = \frac{[HCOO^-](0.020)}{(0.015)} = 1.8 \times 10^{-4} \quad \text{so} \quad [HCOO^-] = 1.4 \times 10^{-4}$$

21.59 What concentration of dichloroacetic acid gives a $[H^+]$ of 8.5×10^{-3}?

$$HC_2HCl_2O_2 + H_2O \rightleftharpoons C_2HCl_2O_2^- + H_3O^+$$

$$K_a = \frac{[H_3O^+][C_2HCl_2O_2^-]}{[HC_2HCl_2O_2]} = \frac{(8.5 \times 10^{-3})^2}{C - (8.5 \times 10^{-3})} = 5.0 \times 10^{-2}$$

or $(8.5 \times 10^{-3})^2 = (0.050)C - (8.5 \times 10^{-3})(0.050)$ and $C = 0.0099$ M

21.60 Calculate $[H^+]$ and $[C_2HCl_2O_2^-]$ in a solution that is 0.0100 M in HCl and 0.0100 M in $HC_2HCl_2O_2$. K_a for $HC_2HCl_2O_2$ (dichloroacetic acid) is 5.0×10^{-2}.

$$HC_2HCl_2O_2 + H_2O \rightleftharpoons C_2HCl_2O_2^- + H_3O^+ \qquad K_a = \frac{[H_3O^+][C_2HCl_2O_2^-]}{[HC_2HCl_2O_2]} = 5.0 \times 10^{-2}$$

	initial	**produced**	**used up**	**equilibrium**
$[HC_2HCl_2O_2]$	0.0100		x	$0.0100 - x$
$[C_2HCl_2O_2^-]$	0	x		x
$[H_3O^+]$	0.0100	x		$0.0100 + x$

$$\frac{x(0.0100 + x)}{0.0100 - x} = 0.050 \quad \text{or} \quad x(0.0100 + x) = 5.0 \times 10^{-4} - 0.050x \quad \text{or} \quad x^2 + 0.0600x - 5 \times 10^{-4} = 0$$

$$x = \frac{-0.060 + \sqrt{(0.060)^2 + 4(5.0 \times 10^{-4})}}{2} = 7.4 \times 10^{-3}$$

$$[H_3O^+] = 0.0100 + (7.4 \times 10^{-3}) = 0.0174 \text{ M}$$

21.61 What is the percent ionization of 0.0065 M chloroacetic acid?

$$HC_2H_2ClO_2 + H_2O \rightleftharpoons C_2H_2ClO_2^- + H_3O^+$$

$$K_a = \frac{[H_3O^+][C_2H_2ClO_2^-]}{[HC_2H_2ClO_2]} = \frac{x^2}{0.0065 - x} = 1.36 \times 10^{-3}$$

$$x^2 = -(1.36 \times 10^{-3})x + 8.84 \times 10^{-6} \quad \text{or} \quad x^2 + (1.36 \times 10^{-3})x - 8.84 \times 10^{-6} = 0$$

$$x = \frac{-1.36 \times 10^{-3} + \sqrt{(1.36 \times 10^{-3})^2 + 4(8.84 \times 10^{-6})}}{2} = 2.37 \times 10^{-3}$$

$$\% \text{ ionization} = \left(\frac{2.37 \times 10^{-3}}{6.5 \times 10^{-3}}\right) \times 100\% = 36\%$$

21.62 Calculate the percent dissociation of acetic acid in a solution 0.200 M in $HC_2H_3O_2$ and 0.100 M in HCl.

Since the HCl is completely dissociated in water, the equilibrium of interest is the ionization of the acetic acid in the presence of 0.100 M H_3O^+:

$$HC_2H_3O_2 + H_2O \rightleftharpoons H_3O^+ + C_2H_3O_2^- \qquad K_a = \frac{[H_3O^+][C_2H_3O_2^-]}{[HC_2H_3O_2]}$$

At equilibrium, let $[C_2H_3O_2^-] = x$ $\quad [HC_2H_3O_2] = 0.200 - x \approx 0.200 \quad [H_3O^+] = 0.100 + x \approx 0.100$

Then $$K_a = \frac{(0.100)x}{0.200} = 1.8 \times 10^{-5} \qquad x = 3.6 \times 10^{-5} \text{ M} = [C_2H_3O_2^-]$$

Since all the acetate ion resulted from the dissociation of acetic acid, the percent dissociation is given by

$$\% \text{ dissociation} = \frac{x}{0.200} \times 100 = \frac{3.6 \times 10^{-5}}{0.200} \times 100 = 0.018\%$$

If there were no HCl in this solution, the acetic acid would have been 0.945% dissociated, as shown in Prob. 21.34. Thus the added strong acid represses the ionization of the acetic acid.

21.63 Consider a solution of a monoprotic weak acid of acidity constant K_a. What is the minimum concentration, C, in terms of K_a, such that the concentration of the undissociated acid can be equated to C within a 10% limit of error? Assume that activity coefficient corrections can be neglected.

▮ Let C = minimum initial concentration

$$HA + H_2O \rightleftharpoons H_3O^+ + A^- \qquad K_a = \frac{[H_3O^+][A^-]}{[HA]} = \frac{x^2}{C - x}$$

The approximation which is usually made is $C - x \approx C$. Within an error of 10%,

$$C - x = 0.90C \quad \text{or} \quad 0.10C = x$$

$$K_a = \frac{x^2}{C - x} = \frac{(0.10C)^2}{0.90C} = \frac{C}{90} \quad \text{so} \quad C = 90K_a$$

21.64 Calculate $[H^+]$, $[C_2H_3O_2^-]$, and $[C_7H_5O_2^-]$ in a solution that is 0.0200 M in $HC_2H_3O_2$ and 0.0100 M in $HC_7H_5O_2$. K_a values for $HC_2H_3O_2$ and $HC_7H_5O_2$ are 1.8×10^{-5} and 6.3×10^{-5}, respectively.

▮ Since the acids are approximately the same strength, we may neglect the ionization of neither.

$$HC_2H_3O_2 + H_2O \rightleftharpoons C_2H_3O_2^- + H_3O^+ \qquad K_a = \frac{[H_3O^+][C_2H_3O_2^-]}{[HC_2H_3O_2]} = 1.8 \times 10^{-5}$$

$$HC_7H_5O_2 + H_2O \rightleftharpoons C_7H_5O_2^- + H_3O^+ \qquad K_a = \frac{[H_3O^+][C_7H_5O_2^-]}{[HC_7H_5O_2]} = 6.3 \times 10^{-5}$$

Assuming a negligible reduction in acid concentration by ionization,

$$[H_3O^+] = \frac{(1.8 \times 10^{-5})(0.0200)}{[C_2H_3O_2^-]} = \frac{(6.3 \times 10^{-5})(0.0100)}{[C_7H_5O_2^-]}$$

By electroneutrality,

$$[H_3O^+] = [C_2H_3O_2^-] + [C_7H_5O_2^-] = \frac{3.6 \times 10^{-7}}{[H_3O^+]} + \frac{6.3 \times 10^{-7}}{[H_3O^+]}$$

$$[H_3O^+]^2 = (3.6 \times 10^{-7}) + (6.3 \times 10^{-7}) = 9.9 \times 10^{-7}$$

$$[H_3O^+] = 1.0 \times 10^{-3}\ \text{M} \qquad [C_2H_3O_2^-] = 3.6 \times 10^{-4}\ \text{M} \qquad [C_7H_5O_2^-] = 6.3 \times 10^{-4}\ \text{M}$$

21.65 Find the value of $[OH^-]$ in a solution made by dissolving 0.0050 mol each of ammonia and pyridine (C_5H_5N) in enough water to make 200 cm^3 of solution. What are the concentrations of ammonium and pyridinium ions?

▮ $[OH^-]$ is not significantly affected by ionization of the weaker base.

$$NH_3 + H_2O \rightleftharpoons NH_4^+ + OH^- \qquad K_b = \frac{[NH_4^+][OH^-]}{[NH_3]} = 1.8 \times 10^{-5}$$

$$\text{Initial } NH_3 \text{ concentration} = \frac{0.0050\ \text{mol}}{0.200\ \text{L}} = 0.025\ \text{M}$$

$$\frac{x^2}{0.025} = 1.8 \times 10^{-5} \quad \text{thus} \quad x = 6.7 \times 10^{-4} = [NH_4^+] = [OH^-]$$

$$C_5H_5N + H_2O \rightleftharpoons C_5H_5NH^+ + OH^- \qquad K_b = \frac{[C_5H_5NH^+][OH^-]}{[C_5H_5N]} = 1.52 \times 10^{-9}$$

$[OH^-]$ was calculated above (both equilibria are in the same solution)

$$K_b = \frac{[C_5H_5NH^+](6.7 \times 10^{-4})}{(0.025)} = 1.52 \times 10^{-9} \quad \text{so} \quad [C_5H_5NH^+] = 5.7 \times 10^{-8}$$

21.66 What are $[H^+]$, $[C_3H_5O_3^-]$, and $[C_6H_5O^-]$ in a solution that is 0.030 M $HC_3H_5O_3$ and 0.100 M HC_6H_5O? K_a values for $HC_3H_5O_3$ and HC_6H_5O are 1.38×10^{-4} and 1.05×10^{-10}, respectively.

▌ The ionization of the weaker acid is neglected in calculating $[H_3O^+]$.

$$HC_3H_5O_3 + H_2O \rightleftharpoons C_3H_5O_3^- + H_3O^+$$

$$K_a = \frac{[H_3O^+][C_3H_5O_3^-]}{[HC_3H_5O_3]} = \frac{x^2}{0.030} = 1.38 \times 10^{-4} \qquad \text{thus} \qquad x = 2.0 \times 10^{-3} = [H_3O^+] = [C_3H_5O_3^-]$$

$$HC_6H_5O + H_2O \rightleftharpoons C_6H_5O^- + H_3O^+ \qquad K_a = \frac{[H_3O^+][C_6H_5O^-]}{[HC_6H_5O]} = 1.05 \times 10^{-10}$$

$[H_3O^+]$ was calculated above. Since both acids are in the same solution, there is only *one* hydrogen ion concentration.

$$K_a = \frac{(2.0 \times 10^{-3})[C_6H_5O^-]}{0.100} = 1.05 \times 10^{-10} \qquad \text{so} \qquad [C_6H_5O^-] = 5.2 \times 10^{-9}\ \text{M}$$

21.67 Calculate $[H^+]$ in a solution that is 0.100 M HCOOH and 0.100 M HOCN.

▌ This is a case in which two weak acids both contribute to $[H^+]$, neither contributing such a preponderant amount that the other's share can be neglected.

	HCOOH ⇌	H^+ +	HCO_2^-	HOCN ⇌	H^+ +	OCN^-
M (initial):	0.100	0	0	0.100	0	0
change by reaction:	$-x$	$+x$	$+x$	$-y$	$+y$	$+y$
M (equilibrium):	$0.100 - x$	$x + y$	x	$0.100 - y$	$x + y$	y
approx. *M* (equilibrium):	0.100	$x + y$	x	0.100	$x + y$	y

The final line in the above tabulation is based on the assumption that x and y are both small compared with 0.100.

$$\frac{x(x+y)}{0.100} = 1.80 \times 10^{-4} \qquad \frac{y(x+y)}{0.100} = 3.3 \times 10^{-4}$$

Dividing the HOCN equation by the HCOOH equation,

$$\frac{y}{x} = \frac{3.3}{1.80} = 1.83 \qquad \text{or} \qquad y = 1.83x$$

Subtracting the HCOOH equation from the HOCN equation,

$$\frac{y(x+y) - x(x+y)}{0.100} = 1.5 \times 10^{-4} \qquad \text{or} \qquad y^2 - x^2 = 1.5 \times 10^{-5}$$

Substitute $y = 1.83x$ into the last equation and solve to obtain $x = 2.5 \times 10^{-3}$. Then

$$y = 1.83x = 4.6 \times 10^{-3} \qquad \text{and} \qquad [H^+] = x + y = 7.1 \times 10^{-3}$$

Check of assumptions: The values of x and y are slightly less than 10% of 0.100.

21.3 IONIZATION OF WATER

21.68 The number 10.92 has how many significant figures? The number 0.92 has how many? If these were pH values, how many significant figures should be reported for the corresponding hydronium ion concentrations?

▌ The number 10.92 has four significant figures; 0.92 has two. Since the integral portion of a logarithm determines the power of 10 only, each hydronium ion concentration should be reported to 2 significant digits—the same number of decimal digits in the pH.

$$\text{antilog}(-10.92) = 10^{-10.92} = 1.2 \times 10^{-11} \qquad \text{antilog}(-0.92) = 10^{-0.92} = 1.2 \times 10^{-1}$$

21.69 What is the pH of 10^{-2} M KOH?

▌ $$[OH^-] = 10^{-2} \qquad [H_3O^+] = 10^{-14}/10^{-2} = 10^{-12} \qquad \text{thus} \qquad \text{pH} = 12$$

21.70 Calculate the pH of a solution which has a hydronium ion concentration of 6.0×10^{-8} M.

▌ $$\text{pH} = -\log(6.0 \times 10^{-8}) = -\log 6.0 - \log 10^{-8} = -0.78 + 8 = 7.22$$

21.71 Calculate the hydronium ion concentration of a solution which has a pH of 11.73.

▌ $$pH = -\log [H_3O^+] = 11.73$$

In order that it may be located in the logarithm table, the part of the logarithm to the right of the decimal point (the mantissa) must be positive. Hence

$$\log [H_3O^+] = -11.73 \qquad \text{thus} \qquad [H_3O^+] = 10^{-11.73} = 10^{0.27} \times 10^{-12}$$

From the logarithm table, $10^{0.27} = 1.9$. Hence $[H_3O^+] = 1.9 \times 10^{-12}$

21.72 In pure water at 25 °C, $[H_3O^+] = 1.0 \times 10^{-7}$. What is the pH of pure water at 25 °C?

▌ $$pH = -\log (1.0 \times 10^{-7}) = 7.00$$

21.73 Calculate the hydronium ion concentration and the hydroxide ion concentration in pure water at 25 °C.

▌ $$2H_2O \rightleftharpoons H_3O^+ + OH^- \qquad K_w = [H_3O^+][OH^-] = 1.0 \times 10^{-14}$$

Let $$x = [H_3O^+] = [OH^-]$$

Hence $$x^2 = 1.0 \times 10^{-14} \qquad \text{and} \qquad x = 1.0 \times 10^{-7}\text{ M} = [H_3O^+] = [OH^-]$$

21.74 Calculate the hydronium ion concentration of a 0.100 M NaOH solution.

▌ The hydroxide ion concentration from the autoionization of water is negligible compared to that provided from the NaOH, and the solution consists of 0.100 M Na^+ and 0.100 M OH^-. Hence $[OH^-] = 0.100$ M. In any dilute aqueous solution $K_w = [H_3O^+][OH^-] = 1.0 \times 10^{-14}$

In this solution, $1.0 \times 10^{-14} = [H_3O^+](0.100)$ thus $[H_3O^+] = 1.0 \times 10^{-13}$

21.75 Express the following H^+ concentrations in terms of pH: (*a*) 1×10^{-3} mol/L (*b*) 5.4×10^{-9} mol/L

▌ (*a*) $pH = -\log [H^+] = -\log 10^{-3} = 3$

(*b*) $pH = -\log [H^+] = -\log (5.4 \times 10^{-9}) = -\log 5.4 + 9 = -0.73 + 9 = 8.27$

21.76 Calculate the pH values, assuming complete ionization, of (*a*) 4.9×10^{-4} M monoprotic acid (*b*) 0.0016 M monoprotic base.

▌ (*a*) Here $[H^+] = 4.9 \times 10^{-4}$

$$pH = -\log [H^+] = -\log (4.9 \times 10^{-4}) = -\log 4.9 + 4 = -0.69 + 4 = 3.31$$

(*b*) $$[H^+] = \frac{10^{-14}}{[OH^-]} = \frac{10^{-14}}{1.6 \times 10^{-3}}$$

$$pH = -\log [H^+] = -\log \frac{10^{-14}}{1.6 \times 10^{-3}} = -(-14 - \log 1.6 + 3) = 14 + 0.20 - 3 = 11.20$$

21.77 Change the following pH values to $[H^+]$ values: (*a*) 4 (*b*) 3.6

▌ (*a*) $[H^+] = 10^{-pH} = 10^{-4}$ (*b*) $[H^+] = 10^{-pH} = 10^{-3.6} = 10^{0.4-4} = 10^{0.4} \times 10^{-4}$

From a table of logarithms,

$$10^{0.4} = \text{antilog } 0.4 = 2.5$$

and so $[H^+] = 2.5 \times 10^{-4}$.

Note that a negative number like -3.6 cannot be found in tables of logarithms, nor can -0.6. Only positive mantissas appear in the printed logarithm tables. The positive mantissa was achieved by adding and subtracting the next higher integer to the negative number: $-3.6 = 4.0 - 3.6 - 4.0 = 0.4 - 4.0$. Then 0.4 can be found to be the logarithm of 2.5.

21.78 Calculate the pH of each of the following solutions from the given hydronium ion molarity: (*a*) 1.0×10^{-3} (*b*) 3.0×10^{-11} (*c*) 3.9×10^{-8} (*d*) 0.15 (*e*) 1.23×10^{-15} (*f*) 1.0 (*g*) 2.0

▌ (*a*) 3.00 (*b*) 10.52 (*c*) 7.41 (*d*) 0.82 (*e*) 14.910 (*f*) 0.00 (*g*) -0.30. For (*c*):

$$pH = -\log (3.9 \times 10^{-8}) = -\log 3.9 - \log 10^{-8} = -(0.59) - (-8) = 7.41$$

21.79 Calculate the hydronium ion concentration of each of the following solutions from the given pH value: (*a*) 5.00 (*b*) 7.572 (*c*) 12.12 (*d*) 0.00 (*e*) 13.85

▌ (*a*) 1.0×10^{-5} (*b*) 2.68×10^{-8} (*c*) 7.6×10^{-13} (*d*) 1.0 (*e*) 1.4×10^{-14}. For (*b*):

$$\log [H_3O^+] = -pH = -7.572 = +0.428 - 8 \quad \text{thus} \quad [H_3O^+] = 2.68 \times 10^{-8}$$

21.80 Calculate the pH of 1.0×10^{-3} M solutions of each of the following: (*a*) HCl (*b*) NaOH (*c*) $Ba(OH)_2$ (*d*) NaCl

▌ (*a*) $[H_3O^+] = 1.0 \times 10^{-3}$ so $pH = 3.00$

(*b*) $[OH^-] = 1.0 \times 10^{-3}$ and $K_w = [H_3O^+][OH^-] = 1.0 \times 10^{-14}$

Thus $[H_3O^+] = 1.0 \times 10^{-11}$ so $pH = 11.00$

(*c*) $[OH^-] = 2.0 \times 10^{-3}$ $[H_3O^+] = 5.0 \times 10^{-12}$ thus $pH = 11.30$

(*d*) $[H_3O^+] = [OH^-] = 1.0 \times 10^{-7}$ so $pH = 7$ (NaCl is neither acidic nor basic.)

21.81 What is the pH of a 500-mL aqueous solution containing 0.050 mol of NaOH?

▌

$$\frac{0.050 \text{ mol}}{0.500 \text{ L}} = 0.100 \text{ M OH}^- \quad \text{so} \quad pH = 13.00$$

21.82 What is the value of K_w (*a*) in 0.10 M NaOH solution? (*b*) in 0.10 M NaCl solution?

▌ (*a*) and (*b*) $K_w = 1.0 \times 10^{-14}$ in dilute aqueous solutions at 25 °C, no matter what the solute.

21.83 Calculate the pH of 0.0030 M $Ba(OH)_2$ solution.

▌

$$[OH^-] = 0.0060 \text{ M} \quad \text{and} \quad K_w = [H_3O^+][OH^-] = 1.0 \times 10^{-14}$$

Thus

$$[H_3O^+] = \frac{1.0 \times 10^{-14}}{0.0060} = 1.7 \times 10^{-12} \quad \text{and} \quad pH = 11.78$$

21.84 Calculate the pH and pOH of the following solutions, assuming complete ionization: (*a*) 0.00345 M nitric acid (*b*) 0.000775 M HCl (*c*) 0.00886 M NaOH

▌ The number of significant decimal places in pH and pOH values is the total number of significant digits in the corresponding concentrations.

$$pH = -\log [H_3O^+] \qquad pH + pOH = 14.000$$

(*a*) $pH = -\log(0.00345) = 2.462$ $\quad pOH = 14.000 - 2.462 = 11.538$

(*b*) $pH = -\log(0.000775) = 3.111$ $\quad pOH = 14.000 - 3.111 = 10.889$

(*c*) $pOH = -\log(0.00886) = 2.053$ $\quad pH = 14.000 - 2.053 = 11.947$

21.85 Convert the following pH values to $[H^+]$ values: (*a*) 4 (*b*) 7 (*c*) 2.50 (*d*) 8.26

▌ $[H_3O^+] = 10^{-pH}$

(*a*) $[H_3O^+] = 10^{-4}$ $\quad$ (*b*) $[H_3O^+] = 10^{-7}$

(*c*) $[H_3O^+] = 10^{2.50} = 3.2 \times 10^{-3}$ $\quad$ (*d*) $[H_3O^+] = 10^{-8.26} = 5.5 \times 10^{-9}$

21.86 The $[H^+]$ of an HNO_3 solution is 1×10^{-3}, and the $[H^+]$ of an NaOH solution is 1×10^{-12}. Find the molar concentration and pH of each solution.

▌ Both HNO_3 and NaOH are completely ionized. Hence, for HNO_3

$$[H_3O^+] = 1 \times 10^{-3} \text{ M} \qquad pH = -\log(1 \times 10^{-3}) = 3.0$$

For NaOH, $[H_3O^+] = 1 \times 10^{-12}$ M so $pH = 12.0$

$$K_w = [H_3O^+][OH^-] = 1.0 \times 10^{-14} \quad \text{thus} \quad [OH^-] = 1 \times 10^{-2} \text{ M}$$

21.87 (*a*) Calculate K_a for an acid whose 0.10 M solution has a pH of 4.50. (*b*) Calculate K_b for a base whose 0.10 M solution has a pH of 10.50.

▮ (*a*) pH = 4.50 so $[H_3O^+] = 3.16 \times 10^{-5}$ $HA + H_2O \longrightarrow H_3O^+ + A^-$

	initial	produced	used up	equilibrium
$[H_3O^+]$	0	3.16×10^{-5}		3.16×10^{-5}
$[A^-]$	0	3.16×10^{-5}		3.16×10^{-5}
$[HA]$	0.10		3.16×10^{-5}	0.10

$$K_a = \frac{[H_3O^+][A^-]}{[HA]} = \frac{(3.16 \times 10^{-5})^2}{0.10} = 1.0 \times 10^{-8}$$

(*b*) pH = 10.50 and pOH = 3.50 so $[OH^-] = 3.16 \times 10^{-4}$

$$B + H_2O \rightleftharpoons BH^+ + OH^- \qquad K_b = \frac{[OH^-][BH^+]}{[B]} = \frac{(3.16 \times 10^{-4})^2}{(0.10)} = 1.0 \times 10^{-6}$$

21.88 Find the pH of the solution resulting when 50 mL of 0.20 M HCl is mixed with 50 mL of 0.20 M $HC_2H_3O_2$.

▮ The diluted HCl is 0.10 M, the pH is 1.00. With added strong acid present, the H^+ from ionization of the weak acid $HC_2H_3O_2$ is insignificant.

21.89 Compute $[H^+]$ and $[OH^-]$ in a 0.0010 M solution of a monobasic acid which is 4.2% ionized. What is the pH of the solution? What are K_a and pK_a for the acid?

▮

$$HA + H_2O \rightleftharpoons H_3O^+ + A^- \qquad K_a = \frac{[H_3O^+][A^-]}{[HA]}$$

$$[H_3O^+] = 0.000042 \text{ M} \qquad \text{thus} \qquad pH = 4.38$$

It follows that pOH = 9.62 so $[OH^-] = 2.40 \times 10^{-10}$ M

$$K_a = \frac{(0.000042)^2}{0.0010 - 0.000042} = 1.8 \times 10^{-6} \qquad \text{and} \qquad pK_a = 5.73$$

21.90 Compute $[OH^-]$ and $[H^+]$ in a 0.10 M solution of a weak monoprotic base which is 1.3% ionized. What is the pH of the solution?

▮

$$B + H_2O \rightleftharpoons BH^+ + OH^-$$

$$[OH^-] = 0.013 \times 0.10 \text{ M} = 1.3 \times 10^{-3} \text{ M}$$

thus $[H_3O^+] = 7.7 \times 10^{-12}$ M and $pH = -\log(7.7 \times 10^{-12}) = 11.11$

21.91 When 0.100 mol of NH_3 is dissolved in sufficient water to make 1.00 L of solution, the solution is found to have a hydroxide ion concentration of 1.33×10^{-3} M. **(*a*)** What is the pH of the solution? **(*b*)** What will be the pH of the solution after 0.100 mol of NaOH is added to it? (Assume no change in volume.) **(*c*)** Calculate K_b for ammonia. **(*d*)** How will NaOH added to the solution affect the extent of dissociation of ammonia?

▮

(*a*) $[OH^-] = 1.33 \times 10^{-3}$ thus $[H_3O^+] = \dfrac{1.00 \times 10^{-14}}{1.33 \times 10^{-3}} = 7.52 \times 10^{-12}$ and pH = 11.12

(*b*) pH = 13.00 (The OH^- concentration from the NH_3 is negligible, since the NaOH represses the dissociation of the weak base.)

(*c*) $NH_3 + H_2O \rightleftharpoons NH_4^+ + OH^-$ $\qquad K_b = \dfrac{[NH_4^+][OH^-]}{[NH_3]} = \dfrac{(1.33 \times 10^{-3})^2}{0.100 - 0.00133} = 1.79 \times 10^{-5}$

(*d*) The OH^- of the NaOH represses the dissociation of NH_3.

21.92 (*a*) What is the pH of a solution containing 0.010 mol HCl/L? (*b*) Calculate the change in pH if 0.020 mol $NaC_2H_3O_2$ is added to 1.0 L of this solution.

▮ (*a*) $HCl + H_2O \longrightarrow H_3O^+ + Cl^-$ (100% ionization)

0.010 M H_3O^+ pH = 2.00

(*b*) The salt contains acetate ions, which react with hydronium ions:

$$H_3O^+ + C_2H_3O_2^- \longrightarrow HC_2H_3O_2 + H_2O$$

Assume complete reaction of the hydronium ion, yielding 0.010 M $HC_2H_3O_2$ and leaving 0.010 M $C_2H_3O_2^-$ in excess:

$$HC_2H_3O_2 + H_2O \rightleftharpoons C_2H_3O_2^- + H_3O^+$$

$$K_a = \frac{[C_2H_3O_2^-][H_3O^+]}{[HC_2H_3O_2]} = \frac{(0.010)[H_3O^+]}{(0.010)} = 1.8 \times 10^{-5}$$

$$[H_3O^+] = 1.8 \times 10^{-5} \quad \text{so} \quad pH = 4.74 \quad \text{and} \quad \Delta pH = 4.74 - 2.00 = 2.74$$

21.93 The value of K_w at the physiological temperature 37 °C is 2.4×10^{-14}. What is the pH at the neutral point of water at this temperature, where there are equal numbers of H^+ and OH^-?

$$[H^+] = \sqrt{2.4 \times 10^{-14}} = 1.55 \times 10^{-7} \quad \text{so} \quad pH = 6.81$$

21.94 Calculate the pH of a 1.0×10^{-8} M solution of HCl.

$$2H_2O \rightleftharpoons H_3O^+ + OH^- \qquad K_w = [OH^-][H_3O^+] = 1.0 \times 10^{-14}$$

Let $x = [OH^-]$. The H_3O^+ concentration is generated (1) from the HCl dissolved, and (2) from the ionization of water. In solutions of 10^{-5} M or greater HCl, the ionization reaction is repressed and negligible. In this very dilute solution, both sources of H_3O^+ must be considered:

$$[H_3O^+] = 10^{-8} + x$$

$$(10^{-8} + x)(x) = 1.0 \times 10^{-14} \quad \text{or} \quad x^2 + 10^{-8}x - 10^{-14} = 0$$

$$[OH^-] = x = \frac{-10^{-8} + \sqrt{10^{-16} + (4 \times 10^{-14})}}{2} = 9.5 \times 10^{-8} \quad \text{so} \quad pOH = 7.02 \quad \text{and} \quad pH = 6.98$$

21.95 Calculate the percent error in the hydronium ion concentration made by neglecting the ionization of water in a 1.0×10^{-6} M NaOH solution.

Neglecting the ionization of water, $[H_3O^+] = 1.0 \times 10^{-8}$. Including the ionization of water (see answer to Prob. 21.94):

Let $$x = [H_3O^+] \quad \text{and} \quad 1.0 \times 10^{-6} + x = [OH^-]$$

$$[(1.0 \times 10^{-6}) + x]x = 1.0 \times 10^{-14} \quad \text{or} \quad x^2 + 1.0 \times 10^{-6}x - 1.0 \times 10^{-14} = 0$$

$$x = \frac{-1.0 \times 10^{-6} + \sqrt{10^{-12} + (4 \times 10^{-14})}}{2} = 9.9 \times 10^{-9}$$

$$\%\ \text{error} = \frac{(10 \times 10^{-9}) - (9.9 \times 10^{-9})}{9.9 \times 10^{-9}} \times 100\% = 1\%$$

21.96 What is the pH of 7.0×10^{-8} M acetic acid? What is the concentration of un-ionized acetic acid?

In such dilute acid solutions, we are able to assume complete ionization of the weak acid, but we must take into account the ionization of the water.

$$2H_2O \rightleftharpoons H_3O^+ + OH^- \qquad K_w = [H_3O^+][OH^-] = 1.0 \times 10^{-14}$$

Let $x = [OH^-]$ then $[H_3O^+] = x + (7.0 \times 10^{-8})$ (from water and acid).

$$K_w = x(x + 7.0 \times 10^{-8}) = 1.0 \times 10^{-14} \quad \text{or} \quad x^2 + (7.0 \times 10^{-8})x - (1.0 \times 10^{-14}) = 0$$

$$x = 7.1 \times 10^{-8}\ M = [OH^-] \quad \text{thus} \quad [H_3O^+] = 1.4 \times 10^{-7}\ M \quad \text{and} \quad pH = 6.85$$

From electroneutrality,

$$[C_2H_3O_2^-] + [OH^-] = [H_3O^+] \qquad [C_2H_3O_2^-] = (1.4 \times 10^{-7}) - (7.1 \times 10^{-8}) = 7 \times 10^{-8}$$

$$HC_2H_3O_2 + H_2O \rightleftharpoons C_2H_3O_2^- + H_3O^+ \qquad K_a = \frac{[C_2H_3O_2^-][H_3O^+]}{[HC_2H_3O_2]} = 1.8 \times 10^{-5}$$

$$[HC_2H_3O_2] = \frac{(1.4 \times 10^{-7})(7 \times 10^{-8})}{1.8 \times 10^{-5}} = 5 \times 10^{-10}\ M$$

21.97 What is the pH of **(a)** 5.0×10^{-8} M HCl **(b)** 5.0×10^{-10} M HCl?

▌ **(a)** If we were to consider only the contribution of the HCl to the acidity of the solution, $[H^+]$ would be 5.0×10^{-8} and the pH would be greater than 7. This obviously cannot be true, because a solution of a pure acid, no matter how dilute, cannot be less acid than pure water alone. It is necessary in this problem to take into account the contribution of water to the total acidity.

	H_2O $\rightleftharpoons$	H^+	+	OH^-
M from HCl:		5.0×10^{-8}		
Change by ionization of H_2O:		x		x
M at equilibrium:		$(5.0 \times 10^{-8}) + x$		x

$$K_w = [H^+][OH^-] = (5.0 \times 10^{-8} + x)x = 1.00 \times 10^{-14} \qquad \text{from which} \quad x = 0.78 \times 10^{-7}.$$

Then $[H^+] = 5.0 \times 10^{-8} + x = 1.28 \times 10^{-7}$ and $pH = -\log(1.28 \times 10^{-7}) = 6.89$

(b) Although the method of (a) could be used here, the problem can be simplified by noting that the HCl is so dilute as to make only a negligible contribution to $[H^+]$ as compared with the ionization of water. Thus $[H^+] = 1.00 \times 10^{-7}$ and $pH = 7.00$.

21.4 BUFFER SOLUTIONS

21.98 Explain why a solution containing a strong base and its salt does not act as a buffer solution.

▌ Addition of OH^- does not shift an equilibrium toward un-ionized base, as it would with a weak base and its conjugate.

21.99 Which of the following combinations of solutes would result in the formation of a buffer solution? **(a)** $NaC_2H_3O_2 + HC_2H_3O_2$ **(b)** $NH_4Cl + NH_3$ **(c)** $HCl + NaCl$ **(d)** $HCl + HC_2H_3O_2$ **(e)** $NaOH + HCl$ **(f)** $NaOH + HC_2H_3O_2$ in a 1:1 mole ratio **(g)** $NH_3 + HCl$ in a 2:1 mole ratio **(h)** $HC_2H_3O_2 + NaOH$ in a 2:1 mole ratio

▌ (a), (b), (g), and (h). [The solution in (g) contains equimolar quantities of NH_4^+ and NH_3. The NH_4^+ resulted from the reaction

$$H^+Cl^- + NH_3 \longrightarrow NH_4^+ + Cl^-$$

The NH_3 was present in excess. A similar situation occurs in (h).

21.100 Calculate the pH of a solution containing 0.10 M H_3BO_3 and 0.18 M NaH_2BO_3.

▌

$$H_3BO_3 + H_2O \longrightarrow H_2BO_3^- + H_3O^+$$

$$K_1 = \frac{[H_2BO_3^-][H_3O^+]}{[H_3BO_3]} = \frac{(0.18)x}{(0.10)} = 7.3 \times 10^{-10} \qquad \text{or} \qquad x = 4.1 \times 10^{-10}$$

$$pH = -\log x = 9.39$$

21.101 **(a)** Calculate the concentration of each ion, Mg^{2+}, H_3O^+, and Br^-, in a solution prepared by dissolving 0.20 mol of $MgBr_2$ and 0.10 mol of HBr in enough water to make 1.0 L of solution. Both substances are completely soluble and ionic in solution. **(b)** Calculate the concentration of each ion, H_3O^+, Mg^{2+}, and $C_2H_3O_2^-$, in a solution prepared by dissolving 0.20 mol of $HC_2H_3O_2$ and 0.050 mol of $Mg(C_2H_3O_2)_2$ in enough water to make 1.0 L of solution.

▌ **(a)** $[Mg^{2+}] = 0.20$ M, $[Br^-] = 0.50$ M, $[H_3O^+] = 0.10$ M

(b) $[Mg^{2+}] = 0.050$ M, $[C_2H_3O_2^-] = 0.10$ M

$$K_a = \frac{[H_3O^+](0.10)}{(0.20)} = 1.8 \times 10^{-5} \qquad \text{thus} \qquad [H_3O^+] = 3.6 \times 10^{-5}\ \text{M}$$

21.102 Calculate the pH of a solution of 0.10 M HA and 0.20 M NaA. $K_a = 1.0 \times 10^{-7}$.

▮

$$HA + H_2O \longrightarrow H_3O^+ + A^- \qquad K_a = \frac{[H_3O^+][A^-]}{[HA]} = 1.0 \times 10^{-7}$$

	initial	produced by reaction	used up by reaction	equilibrium
$[HA]$	0.10		x	$0.10 - x \simeq 0.10$
$[H_3O^+]$	0.00	x		x
$[A^-]$	0.20	x		$0.20 + x \simeq 0.20$

$$\frac{x(0.20)}{(0.10)} = 1.0 \times 10^{-7} \qquad \text{thus} \qquad x = 5.0 \times 10^{-8} = [H_3O^+] \qquad \text{and} \qquad pH = 7.30$$

21.103 It is desired to prepare a buffer solution consisting of 0.10 M $HC_2H_3O_2$ and 0.10 M $NaC_2H_3O_2$. Assuming no volume change upon the addition of the pure compounds, state what reagents and in what quantities should be added to 1.00 L of each of the following solutions to prepare the desired buffer solution: ***(a)*** 0.10 M $HC_2H_3O_2$ ***(b)*** 0.20 M $HC_2H_3O_2$ ***(c)*** 0.20 M $NaC_2H_3O_2$ ***(d)*** 0.10 M $NaC_2H_3O_2$ ***(e)*** 0.10 M NaOH

▮ ***(a)*** 0.10 mol $NaC_2H_3O_2$. ***(b)*** 0.10 mol NaOH. The 0.10 mol of NaOH will react with 0.10 mol of $HC_2H_3O_2$, producing a solution equivalent to that produced in part *(a)*. ***(c)*** 0.10 mol HCl. The HCl will react with $C_2H_3O_2^-$ (producing 0.10 mol of $HC_2H_3O_2$) and yielding the same solution as produced in *(a)* except with NaCl added. ***(d)*** 0.10 mol $HC_2H_3O_2$. ***(e)*** 0.20 mol $HC_2H_3O_2$. Half of this acid will neutralize the NaOH, yielding the same solution as produced in part *(a)*. Note that this combination of reactants is the same as in part *(b)*.

21.104 ***(a)*** What is the hydronium ion concentration in a solution of 0.0200 M $NaC_2H_3O_2$ and 0.0500 M $HC_2H_3O_2$? ($K_a = 1.8 \times 10^{-5}$) ***(b)*** Determine the pH of the solution.

▮

(a) $$K_a = \frac{[C_2H_3O_2^-][H_3O^+]}{[HC_2H_3O_2]} = 1.8 \times 10^{-5}$$

	initial	produced	used up	equilibrium
$[C_2H_3O_2^-]$	0.0200	x		$0.0200 + x \approx 0.0200$
$[H_3O^+]$	0.0000	x		x
$[HC_2H_3O_2]$	0.0500		x	$0.0500 - x \approx 0.0500$

$$\frac{(0.0200)(x)}{0.0500} = 1.8 \times 10^{-5} \qquad \text{thus} \qquad x = 4.5 \times 10^{-5} = [H_3O^+]$$

(b) $pH = -\log[H_3O^+] = -\log(4.5 \times 10^{-5}) = 4.35$

21.105 Calculate the acetic acid to acetate ion concentration ratio in a buffer solution whose pH is 7.00. Explain how it is possible to have *any* acid in a neutral solution.

▮

$$[H_3O^+] = 1.0 \times 10^{-7} \qquad K_a = \frac{[H_3O^+][C_2H_3O_2^-]}{[HC_2H_3O_2]} = 1.8 \times 10^{-5}$$

$$\frac{[HC_2H_3O_2]}{[C_2H_3O_2^-]} = \frac{1.0 \times 10^{-7}}{1.8 \times 10^{-5}} = 5.6 \times 10^{-3}$$

It is possible for an acid to exist at pH 7 if, in the same solution, some base is present—in this case $C_2H_3O_2^-$ (in 180 times the concentration).

21.106 ***(a)*** Determine the pH of a 0.10 M solution of pyridine, C_5H_5N. ***(b)*** Predict the effect of addition of pyridinium ion, $C_5H_5NH^+$, on the position of the equilibrium. Will the pH be raised or lowered? ***(c)*** Calculate the pH of 1.00 L of 0.10 M pyridine solution to which 0.15 mol of pyridinium chloride, $C_5H_5NH^+Cl^-$, has been added, assuming no change in volume.

▮

$$C_5H_5N + H_2O \longrightarrow C_5H_5NH^+ + OH^- \qquad K_b = \frac{[C_5H_5NH^+][OH^-]}{[C_5H_5N]} = 1.52 \times 10^{-9}$$

(*a*) $[C_5H_5NH^+] = [OH^-] = x$ and $[C_5H_5N] = 0.10 - x \simeq 0.10$

$$\frac{x^2}{0.10} = 1.52 \times 10^{-9} \quad \text{thus} \quad x = 1.2 \times 10^{-5} = [OH^-] \quad \text{so} \quad pOH = 4.91 \quad \text{and} \quad pH = 9.09$$

(*b*) According to Le Châtelier's principle, the addition of $C_5H_5NH^+$ will shift the equilibrium to the left, decreasing the OH^- concentration and the pH.

(*c*) $[OH^-] = y \quad [C_5H_5NH^+] = 0.15 + y \simeq 0.15 \quad [C_5H_5N] = 0.10 - y \simeq 0.10$

$$\frac{(0.15)y}{(0.10)} = 1.52 \times 10^{-9} \quad \text{thus} \quad y = 1.01 \times 10^{-9} = [OH^-] \quad \text{so} \quad pOH = 8.99 \quad \text{and} \quad pH = 5.01$$

The prediction of part (*b*) is calculated to be correct; the pH has dropped 4.08 units because of the addition of the pyridinium ion.

21.107 What is the pH of a solution that is 0.0100 M in HCN and 0.0200 M in NaCN?

▮ $$K_a = \frac{[CN^-][H_3O^+]}{[HCN]} = \frac{(0.0200)[H_3O^+]}{(0.0100)} = 6.2 \times 10^{-10} \quad [H_3O^+] = 3.1 \times 10^{-10} \quad \text{thus} \quad pH = 9.51$$

21.108 A solution was made up to be 0.0100 M in chloroacetic acid, $HC_2H_2O_2Cl$, and also 0.0020 M in sodium chloroacetate, $NaC_2H_2O_2Cl$. What is $[H^+]$ in the solution?

▮ Assume that the concentration of chloroacetate ion can be approximated by the molar concentration of the sodium salt, 0.0020, and that the extent of dissociation of the acid is small. Letting $[H^+] = x$,

$$[C_2H_2O_2Cl^-] = 0.0020 + x \approx 0.0020 \qquad [HC_2H_2O_2Cl] = 0.0100 - x \approx 0.0100$$

Then $$x = [H^+] = K_a \frac{[HC_2H_2O_2Cl]}{[C_2H_2O_2Cl^-]} = (1.36 \times 10^{-3})\left(\frac{0.0100}{0.0020}\right) = 6.8 \times 10^{-3}$$

Check: Our assumption is invalid; the value of x obtained, 6.8×10^{-3}, cannot be neglected in comparison with either 0.0020 or 0.0100.

Without making the simplifying assumptions, we have

$$x = [H^+] = (1.36 \times 10^{-3})\left(\frac{0.0100 - x}{0.0020 + x}\right)$$

The solution of the resulting quadratic equation gives the proper value for $[H^+]$, 2.4×10^{-3}.

Chloroacetic acid is apparently strong enough that the common-ion effect does not repress its ionization to a residual value small enough to be neglected. This result might have been guessed from the relatively large value of K_a.

21.109 Calculate $[H^+]$ and $[C_2H_3O_2^-]$ in a solution that is 0.100 M in $HC_2H_3O_2$ and 0.050 M in HCl.

▮ The HCl contributes so much more H^+ than the $HC_2H_3O_2$ that we can take $[H^+]$ as equal to the molar concentration of the HCl, 0.050; this is another example of the common-ion effect.

Then, if $[C_2H_3O_2^-] = x$, we have $[HC_2H_3O_2] = 0.100 - x \approx 0.100$ and

$$[C_2H_3O_2^-] = \frac{[HC_2H_3O_2]K_a}{[H^+]} = \frac{(0.100)(1.8 \times 10^{-5})}{0.050} = 3.6 \times 10^{-5}$$

Check of assumptions: (1) Contribution of acetic acid to $[H^+]$, x, is indeed small compared with 0.050; (2) x is indeed small compared with 0.100.

21.110 Calculate $[H^+]$, $[C_2H_3O_2^-]$, and $[CN^-]$ in a solution that is 0.100 M in $HC_2H_3O_2$ and 0.200 M in HCN.

▮ This problem is similar to Prob. 21.109 in that one of the acids, acetic, completely dominates the other in terms of contribution to the total $[H^+]$ of the solution. We base this assumption on the fact that K_a for $HC_2H_3O_2$ is much greater than for HCN; we will check the assumption after solving the problem. We will proceed by treating the acetic acid as if the HCN were not present.

Let $[H^+] = [C_2H_3O_2^-] = x$; then $[HC_2H_3O_2] = 0.100 - x \approx 0.100$ and

$$\frac{[H^+][C_2H_3O_2^-]}{[HC_2H_3O_2]} = \frac{x^2}{0.100} = 1.8 \times 10^{-5} \quad \text{or} \quad x = 1.3 \times 10^{-3}$$

Check of assumption: x is indeed small compared with 0.100.

Now we treat the HCN equilibrium established at a value of $[H^+]$ determined by the acetic acid, 1.3×10^{-3}. Let $[CN^-] = y$; then

$$[HCN] = 0.200 - y \approx 0.200 \qquad \text{and} \qquad y = [CN^-] = \frac{K_a[HCN]}{[H^+]} = \frac{(6.2 \times 10^{-10})(0.200)}{1.3 \times 10^{-3}} = 9.5 \times 10^{-8}$$

Check of assumptions: (1) y is indeed small compared with 0.200; (2) the amount of H^+ contributed by HCN ionization, equal to the amount of CN^- formed (9.5×10^{-8} mol/L), is indeed small compared with the amount of H^+ contributed by $HC_2H_3O_2$ (1.3×10^{-3} mol/L).

21.111 A buffer solution of pH 8.50 is desired. ***(a)*** Starting with 0.0100 mol of KCN and the usual inorganic reagents of the laboratory, how would you prepare 1.00 L of the buffer solution? ***(b)*** By how much would the pH change after the addition of 5.00×10^{-5} mol HCl to 100 cm^3 of the buffer? ***(c)*** By how much would the pH change after the addition of 5.00×10^{-5} mol NaOH to 100 cm^3 of the buffer? **CAUTION**: NEVER add acid to cyanide solutions.

▌ ***(a)*** To find the desired $[H^+]$:

$$\log [H^+] = -pH = -8.50 = 0.50 - 9.00 \qquad [H^+] = \text{antilog}\, 0.50 \times \text{antilog}\,(-9.00) = 3.2 \times 10^{-9}$$

The buffer solution could be prepared by mixing CN^- (weak base) with HCN (weak acid) in the proper proportions so as to satisfy the ionization constant equilibrium for HCN.

$$HCN \rightleftharpoons H^+ + CN^- \qquad K_a = 6.2 \times 10^{-10} = \frac{[H^+][CN^-]}{[HCN]}$$

Hence
$$\frac{[CN^-]}{[HCN]} = \frac{K_a}{[H^+]} = \frac{6.2 \times 10^{-10}}{3.2 \times 10^{-9}} = 0.194 \qquad (1)$$

This ratio of CN^- to HCN can be attained if some of the CN^- is protonated with a strong acid, like HCl, to form an equivalent amount of HCN. The total cyanide available for both forms is 0.0100 mol. Let $x = [HCN]$; then $[CN^-] = 0.0100 - x$. Substituting in (*1*),

$$\frac{0.0100 - x}{x} = 0.194$$

from which $x = 0.0084$ and $0.0100 - x = 0.0016$. The buffer solution can be prepared by dissolving 0.0100 mol KCN and 0.0084 mol HCl in enough water to make up 1.00 L of solution.

(b) 100 cm^3 of the buffer contains

$$(0.0084 \text{ mol/L})(0.100 \text{ L}) = 8.4 \times 10^{-4} \text{ mol HCN} \qquad \text{and} \qquad (0.0016 \text{ mol/L})(0.100 \text{ L}) = 1.6 \times 10^{-4} \text{ mol } CN^-$$

The addition of 5.00×10^{-5} mol of strong acid will convert more CN^- to HCN. The resulting amount of HCN will be

$$(8.4 \times 10^{-4}) + (0.5 \times 10^{-4}) = 8.9 \times 10^{-4} \text{ mol}$$

and the resulting amount of CN^- will be

$$(1.6 \times 10^{-4}) - (0.5 \times 10^{-4}) = 1.1 \times 10^{-4} \text{ mol}$$

Only the ratio of the two concentrations is needed.

$$[H^+] = K_a \frac{[HCN]}{[CN^-]} = (6.2 \times 10^{-10})\left(\frac{8.9}{1.1}\right) = 5.0 \times 10^{-9} \qquad pH = -\log [H^+] = 9 - 0.70 = 8.30$$

The drop in pH caused by the addition of the acid is $8.50 - 8.30$, or 0.20 pH unit.

(c) The addition of 5.00×10^{-5} mol of strong base will convert an equivalent amount of HCN to CN^-.

HCN: resulting amount $= (8.4 \times 10^{-4}) - (0.5 \times 10^{-4}) = 7.9 \times 10^{-4}$ mol

CN^-: resulting amount $= (1.6 \times 10^{-4}) + (0.5 \times 10^{-4}) = 2.1 \times 10^{-4}$ mol

$$[H^+] = K_a \frac{[HCN]}{[CN^-]} = (6.2 \times 10^{-10})\left(\frac{7.9}{2.1}\right) = 2.3 \times 10^{-9} \qquad pH = -\log [H^+] = 9 - 0.36 = 8.64$$

The rise in pH caused by the addition of base is $8.64 - 8.50$, or 0.14 pH unit.

21.112 Determine $[OH^-]$ of a 0.050 M solution of ammonia to which has been added sufficient NH_4Cl to make the total $[NH_4^+]$ equal to 0.100.

▮

$$NH_3 + H_2O \rightleftharpoons NH_4^+ + OH^-$$

$$K_b = \frac{[NH_4^+][OH^-]}{[NH_3]} = \frac{(0.100)x}{(0.050)} = 1.8 \times 10^{-5} \qquad \text{thus} \qquad x = [OH^-] = 9.0 \times 10^{-6}$$

21.113 Find the value of $[H^+]$ in 1.0 L of solution in which are dissolved 0.080 mol $HC_2H_3O_2$ and 0.100 mol $NaC_2H_3O_2$.

▮

$$HC_2H_3O_2 + H_2O \rightleftharpoons C_2H_3O_2^- + H_3O^+$$

$$K_a = \frac{[H_3O^+][C_2H_3O_2^-]}{[HC_2H_3O_2]} = \frac{x(0.100)}{(0.080)} = 1.8 \times 10^5 \qquad \text{thus} \qquad x = 1.4 \times 10^{-5}$$

21.114 Calculate the pH of a solution prepared by mixing 50.0 mL of 0.200 M $HC_2H_3O_2$ and 50.0 mL of 0.100 M NaOH.

▮ The acid and base react, leaving 0.0500 M $HC_2H_3O_2$ (in 100 mL) and 0.0500 M $C_2H_3O_2^-$.

$$K_a = \frac{[H_3O^+][C_2H_3O_2^-]}{[HC_2H_3O_2]} = 1.8 \times 10^{-5} \qquad [H_3O^+] = 1.8 \times 10^{-5} \qquad \text{thus} \qquad pH = 4.74$$

21.115 Calculate the pH of a solution of 0.10 M acetic acid. Calculate the pH after 50.0 mL of this solution is treated with 25.0 mL of 0.10 M NaOH.

▮

$$HC_2H_3O_2 + H_2O \longrightarrow H_3O^+ + C_2H_3O_2^- \qquad K_a = \frac{[H_3O^+][C_2H_3O_2^-]}{[HC_2H_3O_2]} = 1.8 \times 10^{-5}$$

Before treatment:

$$[H_3O^+] = [C_2H_3O_2^-] = x \qquad [HC_2H_3O_2] = 0.10 - x \simeq 0.10$$

$$\frac{x^2}{0.10} = 1.8 \times 10^{-5} \qquad \text{thus} \qquad x = 1.35 \times 10^{-3} = [H_3O^+] \qquad \text{and} \qquad pH = 2.87$$

After treatment: half the acid is neutralized and half remains. Both solutes are now in 75.0 mL of solution.

$$[H_3O^+] = y \qquad [C_2H_3O_2^-] = \frac{(0.050\ M)(50.0\ mL)}{(75.0\ mL)} + y \approx 0.033\ M$$

$$[HC_2H_3O_2] = \frac{(0.050\ M)(50.0\ mL)}{(75.0\ mL)} - y \approx 0.033\ M \qquad \frac{y(0.033)}{0.033} = 1.8 \times 10^{-5} = y$$

$$[H_3O^+] = 1.8 \times 10^{-5}\ M \qquad \text{thus} \qquad pH = 4.74$$

21.116 Calculate the pH of a solution which results from the mixing of 50.0 mL of 0.300 M HCl with 50.0 mL of 0.400 M NH_3.

▮ The simplest approach to this problem is to consider that the reaction

$$NH_3 + HCl \longrightarrow NH_4^+ + Cl^-$$

goes to completion. The concentrations of all the species which would be present after that reaction are then calculated and used as the initial concentrations for the equilibrium reaction. In this case, on mixing, 15.0 mmol of NH_4^+ is produced and 5.0 mmol of NH_3 remains unreacted, in a total of 100.0 mL of solution. This solution is the same as that of a solution containing 0.050 M NH_3 and 0.150 M NH_4^+. The equilibrium for the ionization of the base is now considered:

$$NH_3 + H_2O \rightleftharpoons NH_4^+ + OH^- \qquad K_b = \frac{[NH_4^+][OH^-]}{[NH_3]} = 1.8 \times 10^{-5}$$

Let $[OH^-] = x \qquad [NH_4^+] = x + 0.150 \approx 0.150 \qquad [NH_3] = 0.050 - x \approx 0.050$

$$K_b = \frac{(0.150)(x)}{0.050} = 1.8 \times 10^{-5} \qquad \text{thus} \qquad x = 6.0 \times 10^{-6} = [OH^-]$$

$$pOH = -(0.78 - 6) = 5.22 \qquad pH = 14.00 - pOH = 8.78$$

21.117 Calculate the pH of a solution prepared by mixing 50.00 mL of 0.0200 M NaOH and 50.00 mL of 0.0400 M $HC_2H_3O_2$.

▮ The combination of 1.00 mmol of OH^- and 2.00 mmol of $HC_2H_3O_2$ yields 1.00 mmol of $C_2H_3O_2^-$, leaving 1.00 mmol of $HC_2H_3O_2$ in excess.

$$K_a = \frac{[H_3O^+][C_2H_3O_2^-]}{[HC_2H_3O_2]} = \frac{[H_3O^+](0.100\ mmol/100\ mL)}{(0.100\ mmol/100\ mL)} = 1.8 \times 10^{-5}$$

$$[H_3O^+] = 1.8 \times 10^{-5} \qquad \text{thus} \qquad pH = 4.74$$

21.118 Calculate the pH of a solution resulting from the addition of 50.0 mL of 0.100 M HCl to 50.0 mL of a solution containing 0.150 M $HC_2H_3O_2$ and 0.200 M $NaC_2H_3O_2$.

Mixing 5.00 mmol of H_3O^+ (from HCl) to the buffer solution of 7.50 mmol of acetic acid and 10.0 mmol of acetate ion results in the reaction

$$H_3O^+ + C_2H_3O_2^- \longrightarrow HC_2H_3O_2 + H_2O$$

In the process, effectively the "initial" concentration (before the equilibrium process is considered) of acetic acid is increased to 12.50 mmol/100 mL and that of acetate ion is reduced to 5.0 mmol/100 mL. Thereafter we have a simple buffer solution problem:

$$HC_2H_3O_2 + H_2O \rightleftharpoons H_3O^+ + C_2H_3O_2^- \qquad K_a = \frac{[H_3O^+][C_2H_3O_2^-]}{[HC_2H_3O_2]} = 1.8 \times 10^{-5}$$

	mmol added	before equilibrium reaction	produced by equilibrium reaction	used up by equilibrium reaction	at equil
$[H_3O^+]$	5.00	0	x		x
$[C_2H_3O_2^-]$	10.0	0.050	x		0.050
$[HC_2H_3O_2]$	7.50	0.125		x	0.125

$$\frac{x(0.050)}{0.125} = 1.8 \times 10^{-5} \quad \text{or} \quad [H_3O^+] = x = 4.5 \times 10^{-5} \quad \text{thus} \quad \text{pH} = 4.35$$

21.119 Calculate the pH of a solution made by mixing 50.0 mL of 0.200 M NH_4Cl and 75.0 mL of 0.100 M NaOH.

Before consideration of the equilibrium reaction, assume that the following reaction goes to completion:

$$\underset{10\text{ mmol}}{NH_4^+} + \underset{7.5\text{ mmol}}{OH^-} \longrightarrow NH_3 + H_2O$$

The equilibrium calculation thus begins with "initial" concentrations

$$\frac{2.5\text{ mmol } NH_4^+}{125\text{ mL}} = 0.020\text{ M } NH_4^+ \qquad \frac{7.5\text{ mmol } NH_3}{125\text{ mL}} = 0.060\text{ M } NH_3$$

$$NH_3 + H_2O \rightleftharpoons NH_4^+ + OH^- \qquad K_b = \frac{[NH_4^+][OH^-]}{[NH_3]} = 1.8 \times 10^{-5}$$

	initial	produced	used up	equilibrium
$[NH_4^+]$	0.020	x		$0.020 + x \simeq 0.020$
$[OH^-]$	0	x		x
$[NH_3]$	0.060		x	$0.060 - x \simeq 0.060$

$$\frac{0.020x}{0.060} = 1.8 \times 10^{-5} \quad \text{or} \quad [OH^-] = x = 5.4 \times 10^{-5} \quad \text{thus} \quad \text{pOH} = 4.27 \quad \text{and} \quad \text{pH} = 9.73$$

21.120 Determine the pH of a solution after 0.10 mol of NaOH is added to 1.00 L of a solution containing 0.15 M $HC_2H_3O_2$ and 0.20 M $NaC_2H_3O_2$. Assume no change in volume.

$$HC_2H_3O_2 + H_2O \rightleftharpoons C_2H_3O_2^- + H_3O^+ \qquad K_a = 1.8 \times 10^{-5}$$

	initial	produced	used up	equilibrium
$[HC_2H_3O_2]$	0.05		x	$0.05 - x \approx 0.05$
$[C_2H_3O_2^-]$	0.30		x	$0.30 - x \approx 0.30$
$[H_3O^+]$	0	x		x

$$K_a = \frac{(0.30)(x)}{0.05} = 1.8 \times 10^{-5} \quad \text{or} \quad x = 3 \times 10^{-6} \quad \text{thus} \quad \text{pH} = 5.5$$

21.121 Assuming that the final volume in each case remains the same, compare the effect of adding 0.010 mol of solid sodium hydroxide to ***(a)*** 1.0 L of a solution 1.8×10^{-5} M in HCl and ***(b)*** 1.0 L of a solution containing 0.10 mol of $NaC_2H_3O_2$ and 0.10 mol of $HC_2H_3O_2$.

▌ The initial hydronium ion concentration of both solutions is 1.8×10^{-5} M. (The reader should verify this statement.)

(a) The 1.8×10^{-5} mol of HCl in the solution will react completely with 1.8×10^{-5} mol of the added NaOH, leaving the excess base in solution.

$$\text{Excess base} = 0.010\text{ mol} - (1.8 \times 10^{-5})\text{ mol} \approx 0.010\text{ mol} \qquad [OH^-] = \frac{0.010\text{ mol}}{1.0\text{ L}} = 0.010\text{ M}$$

The hydroxide ion concentration would be virtually 0.010 M, and the hydronium ion concentration is calculated to be 1.0×10^{-12} M.

(b) The hydroxide ion of the NaOH would react with the 1.8×10^{-5} mol of H_3O^+ present. But as that reaction takes place, more acetic acid ionizes, since removal of the hydronium ion is a stress that shifts the weak acid ionization equilibrium to the right. More and more weak acid dissociates until all the added hydroxide has been used up. Acetate ion is produced in the process. Thus the solution will have a lower acetic acid concentration and a higher acetate ion concentration than it had originally. The solution will be the same as if it had been prepared originally from 0.11 mol of sodium acetate and 0.09 mol of acetic acid in enough water to make 1.0 L of solution.

	initial	**assuming complete chemical reaction with NaOH**	**at equilibrium**
$[H_3O^+]$			x
$[HC_2H_3O_2]$	0.10	0.09	$0.09 - x \cong 0.09$
$[C_2H_3O_2^-]$	0.10	0.11	$0.11 + x \cong 0.11$

$$K_a = \frac{x(0.11)}{(0.09)} = 1.8 \times 10^{-5} \qquad \text{thus} \qquad x = 1.5 \times 10^{-5} = [H_3O^+]$$

The hydronium ion concentration has been reduced only to five-sixths of its original value by the addition of about 500 times as much OH^- as H_3O^+ originally present. In the buffer solution, the hydronium ion concentration remains relatively constant. Added to the unbuffered solution of HCl, the same quantity of base caused an 18 million-fold change in hydronium ion concentration.

21.122 A buffer solution was prepared by dissolving 0.050 mol formic acid and 0.060 mol sodium formate in enough water to make 1.0 L of solution. K_a for formic acid is 1.80×10^{-4}. ***(a)*** Calculate the pH of the solution. ***(b)*** If this solution were diluted to 10 times its volume, what would be the pH? ***(c)*** If the solution in (*b*) were diluted to 10 times *its* volume, what would be the pH?

▌

$$HCHO_2 + H_2O \rightleftharpoons CHO_2^- + H_3O^+ \qquad K_a = \frac{[CHO_2^-][H_3O^+]}{[HCHO_2]} = 1.80 \times 10^{-4}$$

(a) $K_a = \dfrac{(0.060 + x)x}{0.050 - x} = 1.80 \times 10^{-4}$ $\qquad x = 1.5 \times 10^{-4} \quad \text{pH} = 3.83$

(b) $K_a = \dfrac{(0.0060 + x)x}{0.0050 - x} = 1.80 \times 10^{-4}$ $\qquad x = 1.4 \times 10^{-4} \quad \text{pH} = 3.85$

(c) $K_a = \dfrac{(0.00060 + x)x}{(0.00050 - x)} = 1.80 \times 10^{-4}$ $\qquad x = 1.02 \times 10^{-4} \quad \text{pH} = 3.99$

21.123 The base imidazole has a K_b of 9.8×10^{-8}. ***(a)*** In what amounts should 0.0200 M HCl and 0.0200 M imidazole be mixed to make 100 mL of a buffer at pH 7.00? ***(b)*** If the resulting buffer is diluted to 1.00 L, what is the pH of the diluted buffer?

▌ We will represent the base by B:

$$B + H_2O \rightleftharpoons BH^+ + OH^- \qquad K_b = \frac{[BH^+][OH^-]}{[B]} = 9.8 \times 10^{-8}$$

(*a*) At pH = 7.00,

$$\frac{[BH^+]}{[B]} = \frac{9.8 \times 10^{-8}}{1.0 \times 10^{-7}} = 0.98$$

After reaction, we will have 1.00 mol B for every 0.98 mol BH^+, requiring 1.98 mol B initially. Since the acid and base have the same concentration, we will need 0.98 mL HCl for every 1.98 mL B. Let x = volume of B unreacted.

$$0.98x + 1.98x = 100 \text{ mL} \qquad \text{or} \qquad x = 33.8 \text{ mL}$$

$$\text{Volume HCl needed} = 0.98x = 0.98(33.8 \text{ mL}) = 33.1 \text{ mL}$$

$$\text{Volume of base needed} = 1.98(33.8 \text{ mL}) = 66.9 \text{ mL}$$

(*b*) After dilution, the ratio of conjugate acid to base is still the same, so the hydroxide ion concentration and the pH are also the same (pH = 7.00).

21.124 Given 50 mL of 0.20 M solution of the acid HA ($K_a = 1.0 \times 10^{-5}$) and 50 mL of an NaA solution. What should the concentration of the NaA solution be to make a buffer solution with pH = 4.00?

▮

$$HA + H_2O \longrightarrow A^- + H_3O^+$$

The addition of the two solutions halves the concentration of each solute, assuming no reaction at all.

$$K_a = \frac{[A^-][H_3O^+]}{[HA]} = \frac{x(1.0 \times 10^{-4})}{(0.10)} = 1.0 \times 10^{-5} \qquad \text{thus} \qquad x = 1.0 \times 10^{-2} \text{ M}$$

The original NaA solution must be 2.0×10^{-2} M to get this value of x after dilution.

21.125 A buffer solution was prepared by dissolving 0.0200 mol propionic acid and 0.015 mol sodium propionate in enough water to make 1.00 L of solution. (*a*) What is the pH of the buffer? (*b*) What would be the pH change if 1.0×10^{-5} mol HCl were added to 10 mL of the buffer? (*c*) What would be the pH change if 1.0×10^{-5} mol NaOH were added to 10 mL of the buffer? K_a for propionic acid is 1.34×10^{-5}.

▮

$$HC_3H_5O_2 + H_2O \rightleftharpoons C_3H_5O_2^- + H_3O^+ \qquad K_a = \frac{[C_3H_5O_2^-][H_3O^+]}{[HC_3H_5O_2]} = 1.34 \times 10^{-5}$$

(*a*)
$$K_a = \frac{0.0150[H_3O^+]}{0.0200} = 1.34 \times 10^{-5} \qquad \text{thus} \qquad [H_3O^+] = 1.79 \times 10^{-5} \qquad \text{and} \qquad \text{pH} = 4.748$$

(*b*) The effect on 10.0 mL would be the same as adding 1.00×10^{-3} mol HCl to 1.00 L of solution; the concentrations would change to 0.0140 M $C_3H_5O_2^-$ and 0.0210 M $HC_3H_5O_2$

$$K_a = \frac{0.0140[H_3O^+]}{0.0210} = 1.34 \times 10^{-5} \qquad \text{thus} \qquad [H_3O^+] = 2.01 \times 10^{-5} \qquad \text{and} \qquad \text{pH} = 4.697$$

so
$$\Delta\text{pH} = -0.051$$

(*c*) The concentrations would change to 0.0160 M $C_3H_5O_2^-$ and 0.0190 M $HC_3H_5O_2$

$$K_a = \frac{0.0160[H_3O^+]}{0.0190} = 1.34 \times 10^{-5} \qquad \text{thus} \qquad [H_3O^+] = 1.59 \times 10^{-5} \qquad \text{and} \qquad \text{pH} = 4.799$$

so
$$\Delta\text{pH} = +0.051$$

21.126 How many moles of sodium hydroxide can be added to 1.00 L of a solution 0.100 M in NH_3 and 0.100 M in NH_4Cl without changing the pOH by more than 1.00 unit? Assume no change in volume.

▮

$$K_b = 1.8 \times 10^{-5} \qquad \text{so} \qquad pK_b = 4.75 \qquad \frac{[NH_4^+]}{[NH_3]} = 1.00$$

$$\text{pOH} = pK_b + \log\frac{[NH_4^+]}{[NH_3]} = 4.75 + 0.00 = 4.75$$

The original pOH = 4.75. The pOH after addition of NaOH cannot be less than 3.75.

$$\text{pOH} = 3.75 = 4.75 + \log\frac{[NH_4^+]}{[NH_3]}$$

$$\log\frac{[NH_4^+]}{[NH_3]} = -1.00 \qquad \text{so} \qquad \frac{[NH_4^+]}{[NH_3]} = 0.10$$

Hence NaOH can be added until the ratio of $[NH_4^+]$ to $[NH_3]$ is 0.10. Initially $[NH_4^+] + [NH_3] = 0.200$. Although the reaction with OH^- converts NH_4^+ into NH_3, the sum of these two concentrations remains 0.200.

$$[NH_4^+] + [NH_3] = 0.200$$

$$[NH_4^+] = 0.10[NH_3] \qquad \text{(from above)}$$

$$0.10[NH_3] + [NH_3] = 0.200 \qquad \text{so} \qquad 1.10[NH_3] = 0.200 \qquad \text{and} \qquad [NH_3] = 0.182\ \text{M}$$

Hence

$$[NH_4^+] = 0.018\ \text{M}$$

Assuming no change in volume, $0.100 - 0.018 = 0.082$ mol of NaOH can be added without changing the pOH by more than 1.00 pOH unit.

21.127 Using Table 21.1, select a buffer system having a pH of 4.00 and state how you would prepare exactly 1 L of the solution using 0.10 M solutions of the respective conjugates. What practical considerations influence your choice?

▮

For acetic acid: $K_a = 1.8 \times 10^{-5}$ and $pK_a = 4.75$

For nitrous acid: $K_a = 4.5 \times 10^{-4}$ and $pK_a = 3.35$

Either acid would be suitable to make the required buffer having $pH = 4.00$, except that nitrous acid readily undergoes oxidation-reduction reactions and its usefulness is restricted. For acetic acid,

$$pH = 4.00 = 4.75 + \log \frac{[C_2H_3O_2^-]}{[HC_2H_3O_2]}$$

$$\log \frac{[C_2H_3O_2^-]}{[HC_2H_3O_2]} = 4.00 - 4.75 = -0.75 \qquad \text{so} \qquad \frac{[C_2H_3O_2^-]}{[HC_2H_3O_2]} = 0.18$$

Since 1.00 L of the buffer is to be made up from 0.10 M solutions of salt and acid and the volumes of the respective solutions must be in the ratio 0.18:1.00,

let

$$x = \text{the volume of sodium acetate solution in L}$$

then

$$1.00 - x = \text{the volume of acetic acid solution in L}$$

$$\frac{x}{1.00 - x} = 0.18 \qquad \text{so} \qquad x = 0.15\ \text{L}\ NaC_2H_3O_2\ \text{solution} \qquad \text{and} \qquad 1.00 - x = 0.85\ \text{L}\ HC_2H_3O_2\ \text{solution}$$

21.128 Calculate $[H^+]$ in a 0.200 M dichloroacetic acid solution that is also 0.100 M in sodium dichloroacetate.

▮

$$HC_2HCl_2O_2 + H_2O \rightleftharpoons C_2HCl_2O_2^- + H_3O^+ \qquad K_a = \frac{[H_3O^+][C_2HCl_2O_2^-]}{[HC_2HCl_2O_2]} = 5.0 \times 10^{-2}$$

Let $x = [H_3O^+]$

$$\frac{(0.100 + x)x}{(0.200 - x)} = 0.050 \qquad \text{thus} \qquad x^2 + 0.100x = 0.010 - 0.050x \qquad \text{or} \qquad x^2 + 0.150x - 0.010 = 0$$

$$x = 0.050\ \text{M}$$

21.129 How much solid sodium dichloroacetate should be added to a liter of 0.100 M dichloroacetic acid to reduce $[H^+]$ to 0.030? Neglect the increase in volume of the solution on addition of the salt.

▮

$$HC_2HCl_2O_2 + H_2O \rightleftharpoons C_2HCl_2O_2^- + H_3O^+ \qquad K_a = \frac{[H_3O^+][C_2HCl_2O_2^-]}{[HC_2HCl_2O_2]} = 5.0 \times 10^{-2}$$

	initial	produced	used up	equilibrium
$[HC_2HCl_2O_2]$	0.100		0.030	0.070
$[C_2HCl_2O_2^-]$	x	0.030		$x + 0.030$
$[H_3O^+]$	0	0.030		0.030

$$\frac{(x + 0.030)(0.030)}{0.070} = 0.050 \qquad \text{thus} \qquad x = 0.087\ \text{mol/L}$$

$$(1.0\ \text{L})(0.087\ \text{mol/L}) = 0.087\ \text{mol}$$

21.130 Calculate the pH of a solution prepared by addition of sufficient water to make 1.00 L of solution to 0.100 mol of $HC_2H_3O_2$, 0.130 mol of $NaC_2H_3O_2$, and **(a)** 0.090 mol of NaOH **(b)** 0.090 mol of HCl.

▮ **(a)** The solution is the same as that of 0.010 mol of $HC_2H_3O_2$ and 0.220 mol of $Na_2C_2H_3O_2$:

$$HC_2H_3O_2 + H_2O \longrightarrow H_3O^+ + C_2H_3O_2^-$$

$$K_a = \frac{[C_2H_3O_2^-][H_3O^+]}{[HC_2H_3O_2]} = 1.8 \times 10^{-5} = \frac{(0.220)[H_3O^+]}{(0.010)}$$

$$[H_3O^+] = 8.2 \times 10^{-7} \quad \text{so} \quad pH = 6.09$$

(b) The equivalent solution would contain 0.190 mol $HC_2H_3O_2$ and 0.040 mol of $C_2H_3O_2^-$.

$$K_a = \frac{(0.040)[H_3O^+]}{(0.190)} = 1.8 \times 10^{-5} \qquad [H_3O^+] = 8.6 \times 10^{-5} \qquad \text{so} \qquad pH = 4.07$$

21.131 Calculate the mass of $NaC_2H_3O_2(s)$ and the volume of a 5.00 M $HC_2H_3O_2$ solution, plus water as necessary, which are required to prepare 1.00 L of a buffer solution containing 0.200 M $HC_2H_3O_2$ with a pH = 4.00.

▮

$$K_a = \frac{[H_3O^+][C_2H_3O_2^-]}{[HC_2H_3O_2]} = 1.8 \times 10^{-5}$$

Since pH = 4.00, $[H_3O^+] = 1.00 \times 10^{-4}$ and $[HC_2H_3O_2] = 0.200$

$$[C_2H_3O_2^-] = \frac{(1.8 \times 10^{-5})(0.200)}{1.0 \times 10^{-4}} = 0.036\ M$$

Thus 0.200 mol of $HC_2H_3O_2$ and 0.036 mol of $NaC_2H_3O_2$ are necessary to prepare 1.00 L of solution. The quantities of reagents are

$$(0.20\ \text{mol}\ HC_2H_3O_2)\left(\frac{1.00\ L}{5.00\ \text{mol}}\right) = 0.040\ L = 40\ \text{mL}\ HC_2H_3O_2$$

$$(0.036\ \text{mol}\ NaC_2H_3O_2)\left(\frac{82.0\ g}{\text{mol}}\right) = 2.95\ g\ NaC_2H_3O_2$$

21.5 HYDROLYSIS EQUILIBRIUM

The reaction of an ion with water to produce its parent acid or base is sometimes referred to as hydrolysis. The equilibrium constant associated with that reaction can be referred to as K_h. However, the ion is acting as an acid or base in the Brønsted sense, and the equilibrium constant may in some books be referred to as K_a or K_b. Nevertheless, these types of constants are not tabulated, since they may be calculated easily from the K_b or K_a value of their parent base or acid.

21.132 Explain why a solution of NH_4Cl is acidic.

▮ The ammonium ion hydrolyzes somewhat, yielding H_3O^+; the chloride ion, a very weak base, does not hydrolyze.

$$NH_4^+ + H_2O \rightleftharpoons H_3O^+ + NH_3$$

21.133 Calculate the pH of a 0.200 M solution of NH_4Cl.

▮ Cl^-, the conjugate base of a strong acid, does not react with water, but NH_4^+, the conjugate acid of NH_3, does react:

$$NH_4^+ + H_2O \rightleftharpoons H_3O^+ + NH_3 \qquad K_h = \frac{[H_3O^+][NH_3]}{[NH_4^+]} = \frac{K_w}{K_b}$$

Let $[H_3O^+] = x \qquad [NH_3] = x \qquad [NH_4^+] = 0.200 - x \approx 0.200$

$$K_h = \frac{x^2}{0.200} = \frac{1.0 \times 10^{-14}}{1.8 \times 10^{-5}} = 5.6 \times 10^{-10} \qquad \text{thus} \qquad x = 1.1 \times 10^{-5} = [H_3O^+] \qquad \text{so} \qquad pH = 4.96$$

The solution of NH_4Cl in water is acidic, as expected.

21.134 What is the OH^- concentration of a 0.010 M solution of $NaC_2H_3O_2$?

▮ $$C_2H_3O_2^- + H_2O \longrightarrow HC_2H_3O_2 + OH^-$$

$$K_b = \frac{K_w}{K_a} = \frac{[HC_2H_3O_2][OH^-]}{[C_2H_3O_2^-]} = \frac{x^2}{0.010} = \frac{1.0 \times 10^{-14}}{1.8 \times 10^{-5}} \quad \text{so} \quad x = 2.4 \times 10^{-6}\ M = [OH^-]$$

21.135 Which equilibrium constant(s) or ratio of equilibrium constants should be used to calculate the pH of 1.00 L of each of the following solutions? **(a)** KOH **(b)** NH_3 **(c)** $HC_2H_3O_2$ **(d)** $HC_2H_3O_2 + NaC_2H_3O_2$ **(e)** $KC_2H_3O_2$ **(f)** 0.10 mol $HC_2H_3O_2$ + 0.050 mol NaOH **(g)** H_2S **(h)** 0.10 mol NH_4Cl + 0.050 mol NaOH **(i)** 0.10 mol $HC_2H_3O_2$ + 0.10 mol NaOH

▮ **(a)** K_w **(b)** K_b and K_w **(c)** K_a **(d)** K_a **(e)** $K_h = K_w/K_a$ **(f)** K_a. (The solution is the same as 0.05 mol $HC_2H_3O_2$ + 0.05 mol $NaC_2H_3O_2$.) **(g)** K_1 **(h)** K_b and K_w. (The reaction $NH_4^+ + OH^- \rightarrow H_2O + NH_3$ produces a buffer solution of NH_3 with the excess NH_4^+.) **(i)** $K_h = K_w/K_a$. (The neutralized acid results in a solution comparable to $NaC_2H_3O_2$.)

21.136 The salt of which one of the following five weak acids will be the most hydrolyzed?

(a) HA: $K_a = 1 \times 10^{-8}$ **(b)** HB: $K_a = 2 \times 10^{-6}$ **(c)** HC: $K_a = 3 \times 10^{-8}$
(d) HD: $K_a = 4 \times 10^{-10}$ **(e)** HE: $K_a = 1 \times 10^{-7}$

▮ (*d*) $K_b = K_w/K_a$, so the smallest value of K_a will produce the largest value of K_b and hence the most hydrolysis.

21.137 Calculate the pH of 0.10 M NH_4NO_3 solution.

▮ $$NH_4^+ + H_2O \longrightarrow NH_3 + H_3O^+$$

$$K_h = \frac{K_w}{K_b} = \frac{[NH_3][H_3O^+]}{[NH_4^+]} = \frac{x^2}{0.10} = \frac{1.0 \times 10^{-14}}{1.8 \times 10^{-5}} \quad \text{or} \quad x^2 = \frac{1.0 \times 10^{-10}}{1.8} = 0.55 \times 10^{-10}$$

$$[H_3O^+] = x = 0.74 \times 10^{-5} = 7.4 \times 10^{-6} \quad \text{so} \quad pH = 5.13$$

21.138 From the data of Table 21.1, choose an acid and a base, neither of which is strong, which when mixed in equimolar quantities give a solution with pH nearest 7.

▮ $HC_2H_3O_2$ and NH_3, each of which has $K = 1.8 \times 10^{-5}$

21.139 Calculate the extent of hydrolysis in a 0.0100 M solution of NH_4Cl.

▮ $$NH_4^+ \rightleftharpoons NH_3 + H^+ \qquad K_h = \frac{[NH_3][H^+]}{[NH_4^+]} = \frac{K_w}{K_b} = \frac{1.00 \times 10^{-14}}{1.8 \times 10^{-5}} = 5.6 \times 10^{-10}$$

By the reaction equation, equal amounts of NH_3 and H^+ are formed. Let $x = [NH_3] = [H^+]$. Then

$$[NH_4^+] = 0.0100 - x \approx 0.0100$$

and $$K_h = \frac{[NH_3][H^+]}{[NH_4^+]} \quad \text{gives} \quad 5.6 \times 10^{-10} = \frac{x^2}{0.0100}$$

Solving, $x = 2.4 \times 10^{-6}$. *Check of the approximation*: x is very small compared with 0.0100.

$$\%\ \text{hydrolyzed} = \frac{\text{amount hydrolyzed}}{\text{total amount}} = \frac{2.4 \times 10^{-6}\ \text{mol/L}}{0.0100\ \text{mol/L}} = 2.4 \times 10^{-4} = 0.024\%$$

21.140 Calculate $[OH^-]$ in a 1.00 M solution of NaOCN.

▮ $$OCN^- + H_2O \rightleftharpoons HOCN + OH^- \qquad K_h = \frac{[OH^-][HOCN]}{[OCN^-]} = \frac{K_w}{K_a} = \frac{1.00 \times 10^{-14}}{3.3 \times 10^{-4}} = 3.0 \times 10^{-11}$$

Since the source of OH^- and HOCN is the hydrolysis reaction, they must exist in equal concentrations; let $x = [OH^-] = [HOCN]$. Then

$$[OCN^-] = 1.00 - x \approx 1.00 \quad \text{and} \quad K_h = 3.0 \times 10^{-11} = \frac{x^2}{1.00} \quad \text{thus} \quad x = [OH^-] = 5.5 \times 10^{-6}.$$

Check of approximation: x is very small compared with 1.00.

21.141 The acid ionization (hydrolysis) constant of Zn^{2+} is 1.0×10^{-9}. **(a)** Calculate the pH of a 0.0010 M solution of $ZnCl_2$. **(b)** What is the basic dissociation constant of $Zn(OH)^+$?

▮ **(a)** $Zn^{2+} + H_2O \rightleftharpoons Zn(OH)^+ + H^+ \qquad K_a = \dfrac{[Zn(OH)^+][H^+]}{[Zn^{2+}]} = 1.0 \times 10^{-9}$

Let $x = [Zn(OH)^+] = [H^+]$. Then $[Zn^{2+}] = 0.0010 - x \approx 0.0010$ and

$$\frac{x^2}{0.0010} = 1.0 \times 10^{-9} \quad \text{or} \quad x = [H^+] = 1.0 \times 10^{-6}$$

$$pH = -\log(1.0 \times 10^{-6}) = -\log 10^{-6} = +6.0$$

Check of approximation: x is very small compared with 0.0010.

(b) Zn^{2+}, as an acid, is conjugate to the base $Zn(OH)^+$. For the basic dissociation,

$$Zn(OH)^+ \rightleftharpoons Zn^{2+} + OH^- \qquad K_b = \frac{K_w}{K_a(Zn^{2+})} = \frac{1.00 \times 10^{-14}}{1.0 \times 10^{-9}} = 1.0 \times 10^{-5}$$

21.142 Calculate the pH of a solution prepared by mixing 100.0 mL of 0.400 M HCl with 100.0 mL of 0.400 M NH_3.

▮ Mixing the solutions provides 200.0 mL of 0.200 M NH_4Cl, which reacts with water as follows:

$$NH_4^+ + H_2O \longrightarrow NH_3 + H_3O^+ \qquad K = \frac{[NH_3][H_3O^+]}{[NH_4^+]} = \frac{K_w}{K_b} = 5.6 \times 10^{-10}$$

$$\frac{x^2}{0.200 - x} \approx \frac{x^2}{0.200} = 5.6 \times 10^{-10} \quad \text{yielding } x = [H_3O^+] = 1.1 \times 10^{-5} \quad \text{so} \quad pH = 4.96$$

21.143 **(a)** Calculate $[OH^-]$ in a 0.0100 M solution of aniline, $C_6H_5NH_2$. K_b for the basic dissociation of aniline is 4.0×10^{-10}. **(b)** What is the $[OH^-]$ in a 0.0100 M solution of aniline hydrochloride, which contains the ion $C_6H_5NH_3^+$?

▮ **(a)** $C_6H_5NH_2 + H_2O \rightleftharpoons C_6H_5NH_3^+ + OH^-$

$$K_b = \frac{[C_6H_5NH_3^+][OH^-]}{[C_6H_5NH_2]} = \frac{x^2}{0.0100} = 4.0 \times 10^{-10} \quad \text{or} \quad x^2 = 4.0 \times 10^{-12}$$

thus $x = 2.0 \times 10^{-6}$ M

(b) $C_6H_5NH_3^+ + H_2O \rightleftharpoons C_6H_5NH_2 + H_3O^+$

$$K_h = \frac{[C_6H_5NH_2][H_3O^+]}{[C_6H_5NH_3^+]} = \frac{K_w}{K_b} = \frac{x^2}{0.0100} = 2.5 \times 10^{-5}$$

$x^2 = 2.5 \times 10^{-7}$ thus $x = 5.0 \times 10^{-4}\ M = [H_3O^+]$

thus $[OH^-] = \dfrac{1.0 \times 10^{-14}}{5.0 \times 10^{-4}} = 2.0 \times 10^{-11}$ M

21.144 Calculate the percent hydrolysis in a 0.0100 M solution of KCN.

▮

$$CN^- + H_2O \rightleftharpoons HCN + OH^-$$

$$K_h = \frac{[HCN][OH^-]}{[CN^-]} = \frac{K_w}{K_a} = \frac{1.0 \times 10^{-14}}{6.2 \times 10^{-10}} = \frac{x^2}{0.0100} = 1.6 \times 10^{-5} \quad \text{thus} \quad x = 4.0 \times 10^{-4}\ M$$

$$\%\ \text{hydrolysis} = \left(\frac{4.0 \times 10^{-4}}{0.0100}\right) \times 100\% = 4.0\%$$

21.145 Which of the following, when mixed, will give a solution with pH greater than 7?

(a) 0.1 M HCl + 0.2 M NaCl
(b) 100 mL of 0.2 M H_2SO_4 + 100 mL of 0.3 M NaOH
(c) 100 mL of 0.1 M $HC_2H_3O_2$ + 100 mL of 0.1 M KOH
(d) 50 mL of 0.1 M HCl + 50 mL of 0.1 M $NaC_2H_3O_2$
(e) 25 mL of 0.1 M HNO_3 + 25 mL of 0.1 M NH_3

▮ (*c*). In (*a*) the NaCl has no effect. In (*b*), H_2SO_4 requires 0.40 M NaOH for complete neutralization. In (*c*) a weak base, $C_2H_3O_2^-$, is formed. In (*d*), a weak acid is formed. In (*e*) the NH_4NO_3 formed hydrolyzes to yield H_3O^+.

21.146 The basic ionization constant for hydrazine, N_2H_4, is 9.6×10^{-7}. What would be the percent hydrolysis of 0.100 M N_2H_5Cl, a salt containing the acid ion conjugate to hydrazine base?

▍

$$N_2H_5^+ + H_2O \rightleftharpoons N_2H_4 + H_3O^+$$

$$K_h = \frac{[N_2H_4][H_3O^+]}{[N_2H_5^+]} = \frac{K_w}{K_b} = \frac{1.0 \times 10^{-14}}{9.6 \times 10^{-7}} = \frac{x^2}{0.100} = 1.04 \times 10^{-8} \qquad \text{thus} \qquad x = 3.2 \times 10^{-5}$$

$$\%\ \text{hydrolysis} = \frac{3.2 \times 10^{-5}}{0.100} \times 100\% = 0.032\%$$

21.147 A 0.25 M solution of pyridinium chloride, $C_5H_5NH^+Cl^-$, was found to have a pH of 2.89. What is K_b for the basic dissociation of pyridine, C_5H_5N?

▍

$$C_5H_5NH^+ + H_2O \rightleftharpoons C_5H_5N + H_3O^+$$

$$\text{pH} = 2.89 \qquad [H_3O^+] = 1.3 \times 10^{-3} \qquad K_h = \frac{[C_5H_5N][H_3O^+]}{[C_5H_5NH^+]} = \frac{(1.3 \times 10^{-3})^2}{0.25} = 6.6 \times 10^{-6}$$

$$K_b = \frac{K_w}{K_h} = \frac{1.0 \times 10^{-14}}{6.6 \times 10^{-6}} = 1.5 \times 10^{-9}$$

21.148 K_a for the acid ionization of Fe^{3+} to $Fe(OH)^{2+}$ and H^+ is 6.5×10^{-3}. What is the maximum pH value which could be used so that at least 95% of the total iron(III) in a dilute solution exists as Fe^{3+}?

▍

$$Fe^{3+} + H_2O \rightleftharpoons Fe(OH)^{2+} + H_3O^+$$

$$K = \frac{[Fe(OH)^{2+}][H_3O^+]}{[Fe^{3+}]} = \frac{(0.050)[H_3O^+]}{0.95} = 6.5 \times 10^{-3} \qquad [H_3O^+] = 0.12\text{ M} \qquad \text{so} \qquad \text{pH} = 0.91$$

21.149 A 0.010 M solution of $PuO_2(NO_3)_2$ was found to have a pH of 3.80. What is the hydrolysis constant, K_h, for PuO_2^{2+}, and what is K_b for PuO_2OH^+?

▍

$$PuO_2^{2+} + H_2O \rightleftharpoons PuO_2(OH)^+ + H_3O^+$$

$$[H_3O^+] = 10^{-3.80} = 1.6 \times 10^{-4}$$

$$K_h = \frac{(1.6 \times 10^{-4})^2}{0.010} = 2.6 \times 10^{-6} \qquad K_b = \frac{K_w}{K_h} = \frac{1.0 \times 10^{-14}}{2.6 \times 10^{-6}} = 3.8 \times 10^{-9}$$

21.150 Calculate the pH of 1.0×10^{-3} M sodium phenolate, $NaOC_6H_5$. K_a for HOC_6H_5 is 1.05×10^{-10}.

▍

$$C_6H_5O^- + H_2O \rightleftharpoons HC_6H_5O + OH^-$$

$$K_h = \frac{[HC_6H_5O][OH^-]}{[C_6H_5O^-]} = \frac{K_w}{K_a} = \frac{x^2}{(1.0 \times 10^{-3}) - x} = 1.0 \times 10^{-4}$$

$$x^2 + 1.0 \times 10^{-4}x - 1.0 \times 10^{-7} = 0 \qquad \text{thus} \qquad x = 2.7 \times 10^{-4}\text{ M} = [OH^-]$$

$$\text{pOH} = 3.57 \qquad \text{and} \qquad \text{pH} = 14.00 - \text{pOH} = 10.43$$

21.151 Calculate the extent of hydrolysis and the pH of 0.0100 M $NH_4C_2H_3O_2$.

▍ This is a case where both cation and anion hydrolyze.

$$\text{For } NH_4^+ \qquad K_h = \frac{K_w}{K_b(NH_3)} = \frac{1.00 \times 10^{-14}}{1.8 \times 10^{-5}} = 5.6 \times 10^{-10}$$

$$\text{For } C_2H_3O_2^-: \qquad K_h = \frac{K_w}{K_a(HC_2H_3O_2)} = \frac{1.00 \times 10^{-14}}{1.8 \times 10^{-5}} = 5.6 \times 10^{-10}$$

By coincidence, the hydrolysis constants for these two ions are identical. The production of H^+ by NH_4^+ hydrolysis must therefore exactly equal the production of OH^- by $C_2H_3O_2^-$ hydrolysis; and the H^+ and OH^- formed by hydrolysis neutralize each other to maintain the water equilibrium. The solution is thus neutral, $[H^+] = [OH^-] = 1.00 \times 10^{-7}$, and the pH is 7.00.

For NH_4^+ hydrolysis,

$$\frac{[NH_3][H^+]}{[NH_4^+]} = K_h = 5.6 \times 10^{-10}$$

Let $x = [NH_3]$. Then $0.0100 - x = [NH_4^+]$ and

$$\frac{x(1.00 \times 10^{-7})}{0.0100 - x} = 5.6 \times 10^{-10} \qquad \text{or} \qquad x = 5.6 \times 10^{-5}$$

$$\%\ NH_4^+\ \text{hydrolyzed} = \frac{5.6 \times 10^{-5}}{0.0100} \times 100\% = 0.56\%$$

The percent hydrolysis of acetate ion must also be 0.56%, because K_h is the same.

In comparing this result with Prob. 21.139, note that the percent hydrolysis of NH_4^+ is greater in the presence of a hydrolyzing anion. The reason is that the mutual removal of some of the products of the two hydrolyses, H^+ and OH^-, by the water equilibrium reaction allows both hydrolyses to proceed to increasing extent.

21.152 Calculate the pH in a 0.100 M solution of NH_4OCN.

▮ As in Problem 21.151, both cation and anion hydrolyze. Since NH_3 is a weaker base than HOCN is an acid, however, NH_4^+ hydrolyzes more than OCN^-, and the pH of the solution is less than 7. In order to preserve electrical neutrality, there cannot be an appreciable difference between $[NH_4^+]$ and $[OCN^-]$. (The balancing of a slight difference could be accounted for by $[H^+]$ or $[OH^-]$.) Thus $[NH_3]$ must be practically equal to [HOCN], and we will indeed assume that they are equal.

Let $x = [NH_3] = [HOCN]$; then $0.100 - x = [NH_4^+] = [OCN^-]$.

For NH_4^+: $$K_h = \frac{[NH_3][H^+]}{[NH_4^+]} = \frac{K_w}{K_b} = \frac{1.00 \times 10^{-14}}{1.8 \times 10^{-5}} = 5.6 \times 10^{-10}$$

and $$[H^+] = (5.6 \times 10^{-10})\left(\frac{0.100 - x}{x}\right) \qquad (1)$$

For OCN^-: $$K_h = \frac{[HOCN][OH^-]}{[OCN^-]} = \frac{K_w}{K_a} = \frac{1.00 \times 10^{-14}}{3.3 \times 10^{-4}} = 3.0 \times 10^{-11}$$

and $$[OH^-] = (3.0 \times 10^{-11})\left(\frac{0.100 - x}{x}\right) \qquad (2)$$

Dividing (1) by (2), $$\frac{[H^+]}{[OH^-]} = \frac{5.6 \times 10^{-10}}{3.3 \times 10^{-11}} = 17 \qquad (3)$$

Also, $[H^+]$ and $[OH^-]$ must satisfy the K_w relationship:

$$[H^+][OH^-] = K_w = 1.00 \times 10^{-14} \qquad (4)$$

Multiplying (3) by (4), we obtain

$$[H^+]^2 = 17 \times 10^{-14} \qquad [H^+] = 4.1 \times 10^{-7} \qquad pH = -\log[H^+] = 6.39$$

Check of assumption: Our assumption that $[NH_4^+] = [OCN^-]$ was based on the principle of electroneutrality. It is correct if $[H^+]$ and $[OH^-]$ are both small compared with the shift in the concentrations of the other ions. We shall therefore solve for x, the decrease in either $[NH_4^+]$ or $[OCN^-]$ accompanying the hydrolysis. From (1),

$$x = \frac{(5.6 \times 10^{-10})(0.100 - x)}{[H^+]} = \frac{(5.6 \times 10^{-10})(0.100 - x)}{4.2 \times 10^{-7}} \approx 1.4 \times 10^{-4}$$

Both $[H^+]$ and $[OH^-]$ are small compared with x.

21.153 Calculate the ammonia concentration of a solution prepared by dissolving 0.150 mol of $NH_4C_2H_3O_2$ in sufficient water to make 1.0 L of solution.

▮ Since this salt is composed of conjugates of acid and base of equal strength (1.8×10^{-5}), the pH should be 7.0. At any degree of hydrolysis $[NH_3] + [NH_4^+] = 0.150$.

Let $x = [NH_3]$, then $0.150 - x = [NH_4^+]$

$$NH_4^+ + H_2O \longrightarrow NH_3 + H_3O^+$$

$$K_h = \frac{[NH_3][H_3O^+]}{[NH_4^+]} = \frac{1.0 \times 10^{-14}}{1.8 \times 10^{-5}} = \frac{x(1.0 \times 10^{-7})}{(0.150 - x)} = 5.6 \times 10^{-10} \qquad \text{so} \qquad x = 8.3 \times 10^{-4} = [NH_3]$$

21.6 POLYPROTIC ACIDS AND BASES

21.154 Calculate the hydronium ion concentration and the sulfide ion concentration of a 0.100 M solution of H_2S.

▮ The second step in the ionization process is the only source of sulfide ions, but neither the hydronium ion concentration nor the HS^- concentration is significantly changed by this step. Thus $[H_3O^+]$ and $[HS^-]$ can be calculated from K_1 only.

$$K_1 = \frac{[H_3O^+][HS^-]}{[H_2S]} = 1.0 \times 10^{-7}$$

Let $[H_3O^+] = x$ then $[HS^-] = x$ and $[H_2S] = 0.100 - x \approx 0.100$

$$\frac{x^2}{0.100} = 1.0 \times 10^{-7} \qquad \text{so} \qquad x = 1.0 \times 10^{-4} = [H_3O^+] = [HS^-]$$

The sulfide ion concentration is calculated by use of K_2 and the values of $[H_3O^+]$ and $[HS^-]$ just determined.

$$K_2 = \frac{[H_3O^+][S^{2-}]}{[HS^-]} = 1.0 \times 10^{-14} \qquad \text{so} \qquad [H_3O^+] = [HS^-] = 1.0 \times 10^{-4}$$

$$K_2 = \frac{(1.0 \times 10^{-4})[S^{2-}]}{(1.0 \times 10^{-4})} = 1.0 \times 10^{-14} \qquad \text{so} \qquad [S^{2-}] = 1.0 \times 10^{-14}$$

The quantity of additional hydronium ion produced and the quantity of hydrogen sulfide ion used up in the second ionization step are each equal to the quantity of sulfide ion produced and are small compared to the quantities of H_3O^+ and HS^- produced in the first step. As was previously stated, the second ionization does not change the hydronium ion concentration or the hydrogen sulfide ion concentration significantly.

21.155 Determine the $[S^{2-}]$ in a saturated H_2S solution to which enough HCl has been added to produce a $[H^+]$ of 2×10^{-4}.

▮
$$K_1K_2 = \frac{[H^+]^2[S^{2-}]}{[H_2S]} = \frac{(2 \times 10^{-4})^2[S^{2-}]}{0.10} = 1.0 \times 10^{-21} \qquad \text{so} \qquad [S^{2-}] = \frac{1.0 \times 10^{-22}}{4 \times 10^{-8}} = 2.5 \times 10^{-15}$$

Warning: Do not use the K_1K_2 form of the equation unless you have an independent method of calculating $[H^+]$ or $[S^{2-}]$.

21.156 Calculate the values of the equilibrium constants for the reactions with water of $H_2PO_4^-$, HPO_4^{2-}, and PO_4^{3-} as bases. Comparing the relative values of the two equilibrium constants of $H_2PO_4^-$ with water, deduce whether solutions of this ion in water are acidic or basic. Deduce whether solutions of HPO_4^{2-} are acidic or basic.

▮

$$H_2PO_4^- + H_2O \rightleftharpoons H_3PO_4 + OH^- \qquad K_h = \frac{K_w}{K_1} = \frac{1.0 \times 10^{-14}}{7.1 \times 10^{-3}} = 1.4 \times 10^{-12}$$

$$HPO_4^{2-} + H_2O \rightleftharpoons H_2PO_4^- + OH^- \qquad K_h' = \frac{K_w}{K_2} = \frac{1.0 \times 10^{-14}}{6.3 \times 10^{-8}} = 1.6 \times 10^{-7}$$

$$PO_4^{3-} + H_2O \rightleftharpoons HPO_4^{2-} + OH^- \qquad K_h'' = \frac{K_w}{K_3} = \frac{1.0 \times 10^{-14}}{4.5 \times 10^{-13}} = 2.2 \times 10^{-2}$$

$H_2PO_4^-$ can react with water to yield HPO_4^{2-} and H_3O^+, with a constant 6.3×10^{-8}. Its reaction with water to yield H_3PO_4 and OH^- has a much smaller equilibrium constant, 1.4×10^{-12}, and so this ion in water solution is acidic. Comparison of the constants for HPO_4^{2-} (4.5×10^{-13} as an acid, 1.6×10^{-7} as a base) leads to the conclusion that this ion in solution is basic.

21.157 Calculate $[H^+]$, $[H_2PO_4^-]$, $[HPO_4^{2-}]$, and $[PO_4^{3-}]$ in 0.0100 M H_3PO_4.

▮ Begin by assuming that H^+ comes principally from the first stage of dissociation, and that the concentration of any anion formed by one stage of an ionization is not appreciably lowered by the succeeding stage of ionization.

$$H_3PO_4 \rightleftharpoons H^+ + H_2PO_4^- \qquad K_1 = 7.1 \times 10^{-3}$$

Let $[H^+] = [H_2PO_4^-] = x$. Then $[H_3PO_4] = 0.0100 - x$, and

$$\frac{x^2}{0.0100 - x} = 7.1 \times 10^{-3} \qquad \text{or} \qquad x = 0.0056$$

Next take the above value for $[H^+]$ and $[H_2PO_4^-]$ to solve for $[HPO_4^{2-}]$.

$$H_2PO_4^- \rightleftharpoons H^+ + HPO_4^{2-} \qquad K_2 = 6.3 \times 10^{-8}$$

$$[HPO_4^{2-}] = \frac{K_2[H_2PO_4^-]}{[H^+]} = \frac{(6.3 \times 10^{-8})(0.0056)}{0.0056} = 6.3 \times 10^{-8}$$

Check of assumption: The relative extent of the second dissociation $(6.3 \times 10^{-8})/(5.6 \times 10^{-3})$ is indeed small.
Next take the above values for $[H^+]$ and $[HPO_4^{2-}]$ to solve for $[PO_4^{3-}]$.

$$HPO_4^{2-} \rightleftharpoons H^+ + PO_4^{3-} \qquad K_3 = 4.5 \times 10^{-13}$$

$$[PO_4^{3-}] = \frac{K_3[HPO_4^{2-}]}{[H^+]} = \frac{(4.5 \times 10^{-13})(6.3 \times 10^{-8})}{5.6 \times 10^{-3}} = 5.1 \times 10^{-18}$$

Check of assumptions: The depletion of HPO_4^- by the third stage, 5.1×10^{-18} mol/L, is an insignificant fraction of the amount present as a result of the second step, 6.3×10^{-8} mol/L.

21.158 Calculate the pH of a 0.10 M solution of Na_2HPO_4. State any assumptions or approximations used.

▌

$$HPO_4^{2-} + H_2O \rightleftharpoons H_2PO_4^- + OH^- \qquad K_h = \frac{1.0 \times 10^{-14}}{6.3 \times 10^{-8}} = 1.6 \times 10^{-7}$$

(The acid dissociation is negligible, since K_3 is so small.)

$$K_h = \frac{[H_2PO_4^-][OH^-]}{[HPO_4^{2-}]} = \frac{x^2}{0.10 - x} \approx \frac{x^2}{0.10} = 1.6 \times 10^{-7} \qquad \text{so} \qquad x = [OH^-] = 1.26 \times 10^{-4}$$

$$\text{pOH} = 3.90 \qquad \text{and} \qquad \text{pH} = 10.10$$

21.159 Calculate the H^+ concentration of 0.10 M H_2S solution.

▌ Most of the H^+ results from the primary ionization: $H_2S \rightleftharpoons H^+ + HS^-$.
Let $x = [H^+] = [HS^-]$. Then $[H_2S] = 0.10 - x \approx 0.10$.

$$K_1 = \frac{[H^+][HS^-]}{[H_2S]} \qquad \text{or} \qquad 1.0 \times 10^{-7} = \frac{x^2}{0.10} \qquad \text{or} \qquad x = 1.0 \times 10^{-4}$$

Check of assumptions: (1) x is indeed small compared with 0.10. (2) For the above values of $[H^+]$ and $[HS^-]$, the extent of the second dissociation is given by

$$[S^{2-}] = \frac{K_2[HS^-]}{[H^+]} = \frac{(1.0 \times 10^{-14})(1.0 \times 10^{-4})}{1.0 \times 10^{-4}} = 1.0 \times 10^{-14}$$

The second dissociation is so limited that it does not appreciably lower $[HS^-]$ or raise $[H^+]$ as calculated from the first dissociation only. In general, the concentration of the conjugate base resulting from the second dissociation is numerically equal to K_2 whenever the extent of the second dissociation is less than 5%.

21.160 Calculate the concentration of $C_8H_4O_4^{2-}$ (*a*) in a 0.010 M solution of $H_2C_8H_4O_4$, (*b*) in a solution which is 0.010 M with respect to $H_2C_8H_4O_4$ and 0.020 M with respect to HCl. The ionization constants for $H_2C_8H_4O_4$, phthalic acid, are:

$$H_2C_8H_4O_4 \rightleftharpoons H^+ + HC_8H_4O_4^- \qquad K_1 = 1.1 \times 10^{-3}$$
$$HC_8H_4O_4^- \rightleftharpoons H^+ + C_8H_4O_4^{2-} \qquad K_2 = 3.9 \times 10^{-6}$$

▌ (*a*) If there were no second dissociation, the $[H^+]$ could be computed on the basis of the K_1 equation.

$$\frac{x^2}{0.010 - x} = 1.1 \times 10^{-3} \qquad \text{or} \qquad x = [H^+] = [HC_8H_4O_4^-] = 2.8 \times 10^{-3}$$

If we assume that the second dissociation does not appreciably affect $[H^+]$ or $[HC_8H_4O_4^-]$, then

$$[C_8H_4O_4^{2-}] = \frac{K_2[HC_8H_4O_4^-]}{[H^+]} = \frac{(3.9 \times 10^{-6})(2.8 \times 10^{-3})}{2.8 \times 10^{-3}} = 3.9 \times 10^{-6}$$

Check of assumption: The extent of the second dissociation relative to the first,

$$\frac{3.9 \times 10^{-6}}{2.8 \times 10^{-3}} = 1.4 \times 10^{-3} = 0.14\%$$

is sufficiently small to validate the assumption.

(*b*) The $[H^+]$ in solution may be assumed to be essentially that contributed by the HCl. Also, this large common-ion concentration represses the ionization of the phthalic acid, so that we assume that $[H_2C_8H_4O_4] = 0.010$. The most convenient equation to use is the K_1K_2 equation, since all the concentrations for this equation are known but one.

$$\frac{[H^+]^2[C_8H_4O_4^{2-}]}{[H_2C_8H_4O_4]} = K_1K_2 = (1.1 \times 10^{-3})(3.9 \times 10^{-6}) = 4.3 \times 10^{-9}$$

$$[C_8H_4O_4^{2-}] = \frac{(4.3 \times 10^{-9})[H_2C_8H_4O_4]}{[H^+]^2} = \frac{(4.3 \times 10^{-9})(0.010)}{(0.020)^2} = 1.1 \times 10^{-7}$$

Check of assumptions: Solving for the first dissociation,

$$[HC_8H_4O_4^-] = \frac{K_1[H_2C_8H_4O_4]}{[H^+]} = \frac{(1.1 \times 10^{-3})(0.010)}{0.020} = 6 \times 10^{-4}$$

The amount of H^+ contributed by this dissociation, 6×10^{-4} mol/L, is indeed less than 10% of the amount contributed by HCl (0.020 mol/L). The amount of H^+ contributed by the second dissociation is still less.

21.161 What is $[S^{2-}]$ in a solution that is 0.050 M H_2S and 0.0100 M HCl?

▌ The HCl represses the ionization of *both* weak acids, H_2S and HS^-.

$$[H_3O^+] = 0.0100 \text{ M}$$

$$K_1 = \frac{[H_3O^+][HS^-]}{[H_2S]} = \frac{(0.0100)[HS^-]}{0.050} = 1.0 \times 10^{-7} \quad \text{so} \quad [HS^-] = 5.0 \times 10^{-7} \text{ M}$$

$$K_2 = \frac{[H_3O^+][S^{2-}]}{[HS^-]} = \frac{(0.0100)[S^{2-}]}{5.0 \times 10^{-7}} = 1.0 \times 10^{-14} \quad \text{so} \quad [S^{2-}] = 5.0 \times 10^{-19} \text{ M}$$

21.162 K_1 and K_2 for oxalic acid, $H_2C_2O_4$, are 5.6×10^{-2} and 5.4×10^{-5}. What is $[OH^-]$ in a 0.0050 M solution of $Na_2C_2O_4$?

▌ The hydrolysis of $C_2O_4^{2-}$ represses the hydrolysis of $HC_2O_4^-$.

$$C_2O_4^{2-} + H_2O \rightleftharpoons HC_2O_4^- + OH^-$$

$$K_h = \frac{[HC_2O_4^-][OH^-]}{[C_2O_4^{2-}]} = \frac{K_w}{K_2} = \frac{1.0 \times 10^{-14}}{5.4 \times 10^{-5}} = \frac{x^2}{0.0050} = 1.85 \times 10^{-10} \quad \text{so} \quad x = 9.6 \times 10^{-7} \text{ M}$$

21.163 Calculate the sulfide ion concentration of a solution prepared by adding 0.100 mol of H_2S and 0.300 mol of HCl to enough water to make 1.00 L of solution.

▌

$$H_2S + H_2O \xrightarrow{1} HS^- + H_3O^+ \qquad HS^- + H_2O \xrightarrow{2} S^{2-} + H_3O^+$$

$$K_{12} = \frac{[H_3O^+]^2[S^{2-}]}{[H_2S]} = 1.0 \times 10^{-21} \qquad [H_3O^+] = [HCl] = 0.300 \qquad [H_2S] = 0.100$$

$$\frac{(0.300)^2[S^{2-}]}{(0.100)} = 1.0 \times 10^{-21} \quad \text{so} \quad [S^{2-}] = 1.1 \times 10^{-21}$$

21.164 Calculate the sulfate ion concentration in 0.15 M H_2SO_4.

▌ The first proton in H_2SO_4 is ionized completely. For the second step:

$$HSO_4^- + H_2O \longrightarrow SO_4^{2-} + H_3O^+ \qquad K_2 = \frac{[SO_4^{2-}][H_3O^+]}{[HSO_4^-]} = 1.02 \times 10^{-2}$$

	initial	produced	used up	equilibrium
$[H_3O^+]$	0.15 (1st proton)	x		$0.15 + x$
$[SO_4^{2-}]$	0.00	x		x
$[HSO_4^-]$	0.15		x	$0.15 - x$

$$\frac{x(0.15 + x)}{(0.15 - x)} = 0.0102 \quad \text{or} \quad x^2 + 0.160x - 0.0015 = 0$$

$$x = \frac{-0.160 + \sqrt{(0.160)^2 + 0.0060}}{2} = 8.9 \times 10^{-3}$$

21.165 Calculate the pH of a 0.100 M solution of $NH_2CH_2CH_2NH_2$, ethylenediamine (en). Determine the enH_2^{2+} concentration of the solution.

$$en + H_2O \xrightarrow{1} enH^+ + OH^- \qquad K_1 = \frac{[enH^+][OH^-]}{[en]} = 8.5 \times 10^{-5}$$

$$enH^+ + H_2O \xrightarrow{2} enH_2^{2+} + OH^- \qquad K_2 = \frac{[enH_2^{2+}][OH^-]}{[enH^+]} = 7.1 \times 10^{-8}$$

As in the H_2S ionization in pure water, K_1 only is used to calculate $[OH^-]$. The OH^- generated by the first step represses the ionization of the second step.

	initial	produced	used up	equilibrium
$[enH^+]$	0	x		x
$[OH^-]$	0	x		x
$[en]$	0.100		x	$0.100 - x \simeq 0.100$

$$\frac{x^2}{0.100} = 8.5 \times 10^{-5} \qquad \text{so} \qquad x = 2.9 \times 10^{-3} = [OH^-] = [enH^+]$$

thus $\qquad pOH = 2.54 \qquad$ and $\qquad pH = 11.46$

(The approximation made above, $0.100 - x \approx 0.100$, is valid within 2.9%.) The enH_2^{2+} concentration is determined using K_2.

$$\frac{[enH_2^{2+}][2.9 \times 10^{-3}]}{[2.9 \times 10^{-3}]} = 7.1 \times 10^{-8} \qquad \text{so} \qquad [enH_2^{2+}] = 7.1 \times 10^{-8}\ \text{M}$$

21.166 Calculate the pH of a solution containing 0.100 M HCO_3^- and 0.150 M CO_3^{2-}.

This solution is simply a buffer solution involving the second step of the ionization of H_2CO_3.

$$HCO_3^- + H_2O \longrightarrow H_3O^+ + CO_3^{2-}$$

$$K_2 = \frac{[H_3O^+][CO_3^{2-}]}{[HCO_3^-]} = \frac{[H_3O^+](0.15)}{(0.10)} = 4.7 \times 10^{-11} \qquad \text{so} \qquad [H_3O^+] = 3.1 \times 10^{-11} \qquad \text{and} \qquad pH = 10.51$$

21.167 Calculate the OH^- concentration and the H_3PO_4 concentration of a solution prepared by dissolving 0.100 mol of Na_3PO_4 in sufficient water to make 1.00 L of solution.

The PO_4^{3-} hydrolyzes in three steps. K_1, K_2, and K_3 refer to the ionization constants for H_3PO_4.

$$PO_4^{3-} + H_2O \rightleftharpoons HPO_4^{2-} + OH^- \qquad K = \frac{K_w}{K_3} = \frac{1.0 \times 10^{-14}}{4.5 \times 10^{-13}} = 2.22 \times 10^{-2}$$

$$HPO_4^{2-} + H_2O \rightleftharpoons H_2PO_4^- + OH^- \qquad K' = \frac{K_w}{K_2} = \frac{1.0 \times 10^{-14}}{6.3 \times 10^{-8}} = 1.6 \times 10^{-7}$$

$$H_2PO_4^- + H_2O \rightleftharpoons H_3PO_4 + OH^- \qquad K'' = \frac{K_w}{K_1} = \frac{1.0 \times 10^{-14}}{7.1 \times 10^{-3}} = 1.4 \times 10^{-12}$$

Only the first of these reactions is used to calculate the hydroxide ion concentration, since the hydroxide ion from this reaction represses the second and third steps.

$$K = \frac{x^2}{0.100 - x} = 2.22 \times 10^{-2} \qquad \text{or} \qquad x^2 + 2.22 \times 10^{-2}x - 2.22 \times 10^{-3} = 0$$

$$x = \frac{-2.22 \times 10^{-2} + \sqrt{(2.22 \times 10^{-2})^2 + 8.88 \times 10^{-3}}}{2} = 3.73 \times 10^{-2}\ \text{M}$$

The H_3PO_4 concentration is calculated by use of the K' and K'' expressions:

$$K' = \frac{[H_2PO_4^-]x}{x} = 1.6 \times 10^{-7} \qquad \text{so} \qquad [H_2PO_4^-] = 1.6 \times 10^{-7}$$

$$K'' = \frac{[H_3PO_4](3.73 \times 10^{-2})}{(1.6 \times 10^{-7})} = 1.4 \times 10^{-12} \qquad \text{so} \qquad [H_3PO_4] = 6.0 \times 10^{-18}$$

21.168 Calculate the extent of hydrolysis of 0.0050 M K_2CrO_4. The ionization constants of H_2CrO_4 are $K_1 = 1.6$, $K_2 = 3.1 \times 10^{-7}$.

Just as in the ionization of polyprotic acids, so in the hydrolysis of their salts, the reaction proceeds in successive stages. The extent of the second stage is generally very small compared with the first. This is particularly true in this case, where H_2CrO_4 is essentially a strong acid with respect to its first ionization. The equation of interest is

$$CrO_4^{2-} + H_2O \rightleftharpoons HCrO_4^- + OH^-$$

which indicates that the conjugate acid of the hydrolyzing CrO_4^{2-} is $HCrO_4^-$. As the ionization constant for $HCrO_4^-$ is K_2, the hydrolysis constant for the reaction is K_w/K_2.

$$\frac{[OH^-][HCrO_4^-]}{[CrO_4^{2-}]} = K_h = \frac{K_w}{K_2} = \frac{1.0 \times 10^{-14}}{3.1 \times 10^{-7}} = 3.2 \times 10^{-8}$$

Let $x = [OH^-] = [HCrO_4^-]$. Then $[CrO_4^{2-}] = 0.0050 - x \approx 0.0050$ and

$$\frac{x^2}{0.0050} = 3.2 \times 10^{-8} \qquad \text{or} \qquad x = 1.3 \times 10^{-5}$$

$$\%\ \text{hydrolysis} = \frac{1.3 \times 10^{-5}}{0.0050} = 2.6 \times 10^{-3} = 0.26\%$$

Check of assumption: x is indeed small compared with 0.0050.

21.169 What is the pH of a 0.0050 M solution of Na_2S?

As in Prob. 21.168, the first stage of hydrolysis, leading to HS^-, is predominant.

$$S^{2-} + H_2O \rightleftharpoons HS^- + OH^- \qquad K_h = \frac{[HS^-][OH^-]}{[S^{2-}]} = \frac{K_w}{K_2} = \frac{1.00 \times 10^{-14}}{1.0 \times 10^{-14}} = 1.0$$

Because of the large value for K_h, it cannot be assumed that the equilibrium concentration of S^{2-} is approximately 0.0050 mol/L, the stoichiometric concentration. In fact, the hydrolysis is so extensive that the opposite assumption can safely be made; that is, if $x = [S^{2-}]$, then

$$[HS^-] = [OH^-] = 0.0050 - x \approx 0.0050$$

This is tantamount to an assumption of almost 100% hydrolysis. *Check of assumption*:

$$\frac{(0.0050)(0.0050)}{x} = 1.0 \qquad \text{or} \qquad x = 2.5 \times 10^{-5}$$

Thus the residual unhydrolyzed $[S^{2-}]$ is less than 1% of the original.

$$\text{pOH} = -\log 0.0050 = -(0.70 - 3) = 2.30 \qquad \text{pH} = 14.00 - \text{pOH} = 11.70$$

Additional check: Consider the second stage of hydrolysis.

$$HS^- + H_2O \rightleftharpoons H_2S + OH^- \qquad K_h = \frac{K_w}{K_1} = \frac{1.00 \times 10^{-14}}{1.0 \times 10^{-7}} = 1.0 \times 10^{-7}$$

Solve for $[H_2S]$ by assuming the values of $[OH^-]$ and $[HS^-]$ already obtained.

$$[H_2S] = \frac{K_h[HS^-]}{[OH^-]} = \frac{(1.0 \times 10^{-7})(5.0 \times 10^{-3})}{5.0 \times 10^{-3}} = 1.0 \times 10^{-7}$$

The extent of the second hydrolysis compared with the first, $(1.0 \times 10^{-7})/(5.0 \times 10^{-3})$, is indeed small.

21.170 Calculate the $[H^+]$ of a 0.050 M H_2S solution.

Just as in Prob. 21.66, we may neglect the ionization of the weaker acid (HS^-) in the presence of the stronger acid (H_2S—not a strong acid, but much stronger than HS^-).

$$H_2S + H_2O \rightleftharpoons HS^- + H_3O^+$$

$$K_1 = \frac{[HS^-][H_3O^+]}{[H_2S]} = \frac{x^2}{0.050} = 1.0 \times 10^{-7} \qquad \text{so} \qquad x = 7.1 \times 10^{-5}$$

21.171 What is $[S^{2-}]$ in a 0.050 M H_2S solution?

▮ The H_3O^+ and HS^- concentrations were calculated in Prob. 21.66. Since this is in the same solution, these concentrations are the same.

$$HS^- + H_2O \rightleftharpoons S^{2-} + H_3O^+$$

$$K_2 = \frac{[S^{2-}][H_3O^+]}{[HS^-]} = \frac{[S^{2-}](7.1 \times 10^{-5})}{7.1 \times 10^{-5}} = 1.0 \times 10^{-14} \quad \text{so} \quad [S^{2-}] = 1.0 \times 10^{-14}\text{ M}$$

21.172 Malonic acid is a dibasic acid having $K_1 = 1.42 \times 10^{-3}$ and $K_2 = 2.01 \times 10^{-6}$. Compute the concentration of the divalent malonate ion in **(a)** 0.0010 M malonic acid **(b)** a solution that is 0.00010 M in malonic acid and 0.00040 M in HCl.

▮ We will represent the acid as H_2A.

(a) $H_2A + H_2O \rightleftharpoons HA^- + H_3O^+ \qquad HA^- + H_2O \rightleftharpoons A^{2-} + H_3O^+$

$$K_1 = \frac{[HA^-][H_3O^+]}{[H_2A]} = \frac{x^2}{0.0010 - x} = 1.42 \times 10^{-3} \quad \text{so} \quad x = 6.8 \times 10^{-4}$$

$$K_2 = \frac{[A^{2-}][H_3O^+]}{[HA^-]} = \frac{(6.8 \times 10^{-4})[A^{2-}]}{6.8 \times 10^{-4}} = 2.01 \times 10^{-6} \quad \text{so} \quad [A^{2-}] = 2.01 \times 10^{-6}\text{ M}$$

(b) The concentration of HCl is not sufficient to repress the ionization of H_2A entirely.

	initial	produced	used up	equilibrium
$[H_2A]$	0.00010		x	$0.00010 - x$
$[HA^-]$	0	x		x
$[H_3O^+]$	0.00040	x		$0.00040 + x$

$$K_1 = \frac{(x + 0.00040)x}{0.00010 - x} = 1.42 \times 10^{-3}$$

Solving for x using the quadratic equation yields $x = 7.5 \times 10^{-5}$ M HA^-

$$[H_3O^+] = 0.00040 + 0.000075 = 0.000475\text{ M} \quad \text{thus} \quad K_2 = \frac{y(0.000475)}{7.5 \times 10^{-5}} = 2.01 \times 10^{-6}$$

so
$$y = 3.2 \times 10^{-7}\text{ M} = [A^{2-}]$$

21.173 Compute the pH of a 0.010 M solution of H_3PO_4.

▮
$$H_3PO_4 + H_2O \rightleftharpoons H_3O^+ + H_2PO_4^-$$

$$K_1 = \frac{[H_2PO_4^-][H_3O^+]}{[H_3PO_4]} = \frac{x^2}{0.010 - x} = 7.1 \times 10^{-3} \quad \text{or} \quad x^2 + (7.1 \times 10^{-3})x - 7.1 \times 10^{-5} = 0$$

$$x = \frac{-7.1 \times 10^{-3} + \sqrt{(7.1 \times 10^{-3})^2 + 4(7.1 \times 10^{-5})}}{2} = 5.6 \times 10^{-3} \quad \text{so} \quad \text{pH} = -\log x = 2.25$$

21.174 What is $[H^+]$ in a 0.0060 M H_2SO_4 solution? The first ionization of H_2SO_4 is complete and the second ionization has a K_2 of 1.02×10^{-2}. What is $[SO_4^{2-}]$ in the same solution?

▮
$$HSO_4^- + H_2O \rightleftharpoons SO_4^{2-} + H_3O^+ \qquad K_2 = \frac{[H_3O^+][SO_4^{2-}]}{[HSO_4^-]} = 1.02 \times 10^{-2}$$

	concentrations			
	initial	produced	used up	equilibrium
$[HSO_4^-]$	0.0060		x	$0.0060 - x$
$[SO_4^{2-}]$	0	x		x
$[H_3O^+]$	0.0060	x		$0.0060 + x$

The initial concentrations in the table are those after the complete ionization of the first hydrogen atom but before the equilibrium shown in the chemical equation. Substitution of the equilibrium quantities into the equilibrium constant expression allows calculation of the concentrations sought.

$$K_2 = \frac{(0.0060 + x)x}{(0.0060 - x)} = 1.02 \times 10^{-2} \qquad \text{or} \qquad x^2 + 0.0060x = 6.1 \times 10^{-5} - (1.02 \times 10^{-2})x$$

or

$$x^2 + 0.0162x - 6.1 \times 10^{-5} = 0$$

$$x = \frac{-0.0162 + \sqrt{(0.0162)^2 + 4(6.1 \times 10^{-5})}}{2} = 3.2 \times 10^{-3}\ \text{M}\ SO_4^{2-}$$

so

$$[H_3O^+] = 0.0060 + 0.0032 = 0.0092\ \text{M}$$

21.175 Ethylenediamine, $NH_2C_2H_4NH_2$, is a base that can add 1 or 2 protons. The successive pK_b values for the reaction of the neutral base and of the monocation with water are 4.07 and 7.15, respectively. In a 0.010 M solution of ethylenediamine, what are the concentrations of the singly charged cation and of the doubly charged cation?

▌ The base ethylenediamine is represented by B.

$$B + H_2O \rightleftharpoons BH^+ + OH^- \qquad K_1 = \frac{[BH^+][OH^-]}{[B]} = 10^{-4.07} = 8.5 \times 10^{-5}$$

$$BH^+ + H_2O \rightleftharpoons BH_2^{2+} + OH^- \qquad K_2 = \frac{[BH_2^{2+}][OH^-]}{[BH^+]} = 10^{-7.15} = 7.1 \times 10^{-8}$$

$$K_1 = \frac{x^2}{0.010} = 8.5 \times 10^{-5} \qquad \text{so} \qquad x = 9.2 \times 10^{-4}\ \text{M} = [OH^-] = [BH^+]$$

$$K_2 = \frac{(9.2 \times 10^{-4})[BH_2^{2+}]}{9.2 \times 10^{-4}} = 7.1 \times 10^{-8} \qquad \text{so} \qquad [BH_2^{2+}] = 7.1 \times 10^{-8}\ \text{M}$$

21.176 pK_1 and pK_2 for pyrophosphoric acid are 0.80 and 2.20 respectively. Neglecting the third and fourth dissociations of this tetraprotic acid, what would be the concentration of the divalent anion in a 0.050 M solution of the acid?

▌

$$H_4P_2O_7 + H_2O \rightleftharpoons H_3P_2O_7^- + H_3O^+ \qquad K_1 = \frac{[H_3P_2O_7^-][H_3O^+]}{[H_4P_2O_7]} = 10^{-0.80} = 0.16$$

$$H_3P_2O_7^- + H_2O \rightleftharpoons H_2P_2O_7^{2-} + H_3O^+ \qquad K_2 = \frac{[H_2P_2O_7^{2-}][H_3O^+]}{[H_3P_2O_7^-]} = 10^{-2.20} = 6.3 \times 10^{-3}$$

$$K_1 = \frac{x^2}{0.050 - x} = 0.16 \qquad \text{or} \qquad x^2 + 0.16x - 0.0080 = 0 \qquad \text{so} \qquad x = 0.040\ \text{M} = [H_3O^+]$$

$$K_2 = \frac{[H_2P_2O_7^{2-}][H_3O^+]}{[H_3P_2O_7^-]} = \frac{y(0.040 + y)}{0.040 - y} = 10^{-2.20} = 6.3 \times 10^{-3} \qquad \text{so} \qquad y = 4.9 \times 10^{-3}$$

21.177 What is $[CO_3^{2-}]$ in a 0.00100 M Na_2CO_3 solution after the hydrolysis reactions have come to equilibrium?

▌ The first hydrolysis represses the second.

$$CO_3^{2-} + H_2O \rightleftharpoons HCO_3^- + OH^-$$

$$K_h = \frac{[HCO_3^-][OH^-]}{[CO_3^{2-}]} = \frac{K_w}{K_2} = \frac{1.0 \times 10^{-14}}{4.7 \times 10^{-11}} = \frac{x^2}{0.00100 - x} = 2.1 \times 10^{-4}$$

$$x^2 + (2.1 \times 10^{-4})x - 2.1 \times 10^{-7} = 0$$

so

$$x = 3.7 \times 10^{-4} = [OH^-] \qquad \text{and} \qquad [CO_3^{2-}] = 0.00100 - 0.00037 = 0.00063\ \text{M}$$

21.178 Calculate the pH of **(a)** 0.050 M NaH_2PO_4 and **(b)** of 0.0020 M Na_3PO_4.

▌ **(a)** NaH_2PO_4 can react with water in two ways:

$$H_2PO_4^- + H_2O \rightleftharpoons HPO_4^{2-} + H_3O^+ \qquad K_2 = \frac{[HPO_4^{2-}][H_3O^+]}{[H_2PO_4^-]} = 6.3 \times 10^{-8}$$

$$H_2PO_4^- + H_2O \rightleftharpoons H_3PO_4 + OH^- \qquad K_h = \frac{[H_3PO_4][OH^-]}{[H_2PO_4^-]} = \frac{K_w}{K_1} = \frac{1.0 \times 10^{-14}}{7.1 \times 10^{-3}} = 1.4 \times 10^{-12}$$

Assume that H_3PO_4 and HPO_4^{2-} are formed in approximately equal concentrations. Then the ratio of the equilibrium constants above gives

$$\frac{[H_3O^+]}{[OH^-]} = \frac{6.3 \times 10^{-8}}{1.4 \times 10^{-12}} = 4.5 \times 10^4$$

Together with $[H_3O^+][OH^-] = 1.0 \times 10^{-14}$, this yields:

$$[H_3O^+]^2 = 4.5 \times 10^{-10} \qquad [H_3O^+] = 2.1 \times 10^{-5}\ M \qquad pH = 4.68$$

(*b*) $PO_4^{3-} + H_2O \rightleftharpoons HPO_4^{2-} + OH^-$

$$K_h = \frac{[HPO_4^{2-}][OH^-]}{[PO_4^{3-}]} = \frac{x^2}{0.0020 - x} = \frac{K_w}{K_3} = \frac{1.0 \times 10^{-14}}{4.5 \times 10^{-13}} = 2.2 \times 10^{-2}$$

$$x = 1.85 \times 10^{-3}\ M = [OH^-] \quad \text{so} \quad pOH = 2.74 \quad \text{and} \quad pH = 11.26$$

21.179 A buffer solution of pH 6.70 can be prepared by employing solutions of NaH_2PO_4 and Na_2HPO_4. If 0.0050 mol NaH_2PO_4 is weighed out, how much Na_2HPO_4 must be used to make 1.00 L of the solution?

▮

$$H_2PO_4^- + H_2O \rightleftharpoons HPO_4^{2-} + H_3O^+ \qquad K_2 = \frac{[HPO_4^{2-}][H_3O^+]}{[H_2PO_4^-]} = 6.3 \times 10^{-8}$$

At pH = 6.70

$$\frac{[HPO_4^{2-}]}{[H_2PO_4^-]} = \frac{6.3 \times 10^{-8}}{10^{-6.70}} = 0.316$$

$$(0.0050\ \text{mol}\ H_2PO_4^-)\left(\frac{0.316\ \text{mol}\ HPO_4^{2-}}{\text{mol}\ H_2PO_4^-}\right) = 0.0016\ \text{mol}\ HPO_4^{2-}$$

21.180 How much NaOH must be added to 1.0 L of 0.010 M H_3BO_3 to make a buffer solution of pH 10.10?

▮

$$H_3BO_3 + H_2O \rightleftharpoons H_2BO_3^- + H_3O^+ \qquad K_a = \frac{[H_2BO_3^-][H_3O^+]}{[H_3BO_3]} = 7.3 \times 10^{-10}$$

At pH 10.10,

$$\frac{[H_2BO_3^-]}{[H_3BO_3]} = \frac{7.3 \times 10^{-10}}{10^{-10.10}} = 7.3\ \text{antilog}\ 0.10 = 9.2$$

To convert 9.2 parts out of 10.2 parts of H_3BO_3 to its anion requires

$$(0.010\ \text{mol}\ H_3BO_3)\left(\frac{9.2\ \text{mol NaOH}}{10.2\ \text{mol}\ H_3BO_3}\right) = 0.0090\ \text{mol NaOH}$$

21.181 The pH of blood is 7.4. Assuming that the buffer in blood is carbon dioxide, hydrogen carbonate ion, calculate the ratio of conjugate base to acid necessary to maintain blood at its proper pH. What would be the effect of rapid, forced breathing (panting) on the pH of blood?

▮

$$CO_2 + 2H_2O \rightleftharpoons H_3O^+ + HCO_3^- \qquad K_1 = \frac{[H_3O^+][HCO_3^-]}{[CO_2]} = 4.5 \times 10^{-7}$$

Given that pH = 7.4, then $[H_3O^+] = 4.0 \times 10^{-8}$ and $\dfrac{[HCO_3^-]}{[CO_2]} = \dfrac{4.5 \times 10^{-7}}{4.0 \times 10^{-8}} = 11$

Rapid breathing would lower CO_2 concentration, shifting the equilibrium, and changing the pH somewhat. Dizziness might occur.

21.182 If 0.00050 mol $NaHCO_3$ is added to a large volume of a solution buffered at pH 8.00, how much material will exist in each of the three forms H_2CO_3, HCO_3^-, and CO_3^{2-}?

▮

$$HCO_3^- + H_2O \rightleftharpoons H_2CO_3 + OH^- \qquad K_h = \frac{[H_2CO_3][OH^-]}{[HCO_3^-]} = \frac{K_w}{K_1} = \frac{1.0 \times 10^{-14}}{4.5 \times 10^{-7}} = 2.2 \times 10^{-8}$$

$$HCO_3^- + H_2O \rightleftharpoons CO_3^{2-} + H_3O^+ \qquad K_2 = \frac{[CO_3^{2-}][H_3O^+]}{[HCO_3^-]} = 4.7 \times 10^{-11}$$

At pH = 8.00, $[H_3O^+] = 1.00 \times 10^{-8}$ and $[OH^-] = 1.00 \times 10^{-6}$. Thus

$$\frac{[H_2CO_3]}{[HCO_3^-]} = 2.2 \times 10^{-2} \qquad \frac{[CO_3^{2-}]}{[HCO_3^-]} = 4.7 \times 10^{-3}$$

Since all the species are in the same solution, the concentration ratios are equal to the mole ratios.

$$n(H_2CO_3) + n(HCO_3^-) + n(CO_3^{2-}) = 5.0 \times 10^{-4}\ \text{mol}$$
$$0.022n(HCO_3^-) + n(HCO_3^-) + 0.0047n(HCO_3^-) = 0.00050$$
$$n(HCO_3^-) = 4.9 \times 10^{-4}\ \text{mol} \qquad n(H_2CO_3) = (4.9 \times 10^{-4})(2.2 \times 10^{-2}) = 1.1 \times 10^{-5}\ \text{mol}$$
$$n(CO_3^{2-}) = (4.9 \times 10^{-4})(4.7 \times 10^{-3}) = 2.3 \times 10^{-6}\ \text{mol}$$

21.183 Citric acid is a polyprotic acid with pK_1, pK_2, and pK_3 equal to 3.13, 4.76, and 6.40, respectively. Calculate the concentrations of H^+, the monovalent anion, the divalent anion, and the trivalent anion in 0.0100 M citric acid.

▌ The acid will be represented as H_3A:

$$H_3A + H_2O \rightleftharpoons H_2A^- + H_3O^+ \qquad K_1 = \frac{[H_2A^-][H_3O^+]}{[H_3A]} = 10^{-3.13} = 7.4 \times 10^{-4}$$

$$H_2A^- + H_2O \rightleftharpoons HA^{2-} + H_3O^+ \qquad K_2 = \frac{[HA^{2-}][H_3O^+]}{[H_2A^-]} = 10^{-4.76} = 1.7 \times 10^{-5}$$

$$HA^{2-} + H_2O \rightleftharpoons A^{3-} + H_3O^+ \qquad K_3 = \frac{[A^{3-}][H_3O^+]}{[HA^{2-}]} = 10^{-6.40} = 4.0 \times 10^{-7}$$

$$K_1 = \frac{x^2}{0.010 - x} = 7.4 \times 10^{-4} \qquad \text{or} \qquad x^2 + (7.4 \times 10^{-4})x - (7.4 \times 10^{-6}) = 0$$

$$x = \frac{-7.4 \times 10^{-4} + \sqrt{(7.4 \times 10^{-4})^2 + 4(7.4 \times 10^{-6})}}{2} = 2.4 \times 10^{-3}\ \text{M} = [H_3O^+]$$

$$K_2 = \frac{[HA^{2-}]x}{x} = 1.7 \times 10^{-5} \qquad \text{so} \qquad [HA^{2-}] = 1.7 \times 10^{-5}$$

$$K_3 = \frac{[A^{3-}](2.4 \times 10^{-3})}{1.7 \times 10^{-5}} = 4.0 \times 10^{-7} \qquad \text{so} \qquad [A^{3-}] = 2.8 \times 10^{-9}$$

21.184 If 0.0010 mol of citric acid is dissolved in 1.0 L of a solution buffered at pH 5.00 (without changing the volume), what will be the equilibrium concentrations of citric acid, its monovalent anion, the divalent anion, and the trivalent anion? Use pK values from Prob. 21.183.

▌

$$H_3A + H_2O \rightleftharpoons H_2A^- + H_3O^+ \qquad H_2A^- + H_2O \rightleftharpoons HA^{2-} + H_3O^+$$
$$HA^{2-} + H_2O \rightleftharpoons A^{3-} + H_3O^+$$

If one solves the K_1 expression using $[H_3O^+] = 1.0 \times 10^{-5}$, one finds that practically all the acid is ionized at least to the monoanion. Thus we will make a first approximation assuming complete conversion to that anion:

$$K_2 = \frac{[HA^{2-}][H_3O^+]}{[H_2A^-]} = \frac{y(1.0 \times 10^{-5})}{0.0010 - y} = 1.7 \times 10^{-5}$$

thus $\qquad y = 6.3 \times 10^{-4} = [HA^{2-}] \qquad$ and $\qquad [H_2A^-] = 3.7 \times 10^{-4}$

Knowing a value for H_2A^- concentration allows calculation of H_3A concentration:

$$K_1 = \frac{[H_2A^-][H_3O^+]}{[H_3A]} = \frac{(3.7 \times 10^{-4})(1.0 \times 10^{-5})}{[H_3A]} = 7.4 \times 10^{-4} \qquad \text{so} \qquad [H_3A] = 5.0 \times 10^{-6}\ \text{M}$$

Solving for $[A^{3-}]$,

$$K_3 = \frac{[A^{3-}][H_3O^+]}{[HA^{2-}]} = \frac{[A^{3-}](1.0 \times 10^{-5})}{6.3 \times 10^{-4}} = 4.0 \times 10^{-7} \qquad \text{so} \qquad [A^{3-}] = 2.5 \times 10^{-5}$$

The three concentrations calculated so far total more than the initial concentration of the citric acid used. A closer approximation may be made by reducing the H_2A^- and A^{2-} concentrations by the total concentration of H_3A and A^{3-}, in the ratio 1.0:1.7, as given by K_2 at pH 5.00.

$$[H_3A] + [A^{3-}] = 2.5 \times 10^{-5} + 5.0 \times 10^{-6} = 3.0 \times 10^{-5}$$

$$3.0 \times 10^{-5}\left(\frac{1.0}{2.7}\right) = 1.1 \times 10^{-5} \qquad [H_2A^-] = (3.7 \times 10^{-4}) - (1.1 \times 10^{-5}) = 3.6 \times 10^{-4}\ M$$

$$3.0 \times 10^{-5}\left(\frac{1.7}{2.7}\right) = 1.9 \times 10^{-5} \qquad [HA^{2-}] = (6.3 \times 10^{-4}) - (1.9 \times 10^{-5}) = 6.1 \times 10^{-4}\ M$$

These values may now be used with K_1 and K_3 to find more accurate values for $[H_3A]$ and $[A^{3-}]$.

21.185 If 0.00010 mol H_3PO_4 is added to 1.0 L of a solution buffered at pH 7.00, what are the relative proportions of the four forms H_3PO_4, $H_2PO_4^-$, HPO_4^{2-}, PO_4^{3-}?

▌ Since the solution was previously well buffered, we can assume that the pH is not changed by addition of the phosphoric acid. Then, if $[H^+]$ is fixed, the ratio of two of the desired concentrations can be calculated from each of the ionization constant equations.

$$\frac{[H^+][H_2PO_4^-]}{[H_3PO_4]} = K_1 \qquad \frac{[H^+][HPO_4^{2-}]}{[H_2PO_4^-]} = K_2 \qquad \frac{[H^+][PO_4^{3-}]}{[HPO_4^{2-}]} = K_3$$

$$\frac{[H_3PO_4]}{[H_2PO_4^-]} = \frac{[H^+]}{K_1} = \frac{1.00 \times 10^{-7}}{7.1 \times 10^{-3}} = 1.41 \times 10^{-5}$$

$$\frac{[H_2PO_4^-]}{[HPO_4^{2-}]} = \frac{[H^+]}{K_2} = \frac{1.00 \times 10^{-7}}{6.3 \times 10^{-8}} = 1.59$$

$$\frac{[HPO_4^{2-}]}{[PO_4^{3-}]} = \frac{[H^+]}{K_3} = \frac{1.00 \times 10^{-7}}{4.5 \times 10^{-13}} = 2.2 \times 10^5$$

Since the ratio $[H_3PO_4]/[H_2PO_4^-]$ is very small and the ratio $[HPO_4^{2-}]/[PO_4^{3-}]$ is very large, practically all of the material will exist as $H_2PO_4^-$ and HPO_4^{2-}. The sum of the amounts of these two ions will be practically equal to 0.00010 mol; and if the total volume of solution is 1.0 L, the sum of these two ion concentrations will be 0.00010 mol/L.

Let $x = [HPO_4^{2-}]$; then $[H_2PO_4^-] = 0.00010 - x$. Then

$$\frac{[H_2PO_4^-]}{[HPO_4^{2-}]} = 1.59 \quad \text{gives} \quad \frac{0.00010 - x}{x} = 1.59$$

Solving, $\quad x = [HPO_4^{2-}] = 3.9 \times 10^{-5} \qquad [H_2PO_4^-] = 0.00010 - x = 6.1 \times 10^{-5}$

$$[H_3PO_4] = (1.41 \times 10^{-5})[H_2PO_4^-] = 8.6 \times 10^{-10} \qquad [PO_4^{3-}] = \frac{[HPO_4^{2-}]}{2.2 \times 10^5} = 1.8 \times 10^{-10}$$

21.186 What is the pH of 0.0100 M $NaHCO_3$?

▌ This has similarities to Prob. 21.152 in that there is one reaction tending to make the solution acidic (the K_2 acid dissociation of HCO_3^-) and another reaction tending to make the solution basic (the hydrolysis of HCO_3^-).

$$HCO_3^- \rightleftharpoons H^+ + CO_3^{2-} \qquad K_2 = 4.7 \times 10^{-11} \tag{1}$$

$$HCO_3^- + H_2O \rightleftharpoons OH^- + H_2CO_3 \qquad K_h = \frac{K_w}{K_1} = \frac{1.00 \times 10^{-14}}{4.5 \times 10^{-7}} = 2.2 \times 10^{-8} \tag{2}$$

(Note that the hydrolysis constant for reaction (*2*) is related to K_1, because both hydrolysis and the K_1 equilibrium involve H_2CO_3 and HCO_3^-.) We see that the equilibrium constant for (*2*) is greater than that for (*1*); thus the pH is certain to exceed 7.

We assume that after self-neutralization both $[H^+]$ and $[OH^-]$ will be so small as to have no appreciable effect on the ionic charge balance. Therefore electrical neutrality can be preserved only by maintaining a fixed total anionic charge among the various carbonate species, since the cationic charge remains at 0.0100 M, the concentration of Na^+, regardless of the acid and base equilibria. In other words, for every negative charge removed by converting HCO_3^- to H_2CO_3, another negative charge must be created by converting HCO_3^- to CO_3^{2-}. This leads to the following conditions:

$$[H_2CO_3] = [CO_3^{2-}] = x \qquad [HCO_3^-] = 0.0100 - 2x \approx 0.0100$$

$$\frac{[H^+][CO_3^{2-}]}{[HCO_3^-]} = \frac{[H^+]x}{0.0100} = 4.7 \times 10^{-11} \tag{3}$$

$$\frac{[OH^-][H_2CO_3]}{[HCO_3^-]} = \frac{[OH^-]x}{0.0100} = 2.2 \times 10^{-8} \tag{4}$$

Multiplying (3) and (4) and recalling that $[H^+][OH^-] = 1.00 \times 10^{-14}$, we obtain

$$\frac{(1.00 \times 10^{-14})x^2}{(0.0100)^2} = (4.7 \times 10^{-11})(2.2 \times 10^{-8}) \qquad \text{or} \qquad x = 1.02 \times 10^{-4}$$

Provisional check: $2x$ is indeed small compared with 0.0100.

We return now to (3).

$$[H^+] = \frac{(4.7 \times 10^{-11})[HCO_3^-]}{[CO_3^{2-}]} = \frac{(4.7 \times 10^{-11})(0.0100)}{1.02 \times 10^{-4}} = 4.6 \times 10^{-9}$$

$$pH = -\log [H^+] = -\log 4.6 + 9 = -0.66 + 9 = 8.34$$

Final check: Both $[H^+]$ and $[OH^-]$ are small compared with x, the shift in the concentrations of the other ions. Alternatively, $[H^+]$ could have been computed from the K_1 equilibrium; the same result would have been obtained.

21.187 Determine the equilibrium carbonate ion concentration after equal volumes of 1.0 M sodium carbonate and 1.0 M HCl are mixed.

I The solution is the same as that of 0.50 M $NaHCO_3$ (and 0.50 M NaCl). HCO_3^- reacts with water in two ways:

$$HCO_3^- + H_2O \longrightarrow CO_3^{2-} + H_3O^+ \qquad K_a = 4.7 \times 10^{-11}$$

$$HCO_3^- \longrightarrow CO_2 + OH^- \qquad K_h = \frac{K_w}{K_1} = \frac{1.0 \times 10^{-14}}{4.5 \times 10^{-7}} = 2.2 \times 10^{-8}$$

Since the magnitudes of both of these constants are small, the approximation is made that the concentration of HCO_3^- will remain 0.50 M within experimental error. Since the two constants are not too different in magnitude and since each of the reactions aids the progress of the other (because the OH^- produced in one reacts with the H_3O^+ produced in the other), the approximation is made that the carbon dioxide concentration is equal to the carbonate ion concentration.

$$K_2 = \frac{[CO_3^{2-}][H_3O^+]}{[HCO_3^-]} \qquad K_h = \frac{[CO_2][OH^-]}{[HCO_3^-]}$$

Multiplying these expressions together yields

$$K_2 K_h = \frac{[CO_3^{2-}][CO_2]}{[HCO_3^-]^2} K_w$$

Since $K_h = K_w/K_1$,

$$\frac{K_2}{K_1} = \frac{[CO_3^{2-}][CO_2]}{[HCO_3^-]^2}$$

Assuming $[CO_3^{2-}] = [CO_2]$

$$[CO_3^{2-}] \approx \sqrt{K_2/K_1}[HCO_3^-] = \sqrt{\frac{4.7 \times 10^{-11}}{4.5 \times 10^{-7}}}(0.50) = 5 \times 10^{-3}\ M$$

21.188 Calculate the pH of a 0.100 M Na_2HPO_4 solution. State any approximations which are necessary for this calculation.

I See the answer to Prob. 21.156. Assuming no acidic ionization:

$$HPO_4^{2-} + H_2O \longrightarrow H_2PO_4^- + OH^- \qquad K_h' = \frac{[H_2PO_4^-][OH^-]}{[HPO_4^{2-}]} = \frac{x^2}{0.100} = 1.6 \times 10^{-7}$$

$$[OH^-] = x = 1.3 \times 10^{-4} \qquad pOH = 3.90 \qquad pH = 10.10$$

21.189 The amino acid glycine, NH_2CH_2COOH, is basic because of its NH_2— group and acidic because of its —COOH group. By a process equivalent to base dissociation, glycine can acquire an additional proton to form $^+NH_3CH_2COOH$. The resulting cation may be considered to be a diprotic acid, since one proton from the —COOH group and one proton from the $^+NH_3$— group may be lost. The pK_a values for these processes are 2.35 and 9.78, respectively. In a 0.0100 M solution of neutral glycine, what is the pH and what percent of the glycine is in the cationic form at equilibrium?

▌ The abbreviation HGly will be used for glycine.

$$HGly + H_2O \rightleftharpoons H_2Gly^+ + OH^- \qquad K_h = \frac{[H_2Gly^+][OH^-]}{[HGly]} = 10^{-(14.00-2.35)} = 2.2 \times 10^{-12}$$

$$HGly + H_2O \rightleftharpoons Gly^- + H_3O^+ \qquad K_2 = \frac{[Gly^-][H_3O^+]}{[HGly]} = 10^{-9.78} = 1.7 \times 10^{-10}$$

Assuming equal production of H_2Gly^+ and Gly^-,

$$\frac{K_2}{K_h} = \frac{[H_3O^+]}{[OH^-]} = \frac{1.7 \times 10^{-10}}{2.2 \times 10^{-12}} = 75$$

Multiplying this equation by the K_w equation

$$[H_3O^+][OH^-] = 1.0 \times 10^{-14} \quad \text{yields} \quad [H_3O^+]^2 = 75 \times 10^{-14}$$

so $$[H_3O^+] = 8.7 \times 10^{-7} \quad \text{and} \quad pH = 6.06$$

$$K_1 = \frac{(0.0100)(8.7 \times 10^{-7})}{[H_2Gly^+]} = 10^{-2.35} = 4.5 \times 10^{-3} \quad \text{so} \quad [H_2Gly^+] = 1.9 \times 10^{-6}\ M$$

$$\%\ \text{cationic} = \frac{1.9 \times 10^{-6}}{0.0100} \times 100\% = 0.019\%$$

21.7 INDICATORS AND TITRATION

Note: Before working the following problems, you should familiarize yourself with Table 21.2, below.

TABLE 21.2 pH Ranges of Common Indicators

indicator	pH range	indicator	pH range
Methyl violet	−0.3–1.8	Phenol red	6.8–8.6
Methyl orange	2.8–3.8	Cresol red	7.0–9.1
Congo red	2.8–4.8	Thymol blue	0.5–1.8
Methyl red	3.8–6.1	and	7.6–9.2
Bromothymol blue	6.0–7.9	Phenolphthalein	8.0–9.6
Litmus	5.0–8.1	Thymolphthalein	10.2–11.7
		Tropeolin O	11.1–12.6

21.190 Distinguish between the *end point* and the *equivalence point* of a titration. Explain, using these two terms, the importance of choosing the proper indicator for a given titration.

▌ The end point is the point at which the titration is stopped. The equivalence point is the point at which the acid and the base (or oxidizing agent and reducing agent) have been added in equivalent quantities. Since the purpose of the indicator is to stop the titration close to the point at which the acid and base were added in equivalent quantities, it is important that the equivalent point and the end point be as close as possible. The indicator must change color at a pH close to that of a solution of the salt of the acid and base.

21.191 What type of titration curve would be expected for the titration of a weak acid with a weak base? Is such a titration feasible?

▌ Compared to a strong acid, a weak acid titration with base starts at a higher pH. Compared to a strong base, a weak base titration ends at a lower pH. Thus the titration curve is shortened at each end. For titration of a weak acid with a weak base, the nearly vertical portion of the curve would be insufficient for an effective titration.

21.192 Calculate the pH at which an acid indicator with $K_a = 1.0 \times 10^{-5}$ changes color when the indicator is 1.00×10^{-3} M.

▌ The midpoint of the color change range of an indicator is the point at which its acid and conjugate base forms are present in equal concentration, hence

$$K_{ind} = \frac{[H_3O^+][Ind^-]}{[HInd]} = [H_3O^+] = 1.0 \times 10^{-5} \quad \text{and} \quad pH = 5.00$$

21.193 A 40.0-mL sample of 0.0100 M $HC_2H_3O_2$ is titrated with 0.0200 M NaOH. Calculate the pH after the addition of (*a*) 3.0 mL (*b*) 10.0 mL (*c*) 20.0 mL (*d*) 30.0 mL of the NaOH solution.

▮ We can keep track of the changing amounts of the various species and the increasing volume by a scheme such as Table 21.3.

TABLE 21.3

		(*a*)	(*b*)	(*c*)	(*d*)
Amount of base added (L)	0	0.0030	0.0100	0.0200	0.0300
Total volume (L)	0.0400	0.0430	0.0500	0.0600	0.0700
$n(HC_2H_3O_2)$, before neutralization	4.00×10^{-4}	4.00×10^{-4}	4.00×10^{-4}	4.00×10^{-4}	4.00×10^{-4}
$n(OH^-)$ added (Row 1 × 0.0200 M)	0.0×10^{-4}	0.60×10^{-4}	2.00×10^{-4}	4.00×10^{-4}	6.00×10^{-4}
$n(C_2H_3O_2^-)$ formed	0.0×10^{-4}	0.60×10^{-4}	2.00×10^{-4}	4.00×10^{-4}	4.00×10^{-4}
$n(HC_2H_3O_2)$ remaining	4.00×10^{-4}	3.40×10^{-4}	2.00×10^{-4}	x	y
$n(OH^-)$ excess					2.00×10^{-4}

Note that the amount of acetic acid neutralized (amount of $C_2H_3O_2^-$) follows the amount of OH^- added, up to complete neutralization. Additional OH^-, having no more acid to neutralize, accumulates in the solution. Up to the end point, the amount of $HC_2H_3O_2$ remaining is obtained by simply subtracting the amount of $C_2H_3O_2^-$ from the initial amount of $HC_2H_3O_2$. At the end point and beyond, however, $[HC_2H_3O_2]$ cannot be set equal to 0 but must be solved for the ionic equilibrium conditions.

(*a*) and (*b*). Here absolute concentrations of conjugate acid and base are not needed; only their ratio is required.

	(*a*)	(*b*)
$[H^+] = \dfrac{K_a[HC_2H_3O_2]}{[C_2H_3O_2^-]}$	$\dfrac{(1.8 \times 10^{-5})(3.40)}{0.60} = 1.0 \times 10^{-4}$	$\dfrac{(1.8 \times 10^{-5})(2.00)}{2.00} = 1.8 \times 10^{-5}$
$pH = -\log[H^+]$	4.00	4.74

(*c*) $$M(C_2H_3O_2^-) = \frac{4.00 \times 10^{-4}\ \text{mol}}{0.0600\ \text{L}} = 6.7 \times 10^{-3}\ \text{mol/L}$$

The solution at the end point is the same as 6.7×10^{-3} M $NaC_2H_3O_2$. Consider the hydrolysis of $NaC_2H_3O_2$:

$$C_2H_3O_2^- + H_2O \rightleftharpoons HC_2H_3O_2 + OH^-$$

Let $[HC_2H_3O_2] = [OH^-] = x$, $[C_2H_3O_2^-] = 6.7 \times 10^{-3} - x \approx 6.7 \times 10^{-3}$; then

$$\frac{[HC_2H_3O_2][OH^-]}{[C_2H_3O_2^-]} = \frac{x^2}{6.7 \times 10^{-3}} = K_h = \frac{1.00 \times 10^{-14}}{1.8 \times 10^{-5}} \quad \text{or} \quad x = 1.9 \times 10^{-6}$$

Check of assumption: x is indeed small compared with 6.7×10^{-3}.

$$pOH = -\log[OH^-] = -\log(1.9 \times 10^{-6}) = 6 - 0.29 = 5.71 \quad \text{and} \quad pH = 14.00 - 5.71 = 8.29$$

(*d*) From the excess OH^- beyond that needed to neutralize all the acetic acid, we know that

$$M(OH^-) = \frac{2.0 \times 10^{-4}\ \text{mol}}{0.070\ \text{L}} = 2.9 \times 10^{-3}\ \text{mol/L}$$

Hence

$$pOH = -\log[OH^-] = -\log(2.9 \times 10^{-3}) = 2.54 \quad \text{and} \quad pH = 14.00 - 2.54 = 11.46$$

21.194 At what pH will a 1.0×10^{-3} M solution of an indicator with $K_b = 1.0 \times 10^{-10}$ change color?

▮ The indicator changes color when the conjugates are equal in concentration.

$$K_b = \frac{[HInd^+][OH^-]}{[Ind]} = [OH^-] = 1.0 \times 10^{-10} \quad \text{thus} \quad pOH = 10.00 \quad \text{and} \quad pH = 4.00$$

21.195 What indicator should be used for the titration of 0.10 M KH_2BO_3 with 0.10 M HCl?

I

$$H_2BO_3^- + H_3O^+ \longrightarrow H_3BO_3 + H_2O$$

At the equivalence point, 0.050 M H_3BO_3 would be produced. Only the first ionization step of H_3BO_3 is important to the pH.

$$H_3BO_3 + H_2O \longrightarrow H_2BO_3^- + H_3O^+$$

$$K_a = \frac{[H_3O^+][H_2BO_3^-]}{[H_3BO_3]} = \frac{x^2}{0.050} = 7.3 \times 10^{-10} \quad \text{thus} \quad x = 6.0 \times 10^{-6} \quad \text{and} \quad \text{pH} = 5.22$$

pH 5.22 is in the middle of the range of methyl red (Table 21.2), which would therefore be suitable.

21.196 Calculate the pH of 50.0 mL of a 0.100 M acetic acid solution to which each of the following quantities of 0.100 M NaOH has been added: ***(a)*** 15.0 mL ***(b)*** 25.0 mL ***(c)*** 40.0 mL ***(d)*** 49.0 mL ***(e)*** 50.0 mL ***(f)*** 51.0 mL ***(g)*** 60.0 mL ***(h)*** 70.0 mL. Draw a smooth curve through the points which result when pH (on the vertical axis) is plotted against volume (on the horizontal axis).

I A tabulation of variables which assumes complete reaction is useful;

	V, mL	V_{total}, mL	$HC_2H_3O_2$ in excess, mmol	$C_2H_3O_2^-$ produced, mmol	OH^- in excess, mmol
(a)	15.0	65.0	3.50	1.50	
(b)	25.0	75.0	2.50	2.50	
(c)	40.0	90.0	1.00	4.00	
(d)	49.0	99.0	0.10	4.90	
(e)	50.0	100.0	0.00	5.00	0.000
(f)	51.0	101.0		5.00	0.100
(g)	60.0	110.0		5.00	1.00
(h)	70.0	120.0		5.00	2.00

Solutions (*a*) through (*d*) are buffer solutions:

$$K_a = \frac{[H_3O^+][C_2H_3O_2^-]}{[HC_2H_3O_2]} = 1.8 \times 10^{-5}$$

(a) $\frac{[H_3O^+](1.50 \text{ mmol}/65.0 \text{ mL})}{(3.50 \text{ mmol}/65.0 \text{ mL})} = 1.8 \times 10^{-5}$ thus $[H_3O^+] = 4.2 \times 10^{-5}$ and pH = 4.38

(b) $\frac{[H_3O^+](2.50 \text{ mmol}/75.0 \text{ mL})}{(2.50 \text{ mmol}/75.0 \text{ mL})} = 1.8 \times 10^{-5}$ thus $[H_3O^+] = 1.8 \times 10^{-5}$ and pH = 4.74

Similarly:

(c) pH = 5.35 and ***(d)*** pH = 6.43

Solution (*e*), exactly neutralized, acts like a solution of $NaC_2H_3O_2$. The pH is calculated using the hydrolysis equilibrium:

(e) $C_2H_3O_2^- + H_2O \longrightarrow HC_2H_3O_2 + OH^-$

$$K_h = \frac{[HC_2H_3O_2][OH^-]}{[C_2H_3O_2^-]} = \frac{x^2}{0.050} = \frac{1.0 \times 10^{-14}}{1.8 \times 10^{-5}} = 5.6 \times 10^{-10}$$

$x = [OH^-] = 5.3 \times 10^{-6}$ and pOH = 5.28 and pH = 8.72

The solutions of (*f*) through (*h*) have sodium acetate plus excess hydroxide ion. The latter represses the hydrolysis of the acetate, so the pH is governed by the excess OH^- concentration:

(f) $[OH^-] = \frac{0.10 \text{ mmol}}{101.0 \text{ mL}} = 1.0 \times 10^{-3}$ M and pH = 11.00

(g) $[OH^-] = \frac{1.00 \text{ mmol}}{110.0 \text{ mL}} = 9.1 \times 10^{-3}$ M and pH = 11.96

(*h*) $[OH^-] = \dfrac{2.00 \text{ mmol}}{120.0 \text{ mL}} = 1.7 \times 10^{-2}$ M and pH = 12.22

Figure 21.1 shows an almost vertical rise at the equivalence point.

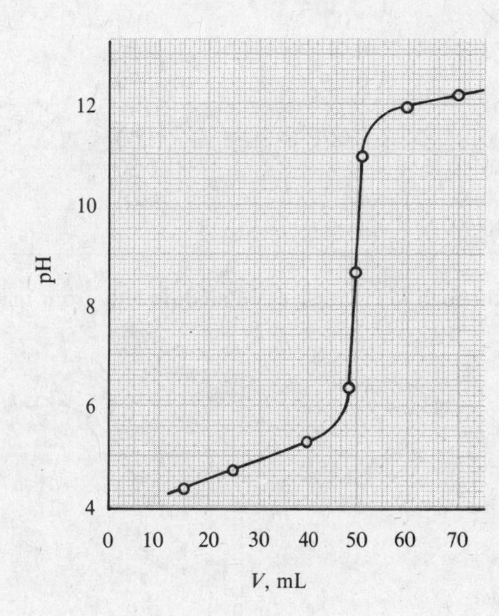

Fig. 21.1

21.197 Calculate the pH at which an indicator with $pK_b = 4.00$ changes color.

▮

$$Ind + H_2O \rightleftharpoons HInd^+ + OH^-$$

At the color change, $[Ind] = [HInd^+]$.

$$K_b = \frac{[OH^-][HInd^+]}{[Ind]} = 1.0 \times 10^{-4} = [OH^-] \quad \text{and} \quad pH = 10.00$$

21.198 At what pH does an indicator change color if the indicator is a weak acid with $K_{Ind} = 4.0 \times 10^{-4}$. For which one(s) of the following neutralizations would the indicator be useful? Explain.

(*a*) $NaOH + HC_2H_3O_2 \longrightarrow$ (*b*) $HCl + NH_3 \longrightarrow$ (*c*) $HCl + NaOH \longrightarrow$

▮

$$HInd + H_2O \rightleftharpoons H_3O^+ + Ind^- \qquad K_{Ind} = \frac{[H_3O^+][Ind^-]}{[HInd]} = 4.0 \times 10^{-4}$$

At the point where the color change occurs, the conjugate acid and base are present in equal concentrations. Hence

$$[H_3O^+] = 4.0 \times 10^{-4}$$

The indicator is suitable for reactions (*b*) and (*c*), but not for (*a*). In (*a*), the equivalence point is the point at which $NaC_2H_3O_2$, a base, is present alone in solution. Its pH is above 7, and an indicator which changes color in acid would not be suitable.

21.199 Construct the curve which would be expected for the titration of 100 mL of 0.010 M H_2S with 1.0 M NaOH.

▮ Compare the process to that of Prob. 21.196.

$$H_2S + H_2O \longrightarrow HS^- + H_3O^+ \qquad K_1 = \frac{[H_3O^+][HS^-]}{[H_2S]} = 1.0 \times 10^{-7}$$

$$HS^- + H_2O \longrightarrow H_3O^+ + S^{2-} \qquad K_2 = \frac{[H_3O^+][S^{2-}]}{[HS^-]} = 1 \times 10^{-14}$$

The critical points to graph are those around the equivalence points and in each buffer region. It is also important to have a point or two past the final equivalence point. With 1.00 mmol of H_2S, the equivalence point will be at 1.00 mL and 2.00 mL. The first buffer region is at 0.50 mL; the second at 1.50 mL. The total volume of the titration mixture will change only slightly. A suitable selection of points for pH calculations might then be

	V, mL	H_2S in excess, mmol	HS^- produced, mmol	S^{2-} produced, mmol	pH
(*a*)	0.00	1.00	0.00		4.50
(*b*)	0.50	0.50	0.50		7.00
(*c*)	0.90	0.10	0.90		7.95
(*d*)	0.99	0.01	0.99		9.00
(*e*)	1.00	0.00	1.00	0.00	9.50
(*f*)	1.50		0.50	0.50	11.7
(*g*)	2.00		0.00	1.00	12.0
(*h*)	2.10			1.00	12.0

See Fig. 21.2. Points (*a*) through (*d*) may be calculated directly from the equation for K_1.

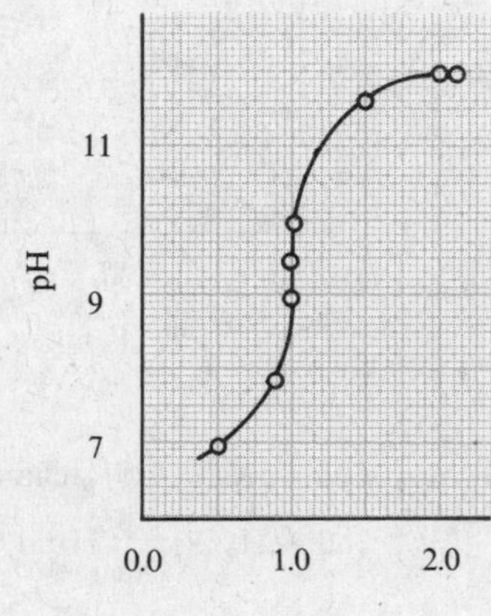

Fig. 21.2

(*a*) $\dfrac{x^2}{0.010} = 1.0 \times 10^{-7}$ so $x = [H_3O^+] = 3.1 \times 10^{-5}$ and $pH = 4.50$

(*b*) $\dfrac{x(0.50)}{(0.50)} = 1.0 \times 10^{-7}$ so $pH = 7.00$

(*c*) $pH = 7.95$ and (*d*) $pH = 9.00$

Point (*e*) is the first equivalence point—the same solution as 1.0 mol of NaHS in 101.0 mL. The HS^- ion can react in water to ionize in either of the two ways:

$$HS^- + H_2O \longrightarrow H_3O^+ + S^{2-} \qquad K_2 = 1 \times 10^{-14}$$

$$HS^- + H_2O \longrightarrow OH^- + H_2S \qquad K_h = \frac{1.0 \times 10^{-14}}{1.0 \times 10^{-7}} = 1.0 \times 10^{-7}$$

The second of these equations will predominate, allowing us to ignore the first.

$$\frac{x^2}{0.010} = 1.0 \times 10^{-7} \quad \text{so} \quad x = 3.1 \times 10^{-5} = [OH^-] \quad \text{thus} \quad pOH = 4.50 \quad \text{and} \quad pH = 9.50$$

With HS^- hydrolyzing much more than ionizing to yield H_3O^+, it might be well to calculate the hydrolysis of S^{2-} at the second equivalence point (point *g*) next.

$$S^{2-} + H_2O \longrightarrow HS^- + OH^- \qquad K_h = \frac{K_w}{K_2} = \frac{x^2}{0.01 - x} = 1 \quad \text{or} \quad x^2 + x - 0.01 = 0$$

$$x = \frac{-1 + \sqrt{1 + 0.04}}{2} = 1 \times 10^{-2} = [OH^-] \quad \text{thus} \quad pOH = 2.0 \quad \text{and} \quad pH = 12.0$$

The extra hydroxide ion added after the first equivalence point has not been neutralized by the very weak HS^- ion. In effect, the titration is somewhat like that of $HC_2H_3O_2$.

(*f*) 1.50 mL added $S^{2-} + H_2O \longrightarrow HS^- + OH^-$

	initial	produced	used	equilibrium
$[S^{2-}]$	0.50×10^{-2}		x	$5 \times 10^{-3} - x$
$[HS^-]$	0.50×10^{-2}	x		$5 \times 10^{-3} + x$
$[OH^-]$	0	x		x

$$\frac{(5 \times 10^{-3} + x)x}{(5 \times 10^{-3} - x)} = 1 \qquad \text{thus} \qquad x = 4.95 \times 10^{-3} \qquad \text{and} \qquad \text{pH} = 11.7$$

21.200 Calculate a point on the titration curve for the addition of 2.0 mL of 0.0100 M NaOH to 50.0 mL of 0.0100 M chloroacetic acid, $HC_2H_2ClO_2$.

▌ If the amount of chloroacetate ion formed were equivalent to the amount of NaOH added, we would have

$$\text{Amount OH}^- \text{ added} = (0.0020\ \text{L})(0.010\ \text{mol/L}) = 2.0 \times 10^{-5}\ \text{mol}$$

$$\text{Total volume} = 0.0520\ \text{L}$$

$$M(C_2H_2ClO_2^-) = \frac{2.0 \times 10^{-5}\ \text{mol}}{0.0520\ \text{L}} = 3.8 \times 10^{-4}\ \text{mol/L} \tag{1}$$

$$M(HC_2H_2ClO_2) = \frac{(0.0500\ \text{L})(0.0100\ \text{mol/L})}{0.0520\ \text{L}} - 3.8 \times 10^{-4}\ \text{mol/L} = 9.2 \times 10^{-3}\ \text{mol/L} \tag{2}$$

$$[H^+] = \frac{K_a[HC_2H_2ClO_2]}{[C_2H_2ClO_2^-]} = \frac{(1.36 \times 10^{-3})(9.2 \times 10^{-3})}{3.8 \times 10^{-4}} = 3.3 \times 10^{-2} \tag{3}$$

But (*3*) is obviously false. $[H^+]$ cannot possibly exceed the initial molar concentration of the acid. Apparently the amount of chloroacetate ion is greater than the equivalent amount of base added. This fact is related to the relatively strong acidity of the acid and to the appreciable ionization of the acid even before the titration begins. Mathematically, this is taken into account by an equation of electroneutrality, according to which there must be equal numbers of cationic and anionic charges in the solution.

$$[H^+] + [Na^+] = [C_2H_2ClO_2^-] + [OH^-] \tag{4}$$

It is safe to drop the $[OH^-]$ term in (*4*) in this case, because it is so much smaller than $[C_2H_2ClO_2^-]$. $[Na^+]$ is obtained from the amount of NaOH added and the total volume of solution.

$$[Na^+] = \frac{2.0 \times 10^{-5}}{0.0520} = 3.8 \times 10^{-4} \tag{5}$$

Chloroacetate ion can be computed from (*4*) and (*5*) with the neglect of $[OH^-]$.

$$[C_2H_2ClO_2^-] = 3.8 \times 10^{-4} + [H^+] \tag{6}$$

The undissociated acid concentration is then the total molar concentration of acid (including its ion) minus $[C_2H_2ClO_2^-]$.

$$[HC_2H_2ClO_2] = \frac{(0.0100)(0.0500)}{0.0520} - (3.8 \times 10^{-4} + [H^+]) = 9.2 \times 10^{-3} - [H^+] \tag{7}$$

Note that (*6*) and (*7*) differ from (*1*) and (*2*) only in the inclusion of the $[H^+]$ terms.

Now we may return to the ionization equilibrium for the acid.

$$[H^+] = \frac{K_a[HC_2H_2ClO_2]}{[C_2H_2ClO_2^-]} = \frac{(1.36 \times 10^{-3})(9.2 \times 10^{-3} - [H^+])}{3.8 \times 10^{-4} + [H^+]}$$

The solution of the quadratic equation is $[H^+] = 2.8 \times 10^{-3}$, and $\text{pH} = -\log[H^+] = 2.55$.

The complication treated in this problem occurs whenever, during partial neutralization, $[H^+]$ or $[OH^-]$ cannot be neglected in comparison with the concentrations of other ions in solution. This is likely to be the case near the beginning of the titration of a moderately weak acid, or near the end of the titration of a strong acid with a moderately weak base.

21.201 In the titration of NH_3 with HCl, explain the pH values in Table 21.4 corresponding to (*a*) 5.0 mL of HCl added (*b*) 10.0 mL of HCl added (*c*) 11.0 mL of HCl added

TABLE 21.4 Titration of 1.000 L of 0.01000 M NH_3 with 1.000 M HCl

HCl added, mL	HCl added, mmol	NH_4^+ formed, mmol	$\frac{[NH_4^+]}{[NH_3]}$	pOH	pH
2.0	2.0	2.0	0.25	4.14	9.86
4.0	4.0	4.0	0.67	4.57	9.43
5.0	5.0	5.0	1.00	4.74	9.26
6.0	6.0	6.0	1.50	4.92	9.08
8.0	8.0	8.0	4.00	5.32	8.68
9.0	9.0	9.0	9.00	5.69	8.31
10.0	10.0	10.0[a]	4.2×10^3	8.38	5.62
11.0	11.0	10.0[a]	1.8×10^6	11.0	3.0
12.0	12.0	10.0[a]	3.6×10^6	11.3	2.7
14.0	14.0	10.0[a]	7.2×10^6	11.6	2.4
16.0	16.0	10.0[a]	1.1×10^7	11.8	2.2
18.0	18.0	10.0[a]	1.4×10^7	11.9	2.1
20.0	20.0	10.0[a]	1.8×10^7	12.0	2.0

[a] The quantity present as NH_3 is negligible.

(*a*) At 5.0 mL of HCl added, half the NH_3 will have been converted to NH_4^+, yielding a buffer solution corresponding to the equation

$$NH_3 + H_2O \rightleftharpoons NH_4^+ + OH^- \qquad K_b = \frac{[NH_4^+][OH^-]}{[NH_3]}$$

A negligible quantity of NH_3 dissociates in the presence of the NH_4^+ produced by the HCl reaction, and the concentrations of NH_3 and NH_4^+ are approximately equal. Thus

$$[OH^-] = K_b = 1.8 \times 10^{-5} \quad \text{thus} \quad pOH = 4.74 \quad \text{and} \quad pH = 9.26$$

(*b*) When 10.0 mL of the HCl has been added, exactly the same number of moles of HCl and NH_3 are present. They will react to yield the same solution as would be produced by 0.0100 mol of NH_4Cl in 1.010 L of solution. A hydrolysis calculation yields a pH value of 5.62.

(*c*) When 11.0 mL of HCl has been added, all the NH_3 has been neutralized and 1.0 mmol of HCl is present in excess. The excess H_3O^+ represses the hydrolysis of the NH_4^+ present, and the pH is determined solely by the 1.0 mmol of H_3O^+ in 1.010 L; that is, pH = 3.0.

21.202 Select an indicator from Table 21.2 which would be suitable to indicate the equivalence point when 50.00 mL of 0.1000 M HCN is titrated with 0.1000 M NaOH.

At the equivalence point there will be equal numbers of mmol of NaOH and HCN, which, if they could react completely, would give 5.000 mmol of NaCN. The solution is the equivalent of a 0.05000 M solution of NaCN, and the following equilibrium will be established:

$$CN^- + H_2O \rightleftharpoons HCN + OH^- \qquad K_h = \frac{K_w}{K_a} = \frac{[HCN][OH^-]}{[CN^-]} = 1.6 \times 10^{-5} \qquad (K_a \text{ from Table 21.1})$$

$$\text{Let} \quad [OH^-] = [HCN] = x \qquad [CN^-] = 0.05000 - x \approx 0.05000$$

$$\frac{x^2}{0.05000} = 1.6 \times 10^{-5} \quad \text{so} \quad x = 8.9 \times 10^{-4} = [OH^-] \quad \text{thus} \quad pOH = 3.05 \quad \text{and} \quad pH = 10.95$$

Thymolphthalein would be a suitable indicator.

21.203 In the titration of HCl with NaOH represented in Fig. 21.3, calculate the pH after the addition of a total of 20.0, 30.0, and 60.0 mL of NaOH.

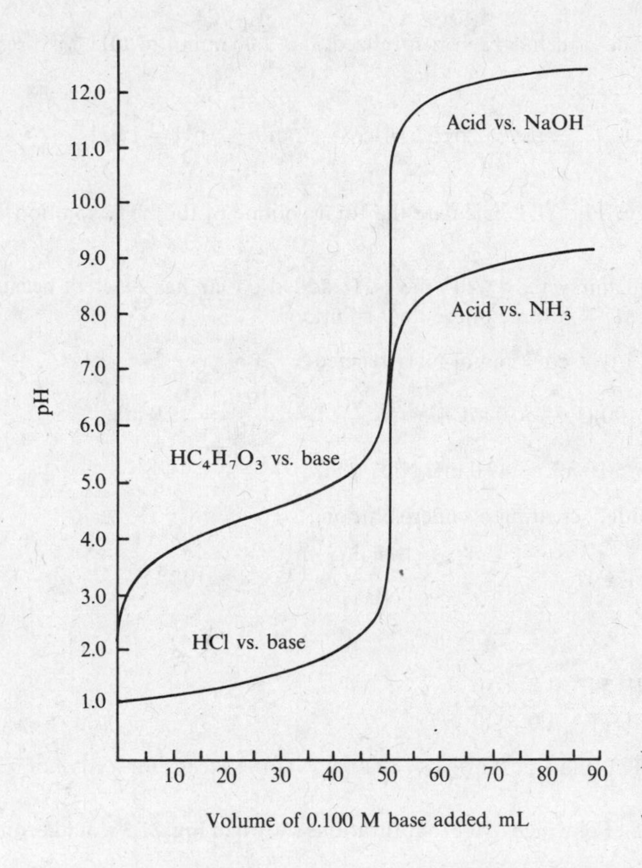

Fig. 21.3 The titration at 25 °C of 50.0 mL of a 0.100 M acid, either HCl or $HC_4H_7O_3$ (β-hydroxybutyric acid), with 0.100 M base, either NaOH or NH_3.

▮ (*a*) After 20.0 mL of NaOH has been added, 2.00 mmol of HCl has been neutralized, and 3.00 mmol remains:

$$\frac{3.00 \text{ mmol}}{70.0 \text{ mL}} = 0.0429 \text{ M } H_3O^+ \qquad pH = 1.368$$

(*b*) After 30.0 mL of NaOH has been added, 2.00 mmol HCl is in excess:

$$\frac{2.00 \text{ mmol}}{80.0 \text{ mL}} = 0.0250 \text{ M } H_3O^+ \qquad pH = 1.602$$

(*c*) After 60.0 mL of NaOH has been added, all the HCl has been neutralized and there is 1.00 mmol of OH^- in excess:

$$\frac{1.00 \text{ mmol } OH^-}{110 \text{ mL}} = 0.00909 \text{ M } OH^- \qquad pOH = 2.041 \qquad \text{and} \qquad pH = 11.959$$

21.204 In the titration of β-hydroxybutyric acid, $HC_4H_7O_3$, with NaOH represented in Fig. 21.3, calculate the pH after the addition of a total of 20.0, 30.0, and 70.0 mL of NaOH. pK'_a for $HC_4H_7O_3$ is 4.39 at the experimental ionic strength.

▮

$$HC_4H_7O_3 + H_2O \rightleftharpoons C_4H_7O_3^- + H_3O^+ \qquad K'_a = \frac{[C_4H_7O_3^-][H_3O^+]}{[HC_4H_7O_3]} = 10^{-4.39} = 4.07 \times 10^{-5}$$

(*a*) After 20.0 mL base added, 2.00 mmol of anion is produced and 3.00 mmol of acid is in excess:

$$K'_a = \frac{(2.00)[H_3O^+]}{3.00} = 4.07 \times 10^{-5} \qquad \text{thus} \qquad [H_3O^+] = 6.1 \times 10^{-5} \qquad \text{and} \qquad pH = 4.21$$

(*b*) After 30.0 mL base added, 3.00 mmol of anion is produced and 2.00 mmol of acid is in excess:

$$K'_a = \frac{(3.00)[H_3O^+]}{2.00} = 4.07 \times 10^{-5} \qquad \text{thus} \qquad [H_3O^+] = 2.7 \times 10^{-5} \qquad \text{and} \qquad pH = 4.57$$

(c) After 70.0 mL of base has been added, all the acid has been neutralized and 2.00 mmol of OH^- is present in excess.

$$\frac{2.00 \text{ mmol } OH^-}{120 \text{ mL}} = 0.0167 \text{ M } OH^- \quad \text{thus} \quad pOH = 1.78 \quad \text{and} \quad pH = 12.22$$

21.205 In the titration of HCl with NH_3 represented in Fig. 21.3, calculate the total volume of the NH_3 solution needed to bring the pH (a) to 3.00 and (b) to 8.00.

▮ **(a)** At pH = 3.00, 0.00100 M HCl is still unreacted. "All" the NH_3 added so far has reacted, because the excess HCl represses hydrolysis of NH_4^+. Let V_t = volume of the final solution:

$$\text{mmol HCl reacted} = \text{mmol } NH_3 \text{ reacted}$$

$$(5.00 \text{ mmol}) - 0.00100V_t = (0.100)(V_t - 50.0 \text{ mL}) \quad \text{so} \quad V_t = \frac{10.0}{0.101} = 99.0 \text{ mL}$$

$$V_{NH_3} = V_t - 50.0 \text{ mL} = 49.0 \text{ mL } NH_3 \text{ added}$$

(b) At pH 8.00, excess NH_3 must have been added, creating a buffer solution.

$$NH_3 + H_2O \rightleftharpoons NH_4^+ + OH^- \qquad K_b = \frac{[NH_4^+][OH^-]}{[NH_3]} = 1.8 \times 10^{-5}$$

Since $[OH^-] = 1.0 \times 10^{-6}$ M,

$$\frac{[NH_4^+]}{[NH_3]} = \frac{1.8 \times 10^{-5}}{1.0 \times 10^{-6}} = 18$$

Thus 50.0 mL is needed just to neutralize the HCl, and 50.0 mL/18 = 2.8 mL is the excess, for a total of 52.8 mL.

21.206 The dye bromcresol green has a pK_a value of 4.95. For which of the four titrations shown in Fig. 21.3 would bromcresol green be a suitable end-point indicator?

▮ The titrations involving strong acid, HCl + NaOH and HCl + NH_3, have steep slopes at pH = 4.95. At that pH, the indicator is half changed from one form to the other and is in the middle of its color change.

21.207 Bromphenol blue is an indicator with a K_a value of 5.84×10^{-5}. What percentage of this indicator is in its basic form at a pH of 4.84?

▮ Representing bromphenol blue as HBb,

$$HBb + H_2O \rightleftharpoons Bb^- + H_3O^+$$

$$K_a = \frac{[Bb^-][H_3O^+]}{[HBb]} = \frac{[Bb^-](10^{-4.84})}{[HBb]} = 5.84 \times 10^{-5} \quad \text{so} \quad \frac{[Bb^-]}{[HBb]} = 4.04$$

$$\% \text{ in basic form} = \frac{[Bb^-]}{[Bb^-] + [HBb]} \times 100\% = \frac{4.04}{1.00 + 4.04} \times 100\% = 0.80 \times 100\% = 80\%$$

21.208 Calculate the pH and $[NH_3]$ at the end point in the titration of β-hydroxybutyric acid with NH_3, at the concentrations indicated for Fig. 21.3. K'_a for the acid is 4.1×10^{-5} at the experimental ionic strength.

▮

$$A^- + H_2O \rightleftharpoons HA + OH^- \qquad NH_4^+ + H_2O \rightleftharpoons NH_3 + H_3O^+$$

Initially, $[NH_4^+] = [A^-] = 0.050$ M; at the end point,

$$\frac{[HA][OH^-]}{[A^-]} = \frac{1.0 \times 10^{-14}}{4.1 \times 10^{-5}} \qquad \frac{[NH_3][H_3O^+]}{[NH_4^+]} = \frac{1.0 \times 10^{-14}}{1.8 \times 10^{-5}} \tag{1}$$

Assuming equal concentration of NH_3 and HA formed, with equal initial concentrations of ions which do not change significantly, we can divide the first equation (1) by the second, yielding

$$\frac{[OH^-]}{[H_3O^+]} = \frac{1.8 \times 10^{-5}}{4.1 \times 10^{-5}} = 0.44$$

Multiplying by

$$[H_3O^+][OH^-] = 1.0 \times 10^{-14}$$

yields

$$[OH^-]^2 = 4.4 \times 10^{-15} \quad \text{so} \quad [OH^-] = 6.6 \times 10^{-8} \quad \text{thus} \quad pOH = 7.18 \quad \text{and} \quad pH = 6.82$$

The second equation (*1*) now gives

$$\frac{[NH_3]\left(\dfrac{1.0 \times 10^{-14}}{6.6 \times 10^{-8}}\right)}{0.050} = \frac{1.0 \times 10^{-14}}{1.8 \times 10^{-5}} \quad \text{so} \quad [NH_3] = 1.8 \times 10^{-4}\ M$$

21.209 An acid-base indicator has a K_a of 3.0×10^{-5}. The acid form of the indicator is red and the basic form is blue. ***(a)*** By how much must the pH change in order to change the indicator from 75% red to 75% blue? ***(b)*** For which of the titrations shown in Fig. 21.3 would this indicator be a suitable choice?

▮ ***(a)***

$$[H^+] = \frac{K_a[\text{acid}]}{[\text{base}]}$$

75% red: $$[H^+] = \frac{(3.0 \times 10^{-5})(75)}{25} = 9.0 \times 10^{-5} \qquad pH = 4.05$$

75% blue: $$[H^+] = \frac{(3.0 \times 10^{-5})(25)}{75} = 1.0 \times 10^{-5} \qquad pH = 5.00$$

The change in pH is $5.00 - 4.05 = 0.95$.

(b) The indicator changes its color in the pH range 4 to 5. The two HCl titration curves in Fig. 21.3 are rising steeply at pH 4 to 5; thus the indicator would be suitable for them. For the $HC_4H_7O_3$ titrations, however, this pH range occurs nowhere near the end point; an indicator is needed which changes its color at higher pH values.

CHAPTER 22
Heterogeneous and Other Equilibria

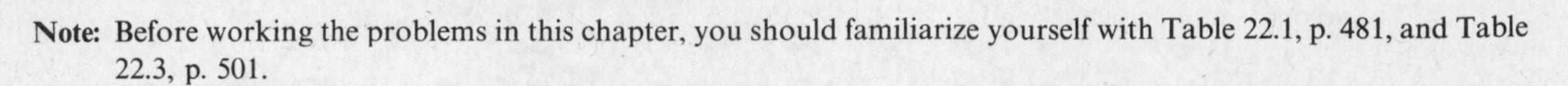

Note: Before working the problems in this chapter, you should familiarize yourself with Table 22.1, p. 481, and Table 22.3, p. 501.

22.1 SOLUBILITY PRODUCT EQUILIBRIA

22.1 When a sample of solid AgCl is shaken with water at 25 °C, a solution containing 1.0×10^{-5} M silver ions is produced. Calculate K_{sp}.

$$AgCl(s) \rightleftharpoons Ag^+ + Cl^-$$

$$K_{sp} = [Ag^+][Cl^-] \qquad [Ag^+] = [Cl^-] = 1.0 \times 10^{-5} \qquad K_{sp} = (1.0 \times 10^{-5})(1.0 \times 10^{-5}) = 1.0 \times 10^{-10}$$

22.2 Calculate the solubility of $Mg(OH)_2$ in water. $K_{sp} = 1.2 \times 10^{-11}$

$$Mg(OH)_2 \rightleftharpoons Mg^{2+} + 2\,OH^- \qquad K_{sp} = [Mg^{2+}][OH^-]^2$$

Let $[Mg^{2+}] = x$; then $[OH^-] = 2x$, and $K_{sp} = (x)(2x)^2 = 4x^3 = 1.2 \times 10^{-11}$. Note that $2x$ (not x) is the concentration of OH^-, and in this example $[OH^-]$ is twice the concentration of magnesium ion. According to the expression for K_{sp}, it is the hydroxide ion concentration, in this case $2x$, which must be squared.

$$x = 1.4 \times 10^{-4} = [Mg^{2+}] \qquad 2x = 2.8 \times 10^{-4} = [OH^-]$$

To check, $K_{sp} = (1.4 \times 10^{-4})(2.8 \times 10^{-4})^2 = 1.1 \times 10^{-11}$

22.3 Calculate the solubility product constants of the following compounds. The solubilities are given in mol/L. (*a*) $BaSO_4$, 1.05×10^{-5} mol/L (*b*) TlBr, 1.9×10^{-3} mol/L (*c*) $Mg(OH)_2$, 1.21×10^{-4} mol/L (*d*) $Ag_2C_2O_4$, 1.15×10^{-4} mol/L (*e*) $La(IO_3)_3$, 7.8×10^{-4} mol/L

(*a*) $K_{sp} = [Ba^{2+}][SO_4^{2-}] = (1.05 \times 10^{-5})^2 = 1.1 \times 10^{-10}$

(*b*) $K_{sp} = [Tl^+][Br^-] = (1.9 \times 10^{-3})^2 = 3.6 \times 10^{-6}$

(*c*) $K_{sp} = [Mg^{2+}][OH^-]^2 = (1.21 \times 10^{-4})(2.42 \times 10^{-4})^2 = 7.09 \times 10^{-12}$

(*d*) $K_{sp} = [Ag^+]^2[C_2O_4^{2-}] = (2.30 \times 10^{-4})^2(1.15 \times 10^{-4}) = 6.1 \times 10^{-12}$ (Assuming no hydrolysis of the $C_2O_4^{2-}$ ion)

(*e*) $K_{sp} = [La^{3+}][IO_3^-]^3 = (7.8 \times 10^{-4})(3 \times 7.8 \times 10^{-4})^3 = 1.0 \times 10^{-11}$

22.4 How many mol CuI ($K_{sp} = 5 \times 10^{-12}$) will dissolve in 1.0 L of 0.10 M NaI solution?

$$K_{sp} = [Cu^+][I^-] = 5 \times 10^{-12} \qquad [I^-] = 0.10\ \text{M} \qquad [Cu^+] = 5 \times 10^{-11}\ \text{M}$$

22.5 Calculate the solubility of A_2X_3 in pure water, assuming that neither kind of ion reacts with water. For A_2X_3, $K_{sp} = 1.1 \times 10^{-23}$.

$$A_2X_3 \longrightarrow 2A^{3+} + 3X^{2-} \qquad K_{sp} = [A^{3+}]^2[X^{2-}]^3 = 1.1 \times 10^{-23}$$

If the solubility of A_2X_3 is y, then $[A^{3+}] = 2y$ and $[X^{2-}] = 3y$.

$$(2y)^2(3y)^3 = 108y^5 = 1.1 \times 10^{-23} \qquad y^5 = 1.0 \times 10^{-25} \qquad y = 1.0 \times 10^{-5}\ \text{mol/L}$$

22.6 A solution of saturated CaF_2 is found to contain 4.1×10^{-4} M fluoride ion. Calculate the K_{sp} of CaF_2. Neglect hydrolysis.

$[Ca^{2+}] = \frac{1}{2}[F^-] = 2.05 \times 10^{-4}$ M $\qquad K_{sp} = [Ca^{2+}][F^-]^2 = (2.05 \times 10^{-4})(4.1 \times 10^{-4})^2 = 3.4 \times 10^{-11}$

22.7 Calculate the nickel ion concentration of a solution prepared by shaking $Ni(OH)_2$ with water until equilibrium is established.

$$Ni(OH)_2 \rightleftharpoons Ni^{2+} + 2OH^- \qquad K_{sp} = [Ni^{2+}][OH^-]^2 = 2 \times 10^{-16} \quad \text{(from Table 22.1)}$$

Let $x = [Ni^{2+}]$; then $[OH^-] = 2x$.

$$x(2x)^2 = 4x^3 = 2 \times 10^{-16} \qquad x = 4 \times 10^{-6}\ \text{M} \quad \text{(one significant figure)}$$

22.8 Calculate the solubility of AgCl in 0.20 M $AgNO_3$ solution.

TABLE 22.1 Solubility Product Constants at 25 °C

$AgBr$	5.0×10^{-13}	CaF_2	3.4×10^{-11}	$Mn(OH)_2$	4×10^{-14}
$AgCN$	2.2×10^{-16}	$CaSO_4$	2×10^{-4}	MnS	2.5×10^{-10}
$Ag_2C_2O_4$	6×10^{-12}	$Cd(OH)_2$	4.5×10^{-15}	$Ni(OH)_2$	2×10^{-16}
$AgCl$	1×10^{-10}	CdS	8×10^{-27}	NiS	2×10^{-21}
Ag_2CrO_4	9×10^{-12}	CoS	3×10^{-26}	$PbCO_3$	3.3×10^{-14}
AgI	1.5×10^{-16}	CuI	5×10^{-12}	$PbCl_2$	1.7×10^{-4}
$AgOCN$	2.3×10^{-7}	$Cu(OH)_2$	1×10^{-19}	PbF_2	3.6×10^{-8}
$AgOH$	1.5×10^{-8}	CuS	8.5×10^{-36}	PbI_2	1.4×10^{-8}
Ag_2S	1.6×10^{-49}	$Fe(OH)_2$	1.6×10^{-14}	$Pb(IO_3)_2$	2.5×10^{-13}
$AgSCN$	1.1×10^{-12}	$Fe(OH)_3$	1.1×10^{-36}	$PbSO_4$	2×10^{-8}
Ag_2SO_4	1.5×10^{-5}	FeS	3.7×10^{-19}	$RaSO_4$	4×10^{-11}
$Al(OH)_3$	2×10^{-33}	Hg_2Cl_2	2×10^{-18}	SrF_2	2.9×10^{-9}
$BaCO_3$	5×10^{-9}	Hg_2I_2	1.2×10^{-28}	$SrSO_4$	2.8×10^{-7}
$BaCrO_4$	2.4×10^{-10}	$MgCO_3$	2.6×10^{-5}	$Zn(OH)_2$	1.8×10^{-14}
BaF_2	1.7×10^{-6}	MgF_2	6.5×10^{-9}	ZnS	1.2×10^{-22}
$BaSO_4$	1×10^{-10}	$Mg(OH)_2$	1.2×10^{-11}		
$CaCO_3$	1×10^{-8}				

▌ $$AgCl \rightleftharpoons Ag^+ + Cl^- \qquad K_{sp} = [Ag^+][Cl^-] = 1 \times 10^{-10} \quad \text{(from Table 22.1)}$$

Let $[Cl^-] = x$; then $[Ag^+] = 0.20 + x \simeq 0.20$.

$$0.20x = 1 \times 10^{-10} \qquad x = 5 \times 10^{-10}\ M$$

22.9 Which has a greater molarity in water, AgCl or $Mg(OH)_2$? Can relative solubilities be predicted on the basis of the relative magnitudes of the K_{sp} values alone? Explain.

▌ See Problem 22.8 for equations for AgCl.

$$Mg(OH)_2 \rightleftharpoons Mg^{2+} + 2OH^- \qquad K_{sp} = [Mg^{2+}][OH^-]^2 = 1.2 \times 10^{-11}$$

Let $[Mg^{2+}] = x$; then $[OH^-] = 2x$.

$$x(2x)^2 = 4x^3 = 1.2 \times 10^{-11} \qquad x = 1.4 \times 10^{-4}\ M$$

The K_{sp} for AgCl is greater than that for $Mg(OH)_2$, but $Mg(OH)_2$ is more soluble. The apparent anomaly results from the fact that the K_{sp} expressions are not comparable—the $Mg(OH)_2$ expression has a square term in it.

22.10 The values of K_{sp} for the slightly soluble salts MX and QX_2 are each equal to 4.0×10^{-18}. Which salt is more soluble? Explain your answer fully.

▌ Proceed as in Problem 22.9.

$$MX \rightleftharpoons M^{n+} + X^{n-} \qquad K_{sp} = [M^{n+}][X^{n-}] = 4.0 \times 10^{-18}$$

$$QX_2 \rightleftharpoons Q^{2n+} + 2X^{n-} \qquad K_{sp} = [Q^{2n+}][X^{n-}]^2 = 4.0 \times 10^{-18}$$

Solving yields $[M^{n+}] = 2.0 \times 10^{-9}$ $\quad [Q^{2n+}] = 1.0 \times 10^{-6}$

QX_2 is more soluble.

22.11 The solubility of $PbSO_4$ in water is 0.038 g/L. Calculate the solubility product constant of $PbSO_4$.

▌ $$PbSO_4(s) \rightleftharpoons Pb^{2+} + SO_4^{2-} \quad \text{(in solution)}$$

The concentrations of the ions must be expressed in mol/L. To convert 0.038 g/L to mol ions per L, divide by the formula weight of $PbSO_4$:

$$0.038\ g/L = \frac{0.038\ g/L}{303\ g/mol} = 1.25 \times 10^{-4}\ mol/L$$

Since 1.25×10^{-4} mol dissolved $PbSO_4$ yields 1.25×10^{-4} mol each Pb^{2+} and SO_4^{2-},

$$K_{sp} = [Pb^{2+}][SO_4^{2-}] = (1.25 \times 10^{-4})(1.25 \times 10^{-4}) = 1.6 \times 10^{-8}$$

This method may be applied to any fairly insoluble salt whose ions do not hydrolyze appreciably or form soluble complexes. (Sulfides, carbonates, and phosphates, and salts of many of the transition metals like iron must be treated by taking into account hydrolysis and, in some cases, complexation.)

22.12 The solubility of Ag_2CrO_4 in water is 0.044 g/L. Determine the solubility product constant.

▮

$$Ag_2CrO_4 \rightleftharpoons 2Ag^+ + CrO_4^{2-}$$

To convert 0.044 g/L to mol/L of the ions, divide by the formula weight of Ag_2CrO_4 (332).

$$0.044 \text{ g/L} = \frac{0.044 \text{ g/L}}{332 \text{ g/mol}} = 1.3 \times 10^{-4} \text{ mol/L}$$

Since 1 mol dissolved Ag_2CrO_4 yields 2 mol Ag^+ and 1 mol CrO_4^{2-},

$$[Ag^+] = 2(1.3 \times 10^{-4}) = 2.6 \times 10^{-4} \qquad [CrO_4^{2-}] = 1.3 \times 10^{-4}$$

and

$$K_{sp} = [Ag^+]^2[CrO_4^{2-}] = (2.6 \times 10^{-4})^2(1.3 \times 10^{-4}) = 8.8 \times 10^{-12}$$

22.13 The concentration of the Ag^+ ion in a saturated solution of $Ag_2C_2O_4$ is 2.3×10^{-4} mol/L. Compute the solubility product constant of $Ag_2C_2O_4$.

▮

$$Ag_2C_2O_4 \rightleftharpoons 2Ag^+ + C_2O_4^{2-}$$

The equation indicates that the concentration of $C_2O_4^{2-}$ is half that of Ag^+. Thus

$$K_{sp} = [Ag^+]^2[C_2O_4^{2-}] = (2.3 \times 10^{-4})^2\left(\frac{2.3 \times 10^{-4}}{2}\right) = 6 \times 10^{-12}$$

22.14 Calculate the solubility product constants of the following salts. Solubilities are given in g/L. (*a*) CaC_2O_4, 0.0055 g/L (*b*) $BaCrO_4$, 0.0037 g/L (*c*) CaF_2, 0.017 g/L

▮ (*a*)

$$\left(\frac{0.0055 \text{ g } CaC_2O_4}{\text{L}}\right)\left(\frac{1 \text{ mol}}{128 \text{ g}}\right) = 4.3 \times 10^{-5} \text{ M}$$

$$K_{sp} = [Ca^{2+}][C_2O_4^{2-}] = (4.3 \times 10^{-5})^2 = 1.8 \times 10^{-9} \qquad \text{(Assuming no hydrolysis of } C_2O_4^{2-} \text{ ion)}$$

(*b*)

$$\left(\frac{0.0037 \text{ g } BaCrO_4}{\text{L}}\right)\left(\frac{1 \text{ mol}}{253 \text{ g}}\right) = 1.46 \times 10^{-5} \text{ M} \qquad K_{sp} = (1.46 \times 10^{-5})^2 = 2.1 \times 10^{-10}$$

(*c*)

$$\left(\frac{0.017 \text{ g } CaF_2}{\text{L}}\right)\left(\frac{1 \text{ mol}}{78.0 \text{ g}}\right) = 2.2 \times 10^{-4} \text{ M}$$

$$K_{sp} = [Ca^{2+}][F^-]^2 = (2.2 \times 10^{-4})(4.4 \times 10^{-4})^2 = 4.3 \times 10^{-11}$$

22.15 The solubility of ML_2 (formula weight, 60 g/mol) in water is 2.4×10^{-5} g/100 mL solution. Calculate the solubility product constant for ML_2.

▮

$$[M^{2+}] = \frac{2.4 \times 10^{-5} \text{ g/0.100 L}}{60 \text{ g/mol}} = 4.0 \times 10^{-6} \text{ M} \qquad [L^-] = 8.0 \times 10^{-6} \text{ M}$$

$$K_{sp} = [M^{2+}][L^-]^2 = (4.0 \times 10^{-6})(8.0 \times 10^{-6})^2 = 2.6 \times 10^{-16}$$

22.16 What is the solubility (in mol/L) of $Fe(OH)_3$ in a solution of pH = 8.0? [K_{sp} for $Fe(OH)_3 = 1.0 \times 10^{-36}$]

▮

$$Fe(OH)_3 \longrightarrow Fe^{3+} + 3OH^-$$

$$K_{sp} = [Fe^{3+}][OH^-]^3 = 1.0 \times 10^{-36} \qquad [Fe^{3+}] = \frac{1.0 \times 10^{-36}}{(1.0 \times 10^{-8})^3} = 1.0 \times 10^{-12} \text{ M}$$

22.17 K_{sp} for $Cu(OH)_2$ equals 1×10^{-19}. What is the solubility (in mol/L) of $Cu(OH)_2$ in water?

▮

$$Cu(OH)_2 \longrightarrow Cu^{2+} + 2OH^-$$

$$K_{sp} = [Cu^{2+}][OH^-]^2 = 1 \times 10^{-19} \qquad (x)(2x)^2 = 1 \times 10^{-19} \qquad \text{Thus } x = 3 \times 10^{-7}$$

22.18 The following three slightly soluble salts have the same solubilities: M_2X, QY_2, PZ_3. How are their K_{sp} values related?

▌ Let w = solubility of each salt.

$$[M^+] = 2w \qquad [X^{2-}] = w \qquad K_{sp} = 4w^3$$
$$[Q^{2+}] = w \qquad [Y^-] = 2w \qquad K_{sp} = 4w^3$$
$$[P^{3+}] = w \qquad [Z^-] = 3w \qquad K_{sp} = 27w^4$$

Since w is small, $K_{sp}(M_2X) = K_{sp}(QY_2) > K_{sp}(PZ_3)$

22.19 (*a*) Determine the solubility of SrF_2 at 25 °C, in mol/L and in mg/cm³. (*b*) What are $[Sr^{2+}]$ and $[F^-]$ (in mol/L) in a saturated solution of SrF_2?

▌ Assuming no hydrolysis of fluoride ion, (*a*) $K_{sp} = [Sr^{2+}][F^-]^2 = 2.9 \times 10^{-9}$. The concentration of F^- is twice that of Sr^{2+}, so

$$(x)(2x)^2 = 2.9 \times 10^{-9} = 4x^3 \qquad x = 9.0 \times 10^{-4} = [Sr^{2+}]$$
$$\left(\frac{9.0 \times 10^{-4} \text{ mol } SrF_2}{\text{L}}\right)\left(\frac{125.6 \text{ g}}{\text{mol}}\right) = 0.11 \text{ g/L} = 0.11 \text{ mg/cm}^3$$

(*b*) $[Sr^{2+}] = 9.0 \times 10^{-4}$ M $\qquad [F^-] = 1.8 \times 10^{-3}$ M

22.20 What is the solubility of $Pb(IO_3)_2$ (*a*) in mol/L (*b*) in g/L?

▌ (*a*) Let the solubility of $Pb(IO_3)_2$ be x mol/L. Then $[Pb^{2+}] = x$ and $[IO_3^-] = 2x$.

$$[Pb^{2+}][IO_3^-]^2 = K_{sp} \quad \text{gives} \quad (x)(2x)^2 = 2.5 \times 10^{-13}$$

Then $4x^3 = 2.5 \times 10^{-13}$, $x^3 = 62 \times 10^{-15}$, and $x = 4.0 \times 10^{-5}$.

(*b*) Solubility = $(4.0 \times 10^{-5}$ mol/L)(557 g/mol) = 0.022 g/L

22.21 The $[Ag^+]$ of a solution is 4×10^{-3}. Calculate the $[Cl^-]$ that must be exceeded before AgCl can precipitate.

▌
$$[Ag^+][Cl^-] = K_{sp} \qquad (4 \times 10^{-3})[Cl^-] = 1.0 \times 10^{-10} \qquad [Cl^-] = 2 \times 10^{-8}$$

Hence a $[Cl^-]$ of 2×10^{-8} M must be exceeded before AgCl precipitates. This problem differs from the previous ones in that the two ions forming the precipitate are furnished to the solution independently. This represents a typical analytical situation, in which some soluble chloride is added to precipitate silver ion present in a solution.

22.22 Determine the solubility of AgCl in 0.10 M $BaCl_2$.

▌
$$[Cl^-] = 0.20 \text{ M from } BaCl_2 \qquad [Ag^+][Cl^-] = 1.0 \times 10^{-10} \qquad [Ag^+] = \frac{1.0 \times 10^{-10}}{0.20} = 5.0 \times 10^{-10} \text{ M}$$

22.23 The solubility of $Fe(OH)_2$ in water is 2×10^{-5} mol/L. Calculate the value of its solubility product constant.

▌
$$Fe(OH)_2 \longrightarrow Fe^{2+} + 2OH^- \qquad K_{sp} = [Fe^{2+}][OH^-]^2 = (2 \times 10^{-5})(4 \times 10^{-5})^2 = 3 \times 10^{-14}$$

22.24 What mass of $BaSO_4$ will dissolve in 450 mL of aqueous solution?

▌
$$BaSO_4 \rightleftharpoons Ba^{2+} + SO_4^{2-} \qquad K_{sp} = [Ba^{2+}][SO_4^{2-}] = 1 \times 10^{-10}$$
$$x^2 = 1 \times 10^{-10} \qquad x = 1 \times 10^{-5} \qquad (1 \times 10^{-5} \text{ mol/L})(0.450 \text{ L})(233 \text{ g/mol}) = 0.001 \text{ g} = 1 \text{ mg}$$

22.25 Determine the mass of PbI_2 that will dissolve in (*a*) 500 mL water (*b*) 500 mL of 0.10 M KI solution (*c*) 500 mL of a solution containing 1.33 g $Pb(NO_3)_2$.

▌
$$PbI_2 \rightleftharpoons Pb^{2+} + 2I^- \qquad K_{sp} = [Pb^{2+}][I^-]^2 = 1.4 \times 10^{-8}$$

(*a*) Let $x = [Pb^{2+}]$, then $[I^-] = 2x$

$4x^3 = 1.4 \times 10^{-8} \qquad x = (1.5 \times 10^{-3}$ mol/L) $\qquad (1.5 \times 10^{-3}$ mol/L)(0.500 L)(461 g/mol) = 0.35 g

(*b*) $[Pb^{2+}](0.10)^2 = 1.4 \times 10^{-8} \qquad [Pb^{2+}] = 1.4 \times 10^{-6}$ M

$(1.4 \times 10^{-6}$ mol/L)(0.500 L)(461 g/mol) = 3.2×10^{-4} g = 0.32 mg

(*c*) $$[1.33 \text{ g } Pb(NO_3)_2]\left(\frac{1 \text{ mol}}{331 \text{ g}}\right)\left(\frac{1}{0.500 \text{ L}}\right) = 8.04 \times 10^{-3} \text{ M}$$

If the concentration of PbI_2 is y, then $[I^-] = 2y$

$(8.04 \times 10^{-3})(2y)^2 = 1.4 \times 10^{-8}$ $y = 6.6 \times 10^{-4}$ M $(6.6 \times 10^{-4}$ mol/L)(0.500 L)(461 g/mol) = 0.15 g

22.26 Should a precipitate of barium fluoride be obtained when 100 mL of 0.25 M NaF and 100 mL of 0.015 M $Ba(NO_3)_2$ are mixed?

▮ (25 mmol F^-)/(200 mL) = 0.125 M (1.5 mmol Ba^{2+})/(200 mL) = 0.0075 M

$$BaF_2 \rightleftharpoons Ba^{2+} + 2F^- \qquad K_{sp} = [Ba^{2+}][F^-]^2 = 1.7 \times 10^{-6}$$

If no precipitation occurs, the product of the molarities, Q, is

$$Q = [Ba^{2+}][F^-]^2 = (7.5 \times 10^{-3})(0.125)^2 = 1.2 \times 10^{-4}$$

Since Q exceeds K_{sp}, precipitation should occur.

22.27 Determine whether a precipitate will form when 100 mL of 0.100 M Pb^{2+} solution is mixed with 100 mL of 0.30 M Cl^- solution.

▮ Assuming that no precipitate forms and the final volume is 200 mL, the lead and chloride ion concentrations would be 0.050 and 0.15 M, respectively,

$$PbCl_2 \rightleftharpoons Pb^{2+} + 2\,Cl^- \qquad K_{sp} = [Pb^{2+}][Cl^-]^2 = 1.7 \times 10^{-4}$$

But if there were no precipitation, $[Pb^{2+}][Cl^-]^2 = (0.050)(0.15)^2 = 1.1 \times 10^{-3}$. Since this product is greater than the value of K_{sp}, the concentrations of ions exceed that required for equilibrium, and precipitation will occur.

22.28 What is the maximum pH of a solution 0.10 M in Mg^{2+} from which $Mg(OH)_2$ will not precipitate?

▮
$$Mg(OH)_2 \rightleftharpoons Mg^{2+} + 2OH^- \qquad K_{sp} = [Mg^{2+}][OH^-]^2 = 1.2 \times 10^{-11}$$

Any OH^- concentration higher than that contained in a saturated solution would cause precipitation. Hence, the solution must be at the point of attaining equilibrium, and the concentrations of ions in solution must be no greater than those required to satisfy the solubility product constant. In this solution $[Mg^{2+}] = 0.10$ M.

$$[Mg^{2+}][OH^-]^2 = 1.2 \times 10^{-11}$$

$$[OH^-]^2 = \frac{1.2 \times 10^{-11}}{0.10} = 1.2 \times 10^{-10} \qquad [OH^-] = 1.1 \times 10^{-5} \qquad pOH = 4.96 \qquad pH = 9.04$$

22.29 If 50.0 mL of a solution containing 0.0010 mol silver ion is mixed with 50.0 mL of 0.100 M HCl solution, how much silver ion remains in solution?

▮ The initial quantity of Cl^- ion is given by (50.0 mL)(0.100 mmol/mL) = 5.00 mmol = 0.00500 mol Cl^-. At equilibrium, let x = number of mol silver ion remaining in solution. Therefore, $0.0010 - x \approx 0.0010$ mol silver ion and an equal fraction of a mol of chloride ion have precipitated, leaving 0.0040 mol chloride ion remaining in solution. The total volume of the solution is 100.0 mL.

$$[Cl^-] = \frac{0.0040 \text{ mol}}{0.100 \text{ L}} = 0.040 \text{ M} \qquad K_{sp} = [Ag^+][Cl^-] = [Ag^+](0.040) = 1 \times 10^{-10}$$

$$[Ag^+] = \frac{1 \times 10^{-10}}{0.040} = 2.5 \times 10^{-9} \text{ M} \qquad x = (2.5 \times 10^{-9} \text{ mol/L})(0.100 \text{ L}) = 2.5 \times 10^{-10} \text{ mol}$$

22.30 What mass of Pb^{2+} ion is left in solution when 50.0 mL of 0.20 M $Pb(NO_3)_2$ is added to 50.0 mL of 1.5 M NaCl?

▮ Assume complete precipitation of the Pb^{2+} followed by solution of the equilibrium concentration to be determined.

(50.0 mL)(0.20 mmol/mL) = 10 mmol Pb^{2+} (50.0 mL)(1.5 mmol/mL) = 75 mmol Cl^-

$$Pb^{2+} + 2Cl^- \longrightarrow PbCl_2$$

What lead ion concentration can exist in a solution of Cl^- containing (75 mmol) − (20 mmol) = 55 mmol in 100 mL? $[Cl^-] = 0.55$ M

$$PbCl_2 \rightleftharpoons Pb^{2+} + 2Cl \qquad K_{sp} = [Pb^{2+}][Cl^-]^2 = 1.7 \times 10^{-4}$$

$$[Pb^{2+}] = \frac{1.7 \times 10^{-4}}{(0.55)^2} = 5.6 \times 10^{-4} \text{ M} \qquad (5.6 \times 10^{-4} \text{ mol/L})(0.100 \text{ L})(208 \text{ g/mol}) = 12 \text{ mg}$$

22.31 Calculate the pH at which $Mg(OH)_2$ will just begin to precipitate from a 0.100 M $Mg(NO_3)_2$ solution by addition of NaOH.

The concentration of Mg^{2+} before any $Mg(OH)_2$ has precipitated is 0.100 M.

$$Mg(OH)_2 \rightleftharpoons Mg^{2+} + 2\,OH^- \qquad K_{sp} = [Mg^{2+}][OH^-]^2 = 1.2 \times 10^{-11}$$

$$0.100(x^2) = 1.2 \times 10^{-11} \qquad x = 1.1 \times 10^{-5}\ M = [OH^-]$$

$$[H_3O^+] = \frac{K_w}{[OH^-]} = \frac{1.0 \times 10^{-14}}{1.1 \times 10^{-5}} = 9.1 \times 10^{-10} \qquad pH = 9.04$$

22.32 Calculate the hydroxide ion concentration of a solution after 100 mL of 0.100 M $MgCl_2$ is added to 100 mL of 0.200 M NaOH.

The 10.0 mmol Mg^{2+} added to 20.0 mmol OH^- will yield 10.0 mmol $Mg(OH)_2$. This $Mg(OH)_2$ dissolves according to K_{sp} values.

$$Mg(OH)_2 \longrightarrow Mg^{2+} + 2OH^- \qquad K_{sp} = [Mg^{2+}][OH^-]^2 = 1.2 \times 10^{-11}$$

At equilibrium, let $[Mg^{2+}] = x$; then $[OH^-] = 2x$.

$$K_{sp} = x(2x)^2 = 4x^3 = 1.2 \times 10^{-11} \qquad x = 1.4 \times 10^{-4} \qquad 2x = 2.8 \times 10^{-4} = [OH^-]$$

22.33 Using the value of K_{sp} for Hg_2I_2 from Table 22.1, calculate the concentrations of cation and anion in a saturated solution of Hg_2I_2 in water.

$$Hg_2I_2 \longrightarrow Hg_2^{2+} + 2I^- \qquad K_{sp} = [Hg_2^{2+}][I^-]^2 = 1.2 \times 10^{-28}$$

Let $[Hg_2^{2+}] = x$; then $[I^-] = 2x$.

$$4x^3 = 1.2 \times 10^{-28} \qquad x = 3.1 \times 10^{-10}\ M = [Hg_2^{2+}] \qquad 2x = 6.2 \times 10^{-10}\ M = [I^-]$$

22.34 What $[SO_4^{2-}]$ must be exceeded to produce a $RaSO_4$ precipitate in 500 cm³ of a solution containing 0.00010 mol Ra^{2+}?

$$[Ra^{2+}] = \frac{0.00010\ \text{mol}}{0.500\ \text{L}} = 0.00020\ M \qquad K_{sp} - [Ra^{2+}][SO_4^{2-}] = 4 \times 10^{-11}$$

$$[SO_4^{2-}] = (4 \times 10^{-11})/(0.00020) = 2 \times 10^{-7}\ M$$

22.35 A solution has a Mg^{2+} concentration of 0.0010 mol/L. Will $Mg(OH)_2$ precipitate if the OH^- concentration of the solution is (*a*) 10^{-5} mol/L (*b*) 10^{-3} mol/L?

$$[Mg^{2+}] = 1.0 \times 10^{-3}\ M \qquad K_{sp} = [Mg^{2+}][OH^-]^2 = 1.2 \times 10^{-11}$$

$$[OH^-]^2 = \frac{1.2 \times 10^{-11}}{1.0 \times 10^{-3}} = 1.2 \times 10^{-8} \text{ at equilibrium} \qquad [OH^-] = 1.1 \times 10^{-4}\ M$$

(*a*) $[OH^-]$ does not exceed $[OH^-]_{sat}$, so no precipitation will occur.
(*b*) $[OH^-]$ exceeds $[OH^-]_{sat}$, so a precipitate will form.

22.36 A mixture of water and AgCl is shaken until a saturated solution is obtained. Then the solid is filtered, and to 100 mL of the filtrate is added 100 mL of 0.030 M NaBr. Should a precipitate be formed?

The silver ion concentration of the AgCl solution is $\sqrt{K_{sp}} = 1 \times 10^{-5}$. When diluted to 200 mL, $[Ag^+] = 0.5 \times 10^{-5}$ and $[Br^-] = 0.015$. The concentration product Q in the second solution is

$$Q = [Ag^+][Br^-] = (0.5 \times 10^{-5})(0.015) = 7.5 \times 10^{-8}$$

Since $Q > K_{sp}$, a precipitate should form. The maximum number of mol AgBr per mL is

$$\frac{0.5 \times 10^{-5}\ \text{mol}}{200\ \text{mL}} = 2.5 \times 10^{-8}\ \text{mol/mL}$$

22.2 COMPETITIVE REACTIONS

22.37 Calculate the simultaneous solubility of CaF_2 and SrF_2.

The two solubilities are not independent of each other because there is a common ion, F^-. We will first assume that most of the F^- in the saturated solution is contributed by the SrF_2, since its K_{sp} is so much larger than that of CaF_2. We can then proceed to solve for the solubility of SrF_2 as if the CaF_2 were not present.

If the solubility of SrF_2 is x mol/L, $x = [Sr^{2+}]$ and $2x = [F^-]$. Then

$$4x^3 = K_{sp} = 2.9 \times 10^{-9} \qquad \text{or} \qquad x = 9 \times 10^{-4}$$

The CaF_2 solubility will have to adapt to the concentration of F^- set by the SrF_2 solubility.

$$[Ca^{2+}] = \frac{K_{sp}}{[F^-]^2} = \frac{3.4 \times 10^{-11}}{(2 \times 9 \times 10^{-4})^2} = 1.0 \times 10^{-5} \qquad \text{(i.e., the solubility of } CaF_2 \text{ is } 1.0 \times 10^{-5} \text{ mol/L)}$$

Check of assumption: The amount of F^- contributed by the solubility of CaF_2 is twice the concentration of Ca^{2+}, or 2.0×10^{-5} mol/L. This is indeed small compared with the amount contributed by SrF_2, $2 \times 9 \times 10^{-4} = 1.8 \times 10^{-3}$ mol/L.

22.38 Calculate the simultaneous solubility of AgSCN and AgBr.

▮ It would be dangerous to proceed as in Problem 22.37 by assuming that the more soluble salt, AgSCN in this case, provides all the common ion, Ag^+. The two K_{sp} values do not differ by very much, so the contribution of AgBr cannot be neglected. Instead we must solve for both equilibria simultaneously.

$$[Ag^+][SCN^-] = 1.1 \times 10^{-12} \qquad (1)$$

$$[Ag^+][Br^-] = 5.0 \times 10^{-13} \qquad (2)$$

Dividing (*1*) by (*2*),

$$\frac{[SCN^-]}{[Br^-]} = 2.2 \qquad (3)$$

Because electric charge must balance,

$$[SCN^-] + [Br^-] = [Ag^+] \qquad (4)$$

Dividing (*4*) by $[Br^-]$,

$$\frac{[SCN^-]}{[Br^-]} + \frac{[Br^-]}{[Br^-]} = \frac{[Ag^+]}{[Br^-]} \qquad (5)$$

Substituting from (*3*),

$$2.2 + 1.0 = 3.2 = \frac{[Ag^+]}{[Br^-]} \qquad (6)$$

Substituting (*6*) into (*2*), $3.2[Br^-]^2 = 5.0 \times 10^{-13}$ or $[Br^-] = 4.0 \times 10^{-7}$. The remaining concentrations can be found by simple substitutions in (*1*) and (*2*): $[Ag^+] = 1.2 \times 10^{-6}$ and $[SCN^-] = 9 \times 10^{-7}$. The solubilities are 4×10^{-7} mol AgBr/L and 9×10^{-7} mol AgSCN/L.

22.39 Ag_2SO_4 and $SrSO_4$ are both shaken up with pure water. Evaluate $[Ag^+]$ and $[Sr^{2+}]$ in the resulting saturated solution.

▮

$$Ag_2SO_4 \longrightarrow 2Ag^+ + SO_4^{2-} \qquad K_{sp} = 1.5 \times 10^{-5} = [Ag^+]^2[SO_4^{2-}]$$

$$SrSO_4 \longrightarrow Sr^{2+} + SO_4^{2-} \qquad K_{sp} = 2.8 \times 10^{-7} = [Sr^{2+}][SO_4^{2-}]$$

The fact that K_{sp} for Ag_2SO_4 is much larger than that for $SrSO_4$ and is the product of three ion concentrations instead of two means that the silver salt is more soluble. Therefore, assume that the $SrSO_4$ provides a negligible concentration of sulfate ion to the solution.

For Ag_2SO_4, let $x = [SO_4^{2-}]$. Then:

$$K_{sp} = (2x)^2(x) = 1.5 \times 10^{-5} \qquad x = 1.55 \times 10^{-2} \qquad [Ag^+] = 2x = 3.1 \times 10^{-2}$$

For $SrSO_4$: $\quad K_{sp} = [Sr^{2+}](1.55 \times 10^{-2}) = 2.8 \times 10^{-7} \qquad [Sr^{2+}] = 1.8 \times 10^{-5}$

Note that the $[SO_4^{2-}]$ provided by solution of $SrSO_4$ (2.8×10^{-5} M, equal to $[Sr^{2+}]$) is negligible when added to that provided by Ag_2SO_4 (1.6×10^{-2} M). The assumption was correct.

22.40 Calculate $[Ag^+]$ in a solution made by dissolving both Ag_2CrO_4 and $Ag_2C_2O_4$ until saturation is reached with respect to both salts.

▮ Assume no hydrolysis of $C_2O_4^{2-}$. The solubilities of the two salts are too similar to make simplifying assumptions about which provides Ag^+ to the solution.

$$Ag_2CrO_4 \rightleftharpoons 2Ag^+ + CrO_4^{2-} \qquad K_{sp} = 9.0 \times 10^{-12} = [Ag^+]^2[CrO_4^{2-}]$$

$$Ag_2C_2O_4 \rightleftharpoons 2Ag^+ + C_2O_4^{2-} \qquad K_{sp} = 6.0 \times 10^{-12} = [Ag^+]^2[C_2O_4^{2-}]$$

By electroneutrality, $0.5[Ag^+] = [CrO_4^{2-}] + [C_2O_4^{2-}] = \dfrac{9.0 \times 10^{-12}}{[Ag^+]^2} + \dfrac{6.0 \times 10^{-12}}{[Ag^+]^2}$

$$0.5[Ag^+]^3 = 9.0 \times 10^{-12} + 6.0 \times 10^{-12} = 1.5 \times 10^{-11} \qquad [Ag^+] = 3.1 \times 10^{-4}\ M$$

22.41 Calculate $[F^-]$ in a solution saturated with respect to both MgF_2 and SrF_2.

I Assume no hydrolysis of F^-.

$$MgF_2 \longrightarrow Mg^{2+} + 2F^- \qquad K_{sp} = [Mg^{2+}][F^-]^2 = 6.5 \times 10^{-9}$$
$$SrF_2 \longrightarrow Sr^{2+} + 2F^- \qquad K_{sp} = [Sr^{2+}][F^-]^2 = 2.9 \times 10^{-9}$$

The solubilities are too similar to neglect the fluoride ion contributed by the less soluble salt. From electroneutrality, $[Mg^{2+}] + [Sr^{2+}] = 0.5[F^-]$. Then

$$\frac{6.5 \times 10^{-9}}{[F^-]^2} + \frac{2.9 \times 10^{-9}}{[F^-]^2} = 0.5[F^-] \qquad 9.4 \times 10^{-9} = 0.5[F^-]^3 \qquad [F^-] = 2.7 \times 10^{-3}\ M$$

22.42 The following solutions were mixed: 500 mL of 0.0100 M $AgNO_3$ and 500 mL of a solution that was both 0.0100 M in NaCl and 0.0100 M in NaBr. Calculate $[Ag^+]$, $[Cl^-]$, and $[Br^-]$ in the equilibrium solution.

I If there were no precipitation, the diluting effect of mixing would make $[Ag^+] = [Cl^-] = [Br^-] = \frac{1}{2}(0.0100) = 0.0050$ M. AgBr is the more insoluble salt and would take precedence in the precipitation process. To find whether AgCl also precipitates, we may assume that it does not. In this case, only Ag^+ and Br^- would be removed by precipitation, and the concentrations of these two ions in solution would remain equal to each other.

$$[Ag^+][Br^-] = [Ag^+]^2 = K_{sp} = 5.0 \times 10^{-13} \quad \text{or} \quad [Ag^+] = [Br^-] = 7.1 \times 10^{-7}\ M$$

We now examine the ion product for AgCl: $[Ag^+][Cl^-] = (7.1 \times 10^{-7})(5.0 \times 10^{-3}) = 3.5 \times 10^{-9}$. Since this ion product exceeds K_{sp} for AgCl, at least some AgCl must also precipitate. In other words, our first assumption was wrong.

Since both halides precipitate, both solubility product requirements must be met simultaneously.

$$[Ag^+][Cl^-] = 1.0 \times 10^{-10} \qquad (1)$$
$$[Ag^+][Br^-] = 5.0 \times 10^{-13} \qquad (2)$$

The third equation needed to define the three unknowns is an equation expressing the balancing of positive and negative charges in solution: $[Na^+] + [Ag^+] = [Cl^-] + [Br^-] + [NO_3^-]$.

$$0.0100 + [Ag^+] = [Cl^-] + [Br^-] + 0.0050 \quad \text{or} \quad [Cl^-] + [Br^-] - [Ag^+] = 0.0050 \qquad (3)$$

Dividing (*1*) by (*2*), $[Cl^-]/[Br^-] = 200$, and we see that Br^- plays a negligible role in the total anion concentration of the solution. Also, $[Ag^+]$ must be negligible in (*3*) because of the insolubility of the two silver salts. We thus assume in (*3*) that $[Cl^-] = 0.0050$. From (*1*),

$$[Ag^+] = \frac{1.0 \times 10^{-10}}{[Cl^-]} = \frac{1.0 \times 10^{-10}}{0.0050} = 2.0 \times 10^{-8}$$

From (*2*), $$[Br^-] = \frac{5.0 \times 10^{-13}}{[Ag^+]} = \frac{5.0 \times 10^{-13}}{2.0 \times 10^{-8}} = 2.5 \times 10^{-5}$$

Check of assumptions: Both $[Ag^+]$ and $[Br^-]$ are negligible compared with 0.0050.

Note: The solubility of saturated H_2S is 0.10 M. The values of K_a for weak acids may be obtained from Table 21.1. When problems involve a precipitation reaction, the reaction should be assumed to go to completion, and then the equilibrium dissolving process should be considered.

22.43 Write equations showing all of the equilibrium reactions occurring in aqueous solutions containing each of the following sets of reagents: (*a*) NaCl (*b*) NaOH (*c*) $NaC_2H_3O_2 + HC_2H_3O_2$ (*d*) $Na_2S + CuS$ (*e*) $NH_4Cl + NH_3 + Mg(OH)_2(s)$

I (*a*) through (*e*) involve $2H_2O \rightarrow H_3O^+ + OH^-$. In addition, (*c*) through (*e*) have the following equilibria:

(*c*) $HC_2H_3O_2 + H_2O \rightleftharpoons H_3O^+ + C_2H_3O_2^-$

(*d*) $CuS \longrightarrow Cu^{2+} + S^{2-}$ $\quad S^{2-} + H_2O \longrightarrow HS^- + OH^-$ $\quad HS^- + H_2O \longrightarrow H_2S + OH^-$

(*e*) $NH_3 + H_2O \longrightarrow NH_4^+ + OH^-$ $\quad Mg(OH)_2 \longrightarrow Mg^{2+} + 2OH^-$

22.44 Calculate the silver ion concentration in a solution prepared by shaking solid Ag_2S with saturated H_2S (0.10 M) in 0.15 M H_3O^+ until equilibrium is established.

▌ $$H_2S + H_2O \rightleftharpoons H_3O^+ + HS^- \qquad K_{12} = \frac{[H_3O^+]^2[S^{2-}]}{[H_2S]} = 1 \times 10^{-21}$$

$$HS^- + H_2O \rightleftharpoons H_3O^+ + S^{2-}$$

In 0.10 M H_2S and 0.15 M H_3O^+, the concentrations are effectively $[H_3O^+] = 0.15$ and $[H_2S] = 0.10$.

$$K_{12} = \frac{(0.15)^2[S^{2-}]}{(0.10)} = 1 \times 10^{-21} \qquad [S^{2-}] = \frac{1 \times 10^{-22}}{0.0225} = 4 \times 10^{-21}$$

$$Ag_2S \rightleftharpoons 2Ag^+ + S^{2-} \qquad K_{sp} = [Ag^+]^2[S^{2-}] = 1.6 \times 10^{-49}$$

$$[Ag^+]^2 = \frac{1.6 \times 10^{-49}}{(4 \times 10^{-21})} = 4 \times 10^{-29} \qquad \text{hence} \quad [Ag^+] = 6 \times 10^{-15}\ \text{M}$$

22.45 Calculate the hydronium ion concentration necessary to just prevent precipitation of ZnS from a solution 0.20 M in Zn^{2+} which is saturated with H_2S.

▌ If precipitation is prevented, $[Zn^{2+}] = 0.20$ M. The sulfide ion concentration is obtained from K_{sp}.

$$[S^{2-}] = \frac{K_{sp}}{[Zn^{2+}]} = \frac{1.2 \times 10^{-22}}{0.20} = 6.0 \times 10^{-22}\ \text{M}$$

See equations in Problem 22.44. $$K_{12} = \frac{[H_3O^+]^2(6.0 \times 10^{-22})}{0.10} = 1 \times 10^{-21} \qquad [H_3O^+] = 0.4\ \text{M}$$

22.46 What is the maximum possible concentration of Ni^{2+} ion in a solution which is also 0.15 M in HCl and 0.10 M in H_2S?

▌ $$[S^{2-}] = 4 \times 10^{-21} \qquad \text{(see Problem 22.44.)}$$

$$K_{sp} = [Ni^{2+}][S^{2-}] = 2 \times 10^{-21} \qquad [Ni^{2+}] = \frac{2 \times 10^{-21}}{4 \times 10^{-21}} = 0.5\ \text{M maximum}$$

22.47 The number of mol CoS which dissolves per L solution exceeds the solubility predicted by the value of K_{sp} for CoS. Explain by means of appropriate chemical equations why this behavior is observed.

▌ Not all of the sulfide which dissolves remains as S^{2-}; most of it hydrolyzes:

$$S^{2-} + H_2O \rightleftharpoons HS^- + OH^-$$

22.48 Calculate the solubility of CoS in 0.10 M H_2S and 0.15 M H_3O^+.

▌ See Problem 22.44 for calculation of $[S^{2-}] = 4 \times 10^{-21}$.

$$K_{sp} = [Co^{2+}][S^{2-}] = 3 \times 10^{-26} \qquad [Co^{2+}] = \frac{3 \times 10^{-26}}{4 \times 10^{-21}} = 7 \times 10^{-6}\ \text{M}$$

22.49 Calculate the copper ion concentration in a solution obtained by shaking CuS with saturated H_2S (0.10 M) in which 0.15 M HCl is present.

▌ $$[S^{2-}] = 4 \times 10^{-21} \qquad \text{(See Problem 22.44.)}$$

$$[Cu^{2+}] = \frac{K_{sp}}{[S^{2-}]} = \frac{8.5 \times 10^{-36}}{4 \times 10^{-21}} = 2 \times 10^{-15}$$

22.50 In an attempted determination of the solubility product constant of Tl_2S, the solubility of this compound in pure CO_2-free water was determined as 6.3×10^{-6} mol/L. Assume that the dissolved sulfide hydrolyzes almost completely to HS^- and that the further hydrolysis to H_2S can be neglected. What is the computed K_{sp}?

▌ $$Tl_2S \rightleftharpoons 2Tl^+ + S^{2-} \qquad K_{sp} = [Tl^+]^2[S^{2-}]$$

$$S^{2-} + H_2O \rightleftharpoons HS^- + OH^- \qquad K_h = \frac{K_w}{K_2} = \frac{1.0 \times 10^{-14}}{1.0 \times 10^{-14}} = 1.0$$

$$[Tl^+] = 2(6.3 \times 10^{-6}) \qquad [HS^-] = 6.3 \times 10^{-6}$$

$$K_h = \frac{(6.3 \times 10^{-6})^2}{[S^{2-}]} = 1.0 \qquad [S^{2-}] = (6.3 \times 10^{-6})^2$$

$$K_{sp} = (6.3 \times 10^{-6})^2[2(6.3 \times 10^{-6})]^2 = 6.3 \times 10^{-21}$$

22.51 The solubility of CuS in pure water is 3.3×10^{-4} g/L at 25 °C. Using this number, calculate the apparent value of K_{sp} for CuS. By precise measurement, the value of K_{sp} for CuS at 25 °C is found to be 8.5×10^{-36}. Explain why CuS is more soluble than predicted by the K_{sp}.

▌

$$CuS \longrightarrow Cu^{2+} + S^{2-}$$

$$(3.3 \times 10^{-4}\ \text{g/L})/(95.6\ \text{g/mol}) = 3.5 \times 10^{-6}\ \text{mol/L} \qquad \text{Apparent } K_{sp} = (3.5 \times 10^{-6})^2 = 1.2 \times 10^{-11}$$

The sulfide ion hydrolyzes extensively. (The amount which dissolves and the amount which exists as S^{2-} in solution are very different.)

22.52 Will FeS precipitate in a saturated H_2S solution if the solution contains 0.010 mol/L of Fe^{2+} and (*a*) 0.20 mol/L of H^+ (*b*) 0.0010 mol/L of H^+?

▌

$$K_1K_2 = \frac{[H^+]^2[S^{2-}]}{[H_2S]} = 1.0 \times 10^{-21}$$

(*a*) $$\frac{(0.20)^2[S^{2-}]}{0.10} = 1.0 \times 10^{-21} \qquad [S^{2-}] = \frac{1.0 \times 10^{-22}}{0.040} = 2.5 \times 10^{-21}$$

$$K_{sp} = [Fe^{2+}][S^{2-}] = 3.7 \times 10^{-19}$$

Since $(0.010)(2.5 \times 10^{-21}) < 3.7 \times 10^{-19}$, FeS will not precipitate.

(*b*) $$\frac{(0.0010)^2[S^{2-}]}{0.10} = 1.0 \times 10^{-21} \qquad [S^{2-}] = 1 \times 10^{-16}$$

Since $(0.010)(1 \times 10^{-16}) > 3.7 \times 10^{-19}$, FeS will precipitate.

22.53 A saturated solution of silver benzoate, $AgOCOC_6H_5$, has a pH of 8.63. K_a for benzoic acid is 6.5×10^{-5}. Estimate the value of K_{sp} for silver benzoate.

▌

$$\text{pH} = 8.63 \qquad \text{pOH} = 5.37 \qquad [OH^-] = 4.3 \times 10^{-6}$$

$$OCOC_6H_5^- + H_2O \longrightarrow HOCOC_6H_5 + OH^-$$

$$K_h = \frac{1.0 \times 10^{-14}}{6.5 \times 10^{-5}} = \frac{(4.3 \times 10^{-6})^2}{[OCOC_6H_5^-]} \qquad [OCOC_6H_5^-] = 0.12$$

$$AgOCOC_6H_5 \longrightarrow Ag^+ + OCOC_6H_5^- \qquad K_{sp} = (0.12)^2 = 1.4 \times 10^{-2}$$

22.54 What is the maximum possible $[Ag^+]$ in a saturated H_2S solution from which precipitation has not occurred?

▌

$$Ag_2S \rightleftharpoons 2Ag^+ + S^{2-} \qquad K_{sp} = [Ag^+]^2[S^{2-}] = 1.6 \times 10^{-49}$$

$$H_2S + H_2O \rightleftharpoons H_3O^+ + HS^- \qquad K_1 = \frac{[H_3O^+][HS^-]}{[H_2S]} = 1.0 \times 10^{-7}$$

$$HS^- + H_2O \rightleftharpoons H_3O^+ + S^{2-} \qquad K_2 = \frac{[H_3O^+][S^{2-}]}{[HS^-]} = 1.0 \times 10^{-14}$$

The second ionization reaction provides negligible hydronium ion. Hence

$$\frac{x^2}{0.10} = 1.0 \times 10^{-7} \qquad x = 1.0 \times 10^{-4} = [H_3O^+] = [HS^-]$$

$$K_2 = \frac{[H_3O^+][S^{2-}]}{[HS^-]} = \frac{(1.0 \times 10^{-4})[S^{2-}]}{1.0 \times 10^{-4}} = 1.0 \times 10^{-14}$$

$$[S^{2-}] = 1.0 \times 10^{-14} \qquad [Ag^+] = \sqrt{\frac{1.6 \times 10^{-49}}{1.0 \times 10^{-14}}} = 4 \times 10^{-18}$$

22.55 Equal volumes of 0.0200 M $AgNO_3$ and 0.0200 M HCN were mixed. Calculate $[Ag^+]$ at equilibrium.

▌ Initially, assume complete precipitation. $Ag^+ + HCN \rightarrow AgCN + H^+$ Since the solutions were diluted 1:1, $[H^+] = 0.0100$ M. Now consider the equilibria:

$$AgCN \rightleftharpoons Ag^+ + CN^- \qquad K_{sp} = 2.2 \times 10^{-16} = [Ag^+][CN^-]$$

$$HCN \rightleftharpoons H^+ + CN^- \qquad K_a = 6.2 \times 10^{-10} = \frac{[H^+][CN^-]}{[HCN]}$$

Since every dissolved CN^- which hydrolyzes yields one HCN, $[Ag^+] = [CN^-] + [HCN]$

$$\frac{2.2 \times 10^{-16}}{[CN^-]} = [CN^-] + \frac{[CN^-](0.0100)}{6.2 \times 10^{-10}} \qquad [CN^-]^2 = \frac{(2.2 \times 10^{-16})(6.2 \times 10^{-10})}{0.0100} \qquad [CN^-] = 3.7 \times 10^{-12}$$

$$[Ag^+] = \frac{K_{sp}}{[CN^-]} = \frac{2.2 \times 10^{-16}}{3.7 \times 10^{-12}} = 5.9 \times 10^{-5}$$

22.56 After solid $SrCO_3$ was equilibrated with a pH 8.60 buffer, the solution was found to have $[Sr^{2+}] = 2.2 \times 10^{-4}$. What is the solubility product constant for $SrCO_3$?

▮

$$K_{sp} = [Sr^{2+}][CO_3^{2-}]$$

$$CO_3^{2-} + H_2O \rightleftharpoons HCO_3^- + OH^- \qquad K_h = \frac{K_w}{K_2} = \frac{[OH^-][HCO_3^-]}{[CO_3^{2-}]} = \frac{K_w}{4.7 \times 10^{-11}}$$

$$[H_3O^+] = -\log 8.6 = 2.51 \times 10^{-9} \qquad [OH^-] = K_w/(2.51 \times 10^{-9})$$

$$\left(\frac{K_w}{2.51 \times 10^{-9}}\right)\frac{[HCO_3^-]}{[CO_3^{2-}]} = \frac{K_w}{4.7 \times 10^{-11}} \qquad \frac{[HCO_3^-]}{[CO_3^{2-}]} = \frac{2.51 \times 10^{-9}}{4.7 \times 10^{-11}} = 53.4$$

The carbonate ion which dissolves forms HCO_3^- in a 1:1 mol ratio or remains unreacted. Thus,

$$[Sr^{2+}] = [HCO_3^-] + [CO_3^{2-}] = 53.4[CO_3^{2-}] + [CO_3^{2-}] = 54.4[CO_3^{2-}] = 2.2 \times 10^{-4}$$

$$[CO_3^{2-}] = \frac{2.2 \times 10^{-4}}{54.4} \qquad K_{sp} = (2.2 \times 10^{-4})\left(\frac{2.2 \times 10^{-4}}{54.4}\right) = 8.9 \times 10^{-10}$$

22.57 Calculate the solubility at 25 °C of $CaCO_3$ in a closed vessel containing a solution of pH 8.60.

▮ See Problem 22.56 for derivation of the equations.

$$[Ca^{2+}] = [CO_3^{2-}] + [HCO_3^-] \qquad \frac{[HCO_3^-]}{[CO_3^{2-}]} = 53.4$$

Let $x = [Ca^{2+}] = 54.4[CO_3^{2-}]$.

$$K_{sp} = [Ca^{2+}][CO_3^{2-}] = x\left(\frac{x}{54.4}\right) = \frac{x^2}{54.4} = 1.0 \times 10^{-8}$$

$$x = 7.4 \times 10^{-4} = [Ca^{2+}] = \text{solubility}$$

22.58 Calculate the NH_4^+ ion concentration (derived from NH_4Cl) needed to prevent $Mg(OH)_2$ from precipitating in 1.0 L of solution which contains 0.010 mol ammonia and 0.0010 mol Mg^{2+}.

▮ First, find the maximum $[OH^-]$ that can be present in the solution without precipitation of $Mg[OH]_2$.

$$[Mg^{2+}][OH^-]^2 = 1.2 \times 10^{-11} \qquad [OH^-] = \sqrt{\frac{1.2 \times 10^{-11}}{[Mg^{2+}]}} = \sqrt{\frac{1.2 \times 10^{-11}}{0.0010}} = \sqrt{1.2 \times 10^{-8}} = 1.1 \times 10^{-4}$$

Now find the $[NH_4^+]$ (derived from NH_4Cl) needed to repress the ionization of NH_3 so that the $[OH^-]$ will not exceed 1.1×10^{-4}.

$$\frac{[NH_4^+][OH^-]}{[NH_3]} = 1.8 \times 10^{-5} \quad \text{or} \quad \frac{[NH_4^+](1.1 \times 10^{-4})}{0.010} = 1.8 \times 10^{-5}$$

and thus $[NH_4^+] = 1.6 \times 10^{-3}$. (Since 0.010 M ammonia is only slightly ionized, especially in the presence of excess NH_4^+, the $[NH_3]$ may be considered to be 0.010.)

22.59 Calculate the solubility of AgCN in a buffer solution of pH 3.00. Assume that no cyano complex is formed.

▮ In this solution the silver that dissolves remains as Ag^+, but the cyanide that dissolves is converted mostly to HCN on account of the fixed acidity of the buffer. [The complex $Ag(CN)_2^-$ forms appreciably only at higher cyanide ion concentrations.] First calculate the ratio of [HCN] to $[CN^-]$ at this pH.

$$\frac{[H^+][CN^-]}{[HCN]} = K_a \quad \text{or} \quad \frac{[HCN]}{[CN^-]} = \frac{[H^+]}{K_a} = \frac{1.0 \times 10^{-3}}{6.2 \times 10^{-10}} = 1.6 \times 10^6$$

Let the solubility of AgCN be x mol/L; then $x = [Ag^+]$ at equilibrium, and $x = [CN^-] + [HCN]$.

Very little error is made by neglecting $[CN^-]$ in comparison to [HCN] (1 part in 1.6 million) and equating [HCN] to x.

$$[CN^-] = \frac{[HCN]}{1.6 \times 10^6} = \frac{x}{1.6 \times 10^6}$$

Substituting in the solubility product equation, $[Ag^+][CN^-] = 2.2 \times 10^{-16}$,

$$x\left(\frac{x}{1.6 \times 10^6}\right) = 2.2 \times 10^{-16} \qquad \text{from which } x = 1.9 \times 10^{-5}.$$

22.60 Given that 0.0010 mol each of Cd^{2+} and Fe^{2+} is contained in 1.0 L of 0.020 M HCl and that this solution is saturated with H_2S: ***(a)*** Determine whether or not each of these ions will precipitate as the sulfide. ***(b)*** How much Cd^{2+} remains in solution at equilibrium?

▌ If either sulfide precipitates, CdS will because of its lower solubility. In doing so, it produces 0.0020 M more H^+.

$$[S^{2-}] = \frac{K_1K_2[H_2S]}{[H^+]^2} = \frac{(1.0 \times 10^{-21})(0.10)}{(0.022)^2} = 2.07 \times 10^{-19}$$

$$[Cd^{2+}] = \frac{8 \times 10^{-27}}{2.07 \times 10^{-19}} = 4 \times 10^{-8}\ \text{M}$$

The cadmium ion concentration has dropped, confirming its precipitation. If the iron does not precipitate, its maximum concentration will be given by

$$[Fe^{2+}] = \frac{3.7 \times 10^{-19}}{2.07 \times 10^{-19}} = 1.8\ \text{M}$$

Since the actual Fe^{2+} concentration is much less, there will be no FeS precipitation. *Note*: In many cases assumptions as to what is apt to precipitate may be made from the qualitative analysis scheme.

22.61 Given that 2.0×10^{-4} mol each of Mn^{2+} and Cu^{2+} was contained in 1.0 L of a 0.0030 M $HClO_4$ solution, and this solution was saturated with H_2S, determine whether or not each of the ions Mn^{2+} and Cu^{2+} will precipitate as the sulfide. The solubility of H_2S, 0.10 mol/L, is assumed to be independent of the presence of other materials in the solution.

▌ $[H_2S] = 0.10$, since the solution is saturated with H_2S; $[H^+] = 0.0030$, since the H_2S contributes negligible H^+ compared with $HClO_4$. Calculate $[S^{2-}]$ from the combined ionization constants.

$$\frac{[H^+]^2[S^{2-}]}{[H_2S]} = K_1K_2 = (1.0 \times 10^{-7})(1.0 \times 10^{-14}) = 1.0 \times 10^{-21}$$

$$[S^{2-}] = (1.0 \times 10^{-21})\frac{[H_2S]}{[H^+]^2} = (1.0 \times 10^{-21})\frac{0.10}{(0.0030)^2} = 1.1 \times 10^{-17}$$

Whenever the product of the concentrations of two ions (e.g., $[Cu^{2+}][S^{2-}]$) in a solution exceeds the value of K_{sp} (e.g., when $[Cu^{2+}][S^{2-}] > 8.5 \times 10^{-36}$), the corresponding ionic compound (CuS) will be precipitated until the product of the concentrations of these two ions again attains the value K_{sp}.

First let us tabulate numbers that would describe the solution if no precipitation occurred.

$$[Mn^{2+}] = [Cu^{2+}] = 2.0 \times 10^{-4}$$
$$[Mn^{2+}][S^{2-}] = (2.0 \times 10^{-4})(1.1 \times 10^{-17}) = 2.2 \times 10^{-21}$$
$$[Cu^{2+}][S^{2-}] = (2.0 \times 10^{-4})(1.1 \times 10^{-17}) = 2.2 \times 10^{-21}$$

Because 2.2×10^{-21} is less than the solubility product of MnS (2.5×10^{-10}), MnS will not precipitate. The large $[H^+]$ furnished by the $HClO_4$ represses the ionization of the H_2S (common H^+ ion) and thus reduces the $[S^{2-}]$ to so low a value that the solubility product of MnS is not reached. But 2.2×10^{-21} is greater than the solubility product of CuS (8.5×10^{-36}); hence CuS will precipitate. When it does, more H^+ is produced, but not nearly enough to change the outcome.

22.62 In Problem 22.61 how much Cu^{2+} escapes precipitation?

▌ Since the Cu^{2+} is nearly completely precipitated, the hydrogen ion concentration is increased to 0.0034 M.

$$Cu^{2+} + H_2S \longrightarrow CuS + 2H^+ \qquad 0.0030\ \text{M} + 0.00040\ \text{M} = 0.0034\ \text{M}$$

$$[S^{2-}] = (1.0 \times 10^{-21})\left(\frac{0.10}{(0.0034)^2}\right) = 8.7 \times 10^{-18} \qquad [Cu^{2+}] = \frac{8.5 \times 10^{-36}}{8.7 \times 10^{-18}} = 9.8 \times 10^{-19}\ \text{M}$$

Thus, 9.8×10^{-19} mol Cu^{2+} remains in solution. On a percent basis the amount of Cu^{2+} remaining unprecipitated is

$$\left(\frac{9.8 \times 10^{-19}}{2 \times 10^{-4}}\right)(100\%) = 4 \times 10^{-13}\%$$

22.63 If the solution in Problem 22.61 is made nearly neutral by buffering the $[H^+]$ at 10^{-9}, will MnS precipitate?

▮

$$[S^{2-}] = (1.0 \times 10^{-21})\left(\frac{0.10}{(10^{-9})^2}\right) = 1.0 \times 10^{-4}$$

Then, if no precipitation occurred, $[Mn^{2+}][S^{2-}] = (2 \times 10^{-4})(1.0 \times 10^{-4}) = 2 \times 10^{-8}$. But 2×10^{-8} is greater than the K_{sp} of MnS (2.5×10^{-10}). Hence MnS will precipitate.

Lowering the $[H^+]$ to 10^{-9} increases the ionization of H_2S to such an extent that sufficient S^{2-} is furnished for the solubility product of MnS to be exceeded.

22.64 H_2S is bubbled into a solution containing 0.15 M Cu^{2+} until no further change takes place. Calculate the concentrations of H_3O^+ produced and of Cu^{2+} remaining in solution.

▮ The precipitation reaction yields H_3O^+, and the H_2S will become saturated (0.10 M) after the precipitation reaction.

$$H_2S + 2H_2O + Cu^{2+} \longrightarrow CuS + 2H_3O^+ \qquad (0.15\text{ M Cu})\left(\frac{2\text{ mol }H_3O^+}{\text{mol }Cu^{2+}}\right) = 0.30\text{ M }H_3O^+$$

See equations in Problem 22.44.

$$K_{12} = \frac{(0.30)^2[S^{2-}]}{0.10} = 1 \times 10^{-21} \qquad [S^{2-}] = 1 \times 10^{-21} \quad \text{(one significant figure)}$$

$$[Cu^{2+}] = \frac{K_{sp}}{[S^{2-}]} = \frac{8.5 \times 10^{-36}}{1 \times 10^{-21}} = 8 \times 10^{-15}\text{ M}$$

22.65 Will MnS be precipitated from a saturated solution of H_2S (0.10 M) which is maintained at pH = 5.00 by a buffer, if the concentration of Mn^{2+} initially present is 0.010 M?

▮

$$K_1K_2 = \frac{[H_3O^+]^2[S^{2-}]}{[H_2S]} = 1.0 \times 10^{-21} = \frac{(1.0 \times 10^{-5})^2[S^{2-}]}{0.10}$$

$$[S^{2-}] = 1.0 \times 10^{-12} \qquad \text{hence} \qquad [Mn^{2+}][S^{2-}] < 2.5 \times 10^{-10} = K_{sp}$$

MnS will not precipitate, since the sulfide ion concentration is 1×10^{-12}, which is small enough for the K_{sp} not to be exceeded.

22.66 Determine the copper ion concentration resulting from the addition to 500 mL of 0.050 M Cu^{2+} solution of ***(a)*** gaseous H_2S until no further change takes place ***(b)*** 500 mL of saturated H_2S solution.

▮ ***(a)*** See Problem 22.64 for solution procedure. In this part, no dilution of the solution takes place, and the H_2S becomes saturated. Hence

$$[H_3O^+] = 0.10\text{ M}, \qquad [H_2S] = 0.10\text{ M}$$

$$K_{12} = \frac{(0.10)^2[S^{2-}]}{0.10} = 1 \times 10^{-21} \qquad [S^{2-}] = 1 \times 10^{-20}$$

$$[Cu^{2+}] = \frac{8.5 \times 10^{-36}}{1 \times 10^{-20}} = 8 \times 10^{-16}\text{ M}$$

(b) In this part, the solutes are all diluted by the doubling of the volume, and half the H_2S is used up in the precipitation reaction. (See Problem 22.64.)

$$[H_3O^+] = 0.050\text{ M} \qquad [H_2S] = 0.025\text{ M}$$

$$K_{12} = \frac{(0.050)^2[S^{2-}]}{0.025} = 1 \times 10^{-21} \qquad [S^{2-}] = 1 \times 10^{-20}\text{ M}$$

$$[Cu^{2+}] = \frac{8.5 \times 10^{-36}}{1 \times 10^{-20}} = 8 \times 10^{-16}\text{ M}$$

(The sulfide ion and therefore the copper ion concentrations just happen to be the same in these two cases.)

22.67 How much Ag^+ would remain in solution after mixing equal volumes of 0.080 M $AgNO_3$ and 0.080 M HOCN?

▌ We must consider both the solubility and the acid ionization equilibria.

$$HOCN \rightleftharpoons H^+ + OCN^- \qquad Ag^+ + OCN^- \rightleftharpoons AgOCN$$

We can think of precipitation as a process which forces HOCN to ionize. H^+ will thus accumulate in the solution as precipitation occurs. To a first approximation, one H^+ must replace every Ag^+ removed by precipitation, to provide electrical balance with respect to the NO_3^-. (This approximation is good as long as OCN^- does not add appreciably to the total equilibrium anion concentration and require still more H^+.) To this approximation, the following equations hold, where $x = [Ag^+]$:

$$[NO_3^-] = 0.040 \qquad \text{(twofold dilution of 0.080 M solution)}$$

Amount of Ag^+ precipitated $= (0.040 - x)$ mol/L Amount of OCN^- precipitated $= (0.040 - x)$ mol/L

$$[HOCN] = x - [OCN^-] \approx x \qquad [H^+] = 0.040 - x$$

Then
$$\frac{[OCN^-]}{[HOCN]} = \frac{K_a}{[H^+]} = \frac{3.3 \times 10^{-4}}{0.040 - x} \qquad \text{or} \qquad [OCN^-] = \frac{3.3 \times 10^{-4}\, x}{0.040 - x}$$

and
$$[Ag^+][OCN^-] = \frac{3.3 \times 10^{-4}\, x^2}{0.040 - x} = K_{sp} = 2.3 \times 10^{-7} \qquad \text{or} \qquad x = [Ag^+] = 5.0 \times 10^{-3}$$

Check of assumptions:
$$[OCN^-] = \frac{3.3 \times 10^{-4}\, x}{0.040 - x} = 4.7 \times 10^{-5}$$

which is small compared with $[NO_3^-]$ and with x.

22.68 Calculate the solubility of MnS in pure water.

▌ This problem differs from the analogous problems dealing with chromates, oxalates, halides, sulfates, and iodates. The difference lies in the extensive hydrolysis of the sulfide ion.

$$S^{2-} + H_2O \rightleftharpoons HS^- + OH^- \qquad K_h = \frac{K_w}{K_2} = \frac{1.0 \times 10^{-14}}{1.0 \times 10^{-14}} = 1.0$$

If x mol/L is the solubility of MnS, we cannot simply equate x to $[S^{2-}]$. Instead, $x = [S^{2-}] + [HS^-] + [H_2S]$. To simplify, we first assume that the first stage of hydrolysis is almost complete and that the second stage proceeds to only a slight extent. In other words, $x = [Mn^{2+}] = [HS^-] = [OH^-]$.

$$[S^{2-}] = \frac{[HS^-][OH^-]}{K_h} = \frac{x^2}{1.0}$$

At equilibrium,
$$[Mn^{2+}][S^{2-}] = \frac{x(x)^2}{1.0} = K_{sp} = 2.5 \times 10^{-10} \qquad \text{or} \qquad x = 6.3 \times 10^{-4}$$

Check of assumptions:

(1) $[S^{2-}] = \dfrac{x^2}{1.0} = \dfrac{(6.3 \times 10^{-4})^2}{1.0} = 4.0 \times 10^{-7}$ $[S^{2-}]$ is indeed negligible compared with $[HS^-]$.

(2) $[H_2S] = \dfrac{[H^+][HS^-]}{K_1} = \dfrac{K_w[HS^-]}{[OH^-]\, K_1} = \dfrac{10^{-14}x}{10^{-7}x} = 10^{-7}$ $[H_2S]$ is also small compared with $[HS^-]$.

The above approximations would not be valid for sulfides like CuS which are much more insoluble than MnS. First, the water dissociation would begin to play an important role in determining $[OH^-]$. Second, the second stage of hydrolysis, producing $[H_2S]$, would not be negligible compared with the first. Even for MnS, an additional complication arises because of the complexation of Mn^{2+} with OH^-. The full treatment of sulfide solubilities is a complicated problem because of the multiple equilibria which must be considered.

22.69 Equal volumes of 0.0100 M $Sr(NO_3)_2$ and 0.0100 M $NaHSO_4$ were mixed. Calculate $[Sr^{2+}]$ and $[H^+]$ at equilibrium. Take into account the amount of H^+ needed to balance the charge of the SO_4^{2-} remaining in the solution.

▌
$$Sr^{2+} + HSO_4^- \longrightarrow SrSO_4 + H^+$$

Assume complete precipitation, yielding $SrSO_4(s)$ in a solution of 0.0050 M H^+. Then consider the equilibria involved.

$$SrSO_4 \rightleftharpoons Sr^{2+} + SO_4^{2-} \qquad K_{sp} = 2.8 \times 10^{-7} = [Sr^{2+}][SO_4^{2-}]$$

$$HSO_4^- \rightleftharpoons H^+ + SO_4^{2-} \qquad K_2 = \frac{[SO_4^{2-}][H^+]}{[HSO_4^-]} = 1.0 \times 10^{-2}$$

$$[Sr^{2+}] = [SO_4^{2-}] + [HSO_4^-] \qquad [H^+] + [HSO_4^-] = 0.0050$$

$$\frac{2.8 \times 10^{-7}}{[SO_4^{2-}]} = [SO_4^{2-}] + \left(\frac{[SO_4^{2-}](0.0050)}{0.010}\right) = 1.5[SO_4^{2-}]$$

$$[SO_4^{2-}]^2 = (2.8 \times 10^{-7})/1.5 \qquad [SO_4^{2-}] = 4.3 \times 10^{-4}\ \text{M} \qquad [Sr^{2+}] = \frac{2.8 \times 10^{-7}}{4.3 \times 10^{-4}} = 6.5 \times 10^{-4}\ \text{M}$$

$$[H^+] = 0.0050 - [HSO_4^-] = 0.0050 - ([Sr^{2+}] - [SO_4^{2-}]) = 0.0050 - 0.00023 = 0.0048\ \text{M}$$

22.70 How much $SrSO_4$ can dissolve in 0.010 M HNO_3?

▌ By use of the calculations of Problem 22.69, we get $[Sr^{2+}] = [SO_4^{2-}] + [HSO_4^-]$

$$\frac{2.8 \times 10^{-7}}{[SO_4^{2-}]} = [SO_4^{2-}] + \frac{0.010[SO_4^{2-}]}{0.010} = 2[SO_4^{2-}]$$

$$[SO_4^{2-}]^2 = 1.4 \times 10^{-7} \qquad [SO_4^{2-}] = 3.7 \times 10^{-4}\ \text{M} \qquad [Sr^{2+}] = \frac{2.8 \times 10^{-7}}{3.7 \times 10^{-4}} = 7.6 \times 10^{-4}\ \text{M}$$

22.71 Calculate the hydronium ion concentration required to *just* prevent precipitation of As_2S_3 from a solution prepared by saturating a solution of 0.100 M As^{3+} with H_2S gas. $K_{sp} = 1.0 \times 10^{-65}$

▌ $K_{sp} = [As^{3+}]^2[S^{2-}]^3 = 1.0 \times 10^{-65}$. Saturated H_2S is 0.10 M. "Just prevent precipitation" means that the concentrations are the equilibrium concentrations, since any more S^{2-} would cause precipitation. Hence $(0.100)^2[S^{2-}]^3 = 1.0 \times 10^{-65}$, giving $[S^{2-}] = 1.0 \times 10^{-21}$. Substitute this value into

$$K_1K_2 = \frac{[H_3O^+]^2[S^{2-}]}{[H_2S]} = 1.0 \times 10^{-21} = \frac{[H_3O^+]^2(1.0 \times 10^{-21})}{0.10} \qquad [H_3O^+] = 0.31\ \text{M}$$

22.72 Calculate the Cu^{2+} concentration of a solution prepared by bubbling H_2S into 0.10 M Cu^{2+} solution until no further change takes place. (Saturated H_2S is 0.10 M.)

▌

$$Cu^{2+} + 2H_2O + H_2S \longrightarrow 2H_3O^+ + CuS$$

$$[H_3O^+] = 0.20\ \text{M} \qquad [H_2S] = 0.10\ \text{M}$$

$$K_{12} = \frac{[H_3O^+]^2[S^{2-}]}{[H_2S]} = \frac{(0.20)^2[S^{2-}]}{0.10} = 1.0 \times 10^{-21}$$

$$[S^{2-}] = 2.5 \times 10^{-21} \qquad K_{sp} = [Cu^{2+}][S^{2-}] = 8.5 \times 10^{-36} \qquad [Cu^{2+}] = 3.4 \times 10^{-15}\ \text{M}$$

22.73 Calculate what hydronium ion concentration must be maintained in order to just prevent precipitation of ZnS from a solution which contains 0.010 M Zn^{2+} and 0.010 M Cd^{2+} when the solution is saturated with H_2S. What concentration of Cd^{2+} will remain in the solution under these conditions?

▌

$$K_{sp} = [Zn^{2+}][S^{2-}] = 1.2 \times 10^{-22} \qquad [Zn^{2+}] = 0.010\ \text{M} \qquad [S^{2-}] = 1.2 \times 10^{-20}$$

This sulfide ion concentration is the maximum possible without precipitation of ZnS. For that sulfide ion concentration, the hydronium ion concentration can be calculated from the equation for saturated H_2S solutions,

$$[H_3O^+]^2[S^{2-}] = 1 \times 10^{-22} \qquad [H_3O^+]^2 \approx 8 \times 10^{-3} \qquad [H_3O^+] \approx 0.09\ \text{M}$$

For cadmium sulfide, $K_{sp} = 8 \times 10^{-27}$. The product of the sulfide ion concentration and the original cadmium ion concentration is 1.2×10^{-22}, which exceeds the value of K_{sp}, and therefore CdS is expected to precipitate. The cadmium ion left in solution will be determined by K_{sp} for CdS.

$$[Cd^{2+}][S^{2-}] = [Cd^{2+}](1.2 \times 10^{-20}) = 8 \times 10^{-27} \qquad [Cd^{2+}] = 7 \times 10^{-7}$$

Thus in 0.09 M H_3O^+ solution saturated with H_2S, the cadmium ion concentration will have been reduced from 0.010 M to 7×10^{-7} M, a factor of 7×10^{-5}, while the zinc ion concentration will be unchanged. It should be noted that during the precipitation reaction, H_2S is used up and H_3O^+ is generated:

$$Cd^{2+} + H_2S + 2H_2O \rightleftharpoons CdS(s) + 2H_3O^+$$

The hydronium ion concentration may be kept constant by means of a buffer, and the hydrogen sulfide concentration must be kept constant by continuously saturating the solution with H_2S gas.

22.74 Excess solid $Ag_2C_2O_4$ is shaken with (*a*) 0.0010 M $HClO_3$ (*b*) 0.00030 M $HClO_3$. What is the equilibrium value of $[Ag^+]$ in the resulting solution? The concentration of free oxalic acid is of no importance in this problem.

▮
$$Ag_2C_2O_4 \longrightarrow 2Ag^+ + C_2O_4^{2-} \qquad K_{sp} = [Ag^+]^2[C_2O_4^{2-}] = 6 \times 10^{-12}$$

$$HC_2O_4^- \longrightarrow H^+ + C_2O_4^{2-} \qquad K_2 = \frac{[C_2O_4^{2-}][H^+]}{[HC_2O_4^-]} = 5.4 \times 10^{-5}$$

The solution process produces two silver ions for each oxalate ion; the latter may or may not hydrolyze:

$$0.5[Ag^+] = [C_2O_4^{2-}] + [HC_2O_4^-]$$

$$(0.5)\left(\frac{6 \times 10^{-12}}{[C_2O_4^{2-}]}\right)^{1/2} = [C_2O_4^{2-}] + \frac{[C_2O_4^{2-}][H^+]}{K_2} \qquad 0.5(2.45 \times 10^{-6}) = [C_2O_4]^{3/2}\left(1 + \frac{[H^+]}{K_2}\right)$$

(*a*) $$0.5(2.45 \times 10^{-6}) = [C_2O_4^{2-}]^{3/2}\left(1 + \frac{0.0010}{5.4 \times 10^{-5}}\right) = 19.5\,[C_2O_4^{2-}]^{3/2}$$

$$[C_2O_4^{2-}]^{3/2} = 6.2 \times 10^{-8} \qquad [C_2O_4^{2-}] = 1.6 \times 10^{-5} \qquad [Ag^+]^2 = \frac{6 \times 10^{-12}}{1.6 \times 10^{-5}} \qquad [Ag^+] = 6 \times 10^{-4}\text{ M}$$

(*b*) $$0.5(2.45 \times 10^{-6}) = [C_2O_4^{2-}]^{3/2}\left(1 + \frac{0.00030}{5.4 \times 10^{-5}}\right)$$

$$[C_2O_4^{2-}]^{3/2} = 1.8 \times 10^{-7} \qquad [C_2O_4^{2-}] = 3.2 \times 10^{-5}\text{ M}$$

$$[Ag^+]^2 = \frac{6 \times 10^{-12}}{3.2 \times 10^{-5}} = 1.9 \times 10^{-7} \qquad [Ag^+] = 4 \times 10^{-4}\text{ M}$$

22.75 Tenth molar solutions of Cu^{2+} and of Zn^{2+} are each saturated with H_2S to produce CuS and ZnS, respectively. What concentrations of H_3O^+ will cause the precipitates to redissolve completely in the saturated H_2S solution?

▮ If the solids are completely dissolved, assuming no change in volume, the concentrations of the metal will be 0.10 M. Thus, for ZnS,

$$K_{sp} = [Zn^{2+}][S^{2-}] = 1.2 \times 10^{-22} \qquad 0.10[S^{2-}] = 1.2 \times 10^{-22} \qquad [S^{2-}] = 1.2 \times 10^{-21}$$

$$\frac{[H_3O^+]^2[S^{2-}]}{[H_2S]} = 1 \times 10^{-21} \qquad [H_3O^+]^2 = \frac{(1 \times 10^{-21})(0.10)}{1.2 \times 10^{-21}} = 8 \times 10^{-2}$$

$$[H_3O^+] = 0.3\text{ M} \quad \text{(to one significant figure)}$$

But for CuS, $[Cu^{2+}][S^{2-}] = 8.5 \times 10^{-36}$ $\quad 0.10[S^{2-}] = 8.5 \times 10^{-36}$ $\quad [S^{2-}] = 8.5 \times 10^{-35}$

$$\frac{[H_3O^+]^2[S^{2-}]}{[H_2S]} = 1 \times 10^{-21} \qquad \frac{[H_3O^+]^2(8.5 \times 10^{-35})}{0.10} = 1 \times 10^{-21} \qquad [H_3O^+] = 1 \times 10^6\text{ M} = 1\text{ million molar}$$

Since it is impossible to achieve such a large hydronium ion concentration, CuS cannot be dissolved by addition of strong acid alone. To dissolve CuS, the equilibrium sulfide ion concentration must be reduced by some method other than the formation of H_2S. A suitable method is the oxidation of the sulfide ion to sulfur by means of HNO_3:

$$CuS \rightleftharpoons Cu^{2+} + S^{2-}$$

$$8H^+ + 2NO_3^- + 3S^{2-} \qquad 3S + 2NO + 4H_2O$$

22.76 Calculate the solubility of FeS in pure water. (*Hint*: The second stage of hydrolysis, producing H_2S, cannot be neglected.)

▮
$$FeS \rightleftharpoons Fe^{2+} + S^{2-} \qquad K_{sp} = [Fe^{2+}][S^{2-}] = 3.7 \times 10^{-19}$$

$$S^{2-} + H_2O \rightleftharpoons HS^- + OH^- \qquad K_{h1} = \frac{[HS^-][OH^-]}{[S^{2-}]} = \frac{K_w}{K_2} = 1.0$$

$$HS^- + H_2O \rightleftharpoons H_2S + OH^- \qquad K_{h2} = \frac{[H_2S][OH^-]}{[HS^-]} = \frac{K_w}{K_1} = 1.0 \times 10^{-7}$$

Assume that Fe^{2+} does not hydrolyze and that S^{2-} hydrolyzes almost completely. Let $x = [Fe^{2+}]$ and $y = [H_2S]$. Then $[HS^-] = x - y$ and $[OH^-] = x + y$.

Combining the first two equations above yields

$$[Fe^{2+}][HS^-][OH^-] = x(x - y)(x + y) = 3.7 \times 10^{-19}$$

As a first approximation, let $y = 0$. Then $x^3 = 3.7 \times 10^{-19}$ and $y = 7.2 \times 10^{-7}$. Now calculate the value for y, using the third equation:

$$[H_2S] = \frac{[HS^-]K_{h2}}{[OH^-]} = \frac{(x-y)K_{h2}}{(x+y)} = y$$

Solving with the quadratic equation yields $y = 8 \times 10^{-8}$. With $y < x$, $a^2 \ll x^2$, and the equation reduces to $x(x-y)(x+y) = x(x^2 - y^2) \cong x^3$. Hence the value of $[Fe^{2+}]$ does not change significantly from 7.2×10^{-7}. The $[S^{2-}]$, calculated from the second equation, is 5.1×10^{-13}, insignificant compared to $[H_2S] + [HS^-]$.

22.77 Calculate the solubility of AgOCN in 0.0010 M HNO_3.

▌ This is a case where no simple approximation can be justified. The dissolved cyanate is not predominantly in a single species at equilibrium, but both $[OCN^-]$ and [HOCN] are of equal orders of magnitude. The student may verify that an assumption that either of these two species may be neglected in comparison with the other leads to an inconsistency.

Let the solubility of AgOCN be x mol/L. The two equilibria that must be satisfied are:

$$[Ag^+][OCN^-] = K_{sp} = 2.3 \times 10^{-7} \qquad (1)$$

$$\frac{[H^+][OCN^-]}{[HOCN]} = K_a = 3.3 \times 10^{-4} \qquad (2)$$

Mass and charge balance impose the following requirements:

$$x = [Ag^+] = [HOCN] + [OCN^-] \quad \text{(the dissolved anion must remain as } OCN^- \text{ or be converted to HOCN)} \qquad (3)$$

$$[H^+] = 0.0010 - [HOCN] \quad (H^+ \text{ is partly used up to form HOCN}) \qquad (4)$$

Algebraically, the four equations are sufficient for determination of the four unknowns. By substitution, a cubic equation could be obtained in one unknown. Rather than proceed rigorously to the cubic, we approach the problem by a method of successive approximations. We employ the following steps:

(i) Estimate x by (3): $x = [HOCN] + [OCN^-]$.

(ii) Estimate $[OCN^-]$ by (1): $[OCN^-] = \dfrac{2.3 \times 10^{-7}}{x}$

(iii) Estimate [HOCN] by (2) with the substitution indicated in (4):

$$[HOCN] = \frac{(0.0010 - [HOCN])[OCN^-]}{3.3 \times 10^{-4}} \quad \text{Solving, } [HOCN] = \frac{0.0010[OCN^-]}{(3.3 \times 10^{-4}) + [OCN^-]}$$

(iv) Return to step (i) and repeat the cycle, each time using the most recent approximate values of the variables. Stop the process when the x values determined in step (i) for two successive cycles are sufficiently close to each other.

For the first cycle of approximations we must have some provisional values of [HOCN] and $[OCN^-]$. A possible first assumption is that $[HOCN] = 0$, that $[OCN^-] = x$, and that, from (1), $x = \sqrt{2.3 \times 10^{-7}} = 4.8 \times 10^{-4}$. Table 22.2 gives the calculations based on this assumption.

TABLE 22.2

approximation no.	step (i) 10^4x	step (ii) $10^4[OCN^-]$	step (iii) $10^4[HOCN]$
1	4.8	4.8	5.9
2	10.7	2.1	3.9
3	6.0	3.8	5.4
4	9.2	2.5	4.3
5	6.8	3.4	5.1
6	8.5	2.7	4.5
7	7.2	3.3	5.0
8	8.3	2.8	4.6
9	7.4	3.1	4.8
10	7.9	2.9	4.7
11	7.6	3.0	4.8

Note the oscillating nature of the approximate x values and the decreasing amplitude of the oscillations. This is a good sign that we have "zeroed in" on the correct x value. We can take as the solution the average of the last two values, between which, according to the above method, the true solution lies:

$$x = 7.8 \times 10^{-4} = [\mathrm{Ag^+}] \qquad [\mathrm{OCN^-}] = 2.9 \times 10^{-4} \qquad [\mathrm{HOCN}] = 4.7 \times 10^{-4}$$

A final check is obtained by substituting these values into (*1*) through (*4*). Or, we could have stopped after approximation 9 by noting that the results in successive approximations were not changing by more than 10%, taking the averages of the results in approximations 8 and 9.

The choice of the first approximate value of x was somewhat arbitrary. Because of the convergence of the procedure, it would not have mattered if some other value of x had been used, even a guessed value of the right order of magnitude.

22.3 COORDINATION EQUILIBRIA

Caution: Watch out for the difference between the equilibrium constant for the formation of a complex ion, K_f, and that for dissociation of a complex, K_d. Selected values of K_f are given in Table 22-3, p. 501.

In the following problems, square brackets are used to denote concentrations and also to indicate the extent of coordination spheres. The latter use is optional, and the concentrations of complex ions are indicated using only one set of square brackets in each case. The charge on the complex ion is inside the square brackets which refer to concentrations of complex ions.

22.78 Assuming no change in volume, calculate the minimum mass of NaCl necessary to dissolve 0.010 mol AgCl in 100 L solution.

$$\mathrm{AgCl + Cl^- \rightleftharpoons AgCl_2^-}$$

The concentration of $AgCl_2^-$ is 0.010 mol/100 L = 1.0×10^{-4} M.

$$K_f = \frac{[\mathrm{AgCl_2^-}]}{[\mathrm{Ag^+}][\mathrm{Cl^-}]^2} = 3 \times 10^5 \qquad [\mathrm{Ag^+}] = \frac{K_{sp}}{[\mathrm{Cl^-}]} \qquad K_f = \frac{[\mathrm{AgCl_2^-}]}{K_{sp}[\mathrm{Cl^-}]}$$

$$[\mathrm{Cl^-}] = \frac{[\mathrm{AgCl_2^-}]}{K_{sp}K_f} = \frac{1.0 \times 10^{-4}}{(1.0 \times 10^{-10})(3 \times 10^5)} = 3\ \mathrm{M}$$

$$(3\ \mathrm{mol/L})(100\ \mathrm{L})(58.5\ \mathrm{g/mol}) = 17\ \mathrm{kg}$$

22.79 What is the concentration of free Cd^{2+} in 0.0050 M $CdCl_2$? K_1 for chloride complexation of Cd^{2-} is 100; K_2 need not be considered.

Assume that the $CdCl_2$ is fully ionized and then undergoes the reaction $Cd^{2+} + Cl^- \rightarrow CdCl^+$.

$$K_1 = \frac{[\mathrm{CdCl^+}]}{[\mathrm{Cd^{2+}}][\mathrm{Cl^-}]} = 100$$

	initial	produced	used up	equilibrium
$[Cd^{2+}]$	0.0050		$0.0050 - x$	x
$[Cl^-]$	0.0100		$0.0050 - x$	$0.0050 + x$
$[CdCl^+]$	0	$0.0050 - x$		$0.0050 - x$

$$K_1 = \frac{0.0050 - x}{x(0.0050 + x)} = 100 \qquad 100x^2 + 1.5x - 0.0050 = 0 \qquad x = 2.8 \times 10^{-3}\ \mathrm{M} = [\mathrm{Cd^{2+}}]$$

22.80 A solution made up to be 0.0100 M $Co(NO_3)_2$ and 0.0200 M N_2H_4 at a total ionic strength of 1 was found to have an equilibrium $[Co^{2+}]$ of 6.2×10^{-3}. Assuming that the only complex formed was $Co(N_2H_4)^{2+}$, what is the apparent K_1 for complex formation at this ionic strength?

$$\mathrm{Co^{2+} + N_2H_4 \longrightarrow Co(N_2H_4)^{2+}} \qquad K_1 = \frac{[\mathrm{Co(N_2H_4)^{2+}}]}{[\mathrm{Co^{2+}}][\mathrm{N_2H_4}]}$$

	initial	produced	used up	equilibrium
$[Co^{2+}]$	0.0100		0.0038	0.0062
$[N_2H_4]$	0.0200		0.0038	0.0162
$[Co(N_2H_4)^{2+}]$	0	0.0038		0.0038

$$K_1 = \frac{0.0038}{(0.0062)(0.0162)} = 38$$

22.81 Sr^{2+} forms a very unstable complex with NO_3^-. A solution that was nominally 0.00100 M $Sr(ClO_4)_2$ and 0.050 M KNO_3 was found to have only 75% of its strontium in the uncomplexed Sr^{2+} form, the balance being $Sr(NO_3)^+$. What is K_1 for complexation?

▮

$$Sr^{2+} + NO_3^- \longrightarrow Sr(NO_3)^+ \qquad K_1 = \frac{[Sr(NO_3)^+]}{[Sr^{2+}][NO_3^-]}$$

	initial	produced	used up	equilibrium
$[Sr^{2+}]$	0.00100		0.00025	0.00075
$[NO_3^-]$	0.050		0.00025	0.050
$[Sr(NO_3)^+]$	0	0.00025		0.00025

$$K_1 = \frac{0.00025}{(0.00075)(0.050)} = 6.7$$

22.82 A recent investigation of the complexation of SCN^- with Fe^{3+} led to values of 130, 16, and 1.0 for K_1, K_2, and K_3, respectively. What is the overall formation constant of $Fe(SCN)_3$ from its component ions, and what is the dissociation constant of $Fe(SCN)_3$ into its simplest ions on the basis of these data?

▮

$$K_s = K_1K_2K_3 = 2.1 \times 10^3 \qquad K_d = 1/K_s = 4.8 \times 10^{-4}$$

22.83 Calculate the formation constant for the reaction of a tripositive metal ion with thiocyanate ion to form the mono complex if the total metal concentration in the solution is 2.00×10^{-3} M, the total SCN^- concentration is 1.50×10^{-3} M, and the free SCN^- concentration is 1.0×10^{-5} M.

▮

$$M^{3+} + SCN^- \rightleftharpoons MSCN^{2+}$$

The bound SCN^- concentration is $(1.50 \times 10^{-3}\ M) - (1.0 \times 10^{-5}\ M) = 1.49 \times 10^{-3}$ M. That is also the concentration of the $MSCN^{2+}$ ion and hence the bound M^{3+} concentration. The free M^{3+} concentration is therefore $(2.00 \times 10^{-3}\ M) - (1.49 \times 10^{-3}\ M) = 0.51 \times 10^{-3}$ M.

$$K = \frac{[MSCN^{2+}]}{[M^{3+}][SCN^-]} = \frac{1.49 \times 10^{-3}}{(0.51 \times 10^{-3})(1.0 \times 10^{-5})} = 2.9 \times 10^5$$

22.84 How much AgBr could dissolve in 1.0 L of 0.40 M NH_3? Assume that $Ag(NH_3)_2^+$ is the only complex formed.

▮

$$AgBr \rightleftharpoons Ag^+ + Br^- \qquad Ag^+ + 2NH_3 \rightleftharpoons Ag(NH_3)_2^+$$

Let x = solubility. Then $x = [Br^-] = [Ag^+] + [Ag(NH_3)_2^+]$.

$$\frac{[Ag(NH_3)_2^+]}{[Ag^+][NH_3]^2} = 1.0 \times 10^8$$

Because $[NH_3] \gg [Br^-]$, $\dfrac{[Ag(NH_3)_2^+]}{[Ag^+](0.40)^2} = 1.0 \times 10^8$

Since the overwhelming majority of the silver is in the form of the complex, $x = [Br^-] = [Ag(NH_3)_2^+]$ and

$$K_{sp} = [Ag][Br^-] = \left(\frac{[Ag(NH_3)_2^+]}{1.6 \times 10^7}\right)[Br^-] = \frac{x^2}{1.6 \times 10^7} = 5.0 \times 10^{-13}$$

$$x^2 = 8.0 \times 10^{-6} \qquad x = 2.8 \times 10^{-3}\ M$$

22.85 Silver ion forms $Ag(CN)_2^-$ in the presence of excess CN^-. How much KCN should be added to 1.0 L of a 0.00050 M Ag^+ solution in order to reduce $[Ag^+]$ to 1.0×10^{-19}?

$$\mathrm{Ag(CN)_2^- \rightleftharpoons Ag^+ + 2CN^-} \qquad K = \frac{[\mathrm{Ag^+}][\mathrm{CN^-}]^2}{[\mathrm{Ag(CN)_2^-}]} = 1.0 \times 10^{-21}$$

	initial	produced	used up	equilibrium
$[Ag^+]$	5.0×10^{-4}		5.0×10^{-4}	1.0×10^{-19}
$[CN^-]$	x		1.0×10^{-3}	$x - 0.0010$
$[Ag(CN)_2^-]$	0	5.0×10^{-4}		5.0×10^{-4}

$$K = \frac{(1.0 \times 10^{-19})[\mathrm{CN^-}]^2}{5.0 \times 10^{-4}} = 1.0 \times 10^{-21}$$

$$[\mathrm{CN^-}] = 2.2 \times 10^{-3} = x - 0.0010 \qquad x = 3.2 \times 10^{-3} = \text{original concentration of } \mathrm{CN^-}$$

For 1.0 L, 3.2×10^{-3} mol should be added, of which 1.0×10^{-3} mol will coordinate.

22.86 What is the $[Cd^{2+}]$ in 1.0 L of solution prepared by dissolving 0.0010 mol $Cd(NO_3)_2$ and 1.5 mol NH_3? K_d for the dissociation of $Cd(NH_3)_4^{2+}$ into Cd^{2+} and $4NH_3$ is 1.8×10^{-7}. Neglect the amount of cadmium in complexes containing fewer than 4 ammonia molecules.

▌

$$\mathrm{Cd(NH_3)_4^{2+} \rightleftharpoons Cd^{2+} + 4NH_3} \qquad K_d = \frac{[\mathrm{Cd^{2+}}][\mathrm{NH_3}]^4}{[\mathrm{Cd(NH_3)_4^{2+}}]} = 1.8 \times 10^{-7}$$

	initial	produced	used up	equilibrium
$[Cd^{2+}]$	0.0010		$0.0010 - x$	x
$[NH_3]$	1.5		$4(0.0010 - x)$	≈ 1.5
$[Cd(NH_3)_4^{2+}]$	0	$0.0010 - x$		$0.0010 - x \approx 0.0010$

$$K_d = \frac{(x)(1.5)^4}{0.0010} = 1.8 \times 10^{-7} \qquad x = 3.6 \times 10^{-11} = [\mathrm{Cd^{2+}}]$$

22.87 A 0.0010 mol sample of solid NaCl was added to 1.0 L of 0.010 M $Hg(NO_3)_2$ at a total ionic strength of 0.5. Calculate the $[Cl^-]$ equilibrated with the newly formed $HgCl^+$. K_1 for $HgCl^+$ formation at this ionic strength is 5.5×10^6. Neglect the K_2 equilibrium.

▌

$$\mathrm{Hg^{2+} + Cl^- \longrightarrow HgCl^+} \qquad K_1 = \frac{[\mathrm{HgCl^+}]}{[\mathrm{Hg^{2+}}][\mathrm{Cl^-}]} = 5.5 \times 10^6$$

Since the equilibrium constant is so large, very little of one (or both) reactant(s) will remain at equilibrium.

	initial	produced	used up	equilibrium
$[Hg^{2+}]$	0.010		$0.0010 - x$	$0.0090 + x \approx 0.0090$
$[Cl^-]$	0.0010		$0.0010 - x$	x
$[HgCl^+]$	0	$0.0010 - x$		$0.0010 - x \approx 0.0010$

$$K_1 = \frac{0.0010}{(0.0090)x} = 5.5 \times 10^6 \qquad x = 2.0 \times 10^{-8}\ \mathrm{M}$$

22.88 One liter of solution was prepared containing 0.0050 mol of silver in the +1 oxidation state and 1.00 mol NH_3. What is the concentration of free Ag^+ in the solution at equilibrium?

▌ Most of the silver, approximately 0.0050 mol, will be in the form of the complex ion $Ag(NH_3)_2^+$. The concentration of free NH_3 at equilibrium is practically unchanged from 1.00 mol/L, since only 0.0100 mol of NH_3 would be used up to form 0.0050 mol of complex.

$$\mathrm{Ag(NH_3)_2^+ \rightleftharpoons Ag^+ + 2NH_3} \qquad K_d = \frac{[\mathrm{Ag^+}][\mathrm{NH_3}]^2}{[\mathrm{Ag(NH_3)_2^+}]} \quad \text{or} \quad 1.0 \times 10^{-8} = \frac{[\mathrm{Ag^+}](1.00)^2}{0.0050}$$

Solving, $[Ag^+] = 5.0 \times 10^{-11}$; i.e., an equilibrium concentration of 5.0×10^{-11} mol/L.

22.89 K_1 for the complexation of one NH_3 with Ag^+ is 2.0×10^3. ***(a)*** With reference to Problem 22.88, what is the concentration of $Ag(NH_3)^+$? ***(b)*** What is K_2 for this system?

(*a*) K_1 refers to the following: $Ag^+ + NH_3 \rightleftharpoons Ag(NH_3)^+ \qquad K_1 = \frac{[Ag(NH_3)^+]}{[Ag^+][NH_3]}$

Then, from Problem 22.88,

$$[Ag(NH_3)^+] = K_1([Ag^+][NH_3]) = (2.0 \times 10^3)(1.0 \times 10^{-10})(1.00) = 2.0 \times 10^{-7}$$

This problem is actually a check on an assumption made in Problem 22.88; namely, that practically all the dissolved silver was in the complex $Ag(NH_3)_2^+$. If $[Ag(NH_3)^+]$ had turned out to be greater than about 5×10^{-4}, the assumption would have been shown to be incorrect.

(*b*) K_1, K_2, and K_d are interrelated.

$$K_2 = \frac{[Ag(NH_3)_2^+]}{[Ag(NH_3)^+][NH_3]} = \frac{[Ag(NH_3)_2^+]/[Ag^+][NH_3]^2}{[Ag(NH_3)^+]/[Ag^+][NH_3]} = \frac{1/K_d}{K_1} = \frac{1}{K_1K_d} = \frac{1}{(2.0 \times 10^3)(1.0 \times 10^{-8})} = 5.0 \times 10^4$$

22.90 How much NH_3 should be added to a solution of 0.0010 M $Cu(NO_3)_2$ to reduce $[Cu^{2+}]$ to 10^{-13}? Neglect the amount of copper in complexes containing fewer than 4 ammonia molecules per copper atom.

$$Cu(NH_3)_4^{2+} \rightleftharpoons Cu^{2+} + 4NH_3 \qquad K_d = \frac{[Cu^{2+}][NH_3]^4}{[Cu(NH_3)_4^{2+}]} = 1.0 \times 10^{-12}$$

Since the sum of the concentrations of copper in the complex and in the free ionic state must equal 0.0010 mol/L, and since the amount of the free ion is very small, the concentration of the complex is taken to be 0.0010 mol/L.

$$\text{Let } x = [NH_3]. \text{ Then } \frac{(10^{-13})(x^4)}{0.0010} = 1.0 \times 10^{-12} \quad \text{or} \quad x^4 = 1.0 \times 10^{-2} \quad \text{or} \quad x = 0.32$$

The concentration of NH_3 at equilibrium is 0.32 mol/L. The amount of NH_3 used up in forming 0.0010 mol/L of complex is 0.0040 mol/L, an amount negligible compared with the amount remaining at equilibrium. Hence the amount of NH_3 to be added is 0.32 mol/L.

22.91 How many mol NH_3 must be added to 1.0 L of 0.750 M $AgNO_3$ in order to reduce the silver ion concentration to 5.0×10^{-8} M?

$$Ag^+ + 2NH_3 \rightleftharpoons Ag(NH_3)_2^+ \qquad K_f = \frac{[Ag(NH_3)_2^+]}{[Ag^+][NH_3]^2} = 1 \times 10^8$$

The complex ion concentration at equilibrium must be equal to $0.750 - (5.0 \times 10^{-8}) = 0.750$ M.

$$\frac{0.750}{(5.0 \times 10^{-8})[NH_3]^2} = 1 \times 10^8 \qquad [NH_3]^2 = 0.15 \qquad [NH_3] = 0.4 \text{ M in solution}$$

As part of the complex ion there is

$$(0.750 \text{ mol } Ag(NH_3)_2^+)\left(\frac{2 \text{ mol } NH_3}{\text{mol } Ag(NH_3)_2^+}\right) = 1.5 \text{ mol } NH_3$$

$$\text{Total } NH_3 \text{ required} = (0.4 \text{ mol}) + (1.5 \text{ mol}) = 1.9 \text{ mol}$$

22.92 Calculate the concentration of Fe^{2+} in a solution prepared by mixing 75.0 mL of 0.030 M $FeSO_4$ with 125.0 mL of 0.20 M KCN.

$$Fe^{2+} + 6CN^- \rightleftharpoons Fe(CN)_6^{4-} \qquad K_f = \frac{[Fe(CN)_6^{4-}]}{[Fe^{2+}][CN^-]^6} = 1 \times 10^{24}$$

Because of the large value of K_f, the iron is essentially all in the form of its complex ion.

$$2.25 \text{ mmol}/200 \text{ mL} = 0.01125 \text{ M } Fe(CN)_6^{4-}$$

Of the 25.0 mmol CN^- initially present, $(6)(2.25 \text{ mmol}) = 13.5$ mmol is complexed, leaving $25.0 - 13.5 = 11.5$ mmol free in solution.

$$11.5 \text{ mmol}/200 \text{ mL} = 0.0575 \text{ M } CN^-$$

$$K_f = \frac{(0.01125)}{x(0.0575)^6} = 1 \times 10^{-24} \qquad x = [Fe^{2+}] = 3 \times 10^{-19} \text{ M}$$

22.93 In the quantitative determination of silver ion as AgCl, a solution of NaCl is used as the precipitating agent. Why should a large excess of that reagent be avoided? (*Hint*: See Table 22.3.)

TABLE 22.3 Overall Formation Constants, K_f, for Selected Complex Ions in Water at 25° C

ammine complexes		cyano complexes		halo complexes	
$[Ag(NH_3)_2]^+$	1×10^8	$[Ag(CN)_2]^-$	1×10^{21}	$[AgCl_2]^-$	3×10^5
$[Cu(NH_3)_4]^{2+}$	1×10^{12}	$[Fe(CN)_6]^{3-}$	1×10^{31}	$[AlF_6]^{3-}$	7×10^{19}
$[Zn(NH_3)_4]^{2+}$	5×10^8	$[Fe(CN)_6]^{4-}$	1×10^{24}	$[CuCl]^+$	1.0
		$[Ni(CN)_4]^{2-}$	1×10^{30}	$[CuCl_4]^{2-}$	4×10^5
hydroxo complexes		$[Zn(CN)_4]^{2-}$	5×10^{16}	$[CuCl_2]^-$	5×10^4
				$[PbCl_4)]^{2-}$	4×10^2
$[Al(OH)_4]^-$	2×10^{28}			$[SnF_6]^{2-}$	1×10^{25}
$[Zn(OH)_4]^{2-}$	5×10^{14}				

▌ Excess Cl^- must be avoided to prevent complex ion formation:

$$AgCl(s) + Cl^- \rightleftharpoons AgCl_2^-(aq)$$

22.94 Calculate the silver ion concentration in a solution of $[Ag(NH_3)_2]^+$ prepared by adding 1.0×10^{-3} mol $AgNO_3$ to 1.0 L of 0.100 M NH_3 solution.

▌

$$Ag^+ + 2NH_3 \rightleftharpoons Ag(NH_3)_2^+ \qquad K_f = \frac{[Ag(NH_3)_2^+]}{[Ag^+][NH_3]^2} = 1 \times 10^8 \qquad \text{(from Table 22.3)}$$

Since K_f is so large, most of the silver will form complex ions. Let $[Ag^+] = x$; then

$$[Ag(NH_3)_2^+] = (1 \times 10^{-3}) - (x) \approx 1 \times 10^{-3} \qquad [NH_3] = 0.100 - 2(1 \times 10^{-3}) + 2x \simeq 0.098$$

$$\frac{1 \times 10^{-3}}{(x)(0.098)^2} = 1 \times 10^8 \qquad x = 1 \times 10^{-9}\ M = [Ag^+]$$

22.95 Calculate the Fe^{2+} concentration in a solution containing 0.200 M $[Fe(CN)_6]^{4-}$ and 0.100 M CN^-.

▌ K_f is found in Table 22.3.

$$Fe^{2+} + 6CN^- \rightleftharpoons Fe(CN)_6^{4-}$$

$$K_f = \frac{[Fe(CN)_6^{4-}]}{[Fe^{2+}][CN^-]^6} = 1 \times 10^{24} = \frac{0.200}{x(0.100)^6} \qquad x = 2 \times 10^{-19}\ M$$

22.96 Calculate the concentration of silver ion which is in equilibrium with 0.15 M $[Ag(NH_3)_2]^+$ and 1.5 M NH_3.

▌ See Problem 22.94.

$$\frac{0.15}{x(1.5)^2} = 1 \times 10^8 \qquad x = 7 \times 10^{-10}\ M$$

22.97 What is the minimum quantity of ammonia which must be added to 1.0 L of solution in order to dissolve 0.10 mol silver chloride by forming $[Ag(NH_3)_2]^+$? $AgCl(s) + 2NH_3 \rightleftharpoons [Ag(NH_3)_2]^+ + Cl^-$

▌ When 0.10 mol AgCl is converted to a complex ion in 1.0 L solution, 0.10 M chloride ion is also formed. The maximum concentration of silver ion which can coexist with that concentration of chloride ion is given by the K_{sp} expression:

$$K_{sp} = [Ag^+][Cl^-] = 1 \times 10^{-10} \qquad [Ag^+] = 1 \times 10^{-9}$$

Therefore the concentration of ammonia in the solution must be sufficient to prevent the silver ion concentration from exceeding 1×10^{-9} M.

$$Ag^+ + 2NH_3 \rightleftharpoons Ag(NH_3)_2^+ \qquad K_f = \frac{[Ag(NH_3)_2^+]}{[Ag^+][NH_3]^2} = 1 \times 10^8$$

Since the complex ion is practically 0.10 M and the silver ion is 1×10^{-9} M, the ammonia concentration can be determined to be

$$[NH_3]^2 = \frac{0.10}{(1 \times 10^{-9})(1 \times 10^8)} = 1.0 \qquad [NH_3] = 1.0\ M$$

Therefore the minimum quantity of ammonia which must be added to 1.0 L solution is at least 1.2 mol: 0.20 mol for the formation of the 0.10 mol of $Ag(NH_3)_2^+$ and 1.0 mol in excess to maintain an ammonia concentration such that the equilibrium silver ion concentration is held at 1×10^{-9} M.

22.98 A certain insoluble compound of M^{2+}, when shaken with water, provides an M^{2+} concentration of 1.0×10^{-4} M. A ligand is added to the system in a quantity which forms a soluble complex with M^{2+} and leaves 1.0×10^{-6} M M^{2+} in solution. Will the insoluble compound tend to dissolve? Explain.

The ligand will tend to dissolve the insoluble compound. Whichever reagent leaves the least amount of common ion in solution is the one which will cause greater reaction.

22.99 Arrange the following solutions in order of decreasing silver ion concentration: ***(a)*** 1 M $[Ag(CN)_2]^-$ ***(b)*** saturated AgCl ***(c)*** 1 M $[Ag(NH_3)_2]^+$ in 0.10 M NH_3 ***(d)*** saturated AgI

From the values of K_{sp} and K_f, it is immediately apparent that AgCl is more soluble than AgI and $Ag(CN)_2^-$ is more stable than $Ag(NH_3)_2^+$. Calculation of $[Ag^+]$ in each case reveals (*b*) > (*c*) > (*d*) > (*a*).

(a) $Ag(CN)_2^- \longrightarrow Ag^+ + 2CN^- \qquad K = \frac{1}{K_f} = \frac{[Ag^+][CN^-]^2}{[Ag(CN)_2^-]} = 1 \times 10^{-21}$

Let $[Ag^+] = x, \qquad [CN^-] = 2x, \qquad [Ag(CN)_2^-] = 1.0 - x \simeq 1.0$

$$K = \frac{(x)(2x)^2}{1.0} = 1 \times 10^{-21} = 4x^3 \qquad x = 6.3 \times 10^{-8}\ M$$

(b) $AgCl \rightleftharpoons Ag^+ + Cl^- \qquad K_{sp} = [Ag^+][Cl^-] = 1 \times 10^{-10}$

$x^2 = 1 \times 10^{-10} \qquad x = 1 \times 10^{-5}$ M

(c) and ***(d)*** are done like (*a*) and (*b*), respectively.

22.100 Calculate the equilibrium concentrations of each of the indicated species necessary to reduce an initial 0.20 M Zn^{2+} solution to 1.0×10^{-4} M Zn^{2+}. In each case, repeat the calculation to determine the concentrations necessary to reduce the 0.20 M Zn^{2+} to 1.0×10^{-10} M Zn^{2+}. ***(a)*** NH_3 and $Zn(NH_3)_4^{2+}$ (assume no partial complexation) ***(b)*** OH^- in equilibrium with $Zn(OH)_2(s)$ ***(c)*** OH and $Zn(OH)_4^{2-}$ ***(d)*** Calculate the OH^- concentration which would be produced by each equilibrium concentration of NH_3 in part (*a*). Using these values, predict whether $Zn(OH)_2$ or $Zn(OH)_4^{2-}$ would form in preference to $Zn(NH_3)_4^{2+}$ upon addition of sufficient ammonia to produce the equilibrium concentrations calculated in part (*a*). ***(e)*** Describe what would be observed if concentrated ammonia solution were added slowly to a 0.20 M solution of Zn^{2+}.

(a) $Zn^{2+} + 4NH_3 \rightleftharpoons Zn(NH_3)_4^{2+} \qquad K_f = \frac{[Zn(NH_3)_4^{2+}]}{[Zn^{2+}][NH_3]^4} = 5 \times 10^8$

For 1×10^{-4} M Zn^{2+}: At equilibrium, $\quad [Zn(NH_3)_4^{2+}] = (0.20) - (1 \times 10^{-4}) = 0.20$

$$K_f = \frac{0.20}{(1.0 \times 10^{-4})[NH_3]^4} = 5 \times 10^8$$

$[NH_3]^4 = 4 \times 10^{-6} = 400 \times 10^{-8} \qquad [NH_3] = 4.5 \times 10^{-2}$ M

(b) $Zn(OH)_2 \longrightarrow Zn^{2+} + 2OH^- \qquad K_{sp} = [Zn^{2+}][OH^-]^2$

$$[OH^-]^2 = \frac{1.8 \times 10^{-14}}{1.0 \times 10^{-4}} = 1.8 \times 10^{-10} \qquad [OH^-] = 1.3 \times 10^{-5}\ M$$

(c) $Zn^{2+} + 4OH^- \longrightarrow Zn(OH)_4^{2-}$

$$\frac{[Zn(OH)_4^{2-}]}{[Zn^{2+}][OH^-]^4} = 5 \times 10^{14} \qquad \frac{0.20}{(1.0 \times 10^{-4})[OH^-]^4} = 5 \times 10^{14}$$

$[OH^-]^4 = 4 \times 10^{-12} \qquad [OH^-] = 1.4 \times 10^{-3}$

(d) The concentration of OH^- in equilibrium with the NH_3 concentration of part (*a*) is given by $NH_3 + H_2O \rightleftharpoons NH_4^+ + OH^-$.

$$K_b = \frac{[NH_4^+][OH^-]}{[NH_3]} = 1.8 \times 10^{-5}$$

Let $x = [NH_4^+] = [OH^-]$.

$$\frac{x^2}{4.5 \times 10^{-2}} = 1.8 \times 10^{-5} \qquad x^2 = 8.1 \times 10^{-7} \qquad x = 9.0 \times 10^{-4} = [OH^-]$$

The hydroxide concentration (9.0×10^{-4}) from the 4.5×10^{-2} M NH_3 is sufficient to precipitate $Zn(OH)_2$ [part (*b*)] but not sufficient to form the $Zn(OH)_4^{2-}$ [part (*c*)].

For the analogous situations in which the $[Zn^{2+}]$ is to be reduced to 1.0×10^{-10} M the answers, obtained in analogous calculations, are (*a*) 1.4 M NH_3 (*b*) 1.3×10^{-2} M OH^- and (*c*) 4.5×10^{-2} M OH^-. The concentration of OH^- needed in (*d*), 5.0×10^{-3} M, is not sufficient to cause precipitation of $Zn(OH)_2$, so the ammonia complex is stable.

(*e*) On addition of NH_3 solution to Zn^{2+} solution, a precipitate of $Zn(OH)_2$ will form, which will later dissolve to yield a clear solution containing $[Zn(NH_3)_4]^{2+}$.

22.101 Explain why 0.10 M NH_3 solution **(*a*)** will precipitate $Fe(OH)_2$ from a 0.10 M solution Fe^{2+} **(*b*)** will not precipitate $Mg(OH)_2$ from a solution which is 0.20 M in NH_4^+ and 0.10 M in Mg^{2+} **(*c*)** will not precipitate AgOH from a solution which is 0.010 M in Ag^+.

▌ **(*a*)** The NH_3 provides sufficient OH^- to exceed K_{sp} for $Fe(OH)_2$. **(*b*)** The NH_4^+ creates a buffer solution, which limits $[OH^-]$ to a level insufficient to cause precipitation of $Mg(OH)_2$. **(*c*)** The NH_3 forms the $Ag(NH_3)_2^+$ complex ion, which is soluble.

22.102 If 0.10 M Ag^+ and 0.10 M H_3O^+ are present in 1.0 L of solution to which 0.010 mol of gaseous NH_3 is added, with which cation will the ammonia react to a greater extent? Support your answer by calculation.

▌

$$Ag^+ + 2NH_3 \rightleftharpoons Ag(NH_3)_2^+ \qquad K_f = \frac{[Ag(NH_3)_2^+]}{[Ag^+][NH_3]^2} = 1 \times 10^8$$

$$H_3O^+ + NH_3 \rightleftharpoons NH_4^+ + H_2O \qquad K = \frac{K_b}{K_w} = \frac{1.8 \times 10^{-5}}{1.0 \times 10^{-14}} = 1.8 \times 10^9$$

Assuming that all the ammonia reacts with Ag^+ yields

$$[Ag(NH_3)_2^+] = 0.0050 \text{ M} \qquad [Ag^+] = 0.095 \text{ M} \qquad [NH_3] = x$$

$$\frac{0.0050}{(0.095)x^2} = 1 \times 10^8 \qquad x = 2.3 \times 10^{-5} \text{ M}$$

In contrast, assuming that all the ammonia reacts with H_3O^+ yields

$$K = \frac{[NH_4^+]}{[H_3O^+][NH_3]} = \frac{0.010}{(0.090)x} = 1.8 \times 10^9 \qquad x = 6.2 \times 10^{-11} \text{ M}$$

Since the H_3O^+ leaves a lower NH_3 concentration, that reaction will predominate.

22.103 How much solid $Na_2S_2O_3$ should be added to 1.0 L of water so that 0.00050 mol $Cd(OH)_2$ could just barely dissolve? K_1 and K_2 for $S_2O_3^{2-}$ complexation with Cd^{2+} are 8.3×10^3 and 2.5×10^2, respectively. (As part of the problem, determine whether CdS_2O_3 or $Cd(S_2O_3)_2^{2-}$ is the predominant species in solution.)

▌

$$Cd(OH)_2 \rightleftharpoons Cd^{2+} + 2OH^- \qquad K_{sp} = [Cd^{2+}][OH^-]^2 = 4.5 \times 10^{-15}$$

$$Cd^{2+} + S_2O_3^{2-} \rightleftharpoons Cd(S_2O_3) \qquad K_1 = \frac{[Cd(S_2O_3)]}{[Cd^{2+}][S_2O_3^{2-}]} = 8.3 \times 10^3$$

$$Cd(S_2O_3) + S_2O_3^{2-} \rightleftharpoons Cd(S_2O_3)_2^{2-} \qquad K_2 = \frac{[Cd(S_2O_3)_2^{2-}]}{[Cd(S_2O_3)][S_2O_3^{2-}]} = 2.5 \times 10^2$$

Assume that $S_2O_3^{2-}$ does not hydrolyze.

$$[Cd^{2+}] + [Cd(S_2O_3)] + [Cd(S_2O_3)_2^{2-}] = 0.00050$$

$$[Cd^{2+}] + K_1[Cd^{2+}][S_2O_3^{2-}] + K_1K_2[Cd^{2+}][S_2O_3^{2-}]^2 = 0.00050$$

$$[Cd^{2+}] = \frac{K_{sp}}{[OH^-]^2} = \frac{4.5 \times 10^{-15}}{(0.0010)^2} = 4.5 \times 10^{-9} \text{ M}$$

Let $x = [S_2O_3^{2-}]$, then $\qquad 1 + K_1x + K_1K_2x^2 = \dfrac{5.0 \times 10^{-4}}{4.5 \times 10^{-9}} = 1.1 \times 10^5$

Assume that the anionic complex prevails:

$$(2.1 \times 10^6)(x^2) = 1.1 \times 10^5 \qquad x = 0.23\ \text{M}$$

The number of mol $S_2O_3^{2-}$ needed to make the complexes is negligible, hence 0.23 mol $S_2O_3^{2-}$ is needed. *Check:* $1 + (8.3 \times 10^3)(0.23) + (2.1 \times 10^6)(0.23)^2 = 1.1 \times 10^5$ The second term is only 1.9×10^3.

22.104 What is the solubility of AgSCN in 0.0030 M NH_3?

❙ We may assume that practically all the dissolved silver will exist as the complex ion $Ag(NH_3)_2^+$. Then, if the solubility of AgSCN is x mol/L, $x = [SCN^-] = [Ag(NH_3)_2^+]$. The concentration of uncomplexed Ag^+ must be computed from the equilibrium constant for the complex ion dissociation.

$$\frac{[Ag^+][NH_3]^2}{[Ag(NH_3)_2^+]} = K_d \qquad \text{or} \qquad [Ag^+] = \frac{K_d[Ag(NH_3)_2^+]}{[NH_3]^2} = \frac{(1.0 \times 10^{-8})x}{(0.0030)^2} = 1.1 \times 10^{-3}\ x$$

This result validates our assumption: the ratio of uncomplexed to complexed silver in solution is only 1.1×10^{-3}. Then, from $[Ag^+][SCN^-] = K_{sp}$,

$$(1.1 \times 10^{-3}\ x)x = 1.1 \times 10^{-12} \qquad \text{or} \qquad x = 3.2 \times 10^{-5}$$

The validity of another assumption is confirmed by the answer. If 3.2×10^{-5} mol of complex is formed per liter, the amount of NH_3 used up for complex formation is $2(3.2 \times 10^{-5}) = 6.4 \times 10^{-5}$ mol/L. The concentration of the remaining free NH_3 in solution is practically unchanged from its initial value, 0.0030 mol/L.

22.105 How much NH_3 must be added to a 0.0040 M Ag^+ solution to just prevent the precipitation of AgCl when $[Cl^-]$ reaches 0.0010?

❙ Just as acids may be used to lower the concentration of anions in solution, so complexing agents may be used in some cases to lower the concentration of cations. In this problem, the addition of NH_3 converts most of the silver to the complex ion $Ag(NH_3)_2^+$. The upper limit for the uncomplexed $[Ag^+]$ without formation of a precipitate can be calculated from the solubility product.

$$[Ag^+][Cl^-] = 1.0 \times 10^{-10} \qquad \text{or} \qquad [Ag^+] = \frac{1.0 \times 10^{-10}}{[Cl^-]} = \frac{1.0 \times 10^{-10}}{0.0010} = 1.0 \times 10^{-7}$$

Enough NH_3 must be added to keep the $[Ag^+]$ below 1.0×10^{-7}. The concentration of $Ag(NH_3)_2^+$ at this limit would then be $0.0040 - (1.0 \times 10^{-7})$, or 0.0040.

$$\frac{[Ag^+][NH_3]^2}{[Ag(NH_3)_2^+]} = K_d \qquad \text{or} \qquad [NH_3]^2 = \frac{K_d[Ag(NH_3)_2^+]}{[Ag^+]} = \frac{(1.0 \times 10^{-8})(0.0040)}{1.0 \times 10^{-7}} = 4.0 \times 10^{-4}$$

and $[NH_3] = 0.020$. The amount of NH_3 that must be added is equal to the sum of the amount of free NH_3 remaining in the solution and the amount of NH_3 used up in forming 0.0040 mol/L of the complex ion, $Ag(NH_3)_2^+$. This sum is $0.020 + 2(0.0040) = 0.028$ mol NH_3 to be added per liter.

22.106 Calculate the concentrations of silver ion, bromide ion, chloride ion, $[Ag(NH_3)_2]^+$ ion, ammonium ion, and hydroxide ion in a solution which results from shaking excess AgCl and AgBr with 0.0200 M ammonia solution. Assume that no monoammine complex is formed.

❙

$$AgCl \rightleftharpoons Ag^+ + Cl^- \qquad K_{sp} = [Ag^+][Cl^-] = 1 \times 10^{-10}$$

$$AgBr \rightleftharpoons Ag^+ + Br^- \qquad K_{sp} = [Ag^+][Br^-] = 5 \times 10^{-13}$$

$$Ag^+ + 2NH_3 \rightleftharpoons Ag(NH_3)_2^+ \qquad K_f = \frac{[Ag(NH_3)_2^+]}{[Ag^+][NH_3]^2} = 1 \times 10^8$$

$$NH_3 + H_2O \rightleftharpoons NH_4^+ + OH^- \qquad K_b = \frac{[NH_4^+][OH^-]}{[NH_3]} = 1.8 \times 10^{-5}$$

K_{sp} for AgBr is 200 times smaller than K_{sp} for AgCl, and the second equation will not contribute significant concentrations of Ag^+ to the solution. The base ionization of NH_3 does not reduce the NH_3 concentration much. Hence, solving for the silver ion concentration from the first and third equations,

$$[Ag^+] = \frac{K_{sp}}{[Cl^-]} = \frac{[Ag(NH_3)_2^+]}{K_f[NH_3]^2}$$

Since most of the silver ion which dissolves must be transformed into the complex, and since each silver ion dissolved also necessitates a chloride ion dissolved,

$$[Cl^-] \approx [Ag(NH_3)_2^+] = c \qquad \frac{K_{sp}}{c} = \frac{c}{K_f[NH_3]^2} \qquad \text{Rearranging: } c^2 = K_{sp}K_f[NH_3]^2$$

The ammonia concentration is approximately $0.0200 - 2c$. Hence

$$c = (\sqrt{K_{sp}K_f})(0.0200 - 2c) = (10^{-1})(0.0200 - 2c) = (2 \times 10^{-3}) - (0.2c)$$

$$1.2c = 2 \times 10^{-3} \qquad c = 1.7 \times 10^{-3} = [Ag(NH_3)_2^+] = [Cl^-]$$

$$[NH_3] = 0.0200 - 2c = 0.0200 - (2)(1.7 \times 10^{-3}) = 1.7 \times 10^{-2}$$

$$[Ag^+] = \underset{(\text{from } K_{sp})}{\frac{1 \times 10^{-10}}{1.7 \times 10^{-3}}} = 6 \times 10^{-8} = \underset{(\text{from } K_f)}{\frac{1.7 \times 10^{-3}}{(1 \times 10^8)(1.7 \times 10^{-2})^2}}$$

$$[Br^-] = \frac{5 \times 10^{-13}}{6 \times 10^{-8}} = 8 \times 10^{-6} \qquad (\text{yielding negligible } Ag(NH_3)_2^+)$$

$$K_b = \frac{[NH_4^+][OH^-]}{[NH_3]} = 1.8 \times 10^{-5} = \frac{x^2}{1.7 \times 10^{-2}}$$

$$x = \sqrt{(1.8 \times 10^{-5})(1.7 \times 10^{-2})} = 6 \times 10^{-4} = [NH_4^+] = [OH^-]$$

The quantity of NH_3 ionized reduces the NH_3 present by about 3.5%, a negligible quantity.

22.107 The stability constants for complexation of Cl^- with Fe^{3+} are 30, 4.5, and 0.10 for K_1, K_2, and K_3, respectively. What is the principal iron-containing species in 0.0100 M $FeCl_3$? Assume that the solution has been mildly acidified to prevent hydrolysis.

▮ There does not exist in this problem a large excess of the complexing agent which would drive the complexation reactions to completion. We may start by imagining that the salt is completely ionized, giving 0.0100 M Fe^{3+} and 0.0300 M Cl^-. Then we assume that only the first stage of complexation is significant.

	Fe^{3+}	+ Cl^-	$\rightleftharpoons FeCl^{2+}$
Conc. if there is no complex	0.0100	0.0300	0.000
Change by complexation	$-x$	$-x$	$+x$
Conc. at equilibrium	$0.0100 - x$	$0.0300 - x$	x

$$\frac{x}{(0.0100 - x)(0.0300 - x)} = K_1 = 30$$

Solving, $x = 0.0043 = [FeCl^{2+}]$, $0.0300 - x = 0.0257 = [Cl^-]$, $0.0100 - x = 0.0057 = [Fe^{3+}]$.
Check of assumption: We must now show that the complexation of a second Cl^- is not extensive enough to modify the above values significantly. We substitute these values into the K_2 equation.

	$FeCl^{2+}$	+ Cl^-	$\rightleftharpoons FeCl_2^+$
Conc. without a 2nd complexation	0.0043	0.0257	0
Change by 2nd complexation	$-y$	$-y$	$+y$
Conc. at equilibrium	$0.0043 - y$	$0.0257 - y$	y

$$\frac{y}{(0.0043 - y)(0.0257 - y)} = K_2 = 4.5 \qquad y = 0.0004 = [FeCl_2^+]$$

This value for $[FeCl_2^+]$ is just barely less than 10% of the previous value for $[FeCl^{2+}]$. Within our allowed error, then, we need not go back to the K_1 equation to correct the extent of the first complexation, in which we had assumed that $[Fe^{3+}] + [FeCl^{2+}] = 0.0100$. (A simultaneous treatment of the first two stages of complexation would give the following values: $[Fe^{3+}] = 0.0054$, $[FeCl^{2+}] = 0.0041$, $[FeCl_2^+] = 0.00046$, $[Cl^-] = 0.0250$. Usually it is not worthwhile to make such a small correction.)

We have thus proved that most of the iron is in the form of Fe^{3+}, almost as much is in $FeCl^{2+}$, and an order of magnitude less is in $FeCl_2^+$. It can be shown similarly that the amount in un-ionized $FeCl_3$ is still less.

22.108 Write stepwise equations for the formation of tetrahydroxodiaquochromate(III) ion, $[Cr(H_2O)_2(OH)_4]^-$, by addition of OH^- to $[Cr(H_2O)_6]^{3+}$ in aqueous solution.

▮

$$Cr(H_2O)_6^{3+} + OH^- \rightleftharpoons [Cr(H_2O)_5(OH)]^{2+}$$
$$[Cr(H_2O)_5(OH)]^{2+} + OH^- \rightleftharpoons [Cr(H_2O)_4(OH)_2]^+$$
$$[Cr(H_2O)_4(OH)_2]^+ + OH^- \rightleftharpoons [Cr(H_2O)_3(OH)_3](s)$$
$$[Cr(H_2O)_3(OH)_3] + OH^- \rightleftharpoons [Cr(H_2O)_2(OH)_4]^-$$

22.109 From the data of Table 22.3, calculate the concentration of each of the ions produced when 1.5 mol of $CuCl_2 \cdot 2H_2O$ is dissolved in enough water to make 1.0 L of solution.

▮

$$Cu^{2+} + Cl^- \rightleftharpoons CuCl^+ \qquad K_f = \frac{[CuCl^+]}{[Cu^{2+}][Cl^-]} = 1.0$$

	initial	produced	used up	equilibrium
$[Cu^{2+}]$	1.5		x	$1.5 - x$
$[Cl^-]$	3.0		x	$3.0 - x$
$[CuCl^+]$	0.0	x		x

$$K_f = \frac{x}{(1.5-x)(3.0-x)} = 1.0 \qquad x = 4.5 - 4.5x + x^2 \qquad 0 = x^2 - 5.5x + 4.5$$

$$x = \frac{5.5 \pm \sqrt{(5.5)^2 - 4(4.5)}}{2} \quad \text{(quadratic formula)} \qquad x = 1.0 \text{ or } \cancel{4.5}$$

$$[CuCl^+] = 1.0\ M \qquad [Cu^{2+}] = 0.5\ M \qquad [Cl^-] = 2.0\ M$$

22.4 MISCELLANEOUS APPLICATIONS OF EQUILIBRIA

22.110 Using appropriate data, determine the entropy of the OH^- ion at 298 K and unit activity.

▮ The standard enthalpy change (Tables 18.2 and 18.3) for the reaction $H_2O \rightleftharpoons H^+ + OH^-$ is 13.4 kcal. The equilibrium constant for this reaction is $K_w = 1.0 \times 10^{-14}$. Hence ΔG° and ΔS° can be calculated:

$$\Delta G^\circ = -2.30RT \log K = 19.1 \text{ kcal/mol} \qquad \Delta S^\circ = \frac{\Delta H^\circ - \Delta G^\circ}{T} = \frac{(13.4 - 19.1)(10^3)}{298} = -19.2 \text{ cal/mol}\cdot\text{K}$$

But
$$\Delta S^\circ = S^\circ(H^+) + S^\circ(OH^-) - S^\circ(H_2O)$$

From Table 18.5, $S^\circ(H_2O) = 70.0$ J/mol·K = 16.7 cal/mol·K

$$-19.2 = 0.0 + S^\circ(OH^-) - 16.7 \qquad S^\circ(OH^-) = -2.5 \text{ cal/mol}\cdot\text{K}$$

22.111 Using data from Tables 18.2, 18.3, 18.5, and 18.6, estimate K_{sp} for AgCl and compare the value obtained with the tabulated value, 1×10^{-10}.

▮ The process is represented by $AgCl(s) \rightleftharpoons Ag^+ + Cl^-$.

$$\Delta H^\circ = \Delta H_f^\circ(Ag^+) + \Delta H_f^\circ(Cl^-) - \Delta H_f^\circ(AgCl) = 25.3 + (-40.0) - (-30.36) = 15.7 \text{ kcal/mol}$$
$$\Delta S^\circ = S^\circ(Ag^+) + S^\circ(Cl^-) - S^\circ(AgCl) = 17.7 + 13.2 - 23.0 = 7.9 \text{ cal/mol}\cdot\text{K}$$
$$\Delta G^\circ = \Delta H^\circ - T\,\Delta S^\circ = 15\,700 - (298)(7.9) = 13\,350 \text{ cal}$$
$$-\log K = \frac{\Delta G^\circ}{2.30RT} = \frac{13\,350}{(2.30)(1.99)(298)} = 9.79 \qquad K = 10^{-9.79} = 1.6 \times 10^{-10}$$

For this type of data, the agreement between the values determined by two different methods is acceptable. The order of magnitude of the number is established, but its precision is poor. It is apparent from this result why many equilibrium constants are given to only one significant figure.

22.112 One liter of a certain solution contained 0.15 mol Cu^{2+} and 0.15 mol Fe^{2+}. The solution was treated with H_2S until no further change occurred and the solution was saturated. Calculate the concentration of Cu^{2+} and Fe^{2+} in the resulting solution.

▮ When H_2S reacts with Cu^{2+}, it generates H_3O^+. In H_3O^+, the Fe^{2+} does not react with H_2S, as deduced from the qualitative analysis scheme. Hence, the solution is the same as that in Problem 22.64.

$$[Cu^{2+}] = 8 \times 10^{-15}\ M \qquad [Fe^{2+}] = 0.15\ M$$

22.113 Using CO_2, NH_3, NH_4NO_3, and K_2CrO_4 as the only reagents, devise a qualitative analysis scheme for separating and identifying the following ions, which might all be present in the same mixture: Ba^{2+}, Ca^{2+}, Mg^{2+}, Na^+, Pb^{2+}. Assume that each cation present is 0.10 M. State the conditions of pH and the reagent concentrations which are required in each step.

$$\left.\begin{matrix}Ba^{2+}\\ Ca^{2+}\\ Mg^{2+}\\ Na^{+}\\ Pb^{2+}\end{matrix}\right\} \xrightarrow{NH_3 + NH_4^+} \begin{cases}\left.\begin{matrix}Ba^{2+}\\ Ca^{2+}\\ Mg^{2+}\\ Na^{+}\end{matrix}\right\} \xrightarrow{NH_3} \begin{cases}\left.\begin{matrix}Ba^{2+}\\ Ca^{2+}\\ Na^{+}\end{matrix}\right\} \xrightarrow[NH_3]{CO_2} \begin{cases}BaCO_3(s)\\ CaCO_3(s)\\ Na^+\end{cases}\\ Mg(OH)_2(s)\end{cases}\\ Pb(OH)_2(s)\end{cases}$$

$$\left.\begin{matrix}BaCO_3\\ CaCO_3\end{matrix}\right\} \xrightarrow{\text{acidify}} \xrightarrow{CrO_4^{2-}} \begin{matrix}BaCrO_4(s)\\ Ca^{2+}(aq)\end{matrix}$$

The Pb^{2+} ion is identified by acidification followed by treatment with chromate ion, upon which it forms a yellow solid precipitate. The Na^+ is identified by the intense yellow flame it yields.

22.114 In the standard qualitative analysis scheme, after removal of the group I cations, the group II cations are separated from the remaining groups by precipitation as sulfides from acid solution. Would these cations precipitate as sulfides from neutral solution? from basic solution? Explain why acid solution is used.

Except for those which form soluble sulfide complexes, the cations would precipitate from neutral or basic solution. The main reason that acid solution is used is to prevent precipitation of group III cations.

22.115 Assuming that the only source of periodic group IIA metals is an equimolar mixture of NaCl, $BaCl_2$, and $MgCl_2$, suggest ways of preparing pure samples of **(a)** $MgSO_4$ **(b)** Ba metal **(c)** $Ba(C_2H_3O_2)_2$.

One possible method is given for each part. **(a)** Add NaOH, precipitating $Mg(OH)_2$. The Na^+ and Ba^{2+} remain in solution. After filtering the $Mg(OH)_2$, add H_2SO_4 to it, yielding $MgSO_4$ in solution. Evaporate most of the water, causing $MgSO_4$ to crystallize. **(b)** Add Na_2CO_3 to the solution remaining from part (*a*), precipitating $BaCO_3$. Add HCl.

$$BaCO_3 + 2HCl \longrightarrow BaCl_2 + H_2O + CO_2$$

Evaporate, melt the solid $BaCl_2$, and electrolyze. **(c)** To the $BaCO_3$ obtained in part (*b*), add $HC_2H_3O_2$, and evaporate.

$$2HC_2H_3O_2 + BaCO_3 \longrightarrow Ba(C_2H_3O_2)_2 + H_2O + CO_2$$

22.116 Aluminum sulfide reacts with water to form hydrogen sulfide and a white precipitate. Identify the latter substance and write equations accounting for its formation.

$$Al_2S_3 + 6H_2O \longrightarrow 2Al(OH)_3 + 3H_2S \qquad Al(OH)_3 \text{ is the white precipitate.}$$

22.117 Explain the following observations: **(a)** When a sample of H_2SO_4 is treated with an equal number of mol of a base such as NaOH, the entire sample of acid is half-neutralized. **(b)** When a sample of $Ba(OH)_2$ is treated with an equal number of mol of an acid such as HCl, half of the sample of base is completely neutralized. **(c)** When a given number of mol of H_2SO_4 is treated with half the number of mol $Ba(OH)_2$, half of the acid is completely neutralized [and the other half does not react at all, in contrast to the situation in part (*a*)].

(a) HSO_4^- is a weak base ($K_2 = 1.2 \times 10^{-2}$). **(b)** $Ba(OH)_2$ is a strong base, which exists in solution entirely as Ba^{2+} and OH^- ions. **(c)** $BaSO_4$ precipitates, eliminating the possibility that the neutralization could stop at HSO_4^-.

22.118 When a solution of Zn^{2+} was added to a solution of NaOH, a clear solution was obtained. When NH_4Cl was added to the clear solution, $Zn(OH)_2$ precipitated. Using balanced chemical equations, explain these observations.

$$Zn^{2+} + 4OH^- \rightleftharpoons Zn(OH)_4^{2-}(aq) \qquad Zn(OH)_4^{2-} + 2NH_4^+ \rightleftharpoons Zn(OH)_2(s) + 2NH_3 + 2H_2O$$

The acid NH_4^+ reacts with $Zn(OH)_4^{2-}$ to cause $Zn(OH)_2$ to precipitate.

22.119 Given the reagents NH_3, NaOH, HCl, and H_2S, which one could be used to separate the ions in each of the following mixtures? **(a)** Cu^{2+} and Zn^{2+} **(b)** Cu^{2+} and Al^{3+} **(c)** Zn^{2+} and Al^{3+}

(a) NaOH in excess [precipitates $Cu(OH)_2$ but forms soluble $Zn(OH)_4^{2-}$] or H_2S in HCl (precipitates CuS but not ZnS). **(b)** NH_3 [forming $Al(OH)_3(s)$ and $Cu(NH_3)_4^{2+}(aq)$], NaOH excess [forms $Al(OH)_4^-(aq)$ and $Cu(OH)_2(s)$], or H_2S + HCl (forms CuS but does not react with Al^{3+}). **(c)** NH_3 [forms $Al(OH)_3(s)$ and $Zn(NH_3)_4^{2+}(aq)$].

22.120 Complete the following equations:

(*a*) $Al(OH)_4^- + H_3O^+ \longrightarrow$ (*b*) $Al(OH)_4^- + 4H_3O^+ \longrightarrow$ (*c*) $Ag(NH_3)_2^+ + Cl^- + 2H_3O^+ \longrightarrow$

▮ (*a*) $Al(OH)_4^- + H_3O^+ \rightleftharpoons Al(OH)_3(s) + 2H_2O$ (*b*) $Al(OH)_4^- + 4H_3O^+ \rightleftharpoons Al^{3+} + 8H_2O$

(*c*) $Ag(NH_3)_2^+ + Cl^- + 2H_3O^+ \rightleftharpoons 2NH_4^+ + AgCl + 2H_2O$

22.121 Using appropriate thermodynamic data, compare $\Delta H°$, $\Delta S°$, and $\Delta G°$ for the precipitations of $CaCO_3$ and $BaCO_3$ from aqueous solution at 25 °C. Show the contributions of $\Delta H°$ and $T\,\Delta S°$ to $\Delta G°$ by plotting the results on an energy diagram (analogous to the enthalpy diagrams used in Chapter 18). Discuss the factors which make the solubilities of $CaCO_3$ and $BaCO_3$ so similar. Suggest an explanation of why the enthalpy change is greater for the precipitation of $CaCO_3$.

▮

$$Ca^{2+} + CO_3^{2-} \xrightarrow{1} CaCO_3 \qquad Ba^{2+} + CO_3^{2-} \xrightarrow{2} BaCO_3$$

$$\Delta H_1 = \Delta H_f^\circ(CaCO_3) - \Delta H_f^\circ(Ca^{2+}) - \Delta H_f^\circ(CO_3^{2-}) = (-288.5) - (-129.77) - (-161.63) = 2.9 \text{ kcal}$$

(from Tables 18.2 and 18.3)

$$\Delta H_2 = \Delta H_f^\circ(BaCO_3) - \Delta H_f^\circ(Ba^{2+}) - \Delta H_f^\circ(CO_3^{2-}) = (-290.8) - (-128.67) - (-161.63) = -0.5 \text{ kcal}$$

$$\Delta G = (-2.30RT)[\log(1/K_{sp})] = (+2.30RT)(\log K_{sp})$$

$$\Delta G_1^\circ = (+2.30)(1.99)(298)[\log(1 \times 10^{-8})] = -10.9 \text{ kcal}$$

$$\Delta G_2^\circ = (+2.30)(1.99)(298)[\log(5 \times 10^{-9})] = -11.3 \text{ kcal}$$

$$\Delta G = \Delta H - T\,\Delta S \qquad \Delta S = (\Delta H - \Delta G)/T$$

$$\Delta S_1 = [2900 - (-10\,900)]/298 = 46.3 \text{ cal/mol}\cdot\text{K}$$

$$\Delta S_2 = [-500 - (-11\,300)]/298 = 36.2 \text{ cal/mol}\cdot\text{K}$$

Figure 22.1 is the energy diagram. The values of ΔG are comparable despite the fact that ΔH values are quite different, because the $T\,\Delta S$ values are about equally different.

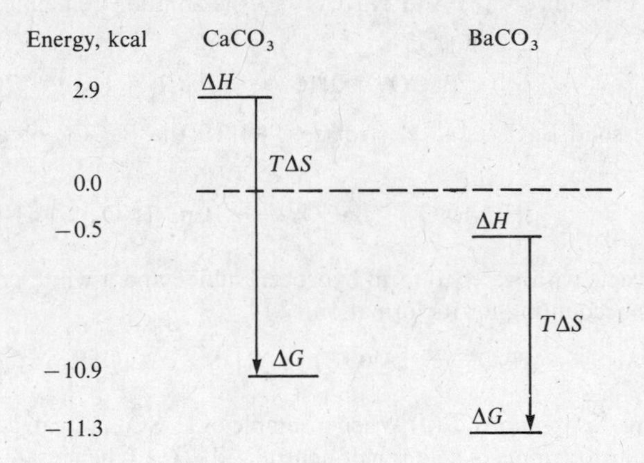

Figure 22.1

22.122 Calculate the enthalpy of solution, ΔH_{298}°, of AB, which dissolves as follows: $AB(s) \rightleftharpoons A^+(aq) + B^-(aq)$ given that $K_{sp} = 8 \times 10^{-12}$ and that $\Delta S° = 25.3$ cal/mol·K.

▮

$$\Delta G° = (-2.30RT)(\log K) = (-2.30)(1.99 \text{ cal/mol}\cdot\text{K})(298 \text{ K})[\log(8 \times 10^{-12})] = +15 \text{ kcal/mol}$$

$$\Delta H° = \Delta G° + T\,\Delta S = (15\,000 \text{ cal/mol}) + (25.3 \text{ cal/mol}\cdot\text{K})(298 \text{ K}) = +22.5 \text{ kcal/mol}$$

22.123 Using thermodynamic data from Tables 18.2, 18.3, 18.5, and 18.6, estimate the K_{sp} of AgBr.

▮ $AgBr(s) \rightleftharpoons Ag^+ + Br^-$. To determine K_{sp}, one can calculate $\Delta G°$ from $\Delta H°$ and $S°$ data. Per mole:

$$\Delta H° = \Delta H_f^\circ(Ag^+) + \Delta H_f^\circ(Br^-) - \Delta H_f^\circ(AgBr) = (25.31 \text{ kcal}) + (-28.9 \text{ kcal}) - (-23.8 \text{ kcal}) = 20.2 \text{ kcal}$$

$$\Delta S° = S°(Ag^+) + S°(Br^-) - S°(AgBr) = (17.7 \text{ cal/K}) + (19.3 \text{ cal/K}) - (25.6 \text{ cal/K}) = 11.4 \text{ cal/K}$$

$$\Delta G° = \Delta H° - T\,\Delta S° = (20\,200 \text{ cal}) - (298 \text{ K})(11.4 \text{ cal/K}) = 16\,800 \text{ cal}$$

$$\Delta G° = (-2.30RT)(\log K)$$

$$\log K = \frac{-\Delta G°}{2.30RT} = \frac{-16\,800 \text{ cal}}{(2.30)(1.99 \text{ cal/K})(298 \text{ K})} = -12.3 \qquad K = 4.8 \times 10^{-13} = K_{sp}$$

22.124 Calculate $\Delta H°$ for the reaction $Ag^+(aq) + Br^-(aq) \rightleftharpoons AgBr(s)$ **(*a*)** directly from the data of Tables 18.2 and 18.3 and **(*b*)** using data of Tables 18.5 and 18.6. **(*c*)** Compare the results.

▮ **(*a*)** $\Delta H° = \Delta H_f°(AgBr) - \Delta H_f°(Ag^+) - \Delta H_f°(Br^-) = (-23.8 \text{ kcal}) - (25.31 \text{ kcal}) - (-28.9 \text{ kcal})$
$= -20.2 \text{ kcal}$

(*b*) $\Delta G° = (-2.30RT)[\log(1/K_{sp})] = (+2.30)(1.99)(298)[\log(5 \times 10^{-13})] = -16.8 \text{ kcal}$

(K_{sp} refers to dissolving, the opposite of this reaction.)

$\Delta S° = S°(AgBr) - S°(Ag^+) - S°(Br^-)$ (from Tables 18.5 and 18.6)
$= (25.6 \text{ cal/K}) - (17.7 \text{ cal/K}) - (19.3 \text{ cal/K}) = -11.4 \text{ cal/K}$

$\Delta H = \Delta G + T\Delta S = (-16\,800 \text{ cal}) + (298 \text{ K})(-11.4 \text{ cal/K}) = -20.2 \text{ kcal}$

(*c*) The results are the same.

CHAPTER 23
Electrochemistry

Note: Before working the problems of this chapter, you should familiarize yourself with Table 23.1, p. 518.

23.1 ELECTRICAL UNITS

23.1 Express each of the following combinations of electrical units as a single unit: **(*a*)** volt·ampere **(*b*)** ampere·second **(*c*)** volt/ampere **(*d*)** joule/volt **(*e*)** watt/ampere·ohm **(*f*)** joule/second **(*g*)** joule/ampere·second **(*h*)** joule/ampere²·second

I (*a*) watt (*b*) coulomb (*c*) ohm (*d*) coulomb (*e*) ampere (*f*) watt (*g*) volt (*h*) ohm

23.2 How many kJ of energy is expended when a current of 1.00 A passes for 100 s under a potential of 115 V?

I Energy is equal to the product of potential times charge or potential times current times time.

$$(115\ \text{V})(1.00\ \text{A})(100\ \text{s}) = 11\,500\ \text{J} = 11.5\ \text{kJ}$$

23.3 From the definition of a faraday, calculate the charge on one electron.

I

$$\left(\frac{96\,500\ \text{C}}{\text{mol}\ e^-}\right)\left(\frac{1\ \text{mol}\ e^-}{6.02\times 10^{23}\ e^-}\right) = 1.60\times 10^{-19}\ \text{C}/e^-$$

23.4 A lamp draws a current of 2.0 A. Find the charge in coulombs used by the lamp in 30 s.

I

$$\text{Charge} = \text{current}\times\text{time} = (2.0\ \text{A})(30\ \text{s}) = 60\ \text{C}$$

23.5 Compute the time required to pass 36 000 C through an electroplating bath using a current of 5 A.

I

$$\text{Time in seconds} = \frac{\text{charge}}{\text{current}} = \frac{36\,000\ \text{C}}{5\ \text{A}} = 7200\ \text{s} = 2\ \text{h}$$

23.6 How many electrons per second pass through a cross section of a copper wire carrying 10^{-16} A?

I

$$\text{Since}\quad 1\ \text{A} = 1\ \text{C/s},\qquad \text{Rate} = \frac{10^{-16}\ \text{C/s}}{1.6\times 10^{-19}\ \text{C}/e} = 600\ e/\text{s}$$

23.7 A galvanic cell was operated under almost ideally reversible conditions at a current of 10^{-16} A. **(*a*)** At this current, how long would it take to deliver 1 mol electrons? **(*b*)** How many electrons would be delivered by the cell to a pulsed measuring circuit in 10 ms of operation?

I

$$(a)\quad (1\ \text{mol}\ e^-)\left(\frac{96\,500\ \text{C}}{\text{mol}\ e^-}\right)\left(\frac{1\ \text{s}}{10^{-16}\ \text{C}}\right) = 9.65\times 10^{20}\ \text{s}$$

$$(9.65\times 10^{20}\ \text{s})\left(\frac{1\ \text{h}}{3600\ \text{s}}\right)\left(\frac{1\ \text{day}}{24\ \text{h}}\right)\left(\frac{1\ \text{year}}{365\ \text{day}}\right) = 3.06\times 10^{13}\ \text{year}$$

$$(b)\quad 10\ \text{ms} = (10\times 10^{-3}\ \text{s})\left(\frac{1\ \text{mol}\ e^-}{9.65\times 10^{20}\ \text{s}}\right)\left(\frac{6.02\times 10^{23}\ e^-}{\text{mol}\ e^-}\right) = 6\,e^-$$

23.8 A resistance heater was wound around a 50 g metallic cylinder. A current of 0.65 A was passed through the heater for 24 s while the measured voltage drop across the heater was 5.4 V. The temperature of the cylinder was 22.5 °C before the heating period and 29.8 °C at the end. If heat losses to the environment can be neglected, what is the specific heat of the cylinder metal in cal/g·K?

I

$$\text{Energy input} = IVt = (0.65\ \text{A})(5.4\ \text{V})(24\ \text{s}) = 84\ \text{J} = \frac{84\ \text{J}}{4.18\ \text{J/cal}} = 20\ \text{cal}$$

But, also, Energy input = (mass) × (specific heat) × (temperature rise)

Therefore, 20 cal = (50 g)(specific heat)[(29.8 − 22.5) K] from which Specific heat = 0.055 cal/g·K.

23.9 A dynamo delivers 15 A at 120 V. **(a)** Compute the power in kW supplied by the dynamo. **(b)** How much electrical energy, in kW·h, is supplied by the dynamo in 2 h? **(c)** What is the cost of this energy at 2¢/kW·h?

I **(a)** Power = (15 A)(120 V) = 1800 W = 1.8 kW **(b)** Energy = (1.8 kW)(2 h) = 3.6 kW·h

(c) Cost = (3.6 kW·h)(2¢/kW·h) = 7.2¢

23.10 How many coulombs per hour pass through an electroplating bath which uses a current of 5.00 A?

I

$$5.0\text{ A} = \left(\frac{5.0\text{ C}}{\text{s}}\right)\left(\frac{3600\text{ s}}{\text{h}}\right) = 1.8 \times 10^4\text{ C/h}$$

23.11 Compute the cost at 5.0¢/kW·h of operating for 8.0 h an electric motor which takes 15 A at 110 V.

I By Problem 23.1(*a*), (110 V)(15 A) = 1650 W = 1.65 kW

$$(1.65\text{ kW})(8.0\text{ h})(5.0\text{ ¢/kW}\cdot\text{h}) = 66\text{¢}$$

23.12 A tank containing 0.20 m^3 of water was used as a constant-temperature bath. How long would it take to heat the bath from 20.0 °C to 25.0 °C with a 250 W immersion heater? Neglect the heat capacity of the tank frame and any heat losses to the air. Assume that the density of water is 1000 kg/m^3.

I

$$\text{Heat} = mc\,\Delta t = (0.20 \times 10^6\text{ g})\left(\frac{4.18\text{ J}}{\text{g}\cdot{}^\circ\text{C}}\right)(25.0\text{ °C} - 20.0\text{ °C}) = (4.18 \times 10^6\text{ J})\left(\frac{1\text{ s}}{250\text{ J}}\right)\left(\frac{1\text{ h}}{3600\text{ s}}\right) = 4.6\text{ h}$$

23.13 The specific heat of a liquid was measured by placing 100 g of the liquid in a calorimeter. The liquid was heated by an electric immersion coil. The heat capacity of the calorimeter together with the coil was previously determined to be 31.4 J/K. With the 100 g sample in place in the calorimeter, a current of 0.500 A was passed through the immersion coil for exactly 3 min. The voltage across the terminals of the coil was measured to be 1.50 V. The temperature of the sample rose by 0.800 °C. Find the specific heat capacity of the liquid.

I

$$(3.000\text{ min})\left(\frac{60\text{ s}}{\text{min}}\right)\left(\frac{0.500\text{ C}}{\text{s}}\right)(1.50\text{ V}) = 135\text{ J}$$

$$\text{Heat} = \underbrace{mc\,\Delta t}_{\text{Liquid}} + \underbrace{(31.4\text{ J/K})\,\Delta t}_{\text{Calorimeter}}$$

$$135\text{ J} = \left((100\text{ g})\,c + \frac{31.4\text{ J}}{\text{g K}}\right)(0.800\text{ K}) \qquad c = 1.38\text{ J/g}\cdot\text{K}$$

23.14 The heat of solution of NH_4NO_3 in water was determined by measuring the amount of electrical work needed to compensate for the cooling which would otherwise occur when the salt dissolves. After the NH_4NO_3 was added to the water, electric energy was provided by passage of a current through a resistance coil until the temperature of the solution reached the value it had prior to the addition of the salt. In a typical experiment, 4.4 g NH_4NO_3 was added to 200 g water. A current of 0.75 A was provided through the heater coil, and the voltage across the terminals was 6.0 V. The current was applied for 5.2 min. Calculate ΔH for the solution of 1.0 mol NH_4NO_3 in enough water to give the same concentration as was attained in the above experiment.

I

$$(5.2\text{ min})\left(\frac{60\text{ s}}{\text{min}}\right)\left(\frac{0.75\text{ C}}{\text{s}}\right)(6.0\text{ V}) = 1.4 \times 10^3\text{ J} = 1.4\text{ kJ}$$

$$(4.4\text{ g NH}_4\text{NO}_3)\left(\frac{1\text{ mol NH}_4\text{NO}_3}{80.0\text{ g NH}_4\text{NO}_3}\right) = 0.055\text{ mol NH}_4\text{NO}_3 \qquad (1.0\text{ mol})\left(\frac{1.4\text{ kJ}}{0.055\text{ mol}}\right) = 25\text{ kJ}$$

23.2 ELECTROLYSIS

23.15 How many faradays of electricity are required to electrolyze 1 mol $CuCl_2$ to copper metal and chlorine gas?

I

$$\text{Cu}^{2+} + 2e^- \longrightarrow \text{Cu} \qquad 2\text{Cl}^- \longrightarrow \text{Cl}_2 + 2e^-$$

Two moles of electronic charge ($2\mathscr{F}$) is required per mol $CuCl_2$.

23.16 A solution of copper(II) sulfate is electrolyzed between copper electrodes by a current of 10.0 A for exactly 1 h. What changes occur at the electrodes and in the solution?

▮ The electrode reactions are

$$\text{Anode:} \qquad Cu \longrightarrow Cu^{2+} + 2e^-$$

$$\text{Cathode:} \qquad Cu^{2+} + 2e^- \longrightarrow Cu$$

$$\text{Number of mol electrons} = (10.0\ \text{A})(3600\ \text{s})\left(\frac{1\ \text{C}}{1\ \text{A}\cdot\text{s}}\right)\left(\frac{1\ \text{mol}\ e^-}{96\,500\ \text{C}}\right) = 0.373\ \text{mol}\ e^-$$

Since it takes 2 mol electrons to react with or produce each mole of copper metal, the number of mol of copper dissolved or deposited is

$$(0.373\ \text{mol}\ e^-)\left(\frac{1\ \text{mol Cu}}{2\ \text{mol}\ e^-}\right) = 0.186\ \text{mol Cu}$$

Thus 0.186 mol copper is dissolved from the anode, 0.186 mol copper is deposited onto the cathode, and the original copper(II) ion concentration of the solution remains unchanged.

23.17 A constant current was passed through a solution of $AuCl_4^-$ ions between gold electrodes. After a period of 10.0 min, the cathode increased in mass by 1.314 g. How much charge, q, was passed? What was the current, I?

▮ The reaction at the cathode is the reduction of the gold(III) to gold metal:

$$AuCl_4^- + 3e^- \longrightarrow Au + 4Cl^- \qquad n(\text{Au}) = \frac{1.314\ \text{g Au}}{197\ \text{g/mol Au}} = 6.67 \times 10^{-3}\ \text{mol Au}$$

$$q = (6.67 \times 10^{-3}\ \text{mol Au})\left(\frac{3\ \text{mol}\ e^-}{\text{mol Au}}\right) = 2.00 \times 10^{-2}\ \mathcal{F} \qquad I = \frac{q}{t} = \frac{(2.00 \times 10^{-2}\ \mathcal{F})(96\,500\ \text{C}/\mathcal{F})}{600\ \text{s}} = 3.22\ \text{A}$$

23.18 What mass of copper is deposited by the passage of 2.0 A of current for 482 s through a 2.0 M solution of $CuSO_4$?

▮

$$(482\ \text{s})\left(\frac{2.0\ \text{C}}{\text{s}}\right)\left(\frac{1\ \text{mol}\ e^-}{96\,500\ \text{C}}\right)\left(\frac{1\ \text{mol Cu}}{2\ \text{mol}\ e^-}\right)\left(\frac{63.5\ \text{g Cu}}{\text{mol Cu}}\right) = 0.32\ \text{g Cu}$$

23.19 A current of 15.0 A is passed through a solution of $CrCl_2$ for 45 min. (*a*) How many g Cr is deposited on the cathode? (*b*) Calculate the volume of $Cl_2(g)$ that is obtained at the anode at 1.00 atm and 273 K.

▮

(*a*) $$(2700\ \text{s})\left(\frac{15.0\ \text{C}}{\text{s}}\right)\left(\frac{1\ \text{mol}\ e^-}{96\,500\ \text{C}}\right) = 0.420\ \text{mol}\ e^- \qquad (0.420\ \text{mol}\ e^-)\left(\frac{1\ \text{mol Cr}}{2\ \text{mol}\ e^-}\right)\left(\frac{52.0\ \text{g Cr}}{\text{mol Cr}}\right) = 10.9\ \text{g Cr}$$

(*b*) $$(0.420\ \text{mol}\ e^-)\left(\frac{1\ \text{mol}\ Cl_2}{2\ \text{mol}\ e^-}\right) = 0.210\ \text{mol}\ Cl_2$$

$$V = \frac{nRT}{P} = \frac{(0.210\ \text{mol}\ Cl_2)(0.0821\ \text{L}\cdot\text{atm/mol}\cdot\text{K})(273\ \text{K})}{1.00\ \text{atm}} = 4.71\ \text{L}$$

23.20 (*a*) Calculate the mass of mercury produced by the reduction of $Hg(NO_3)_2$ by the passage of 19 300 C of charge. (*b*) If $Hg_2(NO_3)_2$ were electrolyzed under the same conditions as in part (*a*), what mass of mercury would be produced?

▮

(*a*) $$(19\,300\ \text{C})\left(\frac{1\ \text{mol}\ e^-}{96\,500\ \text{C}}\right)\left(\frac{1\ \text{mol Hg}}{2\ \text{mol}\ e^-}\right)\left(\frac{201\ \text{g Hg}}{\text{mol Hg}}\right) = 20.1\ \text{g}$$

(*b*) $$Hg_2^{2+} + 2e^- \longrightarrow 2Hg \qquad (19\,300\ \text{C})\left(\frac{1\ \text{mol}\ e^-}{96\,500\ \text{C}}\right)\left(\frac{2\ \text{mol Hg}}{2\ \text{mol}\ e^-}\right)\left(\frac{201\ \text{g Hg}}{\text{mol Hg}}\right) = 40.2\ \text{g}$$

23.21 Determine the mass of copper deposited from a solution of $CuSO_4$ by passage of a 10.0 A current for 965 s.

▮

$$(965\ \text{s})\left(\frac{10.0\ \text{C}}{\text{s}}\right)\left(\frac{1\ \text{mol}\ e^-}{96\,500\ \text{C}}\right)\left(\frac{1\ \text{mol Cu}}{2\ \text{mol}\ e^-}\right)\left(\frac{63.5\ \text{g}}{\text{mol Cu}}\right) = 3.18\ \text{g}$$

23.22 Calculate the mass of copper which can be deposited by the passage of 10.0 A for 20.0 min through a solution of copper(II) sulfate.

▮

$$Cu^{2+} + 2e^- \longrightarrow Cu$$

$$(20.0\ \text{min})\left(\frac{60.0\ \text{s}}{\text{min}}\right)\left(\frac{10.0\ \text{C}}{\text{s}}\right)\left(\frac{1\ \text{mol}\ e^-}{96\,500\ \text{C}}\right)\left(\frac{1\ \text{mol}\ Cu^{2+}}{2\ \text{mol}\ e^-}\right)\left(\frac{63.55\ \text{g Cu}}{\text{mol Cu}}\right) = 3.95\ \text{g Cu}$$

23.23 What is the average current (in A) if 100 g nickel is deposited from a nickel(II) ion solution in 3 h 20 min?

▮

$$(100\text{ g Ni})\left(\frac{1\text{ mol Ni}}{58.7\text{ g Ni}}\right)\left(\frac{2\text{ mol }e^-}{\text{mol Ni}}\right)\left(\frac{96\,500\text{ C}}{\text{mol }e^-}\right) = 3.29\times10^5\text{ C}$$

$$(200\text{ min})(60\text{ s/min}) = 1.20\times10^4\text{ s}\qquad \frac{3.29\times10^5\text{ C}}{1.20\times10^4\text{ s}} = 27.4\text{ A}$$

23.24 Calculate the mass of Hg_2Cl_2 which can be prepared by the reduction of mercury(II) ion in the presence of chloride ion by the passage of a 5.00 A current for 3.00 h.

▮

$$2Hg^{2+} + 2Cl^- + 2e^- \longrightarrow Hg_2Cl_2$$

$$(3.00\text{ h})\left(\frac{3600\text{ s}}{\text{h}}\right)\left(\frac{5.00\text{ C}}{\text{s}}\right)\left(\frac{1\text{ mol }e^-}{96\,500\text{ C}}\right)\left(\frac{1\text{ mol Hg}_2\text{Cl}_2}{2\text{ mol }e^-}\right)\left(\frac{472.09\text{ g Hg}_2\text{Cl}_2}{\text{mol Hg}_2\text{Cl}_2}\right) = 132\text{ g}$$

23.25 A current of 5.00 A flowing for 30.00 min deposits 3.048 g of zinc at the cathode. Calculate the equivalent weight of zinc from this information.

▮

$$\text{Number of coulombs used} = (5.00\text{ A})[(30.00\times60)\text{ s}] = 9.00\times10^3\text{ C}$$

$$n(e^-)\text{ used} = \frac{9.00\times10^3\text{ C}}{9.65\times10^4\text{ C/mol }e^-} = 0.0933\text{ mol }e^-$$

$$\text{Equivalent weight} = \text{mass deposited by 1 mol }e^- = \frac{3.048\text{ g}}{0.0933\text{ mol }e^-} = 32.7\text{ g/eq}$$

23.26 How long would it take to deposit 100 g Al from an electrolytic cell containing Al_2O_3 at a current of 125 A? Assume that Al formation is the only cathode reaction. The solvent is Na_3AlF_6.

▮

$$\text{Equivalent weight of Al} = \tfrac{1}{3}(\text{atomic weight}) = \tfrac{1}{3}(27.0) = 9.00\text{ g Al/mol }e^-$$

$$n(e^-) = \frac{100\text{ g Al}}{9.00\text{ g Al/mol }e^-} = 11.1\text{ mol }e^-$$

$$\text{Time} = \frac{\text{charge}}{\text{current}} = \frac{(11.1\text{ mol }e^-)(9.65\times10^4\text{ C/mol }e^-)}{125\text{ A}} = 8.6\times10^3\text{ s} = 2.4\text{ h}$$

23.27 What mass of bismuth metal is deposited electrolytically from bismuth(III) solution in 30.0 min by a current of 40.0 A?

▮

$$(30.0\text{ min})\left(\frac{60\text{ s}}{\text{min}}\right)\left(\frac{40\text{ C}}{\text{s}}\right)\left(\frac{1\text{ mol }e^-}{96\,500\text{ C}}\right)\left(\frac{1\text{ mol Bi}}{3\text{ mol }e^-}\right)\left(\frac{209\text{ g Bi}}{\text{mol Bi}}\right) = 52.0\text{ g Bi}$$

23.28 What current is required to deposit on the cathode 5.00 g gold/h from a solution containing a salt of gold(III)?

▮

$$\left(\frac{5.00\text{ g Au}}{\text{h}}\right)\left(\frac{1\text{ mol Au}}{197\text{ g Au}}\right)\left(\frac{3\text{ mol }e^-}{\text{mol Au}}\right)\left(\frac{96\,500\text{ C}}{\text{mol }e^-}\right)\left(\frac{1\text{ h}}{3600\text{ s}}\right) = 2.04\text{ A}$$

23.29 An electrolytic cell contains a solution of $CuSO_4$ and an anode of impure copper. How many kg copper will be refined (deposited on the cathode) by 150 A maintained for 12 h?

▮

$$Cu^{2+} + 2e^- \longrightarrow Cu$$

$$(12\text{ h})\left(\frac{3600\text{ s}}{\text{h}}\right)\left(\frac{150\text{ C}}{\text{s}}\right)\left(\frac{1\text{ mol }e^-}{96\,500\text{ C}}\right)\left(\frac{1\text{ mol Cu}}{2\text{ mol }e^-}\right)\left(\frac{63.5\text{ g Cu}}{\text{mol Cu}}\right)\left(\frac{10^{-3}\text{ kg}}{\text{g}}\right) = 2.1\text{ kg Cu}$$

23.30 How many hours is required for a current of 3.0 A to decompose electrolytically 18 g water?

▮

$$H_2O \longrightarrow H_2 + \tfrac{1}{2}O_2$$

$$(18\text{ g H}_2\text{O})\left(\frac{1\text{ mol H}_2\text{O}}{18\text{ g H}_2\text{O}}\right)\left(\frac{2\text{ mol }e^-}{\text{mol H}_2\text{O}}\right)\left(\frac{96\,500\text{ C}}{\text{mol }e^-}\right)\left(\frac{1\text{ s}}{3.0\text{ C}}\right)\left(\frac{1\text{ h}}{3600\text{ s}}\right) = 18\text{ h}$$

(Note that the same 2 mol electrons which produces the 1 mol H_2 also produces the 0.5 mol O_2.)

23.31 The electrodes in a lead storage battery are made of Pb and PbO_2. The overall reaction during discharge is $Pb + PbO_2 + 2H_2SO_4 \rightarrow 2PbSO_4 + 2H_2O$. **(a)** What is the minimum amount (mass in lb) of lead (counting the lead in both free and combined forms) in a battery if the battery is designed to deliver 100 A·h? Assume a 25% "coefficient of use"; this is the percent of the Pb and PbO_2 in the battery case that actually is available for the electrode reactions. **(b)** If the average voltage of a storage battery is 2.00 V under zero load, what is the approximate free-energy change for the reaction as written?

▮ (*a*) $(100\ \text{A}\cdot\text{h}) = \left(\frac{100\ \text{C}}{\text{s}}\right)(3600\ \text{s}) = 3.60 \times 10^5\ \text{C} \qquad (3.60 \times 10^5\ \text{C})\left(\frac{1\ \text{mol}\ e^-}{96\,500\ \text{C}}\right)\left(\frac{1\ \text{mol Pb}}{2\ \text{mol}\ e^-}\right) = 1.87\ \text{mol Pb}$

Also needed is 1.87 mol PbO_2, containing 1.87 mol Pb:

$$\left(\frac{3.74\ \text{mol Pb used}}{0.25\ \text{coefficient}}\right) = (14.9\ \text{mol Pb needed})\left(\frac{207\ \text{g Pb}}{\text{mol Pb}}\right) = 3.09\ \text{kg Pb} \qquad (3.09\ \text{kg Pb})\left(\frac{2.2\ \text{lb}}{\text{kg}}\right) = 6.8\ \text{lb Pb}$$

(*b*) $\Delta G° = -n\mathcal{F}E° = -(2\ \text{mol}\ e^-)(96\,500\ \text{C/mol}\ e^-)(2.00\ \text{V}) = -3.86 \times 10^5\ \text{J} = -386\ \text{kJ}$

23.32 How long will it take for a uniform current of 6.00 A to deposit 78.0 g gold from a solution of $AuCl_4^-$? What mass of chlorine gas will be formed simultaneously at the anode of the electrolytic cell?

▮

$$AuCl_4^- + 3e^- \longrightarrow Au + 4Cl^- \qquad Cl^- \longrightarrow \tfrac{1}{2}Cl_2 + e^-$$

$$(78.0\ \text{g Au})\left(\frac{1\ \text{mol Au}}{197\ \text{g}}\right)\left(\frac{3\ \text{mol}\ e^-}{\text{mol Au}}\right) = 1.19\ \text{mol}\ e^- \qquad (1.19\ \text{mol}\ e^-)\left(\frac{96\,500\ \text{C}}{\text{mol}\ e^-}\right)\left(\frac{1\ \text{s}}{6.00\ \text{C}}\right) = 19\,100\ \text{s} = 5.31\ \text{h}$$

$$(1.19\ \text{mol}\ e^-)\left(\frac{\frac{1}{2}\ \text{mol}\ Cl_2}{\text{mol}\ e^-}\right)\left(\frac{70.9\ \text{g}\ Cl_2}{\text{mol}\ Cl_2}\right) = 42.2\ \text{g}\ Cl_2$$

23.33 Determine a value for Avogadro's number, using the charge on the electron, 1.60×10^{-19} C, and the fact that 96 500 C deposits 107.9 g silver from its solution.

▮

$$(107.9\ \text{g Ag})\left(\frac{1\ \text{mol Ag}}{107.9\ \text{g}}\right)\left(\frac{1\ \text{mol}\ e^-}{\text{mol Ag}}\right) = 1.00\ \text{mol}\ e^-$$

$$(1\ \text{mol}\ e^-) = (96\,500\ \text{C})\left(\frac{1\ \text{electron}}{1.60 \times 10^{-19}\ \text{C}}\right) = 6.02 \times 10^{23}\ e^-$$

Avogadro's number 6.02×10^{23} is equal to the number of electrons per mol electrons.

23.34 A current of 2.00 A passing for 5.00 h through a molten tin salt deposits 22.2 g tin. What is the oxidation state of the tin in the salt?

▮ The number of mol tin and the number of mol electrons are determined.

$$(5.00\ \text{h})\left(\frac{3600\ \text{s}}{\text{h}}\right)\left(\frac{2.00\ \text{C}}{\text{s}}\right)\left(\frac{1\ \text{mol}\ e^-}{96\,500\ \text{C}}\right) = 0.373\ \text{mol}\ e^- \qquad (22.2\ \text{g Sn})\left(\frac{1\ \text{mol Sn}}{118.69\ \text{g Sn}}\right) = 0.187\ \text{mol Sn}$$

$$\frac{0.373\ \text{mol}\ e^-}{0.187\ \text{mol Sn}} = \frac{2\ \text{mol}\ e^-}{\text{mol Sn}}$$

Since the number of mol electrons per mole of tin is equal to the change in oxidation number for each tin atom, there must be a loss of two oxidation numbers. $Sn^{2+} + 2e^- \rightarrow Sn$ The tin initially was tin(II).

23.35 Calculate the number of kW·h of electricity necessary to produce 1.00 metric ton (1000 kg) of aluminum by the Hall process in a cell operating at 15.0 V.

▮

$$Al^{3+} + 3e^- \longrightarrow Al \qquad (1.00 \times 10^6\ \text{g})\left(\frac{1\ \text{mol Al}}{27.0\ \text{g Al}}\right)\left(\frac{3\ \text{mol}\ e^-}{\text{mol Al}}\right)\left(\frac{96\,500\ \text{C}}{\text{mol}\ e^-}\right) = 1.07 \times 10^{10}\ \text{C}$$

$$(1.07 \times 10^{10}\ \text{C})(15.0\ \text{V}) = 1.61 \times 10^{11}\ \text{J}$$

$$(1.61 \times 10^{11}\ \text{W}\cdot\text{s})\left(\frac{1\ \text{kW}}{10^3\ \text{W}}\right)\left(\frac{1\ \text{h}}{3600\ \text{s}}\right) = 4.47 \times 10^4\ \text{kW}\cdot\text{h}$$

23.36 In the reduction of gold(III) from $AuCl_4^-$ ion solution, the solution must be stirred rapidly during electrolysis. Explain concisely why stirring is so important in this process.

▮ The $AuCl_4^-$ anion, which is negatively charged, must be reduced at an electrode with excess electrons, also negatively charged. Since like charges repel, stirring is essential to get the anions to the electrode.

23.37 Two cells containing silver electrodes in aqueous silver nitrate and copper electrodes in aqueous copper(II) sulfate, respectively, are connected in series so that the same current must pass through each of the cells. In a given experiment, 2.000 g silver metal is deposited on the cathode of the first cell. Calculate the mass of silver dissolved from the anode of the first cell and the mass of copper deposited at the cathode of the second cell.

▮

$$Ag^+ + e^- \longrightarrow Ag \qquad Ag \longrightarrow Ag^+ + e^- \qquad Cu^{2+} + 2e^- \longrightarrow Cu$$

The same number of coulombs must pass through each electrode of each cell in the series. If 2.00 g Ag is deposited at one electrode, 2.00 g Ag will be dissolved from the other.

$$(2.00 \text{ g Ag})\left(\frac{1 \text{ mol Ag}}{107.9 \text{ g Ag}}\right)\left(\frac{1 \text{ mol } e^-}{1 \text{ mol Ag}}\right) = 0.0185 \text{ mol } e^- \qquad (0.0185 \text{ mol } e^-)\left(\frac{1 \text{ mol Cu}}{2 \text{ mol } e^-}\right)\left(\frac{63.55 \text{ g Cu}}{\text{mol Cu}}\right) = 0.589 \text{ g Cu}$$

23.38 (*a*) What current is required to pass 1.00 mol electrons per hour through an electroplating bath? How many g of (*b*) bismuth and of (*c*) cadmium will be liberated by 1.00 mol of electrons?

▮

(*a*) $$\left(\frac{1.00 \text{ mol } e^-}{\text{h}}\right)\left(\frac{1 \text{ h}}{3600 \text{ s}}\right)\left(\frac{96\,500 \text{ C}}{\text{mol } e^-}\right) = 26.8 \text{ A}$$

(*b*) $Bi^{3+} + 3e^- \longrightarrow Bi$ $$(1.00 \text{ mol } e^-)\left(\frac{1 \text{ mol Bi}}{3 \text{ mol } e^-}\right)\left(\frac{209.0 \text{ g Bi}}{\text{mol Bi}}\right) = 69.7 \text{ g Bi}$$

(*c*) $Cd^{2+} + 2e^- \longrightarrow Cd$ $$(1.00 \text{ mol } e^-)\left(\frac{1 \text{ mol Cd}}{2 \text{ mol } e^-}\right)\left(\frac{112.4 \text{ g Cd}}{\text{mol Cd}}\right) = 56.2 \text{ g Cd}$$

23.39 The same quantity of electricity that liberated 2.158 g silver was passed through a solution of a gold salt, and 1.314 g gold was deposited. The equivalent weight of silver is 107.9. Calculate the equivalent weight of gold. What is the oxidation state of gold in this gold salt?

▮

$$\text{Number of eq Ag in 2.158 g} = \frac{2.158 \text{ g}}{107.9 \text{ g/eq}} = 0.02000 \text{ eq Ag}$$

Then 1.314 g Au must represent 0.02000 eq, and so

$$\text{Equivalent weight of Au} = \frac{1.314 \text{ g}}{0.02000 \text{ eq}} = 65.70 \text{ g/eq}$$

$$\text{Oxidation state} = \text{number of electrons needed to form one gold atom by reduction}$$

$$= \frac{\text{atomic weight of Au}}{\text{equivalent weight of Au}} = \frac{197.0}{65.7} = 3$$

23.40 A certain current liberates 0.504 g hydrogen in 2.00 h. How many g of oxygen and of copper (from Cu^{2+} solution) can be liberated by the same current in the same time?

▮ Masses of different substances liberated by the same number of coulombs are proportional to their equivalent weights. Equivalent weight of hydrogen is 1.008; of oxygen, 8.00; of copper, 31.8.

$$\text{Number of eq hydrogen in 0.504 g} = \frac{0.504 \text{ g}}{1.008 \text{ g/eq}} = 0.500 \text{ eq}$$

Then 0.500 eq each of oxygen and copper can be liberated.

$$\text{Mass of oxygen liberated} = (0.500 \text{ eq})(8.00 \text{ g/eq}) = 4.00 \text{ g}$$

$$\text{Mass of copper liberated} = (0.500 \text{ eq})(31.8 \text{ g/eq}) = 15.9 \text{ g}$$

23.41 Exactly 0.2000 mol electrons is passed through two electrolytic cells in series. One contains silver ion, and the other, zinc ion. Assume that the only cathode reaction in each cell is the reduction of the ion to the metal. How many g of each metal will be deposited?

▮ One mole of electrons liberates 1 eq of an element. Hence 0.200 mol liberates 0.200 eq. Equivalent weights of Ag^+ and Zn^{2+} are

$$Ag^+: \frac{107.9}{1} = 107.9 \qquad Zn^{2+}: \frac{65.38}{2} = 32.69$$

$$Ag^+ \text{ deposited} = (0.2000 \text{ mol } e^-)(107.9 \text{ g/mol } e^-) = 21.58 \text{ g}$$

$$Zn^{2+} \text{ deposited} = (0.2000 \text{ mol } e^-)(32.69 \text{ g/mol } e^-) = 6.54 \text{ g}$$

23.42 A current of 0.200 A is passed for 600 s through 50.00 mL of 0.1000 M NaCl. If only chlorine gas is produced at the anode and if water is reduced to hydrogen gas at the cathode, what will the hydroxide ion concentration in the solution be after the electrolysis?

▎

$$(600\ \text{s})(0.200\ \text{A})\left(\frac{1\ \text{mol}\ e^-}{96\ 500\ \text{C}}\right) = 1.24 \times 10^{-3}\ \text{mol}\ e^-$$

$$Cl^- \longrightarrow \tfrac{1}{2}Cl_2 + e^- \qquad H_2O + e^- \longrightarrow \tfrac{1}{2}H_2 + OH^-$$

The oxidation does not affect the OH^- concentration. The reduction produces

$$(1.24 \times 10^{-3}\ \text{mol}\ e^-)\left(\frac{1\ \text{mol}\ OH^-}{\text{mol}\ e^-}\right) = 1.24 \times 10^{-3}\ \text{mol}\ OH^- \qquad \frac{1.24\ \text{mmol}\ OH^-}{50.00\ \text{mL}} = 0.0248\ \text{M}\ OH^-$$

23.43 Current efficiency is defined as the extent of a desired electrochemical reaction divided by the theoretical extent of the reaction, times 100%. What is the current efficiency of an electrodeposition of copper metal in which 9.80 g copper is deposited by passage of a 3.00 A current for 10 000 s?

▎

$$(9.80\ \text{g Cu})\left(\frac{1\ \text{mol Cu}}{63.546\ \text{g}}\right)\left(\frac{2\ \text{mol}\ e^-}{\text{mol Cu}}\right) = 0.308\ \text{mol}\ e^- \qquad (10\ 000\ \text{s})\left(\frac{3.00\ \text{C}}{\text{s}}\right)\left(\frac{1\ \text{mol}\ e^-}{96\ 500\ \text{C}}\right) = 0.311\ \text{mol}\ e^-$$

Of 0.311 mol electrons provided by the current, 0.308 mol was used to deposit copper. The current efficiency is thus

$$\text{Current efficiency} = \left(\frac{0.308}{0.311}\right)(100\%) = 99.1\%$$

23.44 The cell of Fig. 23.1 consists of three compartments separated by porous barriers. The first contains a cobalt electrode in 5.00 L of 0.100 M cobalt(II) nitrate; the second contains 5.00 L of 0.100 M KNO_3; the third contains a silver electrode in 5.00 L of 0.100 M $AgNO_3$. Assuming that the current within the cell is carried equally by the positive and negative ions, tabulate the concentrations of ions of each type in each compartment of the cell after the passage of 0.100 mol electrons.

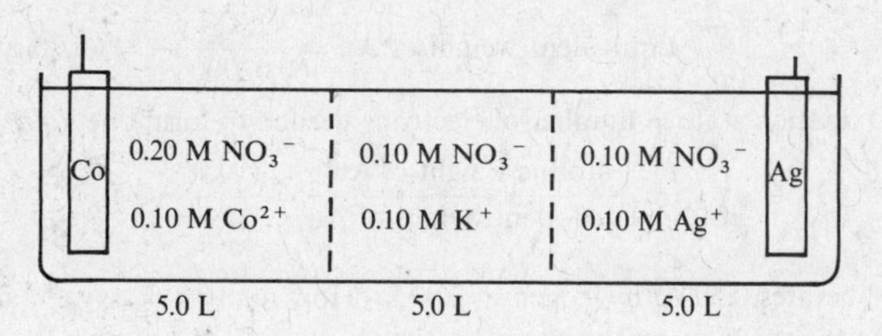

Fig. 23.1

▎

$$Co^{2+} + 2e^- \longrightarrow Co \qquad E° = -0.28\ \text{V}$$
$$Ag^+ + e^- \longrightarrow Ag \qquad E° = 0.80\ \text{V}$$
$$2Ag^+ + Co \longrightarrow Co^{2+} + 2Ag \qquad E° = 1.08\ \text{V}$$

In the left compartment, cobalt will be oxidized to cobalt(II) ion. In the right compartment, silver ion will be reduced to silver. The passage of 0.100 mol electrons will cause the following quantities of change:

	compartment 1	compartment 2	compartment 3
Effect of electrode	+0.0500 mol Co^{2+}		−0.100 mol Ag^+
Positive ion movement	−0.0250 mol Co^{2+}	+0.0250 mol Co^{2+} −0.0500 mol K^+	+0.0500 mol K^+
Negative ion movement	+0.0500 mol NO_3^-		−0.0500 mol NO_3^-

Changing the numbers of mol to concentrations in 5.0 L compartments and adding or subtracting yields the following results:

	compartment 1	compartment 2	compartment 3
Final concentrations (M)	Co^{2+} 0.105 NO_3^- 0.210	Co^{2+} 0.00500 K^+ 0.090 NO_3^- 0.100	Ag^+ 0.0800 NO_3^- 0.0900 K^+ 0.0100

23.45 Hydrogen peroxide can be prepared by the successive reactions

$$2NH_4HSO_4 \longrightarrow H_2 + (NH_4)_2S_2O_8 \qquad (NH_4)_2S_2O_8 + 2H_2O \longrightarrow 2NH_4HSO_4 + H_2O_2$$

The first is an electrolytic reaction, and the second a steam distillation. What current would have to be used in the first reaction to produce enough intermediate to yield 100 g pure H_2O_2 per hour? Assume 50.0% anode current efficiency.

▌

$$\left(\frac{100\text{ g }H_2O_2}{\text{h}}\right)\left(\frac{1\text{ mol }H_2O_2}{34.0\text{ g }H_2O_2}\right)\left(\frac{1\text{ mol }H_2}{\text{mol }H_2O_2}\right)\left(\frac{2\text{ mol }e^-}{\text{mol }H_2}\right)\left(\frac{96\,500\text{ C}}{\text{mol }e^-}\right)\left(\frac{1\text{ h}}{3600\text{ s}}\right) = 158\text{ A at }100\%\text{ efficiency}$$

$$\frac{158\text{ A}}{0.500} = 315\text{ A at }50\%\text{ efficiency}$$

23.46 How many hours will it take to produce 100 lb electrolytic chlorine from NaCl in a cell that carries 1000 A? The anode efficiency for the chlorine reaction is 85%.

▌

$$2Cl^- \longrightarrow Cl_2 + 2e^-$$

$$(100\text{ lb }Cl_2)\left(\frac{454\text{ g}}{\text{lb}}\right)\left(\frac{1\text{ mol }Cl_2}{71.0\text{ g }Cl_2}\right)\left(\frac{2\text{ mol }e^-}{\text{mol }Cl_2}\right)\left(\frac{96\,500\text{ C}}{\text{mol }e^-}\right)\left(\frac{1\text{ s}}{1000\text{ C}}\right)\left(\frac{1\text{ h}}{3600\text{ s}}\right) = 34.3\text{ h} \qquad \text{at }100\%\text{ efficiency}$$

Divide by 0.85 to get the time at 85% efficiency. (Obviously, the time must be longer.)

$$\frac{34.3\text{ h}}{0.85} = 40.3\text{ h at }85\%\text{ efficiency}$$

23.47 A given quantity of electricity passes through two separate electrolytic cells containing solutions of $AgNO_3$ and $SnCl_2$, respectively. If 2.00 g silver is deposited in one cell, how many g tin are deposited in the other cell?

▌

$$(2.00\text{ g Ag})\left(\frac{1\text{ mol Ag}}{107.9\text{ g Ag}}\right)\left(\frac{1\text{ mol }e^-}{1\text{ mol Ag}}\right)\left(\frac{1\text{ mol Sn}}{2\text{ mol }e^-}\right)\left(\frac{118.7\text{ g Sn}}{\text{mol Sn}}\right) = 1.10\text{ g Sn}$$

23.48 How many coulombs must be supplied to a cell for the electrolytic production of 245 g $NaClO_4$ from $NaClO_3$? Because of side reactions, the anode efficiency for the desired reaction is 60%.

▌ First it is necessary to know the equivalent weight of $NaClO_4$ for this reaction. The balanced anode reaction equation is $ClO_3^- + H_2O \rightarrow ClO_4^- + 2H^+ + 2e^-$.

$$\text{Equivalent weight of }NaClO_4 = \frac{\text{formula weight}}{\text{number of electrons transferred}} = \frac{122.4}{2} = 61.2$$

$$\text{Number of eq }NaClO_4 = \frac{245\text{ g}}{61.2\text{ g/eq}} = 4.00\text{ eq }NaClO_4$$

$$n(e^-)\text{ required} = \frac{4.00\text{ eq}}{0.60\text{ eq anode product/mol }e^-} = 6.7\text{ mol }e^-$$

$$\text{Number of coulombs required} = (6.7\text{ mol }e^-)(9.6 \times 10^4\text{ C/mol }e^-) = 6.4 \times 10^5\text{ C}$$

23.49 A current of 15.0 A is employed to plate nickel in a $NiSO_4$ bath. Both Ni and H_2 are formed at the cathode. The current efficiency with respect to formation of Ni is 60.0%. **(*a*)** How many g of nickel is plated on the cathode per hour? **(*b*)** What is the thickness of the plating if the cathode consists of a sheet of metal 4.0 cm square which is coated on both faces? The density of nickel is 8.9 g/cm³. **(*c*)** What volume of H_2 (STP) is formed per hour?

▌ **(*a*)** Total number of coulombs used $= (15.0\text{ A})(3600\text{ s}) = 5.40 \times 10^4\text{ C}$

$$\text{Moles of }e^-\text{ used} = \frac{5.40 \times 10^4\text{ C}}{9.65 \times 10^4\text{ C/mol }e^-} = 0.560\text{ mol }e^-$$

Number of eq Ni deposited $= (0.600)(0.560\text{ mol }e^-)(1\text{ eq Ni/mol }e^-) = 0.336\text{ eq Ni}$

Equivalent weight of Ni $= \frac{1}{2}(\text{atomic weight}) = \frac{1}{2}(58.70) = 29.4\text{ g/eq}$

Mass of Ni deposited $= (0.336\text{ eq})(29.4\text{ g/eq}) = 9.9\text{ g}$

(*b*) Area of two faces $= 2(4.0\text{ cm})(4.0\text{ cm}) = 32\text{ cm}^2$

$$\text{Volume of 9.9 g Ni} = \frac{\text{mass}}{\text{density}} = \frac{9.9\text{ g}}{8.9\text{ g/cm}^3} = 1.11\text{ cm}^3$$

$$\text{Thickness of plating} = \frac{\text{volume}}{\text{area}} = \frac{1.11\text{ cm}^3}{32\text{ cm}^2} = 0.035\text{ cm}$$

(*c*) Number of eq H_2 liberated $= (0.400)(0.560\ \mathscr{F})(1\ \text{eq}\ H_2/\mathscr{F}) = 0.224$ eq H_2
Volume of 1 eq ($\frac{1}{2}$ mol) $H_2 = \frac{1}{2}(22.4\ \text{L}) = 11.2\ \text{L}\ H_2$
Volume of H_2 liberated $= (0.224\ \text{eq})(11.2\ \text{L/eq}) = 2.51\ \text{L}\ H_2$

23.50 Assume that impure copper contains only iron, silver, and gold as impurities. After passage of 140 A for 482.5 s, the mass of the anode decreased by 22.260 g and the cathode increased in mass by 22.011 g. Estimate the % iron and % copper originally present.

▎ The increase in mass at the cathode is due solely to copper. Hence there is 22.011 g of copper (equivalent to 0.3464 mol) and therefore a total of 0.249 g of iron, gold, and silver. Only the iron and copper are oxidized; the gold and silver fall to the bottom in the anode mud. Since each of the active metals requires 2 mol electrons per mol metal, there must be

$$(482\ \text{s})\left(\frac{140\ \text{C}}{\text{s}}\right)\left(\frac{1\ \text{mol}\ e^-}{96\ 500\ \text{C}}\right)\left(\frac{1\ \text{mol}\ M^{2+}}{2\ \text{mol}\ e^-}\right) = 0.3500\ \text{mol}\ M^{2+}$$

The number of mol iron is therefore 0.0036, and the mass of iron is 0.20 g. The metal is 98.88% copper and 0.90% iron.

23.3 GALVANIC CELLS

23.51 What substance can be used to oxidize fluorides to fluorine?

▎ Fluorides may be oxidized electrolytically but not chemically by any substance listed in Table 23.1.

TABLE 23.1 Standard Reduction Potentials at 25 °C

	$E°$, V		$E°$, V
$F_2 + 2e^- \longrightarrow 2F^-$	2.87	$AgCl + e^- \longrightarrow Ag + Cl^-$	0.222
$S_2O_8^{2-} + 2e^- \longrightarrow 2SO_4^{2-}$	2.0	$PdI_4^{2-} + 2e^- \longrightarrow Pd + 4I^-$	0.18
$Co^{3+} + e^- \longrightarrow Co^{2+}$	1.82	$Cu^{2+} + e^- \longrightarrow Cu^+$	0.15
$H_2O_2 + 2H^+ + 2e^- \longrightarrow 2H_2O$	1.77	$Sn^{4+} + 2e^- \longrightarrow Sn^{2+}$	0.13
$MnO_4^- + 4H^+ + 3e^- \longrightarrow MnO_2 + 2H_2O$	1.70	$Ag(S_2O_3)_2^{3-} + e^- \longrightarrow Ag + 2S_2O_3^{2-}$	0.017
$PbO_2 + 4H^+ + SO_4^{2-} + 2e^- \longrightarrow PbSO_4 + 2H_2O$	1.70	$2H^+ + 2e^- \longrightarrow H_2$	0.000
$Ce^{4+} + e^- \longrightarrow Ce^{3+}$	1.70	$Ge^{4+} + 2e^- \longrightarrow Ge^{2+}$	0.0
$MnO_4^- + 8H^+ + 5e^- \longrightarrow Mn^{2+} + 4H_2O$	1.51	$Pb^{2+} + 2e^- \longrightarrow Pb$	−0.126
$Au^{3+} + 3e^- \longrightarrow Au$	1.50	$Sn^{2+} + 2e^- \longrightarrow Sn$	−0.14
$Cl_2 + 2e^- \longrightarrow 2Cl^-$	1.36	$2CuO + H_2O + 2e^- \longrightarrow Cu_2O + 2OH^-$	−0.15
$Cr_2O_7^{2-} + 14H^+ + 6e^- \longrightarrow 2Cr^{3+} + 7H_2O$	1.33	$AgI + e^- \longrightarrow Ag + I^-$	−0.151
$Tl^{3+} + 2e^- \longrightarrow Tl^+$	1.26	$CuI + e^- \longrightarrow Cu + I^-$	−0.17
$MnO_2 + 4H^+ + 2e^- \longrightarrow Mn^{2+} + 2H_2O$	1.23	$Ni^{2+} + 2e^- \longrightarrow Ni$	−0.25
$O_2 + 4H^+ + 4e^- \longrightarrow 2H_2O$	1.229	$Co^{2+} + 2e^- \longrightarrow Co$	−0.28
$2IO_3^- + 12H^+ + 10e^- \longrightarrow I_2 + 6H_2O$	1.20	$PbSO_4 + 2e^- \longrightarrow Pb + SO_4^{2-}$	−0.31
$Br_2 + 2e^- \longrightarrow 2Br^-$	1.09	$Tl^+ + e^- \longrightarrow Tl$	−0.336
$AuCl_4^- + 3e^- \longrightarrow Au + 4Cl^-$	1.00	$Cu_2O + H_2O + 2e^- \longrightarrow 2Cu + 2OH^-$	−0.34
$OCl^- + H_2O + 2e^- \longrightarrow Cl^- + 2OH^-$	0.94	$Cd^{2+} + 2e^- \longrightarrow Cd$	−0.403
$Pd^{2+} + 2e^- \longrightarrow Pd$	0.92	$Fe^{2+} + 2e^- \longrightarrow Fe$	−0.44
$2Hg^{2+} + 2e^- \longrightarrow Hg_2^{2+}$	0.92	$Cr^{3+} + 3e^- \longrightarrow Cr$	−0.74
$Cu^{2+} + I^- + e^- \longrightarrow CuI$	0.85	$Zn^{2+} + 2e^- \longrightarrow Zn$	−0.7628
$Ag^+ + e^- \longrightarrow Ag$	0.799	$2H_2O + 2e^- \longrightarrow H_2 + 2OH^-$	−0.828
$Hg_2^{2+} + 2e^- \longrightarrow 2Hg$	0.79	$Mn^{2+} + 2e^- \longrightarrow Mn$	−1.18
$Fe^{3+} + e^- \longrightarrow Fe^{2+}$	0.771	$Al^{3+} + 3e^- \longrightarrow Al$	−1.66
$O_2 + 2H^+ + 2e^- \longrightarrow H_2O_2$	0.69	$H_2 + 2e^- \longrightarrow 2H^-$	−2.25
$Cu^{2+} + Cl^- + e^- \longrightarrow CuCl$	0.566	$Mg^{2+} + 2e^- \longrightarrow Mg$	−2.37
$I_2 + 2e^- \longrightarrow 2I^-$	0.535	$Ce^{3+} + 3e^- \longrightarrow Ce$	−2.48
$Cu^+ + e^- \longrightarrow Cu$	0.52	$Na^+ + e^- \longrightarrow Na$	−2.713
$Co(dip)_3^{3+} + e^- \longrightarrow Co(dip)_3^{2+}$	0.370	$Ca^{2+} + 2e^- \longrightarrow Ca$	−2.87
$Fe(CN)_6^{3-} + e^- \longrightarrow Fe(CN)_6^{4-}$	0.355	$Ba^{2+} + 2e^- \longrightarrow Ba$	−2.90
$Cu^{2+} + 2e^- \longrightarrow Cu$	0.34	$Cs^+ + e^- \longrightarrow Cs$	−2.92
$Hg_2Cl_2 + 2e^- \longrightarrow 2Hg + 2Cl^-$	0.270	$K^+ + e^- \longrightarrow K$	−2.93
$Hg_2Cl_2 + 2e^- \longrightarrow 2Hg + 2Cl^-$ (satd KCl)	0.244	$Li^+ + e^- \longrightarrow Li$	−3.03
$Ge^{2+} + 2e^- \longrightarrow Ge$	0.23		

23.52 Write the cell reaction and calculate the value of E°_{cell} for the cell: $Zn|Zn^{2+}(1\ M)||Fe^{2+}(1\ M), Fe^{3+}(1\ M)|Pt$

▮ The equations for the half-reactions, with the corresponding standard potentials, are

$$Fe^{3+} + e^- \longrightarrow Fe^{2+} \qquad E^\circ = +0.77\ V$$
$$Zn \longrightarrow Zn^{2+} + 2e^- \qquad E^\circ = +0.76\ V$$

To obtain the equation for the cell reaction, the equation for the iron(II)/iron(III) half-reaction must be multiplied by 2 (to obtain 2 mol electrons) and combined with the zinc/zinc ion half-reaction. The value of the half-cell potential is *not* multiplied by 2, however, because potential does *not* depend on the *quantity* of substance undergoing reaction. Addition of the resulting equations and the two potentials results in the following cell reaction:

$$Zn + 2Fe^{3+} \longrightarrow 2Fe^{2+} + Zn^{2+} \qquad E^\circ = 1.53\ V$$

The sign of E_{cell}, which is proportional to $-\Delta G$, determines the direction of reaction. A positive sign means that the reaction is spontaneous as written, and a negative sign means that the reverse reaction is spontaneous.

23.53 Is 1.0 M H^+ solution under hydrogen gas at 1.0 atm capable of oxidizing silver metal in the presence of 1.0 M silver ion?

▮ The desired reaction is $2Ag + 2H^+\ (1.0\ M) \rightarrow 2Ag^+\ (1.0\ M) + H_2\ (1\ atm)$. Combining the two half-reactions gives

$$2Ag \longrightarrow 2Ag^+ + 2e^- \qquad E^\circ = -0.80\ V$$
$$2H^+ + 2e^- \longrightarrow H_2 \qquad E^\circ = 0.00\ V$$
$$2Ag + 2H^+ \longrightarrow H_2 + 2Ag^+ \qquad E^\circ(\text{cell}) = -0.80\ V$$

The negative value of the standard cell potential indicates that H^+ solution does not oxidize silver metal under these conditions.

23.54 A hydrogen electrode with hydrogen gas at 1 atm and hydrogen ion at unit activity is suitably connected to a half-cell consisting of copper metal immersed in 1.00 M copper(II) sulfate solution. The measured potential for the cell $Pt|H_2\ (1\ atm)|H^+\ (1\ M)||Cu^{2+}\ (1\ M)|Cu$ is 0.34 V. What is the standard electrode potential for the copper/copper(II) ion half-cell?

▮ Since all species are in their standard states, and since the potential of the hydrogen half-cell is zero, the total potential of the cell may be attributed to the copper half-cell; therefore, the standard electrode potential of the copper/copper(II) ion half-cell is +0.34 V; that is, $E^\circ = +0.34$ V.

23.55 The following half-cell reactions are given:

$$Cu^{2+}(aq) + e^- \longrightarrow Cu^+(aq) \qquad E^\circ = +0.15\ V$$
$$2H^+(aq) + 2e^- \longrightarrow H_2(g) \qquad E^\circ = 0.00\ V$$
$$CuI(s) + e^- \longrightarrow Cu(s) + I^-(aq) \qquad E^\circ = -0.17\ V$$
$$Zn^{2+}(aq) + 2e^- \longrightarrow Zn(s) \qquad E^\circ = -0.76\ V$$

Calculate the voltage of the cell $Zn|Zn^{2+}\ (1\ M)||I^-\ (1\ M)|CuI|Cu$.

▮

$$CuI + e^- \longrightarrow Cu + I^- \qquad E^\circ = -0.17\ V$$
$$Zn \longrightarrow Zn^{2+} + 2e^- \qquad E^\circ = +0.76\ V$$
$$2CuI + Zn \longrightarrow Zn^{2+} + 2Cu + 2I^- \qquad E^\circ = +0.59\ V$$

(Do not double E° when you double the number of moles in a half-reaction.)

23.56 Given the following half-cell reactions and corresponding reduction potentials:

(1) $A + e^- \longrightarrow A^- \qquad E^\circ = -0.24\ V$

(2) $B^- + e^- \longrightarrow B^{2-} \qquad E^\circ = 1.25\ V$

(3) $C^- + 2e^- \longrightarrow C^{3-} \qquad E^\circ = -1.25\ V$

(4) $D + 2e^- \longrightarrow D^{2-} \qquad E^\circ = 0.68\ V$

(5) $E + 4e^- \longrightarrow E^{4-} \qquad E^\circ = 0.38\ V$

What combination of two half-cells would result in a cell with the largest potential?

▮ (2) and (3). The two half-cells with the greatest difference in potential yield the cell with the largest potential.

23.57 Make a schematic diagram of a cell for each of the following cell reactions. Indicate on the diagram the anode, the cathode, the directions in which the anions and cations migrate through the cell and the salt bridge, if any, and the direction in which the electrons migrate through the external circuit. Calculate $E°$ for each cell.

(*a*) $Sn + 2Ag^+ \longrightarrow Sn^{2+} + 2Ag$ (*b*) $2Cr + 3H_2O + 3OCl^- \longrightarrow 2Cr^{3+} + 3Cl^- + 6OH^-$

(*c*) $H_2 + I_2 \longrightarrow 2H^+ + 2I^-$

▮ (*a*) The data are from Table 23.1:

$$Sn^{2+} + 2e^- \longrightarrow Sn \qquad E° = -0.14\text{ V}$$

Reversing the equation changes the sign of $E°$:

$$Sn \longrightarrow Sn^{2+} + 2e^- \qquad E° = +0.14\text{ V} \qquad Ag^+ + e^- \longrightarrow Ag \qquad E° = 0.80\text{ V}$$

Doubling each quantity in the chemical equation does not affect $E°$.

$$2Ag^+ + 2e^- \longrightarrow 2Ag \qquad E° = 0.80\text{ V}$$

The cell reaction is $2Ag^+ + Sn \longrightarrow Sn^{2+} + 2Ag \qquad E° = 0.94\text{ V}$ See Fig. 23.2.

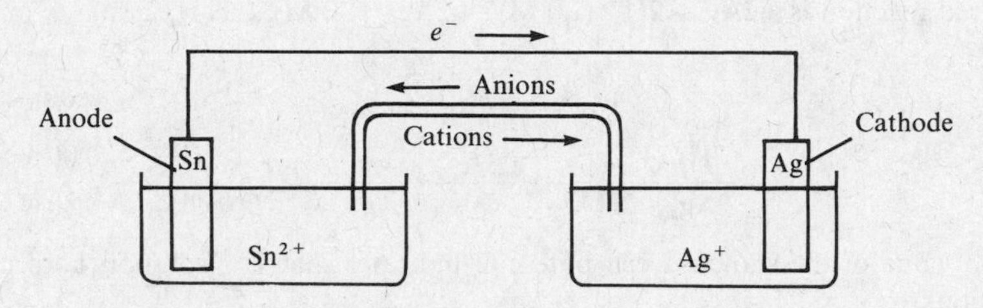

Fig. 23.2

(*b*)

$$Cr^{3+} + 3e^- \longrightarrow Cr \qquad E° = -0.74\text{ V}$$
$$OCl^- + H_2O + 2e^- \longrightarrow Cl^- + 2OH^- \qquad E° = 0.94\text{ V}$$
$$2Cr \longrightarrow 2Cr^{3+} + 6e^- \qquad E° = 0.74\text{ V}$$
$$3OCl^- + 3H_2O + 6e^- \longrightarrow 3Cl^- + 6OH^- \qquad E° = 0.94\text{ V}$$
$$3OCl^- + 2Cr + 3H_2O \longrightarrow 2Cr^{3+} + 3Cl^- + 6OH^- \qquad E° = 1.68\text{ V}$$

See Fig. 23.3.

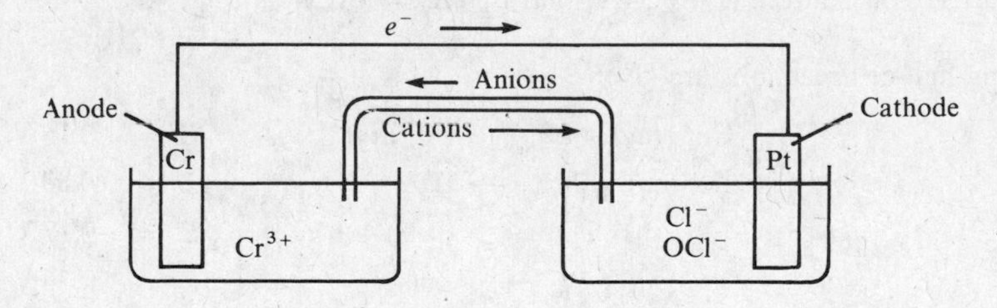

Fig. 23.3

(*c*)

$$2H^+ + 2e^- \longrightarrow H_2 \qquad E° = 0.00\text{ V}$$
$$I_2 + 2e^- \longrightarrow 2I^- \qquad E° = 0.54\text{ V}$$
$$H_2 \longrightarrow 2H^+ + 2e^- \qquad E° = 0.00\text{ V}$$
$$H_2 + I_2 \longrightarrow 2H^+ + 2I^- \qquad E° = 0.54\text{ V}$$

See Fig. 23.4.

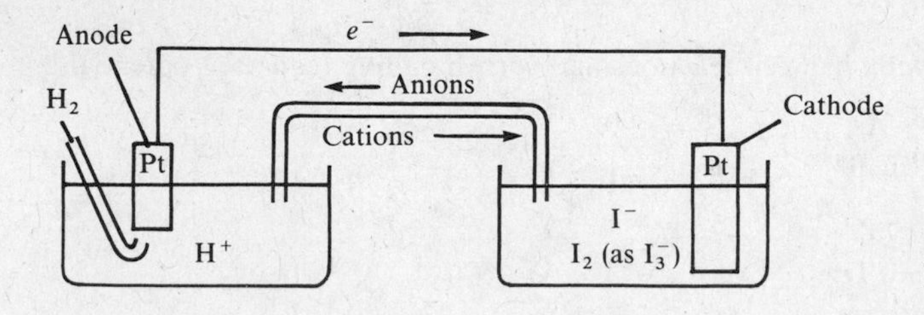

Fig. 23.4

23.58 Calculate the standard potential for each of the cells in which the following reactions occur. State which reactions will proceed spontaneously as written.

(*a*) $2Fe^{2+} + Zn^{2+} \longrightarrow Zn + 2Fe^{3+}$ (*b*) $Fe + 2Fe^{3+} \longrightarrow 3Fe^{2+}$
(*c*) $Cd + 2CuI \longrightarrow Cd^{2+} + 2Cu + 2I^-$ (*d*) $Ag + Cu^{2+} + 2Cl^- \longrightarrow AgCl + CuCl$

▮ (*a*)

$$2Fe^{2+} \longrightarrow 2Fe^{3+} + 2e^- \qquad E^\circ = -0.771 \text{ V}$$
$$Zn^{2+} + 2e^- \longrightarrow Zn \qquad E^\circ = -0.763 \text{ V}$$
$$2Fe^{2+} + Zn^{2+} \longrightarrow 2Fe^{3+} + Zn \qquad E^\circ = -1.534 \text{ V}$$

This reaction is not spontaneous.

(*b*)

$$2Fe^{3+} + 2e^- \longrightarrow 2Fe^{2+} \qquad E^\circ = 0.771 \text{ V}$$
$$Fe \longrightarrow Fe^{2+} + 2e^- \qquad E^\circ = 0.44 \text{ V}$$
$$2Fe^{3+} + Fe \longrightarrow 3Fe^{2+} \qquad E^\circ = 1.21 \text{ V}$$

Since E° is positive, this reaction is spontaneous. Note that Fe^{2+} does *not* disproportionate in solution.

(*c*)

$$Cd \longrightarrow Cd^{2+} + 2e^- \qquad E^\circ = 0.40 \text{ V}$$
$$2CuI + 2e^- \longrightarrow 2Cu + 2I^- \qquad E^\circ = -0.17 \text{ V}$$
$$2CuI + Cd \longrightarrow Cd^{2+} + 2Cu + 2I^- \qquad E^\circ = 0.23 \text{ V}$$

This reaction is spontaneous.

(*d*)

$$Ag + Cl^- \longrightarrow AgCl + e^- \qquad E^\circ = -0.222 \text{ V}$$
$$Cu^{2+} + Cl^- + e^- \longrightarrow CuCl \qquad E^\circ = 0.566 \text{ V}$$
$$Ag + Cu^{2+} + 2Cl^- \longrightarrow AgCl + CuCl \qquad E^\circ = 0.344 \text{ V}$$

The reaction is spontaneous.

23.59 What is the standard potential of a cell that uses the Zn^{2+}/Zn and Ag^+/Ag couples? Which couple is negative? Write the equation for the cell reaction occurring at unit relative concentrations (the standard states).

▮ The standard electrode potentials for Zn^{2+}/Zn and Ag^+/Ag are, from Table 23.1, -0.763 V and $+0.799$ V, respectively. The standard potential of the cell is the difference between these two numbers, $0.799 - (-0.763) = 1.562$ V. The silver electrode potential is higher, and thus silver ion is the oxidizing agent. The zinc couple provides the reducing agent and is the negative electrode. The equation is $Zn + 2Ag^+ \rightarrow Zn^{2+} + 2Ag$

23.60 Can Fe^{3+} oxidize Br^- to Br_2 at unit relative concentrations?

▮ From Table 23.1, the Fe^{3+}/Fe^{2+} couple has a lower standard electrode potential, 0.771 V, than the Br_2/Br^- couple, 1.09 V. Therefore Fe^{2+} can reduce Br_2, but Br^- cannot reduce Fe^{3+}. (On the other hand, I^-, occurring at a much lower standard electrode potential, 0.535 V, is easily oxidized by Fe^{3+} to I_2.)

23.61 (*a*) What is the standard potential of the cell made up of the Cd^{2+}/Cd and Cu^{2+}/Cu couples? (*b*) Which couple is positive?

▮ (*a*)

$$Cd^{2+} + 2e^- \longrightarrow Cd \qquad E^\circ = -0.403 \text{ V}$$
$$Cu^{2+} + 2e^- \longrightarrow Cu \qquad E^\circ = +0.34 \text{ V}$$
$$Cd \longrightarrow Cd^{2+} + 2e^- \qquad E^\circ = +0.403 \text{ V}$$
$$Cu^{2+} + Cd \longrightarrow Cd^{2+} + Cu \qquad E^\circ = +0.74 \text{ V}$$

(*b*) Cu^{2+} is the oxidizing agent, and Cu/Cu^{2+} is the positive couple.

23.62 What is the standard potential of the cell containing the Sn^{2+}/Sn and Br_2/Br^- couples?

▮

$$E^\circ(\text{cell}) = E^\circ(\text{Br}) - E^\circ(\text{Sn}) = (1.08 \text{ V}) - (-0.14 \text{ V}) = 1.22 \text{ V}$$

23.63 Given $2Ce^{4+} + Co \rightarrow 2Ce^{3+} + Co^{2+}$ $E^\circ = 1.89$ V. The standard reduction potential for Co^{2+} is -0.28 V. What is the standard reduction potential for Ce^{4+} to Ce^{3+}?

▮

$$E^\circ(\text{cell}) = E^\circ(\text{Ce}) - E^\circ(\text{Co}) = E^\circ(\text{Ce}) - (-0.28 \text{ V}) = 1.89 \text{ V} \qquad E^\circ(\text{Ce}) = 1.61 \text{ V}$$

23.64 The potential of a silver/silver chloride electrode measured with respect to a saturated calomel electrode is -0.022 V. What is the value of the standard reduction potential for the silver/silver chloride electrode?

▮

$$E^\circ(\text{cell}) = E^\circ(\text{Ag/AgCl}) - E^\circ(\text{cal}) \qquad E^\circ(\text{Ag/AgCl}) = -0.022 \text{ V} + 0.244 \text{ V} = 0.222 \text{ V}$$

23.65 Would H_2O_2 behave as oxidant or reductant with respect to the following couples at standard concentrations? (*a*) I_2/I^- (*b*) $S_2O_8^{2-}/SO_4^{2-}$ (*c*) Fe^{3+}/Fe^{2+}

▮

$$H_2O_2 \longrightarrow O_2 + 2H^+ + 2e^- \qquad E^\circ = -0.69 \text{ V}$$
$$H_2O_2 + 2H^+ + 2e^- \longrightarrow 2H_2O \qquad E^\circ = 1.77 \text{ V}$$

(a)

$$2e^- + I_2 \longrightarrow 2I^- \qquad E^\circ = 0.535\text{ V}$$
$$I_2 + H_2O_2 \longrightarrow O_2 + 2H^+ + 2I^- \qquad E^\circ = -0.15\text{ V}$$
$$2I^- \longrightarrow I_2 + 2e^- \qquad E^\circ = -0.535\text{ V}$$
$$H_2O_2 + 2H^+ + 2I^- \longrightarrow 2H_2O + I_2 \qquad E^\circ = 1.23\text{ V}$$

H_2O_2 acts as an oxidizing agent in the reaction for which its E° value is positive.

(b)

$$S_2O_8^{2-} + 2e^- \longrightarrow 2SO_4^{2-} \qquad E^\circ = 2.0\text{ V}$$
$$H_2O_2 + S_2O_8^{2-} \longrightarrow 2SO_4^{2-} + O_2 + 2H^+ \qquad E^\circ = 1.3\text{ V}$$
$$2SO_4^{2-} \longrightarrow S_2O_8^{2-} + 2e^- \qquad E^\circ = -2.0\text{ V}$$
$$2SO_4^{2-} + H_2O_2 + 2H^+ \longrightarrow S_2O_8^{2-} + 2H_2O \qquad E^\circ = -0.2\text{ V}$$

H_2O_2 is the reducing agent in the reaction with a positive E° value.

(c)

$$Fe^{3+} + e^- \longrightarrow Fe^{2+} \qquad E^\circ = 0.771\text{ V}$$
$$2Fe^{3+} + H_2O_2 \longrightarrow 2Fe^{2+} + O_2 + 2H^+ \qquad E^\circ = 0.02\text{ V}$$
$$Fe^{2+} \longrightarrow e^- + Fe^{3+} \qquad E^\circ = -0.771\text{ V}$$
$$2Fe^{2+} + H_2O_2 + 2H^{2+} \longrightarrow 2Fe^{3+} + 2H_2O \qquad E^\circ = 1.00\text{ V}$$

Since both potentials are positive, H_2O_2 will act as an oxidizing agent and a reducing agent. In fact, iron(II) or iron(III) salts catalyze the self-oxidation-reduction of H_2O_2.

23.66 In the continued electrolysis of each of the following solutions at pH 7.0 and 25 °C, predict the main product at each electrode if there are no (irreversible) electrode polarization effects: **(a)** 1 M $NiSO_4$ with palladium electrodes **(b)** 1 M $NiBr_2$ with inert electrodes **(c)** 1 M Na_2SO_4 with Cu electrodes.

▌ **(a)** *Cathode reaction*: The following two possible reduction processes may be considered:

$$Ni^{2+} + 2e^- \longrightarrow Ni \qquad E^\circ = -0.25\text{ V} \tag{1}$$
$$2H^+ + 2e^- \longrightarrow H_2 \qquad E^\circ = 0.00\text{ V} \tag{2}$$

By the rule that the most probable cathode process is that for which the corresponding electrode potential is algebraically greatest, the hydrogen couple is favored at unit concentrations. Allowing for the effect of the pH 7.0 buffer, however, E for (*2*) is lowered to -0.41 V. The reduction of nickel then becomes the favored process.

Anode reaction: Three possible oxidation processes may be considered, the reverses of the following reduction half-reactions:

$$O_2 + 4H^+ + 4e^- \longrightarrow 2H_2O \qquad E^\circ = 1.23\text{ V} \tag{3}$$
$$Pd^{2+} + 2e^- \longrightarrow Pd \qquad E^\circ = 0.92\text{ V} \tag{4}$$
$$S_2O_8^{2-} + 2e^- \longrightarrow 2SO_4^{2-} \qquad E^\circ = 2.0\text{ V} \tag{5}$$

The standard potentials are reasonable values to take in considering (*4*) and (*5*). Although the initial concentrations of Pd^{2+} and $S_2O_8^{2-}$ are zero, they would increase during prolonged electrolysis if these species were the principal products. In the case of (*3*), however, the buffering of the solution prevents the buildup of $[H^+]$, and it would be more appropriate to take the E value for pH 7.0, 0.82 V. It is apparent that of the three possible anode reactions, (*3*) has the smallest E value and would thus occur most readily.

In conclusion, the electrode processes to be expected are:

$$\text{Anode: } 2H_2O \longrightarrow O_2 + 4H^+ + 4e^- \qquad \text{Cathode: } Ni^{2+} + 2e^- \longrightarrow Ni$$
$$\text{Overall: } 2H_2O + 2Ni^{2+} \longrightarrow O_2 + 4H^+ + 2Ni$$

(b) *Cathode reaction*: As in (*a*), Ni reduction would occur.

Anode reaction: The expression "inert electrode" is often used to indicate that we may neglect reaction of the electrode itself, either by virtue of the intrinsically high value of its electrode potential or because of polarization effects related to the preparation of the electrode surface. The remaining possible anode reactions are the reverses of the following:

$$Br_2 + 2e^- \longrightarrow 2Br^- \qquad E^\circ = 1.08\text{ V} \tag{6}$$
$$O_2 + 4H^+ + 4e^- \longrightarrow 2H_2O \qquad E^\circ = 1.23\text{ V} \tag{3}$$

The E value for (*3*) is 0.82 V for pH 7.0, so oxygen evolution takes precedence. [In practice, "overvoltage" or polarization is more difficult to avoid in the case of reactions involving gases (O_2) with liquids and dissolved solutes, so that electrolysis of $NiBr_2$ at most electrodes would probably lead to Br_2 formation.]

(*c*) *Cathode reaction*: The new couple to be considered is the sodium couple, the reduction reaction for which is:

$$Na^+ + e^- \longrightarrow Na \qquad E^\circ = -2.71\ V \tag{7}$$

This E value is much lower than that of (*2*), the evolution of H_2 at pH 7.0, −0.41 V. Therefore, hydrogen evolution will occur at the cathode.

Anode reaction: In addition to (*3*) and (*5*), the reaction of the Cu anode must be considered, the reverse of which is:

$$Cu^{2+} + 2e^- \longrightarrow Cu \qquad E^\circ = 0.34\ V \tag{8}$$

Process (*8*) has the lowest E value, and copper dissolution would thus take precedence over oxygen evolution.

23.67 What reaction, if any, would zinc(II) ion undergo in the copper(II) half of a Daniell cell? What reaction, if any, would copper(II) ion undergo in the zinc half of a Daniell cell? Which of these ions might actually get into the other half-cell during discharge of the cell? during recharge? Explain why a Daniell cell cannot be fully recharged.

▌ Zinc ion would not undergo any reaction in the copper half-cell. Copper(II) ion, however, would be reduced to copper on contact with the zinc electrode in the Zn^{2+}/Zn half-cell. During discharge, zinc ion might get into the Cu^{2+}/Cu half-cell, which is all right. During recharge, copper(II) ion might get into the Zn^{2+}/Zn half-cell, where it would plate out on the electrode.

23.68 Is tin(II) stable toward disproportionation in noncomplexing media?

▌

$$\begin{array}{lll} 2e^- + Sn^{2+} \longrightarrow Sn & & E^\circ = -0.14\ V \\ Sn^{2+} \longrightarrow Sn^{4+} + 2e^- & & E^\circ = -0.13\ V \\ 2Sn^{2+} \longrightarrow Sn + Sn^{4+} & & E^\circ = -0.27\ V \end{array}$$

The disproportionation reaction is not spontaneous, because the E° value is negative. Hence tin(II) is stable.

23.69 Explain why iron electrodes in a solution of $Fe(NO_3)_3$ would not be appropriate for an electrolysis experiment.

▌ The electrolyte would react with the electrode:

$$2Fe^{3+} + Fe \longrightarrow 3Fe^{2+} \qquad E^\circ = 1.21\ V$$

This reaction occurs even in the absence of a second half-cell or a current.

23.70 What is the standard electrode potential for MnO_4^-/MnO_2 in acid solution?

▌ The reduction half-reaction for this couple is $MnO_4^- + 4H^+ + 3e^- \rightarrow MnO_2 + 2H_2O$, which can be written as the difference of two half-reactions whose electrode potentials are listed in Table 23.1. nE° values may be correspondingly subtracted.

	n	E°, V	nE°, V
$MnO_4^- + 8H^+ + 5e^- \longrightarrow Mn^{2+} + 4H_2O$	5	1.51	7.55
$MnO_2 + 4H^+ + 2e^- \longrightarrow Mn^{2+} + 2H_2O$	2	1.23	2.46
Difference: $MnO_4^- - MnO_2 + 4H^+ + 3e^- \longrightarrow 2H_2O$	3		5.09 V

Rearranging, $MnO_4^- + 4H^+ + 3e^- \rightarrow MnO_2 + 2H_2O$, the desired reaction, in which $n = 3$.

$$E^\circ \text{ for the desired reaction} = \frac{5.09}{3} = 1.70\ V$$

23.71 Predict the stabilities at 25 °C of aqueous solutions of the uncomplexed intermediate oxidation states of **(*a*)** thallium and **(*b*)** copper.

▌ **(*a*)** The question is whether the intermediate state, Tl^+, spontaneously decomposes into the lower and higher states, Tl and Tl^{3+}. The supposed disproportionation reaction, $3Tl^+ \rightarrow 2Tl + Tl^{3+}$, could be written in the ion-electron method as

$$2 \times (Tl^+ + e^- \longrightarrow Tl) \tag{1}$$

$$Tl^+ \longrightarrow Tl^{3+} + 2e^- \tag{2}$$

In (*1*), the Tl^+/Tl couple functions as oxidizing agent; in (*2*), the Tl^{3+}/Tl^+ couple functions as reducing agent. The reaction would occur at unit concentrations if $E°$ for the reducing couple were less than $E°$ for the oxidizing couple. Since (Table 23.1) 1.26 V is greater than −0.34 V, the reaction cannot occur as written. We conclude that Tl^+ does not spontaneously decompose into Tl and Tl^{3+}. On the contrary, the reverse reaction is spontaneous: $2Tl + Tl^{3+} \rightarrow 3Tl^+$. This means that Tl(III) salts are unstable in solution in the presence of metallic Tl.

(*b*) The supposed disproportionation of Cu^+ would take the form $2Cu^+ \rightarrow Cu + Cu^{2+}$. The ion-electron partial equations are $Cu^+ + e^- \rightarrow Cu$ and $Cu^+ \rightarrow Cu^{2+} + e^-$. This process could occur if $E°$ for the supposed reducing couple, Cu^{2+}/Cu^+, were less than $E°$ for the oxidizing couple, Cu^+/Cu. Indeed, +0.16 V (computed by the method of Problem 23.70) is less than 0.52 V. Therefore Cu^+ is unstable to disproportionation in solution. Compounds of Cu(I) can exist only as extremely insoluble substances or as such stable complexes that only a very small concentration of free Cu^+ can exist in solution.

23.72 Using the information in Table 23.1, explain why copper(I) sulfate does not exist in aqueous solution.

▮

$$Cu^{2+} + e^- \longrightarrow Cu^+ \qquad E° = 0.15\text{ V} \qquad nE° = 0.15$$
$$Cu^{2+} + 2e^- \longrightarrow Cu \qquad E° = 0.34\text{ V} \qquad nE° = 0.68$$

Subtraction of the first equation from the second yields $Cu^+ + e^- \rightarrow Cu$, for which $nE° = 0.53$. Since $n = 1$, this half-reaction has a standard reduction potential of 0.53 V. Combination of this equation with $Cu^+ \rightarrow Cu^{2+} + e^-$, for which $E° = -0.15$ V yields

$$2Cu^+ \longrightarrow Cu^{2+} + Cu \qquad E° = +0.38\text{ V}$$

Since $E°$ is positive, this reaction proceeds spontaneously. That is, Cu^+ disproportionates in aqueous solution. Since Cu^+ is not stable, the "copper sulfate" in solution must be copper(II) sulfate.

23.73 From the data of Table 23.1, show that neither Cu^+ nor Co^{3+} is stable in aqueous solution, whereas Fe^{2+} is.

▮ Cu^+ disproportionates in aqueous solution (see Problem 23.72). Co^{3+} oxidizes water:

$$Co^{3+} + e^- \longrightarrow Co^{2+} \qquad E° = 1.82\text{ V}$$
$$2H_2O \longrightarrow O_2(g) + 4H^+ + 4e^- \qquad E° = -1.23\text{ V}$$
$$4Co^{3+} + 2H_2O \longrightarrow O_2 + 4H^+ + 4Co^{2+} \qquad E° = 0.59\text{ V}$$

Fe^{2+} does neither; it is stable. [See Problem 23.58(*b*).]

23.74 Explain why aluminum metal cannot be produced by electrolysis of aqueous solutions of aluminum salts. Explain why aluminum is produced by the electrolysis of a molten mixture of Al_2O_3 and Na_3AlF_6 rather than by electrolysis of molten Al_2O_3 alone.

▮ Water is less difficult to reduce ($E° = -0.414$ V) than aluminum ion ($E° = -1.66$ V). Na_3AlF_6 dissolves Al_2O_3 but is not more easily reduced than aluminum ion. Al_2O_3 cannot be melted economically; it is a refractory substance, used for furnace linings, with a melting point of 2045 °C.

23.75 Neglecting electrode polarization effects, predict the principal product at each electrode in the continued electrolysis at 25 °C of each of the following: **(*a*)** 1 M $Fe_2(SO_4)_3$ with inert electrodes in 0.1 M H_2SO_4 **(*b*)** 1 M LiCl with silver electrodes **(*c*)** 1 M $FeSO_4$ with inert electrodes at pH 7.0 **(*d*)** molten NaF

▮ **(*a*)** Fe ($Fe^{3+} \rightarrow Fe^{2+} \rightarrow Fe$) and O_2 (H_2O is oxidized more easily than SO_4^{2-}.) **(*b*)** H_2 and AgCl (H_2O is reduced more easily than Li^+; AgCl forms as Ag is oxidized in the presence of Cl^- ion.) **(*c*)** H_2 and Fe^{3+} [H_2O is reduced (−0.414 V) slightly in preference to Fe^{2+} (−0.44 V), even with 1 atm H_2 present.] **(*d*)** Na and F_2

23.76 Why are Co^{3+} salts unstable in water?

▮ $e^- + Co^{3+} \longrightarrow Co^{2+} \qquad E° = 1.82$ V That potential is sufficient to oxidize water.

23.77 If H_2O_2 is mixed with Fe^{2+}, which reaction is more likely, the oxidation of Fe^{2+} to Fe^{3+} or the reduction of Fe^{2+} to Fe? Write the reaction for each possibility and compute the standard potential of the equivalent electrochemical cell.

$$\begin{array}{rll}
H_2O_2 + 2H^+ + 2e^- \longrightarrow 2\,H_2O & E^\circ = & 1.77\text{ V} \\
Fe^{2+} \longrightarrow Fe^{3+} + e^- & E^\circ = & -0.77\text{ V} \\
H_2O_2 + 2H^+ + 2Fe^{2+} \longrightarrow 2H_2O + 2Fe^{3+} & E^\circ = & 1.00\text{ V} \\
H_2O_2 \longrightarrow O_2 + 2H^+ + 2e^- & E^\circ = & -0.69\text{ V} \\
2e^- + Fe^{2+} \longrightarrow Fe & E^\circ = & -0.44\text{ V} \\
Fe^{2+} + H_2O_2 \longrightarrow O_2 + 2H^+ + Fe & E^\circ = & -1.13\text{ V}
\end{array}$$

The first reaction is spontaneous; the second is not.

23.78 A certain electrode has a reduction potential of 0.140 V when measured against a saturated calomel electrode. Calculate its potential versus a standard hydrogen electrode.

$$\begin{array}{lll}
M^{n+} + ne^- \longrightarrow M & E^\circ = x & \\
2Hg + 2Cl^-(\text{sat KCl}) \longrightarrow Hg_2Cl_2 + 2e^- & E^\circ = -0.244\text{ V} & E^\circ(cell) = 0.140\text{ V} \\
x - 0.244\text{ V} = 0.140\text{ V} & x = 0.384\text{ V} &
\end{array}$$

23.79 A solution containing Na^+, Sn^{2+}, NO_3^-, Cl^-, and SO_4^{2-} ions, all at unit activity, is electrolyzed between a silver anode and a platinum cathode. What changes occur at the electrodes when current is passed through the cell?

Sn^{2+} will be oxidized to Sn^{4+} first, because it has the highest *oxidation* potential. The silver electrode is oxidized next, followed by water (which has somewhat of an overvoltage). Sn^{2+} is reduced to Sn at the cathode.

23.80 Predict whether 1.0 M Mn^{2+} and 1.0 M MnO_4^- will react in acid solution to produce MnO_2.

Determine if the reduction of MnO_4^- to MnO_2 by Mn^{2+}, which also produces MnO_2, will have a positive E° value in acid solution.

$$\begin{array}{ll}
Mn^{2+} + 2H_2O \longrightarrow MnO_2 + 4H^+ + 2e^- & E^\circ = -1.23\text{ V} \\
MnO_4^- + 4H^+ + 3e^- \longrightarrow MnO_2 + 2H_2O & E^\circ = 1.70\text{ V}
\end{array}$$

Adjusting the numbers of mol of reactants and products would not change the E° value, +0.47 V. Since the value is positive, Mn^{2+} will react with MnO_4^- in acid solution, so $Mn(MnO_4)_2$ would not be stable.

23.81 Describe the products formed when an aqueous solution of aluminum sulfate is electrolyzed between aluminum electrodes.

Aluminum is oxidized to aluminum ion at the anode. Water, more easily reduced than either aluminum ion or sulfate ion, is reduced to hydrogen at the cathode: $H_2O + e^- \rightarrow \frac{1}{2}H_2 + OH^-$

23.82 What fraction of a mole of iron metal will be produced by passage of 4.00 A of current through 1.00 L of 0.100 M Fe^{3+} solution for 1.00 h? Assume that only iron is reduced.

$$(1.00\text{ h})\left(\frac{3600\text{ s}}{\text{h}}\right)\left(\frac{4.00\text{ C}}{\text{s}}\right)\left(\frac{1\text{ mol }e^-}{96\,500\text{ C}}\right) = 0.149\text{ mol }e^- \qquad \left(\frac{0.100\text{ mol }Fe^{3+}}{\text{L}}\right)(1.00\text{ L}) = 0.100\text{ mol }Fe^{3+}$$

$$Fe^{3+} + e^- \longrightarrow Fe^{2+}$$

0.100 mol e^- is required to reduce all the Fe^{3+} to Fe^{2+} and leave 0.049 mol e^- to reduce the Fe^{2+} to Fe.

$$Fe^{2+} + 2e^- \longrightarrow Fe \qquad (0.049\text{ mol }e^-)\left(\frac{1\text{ mol Fe}}{2\text{ mol }e^-}\right) = 0.025\text{ mol Fe}$$

23.83 Prove that for two half-reactions having potentials E_1 and E_2 which are combined to yield a third half-reaction, having a potential E_3,

$$E_3 = \frac{n_1E_1 + n_2E_2}{n_3}$$

$$\Delta G_3 = \Delta G_1 + \Delta G_2$$

$$-E_3n_3\mathscr{F} = -E_1n_1\mathscr{F} - E_2n_2\mathscr{F} \qquad \text{Division by } -n_3\mathscr{F} \text{ yields} \qquad E_3 = \frac{E_1n_1 + E_2n_2}{n_3}$$

23.84 What reaction, if any, would be expected in the following experiments? **(a)** Hg metal is shaken with 1.0 M $AgNO_3$ solution. **(b)** Solid AgCl is shaken with 1.0 M $FeCl_2$ solution. **(c)** 1.0 M $Cr_2O_7^{2-}$ solution is added to 1.0 M HBr solution.

▮ **(a)**

$$2Hg^{2+} + 2e^- \longrightarrow Hg_2^{2+} \qquad E° = 0.92 \qquad nE° = 1.84$$
$$Hg_2^{2+} + 2e^- \longrightarrow 2Hg \qquad E° = 0.79 \qquad nE° = 1.58$$
$$2Hg^{2+} + 4e^- \longrightarrow 2Hg \qquad nE° = 3.42 \qquad E° = 3.42/4 = 0.86\ V$$
$$Hg \longrightarrow Hg^{2+} + 2e^- \qquad E° = -0.86$$
$$Ag^+ + e^- \longrightarrow Ag \qquad E° = 0.80$$
$$Hg + 2Ag^+ \longrightarrow Hg^{2+} + 2Ag \qquad E° = -0.06$$

Ag^+ will oxidize Hg to Hg_2^{2+} but not to Hg^{2+}.

(b)

$$AgCl + e^- \longrightarrow Ag + Cl^- \qquad E° = 0.222\ V$$
$$Fe^{2+} \longrightarrow Fe^{3+} + e^- \qquad E° = -0.771\ V$$
$$AgCl + Fe^{2+} \longrightarrow Ag + Fe^{3+} + Cl^- \qquad E° = -0.549\ V$$

No reaction is expected, since $E°$ is highly negative.

(c)

$$2Br^- \longrightarrow Br_2 + 2e^- \qquad E° = -1.09$$
$$Cr_2O_7^{2-} + 14H^+ + 6e^- \longrightarrow 2Cr^{3+} + 7H_2O \qquad E° = 1.33$$
$$Cr_2O_7^{2-} + 14H^+ + 6Br^- \longrightarrow 2Cr^{3+} + 7H_2O + 3Br_2 \qquad E° = 0.24$$

Since the standard potential is positive, dichromate ion is expected to oxidize bromide ion in acid solution.

23.85 Calculate the standard potential for the reaction $Hg_2Cl_2 + Cl_2 \longrightarrow 2Hg^{2+} + 4Cl^-$.

▮ Just as two half-reactions can be combined to yield a third, three half-reactions can be combined to yield a fourth.

$$Hg_2Cl_2 + 2e^- \longrightarrow 2Hg + 2Cl^- \qquad E° = 0.270 \qquad nE° = 0.540$$
$$Hg_2^{2+} \longrightarrow 2Hg^{2+} + 2e^- \qquad E° = -0.92 \qquad nE° = -1.84$$
$$2Hg \longrightarrow Hg_2^{2+} + 2e^- \qquad E° = -0.79 \qquad nE° = -1.58$$

Adding these three half-reactions yields

$$Hg_2Cl_2 \longrightarrow 2Hg^{2+} + 2Cl^- + 2e^- \qquad nE° = -2.88$$

The half-cell potential is $(-2.88\ V)/2 = -1.44\ V$

Combination with $\quad Cl_2 + 2e^- \longrightarrow 2Cl^- \qquad E° = 1.36\ V$

yields $\quad Hg_2Cl_2 + Cl_2 \longrightarrow 2Hg^{2+} + 4Cl^- \qquad E = -0.08\ V$

23.86 From the data of Table 23.1, compute the standard reduction potentials for **(a)** $Cu^+ + e^- \longrightarrow Cu$ **(b)** $2Cu^+ + 2e^- \longrightarrow 2Cu$

▮ **(a)** 0.53 V (See Problem 23.72.) **(b)** 0.53 V (the same potential)

23.87 Are Fe^{2+} solutions stable in air? Why can such solutions be preserved by the presence of iron nails?

▮

$$Fe^{2+} \longrightarrow Fe^{3+} + e^- \qquad E° = -0.77\ V$$
$$O_2 + 4H^+ + 4e^- \longrightarrow 2H_2O \qquad E° = 1.229\ V$$
$$Fe \longrightarrow Fe^{2+} + 2e^- \qquad E° = 0.44\ V$$

The first two half-cells combine to yield a cell which proceeds spontaneously (even in solutions of $H^+ < 1$ M). The reverse of the first half-cell and the third half-cell combine to give a cell which is highly spontaneous. Fe reduces Fe^{3+} to Fe^{2+}.

23.88 What is the standard potential of the Tl^{3+}/Tl electrode?

▮

	$E°$, V	n	$nE°$, V·mol
$Tl^{3+} + 2e^- \longrightarrow Tl^+$	1.26	2	2.52
$Tl^+ + e^- \longrightarrow Tl$	−0.336	1	−0.336
$Tl^{3+} + 3e^- \longrightarrow Tl$		3	2.18

$$(2.18\ V \cdot mol)/(3\ mol) = 0.73\ V$$

23.4 NERNST EQUATION

23.89 What is the potential of an electrode consisting of zinc metal in a solution in which the zinc ion concentration is 0.0100 M?

▮

$$Zn^{2+} + 2e^- \longrightarrow Zn \qquad E^\circ = -0.763\ \text{V} \qquad E = E^\circ - \left(\frac{0.0592}{n}\right)\left(\log\frac{a(\text{Zn})}{[\text{Zn}^{2+}]}\right)$$

The activity of pure zinc metal, $a(\text{Zn})$, is 1.00, and the Zn^{2+} concentration is 0.0100 M.

$$E = (-0.763) - \left(\frac{0.0592}{2}\right)\left(\log\frac{1.00}{0.0100}\right) = (-0.763) - \left(\frac{0.0592}{2}\right)[\log(100)] = -0.763 - 0.0592 = -0.822\ \text{V}$$

23.90 Calculate the potential of the cell

$$\text{Pt}\,|\,H_2\,(0.50\ \text{atm})\,|\,H^+\,(0.10\ \text{M})\,||\,MnO_4^-\,(0.10\ \text{M}),\ Mn^{2+}\,(1.0\ \text{M}),\ H^+\,(0.10\ \text{M})\,|\,\text{Pt}$$

▮ The half-reactions are combined as follows:

$$\begin{array}{rlr} 5H_2 \longrightarrow & 10H^+ + 10e^- & E^\circ = \ \ 0.00\ \text{V} \\ 16H^+ + 2MnO_4^- + 10e^- \longrightarrow & 2Mn^{2+} + 8H_2O & E^\circ = \ \ 1.51\ \text{V} \\ \hline 6H^+ + 2MnO_4^- + 5H_2 \longrightarrow & 2Mn^{2+} + 8H_2O & E^\circ(\text{cell}) = +1.51\ \text{V} \end{array}$$

$$E(\text{cell}) = E^\circ(\text{cell}) - \left(\frac{0.0592}{10}\right)\left(\log\frac{[Mn^{2+}]^2 a(H_2O)^8}{[H^+]^6[MnO_4^-]^2 P(H_2)^5}\right)$$

In dilute aqueous solutions, water is present in large excess, and its activity is assumed to be 1.00. The activity of the hydrogen gas is assumed to be equal to its pressure in atm, and the activities of the aqueous ions are assumed to be equal to their molar concentrations:

$$E(\text{cell}) = +1.51 - \left(\frac{0.0592}{10}\right)\left(\log\frac{(1.0)^2(1.00)^8}{(0.10)^6(0.10)^2(0.50)^5}\right) = +1.51 - (5.92 \times 10^{-3})[\log(3.2 \times 10^9)]$$
$$= +1.51 - (5.92 \times 10^{-3})(+9.505) = +1.45\ \text{V}$$

23.91 Write the equation for the half-reaction in which HNO_3 is reduced to NO. Under which of the following sets of conditions does the half-cell potential equal the standard reduction potential? ***(a)*** 1 M NO_3^-, 1 M H^+, 1 atm NO ***(b)*** 1 M NO_3^-, 4 M H^+, 1 atm NO ***(c)*** 1 M NO_3^-, 1 M H^+, 1 atm air ***(d)*** 1 M NO_3^-, 4 M H^+, 1 atm air

▮

$$NO_3^- + 4H^+ + 3e^- \longrightarrow NO + 2H_2O$$

Set *a*. By definition, E° refers to a reaction in which all reactants and products are at unit activity (1 M solutes, 1 atm gases).

23.92 ***(a)*** Using the Nernst equation for the cell reaction: $Pb + Sn^{2+} \rightarrow Pb^{2+} + Sn$, calculate the ratio of cation concentrations for which $E = 0$. ***(b)*** Distinguish clearly between the meaning of $E = 0$ and $E^\circ = 0$. Give an example of a reaction in which $E^\circ = 0$.

▮ ***(a)***

$$\begin{array}{rlr} Pb \longrightarrow & Pb^{2+} + 2e^- & E^\circ = \ \ 0.13\ \text{V} \\ Sn^{2+} + 2e^- \longrightarrow & Sn & E^\circ = -0.14\ \text{V} \\ Sn^{2+} + Pb \longrightarrow & Pb^{2+} + Sn & E^\circ = -0.01\ \text{V} \end{array}$$

Sn and Pb are each at unity activity, since each is a pure metal.

$$E = E^\circ - \left(\frac{0.0592}{2}\right)\left(\log\frac{[Pb^{2+}]}{[Sn^{2+}]}\right) = -0.01 - (0.0296)\left(\log\frac{[Pb^{2+}]}{[Sn^{2+}]}\right) = 0$$

$$\log\frac{[Pb^{2+}]}{[Sn^{2+}]} = \frac{0.01}{-0.0296} = -0.3 \qquad \frac{[Pb^{2+}]}{[Sn^{2+}]} = 0.5$$

(b) $E = 0$ is for a cell in which the concentration ratio(s) reduce the E° value to zero. $E^\circ = 0$ indicates a concentration cell.

23.93 Calculate the reduction potential of each of the following half-cells with each metal immersed in 1.00×10^{-6} M solution of the cation. Arrange the half-cells in order of decreasing reduction potential, and compare this list with the relative positions of these half-cells in Table 23.1. ***(a)*** Cu^{2+}/Cu ***(b)*** Cd^{2+}/Cd ***(c)*** Cr^{3+}/Cr ***(d)*** H^+/H_2 (1 atm)(Pt), ***(e)*** Pb^{2+}/Pb ***(f)*** Al^{3+}/Al

▮

$$Cu^{2+} + 2e^- \longrightarrow Cu$$

$$E = E° - \left(\frac{0.0592}{2}\right)\left(\log\frac{1}{[Cu^{2+}]}\right) = 0.34 - (0.0296)\left(\log\frac{1}{10^{-6}}\right) = 0.34 - (0.0296)(6) = 0.16 \text{ V}$$

$$Cr^{3+} + 3e^- \longrightarrow Cr$$

$$E = E° - \left(\frac{0.0592}{3}\right)\left(\log\frac{1}{[Cr^{3+}]}\right) = (-0.74) - (0.0197)\left(\log\frac{1}{10^{-6}}\right) = (-0.74) - (0.0197)(6) = -0.86 \text{ V}$$

Similar operations with the other half-cells yield

		E	$E°$
(*a*)	Cu	0.16	0.34
(*e*)	Pb	−0.31	−0.13
(*d*)	H	−0.355	0.000
(*b*)	Cd	−0.58	−0.40
(*c*)	Cr	−0.86	−0.74
(*f*)	Al	−1.78	−1.66

Only the H_2/H^+ half-cell changed its relative position in this abbreviated listing of Table 23.1

23.94 Calculate the reduction potential of a half-cell consisting of a platinum electrode immersed in 2.0 M Fe^{2+} and 0.020 M Fe^{3+} solution.

▮

$$Fe^{3+} + e^- \longrightarrow Fe^{2+}$$

$$E = E° - \left(\frac{0.0592}{1}\right)\left(\log\frac{[Fe^{2+}]}{[Fe^{3+}]}\right) = 0.771 - (0.0592)\left(\log\frac{2.0}{0.020}\right) = 0.771 - (0.0592)(\log 100) = 0.653 \text{ V}$$

23.95 Determine the potential of the following cell:

▮

$$Pt\,|\,Mn^{2+}\ (0.100\text{ M}),\ MnO_4^-\ (1.00\text{ M}),\ H^+\ (0.500\text{ M})\,||\,Cu^{2+}(0.100\text{ M})\,|\,Cu$$

$$MnO_4^- + 5e^- + 8H^+ \longrightarrow Mn^{2+} + 4H_2O \qquad E° = 1.51 \text{ V}$$

$$Cu \longrightarrow Cu^{2+} + 2e^- \qquad E° = -0.34 \text{ V}$$

$$2MnO_4^- + 5Cu + 16H^+ \longrightarrow 2Mn^{2+} + 5Cu^{2+} + 8H_2O \qquad E° = 1.17 \text{ V}$$

$$E = E° - \left(\frac{0.0592}{10}\right)\left(\log\frac{(0.100)^2(0.100)^5}{(1.00)^2(0.500)^{16}}\right) = 1.17 + 0.01 = 1.18 \text{ V}$$

23.96 For each of the following cell reactions, write each half-cell reaction, and calculate E and $E°$ for the cell.

(*a*) $Hg_2Cl_2(s) \longrightarrow 2Hg(l) + Cl_2\ (0.80\text{ atm})$

(*b*) $6Fe^{2+}\ (1.0\text{ M}) + Cr_2O_7^{2-}\ (0.50\text{ M}) + 14H^+\ (3.0\text{ M}) \longrightarrow 2Cr^{3+}\ (0.71\text{ M}) + 6Fe^{3+}\ (2.0\text{ M}) + 7H_2O(l)$

▮ (*a*)

$$Hg_2Cl_2 + 2e^- \longrightarrow 2Hg + 2Cl^- \qquad E° = 0.270 \text{ V}$$

$$2Cl^- \longrightarrow Cl_2 + 2e^- \qquad E° = -1.36 \text{ V}$$

$$Hg_2Cl_2(s) \longrightarrow 2Hg(l) + Cl_2(0.80\text{ atm}) \qquad E° = -1.09 \text{ V}$$

$$E = E° - \left(\frac{0.0592}{2}\right)[\log P(Cl_2)] = (-1.09) - \left(\frac{0.0592}{2}\right)(\log 0.80) = -1.09 \text{ V}$$

(*b*)

$$6Fe^{2+} \longrightarrow 6Fe^{3+} + 6e^- \qquad E° = -0.77$$

$$Cr_2O_7^{2-} + 14H^+ + 6e^- \longrightarrow 2Cr^{3+} + 7H_2O \qquad E° = 1.33$$

$$6Fe^{2+} + Cr_2O_7^{2-} + 14H^+ \longrightarrow 6Fe^{3+} + 2Cr^{3+} + 7H_2O \qquad E° = +0.56$$

$$E = E° - \left(\frac{0.0592}{6}\right)\left(\log\frac{[Fe^{3+}]^6[Cr^{3+}]^2}{[Fe^{2+}]^6[Cr_2O_7^{2-}][H^+]^{14}}\right) = 0.56 - (0.00987)\left(\log\frac{(2.0)^6(0.71)^2}{(0.50)(3.0)^{14}}\right) = 0.61 \text{ V}$$

23.97 Calculate the potential of a cell consisting of an anode of silver in 0.10 M silver nitrate solution and a cathode of platinum immersed in a solution containing 1.5 M $Cr_2O_7^{2-}$, 0.75 M Cr^{3+}, and 0.25 M H^+.

▮

$$6Ag \longrightarrow 6Ag^+ + 6e^- \qquad E° = -0.80 \text{ V}$$

$$Cr_2O_7^{2-} + 14H^+ + 6e^- \longrightarrow 2Cr^{3+} + 7H_2O \qquad E° = 1.33 \text{ V}$$

$$Cr_2O_7^{2-} + 14H^+ + 6Ag \longrightarrow 2Cr^{3+} + 7H_2O + 6Ag^+ \qquad E° = +0.53$$

$$E = 0.53 - \left(\frac{0.0592}{6}\right)\left(\log\frac{(0.10)^6(0.75)^2}{(1.5)(0.25)^{14}}\right)$$

$$= 0.53 - \left(\frac{0.0592}{6}\right)[6(\log 0.10) + 2(\log 0.75) - (\log 1.5) - 14(\log 0.25)] = 0.51\ \text{V}$$

23.98 Calculate the concentration of Sn^{4+} ion in solution with 1.00 M Sn^{2+} ion in a half-cell which would have a zero potential when suitably connected to a standard hydrogen/hydrogen ion half-cell. Would Sn^{2+} ion tend to be oxidized or would Sn^{4+} ion tend to be reduced under these conditions?

▮

$$Sn^{4+} + 2e^- \longrightarrow Sn^{2+} \qquad E° = 0.13\ \text{V}$$
$$H_2 \longrightarrow 2H^+ + 2e^- \qquad E° = 0.00\ \text{V}$$
$$Sn^{4+} + H_2 \longrightarrow Sn^{2+} + 2H^+ \qquad E° = 0.13\ \text{V}$$

$$E = 0 = E° - \left(\frac{0.0592}{2}\right)\left(\log\frac{[H^+]^2[Sn^{2+}]}{[Sn^{4+}]P(H_2)}\right) = 0.13 - \left(\frac{0.0592}{2}\right)\left(\log\frac{1}{[Sn^{4+}]}\right)$$

$$\log\frac{1}{[Sn^{4+}]} = \frac{2(0.13)}{0.0592} = 4.4 \qquad \log[Sn^{4+}] = -4.4 \qquad [Sn^{4+}] = 4 \times 10^{-5}\ \text{M}$$

Neither, since $E = 0$.

23.99 Given the concentration cell $Zn\,|\,Zn^{2+}\ (1.0\ \text{M})\,||\,Zn^{2+}\ (0.15\ \text{M})\,|\,Zn$ write equations for each half-reaction. Calculate E. As the cell discharges, does the difference in the concentrations of the two solutions becomes smaller or larger?

▮

$$Zn \longrightarrow Zn^{2+}\ (1.0\ \text{M}) + 2e^- \qquad E° = 0.763\ \text{V}$$
$$Zn^{2+}\ (0.15\ \text{M}) + 2e^- \longrightarrow Zn \qquad E° = -0.763\ \text{V}$$
$$Zn^{2+}\ (0.15\ \text{M}) \longrightarrow Zn^{2+}\ (1.0\ \text{M}) \qquad E° = 0.000\ \text{V}$$

$$E = 0.000 - \left(\frac{0.0592}{2}\right)\left(\log\frac{1}{0.15}\right) = -(0.0296)(0.824) = -0.0244\ \text{V}$$

As the cell discharges (the reaction proceeds to the left), the 1.0 M zinc ion is used up and the 0.15 M zinc ion is produced. Thus the two solutions approach each other in concentration (just as they would if the solutions were mixed directly).

23.100 Calculate the potential corresponding to the following cell:

$$Pt\,|\,Co^{2+}\ (2.0\ \text{M}),\ Co^{3+}\ (0.010\ \text{M})\,||\,Cr^{3+}\ (0.50\ \text{M}),\ Cr_2O_7^{2-}\ (4.0\ \text{M}),\ H^+\ (1.5\ \text{M})\,|\,Pt$$

▮

$$Co^{2+} \longrightarrow Co^{3+} + e^- \qquad E° = -1.82\ \text{V}$$
$$14H^+ + 6e^- + Cr_2O_7^{2-} \longrightarrow 2Cr^{3+} + 7H_2O \qquad E° = +1.33\ \text{V}$$
$$6Co^{2+} + 14H^+ + Cr_2O_7^{2-} \longrightarrow 6Co^{3+} + 2Cr^{3+} + 7H_2O \qquad E° = -0.49\ \text{V}$$

$$E = E° - \left(\frac{0.0592}{n}\right)\left(\log\frac{[Co^{3+}]^6[Cr^{3+}]^2}{[Co^{2+}]^6[H^+]^{14}[Cr_2O_7^{2-}]}\right)$$

$$= (-0.49) - \left(\frac{0.0592}{6}\right)\left(\log\frac{(10^{-2})^6(0.50)^2}{(2.0)^6(4.0)(1.5)^{14}}\right) = -0.32\ \text{V}$$

23.101 Determine the potential of the cell: $Pt\,|\,Fe^{2+}, Fe^{3+}\,||\,Cr_2O_7^{2-}, Cr^{3+}, H^+\,|\,Pt$ in which the concentrations of Fe^{2+} and Fe^{3+} are 0.50 and 0.75 M, respectively, and the concentrations of $Cr_2O_7^{2-}$, Cr^{3+}, and H^+ are 2.0, 4.0, and 1.0 M, respectively.

▮

$$6Fe^{2+} \longrightarrow 6Fe^{3+} + 6e^- \qquad E° = -0.77\ \text{V}$$
$$14H^+ + 6e^- + Cr_2O_7^{2-} \longrightarrow 2Cr^{3+} + 7H_2O \qquad E° = 1.33\ \text{V}$$

$$14H^+ + 6Fe^{2+} + Cr_2O_7^{2-} \longrightarrow 6Fe^{3+} + 2Cr^{3+} + 7H_2O \qquad E° = 0.56\ \text{V}$$

$$E = E° - \left(\frac{0.0592}{6}\right)\left(\log\frac{(0.75)^6(4.0)^2}{(1.0)^{14}(0.50)^6(2.0)}\right) = 0.56 - 0.02 = 0.54\ \text{V}$$

23.102 The reversible reduction potential of pure water is -0.414 V under 1.00 atm H_2 pressure. If the reduction is considered to be $2H^+ + 2e^- \rightarrow H_2$, calculate the hydrogen ion concentration of pure water.

▮
$$2H^+ + 2e^- \longrightarrow H_2 \qquad E^\circ = 0.000\ V$$
$$E = -0.414 = (0.000) - \left(\frac{0.0592}{2}\right)\left(\log\frac{1}{[H^+]^2}\right) = (0.0592)(\log [H^+])$$
$$\log [H^+] = -\frac{0.414}{0.0592} = -6.99 \qquad [H^+] = 1.02 \times 10^{-7}\ M$$

23.103 Determine the potential of a Daniell cell, initially containing 1.00 L each of 1.00 M copper(II) ion and 1.00 M zinc(II) ion, after passage of 0.100 MC of charge.

▮ Passage of 0.100×10^6 C causes production of zinc ion and consumption of copper(II) ion. The quantities are

$$(0.100 \times 10^6\ C)\left(\frac{1\ mol\ e^-}{96\ 500\ C}\right)\left(\frac{1\ mol\ M^{2+}}{2\ mol\ e^-}\right) = 0.518\ mol\ M^{2+}$$

Thus $$[Zn^{2+}] = 1.52\ M \quad \text{and} \quad [Cu^{2+}] = 0.48\ M$$
$$E = 1.10 - \left(\frac{0.0592}{2}\right)\left(\log\frac{1.52}{0.48}\right) = 1.09\ V$$

23.104 By how much is the oxidizing power of the MnO_4^-/Mn^{2+} couple decreased if the H^+ concentration is decreased from 1 M to 10^{-4} M at 25 °C?

▮ The half-cell reaction for the reduction is $MnO_4^- + 8H^+ + 5e^- \rightarrow Mn^{2+} + 4H_2O$ with $n = 5$. Assume that only the H^+ concentration deviates from 1 mol/L.

$$E - E^\circ = \left(-\frac{0.0592}{5}\right)\left(\log\frac{[Mn^{2+}]}{[MnO_4^-][H^+]^8}\right) = (-0.0118)\left(\log\frac{1}{(1)(10^{-4})^8}\right) = (-0.0118)(32) = -0.38\ V$$

The couple has moved down the table 0.38 V (to a position of less oxidizing power) from its standard value.

23.105 What is the potential of a cell containing two hydrogen electrodes, the negative one in contact with 10^{-8} molar H^+ and the positive one in contact with 0.025 molar H^+?

▮
$$E = E^\circ - \left(\frac{0.0592}{2}\right)\left(\log\frac{(10^{-8})^2}{(0.025)^2}\right) = 0.000 - (0.0592)[\log (4 \times 10^{-7})] = 0.38\ V$$

23.106 MnO_4^- is reduced to Mn^{2+} in solutions of low pH, and to MnO_2 in solutions of intermediate to high pH. Explain why a potentiometric titration of 0.100 M MnO_4^- should be performed in 2.00 M H^+ rather than in 0.800 M H^+.

▮ Near the end of the titration, most of the hydrogen ion will have been consumed by the reaction $MnO_4^- + 8H^+ + 5e^- \rightarrow Mn^{2+} + 4H_2O$. However, when the pH gets high enough, the precipitation of MnO_2 will start to occur, which causes two difficulties. First, two different reductions will be involved, with no measure of the extent of either. Second, the formation of a precipitate in a titration presents the possibility of occlusion of a reagent inside the solid.

23.107 For a cell consisting of an inert electrode in a solution containing 0.10 M $KMnO_4$, 0.20 M $MnCl_2$, and 1.0 M HCl suitably connected to another inert electrode in a solution containing 0.10 M $K_2Cr_2O_7$, 0.20 M $CrCl_3$, and 0.70 M HCl, calculate E (*a*) by combining E values calculated separately for each half-cell and (*b*) by combining E° values from the half-cells and using the Nernst equation for the overall cell. [*Note*: When performed correctly, procedures (*a*) and (*b*) must give the same result.]

▮ **(*a*)** $MnO_4^- + 8H^+ + 5e^- \longrightarrow Mn^{2+} + 4H_2O \qquad E^\circ = 1.51\ V$

$$E = E^\circ - \left(\frac{0.0592}{5}\right)\left(\log\frac{[Mn^{2+}]}{[MnO_4^-][H^+]^8}\right) = 1.51 - \left(\frac{0.0592}{5}\right)\left(\log\frac{0.20}{0.10}\right) = 1.51\ V$$

$Cr_2O_7^{2-} + 14H^+ + 6e^- \longrightarrow 2Cr^{3+} + 7H_2O \qquad E^\circ = 1.33\ V$

$$E = E^\circ - \left(\frac{0.0592}{6}\right)\left(\log\frac{[Cr^{3+}]^2}{[Cr_2O_7^{2-}][H^+]^{14}}\right) = 1.33 - \left(\frac{0.0592}{6}\right)\left(\log\frac{(0.20)^2}{(0.10)(0.70)^{14}}\right) = 1.31\ V$$

$$E(\text{cell}) = E(MnO_4^-/Mn^{2+}) - E(Cr_2O_7^{2-}/Cr^{3+}) = 0.20\ V$$

(*b*) The overall reaction is the following. The hydronium ions cannot be "canceled" from the two sides, since the concentration is different in each solution, and the concentration affects the terms in the Nernst equation.

$$6MnO_4^- + 48H^+ (1.00\text{ M}) + 10Cr^{3+} + 35H_2O \longrightarrow 6Mn^{2+} + 24H_2O + 5Cr_2O_7^{2-} + 70H^+ (0.70\text{ M})$$

$$E = 0.18 - \left(\frac{0.0592}{30}\right)\left(\log \frac{(0.20)^6(0.10)^5(0.70)^{70}}{(0.10)^6(0.20)^{10}}\right) = 0.19\text{ V}$$

The results would have been identical if more significant figures had been used.

23.108 Calculate the potential of an indicator electrode, versus the standard hydrogen electrode, which originally contained 0.100 M MnO_4^- and 0.800 M H^+ and which has been treated with 90% of the Fe^{2+} necessary to reduce all the MnO_4^- to Mn^{2+}.

▌

$$MnO_4^- + 8H^+ + 5e^- \longrightarrow Mn^{2+} + 4H_2O \qquad E^\circ = 1.51\text{ V}$$
$$5Fe^{2+} \longrightarrow 5Fe^{3+} + 5e^-$$
$$MnO_4^- + 8H^+ + 5Fe^{2+} \longrightarrow 5Fe^{3+} + Mn^{2+} + 4H_2O$$
$$[MnO_4^-] = 0.010 \qquad [Fe^{3+}] = 0.450 \qquad [H^+] = 0.0800 \qquad [Mn^{2+}] = 0.0900$$

Since the Mn^{VII}/Mn^{II} and Fe^{III}/Fe^{II} systems are both in the same vessel, they must both experience the same half-cell potential.

$$E = 1.51 - \left(\frac{0.0592}{5}\right)\left(\log \frac{[Mn^{2+}]}{[MnO_4^-][H^+]^8}\right) = 1.51 - \left(\frac{0.0592}{5}\right)\left(\log \frac{(0.0900)}{(0.0100)(0.0800)^8}\right) = 1.39\text{ V}$$

23.109 To perform an analysis of a mixture of metal ions by electrodeposition, the second metal to be deposited must not begin plating out until the concentration ratio of the second to the first is about 10^6. What must be the minimum difference in standard potential of two metals which form dipositive ions in order for such an analysis to be feasible?

▌ The E value is the same for both metals, since both are present in the same solution.

$$0 = E_1 - E_2 = E_1^\circ - E_2^\circ - \left(\frac{0.0592}{2}\right)(\log 10^6) \qquad E_1^\circ - E_2^\circ = \left(\frac{0.0592}{2}\right)(6.00) = 0.18\text{ V}$$

23.110 At what potential should a solution containing 1 M $CuSO_4$, 1 M $NiSO_4$, and 2 M H_2SO_4 be electrolyzed so as to deposit essentially none of the nickel and all of the copper, leaving 1.0×10^{-9} M Cu^{2+}?

▌

$$Cu^{2+} + 2e^- \longrightarrow Cu \qquad E^\circ = 0.34\text{ V}$$

$$E = 0.34 - \left(\frac{0.0592}{2}\right)\left(\log \frac{1}{[Cu^{2+}]}\right) = 0.34 - (0.0296)(9.00) = 0.34 - 0.266 = 0.07\text{ V}$$

Electrolysis at 0.07 V would leave 1.00×10^{-9} M Cu^{2+} without reducing the Ni^{2+} concentration. [At 0.07 V, the concentration of nickel(II) ion could theoretically be as high as 6×10^{10} M. The calculation follows.]

$$Ni^{2+} + 2e^- \longrightarrow Ni \qquad E^\circ = -0.25\text{ V}$$

$$0.07 = -0.25 - \left(\frac{0.0592}{2}\right)\left(\log \frac{1}{[Ni^{2+}]}\right) \qquad 0.32 = (0.0296)(\log [Ni^{2+}]) \qquad [Ni^{2+}] = 6 \times 10^{10}\text{ M}$$

23.111 From the data for the two calomel, Hg_2Cl_2, half-reactions in Table 23.1, calculate the concentration of saturated KCl at 25 °C.

▌ Treat the E° value for the saturated calomel half-cell as an E value for the standard calomel half-cell.

$$Hg_2Cl_2 + 2e^- \longrightarrow 2Hg + 2Cl^- (\text{sat KCl}) \qquad E^\circ = 0.270\text{ V} \qquad E = 0.244\text{ V}$$

$$E = E^\circ - \left(\frac{0.0592}{2}\right)(\log [Cl^-]^2) \qquad 0.244 = 0.270 - (0.0592)(\log [Cl^-]) \qquad [Cl^-] = 2.8\text{ M}$$

23.112 For the cell $Pt|H_2\,(0.75\text{ atm})|HCl\,(0.25\text{ M})||Sn^{2+}\,(1.50\text{ M}), Sn^{4+}\,(0.60\text{ M})|Pt$ (*a*) write the half-cell reactions (*b*) write the cell reaction (*c*) calculate the cell potential (*d*) calculate the ratio of concentrations of tin(II) to tin(IV) which would cause the potential to be zero.

▌ (*a*)

$$H_2 \longrightarrow 2H^+ + 2e^- \qquad E^\circ = 0.00\text{ V}$$
$$Sn^{4+} + 2e^- \longrightarrow Sn^{2+} \qquad E^\circ = 0.13\text{ V}$$

(*b*) $H_2 + Sn^{4+} \longrightarrow 2H^+ + Sn^{2+} \qquad E^\circ = 0.13\text{ V}$

(*c*) $$E = 0.13 - \left(\frac{0.0592}{2}\right)\left(\log \frac{(1.50)(0.25)^2}{(0.60)(0.75)}\right) = 0.15\text{ V}$$

(*d*) $E = 0.00 = 0.13 - \left(\frac{0.0592}{2}\right)\left(\log\frac{[Sn^{2+}]}{[Sn^{4+}]} + \log\frac{(0.25)^2}{(0.75)}\right)$

$$\frac{2(0.13)}{0.0592} = \log\frac{[Sn^{2+}]}{[Sn^{4+}]} + 0.032 \qquad \log\frac{[Sn^{2+}]}{[Sn^{4+}]} = 4.36 \qquad \frac{[Sn^{2+}]}{[Sn^{4+}]} = 2.3 \times 10^4$$

23.113 Standard reduction potentials for two elements, X and Y, in various oxidation states are as follows:

$$X^{4+} + e^- \longrightarrow X^{3+} \quad E° = +0.6\text{ V} \qquad Y^{3+} + e^- \longrightarrow Y^{2+} \quad E° = +0.6\text{ V}$$
$$X^{3+} + e^- \longrightarrow X^{2+} \quad E° = -0.1\text{ V} \qquad Y^{2+} + e^- \longrightarrow Y^{+} \quad E° = +0.1\text{ V}$$
$$X^{2+} + 2e^- \longrightarrow X \quad E° = -1.0\text{ V} \qquad Y^{+} + e^- \longrightarrow Y \quad E° = +1.0\text{ V}$$

Predict the results of each of the following experiments: (*a*) X^{2+} is added to 1 M H^+ solution. (*b*) Y^+ is added to water. (*c*) 1 M Y^{2+} in 1 M H^+ solution is treated with O_2. (*d*) Y is added in excess to 1 M X^{3+}. (*e*) 25 mL of 0.14 M X^{4+} solution is added to 75 mL of 0.14 M Y^{2+} solution.

▮ Here, one must consider reactions of each substance with every substance present, including itself.

(*a*) $X^{2+} + H^+ \longrightarrow X^{3+} + \frac{1}{2}H_2$, $E° = +0.1$ V (*b*) $2Y^+ \longrightarrow Y^{2+} + Y$, $E° = +0.9$ V

(*c*) $4Y^{2+} + O_2 + 4H^+ \longrightarrow 2H_2O + 4Y^{3+}$, $E° = +0.6$ V

(*d*) No reaction (*e*) The reaction $X^{4+} + Y^{2+} \rightarrow X^{3+} + Y^{3+}$ has an $E°$ value = 0. The reaction will proceed until the concentration ratio of products to reactants adjusts itself so that E becomes 0; the reaction will then stop before all the reactants are used up.

23.114 Calculate the number of coulombs delivered by a Daniell cell, initially containing 1.00 L each of 1.00 M copper(II) ion and 1.00 M zinc(II) ion, which is operated until its potential drops to 1.00 V.

▮

$$E = E° - \left(\frac{0.0592}{2}\right)\left(\log\frac{[Zn^{2+}]}{[Cu^{2+}]}\right) \qquad 1.00 = 1.10 - (0.0296)\left(\log\frac{[Zn^{2+}]}{[Cu^{2+}]}\right)$$

$$\log\frac{[Zn^{2+}]}{[Cu^{2+}]} = 3.38 \qquad \text{hence} \qquad \frac{[Zn^{2+}]}{[Cu^{2+}]} = 2390$$

Since the zinc ion concentration plus the copper ion concentration must always total 2.00 M, $[Zn^{2+}] = 2.00$ and $[Cu^{2+}] = 0.00084$. Essentially all the copper has been consumed; thus 2.00 mol of electrons or 193 000 C has been delivered.

23.5 PRACTICAL APPLICATIONS

23.115 What desirable features are characteristic of a lead storage battery? What undesirable features, if any, are characteristic of a lead storage battery?

▮ The desirable features include rechargeability, relative inexpensiveness, portability, relatively constant potential, and some salvage value. Undesirable features include the use of (dangerous) concentrated sulfuric acid, a somewhat fragile nature, and the fact that lead is becoming more scarce.

23.116 Explain why blocks of magnesium are often strapped to the steel hulls of ocean-going ships.

▮ Magnesium acts, by cathodic protection, to prevent oxidation of the steel by transferring an excess of electrons to the steel.

23.117 What purpose(s) does chrome plating of steel serve? nickel plating?

▮ Chrome plating protects by cathodic action (see Problem 23.116); nickel protects by covering the entire surface with a less active metal (as long as no scratches or imperfections develop).

23.118 Explain why a homeowner should avoid attaching aluminum downspouts to galvanized steel gutters.

▮ The aluminum, while protecting the steel by cathodic protection, would itself be more susceptible to oxidation.

23.119 Explain concisely why a porous plate or a salt bridge is not required in a lead storage cell.

▮ The half-cells in a cell must be separated only if the oxidizing and/or reducing agent can migrate to the other half-cell. In the lead storage cell, the oxidizing agent, PbO_2, and the reducing agent, Pb, as well as their oxidation and reduction product, $PbSO_4$, are solids. That being the case, the half-cells do not need to be in separate vessels; they can be put into the same vessel, eliminating the necessity for a salt bridge or a porous partition.

23.120 Explain why the lead storage cell (*a*) has a relatively constant potential (*b*) has its state of charge signaled by its electrolyte density (*c*) needs no salt bridge (*d*) can be recharged.

▮ (*a*) The reagents are solids or concentrated solutes. (*b*) Discharge is accompanied by conversion of (dense) H_2SO_4 to (less dense) H_2O. (*c*) The oxidizing and reducing agents, in both discharge and recharge, are solids. Thus all the reagents can be placed in the same vessel. (*d*) The solid nature of each oxidizing and reducing agent prevents direct contact no matter which way the reactions are run. [Or same answer as (*c*).]

23.121 $Ni(OH)_2$ and NiO_2 are insoluble in NaOH solution. Design a practical rechargeable cell using these materials. Include in your description all chemical equations for each electrode reaction during charging and discharging. Could the state of charge of your cell be determined easily?

▮

$$NiO_2 + 2H_2O + 2e^- \longrightarrow Ni(OH)_2 + 2OH^- \qquad Ni + 2OH^- \longrightarrow Ni(OH)_2 + 2e^-$$

$$NiO_2 + Ni + 2H_2O \longrightarrow 2Ni(OH)_2$$

Plates of Ni and NiO_2 (inserted in an inert electrode) can be immersed in NaOH solution. The discharge will deposit $Ni(OH)_2$ on each electrode. Recharging will re-form the original electrode materials. Like those of the lead storage cell, the oxidizing agent, the reducing agent, and the oxidation and reduction products are all solids. Unlike the lead storage cell, the electrolyte concentration is affected only by a volume effect. The state of charge cannot be measured very accurately by the OH^- concentration.

23.122 Calculate the energy obtainable from a lead storage battery in which 0.100 mol lead is consumed. Assume a constant concentration of 10.0 M H_2SO_4.

▮

$$PbO_2 + 4H^+ + SO_4^{2-} + 2e^- \longrightarrow PbSO_4 + 2H_2O \qquad E^\circ = 1.70$$

$$Pb + SO_4^{2-} \longrightarrow PbSO_4 + 2e^- \qquad E^\circ = 0.31$$

$$PbO_2 + Pb + 4H^+ + 2SO_4^{2-} \longrightarrow 2PbSO_4 + 2H_2O \qquad E^\circ = 2.01$$

$$E = E^\circ - \left(\frac{0.0592}{2}\right)\left(\log\frac{1}{[H^+]^4[SO_4^{2-}]^2}\right) = 2.01 - \left(\frac{0.0592}{2}\right)\left(\log\frac{1}{(20)^4(10)^2}\right) = 2.22\text{ V}$$

$$(0.100\text{ mol Pb})\left(\frac{2\text{ mol }e^-}{\text{mol Pb}}\right)\left(\frac{96\,500\text{ C}}{\text{mol }e^-}\right) = 19\,300\text{ C}$$

$$\text{Energy} = qE = (19\,300\text{ C})(2.22\text{ V}) = 42.8\text{ kJ}$$

23.6 ELECTROCHEMICAL EQUILIBRIUM AND THERMODYNAMICS

23.123 (*a*) At equimolar concentrations of Fe^{2+} and Fe^{3+}, what must $[Ag^+]$ be so that the voltage of the galvanic cell made from the Ag^+/Ag and Fe^{3+}/Fe^{2+} electrodes equals zero? The reaction is $Fe^{2+} + Ag^+ \rightleftharpoons Fe^{3+} + Ag$. (*b*) Determine the equilibrium constant at 25 °C for the reaction.

▮ (*a*) For the reaction as written, $E^\circ = E^\circ(Ag^+/Ag) - E^\circ(Fe^{3+}/Fe^{2+}) = 0.799 - 0.771 = 0.028$ V. From the Nernst equation,

$$E = E^\circ - (0.0592)\left(\log\frac{[Fe^{3+}]}{[Fe^{2+}][Ag^+]}\right) \qquad 0 = 0.028 - (0.0592)\left(\log\frac{1}{[Ag^+]}\right) = 0.028 + (0.0592)(\log[Ag^+])$$

$$\log[Ag^+] = -\frac{0.028}{0.0592} = -0.47 \qquad [Ag^+] = 0.34$$

(*b*) To find the equilibrium constant, we must combine the relationship between K and ΔG°, with the relationship between ΔG° and E°, $\Delta G^\circ = -n\mathcal{F}E^\circ$.

$$\log K = -\frac{\Delta G^\circ}{2.303RT} = \frac{n\mathcal{F}E^\circ}{2.303RT} = \frac{nE^\circ}{0.0592}$$

Note that the same combination of constants, of value 0.0592 at 25 °C, occurs here as in the Nernst equation.

$$\log K = \frac{0.028}{0.0592} = 0.47 \qquad \text{from which} \qquad K = \frac{[Fe^{3+}]}{[Fe^{2+}][Ag^+]} = 3.0$$

Part (*a*) could have been solved alternatively by using the equilibrium constant, noting that $[Fe^{2+}] = [Fe^{3+}]$ and solving for $[Ag^+]$:

$$[Ag^+] = \frac{[Fe^{3+}]}{[Fe^{2+}] \times 3.0} = \frac{1}{3.0} = 0.33$$

The two solutions must be equivalent because the voltage of a galvanic cell becomes zero when the two couples are at equilibrium with each other.

23.124 An excess of liquid mercury was added to an acidified solution of 1.00×10^{-3} M Fe^{3+}. It was found that only 4.6% of the iron remained as Fe^{3+} at equilibrium at 25 °C. Calculate $E°(Hg_2^{2+}/Hg)$, assuming that the only reaction that occurred was $2Hg + 2Fe^{3+} \rightarrow Hg_2^{2+} + 2Fe^{2+}$.

▮ First we calculate the equilibrium constant for the reaction. At equilibrium,

$$[Fe^{3+}] = (0.046)(1.00 \times 10^{-3}) = 4.6 \times 10^{-5} \qquad [Fe^{2+}] = (1 - 0.046)(1.00 \times 10^{-3}) = 9.5 \times 10^{-4}$$

$$[Hg_2^{2+}] = \tfrac{1}{2}[Fe^{2+}] = 4.8 \times 10^{-4}$$

Liquid Hg is in excess and is in its standard state.

$$K = \frac{[Hg_2^{2+}][Fe^{2+}]^2}{[Fe^{3+}]^2} = \frac{(4.8 \times 10^{-4})(9.5 \times 10^{-4})^2}{(4.6 \times 10^{-5})^2} = 0.205$$

The standard potential of the cell corresponding to the reaction may be computed from the relation found in Problem 23.123(*b*):

$$E° = \left(\frac{0.0592}{n}\right)(\log K) = \frac{(0.0592)(-0.69)}{2} = -0.020\text{ V}$$

For the reaction as written,

$$E° = E°(Fe^{3+}/Fe^{2+}) - E°(Hg_2^{2+}/Hg)$$

or

$$E°(Hg_2^{2+}/Hg) = E°(Fe^{3+}/Fe^{2+}) - E° = 0.771 - (-0.020) = 0.791\text{ V}$$

23.125 In an electrochemical cell, what value does the cell potential have when the reaction is at equilibrium? What is the value at equilibrium of the ratio of concentrations which is part of the Nernst equation?

▮ $E = 0$ at equilibrium. At equilibrium, the concentration ratio equals K.

23.126 Estimate the standard reduction potential for the copper/copper sulfide electrode. For CuS, $K_{sp} = 8.5 \times 10^{-36}$.

▮ One can design a cell in which the half-reaction $CuS + 2e^- \rightleftharpoons Cu + S^{2-}$ occurs. The following cell is suitable: $Cu|Cu^{2+}||S^{2-}|CuS|Cu$. For this cell,

$$CuS + 2e^- \rightleftharpoons Cu + S^{2-} \qquad E°(Cu/CuS) = ?$$

$$Cu \rightleftharpoons Cu^{2+} + 2e^- \qquad E°(Cu/Cu^{2+}) = -0.34\text{ V}$$

$$\text{Cell:} \quad CuS \rightleftharpoons Cu^{2+} + S^{2-} \qquad E°(\text{cell}) = \left(\frac{0.0592}{2}\right)(\log K)$$

$$E°(\text{cell}) = E°(Cu/Cu^{2+}) + E°(Cu/CuS) = \left(\frac{0.0592}{2}\right)[\log(8.5 \times 10^{-36})] = -1.04\text{ V}$$

$$E°(Cu/CuS) = (-1.04\text{ V}) - (-0.34\text{ V}) = -0.70\text{ V}$$

23.127 Calculate $\Delta S°$ for the oxidation of 1.0 mol Zn in the presence of 1.00 M Zn^{2+} by 1.00 M Cu^{2+} in a Daniell cell at 298 K. Assume a large enough volume that there is no change in the concentrations of reactants or products.

▮

$$\Delta G° = -E°n\mathcal{F} = -(1.10)(2)(96\,500) = -212\text{ kJ} \qquad Zn + Cu^{2+} \longrightarrow Cu + Zn^{2+}$$

$$\Delta H° = (-36.34\text{ kcal/mol}) - (15.39\text{ kcal/mol}) = -51.73\text{ kcal} = -216\text{ kJ} \qquad \Delta G° = \Delta H° - T\Delta S°$$

$$\Delta S° = (\Delta H° - \Delta G°)/T = [-216\text{ kJ} - (-212\text{ kJ})]/298\text{ K} = -4000\text{ J}/298\text{ K} = -13\text{ J/K}$$

23.128 Estimate the cell potential of a Daniell cell having 1.00 M Zn^{2+} and originally having 1.00 M Cu^{2+} after sufficient ammonia has been added to the cathode compartment to make the NH_3 concentration 2.00 M. Does that half-cell continue to function as the cathode?

▮ The copper(II) ion concentration is calculated first:

$$K_f = \frac{[Cu(NH_3)_4^{2+}]}{[Cu^{2+}][NH_3]^4} = 1 \times 10^{12} = \frac{1.00}{x(2.00)^4} \qquad x = 6 \times 10^{-14}$$

$$Zn + Cu^{2+} \longrightarrow Cu + Zn^{2+}$$

$$E = E° - \left(\frac{0.0592}{2}\right)\left(\log\frac{[Zn^{2+}]}{[Cu^{2+}]}\right) = 1.10 - \left(\frac{0.0592}{2}\right)[\log(1.6 \times 10^{13})] = 0.71\text{ V}$$

Since the potential is positive, the copper electrode is still the cathode.

23.129 Using data from Tables 23.1 and 22.1, estimate the standard electrode potential for a cobalt/cobalt(II) sulfide electrode.

▌

$$CoS \rightleftharpoons Co^{2+} + S^{2-} \qquad K_{sp} = [Co^{2+}][S^{2-}] = 3 \times 10^{-26}$$

$$Co^{2+} + 2e^- \longrightarrow Co \qquad E^\circ = -0.28 \qquad CoS + 2e^- \longrightarrow Co + S^{2-} \qquad E^\circ = x$$

Subtraction of the first equation from the second yields

$$CoS \longrightarrow Co^{2+} + S^{2-} \qquad E^\circ = x + 0.28$$

$$E^\circ = \frac{(2.30RT)(\log K)}{n\mathscr{F}} = \left(\frac{0.0592}{2}\right)[\log(3 \times 10^{-26})] = x + 0.28$$

$$x = \left(\frac{0.0592}{2}\right)(-25.5) - 0.28 = -1.03 \text{ V}$$

23.130 Using data from Table 23.1, calculate the standard reduction potential for the half-reaction

$$2OCl^- + 2H_2O + 2e^- \longrightarrow Cl_2(g) + 4OH^-$$

▌ Appropriate half-reactions from the table must involve OCl^- and Cl_2 along with a reduction product common to them. The couples OCl^-/Cl^- and Cl_2/Cl^- are selected. Doubling the first and reversing the second yields

	E°	n	nE°
$2OCl^- + 2H_2O + 4e^- \longrightarrow 2Cl^- + 4OH^-$	0.94 V	4	3.76
$2Cl^- \longrightarrow Cl_2(g) + 2e^-$	−1.36 V	2	−2.72

Adding the chemical equations and the nE° values yields

$2OCl^- + 2H_2O + 2e^- \longrightarrow Cl_2(g) + 4OH^-$		2	1.04

$$E^\circ = \frac{nE^\circ}{n} = \frac{1.04}{2} = 0.52 \text{ V}$$

23.131 Calculate ΔG°_{298} and ΔS°_{298} for the reaction $2H_2(g) + O_2(g) \rightarrow 2H_2O(l)$.

▌ This reaction does not occur reversibly under ordinary laboratory conditions; however, a galvanic cell can be constructed for which it is the cell reaction. The appropriate half-reactions are combined as follows:

$$O_2 + 4H^+ + 4e^- \longrightarrow 2H_2O \qquad E^\circ = 1.23 \text{ V}$$

$$2H_2 \longrightarrow 4H^+ + 4e^- \qquad E^\circ = 0.00 \text{ V}$$

$$O_2 + 2H_2 \longrightarrow 2H_2O \qquad E^\circ(\text{cell}) = 1.23 \text{ V}$$

$$\Delta G^\circ_{298} = -n\mathscr{F}E^\circ = -(4 \text{ mol } e^-)(96\,500 \text{ C/mol } e^-)(1.23 \text{ V}) = -475\,000 \text{ J} = -114 \text{ kcal}$$

From Table 18.2, for the formation of water from hydrogen and oxygen, $\Delta H^\circ_f = -68.32$ kcal/mol. Therefore, ΔS° for the above reaction can be calculated as follows:

$$\Delta S^\circ_{298} = \frac{\Delta H^\circ - \Delta G^\circ}{T} = \frac{2(-68.32) - (-114)}{298 \text{ K}} = -0.076 \text{ kcal/K} = -76 \text{ cal/K}$$

23.132 Calculate the standard free energy change for the reaction $Zn + Cu^{2+} \rightarrow Cu + Zn^{2+}$.

▌ E°(cell) is obtained as follows:

$$Zn \longrightarrow Zn^{2+} + 2e^- \qquad E^\circ = 0.76 \text{ V}$$

$$Cu^{2+} + 2e^- \longrightarrow Cu \qquad E^\circ = 0.34 \text{ V}$$

$$Zn + Cu^{2+} \longrightarrow Zn^{2+} + Cu \qquad E^\circ(\text{cell}) = 1.10 \text{ V}$$

Since 2 mol electrons is associated with 1 mol of chemical reaction,

$$\Delta G^\circ = -n\mathscr{F}E^\circ(\text{cell}) = -(2 \text{ mol } e^-)\left(\frac{96\,500 \text{ C}}{\text{mol } e^-}\right)(1.10 \text{ V}) = -212\,000 \text{ J} = -212 \text{ kJ}$$

23.133 Given the following half-reactions and corresponding standard reduction potentials:

$$I_2 + 2e^- \longrightarrow 2I^- \qquad E^\circ = 0.54 \text{ V} \qquad Br_2 + 2e^- \longrightarrow 2Br^- \qquad E^\circ = 1.09 \text{ V}$$

(*a*) If I_2 and Br_2 are added to solutions containing I^- and Br^-, respectively, at unit concentrations, write an equation for the overall cell reaction. (*b*) Give the line notation for the galvanic cell in which the reaction will occur. Use Pt as the inert metal for each electrode. (*c*) Evaluate $\Delta G°$ in J for the overall cell reaciton. (*d*) How will an increase in the concentration of Br^- affect *E*(cell)?

▌ (*a*) $Br_2 + 2I^- \longrightarrow 2Br^- + I_2$ (*b*) $Pt|I_2, I^- \| Br^-, Br_2|Pt$

(*c*) $E° = 1.09 - 0.54 = 0.55$ V $\Delta G° = -n\mathscr{F}E° = -(2\text{ mol})(96\,500\text{ C/mol})(0.55\text{ V}) = 106\text{ kJ}$

(*d*) The value of *E*(ccll) will be reduced.

23.134 Consider the following equations for a cell reaction:

$$A + B \overset{1}{\rightleftharpoons} C + D \qquad 2A + 2B \overset{2}{\rightleftharpoons} 2C + 2D$$

How are $E°$ and K for the two reactions related?

▌ The $E°$ values are equal. $K_2 = \dfrac{[C]^2[D]^2}{[A]^2[B]^2} = (K_1)^2$

23.135 Given the following standard reduction potentials:

$$Sn^{4+} + 2e^- \longrightarrow Sn^{2+} \quad E° = 0.13\text{ V} \qquad Ag^+ + e^- \longrightarrow Ag \quad E° = 0.80\text{ V}$$

(*a*) Calculate the potential of the cell $Pt|Sn^{2+}$ (1.0 M), Sn^{4+} (1.0 M)$\|Ag^+$ (1.0 M)$|Ag$. (*b*) Write equations for both half-cell reactions and the overall cell reaction. (*c*) Calculate the value of $\Delta G°$ for the reaction. (*d*) In what way would an increase in the concentration of silver ion affect the potential of the cell?

▌ (*a*) $E° = -0.13 + 0.80 = 0.67$ V

(*b*) $Sn^{2+} \longrightarrow Sn^{4+} + 2e^-$ $e^- + Ag^+ \longrightarrow Ag$ $Sn^{2+} + 2Ag^+ \longrightarrow Sn^{4+} + 2Ag$

(*c*) $\Delta G° = -n\mathscr{F}E° = -(2)(96\,500\text{ C})(0.67\text{ V}) = -1.3 \times 10^5\text{ J}$

(*d*) The potential would be increased.

23.136 Given the reaction $2M + 6H^+ \rightarrow 2M^{3+} + 3H_2$, for which $\Delta H°_{298} = -3.00$ kcal, the entropies are 6.5 cal/K for M, −22.2 cal/K for M^{3+}, 31.2 cal/K for H_2, and −10.0 cal/K for H^+. $\Delta G°_f$ for H^+ is 0.00. Calculate (*a*) the standard free energy of formation of M^{3+} (*b*) $E°$ for the half-reaction $M^{3+} + 3e^- \rightarrow M$.

▌ (*a*) $\Delta S° = 2(-22.2\text{ cal/K}) + 3(31.2\text{ cal/K}) - 2(6.5\text{ cal/K}) - 6(-10.0\text{ cal/K}) = 96.2\text{ cal/K}$

$\Delta G° = \Delta H° - T\Delta S° = -3000\text{ cal} - (298\text{ K})(96.2\text{ cal/K}) = -31.7\text{ kcal}$

$\Delta G° = 2\,\Delta G°_f(M^{3+}) - 6\,\Delta G°_f(H^+) = 2\,\Delta G°_f(M^{3+})$

$\Delta G°_f(M^{3+}) = (-31.7\text{ kcal})/(2\text{ mol}) = -15.8\text{ kcal/mol}$

(*b*) $E° = -\dfrac{\Delta G°}{n\mathscr{F}} = -\dfrac{-31.7\text{ kcal}}{(6\text{ mol})(96\,500\text{ C/mol})} = \left(\dfrac{5.48 \times 10^{-5}\text{ kcal}}{\text{C}}\right)\left(\dfrac{10^3\text{ cal}}{\text{kcal}}\right)\left(\dfrac{4.184\text{ J}}{\text{cal}}\right) = 0.229\text{ V}$

23.137 From the data of Table 23.1, calculate the standard free energy change for the reaction $Cu^+ + I^- \rightarrow CuI$.

▌ $CuI + e^- \longrightarrow Cu + I^-$ $E° = -0.17$ V (from Table 23.1)

$Cu^+ + e^- \longrightarrow Cu$ $E° = +0.53$ V (from Problem 23.72)

Subtracting the second equation from the first:

$$CuI \longrightarrow Cu^+ + I^- \qquad \Delta G°_1 = -E°_1 n\mathscr{F} \qquad \Delta G°_2 = -E°_2 n\mathscr{F}$$

$$\Delta G° = \Delta G°_1 - \Delta G°_2 = n\mathscr{F}(-E°_1 + E°_2) = (1\text{ mol }e^-)(96\,590\text{ C/mol }e^-)(0.70\text{ V}) = 68\text{ kJ}$$

23.138 Given 1 mol copper atoms and 2 mol iodine atoms, calculate, from the data of Table 23.1, which of the following systems is lowest in free energy: (*a*) $Cu + I_2$ (*b*) $CuI + \frac{1}{2}I_2$ (*c*) CuI_2

▌ From the data of Table 23.1, it may be concluded that CuI_2 is not stable, at least in aqueous solution.

$Cu^{2+} + I^- + e^- \longrightarrow CuI$	$E° = 0.85$ V
$I^- \longrightarrow \frac{1}{2}I_2 + e^-$	$E° = -0.54$ V
$Cu^{2+} + 2I^- \longrightarrow \frac{1}{2}I_2 + CuI$	$E° = 0.31$ V

It also may be concluded that I^- does not spontaneously reduce CuI to Cu.

$$\begin{array}{ll} CuI + e^- \longrightarrow Cu + I^- & E^\circ = -0.17\ V \\ I^- \longrightarrow \tfrac{1}{2}I_2 + e^- & E^\circ = -0.54\ V \\ \hline CuI \longrightarrow Cu + \tfrac{1}{2}I_2 & E^\circ = -0.71\ V \end{array}$$

Thus $CuI + \frac{1}{2}I_2$ is the stable system—the lowest in free energy.

23.139 Calculate the free energy change per mol copper(II) ion formed in a cell consisting of a copper/copper(II) ion half-cell suitably connected to a silver/silver ion half-cell of sufficient size that the concentration of the ions is not changed from 1.00 M.

▌

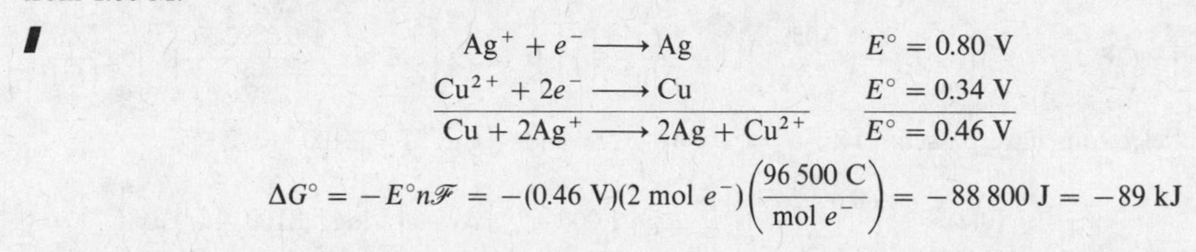

$$\begin{array}{ll} Ag^+ + e^- \longrightarrow Ag & E^\circ = 0.80\ V \\ Cu^{2+} + 2e^- \longrightarrow Cu & E^\circ = 0.34\ V \\ \hline Cu + 2Ag^+ \longrightarrow 2Ag + Cu^{2+} & E^\circ = 0.46\ V \end{array}$$

$$\Delta G^\circ = -E^\circ n\mathscr{F} = -(0.46\ V)(2\ mol\ e^-)\left(\frac{96\,500\ C}{mol\ e^-}\right) = -88\,800\ J = -89\ kJ$$

23.140 When half-reactions are added, the change in free energy of the total is merely the sum of the changes in free energy of the two halves. Show that this statement implies that the potentials are additive for the process in which half-reactions are added to yield an overall reaction, but that they are not additive when added to yield a third half-reaction.

▌ When two half-reactions are added to yield an overall reaction, the numbers of mol electrons involved in each half-reaction and in the overall reaction are necessarily the same.

$$n_1 = n_2 = n_3 = n \qquad E_3 = \frac{n_1E_1 + n_2E_2}{n_3} = \frac{nE_1 + nE_2}{n} = E_1 + E_2$$

However, when two half-reactions are added to yield a third half-reaction, the numbers of mol electrons in all three half-reactions cannot be the same.

23.141 Write the Nernst equation in terms of free energy change instead of potential.

▌ $E = E^\circ - \left(\frac{RT}{n\mathscr{F}}\right)(\ln Q)$ where ln means "natural logarithm" and Q represents the ratio of concentrations.

$$E = \frac{-\Delta G}{n\mathscr{F}} \qquad \frac{-\Delta G}{n\mathscr{F}} = \frac{-\Delta G^\circ}{n\mathscr{F}} - \left(\frac{RT}{n\mathscr{F}}\right)(\ln Q)$$

$$\Delta G = \Delta G^\circ + (RT)(\ln Q) = \Delta G^\circ + (2.303RT)(\log Q)$$

[We have reversed the derivation of the Nernst equation.]

23.142 Assuming that a constant current is delivered, how many kW·h of electricity can be produced by the reaction of 1.00 mol zinc with copper(II) ion in a Daniell cell in which all concentrations remain 1.00 M?

▌

$$E^\circ = E = 1.10\ V \qquad 2\mathscr{F}\ \text{of electricity will be used.}$$

$$-\Delta G = En\mathscr{F} = (1.10\ V)(2\ mol\ e^-)(96\,500\ C/mol\ e^-)$$

$$= 212\,300\ J = (212\,300\ W\cdot s)\left(\frac{1\ kW}{10^3\ W}\right)\left(\frac{1\ h}{3600\ s}\right) = 0.0590\ kW\cdot h$$

23.143 (*a*) What is the potential of the cell containing the Zn^{2+}/Zn and Cu^{2+}/Cu couples if the Zn^{2+} and Cu^{2+} concentrations are 0.10 M and 1.0×10^{-9} M, respectively, at 25 °C? (*b*) What is ΔG for the reduction of 1.0 mol of Cu^{2+} by Zn at the indicated concentrations of the ions, and what is ΔG° for the reaction, both at 25 °C?

▌ (*a*) The cell reaction is $Zn + Cu^{2+} \rightarrow Zn^{2+} + Cu$, with $n = 2$.

$$E = E^\circ - \frac{(0.0592)(\log Q)}{n}$$

E°, the standard cell potential, is equal to the difference between the standard electrode potentials, $0.34 - (-0.76) = 1.10$ V. Q, the concentration function, does not include terms for the solid metals, because the metals are in their standard states.

$$E = 1.10 - \left(\frac{0.0592}{2}\right)\left(\log \frac{[Zn^{2+}]}{[Cu^{2+}]}\right) = 1.10 - (0.0296)\left(\log \frac{1.0 \times 10^{-1}}{1.0 \times 10^{-9}}\right) = 1.10 - (0.0296)(8) = 0.86 \text{ V}$$

(*b*) $\Delta G = -n\mathscr{F}E = -(2 \text{ mol } e^-)(9.65 \times 10^4 \text{ C/mol } e^-)(0.86 \text{ V}) = -1.7 \times 10^5 \text{ J} = -170 \text{ kJ}$

$\Delta G° = -n\mathscr{F}E° = -(2 \times 9.65 \times 10^4 \times 1.10) \text{ J} = -212 \text{ kJ}$

23.144 By means of integral calculus, calculate the total energy theoretically obtainable from a Daniell cell which has a zinc electrode weighing 65.37 g immersed in 1.000 L of 1.000 M zinc ion and a copper electrode in 1.000 L of 1.00 M copper(II) ion.

The electrical work, w_{el}, is equal to $n\mathscr{F}E$. Let x equal the number of mol of electrons which have passed.

$$w_{el} = \int_0^2 E\mathscr{F}\,dx = \mathscr{F}\int_0^2 \left[E° - \left(\frac{0.0592}{2}\right)\left(\log \frac{[Zn^{2+}]}{[Cu^{2+}]}\right)\right] dx = \mathscr{F}\int_0^2 E°\,dx - \frac{0.0592\mathscr{F}}{2}\int_0^2 \log \frac{[Zn^{2+}]}{[Cu^{2+}]}\,dx$$

After x mol electrons have passed, $[Zn^{2+}] = 1.000 + x/2$ and $[Cu^{2+}] = 1.100 - x/2$

$$w_{el} = 1.10(2)\mathscr{F} - \left(\frac{0.0592\mathscr{F}}{2}\right)\left[\int_0^2 \log\left(1.000 + \frac{x}{2}\right) dx - \int_0^2 \log\left(1.100 - \frac{x}{2}\right) dx\right]$$

Integrating by substitution of variables,

$$w_{el} = 2.20\mathscr{F} - (0.0296\mathscr{F})\left[\left(\frac{0.5x + 1}{0.5}\right)\left[\log\left(1.000 + \frac{x}{2}\right)\right] - x - \left(\frac{-0.5x + 1.100}{-0.5}\right)\left[\log\left(1.100 - \frac{x}{2}\right)\right] + x\right]_0^2$$

$$= 2.20\mathscr{F} - (0.0296\mathscr{F})[(4 \log 2) - 0.2 - (2.2 \log 1.100)] = 2.17\mathscr{F} = 2.17(96\,500 \text{ C}) = 209 \text{ kJ}$$

23.145 Compute the potential of the Ag^+/Ag couple with respect to Cu^{2+}/Cu if the concentration of Ag^+ and Cu^{2+} are 4.2×10^{-6} M and 1.3×10^{-3} M, respectively. What is the value of ΔG for the reduction of 1.00 mol of Cu^{2+} by Ag at the indicated ion concentrations?

(*a*) $2Ag^+ + Cu \longrightarrow Cu^{2+} + 2Ag$

$$E = E° - \left(\frac{0.0592}{2}\right)\left(\log \frac{[Cu^{2+}]}{[Ag^+]^2}\right) = 0.46 - \left(\frac{0.0592}{2}\right)\left(\log \frac{1.3 \times 10^{-3}}{(4.2 \times 10^{-6})^2}\right) = 0.23 \text{ V}$$

(*b*) $\Delta G = -n\mathscr{F}E = -2(96500)(0.23) = 44 \text{ kJ}$

23.146 Copper could reduce zinc ions if the resultant copper ions can be kept at a sufficiently low concentration by the formation of an insoluble salt. What is the maximum concentration of Cu^{2+} in solution if this reaction is to occur when Zn^{2+} is 1.00 molar?

$$Cu + Zn^{2+} \longrightarrow Zn + Cu^{2+}$$

$$E = E° - \left(\frac{0.0592}{2}\right)\left(\log \frac{[Cu^{2+}]}{[Zn^{2+}]}\right) \qquad 0 = (-1.10) - (0.0296)\left(\log \frac{[Cu^{2+}]}{1.00}\right)$$

$$\log [Cu^{2+}] = \frac{-1.10}{0.0296} = -37.2 \qquad [Cu^{2+}] = 7 \times 10^{-38} \text{ M}$$

23.147 Evaluate the equilibrium constant for the reaction $Fe(CN)_6^{4-} + Co(dip)_3^{3+} \rightleftharpoons Fe(CN)_6^{3-} + Co(dip)_3^{2+}$.

$$Fe(CN)_6^{4-} \longrightarrow Fe(CN)_6^{3-} + e^- \qquad E° = -0.355 \text{ V}$$
$$e^- + Co(dip)_3^{3+} \longrightarrow Co(dip)_3^{2+} \qquad E° = 0.370 \text{ V}$$
$$E° = 0.015 \text{ V}$$

$$\Delta G° = -E°n\mathscr{F} = -(0.015)(1)(96\,500) = -1.45 \times 10^3 \text{ J} = (2.303RT)(\log K)$$

$$\log K = \frac{-\Delta G°}{2.303RT} = \frac{+1.45 \times 10^3 \text{ J}}{(2.303)(8.31 \text{ J/K})(300 \text{ K})} = 0.252$$

$$K = 1.8$$

23.148 When a rod of metallic lead was added to a 0.0100 M solution of $Co(en)_3^{3+}$, it was found that 68% of the cobalt complex was reduced to $Co(en)_3^{2+}$ by the lead.

(*a*) Find the value of K for $Pb + 2Co(en)_3^{3+} \rightleftharpoons Pb^{2+} + 2Co(en)_3^{2+}$

(*b*) What is the value of $E°(Co(en)_3^{3+})/(Co(en)_3^{2+})$?

▮ (*a*) $[\mathrm{Co(en)_3^{3+}}] = 0.0032 \qquad [\mathrm{Co(en)_3^{2+}}] = 0.0068 \qquad [\mathrm{Pb^{2+}}] = 0.0034$

$$K = \frac{[\mathrm{Pb^{2+}}][\mathrm{Co(en)_3^{2+}}]^2}{[\mathrm{Co(en)_3^{3+}}]^2} = \frac{(0.0068)^2(0.0034)}{(0.0032)^2} = 0.0154$$

(*b*) $\Delta G° = -E°n\mathcal{F} = (-2.303RT)(\log K)$

$$E°(\text{cell}) = \left(\frac{2.303RT}{n\mathcal{F}}\right)(\log K) = \left(\frac{0.0592}{n}\right)(\log K) = \left(\frac{0.0592}{2}\right)(\log 0.0154) = -0.0536\ \text{V}$$

$$E°(\text{Co(II)/Co(III)}) = E°(\text{cell}) + E°(\text{Pb}) = (-0.0536) + (-0.126) = -0.18\ \text{V}$$

23.149 A Tl^+/Tl couple was prepared by saturating 0.100 M KBr with TlBr and allowing the Tl^+ from the relatively insoluble bromide to equilibrate. This couple was observed to have a potential of -0.443 V with respect to a Pb^{2+}/Pb couple in which Pb^{2+} was 0.100 M. What is the solubility product constant of TlBr?

▮ The reduction potential of the lead half-cell is given by

$$E(\text{Pb}) = E°(\text{Pb}) - \left(\frac{0.0592}{2}\right)\left(\log\frac{1}{0.100}\right) = -0.126 - 0.0296 = -0.156\ \text{V}$$

The potential of the thallium half-cell is 0.443 V more negative:

$$E(\text{Tl}) = E(\text{Pb}) - 0.443 = -0.156 - 0.443 = -0.599\ \text{V}$$

For $Tl^+ + e^- \rightarrow Tl$ ($E° = -0.336$ V), the concentration of Tl^+ is given by

$$E = E° - (0.0592)\left(\log\frac{1}{[\mathrm{Tl^+}]}\right) \qquad -0.599 = -0.336 + (0.0592)(\log[\mathrm{Tl^+}])$$

$$[\mathrm{Tl^+}] = 3.6 \times 10^{-5} \qquad K_{sp} = [\mathrm{Tl^+}][\mathrm{Br^-}] = 3.6 \times 10^{-6}$$

23.150 K_d for complete dissociation of $Ag(NH_3)_2^+$ into Ag^+ and NH_3 is 6.0×10^{-8}. Calculate $E°$ for the following half-reaction by reference to Table 23.1: $Ag(NH_3)_2^+ + e^- \rightarrow Ag + 2NH_3$.

▮

$$\mathrm{Ag^+} + e \longrightarrow \mathrm{Ag} \qquad E° = 0.799\ \text{V} \qquad \Delta G° = -n\mathcal{F}E° = -7.71 \times 10^4\ \text{J}$$

$$\mathrm{Ag(NH_3)_2^+} \longrightarrow \mathrm{Ag^+} + 2\mathrm{NH_3} \qquad K = 6.0 \times 10^{-8}$$

$$\Delta G° = (-2.303RT)(\log K) = 4.12 \times 10^4\ \text{J} \qquad \mathrm{Ag(NH_3)_2^+} + e^- \longrightarrow \mathrm{Ag} + 2\mathrm{NH_3}$$

$$\Delta G° = (-7.71 \times 10^4\ \text{J}) + (4.12 \times 10^4\ \text{J}) = -3.59 \times 10^4\ \text{J} \qquad E° = -\frac{\Delta G°}{n\mathcal{F}} = 0.372\ \text{V}$$

23.151 Calculate K_s for formation of PdI_4^{2-} from Pd^{2+} and I^-.

▮

		$E°$, V	$\Delta G° = -E°n\mathcal{F}$, kJ
(*1*)	$PdI_4^{2-} + 2e^- \longrightarrow Pd + 4I^-$	0.18	−34.7
(*2*)	$Pd^{2+} + 2e^- \longrightarrow Pd$	0.92	−177.6
(*3*) = (*2*) − (*1*)	$Pd^{2+} + 4I^- \longrightarrow PdI_4^{2-}$		−142.9

$$\Delta G° = (-2.303RT)(\log K) \qquad \log K = \frac{+142.9 \times 10^3\ \text{J}}{2.303RT} = 25.1 \qquad K = 1.1 \times 10^{25}$$

23.152 Reference tables give the following entry:

$$\mathrm{HO_2^-} + \mathrm{H_2O} + 2e^- \longrightarrow 3\mathrm{OH^-} \qquad E° = 0.88\ \text{V}$$

Combining this information with relevant entries in Table 23.1, find K_1 for the acid dissociation of H_2O_2.

▮ $H_2O_2 \longrightarrow H^+ + HO_2^-$

		$E°$, V	$\Delta G° = -E°n\mathcal{F}$, kJ
(*1*)	$HO_2^- + H_2O + 2e^- \longrightarrow 3OH^-$	0.88	−170
(*2*)	$H_2O_2 + 2H^+ + 2e^- \longrightarrow 2H_2O$	1.77	−342
(*3*) = (*2*)−(*1*)	$H_2O_2 + 2H^+ + 3OH^- \longrightarrow 3H_2O + HO_2^-$		−172

Add one hydrogen ion to each side:

$$(4)\quad H_2O_2 + 3H^+ + 3OH^- \longrightarrow 3H_2O + HO_2^- + H^+ \qquad \Delta G^\circ = -172\text{ kJ}$$

$$K_4 = \frac{[HO_2^-][H^+]}{[H_2O_2][H^+]^3[OH^-]^3} = \frac{K_1}{(K_w)^3}$$

$$\log K_4 = -\frac{\Delta G^\circ}{2.303RT} = 30.1 \qquad K_4 = 1.3 \times 10^{30}$$

$$K_1 = K_4(K_w)^3 = (1.3 \times 10^{30})(1.0 \times 10^{-14})^3 = 1.3 \times 10^{-12}$$

23.153 From data in Table 23.1, calculate the overall stability constant, K_s, of $Ag(S_2O_3)_2^{3-}$ at 25 °C.

▌ There are two entries in the table for couples connecting the zero and +1 oxidation states of silver.

$$(1)\quad Ag^+ + e^- \longrightarrow Ag \qquad E^\circ = 0.799\text{ V}$$

$$(2)\quad Ag(S_2O_3)_2^{3-} + e^- \longrightarrow Ag + 2S_2O_3^{2-} \qquad E^\circ = 0.017\text{ V}$$

Process (*1*) refers to the couple in which Ag^+ is at unit concentration; the Ag^+ concentration to which the E° for (*2*) refers is that value which satisfies the complex ion equilibrium when the other species are at unit concentration.

$$[Ag^+] = \frac{[Ag(S_2O_3)_2^{3-}]}{K_s[S_2O_3^{2-}]^2} = \frac{1}{K_s}$$

In other words, the standard conditions for couple (*2*) may be thought of as a nonstandard condition, $[Ag^+] = 1/K_s$, for couple (*1*). Then the Nernst equation for (*1*) gives

$$E = E^\circ - (0.0592)\left(\log\frac{1}{[Ag^+]}\right) \qquad \text{or} \qquad 0.017 = 0.799 - (0.0592)(\log K_s)$$

from which

$$\log K_s = \frac{0.799 - 0.017}{0.0592} = 13 \qquad \text{and} \qquad K_s = 10^{13}$$

23.154 Knowing that K_{sp} for AgCl is 1.0×10^{-10}, calculate E for a silver/silver chloride electrode immersed in 1.00 M KCl at 25 °C.

▌ The electrode process is a special case of the Ag^+/Ag couple, except that silver in the +1 oxidation state collects as solid AgCl on the electrode itself. Even solid AgCl, however, has some Ag^+ in equilibrium with it in solution. This $[Ag^+]$ can be computed from the K_{sp} equation:

$$[Ag^+] = \frac{K_{sp}}{[Cl^-]} = \frac{1.0 \times 10^{-10}}{1.00} = 1.0 \times 10^{-10}$$

This value for $[Ag^+]$ can be inserted into the Nernst equation for the Ag^+/Ag half-reaction.

$$Ag^+ + e^- \longrightarrow Ag \qquad E^\circ = 0.799\text{ V}$$

$$E = E^\circ - \left(\frac{0.0592}{1}\right)\left(\log\frac{1}{[Ag^+]}\right) = 0.799 - (0.0592)\left(\log\frac{1}{1.0 \times 10^{-10}}\right) = 0.799 - 0.592 = 0.207\text{ V}$$

23.155 Consider the following half-reactions:

$$PbO_2(s) + 4H^+(aq) + SO_4^{2-}(aq) + 2e^- \longrightarrow PbSO_4(s) + 2H_2O \qquad E^\circ = 1.70\text{ V}$$

$$PbSO_4(s) + 2e^- \longrightarrow Pb(s) + SO_4^{2-}(aq) \qquad E^\circ = -0.31\text{ V}$$

(*a*) Calculate the value of E° for the cell. (*b*) Calculate the voltage generated by the cell if $[H^+] = 0.10$ M and $[SO_4^{2-}] = 2.0$ M. (*c*) What voltage is generated by the cell when it is at chemical equilibrium?

▌ (*a*) (1.70 V) − (−0.31 V) = 2.01 V

(*b*) $E = E^\circ - \left(\frac{0.0592}{n}\right)\left(\log\frac{1}{[H^+]^4[SO_4^{2-}]^2}\right) = 2.01 - \left(\frac{0.0592}{2}\right)\left(\log\frac{1}{(0.10)^4(2.0)^2}\right) = 1.91\text{ V}$

(*c*) $E = 0$ at equilibrium

23.156 The overall formation constant for the reaction of 6 mol CN^- with cobalt(II) is 1×10^{19}. The standard reduction potential for the reaction $Co(CN)_6^{3-} + e^- \rightarrow Co(CN)_6^{4-}$ is −0.83 V. Using the data of Table 23.1, calculate the overall formation constant of $Co(CN)_6^{3-}$.

▌ For the reaction $Co^{3+} + e^- \rightarrow Co^{2+}$, $E° = 1.82$ V and $E = E° - (0.0592)\left(\log \frac{[Co^{2+}]}{[Co^{3+}]}\right)$.

$$Co^{2+} + 6CN^- \longrightarrow Co(CN)_6^{4-} \qquad 1 \times 10^{19} = \frac{[Co(CN)_6^{4-}]}{[Co^{2+}][CN^-]^6}$$

$$Co^{3+} + 6CN^- \longrightarrow Co(CN)_6^{3-} \qquad K_f = \frac{[Co(CN)_6^{3-}]}{[Co^{3+}][CN^-]^6}$$

$$\frac{[Co^{2+}]}{[Co^{3+}]} = \frac{[Co(CN)_6^{4-}]/[CN^-]^6(1 \times 10^{19})}{[Co(CN)_6^{3-}]/[CN^-]^6 K_f}$$

The *standard* reduction potential (referring to 1 M concentration of each *complex*) can be regarded as the *actual* potential, E, of the Co^{3+}/Co^{2+} couple.

$$-0.83 = 1.82 - (0.0592)\left(\log \frac{1/(1 \times 10^{19})}{1/K_f}\right) \qquad -2.65 = (-0.0592)\left(\log \frac{K_f}{1 \times 10^{19}}\right)$$

$$\log \frac{K_f}{10^{19}} = 44.8 \qquad \frac{K_f}{10^{19}} = 6 \times 10^{44} \qquad K_f = 6 \times 10^{63}$$

23.157 In aqueous solution, cobalt(III) ion is able to oxidize water. $Co^{3+} + e^- \rightarrow Co^{2+}$, $E° = 1.842$ V. The formation constant for $[Co(NH_3)_6]^{3+}$ is 5×10^{33} and for $[Co(NH_3)_6]^{2+}$, 1×10^5. Show that an aqueous solution of $[Co(NH_3)_6]^{3+}$ in 1 M NH_3 does not oxidize water.

▌

$$Co(NH_3)_6^{2+} \rightleftharpoons 6NH_3 + Co^{2+} \qquad K = 1 \times 10^{-5} = \frac{[Co^{2+}][NH_3]^6}{[Co(NH_3)_6^{2+}]}$$

$$[[Co(NH_3)_6]^{2+}] = [Co^{2+}] \times 10^5 \qquad [Co(NH_3)_6^{3+}] = [Co^{3+}](5 \times 10^{33})$$

$$Co^{3+} + e^- \longrightarrow Co^{2+}$$

$$E = E° - \left(\frac{0.0592}{n}\right)\left(\log \frac{[Co^{2+}]}{[Co^{3+}]}\right) = 1.842 - \left(\frac{0.0592}{1}\right)\left(\log \frac{[Co(NH_3)_6^{2+}]/10^5}{[Co(NH_3)_6^{3+}]/(5 \times 10^{33})}\right)$$

$$= 1.842 - (0.0592)[\log (5 \times 10^{28})] - (0.0592)\left(\log \frac{[Co(NH_3)_6^{2+}]}{[Co(NH_3)_6^{3+}]}\right)$$

$$= 1.842 - (0.0592)(28.7) - (0.0592)\left(\log \frac{[Co(NH_3)_6^{2+}]}{[Co(NH_3)_6^{3+}]}\right)$$

$$= 1.842 - 1.699 - (0.0592)\left(\log \frac{[Co(NH_3)_6^{2+}]}{[Co(NH_3)_6^{3+}]}\right)$$

$$E = 0.143 - (0.0592)\left(\log \frac{[Co(NH_3)_6^{2+}]}{[Co(NH_3)_6^{3+}]}\right)$$

This equation is exactly in the form of the Nernst equation for the reduction of $Co(NH_3)_6^{3+}$ to $Co(NH_3)_6^{2+}$ with an $E°$ value of 0.143 V. This small $E°$ value is not sufficiently positive for the overall cell reaction with water to have a positive E value.

CHAPTER 24

Nuclear and Radiochemistry

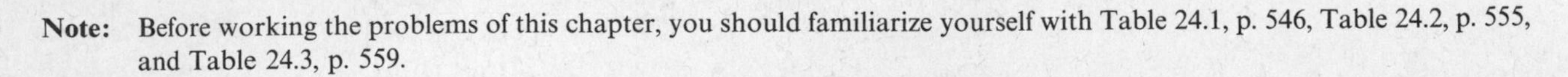

Note: Before working the problems of this chapter, you should familiarize yourself with Table 24.1, p. 546, Table 24.2, p. 555, and Table 24.3, p. 559.

24.1 NUCLEAR PARTICLES AND NUCLEAR REACTIONS

The designations for small particles are:

p	proton	${}_{1}^{1}\mathrm{H}$	γ	gamma ray (photon)	${}_{0}^{0}\gamma$
d	deuteron	${}_{1}^{2}\mathrm{H}$	n	neutron	${}_{0}^{1}\mathrm{n}$
α	alpha particle	${}_{2}^{4}\mathrm{He}$	β^{+}	positron	${}_{+1}^{0}\beta$
β^{-}	beta particle (electron)	${}_{-1}^{0}\beta$			

24.1 Select from the following list of nuclides (*a*) the isotopes (*b*) the isobars (*c*) the isotones: ${}_{18}^{40}\mathrm{Ar}$, ${}_{19}^{41}\mathrm{K}$, ${}_{21}^{40}\mathrm{Sc}$, ${}_{21}^{42}\mathrm{Sc}$, ${}_{40}^{90}\mathrm{Zr}$

▮ (*a*) Isotopes have the same atomic numbers: ${}_{21}^{40}\mathrm{Sc}$ and ${}_{21}^{42}\mathrm{Sc}$. (*b*) Isobars have the same mass numbers: ${}_{18}^{40}\mathrm{Ar}$ and ${}_{21}^{40}\mathrm{Sc}$. (*c*) Isotones have the same numbers of neutrons: ${}_{18}^{40}\mathrm{Ar}$ and ${}_{19}^{41}\mathrm{K}$.

24.2 Show that a mass of 1.00 u is equivalent to 932 MeV of energy.

▮ Since

$$1\ \mathrm{u} = \frac{1}{6.02\times 10^{23}}\ \mathrm{g} = \frac{1}{6.02\times 10^{26}}\ \mathrm{kg}$$

$$E = mc^2 = \left(\frac{1\ \mathrm{kg}}{6.02\times 10^{26}}\right)(3\times 10^{8}\ \mathrm{m/s})^2 = 1.49\times 10^{-10}\ \mathrm{J}$$

But

$$1\ \mathrm{MeV} = 10^{6}\ \mathrm{eV} = (10^{6})(1.6\times 10^{-19}\ \mathrm{C})(1.0\ \mathrm{V}) = 1.6\times 10^{-13}\ \mathrm{J}$$

and so

$$E = (1.49\times 10^{-10}\ \mathrm{J})\left(\frac{1\ \mathrm{MeV}}{1.6\times 10^{-13}\ \mathrm{J}}\right) = 932\ \mathrm{MeV}$$

[A more precise value is 931.5 MeV/u.]

24.3 Do the designations ${}_{1}^{1}\mathrm{H}$ and ${}_{0}^{1}n$ for the proton and neutron imply that these two particles are of equal mass (1 u)?

▮ No. Mass numbers reflect actual masses only to the nearest u. The precise masses are 1.007 276 5 u for the proton and 1.008 665 0 u for the neutron.

24.4 What are the numbers of protons, electrons, and neutrons in ${}^{239}\mathrm{Pu}$?

▮ Protons: 94, from the periodic table. Electrons: 94, since the atom is uncharged. Neutrons: $239 - 94 = 145$.

24.5 Write the complete nuclear symbols for natural fluorine and natural arsenic. Each has only one naturally occurring isotope.

▮ Since each element has only one naturally occurring isotope, the mass number is the integer closest to the atomic weight. ${}_{9}^{19}\mathrm{F}$, ${}_{33}^{75}\mathrm{As}$

24.6 How many protons, neutrons, and electrons are there in each of the following atoms: (*a*) ${}^{3}\mathrm{He}$ (*b*) ${}^{12}\mathrm{C}$ (*c*) ${}^{206}\mathrm{Pb}$?

▮ (*a*) From the periodic table, we see that the atomic number of He is 2; therefore the nucleus must contain 2 protons. Since the mass number of this isotope is 3, the sum of the numbers of protons and neutrons is 3; therefore there is 1 neutron. The number of electrons in the atom is the same as the atomic number, 2. (*b*) The atomic number of carbon is 6; hence the nucleus must contain 6 protons. The number of neutrons is $12 - 6 = 6$. The number of electrons is the same as the atomic number, 6. (*c*) The atomic number of lead is 82; hence there are 82 protons in the nucleus. The number of neutrons is $206 - 82 = 124$. There are 82 electrons.

24.7 Determine the number of (*a*) protons, (*b*) neutrons, (*c*) electrons, in each of the following atoms: (1) ^{70}Ge, (2) ^{72}Ge, (3) ^{9}Be, (4) ^{235}U.

▮ (1) (*a*) 32 (*b*) 38 (*c*) 32 (2) (*a*) 32 (*b*) 40 (*c*) 32 (3) (*a*) 4 (*b*) 5 (*c*) 4 (4) (*a*) 92 (*b*) 143 (*c*) 92

24.8 Classify each of the following nuclides as "probably stable," "beta emitter," or "positron emitter": $^{49}_{20}Ca$ $^{195}_{80}Hg$ $^{208}_{82}Pb$ $^{8}_{5}B$ $^{150}_{67}Ho$ $^{30}_{13}Al$ $^{120}_{50}Sn$ $^{94}_{36}Kr$

▮ The nuclides near the "belt of stability" are probably stable; those above that belt are beta emitters; those below, positron emitters. Thus the stable nuclides are ^{208}Pb and ^{120}Sn; the beta emitters are ^{49}Ca, ^{30}Al, and ^{94}Kr; the positron emitters are ^{195}Hg, ^{8}B, and ^{150}Ho.

24.9 Of the three isobars $^{114}_{48}Cd$, $^{114}_{49}In$, and $^{114}_{50}Sn$, which is likely to be radioactive? Explain your choice.

▮ $^{114}_{49}In$, which has an odd number of protons and an odd number of neutrons.

24.10 Which of the following nuclides is least likely to be stable? (*a*) $^{40}_{20}Ca$ (*b*) $^{30}_{13}Al$ (*c*) $^{119}_{50}Sn$ (*d*) $^{55}_{25}Mn$ (*e*) $^{32}_{16}S$

▮ ^{30}Al, which has an odd number of protons and an odd number of neutrons.

24.11 Using the approximate equation for the radius of the nucleus, $r = 1.4A^{1/3} \times 10^{-13}$ cm, calculate the density of the nucleus of $^{107}_{47}Ag$ and compare it with the density of metallic silver (10.5 g/cm³).

▮
$$r = 1.4(107)^{1/3} \times 10^{-13}\ \text{cm} = 6.65 \times 10^{-13}\ \text{cm}$$

The volume, mass, and density of a single nucleus are calculated from the radius by the following equations:

$$V = \tfrac{4}{3}\pi r^3 = (\tfrac{4}{3}\pi)(6.65 \times 10^{-13}\ \text{cm})^3 = 1.23 \times 10^{-36}\ \text{cm}^3$$

$$m = \frac{107\ \text{g/mol}}{6.02 \times 10^{23}\ \text{atoms/mol}} = 1.78 \times 10^{-22}\ \text{g}$$

$$d = \frac{m}{V} = \frac{1.78 \times 10^{-22}\ \text{g}}{1.23 \times 10^{-36}\ \text{cm}^3} = 1.4 \times 10^{14}\ \text{g/cm}^3$$

The nuclear density is 1.4×10^{13} times that of silver metal.

24.12 Antimatter consists of particles which have properties and characteristics similar to particles in known atoms but opposite charges. For example, an electron and a positron are two particles identical in every respect except for the sign of their charges. Speculate on the existence of a world composed of negative antiprotons and positrons (along with an array of other antiparticles) and neutrons. Would the chemistry in such a world have formulas similar to those of compounds on earth? What would happen if material from earth encountered material from the "antiworld"?

▮ One can imagine that everything would seem "natural," with "hydrogen" atoms consisting of one antiproton and one positron, "carbon" atoms having six antiprotons and six neutrons in the nucleus and six positrons outside the nucleus, etc. The "carbon" and "hydrogen" could form "CH_4" by the "usual" rules of covalent bond formation. Nuclear reactions in which electrons were produced would lead to annihilation reactions: $\beta^- + \beta^+ \rightarrow$ energy. Similarly, reaction of material from the two worlds, e.g., ordinary hydrogen and "hydrogen" from the antiworld, would also lead to annihilation reactions: Proton + antiproton → energy

24.13 When the pure nuclide $^{50}_{24}Cr$ is bombarded with alpha particles, two reactions occur and neutrons and deuterons are observed as product particles. Write equations showing the formation of the possible product nuclides.

▮ $^{50}_{24}Cr(\alpha, n)^{53}_{26}Fe$ $^{50}_{24}Cr(\alpha, d)^{52}_{25}Mn$ (The sum of the subscripts must be the same on both sides of the equation, as must the sum of the superscripts.)

24.14 Complete the following nuclear equations:
(*a*) $^{14}_{7}N + ^{4}_{2}He \longrightarrow ^{17}_{8}O + \cdots$ (*b*) $^{9}_{4}Be + ^{4}_{2}He \longrightarrow ^{12}_{6}C + \cdots$
(*c*) $^{9}_{4}Be(p, \alpha)\cdots$ (*d*) $^{30}_{15}P \longrightarrow ^{30}_{14}S + \cdots$
(*e*) $^{3}_{1}H \longrightarrow ^{3}_{2}He + \cdots$ (*f*) $^{43}_{20}Ca(\alpha, \ldots)^{46}_{21}Sc$

▮ (*a*) The sum of the subscripts on the left is $7 + 2 = 9$. The subscript of the first product on the right is 8. Hence the second product on the right must have a subscript (nuclear charge) of 1. The sum of the superscripts on the left is $14 + 4 = 18$. The superscript of the first product on the right is 17. Hence the second product on the right must have a superscript (mass number) of 1. The particle with nuclear charge 1 and mass number 1 is the proton, ${}^{1}_{1}H$.

(*b*) The charge of the second product particle (its subscript) is $(4+2)-6=0$. The mass number of the particle (its superscript) is $(9+4)-12=1$. Hence the particle must be the neutron, ${}^{1}_{0}n$.

(*c*) The reactants, ${}^{9}_{4}Be$ and ${}^{1}_{1}H$, have a combined nuclear charge of 5 and mass numbers of 10. In addition to the α particle, a product will be formed of charge $5-2=3$, and mass $10-4=6$. This is ${}^{6}_{3}Li$, since lithium is the element of atomic number 3.

(*d*) The charge of the second particle is $15-14=+1$. The mass number is $30-30=0$. Hence the particle must be the positron, ${}^{0}_{+1}e$.

(*e*) The charge of the second particle is $1-2=-1$. Its mass number is $3-3=0$. Hence the particle must be an electron, ${}^{0}_{-1}e$.

(*f*) The reactants, ${}^{43}_{20}Ca$ and ${}^{4}_{2}He$, have a combined nuclear charge of 22 and mass number of 47. The ejected product will have a charge $22-21=1$, and mass $47-46=1$. This is a proton and should be represented within the parentheses by *p*.

24.15 An alkaline earth element is radioactive. It and its daughter elements decay by emitting three alpha particles in succession. In what group should the resulting element be found?

▮ In emitting three alpha particles, six protons (as well as six neutrons) are lost from the nucleus. The atomic number is reduced by 6, and the resulting element is in group IV of the preceding period.

24.16 By natural radioactivity, ${}^{238}U$ emits an α particle. The heavy residual nucleus is called UX_1. UX_1 in turn emits a β^- particle. The heavy residual nucleus from this radioactive process is called UX_2. Determine the atomic numbers and mass numbers of (*a*) UX_1 (*b*) UX_2.

▮ The subscripts and superscripts must total to the same value on each side of the equation.

${}^{238}_{92}U \longrightarrow {}^{4}_{2}He + {}^{238}_{90}UX_1$ With atomic number 90 and mass number 234, the nuclide is ${}^{234}_{90}Th$.

${}^{234}_{90}Th \longrightarrow {}^{0}_{-1}\beta + {}^{234}_{91}UX_2$ With atomic number 91 and mass number 234, UX2 is ${}^{234}_{91}Pa$.

24.17 ${}^{213}_{83}Bi$ decays with emission of an α particle. The resulting nuclide is unstable and emits a β particle. What is the final nuclide formed?

▮ $${}^{213}_{83}Bi \longrightarrow {}^{4}_{2}He + {}^{209}_{81}Tl \qquad {}^{209}_{81}Tl \longrightarrow {}^{0}_{-1}\beta + {}^{209}_{82}Pb$$

24.18 What symbol is needed to complete the nuclear equation ${}^{63}_{29}Cu(p, \ldots){}^{62}_{29}Cu$?

▮ ${}^{63}_{29}Cu + {}^{1}_{1}H \longrightarrow {}^{62}_{29}Cu + {}^{2}_{1}H$ The ejected particle is a deuteron, *d*.

24.19 What symbol is missing in the following representation of a nuclear reaction? ${}^{9}_{4}Be\ (\alpha, \ldots){}^{12}_{6}C$

▮ *n* (neutron) ${}^{9}_{4}Be + {}^{4}_{2}He \longrightarrow {}^{1}_{0}n + {}^{12}_{6}C$

24.20 ${}^{140}_{56}Ba$ decays by β^- emission. Which nuclide would be the product of the decay process?

▮ $${}^{140}_{56}Ba \longrightarrow {}^{0}_{-1}\beta + {}^{140}_{57}La$$

24.21 Which of the following nuclides is the terminal member of the naturally occurring radioactive series which begins with ${}^{232}_{90}Th$?

(*a*) ${}^{209}_{83}Bi$ (*b*) ${}^{208}_{82}Pb$ (*c*) ${}^{206}_{82}Pb$ (*d*) ${}^{207}_{82}Pb$ (*e*) ${}^{210}_{83}Bi$

▮ ${}^{208}_{82}Pb$ (It is a member of the 4*n* series, as is ${}^{232}Th$.)

24.22 By radioactivity, ${}^{239}_{93}Np$ emits a β^- particle. The residual heavy nucleus is also radioactive and gives rise to ${}^{235}U$ by its radioactive process. What small particle is emitted simultaneously with the formation of ${}^{235}U$?

▮ ${}^{239}_{93}Np \longrightarrow {}^{0}_{-1}\beta + {}^{239}_{94}Pa \longrightarrow {}^{235}_{92}U + {}^{4}_{2}He$ An α particle is emitted.

24.23 Complete the following equations.

(*a*) ${}^{23}_{11}Na + {}^{4}_{2}He \longrightarrow {}^{26}_{12}Mg + ?$ (*b*) ${}^{64}_{29}Cu \longrightarrow \beta^+ + ?$

(*c*) ${}^{106}Ag \longrightarrow {}^{106}Cd + ?$ (*d*) ${}^{10}_{5}B + {}^{4}_{2}He \longrightarrow {}^{13}_{7}N + ?$

▌ *(a)* $^{23}_{11}Na + ^{4}_{2}He \rightarrow ^{26}_{12}Mg + ^{1}_{1}H$ *(b)* $^{64}_{29}Cu \rightarrow ^{0}_{1}\beta$(positron) $+ ^{64}_{28}Ni$ *(c)* $^{106}_{47}Ag \rightarrow ^{106}_{48}Cd + ^{0}_{-1}\beta$. The atomic numbers may be read from the periodic table. (*Note*: In general, mass numbers cannot be found on the periodic table; atomic weights and atomic numbers are there.) *(d)* $^{10}_{5}B + ^{4}_{2}He \rightarrow ^{13}_{7}N + ^{1}_{0}n$

24.24 Complete the notations for the following nuclear processes. *(a)* $^{24}Mg(d, \alpha)$? *(b)* $^{26}Mg(d, p)$? *(c)* $^{40}Ar(\alpha, p)$? *(d)* $^{12}C(d, n)$? *(e)* $^{130}Te(d, 2n)$? *(f)* $^{55}Mn(n, \gamma)$? *(g)* $^{59}Co(n, \alpha)$?

▌ The condensed notation includes in parentheses the small particle used as a projectile followed by the small particle which is a product. This type of notation can be expanded. Atomic numbers (but not mass numbers) may be read from the periodic table.

(a) $^{24}Mg(d, \alpha)$? $\quad ^{24}Mg + d \longrightarrow \alpha + ? \quad ^{24}_{12}Mg + ^{2}_{1}H \longrightarrow ^{4}_{2}He + ^{22}_{11}Na$

(b) $^{26}Mg + d \longrightarrow p + ? \quad ^{26}_{12}Mg + ^{2}_{1}H \longrightarrow ^{1}_{1}H + ^{27}_{12}Mg$

(c) $^{40}_{18}Ar + ^{4}_{2}He \longrightarrow ^{1}_{1}H + ^{43}_{19}K$ $\quad$ *(d)* $^{12}_{6}C + ^{2}_{1}H \longrightarrow ^{1}_{0}n + ^{13}_{7}N$

(e) $^{130}_{52}Te + ^{2}_{1}H \longrightarrow 2\,^{1}_{0}n + ^{130}_{53}I$ $\quad$ *(f)* $^{55}_{25}Mn + ^{1}_{0}n \longrightarrow ^{0}_{0}\gamma + ^{56}_{25}Mn$

(g) $^{59}_{27}Co + ^{1}_{0}n \longrightarrow ^{4}_{2}He + ^{56}_{25}Mn$

24.25 If a nuclide of an element in group IA of the periodic table undergoes radioactive decay by emitting positrons, what is the periodic group of the expected resulting element?

▌ When an atom emits a positron, its atomic number decreases by 1. An alkali metal would be converted to a noble gas by such a process.

24.26 If an atom of ^{235}U, after absorption of a slow neutron, undergoes fission to form an atom of ^{139}Xe and an atom of ^{94}Sr, what other particles are produced, and how many?

▌

$$^{235}_{92}U + ^{1}_{0}n \longrightarrow ^{139}_{54}Xe + ^{94}_{38}Sr + ?$$

The subscripts already balance, so the subscript of the unknown particle(s) must be zero. The superscripts on the right total 3 less than those on the left. There must be three neutrons ($^{1}_{0}n$) also produced.

$$^{235}_{92}U + ^{1}_{0}n \longrightarrow ^{139}_{54}Xe + ^{94}_{38}Sr + 3\,^{1}_{0}n$$

Note that gamma particles ($^{0}_{0}\gamma$) might also be produced, but since they have both zero subscript and zero superscript, their presence cannot be deduced by this kind of analysis.

24.27 Which is the more unstable of each of the following pairs, and in each case what type of process could the unstable nucleus undergo? *(a)* ^{16}C, ^{16}N *(b)* ^{18}F, ^{18}Ne

▌ *(a)* $^{16}_{6}C$ has a ratio of neutrons to protons farther above the belt of stability. It would emit a $^{0}_{-1}\beta$ particle to get the neutron/proton ratio back to the stable range. *(b)* $^{18}_{10}Ne$ has a lower neutron/proton ratio than $^{18}_{9}F$. It could emit a $^{0}_{+1}\beta$ (positron) or capture a *K* electron to get that ratio into the range of stability. (Insufficient information is given in the problem to enable prediction of which of these processes will occur, if not both.)

24.28 A typical neutron-initiated fission of ^{235}U yields $^{97}_{42}Mo$, two neutrons, and an isotope of what element?

▌

$$^{235}_{92}U + ^{1}_{0}n \longrightarrow ^{97}_{42}Mo + ^{137}_{50}Sn + 2\,^{1}_{0}n$$

The conservation of charge requires the unknown isotope to have atomic number 50. Thus the product is an isotope of Sn.

24.29 Without consulting tables or other sources of information, explain how one can determine which of the following nuclides is the terminal member of the naturally occurring radioactive series which begins with $^{235}_{92}U$: ^{206}Pb, ^{207}Pb, ^{208}Pb, ^{209}Bi

▌ Only ^{207}Pb is a member of the $(4n + 3)$ series. Emission of α, β, or γ particles changes the mass number by either 0 or 4; hence each daughter must be in the same series as its parent.

24.30 Russian and American scientists have artificially prepared elements with atomic numbers above 100. To what stable isotope would $^{257}_{103}Lr$ decay after having been produced by artificial means? On what basis can such a prediction be made?

▌ $^{257}_{103}Lr$ is a member of the $(4n + 1)$ series and would probably decay to $^{209}_{83}Bi$.

24.31 Which one(s) of the following processes—alpha emission, beta emission, positron emission, electron capture—cause (*a*) an increase in atomic number (*b*) a decrease in atomic number (*c*) emission of an x-ray in every case?

▌ (*a*) Beta emission (*b*) Alpha emission, positron emission, and electron capture (*c*) Electron capture. (The capture of a *K* shell electron leaves a vacancy, which is filled by a higher-energy electron. This transfer is accompanied by x-ray emission.)

24.32 Devise a set of permitted values for the four nuclear quantum numbers—*n*, *l*, *j*, and *m*—which describe the shells of nucleons and the stabilities of nuclei with the magic numbers. Compare your values to those of actual nuclear quantum numbers listed in Problem 6.15.

▌ The quantum numbers must be such that completed shells have 2, 6, 12, etc., protons and the same numbers of neutrons, corresponding to the very stable $^{4}_{2}He$, $^{16}_{8}O$, $^{40}_{20}Ca$ nuclides. (Eight protons corresponds to completed first plus second shells, etc.) The principal quantum number will be the shell number. The others, with the designations and limitations indicated, are

n	1, 2, 3, ...
l	$(n-1), (n-3), (n-5), \ldots$, but no negative number
j	$(l+\frac{1}{2})$ or $(l-\frac{1}{2})$, but no negative number
m	from $-j$ in integral steps to $+j$

Thus for the third shell, for example:

n	3		
l	2		0
j	$2\frac{1}{2}$	$1\frac{1}{2}$	$\frac{1}{2}$ (not $-\frac{1}{2}$)
m	$-2\frac{1}{2}$ $-1\frac{1}{2}$ $-\frac{1}{2}$ $+\frac{1}{2}$ $+1\frac{1}{2}$ $+2\frac{1}{2}$	$-1\frac{1}{2}$ $-\frac{1}{2}$ $+\frac{1}{2}$ $+1\frac{1}{2}$	$-\frac{1}{2}$ $+\frac{1}{2}$

Twelve protons can occupy the third shell, corresponding to the 12 different sets of four quantum numbers.

24.2 HALF-LIFE

TABLE 24.1 Typical Half-Lives

nuclide	half-life	radiation*
$^{238}_{92}U$	4.5×10^{9} y	alpha
$^{237}_{93}Np$	2.2×10^{6} y	alpha
$^{14}_{6}C$	5730 y	beta
$^{90}_{38}Sr$	19.9 y	beta
$^{3}_{1}H$	12.3 y	beta
$^{140}_{56}Ba$	12.5 d	beta
$^{131}_{53}I$	8.0 d	beta
$^{140}_{57}La$	40 h	beta
$^{15}_{8}O$	118 s	beta
$^{94}_{36}Kr$	1.4 s	beta

* In most of these decay processes, gamma radiations are also emitted.

24.33 A radioactive sample has an initial activity of 28 dis/min; $\frac{1}{2}$ h later the activity is 14 dis/min. How many atoms of the radioactive nuclide were there originally? ($\lambda t_{1/2} = 0.693$)

▌ $$t_{1/2} = 30 \text{ min} \qquad A = \lambda N \qquad N = \frac{A}{\lambda} = \frac{28 \text{ dis/min}}{0.693/30 \text{ min}} = 1200$$

24.34 The half-life of $^{90}_{38}Sr$ is 20 y. If a sample of this nuclide has an initial activity of 8000 dis/min today, what will be its activity after 80 y?

▌ 80 y is four half-lives; it will decay to $(\frac{1}{2})^4(8000 \text{ dis/min}) = 500 \text{ dis/min}$

24.35 Prove that $A/A_0 = N/N_0 = m/m_0$, where A and A_0 are activities, N and N_0 are numbers of atoms, and m and m_0 are masses, all of the same decaying nuclide.

▌ $A = \lambda N \qquad \dfrac{A}{A_0} = \dfrac{\lambda N}{\lambda N_0} = \dfrac{N}{N_0} = \dfrac{m}{m_0}$, the last equality because of proportionality of mass and number of atoms.

24.36 ^{18}F is found to undergo 90% radioactive decay in 366 min. What is its computed half-life from this observation?

▌ 90% decay corresponds to 10%, or 0.10, survival.

$$k = -\frac{(2.303)(\log [A]/[A]_0)}{t} = -\frac{(2.303)(\log 0.10)}{366 \text{ min}} = 6.29 \times 10^{-3} \text{ min}^{-1}$$

Then the half-life can be computed:

$$t_{1/2} = \frac{0.693}{k} = \frac{0.693}{6.29 \times 10^{-3} \text{ min}^{-1}} = 110 \text{ min}$$

24.37 A sample of $^{90}_{38}Sr$ originally had an activity of 0.500 mCi. (*a*) What is the specific activity of the sample? (*b*) What is the activity of the sample after 30.0 y?

▌ (*a*) $A_0 = \lambda N_0 = 0.500 \text{ mCi} \qquad \dfrac{0.693}{19.9 \times 365 \times 24.0 \times 3600} N_0 = (0.500)(3.7 \times 10^7 \text{ dis/s})$

$N_0 = 1.68 \times 10^{16}$ atoms

$$\text{Mass of strontium} = \frac{(1.68 \times 10^{16} \text{ atoms})(90.0 \text{ g/mol})}{6.02 \times 10^{23} \text{ atoms/mol}} = 2.50 \times 10^{-6} \text{ g}$$

$$\text{Specific activity} = \frac{(0.500)(3.7 \times 10^7)}{2.50 \times 10^{-6}} = 7.4 \times 10^{12} \text{ dis/g·s}$$

(*b*) To determine the activity after 30.0 y: $(2.30)\left(\log \dfrac{N}{N_0}\right) = -\lambda t = -\left(\dfrac{0.693}{19.9 \text{ y}}\right)(30.0 \text{ y}) = -1.04$

$$\frac{N}{N_0} = \frac{A}{A_0} = 0.353 \qquad A = 0.353A_0 = 0.177 \text{ mCi} = 6.55 \times 10^6 \text{ dis/s}$$

24.38 A sample of river water was found to contain 8×10^{-18} tritium atom, ^3_1H, per atom of ordinary hydrogen. Tritium decomposes radioactively with a half-life of 12.3 y. (*a*) What will be the ratio of tritium to normal hydrogen atoms 49 y after the original sample was taken if the sample is stored in a place where additional tritium atoms cannot be formed? (*b*) How many tritium atoms would 10 g of such a sample contain 40 y after the initial sampling?

▌ (*a*) 49 years is almost exactly four half-lives $(4 \times 12.3 = 49.2 \text{ y})$. It is simpler in this case to avoid the logarithmic formula. The fraction of the tritium atoms remaining after four half-lives would be $(1/2)^4 = 1/16$. The normal hydrogen atoms are not radioactive, and therefore their number does not change. Thus the final ratio of tritium to normal hydrogen will be $(\frac{1}{16})(8 \times 10^{-18}) = 5 \times 10^{-19}$.

(*b*) 10 g of water contains (10 g)/(18 g/mol) or 10/18 mol of water or 10/9 mol of hydrogen atoms. The total number of hydrogen atoms is $(\frac{10}{9})(6 \times 10^{23}) = 6.7 \times 10^{23}$. Of these, the original number of tritium atoms is $(8 \times 10^{-18}) \times (6.7 \times 10^{23}) = 5 \times 10^6$. Only a fraction of this number, $[A]/[A]_0$, remains after 40 years.

$$k = \frac{0.693}{t_{1/2}} = \frac{0.693}{12.3 \text{ y}} = 5.6 \times 10^{-2} \text{ y}^{-1} \qquad \log \frac{[A]}{[A]_0} = -\frac{kt}{2.303} = -\frac{(5.6 \times 10^{-2} \text{ y}^{-1})(40 \text{ y})}{2.303} = -0.973$$

from which $[A]/[A]_0 = 0.106$. The number of remaining tritium atoms is $(0.106)(5 \times 10^6) = 5 \times 10^5$.

24.39 A piece of wood, reportedly from King Tut's tomb, was burned, and 7.32 g CO_2 was collected. The total radioactivity in the CO_2 was 10.8 dis/min. How old was the wood sample? Is there a possibility that it is authentic?

▌ The mass of carbon in the sample is

$$(7.32 \text{ g } CO_2)\left(\frac{1 \text{ mol } CO_2}{44.0 \text{ g}}\right)\left(\frac{1 \text{ mol C}}{\text{mol } CO_2}\right)\left(\frac{12.0 \text{ g C}}{\text{mol C}}\right) = 2.00 \text{ g C}$$

The specific activity is therefore $\frac{10.8 \text{ dis/min}}{2.00 \text{ g}} = 5.40 \text{ dis/min}\cdot\text{g}$

$$(2.30)\left(\log\frac{A}{A_0}\right) = -\left(\frac{0.693}{t_{1/2}}\right)t = -\frac{0.693t}{5730 \text{ y}} \qquad t = (-2.30)\left(\log\frac{5.40}{15.3}\right)\left(\frac{5730 \text{ y}}{0.693}\right) = 8610 \text{ y}$$

Since Tutankhamen (King Tut) died in 1352 BC, the artifact could not have been from his tomb. (There is some uncertainty about the existent specific activity of wood in ancient times, but not that much uncertainty.)

24.40 A sample of uraninite, a uranium-containing mineral, was found on analysis to contain 0.214 g of lead for every g uranium. Assuming that the lead all resulted from the radioactive disintegration of the uranium since the geological formation of the uraninite and that all isotopes of uranium other than ^{238}U can be neglected, estimate the date when the mineral was formed in the earth's crust. The half-life of ^{238}U is 4.5×10^9 y.

▮ The radioactive decay of ^{238}U leads, after 14 steps, to the stable lead isotope ^{206}Pb. The first of these steps, the α decay of ^{238}U with a 4.5×10^9 y half-life, is intrinsically more than 10^4 times as slow as any of the subsequent steps. As a result, the time required for the first step accounts for essentially all the time required for the entire 14-step sequence.

In a sample containing 1.00 g U, there is

$$\frac{0.214 \text{ g Pb}}{206 \text{ g/mol}} = 1.04 \times 10^{-3} \text{ mol Pb} \qquad \text{and} \qquad \frac{1.00 \text{ g U}}{238 \text{ g/mol}} = 4.20 \times 10^{-3} \text{ mol U}$$

If each atom of lead in the mineral today is the daughter of a uranium atom that existed at the time of the formation of the mineral, then the original number of moles of uranium in the sample would have been

$$(1.04 + 4.20) \times 10^{-3} = 5.24 \times 10^{-3}$$

Then the fraction remaining is $\frac{[A]}{[A]_0} = \frac{4.20 \times 10^{-3}}{5.24 \times 10^{-3}} = 0.802$

Letting t be the elapsed time from the formation of the mineral in the earth's crust to the present, we have

$$k = \frac{0.693}{t_{1/2}} = \frac{0.693}{4.5 \times 10^9 \text{ y}} = 1.54 \times 10^{-10} \text{ y}^{-1} \qquad t = -\frac{(2.303)(\log [A]/[A]_0)}{k} = -\frac{(2.303)(\log 0.802)}{1.54 \times 10^{-10} \text{ y}^{-1}} = 1.4 \times 10^9 \text{ y}$$

24.41 Given 2.00 kg of $^{238}_{92}U$ (half-life 4.50×10^9 y). The ultimate decay product in this series is $^{206}_{82}Pb$. ***(a)*** What weight of $^{238}_{92}U$ is left after 3.00×10^9 y? ***(b)*** What weight of $^{206}_{82}Pb$ is produced in this time?

▮ ***(a)***
$$(2.30)\left(\log\frac{m}{m_0}\right) = -\left(\frac{0.693}{t_{1/2}}\right)t = -\left(\frac{0.693}{4.50 \times 10^9 \text{ y}}\right)(3.00 \times 10^9 \text{ y}) = -0.462$$

$$\log(m/m_0) = -0.201 \qquad \text{hence } m/m_0 = 0.630 \qquad m = m_0(0.630) = 1.26 \text{ kg left}$$

(b) Of the 2.00 kg ^{238}U initially present, 2.00 kg − 1.26 kg = 0.74 kg disintegrated. For every U atom which disintegrates, one Pb atom is produced. The mass ratio is just 206/238. Hence,

$$(0.74 \text{ kg } ^{238}\text{U})\left(\frac{206 \text{ g Pb}}{238 \text{ g U}}\right) = 0.64 \text{ kg Pb produced}$$

24.42 The activity of 30 μg of ^{247}Cm is 2.8 nCi. Calculate the disintegration rate constant and the half-life of ^{247}Cm.

▮
$$A = 2.8 \text{ nCi} = (2.8 \times 10^{-9} \text{ Ci})\left(\frac{3.7 \times 10^{10} \text{ dis/s}}{\text{Ci}}\right) = 104 \text{ dis/s}$$

$$N = 30\ \mu\text{g} = (30 \times 10^{-6} \text{ g})\left(\frac{1 \text{ mol}}{247 \text{ g}}\right)\left(\frac{6.0 \times 10^{23} \text{ atoms}}{\text{mol}}\right) = 7.3 \times 10^{16} \text{ atoms}$$

$$A = \lambda N \qquad \lambda = \frac{A}{N} = \frac{104 \text{ s}^{-1}}{7.3 \times 10^{16}} = 1.4 \times 10^{-15} \text{ s}^{-1}$$

$$t_{1/2} = \frac{0.693}{\lambda} = (4.9 \times 10^{14} \text{ s})\left(\frac{1 \text{ h}}{3600 \text{ s}}\right)\left(\frac{1 \text{ day}}{24 \text{ h}}\right)\left(\frac{1 \text{ y}}{365 \text{ day}}\right) = 1.5 \times 10^7 \text{ y}$$

24.43 The bones of a prehistoric bison were found to have a ^{14}C activity of 2.80 dis/min·g carbon. Approximately how long ago did the animal live?

▌ $$(2.30)\left(\log\frac{A}{A_0}\right)=\left(\frac{0.693}{t_{1/2}}\right)t=-\left(\frac{0.693}{5730\text{ y}}\right)t \qquad t=(-2.30)\left(\log\frac{2.80}{15.3}\right)\left(\frac{5730\text{ y}}{0.693}\right)=14\,000\text{ y}$$

24.44 Fallout from nuclear explosions contains ^{131}I and ^{90}Sr. Calculate the time required for the activity of each of these isotopes to fall to 1.0% of its initial value. Radioiodine and radiostrontium tend to concentrate in the thyroid and the bones, respectively, of mammals which ingest them. Which isotope is likely to produce the more serious long-term effects?

▌ $$(2.30)\left(\log\frac{A}{A_0}\right)=-\lambda t=(2.30)(\log 0.010)=-4.60 \qquad t=\frac{4.60}{\lambda}=\frac{4.60t_{1/2}}{0.693}=6.64t_{1/2}$$

For ^{131}I: $t=(6.64)(8.0\text{ d})=53.1\text{ d}$ For ^{90}Sr: $t=(6.64)(19.9\text{ y})=132\text{ y}$

^{90}Sr is likely to be serious; the iodine will soon be gone.

24.45 Upon irradiating californium with neutrons, a scientist discovered a new nuclide having mass number of 250 and a half-life of 0.50 h. Three hours after the irradiation, the observed radioactivity due to the nuclide was 10 dis/min. How many atoms of the nuclide were prepared initially?

▌ The activity at the cessation of bombardment is given by

$$(2.30)\left(\log\frac{A}{A_0}\right)=(2.30)\left(\log\frac{10/\text{min}}{A_0}\right)=-\lambda t=-\left(\frac{0.693}{0.50\text{ h}}\right)(3.0\text{ h})$$

$$\log\frac{10/\text{min}}{A_0}=-1.8 \qquad \frac{10/\text{min}}{A_0}=0.0156 \qquad A_0=\frac{10/\text{min}}{0.0156}=640/\text{min}$$

$$N_0=\frac{A_0}{\lambda}=\frac{(640/\text{min})(30\text{ min})}{0.693}=2.8\times10^4$$

24.46 A radionuclide has an initial activity of 2.00×10^6 dis/min, and after 4.0 days its activity is 9.0×10^5 dis/min. Calculate the activity in the sample after 40 days.

▌ $$\left(\frac{1}{t}\right)\left(\log\frac{A}{A_0}\right)=-\frac{\lambda}{2.30} \qquad \left(\frac{1}{4}\right)\left(\log\frac{9.0\times10^5}{2.00\times10^6}\right)=\left(\frac{1}{40}\right)\left(\log\frac{A}{2.00\times10^6}\right)$$

$$\log\frac{A}{2.00\times10^6}=-3.47 \qquad \frac{A}{2.00\times10^6}=3.41\times10^{-4} \qquad A=6.8\times10^2\text{ dis/min}$$

24.47 A sample of radioactive material has an apparently constant activity of 2000 dis/min. By chemical means, the material is separated into two fractions, one of which has an initial activity of 1000 dis/min. The other fraction decays with a 24 h half-life. Estimate the total activity in *both* samples 48 h after the separation. Explain your estimate.

▌ 2000 dis/min. The total activity when the samples are separated will be the same as the total activity when they are mixed; the mixing makes no difference to the activity.

24.48 The half-life of ^{40}K is 1.28×10^9 y. What mass of this nuclide has an activity of 1.00 μCi?

▌ Let us first calculate the rate constant and express it in s^{-1}.

$$k=\frac{0.693}{t_{1/2}}=\frac{0.693}{(1.28\times10^9\text{ y})(365\text{ d/y})(24\text{ h/d})(3.6\times10^3\text{ s/h})}=1.72\times10^{-17}\text{ s}^{-1}$$

The disintegration rate is an instantaneous rate measured under conditions of essential constancy of the concentration (i.e., the population) of ^{40}K atoms.

$$\text{Rate}=-\frac{\Delta[\text{A}]}{\Delta t}=k[\text{A}]=(3.70\times10^{10}\text{ dis/s}\cdot\text{Ci})(10^{-6}\text{ Ci}/\mu\text{Ci})=3.70\times10^4\text{ dis/s}\cdot\mu\text{Ci}$$

Since one disintegration measures one *atomic* event, it is best to express [A] as the number of ^{40}K atoms.

$$[\text{A}]=\frac{\text{rate}}{k}=\frac{3.70\times10^4\text{ atoms/s}\cdot\mu\text{Ci}}{1.72\times10^{-17}/\text{s}}=2.15\times10^{21}\text{ atoms}/\mu\text{Ci}$$

and the corresponding mass is

$$\frac{(2.15\times10^{21}\text{ atoms}/\mu\text{Ci})(40\text{ g }^{40}\text{K/mol})}{6.0\times10^{23}\text{ atoms/mol}}=0.143\text{ g }^{40}\text{K}/\mu\text{Ci}$$

24.49 A sample of $^{14}CO_2$ was to be mixed with ordinary CO_2 for a biological tracer experiment. In order that 10 cm³ (STP) of the diluted gas should have 10^4 dis/min, how many μCi of radioactive carbon are needed to prepare 60 L of the diluted gas?

▮ $$\text{Total activity} = \left(\frac{10^4\ \text{dis/min}}{10\ \text{cm}^3}\right)\left(\frac{(60\ \text{L})(10^3\ \text{cm}^3/\text{L})}{60\ \text{s/min}}\right) = (10^6\ \text{dis/s})\left(\frac{1\ \text{Ci}}{3.7\times 10^{10}\ \text{dis/s}}\right)\left(\frac{10^6\ \mu\text{Ci}}{1\ \text{Ci}}\right) = 27\ \mu\text{Ci}$$

To find the *mass* of ^{14}C needed to provide the 27 μCi, the procedure of Problem 24.48 would be followed.

24.50 Some radioactive material is decaying with a 30 day half-life. Chemical separation procedures yield two fractions. Immediately after separation, one of the fractions decays with a 2 day half-life. Would the other fraction show, immediately after separation, constant activity, increasing activity, decay with a 30 day half-life, or decay with a 28 day half-life? Explain.

▮ Increasing activity is expected as the daughter nuclide builds up again.

24.51 A mixture of ^{239}Pu and ^{240}Pu has a specific activity of 6.0×10^9 dis/s. The half-lives of the isotopes are 2.44×10^4 and 6.58×10^3 y, respectively. Calculate the isotopic composition of this sample.

▮ The specific activity (activity per g) of each isotope is given by

$$A = \lambda N = \frac{0.693N}{t_{1/2}}$$

$$\text{For } ^{239}\text{Pu:}\quad A = \left(\frac{0.693}{2.44\times 10^4\ \text{y}}\right)\left(\frac{6.02\times 10^{23}}{239\ \text{g}}\right) = 7.15\times 10^{16}/\text{y}\cdot\text{g}$$

$$\text{For } ^{240}\text{Pu:}\quad A = \left(\frac{0.693}{6.58\times 10^3\ \text{y}}\right)\left(\frac{6.02\times 10^{23}}{240\ \text{g}}\right) = 2.64\times 10^{17}/\text{y}\cdot\text{g}$$

The number of seconds in a year is

$$(365\ \text{d})\left(\frac{24\ \text{h}}{\text{d}}\right)\left(\frac{60\ \text{min}}{\text{h}}\right)\left(\frac{60\ \text{s}}{\text{min}}\right) = 3.15\times 10^7\ \text{s}$$

$$\text{For } ^{239}\text{Pu:}\quad A = 2.27\times 10^9/\text{s}\cdot\text{g} \qquad \text{For } ^{240}\text{Pu:}\quad A = 8.37\times 10^9/\text{s}\cdot\text{g}$$

Let the fraction of $^{239}Pu = x$; then the fraction of $^{240}Pu = 1 - x$.

$$(2.27\times 10^9)x + (1-x)(8.37\times 10^9) = 6.0\times 10^9$$

$$(8.37\times 10^9) - (6.10\times 10^9)x = 6.0\times 10^9 \qquad 2.37\times 10^9 = (6.1\times 10^9)x$$

$$x = 0.39 = 39\%\ ^{239}\text{Pu}$$

24.52 ^{120}Q, ^{119}R, and ^{120}R are the stable nuclides of elements Q and R. When element Q is bombarded with slow neutrons, radioactive Q is produced, which decays by beta emission with a 3 h half-life to an isotope of R, which is also beta active with an 11 day half-life. When Q is bombarded with deuterons, the 11 d activity is observed, but not the 3 h activity. When element R is bombarded with fast neutrons, the only activities observed are the 11 d R activity and a 2 y Q activity, which decays by beta emission to an inactive daughter. Assign mass numbers to **(*a*)** the 3 h Q isotope **(*b*)** the 11 d R isotope **(*c*)** the 2 y Q isotope. Write equations for the reaction of **(*d*)** the Q target with slow neutrons **(*e*)** the Q target with deuterons **(*f*)** the R target with fast neutrons.

▮ **(*a*)** 121 **(*b*)** 121 **(*c*)** 119 **(*d*)** $^{120}Q(n,\gamma)^{121}Q$ **(*e*)** $^{120}Q(d,n)^{121}R$
(*f*) $^{119}R(n,p)^{119}Q \longrightarrow {}^{119}R + {}_{-1}\beta$

24.53 Chlorine, with an atomic weight of 35.453 u, is a mixture of 75.53% ^{35}Cl and 24.47% ^{37}Cl, having atomic masses of 34.96885 and 36.96590 u, respectively. Suppose that instead of ^{37}Cl, "natural" chlorine contained ^{36}Cl, having an atomic mass of 35.9787 u and a half-life of 3.1×10^5 y. Assuming the same atomic weight for this mixture of isotopes, calculate the specific activity of a sample of this "natural" chlorine.

▮ Atomic weight = 35.453 u. Let $x = \%\ ^{35}Cl$; then $100 - x = \%\ ^{36}Cl$.

$$\frac{x(34.96885) + (100-x)(35.9787)}{100} = 35.453 \qquad x = 52.1\%\ ^{35}\text{Cl}$$

Per g sample, there is 0.479 g ^{36}Cl.

$$(0.479\ \text{g}\ ^{36}\text{Cl})\left(\frac{6.02\times10^{23}\ \text{atoms}}{35.9787\ \text{g}}\right)=8.01\times10^{21}\ \text{atoms}$$

$$A=\lambda N=\left(\frac{0.693}{3.1\times10^{5}\ \text{y}}\right)(8.01\times10^{21})=1.79\times10^{16}\ \text{dis/y}$$

$$\left(\frac{1.79\times10^{16}\ \text{dis}}{\text{y}}\right)\left(\frac{1\ \text{y}}{365\ \text{d}}\right)\left(\frac{1\ \text{d}}{24\ \text{h}}\right)\left(\frac{1\ \text{h}}{3600\ \text{s}}\right)=5.68\times10^{8}\ \text{dis/s}$$

24.54 The disintegration of ^{239}Pu is accompanied by the loss of 5.24 MeV/dis. The half-life of ^{239}Pu is 24 400 y. Calculate the energy released per day from a 1.00 g sample of ^{239}Pu, in MeV and in kJ.

▮

$$(1.00\ \text{g}\ ^{239}\text{Pu})\left(\frac{1.00\ \text{mol}}{239\ \text{g}}\right)\left(\frac{6.02\times10^{23}\ \text{atoms}}{\text{mol}}\right)=2.52\times10^{21}\ \text{atoms}$$

$$A=\lambda N=\left(\frac{0.693}{t_{1/2}}\right)N=\frac{0.693}{(24\,400\ \text{y})(365\ \text{d/y})}(2.52\times10^{21}\ \text{atoms})=1.96\times10^{14}\ \text{dis/d}$$

$$\text{Power}=(1.96\times10^{14}\ \text{dis/d})(5.24\ \text{MeV/dis})=1.03\times10^{15}\ \text{MeV/d}$$

Taking the conversion factor from Problem 24.2,

$$\text{Power}=(1.03\times10^{15}\ \text{MeV/d})(1.6\times10^{-16}\ \text{kJ/MeV})=0.165\ \text{kJ/d}$$

24.55 The half-life of ^{212}Pb is 10.6 h; that of its daughter ^{212}Bi is 60.5 min. How long will it take for a maximum daughter activity to grow in freshly separated ^{212}Pb?

▮

$$t_{\max}=\left(\frac{2.30}{\lambda_d-\lambda_p}\right)\left(\log\frac{\lambda_d}{\lambda_p}\right)$$

$$\lambda_d=\frac{0.693}{t_{1/2}}=\frac{0.693}{60.5\ \text{min}}=0.01145/\text{min}\qquad \lambda_p=\frac{0.693}{(10.6)(60.0)\ \text{min}}=0.00109/\text{min}$$

$$t_{\max}=\left(\frac{2.30}{(0.01145-0.00109)/\text{min}}\right)\left(\log\frac{0.01145}{0.00109}\right)=227\ \text{min}=3.78\ \text{h}$$

24.56 ^{227}Ac has a half-life of 21.8 y with respect to radioactive decay. The decay follows two parallel paths, one leading to ^{227}Th and one leading to ^{223}Fr. The percent yields of these two daughter nuclides are 1.2% and 98.8%, respectively. What is the rate constant, in y^{-1}, for each of the separate paths?

▮ The rate constant for the decay of Ac, k_{Ac}, can be computed from the half-life.

$$k_{\text{Ac}}=\frac{0.693}{t_{1/2}}=\frac{0.693}{21.8\ \text{y}}=3.18\times10^{-2}\ \text{y}^{-1}$$

The overall rate constant for a set of parallel first-order reactions is equal to the sum of the separate rate constants, $k_{\text{Ac}}=k_{\text{Th}}+k_{\text{Fr}}$, and the fractional yield of either process is equal to the ratio of the rate constant for that process to the overall rate constant.

$$k_{\text{Th}}=(\text{fractional yield of Th})\times k_{\text{Ac}}=(0.012)(3.18\times10^{-2}\ \text{y}^{-1})=3.8\times10^{-4}\ \text{y}^{-1}$$
$$k_{\text{Fr}}=(\text{fractional yield of Fr})\times k_{\text{Ac}}=(0.988)(3.18\times10^{-2}\ \text{y}^{-1})=3.14\times10^{-2}\ \text{y}^{-1}$$

24.57 Assuming each sample was freshly separated from its decay products, what is the mass of a sample which contains 1.00 mCi of **(*a*)** ^{131}I ($t_{1/2}=8.08$ days) **(*b*)** ^{238}U **(*c*)** $^{140}_{56}$Ba?

▮

$$1.00\ \text{mCi}=3.7\times10^{7}\ \text{dis/s}\qquad \text{(The half-lives are from Table 24.1.)}$$

(*a*) $$\lambda=\frac{0.693}{(8.08\ \text{d})(24\ \text{h/d})(3600\ \text{s/h})}=9.93\times10^{-7}\ \text{s}^{-1}$$

$$A=\lambda N\qquad N=\frac{A}{\lambda}=\left(\frac{3.7\times10^{7}\ \text{s}^{-1}}{9.93\times10^{-7}\ \text{s}^{-1}}\right)=3.73\times10^{13}$$

$$(3.73\times10^{13}\ \text{atoms})\left(\frac{1\ \text{mol}}{6.02\times10^{23}\ \text{atoms}}\right)\left(\frac{131\ \text{g}}{\text{mol}}\right)=8.1\times10^{-9}\ \text{g}$$

(*b*) $\lambda = \dfrac{0.693}{(4.5 \times 10^9\ \text{y})[(365)(24)(3600)\ \text{s/y}]} = 4.88 \times 10^{-18}\ \text{s}^{-1}$

$N = \dfrac{3.7 \times 10^7\ \text{s}^{-1}}{4.88 \times 10^{-18}\ \text{s}^{-1}} = 7.58 \times 10^{24}$ atoms

$(7.58 \times 10^{24}\ \text{atoms})\left(\dfrac{1\ \text{mol}}{6.02 \times 10^{23}\ \text{atoms}}\right)\left(\dfrac{238\ \text{g}}{\text{mol}}\right) = 3.0\ \text{kg}$

(*c*) $\lambda = \dfrac{0.693}{(12.5\ \text{days})(24\ \text{h/d})(3600\ \text{s/h})} = 6.42 \times 10^{-7}\ \text{s}^{-1}$

$N = \dfrac{3.7 \times 10^7\ \text{s}^{-1}}{6.42 \times 10^{-7}\ \text{s}^{-1}} = 5.77 \times 10^{13}$

$(5.77 \times 10^{13}\ \text{atoms})\left(\dfrac{1\ \text{mol}}{6.02 \times 10^{23}\ \text{atoms}}\right)\left(\dfrac{140\ \text{g}}{\text{mol}}\right) = 1.3 \times 10^{-8}\ \text{g}$

24.58 Potassium (atomic weight = 39.102 u) contains 93.10 atom % ^{39}K, having atomic mass 38.96371 u; 0.0118 atom % ^{40}K, which has a mass of 40.0 u and is radioactive with $t_{1/2} = 1.3 \times 10^9$ y; and 6.88 atom % ^{41}K having a mass of 40.96184 u. Estimate the specific activity of naturally occurring potassium.

▌ Specific activity refers to the activity of 1 g of material. One gram of naturally occurring potassium contains 1.18×10^{-4} g ^{40}K plus nonradioactive isotopes.

$$A = \lambda N = \left(\frac{0.693}{t_{1/2}}\right)\left(\frac{1.18 \times 10^{-4}\ \text{g}}{(40.0\ \text{g})/(6.02 \times 10^{23}\ \text{nuclei})}\right)$$

$$= \frac{(0.693)(1.18 \times 10^{-4})(6.02 \times 10^{23})}{(1.3 \times 10^9\ \text{y})(40.0)} = 9.47 \times 10^8\ \text{y}^{-1} = 3.00 \times 10^1\ \text{s}^{-1}$$

24.59 A pure radiochemical preparation was observed to disintegrate at the rate of 4280 counts/min at 1:35 p.m. At 4:55 p.m. of the same day, the disintegration rate of the sample was only 1070 counts/min. The disintegration rate is proportional to the number of radioactive atoms in the sample. What is the half-life of the material?

▌ $(2.303)\left(\log \dfrac{[\text{A}]}{[\text{A}]_0}\right) = -kt = -\left(\dfrac{0.693}{t_{1/2}}\right)t \qquad (2.303)\left(\log \dfrac{1070}{4280}\right) = -\left(\dfrac{0.693}{t_{1/2}}\right)(200\ \text{min}) \qquad t_{1/2} = 100\ \text{min}$

Alternatively, it is seen that the sample decomposed to one-fourth its original activity, which took two half-lives. Since two half-lives are 200 min, one half-life is 100 min.

24.60 An atomic battery for pocket watches has been developed which uses the beta particles from ^{147}Pm as the primary energy source. The half-life of ^{147}Pm is 2.62 y. How long would it take for the rate of beta emission in the battery to be reduced to 10% of its initial value?

▌

$$(2.303)(\log 0.10) = -\left(\frac{0.693}{2.62\ \text{y}}\right)t \qquad t = 8.7\ \text{y}$$

24.61 The half-life of ^{14}C is 5730 y. **(*a*)** What fraction of its original ^{14}C would a sample of $CaCO_3$ have after 11 460 y of storage in a locality where additional radioactivity could not be produced? **(*b*)** What fraction of the original ^{14}C would still remain after 13 000 y?

▌ (*a*) Two half-lives, hence 0.25. (*b*) $(2.303)(\log f) = -\left(\dfrac{0.693}{5730\ \text{y}}\right)(13\,000\ \text{y}) = -1.57 \qquad f = 0.208$

24.62 A charcoal sample taken from a fire pit in an archeologist's excavation of a rock shelter was believed to have been formed when early occupants of the shelter burned wood for cooking. A 100 mg sample of pure carbon from the charcoal was found in 1979 to have a disintegration rate of 0.25 count/min. The activity of a contemporary sample taken from a freshly cut tree was found to be 15.3 counts (disintegrations) per min per g carbon. How long ago did the tree grow from which the archeological sample was taken? Use the data from Problem 24.61. Assume that no new ^{14}C was formed after the tree that supplied the wood was felled and that the ^{14}C level in living trees has been the same over time.

▌

$$15.3/\text{g} = (1.53)/(100\ \text{mg})$$

$$(2.303)\left(\log\frac{0.25}{1.53}\right) = -\left(\frac{0.693}{t_{1/2}}\right)t \qquad -1.81 = -\left(\frac{0.693}{5730\ \text{y}}\right)t \qquad t = 15\,000\ \text{y}$$

The event was 15 000 years ago.

24.63 All naturally occurring rubidium ores contain ^{87}Sr resulting from the beta decay of ^{87}Rb. In naturally occurring rubidium, 278 of every 1000 rubidium atoms are ^{87}Rb. A mineral containing 0.85% rubidium was analyzed and found to contain 0.0089% strontium. Assuming that all of the strontium originated by radioactive decay of ^{87}Rb, estimate the age of the mineral. ^{87}Rb has a half-life of 4.7×10^{10} y.

▌ ^{87}Rb and ^{87}Sr have very nearly the same mass, so their mass ratio is equal to their mol ratio.

The mineral contained $$(0.85\%\ \text{Rb})\left(\frac{278\ ^{87}\text{Rb}}{1000\ \text{Rb}}\right) = 0.236\%\ ^{87}\text{Rb}$$

The original percent ^{87}Rb was $$0.236\% + 0.0089\% = 0.245\%\ ^{87}\text{Rb}$$

The fraction remaining is $$\frac{0.236}{0.245} = 0.963$$

$$(2.303)(\log 0.963) = -\left(\frac{0.693}{4.7 \times 10^{10}\ \text{y}}\right)t \qquad t = 2.5 \times 10^9\ \text{y}$$

Note carefully the distinction between fraction disintegrated and fraction remaining.

24.64 Transuranium elements were originally believed not to occur in nature because of their relatively short half-lives. Then ^{244}Pu was reported in a natural ore. The half-life of ^{244}Pu is 8.3×10^7 y. If this element is more stable than any of its radioactive predecessors and thus has not been produced in this ore in significant amounts since the ore was deposited, what fraction of the original ^{244}Pu content would still be present if the ore is assumed to be 5×10^9 y old?

▌

$$(2.303)(\log f) = -\left(\frac{0.693}{8.3 \times 10^7\ \text{y}}\right)(5 \times 10^9\ \text{y}) = -41.75 \qquad f = 10^{-18}$$

24.65 Prior to the use of nuclear weapons, the specific activity of ^{14}C in soluble ocean carbonates was found to be 16 dis/min·g carbon. The amount of carbon in these carbonates has been estimated as 4.5×10^{16} kg. How many MCi of ^{14}C did the ocean carbonates contain?

▌

$$(4.5 \times 10^{16}\ \text{kg C})\left(\frac{16\ \text{dis/min}}{10^{-3}\ \text{kg C}}\right)\left(\frac{1\ \text{min}}{60\ \text{s}}\right)\left(\frac{1\ \text{Ci}}{3.7 \times 10^{10}\ \text{dis/s}}\right) = 3.2 \times 10^8\ \text{Ci} = 320\ \text{MCi}$$

24.66 If the limit of a particular detection system is 0.002 dis/s for a 1 g sample, what would be the maximum half-life that this system could detect on a 1 g sample of a nuclide whose mass number is around 200?

▌

$$A = \left(\frac{0.693}{t_{1/2}}\right)N$$

$$t_{1/2} = \left(\frac{0.693}{A}\right)N = \left(\frac{0.693}{0.002\ \text{s}^{-1}}\right)\left(\frac{6.02 \times 10^{23}\ \text{atoms/mol}}{200\ \text{g/mol}}\right) = 1 \times 10^{24}\ \text{s}$$

$$= (1 \times 10^{24}\ \text{s})\left(\frac{1\ \text{h}}{3600\ \text{s}}\right)\left(\frac{1\ \text{day}}{24\ \text{h}}\right)\left(\frac{1\ \text{y}}{365\ \text{days}}\right) = 3 \times 10^{16}\ \text{y}$$

24.67 Determine the half-life of a nuclide which was separated from a cyclotron target and which gave the following readings on a Géiger-Müller counter:

time, h	counts/min	time, h	counts/min
0	4000	16	1660
2	3660	24	1070
4	3200	40	450
8	2570	60	150

▌ From Fig. 24.1,

$$\text{Slope} = \frac{3.375 - 2.420}{(10.0 - 50.0)\ \text{h}} = -0.0239\ \text{h}^{-1}$$

$$\text{Slope} = -\frac{\lambda}{2.30} = -\frac{0.693}{2.30t_{1/2}} = -\frac{0.301}{t_{1/2}} = -0.0239\ \text{h}^{-1}$$

$$t_{1/2} = \frac{-0.301}{-0.0239\ \text{h}^{-1}} = 12.6\ \text{h}$$ (Compare this solution to that given in Problem 24.68.)

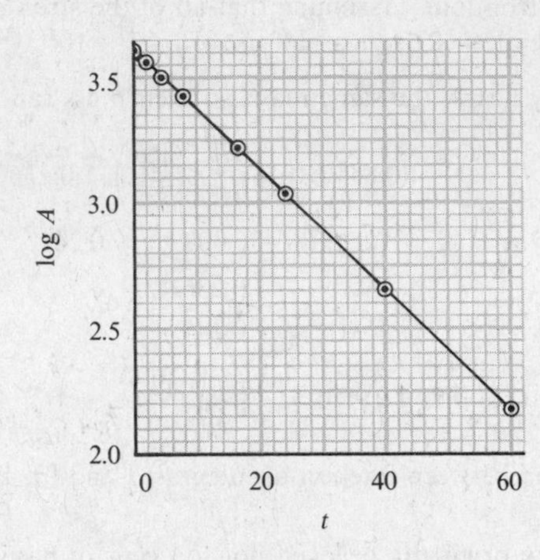

Fig. 24.1

24.68 Determine the half-life of a nuclide which was separated from a cyclotron target and which gave the following readings on a Gieger-Müller counter:

time, min	counts/s	time, min	counts/s
0.0	300	10.0	66.0
2.5	206	12.0	48.9
5.0	141	15.0	31.2
7.5	96.6	20.0	14.7

▌ At $t_{1/2}$ the activity will be 150 counts/s, half the initial activity. Therefore, $\log A = \log 150 = 2.176$. The corresponding value of $t_{1/2}$ can be read from Fig. 24.2 as 4.6 min. (Compare this method of solution to that given in Problem 24.67, a more precise method.)

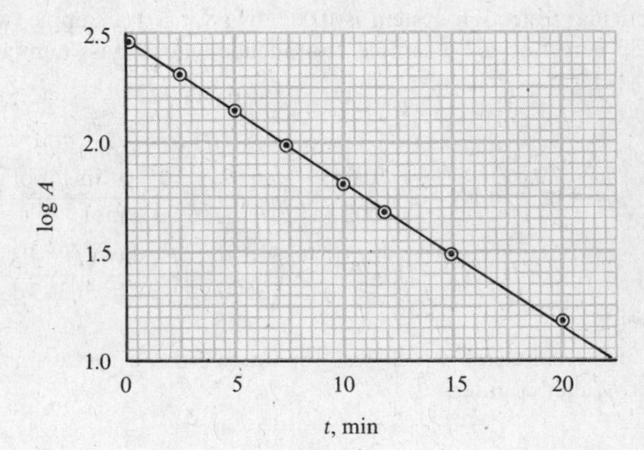

Fig. 24.2

24.3 BINDING ENERGY

24.69 When a ^{238}U nucleus disintegrates spontaneously, it forms a ^{234}Th nucleus and an α particle (^{4}He nucleus). Explain why the change in rest mass during this process can be calculated by subtracting the mass of a ^{234}Th *atom* plus the mass of a ^{4}He *atom* from the mass of the ^{238}U *atom*.

$$\text{Difference} = m(\text{U atom}) - m(\text{He atom}) - m(\text{Th atom})$$
$$= [m(\text{U nucleus}) + 92m_e] - [m(\text{He nucleus}) + 2m_e] - [m(\text{Th nucleus}) + 90m_e]$$
$$= m(\text{U nucleus}) - m(\text{He nucleus}) - m(\text{Th nucleus})$$

(There is a very small difference, due to electronic binding energies, between the rest mass of a neutral atom and the sum of the nuclear and electronic rest masses.)

24.70 How much energy would have to be added to $^4_2$He to make two protons and two neutrons?

▮ To get the $^4_2$He nucleus to come apart into its four nucleons, energy equivalent to 0.0304 u would have to be added. By Problem 24.2, $E = (0.0304\ \text{u})(932\ \text{MeV/u}) = 28.3\ \text{MeV}$

24.71 Refer to Problem 24.70. How much energy is required to dissociate 1 mol helium atoms into protons and neutrons?

$$(28.3\ \text{MeV/atom})(6.02 \times 10^{23}\ \text{atoms/mol})(1.6 \times 10^{-13}\ \text{J/MeV}) = 272.5 \times 10^{10}\ \text{J/mol} = 2725\ \text{GJ/mol}$$

24.72 Red giant stars, which are cooler than the sun, produce energy by means of the reaction $^9_4\text{Be} + ^1_1\text{H} \rightarrow ^6_3\text{Li} + ^4_2\text{He} + \text{energy}$. From the nuclidic masses [^{9}Be(9.01504) and ^{6}Li(6.01702)], calculate the energy released, in MeV, and compare it with the energy released in the carbon cycle (30 MeV) and in the solar helium-hydrogen cycle (26.6 MeV).

▮ The total rest masses of product and reactant nuclei are

$$\text{Products:}\quad (6.01702\ \text{u}) + (4.0015\ \text{u}) = 10.0185\ \text{u}$$
$$\text{Reactants:}\quad (9.01504\ \text{u}) + (1.00728\ \text{u}) = 10.0223\ \text{u}$$
$$\text{Difference} = 0.0038\ \text{u} \qquad (0.0038\ \text{u})\left(\frac{932\ \text{MeV}}{\text{u}}\right) = 3.5\ \text{MeV} \quad \text{(See Problem 24.2.)}$$

This value is much lower than the 30 MeV in the carbon cycle or the 26.6 MeV in the formation of α particles from protons.

24.73 What is the total binding energy of ^{12}C, and what is the average binding energy per nucleon?

▮ Although "binding energy" is a term referring to the nucleus, it is more convenient to use the mass of the whole atom (nuclide) in calculations, since these are the masses that are given in tables. If $M(\text{X})$ is the atomic mass of nuclide X,

$$M(nucleus) = M(x) - ZM(e^-) \tag{1}$$

The nucleus of X consists of Z protons and $A - Z$ neutrons. Hence its binding energy is given by

$$\text{BE} = [ZM(p) + (A - Z)M(n)] - M(nucleus) \tag{2}$$

Applying (1) both to the nucleus of X and to the proton, which is a $^1_1$H nucleus, and substituting in (2),

$$\text{BE} = \{Z[M(^1_1\text{H}) - 1M(e^-)] + (A - Z)M(n)\} - [M(\text{X}) - ZM(e^-)] = \{ZM(^1_1\text{H}) + (A - Z)M(n)\} - M(\text{X})$$

In other words, nuclear masses can be replaced by atomic (nuclidic) masses in calculating the binding energy. Whole atom masses can, in fact, be used for mass-difference calculations in all types of nuclear reactions except for β^+ processes where there is a resulting annihilation of two electron masses (one β^+ and one β^-).

The data needed for ^{12}C can be obtained from Tables 2.1 and 24.2.

TABLE 24.2

particle	symbol	mass, u	charge, e
Proton	p	1.0072765	+1
Neutron	n	1.0086650	0
Electron	$e^-, \beta^-, {}_{-1}\beta$	0.0005486	−1
Positron	$e^+, \beta^+, {}_{+1}\beta$	0.0005486	+1

$$\text{Mass of 6 } {}^1\text{H atoms} = 6 \times 1.00783 = 6.0470 \text{ u}$$
$$\text{Mass of 6 neutrons} = 6 \times 1.00867 = 6.0520 \text{ u}$$
$$\text{Total mass of component particles} = 12.0990 \text{ u}$$
$$\text{Mass of } {}^{12}\text{C} = 12.0000 \text{ u}$$
$$\text{Loss in mass on formation of } {}^{12}\text{C} = 0.0990 \text{ u} \qquad \text{Binding energy} = (932 \text{ MeV})(0.0990 \text{ u}) = 92.3 \text{ MeV}$$

Since there are 12 nucleons (protons and neutrons), the average binding energy per nucleon is (92.3 MeV)/12, or 7.69 MeV.

24.74 Evaluate Q for the $^7\text{Li}(p, n)^7\text{Be}$ reaction.

▌ The change in mass for the reaction must be computed.

Reactants:			Products:		
	^7_3Li	7.01600		^1_0n	1.00867
	^1_1H	1.00783		^7_4Be	7.01693
		8.02383			8.02560

$$\text{Increase of mass} = 8.02560 - 8.02383 = 0.00177 \text{ u}$$

A corresponding net amount of energy must be consumed, equal to (932)(0.00177) MeV, or 1.65 MeV; thus $Q = -1.65$ MeV. This energy is supplied as kinetic energy of the bombarding proton and is part of the acceleration requirement for the proton supplied by the particle accelerator.

24.75 The Q value for the $^3\text{He}(n, p)^3\text{H}$ reaction is 0.76 MeV. What is the nuclidic mass of ^{3}He?

▌ The reaction is $^3_2\text{He} + ^1_0n \rightarrow ^1_1\text{H} + ^3_1\text{H}$. The mass loss must be $0.76/932 = 0.00082$ u. The mass balance can be calculated on the basis of whole atoms.

Reactants:			Products:		
	^3He	x		^1H	1.00783
	^1_0n	1.00867		^3H	3.01605
		$x + 1.00867$			4.02388

Then $\qquad (x + 1.00867) - 4.02388 = 0.00082 \qquad \text{or} \qquad x = 3.01603$ u.

24.76 Calculate the maximum kinetic energy of the β^- emitted in the radioactive decay of ^{6}He. Assume that the β^- has its maximum energy when no other emission accompanies the process.

▌ The process referred to is $^6_2\text{He} \rightarrow ^6_3\text{Li} + \beta^-$. In computing the mass change during this process, only the whole atomic masses of ^{6}He and ^{6}Li need be considered.

$$\text{Mass of } {}^6\text{He} = 6.01889 \text{ u} \qquad \text{Mass of } {}^6\text{Li} = 6.01512 \text{ u} \qquad \text{Loss in mass} = 0.00377 \text{ u}$$
$$\text{Energy equivalent} = (932)(0.00377) \text{ MeV} = 3.51 \text{ MeV}$$

The maximum kinetic energy of the β^- particle is 3.51 MeV.

24.77 ^{13}N decays by β^+ emission. The maximum kinetic energy of the β^+ is 1.20 MeV. What the nuclidic mass of ^{13}N?

▌ The reaction is $^{13}_7\text{N} \rightarrow ^{13}_6\text{C} + \beta^+$. This is the type of process, mentioned in Problem 24.73, in which a simple difference of whole atom masses is not the desired quantity.

$$\begin{aligned}\text{Mass difference} &= [M(\text{nucleus) for } {}^{13}\text{N}] - [M(\text{nucleus) for } {}^{13}\text{C}] - M(e)\\ &= [M({}^{13}\text{N}) - 7M(e)] - [M({}^{13}\text{C}) - 6M(e)] - M(e)\\ &= M({}^{13}\text{N}) - M({}^{13}\text{C}) - 2M(e) = M({}^{13}\text{N}) - 13.00335 - 2(0.00055) = M({}^{13}\text{N}) - 13.00445\end{aligned}$$

This expression must equal the mass equivalent of the maximum kinetic energy of the β^+,

$$\frac{1.20 \text{ MeV}}{932 \text{ MeV/u}} = 0.00129 \text{ u}$$

Then $\qquad 0.00129 = M(^{13}\text{N}) - 13.00445 \qquad \text{or} \qquad M(^{13}\text{N}) = 13.00574$ u

24.78 Consider the two nuclides of mass number 7, ^{7}Li and ^{7}Be. Which of the two is more stable? How does the unstable nuclide decay into the stable one?

▌ Table 2.1 shows that ^7Be has a larger mass than ^7Li. Thus ^7_4Be can decay spontaneously into ^7_3Li, but not vice versa. There are two types of decay process in which Z is decreased by one unit without a change in mass number A: β^+ emission and K capture. These two processes have different mass balance requirements.

Assume that the process is β^+ emission. $^7_4Be \rightarrow \beta^+ + ^7_3Li$ It was shown in Problem 24.77 (third line of mass difference equation) that a positron emission occurs (i.e., Q is positive and the reaction is spontaneous) only if the *nuclidic* mass of the parent species exceeds the *nuclidic* mass of the daughter by at least twice the rest mass of the electron, $2(0.00055) = 0.00110$ u. In the present case the actual mass difference between parent and daughter nuclides is $7.01693 - 7.01600 = 0.00093$ u. We thus see that positron emission in this case is impossible. By elimination, we conclude that ^7Be undergoes K capture.

Note that we have predicted only that ^7Be *should* decay by K capture into ^7Li. We have said nothing about the rate of such a process. *Measurements* show the half-life of the process to be 53 days.

24.79 An isotopic species of lithium hydride, $^6Li^2H$, is a potential nuclear fuel, on the basis of the reaction $^6_3Li + ^2_1H \rightarrow 2\,^4_2He$. Calculate the expected power production, in MW, associated with the consumption of 1.00 g $^6Li^2H$ per day. Assume 100% efficiency in the process.

▌ The change in mass for the reaction is first computed.

Mass of $^6_3Li = 6.01512$ u Mass of $^2_1H = 2.01410$ u Total mass of reactants = 8.02922 u

Mass of products = 2(4.00260) = 8.00520 u Loss in mass = 0.02402 u

$$\text{Energy per atomic event} = (0.02402\text{ u})(932 \times 10^6\text{ eV/u})(1.602 \times 10^{-19}\text{ J/eV}) = 3.59 \times 10^{-12}\text{ J}$$

$$\text{Energy per mol LiH} = (3.59 \times 10^{-12}\text{ J})(6.02 \times 10^{23}\text{ mol}^{-1}) = 2.16 \times 10^{12}\text{ J/mol}$$

$$\text{Power production per g LiH} = \frac{(2.16 \times 10^{12}\text{ J/mol})/(8.02\text{ g/mol})}{(24\text{ h})(3.6 \times 10^3\text{ s/h})} = 3.11 \times 10^6\text{ W/g} = 3.11\text{ MW/g}$$

24.80 One of the most stable nuclei is ^{55}Mn. Its nuclidic mass is 54.938 u. Determine its total binding energy and average binding energy per nucleon.

▌

$^{55}_{25}Mn$:	25 protons + 25 electrons	25(1.00783 u) = 25.19575 u
	30 neutrons	30(1.00867 u) = 30.26010 u

Total mass = 55.45585 u $^{55}_{25}Mn$ mass = 54.938 u Difference = 0.518 u

Then (Problem 24.2)

$$\text{BE} = (0.518\text{ u})(931.5\text{ MeV/u}) = 483\text{ MeV} \quad \text{and} \quad \frac{\text{BE}}{\text{nucleon}} = \frac{483\text{ MeV}}{55\text{ nucleons}} = 8.78\text{ MeV/nucleon}$$

24.81 How much energy is released during each of the following fusion reactions?

(*a*) $^1_1H + ^7_3Li \longrightarrow 2\,^4_2He$ (*b*) $^3_1H + ^2_1H \longrightarrow ^4_2He + ^1_0n$

▌ (*a*) Mass of $^1_1H = 1.00783$ u Mass of $^7_3Li = 7.01600$ u Total = 8.02383 u

Mass of 2 $^4_2He = 8.00520$ u Mass loss = 0.01863 u

$Q = (0.01863\text{ u})(931.5\text{ MeV/u}) = 17.4$ MeV

(*b*)

^3_1H	3.01605 u	^4_2He	4.00260 u
^2_1H	2.01410 u	^1_0n	1.00866 u
Total	5.03015 u	Total	5.01126 u
Less	5.01126 u		
Loss =	0.01889 u		

$Q = (0.01889\text{ u})(931.5\text{ MeV/u}) = 17.6$ MeV

24.82 ^{14}C is believed to be made in the upper atmosphere by an (n, p) process on ^{14}N. What is Q for this reaction?

▌

$$^{14}_7N + ^1_0n \longrightarrow ^1_1H + ^{14}_6C$$

^{14}N	14.00307 u	^1H	1.00783 u
^1_0n	1.008665 u	^{14}C	14.00324 u
Total	15.01174 u	Total	15.01107 u
Less	15.01107 u		
Loss =	0.00067 u		

$Q = (0.00067\text{ u})(931.5\text{ MeV/u}) = 0.62$ MeV

24.83 In the reaction $^{32}S(n, \gamma)^{33}S$ with slow neutrons, the γ is produced with an energy of 8.65 MeV. What is the nuclidic mass of ^{33}S?

Mass of $^{32}S = 31.97207$ u Mass of $^{1}_{0}n = 1.00866$ u Mass Total $= 32.98073$ u

$$(8.65\ \text{MeV})\left(\frac{1\ \text{u}}{931.5\ \text{MeV}}\right) = 0.00929\ \text{u} \qquad (32.98073\ \text{u}) - (0.00929\ \text{u}) = 32.97144\ \text{u}$$

(assuming that all the difference in binding energy is in the γ particle)

24.84 If a β^+ and a β^- annihilate each other and their rest masses are converted into two γ rays of equal energy, what is the energy in MeV of each γ?

Each particle yields one γ ray: $(0.0005486\ \text{u})(931.5\ \text{MeV/u}) = 0.511\ \text{MeV}$

24.85 ΔE for the combustion of 1 mol ethylene in oxygen is -1.4×10^3 kJ. What would be the loss in mass (expressed in u) accompanying the oxidation of one molecule of ethylene?

$$\Delta m = \frac{\Delta E}{c^2} = \frac{(-1.4 \times 10^6\ \text{J/mol})(N_A^{-1}\ \text{mol/molecule})}{(2.998 \times 10^8\ \text{m/s})^2}(10^3 N_A\ \text{u/kg}) = -1.6 \times 10^{-8}\ \text{u/molecule}$$

(This value is so small compared to the molecular mass that the change in mass, as in all chemical reactions, is ordinarily not taken into account.)

24.86 The sun's energy is believed to come from a series of nuclear reactions, the overall result of which is the transformation of four hydrogen atoms into one helium atom. How much energy is released in the formation of one helium atom? (Include the annihilation energy of the two positrons formed in the nuclear reactions with two electrons.)

$4\ ^1H$: $4(1.00783\ \text{u}) = 4.03132$ u ^4He 4.00260 u Difference $= 0.02872$ u

$$(0.02872\ \text{u})(931.5\ \text{MeV/u}) = 26.8\ \text{MeV}$$

[*Note*: If any of the mass difference between ^4He and $4\ ^1H$ had been left in the form of particles (positrons and electrons), their rest masses would have had to be included in the calculation.]

24.87 It is proposed to use the nuclear fusion reaction $2\,^2_1H \longrightarrow {}^3_1H + {}^1_1H + \text{energy}$ to produce industrial electric power. If the output is to be 50 MW and the energy of the reaction is used with 30% efficiency, how many g deuterium fuel will be needed per day?

^3H: 3.01605 u ^1H: 1.00783 u Total $= 4.02388$ u

$2\ ^2H$: $(2)(2.01410\ \text{u}) = 4.02820$ u Loss in mass $= (4.02820 - 4.02388)\ \text{u} = 0.00432$ u

$$(0.00432\ \text{u})\left(\frac{1\ \text{kg}}{6.02 \times 10^{26}\ \text{u}}\right)(2.998 \times 10^8\ \text{m/s})^2 = 6.45 \times 10^{-13}\ \text{J}$$

$$\left(\frac{6.45 \times 10^{-13}\ \text{J}}{2\ \text{atoms}}\right)\left(\frac{6.02 \times 10^{23}\ \text{atoms}}{\text{mol}}\right) = 1.94 \times 10^{11}\ \text{J/mol}\ ^2\text{H}$$

30% efficiency implies that 5.82×10^{10} J/mol ^2H is useful.

$$50\ \text{MW} = 50 \times 10^6\ \text{W} = \left(\frac{50 \times 10^6\ \text{J}}{\text{s}}\right)\left(\frac{3600\ \text{s}}{\text{h}}\right)\left(\frac{24\ \text{h}}{\text{day}}\right)\left(\frac{1\ \text{mol}}{5.82 \times 10^{10}\ \text{J}}\right) = 74.4\ \text{mol/day}$$

$$\left(\frac{74.4\ \text{mol}}{\text{day}}\right)\left(\frac{2.0\ \text{g}}{\text{mol}}\right) = 149\ \text{g}\ ^2\text{H/day}$$

24.88 How much heat would be developed per hour from a 1.0 Ci ^{14}C source if all the energy of the β^- decay were imprisoned?

$$1.0\ \text{Ci} = 3.7 \times 10^{10}\ \text{dis/s} \qquad {}^{14}_{6}\text{C} \longrightarrow {}^{0}_{-1}\beta + {}^{14}_{7}\text{N}$$

^{14}C: 14.00324 u ^{14}N: 14.00307 u $\Delta m = 0.00017$ u

Per atom, $$\Delta E = (\Delta m)c^2 = (0.00017\ \text{u})\left(\frac{1\ \text{kg}}{6.02 \times 10^{26}\ \text{u}}\right)\left(\frac{3.00 \times 10^8\ \text{m}}{\text{s}}\right)^2 = 2.5 \times 10^{-14}\ \text{J}$$

In total, $$(3.7 \times 10^{10}\ \text{dis/s})\left(\frac{3600\ \text{s}}{\text{h}}\right)\left(\frac{2.5 \times 10^{-14}\ \text{J}}{\text{dis}}\right) = 3\ \text{J/h}$$

24.4 NUCLEAR CROSS SECTION

TABLE 24.3 Cross Sections for Low-Energy Neutrons

nuclide	isotopic abundance, %	cross section, barns*	half-life of product
$^{10}_{5}B$	19	3.99×10^3	Stable
$^{23}_{11}Na$	100	5.6×10^{-1}	2.27 min
$^{31}_{15}P$	100	2.3×10^{-1}	14.3 d
$^{55}_{25}Mn$	100	1.34×10^1	2.6 h
$^{107}_{47}Ag$	51.35	4.4×10^1	2.3 min
$^{109}_{47}Ag$	48.65	1.10×10^2	24 s
$^{113}_{48}Cd$	12.26	1.95×10^4	Stable
$_{48}Cd$	All isotopes	2.4×10^3	
$^{115}_{49}In$	95.8	1.45×10^2	54 min
$^{197}_{79}Au$	100	9.8×10^1	2.7 d

* 1 barn = 10^{-24} cm²

24.89 Calculate the effective neutron capture radius of a nucleus having a cross section of 1.0 barn.

▌

$$1.0 \text{ barn} = 1.0 \times 20^{-24} \text{ cm}^2$$

The area of a circle is given by $A = \pi r^2$. Hence $r = \sqrt{A/\pi} = \sqrt{(1.0 \times 10^{-24} \text{ cm}^2)/3.14} = 5.6 \times 10^{-13}$ cm

24.90 For a target containing 1.00 mg of manganese in a nuclear reactor of flux 1.0×10^{13} neutrons/cm²·s, calculate the activity of $^{56}_{25}Mn$ ($t_{1/2} = 2.6$ h) formed in 5.2 h. What would be the activity after 520 h irradiation? The isotopic abundance of $^{55}_{25}Mn$ is 100%, and $\sigma = 13.4$ barns.

▌

$$\lambda_{\text{prod}} = \frac{0.693}{t_{1/2}} = \frac{0.693}{2.6 \text{ h}} = 0.267/\text{h}$$

After 5.2 h,

$$A_{\text{prod}} = N_t \sigma f(1 - 10^{-\lambda_{\text{prod}} t/2.30})$$

$$N_t = \left(\frac{1.00 \times 10^{-3} \, g}{55.0 \text{ g/mol}}\right)(6.02 \times 10^{23} \text{ atoms/mol}) = 1.09 \times 10^{19} \text{ atoms}$$

$$A_{\text{prod}} = (1.09 \times 10^{19})(13.4 \times 10^{-24})(1.00 \times 10^{13})(1 - 10^{-(0.267/\text{h})(5.2 \text{ h})/2.30})$$
$$= (1.46 \times 10^9)(1 - 0.25) = 1.1 \times 10^9 \text{ dis/s}$$

At 520 h, the exponential term will effectively be zero, and the saturation activity will be 1.46×10^9 dis/s.

24.91 The cross section for the reaction $^{127}_{53}I(n,\gamma)^{128}_{53}I$ is 6.3 barns. The half-life of ^{128}I is 25 min. If 12.7 g pure ^{127}I is placed in a reactor in which the neutron flux is 2.0×10^5 neutrons/cm²·s and is bombarded for 25 min, what is the activity of ^{128}I in the sample 100 min *after* bombardment has stopped?

▌

$$A_{\text{prod}} = N_t \sigma f(1 - 10^{-\lambda_{\text{prod}} t/2.30})$$

To solve this equation, one must use units of t and λ which are reciprocals of each other, and the units of σ and f must include units of area which are reciprocals. Thus,

$$N_t = (12.7 \text{ g})\left(\frac{6.02 \times 10^{23} \text{ nuclei}}{127 \text{ g}}\right) = 6.02 \times 10^{22} \text{ nuclei}$$

$$\sigma = 6.3 \text{ barns} = 6.3 \times 10^{-24} \text{ cm}^2 \qquad f = (2.0 \times 10^5)/\text{cm}^2 \cdot \text{s}$$

$$\lambda_{\text{prod}} = \frac{0.693}{25 \text{ min}} = 0.0277 \text{ min}^{-1} \qquad t = 25 \text{ min} \quad (t \text{ is the bombardment time.})$$

The last term of the equation for A_{prod} then is given by

$$1 - 10^{-\lambda_{\text{prod}} t/2.30} = 1 - 10^{-(0.0277/\text{min})(25 \text{ min})/2.30} = 1 - 10^{-0.301} = 1 - 0.500 = 0.500$$

Therefore the activity immediately upon cessation of the bombardment is

$$A_{\text{prod}} = (6.02 \times 10^{22})(6.3 \times 10^{-24} \text{ cm})[(2.0 \times 10^5)/\text{cm}^2 \cdot \text{sec}](0.500) = 3.79 \times 10^4 \text{ s}^{-1}$$

In 100 min, the ^{128}I decays with a 25 min half-life:

$$(2.30)\left(\log\frac{A}{A_0}\right) = -\lambda t = -(0.0277\ \text{min}^{-1})(100\ \text{min}) = -2.77$$

$$\log\frac{A}{A_0} = -1.20 \qquad \frac{A}{A_0} = 0.0625 \qquad A = (0.0625)(3.79 \times 10^4\ \text{s}^{-1}) = 2.4 \times 10^3\ \text{s}^{-1}$$

24.92 Calculate the cross-sectional area of the nucleus of a ^{10}B atom assuming spherical shape. Compare this value to its cross section for neutrons given in Table 24.3. Does the neutron have to make a "direct hit" in order to be captured?

▮ By Problem 24.11,

$$r \approx 1.4A^{1/3} \times 10^{-13} = (1.4)(10)^{1/3} \times 10^{-13} = (1.4)(2.15) \times 10^{-13} = 3.0 \times 10^{-13}\ \text{cm}$$

Thus,

$$\text{Area} = \pi r^2 \approx 2.9 \times 10^{-25}\ \text{cm}^2$$

The cross section, given in Table 24.3, is 3.99×10^3 barns $= 3.99 \times 10^{-21}$ cm^2. In capturing neutrons, ^{10}Be acts as if it has a much larger cross section than its geometric cross section (i.e., a "direct hit" is unnecessary).

24.93 (*a*) Calculate the neutron capture radius of a $^{113}_{48}Cd$ atom from the data of Table 24.3. (*b*) Compare this radius to the "geometric radius" of Problem 24.11.

▮ (*a*) $\sigma = 1.95 \times 10^4$ barns $= 1.95 \times 10^{-20}$ cm$^2 = \pi r^2$

$r = \sqrt{\sigma/\pi} = \sqrt{(1.95 \times 10^{-20}\ \text{cm}^2)/3.14} = 7.88 \times 10^{-11}$ cm

(*b*) $r = 1.4A^{1/3} \times 10^{-13}\ \text{cm} = (1.4)(113)^{1/3} \times 10^{-13}\ \text{cm} = 6.8 \times 10^{-13}$ cm

The actual radius is about 1/100 as great as the neutron capture radius. (Compare Problem 24.92.)

24.94 A 7.00 g sample of nuclidically pure ^{127}I is placed in a nuclear reactor in which the neutron flux is 1.1×10^5 neutrons/cm^2·s and is bombarded for 25 min. The half-life of the product, ^{128}I, is 25 min. Fifty minutes *after* the bombardment has stopped, the activity in the sample is 9000 dis/s. What is the cross section, in barns, for the reaction $^{127}I(n,\gamma)^{128}I$?

▮

$$A_{\text{prod}} = N_t \sigma f(1 - 10^{-\lambda_{\text{prod}} t/2.30}) \qquad N_t = (7.00\ \text{g})\left(\frac{6.02 \times 10^{23}\ \text{nuclei}}{127\ \text{g}}\right) = 3.32 \times 10^{22}$$

$$f = 1.1 \times 10^5/\text{cm}^2\cdot\text{s} \qquad \lambda_{\text{prod}} = \frac{0.693}{25\ \text{min}} = 0.0277\ \text{min}^{-1}$$

$$(2.30)\left(\log\frac{A}{A_{\text{prod}}}\right) = -\lambda t = (2.30)\left(\log\frac{9000\ \text{s}^{-1}}{A_{\text{prod}}}\right) = -(0.0277\ \text{min}^{-1})(50\ \text{min}) = -1.386$$

$$\frac{9000/\text{s}}{A_{\text{prod}}} = 0.250 \qquad A_{\text{prod}} = \frac{9000\ \text{s}^{-1}}{0.250} = 36\,000\ \text{s}^{-1} \qquad \sigma = \frac{A_{\text{prod}}}{N_t f(1 - 10^{-\lambda_{\text{prod}} t/2.30})}$$

$$1 - 10^{-\lambda_{\text{prod}} t/2.30} = 1 - 10^{-(0.0277\ \text{min}^{-1})(25\ \text{min})/2.30} = 0.500$$

Substitution of this expression in the equation for σ yields

$$\sigma = \frac{36000\ \text{s}^{-1}}{(3.32 \times 10^{22})(1.1 \times 10^5/\text{cm}^2\cdot\text{s})(0.500)} = 19.7 \times 10^{-24}\ \text{cm}^2 = 19.7\ \text{barns}$$

24.95 A sample of rock from the Grand Canyon was analyzed for gold content by neutron activation. A 1.00 g sample was irradiated in a flux of 2.0×10^{12} neutrons/cm^2·s for about 2 weeks. Then the sample was removed, dissolved, and some inactive gold carrier was added. Gold was precipitated as the sulfide and was found to have an activity due to ^{198}Au of 50 dis/min, corrected to the time of bombardment. The cross section of ^{197}Au for neutrons is 98 barns. The half-life of ^{198}Au is 2.7 days. What mass of gold is there per g rock sample?

▮

$$A_{\text{prod}} = N_t \sigma f(1 - 10^{-\lambda_{\text{prod}} t/2.30}) = N_t \sigma f(1 - 10^{-(0.693/2.7\ \text{days})(14\ \text{days})/2.30}) = N_t \sigma f(0.973)$$

$$N_t = \frac{A_{\text{prod}}}{\sigma f(0.973)} = \frac{(50/\text{min})(1\ \text{min}/60\ \text{s})}{(98 \times 10^{-24}\ \text{cm}^2)(2.0 \times 10^{12}/\text{cm}^2\cdot\text{s})(0.973)} = 4.4 \times 10^9$$

Per gram of rock:

$$(4.4 \times 10^9\ \text{atoms})\left(\frac{197\ \text{g}}{6.02 \times 10^{23}\ \text{atoms}}\right) = 1.4 \times 10^{-12}\ \text{g}$$

24.96 How long must 5.0 g of the pure nuclide ^{31}P be irradiated with neutron flux 2.0×10^5 neutrons/cm$^2\cdot$s to achieve an activity of ^{32}P of 1.00×10^3 dis/min 100 h after the bombardment has stopped $(t_{1/2} = 14.3 \text{ days})$?

▮

$$(2.30)\left(\log \frac{A}{A_0}\right) = -\left(\frac{0.693}{t_{1/2}}\right)t = -\frac{(0.693)(100 \text{ h})}{(14.3 \text{ days})(24 \text{ h/day})} = -0.202$$

$$\frac{A}{A_0} = 0.817 \qquad A_0 = \frac{A}{0.817} = 1.00 \times 10^3/0.817 = 1.22 \times 10^3$$

The activity at the end of the bombardment must have been 1.22×10^3 counts/min.

$$A_{prod} = N_t \sigma f(1 - 10^{-\lambda_{prod}t/2.30}) = 1.22 \times 10^3 \text{ counts/min}$$

$$N_t = (5.0 \text{ g})\left(\frac{1 \text{ mol}}{31.0 \text{ g}}\right)\left(\frac{6.02 \times 10^{23} \text{ atoms}}{\text{mol}}\right) = 9.7 \times 10^{22} \text{ atoms}$$

$$\sigma = 2.3 \times 10^{-1} \text{ barn} = 2.3 \times 10^{-25} \text{ cm}^2 \qquad f = 2.0 \times 10^5/\text{cm}^2\cdot\text{s} \qquad \lambda_{prod} = 0.693/14.3 \text{ days} = 0.0485/\text{day}$$

$$1 - 10^{-\lambda_{prod}t/2.30} = \frac{A_{prod}}{N_t \sigma f} = \frac{[(1.22 \times 10^3)/\text{min}][(1 \text{ min})/(60 \text{ s})]}{(9.7 \times 10^{22})(2.3 \times 10^{-25} \text{ cm}^2)(2.0 \times 10^5 \text{ cm}^2/\text{s})} = 4.6 \times 10^{-3}$$

$$10^{-\lambda_{prod}t/2.30} = 0.9954 \qquad -\lambda_{prod}t/2.30 = -2.00 \times 10^{-3} \qquad \lambda_{prod}t = 4.60 \times 10^{-3}$$

$$t = \frac{(4.60 \times 10^{-3})(14.3 \text{ d})}{0.693} = 9.5 \times 10^{-2} \text{ d} = 2.3 \text{ h}$$

24.5 RADIOCHEMISTRY

24.97 To 50.00 mL of a solution containing an unknown concentration of zinc ion was added 0.100 μCi of $^{62}Zn^{2+}$ in 10 mL solution, and the total volume was diluted to 100 mL with water. Precipitation of a zinc salt yielded 0.2000 g zinc in the solid phase with an activity of 0.0823 μCi. What was the original concentration of the zinc ion?

▮

$$\% \text{ Zn recovered} = \% \, ^{62}\text{Zn recovered} = \frac{0.0823}{0.100} \times 100 = 82.3\%$$

$$\text{Total zinc} = \frac{0.2000 \text{ g recovered}}{(0.823 \text{ g recovered})/(\text{g total})} = 0.243 \text{ g total}$$

The mass of the added $^{62}Zn^{2+}$ is negligible; hence 0.243 g is the mass of the Zn^{2+} in the original sample.

$$\frac{0.243 \text{ g}}{(65.37 \text{ g/mol})(0.05000 \text{ L})} = 0.0744 \text{ M Zn}^{2+}$$

24.98 When radioactive sulfur is added to alkaline sodium sulfite solution, radioactive thiosulfate ion is formed. Upon adding Ba^{2+}, a precipitate of BaS_2O_3 is obtained. The precipitate is filtered and dried and is then treated with acid, producing solid sulfur, SO_2 gas, and water. The SO_2 is *not* radioactive at all. Write equations for the sequence of reactions, and comment on the structure of the thiosulfate ion as elucidated by this experiment.

▮

$$\tfrac{1}{8}S_8^* + SO_3^{2-} \longrightarrow S^*SO_3^{2-} \quad (S_2O_3^{2-}) \qquad Ba^{2+} + S^*SO_3^{2-} \longrightarrow BaS^*SO_3$$

$$BaS^*SO_3 + 2H_3O^+ \longrightarrow SO_2 + \tfrac{1}{8}S_8^* + 3H_2O + Ba^{2+}$$

The two sulfur atoms in the thiosulfate ion are not equivalent and do not "exchange" with one another. The structure of the thiosulfate ion is

```
      ..       2-
     :O:
  .. .. ..
 :O:S:S*:
  .. .. ..
     :O:
      ..
```

24.99 Describe how radiochemically pure $^{62}_{30}Zn$ $(t_{1/2} = 9 \text{ h})$ can be prepared.

▮ A suitable reaction would be the irradiation of a copper target with 25 MeV protons: $^{63}Cu(p, 2n)^{62}Zn$. Some radioactive ^{62}Cu will also be formed, as well as ^{63}Zn. The ^{63}Zn, having a half-life of 38 min, will decay to ^{63}Cu by electron capture before the separation procedure. The copper produced by these processes will be separated with the bulk of the copper target. About 2 h after the irradiation (the time lapse to allow the ^{63}Zn to decay), the target is disssolved in nitric acid. The pH is adjusted to about 3, copper is precipitated as CuS, and the solid is filtered. Then the filtrate is boiled to concentrate it and to drive off excess H_2S. The only cations remaining in the solution are $^{62}Zn^{2+}$ and H_3O^+. The solution is electrolyzed between platinum electrodes, and radiochemically pure ^{62}Zn is collected at the cathode.

24.100 A 0.20 mL sample of a solution containing 1.0×10^{-7} Ci of ^3_1H is injected into the blood stream of a laboratory animal. After sufficient time for circulatory equilibrium to be established, 0.10 mL blood is found to have an activity of 20 dis/min. Calculate the blood volume of the animal.

/ In the injected sample the activity is

$$(1.0 \times 10^{-7}\ \text{Ci})\left(\frac{3.7 \times 10^{10}\ \text{dis/s}}{\text{Ci}}\right) = 3.7 \times 10^3\ \text{dis/s}$$

In the sample withdrawn: $(20\ \text{dis/min})(1\ \text{min}/60\ \text{s}) = 0.33\ \text{dis/s}$

The total activity of the entire blood volume is equal to the activity of the sample injected. The ratio of total activity to activity of sample withdrawn is equal to the ratio of volumes, where V is the original body blood volume:

$$\frac{3.7 \times 10^3\ \text{dis/s}}{0.33\ \text{dis/s}} = \frac{V + 0.20\ \text{mL}}{0.10\ \text{mL}} \qquad V = 1.1 \times 10^3\ \text{mL} = 1.1\ \text{L}$$

24.101 A sample of $^{131}_{53}I$, as iodide ion, was administered to a patient in a carrier consisting of 0.10 mg of stable iodide ion. After 4.00 days, 67.7% of the initial radioactivity was detected in the thyroid gland of the patient. What mass of the stable iodide ion had migrated to the thyroid gland? Of what diagnostic value is such an experiment?

/ The total activity in the patient after 4.00 d is determined from the half-life (Table 24.1):

$$(2.30)\left(\log\frac{A}{A_0}\right) = -\frac{0.693}{t_{1/2}}t = -\left(\frac{0.693}{8.0\ \text{d}}\right)(4.00\ \text{d}) = -0.3465 \qquad \frac{A}{A_0} = 0.707$$

The total activity is 70.7% of the original activity. Since 67.7% is found in the thyroid, 67.7/70.7 of the radioactive iodide ion is in the thyroid; hence the same percent of stable iodide ion must also be in the thyroid.

$$\left(\frac{67.7}{70.7}\right)(100\%) = 95.8\%$$

CHAPTER 25
Nonmetals

25.1 GENERAL

25.1 Suggest three tests which can be used to distinguish between a metal and a nonmetal.

▌ Among many possible answers are (*a*) physical properties such as metallic luster, heat conductivity, ductility, high density; (*b*) electrical conductivity; and (*c*) reducibility (nonmetallic elements can in general be reduced to negative oxidation states, whereas metallic elements cannot).

25.2 Write equations describing the reaction of freshly prepared aluminum oxide with HCl and with NaOH.

▌
$$Al_2O_3 + 6HCl \longrightarrow 2Al^{3+} + 6Cl^- + 3H_2O$$
$$Al_2O_3 + 2NaOH + 3H_2O \longrightarrow 2Na^+ + 2Al(OH)_4^-$$

25.3 Classify each of the following oxides as strongly acidic, weakly acidic, neutral, amphoteric, weakly basic, or strongly basic: (*a*) SnO_2 (*b*) SnO (*c*) CO (*d*) PbO (*e*) MnO_2 (*f*) RaO (*g*) N_2O (*h*) FeO (*i*) Ag_2O (*j*) OsO_4 (*k*) Al_2O_3 (*l*) Fe_2O_3 (*m*) CeO_2 (*n*) CO_2 (*o*) MgO (*p*) K_2O

▌ (*a*) Amphoteric (*b*) amphoteric (*c*) neutral (*d*) weakly basic (*e*) neutral (*f*) strongly basic (*g*) neutral (*h*) weakly basic (*i*) strongly basic (*j*) weakly acidic (as expected for such a high oxidation state) (*k*) amphoteric (*l*) weakly basic (*m*) strongly basic (*n*) weakly acidic (*o*) moderately strongly basic (*p*) strongly basic

25.4 Select the strongest and the weakest acid in each of the following sets: (*a*) HBr, HF, H_2Te, H_2Se, H_3P, H_2O (*b*) HClO, HIO, H_3PO_3, H_2SO_3, H_3AsO_3

▌ (*a*) HBr, a strong acid, is the strongest in the group; PH_3, which has weakly basic properties, is the least acidic. (*b*) H_2SO_3, which is farthest to the right in the periodic table and has the most oxygen atoms, is the most acidic; HClO, with the fewest oxygen atoms and the farthest up in the periodic table, is the weakest.

25.5 A 0.10 M aqueous solution of which salt in each of the following pairs would have the higher pH? (*a*) $NaNO_2$ or $NaAsO_2$ (*b*) NaF or NaCN (*c*) Na_2SO_3 or Na_2TeO_3 (*d*) NaOCl or NaOBr

▌ Conjugates of stronger acids are weaker bases. The oxyacids are stronger toward the top of the periodic table. By use of these generalizations, one can deduce the substance with the higher pH. (*a*) HNO_2 is a stronger acid; hence NO_2^- is a weaker base. AsO_2^- has a higher pH. (*b*) HF is a stronger acid (see Table 21.1); hence F^- is a weaker base. CN^- has a higher pH. (*c*) H_2SO_3 is a stronger acid; SO_3^{2-} is a weaker base. TeO_3^{2-} has a higher pH. (*d*) HOCl is a stronger acid; OCl^- is a weaker base. OBr^- has a higher pH.

25.6 Identify the good oxidizing agent(s), the good reducing agent(s), the good dehydrating agents, and the strong Brønsted acid(s) from among the following substances: H_2SO_3, HNO_3, P_4O_{10}, H_3PO_4, H_2S, H_2SO_4.

▌ Good oxidizing agents: HNO_3 and H_2SO_4(*conc*) Good reducing agent: H_2S Good dehydrating agents: H_2SO_4 and P_4O_{10}

25.7 State the meaning of each of the following prefixes usually used with oxyacids. Use each one to name a specific compound. (*a*) meta (*b*) pyro (*c*) hypo (*d*) per (*e*) ortho

▌ (*a*) The meta acid has one water molecule fewer than an ortho acid of the same element in the same oxidation state per atom of central element. (*b*) The pyro acid has one-half water molecule fewer than an ortho acid per atom of central element. (*c*) The hypo acid has an oxidation state lower than the ous acid. (*d*) The per acid has an oxidation state higher than the ic acid. (*e*) An ortho acid has as many water molecules added as can fit per atom of central element. Examples of these prefixes include (*a*) HPO_3 (*b*) $H_4P_2O_7$ (*c*) H_3PO_2 (*d*) $HClO_4$ (*e*) H_3PO_4

25.8 Calculate the pH of 1.0 M $NaHCO_3$, using the fact that in such a solution

$$[H_2CO_3] - [CO_3^{2-}] = [OH^-] - [H_3O^+].$$

▮ HCO_3^- reacts with water in two ways:

$$HCO_3^- + H_2O \rightleftharpoons CO_3^{2-} + H_3O^+ \qquad K_2 = 5 \times 10^{-11}$$

$$HCO_3^- + H_2O \rightleftharpoons H_2CO_3 + OH^- \qquad K_h = \frac{K_w}{K_1} = \frac{1.0 \times 10^{-14}}{3.5 \times 10^{-7}} = 2.9 \times 10^{-8}$$

Hence,
$$[CO_3^{2-}] = \frac{K_2[HCO_3^-]}{[H_3O^+]} \qquad [OH^-] = \frac{K_w}{[H_3O^+]}$$

$$[H_2CO_3] = \frac{K_w[HCO_3^-]}{K_1[OH^-]} = \frac{[H_3O^+][HCO_3^-]}{K_1} \qquad [H_3O^+] + [H_2CO_3] = [CO_3^{2-}] + [OH^-]$$

(The last equation results from the combination of the equations describing the distribution of carbon-containing species and the charge balance.)

Rearranging:

$$[H_3O^+] = [CO_3^{2-}] + [OH^-] - [H_2CO_3] = \frac{K_2[HCO_3^-]}{[H_3O^+]} + \frac{K_w}{[H_3O^+]} - \frac{[H_3O^+][HCO_3^-]}{K_1}$$

$$[H_3O^+]^2 = K_2[HCO_3^-] + K_w - [H_3O^+]^2[HCO_3^-]/K_1$$

$$[H_3O^+]^2\left(1 + \frac{[HCO_3^-]}{K_1}\right) = K_2[HCO_3^-] + K_w \qquad [H_3O^+] = \sqrt{\frac{K_2[HCO_3^-] + K_w}{1 + [HCO_3^-]/K_1}}$$

For a solution in which $[HCO_3^-] \approx 1.0$ M,

$$\left(1 + \frac{[HCO_3^-]}{K_1}\right) \simeq \frac{[HCO_3^-]}{K_1} \qquad \text{and} \qquad K_2[HCO_3^-] + K_w \simeq K_2[HCO_3^-]$$

Therefore
$$[H_3O^+] = \sqrt{\frac{K_2[HCO_3^-]}{[HCO_3^-]/K_1}} = \sqrt{K_1K_2}$$

$$pH = -\log[H_3O^+] = -\tfrac{1}{2}(\log K_1 + \log K_2) = \tfrac{1}{2}(pK_1 + pK_2) = \tfrac{1}{2}(6.46 + 10.30) = 8.38$$

At reasonably high $[HCO_3^-]$, the pH is independent of the HCO_3^- concentration.

25.9 Explain why a substance with a formula like $ZrH_{1.9}$ is regarded as a compound rather than a mixture.

▮ It has a definite composition and definite properties distinct from those of its elements, despite being nonstoichiometric.

25.2 HALOGENS

25.10 Write equations showing how each of the following compounds can be prepared, starting with the appropriate elemental halogen: ***(a)*** $HClO_4$ ***(b)*** I_2O_5 ***(c)*** Cl_2O ***(d)*** ClO_2 ***(e)*** $KBrO_3$ ***(f)*** OF_2 ***(g)*** BrO_3 ***(h)*** Br_2O

▮ ***(a)*** $3Cl_2 + 6NaOH \longrightarrow 5NaCl + NaClO_3 + 3H_2O$

$4NaClO_3 \xrightarrow[\text{heat}]{} NaCl + 3NaClO_4 \qquad H_2SO_4 + NaClO_4 \xrightarrow[\text{distill}]{} NaHSO_4 + HClO_4$

(b) $I_2 + 6OH \longrightarrow 5I^- + IO_3^- + 3H_2O$

$IO_3^- + H^+ \longrightarrow HIO_3 \qquad 2HIO_3 \xrightarrow[\text{heat}]{} I_2O_5 + H_2O$

(c) $2Cl_2 + HgO \longrightarrow Cl_2O + HgCl_2$

(d) $Cl_2 + 6OH^- \longrightarrow 5Cl^- + ClO_3^- + 3H_2O$

$2H^+ + 2ClO_3^- + H_2C_2O_4 \longrightarrow 2ClO_2 + 2CO_2 + 2H_2O$

(e) $3Br_2 + 6KOH \longrightarrow 5KBr + KBrO_3$ ***(f)*** $2F_2 + 2OH^- \longrightarrow 2F^- + OF_2 + H_2O$

(g) $\tfrac{1}{2}Br_2 + O_3 \xrightarrow[0^\circ C]{} BrO_3$ ***(h)*** $2Br_2 + HgO \longrightarrow Br_2O + HgBr_2$

25.11 Comparing manufacturing costs and costs of raw materials, explain why it is cheaper to use H_2SO_4 than HCl as an acid for industrial purposes.

▮ HCl is prepared with H_2SO_4 as one of the reagents: $H_2SO_4 + NaCl \rightarrow HCl + NaHSO_4$. Thus the cost of HCl includes the cost of H_2SO_4 plus other costs, so HCl must be more expensive than H_2SO_4.

25.12 Write equations showing a laboratory preparation of each of the following: ***(a)*** HCl ***(b)*** HBr ***(c)*** HI ***(d)*** HIO_4

▮ ***(a)*** $NaCl + H_2SO_4 \longrightarrow HCl + NaHSO_4$

(*b*) $NaBr + H_3PO_4 \longrightarrow HBr + NaH_2PO_4$ (H_2SO_4 is too good an oxidizing agent to use with the easily oxidized Br^- or I^- ions.)

(*c*) $PI_3 + 3H_2O \longrightarrow H_3PO_3 + 3HI$

(*d*) $IO_3^- + ClO^- + H_3O^+ + H_2O \longrightarrow H_5IO_6 + Cl^-$ $\quad H_5IO_6 \xrightarrow[\text{heat}]{} HIO_4 + 2H_2O$

25.13 Pure $HClO_4$ is a liquid which does not conduct electricity. When melted, the solid hydrate, $HClO_4 \cdot H_2O$, does conduct electricity. Draw possible electron dot structures for both the acid and the acid hydrate. Discuss the importance of hydrogen bonding in these examples.

▮

```
        :Ö:        ⎡  H  ⎤+ ⎡   :Ö:   ⎤-
   H:Ö:Cl:Ö:       ⎢H:Ö:H⎥  ⎢:Ö:Cl:Ö: ⎥
        :Ö:        ⎣     ⎦  ⎣   :Ö:   ⎦
```

Hydrogen bonding could be extensive in molecular $HClO_4$. In its hydrate, in which the ions are localized because of the lattice forces, the hydrogen bonding might not be as extensive.

25.14 Suggest an explanation as to why the bond angle in Cl_2O is less than that in ClO_2, and why the O—Cl bond is longer in Cl_2O.

▮

```
     O            Cl
   Cl  Cl       O    O
```

The electron dot structures show two unshared pairs of electrons on the central atom in Cl_2O, which cause the chlorine atoms to be nearly at the tetrahedral angle. Since there are only three nonbonding electrons on the central atom in ClO_2, the repulsion for the bonded pairs is less than that in Cl_2O, and the angle between the bonds is greater in ClO_2. The shorter bond length in ClO_2 results from resonance, with the unpaired electron on the chlorine or the oxygen atoms.

25.15 Tabulate the following data for HCl and HI: (*a*) bond energy (in kcal) (*b*) percent ionic character (*c*) dipole moment (in debyes) (*d*) bond length (in angstroms). Explain why HI is a stronger acid than HCl.

▮

	(*a*)	(*b*)	(*c*)	(*d*)
HCl	103.2 kcal	12%	1.03 D	1.27 Å
HI	71 kcal	5%	0.38 D	1.61 Å

The data in part (*a*) come from Table 7.2, those in part (*b*) from the answer to Problem 7.48, those in part (*c*) from Problem 7.48, and those in part (*d*) from Table 7.1. Of the properties listed, the lower bond energy in HI is the main factor in making HI stronger than HCl, since all the other factors favor HCl as being stronger.

25.16 Draw an electron dot structure for IF_3. Explain why I_4, a molecule which might have a similar electronic and molecular structure, does not exist.

▮

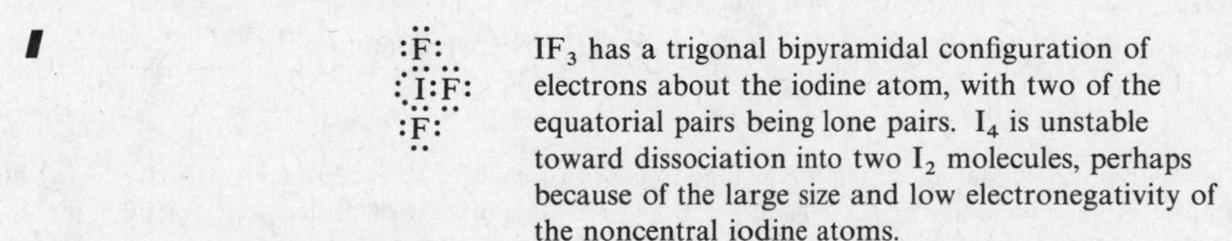

IF_3 has a trigonal bipyramidal configuration of electrons about the iodine atom, with two of the equatorial pairs being lone pairs. I_4 is unstable toward dissociation into two I_2 molecules, perhaps because of the large size and low electronegativity of the noncentral iodine atoms.

25.17 List the properties which are desirable for a liquid to be used as a refrigerant.

▮ High enthalpy of vaporization, boiling point just under the temperature at which it will be used, high critical temperature, nontoxicity, stability, low cost.

25.18 Show in terms of bond energies (*D*) and entropies why Cl_2O would be expected to decompose spontaneously into its elements.

▮

$$Cl_2O \longrightarrow Cl_2 + \tfrac{1}{2}O_2$$

$$\Delta H = 2D(\text{Cl—O}) - D(\text{Cl—Cl}) - \tfrac{1}{2}D(O_2) = 2(50\ \text{kcal/mol}) - (58\ \text{kcal/mol}) - \tfrac{1}{2}(119.2\ \text{kcal/mol}) = -18\ \text{kcal}$$

The decomposition reaction will tend to be spontaneous if the free energy change is negative. The enthalpy portion of the free energy change is negative, and the entropy change is positive, because more mol of gas is produced upon decomposition. Since $\Delta G = \Delta H - T\Delta S$, the positive value of ΔS and the negative value of ΔH ensure that ΔG will be negative.

25.19 Account for the bond angles in each of the following on the basis of electron pair repulsion and on the basis of hybridization of the orbitals on the central atom: ***(a)*** Cl_2O ***(b)*** ClO_2 ***(c)*** Cl_2O_7 ***(d)*** I_3^-

▌ ***(a)*** The tetrahedral orientation of the four electron pairs leads to the bond angle approximating the tetrahedral angle of 109.5°. ***(b)*** Resonance of the single electron among the bonding and nonbonding orbitals on the central atom lead to an angle somewhat greater than the tetrahedral angle. ***(c)*** The tetrahedral angles about the chlorine atoms are as expected. The angle about the central oxygen atom also is expected for an sp^3 hybridized central atom with two of the pairs of electrons bonded to other atoms. ***(d)*** The linear orientation of I_3^- results from the five pairs of electrons being oriented in a trigonal bipyramidal geometry about the central atom. The two bonding pairs repel less than the nonbonding pairs and so are most stable at the apical positions (90° from three other pairs).

The hybridization of the central atoms in (*a*), (*b*), and (*c*) are all sp^3. That in (*d*) is sp^3d. The geometries of the molecules which result from the hybridization argument and the repulsion of electron pair argument are the same.

25.20 The cyanide ion is often referred to as a pseudo-halide ion. Discuss this concept using as illustrations ***(a)*** the cyanogen molecule, C_2N_2 ***(b)*** the reaction of cyanogen with OH^- ***(c)*** the formation of cyanide coordination compounds ***(d)*** the properties of hydrogen cyanide ***(e)*** other evidence, if any.

▌ CN^- resembles Cl^-, Br^-, and I^- in many ways. ***(a)*** Like the halide ions, CN^- can be produced by reduction of an uncharged dimer, C_2N_2. ***(b)*** Cyanogen disproportionates upon treatment with strong base, as do Cl_2, Br_2, and I_2.

$$C_2N_2 + 2OH^- \longrightarrow CN^- + CNO^- + H_2O$$

(c) Cyanide ion forms a wide variety of coordination compounds, as do the halide ions. Because of the possibility of π bonding, the cyanide ion is inherently more strongly bonded. ***(d)*** HCN is an acid. It is weak, in contrast to HCl, HBr, and HI (but not HF). ***(e)*** CN^- is a reasonable reducing agent, like the heavier halide ions. Its electronic structure makes it diamagnetic and very stable toward dissociation into its atoms. Thus it retains its identity through many different types of reactions. Its solubility behavior is very similar to that of the halide ions.

25.3 GROUP VI ELEMENTS

25.21 Diagram the molecular structure of OF_2. What is the oxidation state of oxygen in OF_2?

▌ :F: :F:
  :O:

The oxidation state of oxygen is +2; only with fluorine does oxygen exhibit a positive oxidation state.

25.22 For the reaction (at 25 °C) $3O_2 \rightleftharpoons 2O_3$, $\Delta H° = +68$ kcal; the equilibrium constant is 10^{-54}. Calculate $\Delta G°$ and $\Delta S°$ for the reaction.

▌

$$\Delta G° = (-2.30\,RT)(\log K) = (-2.30)(1.987\text{ cal/mol·K})(298\text{ K})(-54) = +73.5\text{ kcal/mol}$$

$$\Delta G° = \Delta H° - T\Delta S° \qquad \Delta S° = \frac{\Delta H° - \Delta G°}{T} = \frac{(68\,000 - 73\,500)\text{ cal/mol}}{298\text{ K}} = -19\text{ cal/mol·K}$$

25.23 Suggest a practical use for water in which advantage is taken of each of the following properties: ***(a)*** high heat capacity ***(b)*** low molecular weight ***(c)*** high boiling point ***(d)*** polarity of molecules ***(e)*** high melting point ***(f)*** large enthalpy of fusion ***(g)*** large enthalpy of vaporization

▌ One example (of many possible) is given for each part: ***(a)*** heat transfer agent (hot water heating) ***(b)*** steam engine (more gas volume per g substance) ***(c)*** existence of life ***(d)*** solvent for ionic compounds ***(e)*** ice manufacture ***(f)*** cooling power of ice ***(g)*** heating capacity (steam heating)

25.24 Explain why calcium ion makes water "hard," but sodium ion does not.

▌ Calcium forms an insoluble compound with stearate ion (soap anion); sodium stearate is soluble.

25.25 Explain, using equations, how the addition of $Ca(OH)_2$ aids in the removal of calcium ions from temporary hard water. What effect would the addition of too much of the reagent have on the desired process?

▮ Temporary hard water contains $Ca(HCO_3)_2$.

$$Ca^{2+} + 2HCO_3^- + Ca(OH)_2 \longrightarrow 2CaCO_3 + 2H_2O$$ Too much $Ca(OH)_2$ yields more Ca^{2+} to the solution.

25.26 Draw molecular orbital electronic energy level diagrams for the peroxide and the superoxide ions.

▮

O_2^{2-}			O_2^-	
—			—	
↑↓	↑↓		↑↓	↑
↑↓	↑↓		↑↓	↑↓
↑↓			↑↓	

25.27 Account for the fact that the formation of Cs_2O from its elements is less exothermic than the formation of ZnO from its elements.

▮ The lattice energy of ZnO must be greater. Lattice energy is the major contribution to the exothermic nature of each reaction.

25.28 Which of the following should have the greatest enthalpy of combustion at 25 °C: rhombic sulfur, monoclinic sulfur, or plastic sulfur?

▮ Plastic sulfur is the highest energy form listed and should be able to release the most enthalpy per mol upon combustion.

25.29 Determine which of the following has the greatest affinity for water: P_4O_{10}, Cl_2O_7, I_2O_5.

▮ P_4O_{10}. One call tell that P_4O_{10} is more powerful a dehydrating agent than Cl_2O_7 since the latter is the product of dehydration of $HClO_4$ by P_4O_{10}. The reverse reaction cannot be expected to proceed so well. Since I_2O_5 can be prepared by heating HIO_3, it cannot have too great an affinity for water. In fact, P_4O_{10} is one of the most powerful dehydrating agents known.

25.30 Explain why addition of HNO_3 to concentrated H_2SO_4 results in the formation of NO_2^+ and NO_3^- ions. How could one test experimentally to confirm that such ions exist in H_2SO_4 solution?

▮ H_2SO_4 dehydrates HNO_3 to N_2O_5, which in a strongly polar solvent exists in the form of its solvated ions—NO_2^+ and NO_3^-. Conductivity measurements might be easiest to use to confirm the ionic nature of the solute.

25.31 Explain how the freezing of water in the crevices of rocks causes mechanical degradation.

▮ Water expands upon freezing, and the force built up is tremendous. (The expansion on freezing is the reason that antifreeze is used in automobiles in very cold weather.) Thus rocks which experience alternating freezing and thawing are subjected to repeated mechanical degradation.

25.32 Assume that ice has a hexagonal close-packed (hcp) structure with oxygen-oxygen contact and with hydrogen atoms occupying the interstices. **(*a*)** What fraction of the tetrahedral holes would be occupied by hydrogen atoms? **(*b*)** Estimate the density of ice under such conditions.

▮ **(*a*)** All the tetrahedral sites would be occupied for an A_2B-type lattice. **(*b*)** With oxygen-oxygen contact, the density would be much greater than the more open hydrogen-bonded structure which actually exists. The density can be estimated by determining the number of H_2O units per unit cell and the size of such a unit cell. The covalent radius of the oxygen atom, 0.66 Å, Table 7.1, will be used as an estimate for the size of each H_2O unit, since the hydrogen atoms merely fill the interstices.

Since the packing of spheres in hcp and cubic close-packed (ccp) arrangements are equally efficient, the densities of the two are the same. The density of a ccp lattice will be calculated, since it is easier to visualize.. The oxygen atoms can be imagined at the corners of a solid with a square top and rectangular sides, such as shown in Fig. 25.1*a*. Length *C* is to be determined in terms of the radius of one oxygen atom. Figure 25.1*b* depicts the two top rear oxygen atoms, the center one, and the two bottom front ones. It is easy to see that the length marked *A* is merely four times the radius of the oxygen atom. Since the bottom of the triangle is twice the radius of the oxygen atom, the length labeled *B* is $2\sqrt{3}$ times the radius, as determined by the Pythagorean theorem. Figure 25.1*c* allows the determination that length *C* is $2\sqrt{2}$ times the radius of the oxygen atom, again by use of the Pythagorean theorem. The volume of the cube is then $2r \times 2r \times 2\sqrt{2}\,r = 8\sqrt{2}\,r^3$.

$$V = 8\sqrt{2}\,r^3 = (8)(1.414)(0.66\text{ Å})^3 = 3.3\text{ Å}$$

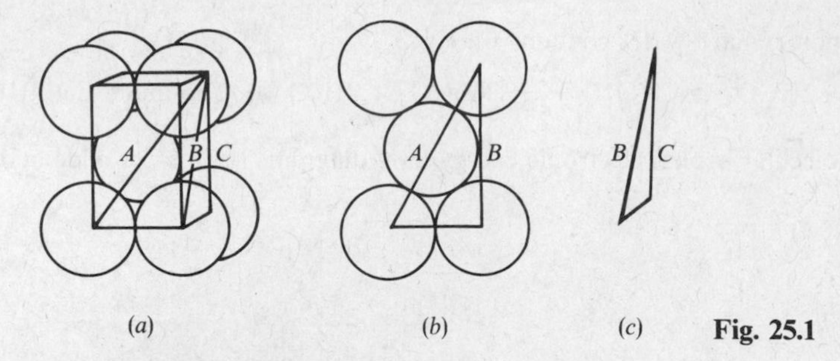

Fig. 25.1

The unit cell contains two molecules of water (one oxygen atom at the eight corners and another one at the center of the cell). The density is then

$$\frac{(36\ \text{u})/(6.02 \times 10^{23}\ \text{u/g})}{(3.3\ \text{Å}^3)(10^{-24}\ \text{cm}^3/\text{Å}^3)} = 18\ \text{g/cm}^3$$

25.4 GROUP V ELEMENTS

25.33 Write equations showing a laboratory preparation of each of the following: **(a)** N_2O **(b)** N_2O_5 **(c)** NO_2 **(d)** H_6TeO_6 **(e)** $COBr_2$

▮ **(a)** $NH_4^+ + NO_3^- \xrightarrow{\text{heat}} N_2O + 2H_2O$ **(b)** $2HNO_3 \xrightarrow{P_4O_{10}} N_2O_5 + H_2O$

(c) $Pb(NO_3)_2 \xrightarrow{\text{heat}} PbO_2 + 2NO_2$

(d) $Te + ClO_3^- + 3H_2O \xrightarrow[\text{regia}]{\text{aqua}} H_6TeO_6 + Cl^-$ The oxidation is done in the presence of aqua regia (HCl + HNO_3), and the reduction products are indeterminate. **(e)** $CO + Br_2 \longrightarrow COBr_2$

25.34 Write an equation for the disproportionation reaction of P_4 in sodium hydroxide.

▮

$$P_4 + 4OH^- + 2H_2O \longrightarrow 2PH_3 + 2HPO_3^{2-}$$

25.35 Suggest tests which could be used to distinguish among NO, N_2O_5, and NO_2.

▮ NO is a colorless gas which turns brown in air. N_2O_5 is a volatile solid which forms only nitric acid upon reaction with water. NO_2 is a brown gas which condenses to a yellow liquid (N_2O_4) below 20 °C.

25.36 **(a)** Write balanced equations for the reactions of arsenic, antimony, and bismuth with concentrated nitric acid to yield H_3AsO_4, Sb_2O_5, and Bi^{3+} ions, respectively. **(b)** Explain why these three elements do not give analogous products upon reaction with HNO_3.

▮ **(a)** $3As + 5HNO_3 + 2H_2O \longrightarrow 3H_3AsO_4 + 5NO$

$6Sb + 10HNO_3 \longrightarrow 3Sb_2O_5 + 10NO + 5H_2O$

$Bi + HNO_3 + 3H^+ \longrightarrow Bi^{3+} + NO + 2H_2O$

(b) As the elements get more metallic (down the periodic table), they tend to form positive ions more and oxyanions less.

25.37 Calculate the equilibrium constant at 25 °C for the disproportionation of 3 mol aqueous HNO_2 to yield gaseous NO and aqueous NO_3^-. The standard potential for the reduction of HNO_2 to NO is 0.99 V; that for the reduction of NO_3^- to HNO_2 is 0.94 V. Comment on the stability of HNO_2.

▮

$$HNO_2 + H_3O^+ + e^- \longrightarrow NO + 2H_2O \qquad E° = 0.99\ \text{V}$$
$$NO_3^- + 3H_3O^+ + 2e^- \longrightarrow HNO_2 + 4H_2O \qquad E° = 0.94\ \text{V}$$

By doubling the first equation and adding the reverse of the second, one obtains

$$3HNO_2 \longrightarrow 2NO + NO_3^- + H_3O^+ \qquad E° = 0.05\ \text{V}$$

$$E°n\mathcal{F} = (2.30RT)(\log K) \qquad \log K = \frac{2.30RT}{E°n\mathcal{F}} = \frac{(2.30)(8.31\ \text{J/K})(298\ \text{K})}{(0.05\ \text{V})(2\mathcal{F})(96\,500\ \text{C}/\mathcal{F})} = 0.6 \qquad K = 4$$

HNO_2 is not very stable; it decomposes on standing.

25.38 Write resonance forms for the N_2O molecule, NNO.

$$:\ddot{N}::N::\ddot{O}: \qquad :\underset{..}{\ddot{N}}:N:::O: \qquad :N:::N:\underset{..}{\ddot{O}}:$$

These elements do not differ much in electronegativity, and the forms with fewer electrons on the oxygen atoms are therefore not much higher in energy than the form shown last.

25.39 In the preparation of P_4O_6, why is a mixture of nitrogen and oxygen used rather than pure oxygen?

Pure oxygen would oxidize the phosphorus to P_4O_{10}, despite an excess of P_4.

23.40 On the basis of appropriate Born-Haber cycles, state what factor(s) is(are) responsible for the fact that lithium nitride is stable while potassium nitride is unstable.

The higher lattice energy of Li_3N is the major difference between the two compounds. The relatively large difference in lattice energy stems from the large difference in ionic size between lithium and potassium ions. The differences in ionization potential, sublimation energy, and the other factors of the Born-Haber cycle are relatively small between the two compounds.

25.41 Look up the appropriate cross sections and describe how astatine might be prepared from bismuth. Which astatine isotopes would be produced? Suggest a method of separating the astatine from the target bismuth.

Astatine is produced by the reaction $^{209}Bi(\alpha, 2n)^{211}At$. To separate the halogen, the bismuth target is dissolved in HNO_3, and the pH is then adjusted to about 3. The I^- and IO_3^- carrier is added to the solution, which is then saturated with H_2S to produce Bi_2S_3. The solid is filtered, after which the solution is treated with zinc metal to reduce all the halogen to X^-. After filtration of excess zinc, the At^- and carrier I^- are precipitated as the silver salts.

25.5 GROUPS IV AND III

25.42 What factors are responsible for the difference in the properties of CO_2 and SiO_2?

Because of the tendency of carbon to form double bonds, CO_2 exists as discrete molecules. In SiO_2 there is an extended network of Si—O single bonds. The existence of *d* orbitals on silicon allows reactions that are impossible with the corresponding carbon compounds.

25.43 $B_{10}C_2H_{12}$ is isostructural and isoelectronic with what borane ion, $B_xH_y^{z-}$?

$B_{12}H_{12}^{2-}$. The boron atoms each have one fewer electron than a carbon atom. To keep the total number of electrons the same in $B_{10}C_2H_{12}$ and the boron hydride ion, the replacement of two carbon atoms with boron atoms must be accompanied by addition of two extra electrons (from two alkali metal ions, for example).

25.44 The formula for the mineral olivine is sometimes written $(Fe,Mg)_2SiO_4$, meaning that 2 mol of any combination of the metal ions is present per mol orthosilicate ion. Does olivine obey the law of definite proportions? Is olivine a compound or a solid solution?

Olivine does not obey the law of definite proportions, since any ratio of iron to magnesium may be found. It therefore is really a solution of Fe_2SiO_4 in Mg_2SiO_4 (or vice versa).

25.45 Explain the structure of C_3O_2 in terms of the bonding in the molecule.

O=C=C=C=O The molecule is linear, as expected for *sp* bonding by each of the carbon atoms. (The bonding is delocalized from one end of the molecule to the other, with each carbon atom involved in two π bonds. The delocalization does not affect the geometry of the molecule, however.)

25.46 What is the empirical formula for the anion of beryl, $Si_6O_{18}^{12-}$? From the number of shared and unshared oxygen atoms in this ion, show how the empirical formula can be deduced without knowing the size of the ring. Determine in this manner the empirical formulas (complete with charges) of the single and double-chain silicate minerals.

The empirical formula is SiO_3^{2-}. If each shared oxygen atom is counted as $\frac{1}{2}$ toward an empirical formula unit, and each unshared oxygen atom is counted fully toward the unit, the number of atoms may be computed as follows:

$$\text{Per Si atom:} \qquad (2 \times 1) + (2 \times \tfrac{1}{2}) = 3 \text{ oxygen atoms}$$

The charge is computed from the oxidation states of Si and O in silicates (always +4 and −2, respectively). Therefore the empirical formula is calculated to be SiO_3^{2-}. Still another method to calculate the empirical formula is to start at any point along the chain and count all the atoms until the next equivalent point along the chain, as shown below.

```
 O  |   O  |   O
 |  |   |  |   |
Si -+O—Si -+O—Si—· · ·
 |  |   |  |   |
 O  |   O  |   O
```

The number of atoms between the vertical lines is easily seen to be one Si and three O atoms.

The single-chain silicate, with each silicon atom sharing two oxygen atoms of the four attached, has an empirical formula SiO_3^{2-} (see discussion above). In the double-chain silicate, half the silicon atoms share two of their four oxygen atoms, and the other half share three of their four. For each two (adjacent) silicon atoms, three oxygen atoms are bonded to only one silicon atom, and the other five are bonded to two.

$(3 \times 1) + (5 \times \frac{1}{2}) = \frac{11}{2}O$ to 2Si, or Si_4O_{11}. The charge must be 6−. The empirical formula is $Si_4O_{11}^{6-}$. The alternative method gives the same results:

```
          |O    O      |O    O
          | \  /       | \  /
        O |  Si    O   |  Si
    O  |  |O    O  |   |O    O     · · ·
      Si  |       Si   |
       |  |        |   |
       O  |        O   |
       |  |        |   |
      Si  |       Si   |
    O  |  |O    O  |   |O    O
       O  |  Si    O   |  Si
          | /  \       | /  \
          |O    O      |O    O
```

25.47 (*a*) In a regular B_{12} icosahedron, how many boron atoms are equidistant from a given boron atom? (*b*) How many edges are there? (*c*) How many valence electrons are there? (*d*) Can each edge line represent an electron pair bond? (*e*) Explain the type of bonding involved in elemental boron.

▌ (*a*) Five (*b*) 30 (5 from the top boron atom, 5 around the upper pentagon, 10 from the upper pentagon to the lower, 5 around the lower pentagon, and 5 to the bottom boron atom) (*c*) $12 \times 3 = 36$ (*d*) Since there are not 60 electrons, each edge line cannot represent an electron pair. (*e*) There must be some three-center "bridge" bonding, just as there is in diborane.

25.48 Assuming that each has the icosahedral structure, determine how many isomers are possible for the $B_{10}C_2H_{12}$ molecule.

▌ Three. The carbon atoms may be at adjacent positions, may have one boron atom between them, or may be on opposite sides of the icosahedron. There are only three because the 12 positions of the icosahedron are all inherently equivalent.

25.49 Borazene, $B_3N_3H_6$, is isoelectronic and isostructural with benzene. (*a*) Diagram the borazene structure. (*b*) Discuss the bonding in borazene. (*c*) How many isotopic disubstituted borazene molecules, $B_3N_3H_4X_2$, are possible without changing the fundamental ring structure?

▌ (*a*)

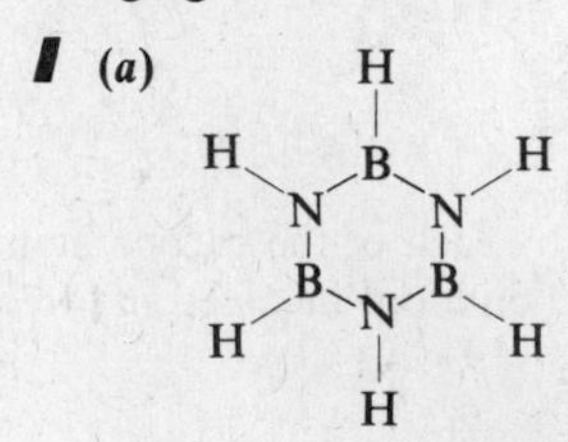

(*b*) There is delocalized π bonding, just as there is in benzene. The number of electrons is the same as in benzene, because each nitrogen atom has one more electron than a carbon atom, but each boron atom has one fewer.

(*c*) There are four such isomers:

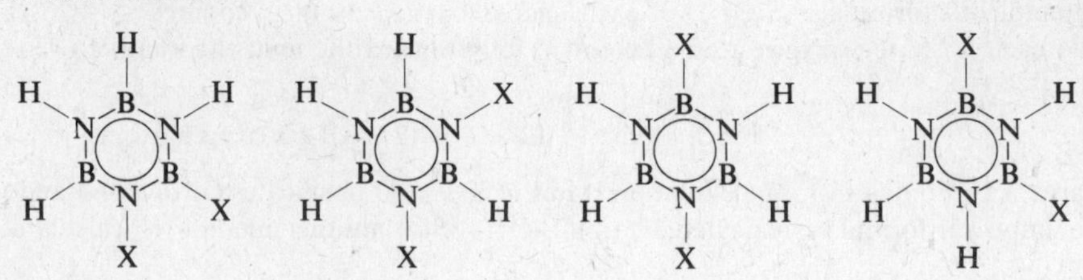

The second and fourth of these are not the same despite the X atoms being one atom apart on each ring. In one case the X atoms are attached to N atoms, and in the other they are attached to B atoms. (Contrast this situation to that in benzene itself.)

25.50 Diagram possible structures of the B_4H_{10} and B_5H_{11} molecules, showing the existence of three-center bridge bonds.

▌

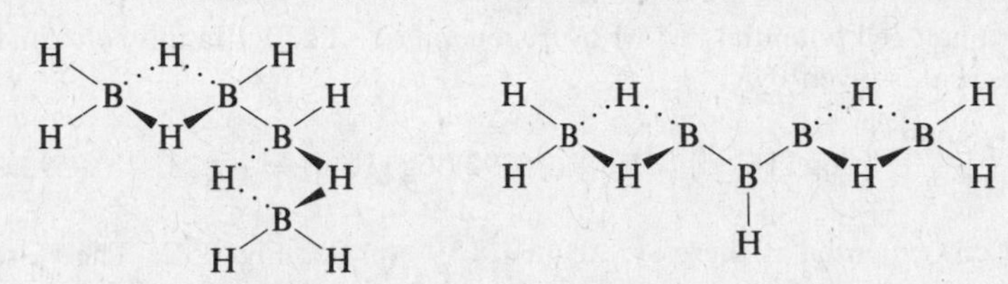

25.51 Diborane, B_2H_6, reacts with water to form boric acid and hydrogen. What is the pH of the solution which results when 1.0 g B_2H_6 reacts with 100 mL water? Assume the final volume to be 100 mL.

▌

$$B_2H_6 + 6H_2O \longrightarrow 2H_3BO_3 + 3H_2$$

$$(1.0\text{ g }B_2H_6)\left(\frac{1\text{ mol }B_2H_6}{27.6\text{ g }B_2H_6}\right)\left(\frac{2\text{ mol }H_3BO_3}{\text{mol }B_2H_6}\right) = 0.072\text{ mol} \qquad \frac{0.072\text{ mol}}{0.10\text{ L}} = 0.72\text{ M }H_3BO_3$$

$$H_2BO_3 + H_2O \longrightarrow H_3O^+ + H_2BO_3^-$$

$$K_a = \frac{[H_2BO_3^-][H_3O^+]}{[H_3BO_3]} = 7.3 \times 10^{-10} \qquad \text{(from Table 21.1)}$$

$$= \frac{x^2}{0.72 - x} \simeq \frac{x^2}{0.72} = 7.3 \times 10^{-10}$$

$$x = 2.3 \times 10^{-5} \qquad \text{pH} = 4.64$$

25.52 Explain why nuclear magnetic resonance experiments show only one type of hydrogen atom in the $B_{12}H_{12}^{2-}$ ion.

▌ Every hydrogen atom is equivalent to every other one, because each is attached to a boron atom, all of which are equivalent. That is, every corner of the icosahedron is equivalent to every other corner.

25.6 NOBLE GASES

25.53 Write electron dot formulas for and explain the geometry of each of the following: (*a*) ICl_3 (*b*) XeF_2

▌

```
 :Cl:
:I:Cl:
 :Cl:
```

T shaped. The five electron pairs are oriented toward the corners of a trigonal bipyramid. The axial positions are occupied by bonding pairs preferentially, since there is less repulsion by bonding pairs than nonbonding pairs.

```
:F:Xe:F:
```

The molecule is linear. The electron pairs are arranged in an octahedral configuration, but only two are bonding pairs, and these are on opposite sides of the molecule.

25.54 Show that it is thermodynamically possible to prepare MnO_4^- by the oxidation of Mn^{2+} with sodium perxenate in acid solution. ($E° = 2.1$ V)

▌ The oxidation of Mn^{2+} in acid solution has a potential of -1.51 V (from Table 23.1). The standard reduction potential of sodium perxenate, $+2.1$ V, is more than adequate to oxidize the Mn^{2+} even when all the substances are present in their standard states. In the absence of MnO_4^- and the reduction products of XeO_6^{4-}, the actual potential for the cell will be even greater. No matter how concentrated the products become, the sign of the potential will not be reversed before the reaction is essentially complete.

25.55 XeF_2 is relatively stable in dilute aqueous HF, despite its reduction potential of approximately 2.2 V. (*a*) Design a galvanic cell in which the following cell reaction occurs:

$$XeF_2 + 2OH^- \longrightarrow Xe + \tfrac{1}{2}O_2 + 2F^- + H_2O$$

(*b*) Predict $E°$ for this cell. (*c*) Sketch the cell, and state what materials would be used for the electrodes, the containers, the salt bridge, etc. (*d*) Will the H_3O^+ from the ionization of HF in the solution have an appreciable effect on $E°$? Explain quantitatively.

▮ (*a*) The cell could have the configuration: $C|OH^-|O_2||XeF_2|Xe|C$ (*b*) The standard reduction potential for the reduction of XeF_2 is about 2.2 V. The half-cell potential for the oxidation half-reaction given may be calculated from the data of Table 23.1 as follows:

$$2OH^- \longrightarrow \tfrac{1}{2}O_2 + H_2O + 2e^-$$

The standard half-cell potential in 1 M hydrogen ion is -1.23 V. In this solution containing 1×10^{-14} M hydrogen ion, the potential is given by

$$E = E^\circ - \left(\frac{0.0592}{2}\right)(\log [H^+]^2) = E^\circ - (0.0592)(\log [H^+]) = -1.23 + (0.0592)(14) = -1.23 + 0.828 = -0.40 \text{ V}$$

The overall cell potential is therefore about 1.8 V. (*c*) See Fig. 25.2. The vessel containing the HF and the salt bridge cannot be made of glass, since glass is attacked by HF. A material such as Teflon, a fluorocarbon plastic, may be used. Inert electrode such as carbon may be used. XeF_2 is a good fluorinating agent and might attack some usually inert materials. (*d*) $XeF_2 + 2e^- \rightarrow Xe + 2F^-$ Since H^+ does not appear in the equation for the reduction half-reaction, the H^+ concentration will not affect the potential. A mixture of HF (1 M) and F^- (1 M) should be used. (Addition of excess strong acid would reduce the fluoride ion concentration and thus affect the potential.)

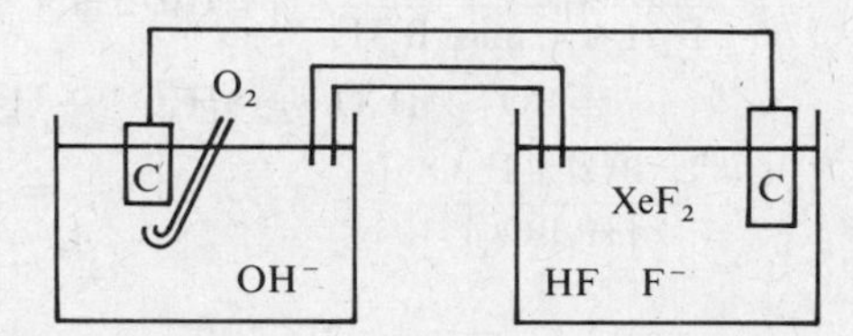

Fig. 25.2.

25.56 Calculate the electronegativities of the noble gases using the method of R. S. Mulliken (Problem 7.25). Are these values reasonable? Explain.

▮ According to Mulliken, $EN = \dfrac{IP + EA}{2}$. Values for fluorine are included in the following tabulation for comparison:

	IP, eV	EA, eV	EN, eV
He	24.8	−0.22	12.3
Ne	21.6	−0.3	10.6
Ar	15.8	−0.36	7.7
Kr	14.0	−0.40	6.8
Xe	12.1	−0.42	5.8
Rn	10.7	−0.42	5.2
F	17.4	3.3	10.4

Comparison of the values of electronegativity calculated for the noble gases with the value for fluorine shows that *in covalent bonds* He and Ne should attract the electrons even more than fluorine does. Since they do not form covalent bonds, however, there is no way to test this hypothesis. The electronegativities of the other noble gases seem to be in a reasonable range. (Note that the values are in electron volts and are not on the usual Pauling scale, where the electronegativity of fluorine is equal to 4.0.)

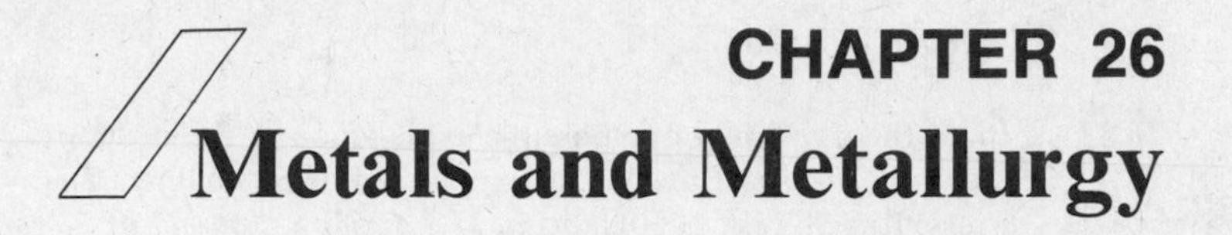

CHAPTER 26
Metals and Metallurgy

26.1 METALLIC BONDING

26.1 Would a sample of mercury vapor appear shiny and metallic? Explain.

▌ Mercury vapor is invisible. Only in a condensed phase does a metal have a metallic luster.

26.2 Are diatomic lithium molecules stable toward (*a*) dissociation into lithium atoms (*b*) condensation into solid lithium metal?

▌ (*a*) They are stable toward dissociation into atoms, because they each have a pair of bonding electrons in excess of the number of pairs of antibonding electrons. (*b*) At room temperature, lithium molecules are unstable toward condensation. The metal is even more stable in the solid state than as diatomic molecules, so the latter are converted to the former under ordinary conditions of temperature and pressure.

26.3 Explain precisely why the two allotropic forms of carbon differ so markedly in their electrical conductance.

▌ The delocalized electrons in the π orbitals of graphite cause this allotropic form to conduct electricity (along the planes of the atoms). There are no delocalized electrons in diamond; hence, diamond does not conduct electricity.

26.4 Suggest how the properties of liquid metals, such as mercury, support the band theory of metallic bonding.

▌ The metallic luster stems from the ability of a solid or liquid metal to absorb and reemit light of all wavelengths because electrons in orbitals have energies so close together that they essentially continuous. The electrical and heat conductivities are due to the very high mobility of the electrons in the bands. The mobility of the liquid stems from the fact that there is only one kind of atom in the lattice and the bonding is nonlocalized.

26.5 What properties of tungsten make it suitable for use as filaments in light bulbs?

▌ High melting point, moderate electrical resistance, low volatility.

26.6 Solutions of alkali metals in liquid ammonia at −33 °C conduct electricity without chemical reaction at the electrodes. Would decreasing the temperature of the solution cause a decrease or increase in conductivity? Explain.

▌ The solution conducts like a metallic conductor (i.e., a wire) with electrons carrying the charge, and so the conductivity should increase as the temperature is lowered.

26.7 W. Hume-Rothery has pointed out that the crystal structures of solution alloys are influenced by the ratio of valence electrons to atoms in the alloy. For example, the crystal structure of solid solutions of zinc in copper have an fcc structure until the composition approaches the "formula" CuZn, at which point the alloys assume a bcc structure known as the beta phase. The valence-electron-to-atom ratio in the beta phase is $\frac{3}{2}$ (or $\frac{21}{14}$). A gamma phase and an epsilon phase are also possible at other characteristic electron-to-atom ratios, where the ratios increase from beta to gamma to epsilon. The following "formulas" of alloys correspond to such phases. Deduce the Hume-Rothery ratios, and classify each alloy in the proper phase: CuZn, Cu_9Al_4, Cu_9Ga_4, Cu_3Sn, Cu_5Zn_8, Cu_5Al_3, Cu_5Sn, Cu_3Ge, $CuZn_3$, Cu_3Al, $Cu_{31}Sn_8$

▌ The Hume-Rothery ratios and the phases of the alloys are tabulated as follows:

formula	valence electrons	atoms	ratio	phase	formula	valence electrons	atoms	ratio	phase
CuZn	3	2	3:2	β	Cu_5Sn	9	6	3:2	β
Cu_9Al_4	21	13	21:13	γ	Cu_3Ge	7	4	7:4	ε
Cu_9Ga_4	21	13	21:13	γ	$CuZn_3$	7	4	7:4	ε
Cu_3Sn	7	4	7:4	ε	Cu_3Al	6	4	3:2	β
Cu_5Zn_8	21	13	21:13	γ	$Cu_{31}Sn_8$	63	39	21:13	γ
Cu_5Al_3	14	8	7:4	ε					

26.8 Band theory is only one possible explanation of the structure and properties of metals. Compare band theory with the *bond theory* of metals, as expounded in L. Pauling, *The Nature of the Chemical Bond*, 3rd ed. (Cornell Univ. Press, Ithaca, NY, 1960), and in J. S. Griffith, *Journal of Inorganic and Nuclear Chemistry*, **3**, 15 (1956).

▮ Pauling empirically assigns "metallic valences" equal to their group numbers to the elements of the first six groups of the periodic table (IA through VIB) on the basis of hardness, strength, and density. Other properties, such as melting point, boiling point, enthalpy of fusion, and enthalpy of vaporization, can be explained with the assigned valences. The metals of the next several groups are assigned "valences" lower than their group numbers on the basis of their properties. He explains the magnetic properties of the elements on the basis of resonance of the electrons in "metallic orbitals" among the atoms, deducing the number of such orbitals empirically once more.

Griffith attempts to explain the variations of the enthalpies of sublimation of transition metals by dividing the sublimation energies into three components. One of these (p) varies smoothly with atomic number, rising to a maximum in the middle of each transition series and then diminishing once more. Another component (P) is a sort of promotional energy term, necessary to get each element into the $d^{n-1}s^1$ state. The third component ($\lambda\Delta$) stems from the splitting of the d orbitals in a field but is judged by Griffith to be of little importance in the cases of these neutral atoms. The theory assumes that electrons in unfilled subshells, which remain as parallel as possible in isolated atoms (Hund's rule), tend to pair up with electrons in adjacent atoms in solid metals. The pairing of electrons between atoms is in essence chemical bonding. Griffith's theory is presented in a highly mathematical form.

In their attempt to explain the properties of metals, the models seek to explain the same kinds of behavior. The band theory focuses on the electrical conductivity of metals, whereas Pauling looks at hardness, density, and strength. Griffith uses enthalpy of sublimation as his major focus. The two bonding theories attempt to explain the magnetic properties of metals, while the band theory as presented in the text does not cover that property of metals. Further exposition of the band theory does attempt to explain magnetic properties, however. Each of the approaches does well in explaining those properties upon which it is based.

26.2 ALLOYS

26.9 A sample of the alloy duralumin contains 4.2% copper by mass. What percent of the alloy consists of the compound $CuAl_2$?

▮ The alloy is composed of $CuAl_2$ plus aluminum. In 100 g alloy there is 4.2 g copper. Hence,

$$(4.2\text{ g Cu})\left(\frac{1\text{ mol Cu}}{63.5\text{ g}}\right)\left(\frac{1\text{ mol CuAl}_2}{1\text{ mol Cu}}\right)\left(\frac{117.5\text{ g CuAl}_2}{\text{mol CuAl}_2}\right) = 7.8\text{ g CuAl}_2$$

In 100 g alloy there is 7.8 g $CuAl_2$; hence the percent $CuAl_2$ is 7.8%.

26.10 Express the composition of the compound alloy Cu_3Au in karats.

▮ Cu_3Au percent composition:

$$(3\text{ mol Cu})(63.5\text{ g/mol}) = 190.5\text{ g} \qquad (1\text{ mol Au})(197\text{ g/mol}) = 197.0\text{ g}$$

$$\text{Formula weight Cu}_3\text{Au} = 190.5 + 197.0 = 387.5\text{ g/mol}$$

$$\%\text{ Cu} = (190.5/387.5)(100\%) = 49.0\% \qquad \%\text{ Au} = (197/387.5)(100\%) = 51.0\%$$

The number of karats in an alloy is the number of parts of gold per 24 parts of the alloy. The alloy is 51.0 parts gold per 100 parts alloy, or

$$\frac{51.0\text{ parts}}{100\text{ parts}} = \frac{x}{24.0\text{ parts}} \qquad x = 12.2\text{ karats}$$

26.11 Name one property of pure metals which makes them more useful than alloys.

▮ The electrical conductivity of pure metals is greater than that of their alloys.

26.12 The compound alloy Cu_3Au crystallizes in a cubic lattice with Cu at the face centers and Au at the corners. How many formula units of the compound are there in each unit cell?

▮ Per unit cell there are $6 \times \frac{1}{2} = 3$ copper atoms and $8 \times \frac{1}{8} = 1$ gold atom. Hence there is only one Cu_3Au formula unit per unit cell.

26.13 What type of alloy would most likely be formed by each of the following pairs? Give an example of each pair to illustrate your choice. Discuss cases in which the choice is not clear-cut, if any. ***(a)*** A pair of metals having similar sized atoms, the same number of valence electrons, and the same type of lattice when pure. (*b*) A pair of metals of vastly different electronegativities and with atoms of widely different sizes. (*c*) A pair of metals having atoms. widely different in size but similar in electronegativities. (*d*) A pair of elements, one of which is a small nonmetal.

▮ (*a*) Substitutional alloy. The metals are so similar that there is little difference in bonding no matter which sites they occupy, so their arrangement is random. The elements copper and nickel are so nearly alike in size that they exhibit complete solubility in each other. Both are fcc. (*b*) Compound alloy. The widely different electronegativities cause compound formation. $MgCu_2$ is an example. (*c*) Eutectic alloy. The widely different sizes tend to make the metals relatively insoluble in each other, so they solidify in separate crystallites. Tin and lead form a eutectic alloy. (*d*) Interstitial alloy. The nonmetal atoms occupy the interstices of the larger, metal atoms. Iron and carbon form an interstitial alloy called steel.

26.14 Predict on the basis of actual calculations which one(s) of the following sets of elements could form an *undistorted* interstitial alloy: Pd-C, La-B, Ba-Be, Pd-H. The metallic radii are Pd, 1.38 Å; La, 1.88 Å; Ba, 2.17 Å. Covalent radii for the nonmetals must be used in lieu of other available data: C, 0.77 Å; B, 0.77 Å; Be, 1.11 Å; H, 0.30 Å. Pd forms a face-centered cubic (fcc) lattice; La forms a hexagonal closest-packed lattice, and Ba forms a body-centered cubic lattice.

▮ The fcc lattice has tetrahedral sites only. The hcp lattice has tetrahedral sites only. The radius ratio for tetrahedral sites varies from 0.414 to 0.225. If this ratio is exceeded, the alloy will be distorted.

The ratio of the radii of carbon and palladium is $0.77/1.38 = 0.56$. Since this ratio is greater than 0.414, the lattice cannot be undistorted.

$B{:}La = 0.77/1.88 = 0.41$ This lattice may be undistorted.

$Be{:}Ba = 1.11/2.17 = 0.511$ This lattice cannot be undistorted.

$H{:}Pd = 0.30/1.38 = 0.22$ This lattice is certainly undistorted.

26.15 Suggest an experimental method of distinguishing between an alloy which is a solid solution from one which is an intermetallic compound.

▮ X-ray diffraction can identify the repeating pattern of an intermetallic compound. Also, the intermetallic compound should have a sharp melting point, whereas the solid solution will melt over a range of temperatures.

26.16 A compound alloy of metals A and B has a unit cell containing A atoms at its corners and B atoms at the face centers. What is the formula of the compound?

▮ $$8 \text{ corners} \times \tfrac{1}{8} \text{ atom each} = 1 \text{ A atom} \qquad 6 \text{ faces} \times \tfrac{1}{2} \text{ atom each} = 3 \text{ B atoms}$$

The compound has a formula AB_3.

26.17 Using Fig. 26.1*a*, draw cooling curves and identify the points at which breaks occur as well as the points at which the temperature is constant during the cooling from the molten state for each of the following Pb-Sn mixtures: (*a*) 25% tin (*b*) 75% tin (*c*) 100% tin (*d*) the eutectic mixture

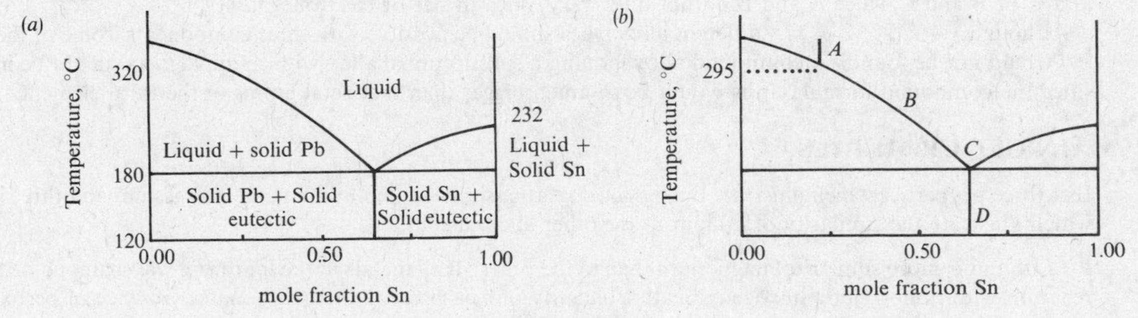

Fig. 26.1

▮ (*a*) At *A* in Fig. 26.1*b* the liquid mixture cools to about 295 °C, where it starts to solidify. *B*, crystallization of Pb occurs, and the percent Sn left in the liquid increases. The temperature falls along the curve to the point *C*. The temperature falls more gradually, however, since solidification is occurring during this phase of the process. At the point at which the composition of the liquid and solid are the same, *C*, no change in composition occurs upon further heat removal, and hence no change in temperature occurs during this phase of the process. *D*, the temperature falls more rapidly once again when all the sample has solidified. The cooling curve is drawn in Fig. 26.2*a*. **(*b*)** The first solidification will occur at about 200 °C. In this case, solid tin will solidify first, until the eutectic temperature is reached. See Fig. 26.2*b*. **(*c*)** Since the sample is pure tin, the temperature will fall to the melting point, hold at that temperature while the entire sample solidifies, and then fall once more (see Fig. 26.2*c*). **(*d*)** The temperature will fall to 181 °C, the eutectic temperature, at which point the eutectic mixture will start to solidify. Since the solid and liquid phases are the same in composition, no change in concentration will occur as the solidification takes place. When the entire sample has solidified, the temperature will again fall. Note the similarity in the cooling curve, Fig. 26.2*d*, to that of pure tin, Fig. 26.2*c*. It would be impossible to distinguish between a pure substance and a eutectic mixture on the basis of the cooling curve alone.

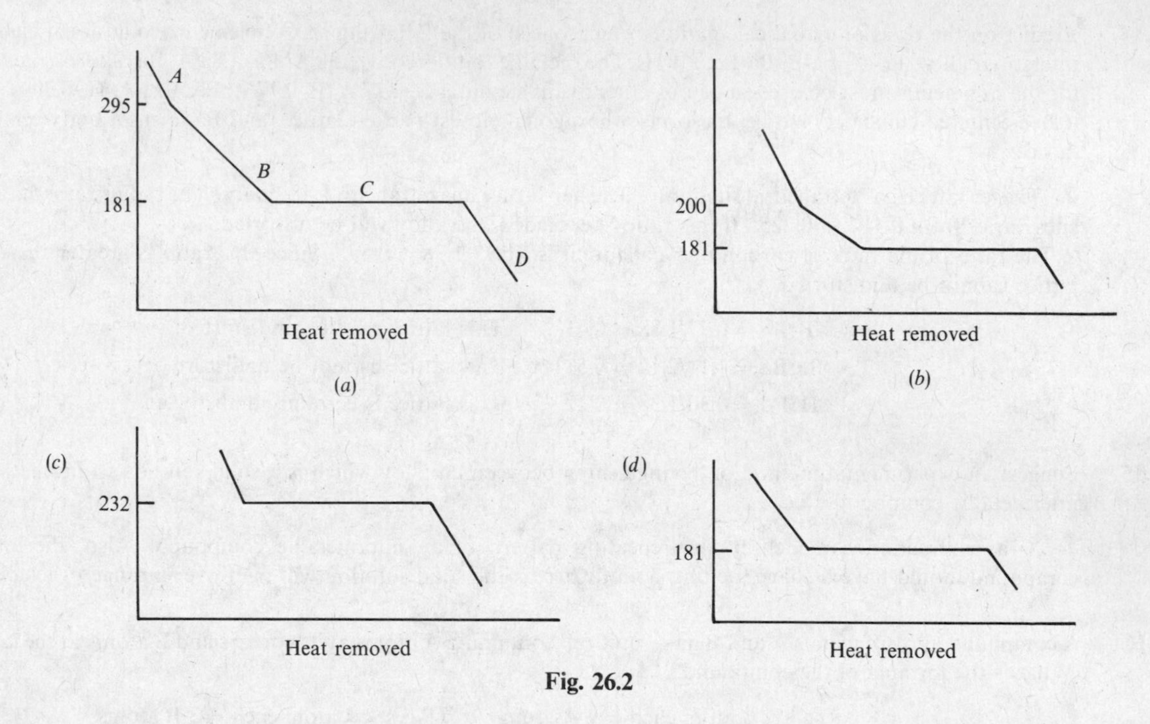

Fig. 26.2

26.18 Speculate on the types of alloys which might be formed in three-component systems. Describe some properties of each type of alloy.

▮ The same types of alloys might be expected in three-component systems as in two-component systems—substitutional, eutectic, compound, and interstitial. But in addition, combinations of these types might be expected. For example, two elements which form a substitutional alloy, A and B, might incorporate a third element, C, in its interstices. Such an interstitial alloy should not differ very much in properties from a two-component alloy of A and C or B and C, since A and B do not differ very much in size or electronegativity.

Although $4 \times 3 \times 2 = 24$ different alloy types should be possible with three components, some of these probably would not be found. A compound alloy forming a substitutional alloy with a simple metal might be impossible, since the compound formula units would be so much bigger than the metal atoms of the third element.

26.3 MAIN GROUP METALS

26.19 List three properties which illustrate the diagonal relationship between lithium and magnesium and three properties which illustrate the similarity of lithium to the other alkali metals.

▮ Lithium is more similar to magnesium than to the other alkali metals in the following ways, among others: direct reaction with atmospheric nitrogen, greater solubility of bicarbonates than carbonates, absence of peroxide in the product of reaction with excess oxygen, solubility of chlorides in organic solvents. It is more similar to the other alkali metals in its electronegativity, its oxidation state in compounds, its basicity, and the solubility of its hydroxide, among others.

26.20 Plot the melting points of NaCl, KCl, RbCl, and CsCl versus their formula weights. By extrapolation of the plot, estimate the melting point of lithium chloride. Explain why LiCl actually melts at a much lower temperature.

▌ When the melting points are plotted against the formula weights, the curve extrapolates to about 840 °C, 230° above its actual melting point. The lattice is comparatively unstable because of the very small size of the Li^+ ion. The radius ratio is 0.33, below the minimum (0.414) for the rock salt structure.

26.21 In each set select the substance which is the **(*a*)** most basic: Al_2O_3, Tl_2O_3, Tl_2O **(*b*)** lowest melting: LiBr, $BeBr_2$, BBr_3 **(*c*)** highest in electronegativity: Li, Be, Mg **(*d*)** most stable toward oxidation: $GeCl_2$, $SnCl_2$, $PbCl_2$, **(*e*)** strongest oxidizing agent: CrO_4^{2-}, MoO_4^{2-}, WO_4^{2-}

▌ **(*a*)** Tl_2O (The farther down in the periodic table and the lower in the oxidation state, the more basic the oxide.) **(*b*)** BBr_3 (The farther to the right in the periodic table, the greater the covalent character of the bonds and the lower the forces holding the parts together.) **(*c*)** Be (The farther to the right and the farther up in the periodic table, the greater the electronegativity.) **(*d*)** $PbCl_2$ (The inert pair of 6th period elements is more stable than those of 5th period elements and far more stable than the analogous pair in 4th period elements.) **(*e*)** CrO_4^{2-} (The high oxidation states of the second and third transition series elements are more stable.)

26.22 Lead(IV) is a powerful oxidizing agent. What can be said about the reducing ability of lead(II)?

▌ Lead(II) is a relatively poor reducing agent.

26.23 Write a balanced chemical equation for the reaction of PbO with NaOH.

▌
$$PbO + 2NaOH \longrightarrow Na_2PbO_2 \text{ (sodium plumbite)} + H_2O$$

26.24 A solution of $BiCl_3$ in aqueous HCl yields a white precipitate when diluted with pure water. By means of a balanced chemical equation, explain this result.

▌
$$BiCl_3 + H_2O \longrightarrow BiOCl \text{ (bismuth oxychloride)} + 2HCl \qquad \text{(See Problem 10.18.)}$$

26.25 Which would you expect to be more soluble in water, LiI or KI? Explain your choice.

▌ LiI. It is composed of a very small cation and a very large anion and so is expected to have a smaller than extrapolated lattice energy. (Compare the answer to Problem 26.3.) Thus the energy required to break up the lattice is lower and is easily provided by the hydration energy. Literature values of the solubilities are LiI, 433 g per 100 g water at 80 °C; KI, 208 g per 100 g water at 100 °C.

26.26 Compounds having the formulas $CsBr_3$ and $CsBrCl_2$ are stable below 100 °C and crystallize in a cubic lattice. Is the existence of these compounds a contradiction of the statement that alkali metals have only one positive oxidation state? Explain.

▌ $CsBr_3$ and $CsBrCl_2$ are compounds containing interhalogen ions, Br_3^- and $BrCl_2^-$, respectively. The alkali metal has a +1 oxidation state.

26.27 Predict the following: **(*a*)** Melting point (°C) of francium **(*b*)** density (g/cm³) of francium **(*c*)** type of lattice of francium **(*d*)** melting point (°C) of FrCl **(*e*)** lattice type and density (g/cm³) of FrI **(*f*)** the product of combustion of francium metal in air

▌ The predictions are made by extrapolation of analogous data for the other alkali metals, which can be obtained, e.g., from the *Handbook of Chemistry and Physics.*

	(*a*)	(*b*)	(*c*)	(*d*)	(*e*)		(*f*)
Na	97.81	0.97	bcc	801	Rock salt	3.667	Na_2O_2
K	63.65	0.86	bcc	770	Rock salt	3.13	KO_2
Rb	38.89	1.532	bcc	718	Rock salt	3.55	RbO_2
Cs	28.40	1.8785	bcc	645	CsCl	4.510	CsO_2
Fr	18	2.2	bcc	550	CsCl	5	FrO_2

26.28 Magnesium metal burns in air to give a white ash. When this material is dissolved in water, the odor of ammonia can be detected. Suggest an explanation for this observation.

▌ Magnesium combines with nitrogen to yield a nitride, which reacts with water to form ammonia:

$$3Mg + N_2 \longrightarrow Mg_3N_2 \qquad Mg_3N_2 + 6H_2O \longrightarrow 3Mg(OH)_2 + 2NH_3$$

26.29 Predict whether Tl^+ will disproportionate in aqueous solution, given the following standard reduction potentials:

$$Tl^+ + e^- \longrightarrow Tl \qquad E° = -0.34\ V \qquad Tl^{3+} + 2e^- \longrightarrow Tl^+ \qquad E° = 1.25\ V$$

▮

$$2Tl^+ + 2e^- \longrightarrow 2Tl \qquad E° = -0.34\ V$$
$$Tl^+ \longrightarrow Tl^{3+} + 2e^- \qquad E° = -1.25\ V$$
$$3Tl^+ \longrightarrow 2Tl + Tl^{3+} \qquad E° = -1.59\ V$$

Since the potential has such a high negative value, Tl^+ will have no tendency to disproportionate; the opposite reaction will tend to proceed spontaneously.

26.30 Explain why aluminium(III) is the only stable oxidation state of aluminium in its compounds, in contrast to thallium, which has states of +1 and +3.

▮ Thallium has an inert pair of electrons ($6s^2$); aluminum has no corresponding inert pair.

26.31 Reduction potential measurements show that, thermodynamically, calcium is a more powerful reducing agent in basic solution than barium, while barium is more powerful than calcium in neutral or acidic solution. Explain these observations.

▮ Calcium hydroxide is much less soluble than barium hydroxide, so in basic solution different species are involved.

26.32 The relative magnitudes of the reduction potentials of metals can be estimated by comparison of the net enthalpy changes for the following processes. If ΔH_{ox} is the enthalpy change for the reaction $M(s) \rightarrow M^{n+}(aq) + ne^-$, then $\Delta H_{ox} = \Delta H_{sub} + \Sigma(IP) + \Delta H_{hyd}$, where ΔH_{sub} is the enthalpy of sublimation of the metal, $\Sigma(IP)$ is the sum of the successive ionization potentials to go from $M(g)$ to $M^{n+}(g)$, and ΔH_{hyd} is the enthalpy of hydration of the gaseous M^{n+} ion. Show what factors are responsible for the following: (*a*) $E°_{red}$ for lithium is more negative than that for rubidium. (*b*) Beryllium and magnesium are poorer reducing agents than barium. (*c*) Silver is not as good a reducing agent as strontium.

▮

$$\begin{array}{ccc} M(s) & \xrightarrow[\Delta H_{ox}]{} & M^{n+}(aq) + ne^- \\ \downarrow \Delta H_{sub} & & \uparrow \Delta H_{hyd} \\ M(g) & \xrightarrow[\Sigma(IP)]{} & M^{n+}(g) + ne^- \end{array}$$

(*a*) ΔH_{hyd} is greater for Li^+ than for Rb^+ because of the small size of the former. (*b*) The first and second ionization potentials for barium are much smaller than the corresponding potentials for beryllium or magnesium. (*c*) The enthalpy of hydration for strontium is much greater than that for silver, because of its double charge. This factor more than offsets the other factors which would have favored the oxidation of silver.

26.4 TRANSITION AND INNER TRANSITION METALS

26.33 Using tables of reduction potentials, predict the products of the reactions of CrO_4^{2-} with excess (*a*) Fe^{2+} (*b*) Zn (*c*) Fe (*d*) H^-

▮ The more powerful reducing agents reduce the chromium below the chromium(III) oxidation state:

(*a*) $3Fe^{2+} + 8H^+ + CrO_4^{2-} \longrightarrow Cr^{3+} + 3Fe^{3+} + 4H_2O$

(*b*) $2Zn + 8H^+ + CrO_4^{2-} \longrightarrow Cr^{2+} + 2Zn^{2+} + 4H_2O$

(*c*) $Fe + 8H^+ + CrO_4^{2-} \longrightarrow Cr^{3+} + Fe^{3+} + 4H_2O$

(*d*) $3H^- + CrO_4^{2-} + H_2O \longrightarrow Cr + 5OH^-$

26.34 Explain why transition group VIIIB includes nine elements, whereas the other transition groups contain three elements each.

▮ The designation of iron, cobalt, nickel, and the six "platinum elements" as one periodic group was a historical accident. The properties of the three elements in each *period* are rather similar, and Mendeleev considered them different from any other group (the noble gases had not yet been discovered). If it were to be done again, there is little reason to suppose that the nine elements would not be put into three groups—"VIII, IX, and X."

26.35 Using complete equations and stating specific conditions where appropriate, tell how each of the following preparations might be performed: (*a*) $AgNO_3$ from AgCl (*b*) Hg from Hg_2Cl_2 (*c*) V from $Pb_5(VO_4)_3Cl$ (*d*) pure Ta and pure Nb from an iron, niobium, tantalum oxide

■ One possible method for each preparation is given:

(*a*) $AgCl(s) + 2NH_3 \longrightarrow Ag(NH_3)_2^+ + Cl^-$

$Ag(NH_3)_2^+ + e^- \xrightarrow{\text{electrolysis}} Ag + 2NH_3 \qquad Ag + 2HNO_3 \longrightarrow AgNO_3 + NO_2 + H_2O$

(*b*) $Hg_2Cl_2 + 6HNO_3 \xrightarrow{\text{heat to dryness}} 2Hg(NO_3)_2 + 2HCl(g) + 2H_2O(g) + 2NO_2(g)$

Dissolve the solid in hot water. $\quad Hg^{2+} + 2e^- \xrightarrow{\text{electrolysis}} Hg$

(*c*) $2Pb_5(VO_4)_3Cl + 18HCl \xrightarrow{\text{hot aqueous solution}} 3V_2O_5 + 10PbCl_2 + 9H_2O$

$3V_2O_5 + 10Al \longrightarrow 5Al_2O_3 + 6V$

Vanadium(V) oxide is not sufficiently basic to form VCl_5 from aqueous solution.

(*d*) $\text{Mixed oxide} \xrightarrow[K_2CO_3]{\text{fuse in}} \xrightarrow[H_2O]{CO_2\text{ in}} \begin{cases} Nb_2O_5 \\ Ta_2O_5 \end{cases}$

$\xrightarrow{HF + KF} \xrightarrow{\text{concentrate}} \begin{cases} K_2NbOF_5(aq) \\ K_2TaF_7(s) \end{cases}$

$K_2NbOF_5 \xrightarrow{\text{heat to dryness}} \xrightarrow{\text{heat with Al}} Nb \qquad K_2TaF_7 \xrightarrow[\text{electrolyze}]{\text{(dilute solution)}} Ta$

26.36 Potassium permanganate can be prepared by melting together equimolar quantities of MnO_2, KOH, and KNO_3. The reaction mixture is allowed to solidify, and then the dark green solid, K_2MnO_4, is crushed and treated with dilute H_2SO_4, whereupon a solution of $KMnO_4$ and a precipitate of MnO_2 are obtained. Write all relevant equations for this preparation.

■

$$3MnO_2 + 4KOH + 2KNO_3 \longrightarrow 3K_2MnO_4 + 2NO + 2H_2O$$

The K_2MnO_4 formed disproportionates in acid solution:

$$3MnO_4^{2-} + 4H^+ \longrightarrow 2MnO_4^- + MnO_2 + 2H_2O$$

26.37 What is the oxidation state of iron in pyrites, FeS_2? Is the anion diatomic? What physical measurements could be performed to test this hypothesis?

■ Iron is in the +2 oxidation state: it is combined with S_2^{2-}. The magnetic moment of the compound would show that the number of unpaired electrons corresponds to iron(II). X-ray diffraction would show that the two sulfur atoms are closer to each other than mere van der Waals forces would place them—they must be bonded. A band in the Raman spectrum corresponding to the sulfur-sulfur bond would confirm this finding.

26.38 Two iron electrodes connected by a wire are placed in a solution of air-free KCl. When oxygen gas is bubbled around one of the electrodes, the other electrode begins to dissolve, and a current is observed in the wire. The solution around the electrode over which oxygen is being bubbled becomes basic. Explain these observations, write appropriate electrode reactions, diagram the cell, label the electrodes as anode and cathode, and indicate the direction of electron flow in the wire.

■ See Fig. 26.3. The oxygen is reduced to hydroxide ion at one electrode, and the iron is oxidized to Fe^{2+} at the other:

$$2H_2O + O_2 + 4e^- \longrightarrow 4OH^- \qquad Fe \longrightarrow Fe^{2+} + 2e^-$$

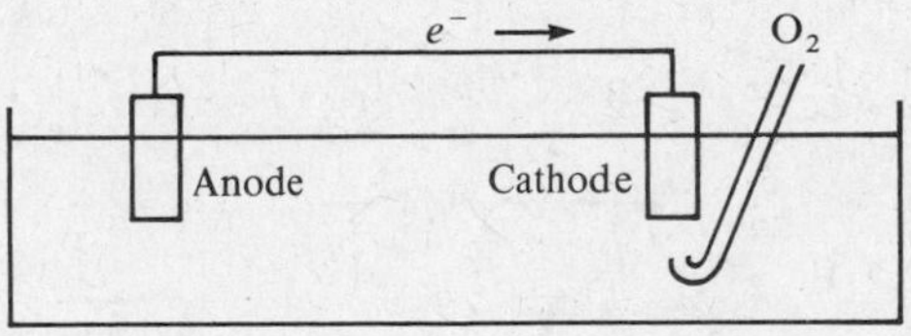

Fig. 26.3

26.39 Osmium forms a +8 oxidation state compound with oxygen but no such compound with fluorine. Explain this behavior.

■ There is space around the osmium atom for four bonded oxygen atoms, but not for eight fluorine atoms.

26.40 Write complete and balanced equations for reactions between the following substances. If no reaction occurs, so indicate.

(*a*) $Cu_2O + MnO_4^-$ (*b*) $Cu_2O + H_2SO_4(dil)$ (*c*) $V^{2+} + I_2$

(*d*) $WO_3 + I_2$ (*e*) $Zn + HCl$ (*f*) $ReO_4^- + Bi^{3+}$

(*g*) $Sn^{2+} + PbO_2$ (*h*) $CrO_4^{2-} + H_3O^+$ (*i*) $Cr_2O_7^{2-} + H_2O_2$
(*j*) $Cr^{2+} + H_2O_2$

▮ (*a*) $3Cu_2O + 2MnO_4^- + 7H_2O \longrightarrow 2MnO_2 + 6Cu(OH)_2 + 2OH^-$
(*b*) $Cu_2O + 2H^+ \longrightarrow Cu^{2+} + Cu + H_2O$ (*c*) $V^{2+} + I_2 + H_2O \longrightarrow VO^{2+} + 2I^- + 2H^+$
(*d*) $WO_3 + I_2 \longrightarrow \text{nr}$ (*e*) $Zn + 2HCl \longrightarrow ZnCl_2 + H_2$
(*f*) $ReO_4^- + Bi^{3+} \longrightarrow \text{nr}$ (*g*) $Sn^{2+} + PbO_2 \longrightarrow Pb^{2+} + SnO_2$
(*h*) $2CrO_4^{2-} + 2H_3O^+ \longrightarrow Cr_2O_7^{2-} + 3H_2O$
(*i*) $Cr_2O_7^{2-} + 3H_2O_2 + 8H_3O^+ \longrightarrow 2Cr^{3+} + 3O_2 + 15H_2O$
(*j*) $2Cr^{2+} + H_2O_2 + 2H_3O^+ \longrightarrow 2Cr^{3+} + 4H_2O$

26.41 Calculate E for the cell $Pt|H_2\ (1.0\ \text{atm})|H^+\ (1.0\ \text{M})||Hg_2^{2+}\ (0.10\ \text{M})|Hg$

▮

$$H_2 \longrightarrow 2H^+ + 2e^- \qquad E^\circ = 0.00\ \text{V} \quad \text{(Table 23.1)}$$
$$2e^- + Hg_2^{2+} \longrightarrow 2Hg \qquad E^\circ = 0.79\ \text{V}$$
$$Hg_2^{2+} + H_2 \longrightarrow 2H^+ + 2Hg \qquad E^\circ = 0.79\ \text{V}$$

$$E = E^\circ - \left(\frac{0.0592}{n}\right)\left(\log \frac{[H^+]^2}{[Hg_2^{2+}]P(H_2)}\right) = 0.79 - \left(\frac{0.0592}{2}\right)\left(\log \frac{1}{0.10}\right) = 0.79 - 0.0296 = 0.76\ \text{V}$$

26.42 After consulting the data of Tables 23.1, 22.1, and 22.3, complete and balance the following equations:
(*a*) $Cu_2O + HI \longrightarrow$ (*b*) $Cu_2O + HCl(\textit{dil}) \longrightarrow$ (*c*) $Cu_2O + HCl(\textit{conc}) \longrightarrow$

▮ (*a*) $Cu_2O + 2HI \longrightarrow 2CuI + H_2O$ (*b*) $Cu_2O + 2HCl(\textit{dil}) \longrightarrow Cu^{2+} + Cu + H_2O$
(*c*) $Cu_2O + 4HCl(\textit{conc}) \longrightarrow 2CuCl_2^- + H_2O + 2H^+$

26.43 Describe, using a diagram, the probable *d* orbital crystal field splittings in the TaF_8^{3-} ion. Assuming only spin effects, would such an ion be high spin or low spin?

▮ A cubic environment (Fig. 26.4*a*) has a crystal field similar to that of a tetrahedral field (Fig. 26.4*c*). Turning one set of four donor atoms through 45° (Fig. 26.4*b*) will not affect the energy of the d_{z^2} orbital, which is symmetrical about the *z* axis. The d_{xy} and $d_{x^2-y^2}$ orbitals will merge to the same energy level in the antiprism field, since each one will be directly over or under one set of four donor atoms and 45° removed from the other set. The d_{xz} and d_{yz} orbitals will be destabilized from the cubic field energies to the same extent, since they will be oriented directly toward one set of four of the donor atoms. The final relative energy levels will be as shown in Fig. 26.4*d*. Ta^V has no *d* electrons left; there is no spin.

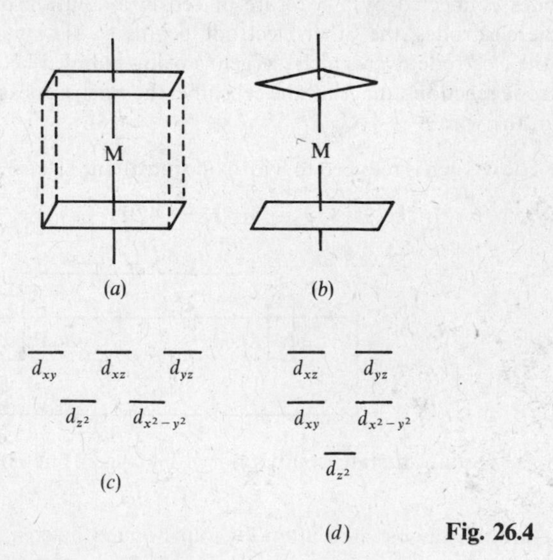

Fig. 26.4

26.44 Calculate the concentration of Au^{3+} in 1.0 L of solution of 0.010 mol gold(III) and 1.0 mol chloride ion.

$$Au^{3+} + 3e^- \longrightarrow Au \qquad E^\circ = 1.50\ \text{V}$$
$$AuCl_4^- + 3e^- \longrightarrow Au + 4Cl^- \qquad E^\circ = 0.994\ \text{V}$$

▌ Treat $E°$ for the second reaction as E for the first:

$$E = E° - \left(\frac{0.0592}{n}\right)\left(\log\frac{1}{[Au^{3+}]}\right) \qquad 0.994 = 1.50 - \left(\frac{0.0592}{3}\right)\left(\log\frac{1}{[Au^{3+}]}\right)$$

$$\log [Au^{3+}] = -25.6 \qquad [Au^{3+}] = 2 \times 10^{-26}\ M$$

26.45 The order of elution of the transuranium elements from an ion exchange resin parallels that of the lanthanoid elements. Suggest an explanation for this observation.

▌ The size variation of the transuranium ions parallels that of the lanthanoids, so the elution behavior does also.

26.46 Explain why none of the lanthanoid sesquioxides, Ln_2O_3, is as acidic as Al_2O_3.

▌ The lanthanoids are lower in the periodic table and are all in group IIIA; hence they are more basic than aluminum. (The added basicity stems from their larger sizes and hence their lower tendency to retain their valence electrons.)

26.47 Suggest a reason why, of the Ln^{3+} ions, the magnetic moment of only Gd^{3+} is in agreement with the moment calculated from the spin-only formula.

▌ The lanthanoid 3+ ions have incomplete $4f$ subshells of electrons. The orbital angular momentum of these electrons contributes to the overall magnetic moment. (In transition metal ions, the orbital angular momentum of the d electrons is more or less "quenched" by the interaction of these electrons and the atoms with which bonding takes place.) Gd^{3+} has a half-filled f subshell, with a total orbital angular momentum of zero. Since there is no orbital angular momentum, it cannot contribute to the moment of the ion, and the "spin only" formula works.

26.48 Which tripositive lanthanoid ion(s) is (are) most likely to react with chromium(II) chloride? What products are expected? Explain.

▌ The lanthanoids which will most likely react with this strong reducing agent are those with stable +2 oxidation states to which they can be reduced—Sm, Eu, and Yb.

26.49 Californium undergoes spontaneous fission as follows: ${}^{252}_{98}Cf \rightarrow {}^{142}_{56}Ba + {}^{106}_{42}Mo + 4\,{}^{1}_{0}n$. The binding energies per nucleon are, respectively, 7.50, 8.25, and 8.50 MeV. Estimate the energy released in this process.

▌

BE/nucleon		no. of nucleons		total BE
7.50	×	252	=	1890
8.25	×	142	=	1171
8.50	×	106	=	901
0	×	1	=	0

The difference in total binding energy between the products and reactant in the nuclear equation gives the approximate energy released, $1890 - 1171 - 901 = -182$ MeV. About 180 MeV of energy will be released.

26.50 What mass of gadolinium will produce the same reduction in neutron flux in a reactor as 1.00 g cadmium? The cross section of gadolinium is 46 kbarns.

▌ The sample will reduce the neutron flux by absorbing neutrons. (The more neutrons absorbed per second, the lower the neutron flux.) The number of neutrons absorbed is equal to the number of nuclei times the cross section of each. The cross section for cadmium (all isotopes) is given in Table 24.3 as 2.4 kbarns. $N\sigma$ for the two must be equal.

$$N(\text{Cd}) = (1.00\ \text{g})\left(\frac{1\ \text{mol}}{112.4\ \text{g}}\right)\left(\frac{6.02 \times 10^{23}\ \text{atoms}}{\text{mol}}\right) = 5.36 \times 10^{21}\ \text{atoms}$$

$$N(\text{Gd}) = (x\ \text{g})\left(\frac{1\ \text{mol}}{157.25\ \text{g}}\right)\left(\frac{6.02 \times 10^{23}\ \text{atoms}}{\text{mol}}\right) = (3.83 \times 10^{21})x\ \text{atoms}$$

$$N(\text{Cd})\sigma(\text{Cd}) = N(\text{Gd})\sigma(\text{Gd}) \qquad x = (1.00\ \text{g})\left(\frac{157.25\ \text{g}}{112.4\ \text{g}}\right)\left(\frac{2.4\ \text{kbarns}}{46\ \text{kbarns}}\right) = 0.073\ \text{g} = 73\ \text{mg}$$

26.5 METALLURGY

26.51 Describe the properties of an ore which is to be concentrated by (*a*) leaching with alkali (*b*) leaching with acid (*c*) flotation (*d*) panning

▌ (*a*) The metal must be amphoteric (or acidic), so that it dissolves in base, leaving behind the gangue and perhaps some other metals. (*b*) The metal must occur as a reasonably basic oxide, hydroxide, or carbonate. (*c*) The mineral desired must be wet by oil more than by soapy water, such as many sulfides are. (*d*) The desired mineral must be much more dense than the gangue [see Problem 26.58(*d*)], which will be washed away by running water while the mineral is left behind.

26.52 There are many minerals in the earth's crust which contain aluminum, but only bauxite is an important ore of this metal. Explain.

▌ Only in bauxite is the aluminum concentrated enough to make its extraction economically feasible.

26.53 Iron oxide is reduced to pig iron with carbon. The excess carbon is oxidized as a part of the steel-making process. Later still, carbon is added to make steel. Explain why the middle step is necessary and why it is not combined with the first or the third step.

▌ The percent carbon in the final product may be controlled more precisely by removing all the carbon initially present and then adding measured quantities.

26.54 In steel manufacture, the word *basic* signifies which of the following? (*a*) fundamental (*b*) alkaline (*c*) unalloyed (*d*) essential

▌ (*b*) Basic furnace linings react with acidic impurities, removing them from the melt.

26.55 What characteristics are desirable for a furnace lining? State under which circumstances SiO_2 would be preferred over CaO or MgO as a furnace lining.

▌ The furnace lining must obviously be very high melting. It should be acidic or basic, depending on the type of impurities which are to be removed. To remove acidic impurities, such as SO_2 or P_4O_{10}, a basic lining is preferable—MgO or CaO. To remove basic impurities, such as metal oxides, an acidic lining—SiO_2—is used.

26.56 Name one metal that is refined by each of the following processes: (*a*) vacuum distillation (*b*) Mond process (*c*) electrolysis (*d*) van Arkel process (*e*) zone refining

▌ (*a*) Mg (*b*) Ni (*c*) Cu (*d*) Zr (*e*) Ga

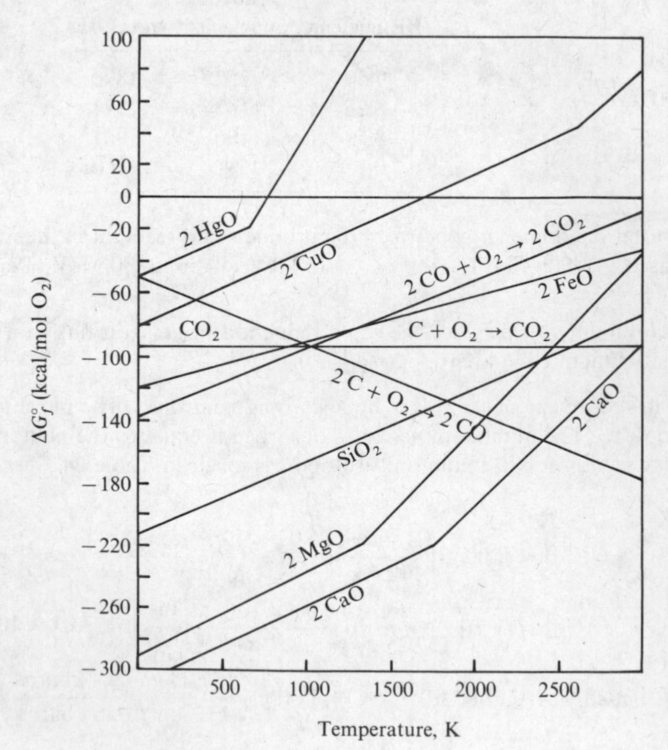

Fig. 26.5

26.57 Using Fig. 26.5, predict the temperature above which CuO would be reduced by roasting in air.

▌ ΔG for the decomposition of CuO is zero at 1700 K. The compound is unstable above that temperature.

26.58 Define each of the following terms: (*a*) karat (*b*) alloy (*c*) ore (*d*) gangue (*e*) slag (*f*) steel (*g*) eutectic point (*h*) eutectic mixture

▌ (*a*) The number of parts of gold per 24 parts of gold alloy. (*b*) A mixture of metals or a metal and a nonmetal which retains metallic characteristics, such as luster and relatively high electrical conductivity. (*c*) A material from which it is economically feasible to extract a metal. (*d*) The portion of an ore other than the desired mineral. (*e*) The by-product of reduction, by which the major portion of the gangue is removed. (*f*) An alloy of iron and carbon with possibly other constituents which has much greater strength than pure iron. (*g*) The temperature at which a eutectic alloy solidifies without change in composition from the liquid state. (*h*) An alloy which solidifies as a mixture of relatively pure crystallites. The solidification from the liquid occurs at constant temperature when the composition of the liquid is the same as that of the solid eutectic mixture.

26.59 Using Figure 26.5, estimate the temperature at which magnesium metal can be prepared by reduction with carbon. What products are expected?

▌ The $2C + O_2 \rightarrow 2CO$ line and the $2MgO \rightarrow 2Mg + O_2$ line cross at about 2050 K. The expected products are CO and Mg. (The $C + O_2 \rightarrow CO_2$ line crosses the MgO line at a higher temperature.)

26.60 Explain how sodium metal may be prepared from NaOH and iron metal (a preparation reported in the chemical literature).

▌ The reaction $6NaOH + 4Fe \rightarrow 2Fe_2O_3 + 6Na + 3H_2$ is not thermodynamically favored at room temperature. The equilibrium lies far to the left. However, if the system is heated enough, the equilibrium shifts to the right, aided by removal of hydrogen and sodium vapor. The reaction is reported to occur in the chemical literature.

26.61 Define each of the following terms: (*a*) refractory (*b*) galvanized steel (*c*) basic metal (*d*) engine knock (*e*) stalactite (*f*) stalagmite

▌ (*a*) Refractory means capable of enduring high temperatures without damage. For example, refractory substances are used in furnace linings because of their high melting points. (*b*) Galvanized steel is steel coated with zinc to improve its corrosion resistance. (*c*) A basic metal is one which can react with mineral acids to form hydrogen and whose oxide reacts as a base but not as an acid. (*d*) Engine knock is the small explosion resulting from premature reaction of fuel in an internal combustion engine, before the spark from the spark plug ignites the mixture of air and fuel for the power stroke of the engine. (*e*) A stalactite is an "icicle-like" calcium carbonate deposit on the ceiling of a limestone cavern formed over long periods of time by the precipitation of $CaCO_3$ from solutions dripping from the ceiling. (*f*) A stalagmite is a deposit similar to a stalactite but on the floor of a limestone cavern.

26.62 In steel manufacture, manganese metal is used as a *scavenger* to reduce traces of iron oxide and iron sulfide. Suggest two reasons why manganese is effective for this purpose.

▌ (1) Manganese is active enough to react with the iron compounds of sulfur and oxygen. (2) Small quantities of the products of these reactions dissolve in the metal without disruption of the lattice. (Large quantities of MnO or MnS would form a slag, which could be skimmed off the molten metal.) Any excess manganese acts as cathodic protection for the iron.

CHAPTER 27

Coordination Compounds

27.1 PROPERTIES OF THE COORDINATION SPHERE

27.1 What is the distinction between a covalent bond and a coordinate covalent bond? How many covalent bonds and how many coordinate covalent bonds are there in the NH_4^+ ion? Is it possible to distinguish between the coordinate covalent bond(s) and the other covalent bond(s) in this ion? Explain.

▌ A coordinate covalent bond is a pair of electrons shared between two atoms which originated on one of the atoms; a covalent bond is a shared pair of electrons. Once formed, there is no difference between them. In an ammonium ion there are four covalent bonds, one of which is a coordinate covalent bond.

27.2 How many faces has an octahedron? How many corners? What is the coordination number of a central ion which has an octahedral coordination sphere?

▌ The octahedron has eight faces, from which it gets its name, but only six corners. Since the ligands are bonded at the corners of the octahedron, the coordination number of a metal with an octahedral coordination sphere is 6.

27.3 The substance $CoBr_3 \cdot 4NH_3 \cdot 2H_2O$ has a molar conductivity of 420 $cm^{-1} \cdot \Omega^{-1}$ (42 kS/m, in SI units) at infinite dilution. Indicate the composition of the coordination sphere.

▌ The conductivity corresponds to that of a (3+, 1−) electrolyte; hence the substance should be represented as $[Co(NH_3)_4(H_2O)_2]^{3+}(Br^-)_3$, or, more simply, $[Co(NH_3)_4(H_2O)_2]Br_3$.

27.4 What is the coordination number of *(a)* molybdenum in $[Mo(CN)_8]^{4-}$? *(b)* copper in $Cu(en)_2^{2+}$ (en = ethylenediamine)?

▌ *(a)* 8 *(b)* 4

27.5 Complete and balance the equation: $AgCl(s) + NH_3 \rightarrow$

▌ $$AgCl(s) + 2NH_3 \longrightarrow Ag(NH_3)_2^+ + Cl^-$$

27.6 Arrange the following compounds in order of increasing molar conductivity: *(a)* $K[Co(NH_3)_2(NO_2)_4]$ *(b)* $[Cr(NH_3)_3(NO_2)_3]$ *(c)* $[Cr(NH_3)_5(NO_2)]_3[Co(NO_2)_6]_2$ *(d)* $Mg[Cr(NH_3)(NO_2)_5]$

▌ The larger the number of ions and the larger the charge on each, the larger the conductivity. The compounds, from lowest conducting to highest, are $b < a < d < c$.

27.7 Give the characteristic coordination number of each of the following central metal ions: *(a)* Cu^{I} *(b)* Cu^{II} *(c)* Co^{III} *(d)* Al^{III} *(e)* Zn^{II} *(f)* Fe^{II} *(g)* Fe^{III} *(h)* Ag^{I}

▌ *(a)* 2 *(b)* 4 *(c)* 6 *(d)* 6, sometimes 4 *(e)* 4 *(f)* 6 *(g)* 6 *(h)* 2

27.8 Indicate the oxidation state of the central metal ion in each of the following complex ions: *(a)* $[Cu(NH_3)_4]^{2+}$ *(b)* $[CuBr_4]^{2-}$ *(c)* $[Cu(CN)_2]^-$ *(d)* $[Cr(NH_3)_4(CO_3)]^+$ *(e)* $[PtCl_4]^{2-}$ *(f)* $[Co(NH_3)_2(NO_2)_4]^-$ *(g)* $Fe(CO)_5$ *(h)* $[ZnCl_4]^{2-}$ *(i)* $[Co(en)_3]^{2+}$

▌ The oxidation state of the central metal ion plus the normal charge on each of the ligands must be equal to the total charge. *(a)* +2 *(b)* +2 *(c)* +1 *(d)* +3 *(e)* +2 *(f)* +3 *(g)* 0 *(h)* +2 *(i)* +2

27.9 Complete the designation of the following coordination spheres by including the charge: *(a)* $[Fe^{III}(CN)_6]$ *(b)* $[Pt^{IV}(NH_3)_3(H_2O)Cl_2]$ *(c)* $[Cr^{III}(NH_3)_2(H_2O)_2Cl_2]$ *(d)* $[Pd^{II}(en)Cl_2]$ *(e)* $[Al(H_2O)_2(OH)_4]$

▌ *(a)* 3− *(b)* 2+ *(c)* 1+ *(d)* 0 *(e)* 1−

27.10 Ethylenediaminetetraacetic acid, EDTA, in the form of its calcium dihydrogen salt, is administered as an antidote for lead poisoning. Explain why this reagent might be an effective medicine. Why is the calcium salt administered rather than the free acid?

▌ EDTA coordinates lead ion in the body, in which form it is passed out of the body without harmful effects. The calcium salt is used so that any excess EDTA will not remove calcium ions from the body.

27.11 Determine the oxidation number of the central metal ion in each of the following: (*a*) $[Co(NH_3)_6]^{3+}$ (*b*) $Ni(CO)_4$ (*c*) $[CuCl_4]^{2-}$ (*d*) $[Ag(CN)_2]^-$ (*e*) $[Co(NH_3)_4(NO_2)_2]^+$

▌ (*a*) +3 (*b*) 0 (*c*) +2 (*d*) +1 (*e*) +3

27.12 The existence of infrared spectral bands corresponding to metal-chloride vibrations in the spectrum of *cis*-$[Cr(NH_3)_4Cl_2]^+$ implies what type of metal-ligand bonding?

▌ The existence of infrared bands implies covalent bonding. (The evidence for covalent bonding makes the crystal field theory, in which only ionic bonding is considered, merely one extreme in the range of theories from pure covalent to pure ionic.)

27.13 Explain why a (2+, 1−) electrolyte is expected to have a higher *molar* conductivity than a (1+, 1−) electrolyte.

▌ The (2+, 1−) electrolyte consists of 3 mol ions per mol, one of which is dipositive. The (1+, 1−) electrolyte has only 2 mol ions per mol, both singly charged.

27.14 A freshly prepared aqueous solution of $Pd(NH_3)_2Cl_2$ does not conduct electricity. Is this compound to be regarded as a strong or weak electrolyte? Explain in terms of its structure.

▌ Since it does not conduct, it is a weak electrolyte, and the structure involves covalent bonding only. Since there are no ions involved in this compound, the chlorine atoms, covalently bonded, must be in the coordination sphere.

27.15 Potassium alum, $KAl(SO_4)_2 \cdot 12H_2O$, is obtained in the form of octahedral crystals when a solution of K_2SO_4 and $Al_2(SO_4)_3$ is concentrated by evaporation. Suggest specific experiments which might be used to determine whether potassium alum is a coordination compound. Explain what results would be expected if it were and if it were not a coordination compound.

▌ One could test the conductivity of the solution of potassium alum. If the conductivity, with equal quantities of each ion total, is equal to that of a solution of K_2SO_4 plus that of a solution of $Al_2(SO_4)_3$, then the compound can be thought of as completely ionic in solution. This is not a coordination compound.

27.16 Explain each of the following phenomena, giving full details, such as pertinent equations and/or numerical justifications: (*a*) The complex $[CuCl_4]^{2-}$ exists, but $[CuI_4]^{2-}$ does not. (*b*) Gold is not attacked by common acids but "dissolves" in aqua regia (HCl + HNO_3). (*c*) Copper metal will dissolve in aqueous KCN solution, with the evolution of hydrogen. (*d*) Since chelate rings containing over six atoms are generally not stable, a single bidentate chelating ligand cannot replace two monodenate ligands which are located in the trans positions of a square planar complex.

▌ (*a*) Cu^{II} is reduced to Cu^{I} by I^- but not by Cl^-. (*b*) The oxidation is aided by the complexation of the product gold(III) as $AuCl_4^-$. (*c*) The stability of $Cu(CN)_4^{2-}$ ion is so great that the E value for oxidation is reduced to a point where H_2O can oxidize the copper. (*d*) The trans positions are too distant for a five-atom chain bidentate ligand to span.

27.17 What is the expected freezing point depression of 0.0100 m $[Co(NH_3)_6]Cl_3$?

▌ The compound dissociates into four ions, one complex ion plus three chloride ions. Neglecting interionic attractions, the molality is 0.0400 m, and the freezing point depression is

$$\Delta t = km = (-1.86\ °C/m)(0.0400\ m) = -0.0744\ °C$$

27.18 The compound with the empirical formula $CsAuCl_3$ is diamagnetic. No metal-to-metal bonds are present in the compound. (*a*) Is there gold(II) in this compound? (*b*) Propose a structure for the compound.

▌ (*a*) The electrons are all paired, as evidenced by the diamagnetism. Any species with a single gold(II) atom would have an odd number of electrons and could not be diamagnetic. The possibility of metal-to-metal bonding to pair up the odd electron is eliminated by the statement in the problem. Hence, the compound cannot contain gold(II). (*b*) The compound must contain at least two oxidation states of gold, the most likely being gold(I) and gold(III), the stable states of gold. The compound actually contains a chain of $AuCl_2^-$ and $AuCl_4^-$ ions with Cs^+ ions to balance the charges.

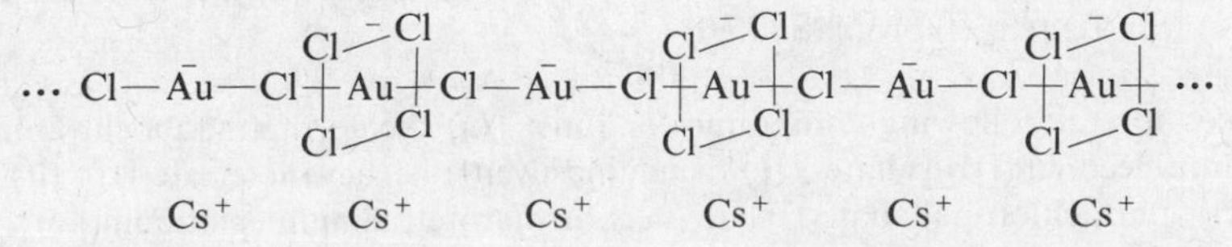

```
                  Cl ⁻ Cl              Cl ⁻ Cl              Cl ⁻ Cl
                    \   |                \   |                \   |
··· Cl—Au⁻—Cl—+—Au—+—Cl—Au⁻—Cl—+—Au—+—Cl—Au⁻—Cl—+—Au—+ ···
                |   \                |   \                |   \
                Cl   Cl              Cl   Cl              Cl   Cl

       Cs⁺      Cs⁺      Cs⁺      Cs⁺      Cs⁺      Cs⁺
```

27.2 NOMENCLATURE OF COORDINATION COMPOUNDS

27.19 Name the following complexes: (*a*) $[FeCl_2(H_2O)_4]^+$ (*b*) $[Pt(NH_3)_2Cl_2]$ (*c*) $[CrCl_4(H_2O)_2]^-$

▮ (*a*) dichlorotetraaquoiron(III) ion (*b*) dichlorodiammineplatinum(II) (*c*) tetrachlorodiaquochromate(III) ion The anionic complex has the suffix *ate* attached to the name of the metal ion.

27.20 Name the following (where en = ethylenediamine, py = pyridine): (*a*) $[Co(NH_3)_5Br]SO_4$ (*b*) $[Cr(en)_2Cl_2]Cl$ (*c*) $[Pt(py)_4][PtCl_4]$ (*d*) $K_2[NiF_6]$ (*e*) $K_3[Fe(CN)_5CO]$ (*f*) $CsTeF_5$

▮ (*a*) bromopentaamminecobalt(III) sulfate
(*b*) dichlorobis(ethylenediamine)chromium(III) chloride
(*c*) tetrapyridineplatinum(II) tetrachloroplatinate(II)
(*d*) potassium hexafluoronickelate(IV)
(*e*) potassium pentacyanocarbonylferrate(II)
(*f*) cesium pentafluorotellurate(IV)

27.21 Write formulas for the following compounds, using brackets to enclose the complex ion portion:

(*a*) bromotriammineplatinum(II) nitrite
(*b*) dichlorobis(ethylenediamine)cobalt(II) monohydrate
(*c*) sulfatopentaamminecobalt(III) bromide
(*d*) potassium hexafluoroplatinate(IV)
(*e*) dibromotetraaquochromium(III) chloride
(*f*) ammonium heptafluorozirconate(IV)

▮ (*a*) $[Pt(NH_3)_3Br]NO_2$ (*b*) $[Co(en)_2Cl_2]\cdot H_2O$ (*c*) $[Co(NH_3)_5SO_4]Br$
(*d*) K_2PtF_6 (*e*) $[Cr(H_2O)_4Br_2]Cl$ (*f*) $(NH_4)_3[ZrF_7]$

27.22 Name the following compounds: (*a*) $[Co(en)_2(CN)_2]ClO_3$ (*b*) $K_4[Co(CN)_6]$ (*c*) $[Ni(NH_3)_6]_3[Co(NO_2)_6]_2$

▮ (*a*) dicyanobis(ethylenediamine)cobalt(III) chlorate (*b*) potassium hexacyanocobaltate(II)
(*c*) hexaamminenickel(II) hexanitrocobaltate(III)

27.23 Write the formula for (*a*) dichlorotetraamminerhodium(III) ion (*b*) tetrahydroxodiaquoaluminate(III) ion (*c*) tetrachlorozincate(II) ion (*d*) aluminum nitrate (*e*) hexaamminecobalt(III) tetrachlorodiamminechromate(III)

▮ (*a*) $[Rh(NH_3)_4Cl_2]^+$ (*b*) $[Al(H_2O)_2(OH)_4]^-$ (*c*) $[ZnCl_4]^{2-}$ (*d*) $Al(NO_3)_3$ (*e*) $[Co(NH_3)_6][Cr(NH_3)_2Cl_4]_3$ There are three anions in part (*e*), just as part (*d*), despite the lack of a *specific* reference to the fact in the name.

27.24 Write formulas for the following compounds: (*a*) hexaamminecobalt(III) bromide (*b*) dibromotetraamminecobalt(III) tetrachlorozincate(II) (*c*) dichlorodiammineplatinum(II)

▮ (*a*) $[Co(NH_3)_6]Br_3$ (*b*) $[Co(NH_3)_4Br_2]_2[ZnCl_4]$ (*c*) $[Pt(NH_3)_2Cl_2]$

27.25 Name the following complex ions: (*a*) $[PdBr_4]^{2-}$ (*b*) $[CuCl_2]^-$ (*c*) $[Au(CN)_4]^-$ (*d*) $[AlF_6]^{3-}$ (*e*) $[Cr(NH_3)_6]^{3+}$ (*f*) $[Zn(NH_3)_4]^{2+}$ (*g*) $[Fe(CN)_6]^{3-}$

▮ (*a*) Tetrabromopalladate(II) ion (*b*) dichlorocuprate(I) ion (*c*) tetracyanoaurate(III) ion (*d*) hexafluoroaluminate(III) ion (*e*) hexaamminechromium(III) ion (*f*) tetraamminezinc(II) ion (*g*) hexacyanoferrate(III) ion

27.26 Name each of the following compounds or ions: (*a*) $[Pt(NH_3)_4Cl_2]^{2+}$ (*b*) $Cr(CO)_6$ (*c*) $[Co(en)Cl_3(H_2O)]$ (*d*) $[Co(NH_3)_5CO_3]_2[CuCl_4]$ (*e*) $Fe[PtCl_4]$

▮ (*a*) Dichlorotetraammineplatinum(IV) ion (*b*) hexacarbonylchromium(0) (*c*) trichloroaquoethylenediaminecobalt(III) (*d*) carbonatopentaamminecobalt(III) tetrachlorocuprate(II) (Note that in the name, no mention is made of the fact that two cations are present per anion in this compound, just as is the case in the name magnesium chloride.) (*e*) iron(II) tetrachloroplatinate(II).

27.27 Write formulas for the following compounds or ions: (*a*) dichlorotetraaquochromium(III) chloride (*b*) bromochlorotetraamminecobalt(III) sulfate (*c*) diamminesilver(I) hexacyanoferrate(II) (*d*) dichlorobis(ethylenediamine) chromium(III) tetrachloropalladate(II) (*e*) *cis*-dichlorotetraammineplatinum(IV) tetrachloroplatinate(II) (*f*) aluminum tetrachloroaurate(III) (*g*) bis(ethylenediamine)copper(II) ion

I (*a*) $[Cr(H_2O)_4Cl_2]Cl$ (*b*) $[Co(NH_3)_4BrCl]_2SO_4$ (Note that the positive charge on the complex ion is 1+, and two complex ions are needed to balance the charge on one sulfate ion.) (*c*) $[Ag(NH_3)_2]_4[Fe(CN)_6]$ (*d*) $[Cr(en)_2Cl_2]_2[PdCl_4]$ (*e*) $[PtCl_2(NH_3)_4]^{2+}$ $[PtCl_4]^{2-}$ (*f*) $Al[AuCl_4]_3$ (*g*) $[Cu(en)_2]^{2+}$

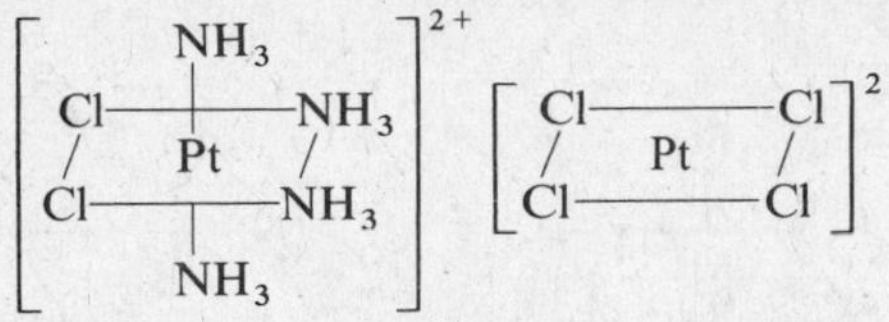

27.3 ISOMERISM OF COORDINATION COMPOUNDS

27.28 Write all geometric isomers for an octahedral complex $[MCl_2(NH_3)_4]$.

I See Fig. 27.1.

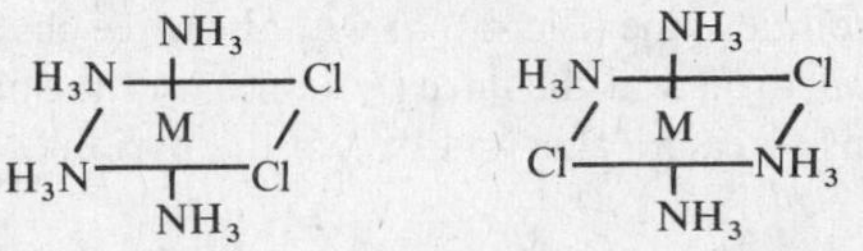

cis *trans* **Fig. 27.1**

27.29 The formula $Co(NH_3)_5CO_3Cl$ could represent a carbonate or a chloride; write the two possible structures and name them.

I

$[Co(NH_3)_5CO_3]Cl$	carbonatopentaamminecobalt(III) chloride
$[Co(NH_3)_5Cl]CO_3$	chloropentaamminecobalt(III) carbonate

27.30 How many isomers are possible for the complex ion $[Cr(NH_3)(OH)_2Cl_3]^{2-}$?

I Three; see Fig. 27.2

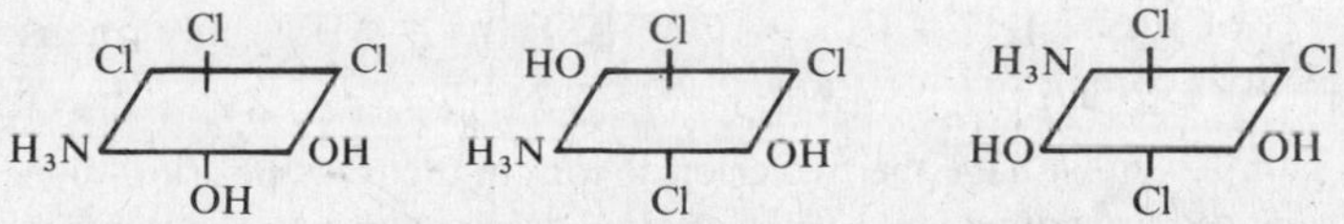

all *cis* Cl, OH *trans* Cl *trans*, OH *cis* **Fig. 27.2**

27.31 The formula $Co(NH_3)_4CO_3Br$ could be that of three isomers. (*a*) Write their possible structures. Tell how you would distinguish the coordination isomers using (*b*) chemical means and (*c*) instrumental methods.

I (*a*) See Fig. 27.3. (*b*) Compound iii can be distinguished from the other two by its ionic nature. The Br^- ion will be precipitated instantaneously by Ag^+, for example. (The coordinated bromine will precipitate somewhat more slowly.) (*c*) The isomers can be distinguished by infrared spectroscopy; the monodentate and bidentate carbonate groups have different absorption patterns. The conductivity is easily determined and distinguishes the nonionic isomers from the ionic one.

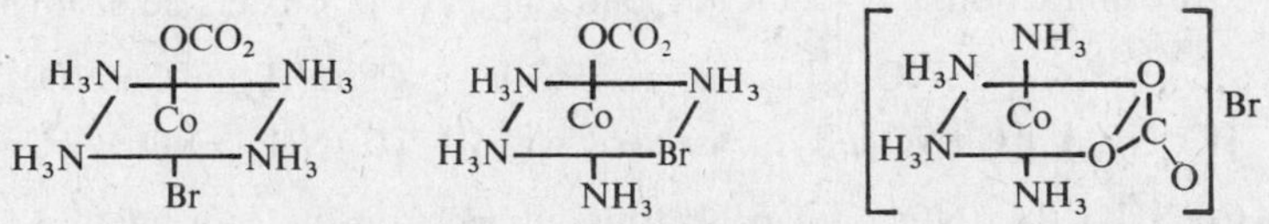

(i) (ii) (iii) **Fig. 27.3**

27.32 For the square coplanar complex $[Pt(NH_3)(NH_2OH)py(NO_2)]^+$, how many geometrical isomers are possible? Draw them.

I Three isomers are possible (Fig. 27.4). Any other possible configuration would merely be a rotation or reflection of one of these.

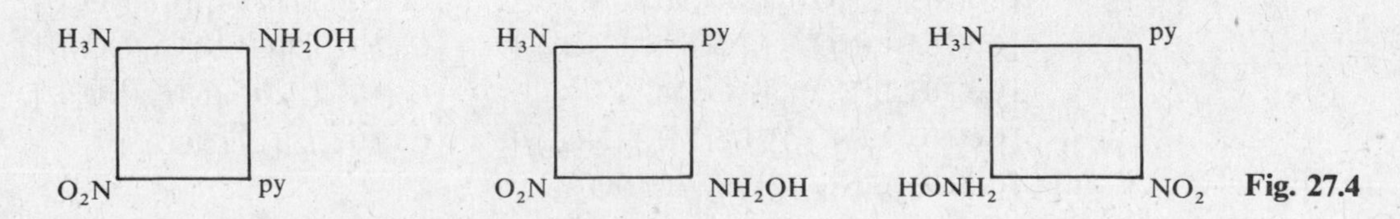

Fig. 27.4

27.33 The ion $Co(en)Cl_2Br_2^-$ has four different isomers (en represents ethylenediamine). Draw the structures of these isomers. *Hint*: Keep the en group in the same position in all the drawings.

▌ See Fig. 27.5.

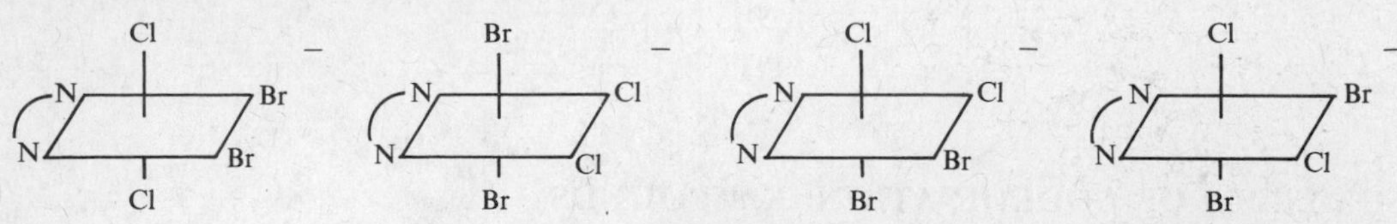

Fig. 27.5

27.34 How many geometrical isomers could $[Rh(py)_3Cl_3]$ have? (py is an abbreviation for the ligand pyridine.)

▌ There are only two possibilities, one (Fig. 27.6*a*) with the three chlorines occupying bonding positions cis to each other on one face of the octahedron and the three pyridines on the opposite face, and the other (Fig. 27.6*b*) in which two of the chlorines are trans to each other and two of the pyridines in turn are trans to each other.

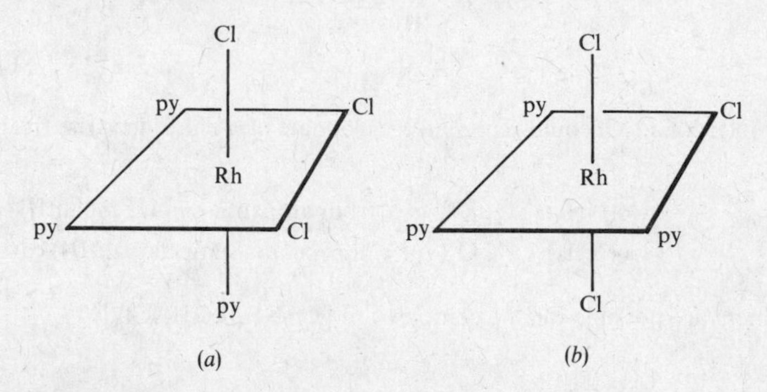

(*a*) (*b*) **Fig. 27.6**

27.35 In the reaction $[CoCl_2(NH_3)_4]^+ + Cl^- \rightarrow [CoCl_3(NH_3)_3] + NH_3$ only one isomer of the complex product is obtained. Is the initial complex cis or trans?

▌ The original complex must have the two chloride ions in the trans position, since then all the other four positions which could be replaced are equivalent. If the original isomer were cis, the third chloride could replace an ammonia cis to both or trans to one, and two isomers would be expected.

27.36 When $[Ni(NH_3)_4]^{2+}$ is treated with concentrated HCl, two compounds having the formula $Ni(NH_3)_2Cl_2$ (designated I and II) are formed. I can be converted into II by boiling in dilute HCl. A solution of I reacts with oxalic acid to form $Ni(NH_3)_2(C_2O_4)$. II does not react with oxalic acid. Deduce the configurations of I and II and the geometry of the nickel(II) complexes.

▌ Compound I is the cis isomer, which can easily form a chelate ring with the oxalate group. The trans isomer cannot form a chelate ring. I and II are square planar.

27.37 Other than by x-ray diffraction, how could the following pairs of isomers be distinguished from one another?

$[Co(NH_3)_6][Cr(NO_2)_6]$ isomeric with $[Cr(NH_3)_6][Co(NO_2)_6]$
$[Cr(NH_3)_6][Cr(NO_2)_6]$ isomeric with $[Cr(NH_3)_4(NO_2)_2][Cr(NH_3)_2(NO_2)_4]$

▌ One way $[Co(NH_3)_6][Cr(NO_2)_6]$ can be distinguished from its coordination isomer is by electrolysis of an aqueous solution. In one case the cobalt(III) complex migrates toward the negative electrode, where cobalt would be deposited. In the other case, chromium would be deposited there. $[Cr(NH_3)_6][Cr(NO_2)_6]$ can be distinguished from $[Cr(NH_3)_4(NO_2)_2][Cr(NH_3)_2(NO_2)_4]$ by conductivity measurements. The former would conduct as a (3+, 3−) electrolyte, while the latter would conduct as a (1+, 1−) electrolyte.

27.38 Write formulas for the nine polymerization isomers of $Co(NH_3)_3(NO_2)_3$

▌

$[Co(NH_3)_4(NO_2)_2][Co(NH_3)_2(NO_2)_4]$
$[Co(NH_3)_4(NO_2)_2]_2[Co(NH_3)(NO_2)_5]$
$[Co(NH_3)_4(NO_2)_2]_3[Co(NO_2)_6]$
$[Co(NH_3)_5(NO_2)][Co(NH_3)_2(NO_2)_4]_2$
$[Co(NH_3)_5(NO_2)][Co(NH_3)(NO_2)_5]$
$[Co(NH_3)_5(NO_2)]_3[Co(NO_2)_6]_2$
$[Co(NH_3)_6][Co(NH_3)_2(NO_2)_4]_3$
$[Co(NH_3)_6]_2[Co(NH_3)(NO_2)_5]_3$
$[Co(NH_3)_6][Co(NO_2)_6]$

27.39 (*a*) Calculate the freezing point of the solution containing 24.8 g solute per kg water for each of the following solutes: (i) $[Co(NH_3)_3(NO_2)_3]$ (ii) $[Co(NH_3)_4(NO_2)_2][Co(NH_3)_2(NO_2)_4]$ (iii) $[Co(NH_3)_5(NO_2)][Co(NH_3)_2(NO_2)_4]_2$ (*b*) Write empirical formulas for the compounds described in (*a*) and in Problem 17.19. (*c*) For which set of compounds is the term polymerization isomerism appropriate? Explain. (*d*) Explain why the compounds of one of these sets cannot be distinguished by freezing point depression methods, while those of the other set can be distinguished in this way. (*e*) Suggest an experimental technique which can be used to distinguish between the polymerization isomers.

(*a*) (i) $(24.8\ \text{g})\left(\frac{1\ \text{mol}}{248\ \text{g}}\right) = 0.100\ \text{mol}$

$\Delta t = (0.100\ \text{m})(1.86\ °\text{C/m}) = 0.186\ °\text{C}; \quad t = -0.186\ °\text{C}$

(ii) $(24.8\ \text{g})\left(\frac{1\ \text{mol compound}}{496\ \text{g}}\right)\left(\frac{2\ \text{mol ions}}{\text{mol compound}}\right) = 0.100\ \text{mol ions}$

$\Delta t = 0.186\ °\text{C}; \quad t = -0.186\ °\text{C}$

(iii) $(24.8\ \text{g})\left(\frac{1\ \text{mol compound}}{744\ \text{g}}\right)\left(\frac{3\ \text{mol ions}}{\text{mol compound}}\right) = 0.100\ \text{mol ions}$

$\Delta t = 0.186\ °\text{C}; \quad t = -0.186\ °\text{C}$

(*b*) $CoN_6H_9O_6$ and CH. (*c*) Polymerization isomerism applies to the coordination compounds, since the freezing points cannot distinguish between the compounds of differing formula weights. (*d*) The value of i is (almost) proportional to the number of ions per formula unit for the complex compounds, whereas $i = 1$ for both of the (covalent) organic compounds. (*e*) Conductance.

27.40 Describe the types of isomerism which can be associated with carbonatoaquotetraamminecobalt(III) chloride monohydrate.

Geometric, hydrate, and coordination isomerism are all possible, including the possibility of the carbonate group being mono- or bidentate. The following compounds results from these types of isomerism: $[Co(NH_3)_4(CO_3)]Cl\cdot 2H_2O$ and cis and trans isomers of each of the following: $[Co(NH_3)_4(CO_3)Cl]\cdot 2H_2O$, $[Co(NH_3)_4(H_2O)_2]CO_3\cdot Cl$, $[Co(NH_3)_4(H_2O)Cl]CO_3\cdot H_2O$, $[Co(NH_3)_4(CO_3)(H_2O)]Cl\cdot H_2O$

27.41 Two compounds have empirical formulas corresponding to $Cr(NH_3)_3(NO_2)_3$. In aqueous solution, one of these is a nonelectrolyte, while the other conducts electricity. What is the lowest possible formula weight of the conducting reagent? What is the highest possible formula weight for the nonconducting reagent?

The lowest conducting polymerization isomer is one of the dimers (compare answer to Problem 27.38), with a formula weight 482 g/mol. The highest nonconducting isomer is the monomer $[Cr(NH_3)_3(NO_2)_3]$, with a formula weight 241 g/mol.

27.42 The compound $Co(NH_3)_3(H_2O)(NO_2)(OCl)Br$ can exist in a variety of isomeric forms. Write structural formulas for at least 10 isomeric forms of this composition, and indicate the types of isomerism shown.

Geometric, optical, linkage, coordination, and hydrate isomerism must be considered. In addition to the five isomers shown in Fig. 27.7, attachment of the nitro group through the oxygen atom instead of the nitrogen atom would yield five additional isomers. Exchange of the NO_2 group with Br yields five more.

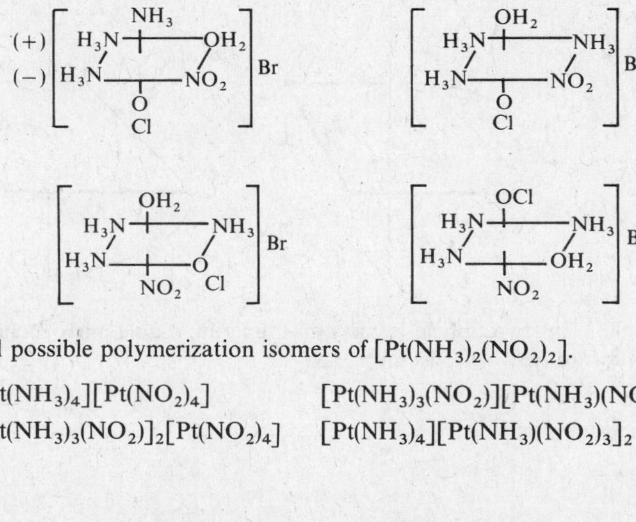

(+) / (−) [NH_3 (top); H_3N, OH_2; H_3N, NO_2; O–Cl (bottom)] Br

[OH_2 (top); H_3N, NH_3; H_3N, NO_2; O–Cl (bottom)] Br

[OH_2 (top); H_3N, NH_3; H_3N, O–Cl; NO_2 (bottom)] Br

[OCl (top); H_3N, NH_3; H_3N, OH_2; NO_2 (bottom)] Br

Fig. 27.7

27.43 Write the formulas for all possible polymerization isomers of $[Pt(NH_3)_2(NO_2)_2]$.

$[Pt(NH_3)_4][Pt(NO_2)_4]$ $\quad$ $[Pt(NH_3)_3(NO_2)][Pt(NH_3)(NO_2)_3]$

$[Pt(NH_3)_3(NO_2)]_2[Pt(NO_2)_4]$ $\quad$ $[Pt(NH_3)_4][Pt(NH_3)(NO_2)_3]_2$

27.44 Sketch the mirror image of the structure indicated in Fig. 27.8 and state if it is optically active.

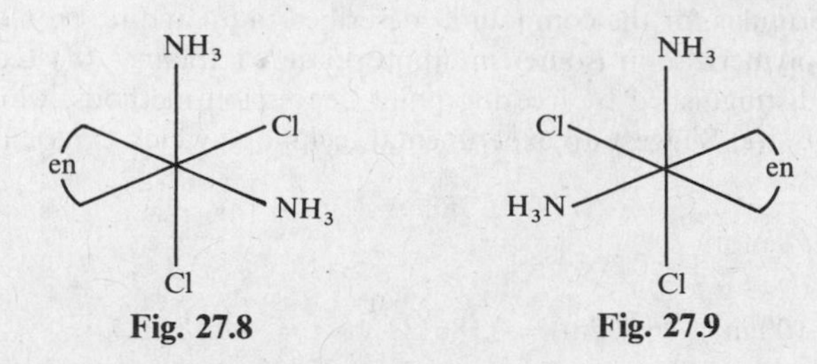

Fig. 27.8 **Fig. 27.9**

▮ See Fig. 27.9. The two isomers are optically active.

27.45 Some ligands are multifunctional; that is, they have two or more atoms which can bind to the central metal atom or ion. Each binding site occupies a different corner on the coordination surface. Ethylenediamine (abbreviated en) is such a ligand; the two binding atoms are nitrogens, and the two binding sites must be cis to each other. How many geometrical isomers of $[Cr(en)_2Cl_2]^+$ should exist, and which isomer(s) might display optical activity?

▮ Two geometrical isomers exist, cis and trans (Fig. 27.10). Each en can be represented by an arc terminating at the two binding sites. By drawing other arrangements of the arcs while preserving the positions of the chlorines, one can see that (*b*) is a distinct mirror image of (*a*), whereas there is no mirror image of (*c*) except (*c*) itself. In other words, only the cis isomer can be optically active; the trans isomer is its own mirror image.

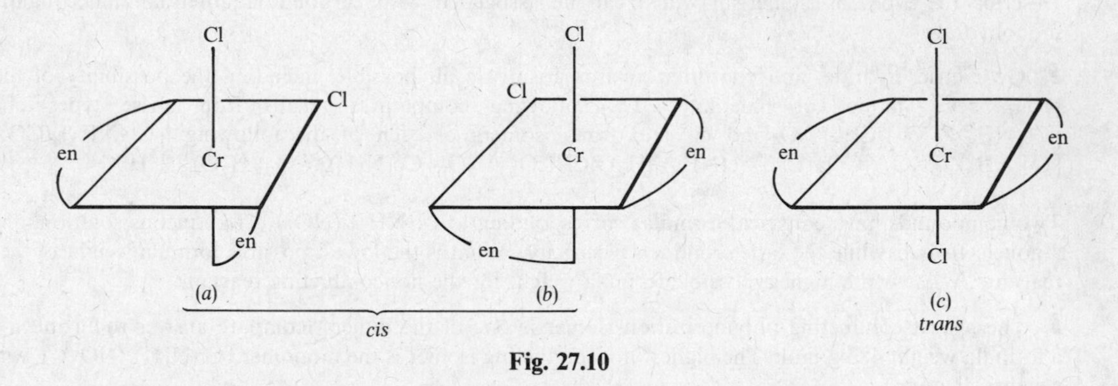

Fig. 27.10

27.46 Predict whether $[Ir(en)_3]^{3+}$ should exhibit optical isomerism. If so, prove by diagrams that the two optical isomers are not simple rotational aspects of the same compound.

▮ There are two optical isomers, which may be represented as in Fig. 27.11. If you make models with cardboard, sticks, and pipe cleaners, you can rotate either one as much as you like and never get the other (without breaking bonds).

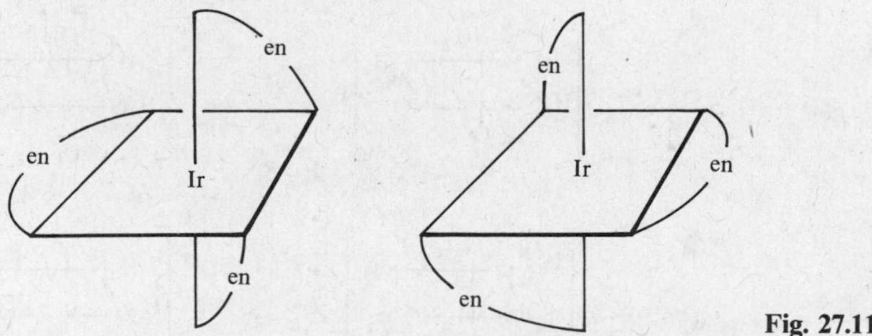

Fig. 27.11

27.47 The $[Co(en)Br_2I_2]^-$ ion has two optical isomers (nonsuperimposable mirror images). Draw the structures of these two isomers. Do *not* draw any other isomers.

▮ See Fig. 27.12.

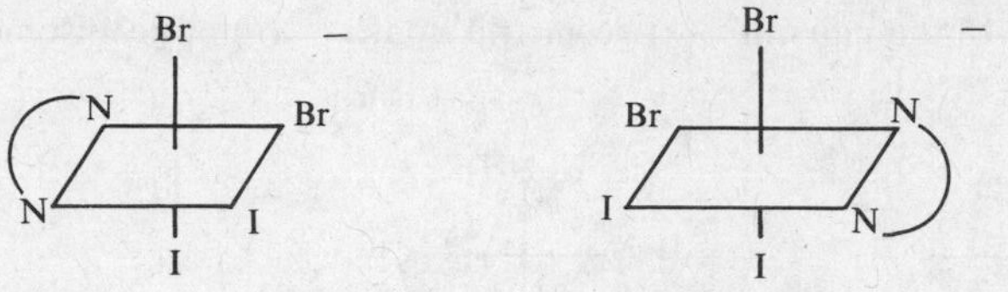

Fig. 27.12

27.48 To demonstrate that optical activity is not related to the presence of carbon atoms in a molecule, Werner prepared $[Co\{(HO)_2Co(NH_3)_4\}_3]^{6+}$. Draw the structure of this complex ion, and show how Werner explained its optical activity.

▮ See Fig. 27.13. This complex, like $Co(en)_3^{3+}$, has a mirror image which is not superimposable; hence it is optically active.

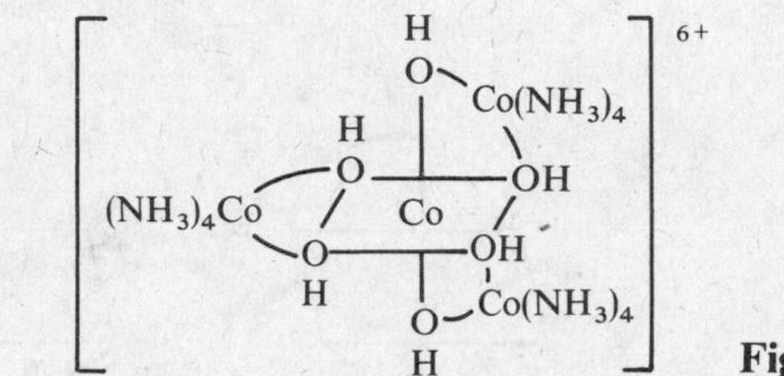

Fig. 27.13

27.49 The complex ion $[M(CN)(NO_2)(H_2O)(NH_3)]^+$ was found to be optically active. What does this finding signify about the configuration of the coordination sphere?

▮ The coordination sphere cannot be planar. By implication it may be tetrahedral, since most simple 4-coordinate complexes are either planar or tetrahedral, but it does not *prove* that the complex is tetrahedral.

27.50 Draw all geometric and optical isomers of $Co(en)Cl_3Br^-$.

▮ See Fig. 27.14. There are no optical isomers, but there are two geometric isomers with two Cl atoms trans to each other or Cl trans to Br.

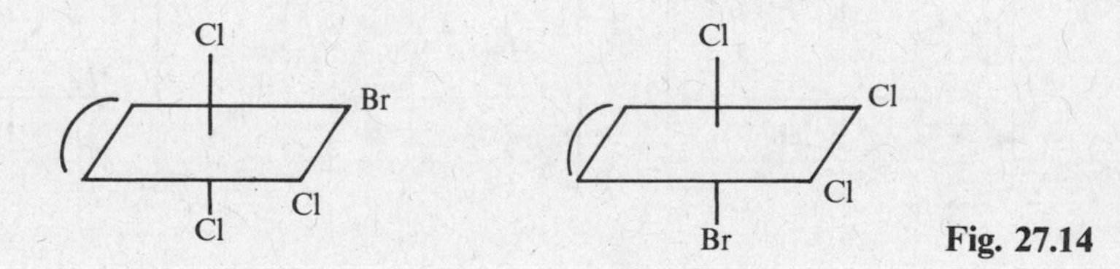

Fig. 27.14

27.51 State the type(s) of isomerism possible for all isomers corresponding to each of the following: **(*a*)** $[Cr(NH_3)_4Br_2]^+$ **(*b*)** $[Cr(NH_3)_4Br_2]NO_2$ **(*c*)** $[Cr(en)_2Cl_2]^+$ **(*d*)** $[Cr(en)_3]^{3+}$ **(*e*)** $[Cr(NH_3)_4ClBr]Br$

▮ **(*a*)** See Fig. 27.15. There is geometric (cis-trans) isomerism. **(*b*)** There is geometric isomerism with two bromine atoms in the coordination sphere, with one bromine atom and a nitro ligand, and with one bromine atom and a nitrito ligand. Thus there are also coordination isomers and linkage isomers. The nitrito ligand is bonded through oxygen; the nitro ligand is bonded through nitrogen. A total of six isomers correspond to the formula. **(*c*)** There is geometric isomerism, and in the cis isomer there also exists the possibility of optical isomerism. A total of three isomers can exist. **(*d*)** Two optical isomers are possible. **(*e*)** Geometric and coordination isomers exist—cis-trans chloro-bromo and cis-trans dibromo isomerism make a total of four possible isomers.

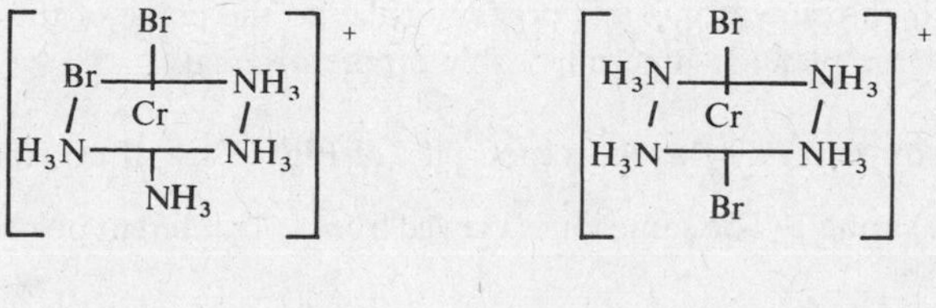

Fig. 27.15

27.52 Select from Fig. 27.16 the pairs of (*a*) geometric isomers (*b*) optical isomers (*c*) identical structures.

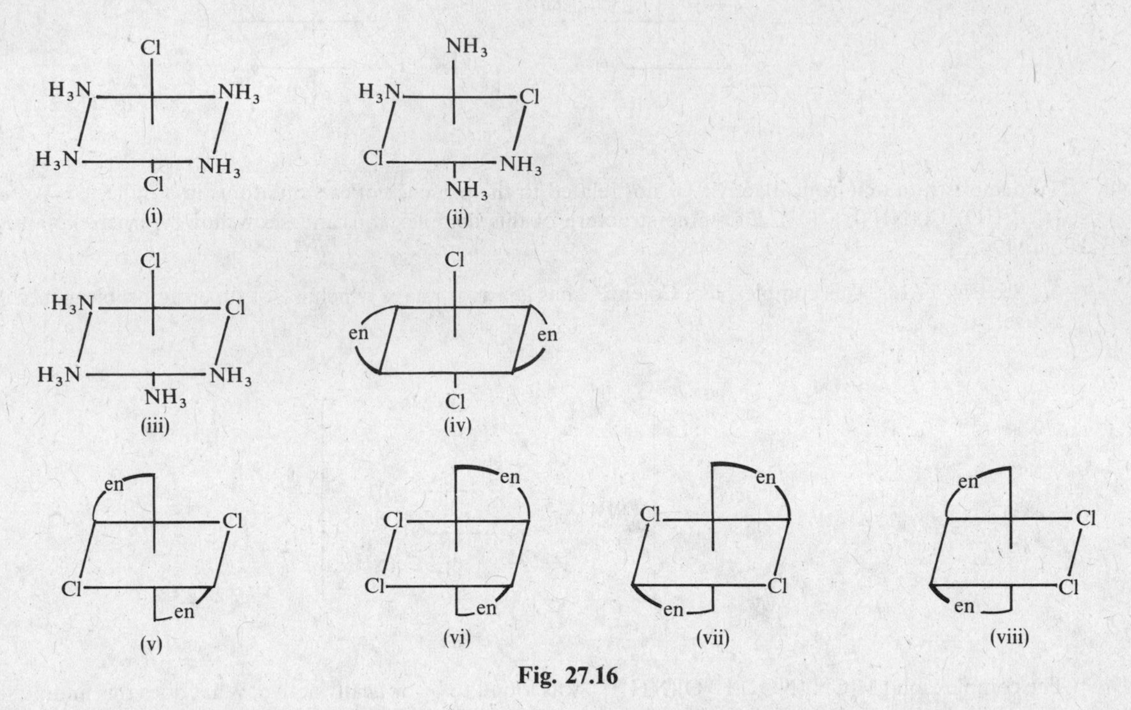

Fig. 27.16

▮ (*a*) Geometric isomers are i and iii; iv with vi and viii. (*b*) Optical isomers are vi and viii. (*c*) Identical structures are i and ii; iv, v, and vii.

27.53 Which one(s) of the ions in Fig. 27.17 would be optically active?

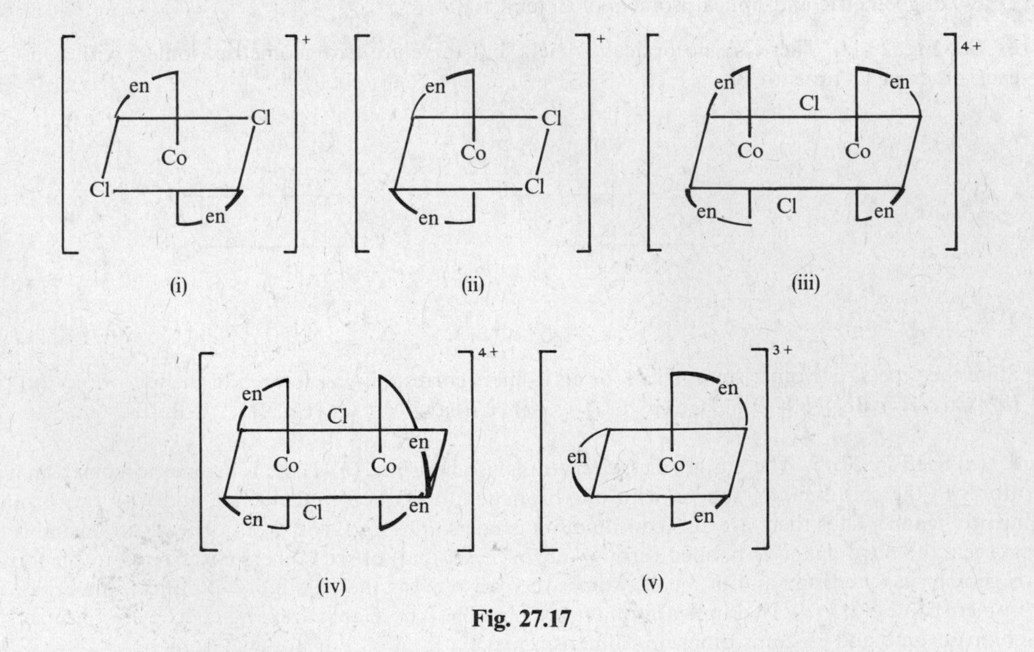

Fig. 27.17

▮ Ions ii and v are optically active. (Ion i is trans, with a mirror plane through the en groups. Ion iii has a mirror plane through the two chlorine atoms and perpendicular to the plane of the Cl and Co atoms. Ion iv has a center of inversion, so its mirror plane is superimposable on the original.)

27.54 Identify the type of isomerism displayed in each part of Fig. 27.18. If no isomerism exists, write "none."

▮ (*a*) Geometric (*b*) none (The same ion is viewed from two different positions.) (*c*) none (*d*) none (*e*) optical

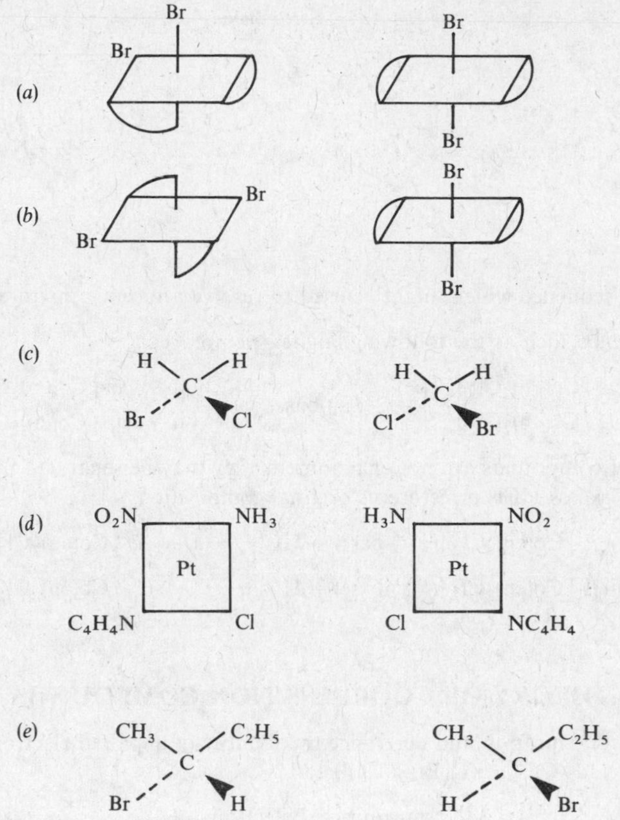

Fig. 27.18

27.55 Neither optical isomers nor geometric isomers can be distinguished by mass spectroscopy. Suggest a reason for this fact.

▌ Since all the atoms in each type of isomer are attached to the same other types of atoms, the same fragments are expected when the molecules are split.

27.56 $[Pt\{NH_2CH(C_6H_5)CH(C_6H_5)NH_2\}\{NH_2C(CH_3)_2CH_2NH_2\}]^{2+}$ was prepared from optically inactive materials and was found to be optically active. Show how this fact proves that the coordination sphere of the platinum is *not* tetrahedral. Does it *prove* that the arrangement is square planar?

▌ If this complex were tetrahedral about the Pt atom, as shown in Fig. 27.19*a*, it would have a plane of symmetry including the left five-membered ring and bisecting the right five-membered ring. Since it is optically active, it cannot be oriented in that manner. The ion actually has a planar coordination sphere, as shown in Fig. 27.19*b*, but its optical activity does not prove that it is planar. It might have a different structure.

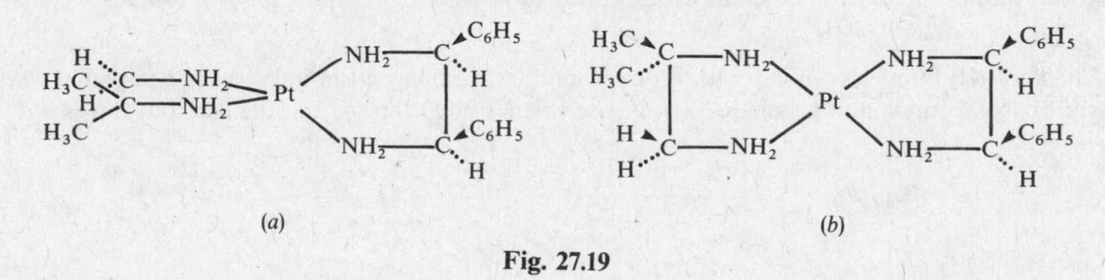

Fig. 27.19

27.57 The ion $[Co(en)_3]^{3+}$ exists in the form of two optical isomers. Show that this fact proves that the complex is not hexagonal or trigonal prismatic. Does it *prove* that the complex is octahedral?

▌ As is shown in Fig. 27.20, both the hexagonal and the prismatic coordination spheres would have a mirror of symmetry within the molecule and thus could not be optically active. The fact that the complex is not hexagonal or trigonal prismatic does not prove that it is octahedral; it might have a still different geometry.

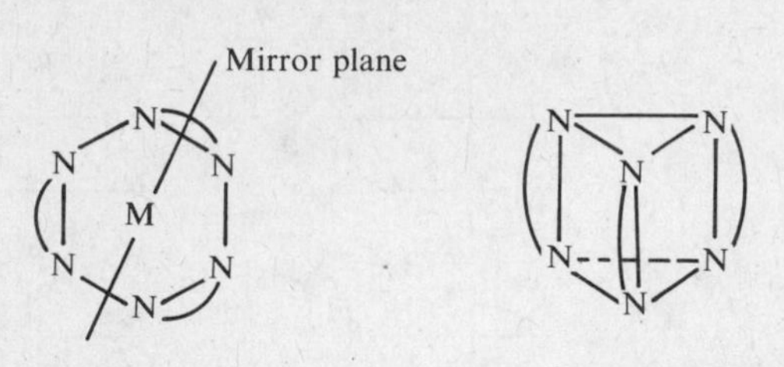

Fig. 27.20

27.58 Describe a reaction sequence which might be used to resolve a racemic mixture of *cis*-$[Co(en)_2Cl_2]^+$.

▌ A reaction sequence such as the following might be employed:

$$\left.\begin{matrix}(+)\text{-}cis\text{-}[Co(en)_2Cl_2]Cl \\ (-)\text{-}cis\text{-}[Co(en)_2Cl_2]Cl\end{matrix}\right\} + 2((-)\text{-base}^-) \longrightarrow \left\{\begin{matrix}(+)\text{-}cis\text{-}[Co(en)_2Cl_2]((-)\text{-base}) + Cl^- \\ (-)\text{-}cis\text{-}[Co(en)_2Cl_2]((-)\text{-base}) + Cl^-\end{matrix}\right.$$

Since the last pair of compounds are not enantiomers, they may be separated by physical means. After separation, each is treated with excess HCl to restore its original composition.

$$(+)\text{-}cis\text{-}[Co(en)_2Cl_2]((-)\text{-base}) + HCl \longrightarrow (+)\text{-}cis\text{-}[Co(en)_2Cl_2]Cl + H((-)\text{-base})$$

$$(-)\text{-}cis\text{-}[Co(en)_2Cl_2]((-)\text{-base}) + HCl \longrightarrow (-)\text{-}cis\text{-}[Co(en)_2Cl_2]Cl + H((-)\text{-base})$$

27.4 VALENCE BOND THEORY OF COORDINATION COMPOUNDS

27.59 Give the electronic configuration and determine the number of unpaired electrons in each of the following: (*a*) Fe (*b*) Fe^{2+} (*c*) Ni^{2+} (*d*) Cu^{2+} (*e*) Pt^{2+} (*f*) Pt^{4+}

▌ (*a*) $1s^2\ 2s^2\ 2p^6\ 3s^2\ 3p^6\ 4s^2\ 3d^6$, four unpaired electrons.

(*b*) $1s^2\ 2s^2\ 2p^6\ 3s^2\ 3p^6\ 3d^6$, four unpaired electrons.

(*c*) $1s^2\ 2s^2\ 2p^6\ 3s^2\ 3p^6\ 3d^8$, two unpaired electrons.

(*d*) $1s^2\ 2s^2\ 2p^6\ 3s^2\ 3p^6\ 3d^9$, one unpaired electron.

(*e*) $1s^2\ 2s^2\ 2p^6\ 3s^2\ 3p^6\ 4s^2\ 3d^{10}\ 4p^6\ 5s^2\ 4d^{10}\ 5p^6\ 5d^8\ 4f^{14}$, two unpaired electrons.

(*f*) $1s^2\ 2s^2\ 2p^6\ 3s^2\ 3p^6\ 4s^2\ 3d^{10}\ 4p^6\ 5s^2\ 4d^{10}\ 5p^6\ 5d^6\ 4f^{14}$, four unpaired electrons.

27.60 Show that all octahedral complexes of nickel(II) must be outer orbital complexes.

▌ The electronic configuration of Ni^{2+} is as follows:

Ni^{2+} ↑↓ ↑↓ ↑↓ ↑ ↑ (3*d*) __ (4*s*) __ __ __ (4*p*) __ __ __ __ __ (4*d*)

Since only one 3*d* orbital can be made available by pairing of the electrons, there cannot be inner orbital d^2sp^3 hybridization. The only octahedral hybridization possible is sp^3d^2, using outer *d* orbitals.

27.61 Manganese(II) forms a complex with bromide ion. Its paramagnetism indicates five unpaired electrons. What are its probable formula and geometry? (*Hint*: Use valence bond theory, and the fact that this is a d^5 ion.)

▌

↑ ↑ ↑ ↑ ↑ (3*d*) | OO (4*s*) OO OO OO (4*p*) — sp^3

$MnBr_4^{2-}$ tetrahedral

27.62 Using valence bond theory, diagram the hybridization of the central atom of the complex ion $[V(H_2O)_6]^{3+}$.

▌

↑ ↑ __ OO OO (3*d*) | OO (4*s*) | OO OO OO (4*p*) — d^2sp^3

27.63 The hybrid orbitals used by manganese in the complex $[MnBr_4]^{2-}$ are sp^3. How many unpaired electrons are there in this complex?

▮

Five electrons are left unpaired: ↑ ↑ ↑ ↑ ↑ (3d) [OO (4s) OO OO OO (4p)] — sp^3

27.64 How many unpaired electrons are there in $[Cr(CN)_6]^{3-}$?

▮

Three electrons are unpaired: ↑ ↑ ↑ [OO OO (3d) OO (4s) OO OO OO (4p)] — d^2sp^3

27.65 In $[ZnBr_4]^{2-}$, electron pairs in sp^3 hybrid orbitals of the zinc atom form bonds to the bromine atoms. Determine the number of unpaired electrons in the complex.

▮ Zinc(II) ion has ten *d* electrons. Since the *d* orbitals are completely filled, there are no unpaired electrons.

27.66 Determine the number of unpaired electrons in $[Cr(NH_3)_6]^{3+}$. Why is it not necessary to designate the complex as inner orbital or outer orbital?

▮ Chromium(III) ion has three 3*d* electrons, all unpaired. The three electrons leave two orbitals vacant for inner orbital bonding without electron pairing; outer orbital bonding is not necessary.

27.67 Write valence bond electronic configurations for each of the following: ***(a)*** $[PtCl_6]^{2-}$ ***(b)*** $Cr(CO)_6$ ***(c)*** $[Ir(NH_3)_6]^{3+}$ ***(d)*** $[Pd(en)_2]^{2+}$

▮

(a) ↑↓ ↑↓ ↑↓ [OO OO (5d) OO (6s) OO OO OO (6p)] — d^2sp^3

(b) ↑↓ ↑↓ ↑↓ [OO OO (3d) OO (4s) OO OO OO (4p)] — d^2sp^3

(c) ↑↓ ↑↓ ↑↓ [OO OO (5d) OO (6s) OO OO OO (6p)] — d^2sp^3

(d) ↑↓ ↑↓ ↑↓ ↑↓ [OO (4d) OO (5s) OO OO (5p)] __ — dsp^2

27.68 *(a)* Explain why a knowledge of the magnetic susceptibility of a complex is often necessary for a correct assignment of the electronic configuration according to valence bond theory. ***(b)*** Draw valence bond representations of the electronic structures of (paramagnetic) $[CoF_6]^{3-}$ and (diamagnetic) $[Co(CN)_6]^{3-}$.

▮ *(a)* Electron pairing is necessary in some cases for inner orbital bonding. The change in magnetic moment is an indication of which orbitals are being used.

(b) CoF_6^{3-} ↑↓ ↑ ↑ ↑ ↑ (3d) [OO (4s) OO OO OO (4p) OO OO (4d)] __ __ __ — sp^3d^2

$Co(CN)_6^{3-}$ ↑↓ ↑↓ ↑↓ [OO OO (3d) OO (4s) OO OO OO (4p)] __ __ __ __ __ (4d) — d^2sp^3

27.69 Using the valence bond method, (*a*) assign electronic configurations to the central metal atoms of the following complex ions, (*b*) predict their geometries, and (*c*) predict their magnetic moments (in Bohr magnetons): (i) $[Ag(CN)_2]^-$ (ii) $[Cu(CN)_4]^{2-}$ (iii) $[Fe(CN)_6]^{3-}$ (iv) $[Zn(CN)_4]^{2-}$

▮

	(*a*)	(*b*)	(*c*)
$Ag(CN)_2^-$	*sp*	Linear	0
$Cu(CN)_4^{2-}$	dsp^2	Square planar	1.73 BM
$Fe(CN)_6^{3-}$	d^2sp^3	Octahedral	1.73 BM
$Zn(CN)_4^{2-}$	sp^3	Tetrahedral	0

27.70 The magnetic moment of $[CoI_4]^{2-}$ is above 3.5 BM. Using the valence bond approach, diagram the electronic configuration of the central metal atom.

▮

$$\underbrace{\uparrow\downarrow\ \ \uparrow\downarrow\ \ \uparrow\ \ \uparrow\ \ \uparrow}_{3d}\ \ \overbrace{\underbrace{\circ\circ}_{4s}\ \ \underbrace{\circ\circ\ \ \circ\circ\ \ \circ\circ}_{4p}}^{sp^3}\qquad \text{(It is high spin.)}$$

27.71 On the basis of valence bond theory, predict whether square planar complexes of palladium(II) are high spin or low spin.

▮

$$\underbrace{\uparrow\downarrow\ \ \uparrow\downarrow\ \ \uparrow\downarrow\ \ \uparrow\downarrow\ \ \circ\circ}_{4d}\ \ \underbrace{\circ\circ}_{5s}\ \ \underbrace{\circ\circ\ \ \circ\circ\ \ \_\_}_{5p}\qquad \text{(low spin)}$$

(dsp^2 spans the last 4d orbital, the 5s orbital and two 5p orbitals.)

27.72 (*a*) A complex of a certain metal ion has a magnetic moment of 4.90 BM; another complex of the same metal ion in the same oxidation state has a zero magnetic moment. The central metal ion could be which one of the following? Cr^{III}, Mn^{II}, Mn^{III}, Fe^{II}, Fe^{III}, Co^{II}. (*b*) If a metal ion had complexes with moments 4.90 and 2.83 BM, which one of these central metal ions could it be?

▮ (*a*) Fe^{II}. (To have either four or zero unpaired electrons, the configuration must be d^6.) (*b*) Mn^{III}. (To have either four or two unpaired electrons, the configuration must be d^4.)

27.5 CRYSTAL FIELD THEORY

27.73 Confirm that the lowering of the stabilized orbitals is 0.4Δ while the raising of the destabilized orbitals is 0.6Δ.

▮ The total energy separation is Δ. Let $x =$ energy of stabilization and $y =$ energy of destabilization. Then $y - x = \Delta$. Since the orbitals yield no net increase in energy when they are equally occupied,

$$3x + 2y = 0 \qquad 3x + 2(\Delta + x) = 0 \qquad 5x = -2\Delta \qquad x = -0.4\Delta$$

27.74 For the $[Cr(H_2O)_6]^{2+}$ ion, the mean pairing energy P is found to be 23 500 cm^{-1}. The magnitude of Δ is 13 900 cm^{-1}. Calculate the crystal field stabilization energy for the complex in configurations corresponding to high-spin and low-spin states. Which is more stable?

▮ For a d^4 ion in a high-spin state,

$$\text{CFSE} = -0.6\Delta = -0.6(13\,900\ \text{cm}^{-1}) = -8340\ \text{cm}^{-1}$$

For a d^4 ion in a low-spin state, the net crystal field stabilization energy is

$$\text{CFSE} = -1.6\Delta + P = -1.6(13\,900\ \text{cm}^{-1}) + 23\,500\ \text{cm}^{-1} = +1260\ \text{cm}^{-1}$$

As is generally the case, the lower energy state will be the more stable. The ligand H_2O does not produce a sufficiently strong crystal field to yield a low-spin cobalt(II) complex. Since $\Delta < P$, it should have been apparent from the outset that the high-spin configuration would be more stable.

27.75 The enthalpy of hydration of Cr^{2+} is -460 kcal/mol. In the absence of crystal field stabilization energy, the value for ΔH would be -435 kcal/mol. Estimate the value of Δ for $Cr(H_2O)_6^{2+}$.

▮ The enthalpy of hydration of Cr^{2+} is 0.6Δ higher for this d^4 weak field ion than it would be in the absence of CFSE. Thus,

$$-0.6\Delta = (-460\ \text{kcal/mol}) - (-435\ \text{kcal/mol}) = -25\ \text{kcal/mol}$$

$$\Delta = \left(\frac{-25\ \text{kcal/mol}}{-0.6}\right)\left(\frac{350\ \text{cm}^{-1}}{1\ \text{kcal/mol}}\right) = 14\,600\ \text{cm}^{-1} \qquad \text{The actual value of } \Delta \text{ is } 13\,900\ \text{cm}^{-1}.$$

27.76 (*a*) Using crystal field theory, depict the electronic configuration of the rhodium(II) ion (Rh^{2+}) in an octahedral field for which the crystal field splitting Δ is greater than the pairing energy P. (*b*) Calculate the crystal field stabilization energy for this configuration (in terms of Δ and P).

▌ (*a*) ↑ ___ ___

↓↑ ↓↑ ↓↑

(*b*) $\text{CFSE} = -1.8\Delta + P$ (Even in a spherical field, the complex must have two pairs of electrons; one *additional* pair is formed in the octahedral field.)

27.77 Which of the following d^n ions will have the smallest crystal field stabilization energy if Δ is greater than P, the pairing energy? d^6, d^7, d^8, d^9, d^{10}

▌ d^{10}. (It has $0\Delta = 0$ CFSE no matter what the value of Δ.)

27.78 A coordination compound in an octahedral field has five electrons in d orbitals. If the mean pairing energy P is 20 500 cm^{-1}, and the $t_{2g} - e_g$ energy gap Δ is 15 500 cm^{-1}, the ground state of the complex will have what spin?

▌ Since $P > \Delta$, the complex is high-spin. (That is the definition of high spin.)

27.79 In terms of valence bond method, cite an example in which the term *inner orbital* does not imply low spin. In terms of crystal field theory, cite an example in which the term *strong field* does not imply low spin.

▌ For cases in which there are three electrons or fewer, the number of unpaired electrons is not affected by inner orbital bonding. For example, chromium(III) complexes (d^3) can be inner orbital with the same number of unpaired electrons as the free metal ion: high spin. This same ion can be referred to as strong field, again without any electron pairing which would be designated as low spin.

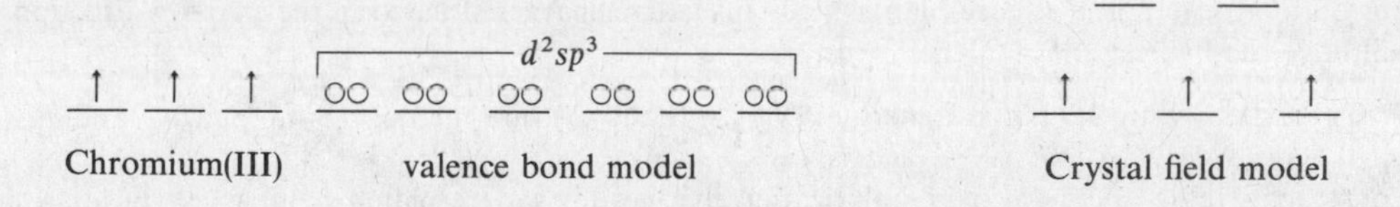

27.80 What factor determines whether the crystal field in an octahedral complex is to be regarded as strong or weak? How many d electrons must be present in orbitals of the central atom for there to be an abrupt change in crystal field stabilization energy between strong and weak fields?

▌ Strong field or weak depends on the relative magnitudes of Δ and P. Abrupt changes in crystal field stabilization energies occur with d^4, d^5, d^6, and d^7 configurations.

27.81 Diagram the electronic configuration of the central metal of each of the following ions from the viewpoint of the crystal field theory (compare Problems 27.67 and 27.68): (*a*) $[Pd(en)_2]^{2+}$ (*b*) $[Co(CN)_6]^{3-}$ (*c*) $[Ir(NH_3)_6]^{3+}$ (*d*) $[PtCl_6]^{2-}$

▌ (*a*) ___ / ↑↓ / ↑↓ / ↑↓ ↑↓ (*b*) ___ ___ / ↑↓ ↑↓ ↑↓ (*c*) ___ ___ / ↑↓ ↑↓ ↑↓ (*d*) ___ ___ / ↑↓ ↑↓ ↑↓

27.82 Determine the crystal field stabilization energy of a d^6 complex having $\Delta = 25\,000\ cm^{-1}$ and $P = 15\,000\ cm^{-1}$.

▌ For a d^6 ion in a strong field ($\Delta > P$ is a strong field):

$$\text{CFSE} = -2.4\Delta + 2P = (-2.4)(25\,000\ cm^{-1}) + (2)(15\,000\ cm^{-1}) = -30\,000\ cm^{-1}$$

27.83 Calculate the crystal field stabilization energy of a d^8 ion in a square planar field for both strong and weak field cases. Are square planar fields likely for complexes of ligands which generate weak fields?

The term "strong field" for square planar complexes means that the splitting of the two highest orbitals is greater than the pairing energy. Thus the strong and weak field configurations are

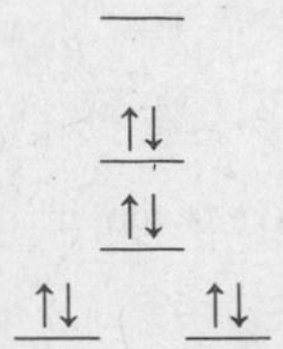

Strong field		Weak field	
—		↑	
↑↓		↑	
↑↓	CFSE = $-2.456\Delta + P$	↑↓	CFSE = -1.456Δ
↑↓ ↑↓		↑↓ ↑↓	

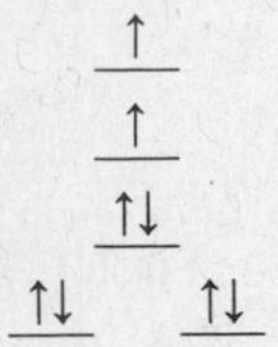

For d^8 ions, the CFSE for octahedral complexes is almost as great as for square planar weak field ions, and the two extra covalent bonds (neglected by crystal field theory) are formed. Thus, square planar weak field complexes are not likely.

27.84 Distinguish between the possibilities in complex ions of $\Delta = 0$ and CFSE = 0. Give an example of each.

$\Delta = 0$ means no field—a free gaseous ion. CFSE = 0 means equal occupancy of the d orbitals—a d^5 weak field ion or a d^{10} ion. See the answer to Problem 27.85.

27.85 What is the crystal field stabilization energy, in terms of Δ, for nickel(II) ion in a square planar field? in a tetrahedral field? in an octahedral field? for zinc(II) ion?

Nickel(II) is d^8. CFSE for square planar field is $-2.456\Delta + P$. (The weak field square planar case would have CFSE = -1.456Δ, if any such ions exist. See Problem 27.83.) For a tetrahedral field, CFSE = -0.356Δ. For an octahedral field, CFSE = -1.20Δ. Zinc(II) is a d^{10} ion and has a zero CFSE for any geometry.

27.86 For an ion located in a square planar field of negative charges situated in the x and y directions, which d orbital(s) will have the highest energy? the lowest?

The orbital with the highest energy is $d_{x^2-y^2}$; the lowest are the degenerate pair d_{xz} and d_{yz}.

27.87 Construct a table of crystal field stabilization energies for square planar complexes. Assume that only the energy difference between the $d_{x^2-y^2}$ and d_{xy} orbitals is sufficiently large to cause electron pairing in some cases.

no. of *d* electrons	CFSE (strong field)
1	-0.514Δ
2	-1.028Δ
3	-1.456Δ
4	-1.228Δ
5	$-1.742\Delta + P$
6	$-2.256\Delta + P$
7	$-2.684\Delta + P$
8	$-2.456\Delta + P$
9	-1.228Δ
10	0

Note that the only splitting large enough to be classified as strong is the Δ difference between the highest and next to highest energy orbitals.

27.88 Explain why crystal field theory is not applied to complexes of main group metals.

The main group metals have no incomplete d subshells. and for empty or completely filled d subshells. CFSE = 0.

27.89 Explain why d^8 complexes are more likely than other complexes to have a square planar geometry.

CFSE is much greater for square planar complexes than for octahedral complexes only in the d^8 and d^9 cases.

27.90 Diagram the electronic configuration of the central metal of each of the following ions from the viewpoint of the crystal field theory: ***(a)*** $[Pt(NH_3)_4]^{2+}$ ***(b)*** $[Cu(NH_3)_4]^{2+}$ ***(c)*** $[Cr(NH_3)_6]^{3+}$

▮ *(a)*

___
↑↓
↑↓
↑↓ ↑↓

(b)

↑
↑↓
↑↓
↑↓ ↑↓

(c)

___ ___
↑ ↑ ↑

27.91 The electronic configuration of a d^9 central ion is analogous to that of a hypothetical ion containing a single positron (e^+) in its d subshell (d^+). ***(a)*** Would the ground state of a d^+ ion be degenerate? ***(b)*** What would be the degeneracy of an *excited* state of the d^+ ion in an octahedral field? ***(c)*** What is the degeneracy of the ground state of a d^1 ion in an octahedral field? ***(d)*** Explain why in an octahedral field the d^9 configuration is regarded as the "inverse" of the d^1 configuration.

▮ ***(a)*** Since the ground state of a d^9 ion is degenerate—the unpaired electron can occupy any of the five orbitals—the ground state of the "positron ion" is also degenerate—the positron can occupy any of the five orbitals. ***(b)*** The excited state would be orbitally threefold degenerate, occupying any of the t_{2g} orbitals. (A positron would be at lower energy in the e_g set.)

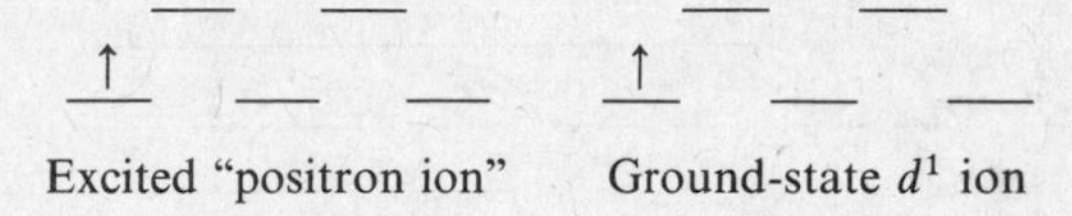

Excited "positron ion" Ground-state d^1 ion

(c) The d^1 ion is orbitally threefold degenerate. ***(d)*** The d^9 configuration, which can be considered equivalent to a d^+ configuration, has a degeneracy in its ground state the same as the d^1 ion in its excited state and vice versa. The two are said to be the inverse of each other.

27.92 Given the following possible conditions with respect to the magnitudes of the crystal field splitting Δ and the electron pairing energy P for an octahedral complex, (1) $\Delta > P$, (2) $\Delta < P$, predict the spin states for the following types of ions: ***(a)*** d^9 ***(b)*** d^3 ***(c)*** d^4 ***(d)*** d^5

▮

	(1) $\Delta > P$	n	x	(2) $\Delta < P$	n	x
(a) d^9	↑↓ ↑ ↑↓ ↑↓ ↑↓	1	2	↑↓ ↑ ↑↓ ↑↓ ↑↓	1	2
(b) d^3	___ ___ ↑ ↑ ↑	3	0	___ ___ ↑ ↑ ↑	3	0
(c) d^4	___ ___ ↑↓ ↑ ↑	2	3	↑ ___ ↑ ↑ ↑	4	2
(d) d^5	___ ___ ↑↓ ↑↓ ↑	1	3	↑ ↑ ↑ ↑ ↑	5	0

n = unpaired electrons; x = degeneracy

27.93 ***(a)*** What is the maximum number of unpaired electrons which a high-spin complex of the first transition series could possess in the ground state? ***(b)*** What elements could show this maximum, and in what oxidation states?

▮ ***(a)*** 5

↑ ↑
↑ ↑ ↑

(b) Mn(II) and Fe(III) are d^5 ions.

27.94 If a complexing metal of the first transition series has a d^i configuration, for what values of i could magnetic properties alone distinguish between strong-field and weak-field ligands in octahedral coordination?

▮ 4, 5, 6, and 7. One, two, or three electrons will occupy the lower orbitals singly, no matter what the value of Δ is. With 8 or more electrons, the lower orbitals will be filled no matter what the value of Δ is.

27.95 Match each of the electronic structures of octahedral complexes, A and B, with one of the numbered structures I through V (which relate to the same complexes using a different theory).

A	B
__ __	↓ ↓
↓↑ ↓↑ ↓↑	↓↑ ↓ ↓

I	↓↑ ↓ ↓ ↓ ↓ [OO OO OO OO OO OO] __ __ __
II	↓↑ ↓↑ ↓↑ OO OO [OO OO OO OO OO OO] __ __ __
III	↓↑ ↓ ↓ [O↓ O↓ OO OO OO OO] __ __ __ __ __
IV	↓↑ ↓↑ ↓↑ [OO OO OO OO OO OO]
V	↓↑ ↓↑ ↓↑ ↓↑ ↓↑ [OO OO OO OO OO OO] __ __ __

▮ A has all six metal electrons paired, so it matches structure IV. B has four electrons unpaired of the six d electrons present, as shown in I.

27.96 What experimental evidence can be cited to prove that the magnitude of Δ in many octahedral complexes is less than the difference in energy between the 1s and 2s orbitals in the hydrogen atom? What orbitals in the hydrogen atom do have energy differences comparable to the magnitude of Δ in octahedral complexes?

▮ The difference in energy between the 1s and 2s orbitals of hydrogen is comparable to the ultraviolet region of the electromagnetic spectrum. The magnitude of Δ of most coordination compounds is comparable to the visible range—less energetic—implying a smaller energy difference. The energy differences between the second shell and higher shells of hydrogen, constituting the Balmer series, are of the same order of magnitude as the Δ values of coordination compounds.

27.97 Draw a crystal field splitting diagram for a complex in a linear field. Assume that the ligands lie on the z axis.

▮ In a linear field with two negative charges on the z axis, the d_{z^2} orbital will be highest in energy. An electron in this orbital will be located in a region of space directly adjacent to the negative charges. The d_{xz} and d_{yz} orbitals are degenerate and are next highest in energy. The d_{xy} and $d_{x^2-y^2}$ orbitals are degenerate (both lying in the xy plane) and lowest in energy.

$$
\begin{array}{cc}
\multicolumn{2}{c}{\underline{\quad\quad}} \\
\multicolumn{2}{c}{d_{z^2}} \\
\underline{\quad\quad} & \underline{\quad\quad} \\
d_{xz} & d_{yz} \\
\underline{\quad\quad} & \underline{\quad\quad} \\
d_{xy} & d_{x^2-y^2}
\end{array}
$$

27.98 Explain why the difference in energy between the $d_{x^2-y^2}$ and d_{xy} orbitals in a square planar field is identical to the difference between the same orbitals in the octahedral field.

▮ Assume that one makes the square planar field by simply elongating two opposite bonds of an octahedral field until the ligands dissociate entirely. As the ligands on the z axis are removed, the effect is the same on the two orbitals in the xy plane. Since each is affected equally, any original difference in energy between these orbitals is maintained as the ligands are withdrawn.

27.99 The enthalpy of hydration of the Fe^{2+} ion is 11.4 kcal/mol higher than would be expected if there were no crystal field stabilization energy. Assuming the aquo complex to be high-spin, estimate the magnitude of Δ for $[Fe(H_2O)_6]^{2+}$.

▮ CFSE for the d^6 weak field ion is -0.40Δ. The ΔH_{hyd} is higher because the energy of the complex is lowered by that much, -11.4 kcal/mol.

$$\text{CFSE} = -11.4 \text{ kcal/mol} = -0.40\Delta \qquad \Delta = 28.5 \text{ kcal/mol}$$

27.100 Using the observed lattice energies tabulated below, estimate the magnitude of Δ in octahedrally coordinated crystals of VO, MnO, and FeO.

Oxide	CaO	TiO	VO	MnO	FeO	CoO	NiO	ZnO
U, kJ/mol	−3465	−3882	−3917	−3813	−3923	−3992	−4076	−4035

▮ Plot (Fig. 27.21) the values of *U* versus atomic number. The Ca^{2+}, Mn^{2+} (weak field), and Zn^{2+} ions, d^0, d^5, and d^{10}, are expected to have CFSE = 0. Draw a smooth curve through these points, and estimate the CFSE of V^{2+} and Fe^{2+} from the difference between the experimental points and those on the smooth curve.

For V^{2+} d^3, $\quad \text{CFSE} = -1.2\Delta = -226 \text{ kJ/mol} \qquad \Delta = 188 \text{ kJ/mol}$

For Fe^{2+} d^6, $\quad \text{CFSE} = -0.4\Delta = -67 \text{ kJ/mol} \qquad \Delta = 167 \text{ kJ/mol}$

There is no way to estimate Δ for Mn^{II}, since CFSE = 0 no matter what the value of Δ, but it probably is in the same range.

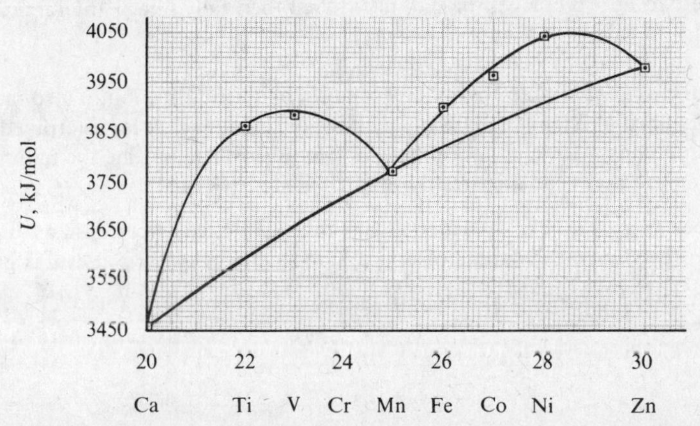

Fig. 27.21

27.101 Explain why platinum(II) and palladium(II) form square planar complexes almost exclusively but only a few nickel(II) complexes are square planar.

▮ Platinum(II) and palladium(II) complexes have greater values of Δ. The greater Δ values imply greater stabilization of the square planar configuration, because the d^8 ions have $\text{CFSE} = -2.456\Delta$ vs. -1.20Δ for octahedral coordination spheres. The greater the magnitude of Δ, the greater is the difference in CFSE. When the difference in CFSE is sufficient to overcome the loss of bonding energy in two bonds (6 to 4), the square planar complex is stabilized.

27.102 The M—O bond distances in a series of 6-coordinate crystal oxides are tabulated below. Plot these data against the atomic numbers of the metals, and explain the observed trends.

Oxide	CaO	TiO	VO	MnO	FeO	NiO	ZnO
d, Å	2.4	2.1	2.05	2.2	2.18	2.08	2.1

▮ See Fig. 27.22. Except for iron(II), the points fall on the now familiar double-humped curve. The curve is inverted, because as the interactions between atoms get stronger the distances get shorter.

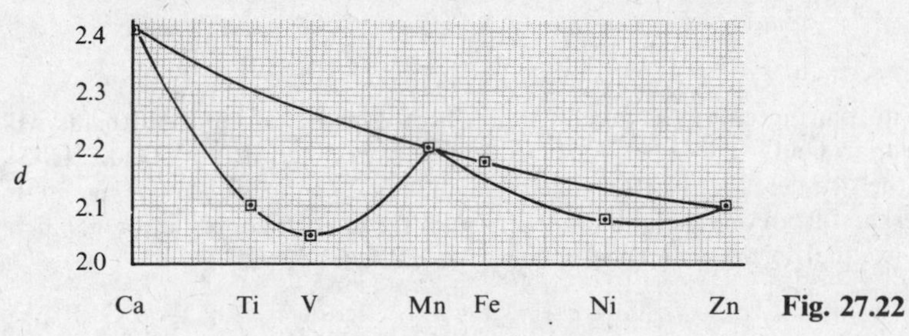

Fig. 27.22

27.103 (*a*) Draw a crystal field splitting diagram for a cubic field (with the ligands at the corners of a cube which has the cartesian axes going through the centers of the cube faces). Compare the magnitude of the splitting(s) with those of the octahedral, square planar, and tetrahedral crystal fields. (*b*) Ions with how many d electrons would give the greatest crystal field stabilization energy in such a cubic field?

▌ (*a*) The diagram is the same as that for a tetrahedral field. The actual splitting is double that of tetrahedral, since the tetrahedral field can be considered to be a cubic field with alternate ligands missing (see Fig. 27.23). The splitting should be 0.889Δ. (*b*) A d^4 strong ion field would be most stable.

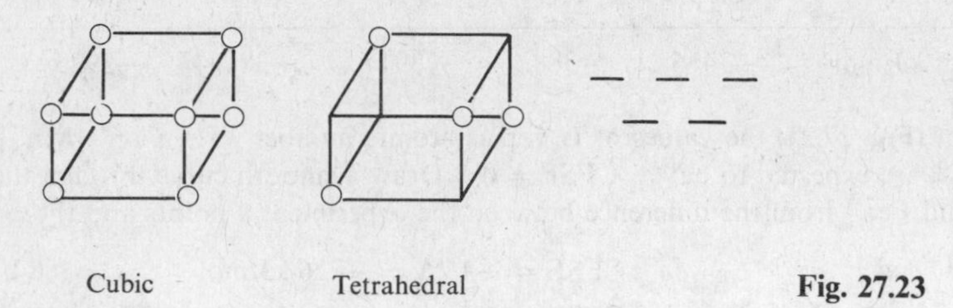

Fig. 27.23

27.104 In terms of crystal field theory, explain why a d^9 octahedral complex with six identical ligands is *not* expected to have all six metal–donor atom distances identical.

▌ One of the two e_g orbitals will be half-filled; the other, fully filled. Interaction with the ligands is thus expected to be different.

27.105 (*a*) Calculate the net coulombic energy of attraction of a dipositive ion in a square planar field of uninegative point charges at a distance of 2.5 Å. (*b*) Calculate the equivalent energy for a tetrahedral field. (*c*) Calculate the crystal field stabilization energy for a d^8 ion in each of these fields. Determine the minimum value of Δ which would favor a square planar geometry for this strictly ionic case.

▌ (*a*) It is seen from Fig. 27.24 that there are four attractive forces and six repulsive forces, two at distances of 5.0 Å and the other four at distances of $2.5\sqrt{2}$ Å. The energy, with q in esu, is given by Coulomb's law as

$$E = \left(4\frac{(2+)(1-)}{2.5} + 4\frac{(1-)^2}{2.5\sqrt{2}} + 2\frac{(1-)^2}{5.0}\right)\left(\frac{(4.8 \times 10^{-10}\ \text{esu})^2}{10^{-8}\ \text{cm}}\right) = -3.84 \times 10^{-11}\ \text{erg}$$

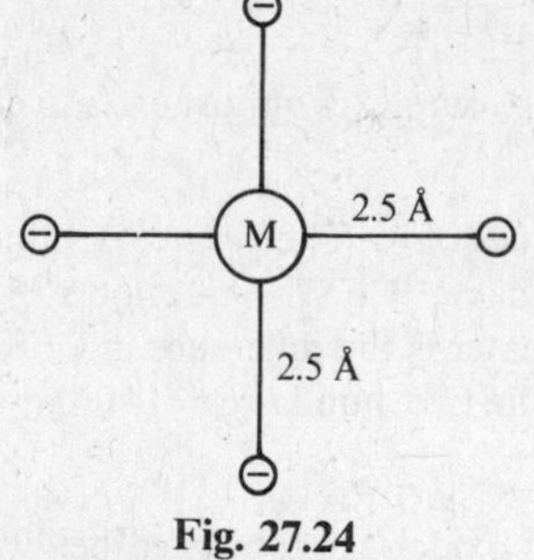

Fig. 27.24

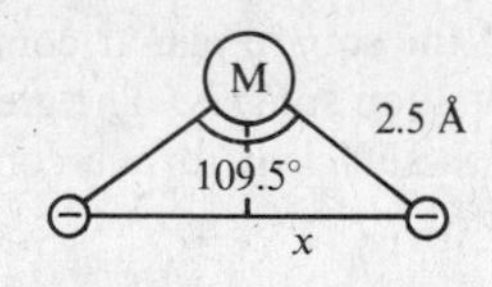

Fig. 27.25

(*b*) For the tetrahedral field (Fig. 27.25), the distance between adjacent negative charges is found by trigonometry to be 4.08 Å.

$$\sin\frac{109.5^\circ}{2} = \frac{x}{2.5\ \text{Å}} \qquad x = 2.04\ \text{Å} \qquad 2x = 4.08\text{Å}$$

There are six such interactions and four attractive forces. The net energy is

$$E = \left(4\frac{(2+)(1-)}{2.5} + 6\frac{(1-)^2}{4.08}\right)\left(\frac{(4.8 \times 10^{-10}\ \text{esu})^2}{10^{-8}\ \text{cm}}\right) = -3.98 \times 10^{-11}\ \text{erg}$$

(*c*) Since no pairing energy is given for this prototype metal ion, both fields will be considered weak. In a weak square planar field, CFSE = −1.456Δ. In a (weak) tetrahedral field, CFSE = −0.356Δ. The difference in CFSE in the two fields is −1.10Δ. The magnitude of Δ changes from one ligand to another, from one oxidation state to another, and from one metal to another, but not from one coordination geometry to another. Thus this difference in CFSE is equal to the difference in energy calculated above.

$$0.14 \times 10^{-11}\ \text{erg} = -1.10\Delta \qquad \Delta = -1.3 \times 10^{-12}\ \text{erg}$$

27.106 Both $Fe(CN)_6^{4-}$ and $Fe(H_2O)_6^{2+}$ appear colorless in dilute solutions. The former ion is low-spin and the latter is high-spin. **(a)** How many unpaired electrons are in each of these ions? **(b)** Why should both ions be colorless, in view of the apparent significant difference in their Δ values?

▮ **(a)**

	—	—		↑	↑
↑↓	↑↓	↑↓	↑↓	↑	↑

0 in $Fe(CN)_6^{4-}$ 4 in $Fe(H_2O)_6^{2+}$

(b) Δ for $Fe(CN)_6^{4-}$ is so large that the absorption peak is in the ultraviolet; Δ for $Fe(H_2O)_6^{2+}$ is so small that its absorption peak is in the infrared. Practically no visible light is absorbed by either.

27.107 Hexaaquoiron(III) ion is practically colorless. Its solutions become red when NCS^- is added. Explain. (Compare with Problem 27.106.)

▮ H_2O is not a strong-field ligand, as noted in Problem 27.106. NCS^-, having vacant π^* orbitals, is a strong-field ligand and causes an increase in Δ and thus a lowering of the wavelength of maximum *d-d* absorption from the near-infrared well into the visible region (actually into the blue-green).

27.108 Which complex of Co^{2+} will have the weakest crystal field splitting? **(a)** $[CoCl_6]^{4-}$ **(b)** $[Co(CN)_6]^{4-}$ **(c)** $[Co(NH_3)_6]^{2+}$ **(d)** $[Co(en)_3]^{2+}$ **(e)** $[Co(H_2O)_6]^{2+}$

▮ $CoCl_6^{4-}$, since Cl^- is lowest in the spectrochemical series of the ligands listed.

27.109 Δ for $IrCl_6^{3-}$ is 27 600 cm^{-1} **(a)** What is the wavelength of maximum absorption? **(b)** What would you predict for the magnetic behavior of this ion?

▮

$$(a)\quad \lambda = \left(\frac{1}{27\,600\ \text{cm}^{-1}}\right)\left(\frac{10^9\ \text{nm}}{10^2\ \text{cm}}\right) = 362\ \text{nm}$$

(b) Since Δ is so high, the Ir(III) would have all electrons paired. The complex is diamagnetic.

	—	—
↑↓	↑↓	↑↓

27.110 A solution of $[Ni(H_2O)_6]^{2+}$ is green, but a solution of $[Ni(CN)_4]^{2-}$ is colorless. Suggest an explanation for these observations.

▮ The value of Δ for the H_2O complex is in the visible region; that for the cyano complex is in the ultraviolet region.

27.111 Calculate the concentration of $[Cu(H_2O)_4]^{2+}$ (represented by $[Cu^{2+}]$) in 1.0 L of a solution made by dissolving 0.10 mol Cu^{2+} in 1.00 M aqueous ammonia. $K_s = 1 \times 10^{12}$

▮

$$Cu^{2+} + 4NH_3 \rightleftharpoons [Cu(NH_3)_4]^{2+} \qquad K = \frac{[Cu(NH_3)_4{}^{2+}]}{[Cu^{2+}][NH_3]^4} = 1 \times 10^{12}$$

The reaction between Cu^{2+} and NH_3 goes almost to completion, as is suggested by the large magnitude of the formation constant. Thus the equilibrium concentration of the tetraamine complex is approximately 0.10 M. (If x mol/L Cu^{2+} ion still remains uncoordinated, the concentration of the complex will be $0.10 - x$.) With the formation of 0.10 M $Cu(NH_3)_4^{2+}$, which requires 4 mol NH_3 per mol complex, the ammonia concentration becomes $1.00 - 4(0.10) = 0.60$ M. (Dissociation of the complex provides $4x$ mol NH_3 per x mol Cu^{2+}.) In summary:

	initial concentration	used up	produced	equilibrium concentration
$[Cu^{2+}]$	0.10	$0.10 - x$		x
$[NH_3]$	1.00	$4(0.10 - x)$		$0.60 + 4x \cong 0.60$
$[Cu(NH_3)_4^{2+}]$	0.00		$0.10 - x$	$0.10 - x \cong 0.10$

$$K = \frac{0.10}{x(0.60)^4} = 1 \times 10^{12} \qquad x = 8 \times 10^{-13}\ \text{M}$$

27.112 An ion, M^{II}, forms the complexes $[M(H_2O)_6]^{2+}$, $[MBr_6]^{4-}$, and $[M(en)_3]^{2+}$. The colors of the complexes, though not necessarily in order, are green, red, and blue. Match the complex with the appropriate color, and explain.

▮ The green complex absorbs red (low-energy) light; the red complex absorbs green (high-energy) light; the blue complex absorbs orange (medium-energy) light. The order of absorption energies should correspond to the order of CFSEs, whence the probable identifications: green complex = $[MBr_6]^{4-}$; blue complex = $[M(H_2O)_6]^{2+}$; red complex = $[M(en)_3]^{2+}$. Only the probability can be expressed, however, because factors other than CFSE, such as electronic interactions, possibly affect the color of a complex.

27.113 $[Ti(H_2O)_6]^{3+}$ absorbs light of wavelength 5000 Å. Name one ligand which would form a titanium(III) complex absorbing light of lower wavelength than 5000 Å and one ligand which would form a complex absorbing light of wavelength higher than 5000 Å.

▮ Any ligand which is below H_2O in the spectrochemical series, for example F^-, would absorb at a lower energy and thus higher wavelength; any ligand above H_2O, for example, CN^-, would absorb at a lower wavelength.

27.114 Predict the colors of $[Co(NH_3)_6]^{3+}$ and $[Cr(H_2O)_6]^{3+}$. The energy difference between the d levels, Δ, in $[Co(NH_3)_6]^{3+}$ is 22 900 cm^{-1}; in $[Cr(H_2O)_6]^{3+}$ it is 17 400 cm^{-1}.

▮

$$\lambda = \frac{1}{\bar{\nu}} = \frac{1}{22\,900\ \text{cm}^{-1}} = 4.37 \times 10^{-5}\ \text{cm} = 4370\ \text{Å}$$

An absorption band is expected at about 4400 Å, which corresponds to the absorption of blue light. The observed color of the complex is orange, as expected. In $[Cr(H_2O)_6]^{3+}$ the value of Δ corresponds to the absorption of light at about 5750 Å. The expected color is violet; however, the actual color is green. In this case an additional absorption occurs at 4050 Å; hence the color cannot be predicted from a consideration of Δ alone.

27.115 Soluble compounds of the complex ion $Co(NH_3)_6^{3+}$ have a maximum in absorption of visible light at 437 nm. (*a*) What is the value of Δ for this complex ion, expressed in cm^{-1}, and what is the ion's color? (*b*) How many unpaired electrons would you expect this ion to have if it is considered low-spin, and how many if it is considered high-spin?

▮

$$(a)\quad \Delta = \left(\frac{1}{437\ \text{nm}}\right)\left(\frac{10^9\ \text{nm}}{10^2\ \text{cm}}\right) = 22\,900\ \text{cm}^{-1}$$

The color of the ion is complementary to the light absorbed. Although the prediction of the color from the data is not absolute, since the color depends not only on the wavelength of the absorption maximum but also on the shape of the whole absorption band and on the color sensitivity of the human eye, fairly reliable conclusions can be drawn. The absorption, which peaks in the blue-violet region of the spectrum, would be expected to cover most of the blue region and part of the green. The color would be expected to be yellow. (*b*) The outer-electron configuration of Co^{3+} is $3s^2\ 3p^6\ 3d^6$. The spin is due to unpaired d electrons. For low spin, the six d electrons would all be paired in the three t_{2g} orbitals, and the spin would be zero. For high spin, the two e_g molecular orbitals would be available, as well as the three t_{2g} orbitals. Four of the available orbitals would be singly occupied and one doubly occupied, in order to maintain the maximum number of unpaired electrons, four in this case. The Δ value in this case is large enough to rule out high spin, and the ion is diamagnetic.

27.116 Predict the magnetic properties of (*a*) $Rh(NH_3)_6^{3+}$ (*b*) CoF_6^{3-}.

▮ (*a*) This problem can be approached by comparison with Problem 27.115. For analogous complexes of two different members of the same group in the periodic table, Δ increases with increasing atomic number. Since Δ for $Co(NH_3)_6^{3+}$ is already so high that the ion is low-spin, $Rh(NH_3)_6^{3+}$ would certainly be low-spin and diamagnetic. [The observed Δ for $Rh(NH_3)_6^{3+}$ is 34 000 cm^{-1} and the ion is diamagnetic.] (*b*) F^- is a weak-field ligand, tending to form complexes with a low Δ value, so that the ion would be expected to be high-spin, with four unpaired and parallel electron spins [compare with Problem 27.115(*b*)]. The measured Δ is 13 000 cm^{-1}, a low figure, and the ion is indeed paramagnetic.

27.117 The magnetic moment of $[Mn(CN)_6]^{3-}$ is 2.8 BM. The magnetic moment of $[MnBr_4]^{2-}$ is 5.9 BM. What are the geometries of these complex ions?

▮ From the spin-only formula it may be calculated that $[Mn(CN)_6]^{3-}$ has two unpaired electrons, while $[MnBr_4]^{2-}$ has five. (Note that the manganese is in different oxidation states.) The electronic configurations corresponding to the simple ions and the complexes with the indicated numbers of unpaired electrons are as follows:

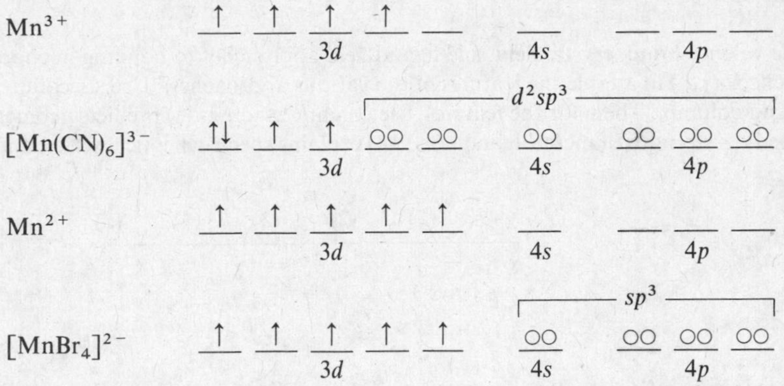

	3d					4s	4p		
Mn^{3+}	↑	↑	↑	↑	___	___	___	___	___
$[Mn(CN)_6]^{3-}$ (d^2sp^3)	↑↓	↑	↑	OO	OO	OO	OO	OO	OO
Mn^{2+}	↑	↑	↑	↑	↑	___	___	___	___
$[MnBr_4]^{2-}$ (sp^3)	↑	↑	↑	↑	↑	OO	OO	OO	OO

The first complex is octahedral and the second is tetrahedral.

27.6 OTHER CONCEPTS

27.118 What are the effective atomic numbers of the metal atoms in $Fe(CO)_5$ and in $Co_2(CO)_8$?

▮ The effective atomic number of iron includes 26 electrons from iron(0) plus 10 electrons from five CO molecules, totaling 36, the atomic number of krypton. Each cobalt atom in $Co_2(CO)_8$ has 27 electrons of its own and shares one with the other cobalt atom in addition to the eight electrons from four CO molecules, again for a total of 36 electrons per Co atom.

27.119 Deduce the value of x in the formula $Mo(CO)_x$.

▮ Since the effective atomic number of molybdenum must be 54 and its atomic number is 42, 12 electrons $(54 - 42)$ must be donated by the CO molecules. Hence, six CO molecules must each donate a pair of electrons, resulting in the formula $Mo(CO)_6$.

27.120 Determine the effective atomic number of iron in $Fe(NO)_2(CO)_2$ and in $Fe(C_5H_5)_2$.

▮ The effective atomic number of iron in $Fe(NO)_2(CO)_2$ includes 26 electrons from Fe, 6 electrons from the two NO molecules, and 4 electrons from the two CO molecules, for a total of 36—the atomic number of krypton. In $Fe(C_5H_5)_2$, the 26 electrons from the Fe are added to the 10 electrons from the two $C_5H_5^-$ ions, again for a total of 36.

27.121 Predict the value of x in each of the following carbonyls: **(a)** $Co_2(CO)_x$ **(b)** $H_xCr(CO)_5$ **(c)** $H_xCo(CO)_4$

▮ The carbonyls in general obey the effective atomic number rule—their effective atomic numbers (EAN) are equal to the atomic numbers of the next noble gases. Each CO molecule contributes a pair of electrons to the metal.

(a) Co electrons = 27 Shared with other Co atom = 1 Total = 28
Needed for EAN of Kr = 36 Shared with CO molecules = 36 − 28 = 8
Four CO molecules are needed per Co atom. For two Co atoms, $x = 8$. The formula is $Co_2(CO)_8$.

(b) Cr electrons = 24 Electrons from 5 CO groups = 10 Needed from H atoms = 2
Total = 36 EAN of Kr Formula: $H_2Cr(CO)_4$.

(c) Co electrons = 27 Electrons from 4 CO groups = 8 Needed from H atom(s) = 1
Total = 36 EAN of Kr Formula: $HCo(CO)_4$.

27.122 The "ferroin" complex of iron(II), $[Fe(o\text{-phen})_3]^{2+}$, is highly colored. The corresponding complex of iron(III) is weakly colored. Suggest how ferroin can be used in an oxidation-reduction titration method.

▮ The ferroin complex can be used as an indicator, which when oxidized loses its color.

27.123 Metal M forms a highly colored complex with ligand A and a colorless complex with ligand B, which has a larger formation constant than the complex of M and A. Suggest a method to determine the concentration of M ions in a solution by titration using A and B as reagents.

▮ One should use B as the titrant and A as the indicator. Titrant B will react with the colored complex of the metal with A when there is no free metal ion left for B to react with. When the complex of the metal and A is dissociated, the color will disappear, signalling an end to the titration. Naturally, only a small quantity of reagent A is used, as with any indicator.

27.124 Compare the valence bond, crystal field, and ligand field approaches to bonding in coordination compounds by indicating with checks (✓) in a table the features of the various approaches. Use six columns. List the three approaches in the left-hand column. Then, for the features, use column headings (1) predicts geometry, (2) predicts magnetic moments, (3) predicts strength of metal-ligand bond, (4) explains spectrochemical series, and (5) explains thermodynamic properties.

▮

	(1)	(2)	(3)	(4)	(5)
VB	✓	✓			
CF	✓	✓	✓	✓	✓
LF	✓	✓	✓	✓	✓

27.125 Draw representations of the electronic structures of the following complex ions according to ligand field theory: (*a*) $[Co(NH_3)_6]^{2+}$ (*b*) $[Pt(NH_3)_6]^{4+}$ (*c*) $[CoF_6]^{4-}$

▮ See Fig. 27.26. The circled orbitals are the ones which are considered in the crystal field theory. Note that the complex in (*b*) is strong-field; the others are weak-field.

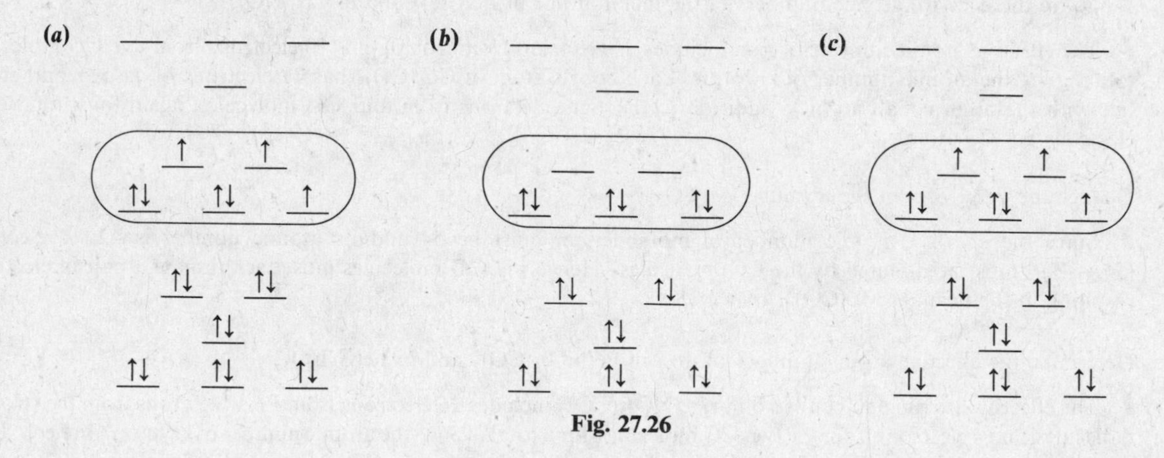

Fig. 27.26

1 H 1.0079	IIA																2 He 4.00260
3 Li 6.941	4 Be 9.01218											5 B 10.81	6 C 12.011	7 N 14.0067	8 O 15.9994	9 F 18.998403	10 Ne 20.179
11 Na 22.98977	12 Mg 24.305											13 Al 26.98154	14 Si 28.0855	15 P 30.97376	16 S 32.06	17 Cl 35.453	18 Ar 39.948
19 K 39.0983	20 Ca 40.08	21 Sc 44.9559	22 Ti 47.90	23 V 50.9415	24 Cr 51.996	25 Mn 54.9380	26 Fe 55.847	27 Co 58.9332	28 Ni 58.70	29 Cu 63.546	30 Zn 65.38	31 Ga 69.72	32 Ge 72.59	33 As 74.9216	34 Se 78.96	35 Br 79.904	36 Kr 83.80
37 Rb 85.4678	38 Sr 87.62	39 Y 88.9059	40 Zr 91.22	41 Nb 92.9064	42 Mo 95.94	43 Tc (98)	44 Ru 101.07	45 Rh 102.9055	46 Pd 106.4	47 Ag 107.868	48 Cd 112.41	49 In 114.82	50 Sn 118.69	51 Sb 121.75	52 Te 127.60	53 I 126.9045	54 Xe 131.30
55 Cs 132.9054	56 Ba 137.33	57 La 138.9055 *	72 Hf 178.49	73 Ta 180.9479	74 W 183.85	75 Re 186.207	76 Os 190.2	77 Ir 192.22	78 Pt 195.09	79 Au 196.9665	80 Hg 200.59	81 Tl 204.37	82 Pb 207.2	83 Bi 208.9804	84 Po (209)	85 At (210)	86 Rn (222)
87 Fr (223)	88 Ra 226.0254	89 Ac 227.0278 †	104 Unq (261)	105 Unp (262)	106 Unh (263)												

*	58 Ce 140.12	59 Pr 140.9077	60 Nd 144.24	61 Pm (145)	62 Sm 150.4	63 Eu 151.96	64 Gd 157.25	65 Tb 158.9254	66 Dy 162.50	67 Ho 164.9304	68 Er 167.26	69 Tm 168.9342	70 Yb 173.04	71 Lu 174.967
†	90 Th 232.0381	91 Pa 231.0359	92 U 238.029	93 Np 237.0482	94 Pu (244)	95 Am (243)	96 Cm (247)	97 Bk (247)	98 Cf (251)	99 Es (252)	100 Fm (257)	101 Md (258)	102 No (259)	103 Lr (260)

INDEX

Reference is made to problem numbers, not page numbers, except for tables.